P9-DFM-082

For All Practical Purposes

Project Director

Solomon Garfunkel
Consortium for Mathematics and Its Applications

Contributing Authors

PART I MANAGEMENT SCIENCE
Joseph Malkevitch
York College, CUNY

PART II STATISTICS: THE SCIENCE OF DATA
Lawrence M. Lesser
The University of Texas at El Paso
[*Founding FAPP author is David S. Moore, Purdue University*]

PART III VOTING AND SOCIAL CHOICE
Alan D. Taylor | **Bruce P. Conrad** | **Steven J. Brams**
Union College | Temple University | New York University

PART IV FAIRNESS AND GAME THEORY
Alan D. Taylor | **Bruce P. Conrad** | **Steven J. Brams**
Union College | Temple University | New York University

PART V THE DIGITAL REVOLUTION
Joseph A. Gallian
University of Minnesota Duluth

PART VI ON SIZE AND GROWTH
Paul J. Campbell
Beloit College

PART VII YOUR MONEY AND RESOURCES
Paul J. Campbell
Beloit College

9th
Edition

For All Practical Purposes

Mathematical Literacy in Today's World

W. H. FREEMAN AND COMPANY
New York

Publisher: Ruth Baruth
Executive Acquisitions Editor: Terri Ward
Executive Marketing Manager: Jennifer Somerville
Developmental Editors: Andrew Sylvester and Lisa Collette
Senior Media Acquisitions Editor: Roland Cheyney
Senior Media Editor: Laura Capuano
Associate Editor: Katrina Wilhelm
Assistant Media Editor: Catriona Kaplan
Editorial Assistant: Tyler Holzer
Photo Editor: Bianca Moscatelli
Photo Researcher: Christina Micek
Cover Designer: Vicki Tomaselli
Text Designer: Marsha Cohen
Project Editor: Vivien Weiss
Director of Production: Ellen Cash
Production Manager: Julia DeRosa
Illustrations: MPS Limited, a Macmillan Company
Composition: MPS Limited, a Macmillan Company
Printing and Binding: QuadGraphics

Library of Congress Control Number: 2011937302

Student Edition (hardcover):
ISBN-13: 978-1-4292-4316-2
ISBN-10: 1-4292-4316-3

Student Edition (paperback):
ISBN-13: 978-1-4292-5482-3
ISBN-10: 1-4292-5482-3

Student Edition (looseleaf):
ISBN-13: 978-1-4641-0157-1
ISBN-10: 1-4641-0157-4

Instructor Comp Copy (hardcover):
ISBN-13: 978-1-4641-0156-4
ISBN-10: 1-4641-0156-6

NASTA-spec version:
ISBN-13: 978-1-4292-5481-6
ISBN-10: 1-4292-5481-5

Printed in the United States of America
First printing

W. H. Freeman and Company
41 Madison Avenue
New York, NY 10010
Houndmills, Basingstoke RG21 6XS, England

www.whfreeman.com

Brief Contents

Contents

PART I Management Science / 3

PART II Statistics: The Science of Data / 163

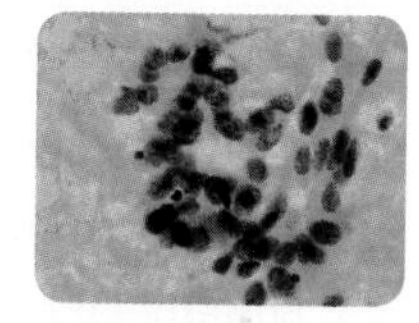

VOTE
VOTE
VOTE

PART V The Digital Revolution / 571

PART VI On Size and Growth / 641

CUSTOM CHAPTERS

The following chapters are available through Freeman's custom publishing (restrictions may apply), as well as through the Online Study Center, MathPortal, and eBook:

Sets
Problem Solving
Logic
Geometry
Counting and Probability
Numeration Systems
Personal Finance

Preface

To the Student

For All Practical Purposes, Ninth Edition, continues our effort to bring the excitement of contemporary mathematical thinking to the nonspecialist. In science and industry, mathematical models are the main tools for analyzing and solving problems that arise. In this book, our goal is to convey the power of mathematics by showing you the wide variety of problems that can be modeled and solved by quantitative means. An extensive package of supplements designed to make study time supremely effective complements the Ninth Edition. Highlights of the supplements package include the *Student Study Guide* and *Student Solutions Manual.* Between the text and the available resources, *For All Practical Purposes* offers you the tools to succeed in the course and apply your new knowledge to daily life experiences.

There are many ways to talk about why mathematics and its applications matter. You will hear expressions such as *mathematical literacy* or *quantitative literacy.* They mean, essentially, that math is important. It is important because knowing it can make your life easier. In other words, it can help to explain how your world works. We created this course and this book because we know that not everyone looks at mathematics in this way.

In school, you spent a great deal of time learning the tools of mathematics—how to manipulate symbols, how to solve equations. In this course, you will spend time learning the uses of mathematics and the power of mathematics to help us to understand so many different parts of our everyday lives and the world itself. We hope that this exploration will give you a broader sense of what our subject is about and why we wanted you to take a math course every year you were in school. It's "for all practical purposes" because, in a sense, you've learned to hammer nails and saw wood. Now we're going to build houses.

Enjoy!

To the Instructor

Because *For All Practical Purposes* stresses the connections between contemporary mathematics and modern society, our text must be flexible enough to accommodate new ideas in mathematics and their new applications to our daily lives. We maintain this flexibility in the ninth edition.

Our primary goal for this edition was to further improve the ease of use for instructors and students alike. An extensive supplements package is available, including the new MathPortal, available packaged with the text or sold separately. This innovative online resource brings together the complete text and its media in one easy-to-use learning space. From the ebook and Assignment Center to the full array of resources, including practice quizzes, exercise solutions, interactive applets, flashcards, video clips, and much more, it's new, it's innovative, it's a must have!

New to the Ninth Edition

Improved Pedagogical Structure

The enhanced pedagogy makes it easier to navigate the text. Chapter openers have been revised to include more guidance for what is to come in the chapter. New boxed "Algebra Review" features have been added throughout the text to provide a basic review of algebra topics for students who need the extra support.

New Examples

New and updated examples are included throughout the text to address new topics and changes to the material. Examples provide new topics for class discussion and new ways of relating to essential concepts.

New Exercises

- All exercise sets have been updated and refreshed.
- The Skills Check questions have also been updated and refreshed, and the number of questions in each chapter has been increased from twenty to thirty.

Part-Specific Content Changes

Part I:

- Revised discussion of graphing (Section 1.1), including an expansion of Example 1: Describing a Graph.
- Revised Spotlight 1.3 on the Chinese Postman Problem.

Part II:

- Throughout Part II, the colloquial use of words (*spread, range, random, normal, correlation,* and *expected value*) has been distinguished from the precise academic usage of these words, in accordance with current statistics education research.
- Improved explanation of standard deviation (Chapter 5), including a new example addressing stock and mutual fund performance over the last decade.
- Expanded explanation of distribution (Chapter 5), which now distinguishes among three forms of distribution: frequency distribution, grouped frequency distribution, and relative frequency distribution,
- A new Figure 7.3 distinguishes the types of designs.
- A new Table 8.1 distinguishes independent and disjoint events.
- New Spotlights and examples relating to current events and student interests (e.g., the 2010 Census, class size, ratemyprofessors.com, and ethics) have been added.

Part III:

- Revised coverage of the Pareto Condition, Hare System, and Approval Voting (Chapter 9).
- New definition of Group Manipulability (Section 10.3) and revised coverage of the chair's paradox (Section 10.5)
- Expanded coverage of weighted voting, the Shapley-Shubik power index, and the Banzhaf power index (including a simplified method for calculation), with new examples (Chapter 11).

Part IV:

- Results from the 2008 election have been added to Chapter 12, and a discussion of the growing support for the National Popular Vote Law has been added.
- New Section 13.8 on Vickrey auctions, including a new example.
- Addition of applications of apportionment methods to parliamentary elections, including new examples (Chapter 14).

Part V:

- A revision of Section 16.1 expands on the content and more clearly divides and summarizes the various methods described.
- New Spotlight 16.5 on census records.
- New coverage of the linear cipher (Section 17.3).

Part VI:

- Revised coverage of the language of growth, enlargement, and decrease, including a new example.
- New Spotlight 18.2: Is There Any Advantage to Being Gigantic?
- Revised Spotlight 18.4: Helping to Find Missing Children
- Revised writing projects that cover projected government spending, federal food subsidies, and marijuana arrests in California.
- New Spotlight 19.4 on the aspect ratio for digital screens.
- New examples and exercises in connection with symmetry groups, featuring mattress turning and tire rotation (Chapter 19).

Part VII:

- Expanded coverage of effective rate and geometric series (Chapter 21).
- New Spotlights covering derivatives in the financial market and using spreadsheets for financial calculations (Chapter 21).
- New Section 23.3, on radioactive decay, featuring half-life, carbon-14 dating, and the isotopes involved in the Japanese nuclear power plant disaster.

Focus on Accuracy

For this edition, we once again implemented a detailed accuracy-checking plan to sustain the quality of the exercises and improve the solutions. To this end, we are very grateful to John Samons of Florida State College at Jacksonville. He tirelessly worked with the authors to ensure accuracy in this edition of the text. John once again collaborated with the supplements author, Heidi A. Howard of Florida State College at Jacksonville, to ensure both accuracy and consistency between the text and supplements package.

We are also grateful to Paul Lorczak for his participation in a detailed line-edit review of the ninth edition.

Custom Options

In addition to the extensive topics covered in the text, more traditional chapters (including "Problem Solving," "Sets," "Logic," "Geometry," "Counting and Probability," "Numeration Systems," and "Personal Finance") are available with *For All Practical Purposes* through custom publishing. For more information, please contact your W. H. Freeman representative or go to www.whfreeman.com/fapp9e. Restrictions apply.

Media and Supplements

The media and supplements package for the ninth edition has been updated to reflect changes in the book. Both instructors and students will benefit from the innovative materials available to them.

Student Resources

MathPortal

http://courses.bfwpub.com/fapp9e (Access code required. Available packaged with *For All Practical Purposes,* Ninth Edition, or for purchase online.)

MathPortal is the digital gateway to *For All Practical Purposes,* Ninth Edition, and it is designed to enrich the course and enhance students' study skills through a collection of Web-based tools. MathPortal integrates a rich suite of diagnostic, assessment, tutorial, and enrichment features, enabling students to master statistics at their own pace. MathPortal is organized around three main teaching and learning components:

1. Interactive ebook offers a complete and customizable online version of the text, fully integrated with all of the media resources available with *For All Practical Purposes,* Ninth Edition. The ebook allows students to quickly search the text, highlight key areas, and add notes about what they're reading. Instructors can customize the ebook to add, hide, and reorder content, add their own material, and highlight key text for students.

2. Resources organizes all student and instructor resources in one easily searchable location. Resources include:

- **Interactive applets** offer a series of interactive applets to help students master key mathematical concepts and work exercises from the text.
- **Student Solutions Manual** provides solutions to the odd-numbered exercises, with stepped out solutions to select problems.
- **NEW! *LEARNING*Curve** is a formative quizzing system that offers immediate feedback at the question level to help students master course material.
- **Math Clips:** These animated whiteboard videos illuminate key concepts in the text by showing students step-by-step solutions to selected exercises.
- **Self-quizzes, flash cards, video clips, news feeds, and other projects**

MathPortal Resources (Instructors)

- **Instructor's Guide with Full Solutions** includes teaching suggestions, chapter comments, and detailed solutions to all exercises.
- **Test Bank** offers thousands of multiple-choice questions.
- **PowerPoint slides** offer a detailed lecture presentation of concepts covered in each chapter of *For All Practical Purposes,* Ninth Edition.
- **NEW! SolutionMaster** is a Web-based version of the solutions in the *Instructor's Guide with Full Solutions*. This easy-to-use tool allows instructors to generate a solution file for any set of homework exercises. Solutions can be downloaded in PDF format for convenient printing and posting. For more information or a demonstration, contact your local W. H. Freeman sales representative.

3. Assignments organizes assignments and guides instructors through an easy-to-create assignment process providing access to questions from the Test Bank and exercises from the text, including many algorithmic problems. The Assignment Center enables instructors to create their own assignments from a variety of question types for machine-gradable assignments. This powerful assignment manager allows instructors to select their preferred policies in regard to scheduling, maximum attempts, time limitations, feedback, and more!

Online Study Center

www.whfreeman.com/osc/fapp9e (Access code or online purchase required.)
The Online Study Center offers all the resources available in MathPortal, except the ebook and Assignment Center.

Companion Web Site

www.whfreeman.com/fapp9e
This open-access Web site provides students with access to study tools and instructors a range of assessment, presentation, and course management resources.

Student Study Guide, ISBN: 1-4292-4365-1

Heidi A. Howard, Florida State College at Jacksonville
Offers study tips and tools to help students gain a better understanding of course material.

Student Solutions Manual, ISBN: 1-4292-4364-3

Heidi A. Howard, Florida State College at Jacksonville
Contains full, worked solutions to the odd-numbered problems in the text.

Instructor Resources

Instructor's Manual with Full Solutions, ISBN: 1-4292-4363-5

Heidi A. Howard, Florida State College at Jacksonville
Includes teaching support for each chapter and full solutions for all problems in the text.

Teaching Guide for First-Time Instructors, ISBN: 1-4292-4358-9

Heidi A. Howard, Florida State College at Jacksonville
This guide for new instructors, adjuncts, and teaching assistants will help make planning your course and teaching with *For All Practical Purposes* easier and more effective. Ideas set forth in this guide also offer fresh perspective and ideas to experienced instructors.

Enhanced Instructor's Resource CD-ROM (IRCD), ISBN: 1-4292-4359-7

Created to help instructors develop lecture presentations, course Web sites, and other resources, this CD-ROM allows instructors to search and export all the resources contained below by key term or chapter:

- All text images.
- Applets, movies, flash cards, spreadsheet projects, and self-quizzes available on the Web site.
- *Instructor's Manual with Full Solutions.*

Assessment

Test Bank

CD-ROM (Windows and Macintosh): 1-4292-4361-9
Printed: 1-4292-4362-7
The Test Bank offers over 75 multiple-choice and fill-in-the-blank questions and 35 short-answer questions per chapter. The easy-to-use CD includes Windows and Macintosh versions on a single disc, in a format that lets you add, edit, and resequence questions to suit your needs.

I>Clicker

i>clicker is a two-way radio-frequency classroom response solution developed by educators for educators. University of Illinois physicists Tim Stelzer, Gary Gladding, Mats Selen, and Benny Brown created the i>clicker system after using competing classroom response solutions and discovering they were neither classroom-appropriate nor student-friendly. Each step of i>clicker's development has been informed by teaching and learning. i>clicker is superior to other systems from both a pedagogical and technical standpoint. To learn more about packaging i>clicker with this textbook, please contact your local sales rep or visit **www.iclicker.com**.

Course Management Systems

W.H. Freeman and Company provides courses for Blackboard, WebCT (Campus Edition and Vista), Angel, Desire2Learn, Moodle, and Sakai course management systems. These are completely integrated solutions that you can customize and adapt easily to meet your teaching goals and course objectives. Visit **www.bfwpub.com/lms** for more information.

Acknowledgments

For All Practical Purposes continues to evolve in great part due to our many friends and colleagues who have offered suggestions, comments, and corrections. We are grateful to them all.

Rosalie Abraham, *Florida State College at Jacksonville*
Alison Ahlgren, *University of Illinois at Urbana-Champaign*
Julian Allagan, *Gainesville State College*
James Baglama, *University of Rhode Island*
Scott Balcomb, *St. Joseph's University*
Nancy Balle, *Ball State University*
Richard Bedient, *Hamilton College*
Rebecca Bergs, *Ball State University*
Terence R. Blows, *Northern Arizona University*
Raouf N. Boules, *Towson University*
Kristina K. Bowers, *Florida State University*
Terry Boyd, *University of Indianapolis*
Linda Braddy, *East Central University*
Barry Brunson, *Western Kentucky University*
Paul Buckelew, *Oklahoma City Community College*
Annette M. Burden, *Youngstown State University*
Shana Calaway, *Shoreline Community College*
Tim Carroll, *Eastern Michigan University*
Christina Cedzo, *Gannon University*
G. Andy Chang, *Youngstown State University*
Yi Cheng, *Indiana University South Bend*
Leo Chouinard, *University of Nebraska–Lincoln*
Hilary Clark, *Virginia Commonwealth University*
Karen Clark, *Tacoma Community College*
Drue Coles, *Bloomsburg University*
John Czaplewski, *Winona State University*
Valerie Morgan-Crick, *Tacoma Community College*
Greg Crow, *Point Loma Nazarene University*
Sloan Despeaux, *Western Carolina University*
Rob Donnelly, *Murray State University*
Daniel Dreibelbis, *University of North Florida*
Gina Poore Dunn, *Lander University*
Nancy Eaton, *University of Rhode Island*
Kristy J. Eisenhart, *Western Michigan University*
John W. Emert, *Ball State University*
Scott Fallstrom, *University of Oregon*
Kevin Ferland, *Bloomsburg University*
Sandra Fillebrown, *Saint Joseph's University*
Joseph Fox, *Salem State College*
W. Bart Frye, *Ball State University*
Martha Gady, *Whitworth College*
Monica Pierri-Galvao, *Gannon University*
Steve Gendler, *Clarion University*
Marty Getz, *University of Alaska, Fairbanks*
Carol E. Gibbons, *Salve Regina University*
T. R. Hamlett, *East Central University*
Geoffrey Hagopian, *College of the Desert*
Mohammad Halim, *Ball State University*
Kathy Hays, *Anne Arundel Community College*
Frederick Hoffman, *Florida Atlantic University*
Michael Hull, *Northern Arizona University*
Scott Inch, *Bloomsburg University of Pennsylvania*
Peter Johnson, *Auburn University*
W. T. Kiley, *George Mason University*
Julie Killingbeck, *Ball State University*
Nancy Kitt, *Ball State University*
Samuel Kohn, *Thomas Edison State College*
Kathy Lewis, *State University of New York, Oswego*
Aihua Li, *Montclair State University*
Monica Liddle, *State University of New York, Delhi*
Anthony Malone, *University of Cincinnati*
Jay Malmstrom, *Oklahoma City Community College*
Barbara Margoulius, *Cleveland State University*
Vania Mascioni, *Ball State University*
Mary T. McMahon, *North Central College*
Christopher McCord, *University of Cincinnati*
Ricardo Moena, *University of Cincinnati*
Steve Morics, *University of Redlands*
Dean Morrow, *Washington and Jefferson College*
Anne Marie Mosher, *St. Louis Community College*
Ellen Mulqueeny, *Cleveland State University*
Mika Munakata, *Montclair State University*
Chris Oehrlein, *Oklahoma City Community College*
Steven Ohs, *Western Michigan University*
Patricia Parkison, *Ball State University*
Deb Pearson, *Ball State University*
Andrew B. Perry, *Springfield College*
Marilyn Reba, *Clemson University*
Leo Robinson, *Ball State University*
Chris Rodger, *Auburn University*
Jennifer Marie Rodin, *University of South Carolina Aiken*
Michael Rosenthal, *Florida International University*
Robin Ruffato, *Ball State University*
Daniel Russow, *Arizona Western University*
Steven Schecter, *North Carolina State*

Art Shindhelm, *Western Kentucky University*
Robert Shih, *North Carolina State University*
Brian Siebenaler, *Ball State University*
Debora J. Simonson, *University of North Florida*
Samuel Bruce Smith, *St. Joseph's University*
Patricia Stanley, *Ball State University*
James D. Stoops, *Ball State University*
William R. Stout, *Salve Regina University*
Tamas Szabo, *Weber State University*
Robert Terrell, *Cornell University*
Helen Thorwarth, *Northern Kentucky University*
Aaron K. Trautwein, *Carthage College*
David Urion, *Winona State University*
Bonnie Wachhaus, *Messiah College*
W. D. Wallis, *Southern Illinois University*
Kim Ward, *Eastern Connecticut State University*
John Weglarz, *Kirkwood Community College*
Gideon Weinstein, *Montclair State University*
Cheryl Whitelaw, *Southern Utah University*
Liz Whittern, *Ball State University*
Scott Wilde, *Baylor University*
Bruce Woodcock, *Central Washington University*
Meredith Wort, *East Central University*
Mingqing Xiao, *Southern Illinois University*
Christian Yankov, *Eastern Connecticut State University*
Janet Yi, *Ball State University*
Laurie Margaret Zack, *High Point University*
John Zerger, *Catawba College*
Cathleen M. Zucco-Teveloff, *Rowan University*

We owe our appreciation to the people at W. H. Freeman and Company who participated in the development and production of this edition. We wish especially to thank the editorial staff for their tireless efforts and support. Among them are Ruth Baruth, Publisher; Terri Ward, Executive Acquisitions Editor; Andrew Sylvester, Developmental Editor; Vivien Weiss, Project Editor; Blake Logan, Cover Designer; Bianca Moscatelli, Photo Editor; Roland Cheyney, Senior Media Editor; and Ellen Cash, Director of Production. We would also like to extend our appreciation to outside Development Editor Lisa Collette.

To the staff of COMAP and everyone who made our purposes practical, we offer our appreciation for an exciting and exhilarating time.

Solomon Garfunkel, COMAP

For All Practical Purposes

NASA

Management Science

Part I

Chapter 1
Urban Services

Chapter 2
Business Efficiency

Chapter 3
Planning and Scheduling

Chapter 4
Linear Programming

Getting through a typical day can be a challenge: getting to or from school or work on time; finding a parking spot when we are late for a date; making sure we have food in the refrigerator or getting to our favorite fast-food restaurant to stay properly fed and "fueled" with coffee; making sure our bodies are fit by getting to the gym or exercising at home, and seeing the doctor for a regular checkup or when we are ill. However, for real complexity, try to put yourself in the position of one of the astronauts in the International Space Station (ISS), which orbits more than 300 miles above the earth. In late July 2010, the cooling system of the ISS failed, and while there was a backup system, had that system failed the situation on the ISS could have been dire. We tend to think of space as a very cold place, but the intensity of the sun provides plenty of heat for the astronauts via solar panels. The ISS has a heat dissipation system to help keep it cool, and it was this system that failed.

So what does this have to do with mathematics? To correct the problem, the astronauts on board the ISS, in conjunction with the ground controllers, had to work out a complex system of space walks that used the ISS robotic arm—developed with input from mathematicians and computer scientists—to make the repairs. The part of mathematics concerned with efficient operations of business and governments is called **operations research (OR)** or **management science**. The domain of OR includes resource allocation, scheduling, queues (waiting lines), inventory analysis, routing problems, and cost minimization, to mention but a few of OR's growing areas of applicability.

Using the skills of the ground team and astronauts, the latter carried out a series of complex space walks. These activities, performed under the harshest physical conditions, required many adaptations to deal with contingencies not originally anticipated. Relying on the systems and tools that mathematics had helped put in place, the ISS was restored to stable condition by an international team of courageous astronauts. Chalk up another triumph for OR! What follows will help you, too, to know about and use such tools.

FEDERAL EXPRESS
WORLDWIDE SERVICE
777-6500
STOP

1

Urban Services

T*he underlying theme of management science,* also called **operations research (OR)**, is finding the best method for solving some problem—what mathematicians call the **optimal solution**. In some cases, the goal may be to finish a job as quickly as possible. In other situations, the objective might be to maximize profit or minimize cost. In this chapter, our goal is to save time (and usually taxpayer money) in traversing a street network while providing services such as checking parking meters, collecting garbage or bottles for recycling, de-icing roads, inspecting for potholes or gas leaks, or delivering mail.

Let's begin by assisting the parking department of a city government. Most cities and many small towns have parking meters that must be regularly checked for parking violations or emptied of coins. We will use an imaginary town to show how management science techniques can help to make parking control more efficient.

1.1 Euler Circuits

The street map in Figure 1.1 is typical of many towns across the United States, with streets, residential blocks, and a town park. Our job, or that of the commissioner of parking, is to find the most efficient route for the parking-control officer, who travels on foot, to check the meters in an area. Efficient routes save money. Our map shows only a small area, allowing us to start with an easy problem. But the problem occurs on a larger scale in all cities and towns and for larger areas. The bigger the region involved, the greater the potential for cost savings.

The commissioner has two goals in mind: (1) The parking-control officer must cover all the sidewalks that have parking meters without retracing any more steps than are necessary; and (2) the route should end at the same point at which it began, perhaps where the officer's patrol car is parked. To be specific, suppose there are only two blocks that have parking meters, the two lightly shaded blocks that are side by side toward the top of Figure 1.1. Suppose further that the parking-control officer must start and end at the upper left corner of the left-hand block. You might enjoy working out some routes by trial and error and evaluating their good and bad features. We are going to leave this problem for the moment and establish some concepts that will give us a better method than trial and error to deal with this problem.

Figure 1.1 A street map for part of a town.

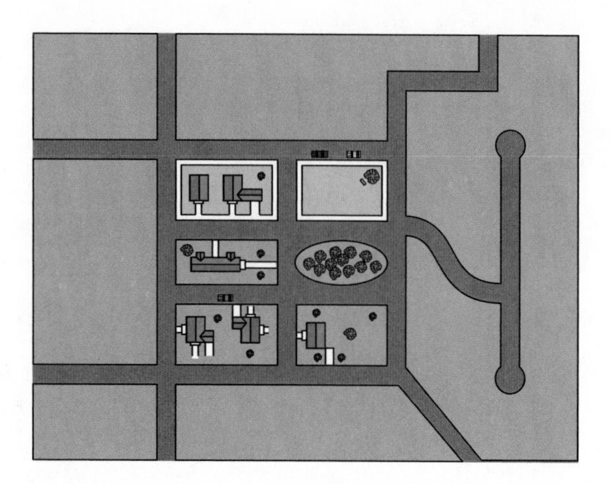

What Is a Graph? DEFINITION

A **graph** is a finite set of dots and connecting links. The dots are called **vertices** (a single dot is called a **vertex**), and the links are called **edges**. Each edge must connect two different vertices. A graph can represent our city map, a communications network, a system of air routes, or electrical power lines.

Path and Circuit DEFINITION

A **path** is a connected sequence of edges showing a route on the graph that starts at a vertex and ends at a vertex; a path is usually described by naming in turn the vertices visited in traversing it. A path that starts and ends at the same vertex is called a **circuit**.

EXAMPLE 1
Parts of a Graph

We can use the graph in Figure 1.2 to help explain these technical terms. The graph shown has 5 vertices and 8 edges. The vertices represent cities, and the edges represent nonstop airline routes between them. We see that there is a nonstop flight between Berlin and Rome, but no such flight between New York and Berlin. There are several paths that describe how a person might travel with this airline from New York to Berlin. The path that seems most direct is New York, London, Berlin, but New York, Miami, Rome, Berlin is also a path. We can describe these two paths as *NLB* and *NMRB*. Another

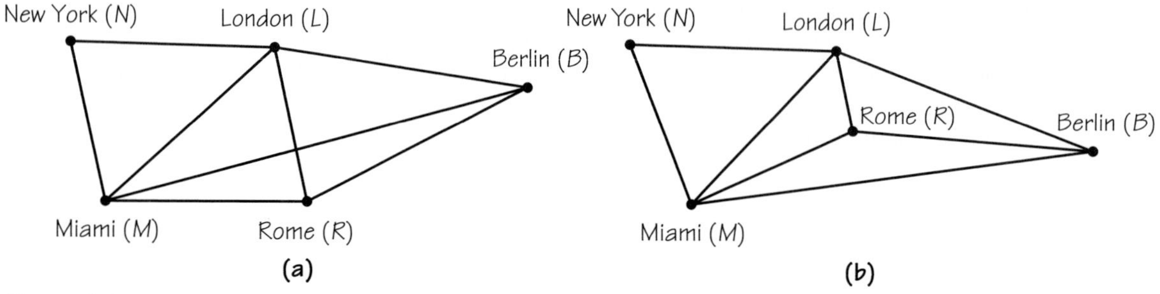

Figure 1.2 (a) The edges of the graph show nonstop routes that an airline might offer. (b) The graph in (a) redrawn without the accidental crossing.

path would be New York, Miami, Rome, London, Berlin, which can be written as *NMRLB*. An example of a circuit is Miami, Rome, London, Miami. It is a circuit because the path starts and ends at the same vertex. This circuit can best be described in symbols by *MRLM.* Another example of a circuit in this graph would be *LRBL*, which is the circuit involving the cities London, Rome, Berlin, and back to London. In this chapter, we are especially interested in circuits, just as we are in real life. Most of us end our day in the same place that we start it—at home!

Notice that the edges *MB* (which could also be denoted *BM*) and *RL* shown in Figure 1.2a meet at a point that has no label. Furthermore, this point does not have a dark dot. This is because this point does not represent a vertex of our graph; it does not represent a city. It arises as an "accidental" consequence of the way this diagram has been drawn. We could join *M* and *B* with a curved line segment so that the edges *LR* and *MB* do not cross, or redraw the diagram so as to avoid a crossing in this case. We will be working often in situations where graphs can be drawn without accidental crossings and we will try to avoid such crossings when it is convenient to do so. However, there are graphs which—when they are drawn on a flat piece of paper—accidental crossings are unavoidable. (The graph in Figure 2.12 on p. 48 is an example of such a graph.)

Returning to the case of parking control in Figure 1.1, we can use a graph to represent the whole territory to be patrolled: Think of each street intersection as a vertex and each sidewalk that contains meters as an edge, as in Figure 1.3. Notice in Figure 1.3b that the width of the street separating the blocks is not explicitly represented; it has been shrunk to nothing. In effect, we are simplifying our problem by ignoring any distance traveled in crossing streets. In drawing graph diagrams such as those in Figure 1.3 or Figure 1.5, we usually use straight line segments to represent edges. However, sometimes we cannot avoid the use of "curves," or we may prefer to use curved edges because they convey aspects of the original problem that we desire to emphasize.

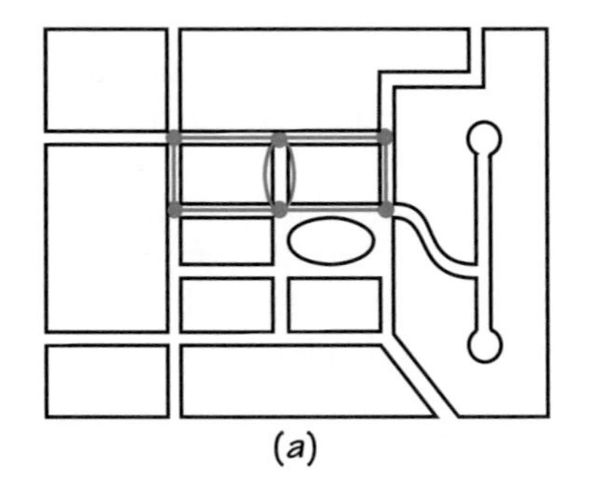

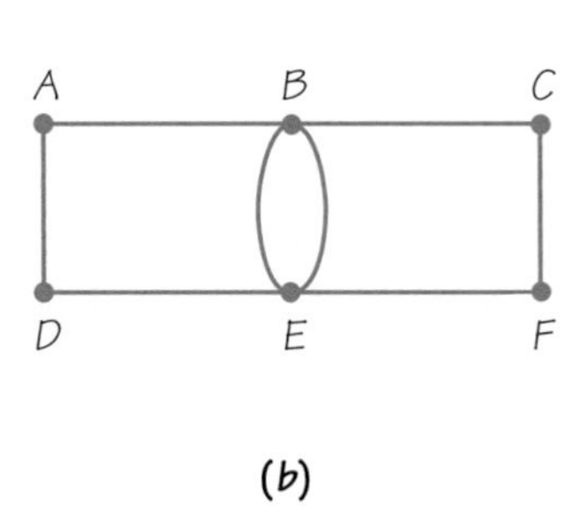

Figure 1.3 (a) A graph superimposed upon a street map. The edges show which sidewalks have parking meters. (b) The same graph enlarged.

The numbered sequence of edges in Figure 1.4a shows one circuit that covers all the meters. (Note that it is a circuit because its path returns to its starting point.) However, one edge is traversed three times. Figure 1.4b shows another solution that is better because its circuit covers every edge (sidewalk) exactly once. In Figure 1.4b, no edge is covered more than once, or *deadheaded* (a term borrowed from shipping, which means making a return trip without a load). When deadheading is required in an applied situation, such as inspecting parking meters or snow removal, typically time and effort is being spent but no productive work is accomplished because the productive work was done the first time the edge was covered (traversed).

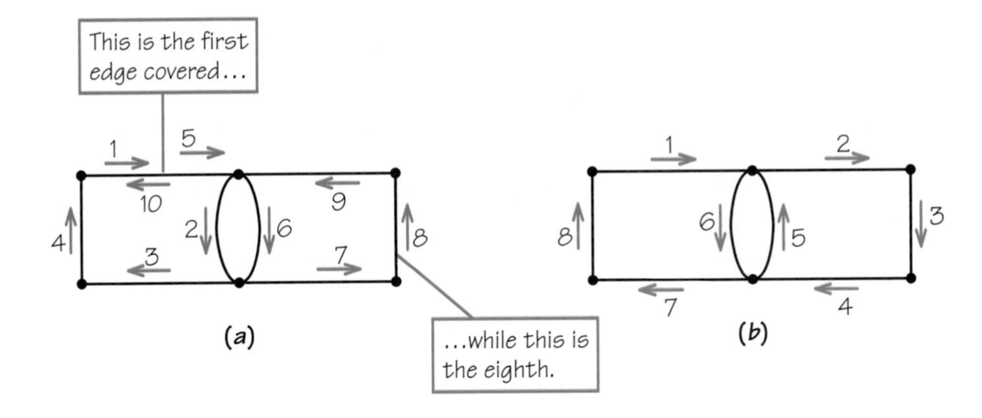

Figure 1.4 (a) A circuit and (b) an Euler circuit.

Euler Circuit DEFINITION

A circuit that covers each edge of a graph once, but not more than once, is called an **Euler circuit**.

Figure 1.4b shows an Euler circuit. These circuits get their name from the great eighteenth-century mathematician Leonhard Euler (pronounced *oy′ lur*), who first studied them (see Spotlight 1.1). Euler was the founder of the theory of graphs, or graph theory. One of his first discoveries was that some graphs have no Euler circuits at all. For example, in the graph in Figure 1.5b, it would be impossible to start at one point, return to that starting vertex and cover all the edges without retracing some steps: If we try to start a circuit at the leftmost vertex, we discover that once we have left the vertex, we have "used up" the only edge meeting it. We have no way to return to our starting point except to reuse that edge. But this is not allowed in an Euler circuit. If we try to start a circuit at one of the other two vertices, we likewise can't complete it to form an Euler circuit.

As mentioned in Spotlight 1.2, realistic problems of this type will involve larger neighborhoods that might require the use of a computer. In addition, there may be other complications that might take us beyond the simple mathematics we want to stick to.

Because we are interested in finding circuits, and Euler circuits are the most efficient ones, we will want to know how to find them. If a graph has no Euler circuit, we will want to develop the next best circuits, those having minimum deadheading. These topics make up the rest of this chapter.

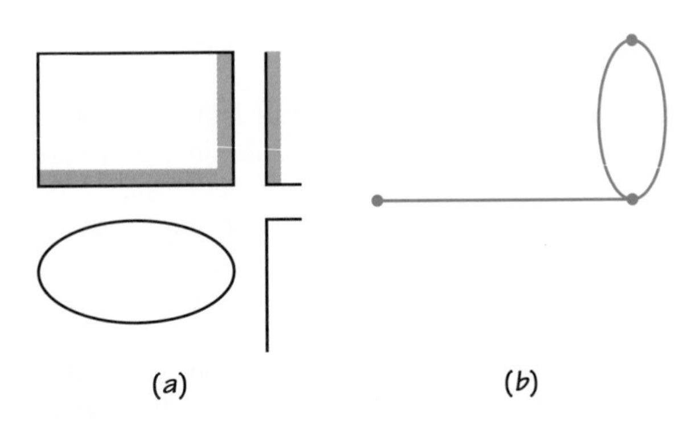

Figure 1.5 (a) The three shaded sidewalks cannot be covered by an Euler circuit. (b) The graph of the shaded sidewalks in part (a).

SPOTLIGHT 1.1
Leonhard Euler

Leonhard Euler (1707–1783) was one of those rare individuals who was remarkable in many ways. He was extremely prolific, publishing over 500 works in his lifetime. But he wasn't devoted just to mathematics; he was a people person, too. He was extremely fond of children and had thirteen of his own, of whom only five survived childhood. It is said that he often wrote difficult mathematical works with a child or two in his lap.

Human interest stories about Euler have been handed down through two centuries. He was a prodigy at doing complex mathematical calculations under less than ideal conditions, and he continued to do them even after he became totally blind later in life. His blindness diminished neither the quantity nor the quality of his output. Throughout his life, he was able to mentally calculate in a short time what would have taken ordinary mathematicians hours of pencil-and-paper work. A contemporary claimed that Euler could calculate effortlessly, "just as men breathe, as eagles sustain themselves in the air." His collected works and numerous letters to other scholars of his day are still being published.

Euler invented the idea of a graph in 1736 when he solved a problem in "recreational mathematics." He showed that it was impossible to stroll a route visiting the seven bridges of the German town of Königsberg exactly once. Ironically, in 1752 he discovered that three-dimensional polyhedra obey the remarkable formula $V - E + F = 2$ (that is, number of vertices − number of edges + number of faces = 2) but failed to give a proof because he did not analyze the situation using graph theory methods.

Leonhard Euler
(Portrait by Emanuel Handmann, Bildnis des Mathematikers, 1753, Oeffentliche Kunstsammlung Basel, Kunstmuseum.)

1.2 Finding Euler Circuits

Now that we know what an Euler circuit is, we are faced with two obvious questions:

1. Is there a way to tell by calculation or logical reasoning, not by trial and error, if a graph has an Euler circuit?
2. Is there a method, other than trial and error, for finding an Euler circuit when one exists and finding it quickly?

Loosely speaking, the first question lies within the concerns of mathematicians because it asks whether or not a certain problem admits a solution. Typically, the second question lies in the domain of computer science because it concerns finding the actual answer to a complex version of a problem in a short enough time to be useful.

Euler investigated these questions in 1735 by using the concepts of **valence** and **connectedness**.

Valence DEFINITION

The **valence** of a vertex in a graph is the number of edges meeting at the vertex.

SPOTLIGHT 1.2

The Human Aspect of Problem Solving

Thomas Magnanti, professor of operations research and management science, is Dean of Engineering at MIT. Here are some of his observations:

> Typically, a management science approach has several different ingredients. One is just structuring the problem—understanding that the problem is an Euler circuit problem or a related management science problem. After that, one has to develop the solution methods.
>
> But one should also recognize that you don't just push a button and get the answer. In using these underlying mathematical tools, we never want to lose sight of our common sense, understanding, intuition, and judgment. The computer provides certain kinds of insights. It deals with some of the combinatorial complexities of these problems very nicely. But a model such as an Euler circuit can never capture the full essence of a decision-making problem.
>
> Typically, when we solve the mathematical problem, we see that it doesn't quite correspond to the real problem we want to solve. So we make modifications in the underlying model. It is an interactive approach, using the best of what computers and mathematics have to offer and the best of what we, as human beings, with our own decision-making capabilities, have to offer.

Thomas Magnanti
(*Courtesy of Thomas Magnanti.*)

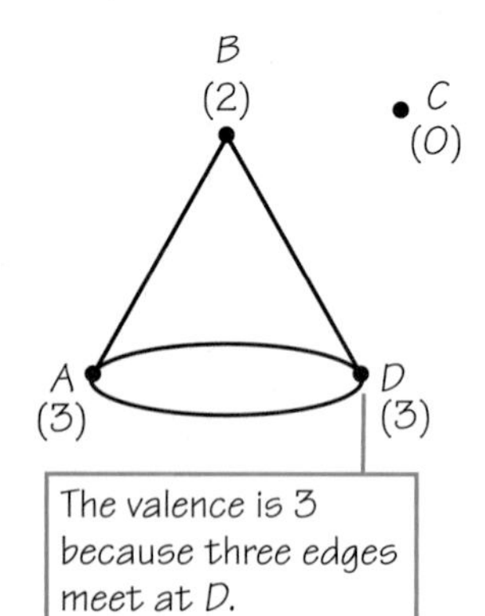

Figure 1.6 Valences of vertices.

Figure 1.6 illustrates the concept of valence, with vertices *A* and *D* having valence 3, vertex *B* having valence 2, and vertex *C* having valence 0. Isolated vertices such as vertex *C* are an annoyance in Euler circuit theory. Because they don't occur in typical applications, we henceforth assume that our graphs have no vertices of valence 0.

Figure 1.3b has four vertices of valence 2, namely, *A, C, F,* and *D*. This graph also has two vertices, *B* and *E,* of valence 4. Notice that each vertex has a valence that is an even number. We'll soon see that this is very significant.

Connected Graph DEFINITION

A graph is said to be **connected** if for every pair of its vertices, there is at least one path connecting the two vertices.

Given a graph, if we can find even one pair of vertices not connected by a path, then we say that the graph is not connected. For example, the graph in Figure 1.7 is not connected because we are unable to join *A* to *D* with a path of edges. However, the graph does consist of two "pieces" or connected components, one containing the vertices *A, B, F,* and *G,* the other containing *C, D,* and *E.* A connected graph will contain a single connected component. Notice that the parking-control graph of Figure 1.3b is connected.

We can now state Euler's theorem, his simple answer to the problem of detecting when a graph *G* has an Euler circuit.

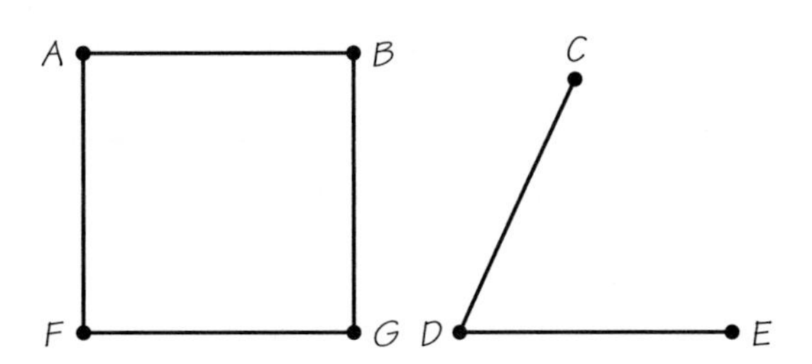

Figure 1.7 A graph that is not connected.

Euler Circuit Theorem THEOREM

1. If G is connected and has all valences even, then G has an Euler circuit.
2. Conversely, if G has an Euler circuit, then G must be connected and all its valences must be even numbers.

Because the parking-control graph of Figure 1.3b conforms to the connectedness and even-valence conditions, Euler's theorem tells us that it has an Euler circuit. We already have found an Euler circuit for the graph shown in Figure 1.4b by trial and error. For a very large graph, however, trial and error may take a long time. It is usually quicker to check whether the graph is connected and even-valent than to find out if it has an Euler circuit.

Once we know there is an Euler circuit in a certain graph, how do we find it? Many people find that, after a little practice, they can find Euler circuits by trial and error, and they don't need detailed instructions on how to proceed. At this point you should see if you can develop this skill by trying to find Euler circuits in Figure 1.8a, Figure 1.9a, and Figure 1.10. In doing your experiments, draw your graph in ink and the circuit in pencil so you can erase if necessary.

If you would like more guidance on how to find an Euler circuit without trial and error, here is a method that works: Never use an edge that is the only link between two parts of the graph that still need to be covered. Figure 1.9b illustrates this. Here we have started the circuit at A and gotten to D via B and C, and we want to know what to do next. Going to E would be a bad idea because the uncovered part of the graph would then be disconnected into left and right portions. You will never be able to get from the left part back to the right part because you have just used the last remaining link between these parts. Therefore, you should stay on the right side and finish that before using the edge from D to E. This kind of thinking needs to be applied every time you need to choose a new edge.

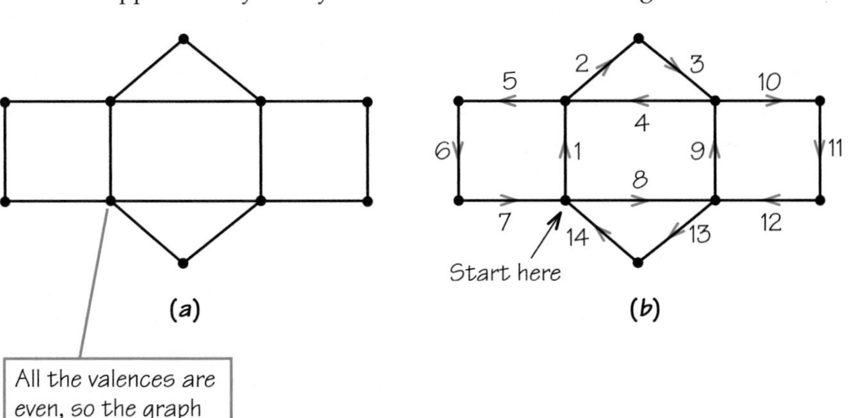

Figure 1.8 (a) A graph having (b) an Euler circuit.

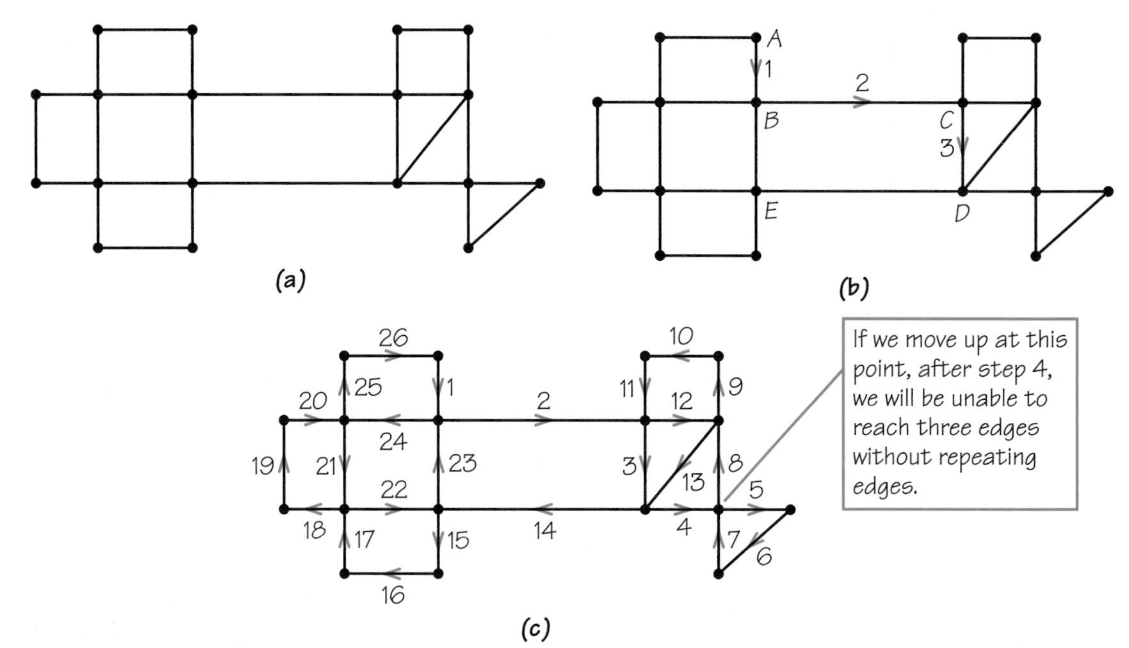

Figure 1.9 (a) A graph that has an Euler circuit. (b) A critical junction in finding an Euler circuit in this graph, starting from vertex *A*. (c) A description of a full Euler circuit for this graph.

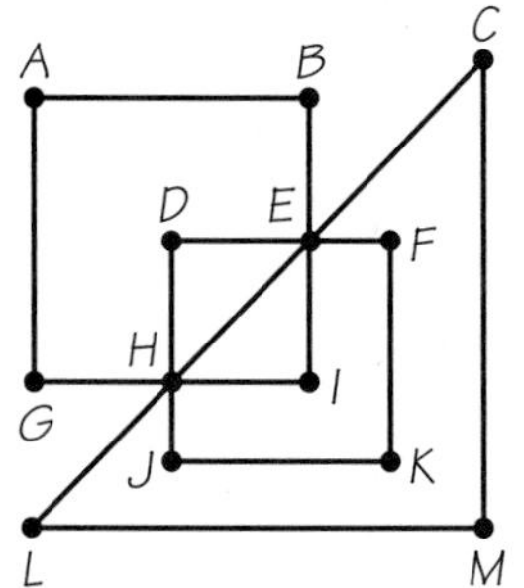

Figure 1.10 A graph with an Euler circuit.

Let's see how this works, starting at the beginning at *A*. From vertex *A* there are two possible edges, and neither of them disconnects the unused portion of the graph. Thus, we could have gone either to the left or down. Having gone down to *B*, we now have three choices, none of which disconnects the unused part of the graph. After choosing to go from *B* to *C*, we find that any of the three choices at *C* is acceptable. Can you complete the Euler circuit? Figure 1.9c shows one of many ways to do this.

The method just described leaves many edge choices up to you. When there are many acceptable edges for your next step, you can pick one at random.

EXAMPLE 2
Finding an Euler Circuit

Check the valences of the vertices and the connectivity of the graph in Figure 1.8a to verify that the graph does have an Euler circuit. Now try to find an Euler circuit for that graph. You can start at any vertex. When you are done, compare your solution with the Euler circuit given in Figure 1.8b. If your path covers each edge exactly once and returns to its original vertex (is a circuit), then it is an Euler circuit, even if it is not the same as the one we give.

Understanding Euler's Theorem

We'll start by showing that if a graph has an Euler circuit *R*, then it must have only even-valent vertices and it must be connected. Let *X* be any vertex of the graph. We will show that the edges at *X* can be paired up, and this will prove that the valence is even. Every edge at *X* is used by *R* as an outgoing edge (leaving from *X*) or an incoming edge (arriving at *X*). If the Euler circuit starts at *X*, then pair up the first edge used by *R* with the last one (when the circuit returns to *X* for the last time). In addition, each other edge at X that is used by the circuit as an incoming edge will be paired with the outgoing edge that is used next. Because all edges at X are used by the Euler circuit, none more than once, this pairs up the edges. Hence, X must be even-valent because we have "organized" the edges of R in pairs.

But what if X is not the start of the Euler circuit? Then do the pairing like this: The first incoming edge at X is paired with the outgoing one used next, the second incoming edge at X is paired with the outgoing one used next, and so on. For example, in Figure 1.11 at vertex B, we would pair up edges 2 and 3 and edges 9 and 10. At vertex C, we would pair up edges 4 and 5 and edges 8 and 9. Can you see how the pairings would work at D? How about vertex A?

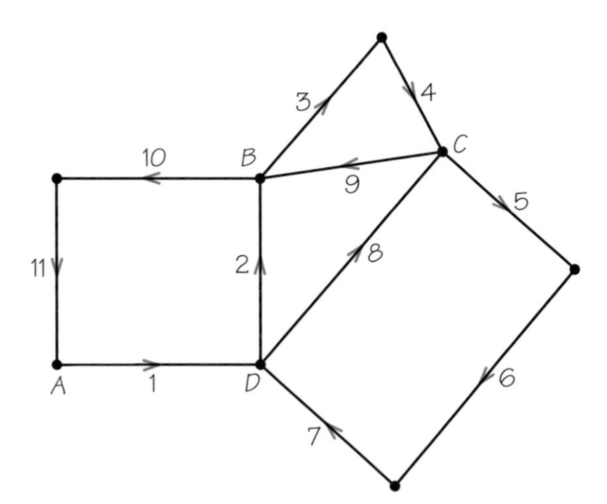

Figure 1.11 An Euler circuit starting and ending at A.

In studying this particular example, you might think it would be simpler to count the edges at a vertex to see that the valence is even. True, but our pairing method works for a graph about which we know nothing except that it has an Euler circuit.

To see that a graph with an Euler circuit is connected, note that by following the Euler circuit around we can get from any edge to any other edge (it covers them all) using a portion of the Euler circuit. Because every vertex is on an edge (there are no vertices of valence 0), we can get from any vertex to any other using a portion of the Euler circuit.

So far, this is not a complete proof of Euler's theorem. To complete the proof, we would need to prove that if a graph has all vertices even-valent and is connected, then an Euler circuit can be found for it.

1.3 Beyond Euler Circuits

Now let's see what Euler's theorem tells us about the three-block neighborhood with parking meters, represented by dots in Figure 1.12a. Figure 1.12b shows the corresponding graph. (Because we use edges to represent only sidewalks along which the officer must walk, the sidewalk with no meters is not represented by any edge in the graph.) This graph has vertices with odd valences (at vertices C and G), so Euler's theorem tells us that there is no Euler circuit for this graph.

Because we must reuse some edges in this graph to cover all edges in a circuit, for efficiency we need to keep the total length of reused edges to a minimum. This type of problem, in which we want to minimize the length of a circuit by carefully choosing which edges to retrace, is often called the **Chinese postman problem**. (Like parking-control routes, mail routes need to be efficient.) The problem was

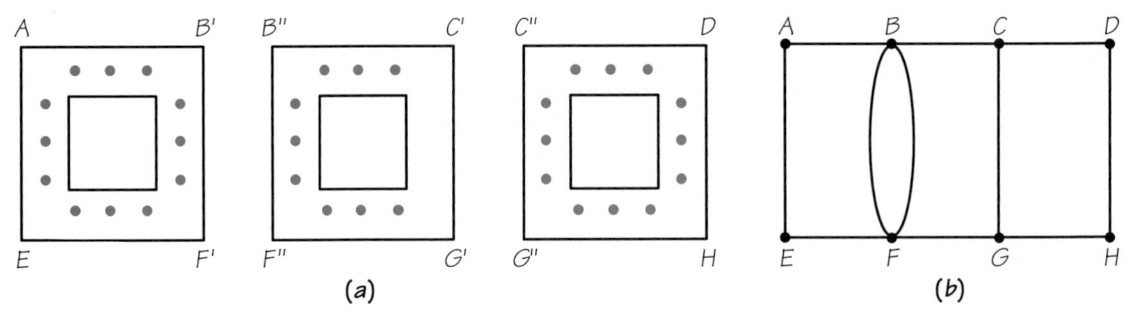

Figure 1.12 (a) A street network and (b) its graphic representation. Locations such as *B'* and *B''*, *C'* and *C''*, *F'* and *F''*, and *G'* and *G''* are merged to form the vertices *B*, *C*, *F*, and *G*. The dots shown represent parking meters.

first studied by the Chinese mathematician Meigu Guan in 1962—hence the name. The remainder of this chapter is dedicated to solving the Chinese postman problem and discussing applications beyond parking control.

Solving the Chinese Postman Problem

In a realistic Chinese postman problem, we need to consider the lengths of the sidewalks, streets, or whatever the edges represent, because we want to minimize the total length of the reused edges. However, to simplify things at the start, we can suppose that all edges represent the same length. (This is often called the *simplified* Chinese postman problem.) In this case, we need only count reused edges and need not add up their lengths. To solve the problem, we want to find a circuit that covers each edge and that has the minimal number of reuses of edges already covered.

To follow the procedure we are going to develop, look at the graph in Figure 1.13a, which is the same graph as in Figure 1.12b. This graph has no Euler circuit, but there is a circuit that has only one reuse of an edge (*CG*), namely, *ABCDHGCGFBFEA*. Let's draw this circuit so that when edge *CG* is about to be reused, we install a new, extra, blue edge in the graph for the circuit to use. By duplicating edge *CG*, we can avoid reusing the edge. To duplicate an edge, we must add an edge that joins the two vertices that are already joined by the edge we want to duplicate. (It makes no sense to join vertices that are not already

Figure 1.13 Making a circuit by reusing an edge.

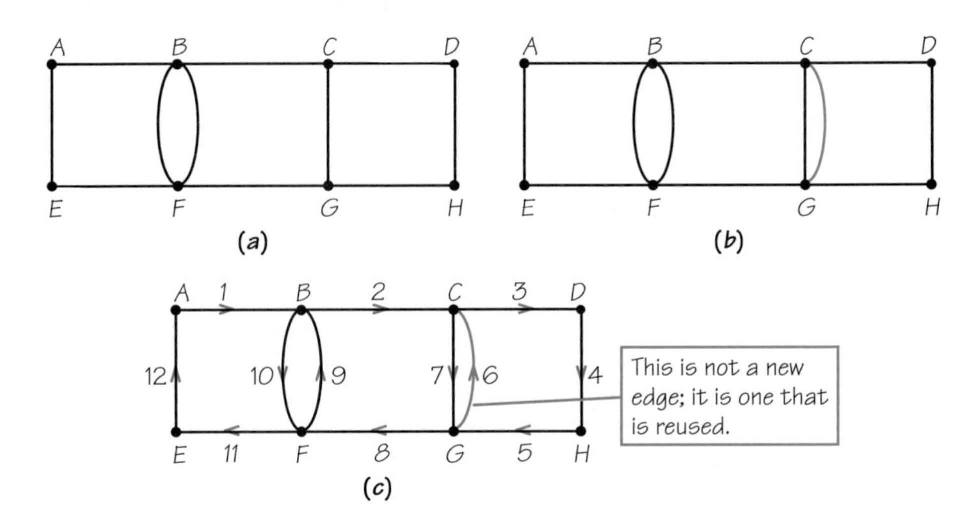

connected by an edge, because such edges would not represent sidewalk sections with meters; see Figure 1.15.)

We have now created the graph of Figure 1.13b. In the graphs we draw, the edges that are added will be shown in color to distinguish them from the original edges, which are shown in black. (In the graphs that you draw, you may want to use a similar scheme to help you remember which edges are the originals and which are duplicated.) In the graph of Figure 1.13b, the original circuit can be traced as an Euler circuit, using the new edge when needed. The circuit is shown in Figure 1.13c. Our theory will be based on using this idea in reverse, as follows:

1. Take the given graph and add edges by duplicating existing edges, until you arrive at a graph that is connected and even-valent. Note that after a graph is *eulerized,* the new graph produced will have an Euler circuit.
2. Find an Euler circuit on the eulerized graph.
3. "Squeeze" this Euler circuit from the eulerized graph onto the original graph by reusing an edge of the original graph each time the circuit on the eulerized graph uses an added edge.

Eulerizing a Graph DEFINITION

Adding edges that duplicate existing edges to a connected graph to make all valences even is called **eulerizing** the graph.

EXAMPLE 3
Eulerizing a Graph

Suppose we want to eulerize the graph of Figure 1.14a. When we eulerize a graph, we first locate the vertices with odd valence. The graph in Figure 1.14a has two, *B* and *C*. Next, we add one end of an edge at each such vertex, matching up the new edge with an existing edge in the original graph. Figure 1.14b shows one way to eulerize the graph. Note that *B* and *C* have even valence in the second graph. After eulerization, each vertex

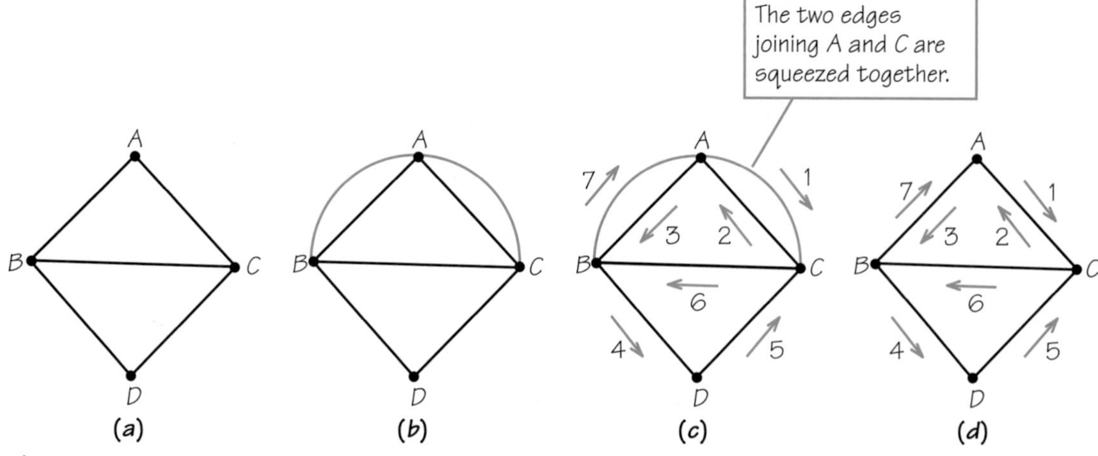

Figure 1.14 Eulerizing a graph.

has even valence. To see an Euler circuit on the eulerized graph in Figure 1.14c, simply follow the edges in numerical order and in the direction of the arrows, beginning and ending at vertex *A*. The final step, shown in Figure 1.14d, is to "squeeze" our Euler circuit onto the original graph. There are two reuses of previously covered edges. Notice that each reuse of an edge corresponds to an added edge.

In the previous example, we noticed that we could count how many reuses we needed by counting added edges. This is generally true in this type of problem: *If you add the new edges correctly, the number of reuses of edges equals the number of edges added during eulerization.*

Adding new edges correctly means adding only edges that are duplicates of existing edges. Doing this makes the rule just stated in italics always true, and so it is easy to count the needed reuses.

To see why we add only duplicate edges, examine Figure 1.15a. We need to alter the valences of vertices *X* and *Y* by adding edges so that they become even-valent. Adding one long edge from *X* to *Y* (Figure 1.15b) might seem like an attractive idea, but adding this edge is equivalent to asking a snowplow, say, to get from *X* to *Y* without moving along existing streets. At times it is necessary to traverse sections of the graph that have been previously traversed. This is the significance of the duplicated edges. Here the structure of the graph forces us to repeat some edges. We cannot get away with fewer than three repeats: the three edges *XU*, *UV*, and *VY* (Figure 1.15c). The duplicated edges are shown in color.

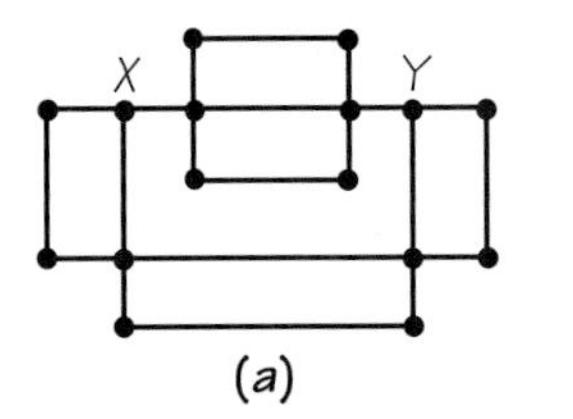

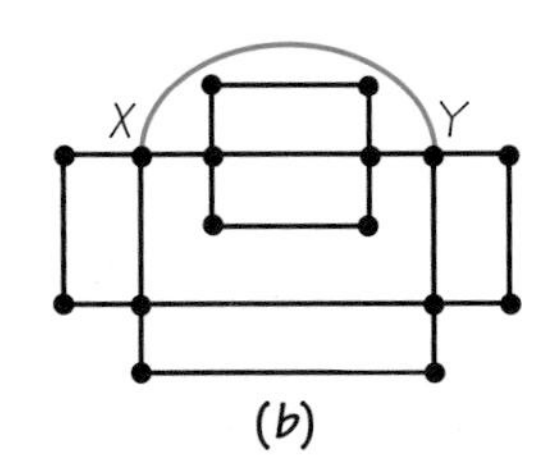

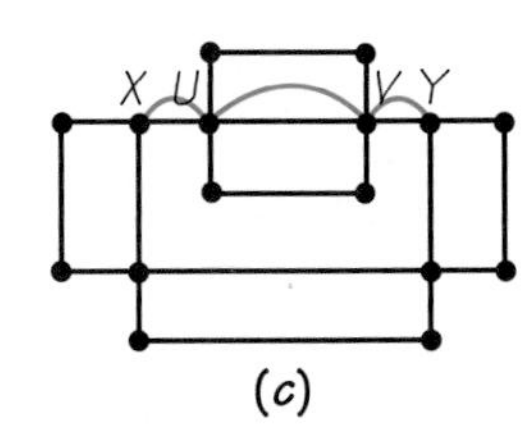

Figure 1.15 Eulerizing when the odd-valent vertices are more than one edge apart.

Now that we have learned to eulerize, the next step is to try to get a best eulerization we can—one with the fewest added edges. It turns out that there are many ways to eulerize a graph. It is even possible that the smallest number of added edges can be achieved with two different eulerizations. This is the reason we use the phrase "a best eulerization" rather than "the best eulerization." Remember, we want a best eulerization because this enables us to find the circuit for the original graph that has the minimum number of reuses of edges, which in typical applications means saving time or money by avoiding retraversing edges where the productive work has already been accomplished (avoiding deadheading).

EXAMPLE 4
A Better Eulerization

In Figure 1.16a, we begin with the same graph as in Figure 1.14, but we eulerize it in a different way—by adding only one edge (see Figure 1.16b). Figure 1.16c shows an Euler circuit on the eulerized graph, and in Figure 1.14d we see how it is squeezed onto the original graph. There is only one reuse of an edge, because we added one edge during eulerization.

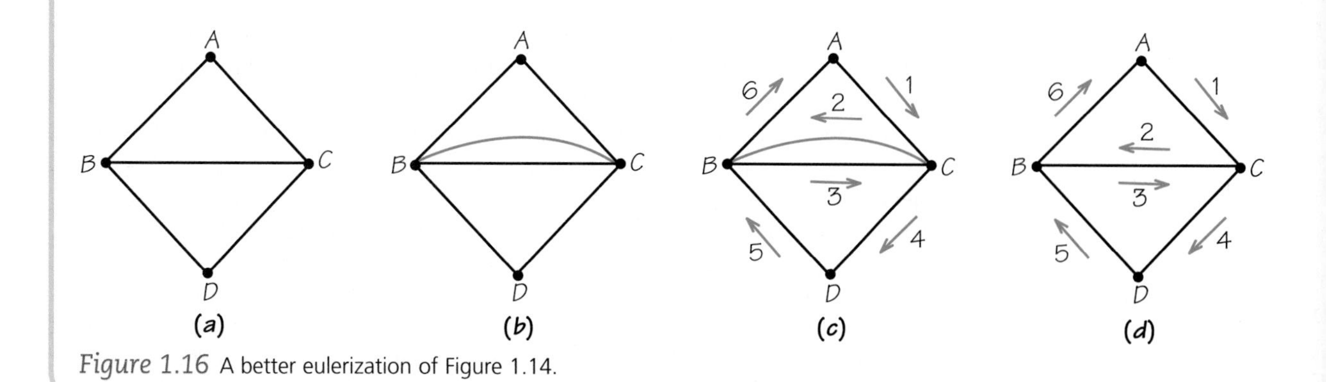

Figure 1.16 A better eulerization of Figure 1.14.

The solution in Figure 1.16 is better than the solution in Figure 1.14 because one reuse is better than two. These examples suggest the following addition to our solution procedure: Try to find an eulerization with the smallest number of added edges. This extra requirement makes the problem both more interesting and more difficult. For large graphs, a best eulerization may not be obvious. We can try out a few and pick the best among the ones we find, but there may be an even better one that our haphazard search does not turn up.

A systematic procedure for finding a best eulerization does exist, but the process is complicated. There is an especially easy technique for eulerizing the following special category of networks often found in our neighborhoods.

Rectangular Network — DEFINITION

If a street network is composed of a series of rectangular blocks that form a large rectangle a certain number of blocks high and a certain number of blocks wide, the network is called **rectangular**.

Examples of rectangular street networks (a 3-by-3, a 3-by-4, and a 4-by-4) are shown in Figure 1.17. The graph on the right in each pair shows a best eulerization for the rectangular street network on the left. There appear to be three different eulerization patterns, depending upon whether the rectangle height and width in

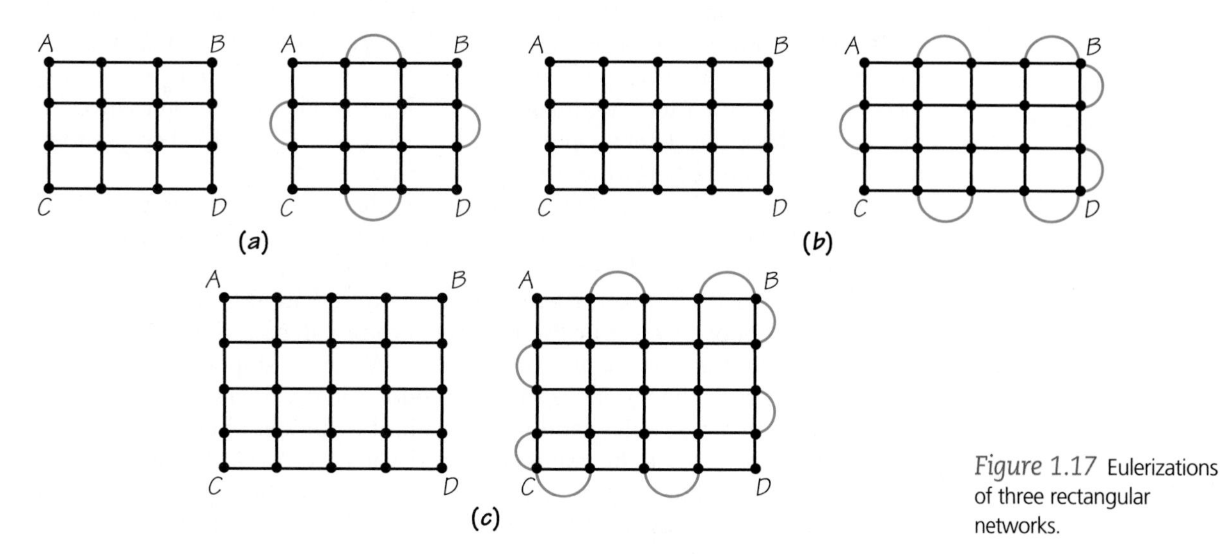

Figure 1.17 Eulerizations of three rectangular networks.

the original graph are odd or even numbers. In Figure 1.17a, both lengths are 3, both odd; in Figure 1.17b, one length is odd (3) and one is even (4); in Figure 1.17c, both lengths are 4, an even number.

Although the patterns appear different, one technique can be used to create all of them. This technique can be thought of as involving an "edge walker" who walks around the outer boundary of the large rectangle in some direction, say, clockwise. He starts at any corner, say, the upper-left corner. As he goes around, he adds edges by the following rules. When he comes to an odd-valent vertex, he links it to the next vertex with an added edge. This next vertex now becomes either even or odd. If it became even, he skips it and continues around, looking for an odd vertex. If it became odd (this could happen only at a corner of the big rectangle), the edge walker links it to the next vertex and then checks this vertex to see whether it is even or odd. Each of the three parts of Figure 1.17 has been eulerized by this method.

In a street network that is not rectangular, the eulerization process is started by locating all the vertices with odd valence and then pairing these vertices with each other and finding the length of the shortest path between each pair. We look for the shortest paths because each edge on the connecting paths will be duplicated. The idea is to choose the pairings cleverly so that the sum of the lengths of those paths is the smallest it can be. With a little practice, most people can find a best or nearly best eulerization using this idea, together with trial and error and some ingenuity.

Finding Good Eulerizations

Suppose we want a perfect procedure for eulerizing a graph. What theoretical ideas and methods could we use to build such a tool?

One building block we could use is a method for finding the shortest path between two given vertices of a graph. For example, let us focus on vertices *X* and *Y* in Figure 1.18a. Both have odd valence. We can connect them with a pattern of duplicate edges, as in Figure 1.18b. The cost of this is the length of the path we duplicated from *X* to *Y*. A shorter path from *X* to *Y*, such as the one shown in Figure 1.18c, would be better. Fortunately, the *shortest-path problem* has been well studied, and we have many good procedures for solving it exactly, even in large, complex graphs.

But there is more to eulerizing the graph in Figure 1.18a than dealing with *X* and *Y*. Notice that we have odd valences at *Z* and *W*. Should we connect *X* and *Y* with a path, and then connect *Z* and *W*, as in Figure 1.18d? Or should we connect *X* to *Z* and *Y* to *W*, as in Figure 1.18e? Another alternative is to use connections *X* to *W* and *Y* to *Z*, as in Figure 1.18f. It turns out that the alternatives in both Figures 1.18e and 1.18f are preferable to the one in Figure 1.18d because they involve seven added edges, whereas Figure 1.18d uses nine.

We know there is a simple way to test whether a connected graph has an Euler circuit: Check to see if the graph is even-valent. Is there a very easy way to compute the number of edges in a best eulerization of a graph? Unfortunately, the answer is "no." However, there is a simple observation that often saves a lot of work. Suppose we count the number of odd-valent vertices in a graph. This number must always be an even number. When we duplicate an existing edge, we can never change more than two odd-valent vertices to even-valent vertices. Thus, in a best eulerization of a graph, the number of edges that must be duplicated is at least the number of odd-valent vertices divided by two. If, for example, we have a graph with 10 odd-valent vertices and we find an eulerization with five added edges, there may be other eulerizations which also

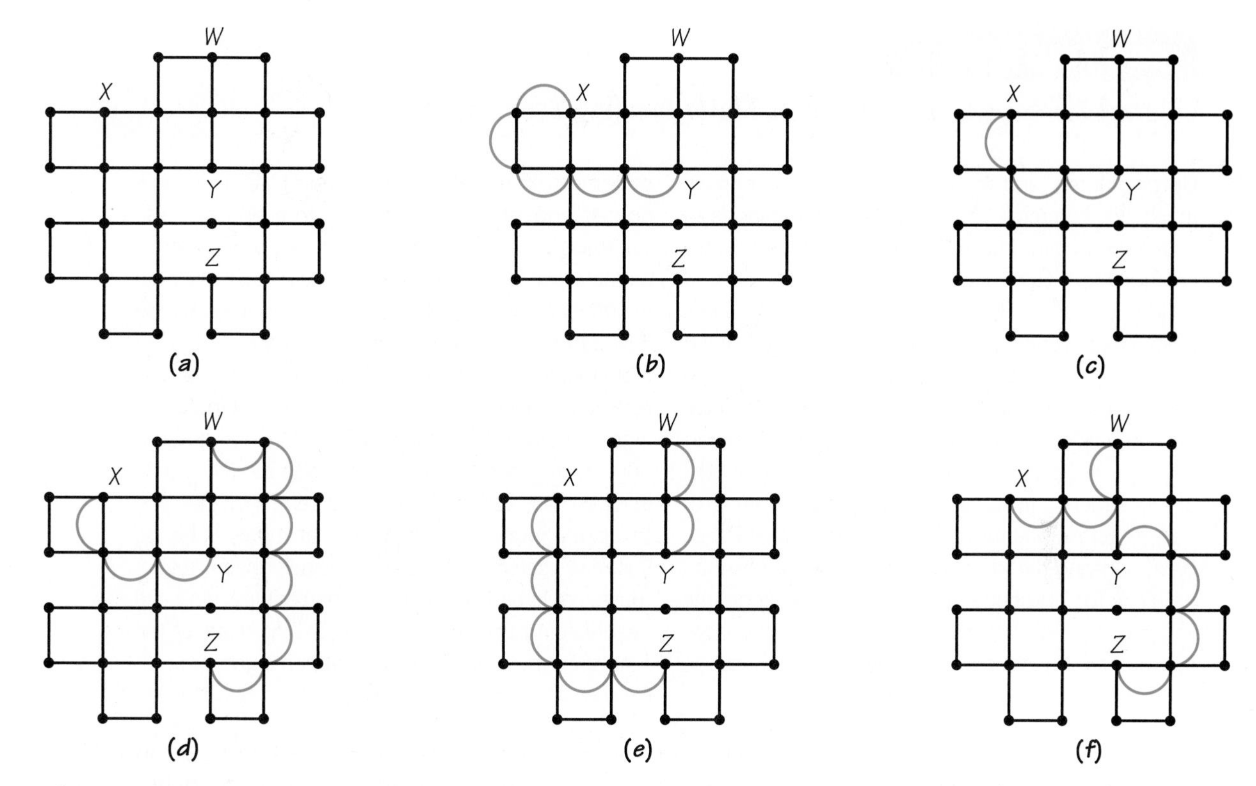

Figure 1.18 Choosing among eulerizations.

have five duplicated edges, but there can be no eulerization with fewer than five duplicated edges.

Remember that when an unweighted graph is eulerized in an optimal way, then the total cost of traversing each edge at least once can be found by adding the total number of edges in the graph to the number of edges that are reused (duplicated). Small problems involving eulerization can be carried out by trial-and-error methods. Unfortunately, although there is an algorithm that can be applied to find the best eulerization for large problems, the details of this algorithm are quite complex. However, the procedure works quickly not only for graphs without weights but also for graphs with weights on the edges.

1.4 Urban Graph Traversal Problems

Euler circuits and eulerizing have many more practical applications than just checking parking meters. Almost anytime services must be delivered along streets or roads, our theory can make the job more efficient. Examples include collecting garbage, salting icy roads, plowing snow, inspecting railroad tracks for flaws, collecting debris or leaves from urban curbs, mowing grass along superhighways, or reading electric meters at private houses (see Spotlight 1.3).

Each of these problems has its own special requirements that may call for modifications in the theory. For example, in the case of garbage collection, the edges of our graph will represent streets, not sidewalks. If some of the streets are one-way, we need to put arrows on the corresponding edges, resulting in a directed graph, or **digraph**. The circuits we seek will have to obey these arrows. In the case of salt spreaders and snowplows, each lane of a street needs to be

SPOTLIGHT 1.3

Israel Electric Company Reduces Meter-Reading Task

Electric meters and icy roads don't seem to have much in common, but private companies and governments can draw on a common source—mathematics—to save customers money and drivers from accidents. Thus, in Israel there was a branch of the major electric company that needed to make the job of reading its electric meters more efficient, while in Minnesota and Ohio, the counties and cities must plow and/or de-ice winter roads. Both problems start in the same place: constructing a graph theory model of the streets and roads the meter readers need to traverse and winter maintenance vehicles must travel. Next, applying the ideas about the Chinese Postman problem for efficiently providing services along the edges of a graph at least once can lead to time saving routes for meter reading or de-icing. Because the customers of the electric company change as new houses and businesses need service and because the roads needing plowing vary with storm patterns, digitized versions of street networks are usually now maintained online; solutions to either the Chinese Postman problem or variants of it can be solved in real time. In Minnesota and Ohio, routes are typically reevaluated every winter season. In one Ohio county where nearly 200 segments of road involving about 300 miles need plowing, the number of trucks involved was reduced by 30 percent and the time to do the work was cut 40 percent when Chinese Postman ideas were used. In Israel, meter reader routes could be traversed in 40 percent less time and deadheading time was reduced to 1 percent of the time spent by using Chinese Postman algorithms.

modeled as a directed edge, as shown in Figure 1.19. Note that the arrows on the map and digraph are not in color because these arrows denote restrictions in traversal possibilities, not parts of circuits.

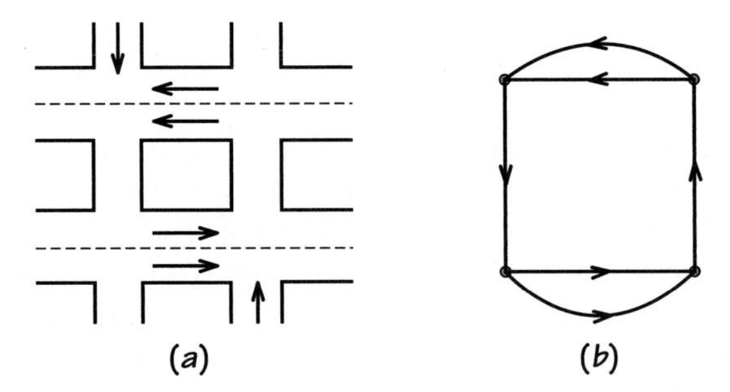

Figure 1.19 (a) Salt-spreading route, where each east–west street has two traffic lanes in the same direction, and (b) an appropriate digraph model.

Like salt spreaders, street-sweeping trucks can travel in only one lane at a time and must obey the direction of traffic. Street sweepers, however, have an additional complication: parked cars. It is very difficult to clean the street if cars are parked along the curb. Yet for overall efficiency, those who are responsible for routing street sweepers want to interfere with parking as little as possible. The common solution is to post signs specifying times when parking is prohibited. Because the parking-time factor is a constraint on street sweeping, it is important to find not only an Euler circuit, or a circuit with very few duplications, but also a circuit that visits streets when they are free of cars. Once again, mathematicians have developed techniques to handle this constraint.

Finally, because towns and cities of any size will have more than one street sweeper, parking officer, or garbage truck, a single best route may not suffice. Instead, it becomes necessary to divide the territory into multiple routes. The general goal is to find optimal solutions while taking into account traffic direction, number of lanes, parking-time restrictions, and divided routes (see Figure 1.20).

Management science makes all this possible. For example, a pilot study done in the 1970s in New York City showed that applying these techniques to street

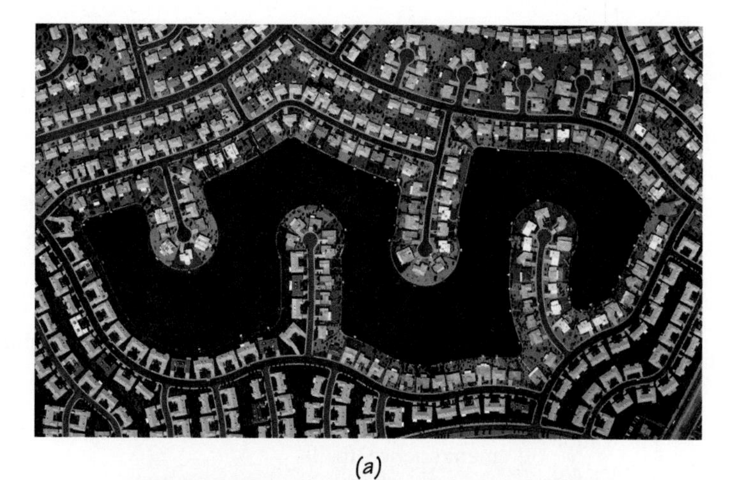

(a)

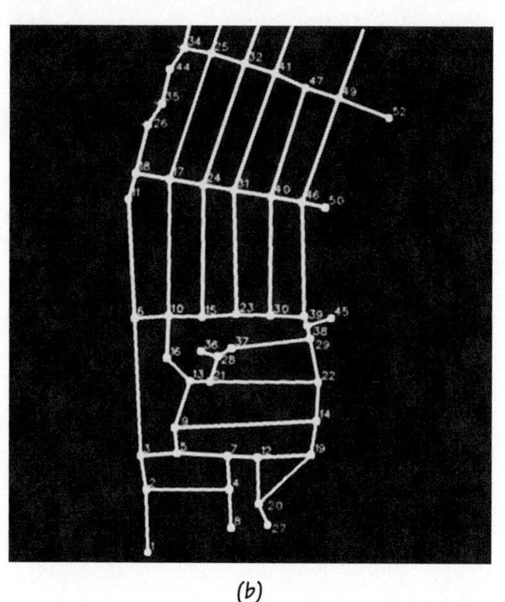

(b)

Figure 1.20
(a) Residential neighborhoods, whether they be in cities or the suburbs, require many services such as mail delivery, garbage collection, street sweeping, meter reading, or sewage systems. The mathematical techniques of operations research make it possible to provide these services as cheaply as possible. When optimal solutions to providing such services can be found, everyone is a winner. *(Brand X Pictures/Picturequest.)*
(b) Computers can be used to extract the essential information needed to solve routing problems from photographs.

sweepers in just one district could save about $30,000 per year. With 57 sanitation districts in New York, this would amount to a savings of more than $1.5 million in a single year. This translates to about $5.5 million in 2010 dollars. In addition, the same principles could be extended to garbage collection, parking control, and other services carried out on street networks.

This plan was not adopted when first proposed. Because city services take place in a political context, several other factors come into play. For example, union leaders try to protect the jobs of city workers, bureaucrats might try to keep their departmental budgets high, and elected politicians rarely want to be accused of cutting the jobs of their constituents. Thus, political obstacles can overrule management science. As mentioned in Spotlight 1.2, such human factors often arise when applying management science. Perhaps a more acceptable street-sweeping plan would have been devised for New York City if more attention had been paid to the human factors earlier.

Despite the complications of real-world problems, management science principles provide ways to understand these problems by using graphs as models. We can reason about the graphs and then return to the real-world problem with a workable solution. The results we get can have a lasting effect on the efficiency and economic well-being of any organization or community.

REVIEW VOCABULARY

Chinese postman problem The problem of finding a circuit on a graph that covers every edge of the graph at least once and that has the shortest possible length. (p. 13)

Circuit A path that starts and ends at the same vertex. (p. 6)

Connected graph A graph is connected if it is possible to reach any vertex from any specified starting vertex by traversing edges. (p. 10)

Digraph A graph in which each edge has an arrow indicating the direction of the edge. Such directed edges are appropriate when the relationship is "one-sided" rather than symmetric (for instance, one-way streets as opposed to regular streets). (p. 19)

Edge A link joining two vertices in a graph. (p. 6)

Euler circuit A circuit that traverses each edge of a graph exactly once. (p. 8)

Eulerizing Adding new edges which duplicate existing edges to a connected graph so as to make a graph that possesses an Euler circuit. (p. 15)

Graph A mathematical structure in which points (called vertices) are used to represent things of interest and in which links (called edges) are used to connect vertices, denoting that the connected vertices have a certain relationship. (p. 6)

Management science A discipline in which mathematical methods are applied to management problems in pursuit of optimal solutions that cannot readily be obtained by common sense. (p. 3)

Operations research (OR) Another name for management science. (p. 3)

Optimal solution When a problem has various solutions that can be ranked in preference order (perhaps according to some numerical measure of "goodness"), the optimal solution is the best-ranking solution. (p. 5)

Path A connected sequence of edges in a graph. (p. 6)

Valence (of a vertex) The number of edges touching that vertex. (p. 9)

Vertex A point in a graph where one or more edges end. (p. 6)

SKILLS CHECK

1. The accompanying graph has

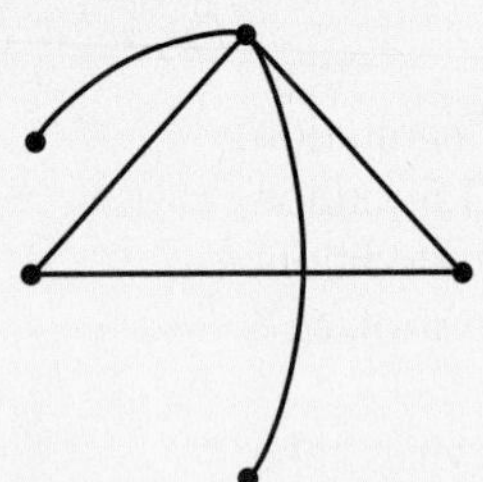

(a) four vertices and six edges.
(b) four vertices and four edges.
(c) five vertices and five edges.

2. The number of vertices in the following graph is _____, while the number of edges in this graph is _________.

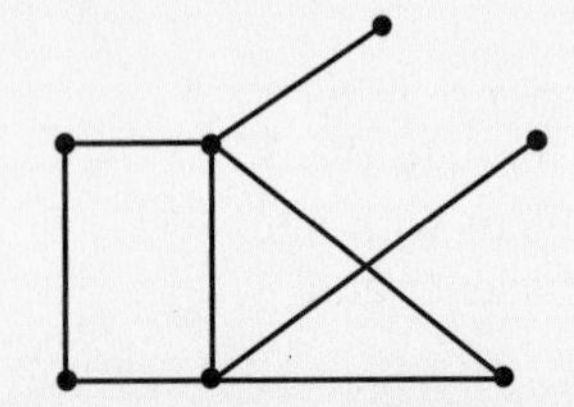

3. The accompanying graph is:

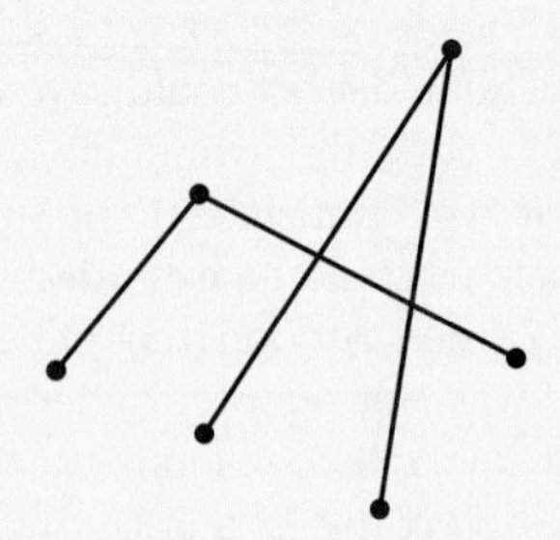

(a) Even-valent
(b) Not connected
(c) Has 6 components

4. The graph shown below is not connected because it consists of ________ parts.

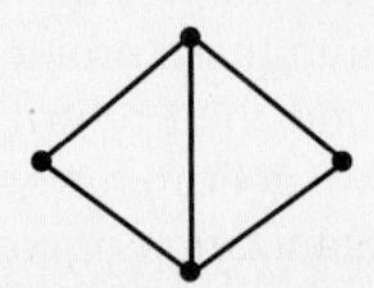
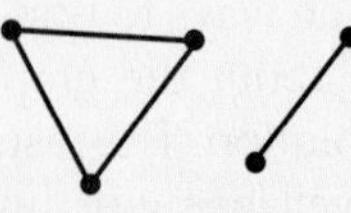
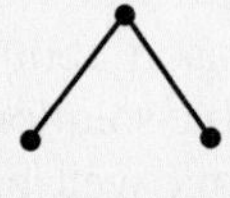

5. What is the valence of vertex *A* in the graph below?

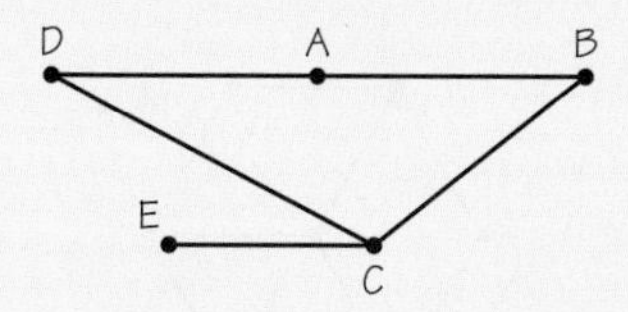

(a) 2 **(b)** 1 **(c)** 3

6. A graph has 30 edges and all the vertices of the graph have valence 3. The number of vertices of the graph must be _______ .

7. The valences of the vertices in the accompanying graph, listed in decreasing order, are

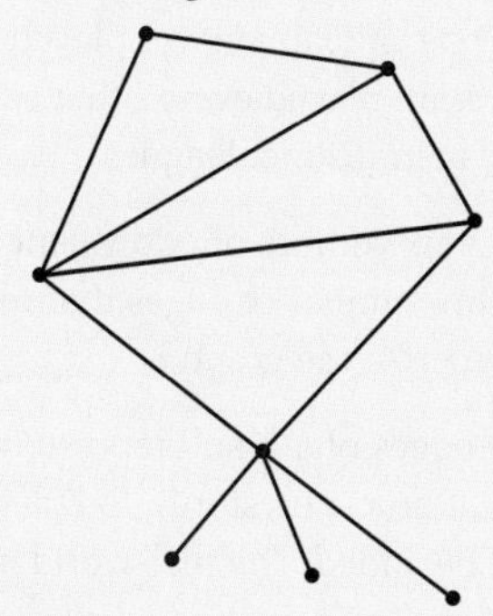

(a) 5, 4, 3, 3, 2, 1, 1, 1. **(b)** 1, 3, 4, 4, 5, 5.
(c) 5, 5, 4, 3, 3, 1

8. The alphabetically ordered list of even-valent vertices of the graph below is ______, ______ .

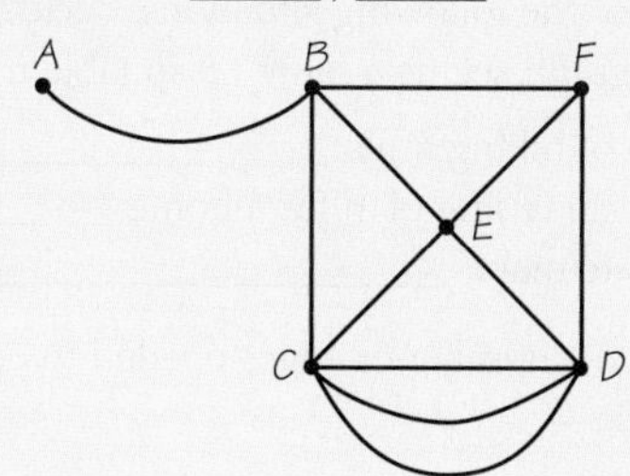

9. Which of the statements below is false for the accompanying graph:

(a) It is connected
(b) It is not a graph because it includes curved edges
(c) It is even-valent

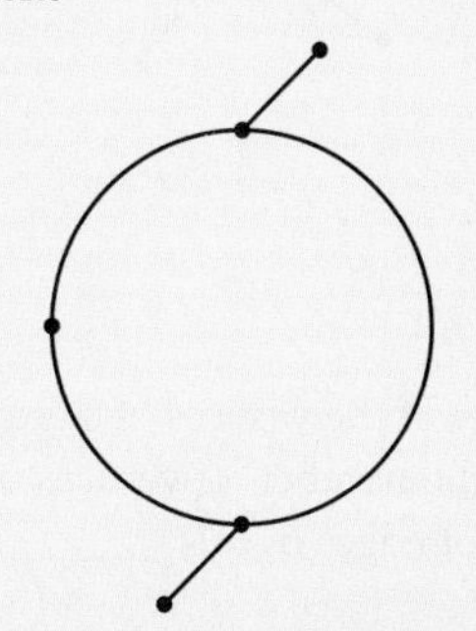

10. The accompanying graph shown has _______ edges and _______ vertices.

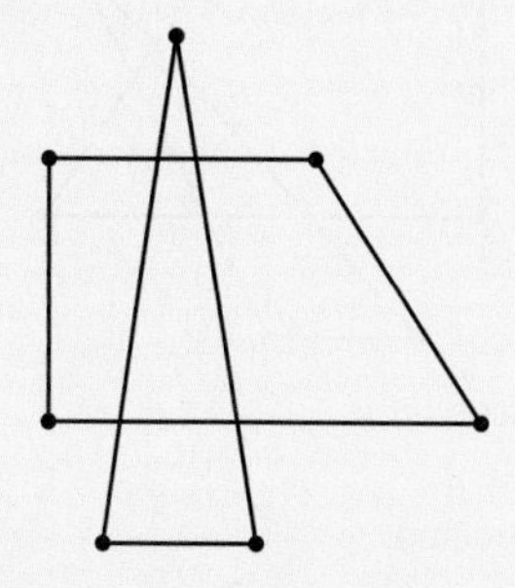

11. Which of the following statements is true about a *path*?

(a) A path always forms a circuit.
(b) A path is always connected.
(c) A path can visit any vertex only once.

12. A graph *G* has 10 edges, and all its vertices have the same valence. The possible valences of the vertices of *G* are ______, ______, ______, ______, ______ .

13. It is not possible for a graph to have five vertices of valence 3 and six vertices of valence 4 because

(a) there are no graphs with exactly 11 vertices.
(b) a graph cannot have an even number of 4-valent vertices.
(c) a graph cannot have an odd number of odd-valent vertices.

14. If a graph consists of four vertices and every pair of vertices is connected by a single edge, the number of edges in the graph is exactly _______ .

15. For which of the situations below is it most desirable to find an Euler circuit or an efficient eulerization of the graph?

(a) Sweeping the sidewalks of a small town
(b) Planning a new highway
(c) Planning a parade route in Muncie, Indiana

16. For the accompanying graph, the vertex that has the largest valence is _______ and the number of edges in the graph is _______ .

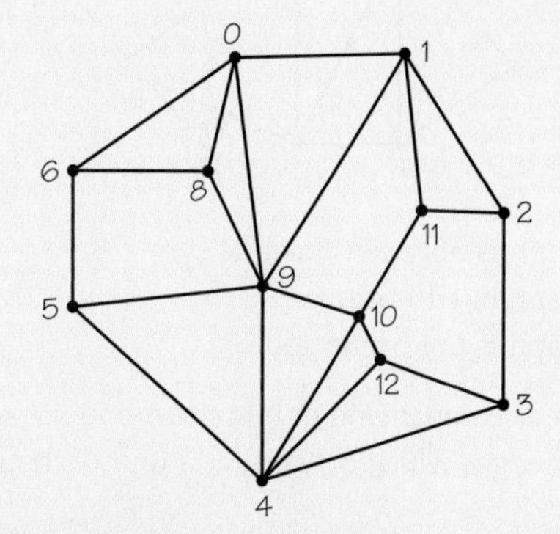

17. The following graph has no Eulerian circuit because

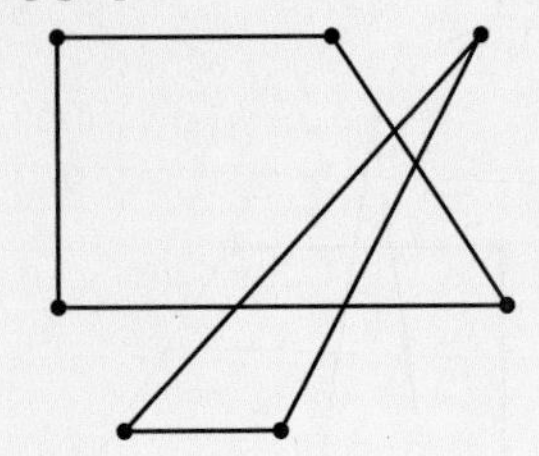

(a) it has 7 vertices.
(b) it is even-valent.
(c) it is not connected.

18. If a graph is connected and has seven vertices, the graph must have at least ________ edges.

19. Consider the path represented by the sequence of numbered edges on the graph below. Which statement is correct?

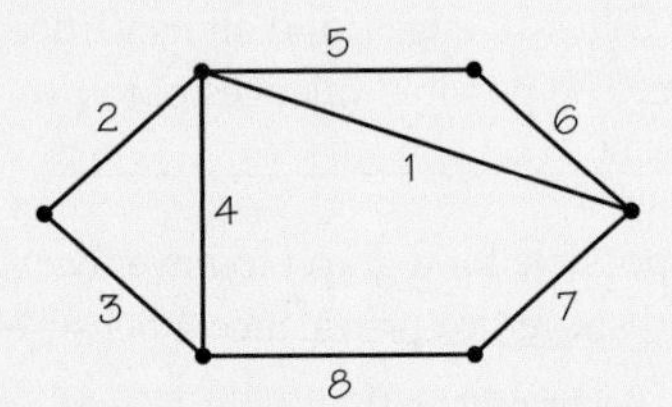

(a) The sequence of numbered edges forms an Euler circuit.
(b) The sequence of numbered edges traverses each edge exactly once but is not an Euler circuit.
(c) The sequence of numbered edges forms a circuit but not an Euler circuit.

20. The minimum number of edges which must be duplicated to create a best possible eulerization of the following graph is ______ .

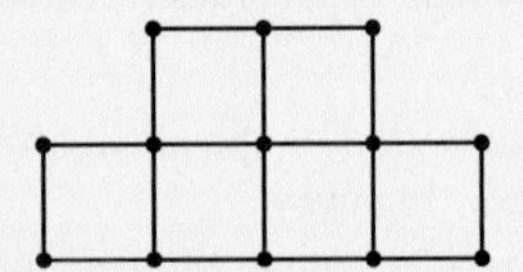

21. The accompanying graph has no Eulerian circuit because

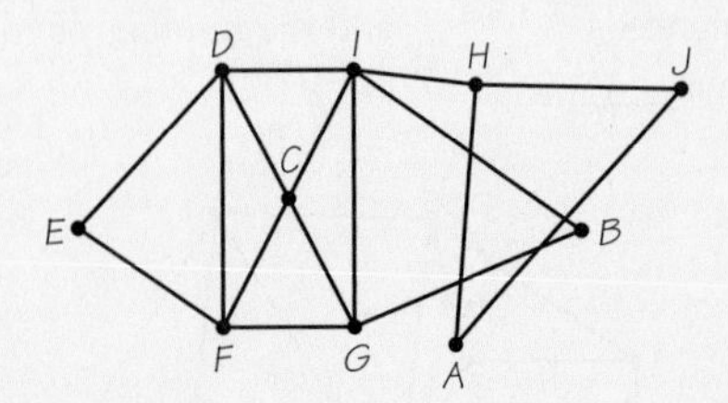

(a) the graph is not even-valent.
(b) the graph has 10 vertices.
(c) the graph is not connected.

22. For the accompanying graph, the minimum *total* number of edges which constitutes a tour of the graph, starting and ending at the same vertex, and which visits each edge at least once, is ______ .

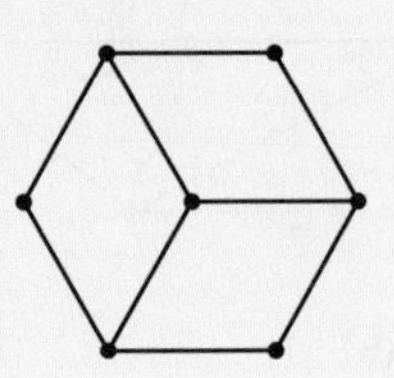

23. Suppose each vertex of a graph represents a baseball team and each edge represents a game played by two baseball teams. If the resulting graph is not connected, which of the following statements must be true?

(a) At least one pair of teams never played a game.
(b) At least one team played every other team.
(c) The teams play in distinct leagues.

24. If a graph has six vertices of odd valence, the absolute minimum number of edges that must be added (duplicated) to eulerize the graph is ______ .

25. Suppose the edges of a graph represent streets that must be plowed after a snowstorm. To eulerize the graph, four edges must be added. The real-world interpretation of this is that

(a) four streets will not be plowed.
(b) four streets will be traversed twice.
(c) four new streets would be built.

26. For each of the following situations, decide whether a graph or a digraph seems a more reasonable model:

(a) A system of hiking trails: ________________ .
(b) An electrical wiring plan for a home: __________ .
(c) A bus route map: ____________________ .

27. The smallest number of edges needed to eulerize the graph below is

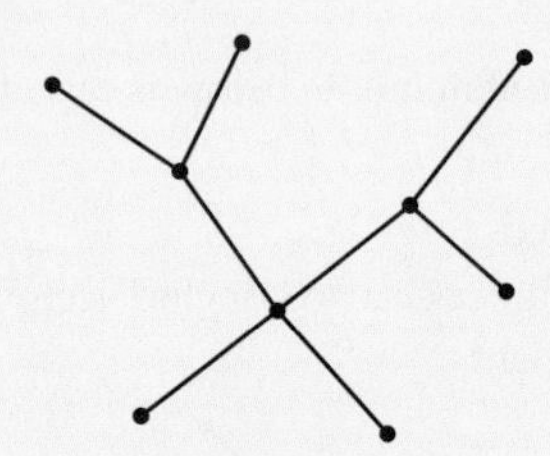

(a) 4.
(b) 6.
(c) 8.

28. If the valences of the vertices of a graph G are: 5, 4, 4, 4, 3, 2, 2, and 2, the number of vertices of G is ______ and the number of edges of G is ______ .

29. The smallest number of edges that must be added to the accompanying graph below for it to have an Euler circuit is

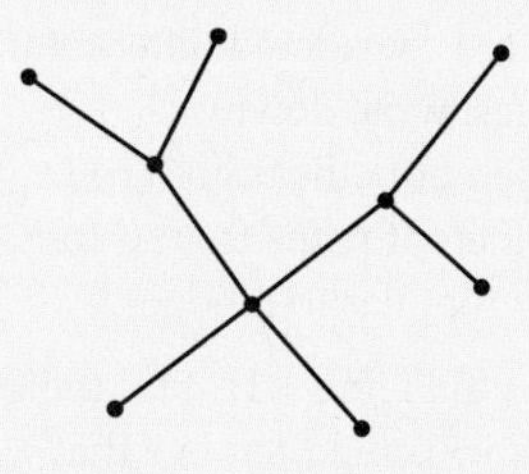

(a) 4. **(b)** 6. **(c)** 8.

30. The number of edges in a Chinese postman tour (i.e. a tour with a minimum number of edges that starts and ends at the same vertex and visits each edge at least once) for the accompanying graph is __________ .

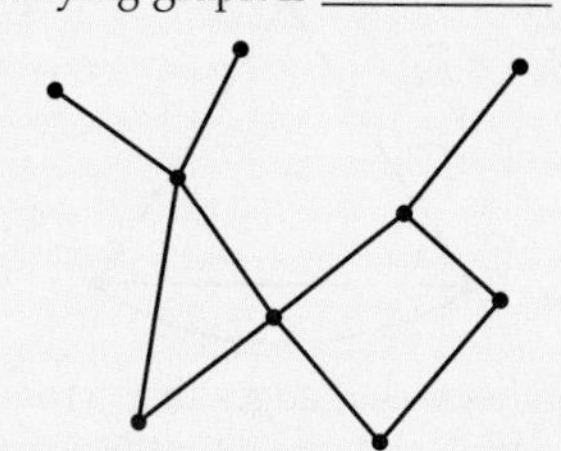

CHAPTER 1 EXERCISES

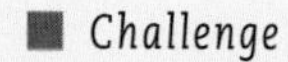 *Challenge* 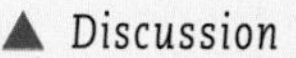 *Discussion*

1.1 Euler Circuits

1.2 Finding Euler Circuits

1. (a) Locate within Figure 1.3(a) a section of the street network shown that indicates how the graph below could be interpreted as meters that require inspection in this urban street layout.

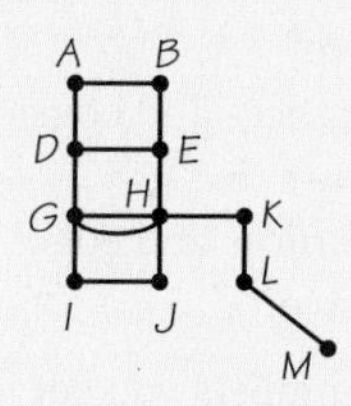

(b) Can you find a tour of the edges in the graph shown above that starts at *A* and allows for inspection of the parking meters with as few deadheaded edges as possible?

2. (a) Determine the number of vertices and edges in the accompanying graph *G*.

(b) What is the smallest number of edges that must be added to graph *G* so that the result is a connected graph?

(c) What are the valences of the vertices in *G*?

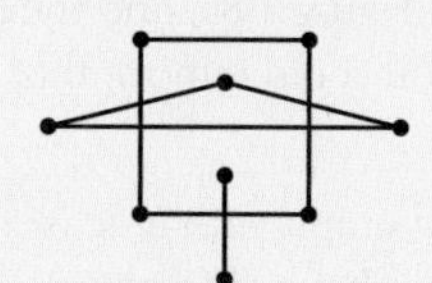

3. In the accompanying graph, the vertices represent houses and two vertices are joined by an edge if it is possible to drive between the two houses in under 10 minutes.

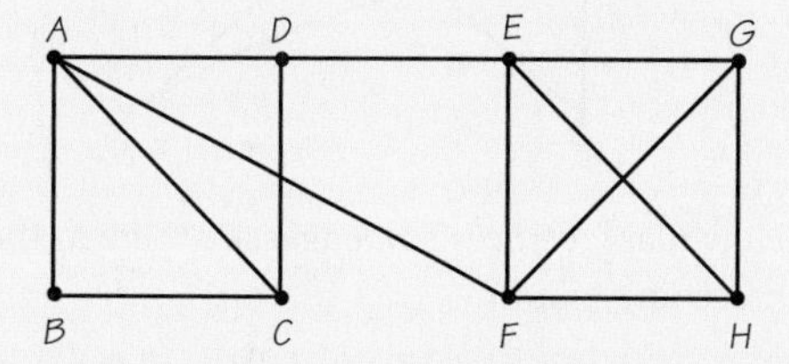

(a) How many vertices does the graph have?

(b) How many edges does the graph have?

(c) What are the valences of the vertices in this graph?

(d) Based on the information given by the accompanying graph, for which houses, if any, is it possible to drive to all the other houses in less than 20 minutes?

(e) Based on the accompanying graph, from house *B* which houses require a trip of longer than 20 minutes?

4. The graph below shows the stores and roads connecting them in a small shopping mall.

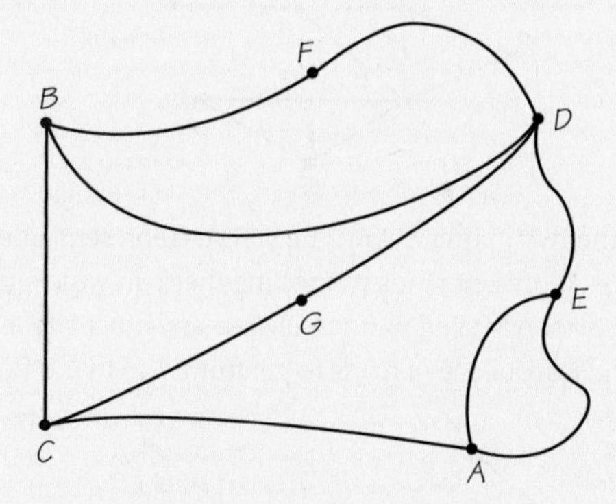

(a) How many stores does the mall have?

(b) How many roads connect up the stores in the mall?

(c) Write down a path from *C* to *F*.

(d) Write down a path from *E* to *B*.

5. **(a)** Is the figure below a graph? Explain your answer.

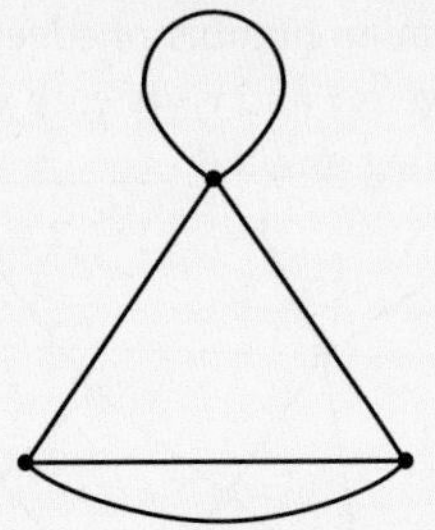

(b) The graph below has edges that "cross" at points that are not vertices of the graph. Which edges are these?

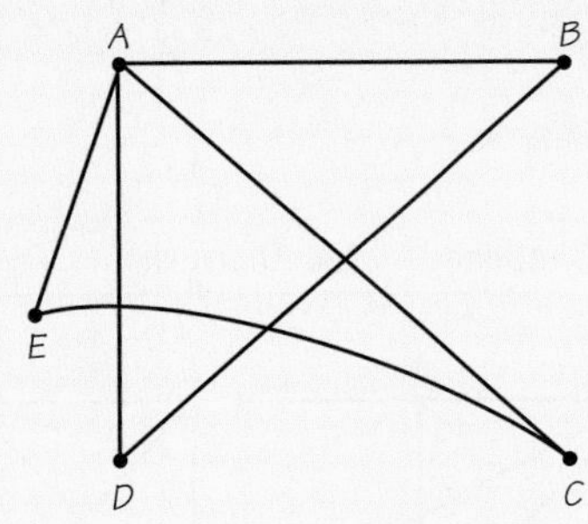

(c) How many vertices and edges are there in the preceding graph?

6. **(a)** Redraw the graph in Figure 1.2a to obtain a graph that has the same information but where the edges only meet other edges at vertices.

(b) List all the routes that start on the U.S. side of the Atlantic Ocean and cross the ocean once and immediately.

7. In the graph below, the vertices represent cities and the edges represent roads connecting them. What are the valences of the vertices in this graph? (Keep in mind that *E* is part of the graph.) What might the valence of city *E* be showing about the geography?

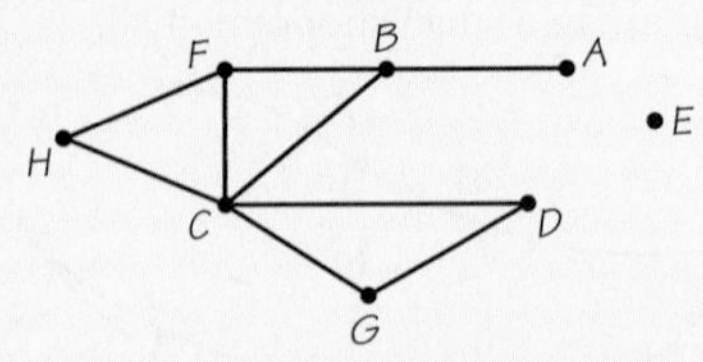

8. In the two graphs below, the vertices represent cities and the edges represent roads connecting them. In which graphs could a person located in city *A* choose any other city and then find a sequence of roads to get from *A* to that other city?

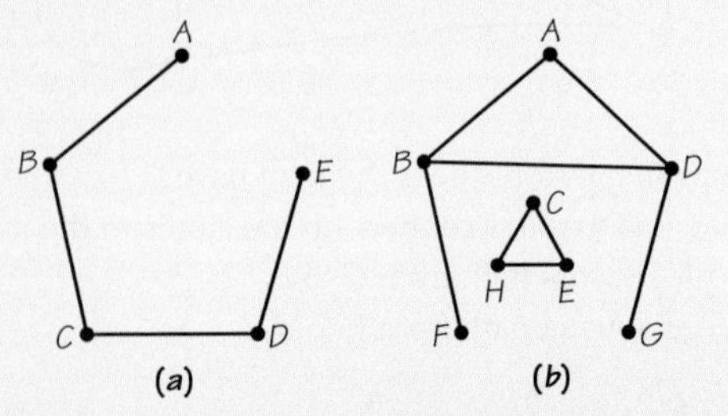

9. Refer to the figure in Exercise 4.

(a) Write down a circuit that includes the vertices *G* and *D* but does not start or end at either of these vertices.

(b) If two paths are considered different if they use different edges, write down:

(i) two different paths from *B* to *D*.

(ii) three different paths from *C* to *F*.

(iii) a circuit that has four edges.

10. In the graphs in Figure 1.17, find the smallest possible number of edges you could remove that would disconnect the graph.

11. Draw a graph with eight vertices that is connected where

(a) each vertex has valence 3.

(b) each vertex has valence 4.

(c) Do all graphs with eight vertices having valence 2 have the same number of edges?

12. Jack and Jill are located in Miami and want to fly to Berlin (see Figure 1.2).

(a) Find three paths for them to carry out this trip.

(b) What is the largest number of paths that can be used to carry out this trip that do not repeat a vertex (city)?

(c) Explain why it is reasonable not to want to repeat a vertex in this situation.

13. **(a)** How many vertices and edges does the graph in Figure 1.6 have?

(b) How many vertices and edges does the graph in Figure 1.7 have?

(c) How many vertices and edges does the graph in Figure 1.8a have?

14. **(a)** Add up the numbers you get for the valences of the vertices in Figure 1.6.

(b) Add up the numbers you get for the valences of the vertices in Figure 1.7.

(c) Add up the numbers you get for the valences of the vertices in Figure 1.8a.

(d) Describe the pattern you see in the answers you got for parts (a) through (c).

(e) Show that the pattern describes a fact that is true for any graph. (*Hint*: How many endpoints does an edge have?)

15. In the graph in Figure 1.8a, find the smallest possible number of edges you could remove that would disconnect the graph.

16. Can you draw a graph with at least two vertices for which all the vertices have different valences?

17. **(a)** Can you draw a graph that is connected and for which the valences of vertices are the consecutive integers from 2 to 6?

(b) If a graph such as the one described in (a) exists, can it have an Euler circuit? (Explain your answer.)

18. (a) What is the number of vertices and edges in the accompanying graph G?

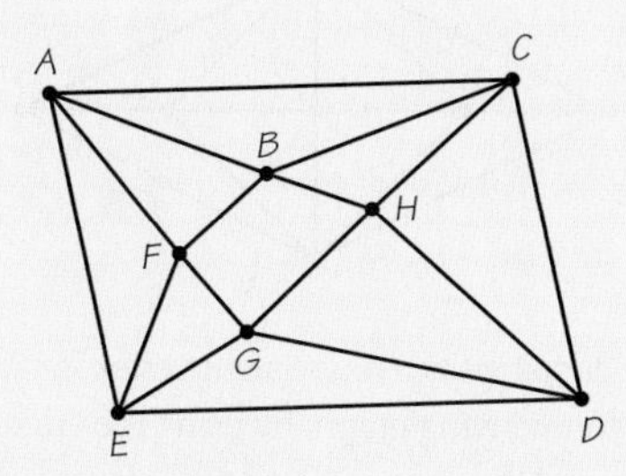

(b) Write down an Euler circuit for the accompanying graph, first by numbering the edges consecutively starting at an edge emanating from vertex D and then using an alternative description that lists the vertices as they are traversed by the Euler circuit.

19. Draw a connected graph with six vertices, all of whose vertices have valence 2.

20. (a) Is it possible that a street network gives rise to a disconnected graph? If so, draw such a network of blocks and streets and parking meters (in the style of Figure 1.12a). Then draw the disconnected graph it gives rise to.

(b) Draw a graph where every vertex has valence of at least 3 but where removing a single edge disconnects the graph.

(c) In what urban settings might a road network be represented by a graph that has an edge whose removal would disconnect the graph?

21. (a) Find a graph where the valences of the seven vertices of the graph are 1, 2, 2, 3, 3, 3, 4.

(b) Find another graph with the same valences as above that is "different" from the one you found for part (a).

22. For some services provided along streets, it may matter whether the roads are one-way or two-way. Give some examples where the street directions do and do not matter for our graph model analysis.

▲ **23.** A postal worker is supposed to deliver mail on all streets represented by edges in the accompanying graph by traversing each edge exactly once. The first day, the worker traverses the numbered edges in the order shown in (a), but the supervisor is not satisfied—why? The second day the worker follows the path indicated in (b), and the worker is unhappy—why? Is the original job description realistic? Why?

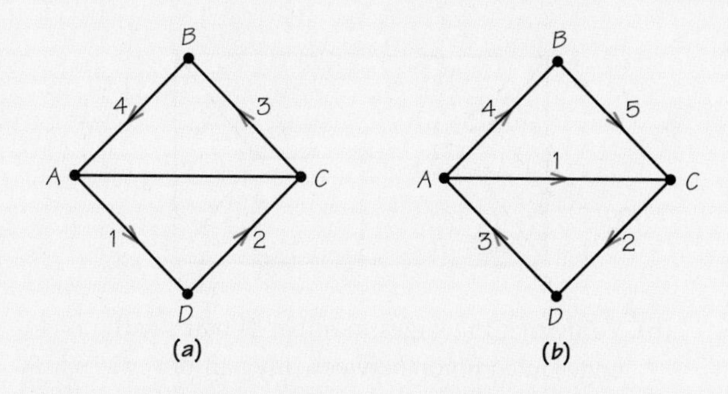

24. For the street network in Exercise 23, draw the graph that would be useful for routing a snowplow. Assume that all streets are two-way, one lane in each direction, and that you need to pass down each lane separately.

25. Find an efficient route for the snowplow to follow in the graph you drew in Exercise 24.

26. (a) Give examples of services that could be performed by a vehicle that moved in the direction of traffic down either lane of a two-way street.

(b) Give examples of services that would probably require a vehicle to travel down each of the lanes of a two-way street (in the direction of traffic for that lane) to perform the service.

27. For the street network shown below, draw the graph that would be useful for finding an efficient route for checking parking meters. (*Hint:* Notice that not every sidewalk has a meter; see Figure 1.12.)

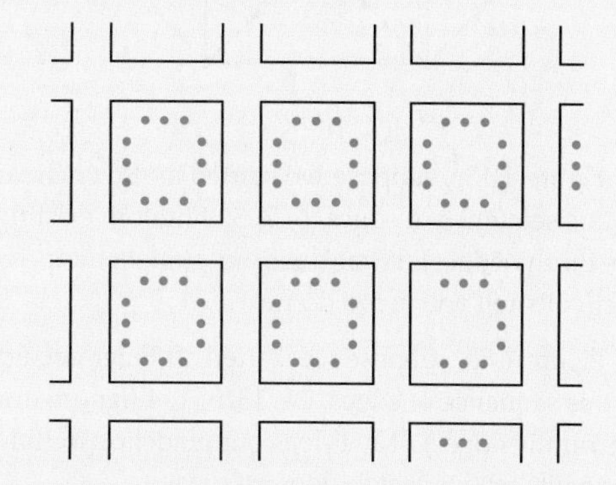

28. (a) For the street network in Exercise 27, draw a graph that would be useful for routing a garbage truck. Assume that all streets are two-way and that passing once down a street suffices to collect from both sides.

(b) Do the same problem using the assumption that one pass down the street suffices to collect from only one side.

29. (a) In the accompanying graph, find the largest number of paths from A to F that do not have any edges in common.

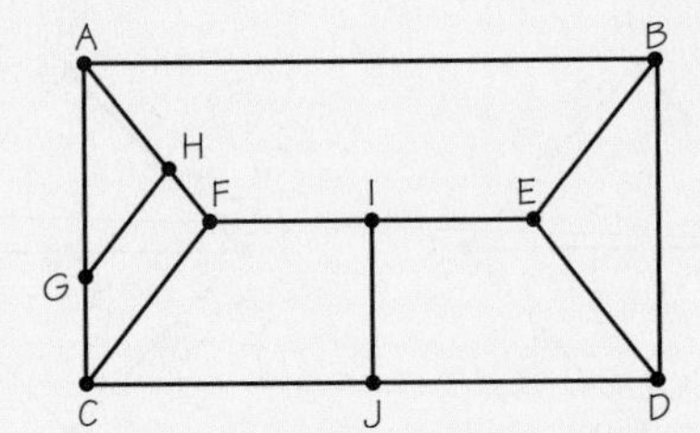

(b) Verify that the largest number of paths with no edges in common between any pair of vertices in this graph is the same.

(c) Why might one want to be able to design graphs such that one can move between two vertices of the graph using paths that have no edges in common?

30. Examine the paths represented by the numbered sequences of edges in both parts of the figure below. Determine whether each path is a circuit. If it is a circuit, determine if it is an Euler circuit.

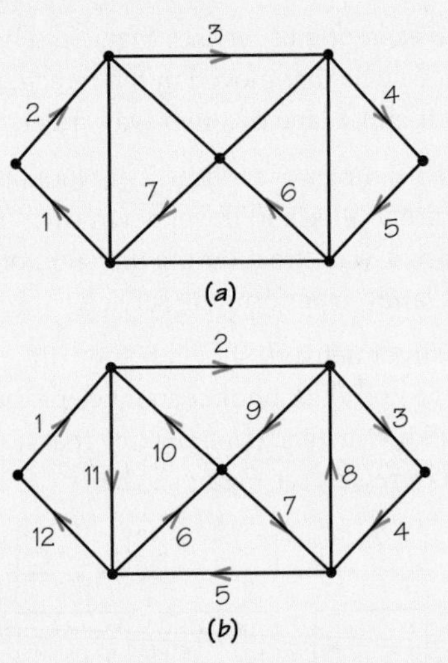

(a)

(b)

31. In Figure 1.13c, suppose we started an Euler circuit using this sequence of edges: 6, 7, 8, 9 (ignore existing arrows on the edges). What does our guideline for finding Euler circuits tell you *not* to do next?

32. In Figure 1.8b, suppose we started an Euler circuit using this sequence of edges: 14, 13, 8, 1, 4 (ignore existing arrows on the edges). What does our guideline for finding Euler circuits tell you *not* to do next?

33. Find an Euler circuit on the graph of Figure 1.15c (including the blue edges).

34. Find Euler circuits in the right-hand graphs in Figures 1.17a and 1.17b.

35. **(a)** Which vertices in the accompanying graph are odd-valent?

(b) In the accompanying graph, we see a territory for a parking-control officer that has no Euler circuit. How many sidewalks (edges) need to be omitted in order to enable us to find an Euler circuit? What effect would this have in the associated real-world situation?

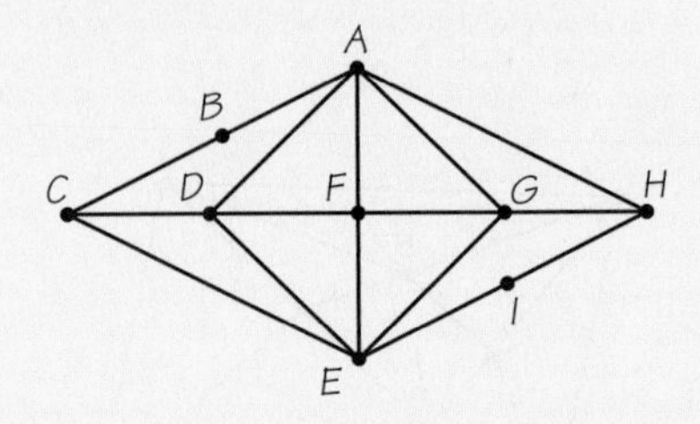

36. An Euler circuit visits a four-valent vertex *X*, such as the one in the accompanying graph, by using the edges *AX* and *XB* consecutively, and then using *CX* and *XD* consecutively. When this happens, we say that the Euler circuit cuts through at *X*.

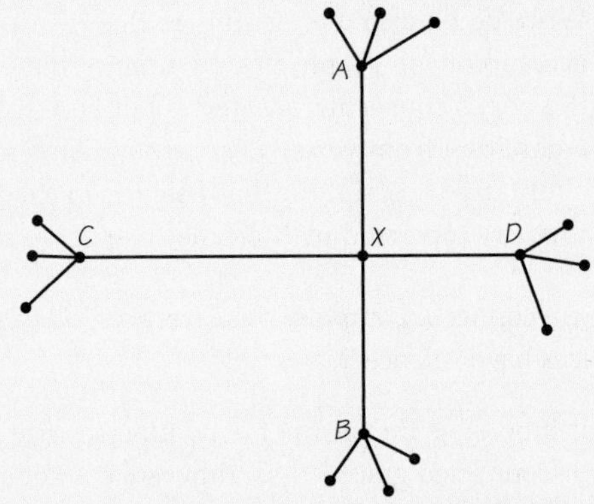

Suppose *G* is a four-valent graph such as that in the diagram below. Is it possible to find an Euler circuit of this graph that never cuts through any vertex? Explain why it might be desirable to find an Euler circuit of this special kind in an applied situation.

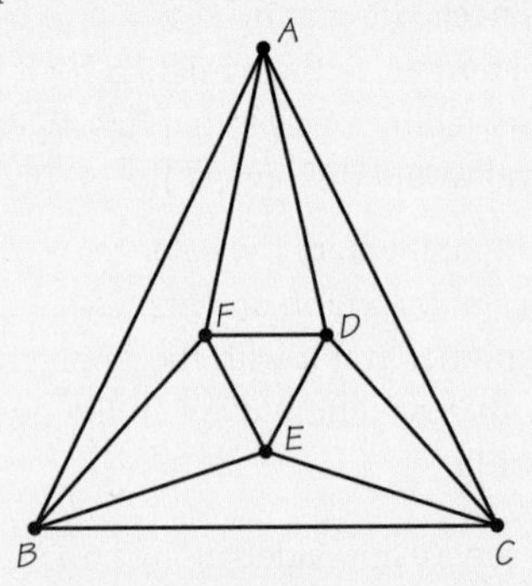

37. The company that distributes natural gas in a small town is located at vertex C in the accompanying graph *G*, which models the two-way streets for the town.

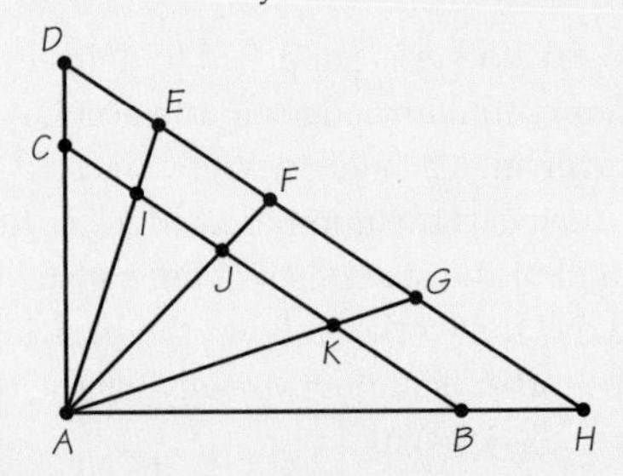

(a) Can an inspector for gas leaks travel along the street network and find a route that starts and ends at C and has no deadheads (repeated edges)?

(b) What is the minimum number of repeated edges in a shortest-length tour T, starting and ending at C, which looks for gas leaks?

(c) By duplicating existing edges in G, use tour T to modify graph G to obtain a new graph G^*, which shows that T can be interpreted as an Euler circuit in G^*.

1.3 Beyond Euler Circuits

1.4 Urban Graph Traversal Problems

38. Squeeze the circuit shown in graph (a) below onto graph (b). Show your answers by writing numbered arrows on the edges and by listing a sequence of vertices (for example, $ABEB \ldots A$).

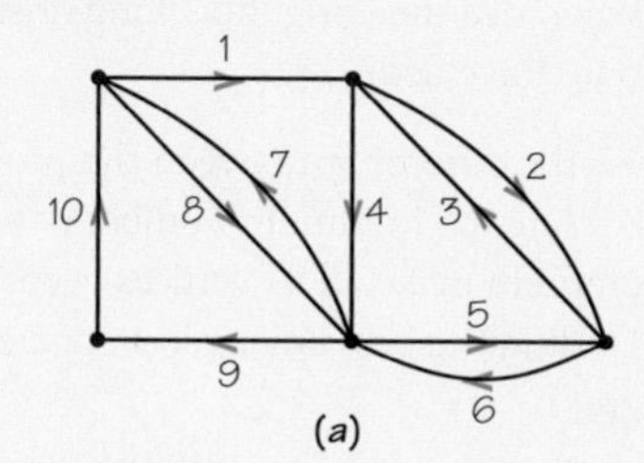

(a)

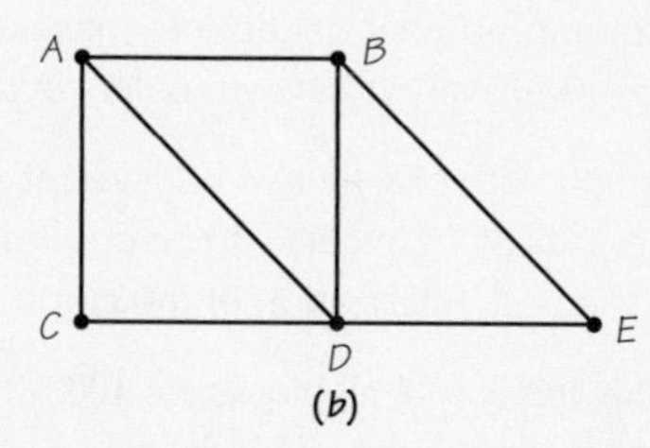

(b)

Then squeeze the circuit shown in graph (c) onto graph (d). Show your answers by writing numbered arrows on the edges and by listing a sequence of vertices.

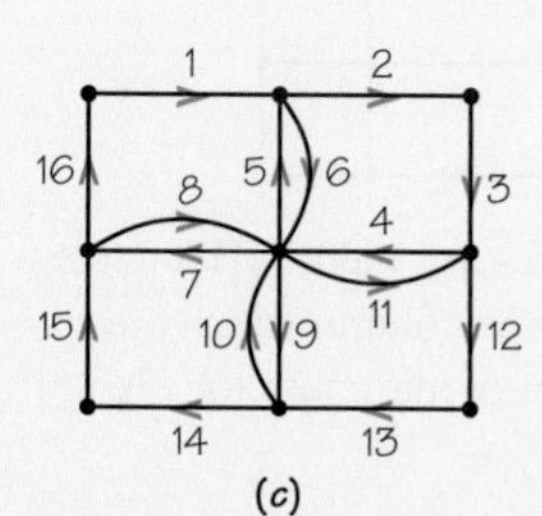

(c)

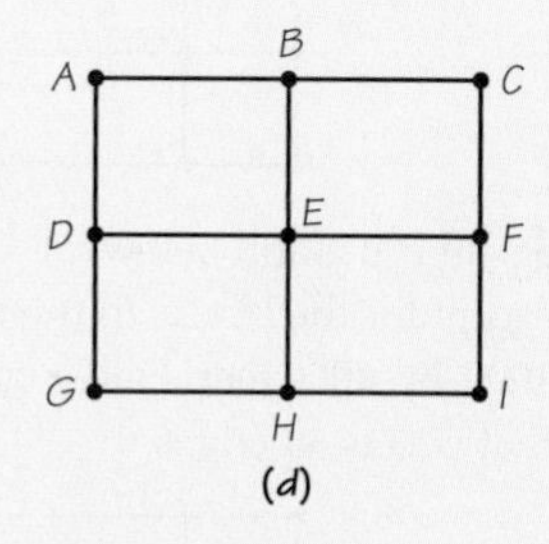

(d)

39. Find an Euler circuit on the eulerized graph (b) of the accompanying figure. Use it to find a circuit on the original graph (a) that covers all edges and reuses edges only five times. Can fewer than five reused edges be achieved?

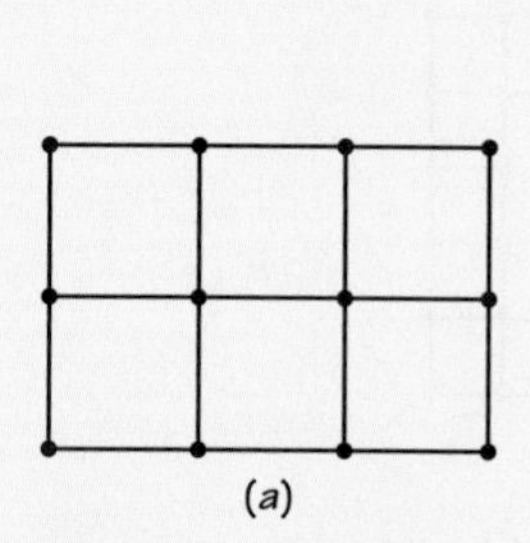
(a)

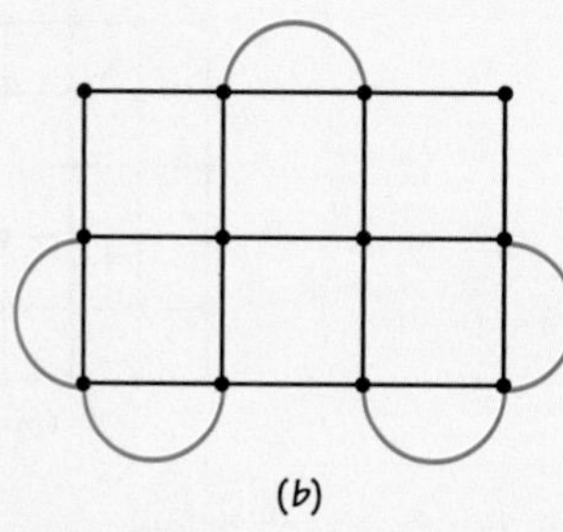
(b)

40. In the accompanying graph, add one or more edges to produce a graph that has an Euler circuit.

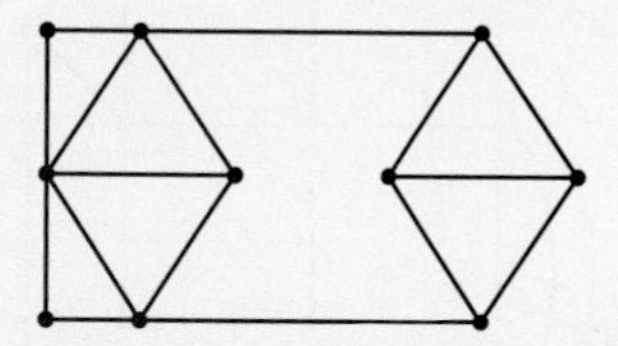

41. A college campus has a central square with sides arranged as shown by the edges in the graph below. Show how all these sidewalks can be traversed at least once in a tour that starts and ends at the same vertex.

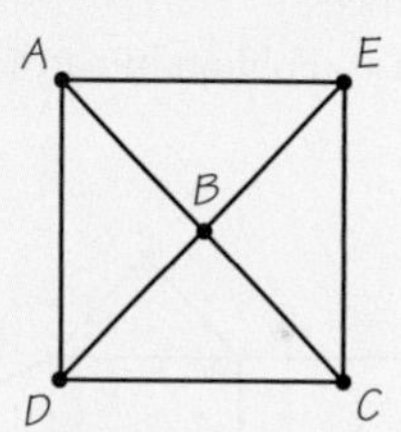

42. Find a circuit in the accompanying graph that covers every edge and has as few reuses as possible.

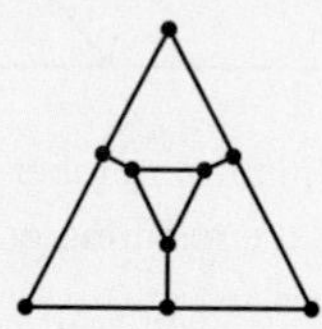

43. Eulerize these rectangular street networks using the same patterns that would be used by the edge walker described in the text.

(a) A 5×5 rectangle **(b)** A 4×5 rectangle

(c) A 6×6 rectangle

(d) Can you find an eulerization with nine added edges for a 2-by-7-block rectangular street network? Can you do better than nine added edges?

44. Find good eulerizations for the accompanying graphs, using as few duplicated edges as you can. See "Finding Good Eulerizations" (on p. 18) for hints.

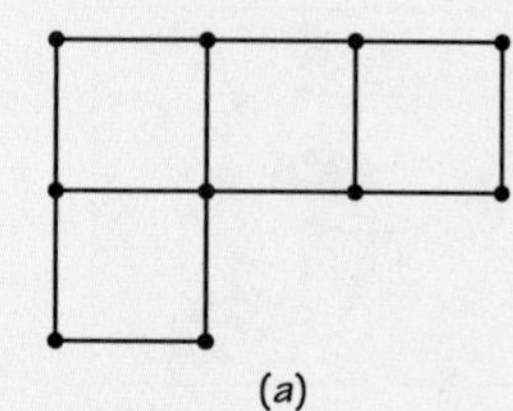
(a)

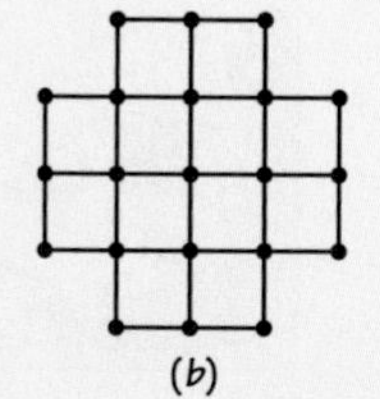
(b)

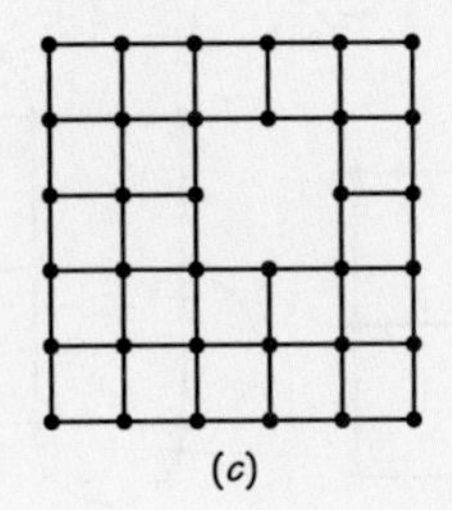

45. For the graph below:

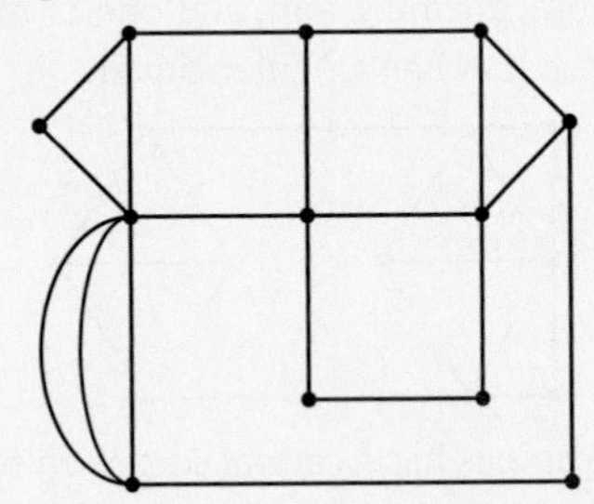

(a) Determine the minimum number of edges that have to be removed for the resulting graph to have all even-valent vertices.

(b) Does the graph you obtain in part (a) have an Euler circuit?

For the graph below:

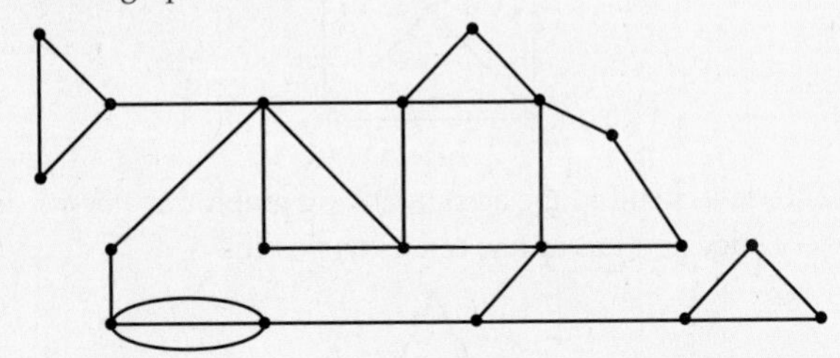

(c) Determine the minimum number of edges that have to be removed for the resulting graph to have all even-valent vertices.

(d) Does the graph you obtain in part (c) have an Euler circuit?

46. The following figure shows a river, some islands, and bridges connecting the islands and riverbanks. A charity is sponsoring a race in which entrants have to start at A, go over each bridge at least once, and end at A. Draw a graph that would be useful for finding a route that requires the least recrossing of bridges. Show what that route would be. (*Historical note:* This situation resembles the one that inspired Leonhard Euler's 1736 "recreational mathematics" problem that resulted in the first work in graph theory.)

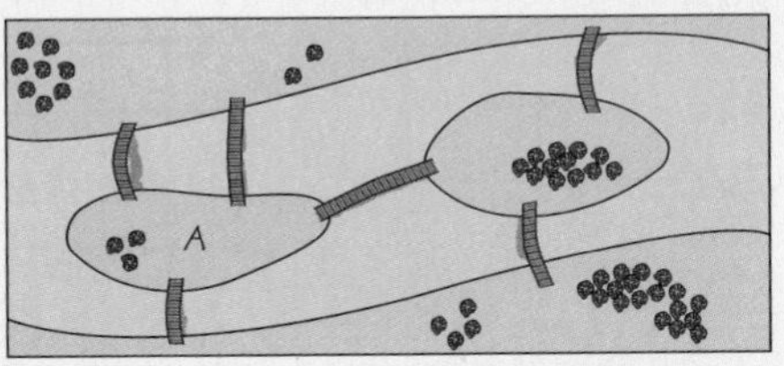

47. On Wednesdays during the summer, a small town that has only two-way streets does not allow on-street parking in the mornings. A truck that flushes water onto the streets to clean off "debris" can, thus, easily perform its work.

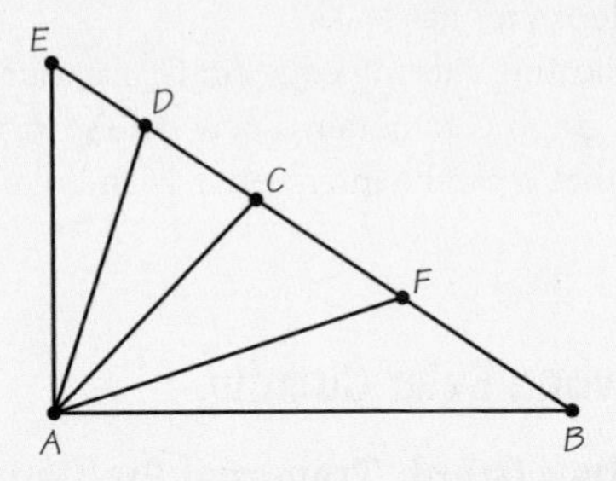

(a) For the street network of the town shown in the accompanying graph, design an efficient route R for the truck starting and ending at vertex A.

(b) If all the edges take equally long and take 9 minutes to cover when flushing is being carried out and 7 minutes when deadheading, how long does it take for route R to be traversed?

48. (a) Discuss the difference between the problem of:

(i) Adding the minimum number of edges to a graph to make all its vertices even-valent, and

(ii) Finding the best eulerization of a connected graph.

(b) In (i) must the graph that results from adding a minimum number of edges to make all the vertices even-valent have an Euler circuit?

49. Draw a graph with exactly two odd-valent vertices which requires exactly nine edges to be duplicated in order to find the best eulerization of the graph.

50. In the following graph, all blocks are 1000-by-1000 feet, except for the middle column of blocks, which are 1000-by-4000 feet. Find a circuit of minimum total length that covers all edges.

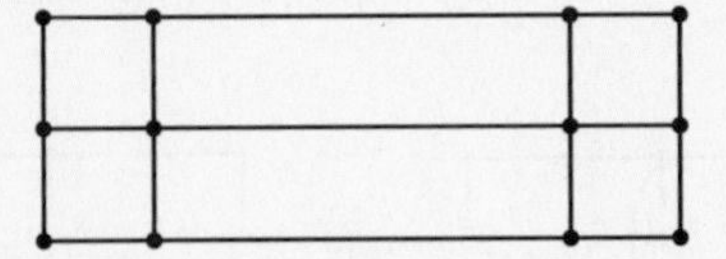

51. In the graph below, all blocks are 1000-by-1000 feet, except for the blocks in the middle column which are 1000-by-4000 feet. Find a circuit of minimum total length that covers all edges.

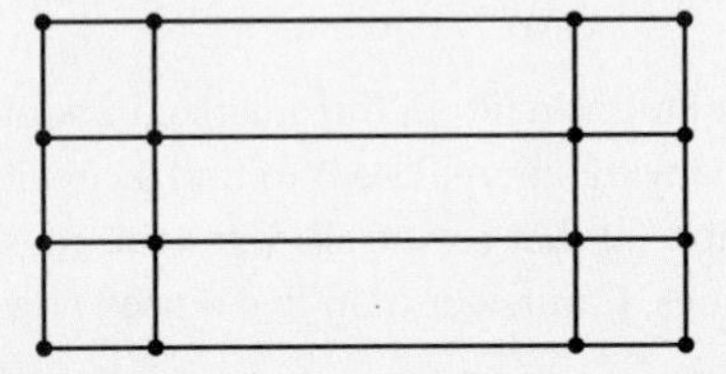

52. (a) Find the cheapest route in the accompanying graph, where one starts at vertex A, finishes at vertex A, and traverses each edge at least once. The cost of a route is computed by summing the numbers along the edges that one uses.

(b) How many edges are repeated in the minimal-cost route?

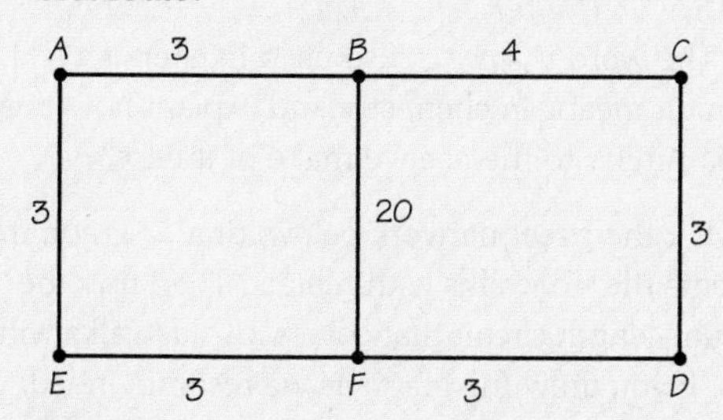

(c) Discuss the implications of this example for the relation between finding good eulerizations of graphs and the problem of finding cheap routes that start and end at the same vertex and traverse each edge at least once.

(d) The physical edge with cost 20 in the diagram is not physically longer than other edges with lower costs attached to them. Explain why in an urban setting it might make sense to assign two stretches of street of similar length very different "costs" for traversing them.

(e) What are some different meanings that "weights" (for example, traffic volume) potentially assigned to edges in a graph might have in an urban setting?

53. Which graphs (see accompanying figures) have Euler circuits? In the ones that do, find the Euler circuits by numbering the edges in the order the Euler circuit uses them. For the ones that don't, explain why no Euler circuit is possible.

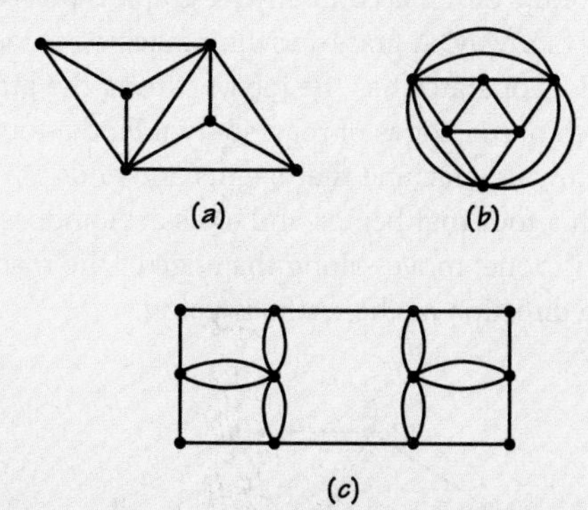

54. Eulerize the accompanying graph by using four new edges. Find an Euler circuit in the eulerized graph and use that circuit to find a circuit of the original graph that covers all edges but reuses edges only four times. How many different ways can the four edges be chosen? What is the total number of edges in the Euler circuit in the eulerized graph?

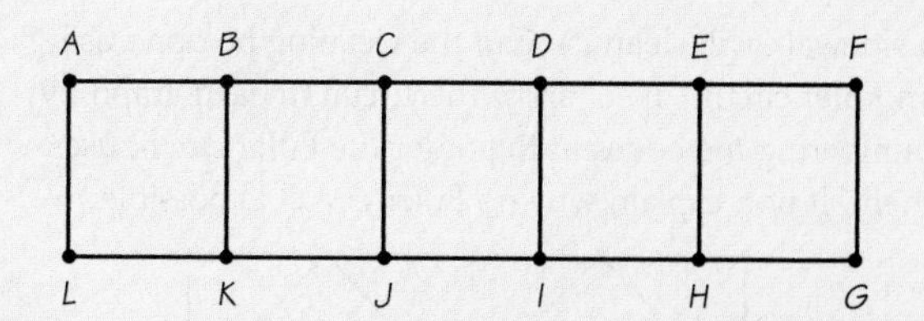

55. A graph G represents a street network to be traveled by a postal worker on foot, who must deliver mail to the houses on both sides of each street. How does one obtain from G a new graph H that represents the sidewalks that must be traversed? Does graph H always have an Euler circuit? Explain your answer.

56. In the following graph, find a circuit that covers every edge and has as few reuses as possible.

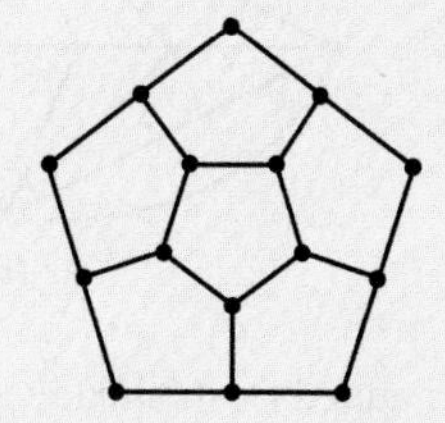

57. (a) Find the best eulerizations you can for the two accompanying graphs.

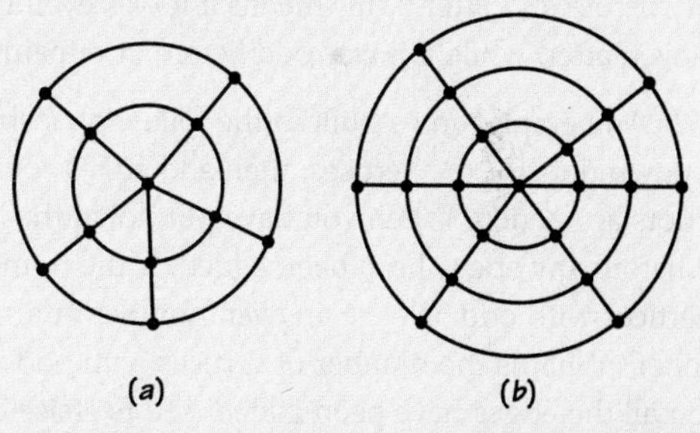

(b) Graph (a) can be thought of as having five rays and two circles, and graph (b) as having six rays and three circles. Draw a graph with four rays and four circles and find the best eulerization you can for this graph.

(c) Find a "formula" involving r and s for the smallest number of edges needed to eulerize a graph of this type having r rays and s circles.

■ **58.** Suppose that for a certain connected graph, it is possible to disconnect it by removing one edge. Explain why such a graph (before the edge is removed) must have at least one vertex of odd valence. (*Hint:* Show that it cannot have an Euler circuit.)

59. (a) Can you draw a graph with six vertices where the valence of each vertex is 5?

(b) If such a graph exists, how many edges must it have?

60. Each of the accompanying graphs represent the sidewalks to be cleaned in a fancy garden (one pass over a sidewalk will clean it). Can the cleaning be done using an Euler circuit? If so, show the circuit in each graph by numbering the edges in the order the Euler circuit uses them. If not, explain why no Euler circuit is possible.

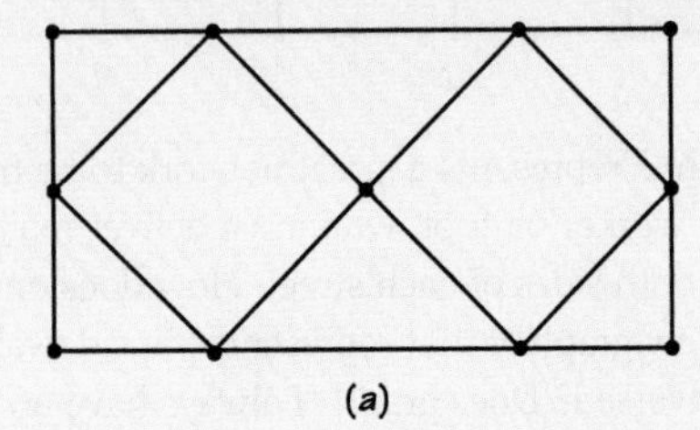

(*a*)

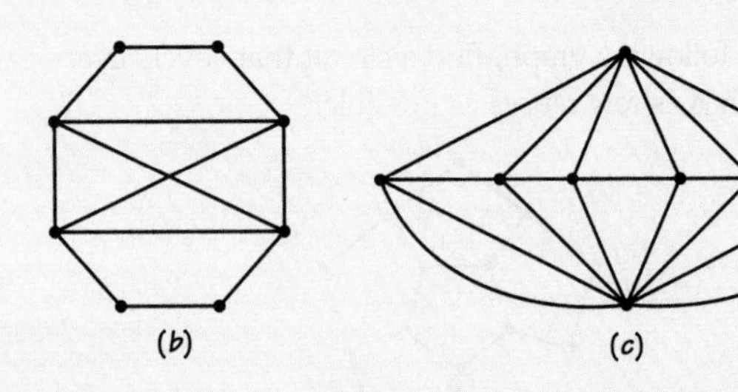

(*b*) (*c*)

■ **61.** If an edge is added to an already existing graph, connecting two vertices already in the graph, explain why the number of vertices with odd valence has the same parity before and after. (This means if it was even before, it is even after, while if it was odd before, it remains odd.)

■ **62.** Any graph can be built in the following fashion: Put down dots for the vertices, then add edges connecting the dots as needed. When you have put down the dots, and before any edges have been added, is the number of vertices with odd valence an even number or an odd number? What is the number of vertices with odd valence when all the edges have been added (see Exercise 61)?

63. Draw the graph for the parking-control territory shown in the figure below. Label each vertex with its valence and determine if the graph is connected.

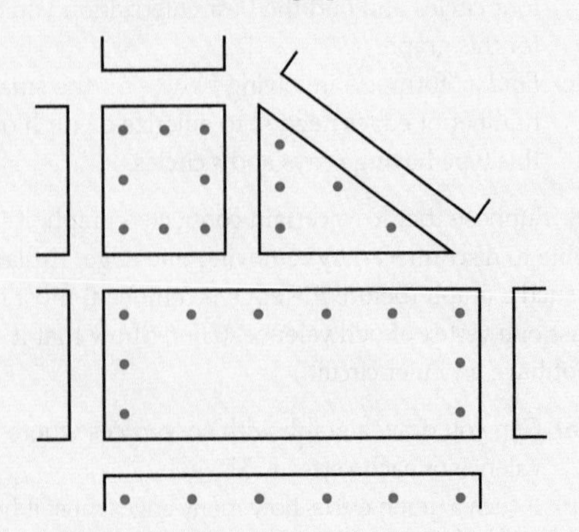

64. If a rectangular street network is r blocks by s blocks, find a formula for the minimum number of edges that must be added to eulerize a graph representing the network in terms of r and s. (*Hint:* Treat the case $r = 1$ separately. Test your formula with the cases 6 blocks by 5 blocks, 6 blocks by 6 blocks, and 5 blocks by 3 blocks.)

▲ **65.** The word *valence* is also used in chemistry. Find out what it means in chemistry and explain how this usage is similar to the use we make of it here.

■ **66.** For the street network below, draw a graph that represents the sidewalks with meters. Then find the minimum-length circuit that covers all sidewalks with meters. If you drew the graph as we recommended, you would find that the shortest circuit has length 18 (it reuses every edge).

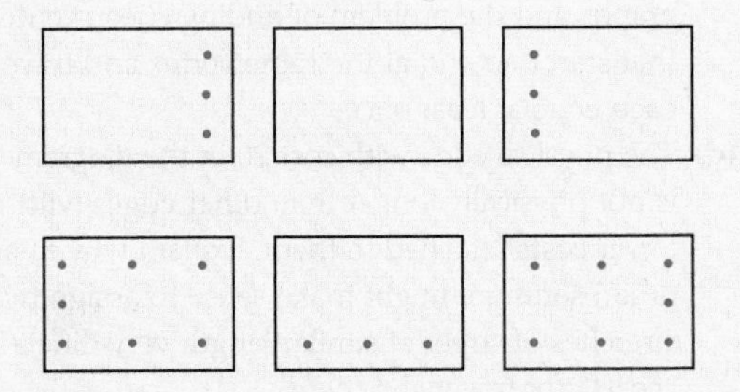

But the meter checker comes to you and says: "I don't know anything about your theories, but I have found a way to cover the sidewalks with meters using a circuit of length 10. My trick is that I don't rule out walking on sidewalks with no meters." Explain what he means and discuss whether his strategy can be used in other problems.

67. Each edge of the accompanying graph represents a two-lane highway. A grass-mowing machine is located at *A*, and its operator has the job of cutting the grass along each of the edges of road shown. Find a tour for the mowing machine that begins and ends at *A*. Find such a tour that begins and ends at *A* and, as the mowing is done, moves along the edge of the road in the same direction as the traffic is going.

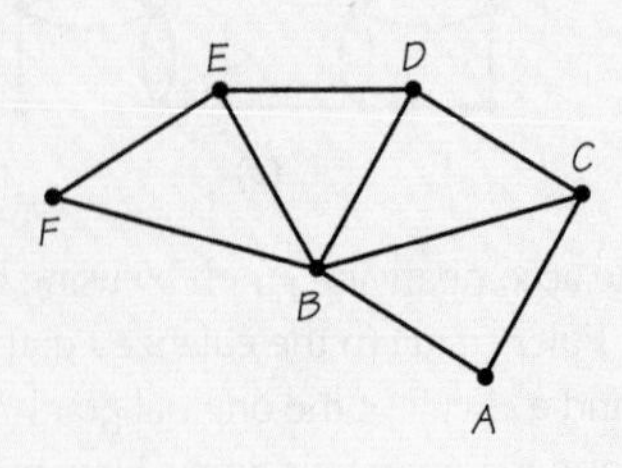

APPLET EXERCISES

To do these exercises, go to www.whfreeman.com/fapp9e.

Eulerizing a Graph

We learned that if a graph has exactly two vertices with odd valences, then an Euler circuit does not exist—but an Euler path does. It is also possible to produce an Euler circuit through the process of eulerization, by duplicating certain edges of the graph. But how many duplications are necessary to obtain an Euler circuit? Investigate this problem and more general related topics using the *Eulerizing a Graph* applet.

Euler Circuits

We know that if all the vertices have even valence, then an Euler circuit exists. Try your hand at finding such circuits in the *Euler Circuit* applet.

WRITING PROJECTS

1. Write a memo to your local department of parking control (or police department) in which you suggest that management science techniques like the ones in this chapter be used to plan routes. Assume that the person to whom you are writing is not extensively trained in mathematics but is willing to read through some technical material, provided you make it seem worth the trouble.

2. Do the same as in Writing Project 1, but to the department in charge of spreading salt on roads after snowstorms.

3. If you were making a recommendation to the mayor of New York City concerning proposed new street-sweeping routes, designed using the theory of this chapter, would you recommend that the changes be adopted or not? Write a memo that outlines the pros and cons as fairly as you can, and then conclude with your recommendation.

Suggested Readings

BELTRAMI, EDWARD J., *Models for Public Systems Analysis*, Academic Press, New York, 1977. This book gives a good overview of the way that operations research has provided and continues to provide new tools for solving societal problems. Among the ideas discussed are police patrol tactics, organization of emergency services, and scheduling. Some of the mathematics used is advanced.

MALKEVITCH, JOSEPH, and WALTER MEYER, *Graphs, Models, and Finite Mathematics*, Prentice-Hall, Englewood Cliffs, NJ, 1974. This introductory book includes much of the same material as presented here but provides more details of the proofs and uses somewhat different algorithms for solving the problems involved.

The following books treat many of the topics discussed here as well as shortest-path problems and matching problems, and they formulate some problems in more realistic terms:

ROBERTS, FRED S., AND BARRY TESMAN, *Applied Combinatorics*, 2nd ed., Pearson Prentice Hall, Upper Saddle River, NJ, 2004.

TUCKER, ALAN. *Applied Combinatorics*, 3rd ed., Wiley, New York, 1995.

Suggested Web Sites

www.hsor.org/what_is_or.cfm This site discusses the history of OR and some of the areas where OR is being applied. Be sure to follow the "Networks Routing" link to see applications of the Chinese postman problem.

www.geom.uiuc.edu/~doty/applications This Web page provides some examples of how to apply Euler circuits.

http://www-history.mcs.st-andrews.ac.uk/Biographies/Euler.html This essay discusses the numerous contributions that Euler made to mathematics, and provides biographical information about him.

http://www.ams.org/samplings/feature-column/fcarc-urban-geom This Web page includes an introduction to how graph theory has provided tools for urban operations research.

Walter Hodges/Stone/Getty Images

2

Business Efficiency

In the previous chapter, we saw that there was an easy way of telling whether a connected graph has a circuit that traverses each of the edges of a graph exactly once—for example, a route for a snowplow that covers the streets of a section of a town. However, the situation changes radically if we make a seemingly small change in the problem: When is it possible to find a route along distinct edges of a graph that visits each *vertex* once and only once in a simple circuit? Perhaps there has been a hurricane and it is important to check whether the storm sewers at every corner in town are clogged. We want all the sewers to be checked. However, this typically will not require that all the streets in the town be traversed.

This problem is called the *Hamiltonian circuit problem,* and, like the Euler circuit problem, it is a graph theory problem. The Hamiltonian circuit problem has many applications—for example, the delivery of parcels or water meter inspections. Suppose inspections or deliveries need to be made at each vertex (rather than along each edge) of a graph. An "efficient" tour of the graph would be a route that started and ended at the same vertex and passed through all the vertices without reuse, or repetition; that is, the route would be a **Hamiltonian circuit**. Such routes would be useful for inspecting traffic signals, for delivering mail to drop-off boxes, which hold heavy loads of mail so that urban postal carriers do not have to carry them long distances, or for delivering Meals on Wheels to the elderly.

Hamiltonian Circuit DEFINITION

A tour that starts at a vertex of a graph and visits each vertex once and only once, returning to where it started, is called a **Hamiltonian circuit**.

For example, the wiggly line in Figure 2.1a shows a circuit we can take to tour that graph, visiting each vertex once and only once. This tour can be written *ABDGIHFECA.* Note that another way of writing the same circuit would be *EFHIGDBACE.* A different circuit visiting each vertex once and only once would be *CDBIGFEHAC* (Figure 2.1b). Do not be confused because *C* is written twice when we write down this list of vertices. We can think of the circuit as starting at any of its vertices, but we do start and end at the same vertex.

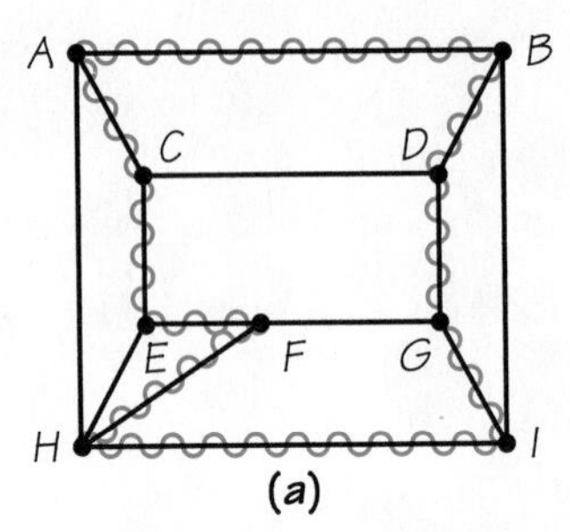

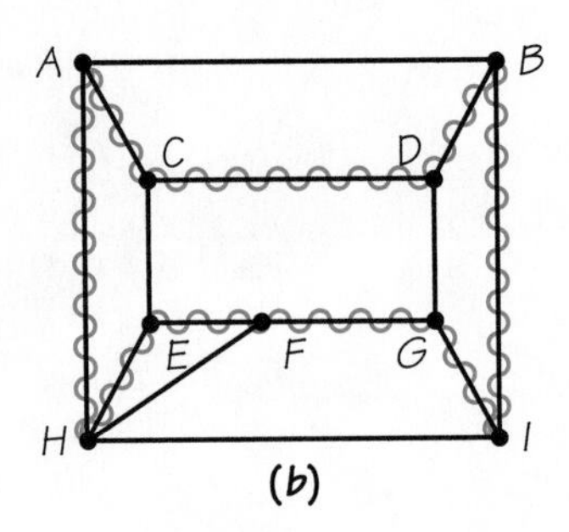

Figure 2.1 Wiggly edges illustrate Hamiltonian circuits.

2.1 Hamiltonian Circuits

The concept is named for the Irish mathematician William Rowan Hamilton (1805–1865), who was one of the first to study it. We now know that the concept was discovered somewhat earlier by Thomas Kirkman (1806–1895), a British minister with a penchant for mathematics.

It may seem that delivering mail in a cheap and timely manner should not be that hard. However, finding the best way to deliver mail over a variety of environments, rural, suburban, and urban, is very complex. How should a large geographic area be divided into smaller sections? Should each mail carrier use a truck as a "depot" to resupply small amounts of mail for delivery or should there be deposit boxes on street corners? Mathematics can be used to find answers to such questions. (*Bill Aron/PhotoEdit*)

The concepts of Euler and Hamiltonian circuits are similar in that both forbid reuse. An Euler circuit forbids the reuse of edges, while a Hamiltonian circuit forbids the reuse of vertices. However, it is far more difficult to determine which connected graphs possess a Hamiltonian circuit than to determine which connected graphs have Euler circuits. Using the concepts of Euler circuit and Hamiltonian circuit in practical applications in the real world involves actually being able to find such circuits in large graphs. As we saw in Chapter 1, looking at the valences of vertices tells us whether a connected graph has an Euler circuit, but we have no such simple method for telling whether or not a graph has a Hamiltonian circuit.

Some special classes of graphs are known to have Hamiltonian circuits, and some special classes of graphs are known to lack them. For example, here is a method for constructing an infinite family of graphs where each graph in the family cannot have a Hamiltonian circuit. Construct a vertical column of m vertices and a parallel column of n vertices, where m is bigger than n, as shown in Figure 2.2a. The figure illustrates a typical case where $m = 4$ and $n = 2$. Now join each vertex on the left in the figure to every vertex on the right. As m and n vary, we get a family of different graphs.

No graph obtained in this manner can have a Hamiltonian circuit. If a Hamiltonian circuit existed, it would have to alternately include vertices on the left and right of the figure. This is not possible because the number of vertices on the left and right are not the same. It is unlikely that a method will ever be found to easily determine whether or not a graph has a Hamiltonian circuit. If Hamiltonian circuits were easy to find in any graph at all, many applied problems could be solved in a less costly way.

In many urban operations research situations, "grid graphs" such as the one in Figure 2.2b are of interest. If we wanted an efficient route (circuit) to inspect traffic surveillance cameras located at urban street intersections, we would need to find a Hamiltonian circuit for the graph in Figure 2.2b. Note that in going from one vertex to another, we move from a vertex of one color to a vertex of the other color. Since colors would alternate in a Hamiltonian circuit, it follows that the number of vertices of each color would have to be the same if there is a Hamiltonian circuit in this graph. Since the number of vertices of the two colors is *not* the same, there is no Hamiltonian circuit and, hence, no fully efficient route for inspecting the traffic control cameras.

The Hamiltonian circuit problem itself has many applications. This is not unusual in mathematics. Often mathematics used to solve a particular real-world problem leads to new mathematics that suggests applications to other real-world situations. One class of problems to which we can apply Hamiltonian circuits is vacation planning.

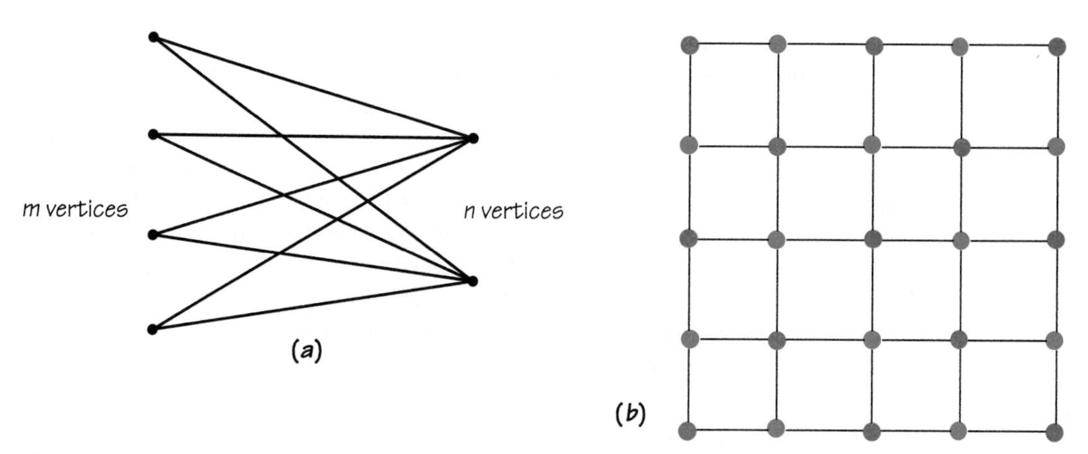

Figure 2.2 (a) An example of one graph from a family of graphs that has no Hamiltonian circuit. The number of vertices m on the left is chosen to be greater than the number of vertices n on the right. The case $m = 4$ and $n = 2$ is shown. (b) A graph used to model a portion of a city. Since the graph reflects the block structure of the city, it is known as a "grid graph."

EXAMPLE 1
Vacation Planning

Let's imagine that you are a college student studying in Chicago. During spring break, you and a group of friends have decided to take a car trip to visit other friends in Minneapolis, Cleveland, and St. Louis. There are many choices as to the order of visiting the cities and returning to Chicago, but you want to design a route that minimizes the distance you have to travel. Presumably, you also want a route that cuts costs, and you know that minimizing distance will minimize the cost of gasoline for the trip. Similar problems with different complications would arise for bus, railroad, or airplane trips.

Express mail and parcel post delivery companies need to make complicated patterns of deliveries and pick-ups. To do this they need to know driving distances between the various geographical locations involved. Using this information, together with driving times, they can use mathematics to cut costs and to make the pick-ups and deliveries on time. (*© Rhoda Sidney/PhotoEdit.*)

Imagine now that the Internet has provided you with the intercity driving distances between Chicago, Minneapolis, Cleveland, and St. Louis. We can construct a graph model with this information, representing each city by a vertex and the legs of the journey between the cities by edges joining the vertices. To construct the model, we add a number called a **weight** to each graph edge, as in Figure 2.3. In this example, the weights represent the distances between the cities, each of which corresponds to one of the endpoints of the edges in the graph. (In other examples the weight might represent a cost, time, satisfaction rating, or profit.) We want to find a minimum-cost tour that starts and ends in Chicago and visits each other city only once. Using our earlier terminology, what we wish to find is a **minimum-cost Hamiltonian circuit**—a Hamiltonian circuit with the lowest possible sum of the weights of its edges.

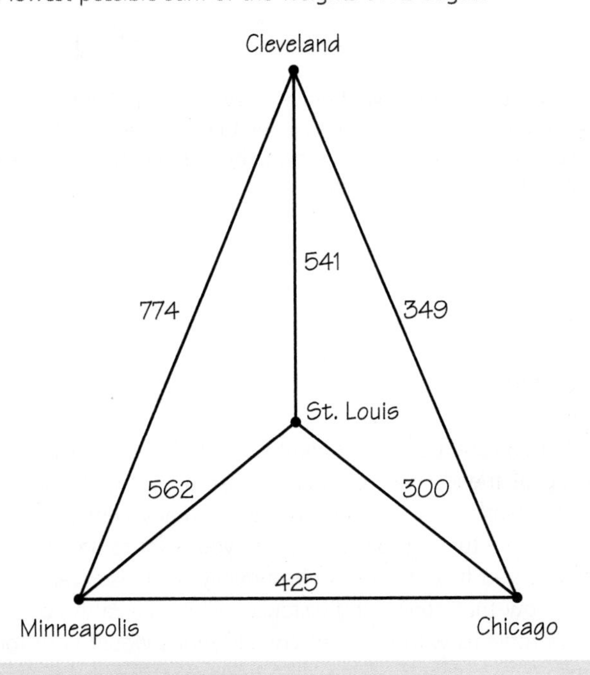

Figure 2.3 Road mileages between four cities.

Finding a Minimum-Cost Hamiltonian Circuit PROCEDURE

How can we determine which Hamiltonian circuit has minimum cost? There is a conceptually easy **algorithm**, or mechanical step-by-step process, for solving this problem:

1. Generate all possible Hamiltonian tours (starting from Chicago).
2. Add up the distances on the edges of each tour.
3. Choose a tour with total distance being a minimum, that is, as small as possible.

Steps 2 and 3 of the algorithm are straightforward. Thus, we need worry only about Step 1, generating all the possible Hamiltonian circuits in a systematic way. To find the Hamiltonian tours, we will use the **method of trees**, as follows: Starting from Chicago, we can choose any of the three cities to visit after leaving Chicago. The first stage of the enumeration tree is shown in Figure 2.4. If Minneapolis is chosen as the first city to visit, then there are two possible cities to visit next, Cleveland and St. Louis. The possible branchings of the **tree** at this stage are shown in Figure 2.5. In this second stage, however, for each choice of first city to visit, there are two choices from this city to the second city to visit. This would lead to the diagram in Figure 2.6.

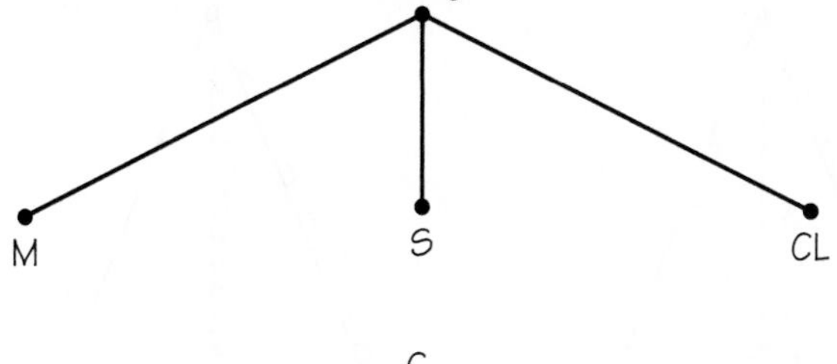

Figure 2.4 First stage in finding vacation-planning routes.

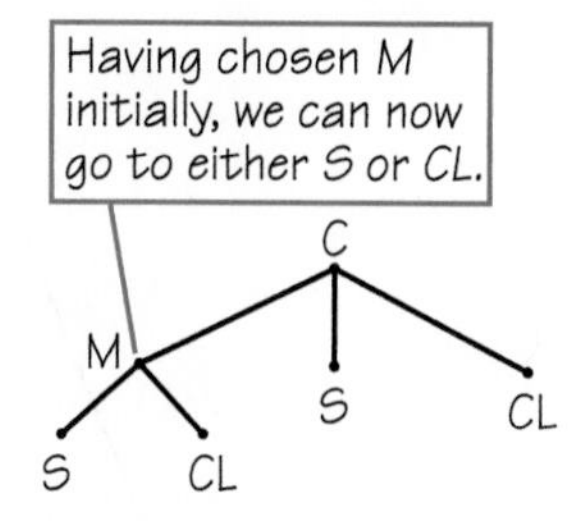

Figure 2.5 Part of the second stage in finding vacation-planning routes.

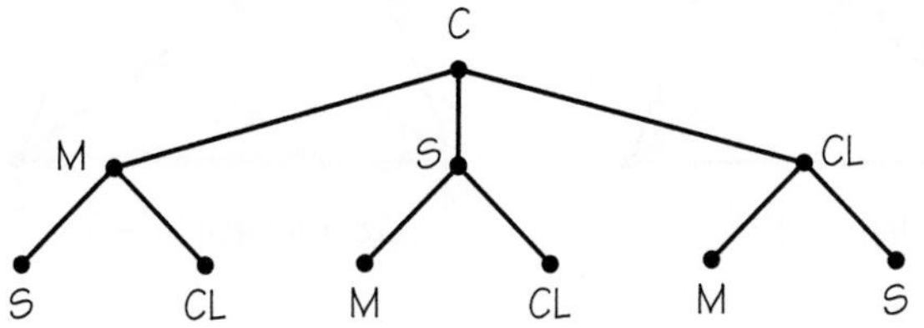

Figure 2.6 Complete second stage in finding vacation-planning routes.

Having chosen the order of the first two cities to visit, and knowing that no revisits (reuses) can occur in a Hamiltonian circuit, we have only one choice left for the next city. From this city, we return to Chicago. The complete tree diagram showing the third and fourth stages for these routes is given in Figure 2.7. Notice, however, that because we can traverse a circular tour in either of two directions, the paths shown in the tree diagram of Figure 2.7 do not correspond to different Hamiltonian circuits. For example, the leftmost path (C–M–S–CL–C) and the rightmost path (C–CL–S–M–C) represent the same Hamiltonian circuit. Thus, among what appear to be six different paths in the tree diagram, only three in fact correspond to different Hamiltonian circuits. These three distinct Hamiltonian circuits are shown in Figure 2.8.

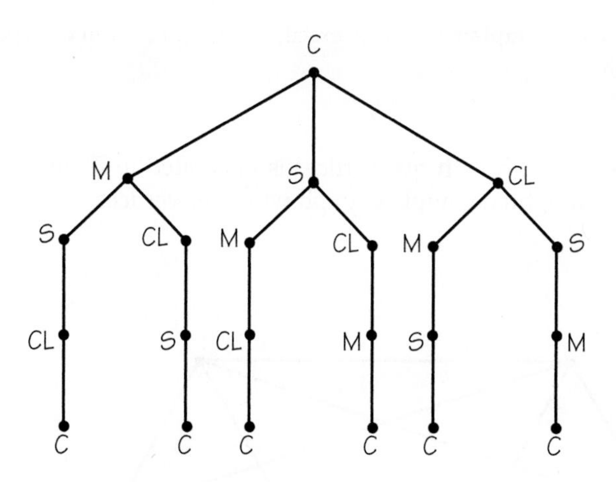

Figure 2.7 Completed enumeration of routes using the method of trees for the vacation-planning problem.

Note that in generating the Hamiltonian circuits, we disregard the distances involved. We are concerned only with the different patterns of carrying out the visits. To find the optimal route, however, we must add up the distances on the edges to get each tour's length and select a tour whose total length is smallest. Figure 2.8 shows that the optimal tour is Chicago, Minneapolis, St. Louis, Cleveland, Chicago. The length of this tour is 1877 miles, which saves 163 miles over the longest choice of tour.

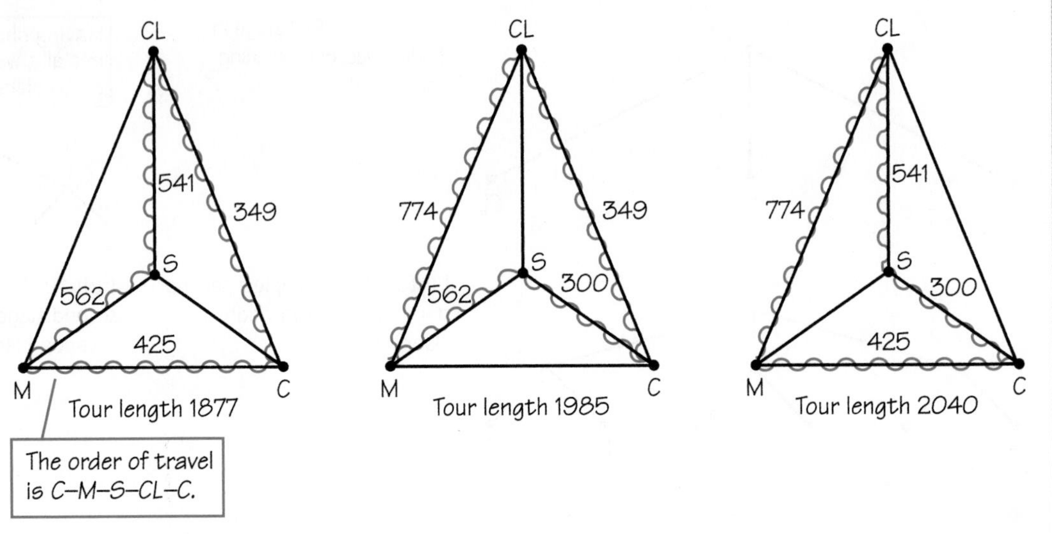

Figure 2.8 The three Hamiltonian circuits for the vacation-planning problem of Figure 2.3.

The method of trees is not always as easy to use as our example suggests. Instead of doing our analysis for four cities, consider the general case of n cities. The graph model similar to that in Figure 2.3 would consist of a weighted graph with n vertices, with every pair of vertices joined by an edge.

Complete Graph DEFINITION

A graph is called **complete** if there is exactly one edge between each pair of vertices in the graph.

A complete graph with five vertices is illustrated in Figure 2.9. The graph in Figure 2.3 is a weighted complete graph with four vertices.

Figure 2.9 A complete graph with five vertices. Every pair of vertices is joined by an edge.

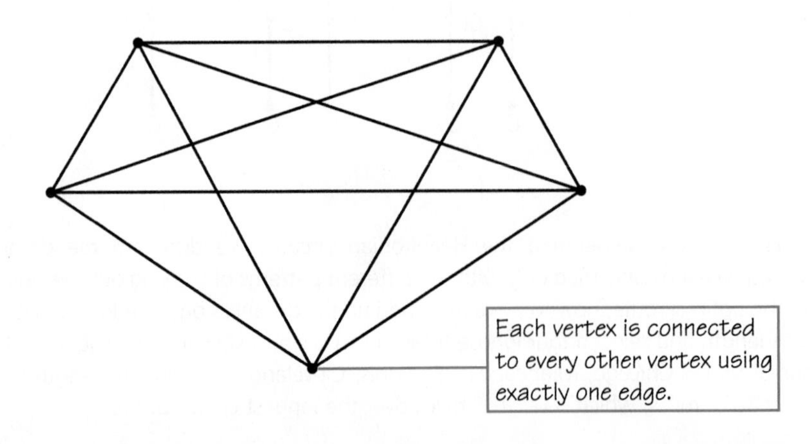

Fundamental Principle of Counting

How many Hamiltonian circuits are in a complete graph of n vertices? We can solve this problem by using the same type of analysis that we used in the method of trees. The method of trees is a visual application of the **fundamental principle of counting,** a procedure for counting outcomes in multistage processes. Using this procedure, we can count how many patterns occur in a situation by looking at the number of ways in which the component parts can occur. For example, if Jack has 10 shirts and 4 pairs of trousers, he can wear $10 \times 4 = 40$ shirt–pants outfits. Each shirt can be worn with any of the pants. (This can be verified by drawing a tree diagram, but such a diagram is cumbersome for big numbers.)

The Fundamental Principle of Counting DEFINITION

In general, the **fundamental principle of counting** can be stated this way: If there are a ways of choosing one thing, b ways of choosing a second after the first is chosen, . . ., and z ways of choosing the last item after the earlier choices, then the total number of choice patterns is $a \times b \times c \times \cdots \times z$.

EXAMPLE 2
Counting

Here are some examples of how to use the fundamental principle of counting:

1. In a restaurant, there are 4 kinds of soup, 12 entrees, 6 desserts, and 3 drinks. How many different four-course meals can a patron choose? The four choices can be made in 4, 12, 6, and 3 ways, respectively. Hence, applying the fundamental principle of counting, there are $4 \times 12 \times 6 \times 3 = 864$ possible meals.
2. In a state lottery, a contestant gets to pick a four-digit number that does not contain a zero followed by an uppercase or lowercase letter. How many such sequences of digits and a letter are there? Each of the four digits can be chosen in 9 ways (that is, 1, 2, . . . , 9), and the letter can be chosen in 52 ways (that is, A, B, . . . , Z plus a, b, . . . , z). Hence, there are $9 \times 9 \times 9 \times 9 \times 52 = 341{,}172$ possible patterns.
3. A corporation is planning a musical logo consisting of four different ordered notes from the scale C, D, E, F, G, A, and B. How many logos are there to choose from? The first note can be chosen in 7 ways, but because reuse is not allowed, the next note can be chosen in only 6 ways. The remaining two notes can be chosen in 5 and 4 ways, respectively. Using the fundamental principle of counting, $7 \times 6 \times 5 \times 4 = 840$ musical logos are possible. If reuse of notes is allowed, $7 \times 7 \times 7 \times 7 = 2401$ logos are possible.

Let's now return to the problem of enumerating Hamiltonian circuits for the complete graph with n vertices. The city visited first after the home city can be chosen in $n - 1$ ways, the next city in $n - 2$ ways, and so on, until only one choice remains. Using the fundamental principle of counting, there are $(n - 1)! = (n - 1)(n - 2) \times \cdots \times 3 \times 2 \times 1$ routes. The exclamation mark in $(n - 1)!$ is read "factorial" and is shorthand notation for the product $(n - 1)(n - 2) \times \cdots \times 3 \times 2 \times 1$. For example, $5! = 5 \times 4 \times 3 \times 2 \times 1 = 120$.

As we saw in Figure 2.7, pairs of routes correspond to the same Hamiltonian circuit because one route can be obtained from the other by traversing the cities in reverse order. Thus, although there are $(n - 1)!$ possible routes, there are only half as many, or $(n - 1)!/2$, different Hamiltonian circuits. Now, if we have only a few cities to visit, $(n - 1)!/2$ Hamiltonian circuits can be listed and examined in a reasonable amount of time. Analysis of a six-city problem would require generation of $(6 - 1)!/2 = 5!/2 = 120/2 = 60$ tours. But for, say, 25 cities, $24!/2$ is approximately 3×10^{23}. Even if these tours could be generated at the rate of 1 million a second, it would take 10 billion years to generate them all. Because it would take so long to solve large vacation-planning problems using this method, it is sometimes referred to as a **brute force method** (that is, trying all the possibilities). Computer scientists and engineers have made it possible to market faster and faster computers. However, governments and businesses need to solve larger scale problems; say, for example, finding a Hamiltonian circuit in a graph with 10,000 vertices. If the methods one knows for solving such problems are not much better than brute force, then it's unlikely that even these faster computers can solve large versions of such problems. Mathematicians and computer scientists are actively seeking procedures that will significantly improve our ability to solve large versions of important problems.

2.2 Traveling Salesman Problem

If the only benefit were saving money and time in vacation planning, the difficulty of finding a minimum-cost Hamiltonian circuit in a complete graph with n vertices for large values of n would not be of great concern. However, the problem we are discussing is one of the most common in *operations research,* the branch of mathematics concerned with getting governments and businesses to operate more efficiently. This problem is usually called the **traveling salesman problem (TSP)** because of its early formulation.

Traveling Salesman Problem (TSP) DEFINITION

The **traveling salesman problem (TSP)** involves finding the trip of minimum cost that a salesman can make to visit the cities in a sales territory once and only once (represented by a complete graph with weights on the edges), starting and ending the trip in the same city.

Many situations require in essence solving a TSP:

1. A lobster fisherman has set out traps at various locations and wishes to pick up his catch.
2. The telephone company wishes to pick up the coins from its pay telephone booths. (To avoid the high cost of picking up these coins, phone companies in many countries have adopted a system that uses prepurchased phone cards to operate phones. This means that there are no coins to collect.) Due to the increased use of cell phones, fewer pay phones are available.
3. The electric (or gas) company needs to design a route for its meter readers.
4. A minibus must pick up six day campers, deliver them to camp, and return them home later in the day.
5. In drilling holes in a series of plates, the drill press operator (perhaps a robot) must drill the holes in a predetermined order.

6. Physical records generated at automated teller machine (ATM) locations—as backup in case of failure of the electronic systems—must be picked up periodically.
7. A limousine service with a van located at an airport must pick up five customers and deliver them to the airport in time to catch their flights.
8. A local ethnic fast-food restaurant must hire someone to deliver the food quickly and while it is still hot, before returning to the restaurant for the next batch of orders to be delivered.

Perhaps surprisingly, TSP problems are also solved regularly in the design of computer chips. The components must be located so that the machines involved in the assembly can insert them on the chips as efficiently as possible. Because many chips are manufactured, even a small improvement in the time needed to make a chip can save a lot of money.

The meaning of *cost* can vary from one formulation of a TSP to another. We can measure cost as distance, airplane ticket prices, time, or any other factor that is to be optimized. In many situations, the TSP arises as a subproblem of a more complicated problem. For example, a supermarket chain may have a very large number of stores to be served from a single large warehouse. If there are fewer trucks than stores, the stores must be grouped into clusters so that one truck can serve each cluster. If we then solve the TSP for every truck, we can minimize total costs for the supermarket chain. Similar vehicle-routing problems—for dial-a-ride services that transport senior citizens to activity centers, for example, or that deliver children to their schools or camps—often involve solving the TSP as a subproblem.

2.3 Helping Traveling Salesmen

Because the traveling salesman problem arises so often in situations where the associated complete graphs would be very large, we must find a faster method than the brute force method we have described. One intuitive idea is to try to visit nearby locations sooner. Recall that our goal is to find the minimum-cost Hamiltonian circuit.

Nearest-Neighbor Algorithm PROCEDURE

Starting from the home city, first visit the nearest city, then visit the nearest city that has not already been visited. We return to the start city when no other choice is available. This approach is called the **nearest-neighbor algorithm**.

EXAMPLE 3
Applying the Nearest-Neighbor Algorithm

Applying this algorithm to the TSP in Figure 2.3 quickly leads to the tour of Chicago, St. Louis, Cleveland, Minneapolis, and Chicago, with a length of 2040 miles. Here is how this tour is determined. Because we are starting in Chicago, there is a choice of going to a city that is 425, 300, or 349 miles away. Because the smallest of these numbers is 300, we next visit St. Louis, which is the nearest neighbor of Chicago not already visited.

At St. Louis, we have a choice of visiting next cities that are 541 or 562 miles away. Hence, Cleveland, which is nearer (541 miles), is visited. To complete the tour, we visit Minneapolis and return to Chicago, thereby adding 774 and 425 miles to the length of the tour. The total length of the tour is 2040 miles.

The nearest-neighbor algorithm is an example of a **greedy algorithm** because at each stage a best (greedy) choice, based on an appropriate criterion, is made. Unfortunately, this is not the optimal tour, which we saw was C–M–S–CL–C, for a total length of 1877 miles. Making the best choice at each stage may not yield the best "global" solution. However, even for a large TSP, one can always find a nearest-neighbor route quickly.

EXAMPLE 4
Applying the Nearest-Neighbor Algorithm Revisited

Figure 2.10 again illustrates the ease of applying the nearest-neighbor algorithm, this time to a weighted complete graph with five vertices. Starting at vertex *A*, we get the tour *ADECBA* (cost 2800) (Figure 2.10a). Note that the nearest-neighbor algorithm starting at vertex *B* yields the tour *BCADEB* (cost 3050) (Figure 2.10b).

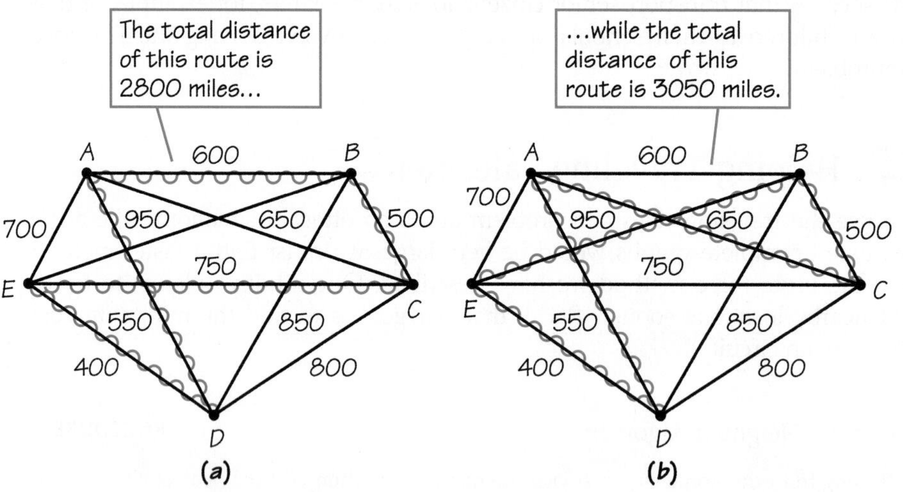

Figure 2.10 (a) A weighted complete graph with five vertices that illustrates the use of the nearest-neighbor algorithm (starting at *A*). (b) TSP tour generated by the nearest-neighbor algorithm (starting at *B*).

This example illustrates that a nearest-neighbor tour can be computed for each vertex of the complete graph being considered and that different nearest-neighbor tours can be obtained starting at different vertices. Thus, even though we may seek a tour starting at a particular vertex—say, *A* in Figure 2.10—because a Hamiltonian circuit can be thought of as starting at any of its vertices, we can just as easily apply the nearest-neighbor procedure starting at vertex *B* (rather than at *A*). The Hamiltonian circuit we get can still be thought of as beginning at vertex *A* rather than *B*. Even for complete graphs with a large number of vertices, it would still be faster to apply the nearest-neighbor algorithm for each vertex and pick the cheapest of the tours generated (though such a tour might not be optimal) than to apply the brute force method. Consider a different approach to the TSP that might find a good solution quickly.

Sorted-Edges Algorithm PROCEDURE

Start by sorting or arranging the edges of the complete graph in order of increasing cost (or, equivalently, arranging the intercity distances in order of increasing distance). Then at each stage select an edge that has not been previously chosen of least cost that (1) never requires that three used edges meet at a vertex (because a Hamiltonian circuit uses up exactly two edges at each vertex) and that (2) never closes up a circular tour that doesn't include all the vertices. This algorithm is called the **sorted-edges algorithm**.

EXAMPLE 5
Applying the Sorted-Edges Algorithm

Applying the sorted-edges algorithm to the TSP in Figure 2.3 works as follows: First, the six weights on the edges listed in increasing order would be 300, 349, 425, 541, 562, and 774. Because the cheapest edge in this sorted list is 300, this is the first edge we select for the tour we are building. Next we add the edge with weight 349 to the tour. The next-cheapest edge would be 425, but using this edge together with those already selected would result in having three edges at a vertex (Figure 2.11a), which is not consistent with having a Hamiltonian circuit. Hence, we do not use this edge. The next-cheapest edge, 541, used together with the edges already selected, would create a circuit (see Figure 2.11b) that does not include all the vertices. Thus, this edge, too, would be skipped over. However, we are able to add the edges 562 and 774 without either creating a circuit shorter than one including all the vertices or having three edges at a vertex. Hence, the tour we arrive at is Chicago, St. Louis, Minneapolis, Cleveland, and Chicago. Again, this solution is not optimal because its length is 1985. Note that this algorithm, like the nearest-neighbor, is greedy.

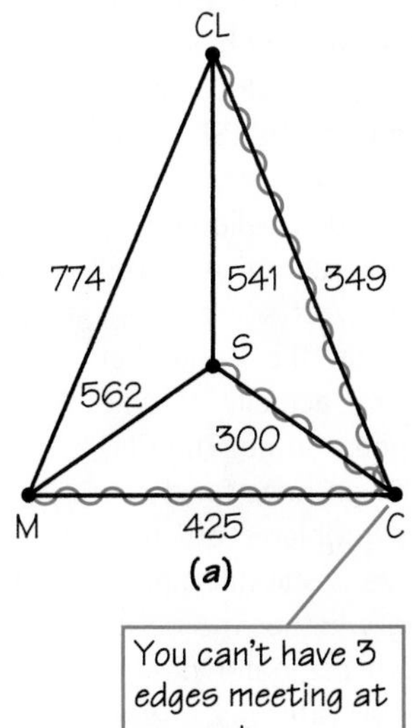

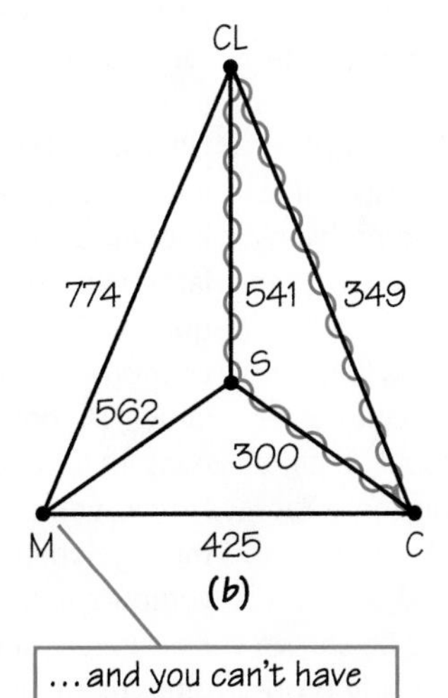

Figure 2.11 (a) When three shortest edges are added in order of increasing distance, three edges at a vertex are selected, which is not allowed as part of a Hamiltonian circuit. (b) When the edges of distances 300, 349, and 541 are selected, a circuit that does not include all vertices results.

EXAMPLE 6
Applying the Sorted-Edges Algorithm Revisited

Although the edges selected by applying the sorted-edges method to the example in Figure 2.3 are connected to each other at every stage, this does not always happen. For example, if we apply the sorted-edges algorithm to the graph in Figure 2.10a, we build up the tour first with edge *ED* (400) and then edge *BC* (500), which do not touch. The edges that are then selected are *AD*, *AB*, and *EC*, giving the circuit *EDABCE*, which is the same as the nearest-neighbor circuit starting at vertex *A*.

Many "quick-and-dirty" methods for solving the TSP have been suggested; while some methods give an optimal solution in some cases, none of these methods guarantees an optimal solution. Surprisingly, most experts believe that no efficient method that guarantees an optimal solution for the TSP will ever be found (see Spotlight 2.1).

Recently, mathematical researchers have adopted a somewhat different strategy for dealing with TSP problems. If finding a fast algorithm to generate optimal solutions for large problems is unlikely, perhaps we can show that the quick-and-dirty methods, usually called **heuristic algorithms**, come close enough to giving optimal solutions to be important for practical use. For example, suppose we could prove that the nearest-neighbor heuristic was never off by more than 25 percent in the worst case or by more than 15 percent in the average case. For a medium-sized TSP, we would then have to choose whether to spend a lot of time (or money) to find an optimal solution or instead to use a heuristic algorithm to obtain a fairly good solution. Investigators at AT&T Research have developed many remarkably

SPOTLIGHT 2.1
NP-Complete Problems

Steven Cook, a computer scientist at the University of Toronto, showed in 1971 that certain computational problems are equivalently difficult. This class of problems, now referred to as **NP-complete problems**, has the following characteristic: If a "fast" algorithm for solving one of these problems could be found, then a fast method would exist for all these problems.

In this context, "fast" means that as the size *n* of the problem grows (the number of cities gives the problem size in the TSP), the amount of time needed to solve the problem grows no more rapidly than a polynomial function in *n*. (A polynomial function has the form $a_k n^k + a_{k-1} n^{k-1} + \cdots + a_1 n + a_0$.) On the other hand, if it could be shown that any problem in the class of NP-complete problems required an amount of time that grows faster than any polynomial (an exponential function, such as 3^n, is an example of a function that grows faster than any polynomial) as the problem size increased, then all problems in the NP-complete class would share this characteristic. The TSP, along with a wide variety of other practical problems, is known to be NP-complete. It is widely believed that large versions of these problems cannot be solved quickly. Furthermore, the security of some recent cryptographical systems relies on the hope that large NP-complete problems are actually as time consuming to solve as they appear to be. The Clay Foundation is offering a $1 million prize for determining whether NP-complete problems are truly computationally hard. The prize is still unclaimed! Researchers are also exploring whether the development of new approaches to computer design, such as quantum computing, will offer faster ways to solve very difficult problems.

good heuristic algorithms. The best-known guarantee for a heuristic algorithm for a TSP is that it yields a cost no worse than one and a half times the optimal cost. Interestingly, this heuristic algorithm involves solving a Chinese postman problem (see Chapter 1), for which a "fast" algorithm is known to exist.

Throughout our discussion of the TSP, we have concentrated on the goal of minimizing the cost (or time) of a tour that visited each of a variety of sites once and only once. However, the subtle issues that arise in specific real-world situations (or that provide a contrast between seemingly similar situations) are the things that make mathematical modeling exciting. For example, suppose the TSP situation is to pick up day campers and take them to and from the camp. The camp wants to minimize the total length of time that the bus needs to pick up the campers. The parents of the campers, however, may want to minimize the time their children spend on the bus. For some problems, the tour that minimizes the mean (average) time that a child spends on the bus may not be the same tour that minimizes the total time of the tour. (Specifically, if the bus first picks up the child who lives the farthest from the camp, and then picks up the other children, this may yield a relatively short time on the bus for the kids but a relatively long time for the tour itself.) Mathematicians return to examine these subtleties between problems at a later time, after the basic structure of the main problem itself is well understood. It is in this way that mathematics continues to grow, explore new ideas, and find new applications.

2.4 Minimum-Cost Spanning Trees

The TSP is but one of many graph theory optimization problems that have grown out of real-world problems in both government and industry. Here is another.

EXAMPLE 7
Pictaphone Service

Imagine that Pictaphone service (telephone service that provides a video image of the callers) will be set up on an experimental basis among five cities. The graph in Figure 2.12 shows the possible links that might be included in the Pictaphone network, with each edge showing the cost in millions of dollars to create that particular link. To send a Pictaphone message between two cities, a direct communication link is not necessary because it is possible to send a message indirectly via another city. Thus, in Figure 2.12, sending a message from *A* to *C* could be achieved by sending the message from *A* to *B*, from *B* to *E*, and

Some videophone (pictaphone) and video conferencing requires the creation of specialized networks for which the techniques of finding a minimum-cost spanning tree are required. (© *POOL/Reuters/Corbis*)

Figure 2.12 Costs (in millions of dollars) of installing Pictaphone service among five cities.

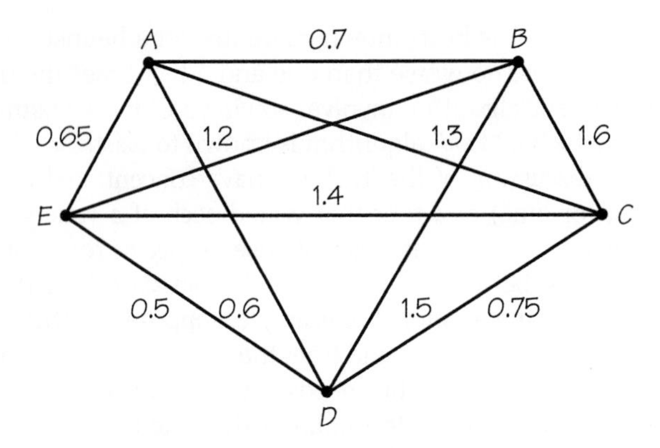

from E to C, provided the links AB, BE, and EC are part of the network. We assume that the cost of relaying a message, compared with the cost of the direct communication link, is so small that we can neglect this amount. The problem that concerns us, therefore, is to provide service between any pair of cities in a way that minimizes the total cost of the links.

Our first guess at a solution is to put in the cheapest possible links between cities first, until all cities could send messages to any other city. Such an approach is analogous to the sorted-edges method that was used to study the traveling salesman problem. In our example, if the cheapest links are added until all cities are joined, we obtain the connections shown in Figure 2.13a.

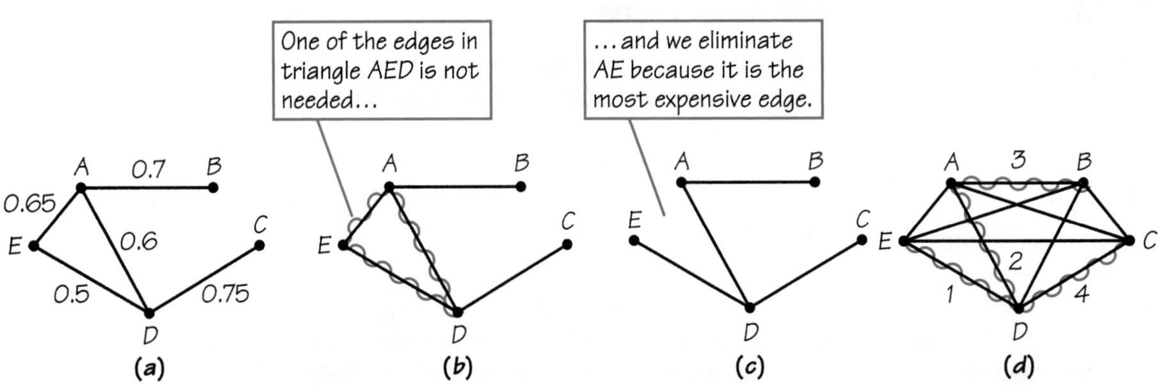

Figure 2.13 (a) Cities are linked in order of increasing cost until all cities are connected. (b) Circuit in part (a) highlighted. (c) Most expensive link in circuit in part (a) deleted. (d) Highlighted edges show, as a subgraph of the original graph, those links connecting the cities with minimum cost, obtained using Kruskal's algorithm. The numbers show the order in which Kruskal's algorithm selects the edges.

The links were added in the order ED, AD, AE, AB, DC. However, because this graph contains the circuit $ADEA$ (wiggly edges in Figure 2.13b), it has redundant edges: We can still send messages between any pair of cities using relays after omitting the most expensive edge in the circuit—AE. After an edge of a circuit is deleted, a message can still be relayed among the cities of the circuit by sending signals the long way around. After AE is deleted, messages from A to E can be sent via D (Figure 2.13c). These ideas constitute a procedure developed by Joseph Kruskal (1928–2010) in 1956.

Kruskal's Algorithm

PROCEDURE

Kruskal's algorithm: Add links in order of cheapest cost so that no circuits form and so that every vertex belongs to some link added (Figure 2.13d).

In Kruskal's procedure, as in the sorted-edges method for the TSP, the edges that are added need not be connected to each other until the end. A subgraph formed in this way will be a **tree;** that is, it will consist of one piece and contain no circuits. It will also include all the vertices of the original graph. A subgraph that is a tree and that contains all the vertices of the original graph is called a **spanning tree** of the original graph.

To understand these concepts better, consider the graph *G* in Figure 2.14a. The wiggly edges in Figure 2.14b would constitute a subgraph of *G* that is a tree (because it is connected and has no circuit), but this tree would not be a spanning tree of *G* because the vertices *D* and *E* would not be included. On the other hand, the wiggly edges in Figure 2.14c and 2.14d show subgraphs of *G* that include all the vertices of *G* but are not trees because the first is not connected and the second contains a circuit. Figure 2.14e shows a spanning tree of *G*; the wiggly edges are connected and contain no circuit, and every vertex of the original graph is an endpoint of some wiggly edge.

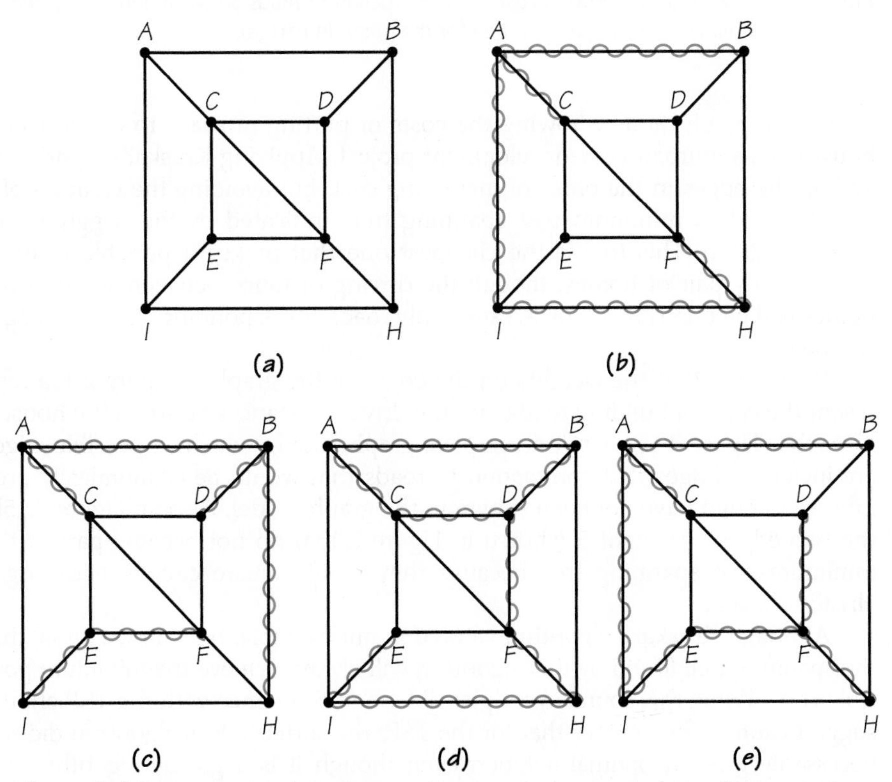

Figure 2.14 (a) A graph to help illustrate the concept of a spanning tree. (b) The wiggly edges are a tree, but not a spanning tree, because vertices *D* and *E* are not part of the tree. (c) The wiggly edges are not a tree, because they are not connected. All of the vertices of the graph are, however, end points of wiggly edges. (d) The wiggly edges are not a tree, because they contain the edges of the circuit *BDCAB*. All the vertices of the graph are, however, endpoints of wiggly edges. (e) The wiggly edges form a tree and include all of the vertices of the graph as endpoints of wiggly edges. Thus, the wiggly edges are a spanning tree.

Finding a **minimum-cost spanning tree**—that is, a spanning tree whose edge weights sum to a minimum value—solves the Pictaphone problem. Note that having a different goal in the Pictaphone problem led to a different mathematical question from that of finding a Chinese postman tour or TSP tour. This application required that we find a minimum-cost spanning tree. In Figure 2.15a,

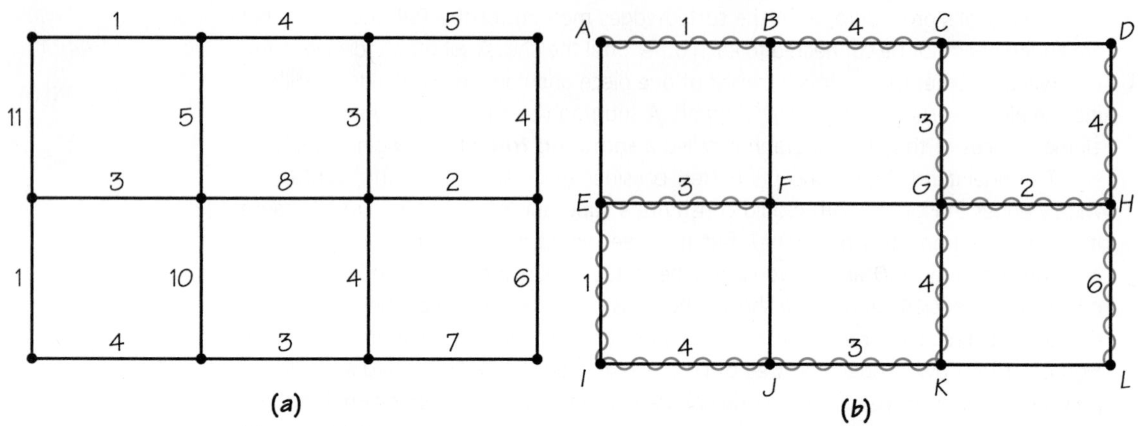

Figure 2.15 (a) A graph showing costs for construction of roads between houses. (b) Wiggly edges show a minimum-cost spanning tree for the graph in part (a).

we have a graph model showing the costs of putting in roads to connect new houses in a suburban land-development project. Applying Kruskal's algorithm–adding the edges in the order of increasing cost, but avoiding the creation of a circuit—yields a minimum-cost spanning tree, indicated by the wiggly edges in Figure 2.15b. This tree is the cheapest one that makes it possible to drive between any pair of homes, though the driving distance between some of the homes will be relatively large, because only roads corresponding to wiggly edges will be built.

Remember that the weights on the edges of the graph in Figure 2.15a represent the costs of building roads, not the driving distances between the houses. Note that Figure 2.15a is not a complete graph, one in which all possible edges are included. Edges that correspond to roads that would be economically prohibitive to build have not been shown in the graph model. Also, in Figure 2.15b, the two edges of weight 5 (shown in Figure 2.15a) do not become part of the minimum-cost spanning tree because they would create circuits with edges already chosen.

Although Kruskal's algorithm worked in our example, how do we know that the spanning tree found by this algorithm will always achieve the minimum possible cost? While this sounds very plausible, our experience with the TSP should suggest caution. Remember that for the TSP, the sorted-edges algorithm did not necessarily give an optimal solution even though it is a greedy algorithm like Kruskal's. On what basis should we have more faith in Kruskal's algorithm?

Kruskal proposed his algorithm as a way to solve a pure mathematics problem put forward by Czechoslovakian mathematician Otakar Borůvka. In mathematics, it is surprising but not uncommon to find that ideas used to solve problems with no apparent application often turn out to have many real-world uses. Kruskal's solution to the problem of finding a minimum-cost spanning tree in a graph with weights is a good example of this phenomenon.

Kruskal showed that the greedy algorithm described does yield the minimum answer, and his work led to applications of these and related ideas in designing minimum-cost computer networks, phone connections, sewer systems, and road and railway systems. For additional discussion of operations research in the communications industry, see Spotlight 2.2. To explore how one can reconstruct full information from partial information using the tree concept, see Spotlight 2.3.

SPOTLIGHT 2.2

AT&T Manager Explains How Long-Distance Calls Run Smoothly

Although long-distance calls are now routine, it takes great expertise and careful planning for a company like AT&T to handle its vast amounts of telephone traffic. Rich Wetmore was district manager of AT&T's Communications Network Operations Center in Bedminster, New Jersey. Here are his responses to questions about how AT&T handles its huge volumes of long-distance traffic and how it tracks its operations to keep things running smoothly.

How do you make sure that a customer doesn't run into a delayed signal when attempting a long-distance call?

We monitor the performance of our AT&T network by displaying data collected from all over the country on a special wallboard. The wallboard is configured to tell us if a customer's call is not going through because the network doesn't have enough capacity to handle it.

That's when we step in and take control to correct the problem. The typical control we use is to reroute the call. Instead of sending the customer's call directly to its destination, we'll route it via a third city—to someplace else in the country that has the capacity to complete the call.

It would seem that routing via another city would take longer. Is the customer aware of this process?

Routing a call via a third city is entirely transparent [imperceptible] to the customer. I'm an expert about the network, and even when I make a phone call, I have no idea how that individual call was routed. It's transparent both in terms of how far away the other person sounds and in how quickly the telephone call gets set up. With the signaling network we use, it takes milliseconds for switching systems to "talk" to each other to set up a call. So the fact that you are involved in a third switch in some distant city is something you would never know.

You want to be sure to keep costs down while supplying enough service to customers. So how do you balance company benefits with customer benefit?

In terms of making the network efficient, we want to do two things. First, we want our customers to be happy with our service and for all their calls to go through, which means we must build enough capacity in the network to allow that to happen. Second, we want to be efficient for stockholders and not spend more money than we need to for the network to be at the optimum size.

There are basically two costs in terms of building the network. There is the cost of switching systems and the cost of the circuits that connect the switching systems. Basically, you can use operations-research techniques and mathematics to determine cost trade-offs. It may make sense to build direct routing between two switching systems and use a lot of circuits, or maybe to involve three switching systems, with fewer circuits between the main two, and so on.

In our discussion of routing problems in graphs, we have not touched on one of the most obvious ones: finding the path between two specified, distinct vertices while keeping the sum of the weights of the edges in the path as small as possible. (Here there is no need to cover all vertices or to cover all edges.) We have seen that the weights on the edges have many possible interpretations, including time, distance, and cost. The following are some of the many possible applications:

1. Design routes to be used by an ambulance, police car, or fire engine to get to an emergency as quickly as possible.
2. Design delivery routes that minimize gasoline use.
3. Design routes to bring soldiers to the front as quickly as possible.
4. Design a route for a truck carrying nuclear or biomedical waste.

Many people find it increasingly convenient to use the Web or software installed in their cars to get driving directions and driving time estimates to a place they

SPOTLIGHT 2.3

Common Ancestors?

In the study of ancient manuscripts, different manuscripts of the same book are available, even though the original manuscript upon which they are based has been lost. Examples of this include Euclid's *Elements* and Chaucer's *Canterbury Tales*. What interests scholars is reconstructing the relationships between the manuscripts and the common ancestors of the manuscripts, even when some of the ancestors are now missing.

Similarly, perceptual psychologists may be interested in which colors people perceive as being closely related and comparing these perceptions with those of people who are color-blind. Linguists are interested in the connections between languages that seem very different today, but have some words that are similar. Finally, in studying different species, biologists are interested in determining which species are more closely related to each other, including species known only in fossil form, and constructing a "tree" of life that shows which species were ancestors of others.

Reconstructions of this kind are made possible by using graph theory, specifically using the graph theory concept of a tree. The value of the graph theory in these and many other situations lies in using the distance between pairs of vertices in the tree as a way of reflecting the closeness of the relationships that pairs of manuscripts, pairs of colors, or pairs of species have. The distance between two vertices in a tree is the sum of the weights along the one path that joins the two vertices. If there are no weights on the edges, the distance is the number of edges in the path. In some reconstruction problems, a special vertex of the tree called the *root* is singled out. This root plays the role of the original common ancestors, and distances to the root are of critical interest.

In the case of species, trees of family relatedness were traditionally constructed based on similarities of bones and physical appearance. With the discovery of molecular biology, many new avenues have been opened up. We can now draw trees of relatedness based on an organism's genetic material, DNA, or the proteins that the DNA codes for. The traditional trees based on physical traits often show different species as being more closely related than trees based on newer molecular biological approaches. These differences cause scholars to focus on how to resolve the discrepancies and thereby reach a deeper understanding of the unity of life.

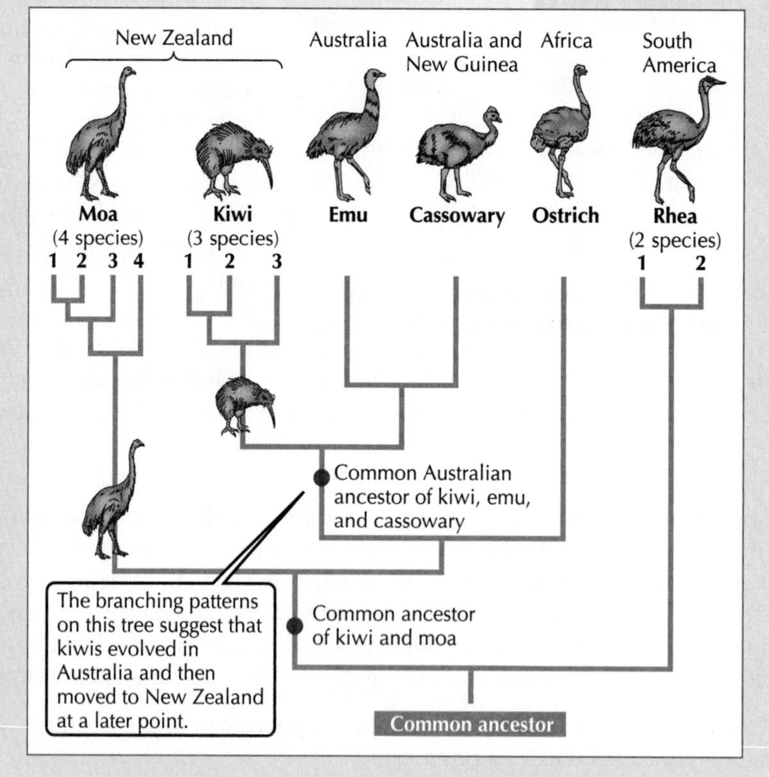

wish to visit. The software that provides this information relies on algorithms that compute the shortest-path route in an appropriate weighted graph, which involves distances or times.

The need to find shortest paths seems natural. Next we investigate a situation in which finding a *longest* path is the right tool.

2.5 Critical-Path Analysis

Mathematics can confirm the obvious in certain situations while showing that our intuition is wrong in other circumstances. Our next group of applications will illustrate this point.

A characteristic of American life is its fast pace. People are interested in getting things done quickly and efficiently. This means that when you take your car in to be repaired before going to work, you want to know for sure that the repairs will be done when you pick the car up. You want the trains and the bus that take you to your doctor's appointment to run on time. When you arrive at the doctor's office, you want a technician to be free to take a blood sample and a throat culture. You want your outpatient appointment for an X-ray at the local hospital to occur on schedule. You want the X-ray to be interpreted quickly and the results reported back to your internist.

Scheduling machines and people is a big part of modern life. It is involved in running a school, a hospital, an airline, or in landing a person on Mars, and modern mathematics plays a big part in solving scheduling problems.

Part of what makes scheduling complicated is that the tasks that make up a job usually cannot be done in a random order. For example, to make Thanksgiving dinner, you must buy and prepare the turkey before putting it in the oven, and you must set the table before serving the food.

If the tasks cannot be performed in a random order, we can specify the order in an **order-requirement digraph**. The term *digraph* is short for "directed graph." A digraph is a geometrical tool similar to a graph except that each edge has an arrow on it to indicate a direction for that edge. Digraphs can be used to illustrate that traffic on a street must go in one direction or that certain tasks in a job must be completed before other tasks. A typical example of an order-requirement digraph is shown in Figure 2.16. There is a vertex in this digraph for each task. If one task must be done immediately before another, we draw a directed edge, or arrow, from the prerequisite task to the subsequent task. The numbers within the circles representing vertices are the times it takes to complete the tasks. In Figure 2.16, there is no arrow from T_1 to T_5 because task T_2 intervenes. Tasks T_7 and T_8 are independent of each other. There are no directed arrows entering or leaving them, and such arrows would force them to be performed in a particular order with respect to the other tasks or to each other. Also, T_1, T_7, and T_8 have no tasks that must precede them. Hence, if there are at least three processors (such as people or machines) available, tasks T_1, T_7, and T_8 can be worked on simultaneously at the start of the job.

Let's investigate a typical scheduling problem faced by a business.

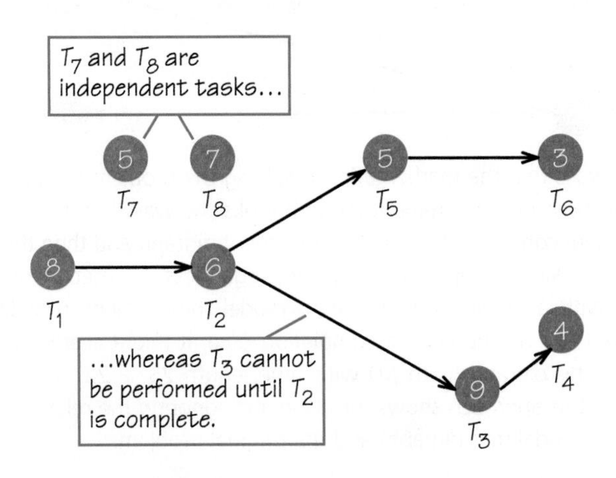

Figure 2.16 A typical order-requirement digraph.

EXAMPLE 8
Turning a Plane Around

Consider an airplane that carries both freight and passengers. The plane must have its passengers and freight unloaded and new passengers and cargo loaded before it can take off again. Also, the cabin must be cleaned before departure can occur. Thus, the job of "turning the plane around" requires completion of five tasks:

TASK *A*	Unload passengers	13 minutes
TASK *B*	Unload cargo	25 minutes
TASK *C*	Clean cabin	15 minutes
TASK *D*	Load new cargo	22 minutes
TASK *E*	Load new passengers	27 minutes

Turning a plane around, which involves such tasks as refueling, unloading, cabin cleanup, and then reloading cargo and passengers, entails very careful scheduling to avoid time slippage. (*David Butow/Corbis Saba.*)

The order-requirement digraph for the problem of turning an airplane around is shown in Figure 2.17. The presence or absence of an edge in the order-requirement digraph depends on the analysis made as part of the modeling process for the problem. It seems natural that we need an arrow between task *A* and task *C*, because the passengers have to be unloaded before the cabin is cleaned. Other arrows may not seem natural—say, perhaps the arrow from task *B* (unload the cargo) to task *E* (load new passengers). This arrow may be due to government rules or union requirements.

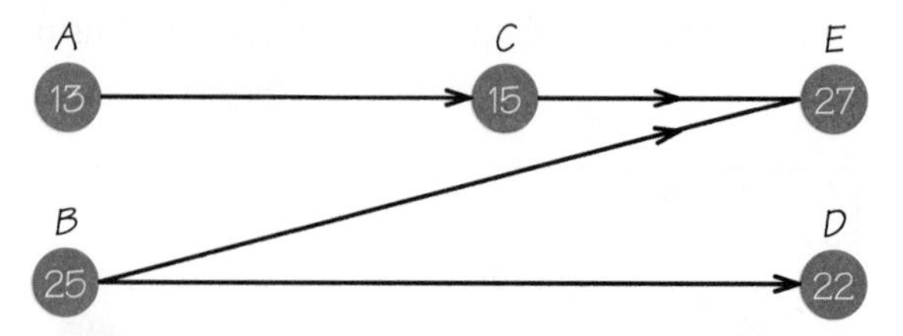

Figure 2.17 An order-requirement digraph for turning an airplane around after landing.

What matters is that the mathematics of solving the problem does not depend on the reason that the order-requirement digraph looks the way that it does. The person solving the problem constructs the order-requirement digraph and then the mathematical techniques we will develop can be applied, regardless of whether or not another business faced with a similar problem might model the problem in a different way. Because we want to find the earliest completion time, it might seem that finding the shortest path in the digraph (path *BD* with time length $25 + 22 = 47$) would solve the problem. But this approach shows the danger of ignoring the relationship between the mathematical model (the digraph) and the original problem.

The time required to complete all the tasks, *A* through *E*, must be at least as long as the time necessary to do the tasks on any particular path. Consider the path *BD*, which has length 25 + 22 = 47. Recall that here *length* of a path refers to the sum of the times of the tasks that lie along the path. Because task *B* must be done before task *D* can begin, the two tasks *B* and *D* cannot be completed before time 47. Hence, even if work on other tasks (such as *A*, *C*, and *E*) proceeds during this period, all the tasks cannot be finished before the tasks on path *BD* are finished. The same statement is true for every other path in the order-requirement digraph. Thus, the earliest completion time actually corresponds to the length of the longest path. In the airplane example, this earliest completion time is 55 (= 13 + 15 + 27) minutes, corresponding to the path *ACE*. We call *ACE* the **critical path** because the times of the tasks on this path determine the earliest completion time.

Critical Path DEFINITION

A **critical path** in an order-requirement digraph is a longest path. The length is measured in terms of summing the task times of the tasks making up the path.

Note that if none of these tasks could go on simultaneously, the time to complete all the tasks would be 13 + 25 + 15 + 22 + 27 = 102 minutes. However, even though some tasks may be performed simultaneously, the fact that the length of the critical path is 55 means that completion of the tasks in less than 55 minutes is not possible. Only by speeding up the times to complete the critical-path tasks themselves can a completion time less than 55 minutes be achieved.

Suppose it were desirable to speed the turnaround of the plane to less than 55 minutes. One way to do this might be to build a second jetway to help unload passengers more quickly. For example, we could unload passengers (task *A*) in 7 minutes instead of 13. However, reducing task *A* to 7 minutes does not reduce the completion time by 6 minutes, because in the new digraph (Figure 2.18) *ACE* is no longer the critical (longest) path. The longest path is now *BE*, which has a length of 52 minutes. Thus, shortening task *A* by 6 minutes results in only a 3-minute saving in completion time. This may mean that building a new jetway is uneconomical.

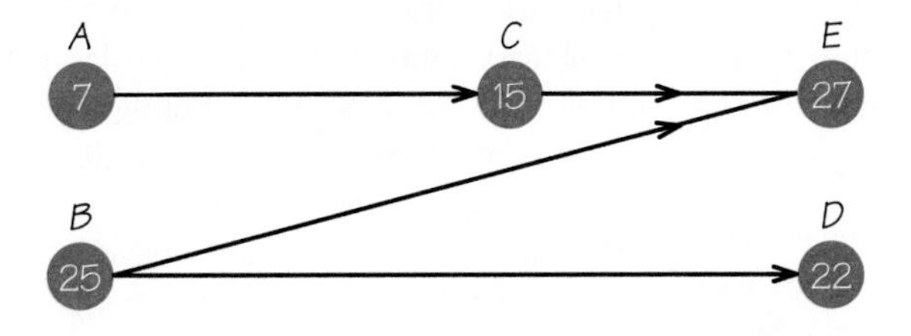

Figure 2.18 An order-requirement digraph for turning an airplane around in reduced time due to construction of a new jetway.

Note also that shortening the time to complete tasks that are not on the original critical path *ACE* will not shorten the completion time at all. Speeding up tasks on the critical path will shorten completion time of the job only up to the point where a new critical path is created. Also note that a digraph may have more than one longest path.

Not all order-requirement digraphs are as simple as the one shown in Figure 2.17. The order-requirement digraph in Figure 2.19 has 12 paths, which can be found by exhaustive search. Examples of such paths are $T_1T_2T_3$, $T_1T_5T_9$, $T_4T_5T_9$, and $T_7T_5T_3$. (Although we have not discussed them here, fast algorithms for finding longest and shortest paths in graphs are known.) The critical path is $T_7T_8T_6$ (length 21), and the earliest completion time for all nine tasks is time 21, though the actual completion time may be later than time 21, depending on the resources available to carry out the tasks. Completing the tasks by time 21 depends on having sufficient resources available so that some of the tasks can be worked on simultaneously.

These examples are typical of many scheduling problems that occur in practice (see Spotlight 2.4). Perhaps the most dramatic use of critical-path analysis is in the construction trades. No major new building project is now carried out without a critical-path analysis first being performed to ensure that the proper personnel and materials are available at the right times in order to have the project finished as quickly as possible. Many such problems are too large and complicated to be solved without the aid of computers.

The critical-path method was popularized and came into wider use as a consequence of the *Apollo* project. This project, which aimed at landing a man on the moon within 10 years of 1960, was one of the most sophisticated projects in planning and scheduling ever attempted. The dramatic success of the project can be attributed partly to the use of critical-path ideas and the related program evaluation and review technique (PERT), which helped keep the project on schedule.

In Chapter 3, we will see how mathematical ideas drawn from outside of graph theory can be used to gain insight into scheduling problems.

SPOTLIGHT 2.4

Every Moment Counts in Rigorous Airline Scheduling

When people think of airline scheduling, the first thing that comes to mind is how quickly a particular plane can safely reach its destination. But using ground time efficiently is just as important to an airline's timetable as the time spent in flight. Bill Rodenhizer was the manager of control operations for an airline that provided shuttle service between Boston and New York. He is considered to be an expert on airplane turnaround time, the process by which an airplane is prepared for almost immediate takeoff once it has landed. He tells us how this well-orchestrated effort works:

> Scheduling, to the airline, is just about the whole ball game. Everything is scheduled right to the minute. The whole fleet operates on a strict schedule. Each of the departments responsible for turning around an aircraft has an allotted period of time in which to perform its function. Manpower is geared to the amount of ground time scheduled for that aircraft. This would be adjusted during off-weather or bad-weather days or during heavy air-traffic delays.
>
> Most of our aircraft in Boston are scheduled for a 42- to 65-minute ground time. Boston is the end of the line, so it is a "terminating and originating station." In plain talk, that means almost every aircraft that comes in must be fully unloaded, refueled, serviced, and dispatched within roughly an hour's time.
>
> This is how the process works: In the larger aircraft, it takes passengers roughly 20 minutes to load and 20 minutes to unload. During this period, we will have completely cleaned the aircraft and unloaded the cargo, and the caterers will have taken care of the food. The ramp service may take 20 to 30 minutes to unload the baggage, mail, and cargo from underneath the plane, and it will take the same amount of time to load it up again. We double-crew those aircraft with heavier weights so that the workload will fit the time it takes passengers to load and unload upstairs.
>
> While this has been going on, the fueler has fueled the aircraft. As to repairs, most major maintenance is done during the midnight shift, when [most of our] several hundred aircraft are inactive. We all work under a very strict time frame.

New security requirements in the wake of the World Trade Center attack (9/11/2001) have increased the difficulty of adhering to timetables in operating shuttle services between East Coast cities such as New York and Boston. This makes it even more important to use analytical tools in keeping operations on schedule.

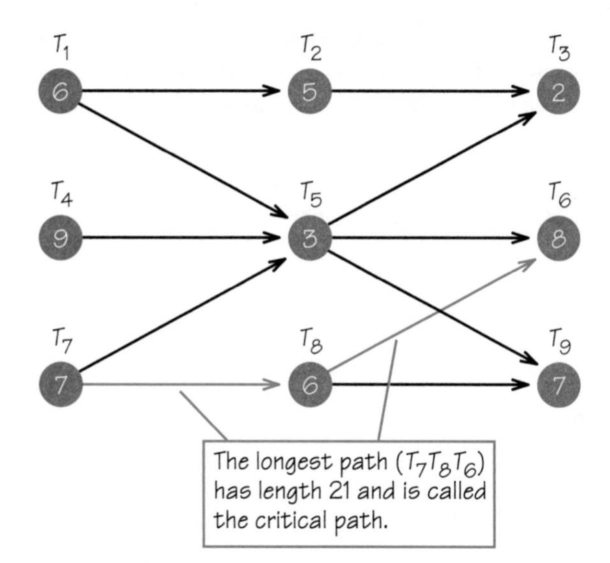

Figure 2.19 An order-requirement digraph with 12 paths, to examine how to find the length of the longest path.

REVIEW VOCABULARY

Algorithm A step-by-step description of how to solve a problem. (p. 38)

Brute force method The method that solves the traveling salesman problem (TSP) by enumerating all the Hamiltonian circuits and then selecting the one with minimum cost. (p. 42)

Complete graph A graph in which every pair of vertices is joined by an edge. (p. 40)

Critical path The longest path in an order-requirement digraph. The length of this path gives the earliest completion time for all the tasks making up the job consisting of the tasks in the digraph. (p. 55)

Fundamental principle of counting A method for counting outcomes of multistage processes. (p. 41)

Greedy algorithm An approach for solving an optimization problem, where at each stage of the algorithm the best (or cheapest) action is taken. Unfortunately, greedy algorithms do not always lead to optimal solutions. (p. 44)

Hamiltonian circuit A circuit using distinct edges of a graph that starts and ends at a particular vertex of the graph and visits each vertex once and only once. A Hamiltonian circuit can start at any one of its vertices. (p. 35)

Heuristic algorithm A method of solving an optimization problem that is "fast" but does not guarantee an optimal answer to the problem. (p. 46)

Kruskal's algorithm An algorithm developed by Joseph Kruskal (AT&T Research) that solves the minimum-cost spanning-tree problem by selecting edges in order of increasing cost, but in such a way that no edge forms a circuit with edges chosen earlier. It can be proved that this algorithm always produces an optimal solution. (p. 48)

Method of trees A visual method of carrying out the fundamental principle of counting. (p. 38)

Minimum-cost Hamiltonian circuit A Hamiltonian circuit in a graph with weights on the edges, for which the sum of the weights of the edges of the Hamiltonian circuit is as small as possible. (p. 38)

Minimum-cost spanning tree A spanning tree of a weighted connected graph having minimum cost. The cost of a tree is the sum of the weights on the edges of the tree. (p. 49)

Nearest-neighbor algorithm An algorithm for attempting to solve the TSP that begins at a "home" vertex and visits next that vertex not already visited that can be reached most cheaply. When all other vertices have been visited, the tour returns to home. This method may not give an optimal answer. (p. 43)

NP-complete problems A collection of problems, which includes the TSP, that appear to be very hard to solve quickly for an optimal solution. (p. 46)

Order-requirement digraph A directed graph that shows which tasks precede other tasks among the collection of tasks making up a job. (p. 53)

Sorted-edges algorithm An algorithm for attempting to solve the TSP where the edges added to the circuit being built up are selected in order of increasing cost, but no edge is chosen that would prevent a Hamiltonian circuit from forming. These edges must all be connected at the end, but not necessarily at earlier stages. The tour obtained may not have the lowest possible cost. (p. 45)

Spanning tree A subgraph of a connected graph that is a tree and includes all the vertices of the original graph. (p. 49)

Traveling salesman problem (TSP) The problem of finding a minimum-cost Hamiltonian circuit in a complete graph where each edge has been assigned a cost (or weight). (p. 42)

Tree A connected graph with no circuits. (p. 38)

Weight A number assigned to an edge of a graph that can be thought of as a cost, distance, or time associated with that edge. (p. 38)

SKILLS CHECK

1. Which of the statements below is false for the graph G shown?

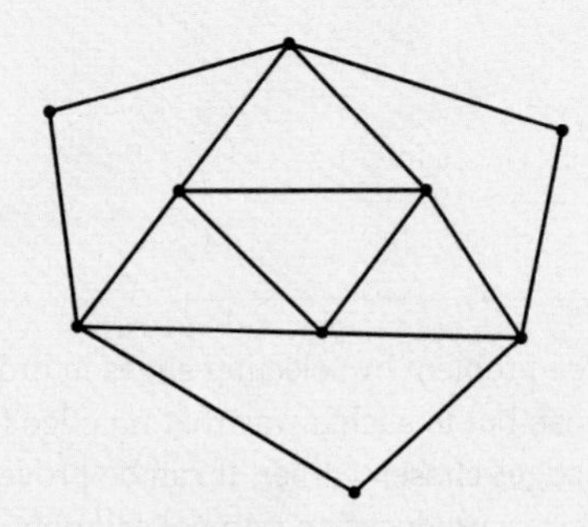

(a) G has an Euler circuit.
(b) G has no spanning tree.
(c) G is not even-valent.
(d) G has no Hamiltonian circuit.

2. The cost of the nearest neighbor tour for the accompanying graph, starting at vertex 3, is ________________.

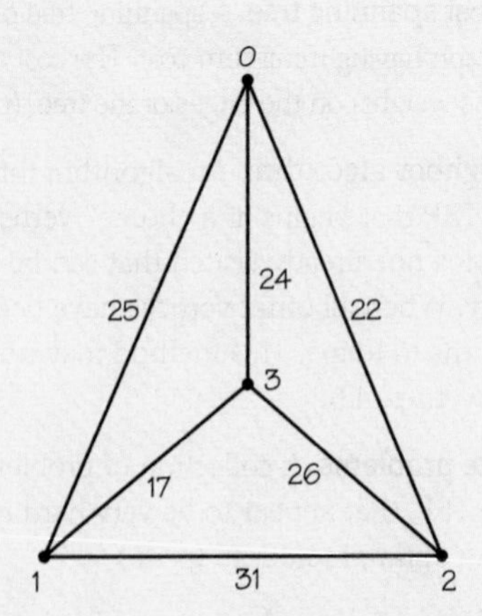

3. The cost associated with the TSP tour 0, 2, 1, 3, 0 in the accompanying graph is

(a) 48.
(b) 47.
(c) −48.

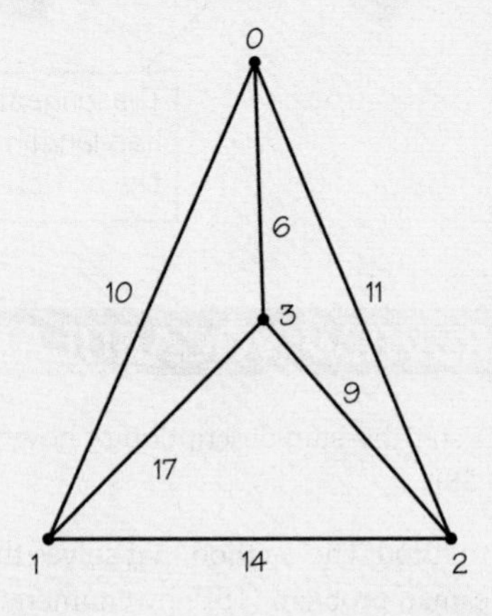

4. The difference between the cost of a nearest-neighbor tour starting at D and the cost of a sorted-edges tour for the accompanying graph is ________.

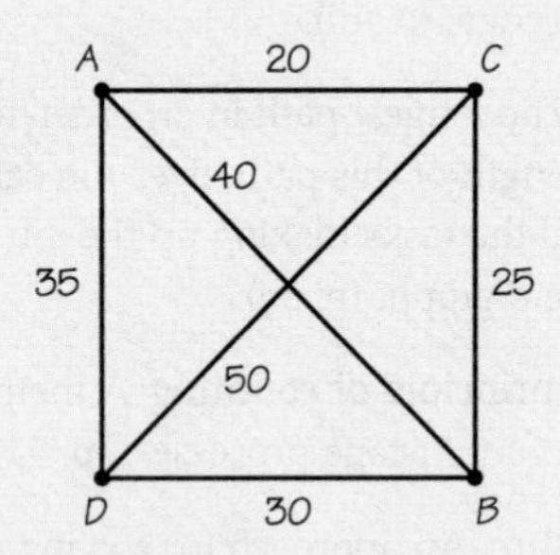

5. The tour 1, 2, 3, 4 is not a Hamiltonian circuit because

(a) it does not include all of the vertices of the graph.
(b) it is not a circuit.
(c) there are other ways to get from 1 to 4 in this graph.

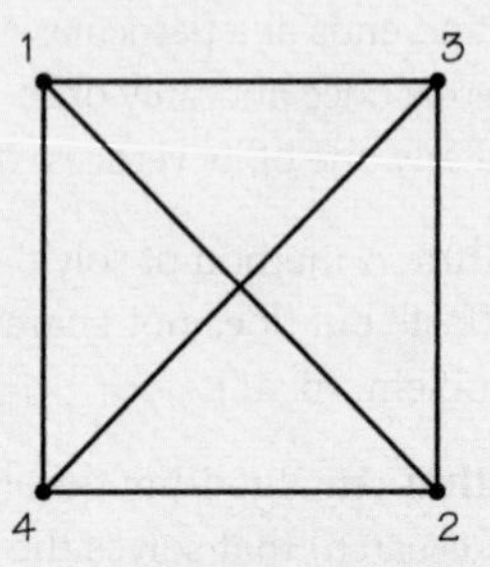

6. The cost of the nearest-neighbor tour (Hamiltonian circuit) that starts at vertex A for the accompanying graph is ____________________ .

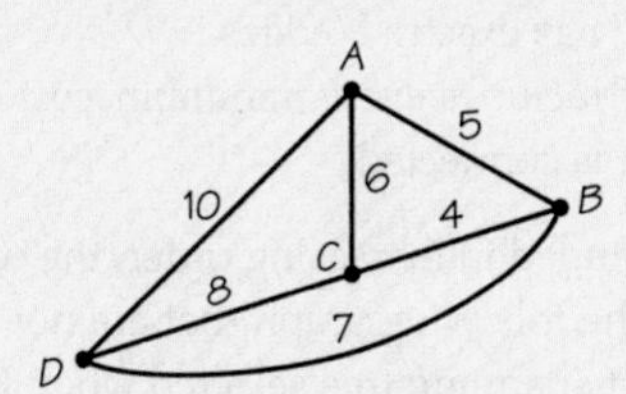

7. The accompanying graph has no Hamiltonian circuit because

(a) it is even-valent.
(b) it has an odd number of vertices.
(c) it is not connected.

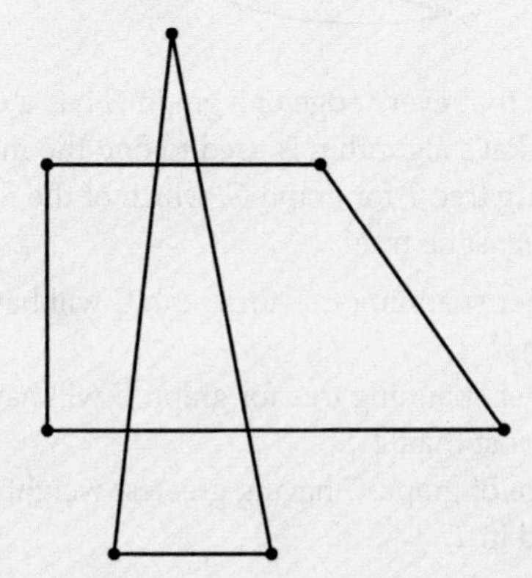

8. The cost of the sorted-edges tour (Hamiltonian circuit) for the accompanying graph is

____________________________ .

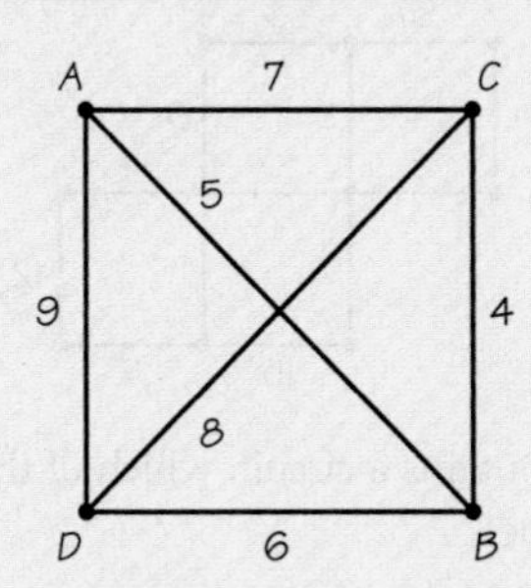

9. Which of the following describes a Hamiltonian circuit for the graph below?

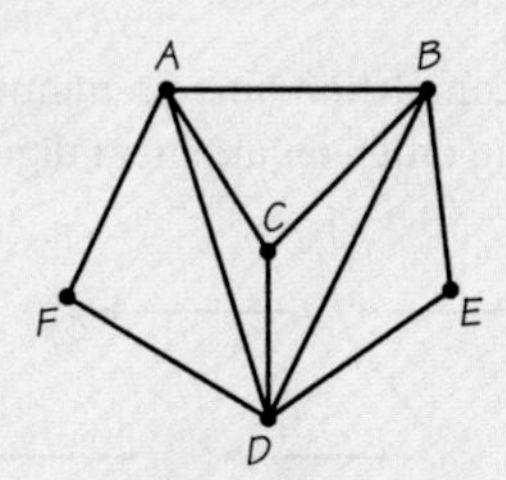

(a) *ABCDFA*
(b) *AFDCBE*
(c) *ACBEDFA*
(d) *ACEBDFA*

10. The cost of the nearest-neighbor traveling salesman tour that starts at B for the graph below is _______ .

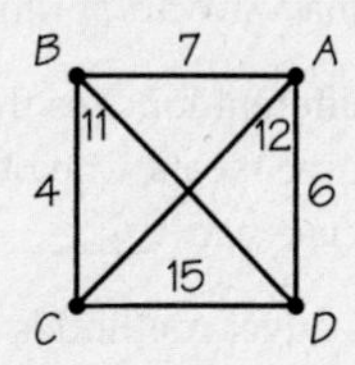

11. Suppose that after a hurricane, a van is dispatched to pick up five nurses at their homes and bring them to work at the local hospital. Which of these techniques is most likely to be useful in solving this problem?

(a) Finding an Euler circuit in a graph
(b) Solving a TSP (traveling salesman problem)
(c) Finding a minimum-cost spanning tree in a graph

12. If a graph has E edges and V vertices as well as a Hamiltonian circuit, then the number of edges in the Hamiltonian circuit is ______________ .

13. The graph shown below has

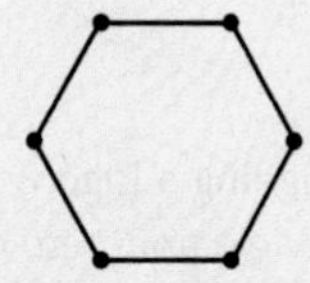

(a) no Hamiltonian circuit and no Euler circuit.
(b) an Euler circuit and a Hamiltonian circuit.
(c) no Hamiltonian circuit, but it has an Euler circuit.

14. The longest-circuit tour of vertices that arises by applying the nearest-neighbor algorithm starting at any of the sites 0, 1, 2, 3 in the graph below has a cost of ________.

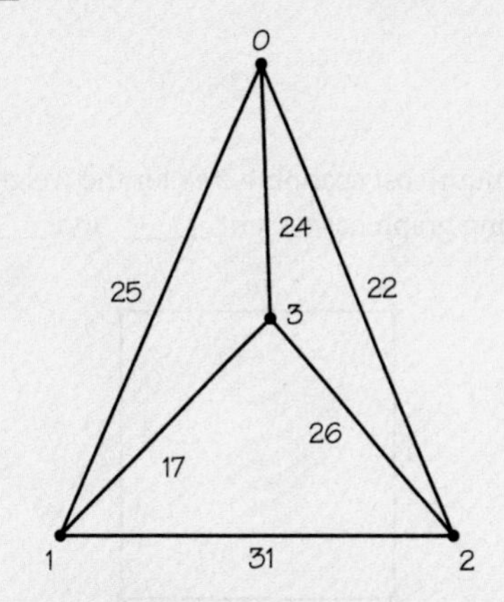

15. When the sorted-edges method and nearest-neighbor method are applied to a complete graph on seven vertices with nonnegative weights,

(a) both methods always give the same optimal answer.
(b) both methods always give the same answer but that answer may not be optimal.
(c) neither method may give an optimal answer.

16. The number of different lunches that Jules can design by selecting one of three meats, one of three salads, and one of six vegetables is exactly ______________ .

17. When the sorted-edges method is applied to the TSP where all the edges have distinct costs, which of the following *must* be true of the tour S obtained?

(a) The largest-weight edge cannot be part of tour S.
(b) The shortest-weight edge must be part of tour S.
(c) The length of the tour S must be even.

18. If a three-character password system must begin with a lowercase letter of the English alphabet followed by two decimal digits that may be repeated, the number of different possible passwords is ________________ .

19. Paul has packed four ties, three shirts, and two pairs of pants for a trip. How many different outfits can he create if he never wears a tie?

(a) Fewer than 10
(b) Between 10 and 25
(c) More than 25

20. A company is designing a logo with two identical capital letters with a single non-zero digit between the letters. A typical possible example would be R7R. The number of such logos is ____________.

21. An ice-cream shop offers 3 types of cones, 20 flavors, and 4 different toppings (crushed peanuts, crushed almonds, chocolate bits, or corn flakes). If a customer is allergic to nuts, how many different choices can she choose from?

(a) 240
(b) 120
(c) 25

22. A minimum-cost spanning tree for the weighted accompanying graph has weight _____ and _____ edges.

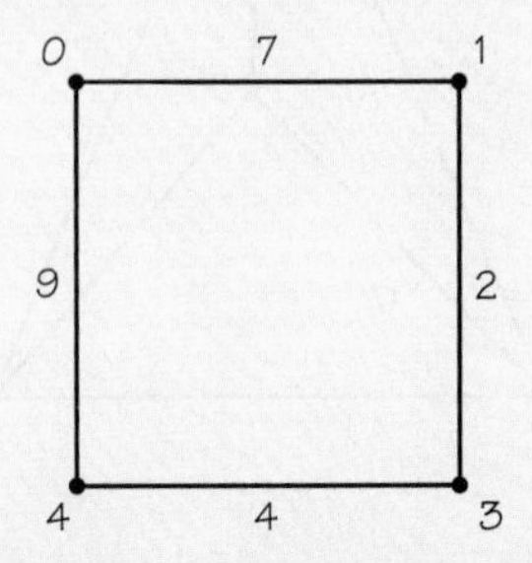

23. Assuming a graph with E edges and V vertices has a minimum-cost spanning tree T, which of the following statements *must* be true?

(a) The tree T has exactly V edges.
(b) The tree T includes every minimum-cost edge.
(c) The graph is connected.

24. When arranged in increasing order, the weights of the edges in the following graph that are not part of the minimum-cost spanning tree selected when Kruskal's algorithm is applied are _______, _______, _______ .

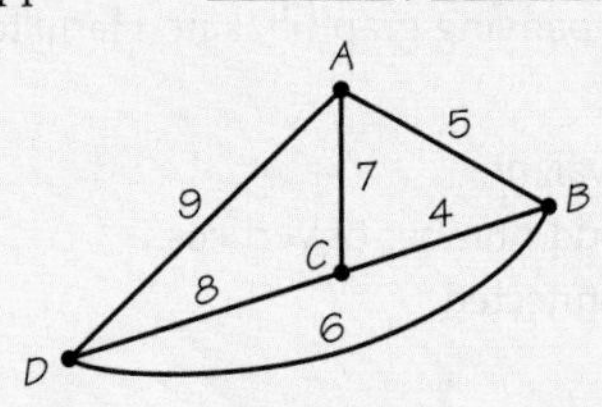

25. Assume that every edge of a graph G has a different cost. If Kruskal's algorithm is used to find the minimum-cost spanning tree T for graph G, which of the following statements *must* be true?

(a) Any other spanning tree for graph G will have more edges than T.
(b) Any other spanning tree for graph G will have a greater cost than T.
(c) The edge of graph G having greatest weight is included in T.

26. The cost of the last edge that Kruskal's algorithm selects to be part of a minimal-cost spanning tree for the weighted graph below is _______.

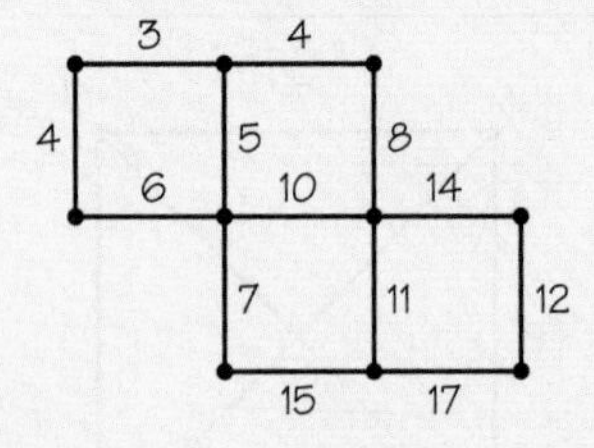

27. If a graph contains a circuit, which of the following statements is true?

(a) The graph cannot be a tree.
(b) The graph must have the same number of vertices as edges.
(c) The graph is not connected.

28. The earliest completion time (in minutes) for a job with the following order-requirement digraph is _____ .

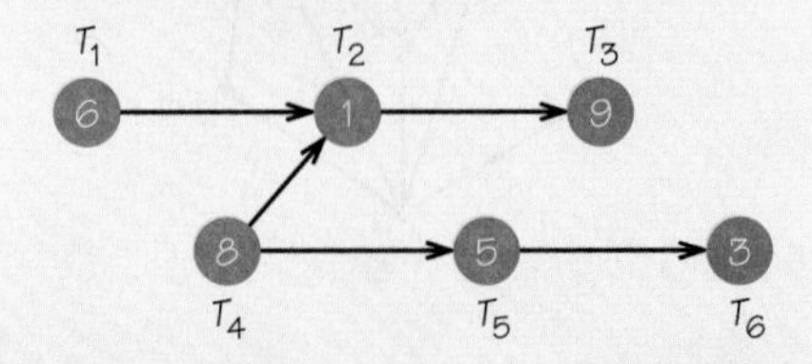

29. Assume a job has an order-requirement digraph with five tasks whose critical path is 25 minutes in length. Based on this information, what can be said about the tasks?

(a) Each task takes exactly 5 minutes.
(b) Some task takes 25 minutes.
(c) The five tasks in total take at least 25 minutes.

30. The length of the critical path in the accompanying order-requirement digraph is ______________ minutes.

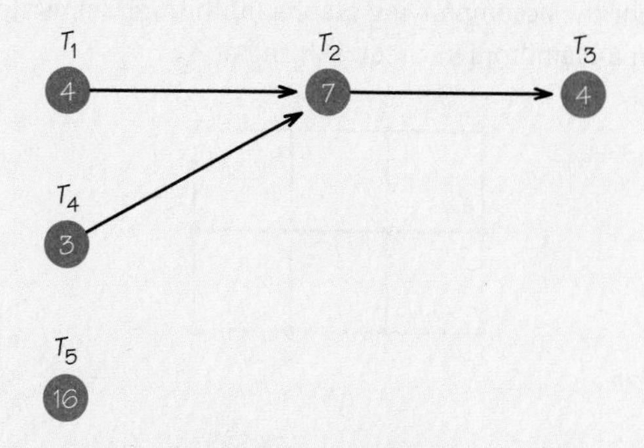

CHAPTER 2 EXERCISES

■ Challenge ▲ Discussion

2.1 Hamiltonian Circuits

1. For the accompanying graphs (a) and (b), write down, if possible, a Hamiltonian circuit starting at X_3.

(a)

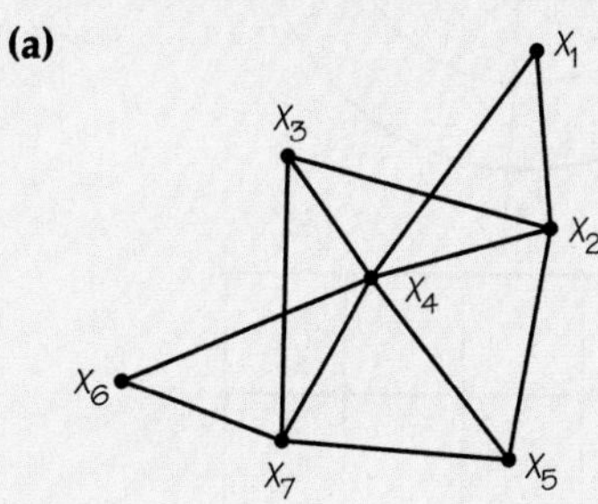

(b)

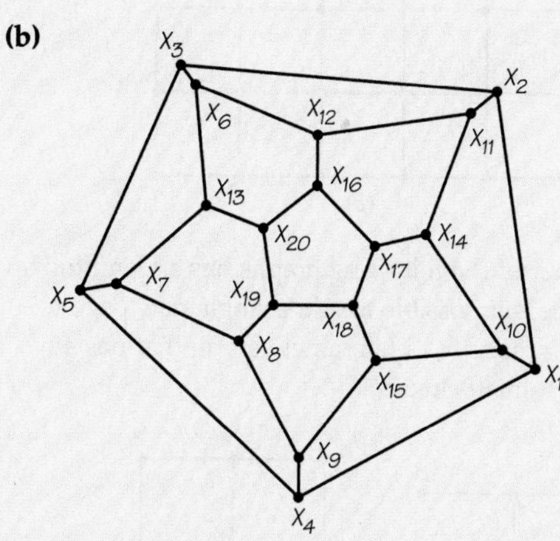

2. **(a)** For the graph below, write down a Hamiltonian circuit that starts at X_3.

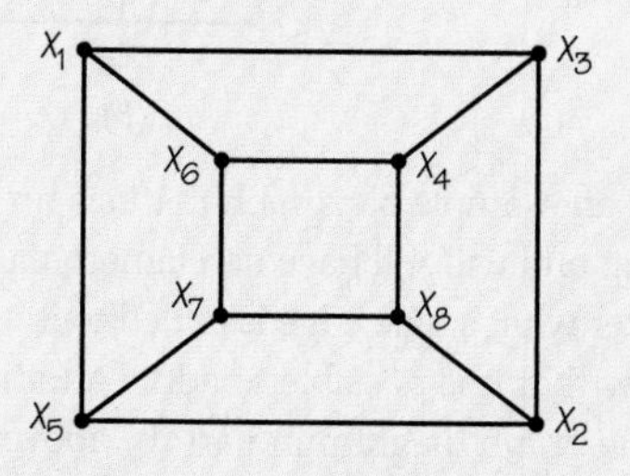

(b) How many vertices are there in the Hamiltonian circuit you found in (a)?

(c) How many edges are there in the Hamiltonian circuit you found in (a)?

(d) What is the largest number of edges you can remove from the graph shown in (a) that will still allow one to find a Hamiltonian circuit in the graph after the edges are removed?

3. For the accompanying graphs (a) through (c), write a Hamiltonian circuit starting at X_3.

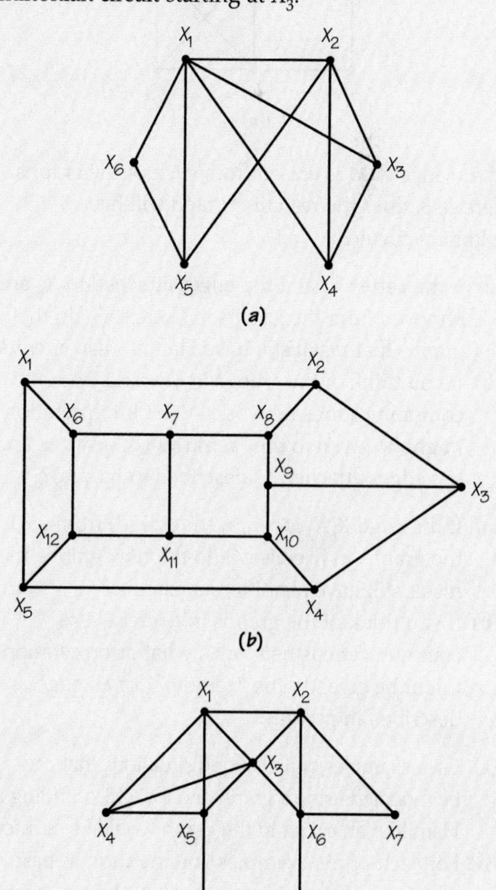

4. For the accompanying graphs (a) through (c), write down a Hamiltonian circuit starting at X_2.

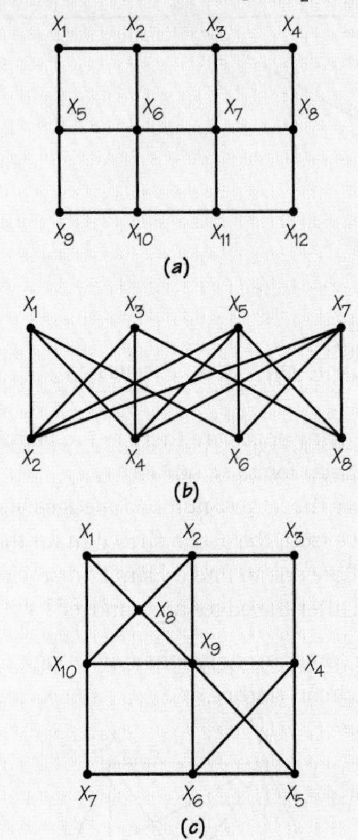

5. If the edge X_2X_3 is erased from each of the graphs in Exercise 3, does the resulting graph still have a Hamiltonian circuit?

6. **(a)** If the vertex X_6 and the edges attached to X_6 are removed from the graphs in Exercise 3, do the new graphs that result still have Hamiltonian circuits?
(b) If you think of the graphs in Exercise 3 as communications networks, what interpretation might be given to the "removal" of a vertex and the edges attached as described in part (a)?

7. **(a)** If the edge X_6X_7 is removed (erased) from each of the graphs in Exercise 4, do the new graphs that result still have Hamiltonian circuits?
(b) If you think of the graphs in Exercise 4 as communications networks, what interpretation might be given to the "removal" of an edge as described in part (a)?

8. **(a)** Give examples of real-world situations that can be modeled using a graph and for which finding a Hamiltonian circuit in the graph would be of interest.
(b) For each of the examples you mention in part (a), can you adapt the question about the real-world situation involved so that finding an Eulerian circuit in the same graph would be of interest?

9. Suppose two Hamiltonian circuits are considered different if the collections of edges that they use are different. How many other Hamiltonian circuits can you find in the graph in Figure 2.1 that are different from the two discussed?

10. **(a)** For each of the following graphs, add wiggly edges to indicate a Hamiltonian circuit.
(b) Can you see how to construct an infinite family of graphs, each with a Hamiltonian circuit, inspired by how you constructed a Hamiltonian circuit in graph (b) of part (a)?

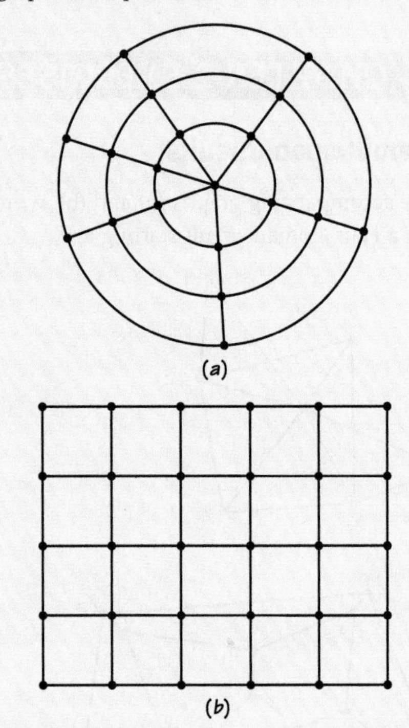

11. **(a)** Neither of the following graphs has a Hamiltonian circuit. Is it possible to add a single new edge to these graphs to obtain a new graph that has a Hamiltonian circuit?

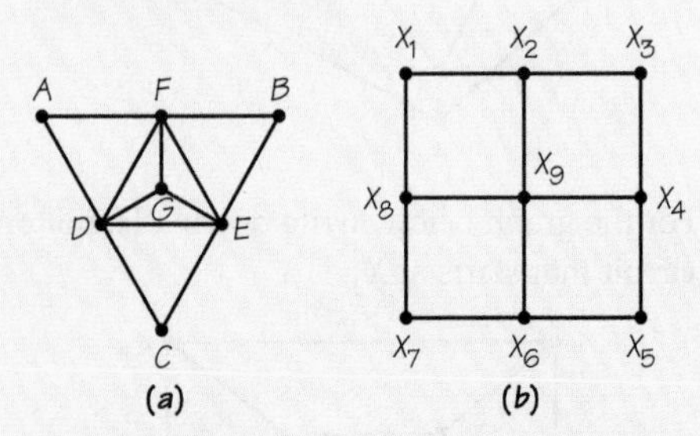

(b) Find an example of a graph that has no Hamiltonian circuit and will still have no Hamiltonian circuit no matter what single edge is added to it.
(c) Show that it is possible to add 4 additional edges to the graph diagram in part (b) above so that the resulting new graph will still have no Hamiltonian circuit.

12. Explain why the graph below has no Hamiltonian circuit.

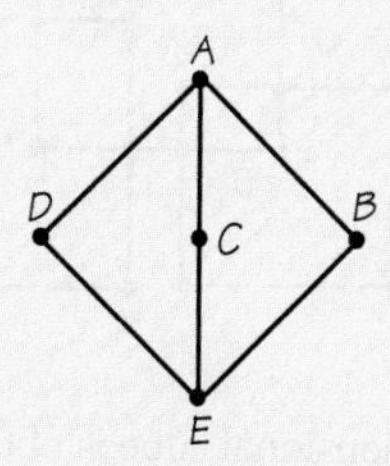

13. Use the graph shown in Exercise 12 to help you construct a connected graph for which every vertex has valence 3 and that does not have a Hamiltonian circuit.

■ **14.** Explain why the tour *BDABCFECB* is not a Hamiltonian circuit for the accompanying graph. Does this graph have a Hamiltonian circuit?

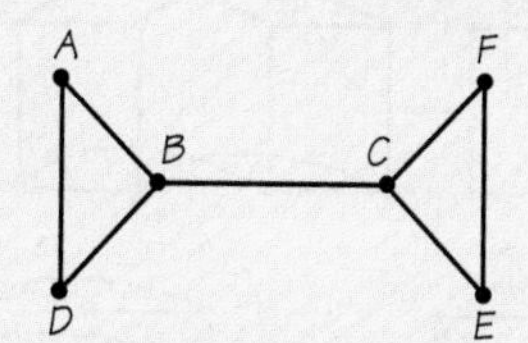

■ **15.** Do the following graphs have Hamiltonian circuits? If not, can you demonstrate why not?

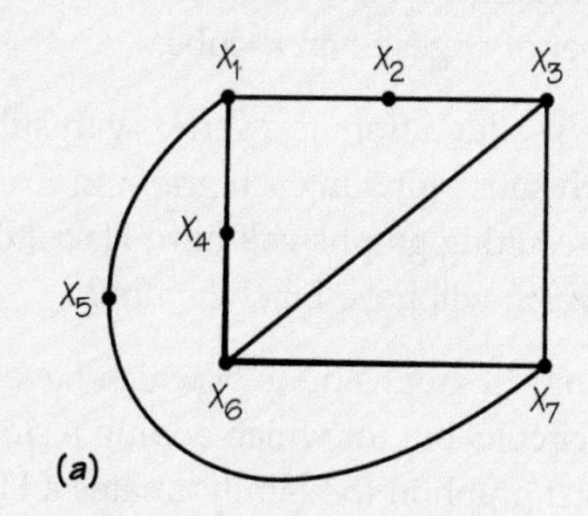

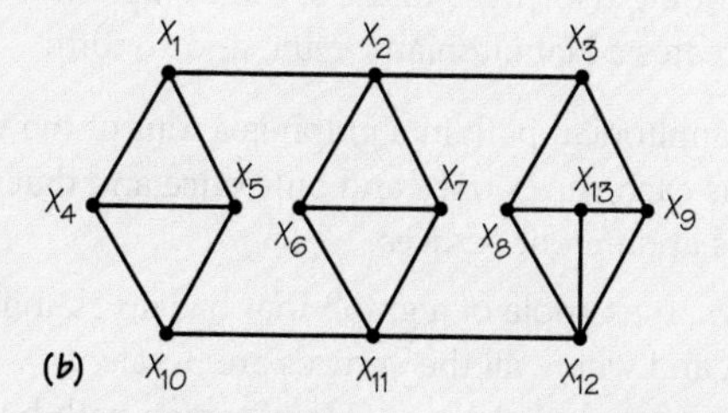

16. If an edge from X_2 to X_5 is added to each graph in Exercise 15, do the new graphs that result have a Hamiltonian circuit?

17. For each of the accompanying graphs, determine whether there is a Hamiltonian circuit.

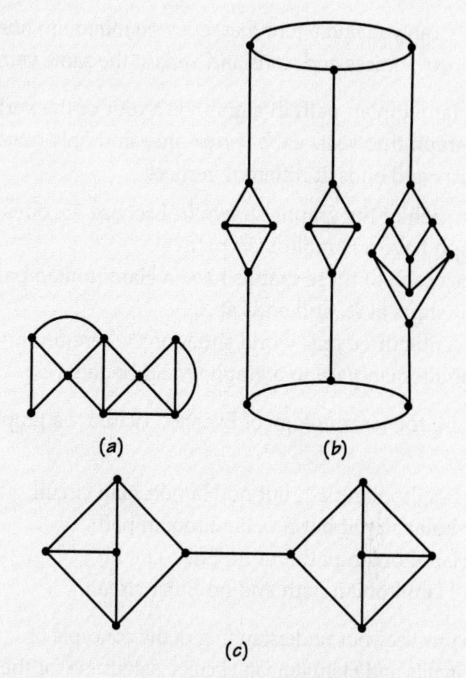

■ **18. (a)** The graph below is known as a four spokes and three concentric circles graph. What conditions on m and n guarantee that an m spokes and n concentric circles graph has a Hamiltonian circuit? (Assume $m \geq 2$, $n \geq 1$.)

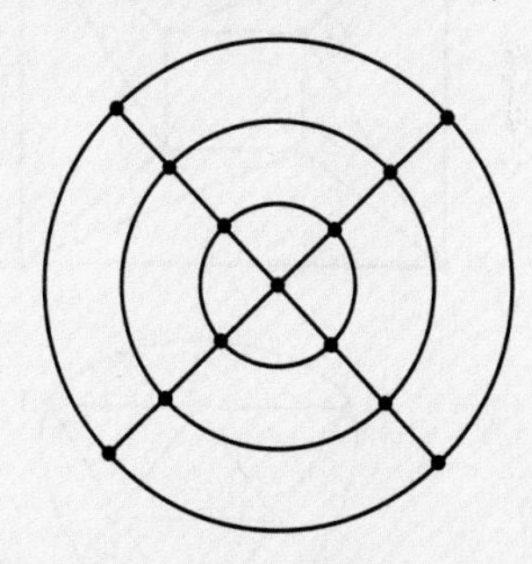

(b) The graph below is known as a 3×4 grid graph. What conditions on m and n guarantee that an $m \times n$ grid graph has a Hamiltonian circuit?

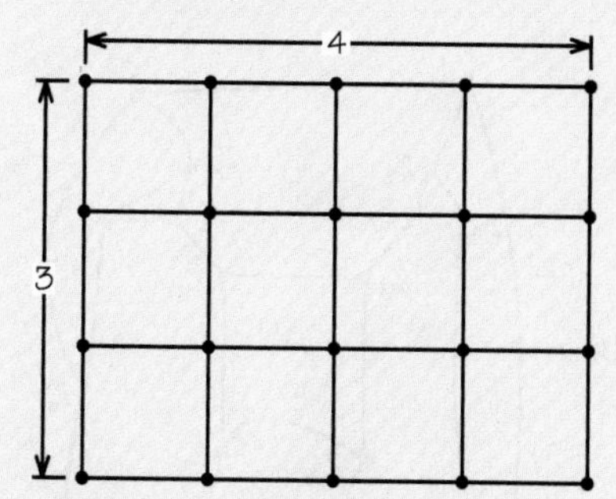

Can you think of a real-world situation in which finding a Hamiltonian circuit in an $m \times n$ grid graph would represent a solution to the problem? If an $m \times n$ grid graph has no Hamiltonian circuit, can you find a tour that repeats a minimum number of vertices and starts and ends at the same vertex?

19. A Hamiltonian path in a graph is a tour of the vertices of the graph that visits each vertex once and only once and starts and ends at different vertices.

(a) For each of the graphs shown in Exercise 15, does the graph have a Hamiltonian path?

(b) Does each of these graphs have a Hamiltonian path that starts at X_1 and ends at X_2?

(c) Describe three real-world situations where finding a Hamiltonian path in a graph would be required.

20. Using the terminology of Exercise 19, draw a graph that has

(a) a Hamiltonian path but no Hamiltonian circuit.

(b) an Euler circuit but no Hamiltonian path.

(c) a Hamiltonian path but no Euler circuit.

(d) no Hamiltonian path and no Euler circuit.

21. To practice your understanding of the concepts of Euler circuits and Hamiltonian circuits, determine for the accompanying graphs (a) through (d) whether there is an Euler circuit and/or a Hamiltonian circuit. If so, write it down.

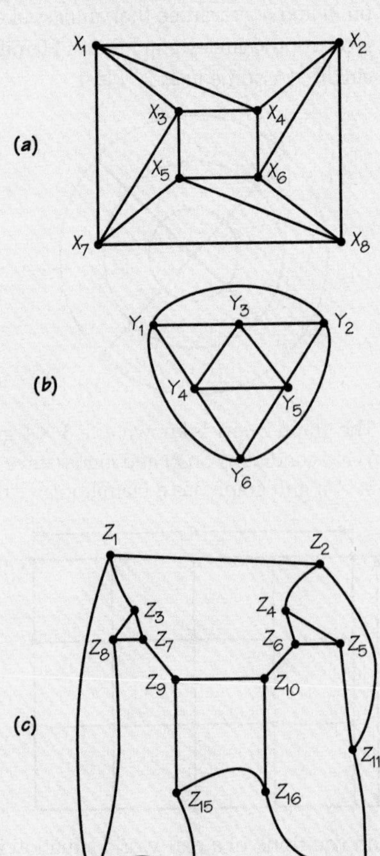

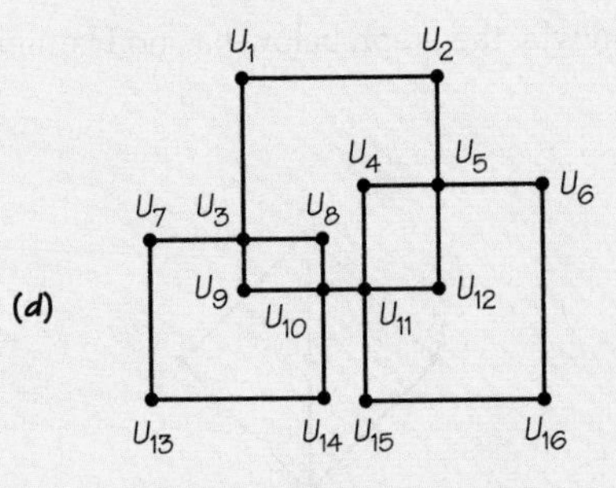

■ **22. (a)** The n-dimensional cube is obtained from two copies of an $(n - 1)$-dimensional cube by joining corresponding vertices. (The process is illustrated for the 3-cube and the 4-cube in the following figure.) Can you show that every n-cube has a Hamiltonian circuit? [*Hint:* Show that if you know how to find a Hamiltonian circuit on an $(n - 1)$-cube, then you can use two copies of this to build a Hamiltonian circuit on an n-cube.]

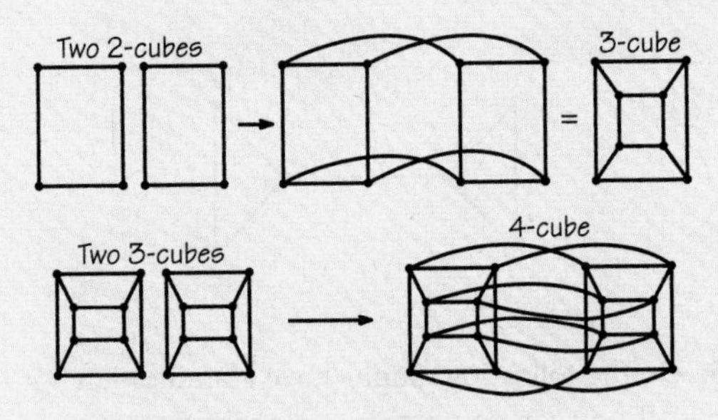

(b) Find formulas for the number of vertices and the number of edges of an n-cube.

23. If an edge is added from the vertex with subscript 4 to the vertex with subscript 5 in each graph in Exercise 21, which of the resulting graphs will have Hamiltonian circuits and which will have Euler circuits?

24. Find a family of graphs none of which have Hamiltonian circuits but for which adding a single edge to the first graph in the family creates a Hamiltonian circuit, adding two edges to the second graph in the family creates a Hamiltonian circuit, and so forth.

25. A Hamiltonian path in a graph is a tour of the vertices that visits each vertex once and only once and that starts and ends at different vertices.

(a) Draw an example of a graph that has no Hamiltonian path and where all the vertices are 3-valent.

(b) Draw a graph that has no Hamiltonian path but that does have an Euler circuit.

(c) By analogy with the Hamiltonian path, develop a definition of "Euler path."

26. (a) When going outside on a cold winter day, Jill can choose from four winter coats (two are red), five wool scarves (one is green), four pairs of boots, and three ski hats (two are blue). How many outfits might her friends see her in?

(b) If Jill always insists on wearing a red coat and a green scarf, how many outfits might her friends see her in?

27. The notes C, D, E, F, G, A, and B are to be used to form an ordered five-note musical logo. In how many ways can this be done if **(a)** no note can be repeated; **(b)** notes can be repeated; **(c)** notes can be repeated but all the notes cannot be the same?

28. A lottery game requires a person to select an uppercase or lowercase letter followed by 3 different odd digits. How many choices can a customer make?

29. (a) In designing a security system for its accounts, a bank asks each customer to choose a five-digit number, all the digits to be distinct and nonzero. How many choices can a customer make?

(b) A suitcase with a liquid-crystal display allows one to unlock it with a specific combination of three capital letters that are not necessarily different. How many choices would a thief have to go through to be sure that all the possibilities had been tried? How does this compare to a "standard" combination lock?

30. To encourage her son to try new things, a mother offers to take him for a dish of ice cream with a topping once a week, for as many weeks as he does not get the same choice as on a previous occasion. If the store offers 12 flavors and six toppings, for how many weeks will she have to do this if her son never picks either of the two types of chocolate ice cream or the three types of nut topping that the store carries?

31. A large corporation has found that it has "outgrown" its current code system for routing interoffice mail. The current system places a code of three ordered, distinct nonzero digits on the mail. The new proposal calls for the use of two ordered capital letters. Does the new system have more code numbers than the old system? If so, how many more locations will the new system enable the company to encode over the current system?

32. Repeat Exercise 27a, except that exactly one of the notes in the musical logo must be a sharp and the note chosen to be sharped cannot appear elsewhere (for example, BCD#AG, where D# denotes D sharp).

▲ **33. (a)** In New York State, one type of license plate has three letters followed by three numbers. Suppose the digits from 0, 1, · · ·, 9 can be used, except that all three digits cannot be zero, and that any letter from A to Z (repeats allowed) can be used. How many plates are possible?

(b) Investigate what schemes for license plates are used in your state and determine how many different plates are possible.

34. A restaurant offers 6 soups, 10 entrees, and 8 desserts. How many different choices for a meal can a customer make if one selection is made from each category? If 3 of the desserts are pies and the customer will never order pie, how many meals can the customer choose?

35. In the last several years, heavily populated regions that previously had only one area code have been divided into service areas with more than one area code. What is the largest number of different phone numbers that can be served using one area code? If an area code cannot begin with a zero, how many different area codes are possible?

36. (a) A credit-card company makes it easier for customers to memorize their PIN (personal identification number) by using a four-digit PIN that consists of three different digits selected from 0, 1, 2, · · ·, 9 where one of the digits must be a zero, another is a nonzero digit that is repeated, and another is a digit different from these two. How many different PINs of this kind are there?

(b) How many PINs are possible if there are no restrictions on repeats of the 10 possible digits that can be used?

2.2 Traveling Salesman Problem

2.3 Helping Traveling Salesmen

37. Draw complete graphs with four, five, and six vertices. How many edges do these graphs have? Can you generalize to n vertices? How many TSP tours would these graphs have? (Tours yielding the same Hamiltonian circuit are considered the same.)

38. Calculate the values of 5!, 6!, 7!, 8!, 9!, and 10!. Then find the number of TSP tours in the complete graph with eight vertices.

39. The following table shows the mileage between four cities: Springfield, Ill. (*S*); Urbana, Ill. (*U*); Effingham, Ill. (*E*); and Indianapolis, Ind. (*I*).

	E	I	S	U
E	—	147	92	79
I	147	—	190	119
S	92	190	—	88
U	79	119	88	—

(a) Represent this information by drawing a weighted complete graph on four vertices.

(b) Use the weighted graph in part (a) to find the cost of the three distinct Hamiltonian circuits in the graph. (List them starting at *U*.)

(c) Which circuit gives the minimum cost?

(d) Would there be any different in parts (b) and (c) if the starting vertex were at *I*?

(e) If one applies the nearest-neighbor method starting at *U*, what circuit would be obtained? Does the answer change if one applies the nearest-neighbor algorithm starting at *S*? At *E*? At *I*?

(f) If one applies the sorted-edges method, what circuit would be obtained? Does one get the optimal answer?

40. After a party at her house, Francine (*F*) has agreed to drive Mary (*M*), Rachel (*R*), and Constance (*C*) home. If the times (in minutes) to drive between her friends' homes are shown below, what route gets Francine back home the quickest?

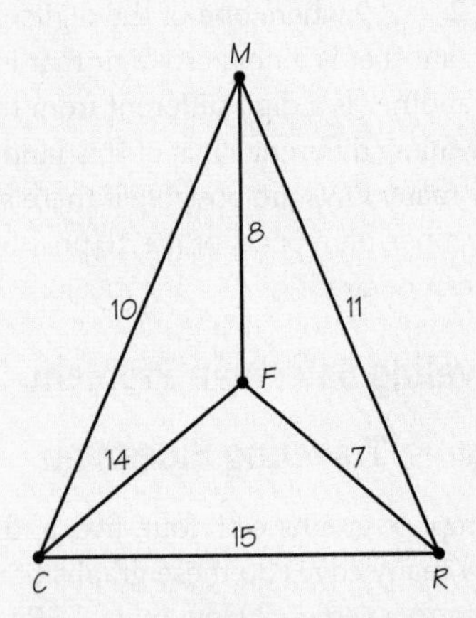

41. In Exercise 40, what route would Francine have to follow to get home as quickly as possible, assuming she promised to drive Mary home first?

42. In Exercise 40, Francine is planning to deliver her friends home and then spend the night at Rachel's house. What would her fastest route be?

43. A religious charity arranges free pickups for donated goods to encourage such donations but would like to keep its pickup costs low. The accompanying diagram shows the estimated amounts of time to get between the charity (*H*) and the locations of the three pickups scheduled for Wednesday.

(a) What route is selected using the nearest-neighbor algorithm starting at *H*?

(b) What route is selected using the sorted-edges algorithm?

(c) Use "brute force" to determine if the solution in either (a) or (b) yields an optimal solution.

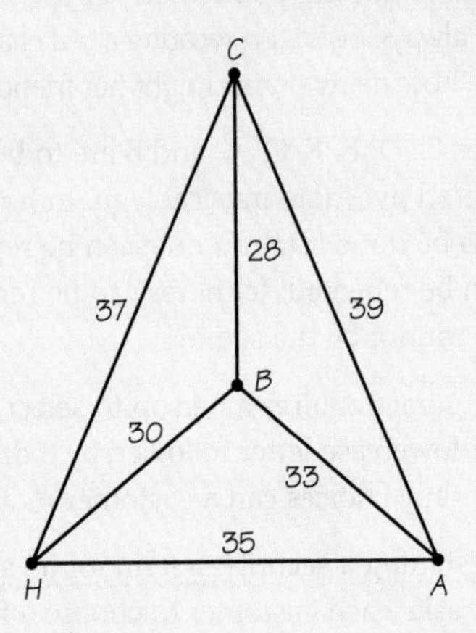

44. Meals on Wheels must deliver food to three clients and wants to keep its costs down. Typical times to get between its home site *H* and the clients are given in the accompanying graph.

(a) What route is selected using the nearest-neighbor algorithm starting at *H*?

(b) What route is selected using the sorted-edges algorithm?

(c) Use "brute force" to determine if the solution in either (a) or (b) yields an optimal solution.

(d) Compare the times in the diagrams in this excercise and Exercise 43. How do they differ? Can you state a general result about solving TSP problems based on what you notice?

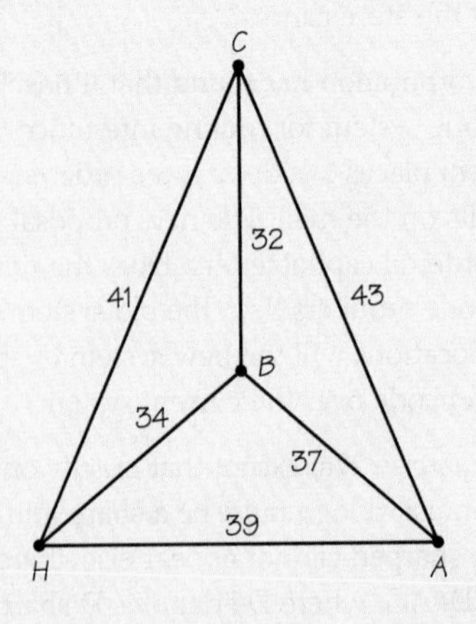

45. Starting from the location where she moors her boat (*M*), a fisherwoman wishes to visit three areas—*A*, *B*, and *C*—where she has set fishing nets. If the times (in minutes) between the locales are given in the accompanying figure, what route to visit the three sites and return to the mooring place would be optimal?

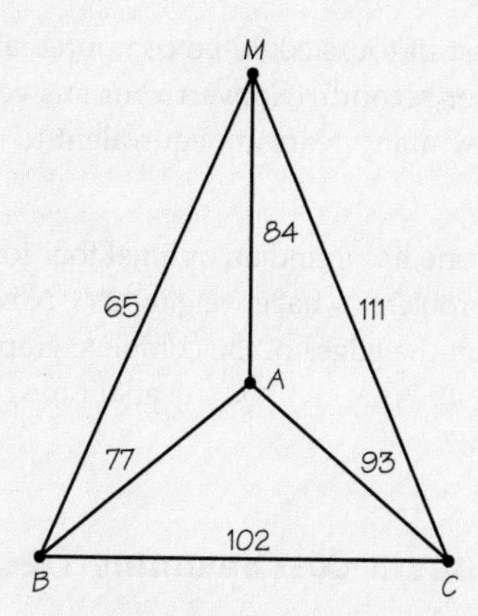

46. (a) For the two complete graphs that follow, find the costs of the nearest-neighbor tour starting at B and of the tour generated by the sorted-edges algorithm.

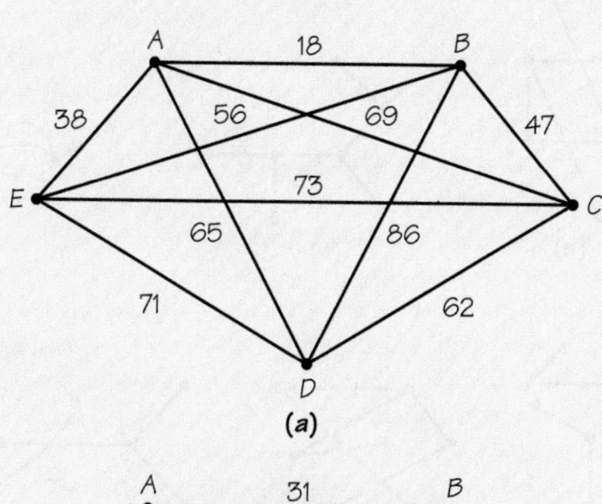

(a)

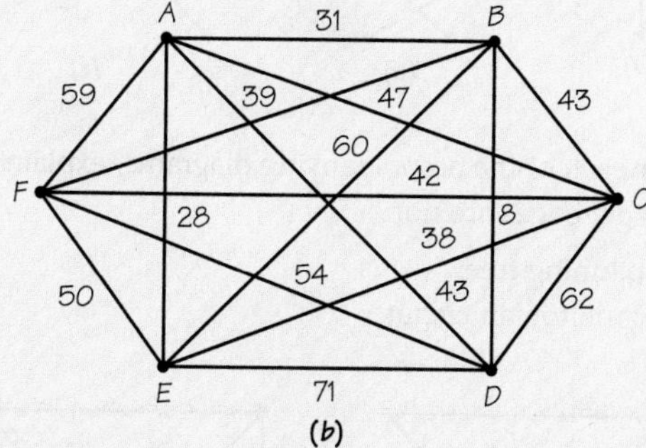

(b)

■ **(b)** How many Hamiltonian circuits would have to be examined to find a shortest route for part (a) by the brute force method?

(c) Invent an algorithm different from the sorted-edges and nearest-neighbor algorithms that is easy to apply for finding TSP solutions.

47. An airport limo must take its five passengers from the airport to different downtown hotels. Is this a traveling salesman problem, a Chinese postman problem, or an Euler circuit problem?

48. For each of the accompanying graphs with weights, apply the nearest-neighbor method (starting at vertex A) and the sorted-edges method to find (it is hoped) a cheap tour.

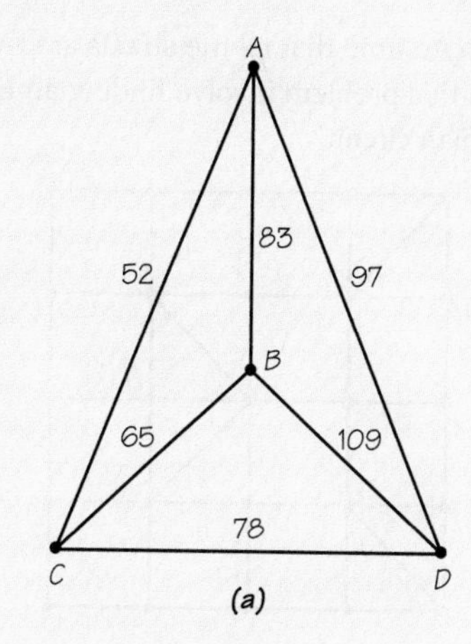

(a)

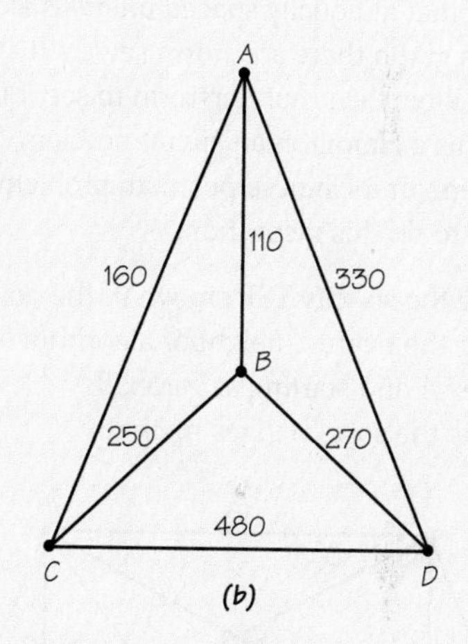

(b)

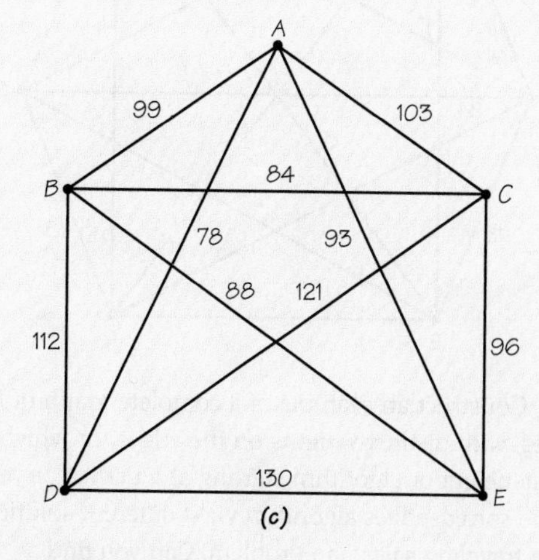

(c)

49. The accompanying figure represents a town where there is a sewer located at each corner (where two or more streets meet). After every thunderstorm, the department of public works wishes to have a truck start at its headquarters (at vertex H) and make an inspection of sewer drains to be sure that leaves are not clogging them. Can a route start and end at H that visits each corner exactly once? (Assume that all the streets are two-way streets.) Does this problem involve finding an Euler circuit or a Hamiltonian circuit?

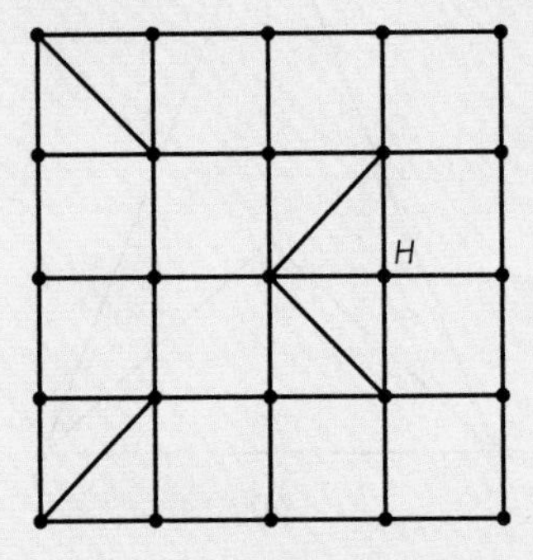

Assume that at equally spaced intervals along the blocks in this graph there are storm sewers that must be inspected after each thunderstorm to see if they are clogged. Is this a Hamiltonian circuit problem, an Euler circuit problem, or a Chinese postman problem? Find an optimal tour to do this inspection.

50. (a) Solve the six-city TSP shown in the diagram using the nearest-neighbor algorithm starting at vertex A and starting at vertex B.

(b) Apply the sorted-edges method.

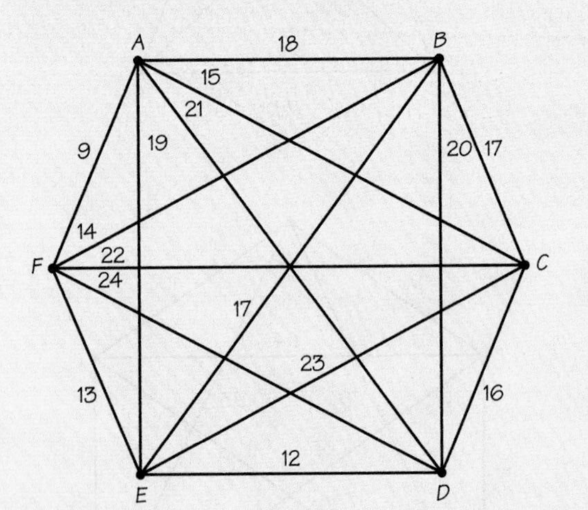

■ **51.** Construct an example of a complete graph of five vertices, with distinct weights on the edges for which the nearest-neighbor algorithm starting at a particular vertex and the sorted-edges algorithm yield different solutions for the traveling salesman problem. Can you find a five-vertex complete graph with weights on the edges in which the optimal solution, the nearest-neighbor solution, and the sorted-edges algorithm solution are all different?

■ **52.** If the brute force method of solving a 20-city TSP is employed, use a calculator to determine how many Hamiltonian circuits must be examined. How long would it take to determine the minimum-cost tour if the cost of tours could be computed at the rate of 1 billion per second? (Convert your answer to years by seeing how many years are equivalent to a billion seconds!)

53. Suppose one has found an optimal tour for a given 10-city TSP problem to have weight 4200. Now suppose the weights on the edges of the complete graph are increased by 50. What can you say about the optimal tour and its weight?

2.4 Minimum-Cost Spanning Trees

54. For each graph below, explain why it is or is not a tree.

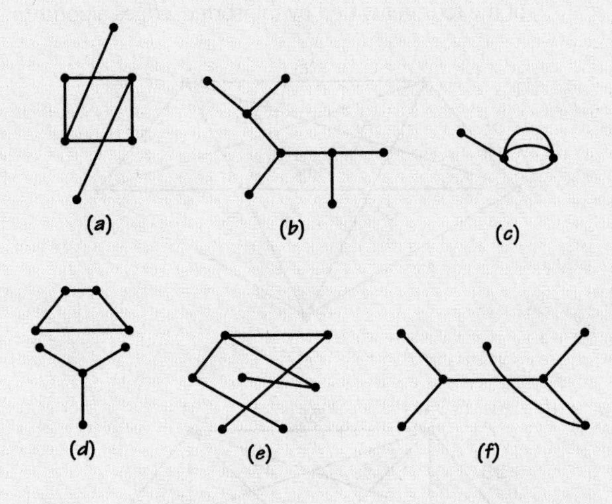

55. For each of the accompanying diagrams, explain why the wiggly edges are not

(a) a spanning tree.

(b) a Hamiltonian circuit.

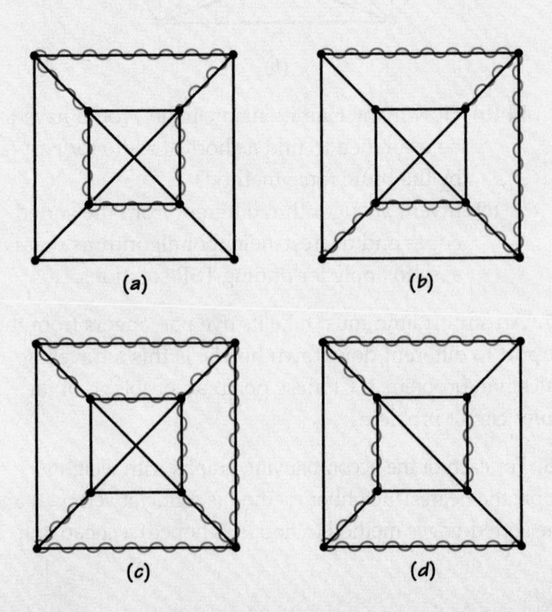

56. Find all the spanning trees in the accompanying graphs.

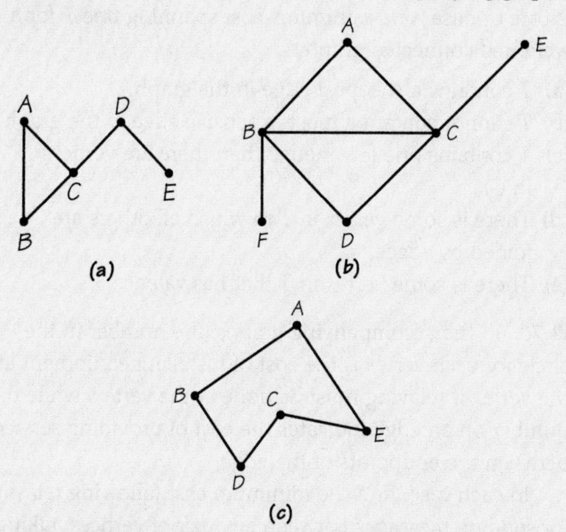

57. Use Kruskal's algorithm to find a minimum-cost spanning tree for the following graphs (a), (b), (c), and (d). In each case, what is the cost associated with the tree?

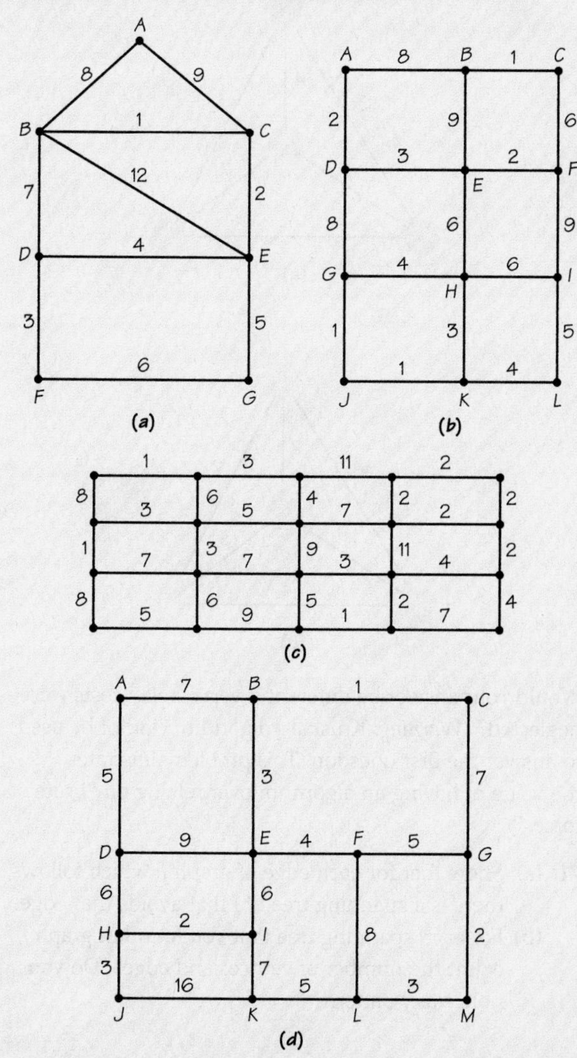

58. A connected graph G has 16 vertices. How many edges does a spanning tree of G have? How many vertices does a spanning tree of G have? What can one say about the number of edges G has?

59. A connected graph H has a spanning tree with 26 edges. How many vertices does the spanning tree have? How many vertices does H have? What can one say about the number of edges H has?

60. A large company wishes to install a pneumatic tube system that would enable small items to be sent between any of 10 locales, possibly by using relay. If the nonprohibitive costs (in $100) are shown in the accompanying graph model, between which sites should the tube be installed to minimize the total cost?

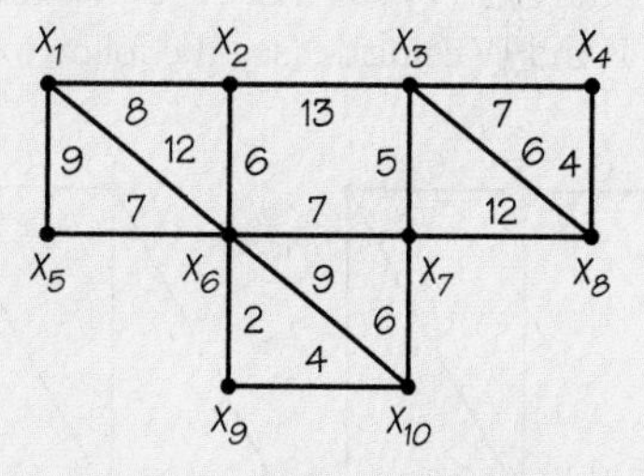

61. If the weight of each edge in Exercise 60 is increased by 3, will the tree that achieves minimum cost for the new collection of weights be the same as the one that achieves minimum cost for the original set of weights?

▲ **62.** Give examples of real-world situations that can be modeled using a weighted graph and for which finding a minimum-cost spanning tree for the graph would be of interest.

■ **63.** Can Kruskal's algorithm be modified to find a maximum-weight spanning tree? Can you think of an application for finding a maximum-weight spanning tree?

▲ **64.** Find the cost of providing a relay network between the six cities with the largest populations in your home state, using the road distances between the cities as costs. Does it follow that the same solution would be obtained if air distances were used instead?

■ **65.** Would there ever be a reason to find a minimum-cost spanning tree for a weighted graph in which the weights on some of the edges were negative? Would Kruskal's algorithm still apply?

■ **66.** Suppose G is a graph such that all the weights on its edges are different numbers. Show that there is a unique minimum-cost spanning tree.

67. Two spanning trees of a (weighted) graph are considered different if they use different edges. Show that the following graph has different minimum-cost spanning trees, though all these different trees have the same cost.

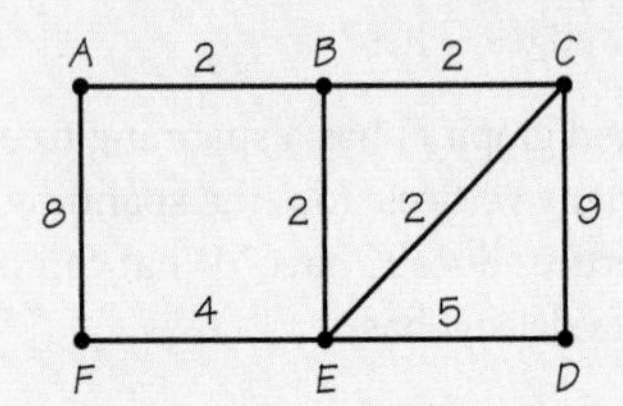

68. Let G be a graph with weights assigned to each edge. Consider the following algorithm:

(a) Pick any vertex V of G.

(b) Select an edge E with a vertex at V that has a minimum weight. Let the other endpoints of E be W.

(c) Contract the edge VW so that edge VW disappears and vertices V and W coincide (see the following figures).

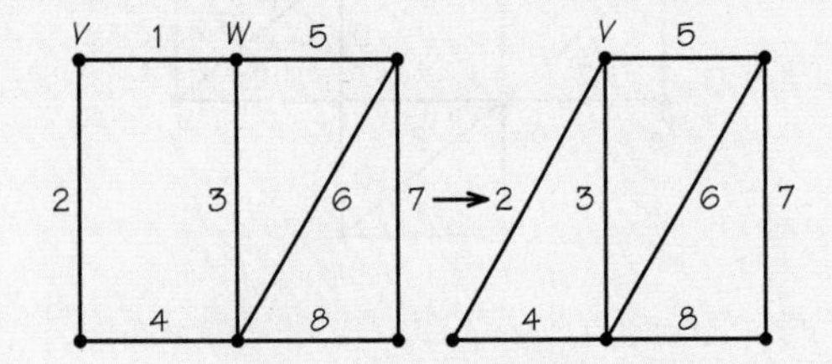

If in the new graph two or more edges join a pair of vertices, delete all but the cheapest. Continue to call the new vertex V.

(d) Repeat steps (b) and (c) until a single point is obtained. The edges selected in the course of this algorithm (called Prim's algorithm) form a minimum-cost spanning tree. Apply this algorithm to the accompanying graphs.

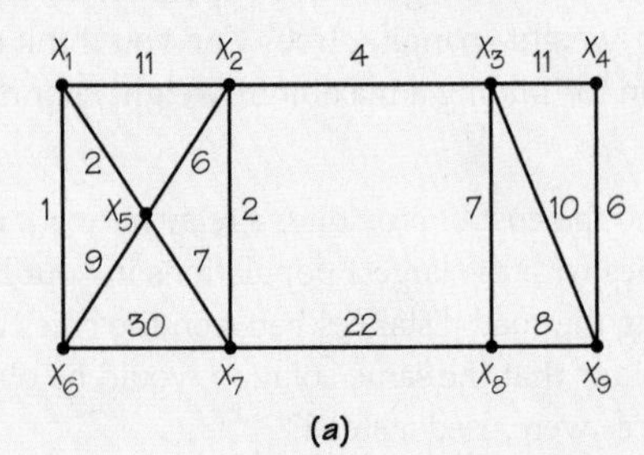

(a)

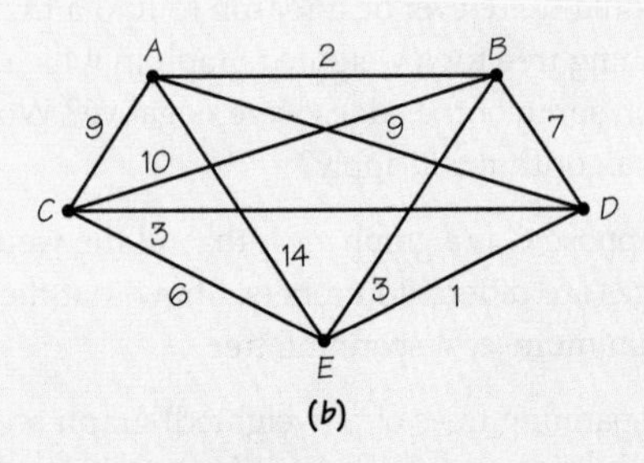

(b)

69. Determine whether each of the following statements is true or false for a minimum-cost spanning tree T for a weighted connected graph G:

(a) T contains a cheapest edge in the graph.

(b) T cannot contain a most expensive edge in the graph.

(c) T contains one fewer edge than there are vertices in G.

(d) There is some vertex in T to which all others are joined by edges.

(e) There is some vertex in T that has valence 3.

■ **70.** In the accompanying graphs, the number in the circle for each vertex is the cost of installing equipment at the vertex if relaying must be done at the vertex, while the number on an edge indicates the cost of providing service between the endpoints of the edge.

In each case, find the minimum cost (allowing relays) for sending messages between any pair of vertices, taking vertex relay costs into account.

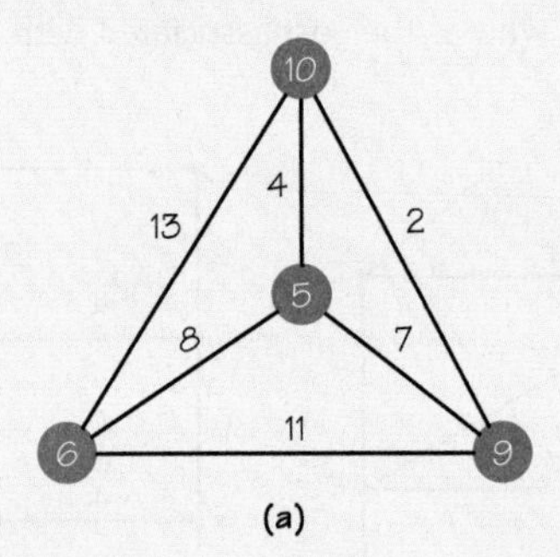

(a)

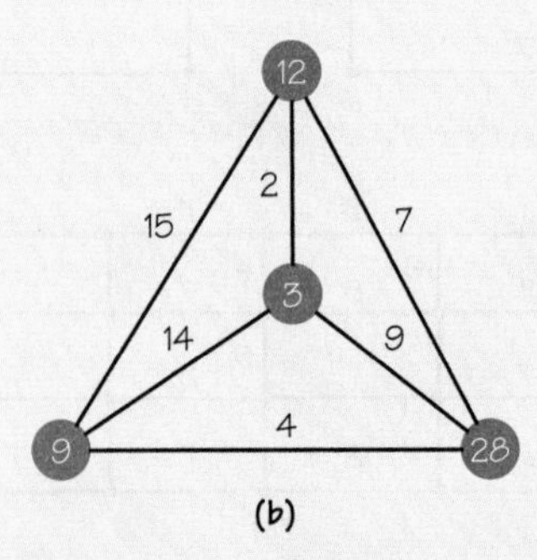

(b)

Would your answer be different if vertex relay costs were neglected? (*Warning:* Kruskal's algorithm cannot be used to answer the first question. This problem illustrates the value of having an algorithm over relying on "brute force.")

71. (a) Show that for each edge of graph J, which follows, there is a spanning tree of J that avoids that edge.

(b) For each spanning tree that you found in graph J, count the number of vertices and edges. Do you notice any pattern?

(c) For graph H, which follows, and each edge in the graph, is there a spanning tree that does not include that edge of H?

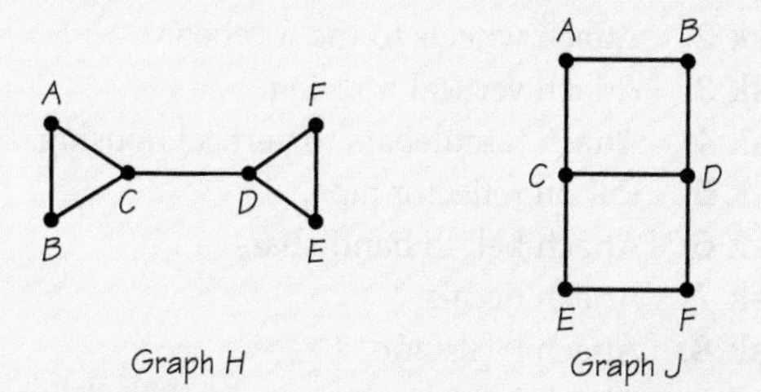

72. (a) The table shown gives the "closeness" or distance values between four objects. Construct a four-vertex tree with weights on its edges such that the distances between pairs of vertices of the tree (as measured by the sum of the weights on the path in the tree between these vertices) give rise to this table.

	A	B	C	D
A	0	3	10	14
B	3	0	7	11
C	10	7	0	4
D	14	11	4	0

(b) Produce several real-world contexts that might give rise to the situation described here.

73. The accompanying figure represents four objects using a tree with weights on the edges. Construct a table with four rows and four columns recording how "close" pairs of vertices in the tree are to each other. To find how close a pair of objects is, add together the weights along the path that joins these two objects.

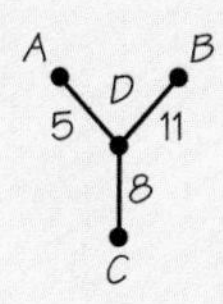

2.5 Critical-Path Analysis

74. Find the earliest completion time and critical paths for the accompanying order-requirement digraphs.

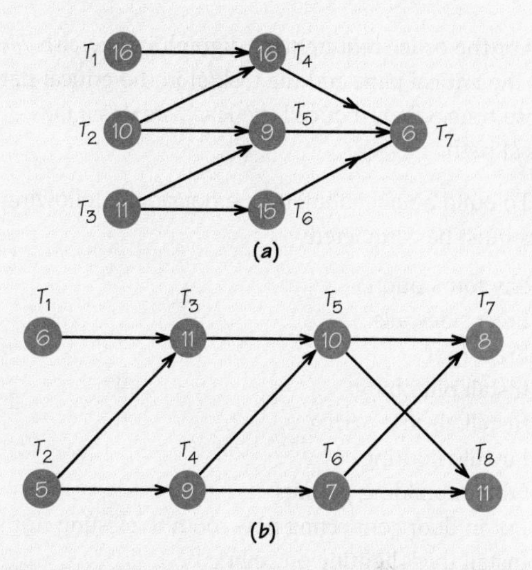

(a)
(b)

75. Find the earliest completion time and critical paths for the following order-requirement digraphs.

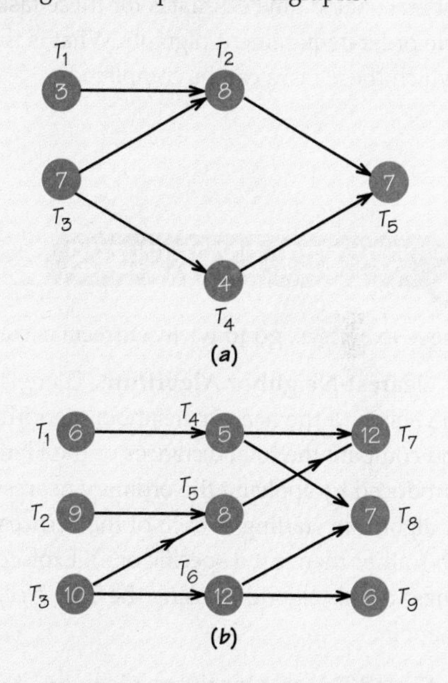

(a)
(b)

76. Construct an example of an order-requirement digraph with exactly three different critical paths.

77. In the order-requirement accompanying digraph, determine which tasks, if shortened, would reduce the earliest completion time and which would not. Then find the earliest completion time if task T_5 is reduced to time length 7. What is the new critical path?

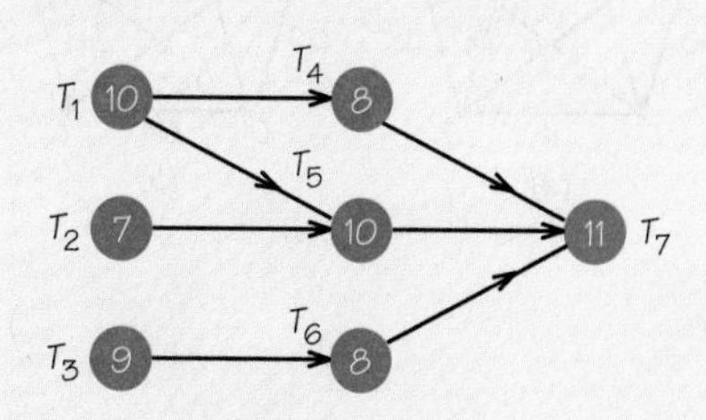

78. For the order-requirement digraph in Exercise 77, find the critical path and the task(s) in the critical path whose time, when reduced the least, creates a new critical path.

79. To build a new addition on a house, the following tasks must be completed:

(a) Lay foundation.
(b) Erect sidewalls.
(c) Erect roof.
(d) Install plumbing.
(e) Install electric wiring.
(f) Lay tile flooring.
(g) Obtain building permits.
(h) Put in door connecting new room to existing house.
(i) Install track lighting on ceiling.
(j) Install wall air-conditioner.

Construct reasonable time estimates for these tasks and a reasonable order-requirement digraph. What is the fastest time in which these tasks can be completed?

80. At a large toy store, scooters arrive unassembled in boxes. To assemble a scooter, the following tasks must be performed:

Task 1. Remove parts from the box.
Task 2. Attach wheels to the footboard.
Task 3. Attach vertical housing.
Task 4. Attach handlebars to vertical housing.
Task 5. Put on reflector tape.
Task 6. Attach bell to handlebars.
Task 7. Attach decals.
Task 8. Attach kickstand.
Task 9. Attach safety instructions to handlebars.

Give reasonable time estimates for these tasks and construct a reasonable order-requirement digraph. What is the earliest time by which these tasks can be completed?

81. Construct an order-requirement digraph with six tasks that has three critical paths of length 26.

APPLET EXERCISES

To do these exercises, go to www.whfreeman.com/fapp9e.

1. TSP: Nearest-Neighbor Algorithm. There is an extended version of the nearest-neighbor algorithm in which you compare the total distances of the Hamiltonian circuits produced by applying the ordinary nearest-neighbor algorithm starting at each of the vertices of the graph (rather than just a specific one). Explore the effectiveness of this algorithm using the *TSP: Nearest-Neighbor* applet.

2. TSP: Sorted-Edges Algorithm. Go to the *TSP: Sorted Edges* applet, where you can apply the sorted-edges algorithm to see if it solves the traveling salesman problem for the following graphs (and others):

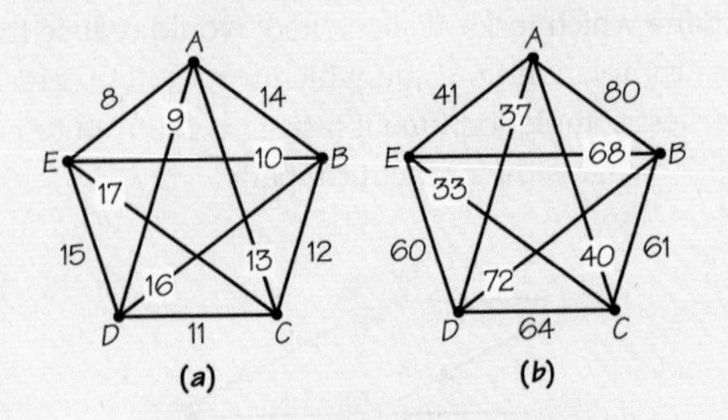

3. Kruskal's Algorithm. Go to the *Kruskal's Algorithm* applet, where you can apply Kruskal's algorithm to find the minimum-cost spanning trees in the following graphs (and others):

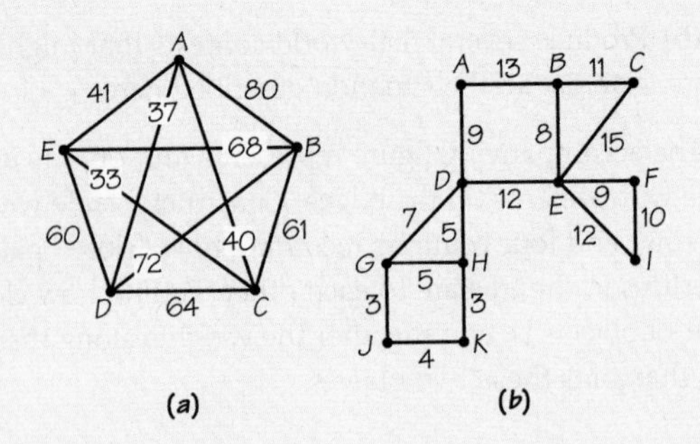

WRITING PROJECTS

1. Write an essay about a variety of situations in which you are personally involved for which a solution of the TSP is (perhaps implicitly) required. Explain under what circumstances it might be valuable to carry out a formal mathematical solution to such TSPs rather than use an ad hoc solution.

2. Construct an example that shows that in a situation where three day campers must be picked up and brought to camp, it may make a difference if the optimization criterion is minimizing distance traveled by the camp bus versus minimizing average time that the children spend on the bus.

3. Determine the six largest cities in the state in which you live. By consulting a road atlas (or by some other means) construct the graph that represents the road distances between your hometown and these six other cities. Now apply (a) the nearest-neighbor method, (b) the sorted-edges method, and (c) the nearest neighbor from each city, and pick the minimum tour method to solve the associated TSP. Do you have reason to believe that the answers you get might include an optimum solution among them?

Suggested Readings

BODIN, LAWRENCE. Twenty years of routing and scheduling, *Operations Research,* 38 (1990): 571–579. A survey of real-world situations where routing and scheduling were used, written by a pioneer in this area.

DOLAN, ALAN, and JOAN ALDUS. *Networks and Algorithms: An Introductory Approach,* Wiley, Chichester, England, 1993. An excellent introduction to graph theory algorithms.

GUSFIELD, DAN. *Algorithms on Strings, Trees, and Sequences,* Cambridge University Press, New York, 1997. Details applications of graph theory in pattern recognition and reconstruction problems.

JONES, NEIL C., and PAVEL A. PEVZNER, *An Introduction to Bioinformatics Algorithms,* MIT Press, Cambridge, MA, 2004. This book has material on how graph theory ideas, particularly those related to Hamiltonian circuits, are being used in molecular genetics and computational biology.

LAWLER, EUGENE, J. LENSTRA, RINNOY KAN, and D. SHMOYS, eds. *The Traveling Salesman Problem,* Prentice-Hall, Englewood Cliffs, NJ, 1985. Includes survey and technical articles on all aspects of the TSP.

LUCAS, WILLIAM, FRED ROBERTS, and ROBERT THRALL, eds. *Discrete and Systems Models,* vol. 3: *Modules in Applied Mathematics,* Springer-Verlag, New York, 1983. Chapter 6, "A Model for Municipal Street Sweeping Operations," by A. Tucker and L. Bodin, describes street-sweeping and related models in detail. Other chapters detail many recent applications of mathematics.

ROBERTS, FRED S., and BARRY TESMAN, *Applied Combinatorics,* 2nd ed., Pearson Prentice Hall, Upper Saddle River, NJ, 2004. The material on network-optimization problems is excellent.

ROBERTS, FRED. *Graph Theory and Its Applications to Problems of Society,* Society for Industrial and Applied Mathematics, Philadelphia, 1978. A very readable account of how graph theory is finding a wide variety of applications.

Suggested Web Sites

www.tsp.gatech.edu This site provides a detailed history and many applications of the TSP.

en.wikipedia.org/wiki/Minimum_cost_spanning_tree This Web page provides basic ideas about minimum-cost spanning trees, their applications, and extensions of this idea.

http://www-gap.dcs.st-and.ac.uk/~history/Biographies/Hamilton.html This site provides biographical information about William Rowan Hamilton, for whom Hamiltonian circuits are named.

http://www.ams.org/samplings/feature-column/fcarc-tsp
http://www.ams.org/samplings/feature-column/fcarc-trees
These sites provide some history and information about applications of the Traveling Salesman Problem and of minimum-cost spanning trees.

AP Photo/Amy Sancetta

3

Planning and Scheduling

In a society as complex as ours, everyday problems such as providing services efficiently and on time require accurate planning of both people and machines. Take the example of a hospital in a major city. Around-the-clock scheduling of nurses, doctors, and emergency room staff must be provided to guarantee that people with particular expertise are available during each shift. The operating rooms must be scheduled in a manner flexible enough to deal with emergencies. Equipment used for X-ray, CT, or MRI scans must be scheduled for maximum efficiency.

Although many scheduling problems are often solved on an ad hoc basis, we can also use mathematical ideas to gain insight into the complications that arise in scheduling. The ideas we develop in this chapter have practical value in a relatively narrow range of applications, but they shed light on many characteristics of more realistic, and hence more complex, scheduling problems.

3.1 Scheduling Tasks

Assume that a certain number of identical **processors** (machines, humans, or robots) work on a series of tasks that make up a job. Associated with each task is a specified amount of time required to complete the task. For simplicity, we assume that any of the processors can work on any of the tasks. Our problem, known as the **machine-scheduling problem**, is to decide how the tasks should be scheduled so that the completion time for the tasks collectively is as early as possible.

Even with these simplifying assumptions, complications in scheduling will arise. Some tasks may be more important than others and perhaps should be scheduled first. When "ties" occur, they must be resolved by special rules. As an example, suppose we are scheduling patients to be seen in a hospital emergency room staffed by one doctor. If two patients arrive simultaneously, one with a bleeding foot, the other with a bleeding arm, which patient should be examined first? Suppose the doctor treats the arm patient first, and while treatment is going on, a person in cardiac arrest or having a stroke arrives. Scheduling rules must establish appropriate priorities for cases such as these.

Another common complication arises with jobs consisting of several tasks that cannot be done in an arbitrary order. For example, if the job of putting up a new house is treated as a scheduling problem, the task of laying the foundation must precede the task of putting up the walls, which in turn must be completed before work on the roof can begin. The plumbing system can be scheduled for installation later.

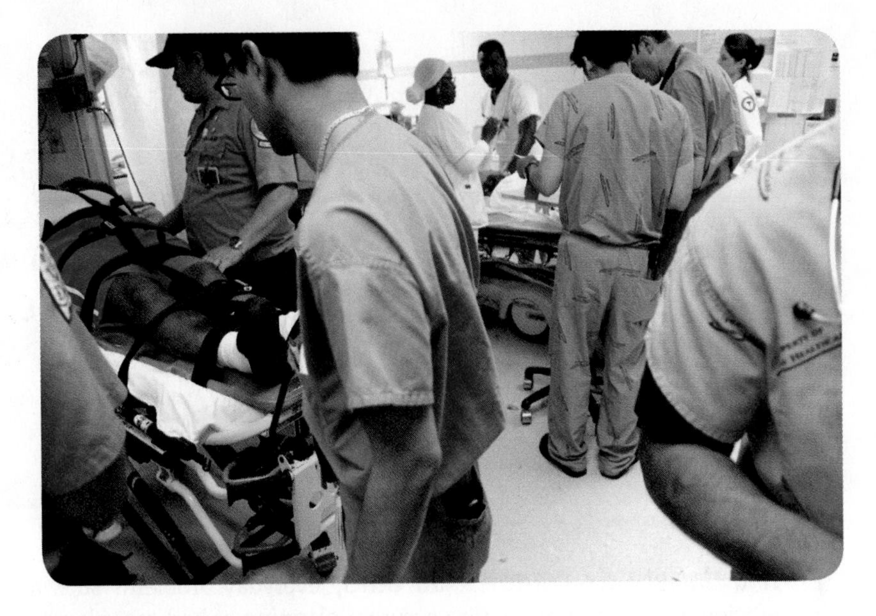

Hospitals are increasingly making use of mathematical techniques applied to scheduling problems. To make efficient use of one or more emergency rooms (operating rooms) requires the complicated assembly of a team of doctors, nurses, equipment, and support staff. Mathematical techniques for scheduling have made it possible to treat more patients in less time. *(Erik S. Lesser/ZUMA Press)*

Assumptions and Goals

To simplify our analysis, we need to make clear and explicit assumptions:

1. If a processor starts work on a task, the work on that task will continue without interruption until the task is completed.
2. No processor stays voluntarily idle. In other words, if there is a processor free and a task available to be worked on, then that processor will immediately begin work on that task.
3. The requirements for ordering the tasks are given by an order-requirement digraph. (A typical example is shown in Figure 3.1, with task times highlighted within each vertex. The ordering of the tasks imposed by the order-requirement digraph often represents constraints of physical reality. For example, you cannot fly a plane until it has taken fuel on board.)
4. The tasks are arranged in a **priority list** that is independent of the order requirements. (The priority list is a ranking of the tasks according to some criterion of "importance.")

Figure 3.1 A typical order-requirement digraph. Tasks with no edges entering or leaving the vertices representing them (T_7, T_8) can be more flexibly scheduled than the other tasks.

EXAMPLE 1

Home Construction

Let's see how these assumptions might work for an example involving a home construction project. In this case, the processors are human workers with identical skills. Assumption 1 means that once a worker begins a task, the work on this task is finished without interruption. Assumption 2 means that no worker will stay idle if there is some task for which the predecessors are finished. Assumption 3 requires that the ordering of the tasks be summarized in an order-requirement digraph. This digraph would code facts such as that the site must be cleared before the task of laying the foundation is begun. Assumption 4 requires that the tasks be ranked in a list from some perspective, perhaps a subjective view.

The task with highest priority rank is listed first in the list, followed left to right by the other tasks in priority rank. The priority list might be based on the size of the payments made to the construction company when a task is completed, even though these payments have no relation to the way the tasks must be done, as indicated in the order-requirement digraph. Alternatively, the priority list might reflect an attempt to find an algorithm to schedule the tasks needed to complete the whole job more quickly.

When considering a scheduling problem, there are various goals we might want to achieve. Among these are:

Goal 1. Minimizing the completion time of the job.

Goal 2. Minimizing the total time that processors are idle.

Goal 3. Finding the minimum number of processors necessary to finish the job by a specified time.

In the context of the construction example, goal 1 would complete the home as quickly as possible. Goal 2 would ensure that workers, who are perhaps paid by the hour, were not paid for doing nothing. One way of accomplishing this would be to hire one fewer worker even if it means the house takes longer to finish. Goal 3 might be reasonable if the family wants the house done before the first day of school, even if they have to pay a lot more workers to get the house done by this time.

For now we will concentrate on goal 1, finishing all the tasks at the earliest possible time. Note, however, that optimizing with respect to one goal may not optimize with respect to another. Our discussion here goes beyond what was discussed in Chapter 2 (see section 2.4) by dealing with how to assign tasks in a job to the processors that do the work. To build a new skyscraper involves designing a schedule for who will do what work when.

List-Processing Algorithm

The scheduling problem we have described sounds more complicated than the traveling salesman problem (TSP). Indeed, like the TSP, it is known to be NP-complete. This means that it is unlikely that anyone will ever find a computationally fast algorithm that can find an optimal solution for scheduling problems involving a very large number of tasks. Thus, we will be content to seek a solution method that is computationally fast and gives only approximately optimal answers.

List-Processing Algorithm: Part I and Ready Task — PROCEDURE

The algorithm we use to schedule tasks is the **list-processing algorithm**. In describing it, we will call a task **ready** at a particular time if all its predecessors as indicated in the order-requirement digraph have been completed at that time. In Figure 3.1 at time 0, the ready tasks are T_1, T_7, and T_8, while T_2 cannot be ready until 8 time units after T_1 is started. The algorithm works as follows: At a given time, assign to the lowest-numbered free processor the first task on the priority list that is *ready* at that time and that hasn't already been assigned to a processor.

In applying this algorithm, we will need to develop skill at coordinating the use of the information in the order-requirement digraph and the priority list. It will be helpful to cross out the tasks in the priority list as they are assigned to a processor to keep track of which tasks remain to be scheduled.

EXAMPLE 2
Applying the List-Processing Algorithm

Let's apply the list-processing algorithm to one possible priority list—$T_8, T_7, T_6, \ldots, T_1$—using two processors and the order-requirement digraph in Figure 3.1. The result is the schedule shown in Figure 3.2, where idle processor time (time during which a processor is not at work on a task) is indicated by white. How does the list-processing algorithm generate this schedule?

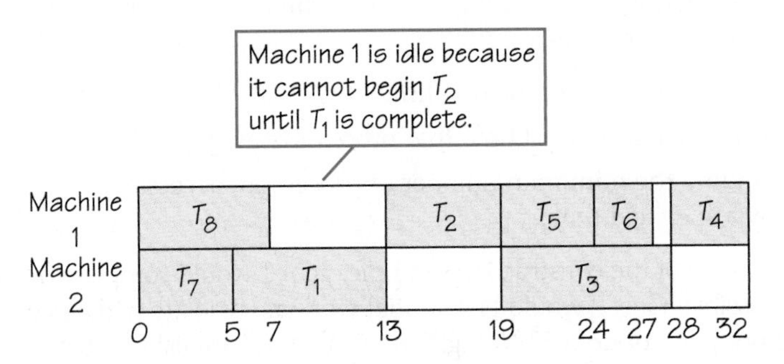

Figure 3.2 The schedule produced by applying the list-processing algorithm to the order-requirement digraph in Figure 3.1 using the list $T_8, T_7, \ldots, T_1$.

T_8 (task 8) is first on the priority list and ready at time 0 since it has no predecessors. It is assigned to the lowest-numbered free processor, processor 1. Task 7, next on the priority list, is also ready at time 0 and thus is assigned to processor 2. The first processor to become free is processor 2 at time 5. Recall that by assumption 1, once a processor starts work on a task, its work cannot be interrupted until the task is complete. Task 6, the next unassigned task on the list, is not ready at time 5, as can be seen by consulting Figure 3.1. The reason task 6 is not ready at time 5 is that task 5 has not been completed by time 5. In fact, at time 5, the only ready task on the list is T_1, so that task is assigned to processor 2. At time 7, processor 1 becomes free, but no task becomes ready until time 13.

Thus, processor 1 stays idle from time 7 to time 13. At this time, because T_2 is the first ready task on the list not already scheduled, it is assigned to processor 1. Processor 2, however, stays idle because no other ready task is available at this time. The remainder of the scheduling shown in Figure 3.2 is completed in this manner.

We can summarize this procedure as follows:

List-Processing Algorithm: Part II PROCEDURE

As the priority list is scanned from left to right to assign a task to a processor at a particular time, we pass over tasks that are not ready to find ones that are ready. If no task can be assigned in this manner, we keep one or more processors idle until such time that, reading the priority list from the left, there is a ready task not already assigned. After a task is assigned to a processor, we resume scanning the priority list for unassigned tasks, starting over at the far left.

When Is a Schedule Optimal?

The schedule in Figure 3.2 has a lot of idle time, so it may not be optimal. Indeed, if we apply the list-processing algorithm for two processors to another possible priority list $T_1, \ldots, T_8$, using the digraph in Figure 3.1, the resulting schedule is that shown in Figure 3.3.

Here are the details of how we arrived at this schedule. Remember that we must coordinate the list $T_1, T_2, \ldots, T_8$ with the information in the order-requirement digraph shown in Figure 3.3a. At time 0, task T_1 is ready, so this task is assigned to processor 1. However, at time 0, tasks $T_2, T_3, \ldots, T_6$ are not ready because their predecessors are not done. For example, T_2 is not ready at time 0 because T_1, which precedes it, is not done at time 0. The first ready task on the list, reading from left to right, that is not already assigned is T_7, so task T_7 gets assigned to processor 2. Both processors are now busy until time 5, at which point processor 2 becomes available to work on another task (Figure 3.3b).

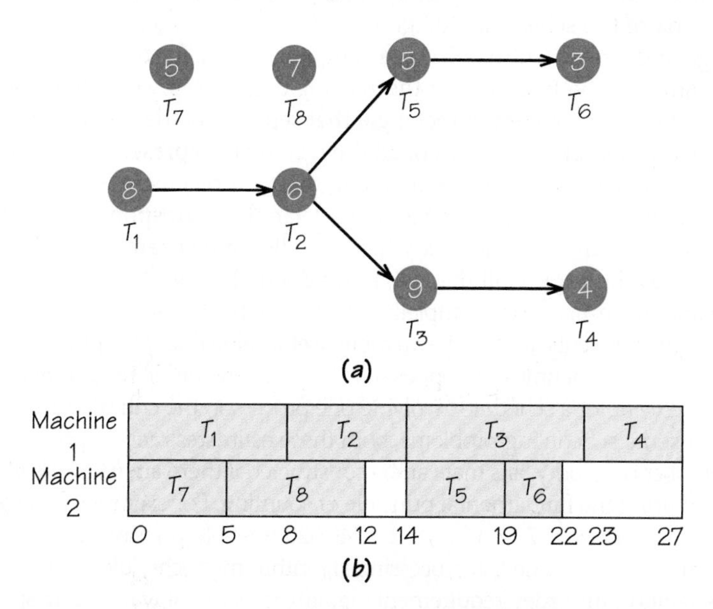

Figure 3.3 (a) A typical order-requirement digraph (repeat of Figure 3.1). (b) The schedule produced by applying the list-processing algorithm to the order-requirement digraph in Figure 3.3a using the list $T_1, T_2, \ldots, T_8$.

Tasks T_1 and T_7 have been assigned. Reading from left to right along the list, the first task not already assigned whose predecessors are done by time 5 is T_8, so this task is started at time 5 on processor 2; processor 2 will continue to work on this task until time 12, because the task time for this task is 7 time units. At time 8, processor 1 becomes free, and reading the list from left to right we find that T_2 is ready (because T_1 has just been completed). Thus, T_2 is assigned to processor 1, which will stay busy on this task until time 14. At time 12, processor 2 becomes free, but the tasks that have not already been assigned from the list, T_3, T_4, T_5, T_6 are not ready, because they depend on T_2 being completed before these tasks can start. Thus, processor 2 must stay idle until time 14. At this time, T_3 and T_5 become ready. Since both processors 1 and 2 are idle at time 14, the lower numbered of the two, processor 1, gets to start on T_3 because it is the first ready task remaining to be assigned on the list scanned from left to right. Task T_5 gets assigned to processor 2 at time 14. The remaining tasks are assigned in a similar manner.

The schedule shown in Figure 3.3b is optimal because the path T_1, T_2, T_3, T_4, with length 27, is the critical path in the order-requirement digraph. As we saw in Chapter 2, the earliest completion time for the job made up of all the tasks is the length of the longest path in the order-requirement digraph. Different lists can give rise to the same or different optimal scheduling diagrams. The optimal schedule may have a completion time equal to or longer than the length of the critical path, but it cannot have a completion time shorter than the length of the critical path.

There is another way of relating optimal completion time to the completion time that is yielded by the list-processing algorithm. Suppose that we add all the task times given in the order-requirement digraph and divide by the number of processors. The completion time using the list-processing algorithm must be at least as large as this number. For example, the task times for the order-requirement digraph in Figure 3.3a sum to 47. Thus, if these tasks are scheduled on two processors, the completion time is at least $\frac{47}{2} = 23.5$ (in fact, 24, because the list-processing algorithm applied to integer task times must yield an integer solution), while for three processors the completion time is at least $\frac{47}{3}$ (in fact, 16).

Why is it helpful to take the total time to do all the tasks in a job and divide this number by the number of processors? Think of each task that must be scheduled as a rectangle that is 1 unit high and t units wide, where t is the time allotted for the task. Think of the scheduling diagram with m processors as a rectangle that is m units high and whose width, W, is the completion time for the tasks. The scheduling diagram is to be filled up by the rectangles that represent the tasks. How small can W be? The area of the rectangle that represents the scheduling diagram must be at least as large as the sum of all the rectangles representing tasks that are "packed" into it. The area of the scheduling diagram rectangle is mW. The combined areas of all the tasks, plus the area of rectangles corresponding to idle time, will equal mW. Width W is smallest when the idle time is zero. Thus, W must be at least as big as the sum of all the task times divided by m.

Sometimes the estimate for completion time given by the list-processing algorithm from the length of the critical path gives a more useful value than the approach based on adding task times. Sometimes the opposite is true. For the order-requirement digraph in Figure 3.1, except for a schedule involving one processor, the critical-path estimate is superior. For some scheduling problems, both these estimates may be poor.

The number of priority lists that can be constructed if there are n tasks is $n!$ and can be computed using the fundamental principle of counting. For example, for eight tasks, $T_1, \ldots, T_8$, there are $8 \times 7 \times 6 \ldots \times 1 = 40{,}320$ possible priority lists. For different choices of the priority list, the list-processing algorithm may schedule the tasks, subject to the constraints of the order-requirement digraph, in different ways. More specifically, two different lists may yield different completion times or the same completion time, but the order in which the tasks are carried out will be different. It is also possible that two different lists produce identical ordering of the assignments of the tasks to processors and completion times. Soon we will see a method that can be used to select a list that, if we are lucky, will give a schedule with a relatively good completion time. In fact, no method is known, except for very specialized cases, of how to choose a list that can be guaranteed to produce an optimal schedule when the list algorithm is applied to it.

Strange Happenings

The list-processing algorithm involves four factors that affect the final schedule. The answer we get depends on the following:

1. The times to carry out the tasks
2. Number of processors
3. Order-requirement digraph
4. Ordering of the tasks on the priority list

To see the interplay of these four factors, consider another scheduling problem, this time asociated with the order-requirement digraph shown in Figure 3.4. (The highlighted

SPOTLIGHT 3.1

Management Science and Disaster Recovery

The city of New York depends on a public transportation system of subways and roads to bring hundreds of thousands of people who live in the four outer boroughs (Queens, Brooklyn, the Bronx, and Staten Island) into Manhattan to work and "play." New York City (NYC) also has a communications system of telephones, radio and television stations, and computer networks. These systems speed information between New York's citizens and people outside the city and around the world. The area in southern Manhattan, in the vicinity of the World Trade Center (WTC), was a center for banking, insurance, financial markets, and domestic and international commerce. The attack on the World Trade Center on September 11, 2001, disrupted these networks and markets but did not destroy them, partly because the principles of operations research and management science were used in the design and development of these systems over a long period of time.

The diagram below shows a very simple subway (train) system between an eastern and a western terminus.

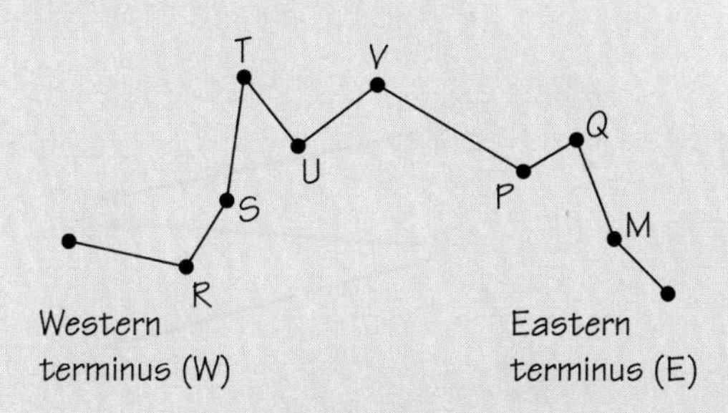

There are two tracks, each dedicated for use by westbound or eastbound trains to run between the two termini. The only place where trains can be turned around is at these termini. Simple graph theory tells us that in such a system, if a vertex is "destroyed" or out of service, or an edge is "destroyed" or out of service, the system totally breaks down. However, the simple provision that trains can be turned around at *U*, even though this is usually only one stop on the way from *W* to *E*, gives much greater flexibility to the system if there is a water main break, or a gas leak, etc. Thanks to simple principles of this kind and the creation of routes that use independent lines with many transfer points, New Yorkers were able to use the subway system in a flexible way after the World Trade Center disaster. In the days right after the WTC collapsed, trains were not allowed past the geographic area near the WTC for fear that the tunnels' structural foundation had been weakened and that subway vibrations could cause the collapse of damaged buildings. After it was ascertained that running the subways was safe, both for partially damaged buildings and for the subways themselves, routes were altered several times to give rescue workers and people returning to their daily routines maximum support. One line's tunnels did collapse, and several stations had to be closed for extended periods, but due to the redundancy and flexibility of the design of the system, a remarkable amount of service was restored quickly. Recent projects to improve the infrastructure of the NYC subway system are also making it more flexible in dealing with potentially crippling events such as a huge snowfall.

(AP/World Wide Photo.)

Good planning and wise application of the principles of management science make it possible to minimize the effects of natural and manmade disasters.

show how to construct a specific priority list based on this principle, to which the list-processing algorithm can then be applied.

Recall from our discussion of critical-path analysis in Chapter 2 that no matter how a schedule is constructed, the finish time cannot be earlier than the length of the longest path in the order-requirement digraph. This suggests that we should try to schedule first those tasks that occur early in long paths, because they might be a bottleneck for the other tasks. This idea leads to **critical-path scheduling**.

EXAMPLE 3
Scheduling Two Processors

To illustrate this method, consider the order-requirement digraph in Figure 3.10a. Suppose we wish to schedule these tasks on two processors. Initially, there are two critical paths of length 64: T_1, T_2, T_3 and T_1, T_4, T_3. Thus, we place T_1 first on the priority list. With T_1 "gone," there is a new critical path of length 60 (T_5, T_6, T_4, T_3) that starts with T_5, so T_5 is placed second on the priority list. At this stage, with T_1 and T_5 removed, we have the residual order-requirement digraph shown in Figure 3.10b. In this diagram, there are paths of length 50 (T_2, T_3), 56 (T_6, T_4, T_3), 36 (T_6, T_4, T_7), and 24 (T_8, T_9, T_{10}). Because T_6 heads the path that is currently longest in length, it is placed third in the priority list. Once T_6 is removed from Figure 3.10b, there is a tie for which is the longest path remaining, because both T_2, T_3 and T_4, T_3 are paths of length 50.

When there is a tie between two longest paths, we place next on the priority list the lowest-numbered task heading a longest path. In the example shown here, this means that T_2 is placed next on the priority list, to be followed by T_4. Continuing in this fashion, we obtain the priority list T_1, T_5, T_6, T_2, T_4, T_3, T_8, T_9, T_7, T_{10}. Note that the order of T_7 and T_{10} was decided using the rule for breaking ties. The list-processing algorithm is now applied using this priority list and the order-requirement digraph in Figure 3.10a. We obtain the schedule in Figure 3.11.

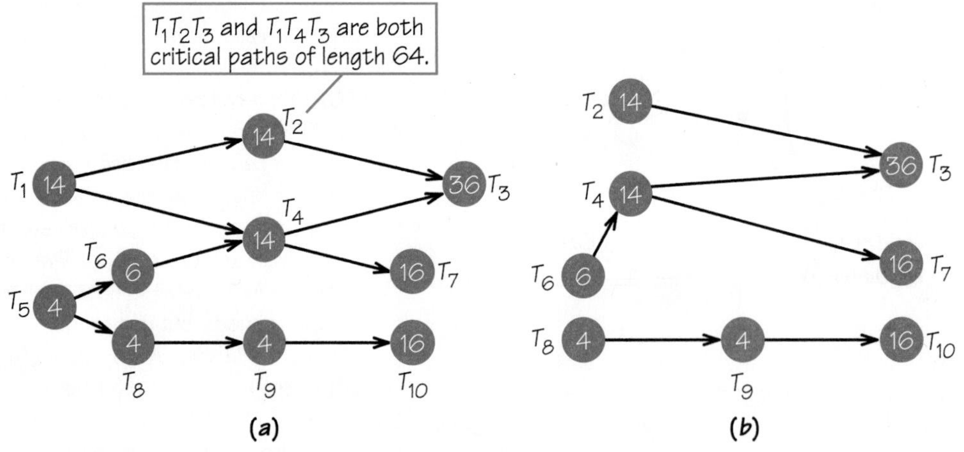

Figure 3.10 (a) An order-requirement digraph used to illustrate the critical-path scheduling method. (b) Residual order-requirement digraph after tasks T_1 and T_5 have been removed.

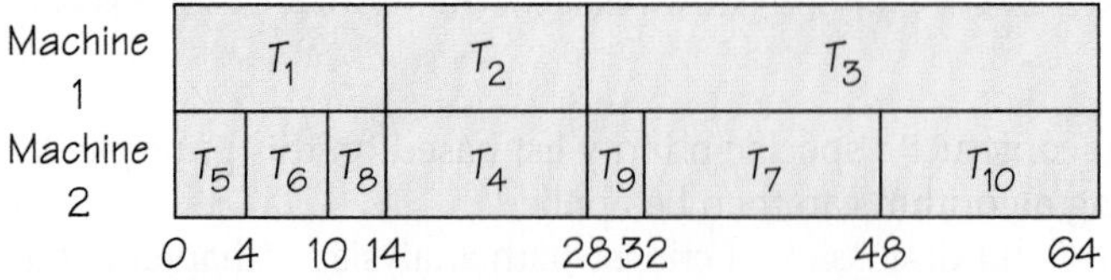

Figure 3.11 The optimal schedule produced by applying the critical-path scheduling method to the order-requirement digraph in Figure 3.10. The list used was T_1, T_5, T_6, T_2, T_4, T_3, T_8, T_9, T_7, T_{10}.

Critical-Path Scheduling

PROCEDURE

The **critical-path scheduling** algorithm applies the list-processing algorithm using the priority list L obtained as follows:

1. Find a task that heads a critical (longest) path in the order-requirement digraph. If there is a tie, choose the task with the lower number.
2. Place the task found in step 1 next on the list L. (The first time through the process, this task will head the list.)
3. Remove the task found in step 1 and the edges attached to it from the current order-requirement digraph, obtaining a new (modified) order-requirement digraph.
4. If there are no vertices left in the new order-requirement digraph, the procedure is complete; if there are vertices left, go to step 1.

This procedure will terminate when all the tasks in the original order-requirement digraph have been placed on the list L.

The preceding example shows that critical-path scheduling can sometimes yield optimal solutions. Unfortunately, this algorithm does not always perform well. For example, the critical-path method employing four processors applied to the order-requirement digraph shown in Figure 3.12 yields the list T_1, T_8, T_9, T_{10}, T_{11}, T_5, T_6, T_7, T_{12}, T_2, T_3, T_4 and then the schedule in Figure 13.3. (Note that T_5, T_6, T_7 are thought of as heading paths of length 10.) In fact, there can be no worse schedule than this one. An optimal schedule is shown in Figure 3.14.

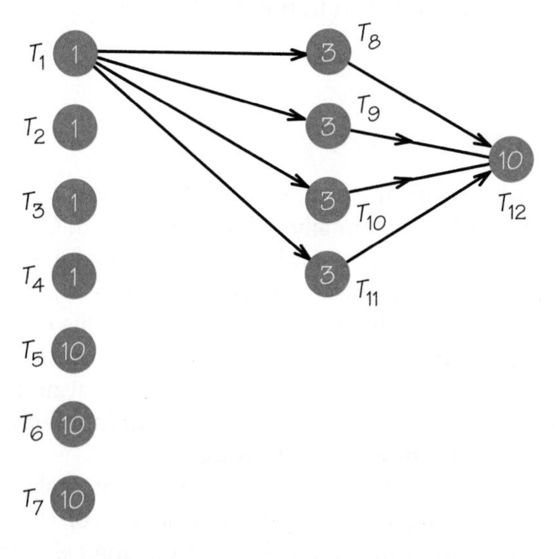

Figure 3.12 An order-requirement digraph used to illustrate how poorly the critical-path scheduling method can sometimes behave.

	0–1	1–4	4–7	7–10	10–11	11–13	13–23
Machine 1	T_1	T_8	T_9	T_{10}	T_{11}		T_{12}
Machine 2		T_5			T_2		
Machine 3		T_6			T_3		
Machine 4		T_7			T_4		

Time axis: 0 1 4 7 10 11 13 23

Figure 3.13 The schedule produced by applying the critical-path scheduling method to the order-requirement digraph in Figure 3.12 using four processors. The list used was T_1, T_8, T_9, T_{10}, T_{11}, T_5, T_6, T_7, T_{12}, T_2, T_3, T_4.

Figure 3.14 An optimal schedule for the order-requirement digraph in Figure 3.12 using four processors.

	0–1	1–4	4–14
Machine 1	T_1	T_8	T_5
Machine 2	T_2	T_9	T_6
Machine 3	T_3	T_{10}	T_7
Machine 4	T_4	T_{11}	T_{12}

Many of the results we have examined so far are negative because we are dealing with a general class of problems that defy our using computationally efficient algorithms to find an optimal schedule. But we can close on a more positive note. Consider an arbitrary order-requirement digraph, but assume all the tasks take equal time. It turns out that we can always construct an optimal schedule using two processors in this situation. Ironically, we can choose among many algorithms to produce these optimal schedules. The algorithms are easy to understand (though not easy to prove optimal) and have all been discovered since 1969! Many people think that mathematics is a subject that is no longer alive, and that all its ideas and methods were discovered hundreds of years ago—but as we have just seen, this is not true. In fact, more new mathematics has been discovered and published in the last 30 years than during any previous 30-year period.

3.3 Independent Tasks

Mathematicians suspect that no computationally efficient algorithm for solving general scheduling problems optimally will ever be found. Owing to our limited success in designing algorithms for finding optimal schedules for general order-requirement digraphs, we will consider a special class of scheduling problems for which the order-requirement digraph has no edges. In this case, we say that the tasks are *independent* of one another, because they can be performed in any order. (No edges in the order-requirement digraph indicates that no tasks need to precede others; that is, the tasks can be done in any order.) In this section, we consider the problem of scheduling **independent tasks**.

Geometrically, we can think of the independent tasks as rectangles of height 1 whose lengths are equal to the time length of the task. Finding an optimal schedule amounts to packing the task rectangles, with no "idle time" gaps between adjacent rectangles, into a longer rectangle whose height equals the number of machines. For example, Figure 3.15 shows two different ways to schedule tasks of length 10, 4, 5, 9, 7, 7 on two machines. (For convenience, the rectangles in the case of independent tasks are labeled with their task times rather than their task numbers.) Scheduling basically means efficiently packing the task rectangles into the machine rectangle. Finding the optimal answer among all possible ways to pack these rectangles is like looking for a needle in a haystack. The list-processing algorithm produces a packing, but it may not be a good one.

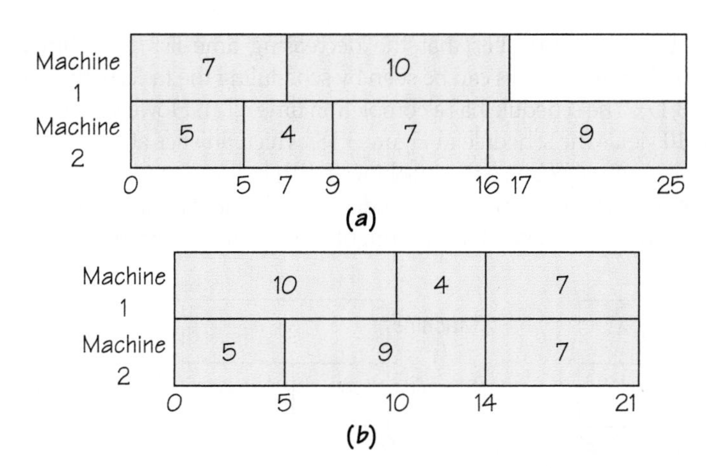

Figure 3.15 (a) A nonoptimal way to schedule independent tasks of time lengths 10, 4, 5, 9, 7, 7 using two processors. (b) An optimal way to schedule independent tasks of time lengths 10, 4, 5, 9, 7, 7 using two processors.

There are two approaches we can consider. To study **average-case analysis**, we might ask: *Is the average (mean) of the completion times arrived at by using the list-processing algorithm with all the possible different lists close to the optimal possible completion time?* To study **worst-case analysis**, we might ask: *How far from optimal is a schedule obtained using the list-processing algorithm with one particular priority list?* What is being contrasted with these two points of view is that an algorithm may work well most of the time (give an answer close to optimal) even though there may be a few cases in which it performs very badly. Average-case analysis is amenable to mathematical solution but requires methods of great sophistication.

Decreasing-Time Lists

Is there some way of choosing a priority list for independent tasks that consistently yields relatively good schedules? The surprising answer is yes! The idea is that when long tasks appear toward the end of the list, they often seem to "stick out" on the right end, as in Figure 3.15a. This suggests that before one tries to schedule a collection of tasks, the tasks should be placed in a list where the longest tasks are listed first.

Decreasing-Time-List Algorithm PROCEDURE

The list-processing algorithm applied to a list of task times arranged in order of non-increasing size is called the **decreasing-time-list algorithm**.

If we apply the **decreasing-time-list algorithm** to the set of tasks listed previously (10, 4, 5, 9, 7, 7), we obtain the times 10, 9, 7, 7, 5, 4 and the schedule (packing) shown in Figure 3.16. This packing is again optimal, but it is different from the optimal scheduling in Figure 3.15b. It is worth noting that the decreasing-time list and the list obtained by the critical-path method discussed earlier will coincide in the case of independent tasks. The decreasing-time list can also be constructed for the case in which the tasks are not independent. For general order-requirement digraphs, the decreasing-time list does not produce particularly good schedules.

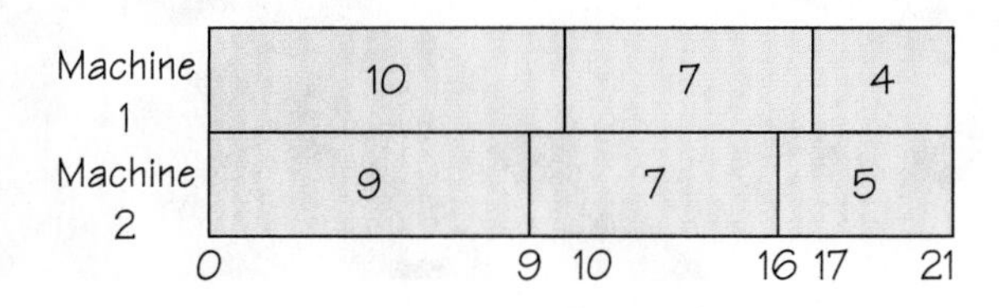

Figure 3.16 The optimal schedule resulting from applying the decreasing-time-list algorithm to a collection of independent tasks. The list used, written in terms of task times only, is 10, 9, 7, 7, 5, 4.

It is important to remember that the decreasing-time-list algorithm does not *guarantee* optimal solutions. This can be seen by scheduling the tasks with times 11, 10, 9, 6, 4 (Figure 3.17). The schedule has a completion time of 21. However, the rearranged list 9, 4, 6, 11, 10 yields the schedule in Figure 3.18, which finishes at time 20. This solution is obviously optimal because the machines finish at the same time and there is no idle time. Note that when tasks are independent, if there are m machines available, the completion time cannot be less than the sum of the task times divided by m.

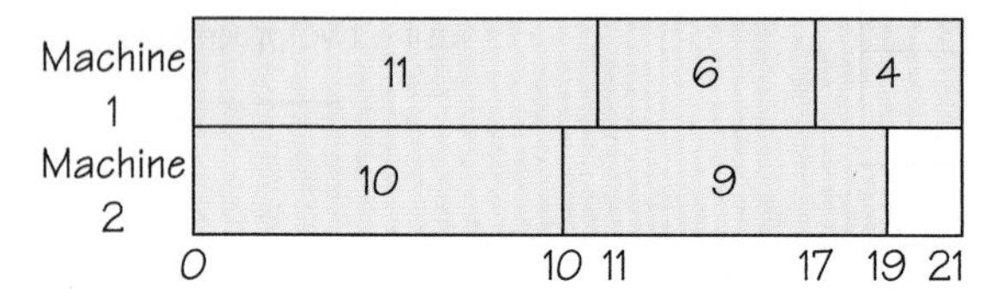

Figure 3.17 The nonoptimal schedule resulting from applying the decreasing-time-list algorithm to a collection of independent tasks. The list used, written in terms of task times only, is 11, 10, 9, 6, 4.

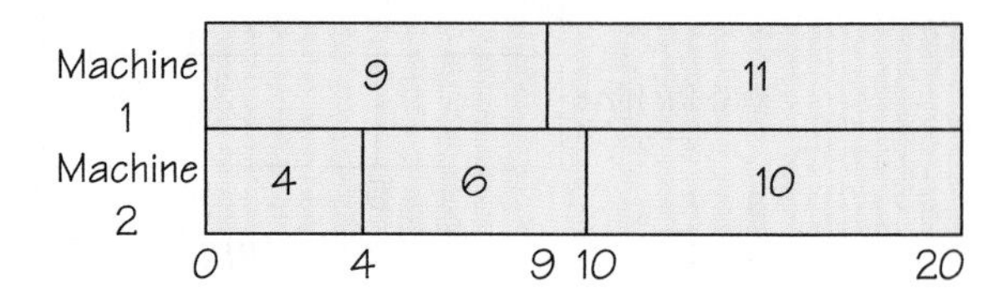

Figure 3.18 The optimal schedule resulting from applying the list-processing algorithm to a collection of independent tasks. The list used, written in terms of task times only, is 9, 4, 6, 11, 10.

EXAMPLE 4
Photocopy Shop and Data Entry Problems

Imagine a photocopy shop with three photocopiers. Photocopying tasks that must be completed overnight are accepted until 5 P.M. The tasks are to be done in any manner that minimizes the finish time for all the work. Because this problem involves scheduling machines for independent tasks, the decreasing-time-list algorithm would be a good heuristic to apply.

For another example, consider a data entry pool at a large corporation or college, where individual entry tasks can be assigned to any data entry specialist. In this setting, however, the assumption that the data entry workers are identical in skill is less likely to be true. Hence, the tasks might have different times with different processors. This phenomenon, which occurs in real-world scheduling problems, violates one of the assumptions of our mathematical model.

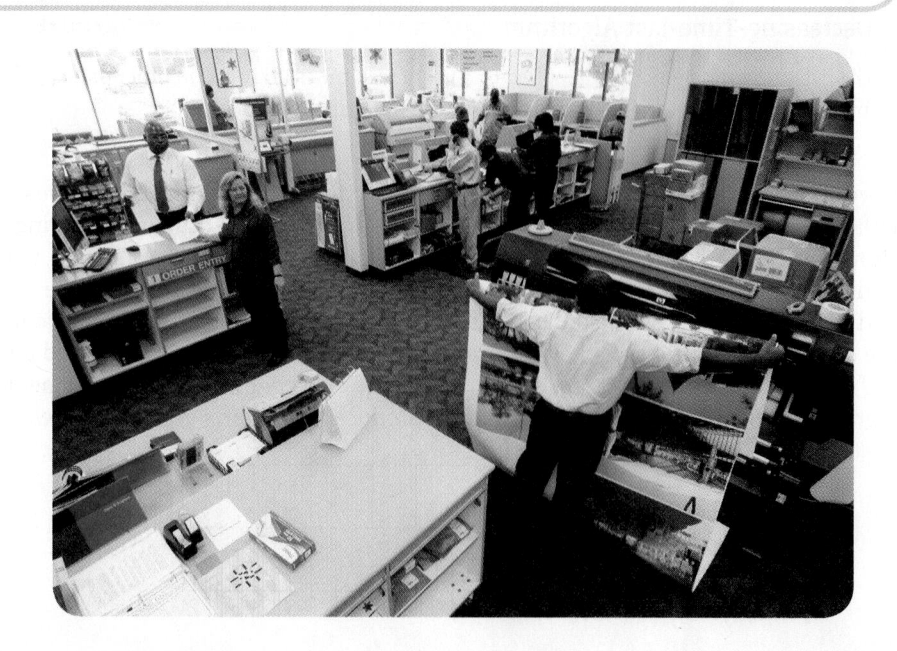

A modern copy shop provides a wide array of services ranging from copying a few sheets for a "drop in" customer, to printing elaborate reports for small businesses, to publishing monographs and advertising flyers. Using mathematical scheduling techniques can save time for the customer and increase profit for the shop owner by ensuring that the many tasks are completed most efficiently. *(The Commercial Appeal/ZUMApress.com)*

3.4 Bin Packing

Suppose you plan to build a wall system for your books, CDs, DVDs, and stereo set. This project requires 24 wooden shelves of various lengths: 6, 6, 5, 5, 5, 4, 4, 4, 4, 2, 2, 2, 2, 3, 3, 7, 7, 5, 5, 8, 8, 4, 4, and 5 feet. The lumberyard, however, sells wood only in boards of length 9 feet. If each board costs $8, what is the minimum cost to buy sufficient wood for this wall system?

Because all shelves required for the wall system are shorter than the boards sold at the lumberyard, the largest number of boards needed is 24, the precise number of shelves needed for the wall system. Buying 24 boards would, of course, be a waste of wood and money because several of the shelves you need could be cut from one board. For example, pieces of length 2, 2, 2, and 3 feet can be cut from one 9-foot board, assuming no loss of wood is created by the cutting process.

To make the ideas we develop more flexible, we think of the boards as bins of capacity W (9 feet in this case) into which we will pack (without overlap) n weights (in this case, lengths) whose values are $w_1, \ldots, w_n$, where each $w_i \leq W$. We wish to find the minimum number of bins into which the weights can be packed. In this formulation, the problem is known as the **bin-packing problem**.

Bin-Packing Problem DEFINITION

The **bin-packing problem** involves finding the minimum number of bins of weight capacity W into which weights $w_1, w_2, \ldots, w_n$ (each less than or equal to W) can be packed without exceeding the capacity of the bins.

At first glance, bin-packing problems may appear unrelated to the machine-scheduling problems we have been studying. However, there is a connection.

Let's suppose we want to schedule independent tasks so that each machine working on the tasks finishes its work by time W. Instead of fixing the number of machines and trying to find the earliest completion time, we must find the minimum number of machines that will guarantee completion by the fixed completion time (W). Despite this similarity between the machine-scheduling problem and the bin-packing problem, the discussion that follows will use the traditional terminology of bin packing.

By now, it should come as no surprise to learn that no one knows a fast algorithm that always picks the optimal (smallest) number of bins (boards). In fact, the bin-packing problem belongs to the class of NP-complete problems (see Spotlight 2.1 on page 46), which means that most experts think it unlikely that any fast optimal algorithm will ever be found. Relatively good algorithms for problems that come up in actual applications are known.

Bin-Packing Heuristics

We will think of the items to be packed, in any particular order, as constituting a list. In what follows we will use the list of 24 shelf lengths given for the wall system. We will consider various **heuristic algorithms**, namely, methods that can be carried out quickly but cannot be guaranteed to produce optimal results. Probably the easiest approach is simply to put the weights into the first bin until the next weight won't fit, and then start a new bin. (Once you open a new bin, don't use leftover space in an earlier, partially filled bin.) Continue in the same way until as many bins as necessary are used.

The resulting solution is shown in Figure 3.19. This algorithm, called **next-fit (NF)**, has the advantage of not requiring knowledge of all the weights in advance. Only the remaining space in the bin currently being packed must be remembered. The disadvantage of this heuristic is that a bin packed early on may have had room for small items that come later in the list.

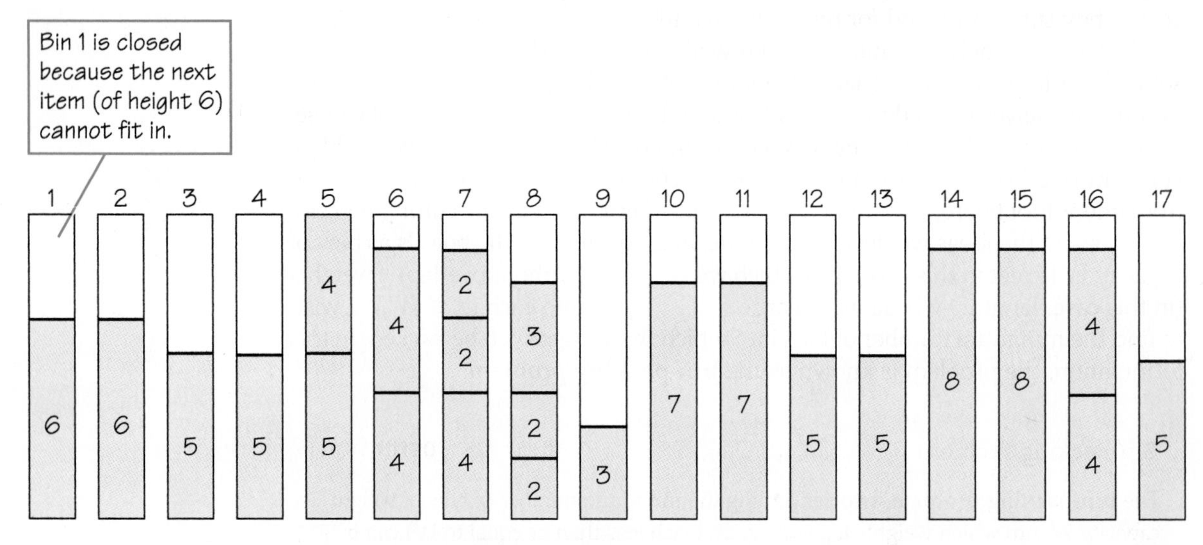

Figure 3.19 The list 6, 6, 5, 5, 5, 4, 4, 4, 4, 2, 2, 2, 2, 3, 3, 7, 7, 5, 5, 8, 8, 4, 4, 5 packed in bins using next fit.

Our wish to avoid permanently closing a bin too early suggests a different heuristic—**first-fit (FF)**: Put the next weight into the first bin already opened that has room for this weight. If no such bin exists, start a new bin. Note that a computer program to carry out first fit would have to keep track of how much room was left in all the previously opened bins. For the 24 wall-system shelves, the FF algorithm would generate a solution that uses only 14 bins (see Figure 3.20) instead of the 17 bins generated by the NF algorithm.

If we are keeping track of how much room remains in each partially filled bin, we can put the next item to be packed into the bin that currently has the most room available. This heuristic will be called **worst-fit (WF)**. The name *worst fit* refers to the fact that an item is packed into a bin with the most room available, that is, into which it fits "worst," rather than into a bin that will leave little room left over after it is placed in that bin ("best fit"). The solution generated by this approach looks the same as that shown in Figure 3.20. Although this heuristic also

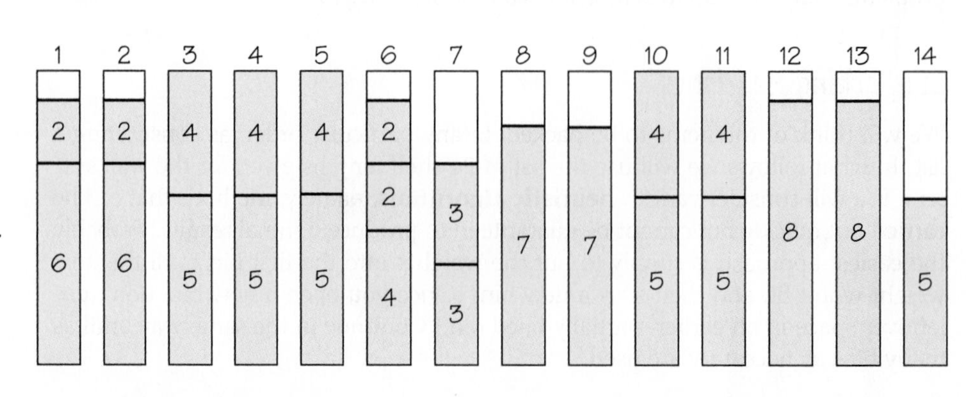

Figure 3.20 The list 6, 6, 5, 5, 5, 4, 4, 4, 4, 2, 2, 2, 2, 3, 3, 7, 7, 5, 5, 8, 8, 4, 4, 5 packed in bins using first fit. Worst fit would yield a packing that would look identical.

leads to 14 bins, the items are packed in a different order. For example, the first item of size 2, the tenth item in the list, is put into bin 6 in worst fit, but into bin 1 in first fit.

Decreasing-Time Heuristics

One difficulty with all three of these heuristics is that large weights that appear late in the list can't be packed efficiently. Therefore, we should first sort the items to be packed in order of decreasing size, assuming that all items are known in advance. We can then pack large items first and the smaller items into leftover spaces. This approach yields three new heuristics: **next-fit decreasing (NFD)**, **first-fit decreasing (FFD)**, and **worst-fit decreasing (WFD)**. Here is the original list sorted by decreasing size: 8, 8, 7, 7, 6, 6, 5, 5, 5, 5, 5, 5, 4, 4, 4, 4, 4, 4, 3, 3, 2, 2, 2, 2. Packing using FFD order yields the solution in Figure 3.21. This solution uses only 13 bins.

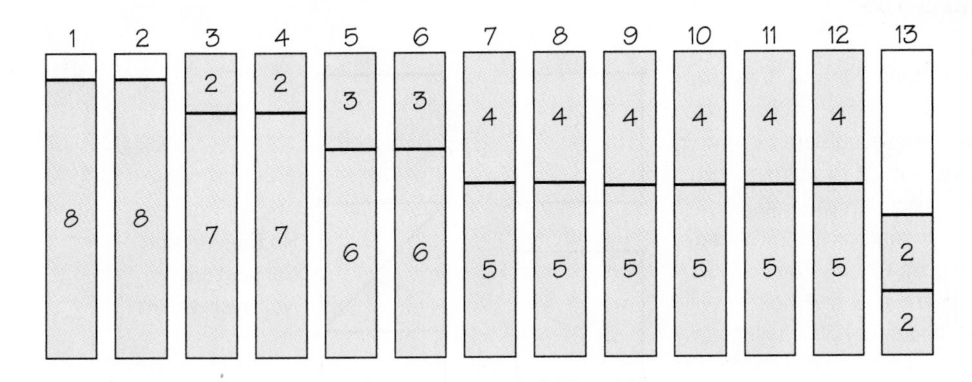

Figure 3.21 The bin packing resulting from applying first-fit decreasing to the wall-system numbers. The list involved, which uses the original list sorted in decreasing order, is 8, 8, 7, 7, 6, 6, 5, 5, 5, 5, 5, 5, 4, 4, 4, 4, 4, 4, 3, 3, 2, 2, 2, 2.

Is there any packing that uses only 12 bins? No. In Figure 3.21, there are only 2 free units (1 unit each in bins 1 and 2) of space in the first 12 bins, but 4 occupied units (two 2's) in bin 13. We could have predicted this by dividing the total length of the shelves (110) by the capacity of each bin (board): $\frac{110}{9} = 12\frac{2}{9}$. Thus, no packing could squeeze these shelves into 12 bins; there would always be at least 2 units left over for the 13th bin. (In Figure 3.21, there are 4 units in bin 13 because of the 2 wasted empty spaces in bins 1 and 2.) Even if this division created a zero remainder, there would still be no guarantee that the items could be packed to fill each bin without wasted space. For example, if the bin capacity is 10 and there are weights of 6, 6, 6, 6, and 6, the total weight is 30; dividing by 10, we get 3 bins as the minimum requirement. Clearly, however, 5 bins are needed to pack the five 6's.

None of the six heuristic methods shown will necessarily find the optimal number of bins for an arbitrary problem. How can we decide which heuristic to use? One approach is to see how far from the optimal solution each method might stray.

Various formulas have been discovered to calculate the maximum discrepancy between what a bin-packing algorithm actually produces and the best possible result. For example, in situations where a large number of bins are to be packed, FF can be off by as much as 70%, but FFD is never off by more than 22%. Of course, FFD doesn't give an answer as quickly as FF, because extra time for sorting a large collection of weights may be considerable. Also, FFD requires knowing the whole list of weights in advance, whereas FF does not. It is important to emphasize that a 22% margin of error is a worst-case figure. In many cases, FFD will perform much better. Results obtained by computer simulation indicate excellent average-case performance for this algorithm.

When solving real-world problems, we always have to look at the relationship between mathematics and the real world. Thus, first-fit decreasing usually results in fewer bins than next fit, but next fit can be used even when all the weights are not known in advance. Next fit also requires much less computer storage than first fit, because once a bin is packed, it need never be looked at again.

Fine-tuning of the conditions of the actual problem often results in better practical solutions and in interesting new mathematics as well. See Spotlight 3.2

SPOTLIGHT 3.2

Using Mathematical Tools

The tools of a carpenter include the saw, T square, level, and hammer. A mathematician also requires tools of the trade. Some of these tools are the proof techniques that enable verification of mathematical truths. Another set of tools consists of strategies to sharpen or extend the mathematical truths already known. For example, suppose that if *A* and *B* hold, then *C* is true. What happens if only *A* holds? Will *C* still be true? Similarly, if only *B* holds, will *C* still be true?

This type of thinking is of value because such questions will result either in more general cases where *C* holds or in examples showing that *B* alone and/or *A* alone can't imply *C*. For example, we saw that if a graph *G* is connected (hypothesis *A*) and even-valent (hypothesis *B*), then *G* has a circuit which uses each edge only once (conclusion *C*). If either hypothesis is omitted, the conclusion fails to hold. The figures illustrate this point. On the left is an even-valent but nonconnected graph; on the right, a connected graph with two odd-valent vertices. Neither graph has an Euler circuit.

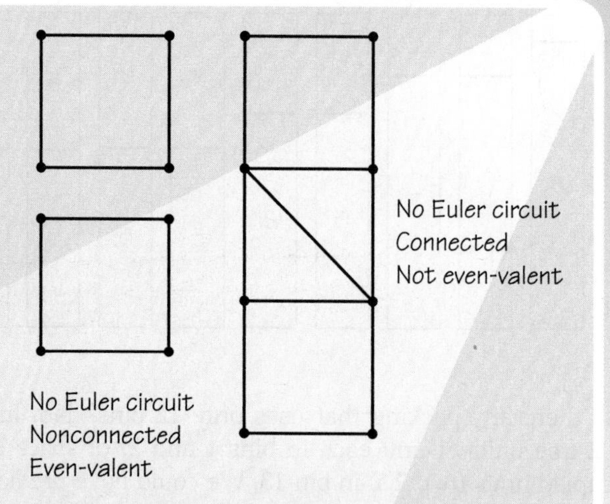

Here is another way that a mathematician might approach extending mathematical knowledge. If *A* and *B* imply *C*, will *A* and *B* imply both *C* and *D*, where *D* extends the conclusion of *C*? For example, not only can we prove that a connected, even-valent (hypotheses *A* and *B*) graph has an Euler circuit, but we can also show that the first edge of the Euler circuit can be chosen arbitrarily (conclusions *C* and *D*). It turns out that being able to specify the first two edges of the Euler circuit may not always be possible. Mathematicians are trained to vary the hypotheses and conclusions of results they prove, in an attempt to clarify and sharpen the range of applicability of the results.

We have seen that machine scheduling and bin packing are probably computationally difficult to solve because they are NP-complete. A mathematician could then try to find the simplest version of a bin-packing problem that would still be NP-complete: What if the items to be packed can have only eight weights? What if the weights are only 1 and 2? Asking questions like these is part of the mathematician's craft. Such questions help to extend the domain of mathematics and hence the applications of mathematics.

for a discussion of some of the tools mathematicians use to verify and even extend mathematical truths by raising new mathematical problems.

3.5 Resolving Conflict via Coloring

In attempting to understand situations that involve scheduling, one might desire to achieve a wide variety of goals. For example, in certain types of scheduling problems, as we have seen here, one is interested in optimization issues. What is the earliest completion time for getting a collection of tasks done on two identical processors? However, in other situations, a different goal may arise. For example, in sports, consider a league of baseball teams. Each team has to play some games during the day, some at night, some at home, and some away from home. In the interests of *equity*, it may be desirable for each team to play the same number of day games and night games both at home and away against each of the other teams in the league. If, for example, team A plays 8 games away against team B and 2 games at home against B, then if A wins both home games but loses 7 out of 8 away games, it may appear that B had an advantage due to the way its games against A were scheduled.

Graph theory can be used to resolve scheduling conflicts that occur in trying to provide students access to limited database or computer resources. *(Bananastock/Picturequest.)*

Another goal of scheduling, other than optimization and equity, may be to prevent conflicts from occurring. We can use our knowledge of graph theory to solve some interesting scheduling problems where the goal is "conflict resolution." For example, at most colleges, final examinations must be scheduled every semester and summer session. From the point of view of students and faculty both, it would be desirable to schedule these examinations so that (1) no two examinations are scheduled at the same time when a student is enrolled in both of the courses and (2) the examinations are scheduled in as "compact" a way as possible; that is, in as few time slots or days as possible. The administration of the college may share the desire for these two features and want still another property for the scheduling: (3) no more than five examinations are scheduled for any time slot. The reason for a condition such as the last might be that during the summer only five rooms with reliable enough air conditioning are available (or there might be only five rooms large enough to hold all the students taking the common final for multiple-section courses).

EXAMPLE 5
Scheduling Examinations

Small State is offering eight courses during its summer session. The table shows with an X which pairs of courses have one or more students in common. Only two air-conditioned lecture halls are available for use at any one time. To design an efficient way to schedule the final examinations, we can represent the information in this table by using a graph, as shown in Figure 3.22a. In the graph, courses are represented by vertices and two courses are joined by an edge if there is any student enrolled in both courses.

	F	M	H	P	E	I	S	C
French (F)		X		X	X	X		X
Mathematics (M)	X				X	X		
History (H)						X	X	X
Philosophy (P)	X							X
English (E)	X	X				X		
Italian (I)	X	X	X		X		X	
Spanish (S)			X			X		
Chemistry (C)	X		X	X				

We are faced with the following graph theory problem: Can we assign labels to the vertices of the graph in such a way that vertices that are joined by an edge get different labels? We think of the labels as the time slots the courses are assigned for final examinations. Traditionally, in graph theory such labels are referred to as *colors*. In this language we seek to color the vertices of the graph so that vertices that are joined by an edge get different colors. Such a coloring is called a **vertex coloring**.

Vertex Coloring — DEFINITION

The **vertex coloring** problem for a graph requires assigning each vertex of the graph a color (label) such that two vertices joined by an edge are assigned different colors.

Figure 3.22b shows one way to color the vertices of the graph so that each vertex gets a different color. Note that numbers are being used to represent the different colors. This solution is not very valuable, however, because it means that each course must be given its own time slot.

Can we improve upon eight colors? Note that *F*, *M*, *I*, *E*, and the edges that join them form a complete graph on four vertices. To have the vertices that are joined by an edge colored differently, we must assign *F*, *M*, *I*, and *E* four different colors because each of these four vertices is joined to the other three (see Figure 3.22c). Thus, any coloring of the graph in Figure 3.22a must use at least four colors. Exactly four colors will do, because the four colors used for *F*, *M*, *I*, and *E* can be used to color the remaining vertices while ensuring that no two vertices joined by an edge have the same color. The improved coloring in Figure 3.22c was found by trial and error.

Chromatic Number — DEFINITION

The **chromatic number** is the minimum number of colors needed to label the vertices of a graph so that no two vertices of the graph joined by an edge get the same color.

The examination graph we have been studying has chromatic number 4; hence, we can schedule the eight examinations in four time slots without a conflict. Notice, however, that the coloring in Figure 3.22c schedules three different courses for the time slot corresponding to color 2. This means that not enough rooms with air conditioning will be available. Is there a way to recolor the graph with four colors so that each of the four colors is used only twice? Figure 3.22d shows that the answer is yes.

Thus, we are able to schedule the eight final examinations in four time slots, using only two air-conditioned rooms, and no student will have a conflict under this schedule!

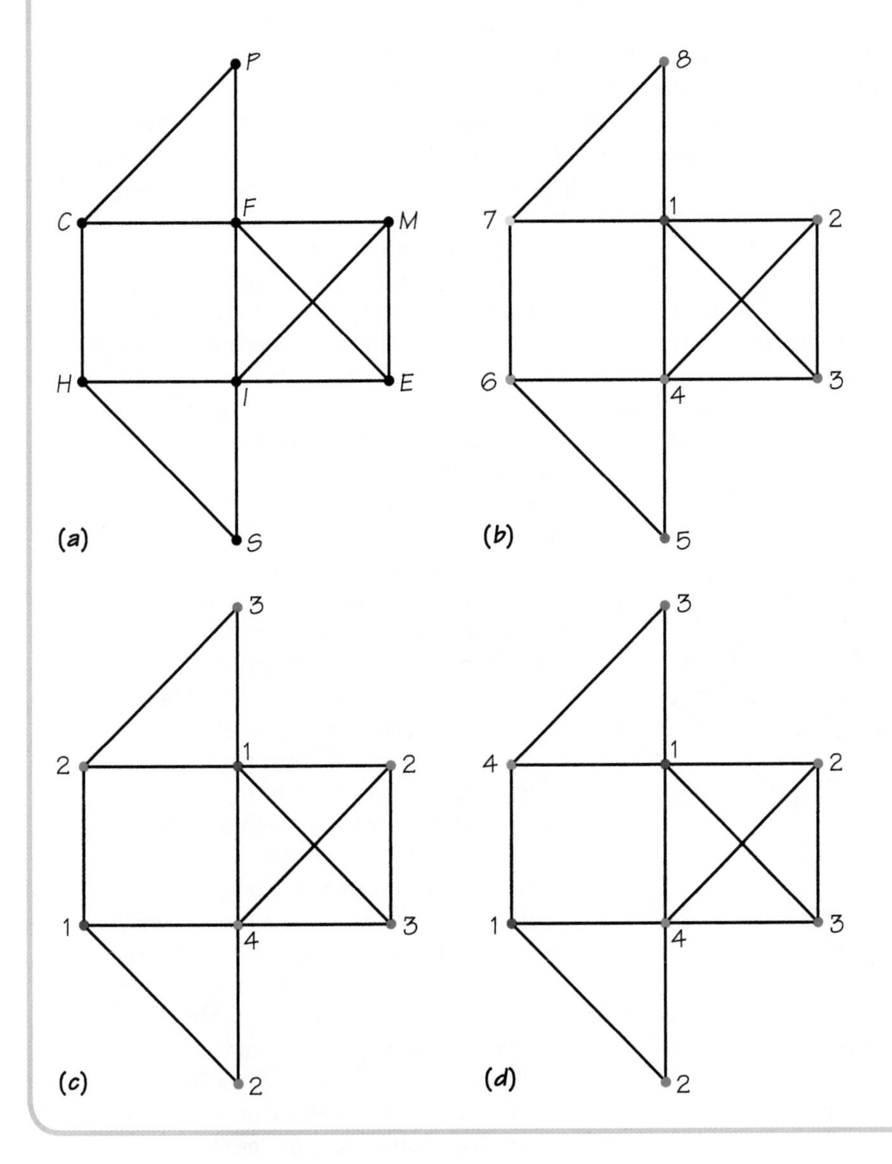

Figure 3.22 (a) A graph used to represent conflict information about courses. When two courses have a common student, an edge is drawn between the vertices that represent these courses. (b) A coloring of the scheduling graph with 8 colors, representing 8 time slots. Using this coloring would lead to a schedule where 8 time slots are used to schedule the examinations. This number is far from optimal. (c) A coloring of the scheduling graph with 4 colors. This translates into a way of scheduling the examinations during 4 time slots, and it is not possible to design a schedule with fewer time slots. However, this schedule calls for the use of three different rooms, because three examinations are scheduled during time slot 2. (d) A coloring of the scheduling graph with 4 colors. This means that the examinations can be scheduled in 4 time slots. However, because each color appears only twice, all the examinations can be scheduled in two air-conditioned rooms.

Realistic problems in scheduling government committees, high school and university final examinations, and job interviews (see Spotlight 3.4) are usually so large that graph coloring algorithms have to be incorporated into elaborate software packages to solve them.

Mathematicians have examined many kinds of coloring problems. Many developments about coloring graphs have been an outgrowth of work on the Four Color Problem (see Spotlight 3.3). One can study problems that involve the coloring of

SPOTLIGHT 3.3

Four Color Problem

Many people perceive mathematics as complex because it often uses strange notations and algebraic symbols. Thus, it may come as a surprise that a problem that is relatively easy to state and understand without complex symbolism eluded solution for about 100 years. When it was finally solved, it set off a "firestorm," with some saying that it had not truly been solved. More importantly, many of the ideas that have been developed in the theory of graphs were expanded or developed in the course of trying to prove this "guess."

When a graph can be drawn on a flat piece of paper so that edges meet only at vertices, we can talk about not only the vertices and edges of the graph, but also about its regions or *faces*. Such graphs are known as *plane graphs*. Two examples of plane graphs are shown in the diagram below.

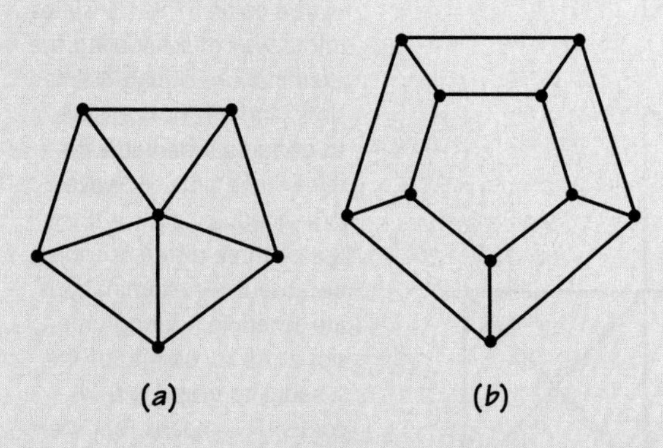

Graph (a) has 6 regions (the area "outside" of the graph is counted as one of the 6 regions), 5 of which have 3 sides and 1 of which has 5 sides, while graph (b) has 7 regions, 2 of which have 5 sides and 5 of which have 4 sides. To count the number of sides of a region, imagine that you are a small ant and are following the edges around the region, starting at some vertex *w*. You count edges until you get back to *w*. Note that for each of these graphs, there is one *unbounded* (goes off to "infinity") region, in addition to the other regions. When you color the regions of a plane map, do not forget to assign a color to the unbounded region.

If you think of the regions of a plane graph as being distinct countries on a page that is to appear in an atlas, it would be nice if countries that share a border got different colors so that they can be distinguished. Countries that meet at a vertex, but do not share an edge representing a common border, can be given the same color. It is convenient to use the term *map* for the regions created by the drawing of a plane graph.

The following provocative question was raised in a letter (1852) from Augustus De Morgan to William Rowan Hamilton that was based on a problem posed to De Morgan by his student Fredrick Guthrie, who heard the question from his brother Francis:

Can the regions of any (plane) map always be colored with four or fewer colors?

A clever approach to proving the "Four Color Conjecture" was suggested by Alfred Kempe. Kempe's "proof" had a subtle error, which defied detection for many years, showing that proofs in mathematics really depend on the community of mathematicians to guarantee their accuracy. The British mathematician Percy Heawood discovered the error Kempe made. Heawood adapted Kempe's proof to show correctly that any map can be colored with five or fewer colors. Approximately 100 years elapsed before a proof that the Four Color Conjecture was true was found. This occurred in 1976, but there was a curious loose end: The proof found by Wolfgang Haken and Kenneth Appel required that a computer verify a large collection of "calculations," which were too numerous to be done by hand. This proof troubled some philosophers and mathematicians, but has been widely accepted by the mathematics community. In 1995, Neil Robertson, Daniel Sanders, Paul Seymour, and Robin Thomas found another proof. This proof, while simpler and shorter than the earlier Haken-Appel proof, also required computer calculations too numerous to be checked by "hand." Though it is possible that some new approach to the Four Color Conjecture will avoid the use of computers, this is not widely thought to be likely. However, human ingenuity sometimes surprises us!

SPOTLIGHT 3.4

Scheduling Job Interviews

A group of companies is coming to campus for job interviews. Different companies may want different numbers of time slots to hold their interviews. In each time slot one student can be interviewed. In the example below, all the companies have requested contiguous time slots for the interviews, but this need not be the case. Due to the fact that classes are going on at the same time, five departmental conference rooms have been made available to the companies to conduct their interviews.

The interviews will follow the school's regular hourly periods, which start at 9 A.M. and end at 4 P.M. (Companies will be scheduled for continuous interviews during lunch-hour times. Interviews cannot be scheduled beyond the end of the period that starts at 4 P.M. and ends at 5 P.M.)

	Company	Time Slot Requested
A	(Apricot Computers)	7
B	(Big Green)	1
C	(Challenge Insurance)	4, 5
D	(Daisy Printers)	7, 8
E	(Earnest Engine)	4, 5, 6
F	(Flexible Systems)	2, 3
G	(Gutter Leaders)	1, 2
H	(Halley's Combs)	6, 7
I	(Indelible Ink Corporation)	7, 8
J	(Jay's Produce)	4, 5
K	(Kelly's Detective Agency)	2, 3
L	(Large Clothes)	4, 5, 6
M	(Metropolitan TV)	1, 2
N	(Nationwide Bank)	4, 5, 6, 7

Look at the list of time blocks that the companies requested (where 1 = 9–10 A.M., . . . , 8 = 4–5 P.M.). Is it possible to accommodate all the companies that wish to do interviewing in the five rooms available while meeting their desired schedule times?

Problems of this kind seem simple enough, and you should try your hand at solving this particular one, for which a schedule does exist! However, this situation is not simple at all. The following facts are known about problems of this kind.

Fact 1. Suppose there are i interviews, p time periods, and r rooms where interviews can be scheduled. Each interviewer has specified periods during which he or she wishes to conduct interviews. Is it possible to design a schedule that meets the desired specifications? It turns out that this problem is NP-complete (see Spotlight 2.1); that is, it belongs to a large group of problems for which, among other things, the fastest known algorithms run very slowly on large-problem versions.

Fact 2. The problem just described remains NP-complete even for the case where there are only three rooms to be scheduled ($p = 3$).

The moral is *surprisingly simple*: Scheduling problems are very hard to solve.

However, the situation is not always as hopeless as it might seem. If you look at the list of time requests for the corporations, you will note again that, not surprisingly, each company has requested a contiguous block of times. It turns out that when this condition holds, it is possible to determine whether there is a feasible schedule using an algorithm that works relatively quickly.

the edges of a graph rather than its vertices. Using techniques that have emerged from the study of coloring problems, problems involving such diverse contexts as scheduling government committees, using runways at airports efficiently, assigning frequencies for use by mobile pagers and cell phones, and designing timetables for public transportation have been solved—all these benefits from a problem that at first glance looks as if it belongs to recreational mathematics!

REVIEW VOCABULARY

Average-case analysis The study of the list-processing algorithm (more generally, any algorithm) from the point of view of how well it performs in all the types of problems it may be used for and seeing on average how well it does. *See also* worst-case analysis. (p. 87)

Bin-packing problem The problem of determining the minimum number of containers of capacity W into which objects of size $w_1, \ldots, w_n$ ($w_i \leq W$) can be packed. (p. 89)

Chromatic number The chromatic number of a graph G is the minimum number of colors (labels) needed in any vertex coloring of G. (p. 94)

Critical-path scheduling A heuristic algorithm for solving scheduling problems where the list-processing algorithm is applied to the priority list obtained by listing next in the priority list a task that heads a longest path in the order-requirement digraph. This task is then deleted from the order-requirement digraph, and the next task placed in the priority list is obtained by repeating the process. (p. 83)

Decreasing-time-list algorithm The heuristic algorithm that applies the list-processing algorithm to the priority list obtained by listing the tasks in decreasing order of their time length. (p. 87)

First-fit (FF) A heuristic algorithm for bin packing in which the next weight to be packed is placed in the lowest-numbered bin already opened into which it will fit. If it fits in no open bin, a new bin is opened. (p. 90)

First-fit decreasing (FFD) A heuristic algorithm for bin packing where the first-fit algorithm is applied to the list of weights sorted so that they appear in decreasing order. (p. 91)

Heuristic algorithm An algorithm that is fast to carry out but that doesn't necessarily give an optimal solution to an optimization problem. (p. 89)

Independent tasks Tasks are independent when there are no edges in the order-requirement digraph. These are tasks that can be performed in any order. (p. 86)

List-processing algorithm A heuristic algorithm for assigning tasks to processors: Assign the first ready task on the priority list that has not already been assigned to the lowest-numbered processor that is not working on a task. (p. 77)

Machine-scheduling problem The problem of assigning tasks to processors so as to complete the tasks by the earliest time possible. (p. 75)

Next-fit (NF) A heuristic algorithm for bin packing in which a new bin is opened if the weight to be packed next will not fit in the bin that is currently being filled; the current bin is then closed. (p. 90)

Next-fit decreasing (NFD) A heuristic algorithm for bin packing where the next-fit algorithm is applied to the list of weights sorted so that they appear in decreasing order. (p. 91)

Priority list An ordering of the collection of tasks to be scheduled for the purpose of attaining a particular scheduling goal. One such goal is minimizing completion time when the list-processing algorithm is applied. (p. 76)

Processor A person, machine, robot, operating room, or runway with time that must be scheduled. (p. 75)

Ready task A task is called ready at a particular time if its predecessors, as given by the order-requirement digraph, have been completed by that time. (p. 77)

Vertex coloring A vertex coloring of a graph G is an assignment of labels, which can be thought of as "colors," to the vertices of G so that vertices joined by an edge get different labels (colors). (p. 94)

Worst-case analysis The study of the list-processing algorithm (more generally, any algorithm) from the point of view of how well it performs on the hardest problems it may be used on. *See also* average-case analysis. (p. 87)

Worst-fit (WF) A heuristic algorithm for bin packing in which the next weight to be packed is placed into the open bin with the largest amount of room remaining. If the weight fits in no open bin, a new bin is opened. (p. 90)

Worst-fit decreasing (WFD) A heuristic algorithm for bin packing where the worst-fit algorithm is applied to the list of weights sorted so that they appear in decreasing order. (p. 91)

SKILLS CHECK

1. What is the minimum time required to complete 8 independent tasks with a total task time of 64 minutes on 4 machines?

(a) Less than 8 minutes
(b) Between 8 and 10 minutes
(c) More than 12 minutes

2. If the list-processing algorithm is applied to the accompanying order-requirement digraph (task time in minutes) using the list T_1, T_2, T_3, T_4, and three machines are available, the earliest completion time for all the tasks will be _____.

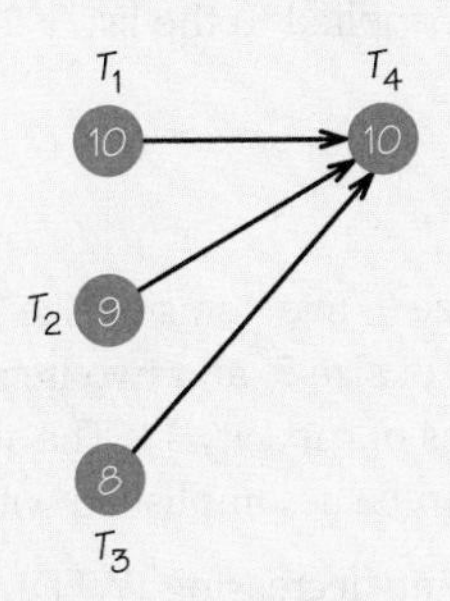

3. The following digraph cannot be an order-requirement digraph because

(a) no vertex has three edges that enter that particular vertex.
(b) it has a directed circuit.
(c) all the tasks require the same time to complete.

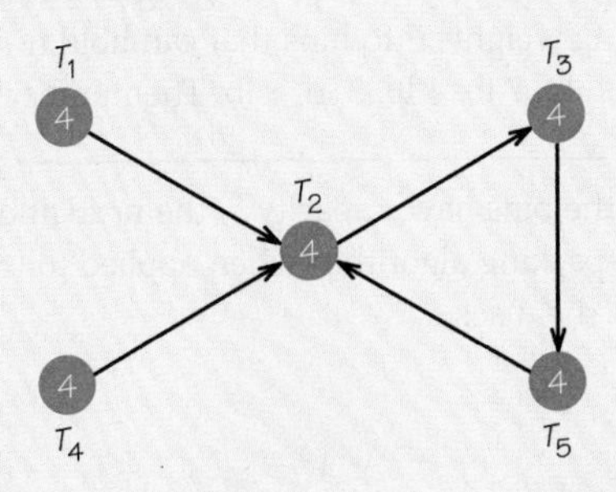

4. Given the accompanying order-requirement digraph (time in minutes) and the priority list T_1, T_2, T_3, T_4, T_5, T_6, apply the list-processing algorithm to construct a schedule using two processors. The completion time of the resulting schedule is _____________________.

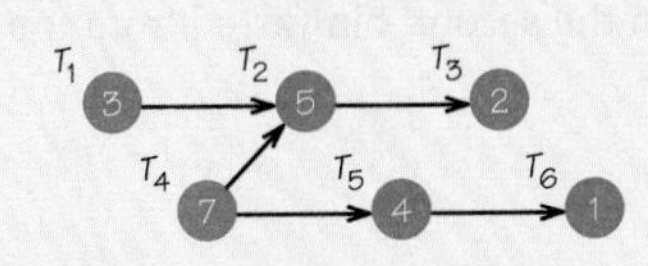

5. The shortest path and longest path, respectively, in the accompanying task-analysis digraph have, respectively, lengths

(a) 1 and 9.
(b) 4 and 14.
(c) 9 and 12.

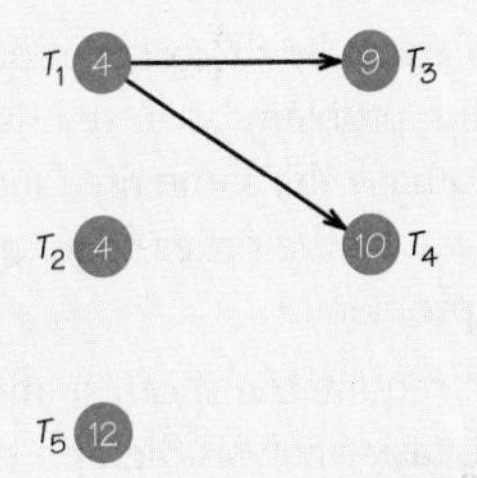

6. Suppose that a crew can complete in a minimum amount of time the job whose order-requirement digraph is shown below. If task T_2 is shortened from 5 minutes to 2 minutes, then the maximum amount by which the completion time for the entire job can be shortened is _____________________.

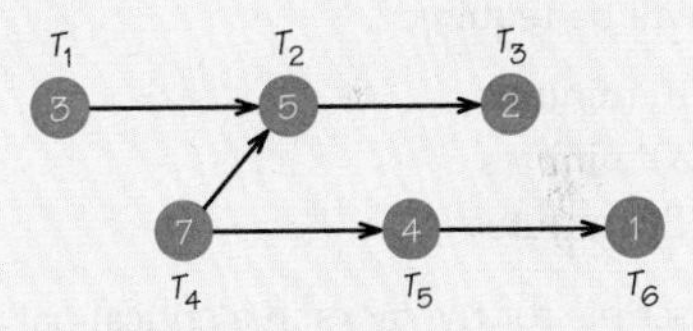

7. Suppose that independent tasks require a total of 30 minutes, while only one task takes as long as 10 minutes. If these tasks are scheduled on two machines,

(a) the tasks might take longer than 16 minutes to complete.
(b) the tasks can never take longer than 15 minutes to complete.
(c) the tasks can always be completed within 16 minutes.

8. The subscripts for the tasks that make up a critical path for the order-requirement digraph below are: ____, ____, ____.

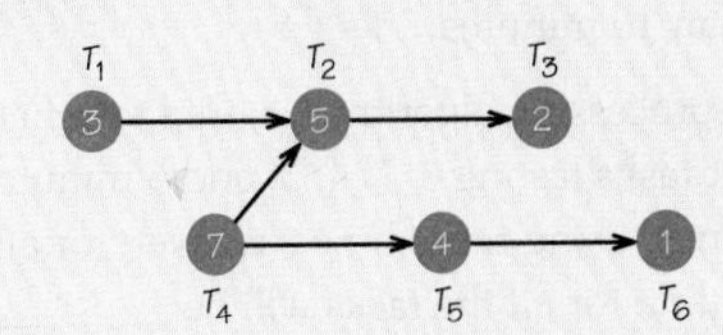

9. Which statement is true for the accompanying digraph?

(a) This digraph cannot be the order-requirement digraph for a scheduling problem because the digraph has no (directed) edges.
(b) This digraph cannot be the order-requirement digraph for a scheduling problem because it is not allowed for all the tasks to have the same time length.
(c) This digraph can be the order-requirement digraph for a scheduling problem.

10. The tasks that require the shortest and longest time to carry out in the task-analysis digraph below are, respectively, _____ and _____.

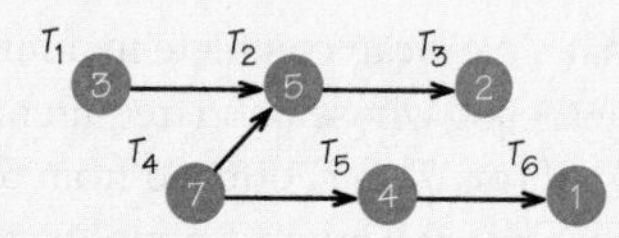

11. Assume an order-requirement digraph has a critical path with length 20 minutes. Based on this information, when the tasks are scheduled on two machines, how much time will be required?

(a) Exactly 10 minutes
(b) Exactly 20 minutes
(c) At least 20 minutes

12. The subscripts for the tasks in a critical path list associated with the following order-requirement digraph are: _____, _____, _____.

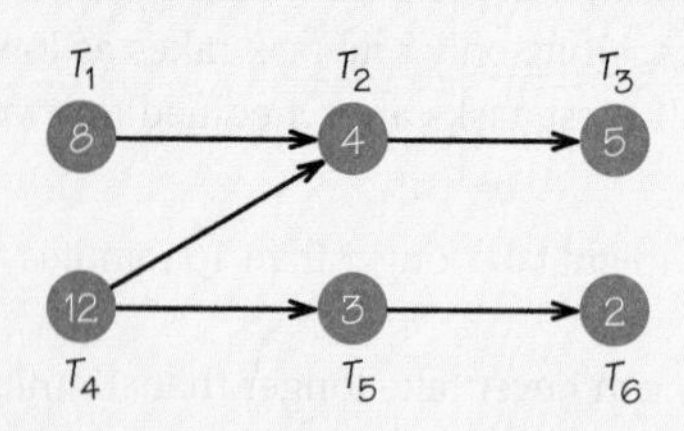

13. Assume that a job consists of six independent tasks ranging in time from 2 to 10 minutes and totaling 27 minutes. Efficiently scheduled on three machines, how much time will the job require?

(a) Exactly 9 minutes
(b) Exactly 10 minutes
(c) More than 10 minutes

14. The list-processing algorithm is used to schedule independent tasks lasting 6, 7, 4, 3, and 6 minutes on three machines, using these times as given for a list. The completion time for all the tasks will be __________.

15. A radio announcer has 10 songs of various lengths to schedule into several segments. The announcer must identify the station at least once every 15 minutes, so the segments cannot be longer than 15 minutes. This job can be solved using the

(a) list-processing algorithm for independent tasks.
(b) critical-path scheduling algorithm.
(c) first-fit algorithm for bin packing.

16. When the decreasing-time-list algorithm is used to schedule independent tasks lasting 6 minutes, 7 minutes, 4 minutes, 3 minutes, and 6 minutes, on two machines, a schedule results where the tasks are completed after _____ minutes.

17. When the bins have capacity 5, the Next Fit bin-packing algorithm applied to the list 3, 2, 4, 1, 1, 4, 4 uses

(a) 7 bins.
(b) 5 bins.
(c) 4 bins.

18. Six items of size 9, five items of size 7, four items of size 6, three items of size 5, and two items of size 9 are to be packed into bins of capacity 11. The smallest number of bins that this can be accomplished with is _____.

19. Use the worst-fit-decreasing (WFD) bin-packing algorithm to pack the following weights into bins that can hold no more than 10 lb: 6 lb, 7 lb, 4 lb, 3 lb, 6 lb. How many bins are holding a full 10 lb?

(a) 0 bins
(b) 1 bin
(c) 2 bins

20. Use the first-fit (FF) bin-packing algorithm to pack the following weights into bins that can hold no more than 10 lb: 6 lb, 7 lb, 4 lb, 3 lb, 6 lb. The number of bins required is __________.

21. When the bins have capacity 5, the next-fit decreasing (NFD) bin-packing algorithm when applied to the list 3, 2, 4, 1, 1, 4, 4 uses

(a) 7 bins.
(b) 5 bins.
(c) 4 bins.

22. When the bins have capacity 5, the items packed in the second bin (listed from top to bottom) using the list 3, 2, 4, 1, 1, 4, 4 by the first-fit (FF) bin-packing algorithm will be _____ and _____.

23. When the bins have capacity 5, the best-fit bin-packing algorithm applied to the list 2, 4, 1, 4, 3 packs which items in the second bin (listed from top to bottom)?

(a) 1, 4
(b) 4
(c) 4, 1

24. The first-fit decreasing (FFD) bin-packing algorithm is applied to the weight list 1, 2, 3, 4, 5, 5, 6, 8 for packing into bins of capacity 10. The item of weight 2 is packed into the bin numbered ________ when the packed bins are numbered from left to right.

25. A vertex coloring seeks to color the vertices of a graph to ensure which of the following traits?

(a) Every color is used.
(b) Every edge connects vertices of the same color.
(c) Vertices of the same color are never connected by an edge.

26. Assume the 8 corners of a cube represent vertices of a graph and the 12 edges of a cube represent the cube's edges. The chromatic number of this graph is ________________.

27. Which of the following statements is true?

(a) The vertices of a graph that can be drawn in the plane so that the edges meet only at the vertices can always be colored with at most three colors.
(b) The number of inequivalent (non-isomorphic) graphs that can be vertex-colored with exactly three colors is finite.
(c) The vertices of a graph that can be drawn in the plane so that the edges meet only at the vertices can be colored with at most four colors.

28. Graphs that have circuits of only even lengths have chromatic number ________.

29. The minimum number of colors needed to color the vertices of the accompanying graph is

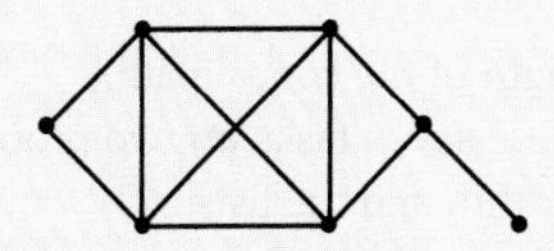

(a) 4.
(b) 2.
(c) 3.

30. A graph that has a circuit of length 3 can always be vertex colored with no fewer than _____ colors.

CHAPTER 3 EXERCISES

■ Challenge ▲ Discussion

3.1 Scheduling Tasks

3.2 Critical-Path Schedules

1. You and your two housemates are planning to have a party this Friday night at your apartment. Eight guests are expected, and you plan to serve a small homemade dinner. List the tasks involved in carrying out such a party and the types of processors to be used to carry out the tasks. Can any of the tasks be done simultaneously?

▲ **2.** Jane is planning a getaway weekend at a ski resort. She plans to leave work in Manhattan at 1 P.M. and must make her way to a local airport for a 5 P.M. shuttle plane to Boston. She then hopes to get a bus to the nearby resort. Discuss the tasks that Jane must complete to be at the resort by 10 P.M. What are the different types of processors involved in getting these tasks done? Can any of these tasks be done simultaneously?

3. List as many scheduling situations as you can for these environments:

(a) Your school
(b) Bus/train terminal
(c) Hospital
(d) Police station
(e) Bookstore
(f) Firehouse
(g) Television studio

4. Compare and contrast the scheduling problems which arise at a

(a) Fast food restaurant.
(b) Standard sit-down restaurant.

5. Use the list-processing algorithm to schedule the tasks in the accompanying order-requirement digraph on

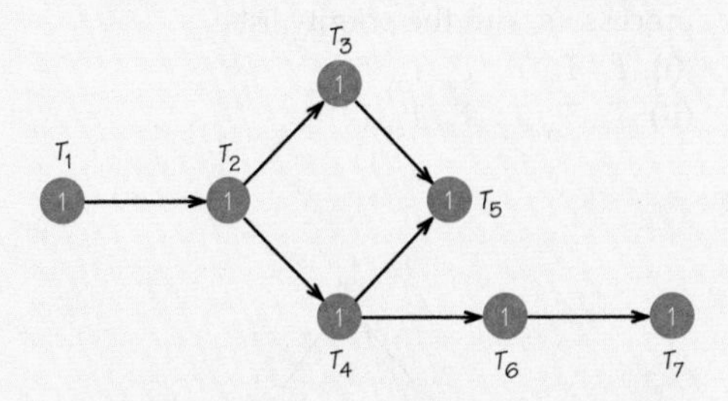

(a) two processors using the list $T_1, \ldots, T_7$.
(b) two processors using the list $T_1, T_2, T_3, T_4, T_6, T_5, T_7$.
(c) Is either of the schedules that you obtain optimal?
(d) Will adding a third processor enable the tasks to be finished earlier?
(e) Which tasks in this order-requirement digraph can be shortened and not affect the completion time of all the tasks?

6. Consider the order-requirement digraph below:

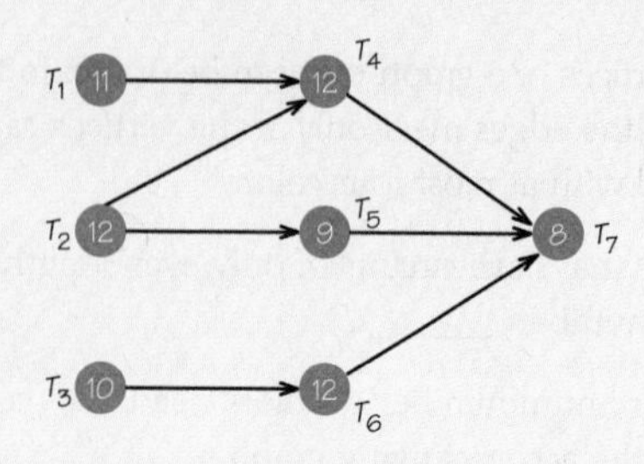

(a) Find the length of the critical path.

(b) Schedule these seven tasks on two processors using the list algorithm and the lists:

(i) $T_1, T_2, T_3, T_4, T_5, T_6, T_7$

(ii) $T_2, T_1, T_3, T_6, T_5, T_4, T_7$

(c) Does either list lead to a completion time that equals the length of the critical path?

(d) Show that no list can ever lead to a completion time equal to the length of the critical path (providing the schedule uses two processors).

7. An order-requirement digraph has exactly two vertices that have no edges coming into these vertices. Explain why, if there are three processors available, one or more of these processors must be idle some of the time.

8. If the tasks subject to an order-requirement digraph are scheduled on only one machine, explain what different goals one might have in choosing to use different lists to schedule the tasks.

9. (a) Use the accompanying order-requirement digraph to schedule the 6 tasks $T_1, T_2, T_3, T_4, T_5, T_6$ on two processors with the priority lists:

(i) $T_1, T_2, T_3, T_4, T_5, T_6$

(ii) $T_1, T_6, T_3, T_5, T_4, T_2$

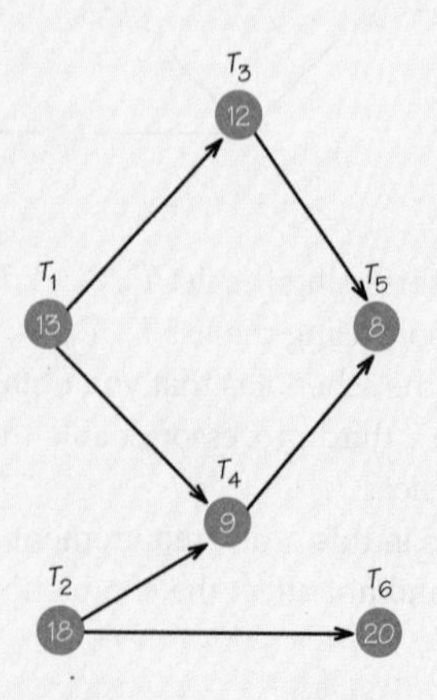

(b) Are either of the schedules produced from these lists optimal? If not, can you find a priority list that will result in an optimal schedule?

(c) Find the critical path and its length. Explain why no schedule has earliest completion time equal to the length of the critical path.

10. (a) Repeat Exercise 9, but interchange the task times of tasks T_2 and T_6.

(b) How does the completion time for an optimum schedule for this situation compare with the optimum schedule for Exercise 9?

11. (a) If one adds a new directed edge to an order-requirement digraph D, can the critical path in the new order-requirement digraph D' have longer length?

(b) If one adds a new directed edge to an order-requirement digraph D, can the critical path in the the new order-requirement digraph D' have shorter length?

▲ **12.** Discuss scheduling problems for which it is not reasonable to assume that once a processor starts a task, it will always complete that task, before it works on any other task. Give examples for which this approach would be reasonable.

13. Can you give examples of scheduling problems for which it seems reasonable to assume that all the task times are the same?

14. Use the list-processing algorithm to schedule the tasks in the below order-requirement digraph on

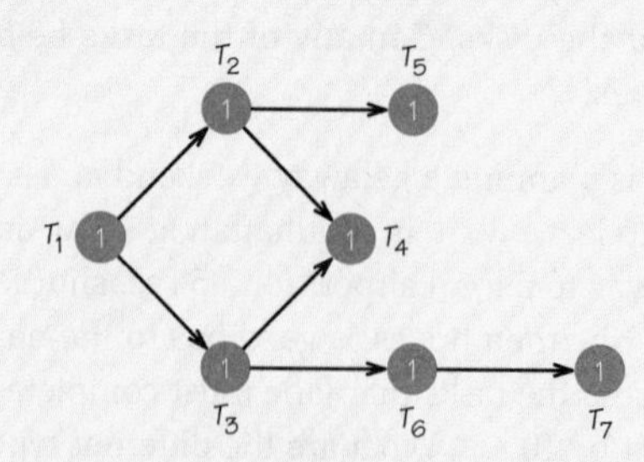

(a) two processors using the list $T_1, \ldots, T_7$.

(b) two processors using the list $T_1, T_2, T_3, T_4, T_6, T_5, T_7$.

(c) Is either of the schedules that you obtain optimal?

15. For the accompanying order-requirement digraph, apply the list-processing algorithm, using three processors for lists (a) through (c). How do the completion times obtained compare with the length of the critical path?

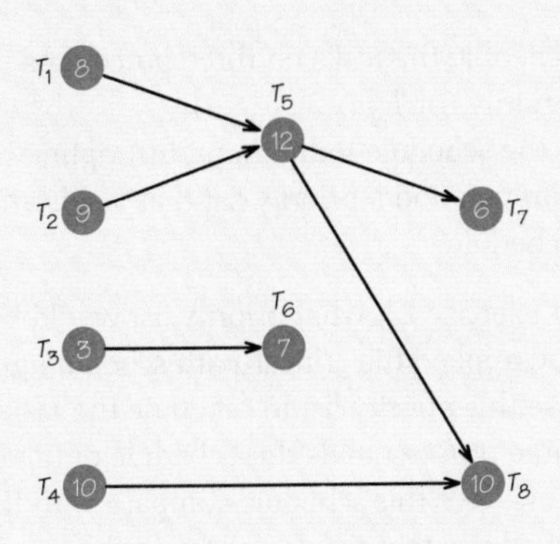

(a) $T_1, T_2, T_3, T_4, T_5, T_6, T_7, T_8$
(b) $T_1, T_3, T_5, T_7, T_2, T_4, T_6, T_8$
(c) $T_8, T_6, T_4, T_2, T_1, T_3, T_5, T_7$

16. (a) Can you find an order-requirement digraph with four tasks for which every priority list used to schedule the tasks on two machines assigns task T_4 to machine 1 at time 0?
(b) Can you choose the order-requirement digraph in part (a) so that machine 2 stays idle for all lists from time 0 to time 3?

17. Can you find a list that gives rise to the optimal schedule shown in Figure 3.14 (on page 86) for the order-requirement digraph in Figure 3.12 (on page 85)?

18. Consider the accompanying order-requirement digraph:

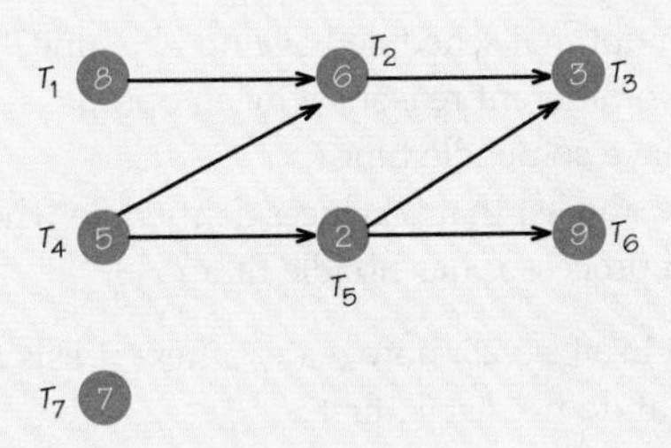

(a) Find the critical path(s).
(b) Schedule these tasks on one processor using the critical-path scheduling method.
(c) Schedule these tasks on one processor using the priority list obtained by listing the tasks in order of decreasing time.
(d) Does either of these schedules have idle time? How do their completion times compare?
(e) If two different schedules have the same completion time, what criteria can be used to say one schedule is superior to the other?
(f) Schedule these tasks on two processors using the order-requirement digraph shown and the priority list from part (b).
(g) Does the schedule produced in part (f) finish in half the time that the schedule in part (b) did, which might be expected, since the number of processors has doubled?
(h) Schedule the tasks on (i) one processor and (ii) two processors (using the decreasing-time list), assuming that each task time has been reduced by one. Do the changes in completion time agree with your expectations?

19. Given the order-requirement digraph below:

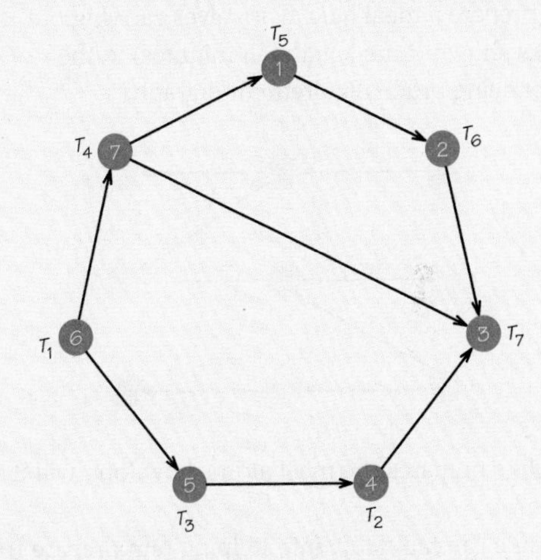

(a) Find the length of the shortest path from T_1 to T_2.
(b) What is the length of the critical path?
(c) Give a schedule that completes the tasks by the time length of the critical path on two machines, or explain why this is not possible. (If it is possible, provide a list that gives rise to this schedule.)

20. The order-requirement diagram accompanying Exercise 19 shows a directed edge from T_4 to T_7. Explain why this edge can be omitted from the order-requirement digraph because it is "redundant."

21. (a) Can all the processors being used to schedule tasks be simultaneously idle at a time before the completion time of a collection of tasks scheduled using the list-processing algorithm?
(b) Explain why the list-processing algorithm cannot give rise to the schedule below, regardless of what priority list was used to schedule the tasks on the three processors.

Machine 1	T_1	T_4	T_6
Machine 2	T_2		T_7
Machine 3	T_3	T_5	

(c) Construct an order-requirement digraph and a priority list that will give rise to the following schedule on two processors.

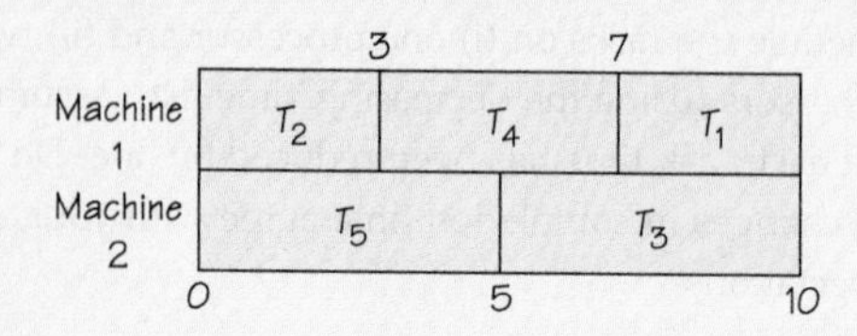

22. To prepare a meal quickly involves carrying out the tasks shown (time lengths in minutes) in the accompanying order-requirement digraph:

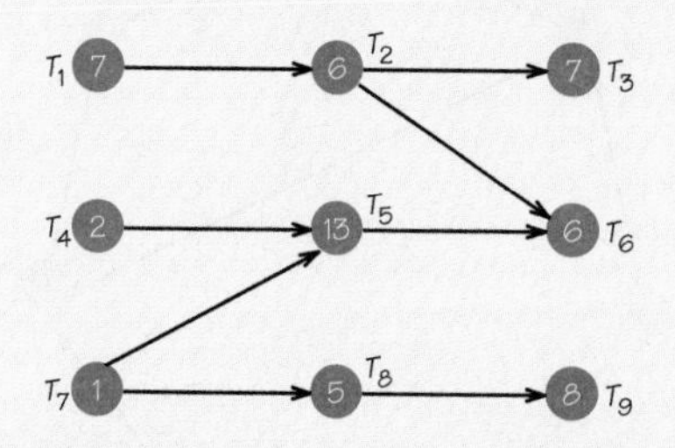

(a) If Mike prepares the meal alone, how long will it take?

(b) If Mike can talk Mary into helping him prepare the meal, how long will it take them if the tasks are scheduled using the list $T_5, T_9, T_1, T_3, T_2, T_6, T_8, T_4, T_7$ and the list-processing algorithm?

(c) If Mike can talk Mary and Jack into helping him prepare the meal, how long will it take if the tasks are scheduled using the same list as in part (b)?

(d) What would be a reasonable set of criteria for choosing a priority list in this situation?

23. (a) Making use of the order-requirement digraph below, determine at time 0 which tasks are ready.

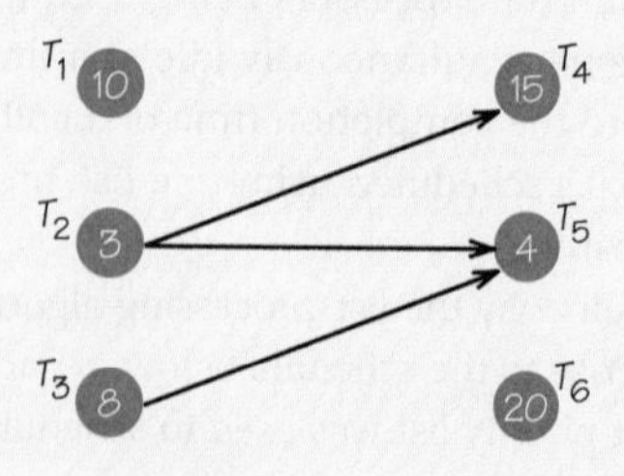

(b) What is special about tasks T_1 and T_6?

(c) What is the critical path, and what is its length?

(d) Schedule the tasks on three processors with the priority list $T_1, \ldots, T_6$.

(e) Is the schedule found in part (d) optimal?

(f) Schedule the tasks on three processors using the priority list $T_6, \ldots, T_1$.

(g) Is the schedule found in part (f) optimal?

(h) Can you find a priority list that yields an optimal schedule?

24. (a) In Exercise 23, what priority list would be used if you applied the critical-path scheduling method?

(b) Use this priority list to schedule the tasks on three processors. Is this schedule optimal?

(c) How does this schedule compare with the schedules that you found using the lists in Exercise 23?

25. Consider the order-requirement digraph below. Suppose one plans to schedule these tasks on two identical processors.

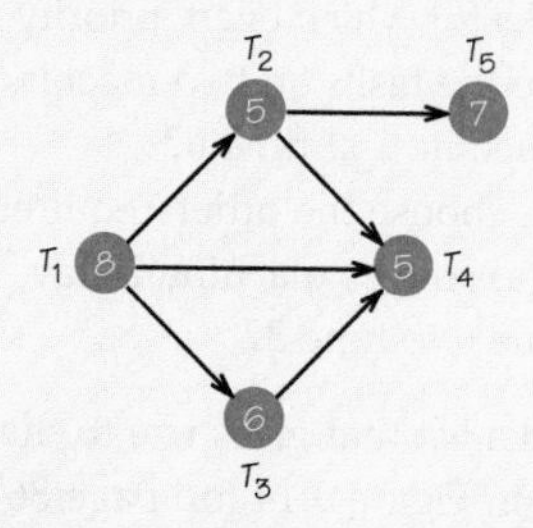

(a) How many different priority lists can be used to schedule the tasks?

(b) Can all these priority lists lead to different schedules? If not, why not?

(c) Can an optimal schedule have no idle time? Can you give two different reasons why an optimal schedule must have some idle time?

(d) Is there any list that produces a schedule where the second processor has no idle time?

26. (a) In Exercise 25, how many different lists are there that do not list T_1 first?

(b) Would it make any sense not to list T_1 first in a list?

(c) Construct a list and schedule the tasks on two processors.

(d) Can you find another list that leads to a different completion time than the schedule you found for part (c)?

(e) Find a list that leads to an optimal schedule.

27. Can you find an order-requirement digraph with five tasks for which every possible list yields exactly the same schedule?

28. Can you find an order-requirement digraph involving three tasks such that the schedule corresponding to every list is different?

29. At a large toy store, scooters arrive unassembled in boxes. To assemble a scooter, the following tasks must be performed:

Task 1. Remove parts from the box.
Task 2. Attach wheels to the footboard.
Task 3. Attach vertical housing.
Task 4. Attach handlebars to vertical housing.
Task 5. Put on reflector tape.
Task 6. Attach bell to handlebars.
Task 7. Attach decals.
Task 8. Attach kickstand.
Task 9. Attach safety instructions to handlebars.

(a) Give reasonable time estimates for these tasks and construct a reasonable order-requirement digraph. What is the earliest time by which these tasks can be completed?
(b) Schedule this job on two processors (humans) using the decreasing-time-list algorithm.

30. If two schedules for the same number of processors have the same completion time, can one schedule have more idle time than the other?

31. Could the schedule below be obtained by applying the list-scheduling algorithm to some order-requirement digraph?

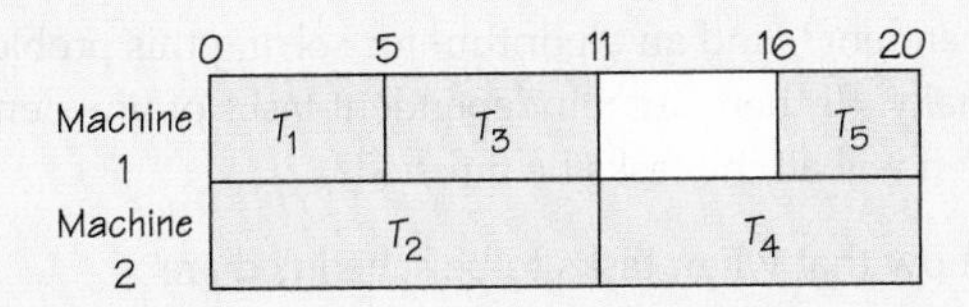

3.3 Independent Tasks

32. Could the following schedule be obtained by applying the list-scheduling algorithm to some order-requirement digraph?

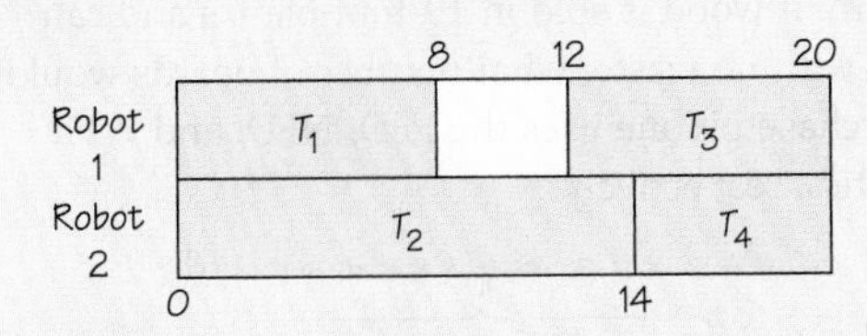

33. For the accompanying schedules, can you produce a list so that the list-processing algorithm produces the schedule shown when the tasks are independent? What are the times for each task?

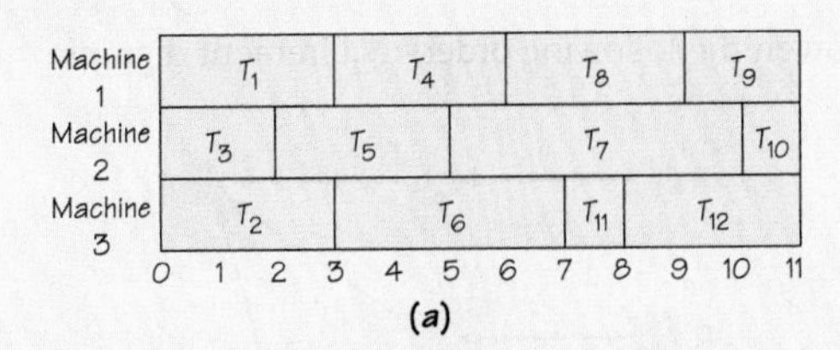

(*a*)

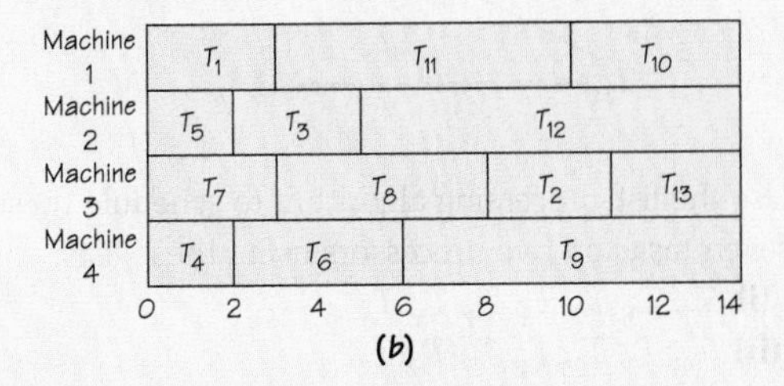

(*b*)

▲ **34.** Once an optimal schedule has been found for independent tasks (see diagrams in Exercise 33), usually the scheduling of the tasks can be rearranged and the same optimal time achieved.

One can, among other things, reorder the tasks done by a particular processor. Discuss criteria that might be used to implement the rearrangement process.

35. The task times of eight independent tasks T_1 to T_8 are 1, 2, 3, 4, 5, 6, 7, 8.

(a) Schedule the tasks on two processors using the lists (i) $T_1, T_2, \ldots, T_8$ and (ii) $T_8, T_7, \ldots, T_1$.
(b) Is either of the schedules you get in part (a) optimal? If not, find a list that gives an optimal schedule.

36. Repeat Exercise 35, but schedule the tasks (with the same lists) on three processors. If the schedules you get are not optimal, find a list that gives an optimal schedule.

▲ **37.** Discuss different criteria that might be used to construct a priority list for a scheduling problem.

▲ **38.** Some scheduling projects have due dates for tasks (times by which a given task should be completed) and release dates (times before which a task cannot have work begun on it). Give examples of circumstances where these situations might arise.

39. Using the lists you found in Exercise 33 and the task times you computed for those independent tasks, schedule the tasks for (a) on four processors and the tasks for (b) on five processors. Can you see why for any schedule you may produce for (a) on four processors and (b) on five processors there must be some idle time for one or more processors?

40. Given the following order-requirement digraph:

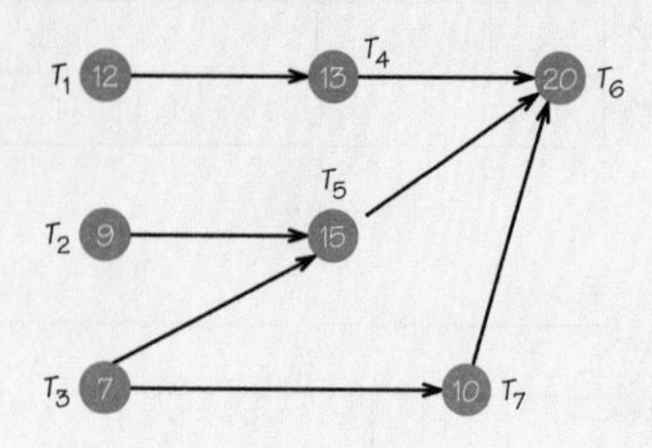

(a) Use the list-processing algorithm to schedule these seven tasks on two processors using these lists:

(i) $T_1, T_3, T_7, T_2, T_4, T_5, T_6$

(ii) $T_1, T_3, T_2, T_4, T_5, T_6, T_7$

(iii) The list obtained by listing the tasks in order of decreasing time

(b) Try to determine if any of the resulting schedules are optimal.

(c) Schedule the tasks using the critical-path scheduling method. Try to determine if this schedule is optimal.

41. Repeat the questions in Exercise 40 using the order-requirement digraph obtained by erasing all the (directed) edges shown there. How do the schedules you get compare with the ones you originally got?

42. (a) Find the completion time for independent tasks of length 8, 11, 17, 14, 16, 9, 2, 1, 18, 5, 3, 7, 6, 2, 1 on two processors, using the list-processing algorithm.

(b) Find the completion time for the tasks in part (a) on two processors, using the decreasing-time-list algorithm.

(c) Does either algorithm give rise to an optimal schedule?

(d) Repeat for tasks of lengths 19, 19, 20, 20, 1, 1, 2, 2, 3, 3, 5, 5, 11, 11, 17, 18, 18, 17, 2, 16, 16, 2.

43. Repeat parts (a)–(c) of Exercise 42 for independent tasks of lengths 19, 19, 20, 20, 1, 1, 2, 2, 3, 3, 5, 5, 11, 11, 17, 17, 18, 18, 17, 2, 16, 16, 2.

44. Suppose that independent tasks require a total of 36 minutes, while only one of the tasks takes as long as 12 minutes. If these tasks are scheduled on two machines, show by an example that the earliest completion time may be as long as 22 minutes.

45. A photocopy shop must schedule independent batches of documents to be copied. The times for the different sets of documents are (in minutes): 12, 23, 32, 13, 24, 45, 23, 23, 14, 21, 34, 53, 18, 63, 47, 25, 74, 23, 43, 43, 16, 16, 76.

(a) Construct a schedule using the list-processing algorithm on three machines.

(b) Construct a schedule using the list-processing algorithm on four machines.

(c) Repeat parts (a) and (b), but use the decreasing-time-list algorithm.

(d) Suppose union regulations require that an 8-minute rest period be allowed for any photocopy task over 45 minutes. Use the decreasing-time-list algorithm, with the preceding times modified to take into account the union requirement, to schedule the tasks on three human-operated machines.

46. Find a list that produces the following optimal schedule when the list-processing algorithm is applied to this list. (Assume that the tasks are independent.)

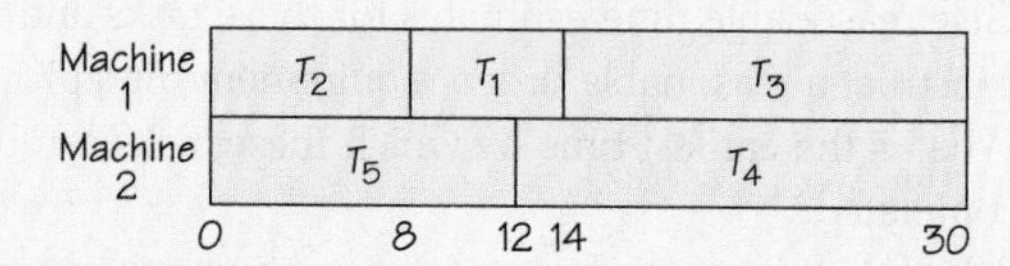

What completion time and schedule are obtained when the decreasing-time-list algorithm is applied to this list?

47. Can you think of situations other than those mentioned in the text where scheduling independent tasks on processors occurs?

48. Can you think of real-world scheduling situations in which all the tasks have the same time and are independent? Find an algorithm for solving this problem optimally. (If there are n independent tasks of time length k, when will all the tasks be finished?)

49. Show that when tasks to be scheduled are independent, the critical-path method and the decreasing-time-list method are identical.

3.4 Bin Packing

50. Two wooden wall systems are to be made of pieces of wood with lengths shown in the accompanying diagram. If wood is sold in 10-foot planks and can be cut with no waste, what number of boards would be purchased if one uses the FFD, NFD, and WFD heuristics, respectively?

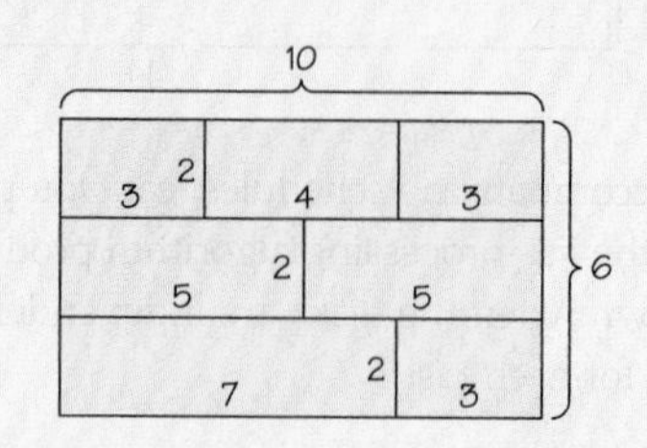

In solving this problem, does it make a difference if the 10-foot horizontal shelves and 6-foot vertical boards employ single-length pieces, as compared with using pieces of boards that add up to 10- and 6-foot lengths?

51. It takes 4 seconds to photocopy one page. Manuscripts of 10, 8, 15, 24, 22, 24, 20, 14, 19, 12, 16, 30, 15, and 16 pages are to be photocopied. How many photocopy machines would be required, using the FFD algorithm, to guarantee that all manuscripts are photocopied in 2 minutes or less? Would the solution differ if WFD were used?

52. A radio station's policy allows advertising breaks of no longer than 2 minutes, 15 seconds. Using FF and FFD algorithms, determine the minimum number of breaks into which the following ads will fit (lengths given in seconds): 80, 90, 130, 50, 60, 20, 90, 30, 30, 40. Can you find the optimal solution? Do the same for these ad lengths: 60, 50, 40, 40, 60, 90, 90, 50, 20, 30, 30, 50.

53. Fiberglass insulation comes in 36-inch precut sections. A plumber must install insulation in a basement on piping that is interrupted often by joints. The distances between the joints on the stretches of pipe that must be insulated are 12, 15, 16, 12, 9, 11, 15, 17, 12, 14, 17, 18, 19, 21, 31, 7, 21, 9, 23, 24, 15, 16, 12, 9, 8, 27, 22, 18 inches. How many precut sections would he have to use to provide the insulation if he bases his decision on

(a) next-fit?
(b) next-fit decreasing?
(c) worst-fit?
(d) worst-fit decreasing?

54. The files that a company has for its employees dealing with utilities occupy 100, 120, 60, 90, 110, 45, 30, 70, 60, 50, 40, 25, 65, 25, 55, 35, 45, 60, 75, 30, 120, 100, 60, 90, 85 sectors. If, after operating systems are installed, a disk can store up to 480 sectors, determine the number of disks needed to store the utilities if each of these heuristics is used to pack the disk with files:

(a) next-fit
(b) next-fit decreasing
(c) first-fit
(d) first-fit decreasing

55. Advertisements for the TV show *Q* are permitted to last up to a total of 8 minutes, and each group of ads can last up to 2 minutes. If the ads slated for *Q* last 63, 32, 11, 19, 24, 87, 64, 36, 27, 42, 63 seconds, determine if FF and FFD yield acceptable configurations for the ads.

56. Consider the heuristic for packing bins known as *best-fit* described as follows: Keep track of how much room remains in each unfilled bin and put the next item to be packed into that bin that would leave the least room left over after the item is put into the bin. (For example, suppose that bin 4 had 6 units left, bin 7 had 5 units left, and bin 9 had 8 units left. If the next item in the list had size 5, then first-fit would place this item in bin 4, worst-fit would place the item in bin 9, while best-fit would place the item in bin 7.) If there is a tie, place the item into the bin with the lowest number. Apply this heuristic to the list 8, 7, 1, 9, 2, 5, 7, 3, 6, 4, where the bins have capacity 10.

▲ **57.** We have described two algorithms for bin packing called worst-fit and best-fit (see page 90 and Exercise 56). The words *best* and *worst* have connotations in English. However, the performance of algorithms depends on their merits as algorithms, not on the names we give them.

(a) On the basis of experiments you perform with the best-fit and worst-fit algorithms, which one do you think is the "better" of the two?
(b) Can you construct an example where worst-fit uses fewer bins than best-fit?

58. The best-fit heuristic (see Exercise 56) also has a "decreasing" version, where the list is first sorted in decreasing order. Using bins of capacity 10, apply the best-fit heuristic and its decreasing version to the following list: 6, 9, 5, 8, 3, 2, 1, 9, 2, 7, 2, 5, 4, 3, 7, 6, 2, 8, 3, 7, 1, 6, 4, 2, 5, 3, 7, 2, 5, 2, 3, 6, 2, 7, 1, 3, 5, 4, 2, 6.

■ **59.** One pianist's recording of the complete Mozart piano sonatas takes the following times (given in minutes and seconds): 13:46, 6:15, 3:29, 5:37, 7:52, 2:55, 5:00, 4:28, 4:21, 7:39, 7:55, 6:42, 4:23, 3:52, 4:21, 4:20, 5:46, 6:29, 5:34, 6:23, 6:39, 7:19, 5:54, 6:54, 2:58, 5:22, 1:42, 5:00, 1:29, 5:47, 7:30, 8:19, 4:44, 4:57, 4:09, 14:31, 3:55, 4:04, 4:01, 6:06, 6:50, 5:27, 4:28, 5:40, 2:52, 5:16, 5:34, 3:10, 7:22, 4:40, 3:08, 6:32, 4:47, 6:59, 5:38, 7:57, 3:38. If the maximum time that can be recorded on a compact disc is 70:30, can all the music be performed on four compact discs? Can all the music be performed on five compact discs?

■ **60.** In the wall-system example in the text, first-fit and worst-fit required equal numbers of bins (see Figure 3.20 on page 90). Can you find an example where first-fit and worst-fit yield different numbers of bins? Can you find an example where first-fit, worst-fit, and next-fit yield answers with different numbers of bins?

▲ **61.** A common suggestion for heuristics for the bin-packing problem with bins of capacity *W* involves finding weights that sum to exactly W. Discuss the pros and cons of a heuristic of this type.

■ **62.** A recording company wishes to record all the Beethoven string quartets (16 quartets, each consisting of several consecutive parts called movements) on LPs. It wishes to complete the project on as few records as possible. Recording can be done on two sides as long as the movements are consecutive. Is this an example of a bin-packing problem? (Defend your answer.) If the project were to record the quartets on (standard) tape cassettes or compact discs, would your answer be different?

63. Give examples where it would be realistic to keep bins open as more items "arrive" to be packed, rather than to close a bin permanently based on some criterion.

64. Give examples where it would be unrealistic to keep bins open as more items "arrive" to be packed, rather than to close a bin permanently based on some criterion.

65. A data entry group must handle 30 (independent) tasks that will take the following amounts of time (in minutes) to type: 25, 18, 13, 19, 30, 32, 12, 36, 25, 17, 18, 26, 12, 15, 31, 18, 15, 18, 16, 19, 30, 12, 16, 15, 24, 16, 27, 18, 9, 14. Using these times as a priority list:

(a) Use the list-processing algorithm to find the completion time for scheduling tasks with four secretaries. Also, solve with five secretaries.

(b) Repeat the scheduling using the decreasing-time-list algorithm.

(c) Can you show that any of the schedules that you get are optimal?

If one needs to finish the typing in one hour:

(d) Use the FFD heuristic to find how many typists would be needed.

(e) Repeat for the NFD and WFD heuristics.

(f) Can you show that any of the solutions you get are optimal?

66. Find the minimum number of bins necessary to pack items of size 8, 5, 3, 4, 3, 7, 8, 8, 6, 5, 3, 2, 1, 2, 1, 2, 1, 3, 5, 2, 4, 2, 6, 5, 3, 4, 2, 6, 7, 7, 8, 6, 5, 4, 6, 1, 4, 7, 5, 1, 2, 4 in bins of capacity (a) through (d) using the FF and FFD algorithms. Can you determine if any of the packings you get are optimal?

(a) 9
(b) 10
(c) 11
(d) 12

■ **67.** Two-dimensional bin packing refers to the problem of packing rectangles of various sizes into a minimum number of $m \times n$ rectangles, with the sides of the packed rectangles parallel to those of the containing rectangle.

(a) Suggest some possible real-world applications of this problem.

(b) Devise a heuristic algorithm for this problem.

(c) Give an argument to show that the problem is at least as hard to solve as the usual bin-packing problem.

(d) If you have $1 \times m$ rectangles with total area W to be packed into a single rectangle of area $p \times q = W$, can the packing always be accomplished?

▲ **68.** In what situations would packing bins of different capacities be the appropriate model for real-world situations? Suggest some possible algorithms for this type of problem.

■ **69.** Find an example of weights that, when packed into bins using first-fit, use fewer bins than the number of bins used when the first-fit algorithm is applied with the first weight on the list removed.

▲ **70.** Formulate "paradoxical" situations for bin packing that are analogous to those we found for scheduling processors.

3.5 Resolving Conflict via Coloring

71. For each of the graphs below:

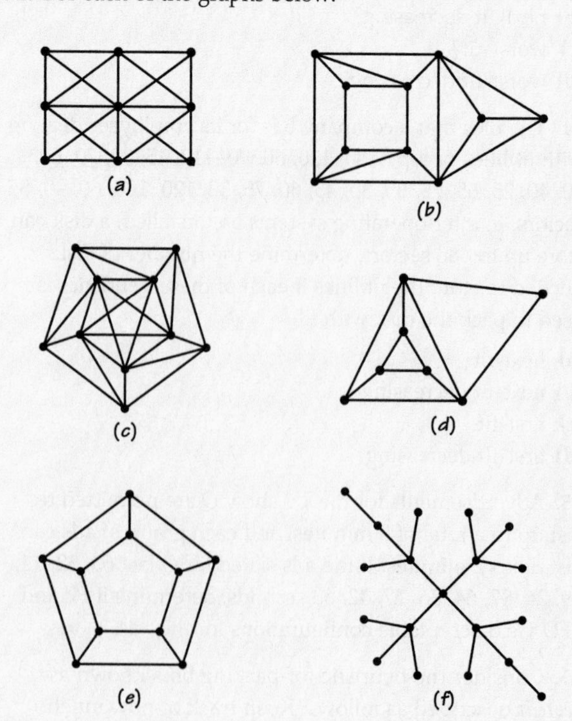

(a) Color the vertices (if possible) with three different colors.

(b) Color the vertices (if possible) with four different colors.

(c) Find the chromatic number of the graph.

72. For each of the accompanying graphs:

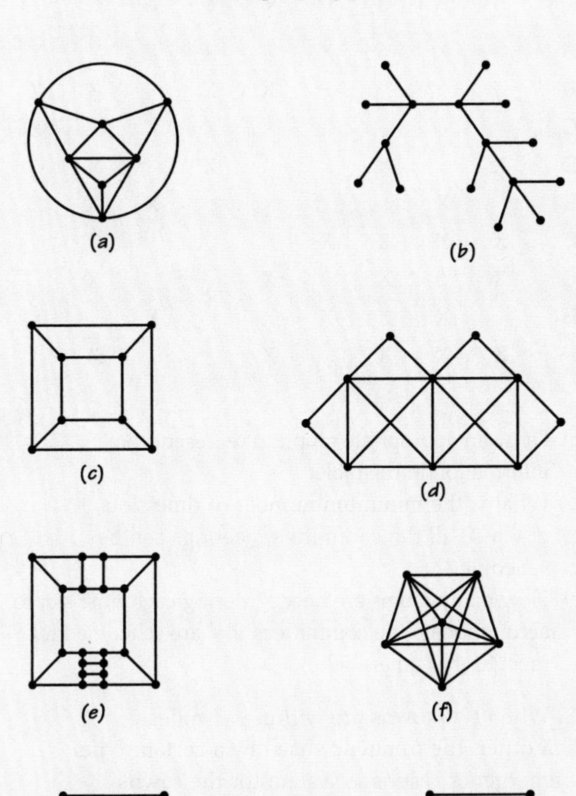

(a) (b) (c) (d) (e) (f) (g) (h)

(a) Color the vertices (if possible) with two different colors.

(b) Color the vertices (if possible) with three different colors.

(c) Find the chromatic number of the graph.

73. The owner of a new pet store wishes to display tropical fish in display tanks. The accompanying table shows the incompatibilities between the species, in the sense that an X indicates that it is unwise to allow those species in the row and column that meet at the X to be in the same tank.

	A	B	C	D	E	F	G	H	I
A						X	X		X
B			X					X	
C		X			X			X	
D					X	X		X	
E			X	X			X		
F	X			X			X		X
G	X				X	X		X	X
H		X	X	X			X		
I	X					X	X		

(a) Draw an appropriate graph to represent the information in the table.

(b) What is the minimum number of tanks needed to display all the fish she wishes to sell?

(c) Display the species so that the number of species in each tank is as nearly equal as possible.

74. The managers of a zoo are planning to open a small satellite branch. The animals are to be in enclosures in which compatible animals are displayed together. The accompanying table indicates those pairs of animals that are compatible. (Thus, an X in a particular row and column means that the animals that label this row and column *can* share an enclosure.)

	A	B	C	D	E	F	G	H	I	J
A	X	X		X	X	X	X			
B	X	X			X	X	X		X	X
C			X		X	X	X			
D	X			X	X	X	X		X	X
E	X	X	X	X	X			X	X	
F	X	X	X	X		X	X	X	X	
G	X	X	X	X		X	X	X		
H					X	X	X	X		
I		X		X	X	X			X	
J		X		X						X

(a) Draw an appropriate graph to represent the information in the table.

(b) What is the minimum number of enclosures needed to avoid housing incompatible animals in the same enclosure?

(c) Is it possible to enclose the animals in such a way that each enclosure contains the same number of animals?

(d) Why might that be desirable? Why might this approach to grouping the animals not be ideal?

75. The nine standing committees of a state legislature are designing a schedule for when the committees can meet. The matrix shown in the following table has an X in a position where the committees corresponding to the row and column have a common member and, hence, should not be scheduled to meet at the same hour. The committees involved are Agriculture (A), Commerce (C), Consumer Affairs (CA), Education (E), Forests (F), Health (H), Justice (J), Labor (L), and Rules (R).

	A	C	CA	E	F	H	J	L	R
A		X	X			X			
C	X		X	X	X				
CA	X	X					X		X
E		X			X	X			
F		X		X		X	X		
H	X			X	X			X	
J			X		X			X	X
L						X	X		X
R			X				X	X	

(a) Draw a graph that will be of value in determining the minimum number of time slots the committees can meet in without any legislator having to be in two places at one time.

(b) What is the minimum number of time slots in which the committees can be scheduled without a conflict?

(c) How many different rooms are needed at any time that a committee is scheduled to meet? (Why might this issue matter?)

76. Determine the minimum number of colors, and how often each color is used, in a vertex coloring of the graphs below.

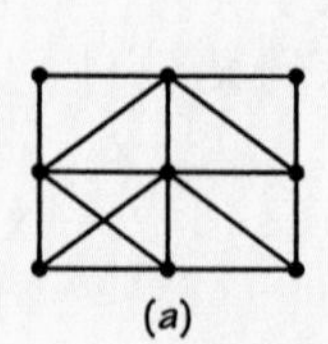
(a)

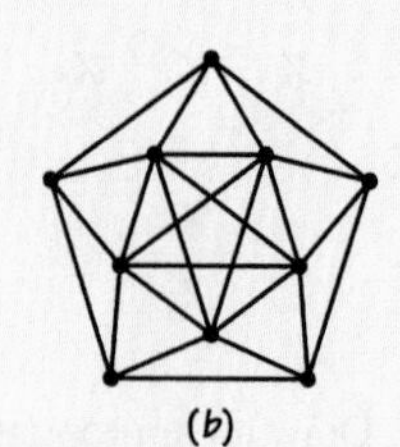
(b)

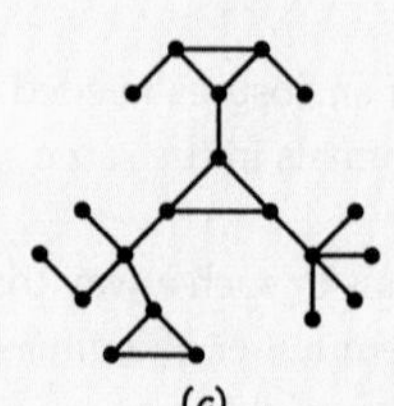
(c)

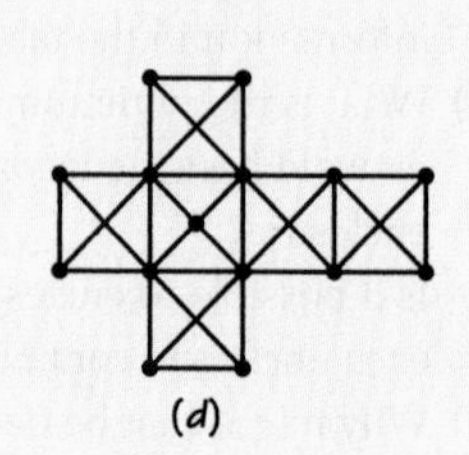
(d)

77. The faculty–student governing council at All State College has nine standing committees (such as Curriculum, Academic Standards, Campus Life) that are designated A, B, C, D, …, I for convenience. The following table shows which committees have no member in common.

	A	B	C	D	E	F	G	H	I
A		X		X		X	X		X
B	X				X	X		X	X
C				X		X	X	X	X
D	X		X			X		X	
E		X					X	X	X
F	X	X	X	X					
G	X		X		X			X	
H		X	X	X	X		X		X
I	X	X	X		X			X	

(a) Draw an appropriate graph to represent the information in the table.

(b) What is the minimum number of time slots in which all the committee meetings can be scheduled?

(c) How many rooms are needed during each time slot to accommodate the committees that are scheduled to meet in that time slot?

78. When two towns are within 145 miles of each other, the frequency used by a certain type of emergency response system for the towns requires that they be on different frequencies to avoid possible interference with each other. The following table shows the mileage distances between six towns.

	E	F	G	I	S	T
Evansville (E)		290	277	168	303	133
Ft. Wayne (F)	290		132	83	79	201
Gary (G)	277	132		153	58	164
Indianapolis (I)	168	83	153		140	71
South Bend (S)	303	79	50	140		196
Terre Haute (T)	113	201	164	71	196	

(a) What would be the minimum number of frequencies that are needed for each town to have its emergency broadcasts not conflict with those of any other town using this system?

(b) How many different towns would be assigned to each frequency used?

79. The legislature of a city has committees devoted to the following governmental areas:
A = agriculture; P = planning; D = districting; F = finance; E = education; T = transportation; H = housing, C = courts. To schedule meetings for these committees in as few time slots as possible, consider the graph model below.

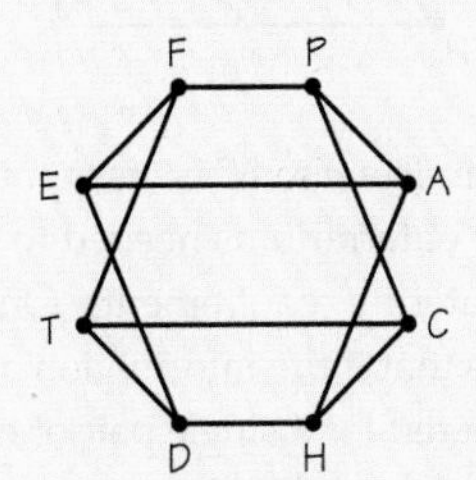

(a) Complete the entries in the table below that would display the "conflict" information represented by the graph, where an X indicates two committees that cannot meet at the same time because they have at least one common member.

	F	E	H	P	T	A	D	C
F								
E								
H								
P								
T								
A								
D								
C								

(b) What is the minimum number of time slots in which it is possible to schedule these committees?

(c) If there are only three rooms with the audio and video setups needed for the committees to meet, what is the minimum number of time slots during which the committees can meet?

80. A small college has a mini-session in January (between its regular semesters) in which nine classes are offered: A = Art; B = Biology, C = Chemistry; D = Diversity; E = English, F = French, G = German, H = Health, I = Italian. The accompanying table indicates those courses that have students in common.

	A	B	C	D	E	F	G	H	I
A		X	X	X					
B	X				X	X			
C	X			X			X		X
D	X		X				X		
E		X				X		X	
F		X			X			X	X
G			X	X				X	X
H					X	X	X		X
I			X			X	X	X	

(a) Draw a graph model for the information in this table.

(b) What is the minimum number of time slots in which final examinations can be scheduled for these nine courses?

(c) Can this minimum be achieved if only three rooms are large enough to hold the finals?

81. Show that the vertices of any tree can be colored with two colors.

82. Can you find a family of graphs H_n ($n \geq 1$) that require n colors to color their vertices?

83. The edge-coloring number of a graph G is the minimum number of colors needed to color the edges of G so that edges that share a common vertex get different colors. Determine the edge-coloring number for each of the graphs in Exercise 71. Can you make a conjecture about the value of the minimum number of colors needed to color the edges of any graph?

84. Can you think of any applications that require determining the minimum number of colors needed to color the edges of a graph?

85. When a graph has been drawn on a piece of paper so that edges meet only at vertices, the graph divides the paper up into regions called *faces*. The faces include one called the "infinite" face, which surrounds the whole graph. The face-coloring number of a graph G (which can be drawn in this special way) is the minimum number of colors needed to color the faces of G so that two faces that share an edge receive different colors. (Note that if two faces meet only at a vertex, they can be colored the same color.)

(a) Determine the minimum number of colors needed to color the faces of the accompanying graphs. In each case, remember to color the infinite face, which is labeled I (for "infinite").

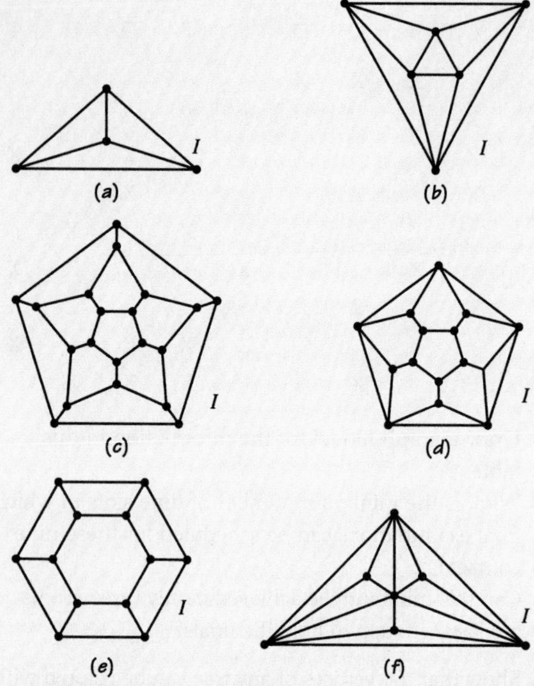

(b) Can you think of an application of the problem of coloring the faces of a graph with a minimum number of colors?

86. For each of the graphs in Exercise 72 where the graph shown has edges that meet only at vertices, verify that the Four Color Theorem holds by showing that the regions (faces) of the graph can be colored with four or fewer colors so that regions that share an edge get different colors. (Remember to assign a color to the unbounded, so-called infinite region.)

87. A company sells herbs, each of which requires a certain level of proper watering. The accompanying graph is constructed by having one vertex for each type of herb. The vertices representing two herbs are joined by an edge if they must have different levels of watering. What is the minimum number of terrariums that the herbs can be displayed in so that herbs in the same terrarium can be watered at the same level?

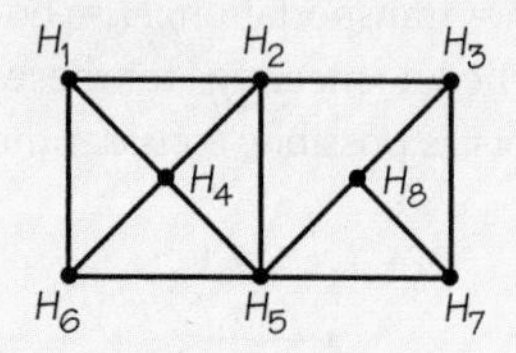

88. The company in Exercise 87 is disappointed by the minimum number of terrariums needed to display the herbs with the proper watering requirements. One company employee suggests that if the information about watering requirements is altered for a single pair of herbs (e.g., a single edge is erased from the diagram), then the number of terrariums needed will be reduced by 1. Is this true?

89. Each vertex in the graph below represents a child who attends a day care center. An edge between two children indicates these children tend to cause problems when they are in the same play group. What is the minimum number of play groups that will ensure that no conflicts arise? Can conflict-free play groups with the same number of children in each group be formed?

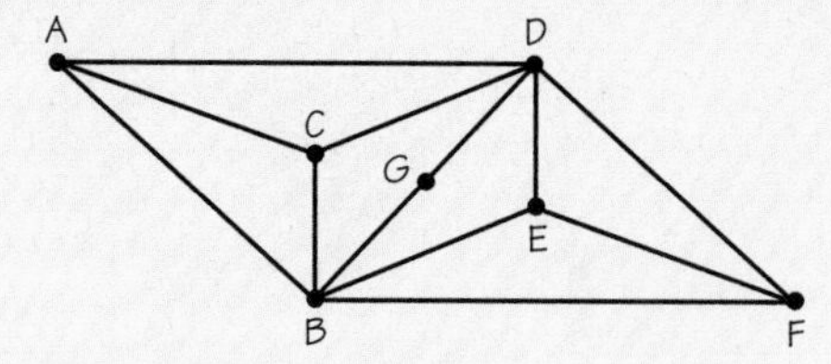

APPLET EXERCISES

To do these exercises, go to www.whfreeman.com/fapp9e.

Graph Coloring

Solving a scheduling problem such as the one below can be accomplished by constructing a related graph and then coloring it in a way that adjacent vertices have different colors. Explore the problem of graph coloring in the *Graph Coloring* applet.

Scheduling

A mathematics department has seven faculty committees—A, B, C, D, E, F, and G. Because there is overlap in the composition of the committees, the chairman of the department is attempting to work out a schedule that will avoid conflicts among the committees.

The following chart indicates the overlapping committee structure:

	A	B	C	D	E	F	G
A		X		X		X	
B	X		X			X	
C		X			X		X
D	X						X
E			X			X	X
F	X	X			X		
G			X	X	X		

Help the chairman arrange a schedule without conflicts in the *Scheduling* applet.

WRITING PROJECTS

1. Scheduling is important for hospitals, schools, transportation systems, police services, and fire services. Pick one of these areas and write about the different scheduling situations that come up, the types of processors, and the extent to which the assumptions of the list-processing model hold for the area you pick.

2. Compare and contrast the basic scheduling problem we investigated with the scheduling version of the bin-packing problem.

3. One of the oversimplifications made in our discussion of scheduling was that there were no "due dates" involved for the tasks making up a job. Develop an algorithm for solving a scheduling problem under the assumption that each task has a due date as well as a time length. You will probably want to decide on a penalty amount that will occur when a due date is exceeded.

4. Consider the problem of scheduling tasks on a single machine. Design different algorithms for achieving different goals. You will probably wish to assume that each task has a due date such that if the task is not finished by this date, some penalty payment must be made.

5. Discuss the role of graph colorings for scheduling committee meetings so as to avoid conflicts. Research whether or not these ideas are used in the legislature of your home state.

6. In choosing a location (vertex) for trains to turn around in the graph shown in Spotlight 3.1 (on page 83), explain why it seems to be a much better choice to use V as a place to allow the turnarounds, rather than at M or at R.

Suggested Readings

BRUCKER, P. *Scheduling Algorithms*, 4th ed., Springer-Verlag, Heidelberg, Germany, 2004. A detailed mathematical look at scheduling.

GRAHAM, RONALD. Combinatorial scheduling theory, in Lynn Steen (ed.), *Mathematics Today*, Springer-Verlag, New York, 1978, pp. 183–211. This essay on scheduling is one of many excellent accounts of recent developments in mathematics in this book.

GRAHAM, RONALD. The combinatorial mathematics of scheduling. *Scientific American*, March 1978, pp. 124–32. A very readable introduction to scheduling and bin packing.

JENSEN, T. R., and BJARNE TOFT. *Graph Coloring Problems*, Wiley, New York, 1995. A detailed summary of what is known about coloring problems and many questions that await answering.

LAWLER, E., et al. Sequencing and scheduling algorithms and complexity, in S. C. Graves et al. (eds.), *Handbooks in OR and MS*, vol. 4, Elsevier, New York, 1993, pp. 445–522. A recent survey of results about scheduling.

LEUNG, JOSEPH Y-T., *Handbook of Scheduling*, Chapman & Hall/CRC, Boca Raton, Florida, 2004. This book has an encyclopedic treatment of scheduling algorithms and the great variety of situations where mathematical analysis has assisted schedulers, ranging from sports to hospitals.

PARKER, R. GARY, *Deterministic Scheduling Theory*, Chapman & Hall, London, 1995. A wide-ranging look at scheduling methods and their applications.

Suggested Web Sites

www.ctl.ua.edu/math103/scheduling/schedmnu.htm This site provides an overview of scheduling as discussed in this chapter.

http://www.ams.org/samplings/feature-column/fcarc-machines1
http://www.ams.org/samplings/feature-column/fcarc-packings1
http://www.ams.org/samplings/feature-column/fcarc-bins1 These Web pages describe mathematical aspects of machine scheduling and bin packing and give a discussion of the relationship between these two mathematical problems.

www.ie.bilkent.edu.tr/~ie672/docs/resources.html This Web page contains links to many aspects of scheduling theory, including research on the frontier.

Jeremy Walker/Stone/Getty Images

4

Linear Programming

A manager's job often calls for making very complicated decisions. One set of decisions involves planning what products the business is to make and determining what resources are needed to make these products. In the modern business world, diversification of products provides a company with stability in a climate of changing tastes and needs. So it is not surprising that companies would produce many products, some of which share resource needs. For example, any bakery uses many resources—like butter, sugar, eggs, and flour—to make its products such as cookies, cakes, pies, and breads. Similarly, car manufacturers use many kinds of metals in the different models of cars they make, and manufacturers of gasoline use different kinds of crude oils to make their product.

Resources can include more than just raw materials. Farmland, time, machinery, and a labor force with appropriate skills are also resources. Typically, resources are limited: A farmer owns only so much land; there are only so many hours in a day; in a year of drought the wheat crop is very small; a winter freeze may damage an orange crop. Resource availability is also limited by location and competition.

Because resources are limited, management faces important questions: How should the available resources be shared among the possible products? One goal of management is to maximize profit. How can that determine how much of each product should be produced? There are usually so many alternative product mixes that it is impossible to evaluate them all individually. Despite this complexity, millions of dollars may ride on management's decision.

Many business and government agencies must deal with supply-and-demand problems. The general idea is that goods or services can be provided by different providers to individuals or businesses who need these goods or services. There are varying costs to the suppliers to provide different recipients with these goods or services. The goal is to find how to meet the demands from the supplies as cheaply as possible. For example, what is the cheapest way for a company with several oil refineries to provide oil distributors, in many different geographical locations, with the oil they need?

4.1 Linear Programming and Mixture Problems: Combining Resources to Maximize Profit

Here, we learn about **linear programming**, a management science technique that helps a business allocate the resources it has on hand or can purchase to make a particular mix of products that will maximize profit.

Linear Programming

DEFINITION

Linear programming is a tool for maximizing or minimizing a quantity, typically a profit or a cost, subject to a set of constraints.

The technique is so powerful that it is estimated that much more computer time is used solving linear programming types of problems than for any other purpose for which business managers and decision makers use computers.

Linear programming is an example of "new" mathematics. It came into being, along with many other management science techniques, during and shortly after World War II, in the 1940s. It is quite young as intellectual ideas go. Yet, during its short history, linear programming has changed the way businesses and governments make decisions, from "seat-of-the-pants" methods based on guesswork and intuition to using an algorithm based on available data and guaranteed to produce an optimal decision.

Linear programming is but one operations research tool belonging to a family of tools known as mathematical programming. Another such tool is integer programming. The difference between linear programming and integer programming is that for linear programming, the quantities being studied can take on values such as $\pi = 3.14159\ldots$ or $7\frac{1}{8}$; in integer programming, the values are confined to whole numbers such as 8, 50, or 1,102,362. Whole numbers are conceptually easier than the broader group consisting of all numbers that can be represented by decimals (1.32, 1.455555 . . .), yet integer-programming problems have proved much harder to solve.

The assembly of an automobile requires many complicated steps and processes. The use of linear-programming techniques enables the robots and humans to carry out their tasks faster and more accurately than would be possible without the use of mathematics. This makes American cars more competitive and of a higher quality than otherwise would be the case. *(Age fotostock/Photolibrary)*

In the discussions that follow, we often describe "relaxed" versions of integer-programming problems as linear-programming problems. For example, it would make no sense to produce 3.24 dolls to sell. So, strictly speaking, we must find an optimum whole number of dolls to produce. If we are "lucky," the linear-programming problem associated with an integer-programming problem has an integer solution. In this case, we have also found the correct answer to the integer-programming problem. Some other examples that fall into this category are discussed below.

Linear programming has saved businesses and governments billions of dollars. Of all the management science techniques presented in this book, linear programming is far and away the most frequently used. It can be applied in a variety of situations, in addition to the one we study in this chapter. Some of the problems studied in Chapters 1, 2, and 3—for example, the traveling salesman (TSP) and scheduling problems—can be viewed as linear-programming problems. Linear programming is an excellent example of a mathematical technique useful for solving many different kinds of problems that at first do not seem to be similar problems at all. It has been suggested that without linear programming, management science would not exist.

Next, we study how to use linear programming to solve a special kind of problem—a **mixture problem**. Realistic versions of such problems would be much more involved. Our discussion is designed to give you the flavor of what is actually done. Realistic examples of what follows are commonly used in the manufacture of different kinds of breads from the grain flours available, and in the making of different kinds of sausages from meats such as beef and pork.

Mixture Problem DEFINITION

In a **mixture problem**, limited resources are combined into products so that the profit from selling those products is a maximum.

Mixture problems are widespread because nearly every product in our economy is created by combining resources. A typical example would be how different kinds of aviation fuel are manufactured using different kinds of crude oil.

Let's analyze small versions of the kinds of problems that might confront a toy or a beverage manufacturer. Both manufacturers can sell many different products on which each company can make a profit. There could be dozens of possible products and many resources. A manufacturer must periodically look at the quantities and prices of resources and then determine which products should be produced in which quantities in order to gain the greatest, or optimum, profit. This is an enormous task that usually requires a computer to solve.

What does it mean to find a solution to a linear-programming mixture problem? A solution to a mixture problem is a production policy that tells us how many units of each product to make.

Optimal Production Policy DEFINITION

An **optimal production policy** has two properties:

1. It is possible; that is, it does not violate any of the limitations under which the manufacturer operates, such as availability of resources.
2. It gives the maximum profit.

SPOTLIGHT 4.1

Case Studies in Linear Programming

Linear programming is not limited to mixture problems. Here are two case studies that do not involve mixture problems, yet where applying linear-programming techniques produced impressive savings:

- The Exxon Corporation spends several million dollars per day running refineries in the United States. Because running a refinery takes a lot of energy, energy-saving measures can have a large effect. Managers at Exxon's Baton Rouge plant had over 600 energy-saving projects under consideration. They couldn't implement them all because some conflicted with others, and there were so many ways of making a selection from the 600 that it was impossible to evaluate all selections individually.

  Exxon used linear programming to select an optimal configuration of about 200 projects, resulting in millions of dollars in savings.
- Edwards Lifesciences uses heart valves from pigs to produce artificial heart valves for human beings. Pig heart valves come in different sizes. Shipments of pig heart valves often contain too many of some sizes and too few of others. However, each supplier tends to ship roughly the same imbalance of valve sizes in every order, so the company can expect consistently different imbalances from the different suppliers. Thus, if they order shipments from all the suppliers, the imbalances could cancel each other out in a fairly predictable way. The amount of cancellation will depend on the sizes of the individual shipments. Unfortunately, there are too many combinations of shipment sizes to consider all combinations individually.

  Edwards Lifesciences used linear programming to figure out which combination of shipment sizes would give the best cancellation effect. This reduced the company's annual cost by $1.5 million.

Common Features of Mixture Problems

Although our first mixture problem (Example 1) has only two products and one resource, it does contain the essential features that are common to *all* mixture problems:

- *Resources.* Definite resources are available in limited, known quantities for the time period in question. The resource in Example 1 is containers of plastic.
- *Products.* Definite products can be made by combining, or mixing, the resources. In Example 1, the products are skateboards and dolls.
- *Recipes.* A recipe for each product specifies how many units of each resource are needed to make one unit of that product. Each skateboard in Example 1 uses five units of plastic, and each doll uses two units.
- *Profits.* Each product earns a known profit per unit. (We assume that every unit produced can be sold. More complicated mathematical models, which we will not discuss here, are needed if we want to consider the possibility of items being produced but not sold.)
- *Objective.* The objective in a mixture problem is to find how much of each product to make so as to maximize the profit without exceeding any of the resource limitations.

The examples we show are not designed to be realistic. Rather, our goal is to demonstrate how ideas whose roots are in basic algebra and geometry can solve, when scaled up to realistic versions, problems that save Americans much time and money and make our government and American businesses more efficient.

EXAMPLE 1
Making Skateboards and Dolls

A toy manufacturer can manufacture only skateboards, only dolls, or some mixture of skateboards and dolls. Skateboards require 5 units of plastic and can be sold for a profit of $1, while dolls require 2 units of plastic and can be sold for a $0.55 profit. If 60 units of plastic are available, what numbers of skateboards and/or dolls should be manufactured for the company to maximize its profit?

Attacking this and other mixture problems requires carrying out a series of steps that determine the essence of the problem.

As a first step, we need to take the "verbal" information that we have been given and display it in a form that makes it easier to convert into the mathematics necessary to solve the problem. This is done by making a **mixture chart** for the information we are given (see Figure 4.1).

In the rows of this chart, we display the products we want to make, and in the column of the chart, we display the resources and the profit margin information that is available. In this case, we have two products, so we have two rows. We have one resource, which accounts for there being one column.

The other column is reserved for profit information. Since this information is used in a somewhat different way from the information about the resources, we will separate the resource column(s) from the profit column by a double bar.

Because we do not know the number of skateboards the company should make, we will use a letter x to represent the unknown number of skateboard units that the company might manufacture. Similarly, y will represent the number of dolls that the

(Patrik Giardino/Corbis.)

company might manufacture. We enter these letters as part of the labels of the rows of our table.

		RESOURCE(S) Containers of Plastic 60	PROFIT
PRODUCTS	**Skateboards** (x units)	5	$1.00
	Dolls (y units)	2	$0.55

Figure 4.1 Mixture chart for Example 1.

We can now enter the numbers about resources in the columns based on the information we have been given. In this case, there is one resource: plastic. Thus, for the 5 units of plastic needed for a skateboard, we record a 5 in the skateboards row and the containers of plastic column. Similarly, we enter a 2 in the second row and first column because dolls require 2 units of plastic each. Because we have 60 units of plastic available, we display this fact by placing the number 60 at the top of this column. We complete the table with the information about profit. We enter $1 in the skateboards row and profit column and $0.55 in the dolls row and profit column.

EXAMPLE 2
Making a Mixture Chart

Make a mixture chart to display this situation: A clothing manufacturer has 60 yards of cloth available to make shirts and decorated vests. Each shirt requires 3 yards of cloth and provides a profit of $5. Each vest requires 2 yards of cloth and provides a profit of $3.

SOLUTION: See the mixture chart in Figure 4.2.

		RESOURCE(S) Yards of Cloth 60	PROFIT
PRODUCTS	**Shirts** (x units)	3	$5
	Vests (y units)	2	$3

Figure 4.2 Mixture chart for the clothing manufacturer.

Translating Mixture Charts into Mathematical Form

Consider again the mixture chart in Figure 4.1. What can we say about the numbers of skateboards and dolls that might be manufactured? Clearly, we cannot make negative numbers of skateboards or dolls. Because we are using the letter x to represent the number of skateboards we plan to make, we can write down the algebraic expression that $x \geq 0$. Here, we are using the standard symbol $\geq$ for "greater than or equal to."

Algebraic expressions that involve the symbol $\geq$ or its companion symbols $\leq$ (less than or equal to), $>$ (greater than), and $<$ (less than) are known as *inequalities*. We can also write down an inequality for the y number of dolls we plan to make, based on the fact that we cannot make a negative number of dolls. Thus, we must have $y \geq 0$. We will use the phrase **minimum constraints** for these two inequalities, $x \geq 0$ and $y \geq 0$, which say simply that one cannot manufacture negative numbers of objects.

However, we also have only a limited number of units of plastic available. How can we represent this information? Consulting the mixture table, we see that we need 5 units of plastic for every skateboard we make. Thus, we will need $5x$ (5 times x) units of plastic for the x skateboards we make. Similarly, we will need $2y$ (2 times y) units of plastic for the y dolls we make. Hence we will need $5x + 2y$ units of plastic for the mixture of skateboards and dolls we make. We added the $5x$ and $2y$ because we need to find the total plastic used when we make a mixture of skateboards and dolls.

Reading from the table, we see that we are limited by having only 60 units of plastic. So we can express the **resource constraint** imposed by the limited number of units of plastic by writing that $5x + 2y \leq 60$. Here, we use the symbol for less than or equal to, $\leq$, to express the fact that we cannot use more than the amount of plastic we have available.

Notice that all the numbers in this inequality can be obtained from a column of the mixture chart. One of the reasons that we construct a mixture chart is that it helps speed up the conversion of the information about the problem we wish to solve into inequalities. The setup phase of a realistic linear programming problem, constructing the mathematical model of a manufacturing situation, is often the hardest and most complex step in getting an answer.

In addition to the resource inequalities (of which realistic problems will often have hundreds), there is one additional algebraic expression, this time an equality, that the mixture table allows us to create. Using the mixture table, we can compute the profit that will be produced when we manufacture different mixtures. For each skateboard, we make a profit of \$1, so if x skateboards are made, the profit is $1x$ (1 times x). For each doll made, the profit is \$0.55. So if y dolls are made, the profit is $0.55y$ (0.55 times y). Denoting by P the total profit from making x skateboards and y dolls, we get the equation

$$P = 1x + 0.55y$$

Note that unlike the situation for the resources where we got an inequality, here we get an expression for what the profit will be as we vary the numbers of skateboards and dolls manufactured. Our goal is to find which values of x and y (skateboards and dolls) make this profit as large as possible.

Algebra Review — Linear Equations and Intercepts

A linear equation in two variables (typically x and y) can be written in the form $ax + by = c$. A solution to such an equation is an ordered pair, (x, y), that satisfies the equation. For example, $(4, -3)$ is a solution to $3x + 2y = 6$ and $(1, 1)$ is not.

check $(4, -3)$	check $(1, 1)$
$3(4) + 2(-3) \stackrel{?}{=} 6$	$3(1) + 2(1) \stackrel{?}{=} 6$
$12 + (-6) \stackrel{?}{=} 6$	$3 + 2 \stackrel{?}{=} 6$
$6 = 6$ True	$5 = 6$ False

A line is a visual representation of the linear equation's solution set and can be uniquely graphed by two points. When the linear equation is in the form $ax + by = c$, an efficient method of graphing is to graph by intercepts. The y-intercept can be found by substituting $x = 0$. The x-intercept can be found by substituting $y = 0$. In the equation $3x + 2y = 6$, the y-intercept is $(0, 3)$ and the x-intercept is $(2, 0)$.

substitute $x = 0$	substitute $y = 0$
$3(0) + 2y = 6$	$3x + 2(0) = 6$
$0 + 2y = 6$	$3x + 0 = 6$
$2y = 6$	$3x = 6$
$y = 3$	$x = 2$

The graph of $3x + 2y = 6$ can be found by plotting the intercepts and drawing the line. Notice that $(4, -3)$ is on the line but $(1, 1)$ is not.

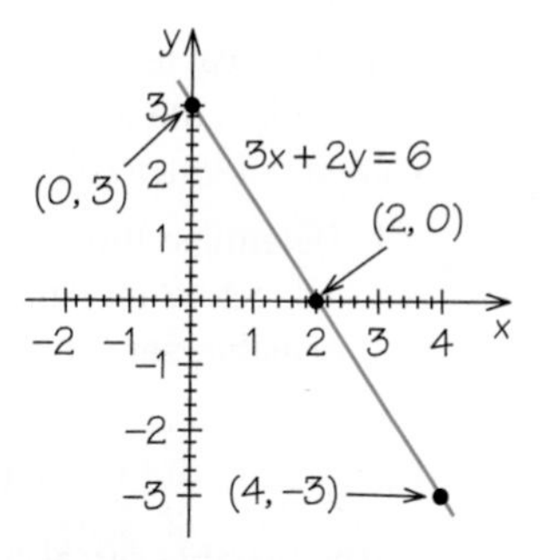

Two special kinds of lines of interest are horizontal and vertical lines. A horizontal line can be written in the form $y = b$ (b is where the line crosses the y-axis). A vertical line can be written in the form $x = a$ (a is where the line crosses the x-axis). The graphs of $y = 2$ and $x = 1$ are below.

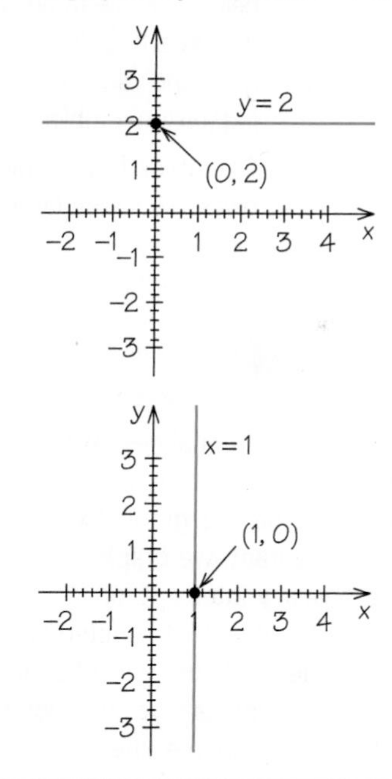

EXAMPLE 3
Revisiting Our Clothing Manufacturer

We can also translate the information in the mixture chart shown in Figure 4.2 into inequalities and an equation for expressing the profit in terms of how many shirts and vests are produced. Using the first column of the mixture chart and the fact that only 60 units of cloth are available, we can write

$$3x + 2y \leq 60$$

And using the last column, we get the following expression for the profit P:

$$P = 5x + 3y$$

Now that we have the information from the original problems represented in mathematical terms, we will return our attention to finding a solution to the problems.

Finding the best (largest profit) mixture of skateboards and/or dolls to make can be carried out in two phases.

1. Determine those mixtures of skateboards and/or dolls that can be manufactured subject to the limited resources that are available. This step involves finding the **feasible set** for the mixture problem.

Feasible Set or Feasible Region DEFINITION

The **feasible set,** also called the **feasible region,** for a linear-programming problem is the collection of all physically possible solution choices that can be made.

We can use a geometric diagram such as the one in Figure 4.3 to help us understand the feasible set of options that the manufacturer of skateboards and dolls has available. The geometric diagram we draw will have as many "dimensions" as there are products being manufactured. We have two products represented by the variables x and y, so we use a two-dimensional picture. Even diagrams involving three variables are hard to draw and visualize. Though these diagrams helped with developing algorithms for solving mixture problems, they are of little practical use for realistic problems.

2. Determine how to pick out, from the feasible set, the mixture (or mixtures) that gives rise to the largest profit.

Algebra Review Linear Inequalities

To graph linear inequalities, we first graph the line and then determine which side of the line to shade. To graph $3x + 2y \leq 12$ in the first quadrant, we graph $3x + 2y = 12$. This line has a y-intercept of (0, 6) and an x-intercept of (4, 0). To determine which side of the line to shade, we substitute a test point. A test point can be any point in the plane that is not on the line. In most cases, choosing the origin as a test point allows for convenient calculation. Testing (0, 0), we have the statement $3(0) + 2(0) \leq 12$ or $0 \leq 12$. This is a true statement, so we shade the side containing our test point.

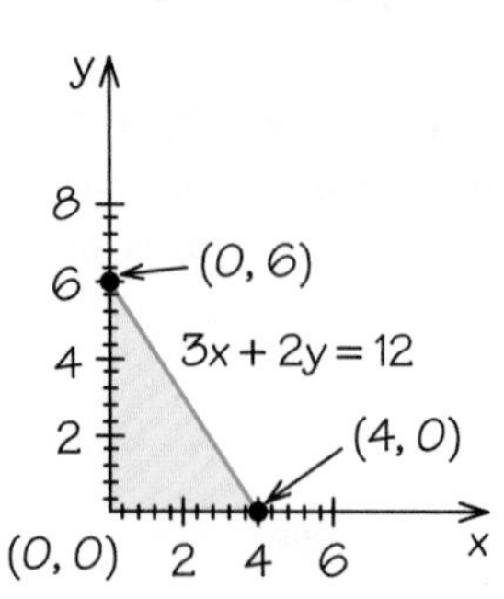

Representing the Feasible Region with a Picture

After we have constructed inequalities using a mixture chart or have the inequalities that must be obeyed for a more general linear-programming problem, we can draw a helpful picture to visualize the choices to be made in solving the problem. This picture will show in a convenient way the different choices that are available in solving the linear-programming problem at hand. To get the picture that will help us, we need to draw graphs of the inequalities associated with the linear-programming problem.

To draw the graph of an inequality, let's first review how to draw the graph of the equation of a straight line. Remember that two points can be used to uniquely determine a straight line. Let's use the equation associated with the less-than-or-equal-to inequality

$$5x + 2y \leq 60$$

namely,

$$5x + 2y = 60$$

There are two points that are easy to find on this line: the x- and y-intercepts. When $x = 0$, this gives rise to one point on the line, and when $y = 0$, we can find another point. (See Figure 4.3.) When $x = 0$, if we substitute this value in the equation $5x + 2y = 60$, we get $5(0) + 2y = 60$. Solving this equation, we discover that $y = 30$. Similarly, if we substitute $y = 0$ in the equation $5x + 2y = 60$, we get $5x + 2(0) = 60$, from which we conclude that $x = 12$. We now have two points (0, 30) and (12, 0) that lie on the line $5x + 2y = 60$.

We are using the usual convention that when we write a pair such as (3, 10), we are describing a point that has $x = 3$ and $y = 10$. We always list the x-value (x-coordinate) first in such a pair and the y-value (y-coordinate) second. Furthermore, when a point (x, y) is represented in a diagram, larger values of x are shown farther to the right (east) and larger values of y are shown farther up (north).

Using the two points we found on the line $5x + 2y = 60$, we can draw the graph shown in Figure 4.3a, where we have also displayed the point (3, 10). How do we know that (3, 10) is not on the line? We can see this by replacing x by 3 and y by 10 in the equation $5x + 2y = 60$, getting $5(3) + 2(10) = 15 + 20 = 35$. To have been on the line, we would have to have had the value 60. Furthermore, we know that the point (3, 10) is below the line $5x + 2y = 60$ because when $x = 3$, and we replace x by this value in the equation $5x + 2y = 60$, we get $5(3) + 2y = 60$, which means that $y = 45/2$. Since 45/2 is greater than 10, the y-value for the point (3, 10), we conclude that (3, 10) is below the line.

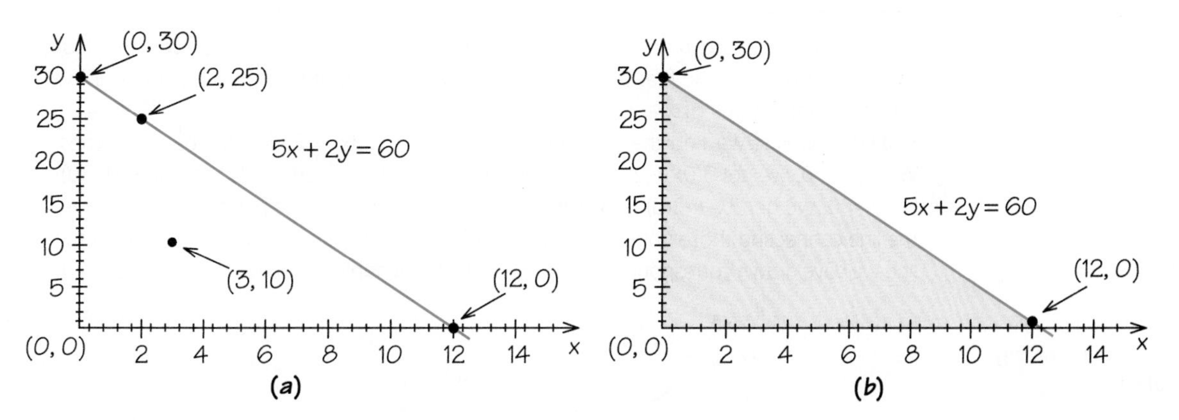

Figure 4.3 The feasible region for Example 3. (a) Graph of $5x + 2y = 60$. (b) Shading of the half-plane $5x + 2y < 60$, and where $x \geq 0$, $y \geq 0$.

Now that we know what the graph of the equation $5x + 2y = 60$ looks like, we can think through where points (x, y) that satisfy $5x + 2y < 60$ are located. The points that are either on the line $5x + 2y = 60$ or satisfy $5x + 2y < 60$ will satisfy $5x + 2y \leq 60$.

Any line, for example, $5x + 2y = 60$, divides the xy-plane into three parts: those points on the line, and the points in one of two half-planes. In one of these half-planes, we have the points for which $5x + 2y < 60$, and in the other, we have the points for which $5x + 2y > 60$. How can we tell which of the two half-planes is above the line $5x + 2y = 60$ and which is below?

The key is the use of a test point (x, y) that is not on the line and whose half-planes we wish to distinguish. We saw above that (3, 10) is not on the line $5x + 2y = 60$ and is below the line. This enables us to see that the half-plane for which $5x + 2y < 60$ consists of the points below the line $5x + 2y = 60$.

To complete the drawing of the points that are feasible for the skateboard and dolls manufacturing problem, we also have to know which points satisfy the constraints that state that the number of skateboards produced x cannot be negative ($x \geq 0$) and the number of dolls produced y cannot be negative ($y \geq 0$). Each of these inequalities corresponds to a half-plane, and we can again test which of the half-planes associated with the line $x = 0$ is determined by $x \geq 0$.

This can be done using the point (3, 10) as a test point again. Because $x = 3$ is greater than 0, $x \geq 0$ determines the half-plane to the right of the line $x = 0$ (the y-axis). Similarly, using the point (3, 10), we see that because $y = 10$ is greater than 0, $y \geq 0$ determines the half-plane above the line $y = 0$ (the x-axis).

Putting this information together leads us to the conclusion that the collection of points (x, y) that meets the three inequalities involved ($x \geq 0$, $y \geq 0$, $5x + 2y \leq 60$) corresponds to the shaded region in Figure 4.3b. Remember that when graphing equations, lines with equations like $x = 2$ are vertical lines, whereas those like $y = 4$ are horizontal lines.

Note that since the minimality conditions are always present in the kind of linear-programming problems we are dealing with, the points that are feasible for these problems are always in the upper right region (quadrant) that is created by the x-axis and y-axis. Next, we draw the feasible region for the clothing manufacturing problem.

EXAMPLE 4
Drawing a Feasible Region

In the earlier clothing manufacturer example, we developed a resource constraint of $3x + 2y \leq 60$. Draw the feasible region corresponding to that resource constraint, using the reality minimums of $x \geq 0$ and $y \geq 0$.

SOLUTION: First we find the two points where the line $3x + 2y = 60$ crosses the axes. When $x = 0$, we get $3(0) + 2y = 60$, giving $y = \frac{60}{2} = 30$, which yields the point (0, 30). For $y = 0$, we get $3x + 2(0) = 60$, or $x = \frac{60}{3} = 20$, so we have the point (20, 0). We draw the line connecting those points. Testing the point (0, 0), we find that the down side of the line we have drawn corresponds to $3x + 2y < 60$. The feasible region is shown in Figure 4.4.

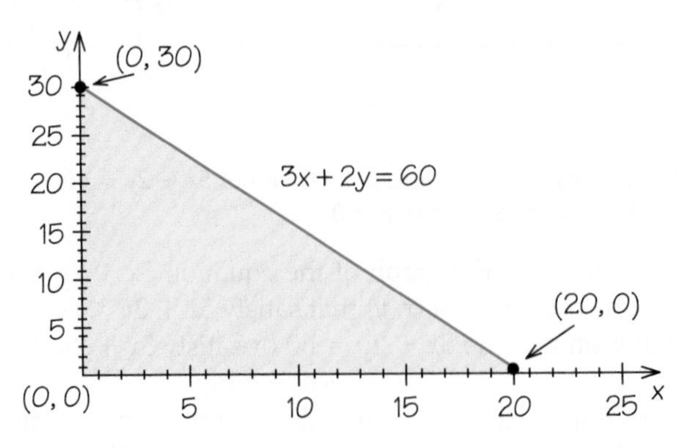

Figure 4.4 Feasible region for the clothing manufacturer.

4.2 Finding the Optimal Production Policy

Our next step is that we still must find the *optimal production policy*, a point within the feasible region that gives a maximum profit. There are a lot of points in that region. If you consider points with only whole numbers as values for x or y, there are many points, but, in fact, either x or y or both could be some fractional number. There are so many points in this feasible region that to consider the profit at each one of them would require us to calculate profits from now until we grow very old, and still the calculations would not be done. Here is where the genius of the linear-programming technique comes in, with the **corner point principle**, which we define in terms of our mixture problems.

Corner Point Principle THEOREM

The **corner point principle** states that in a linear-programming problem, the maximum value for the profit formula always corresponds to a corner point of the feasible region.

The corner point principle is probably the most important insight into the theory of linear programming. Later in this chapter, we will explain why this principle works. The geometric nature of this principle explains the value of creating a geometric model from the data in a mixture chart.

The corner point principle gives us the following method to solve a linear-programming problem:

1. Determine the corner points of the feasible region.
2. Evaluate the profit at each corner point of the feasible region.
3. Choose the corner point with the highest profit as the production policy.

Table 4.1

Calculation of the Profit Formula for Skateboards and Dolls

Corner Point	Value of the Profit Formula: $1.00x + $0.55y
(0, 0)	$1.00(0) + $0.55(0) = $0.00 + $0.00 = $0.00
(0, 30)	$1.00(0) + $0.55(30) = $0.00 + $16.50 = $16.50
(12, 0)	$1.00(12) + $0.55(0) = $12.00 + $0.00 = $12.00

Let's look at the feasible region that we drew in Figure 4.3. It is a triangle having three corners; namely, (0, 0), (0, 30), and (12, 0). Now all we need to do is find out which of these three points gives us the highest value for the profit formula, which in this problem is \$1.00x + \$0.55y. We display our calculations in Table 4.1. The maximum profit for the toy manufacturer is \$16.50, and that happens if the manufacturer makes 0 skateboards and 30 dolls. The point (0, 30) is called the *optimal production policy*.

Optimal Production Policy THEOREM

An **optimal production policy** corresponds to a corner point of the feasible region where the profit formula has a maximum value.

EXAMPLE 5
Finding the Optimal Production Policy

Our analysis of the clothing manufacturer problem resulted in a feasible region with three corner points: (0, 0), (0, 30), and (20, 0). Which of these maximizes the profit formula, \$5*x* + \$3*y*, and what does that corner represent in terms of how many shirts and vests to manufacture?

SOLUTION: The evaluation of the profit formula at the corner points is shown in Table 4.2. The maximum profit of \$100 occurs at the corner point (20, 0), which represents making 20 shirts and no vests.

Table 4.2

Evaluating the Profit Formula in the Clothing Example

Corner Point	Value of the Profit Formula: \$5x + \$3y
(0, 0)	\$5(0) + \$3(0) = \$0 + \$0 = \$0
(0, 30)	\$5(0) + \$3(30) = \$0 + \$90 = \$90
(20, 0)	\$5(20) + \$3(0) = \$100 + \$0 = \$100

General Shape of Feasible Regions

The shape of a feasible region for a linear-programming mixture problem has some important characteristics, without which the corner point principle would not work:

1. The feasible region is a polygon in the first quadrant, where both $x \geq 0$ and $y \geq 0$. This is because the minimum constraints require that both x and y be nonnegative.
2. The region is a polygon that has neither dents (as in Figure 4.5a) nor holes (as in Figure 4.5b). Figure 4.5c is a typical example. Such polygons are called *convex*.

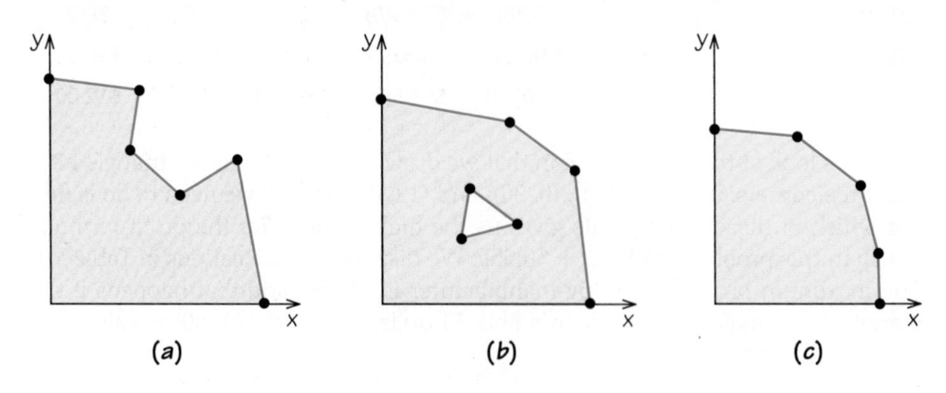

Figure 4.5 A feasible region may not have (a) dents or (b) holes. Graph (c) shows a typical feasible region.

The Role of the Profit Formula: Skateboards and Dolls

In practice, there are often different amounts of resources available in different time periods. The selling price for the products can also change. For example, if competition forces us to cut our selling price, the profit per unit can decrease.

To maximize profit, it is usually necessary for a manufacturer to redo the mixture problem calculations whenever any of the numbers change.

Suppose that business conditions change, and now the profits per skateboard and doll are, respectively, \$1.05 and \$0.40. Let us keep everything else about the skateboards and dolls problem the same. The change in profits would give us a new profit formula of $\$1.05x + \$0.40y$. When we evaluate the new profit formula at the corner points, we get the results shown in Table 4.3. This time, the optimal production policy, the point that gives the maximum value for the profit formula, is the point (12, 0). To get the maximum profit of \$12.60, the toy manufacturer should now make 12 skateboards and 0 dolls.

Table 4.3

A Different Profit Formula: Skateboards and Dolls

Corner Point	Value of the Profit Formula: \$1.05x + \$0.40y
(0, 0)	\$1.05(0) + \$0.40(0) = \$0.00 + \$0.00 = \$0.00
(0, 30)	\$1.05(0) + \$0.40(30) = \$0.00 + \$12.00 = \$12.00
(12, 0)	\$1.05(12) + \$0.40(0) = \$12.60 + \$0.00 = \$12.60

We see from this example that the shape of the feasible region, and thus the corner points we test, are determined by the constraint inequalities. The profit formula is used to choose an optimal point from among the corner points, so it is not surprising that different profit formulas might give us different optimal production policies.

We started the exploration of skateboard and doll production with the idea that a toy manufacturer has a product line with either one to two products. But both linear-programming solutions we have found tell the manufacturer that to maximize profit, make just one product. This is probably not an acceptable result for the manufacturer, who might want to produce both products for business reasons other than profit, such as establishing brand loyalty. And it certainly would be very difficult for the manufacturer to be ready to switch back and forth between producing either skateboards or dolls every time the profit formula changed. Linear programming is a flexible enough technique that it can accommodate the desire for there to be both products in the optimal production policy. This is done by specifying that there be nonzero minimum quantities for each period.

Summary of the Pictorial Method Using a Feasible Region

Let's stop and summarize the steps we are following to find the optimal production policy in a mixture problem:

1. Read the problem carefully to identify the resources and the products.
2. Make a mixture chart showing the resources (associated with limited quantities), the products (associated with profits), the recipes for creating the products from the resources, the profit from each product, and the amount of each resource on hand. If the problem has nonzero minimums, include a column for those as well.
3. Assign an unknown quantity, x or y, to each product. Use the mixture chart to write down the resource constraints, the minimum constraints, and the profit formula.

4. Graph the line corresponding to each resource constraint and determine which side of the line is in the feasible region. If there are nonzero minimum constraints, graph lines for them also, and determine which side of each is in the feasible region. Sketch the feasible region by finding the common points in the half-planes from all the resource constraints plus the minimum constraints. (This process is called finding the "intersection" of the half-planes.)
5. Find the coordinates of all the corner points of the feasible region. Some of these may have been calculated so that you can graph the individual lines. Proceed in order around the boundary of the feasible region. Be sure that every point you consider is part of the feasible region.
6. Evaluate the profit formula for each of the corner points. The production policy that maximizes profit is the one that gives the biggest value to the profit formula.

Algebra Review: Systems of Linear Equations and Inequalities

When we graph more than one line in a plane, we are often interested in locating where two lines intersect. If the two lines are parallel, there will be no point of intersection. Suppose that we wish to graph $x + y = 5$ and $x + 2y = 7$ in the first quadrant (upper-right quadrant). By finding the intercepts of each line, we get the following:

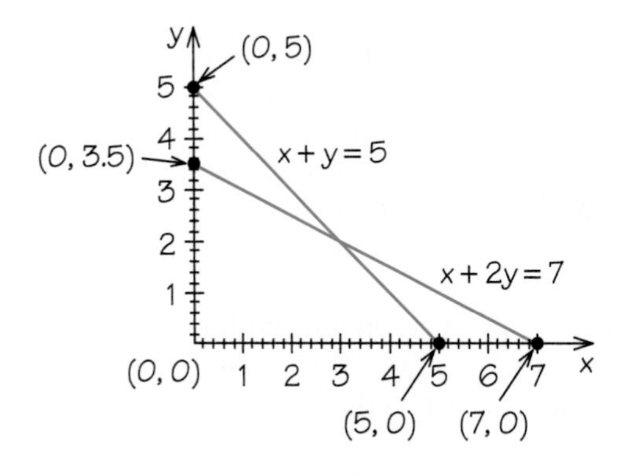

It appears from the graph that (3, 2) is the point of intersection. This can be verified by substituting this point into both equations.

check (3, 2) in $x + y = 5$	check (3, 2) in $x + 2y = 7$
$3 + 2 \stackrel{?}{=} 5$	$3 + 2(2) \stackrel{?}{=} 7$
$5 = 5$ True	$7 = 7$ True

To find the point of intersection algebraically, we can eliminate one of the variables by multiplying both sides of $x + y = 5$ by -1, and adding the result to $x + 2y = 7$.

$$\begin{aligned} -x - y &= -5 \\ x + 2y &= 7 \\ \hline y &= 2 \end{aligned}$$

By inserting $y = 2$ into $x + y = 5$, we find $x = 3$. Thus, the point of intersection is (3, 2).

If you are given more than one linear inequality, then we have a system of linear inequalities. In graphing a system of linear inequalities, you look for the region that is common to all the individual graphs. For example, given $x \geq 1$, $y \geq 2$, and $3x + 2y \leq 12$, we can see that the following region satisfies all three inequalities. In addition, the three points of intersection are shown. These were found by examining the three lines, two at a time.

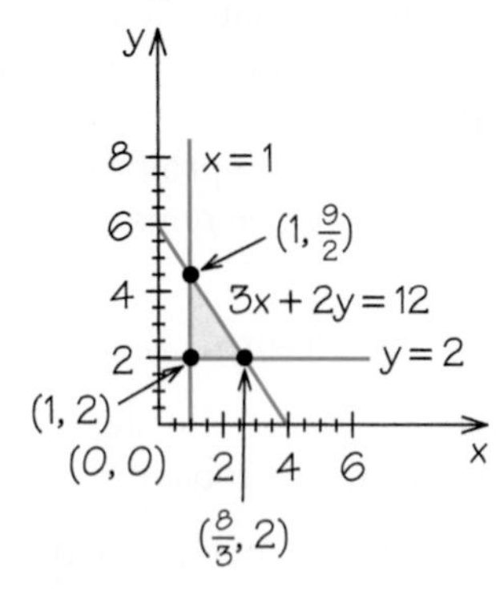

In applications that involve setting up a system of inequalities, two special inequalities may not be stated but are generally implied. These two inequalities are $x \geq 0$ and $y \geq 0$. When these two inequalities occur, the meaning is that your graph is restricted to the first quadrant.

Two Products and Two Resources: Skateboards and Dolls

We return to the toy manufacturer now to consider two limited resources instead of one. The second limited resource will be time, the number of person-minutes available to prepare the products. Suppose that there are 360 person-minutes of labor available and that making one skateboard requires 15 person-minutes and making one doll requires 18 person-minutes. We will continue to use the original values regarding containers of plastic, the first of our two profit formulas, and to keep the problem relatively simple, we use the zero minimum constraints: $x \geq 0$ and $y \geq 0$. We need a new mixture chart. In general, we will include a column for minimums in a mixture chart only if there are any nonzero minimum constraints. In Figure 4.6, we have the mixture chart for this problem. Using the mixture chart, we can write the two resource constraints:

$$5x + 2y \leq 60 \qquad \text{for containers of plastic}$$

and

$$15x + 18y \leq 360 \qquad \text{for person-minutes}$$

We can also write the profit formula: $\$1.00x + \$0.55y$.

PRODUCTS	RESOURCE(S): Containers of Plastic 60	RESOURCE(S): Person-minutes 360	PROFIT
Skateboards (x units)	5	15	$1.00
Dolls (y units)	2	18	$0.55

Figure 4.6 Mixture chart for Skateboards and Dolls (two resources).

The half-plane corresponding to the plastic resource is shown in Figure 4.7a. We now need to graph the half-plane corresponding to the time constraint. We find where the line $15x + 18y = 360$ intersects the two axes by substituting first $x = 0$ and then $y = 0$ into that equation.

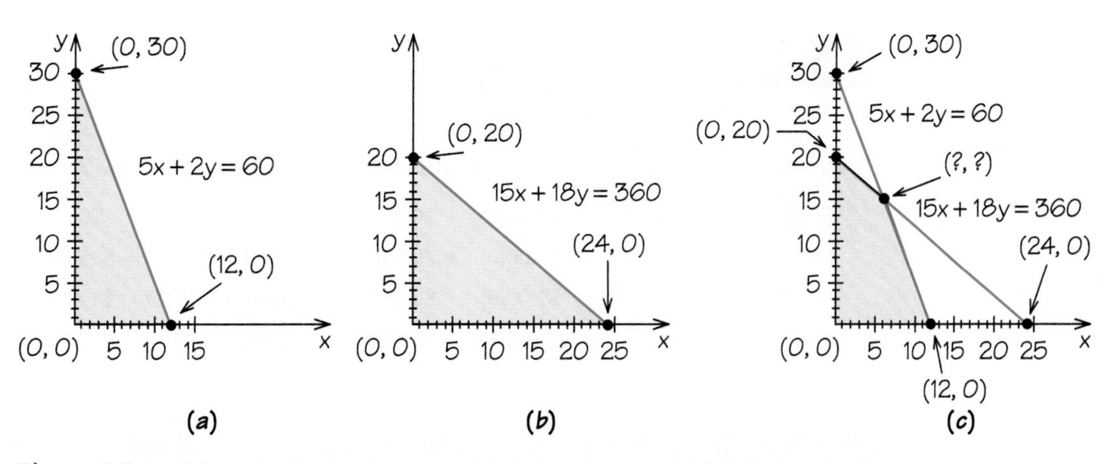

Figure 4.7 Feasible region for skateboards and dolls (two resources). (a) Half-plane for the plastic resource constraint. (b) Half-plane for the time resource constraint. (c) Intersection of the two half-planes.

The line corresponding to the time constraint contains the two points (0, 20) and (24, 0). When we insert the point (0, 0) into the inequality $15x + 18y < 360$, we get $15(0) + 18(0) < 360$, or $0 < 360$, which is true, so (0, 0) is on the side of the line that we shade. Putting all this together, we get the half-plane in Figure 4.7b as the correct half-plane for the time resource constraint.

We are not permitted to exceed the supply of even a single resource. Therefore, the feasible region must be made up of points that are shaded twice—both in the half-plane for the plastic resource constraint, shown in Figure 4.7a, and in the half-plane for the time resource constraint in Figure 4.7b. The procedure with several half-plane constraints is that we build our feasible region by finding the intersection, or overlap, of the individual half-planes in the problem. In Figure 4.7c, we show the result of intersecting the half-planes from the two resource constraints. Because this problem has minimums that are zeroes, the shaded region in Figure 4.7c is in fact the feasible region for the problem.

The next step that we need to carry out to use the pictorial method for solving this problem is to find the corner points of the feasible region. This is carried out by doing the algebra necessary to solve two linear equations in two unknowns. This leads to the points (0, 0), (0, 20), (6, 15), and (12, 0). Three of these points have a zero value for one or more of the unknowns, so the calculations are easy.

To find the coordinates of the point (6, 15), it is necessary to solve for the point that satisfies both of the equations $5x + 2y = 60$ and $15x + 18y = 360$. To solve these two equations simultaneously, you must multiply one or both of the equations by a number to create equivalent equations, so that when added together, one of the variables will cancel out. One way to do this is to multiply the first of these equations by -3, obtaining the equation $-15x - 6y = -180$. When this is added to the second equation, the x term "drops out" and we can solve $12y = 180$, to get $y = 15$. Now it is an easy matter to substitute this value into either of the original equations to get the x value of 6.

We are ready to finish the problem. In Table 4.4 we have evaluated the profit formula at the four corner points of the feasible region. The optimal production policy for the toy manufacturer would be to make 6 skateboards and 15 dolls, for a maximum profit of \$14.25.

Table 4.4

The Profit at the Four Corner Points

Corner Point	Value of the Profit Formula: \$1.00x + \$0.55y
(0, 0)	\$1.00(0) + \$0.55(0) = \$0.00 + \$0.00 = \$0.00
(0, 20)	\$1.00(0) + \$0.55(20) = \$0.00 + \$11.00 = \$11.00
(6, 15)	\$1.00(6) + \$0.55(15) = \$6.00 + \$8.25 = \$14.25
(12, 0)	\$1.00(12) + \$0.55(0) = \$12.00 + \$0.00 = \$12.00

Here is another mixture problem example of how the pictorial method using a feasible region works from start to finish.

EXAMPLE 6
Mixtures of Two Fruit Juices: Beverages

A juice manufacturer produces and sells two fruit beverages: 1 gallon of cranapple is made from 3 quarts of cranberry juice and 1 quart of apple juice; and 1 gallon of appleberry is made from 2 quarts of apple juice and 2 quarts of cranberry juice. The manufacturer makes a profit of 3 cents on a gallon of cranapple and 4 cents on a gallon of appleberry. Today, there are 200 quarts of cranberry juice and 100 quarts of apple juice available. How many gallons of cranapple and how many gallons of appleberry should be produced to obtain the highest profit without exceeding available supplies? We use zeroes as "reality minimums." The mixture chart for this problem is shown in Figure 4.8.

		RESOURCE(S)		
		Cranberry 200 quarts	Apple 100 quarts	PROFIT
PRODUCTS	**Cranapple** (x gallons)	3 quarts	1 quart	3 cents/gallon
	Appleberry (y gallons)	2 quarts	2 quarts	4 cents/gallon

Figure 4.8 A mixture chart for Example 6.

For each resource, we develop a resource constraint reflecting the fact that the manufacturer cannot use more of that resource than is available. The number of quarts of cranberry juice needed for x gallons of cranapple is $3x$. Similarly, $2y$ quarts of cranberry are needed for making y gallons of appleberry. So if the manufacturer makes x gallons of cranapple and y gallons of appleberry, then $3x + 2y$ quarts of cranberry juice will be used. Because there are only 200 quarts of cranberry available, we get the cranberry resource constraint $3x + 2y \leq 200$. Note that the numbers 3, 2, and 200 are all in the cranberry column. We get another resource constraint from the column for the apple juice resource: $1x + 2y \leq 100$. We also have these minimum constraints: $x \geq 0$ and $y \geq 0$.

Finally, we have the profit formula. Because $3x$ is the profit from making x units of cranapple and $4y$ is the profit from making y units of appleberry, we get the profit formula $3x + 4y$.

We summarize our analysis of the juice mixture problem. Maximize the profit formula, $3x + 4y$, given these constraints:

cranberry:	$3x + 2y \leq 200$
apple:	$1x + 2y \leq 100$
minimums:	$x \geq 0$ and $y \geq 0$

Remember, in a mixture problem, our job is to find a production policy (x, y), that makes all the constraints true and maximizes the profit.

Figure 4.9a shows the result of graphing the constraint associated with the cranberry resource, while Figure 4.9b shows the result of graphing the constraint associated with the apple resource, taking into account that the amounts of these resources used cannot be negative. When these two diagrams are superimposed, we get the diagram in

Figure 4.9c. Now, to carry out the pictorial method, we need to find the profits associated with the four corner points shown. This is done in Table 4.5.

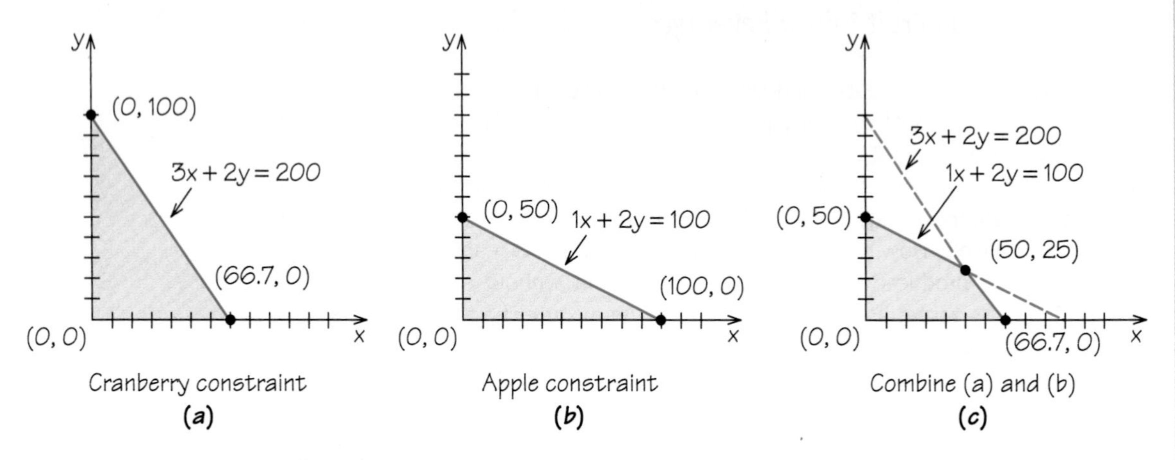

Figure 4.9 Feasible region for Example 6.

When we evaluate the profit formula at the four corner points, we see that the optimal production policy is to make 50 gallons of cranapple and 25 gallons of appleberry, for a profit of 250 cents.

Table 4.5

Finding the Optimal Production Policy for Beverages

Corner Point	Value of the Profit Formula: 3x + 4y cents
(0, 0)	3(0) + 4(0) = 0 cents
(0, 50)	3(0) + 4(50) = 200 cents
(50, 25)	3(50) + 4(25) = 250 cents
(66.7, 0)	3(66.7) + 4(0) = 200 cents (rounded)

4.3 Why the Corner Point Principle Works

In finding solutions to our mixture problems, we have been using the corner point principle, which says that the highest profit value on a convex polygonal feasible region is always at a corner point. A feasible region has infinitely many points, making it impossible to compute the profit for each point. The corner point principle gives us a finite set of points, making the calculation possible.

You can visualize a mathematical proof of the corner point principle by imagining that each point of the plane is a tiny light bulb that is capable of lighting up. For the juice mixture example, whose feasible region is shown in Figure 4.9c, imagine what would happen if we ask this question: Will all points with profit = 360 please light up? What geometric figure do these lit-up points form?

In algebraic terms, we can restate the profit question in this way: Will all points (x, y) with $3x + 4y = 360$ please light up? As it happens, this version of the profit question is one that mathematicians learned to answer hundreds of years before linear programming was born.

The points that light up make a straight line because $3x + 4y = 360$ is the equation of a straight line. Furthermore, it is a routine matter to determine the exact position of the line. We call this line the **profit line** for 360; it is shown in Figure 4.10. For numbers other than 360, we would get different profit lines. Unfortunately, there are no points on the profit line for 360 that are feasible, that is, which lie in the feasible region. Therefore, the profit of 360 is impossible. *If the profit line corresponding to a certain profit value doesn't touch the feasible region, then that profit value isn't possible.*

Because 360 is too big, perhaps we should ask the profit line for a more modest amount, say, 160, to light up. You can see that the new profit line of 160 in Figure 4.10 is parallel to the first profit line and closer to the origin. This is no accident: All profit lines for the profit formula $3x + 4y$ have the same coefficients for x and y—namely, 3 for x and 4 for y. Because the slope of the line is determined by those coefficients, they all have the same slope. Changing the profit value from 360 to 160 has the effect of changing where the line intersects the y-axis, but it does not affect the slope. These different profit lines are parallel to each other.

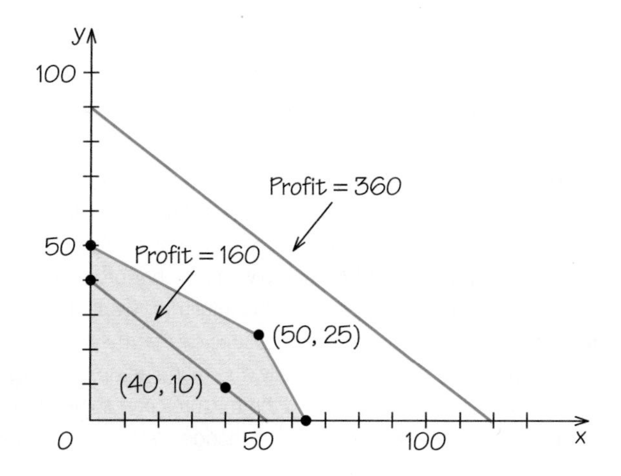

Figure 4.10 The profit line for 360 lies outside the feasible region, whereas the profit line for 160 passes through the region.

The most important feature of the profit line for 160 is that it has points in common with the interior of the feasible region. For example, (40, 10) is on that profit line because $3(40) + 4(10) = 160$; in addition, (40, 10) is a feasible point. This means that it is possible to make 40 gallons of cranapple and 10 gallons of appleberry and that if we do so, we will have a profit of 160.

Can we do better than a 160 profit? As we slowly increase our desired profit from 160 toward 360, the location of the profit line that lights up shifts smoothly upward away from the origin. So long as the line continues to cross the feasible region, we are happy to see it move away from the origin because the more it moves, the higher the profit represented by the line. We would like to stop the movement of the line at the last possible instant, while the line still has one or more points in common with the feasible region. It should be obvious that this will occur when the line is just touching the feasible region either at a corner point (Figure 4.11a) or along a line segment joining two corners (Figure 4.11b). That point or line segment corresponds to the production policy or policies with the maximum achievable profit. This is just what the corner point principle says: The maximum profit always occurs at a corner or along an edge of the feasible region.

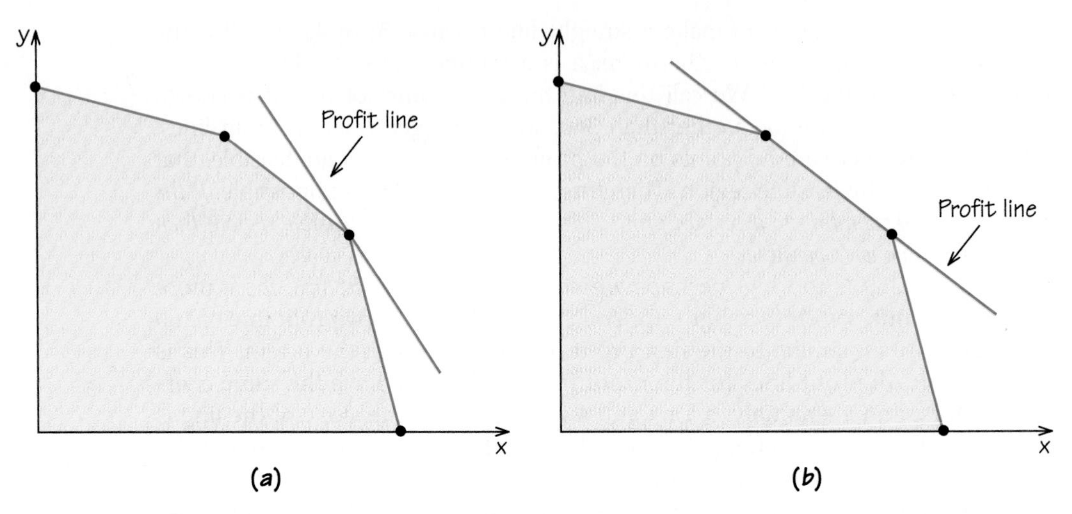

Figure 4.11 The highest profit will occur when the profit is just touching the feasible region, either (a) at the corner point or (b) along a line segment, which will include corner points.

EXAMPLE 7
Adding Nonzero Minimums: Beverages

Suppose that in Example 6, the profit for cranapple changes from 3 cents per gallon to 2 cents and the profit for appleberry changes from 4 cents per gallon to 5 cents. You can verify that this change moves the optimal production policy to the point (0, 50); no cranapple is produced. This result is not surprising: Appleberry is giving a higher profit and the policy is to produce as much of it as possible. But suppose the manufacturer wants to incorporate nonzero minimums into the linear-programming specifications so that there will always be both cranapple, x, and appleberry, y, produced. Specifically, the manufacturer decides that $x \geq 20$ and $y \geq 10$ are desirable minimums. Figure 4.12 is the mixture chart showing the new profit formula and the nonzero minimums, along with the unchanged rest of the beverage problem.

Figure 4.12 Mixture chart for Example 7.

PRODUCTS	RESOURCE(S): Cranberry Juice 200 quarts	Apple Juice 100 quarts	MINIMUMS	PROFIT
Cranapple (x gallons)	3	1	20	2 cents
Appleberry (y gallons)	2	2	10	5 cents

The feasible region for beverages, Example 6, is shown in Figure 4.13a. The feasible region for beverages, Example 7, is shown in Figure 4.13b. You can verify that, starting at the lower-left corner of the new feasible region and moving clockwise around its boundary, we have corner points (20, 10), (20, 40), (50, 25), and (60, 10). (One of those points was also a corner point of the old feasible region. Can you explain why?) Table 4.6 shows the evaluation of the profit formula at these corner points. For this modified problem, the optimal production policy is to produce 20 gallons of cranapple and 40 of appleberry, for a maximum profit of 240 cents.

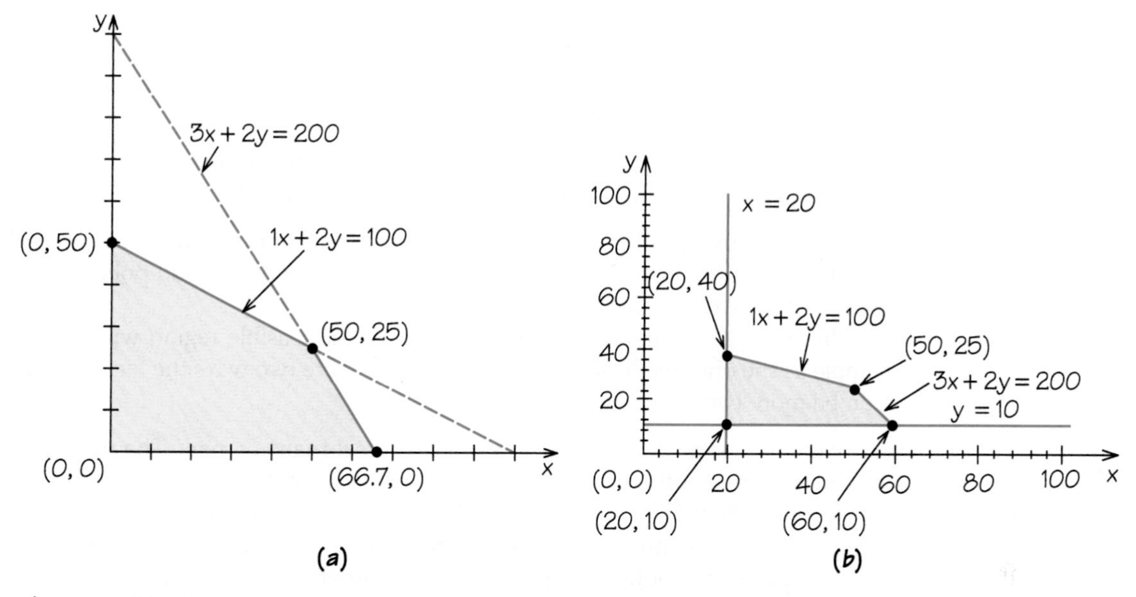

Figure 4.13 Feasible region for Examples 6 and 7. (a) Zero minimums. (b) Nonzero minimums.

Table 4.6

Profit Evaluation for Beverages

Corner Point	Value of the Profit Formula: 2x + 5y
(20, 10)	2(20) + 5(10) = 40 + 50 = 90 cents
(20, 40)	2(20) + 5(40) = 40 + 200 = 240 cents
(50, 25)	2(50) + 5(25) = 100 + 125 = 225 cents
(60, 10)	2(60) + 5(10) = 120 + 50 = 170 cents

One final note about this solution concerns the resources. The point (20, 40) is on the resource constraint line for the apple juice resource, so it represents using up all the available apple juice. We can see this by inserting the apple juice resource constraint: $1(20) + 2(40) = 100$ is true. However, (20, 40) is *below* the line for the cranberry juice resource, indicating that there will be *slack*, or leftover, amounts of cranberry juice. Specifically, substituting (20, 40) into the cranberry juice constraint gives $3x + 2y = 3(20) + 2(40) = 60 + 80 = 140$, which is 60 quarts less than the 200 quarts available. The slack is 60 quarts of cranberry juice. Dealing with slack can be an important consideration for manufacturers. Can you see why?

4.4 Linear Programming: Life Is Complicated

Every algorithm for solving a linear-programming problem has the following three characteristics, which hold true regardless of the number of products or the number of resources in the problem:

1. The algorithm can distinguish between "good" production policies—those in the feasible set that satisfy all the constraints—and those that violate some constraint(s) and are thus not feasible. There are usually many good points,

each of which corresponds to some production policy; for example, "Make x units of product 1 and y units of product 2."

2. The algorithm uses some geometric principles—one of which is the corner point principle—to select a special subset of the feasible set.
3. The algorithm evaluates the profit formula at points in the special subset to find which corner point actually gives the maximum profit.

The various algorithms for linear programming differ in how they process the feasible set and in how quickly the algorithm finds the production policy—corner point—that gives the optimal profit.

In practical linear-programming problems, the feasible region will not be as simple as the ones we have examined here. There are two ways the feasible region can be more complex:

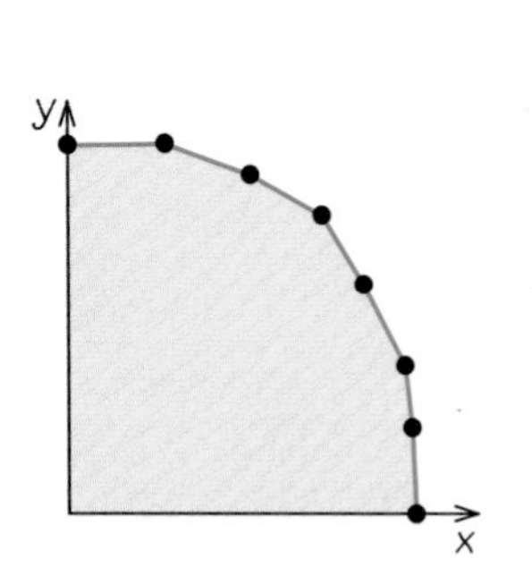

Figure 4.14 A feasible region with many corners.

1. Sometimes, as in Figure 4.14, we have a great many corners. The more corners there are, the more calculations we need to determine the coordinates of all of them and the profit at each one. The number of corners literally can exceed the number of grains of sand on the earth. Even with the fastest computer, computing the profit of every corner is impossible.
2. It is not possible to visualize the feasible region as a part of two-dimensional space when there are more than two products. Each product is represented by an unknown, and each unknown is represented by a dimension of space. If we have 50 products, we would need 50 dimensions and couldn't visualize the feasible region.

Another type of complication can occur even in simple two-dimensional regions: Corner points can have fractional coordinates, not the integer ones we see in the specially constructed problems in this text. Making 3.75 skateboards and 5.45 dolls is not possible. As mentioned earlier, integer programming, a special type of linear programming, is used when it is not possible to use fractional answers.

We have looked at some simple mixture problems to illustrate the power of linear programming to solve optimization problems. Solving similar problems scaled up to a realistic level is widely applied by governments, businesses, and humanitarian organizations to make the world a better place that runs more smoothly.

The Simplex Method

Several methods are used for the typically large linear-programming problems solved in practice. The oldest method is the **simplex method**, which is still the most commonly used. Devised by the American mathematician George Dantzig (see Spotlight 4.2), this ingenious mathematical invention makes it possible to find the best corner point by evaluating only a tiny fraction of all the corners. With the use of the simplex method, a problem that might be impossible to solve if each corner point had to be checked can be solved in a few minutes or even a few seconds on a typical business computer.

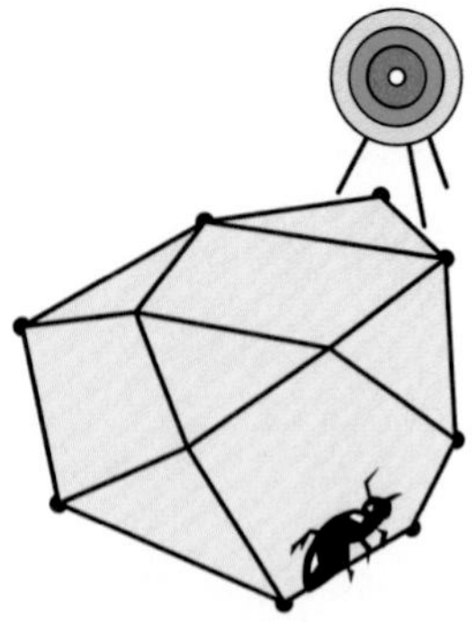

Figure 4.15 The simplex method can be compared to an ant crawling along the edges of a polyhedron, looking for the "target"—the optimal corner point.

The operation of the simplex method may be likened to the behavior of an ant crawling on the edges of a polyhedron (a solid with flat sides) looking for an optimal corner point—one that gives the highest profit (Figure 4.15). The ant cannot see where the optimal corner is. As a result, if it were to wander along the edges randomly, it might take a long time to reach that corner. The ant will do much better if it has a temperature clue to let it know it is getting warmer (closer to the optimal corner) or colder (farther from the optimal corner).

SPOTLIGHT 4.2

The Father of Linear Programming Recalls Its Origins

George Dantzig, who died in 2005, spent most of his career as a professor of operations research and computer science at Stanford University. He is credited with inventing the linear-programming technique called the simplex method. Since its invention in the 1940s, the simplex method has provided solutions to linear-programming problems that have saved both industry and the military time and money. Here Dantzig talks about the background of his famous technique:

George Dantzig
George Dantzig (left), sometimes referred to as the "father" of linear programming, shown with Leonid Khaciyan (right) who developed an important new approach to solving linear programming problems. *(Kees Roos.)*

> Initially, all the work we did had to do with military planning. During World War II, we were planning on a very extensive scale. The civilian population and the military were all performing scheduling and planning tasks, perhaps on a larger scale than at any time in history. And this was the case up until about 1950. From 1950 on, the whole emphasis shifted from military planning to practical planning for the civilian population, and industry picked it up.
>
> The first areas of industry to use linear programming were the petroleum refineries. They used it for blending gasoline. Nowadays, all of the refineries in the world (except for one) use linear programming methods. They are one of the biggest users of it, and it's been picked up by every other industry you can think of—the forestry industry, the steel industry. You could fill up a book with all the different places it's used.
>
> The question of why linear programming wasn't invented before World War II is an interesting one. In the postwar period, various technologies just evolved that had never been there before. Computers were one example. These technologies were talked about before. You can go back in history and you'll find papers on them, but these were isolated cases that never went anywhere. . . .
>
> The problems we solve nowadays have thousands of equations, sometimes a million variables. One of the things that still amazes me is to see a program run on the computer—and to see the answer come out. If we think of the number of combinations of different solutions that we're trying to choose the best of, it's akin to the stars in the heavens. Yet we solve them in a matter of moments. This, to me, is staggering. Not that we can solve them—but that we can solve them so rapidly and efficiently.
>
> The simplex method has been used now for roughly 70 years. There has been steady work going on trying to use different versions of the simplex method, nonlinear methods, and interior methods. It has been recognized that certain classes of problems can be solved much more rapidly by special algorithms than by using the simplex method. If I were to say what my field of specialty is, it is in looking at these different methods and seeing which are more promising than others. There's a lot of promise in this—there's always something new to be looked at.

Think of the simplex method as a way of calculating these temperature hints. We begin at any corner. All neighboring corners are evaluated to see which ones are warmer and which are colder. A new corner is chosen from among the warmer ones, and the evaluation of neighbors is repeated—this time checking neighbors of the new corner. The process ends when we arrive at a corner point, all of whose neighbors are colder than it is.

Part of what the simplex method has going for it is that it works faster in practice than its worst-case behavior would lead us to believe. Although mathematicians have devised artificial cases for which the simplex method bogs down in unacceptable amounts of arithmetic, the examples arising from real applications are never like that. This may be the world's most impressive counterexample to Murphy's law, which says that if something can go wrong, it will.

Although the simplex method usually avoids visiting every corner, it may require visiting many intermediate ones as it moves from the starting corner to the optimal one. The simplex method has to search along edges on the boundary of the polyhedron. If it happens that there are a great many small edges lying between the starting corner and the optimal one, the simplex method must operate like a slow-moving bus that stops on every block.

Many computer programs are available that will use the simplex method to produce an optimal production policy if we just supply the computer with the constraint inequalities and profit formula. Simplex method programs can be found in a variety of places, including electronic spreadsheets, packages of mathematics programs designed for business applications or finite mathematics courses, and large "all-purpose" mathematics packages. A graphical solution is possible only for problems limited to two products; these special exercises involve more than two products.

An Alternative to the Simplex Method

In 1984, Narendra Karmarkar (see Figure 4.16), a mathematician working at Bell Laboratories, devised an alternative method for linear programming that finds the optimal corner point in fewer steps than the simplex algorithm by making use of search routes through the interior of the feasible region. The applications of Karmarkar's algorithm are important to a lot of industries, including telephone communications and the airlines (see Spotlight 4.3). Routing millions of long-distance calls, for example, means deciding how to use the resources of long-distance landlines, repeater amplifiers, and satellite terminals to best advantage. The problem is similar to the juice company's need to find the best use of its stocks of juice to create the most profitable mix of products.

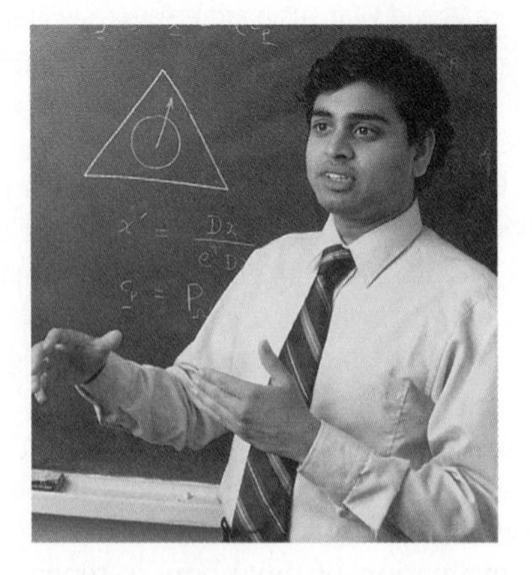

Figure 4.16 Narendra Karmarkar, a researcher at AT&T Bell Laboratories, invented a powerful new linear-programming algorithm that solves many complex linear-programming problems faster and more efficiently than any previous method. (*Courtesy of AT&T Labs.*)

Many airlines use software based on Karmarkar's algorithm to reduce fuel costs and deal with delays caused by storms.

In the 1980s, scientists at Bell Labs applied Karmarkar's algorithm to a problem of unprecedented complexity: deciding how to economically build telephone links between cities so that calls can get from any city to any other, possibly being

SPOTLIGHT 4.3

Finding Fast Algorithms Means Better Airline Service

Linear-programming techniques have a direct impact on the efficiency and profitability of major airlines. Thomas Cook, once director of operations research at American Airlines, made these comments concerning why optimal solutions are essential to the airline business:

> Finding an optimal solution means finding the best solution. Let's say you are trying to minimize a cost function of some kind. For example, we may want to minimize the excess costs related to scheduling crews, hotels, and other costs that are not associated with flight time. So we try to minimize that excess cost, subject to a lot of constraints, such as the amount of time a pilot can fly, how much rest time is needed, and so forth.
>
> An optimal solution, then, is either a minimum-cost solution or a maximizing solution. For example, we might want to maximize the profit associated with assigning aircrafts to the schedule; so we assign large aircraft to high-need segments and small aircraft to low-load segments.
>
> The simplex method, which was developed some 50 years ago by George Dantzig, has been very useful at American Airlines and, indeed, at a lot of large businesses. The difference between his method and Narendra Karmarkar's is speed. Finding fast solutions to linear-programming problems is also essential. With an algorithm like Karmarkar's, which is 50 to 100 times faster than the simplex method, we could do a lot of things that we couldn't do otherwise. For example, some applications could be real-time applications, as opposed to batch applications. So instead of running a job overnight and getting an answer the next morning, we could actually key in the data or access the database, generate the matrix, and come up with a solution that could be implemented a few minutes after keying in the data.
>
> A good example of this kind of application is what we call a major weather disruption. If we get a major weather disruption at one of the hubs, such as Dallas or Chicago, then a lot of flights may get canceled, which means we have a lot of crews and airplanes in the wrong places. What we need is a way to put that whole operation back together again so that the crews and airplanes are in the right places. That way, we minimize the cost of the disruption, as well as passenger inconvenience.

relayed through intermediate cities. Figure 4.17 shows a graph theory model (with color coding to show the intensity of traffic) for a portion of the internet as it existed in the 1990's. Now, the number of ways to route internet traffic has become unimaginably large, so picking the most efficient way to route email or other data packages is very difficult without using OR techniques.

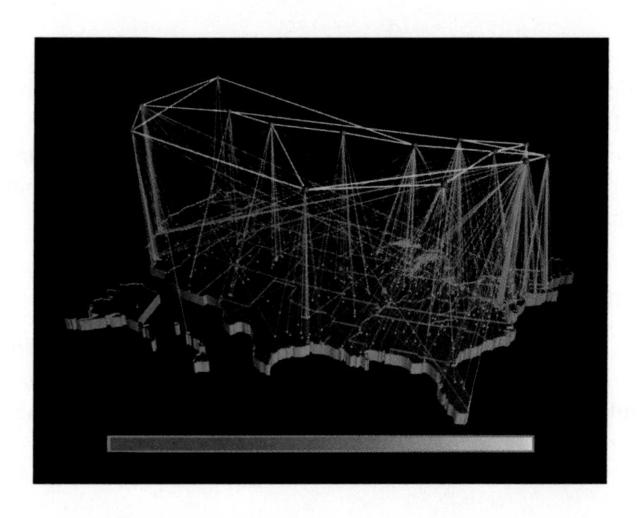

Figure 4.17 A map of the United States showing internet traffic among a collection of major sites, with the volume of traffic color coded. Routing traffic (email, data packets, streaming video) over immense networks such as these can sometimes benefit from sophisticated linear programming techniques and high speed computers. (*Donna Cox and Robert Patterson, courtesy of the National Center for Supercomputing Applications (NCSA) and the Board of Trustees of the University of Illinois.*)

Now that cell phones and Internet phone services are increasingly being used instead of landlines, new kinds of linear and integer programming problems and solutions have emerged to help provide better communications services. A typical example is using linear programming to determine the optimal locations for cell phone tower placement to achieve the best service.

4.5 A Transportation Problem: Delivering Perishables

A supermarket chain gets bread deliveries from a bakery chain that does its baking in different places. Each supermarket store needs a certain number of loaves each day, and the supplier bakes, in total, enough breads to exactly meet the demands. Figure 4.18 shows the cost to ship a loaf from a particular baking location to the store involved. How many breads should be shipped from each locale to each of the stores to stay within the demands and to minimize the cost?

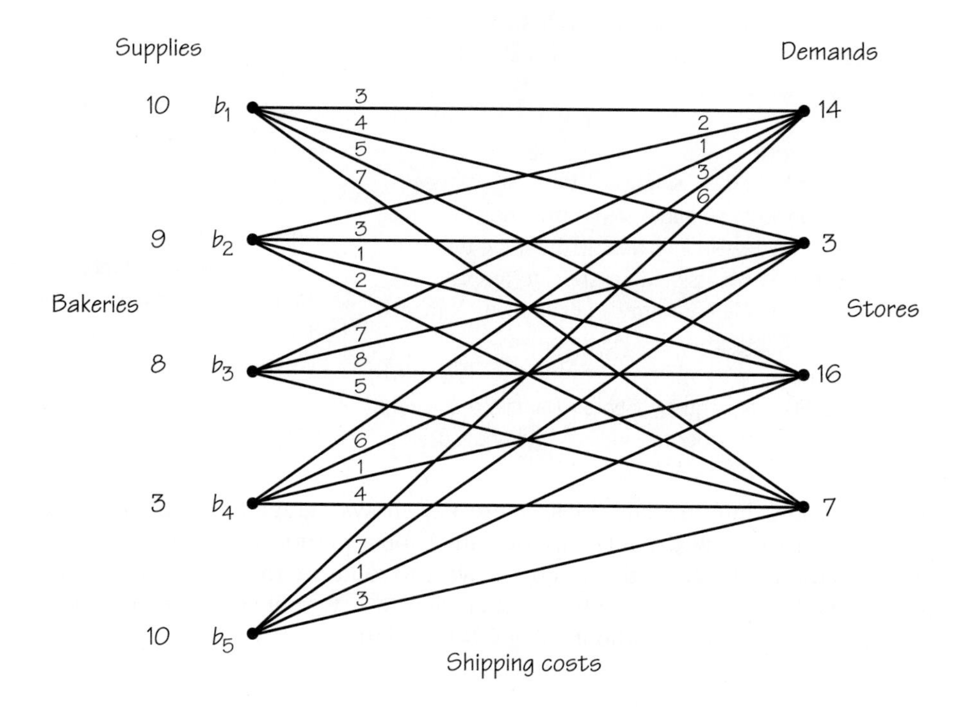

Figure 4.18 A graph theory representation of a supply-and-demand transportation problem that involves shipping breads from bakeries to stores.

Similarly, after a long holiday weekend, a car rental company will have extra cars in some cities and too few cars in other cities. It is faced with the problem of reshuffling the cars at minimal cost so that each city has the right number of cars. Problems such as these go under the general name of **transportation problems** and they form a special class of linear-programming problems that can be solved by a specialized method.

Transportation Problem DEFINITION

A group of suppliers must meet the needs of users of these supplies. There is a cost for shipping from a particular supplier to a particular user (demander). The **transportation problem** involves minimizing the total shipping cost of meeting the required demands from the supplies available.

EXAMPLE 8
Delivering Bread

Imagine that we have three bakeries and three stores, though the ideas we develop will also solve problems where the number of stores and bakeries are not the same. The three stores require 3 dozen, 7 dozen, and 1 dozen loaves of bread, respectively, while the three bakeries can supply 8 dozen, 1 dozen, and 2 dozen loaves, respectively. The information given so far can be displayed in Figure 4.19, where the "suppliers" are represented by the rows of the table (labeled with Roman numerals) and the "demanders" are represented by the columns (labeled with Hindu-Arabic numerals).

The numbers of breads available and the numbers being required are shown on the right side and bottom of the table and will be referred to as **rim conditions**. Each entry of the table shown in Figure 4.19 is known as a cell. It is convenient to have a name for each of these cells. For example, the cell in the third row and second column will be denoted (III, 2). The first number always corresponds to a row, the second to a column. Thus cell (I, 2) refers to bakery I and store 2.

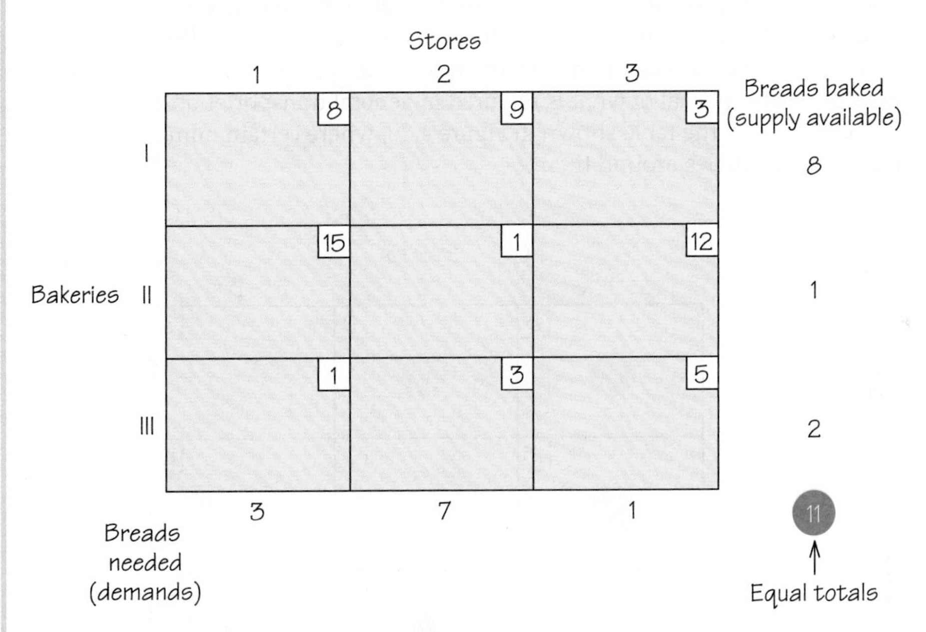

Figure 4.19 A representation of a specific problem involving meeting the demands of three stores for breads from the supplies available at three bakeries. Shipping costs between bakeries and stores are also shown.

In deciding which bakeries should ship to which stores, it seems natural to take into account the costs of shipping a dozen breads from a particular bakery to a particular store. If bakery I is farther from store 2 than is bakery II, it seems reasonable that the shipping cost for I will be higher than for II when shipping to that particular store.

However, the costs of shipping may also involve time considerations. (The distance to a store may be shorter, but it may be that this route is a very slow one.) Also, it may take extra time for a truck coming from I to park when making the next delivery.

The numbers we use in our diagrams are "aggregate" costs. The nice thing about what we are doing is that the solution method works independently of the way the costs are arrived at or computed. These costs (see Figure 4.19) are shown in the upper-right-hand corner of a cell. Thus the number 9 shown in cell (I, 2) means that it costs nine units to ship a bread from bakery I to store 2. Our goal will be to supply the stores with the breads they require from the supplies available at the bakeries so that the total cost of providing the breads to the stores is as small as possible (a minimum).

The tools for solving transportation problems like these were developed during World War II in conjunction with getting supplies from different ports in the United States to different ports in Europe (mostly the United Kingdom) as efficiently as possible. (The U.S. ports were like the bakeries, and the British ports were like the stores that needed the breads.)

We can think of finding a solution to a problem like this as a special kind of linear-programming problem because we can express the objective of minimizing the cost using a linear relationship. The constraints that express that the rim conditions are met can also be expressed using linear equations. However, it turns out there are algorithms that make it possible to solve problems of this kind that are rather larger than general linear-programming problems which can be solved by hand. These algorithms are intuitively appealing.

We can divide the problem of finding a solution to a transportation problem into two phases, as we did for general linear-programming problems. First, find a solution that is feasible (that is, a solution that does not violate any of the constraints of the problem). Second, if the current solution is not optimal, we move to a better one. Thus, we will first find a solution that meets the constraints and then try to find an improved solution. If there is no better solution than the one we have, under suitable circumstances we show that there is never a better one. Thus, the solution that we have found is an optimal solution. We will work our way through a simple example that is typical of what is required in general transportation problems.

Let's turn to the table shown in Figure 4.20, where certain numbers have been inserted with circles around them.

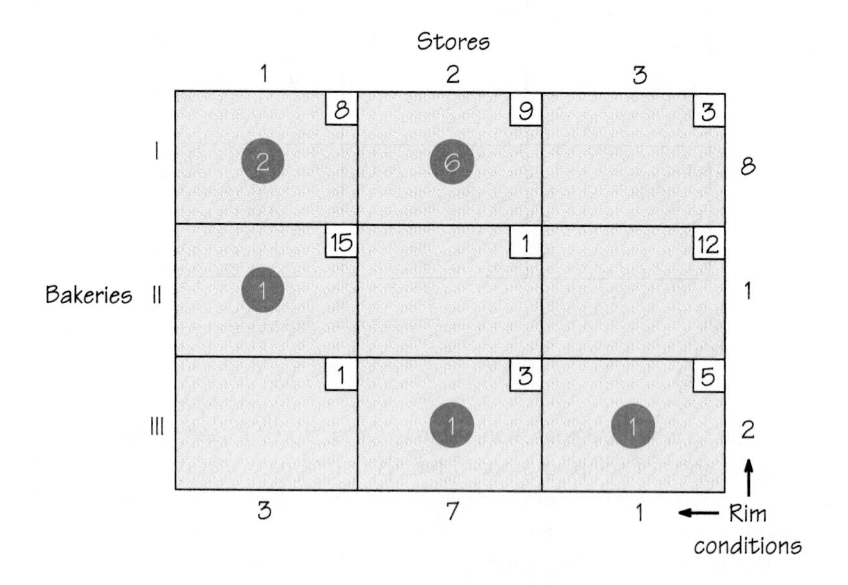

Figure 4.20 A possible solution to meeting the needs of three stores for bread from supplies available at three bakeries. The circled numbers show the amounts shipped from the bakeries to the stores.

Tableau DEFINITION

A table showing costs and rim conditions for a transportation problem is known as a **tableau**.

When we see a circled number such as the 6 in row I, column 2, this means that we plan to ship six breads to store 2 from bakery I. Similarly, the circled number 1 in row III and column 3 means we plan to ship one bread from bakery III to store 3. The cells that have no circled numbers are thought of as having zero entries; no breads are being shipped between these stores and these bakeries. Note, for example, that the row sum of the circled numbers in the first row is 8. This means that all the breads available at bakery I are being shipped to some store.

Similarly, the fact that the circled entries in column 2 add up to 7 means that all the breads needed by store 2 are being supplied to it. You can verify that all the row sums and column sums add to exactly the numbers that we want to ship from each bakery to each store. Note that 11 breads have been shipped by the bakeries and received by the stores. When this happens, the circled numbers are said to be a *feasible* solution to the problem.

How much will it cost to ship these amounts of breads (see Figure 4.20) to the stores? The number can be computed by multiplying the circled numbers by the cost shown in the associated cell. For example, to ship two breads from bakery I to store 1 costs $2(8) = 16$ because the cost associated with the cell in which the 2 appears is 8. The cost of shipping six breads from bakery I to store 2 is $6(9) = 54$. To get the total cost of this "shipment plan," we sum all the shipped amounts by the associated costs to get

$$2(8) + 6(9) + 1(15) + 1(3) + 1(5) = 16 + 54 + 15 + 3 + 5 = 93$$

However, at this point we do not know if there is a cheaper way to ship the breads to the stores. Notice that the number of cells with circled numbers is exactly equal to the number of rows m plus the number of columns n minus 1. This is the general pattern with transportation problems. Cells that are used for shipping are circled. On occasion, we ship a zero amount because the procedure works only when $m + n - 1$ cells are circled.

If we look at the pattern of circled numbers in the tableau in Figure 4.21, we see that there is a difficulty even though 11 breads are involved (the sum of all the circled numbers).

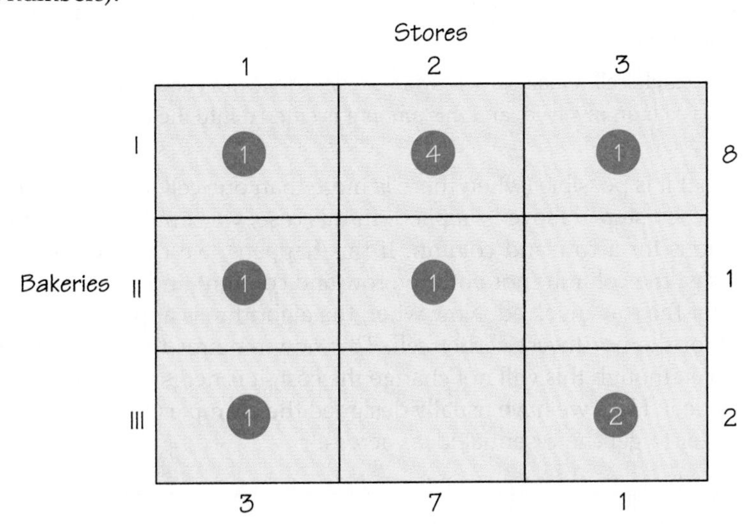

Figure 4.21 The circled numbers are not a possible solution to this transportation problem because the rim conditions are not satisfied.

The numbers in the first row add to 6, which means that at bakery I there will be breads left over that have not been shipped. In row 2, the sum of the circled numbers is 2, but this means that something is wrong. How can bakery II, which has a supply of only 1 bread, ship 2 breads? Furthermore, column 3 sums to 3, which means that 3 breads have been shipped to store 3 despite the fact that it only requested 1 bread! These facts add up to the realization that this assignment of numbers to the cells violates the rules we are requiring. This proposed shipment plan also violates our rule that we are not allowed to circle more than five cells.

How can we find a solution that meets the constraints of the problem (the rim conditions)? We will show two ways to do this. The first is "fast and dirty" but typically does not find a very good solution with which to start. The second (developed in the exercises) usually gives a better "initial" solution but is a little harder to carry out.

This pair of approaches displays a common tension in problem solving: the ease of getting started but requiring more work later on, or more work at the start, which often proves to be a good investment of extra effort because less work is needed to find an optimal solution. If we know in advance the method being chosen to solve a problem, we can often find an example where this particular method does poorly. Mathematicians work hard to find methods that work well on the kinds of problems that come up in genuine applications.

Northwest Corner Rule

The easier approach involves what is called the **Northwest Corner Rule.** This rule is simple because it is based on the geometry of the table that is involved and does not even look at the costs associated with the cells in the table, which in the long run cannot be a good idea, because these costs come into play when trying to get an optimal solution.

How does the Northwest Corner Rule work? The algorithm carries out the following procedure until exactly one cell remains in the "altered tableau."

Northwest Corner Rule PROCEDURE

1. Locate that cell of the current tableau that is as far to the top and to the left as possible (that is, in the northwest corner). Ship via this cell the smaller of the two rim values (call the value s) associated with the row and column of this cell. (Indicate that this cell is being used by putting a circle around the entry in the tableau.)
2. Cross out the row or column that had rim value s and reduce the other rim value for this cell by s.
3. When a single cell remains, there will be a tie for the rim conditions of both the row and column involved, and this amount is entered into the cell and circled.

Note that it is possible (when there is more than one cell at the start) for there to be a tie when step 1 above is applied. In this case, we simultaneously fulfill the rim conditions for a row and column. If this happens, we can always choose to cross out, say, the column (not both the row and column) and reduce to 0 the rim condition for the row involved. Now when the algorithm is applied, one has a rim value of 0 for the northwest corner cell. This now requires that 0 be shipped via that cell. Even though this will not change the cost, it is necessary to put a 0 in this cell and circle it. Here we have usually designed the examples to avoid ties so as to make it easier to get the essential ideas across.

EXAMPLE 9
Using the Northwest Corner Rule

Applying the Northwest Corner Rule to our original tableau (see Figure 4.19), we get the sequence of tableaux in Figure 4.22 as we cross out the rows or columns, where for clarity the costs associated with the cells are suppressed. The last diagram in the sequence shows the results on the original tableau, with the cost restored. Note that, at the steps in between, the costs played no role. It is a good idea to check that the circled numbers in each row and column really add up to the rim value for that row and column and that exactly $m + n - 1$ cells are filled.

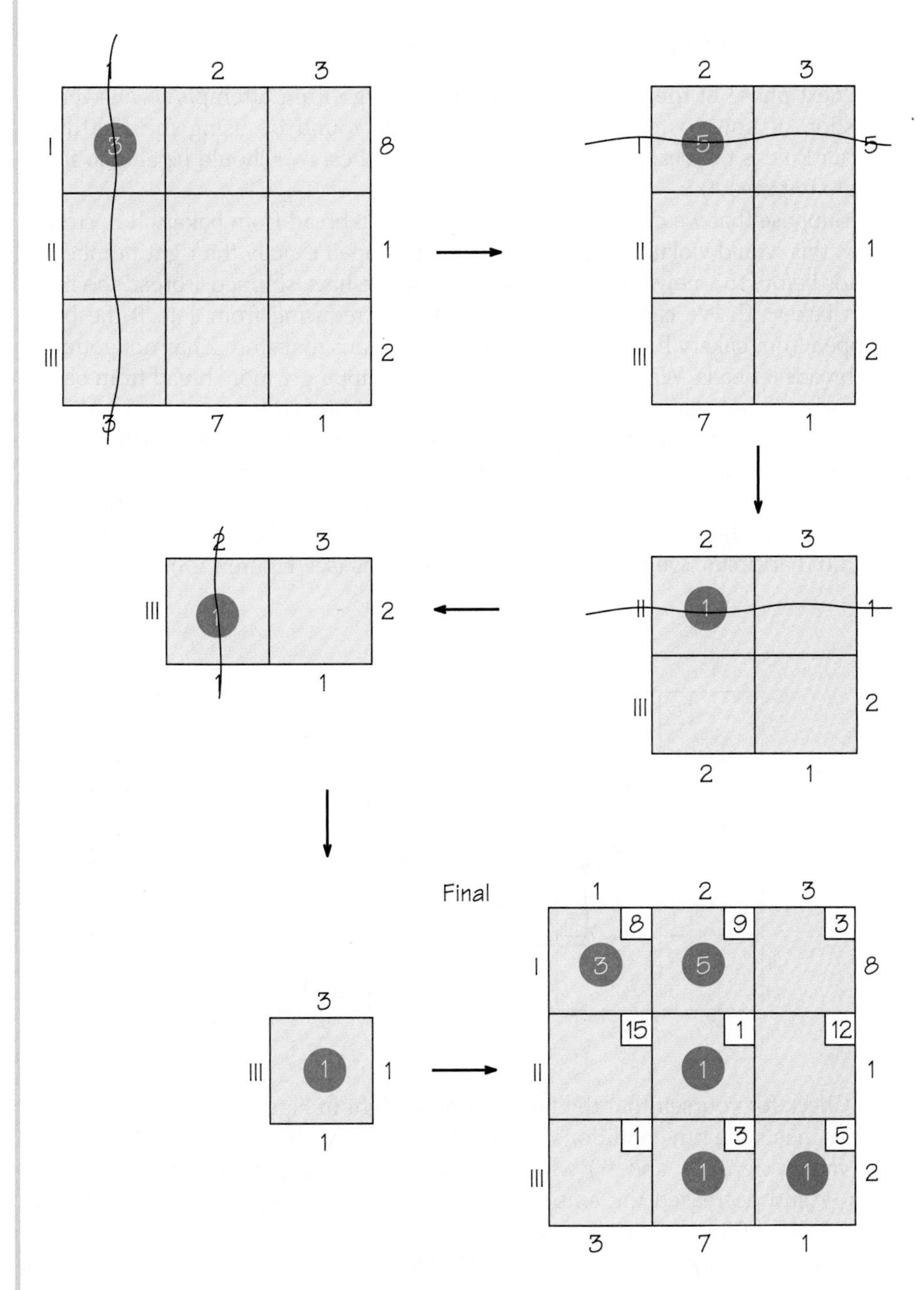

Figure 4.22 The construction of an initial solution to a transportation problem using the Northwest Corner Rule.

We can now compute the cost of the associated solution that we have found (feasible solution), which obeys the rim conditions. As we did previously, we add up the cost multiplied by the amount shipped for each cell with a circled entry. We get the following calculation:

$$3(8) + 5(9) + 1(1) + 1(3) + 1(5) = 78$$

This shows a cost that is smaller than the solution we found earlier. That solution involved a cost of 93. But is this solution the cheapest one? That finding this feasible solution did not use the costs on the cells suggests that it is not very likely.

Improving the Feasible Solution

The next phase of the transportation problem algorithm attempts to answer the question of how to tell if the feasible solution found by using the Northwest Corner Rule is the best. If this solution is not the best, we should be able to find a way to improve it.

Suppose that we decided to ship an additional bread from bakery II to store 3. Now, this would violate the fact that we had shipped exactly the right numbers of breads before this new additional shipment, so we have shipped 1 bread too many from bakery II. We can compensate for this by reducing from 1 to 0 the bread shipped from bakery II to store 2. But this now means that store 2 has not gotten all the breads it needs. We can take care of this by shipping 1 more bread from bakery III to store 2, but again we now have 1 extra bread shipped from bakery III. We can compensate for this by reducing the number of breads shipped via cell (III, 3)—that is, from bakery III to store 3. This step will ensure that the rim conditions will hold for the circled numbers. This is because we have located a circuit—(II, 3), (II, 2), (III, 2), (III, 3), (II, 3)—where, if we increase and decrease the breads alternately going around that circuit, we maintain the rim conditions (see Figure 4.23).

Figure 4.23 An illustration of how to take a current solution to a transportation problem and try to get an improved cheaper solution that still meets the rim conditions.

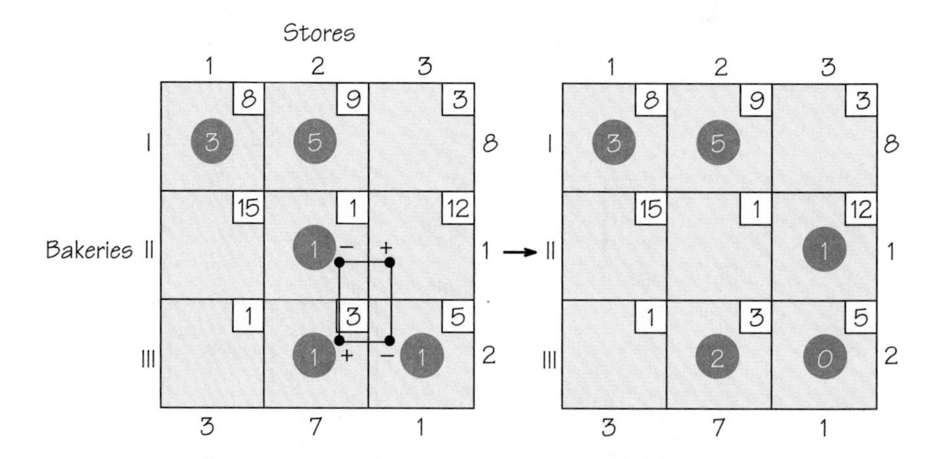

Check for yourself that the tableau on the right in Figure 4.23 with the circled entries meets the rim conditions. On the left, we show the circuit of cells with plus and minus signs (+ and −) where we have increased the amounts in the cells with + and decreased the amounts in the cells with − by 1 unit. (Note that to keep a total of five cells circled, we have set one of the circled cells to 0, because when we reduce the amounts of bread shipped in cells (II, 2) and (III, 3), we get a tie of value 0.)

We now have to ask whether this new solution is cheaper or more expensive than the one we started with. We can figure out whether this is a better or worse solution by tracking the costs of moving from the previous solution to the new one.

We went around a circuit where we increased a cost, decreased a cost (because we reduced the number of breads in that cell), increased a cost, and then decreased a cost before coming back to where we started, having traversed a circuit of length 4. The net effect of this collection of cost changes is $+12 - 1 + 3 - 5 = +9$. Thus, these changes, while producing a new feasible solution, give a more costly solution!

Perhaps increasing the amount shipped via a different circuit of cells would be better. Suppose we try the same process for cell (I, 3) (Figure 4.24)—that is, increase the shipping of breads from bakery I to store 3. To see if this is worthwhile, check the circuit formed by shipping more via cell (I, 3). It takes a bit of practice to find the circuit that this cell forms. In the case of cell (I, 3), the circuit we get is (I, 3), (I, 2), (III, 2), (III, 3), (I, 3). The cost of moving around this circuit is $+3 - 9 + 3 - 5 = -8$. We will refer to this number as the **indicator value** for this cell.

Indicator Value DEFINITION

The **indicator value of a cell** C (not currently a circled cell) is the cost change associated with increasing or decreasing the amounts shipped in a circuit of cells starting at C. It is computed by summing with alternating signs the costs of the cells in the circuit.

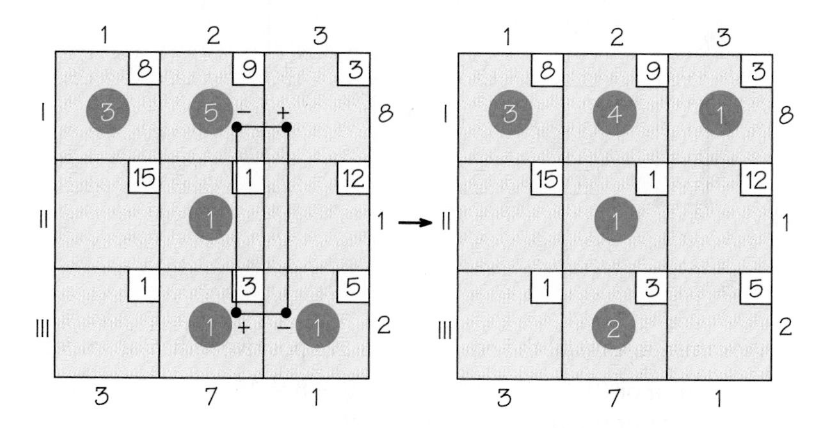

Figure 4.24 Using a cell with a negative indicator value, we can find a cheaper way of meeting the demands from the supplies.

The -8 means that we can lower the cost of shipping breads by using a different pattern of meeting the demands from the supplies. We have computed the saving for shipping one bread more via cell (I, 3), but perhaps we might be able to save even more by shipping even more breads via this cell. To determine whether we could, we look at the circuit that begins at cell (I, 3). To maintain a feasible solution, we have to increase the amounts shipped via some cells of this circuit and decrease the amounts shipped via others. Because we cannot decrease the amount shipped via any cell below zero, the minimum value of any cell that must be reduced is the maximum amount that can be shipped via cell (I, 3). In this case, it means that only one bread can be shipped via cell (I, 3), thereby lowering the cost from the previous solution by 8.

When we looked to improve the solution shown in Figure 4.22, we have now seen that by shipping via cell (I, 3), we can get a better solution. However, there might be several cells in the solution shown in Figure 4.22 that would lead to improvement. Which one should we choose? The answer is that we should adopt a greedy point of view. If there are several cells with a negative indicator value, pick the one that is "most negative" to improve the solution.

Given a current feasible solution (one that satisfies the rim condition), we check each cell that does not have a circled number for improvement if we ship via that cell. If a cell leads to a positive indicator value with the circuit associated with it, no improvement is possible. If a cell has a negative indicator value associated with the circuit for that cell, we can get an improvement. We select as the cell to increase that cell with the largest negative indicator value. We now have a new feasible solution that is cheaper than the one we started with and can repeat our procedure just described starting from this new feasible solution.

It turns out that there was no better cell than (I, 3) (using this greedy approach) to get an improved solution. We will take the current best solution and see if we can improve it more. It turns out that for the current tableau (Figure 4.24), all the cells have a positive indicator except for cell (III, 1):

$$\text{Indicator for cell (III, 1):} \qquad +1 - 3 + 9 - 8 = -1$$

Because the minimum of the circled numbers in the cell with a negative label is 2 in cell (III, 2), we can increase by 2 the amount shipped in cell (III, 1) and get a new solution, as shown in Figure 4.25.

Figure 4.25 We can find an even cheaper way of meeting the demands from the supplies available using a cell with a negative indicator value.

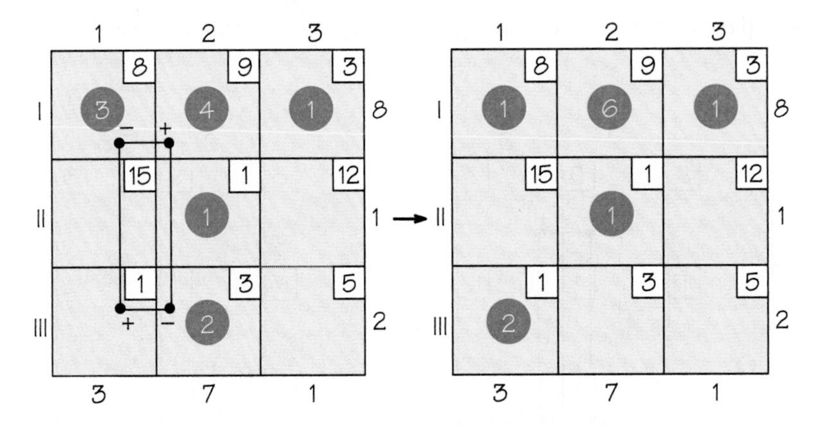

Now, for this tableau, all the empty cells have positive indicator values.

$$\begin{aligned}
&\text{Indicator (II, 1):} && +15 - 1 + 9 - 8 = 15\\
&\text{Indicator (II, 3):} && +12 - 3 + 9 - 1 = 17\\
&\text{Indicator (III, 2):} && +3 \ - 9 + 8 - 1 = 1\\
&\text{Indicator (III, 3):} && +5 \ - 3 + 8 - 1 = 9
\end{aligned}$$

This means that the current solution is optimal. The cost of this solution is

$$1(8) + 6(9) + 1(3) + 1(1) + 2(1) = 8 + 54 + 3 + 1 + 2 = 68$$

It turns out that if all the cells associated with a feasible solution have positive indicator values, then the solution that one has reached is optimal. (Cells with a zero indicator value show that there are other solutions that achieve the same optimal value.)

How to Recognize an Optimal Solution THEOREM

We are given a transportation problem with m suppliers and n demanders where the amount of the supplies equals the amount of demands. A collection of $m + n - 1$ circled cells is optimal (that is, the circled cells determine a minimum cost solution) if the indicator value associated with the empty cells is positive. If some indicator cells are positive and some are zero, there are multiple solutions for an optimal value.

This theorem is the analog of the result for linear programming that states that if a corner point is feasible, and if no neighbor of the corner point has a better value of the objective function, then the corner point we are at is already an optimal one. Note that there may be other optimal solutions that use a different number of cells than $m + n - 1$, but we can never do any better in terms of the cheapness of a solution than what we have described above.

For those interested in the exciting fact that one piece of mathematics is often useful for other mathematics, we see an example of that here. The reason that an empty cell gives rise to a unique circuit with which we can try to improve the current solution of a transportation problem is the fact that when an edge not in a tree is added to a tree, it creates a unique circuit (see Chapter 3). Because we have m rows and n columns, a tree associated with a graph on $m + n$ vertices has $m + n - 1$ edges, exactly the number of cells we need to fill in a transportation problem!

4.6 Improving on the Current Solution

Now we will describe a method guaranteed to find an optimal solution to a transportation problem.

The Stepping Stone Method DEFINITION

The **stepping stone method** consists of taking some feasible solution of a transportation problem and improving this solution, if it is not optimal, by shipping an additional amount using a cell with a negative indicator value.

EXAMPLE 10
Applying the Stepping Stone Method

We will work out another small example to illustrate the technique of applying the Northwest Corner Rule to get an initial solution, and then improving this solution if it is not optimal. Again, we do so by computing the indicator values of the cells and improving the current solution by shipping using a cell with a negative indicator value.

We start with an initial tableau where there are two mines that can supply ore to three companies that extract ore. There are 10 units of ore being mined and the extractors need 10 units to run at full capacity. The initial tableau for the problem is displayed in Figure 4.26.

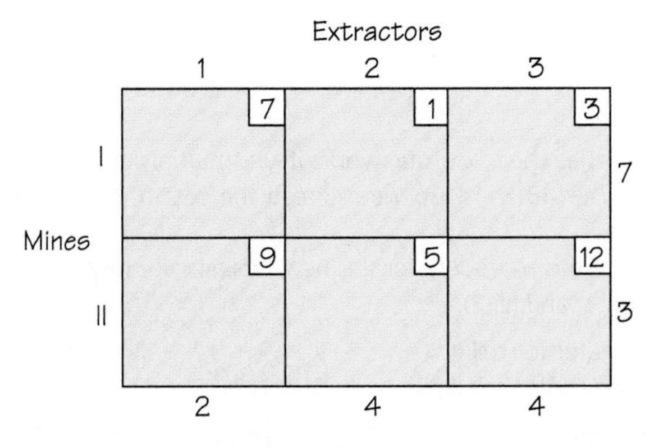

Figure 4.26 A transportation problem where two mines supply ore to three companies that extract metal from the ore. The shipping costs are indicated.

Using the Northwest Corner Rule, we find an initial feasible solution as shown in Figure 4.27. When applying the Northwest Corner Rule, we eliminate a row or column as follows: first column 1, then column 2, then row I, and we are now left with a single cell. The cost of the feasible solution shown is

$$2(7) + 4(1) + 1(3) + 3(12) = 14 + 4 + 3 + 36 = 57$$

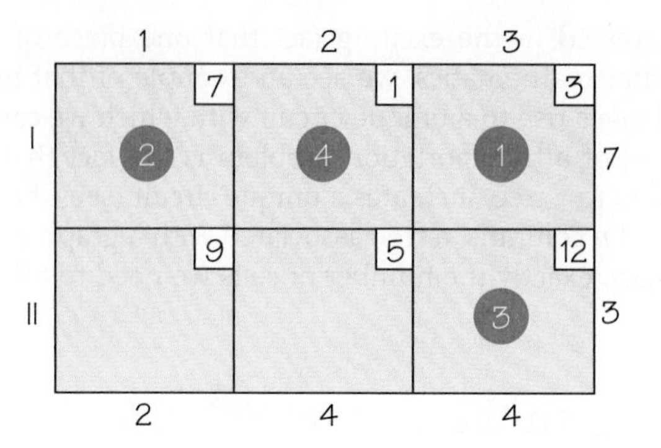

Figure 4.27 The Northwest Corner Rule has been used to find a possible way to meet the demands from the supplies for the tableau in Figure 4.26.

The two empty cells we have are (II, 1) and (II, 2). We compute the indicator value for each of these cells:

Indicator for cell (II, 1): $+9 - 12 + 3 - 7 = -7$
Indicator for cell (II, 2): $+5 - 12 + 3 - 1 = -5$

Because cell (II, 1) has a more negative indicator value, we can reduce the cost more by using that cell. Increasing by 2 (because this is the minimum of circled numbers with negative signs in the computation of the indicator) the amount of metal shipped via cell (II, 1) and cell (I, 3) and reducing by 2 the amount in cells (I, 1) and (II, 3), we obtain the new tableau in Figure 4.28. This has cost

$$4(1) + 3(3) + 2(9) + 1(12) = 4 + 9 + 18 + 12 = 43$$

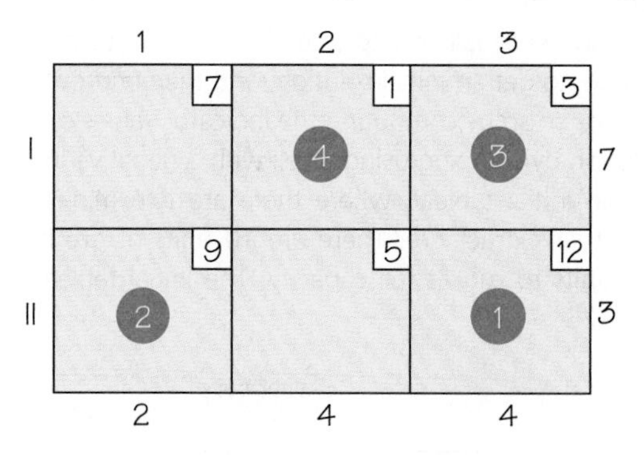

Figure 4.28 An improved solution based on the negative indicator value for cell (II, 1) in Figure 4.27.

Note that as a partial check on our work, if we multiply the indicator (−7) by 2, this is −14 and 57 − 43 = 14, so we reduced the cost of our first solution by 14, as expected.

We now repeat this procedure for this new tableau. We must compute the indicator value of cells (I, 1) and (II, 2).

Indicator for cell (I, 1): $+7 - 9 + 12 - 3 = +7$
Indicator for cell (II, 2): $+5 - 12 + 3 - 1 = -5$

Thus, it turns out that we can increase by 1 the amount shipped by (II, 2), and get the tableau in Figure 4.29.

From this tableau, we need to compute the indicator values for the cells (I, 1) and (II, 3). We obtain

Indicator for cell (I, 1): $+7 - 9 + 5 - 1 = +2$
Indicator for cell (II, 3): $+12 - 3 + 1 - 5 = +5$

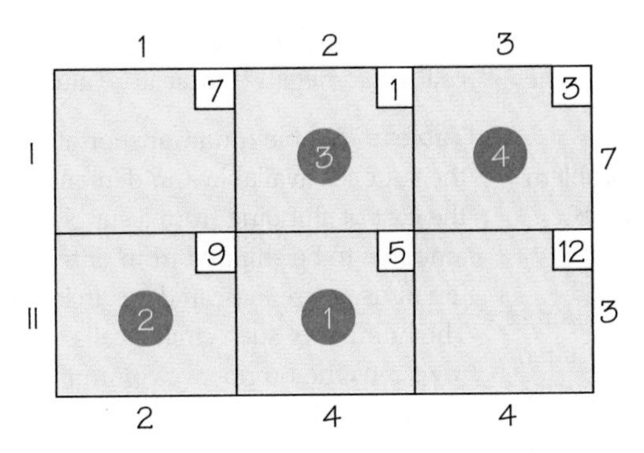

Figure 4.29 An improved solution, which turns out to be optimal, based on the negative indicator value for cell (II, 2) in Figure 4.28.

Not surprisingly, the cell (II, 2) has a positive indicator value because in the previous tableau, that cell was the one that, when we shipped less via it, enabled us to reduce the cost. The fact that both of these indicator values are positive means that the current shipping schedule is an optimal one; that is, using a shipping schedule that ships via only four cells, we cannot find any other solution with the same value.

Transportation problems arise in a very large range of situations including shipping milk from dairies to supermarkets, vegetables to health food stores, and vitamins to your local drug store. The next time you sit down to breakfast, think about how many mathematics problems were solved for you to have a healthy breakfast!

REVIEW VOCABULARY

Corner point principle This principle states that there is a corner point of the feasible region that yields the optimal solution. (p. 125)

Feasible points A possible solution (but not necessarily the best one) to a linear-programming problem. With just two products, we can think of a feasible point as a point on the plane. (p. 151)

Feasible region The set of all **feasible points**, that is, possible solutions to a linear-programming problem. For problems with just two products, the feasible region is a part of the plane. Also called **feasible set**. (p. 122)

Indicator value of a cell The change in cost due to shipping an increased or decreased amount, using the cells in a transportation tableau that form a circuit consisting of circled cells together with a selected cell that is not circled. When an indicator value is negative, a cheaper solution can be found by shipping using this cell. (p. 147)

Linear programming A set of organized methods of management science used to solve problems of finding optimal solutions, while at the same time respecting certain important constraints. The mathematical formulations of the constraints in linear-programming problems are linear equations and inequalities. Mixture problems usually are solved by some type of linear programming. (p. 115)

Minimum constraint An inequality in a mixture problem that gives a minimum quantity of a product. Negative quantities can never be produced. (p. 120)

Mixture chart A table displaying the relevant data in a linear-programming mixture problem. The table has a row for each product and a column for each resource, for any nonzero minimums, and for the profit. (p. 118)

Mixture problem A problem in which a variety of resources available in limited quantities can be combined in different ways to make different products. It usually is desirable to find the way of combining the resources that produces the most profit. (p. 116)

Northwest Corner Rule A method for finding an initial, but rarely optimal, solution to a transportation problem starting from a tableau with rim conditions. The amounts to be shipped between the suppliers and demanders are indicated by circling numbers in the cells in the tableau. The number of cells circled after applying the method will equal the number of rows plus the number of columns minus 1. The method depends on locating at each stage the "northwest corner" of the original tableau or a part of it. (p. 144)

Optimal production policy A corner point of the feasible region where the profit formula has a maximum value. (p. 117)

Profit line In a two-dimensional, two-product, linear-programming problem, the set of all points that yield the same profit. (p. 133)

Resource constraint An inequality in a mixture problem that reflects the fact that no more of a resource can be used than what is available. (p. 120)

Rim conditions The supplies available (listed in a column at the right of a transportation tableau) and demands required (listed in a row at the bottom of a transportation tableau) in a transportation problem. The supplies available are usually taken to meet exactly the demands required. (p. 141)

Simplex method One of a number of algorithms for solving linear-programming problems. (p. 136)

Stepping stone method A method for solving a transportation problem that improves the current solution, when it is not optimal, by increasing the amount shipped using a cell with a negative indicator value. (p. 149)

Tableau A table for a transportation problem indicating the supplies available and demands required, as well as the cost of shipping from a supplier to a demander. The amounts to be shipped from different suppliers to different users are indicated by circled cells in the tableau. The number of such circled cells is always the number of rows plus the number of columns diminished by 1 for the tableau. (p. 142)

Transportation problem A special type of linear-programming problem where we have sources of supplies and users of, or demand for, these supplies. There is a cost to ship an item from a supplier to a demander. The goal is to minimize the total shipping cost to meet the demands from the supplies. (p. 140)

SKILLS CHECK

1. Where do the lines $6x + 2y = 26$ and $2x + 3y = 18$ intersect?

(a) At the point (3, 4)
(b) At the point (6, 2)
(c) At the point (3, 2)

2. The x-coordinate and y-coordinate of the point where the line $x = -2$ intersects the line $5y - 2x = -11$ are _____ and _____, respectively.

3. The y-coordinate of the point whose x value is 3 on the line $3x + 2y = 12$ is

(a) 3/2.
(b) 0.
(c) 2.

4. The lines $x + 3y = 12$ and $y = 2$ intersect at the point with x-coordinate _____ and y-coordinate _____.

5. The two lines $2x + 3y = 12$ and $6x + 9y = 7$

(a) intersect at (0, 0).
(b) intersect at (−3, −2).
(c) are parallel.

6. The x-coordinate and y-coordinate of the points where the line $2x + 7y = 28$ crosses the x-axis and y-axis, respectively, are _____ and _____.

7. Which of these points lie in the region $4x + 3y \geq 24$, $x \geq 0$, $y \geq 0$?

(a) Points (5, 2) and (3, 4)
(b) Points (2, 5) and (3, 4)
(c) Points (5, 2) and (2, 5)

8. The difference between the set of those pairs (x, y) that satisfy $x + 2y < 8$ and $x + 2y \leq 8$ is that the second inequality holds for points on the line _____, whereas the first inequality does not hold for points on this line.

9. A tart requires 3 oz of fruit and 2 oz of dough; a pie requires 13 oz of fruit and 7 oz of dough. There are 140 oz of fruit and 90 oz of dough available. Each tart earns 6 cents profit; each pie earns 25 cents profit. What are the resource inequalities of this situation?

(a) $3x + 2y \leq 140$
$13x + 7y \leq 90$
$x \geq 0,\ y \geq 0$

(b) $3x + 13y \leq 140$
$2x + 7y \leq 90$
$x \geq 0,\ y \geq 0$

(c) $3x + 2y \leq 6$
$13x + 7y \leq 25$
$x \geq 0,\ y \geq 0$

10. If the profit P for making x large and y small shovels is ($P = 7x + 6y$), then the profit made if it were feasible to manufacture 6 small shovels and 8 large shovels would be _____.

11. The cost C of manufacturing x pounds of flour blend X and y pounds of flour blend Y is given as $C = 9x + 4y$. If a company is using linear programming to minimize the cost of making the blends X and Y and this happens when $x = 11$ and $y = 9$, which of the following statements must hold?

(a) $x = 11$ and $y = 9$ cannot be a corner point.
(b) $x = 11$ and $y = 9$ cannot be an interior point of the feasible region.
(c) $x = 11$ and $y = 9$ is the only point where the minimum can occur.

12. Producing a bench (x) requires 2 boards, and producing a table (y) requires 5 boards. There are 25 boards available. The resource constraint associated with this situation is _____ x + _____ $y \le 25$.

13. For the feasible region of a linear-programming problem defined by the inequalities $x \ge 0$, $y \ge 0$, and $2x + 5y \le 10$, which of the following pairs of points lies within the feasible region?

(a) (0, 2) and (0, 0)
(b) (2, 1) and (3, 1)
(c) (5, 0) and (1/2, 2)

14. A tart requires 3 oz of fruit and 2 oz of dough; a pie requires 13 oz of fruit and 7 oz of dough. There are 140 oz of fruit and 90 oz of dough available. Each tart earns 6 cents profit; each pie earns 25 cents profit. The profit formula for this situation, if x represents the numbers of pies produced and y represents the number of tarts produced, is given by P (in cents) = ____ x + ____ y.

15. Graph the feasible region identified by the following inequalities:

$2x + 4y \le 20$
$4x + 2y \le 16$
$x \ge 0, y \ge 0$

Which of these points is *not* in the feasible region of the graph drawn?

(a) (2, 4)
(b) (1, 1)
(c) (10, 0)

16. Suppose the feasible region has four corners, at points (0, 0), (4, 0), (0, 3), and (3, 2). If the profit formula is \$3x − \$2y, the maximum value for the profit is ____________________.

17. Suppose that the feasible region has four corners, at points (0, 0), (4, 0), (0, 3), and (3, 2). For which of these profit formulas is the profit maximized by producing a mix of products?

(a) \$2x − \$2y
(b) \$$x$ + \$2y
(c) \$2x − \$$y$

18. The corner point principle cannot be applied to find the optimal answer for the value of the profit $P = 3x + 7y$, where the feasible region is shown in the diagram, because the feasible region is not ______________.

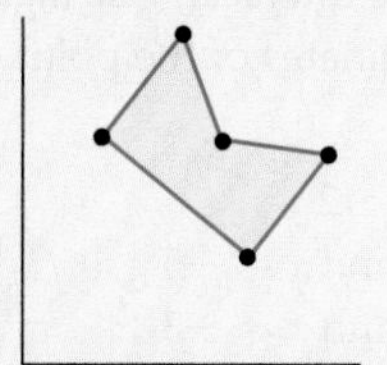

19. The shaded region in the accompanying diagram is an example of a region

(a) whose area is not bounded.
(b) that is not convex.
(c) that is not bounded by straight-line segments.

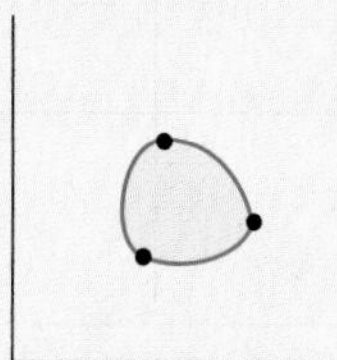

20. Given the feasible region for a linear-programming problem defined by the inequalities $x \ge 0$, $y \ge 0$, $2x + y \le 12$, the feasible point where $x = y$ with the largest possible x-coordinate has x-coordinate ______ and y-coordinate ______.

21. Suppose that the feasible region has five corners, at points (1, 1), (2, 1), (3, 2), (2, 4), (1, 5). Which of these points is *not* in the feasible region?

(a) (1, 3)
(b) (2, 2)
(c) (0, 0)

22. Suppose the feasible region has five corners, at points (1, 1), (2, 1), (3, 2), (2, 4), (1, 5). If the profit formula is \$5x − \$3y, the corner point which maximizes the profit has x-coordinate ____ and y-coordinate ____.

23. How does the line representing the maximum feasible profit intersect the feasible region?

(a) No points of intersection
(b) Only one point of intersection
(c) At least one point of intersection, and sometimes more than one point of intersection

24. Consider the feasible region identified by the inequalities $x \ge 0$, $y \ge 0$, $3x + y \le 10$, $x + 2y \le 6$. The corner point of this region, which is not (0, 0), that has x-coordinate 0 has y-coordinate ________________.

25. Given the feasible region for a linear-programming problem defined by the inequalities $x \ge 0$, $y \ge 0$, $2x + y \le 15$, the feasible point where $x = y$ with the largest possible value for a y-coordinate has coordinates

(a) (5, 5).
(b) (3, 3).
(c) (0, 15).

26. Consider the feasible region for a linear programming problem involving the inequalities $x \ge 0$, $y \ge 0$, $3x + y \le 10$, $x + 2y \le 5$. The corner point for this feasible region that has no zero coordinates has x-coordinate ______ and y-coordinate ______.

27. When the Northwest Corner Rule is applied to the accompanying transportation problem tableau, the cells that remain empty are

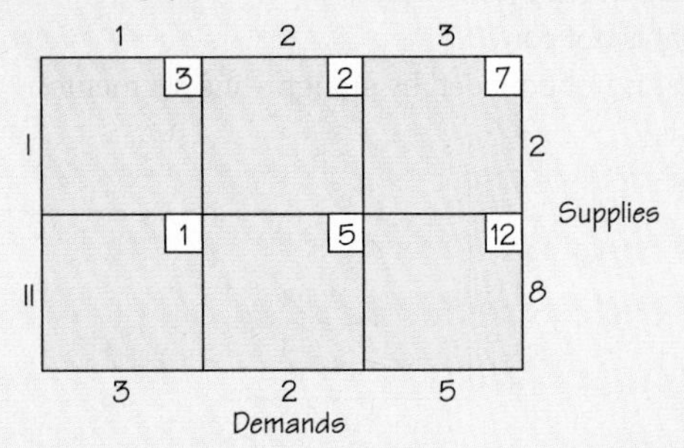

(a) cell (II, 2) and cell (I, 2).
(b) cell (I, 2) and cell (II, 3).
(c) cell (I, 2) and cell (I, 3).

28. The circled cells in the accompanying tableau give a solution that satisfies the rim conditions. The cost associated with this solution is __________.

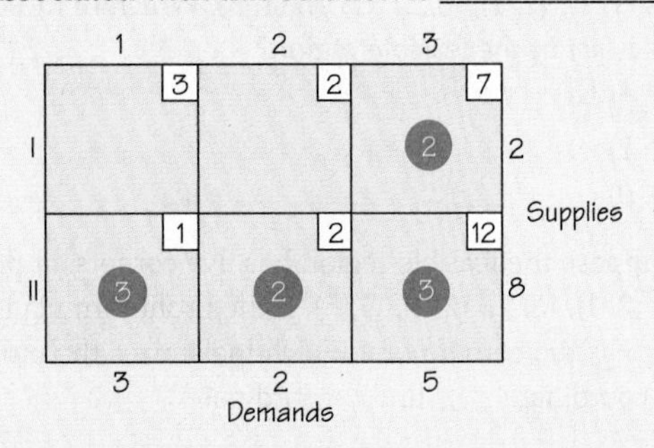

29. The circled cells in the accompanying tableau satisfy the rim conditions. When the indicator value for cell (I, 2) is computed,

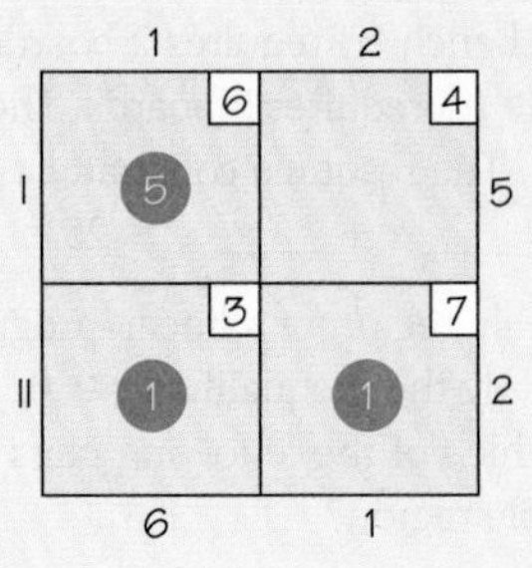

(a) the result being positive, the current solution is optimal.
(b) the result being positive, this tableau has no minimal cost solution.
(c) the result being negative, this tableau does not give a minimal cost solution.

30. The indicator value associated with cell (I, 2) of the accompanying tableau is __________.

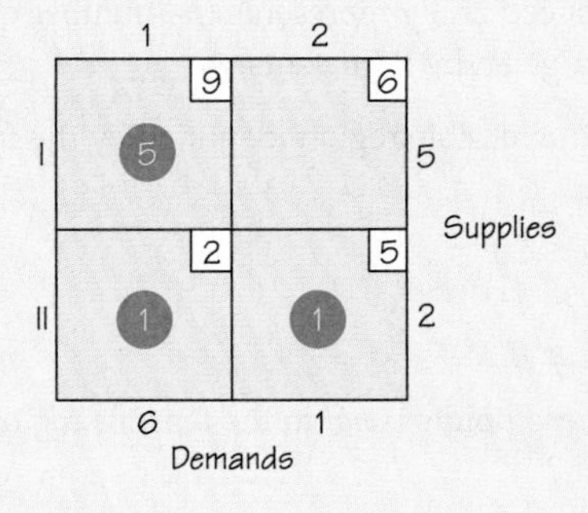

CHAPTER 4 EXERCISES

■ Challenge ▲ Discussion

4.1 Linear Programming and Mixture Problems: Combining Resources to Maximize Profit

1. Find the coordinates of all points where the lines $x + y = 8$, $y = 4$, and $x = 2$ intersect.

2. Find the x- and y-intercepts of the line $3y + 5x = 30$.

3. Using intercepts, the points where the lines cross the axes, graph each line.

(a) $2x + 3y = 12$
(b) $3x + 5y = 30$
(c) $4x + 3y = 24$

4. Using intercepts, the points where the lines cross the axes, graph each line.

(a) $7x + 4y = 42$
(b) $x = -3$
(c) $y = 6$

5. Graph both lines on the same axes. Put a dot where the lines intersect. Use algebra to find the x- and y-coordinates of the point of intersection.

(a) $4x + 9y = 18$ and $x = 0$
(b) $5x + 3y = 45$ and $y = -5$
(c) $5x + 3y = 45$ and $x = 3$

6. Graph both lines on the same axes. Put a dot where the lines intersect. Use algebra to find the x- and y-coordinates of the point of intersection.

(a) $x = 3$ and $y = -2$
(b) $3x + 5y = 45$ and $x = -5$
(c) $5x + 3y = 45$ and $x = -3$

7. Graph both lines on the same axes. Put a dot where the lines intersect. Use algebra to find the x- and y-coordinates of the point of intersection.

(a) $x + y = 10$ and $x + 2y = 14$
(b) $y - 2x = 0$ and $x = 2$

8. Graph the line and half-plane corresponding to the inequality, a typical constraint from a mixture problem.

(a) $x \geq 6$
(b) $y \geq 4$
(c) $5x + 3y \leq 15$
(d) $4x + 5y \leq 30$

9. Graph the line and half-plane corresponding to the inequality, a typical constraint from a mixture problem.

(a) $x \geq 2$
(b) $y \geq 8$
(c) $3x + 2y \leq 18$
(d) $7x + 2y \leq 42$

In Exercises 10–12, for each description, write one or more suitable resource-constraint inequalities. The unknown to use for each product is given in parentheses.

10. (a) One bridesmaid's bouquet (x) requires 2 roses, and one corsage (y) requires 4 roses. There are 28 roses available.
(b) Maintaining a large tree (x) takes 1 hour of pruning time and 30 minutes of shredder time; maintaining a small tree (y) takes 30 minutes of pruning time and 15 minutes of shredder time. There are 40 hours of pruning time and 2 hours of shredder time available.

11. (a) Manufacturing one package of hot dogs (x) requires 6 oz of beef, and manufacturing one package of bologna (y) requires 4 oz of beef. There are 300 oz of beef available.
(b) It takes 30 ft of 12-in. board to make one bookcase (x); it takes 72 ft of 12-in. board to make one table (y). There are 420 ft of 12-in. board available.

12. Manufacturing one salami (x) requires 12 oz of beef and 4 oz of pork. Manufacturing one bologna (y) requires 10 oz of beef and 3 oz of pork. There are 40 lb of beef and 480 oz of pork available.

In Exercises 13–18, graph the feasible region, label each line segment bounding it with the appropriate equation, and give the coordinates of every corner point.

13. $x \geq 0;\ y \geq 0;\ 2x + y \leq 10$

14. $x \geq 0;\ y \geq 0;\ x + 2y \leq 12$

15. $x \geq 0;\ y \geq 0;\ 2x + 5y \leq 60$

16. $x \geq 10;\ y \geq 0;\ 3x + 5y \leq 120$

17. $x \geq 0;\ y \geq 4;\ x + y \leq 20$

18. $x \geq 2;\ y \geq 6;\ 3x + 2y \leq 30$

In Exercises 19–20, determine whether the points (2, 4) and/or (10, 6) are points of the given feasible regions of:

19. Exercises 13, 15, and 17.

20. Exercises 14, 16, and 18.

21. In the toy problem, x represents the number of skateboards and y the number of dolls. Using the version of that problem whose feasible region is presented in Figure 4.3b, with the profit formula $\$2.30x + \$3.70y$, write a sentence giving the maximum profit and describing the production policy that gives that profit.

22. In the toy problem, x represents the number of skateboards and y the number of dolls. Using the version of that problem whose feasible region is presented in Figure 4.3b, with the profit formula $\$5.50x + \$1.80y$, write a sentence giving the maximum profit and describing the production policy that gives that profit.

23. Graph both lines on the same axes. Put a dot where the lines intersect. Use algebra to find the x- and y-coordinates of the point of intersection.

(a) $5x + 4y = 22$ and $5x + 10y = 40$
(b) $x + y = 7$ and $3x + 4y = 24$

In Exercises 24–27, graph the feasible region, label each line segment bounding it with the appropriate equation, and give the coordinates of every corner point.

24. $x \geq 0;\ y \geq 0;\ 3x + y \leq 9;\ x + y \leq 7$

25. $x \geq 0;\ y \geq 0;\ 2x + y \leq 4;\ 4x + 4y \leq 12$

26. $x \geq 0;\ y \geq 2;\ 5x + y \leq 14;\ x + 2y \leq 10$

27. $x \geq 4;\ y \geq 0;\ 5x + 4y \leq 60;\ x + y \leq 13$

28. Determine whether the points (4, 2) and/or (1, 3) are points of the given feasible regions of Exercises 25 and 27.

29. Determine the maximum value of P given by $P = 3x + 2y$ subject to the constraints:
$x \geq 0$, $y \geq 0$, $x \leq 7$, and $y \leq 4$

30. A linear-programming problem has constraints given by:
$x \geq 0$, $y \geq 0$, $5x - y \leq 15$ and $4y + x \leq 24$

(a) What are the corner points of the feasible region for this LP problem?
(b) Sketch a graph of the feasible region.

4.2 Finding the Optimal Production Policy

4.3 Why the Corner Point Principle Works

4.4 Linear Programming: Life Is Complicated

31. A linear-programming problem has constraints given by:
$x \geq 0$, $y \geq 0$, $x \leq 4$ and $4y + 13x \leq 60$

(a) Which, if any, of the constraints involve vertical and horizontal lines?

(b) Sketch a graph of the feasible region.
(c) What are the corner points of the feasible region?
(d) If profit is given by the expression $P = 3x + 7y$,
i. What is the profit associated with the point (3, 1)?
ii. Which corner points have higher profit than (3, 1)?

32. Nuts Galore sells two spiced nut mixtures: Grade A and Grade B. Grade A requires 7 oz of peanuts for every 8 oz of almonds. Grade B requires 9 oz of peanuts for every 8 oz of almonds. There are 630 oz of peanuts and 640 oz of almonds available. Grade A makes Nuts Galore a profit of \$1.70 and Grade B makes a profit of \$2.40 per unit assembled. How many units of Grade A and Grade B nut mixtures should be made to maximize the company's profit, assuming that all units made can be sold?

33. Find the maximum value of P where $P = 3x + 2y$ subject to the constraints $x \geq 3$, $y \geq 2$, $x + y \leq 10$, $2x + 3y \leq 24$.

34. Find the maximum value of P where $P = 3x - 2y$ subject to the constraints $x \geq 2$, $y \geq 3$, $3x + y \leq 18$, $6x + 4y \leq 48$.

35. Find the maximum value of P where $P = 5x + 2y$ subject to the constraints $x \geq 2$, $y \geq 4$, $x + y \leq 10$.

36. Given profit $P = 21x + 11y$ subject to the constraints $x \geq 0$, $y \geq 0$, $7x + 4y \leq 13$:

(a) Graph the feasible region.
(b) Determine a corner point where there is an optimal solution.

(*Warning:* The corner point where the optimal solution occurs may not have integer values for both x and y.)

37. (a) Referring to Exercise 36, use the usual rounding rule to round the x-coordinate and the y-coordinate of the point where the optimal linear-programming solution occurs. Call the point with these coordinates Q.
(b) Determine if Q's coordinates define a feasible point by checking them against the constraints.
(c) Evaluate the profit value P at point Q. How does the profit value compare with the point where the optimal value occurred in Exercise 36?
(d) Let R be the point with coordinates (0, 3). Is R in the feasible region? Evaluate P at point R and compare the result with the answer at Q and where the optimum linear-programming value occurred.
(e) Explain the significance of the situation here for solving maximization problems where $P = ax + by$ (with a and b known in advance) is subject to linear constraints but where the variables must be nonnegative integers rather than arbitrary nonnegative decimal numbers.

Exercises 38–49 each have several steps leading to a complete solution to a mixture problem. Practice a specific step of the solution algorithm by working out just that step for several problems. The steps are:

(a) Make a mixture chart for the problem.
(b) Using the mixture chart, write the profit formula and the resource- and minimum-constraint inequalities.
(c) Draw the feasible region for those constraints and find the coordinates of the corner points.
(d) Evaluate the profit information at the corner points to determine the production policy that best answers the question.
(e) (Requires technology) Compare your answer with the one you get from running the same problem on a simplex algorithm computer program.

38. A clothing manufacturer has 600 yd of cloth available to make shirts and decorated vests. Each shirt requires 3 yd of material and provides a profit of \$5. Each vest requires 2 yd of material and provides a profit of \$2. The manufacturer wants to guarantee that under all circumstances, there are minimums of 100 shirts and 30 vests produced. How many of each garment should be made to maximize profit? If there are no minimum quantities, how, if at all, does the optimal production policy change?

39. A car maintenance shop must decide how many oil changes and how many tune-ups can be scheduled in a typical week. The oil change takes 20 min, and the tune-up requires 100 min. The maintenance shop makes a profit of \$15 on an oil change and \$65 on a tune-up. What mix of services should the shop schedule if the typical week has 8000 min available for these two types of services? How, if at all, do the maximum profit and optimal production policy change if the shop is required to schedule at least 50 oil changes and 20 tune-ups?

40. A clerk in a bookstore has 90 min at the end of each workday to process orders received by mail or on voice mail. The store has found that a typical mail order brings in a profit of \$30 and a typical voice-mail order brings in a profit of \$40. Each mail order takes 10 min to process and each voice-mail order takes 15 min. How many of each type of order should the clerk process? How, if at all, do the maximum profit and optimal processing policy change if the clerk must process at least three mail orders and two voice-mail orders?

41. In a certain medical office, a routine office visit requires 5 min of doctors' time and a comprehensive office visit requires 25 min of doctors' time. In a typical week, there are 1800 min of doctors' time available. If the medical office clears \$30 from a routine visit and \$50 from a comprehensive visit, how many of each should be scheduled per week? How, if at all, do the maximum profit and optimal production policy change if the office is required to schedule at least 20 routine visits and 30 comprehensive ones?

42. A bakery makes 600 specialty breads—multigrain or herb—each week. Standing orders from restaurants are for 100 multigrain breads and 200 herb breads. The profit on each multigrain bread is $8 and on herb bread, $10. How many breads of each type should the bakery make to maximize profit? How, if at all, do the maximum profit and optimal production policy change if the bakery has no standing orders?

43. A student has decided that passing a mathematics course will, in the long run, be twice as valuable as passing any other kind of course. The student estimates that passing a typical math course will require 12 hr a week to study and do homework. The student estimates that any other course will require only 8 hr a week. The student has 48 hr available for study per week. How many of each kind of course should the student take? (*Hint:* The profit could be viewed as 2 "value points" for passing a math course and 1 "value point" for passing any other course.) How, if at all, do the maximum value and optimal course mix change if the student decides to take at least two math courses and two other courses?

Exercises 44–47 require finding the point of intersection of two lines, each corresponding to a resource constraint.

44. The firm WebsAreUs creates and maintains Web sites for client companies. There are two types of Web sites: "Hot" sites change their layout frequently but keep their content for long times; "cool" sites keep their layout for a while but frequently change their content. Maintaining a hot site requires 1.5 hr of layout time and 1 hr for content changes. Maintaining a cool site requires 1 hr of layout time and 2 hr for content changes. Every day, WebsAreUs has 12 hr available for layout changes and 16 hr for content changes. Net profit is $50 for a set of changes on a hot site and $250 for a set of changes on a cool site. To maximize profit, how many of each type of site should WebsAreUs maintain daily? How, if at all, do the maximum profit and optimal policy change if the company must maintain at least two hot and three cool sites daily?

45. A paper recycling company uses scrap cloth and scrap paper to make two different grades of recycled paper. A single batch of grade A recycled paper is made from 25 lb of scrap cloth and 10 lb of scrap paper, whereas one batch of grade B recycle paper is made from 10 lb of scrap cloth and 20 lb of scrap paper. The company has 100 lb of scrap cloth and 120 lb of scrap paper on hand. A batch of grade A paper brings a profit of $500, whereas a batch of grade B paper brings a profit of $250. What amounts of each grade should be made? How, if at all, do the maximum profit and optimal production policy change if the company is required to produce at least one batch of each type?

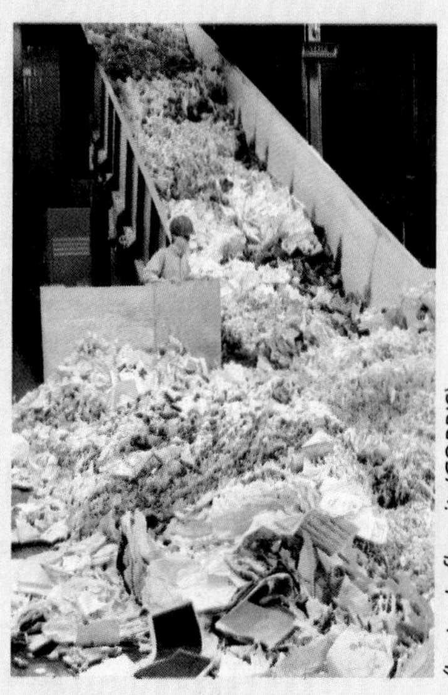
(Lester Lefkowitz/CORBIS)

46. Jerry Wolfe has a 100-acre farm that he is dividing into one-acre plots, on each of which he builds a house. He then sells the house and land. It costs him $20,000 to build a modest house and $40,000 to build a deluxe house. He has $2,600,000 to cover these costs. The profits are $25,000 for a modest house and $60,000 for a deluxe house. How many of each type of house should he build to maximize profit? How, if at all, do the maximum profit and optimal production policy change if Wolfe is required to build at least 20 of each type of house?

47. The maximum production of a soft-drink bottling company is 5000 cartons per day. The company produces regular and diet drinks and must make at least 600 cartons of regular and 1000 cartons of diet per day. Production costs are $1.00 per carton of regular and $1.20 per carton of diet. The daily operating budget is $5400. How many cartons of each type of drink should be produced if the profit is $0.10 per regular and $0.11 per diet? How, if at all, do the maximum profit and optimal bottling policy change if the company has no minimum required production?

48. Wild Things raises pheasants and partridges to restock the woodlands and has room to raise 100 birds during the season. The cost of raising one bird is $20 per pheasant and $30 per partridge. The Wildlife Foundation pays Wild Things for the birds; the latter clears a profit of $14 per pheasant and $16 per partridge. Wild Things has $2400 available to cover costs. How many of each type of bird should they raise? How, if at all, do the maximum profit and optimal restocking policy change if Wild Things is required to raise at least 20 pheasants and 10 partridges?

49. Lights Afire makes desk lamps and floor lamps, on which the profits are $2.65 and $4.67, respectively. The company has 1200 hr of labor and $4200 for materials each week. A desk lamp takes 0.8 hr of labor and $4 for materials; a floor lamp takes 1.0 hr of labor and $3 for materials. What production policy maximizes profit? How, if at all, do the maximum profit and optimal production policy change if Lights Afire wants to produce at least 150 desk lamps and 200 floor lamps per week?

In Exercises 50–53, there are more than two products in the problem. Although you cannot solve these problems using the two-dimensional graphical method, you can follow these steps:

(a) Make a mixture chart for each problem.

(b) Using the mixture chart, write the resource- and minimum-constraint inequalities. Also write the profit formula.

(c) (Requires software) If you have a simplex method program available, run the program to obtain the optimal production policy.

50. A toy company makes three types of toys, each of which must be processed by three machines: a shaper, a smoother, and a painter. Each Toy A requires 1 hr in the shaper, 2 hr in the smoother, and 1 hr in the painter, and brings in a $4 profit. Each Toy B requires 2 hr in the shaper, 1 hr in the smoother, and 3 hr in the painter, and brings in a $5 profit. Each Toy C requires 3 hr in the shaper, 2 hr in the smoother, and 1 hr in the painter, and brings in a $9 profit. The shaper can work at most 50 hr per week, the smoother 40 hr, and the painter 60 hr. What production policy would maximize the toy company's profit?

51. A rustic furniture company hand-crafts chairs, tables, and beds. It has three workers, Chris, Sue, and Juan. Chris can work only 80 hr per month, but Sue and Juan can each put in 200 hr. Each of these artisans has special skills. To make a chair takes 1 hr of Chris's time, 3 from Sue, and 2 from Juan. A table needs 3 hr from Chris, 5 from Sue, and 4 from Juan. A bed requires 5 hr from Chris, 4 from Sue, and 8 from Juan. Even artisans are concerned about maximizing their profit, so what product mix should the company stick with if it gets $100 profit per chair, $250 per table, and $350 per bed?

52. A candy manufacturer has 1000 lb of chocolate, 200 lb of nuts, and 100 lb of fruit in stock. The Special Mix requires 3 lb of chocolate, 1 lb each of nuts and fruit, and it brings in $10. The Regular Mix requires 4 lb of chocolate, 0.5 lb of nuts, and no fruit, and brings in $6. The Purist Mix requires 5 lb of chocolate, no nuts or fruit, and brings in $4. How many boxes of each type should be produced to maximize profit?

(Atlantide Phototravel/Corbis.)

53. A gourmet coffee distributor has on hand 17,600 lb of African coffee, 21,120 oz of Brazilian coffee, and 12,320 oz of Colombian coffee. It sells four blends—Excellent, Southern, World, and Special—on which it makes these per-pound profits, respectively: $1.80, $1.40, $1.20, and $1.00. One pound of Excellent is 16 oz of Colombian; it is not a blend at all. One pound of Southern consists of 12 oz of Brazilian and 4 oz of Colombian. One pound of World requires 6 oz of African, 8 oz of Brazilian, and 2 oz of Colombian. One pound of Special is made up of 10 oz of African and 6 oz of Brazilian. What product mix should the gourmet coffee distributor prepare to maximize profit?

In Exercises 54 and 55, use the fact that the corner point approach can also solve minimization problems to minimize the given expression for cost C.

54. Minimize C given by $C = 7x + 8y$ over the feasible region for Exercise 33.

55. Minimize C given by $C = 5x + 11y$ over the feasible region for Exercise 34.

56. Show by example that a feasible region that has the nonnegativity constraints $x \geq 0$, $y \geq 0$, and $x + y \leq 0.5$ can have no feasible points with integer coordinates other than (0, 0).

57. Courtesy Calls makes telephone calls for businesses and charities. A profit of $0.50 is made for each business call and $0.40 for each charity call. It takes 4 min (on average) to make a business call and 6 min (on average) to make a charity call. If there are 240 min of calling time to be distributed each day, how should that time be spent so that Courtesy Calls makes a maximum profit? What changes, if any, occur in the maximum profit and optimal production policy if they must make at least 12 business and 10 charity calls every day?

58. A refinery mixes high-octane and low-octane fuels to produce regular and premium gasolines. The profits per gallon on the two gasolines are $0.30 and $0.40, respectively. One gallon of premium gasoline is produced by mixing 0.5 gal of each of the fuels. One gallon of regular gasoline is produced by mixing 0.25 gal of high octane with 0.75 gal of low octane. If there are 500 gal of high octane and 600 gal of low octane available, how many gallons of each gasoline should the refinery make? How, if at all, do the maximum profit and optimal production policy change if the refinery is required to produce at least 100 gal of each gasoline?

59. A toy manufacturer makes bikes, for a profit of $12, and wagons, for a profit of $10. To produce a bike requires 2 hr of machine time and 4 hr painting time. To produce a wagon requires 3 hr machine time and 2 hr painting time. There are 12 hr of machine time and 16 hr of painting time available per day. How many of each toy should be produced to maximize profit? How, if at all, do the maximum profit and optimal production policy change if the manufacturer must produce at least two bikes and two wagons daily?

4.5 A Transportation Problem: Delivering Perishables

4.6 Improving on the Current Solution

60. Apply the Northwest Corner Rule, thereby finding a feasible solution that obeys the rim conditions, to the following transportation problem tableaux.

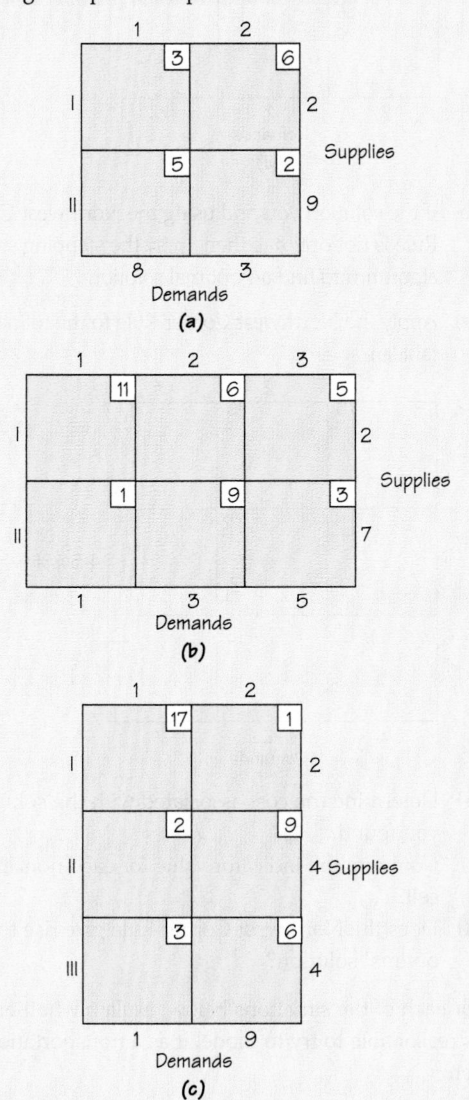

(d) For each tableau, give a possible real-world setting for the problem.

(e) For each tableau, find the cost of shipping using the cells that were circled when you used the Northwest Corner Rule.

61. (a) Apply the Northwest Corner Rule, thereby finding a feasible solution that obeys the rim conditions, to the accompanying tableau which arose from meeting the demands of fruit stands for peaches from supplies available from local orchards.

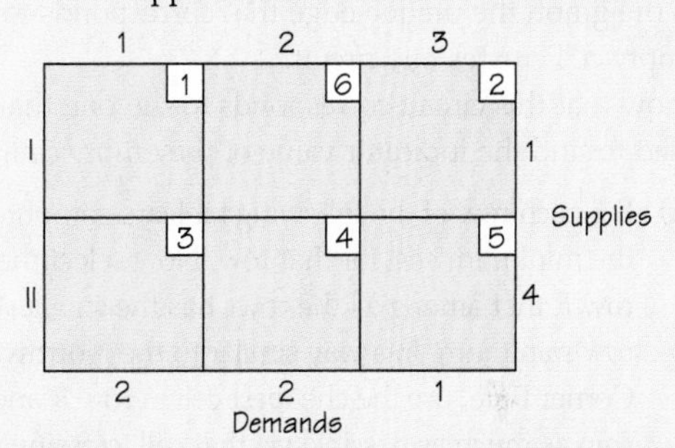

(b) Determine the cost associated with the solution that you found.

(c) Compute the indicator value for each noncircled cell.

62. The accompanying tableau represents the shipping costs and supply-and-demand constraints for supplies of purified water to be shipped to companies that resell the water to office buildings.

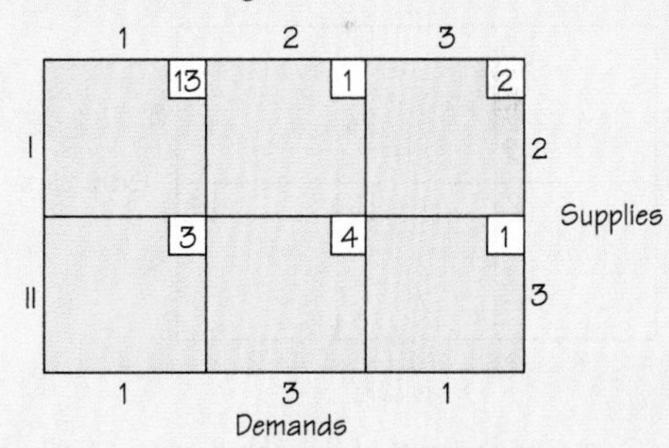

(a) Find the Northwest Corner Rule initial solution.

(b) Determine the indicator value for each noncircled cell.

(c) Is the current solution optimal? If not, find a cheaper solution.

63. The accompanying tableau arose by applying the Northwest Corner Rule.

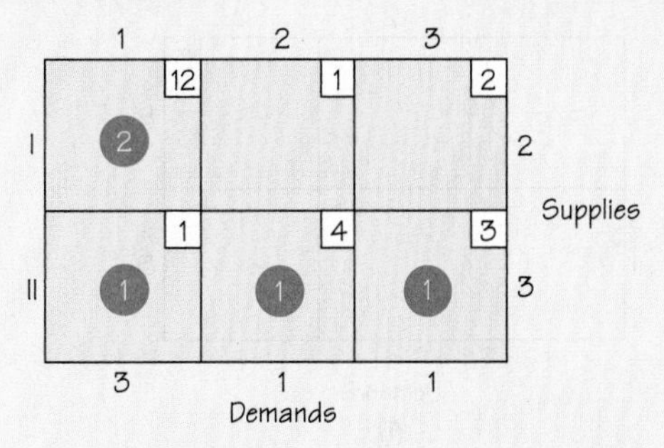

The accompanying graph was constructed so that there is one edge for each circled vertex in the tableau above.

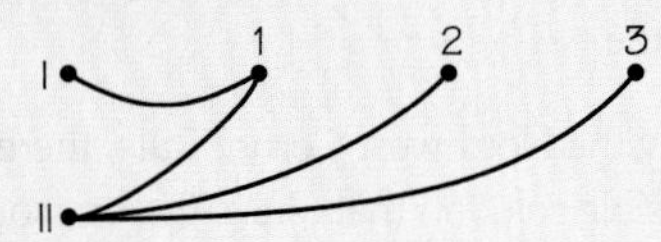

(a) Verify that the graph is a tree.

(b) Show that for each empty cell in the tableau, adding to the graph the unique edge that corresponds to the empty cell creates one circuit.

(c) Show that this circuit corresponds to the one that is used to find the indicator value of the empty cell.

64. **(a)** For each row of the following tableau, compute the minimum cost for that row. Now select the row R that among all the rows has the smallest row minimum. In a way similar to the Northwest Corner Rule, use the cheapest cell in row R and ship as much as possible via that cell, crossing out a row or a column, and adjust the rim conditions and repeat the process. This is known as the *minimum row entry method*. Use the minimum row entry method to find an initial solution to the following transportation problem, which shows the costs of returning rental cars from cities that have more cars than necessary to cities that have too few cars.

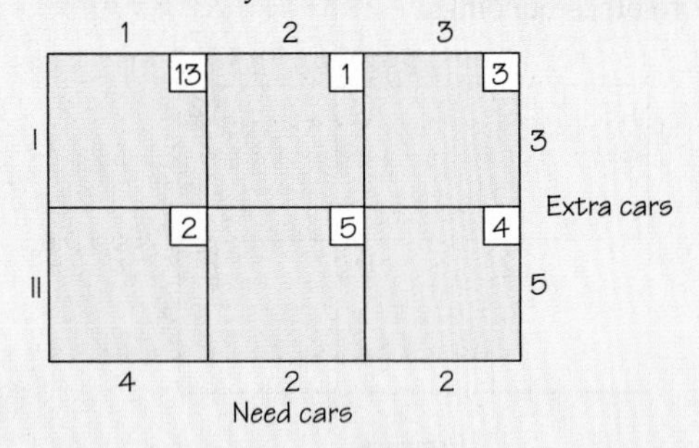

(b) Compute the cost of the solution you find using the minimum row entry method.

(c) Compare the cost found in part (b) with the cost of the initial solution obtained using the Northwest Corner Rule.

65. **(a)** For each of the following tableaux, find an initial solution using the Northwest Corner Rule.

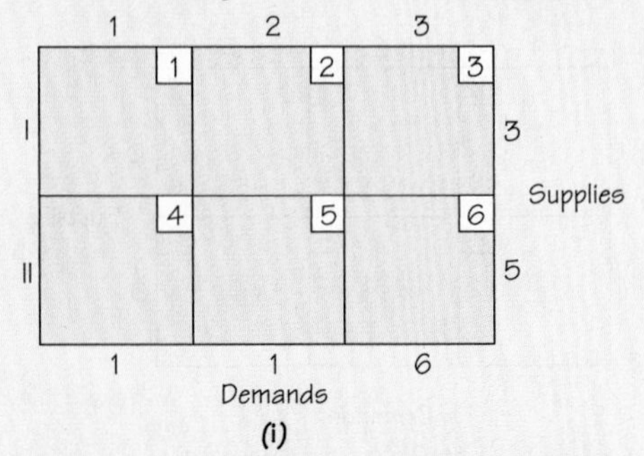

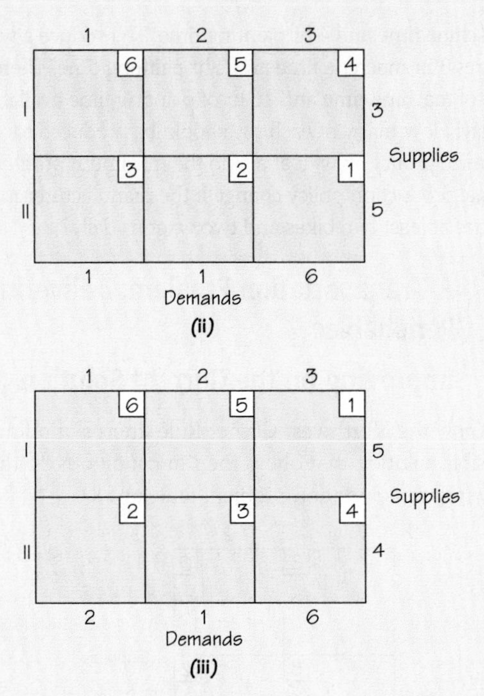

(b) If the solution you find using the Northwest Corner Rule is not optimal, then apply the stepping stone algorithm to find an optimal solution.

66. **(a)** Apply the Northwest Corner Rule to the following tableau.

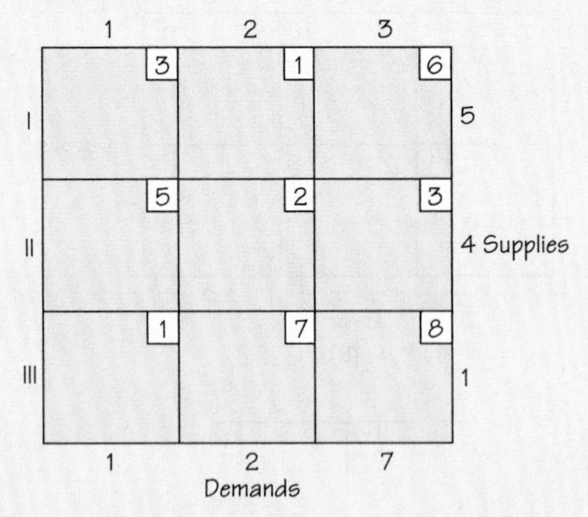

(b) Determine the cost associated with the solution you found.

(c) Compute the indicator value for each noncircled cell.

(d) Does the Northwest Corner Rule give rise to an optimal solution?

67. For each of the situations below, explain whether it seems reasonable to try to model it as a transportation problem.

(a) A supermarket chain is arranging to control costs in supplying the delivery of vegetables from its suppliers to its many branch stores.

(b) A mining company is trying to control the costs of repaving the roads that form the road network within the mine premises.

(c) A company is operating oil refineries to produce gasoline, as well as gasoline stations, to keep the cost of gasoline at the pump down.

WRITING PROJECTS

1. Interview a local businessperson who is in charge of deciding the product mix for a business. Must this business take into consideration situations other than minimum and resource constraints? If so, what are these considerations? Find out what methods the person uses to make production policy decisions. Is linear programming used? Are other methods used? If so, what are they? Write a report of your findings, and add some of your own conclusions about the usefulness of linear programming for this business.

2. In economics, it is often useful to distinguish between a firm that has a monopoly (for example, is the only supplier of a product) and firms that supply only a small share of the market. How would the presence of a monopoly affect the relation between production and price? Would the presence of a monopoly tend to ensure the fixed-profit assumption of linear programming, or would it make it more likely that the interplay of supply and demand would have to be considered in order to have a truly realistic model? Write an essay addressing these issues.

Suggested Readings

ANDERSON, DAVID R., DENNIS J. SWEENEY, and THOMAS A. WILLIAMS. *An Introduction to Management Science: Quantitative Approaches to Decision Making*, West, St. Paul, MN 1985. A business management text with seven chapters on linear programming.

DOLAN, ALAN, and JOAN ALDUS, *Networks and Algorithms: An Introductory Approach*, Wiley, NY 1993. A graph theoretical approach to network optimization problems, including the transportation problem.

GASS, SAUL I. *Decision Making, Models, and Algorithms*, Krieger, Melbourne, FL 1991. This book demonstrates how to use linear programming and related ideas to solve a variety of industrial and governmental problems.

HARDWICK, I., *Decision and Discrete Mathematics*, Albion Publishing, Chichester, England, 1996. A survey of situations that can be modeled using graphs in the area of operations research. It treats both the simplex method for solving linear-programming problems and the transportation problem.

Note: Simplex software can be found in *Maple* (keyword is *simplex*), *Mathematica* (keyword is *Linear Programming*), in both *Lotus 1-2-3* and *MSExcel* via *Solver*, and in other software packages, especially those intended for quantitative mathematics courses focusing on business applications.

Suggested Web Sites

www.informs.org This Web site is maintained by the Institute for Operations Research and the Management Sciences, the main professional organization in these fields in the United States. It contains information on (and/or links to) news items about operations research and management science and employment opportunities and summer internships; it also has a student newsletter. Much of the material is written in a nontechnical style.

www.hsor.org/what_is_or.cfm?name=linear_programming This Web page discusses how linear programming fits into the broader subject of operations research.

http://www-gap.dcs.st-and.ac.uk/~history/Biographies/Dantzig_George.html This site contains biographical information about George Dantzig, who, by developing the simplex method, greatly expanded the use and applicability of linear programming.

http://www.neos-guide.org/NEOS/index.php/Linear_Programming_FAQ This site is the "frequently asked questions" section of an online newsgroup for people interested in linear programming.

en.wikipedia.org/wiki/Linear_programming This Web page outlines the theory of linear programming.

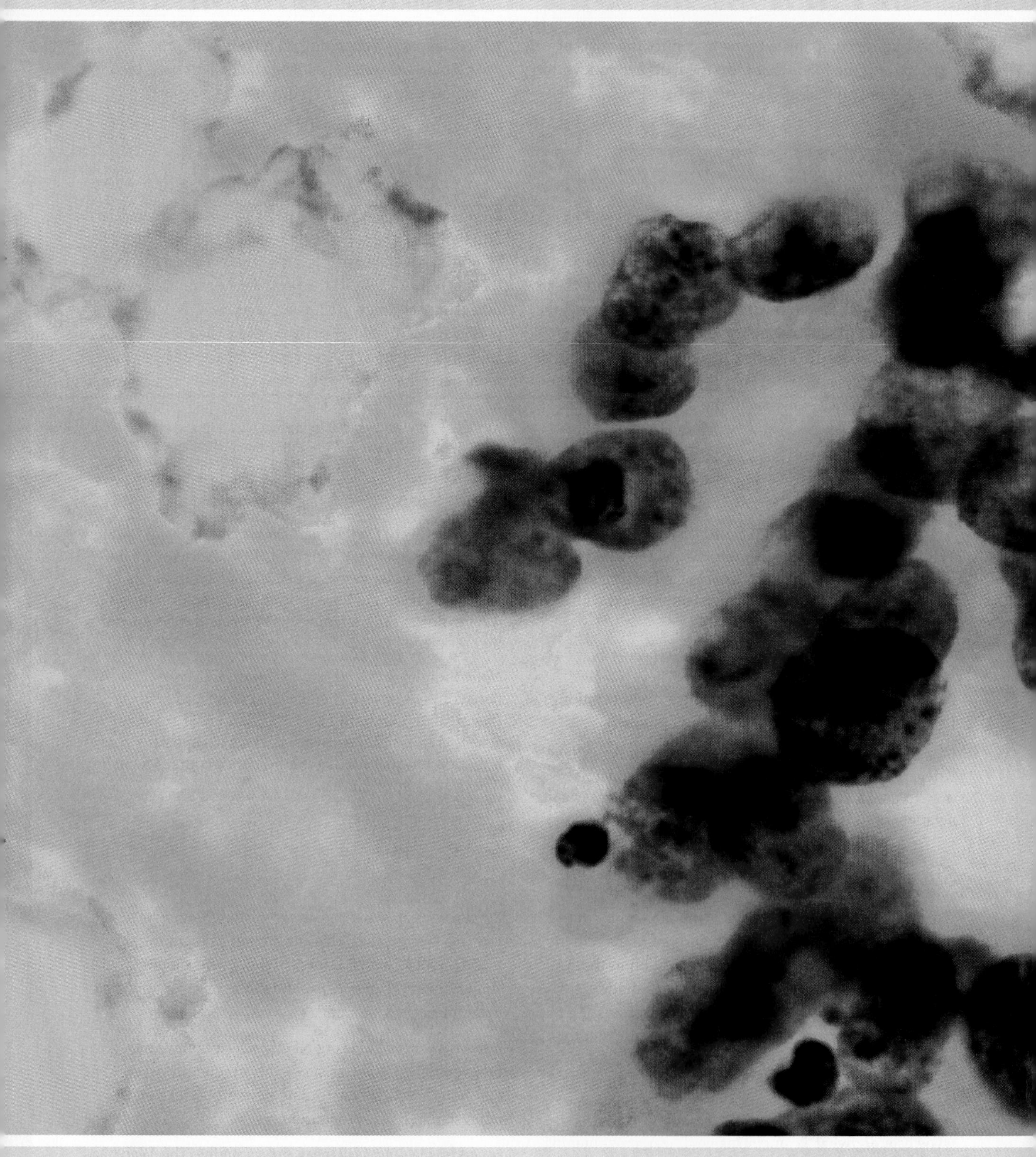

Spike Walker/Getty Images

Statistics: The Science of Data

Part II

Chapter 5
Exploring Data: Distributions

Chapter 6
Exploring Data: Relationships

Chapter 7
Data for Decisions

Chapter 8
Probability: The Mathematics of Chance

What books are big sellers this week? When you buy a book, the checkout scanner probably reports your choice to a company that tallies sales and reports the winners. Are there genetic differences between two related types of cancer? To find out, biologists use microarrays to report the activity of thousands of genes at once. Checkout scanners and microarrays produce immense amounts of *data*, numerical facts. So do opinion polls, medical studies, and even the sports pages. *Statistics* is the science of collecting, organizing, and interpreting data.

Chapters 5 and 6 concern *data analysis*, the art of studying what data say. We learn from data by making graphs and doing calculations, guided by principles that help us decide what graphs to make, what to look for in our graphs, and what calculations are helpful based on what we see. Sometimes we want to know more: An opinion poll or a medical study looks at only some people, but we want conclusions that apply to all voters or all patients. This is called *statistical inference* because we infer conclusions about a large group from data on a small part of the group. Chapter 7 discusses inference from beginning to end, from how to produce data when we have inference in mind to how to say just how much confidence we can have in our conclusions. Confidence, uncertainty, risk, chance—the mathematics that describes all these ideas is *probability theory*, the topic of Chapter 8. Probability is the mathematics behind statistical inference, but that's just a small part of its usefulness.

Topham/The Image Works

5

Exploring Data: Distributions

If a map included every pothole, traffic sign, and store throughout our road trip, it would be far too cluttered to be used or read easily. But if the map included only a couple of reference points, it would be too easy to get lost and miss our destination. So a good map gives us just the right level of detail, calling our attention to special features and main roads.

The undigested blizzard of data we encounter in modern society can feel overwhelming, like that first type of map. But if we simply ignore data, we risk the pitfall of the second type of map. Failure to detect patterns in data in a timely manner has had serious consequences, ranging from the loss of a NASA spacecraft to large-scale inappropriate financial trading practices that caused billions of dollars of losses. And so we need to develop good skills to "read" and appropriately summarize data so that we can navigate the terrain of information and numbers where we live and travel. Just as it helps for directions to have both numerical information (e.g., "3.2 miles on Gluckin Avenue") and visual diagrams or landmarks (e.g., "the right turn comes just after you pass the water tower"), it is important for data analysis to have both numerical and graphical techniques as well.

This chapter starts with an introduction to the concepts of exploring data from one quantitative variable. (Chapter 6 will look at two such variables at a time.) There is emphasis on the importance of graphical (not just numerical) techniques: histograms, stemplots, dotplots, and boxplots. The sequence of sections shows how the pattern of data values can be described by features such as its shape (e.g., skewed or symmetric distribution), any outliers, center (e.g., mean or median in Section 5.4), and variability (e.g., range and quartiles in Section 5.5, five-number summary and boxplots in Section 5.6, and standard deviation in Section 5.7). The chapter's concluding sections (5.8 and 5.9) highlight a particular shape for a data distribution known as a normal distribution. It is special because of its pervasiveness in statistics and its presence in the real world. Happy travels through statistics!

5.1 Displaying Distributions: Histograms

Any set of data contains information about some group of **individuals**. The information is organized in **variables**. We will briefly note some basics about data before we learn tools for exploring and summarizing.

Individuals DEFINITION

Individuals are the entities described by a set of data. Individuals may be people, but they may also be groups, animals, or things.

Variable DEFINITION

A **variable** is a particular characteristic or trait that can take on different values for different individuals. A particular variable may be either qualitative (e.g., gender) or quantitative (e.g., age), though our focus here is the latter.

Note that in statistics class, the term *variable* means something different from what it means in algebra class, where a variable x typically has a single fixed value to solve for in an equation such as $3x + 5 = 11$, or has a value that is determined exactly when another variable's value is given (such as $y = 3x + 5$ or $C = 2\pi r$).

EXAMPLE 1
Data from a Student Questionnaire

Figure 5.1 is a small part of a dataset that describes the students in a large statistics class. The data come from anonymous responses to a class questionnaire. Most data tables follow this format: each row records data on one individual (that is, one student) and each column contains values of one variable for all the individuals. This dataset appears in a *spreadsheet* program that has rows and columns ready for your use. Spreadsheets are commonly used to enter and transmit data, and spreadsheet programs also have functions for basic statistics.

Figure 5.1 Part of a dataset as displayed by the Microsoft Excel spreadsheet program.

	A	B	C	D	E	F
1	SEX	HAND	HEIGHT	STUDY	COINS	
2	F	L	65	200	50	
3	M	L	72	30	35	
4	M	R	62	95	35	
5	F	L	64	120	0	
6	M	R	63	220	0	
7	F	R	58	60	76	
8	F	R	67	150	215	
9						

Sheet1 / Sheet2 / Sheet3

There are seven individuals and five variables. Sex (female or male) and handedness (left-handed or right-handed) are variables that are usually described as *categorical* or *qualitative* because they categorize individuals by traits and do not take numerical values. The remaining three variables are *measurement* or *quantitative* variables because they do take numerical values. They are: height (inches), time spent studying (in minutes per weeknight), and "How many cents in coins are you carrying?" Our main focus will be on variables involving numerical data, because you have probably already had much experience with the usual ways to summarize categorical data (proportions, pie charts, and bar graphs).

Knowing the context of the data—that these are student responses to a class questionnaire—helps us make sense of them. For example, one student claimed to study 1500 minutes on a typical night. We know that this is impossible!

Statistical tools and ideas help us examine data to describe their main features. This examination is called **exploratory data analysis**. Like an explorer crossing unknown lands, we first want to describe simply what we see so that we can start to draft our map of the terrain. In this chapter and the next, we use both numbers and graphs to explore data. Here is a process for exploratory analysis of data.

Exploring Data PROCEDURE

1. Begin by examining each variable by itself, starting with one or more graphical summaries and then making numerical summaries of specific aspects of the data.
2. Explore possible relationships among variables, using graphical and then numerical summaries.

These principles also organize the material in Chapters 5 and 6. In this chapter, we look at data on a single variable. Chapter 6 moves on to relations among several variables. In each chapter, we first display data in graphs and then add numerical summaries.

Data analysis begins with graphical displays of the values of a single variable. For example, you may want to compare the study times claimed by female and male students. Because individual study times vary so much, we are interested in the **distribution** of study time for female and male students.

Distribution DEFINITION

The **distribution** of a variable gives information (as a table, graph, or formula) about how often the variable takes certain values or intervals of values.

The above definition has several different manifestations. The simplest one is like making a "tally chart" and is called a **frequency distribution**.

Frequency Distribution DEFINITION

The **frequency distribution** of a variable states all observed values of the variable and how many times the variable takes on each of these values.

A frequency distribution can be represented graphically with stemplots and dotplots (e.g., Figures 5.5 and 5.6 on pages 175 and 176, respectively) or in a table, such as this one for the variable "COINS" in the dataset excerpt of Figure 5.1:

Value	0	35	50	76	215
Frequency	2	2	1	1	1

A frequency distribution is most often used when full detail in the statistical "map" is needed so that every single value of the dataset remains in view.

However, sometimes that much detail is too distracting or time-consuming (especially with large datasets), and we look for a way to summarize the dataset. One way to reduce detail is to display not the absolute count (or raw frequency) of how many times each value occurs, but simply how often on a relative basis. This part-to-whole relationship can be expressed by a **relative frequency distribution**.

Relative Frequency Distribution

DEFINITION

The **relative frequency distribution** of a variable states all observed values of the variable and what fraction (or percentage) of the time each value occurs.

The easy way to convert a frequency distribution into a relative frequency distribution is to divide each frequency by the total number of values in the dataset. So, for the "COINS" example just mentioned (which has a sample size of 7), the relative frequency distribution could be shown with this table:

Value	0	35	50	76	215
Relative Frequency	$\frac{2}{7} \approx 29\%$	$\frac{2}{7} \approx 29\%$	$\frac{1}{7} \approx 14\%$	$\frac{1}{7} \approx 14\%$	$\frac{1}{7} \approx 14\%$

A relative frequency distribution can also be represented graphically, such as the one in Figure 5.17 on page 199.

Another way to reduce the full detail of a frequency distribution (to make it easier to see patterns) is to group neighboring data values into consecutive nonoverlapping intervals. This results in a **grouped frequency distribution**.

Grouped Frequency Distribution

DEFINITION

A **grouped frequency distribution** of a variable gives how many times in each interval of values the variable takes on a particular value.

A grouped frequency distribution can be displayed as either a table or as a graph known as a **histogram**.

Although a histogram and a bar graph both use rectangular bars, they are different because they display one quantitative variable and one qualitative variable, respectively. Only a histogram's bars start from an axis that represents a numerical scale.

Histogram

DEFINITION

A **histogram** is a graph of the distribution of outcomes (divided into consecutive nonoverlapping classes) for a single numerical variable. The height of each bar is the number of observations in the class of outcomes covered by the base of the bar. All classes should have the same width so that bars can be compared to each other accurately and readily.

We will now explore Example 2, where there is too much detail in Table 5.1 to find patterns and trends easily. Therefore, we will group the data into convenient intervals (or "classes") to make a grouped frequency distribution in Step 2 and its associated histogram in Figure 5.2.

EXAMPLE 2
Population Distribution

Every 10 years, the Census Bureau (www.census.gov) tries to contact every household in the United States. One finding of the 2010 Census was that the Hispanic population (which is now over 50 million) accounted for most of the nation's growth in the past decade. Table 5.1 presents the percent of adult residents (age 18 and over) in each of the 50 states who identified themselves in the 2010 Census as "Hispanic, Latino, or Spanish origin." Because we are interested in patterns at the state level, the *individuals* in this dataset are not the millions of Americans but the 50 states. The *variable* is the percent of Hispanics in a state's adult population. To make a histogram of the distribution of this variable, proceed as follows:

Making a Histogram

Step 1. Choose the classes. Divide the range of the data into some reasonable number of classes of equal width. The data in Table 5.1 range from 1.0 to 42.3, so here's one way to choose classes:

$$0.0 \leq \text{percent Hispanic} < 5.0$$
$$5.0 \leq \text{percent Hispanic} < 10.0$$
$$\vdots$$
$$40.0 \leq \text{percent Hispanic} < 45.0$$

Be sure to specify the classes precisely so that each individual falls into exactly one class. A state with 4.9% Hispanic residents would fall into the first class, but a state (like Arkansas) with 5.0% falls into the second.

Table 5.1

Percent of Adult Population of Hispanic Origin, by State (2010 Census)

State	Percent	State	Percent	State	Percent
Alabama	3.2	Louisiana	4.0	Ohio	2.5
Alaska	4.7	Maine	1.0	Oklahoma	7.1
Arizona	25.0	Maryland	7.3	Oregon	9.1
Arkansas	5.0	Massachusetts	8.1	Pennsylvania	4.6
California	33.1	Michigan	3.5	Rhode Island	10.2
Colorado	17.5	Minnesota	3.7	South Carolina	4.3
Connecticut	11.6	Mississippi	2.5	South Dakota	2.1
Delaware	6.7	Missouri	2.9	Tennessee	3.8
Florida	21.1	Montana	2.3	Texas	33.6
Georgia	7.5	Nebraska	7.2	Utah	11.3
Hawaii	7.2	Nevada	22.3	Vermont	1.3
Idaho	9.0	New Hampshire	2.2	Virginia	6.9
Illinois	13.4	New Jersey	16.3	Washington	8.9
Indiana	4.8	New Mexico	42.3	West Virginia	1.0
Iowa	3.8	New York	16.2	Wisconsin	4.6
Kansas	8.4	North Carolina	6.8	Wyoming	7.5
Kentucky	2.5	North Dakota	1.5		

Step 2. Count the individuals in each class to obtain the grouped frequency distribution:

Class	Count	Class	Count	Class	Count
0.0 to 4.9	22	15.0 to 19.9	3	30.0 to 34.9	2
5.0 to 9.9	15	20.0 to 24.9	2	35.0 to 39.9	0
10.0 to 14.9	4	25.0 to 29.9	1	40.0 to 44.9	1

Step 3. Draw the histogram. First, mark the scale for the variable whose distribution you are displaying on the horizontal axis. That's the percentage of a state's population who are Hispanic. The scale runs from 0 to 45 because that is the span of the classes we chose. The vertical axis contains the scale of counts. Each bar represents a class. The base of the bar covers the class, and the bar height is the class count. There is no horizontal space between the bars unless a class is empty, so that its bar has height zero. Figure 5.2 is our histogram.

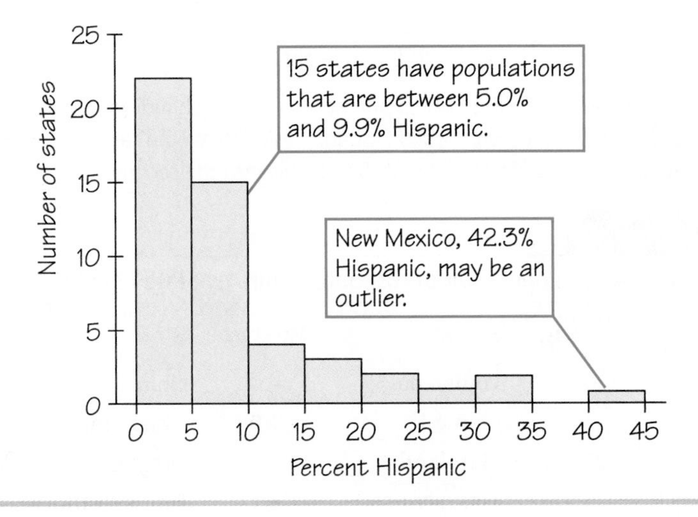

Figure 5.2 Histogram of the percentage of Hispanics among the adult residents of the states.

Our eyes respond to the *area* of the bars in a histogram. Because the classes are all the same width, area is determined by height and all classes are fairly represented. While something between 5 and 20 classes usually works for all real-world datasets, there is no one right choice for the number of the classes in a histogram. Subtle variation or the presence of more than one "hill" in the data may be hard to see in a graph where there is nothing but two or three huge bars. Similarly, it will be hard to see the main patterns in a graph where there are so many classes that most classes have one or zero observations. Neither choice will give a good picture of the shape of the distribution. You must use your judgment in choosing a number of classes that best communicates the shape of the distribution. Statistics software will choose (usually reasonably) the number of classes for you (using any of several methods, such as the square root of the sample size), but you can change it if you want. Sometimes it is more convenient to start by choosing a convenient class width (e.g., the width of 5 in Example 2) and seeing if that yields a number of classes that reveals a useful degree of detail in the data (and being prepared to increase or decrease this if it can be improved).

5.2 Interpreting Histograms

Making a statistical graph is not an end in itself. The purpose of the graph is to help us understand the data. After you make a graph, always ask, "What do I see?" Once you have displayed a distribution, you can see its important features as follows.

Outlier DEFINITION

In any graph of data, look for the overall pattern and for striking deviations from that pattern. You can describe the overall pattern of a distribution by its shape, center, and variability. An important kind of deviation is an **outlier,** an individual value that falls outside the overall pattern.

We will soon learn how to describe center and variability numerically. For now, you can describe the center of a distribution by its middle value, with roughly half the observations taking smaller values and half taking larger values. You can give a rough description of the variability of a distribution by giving the *smallest and largest values*.

EXAMPLE 3
Describing a Distribution

Look again at the histogram in Figure 5.2. **Shape**: The distribution has a *single peak,* which represents states in which less than 5 percent of adults are Hispanic. Most states have no more than 10 percent Hispanics, but some states have much higher percentages, so that the graph trails off to the right. **Center**: Table 5.1 shows that about half the states have less than 7 percent Hispanics among their adult residents and the rest have more. So the middle of the distribution is almost 7 percent. **Variability**: The data's span is from about 1 to 42 percent, but only six states exceed 20 percent. **Outliers**: California, New Mexico, and Texas stand out. Whether these are outliers or just part of the long right tail of the distribution is a matter of judgment, perhaps taking into account that those three states are heavily Hispanic by history and location.

Some statistical software packages "flag" outlier values using methods such as those in Exercises 31 or 54, but there is no one universal rule for calling an observation an outlier. Once you have spotted possible outliers, look for an explanation. Some outliers are due to mistakes, such as typing 1.8 as 18 or as 8.1. Other outliers point to the special nature of some observations.

When you describe a distribution, concentrate on the main features. Look for major peaks, not for minor ups and downs, in the bars of the histogram. Look for clear outliers, not just for the smallest and largest observations. Look for rough *symmetry* or clear departures from it.

Some variables have distributions with predictable shapes. For example, some distributions may have a shape in which the bulk of the values form a heap on one side, close to the distribution's balance point, and the rest of the values stretch out into a long tail on the other side of the balance point. This trait is known as **skewness,** and the direction of skewness may be either to the right or to the left:

Right-Skewed Distribution

DEFINITION

A **right-skewed distribution** is a distribution in which the longer tail of the histogram is on the right side. (Because positive numbers lie on the right side of a number line, such a distribution is also called "positively skewed.") For example, see Figure 5.2 or Figure 5.7.

Left-Skewed Distribution

DEFINITION

A **left-skewed distribution** is a distribution in which the longer tail of the histogram is on the left side. (Because negative numbers lie on the left side of a number line, such a distribution is also called "negatively skewed.") For example, see Figure 5.21.

Let's consider examples of each. Incomes, house prices, and other money amounts usually have right-skewed distributions because there are always a few CEOs and celebrities well to the right of most of us! On the other hand, a simple test designed to measure basic achievement may yield a left-skewed distribution because most students will cluster together with high scores, but there are usually still a few people who perform low (e.g., due to lack of attendance or effort) and give the distribution a tail stretching out to the left.

Other distributions may have little or no skewness. For example, the distribution of heights (or handspans) in an adult population may look like two hills of equal size (picture a two-humped camel) if males cluster around one value and females cluster around another. A more common and more important shape without skewness is the bell-shaped histogram, which will be revisited in Section 5.8. The bell shape is yielded by most standardized tests (e.g., Figure 5.3), as well as by many biological measurements (such as height, length of thigh bone, and so on) on specimens from the same species and sex. Distributions without much skewness where values are distributed similarly on both sides of the distribution can typically be described as **symmetric**:

Symmetric Distribution DEFINITION

A distribution is **symmetric** if a vertical line could be superimposed on the histogram and have the left and right sides be approximate mirror images of each other.

EXAMPLE 4
Iowa Test Scores

Figure 5.3 displays the scores of all 947 seventh-grade students in the public schools of Gary, Indiana, on the vocabulary part of the Iowa Test of Basic Skills. The distribution is *single-peaked* and *symmetric*. In mathematics, the two sides of symmetric patterns are exact mirror images, but real-life data are almost never exactly symmetric. We are content to describe Figure 5.3 as symmetric. The center (half above, half below) is close to 7. This is a seventh-grade reading level. The scores range from 2.0 (second-grade level) to 12.1 (twelfth-grade level).

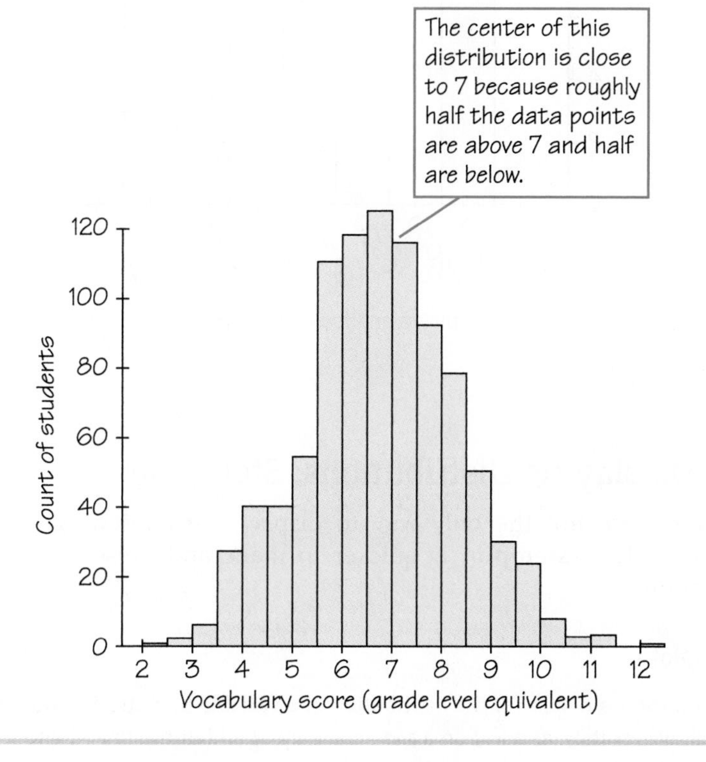

Figure 5.3 Histogram of the Iowa Test of Basic Skills vocabulary scores for all 947 seventh-grade students from Gary, Indiana.

EXAMPLE 5
College Tuition

Jeanna plans to attend college in her home state of Massachusetts. She looks up the tuition and fees for all 55 four-year colleges in Massachusetts (omitting art schools and other special colleges). The data for a recent academic year (see Exercise 23 on page 202) are displayed in a histogram (Figure 5.4). For example, the tallest bar tells us that there are 16 colleges charging between $34,000 and $38,000. As is often the case, we can't call this irregular distribution either symmetric or skewed. It does show two separate *clusters* of colleges, 11 with tuition less than $10,000 and the remaining 44 costing more than $14,000. Clusters suggest that two types of individuals are mixed in the dataset. In fact, the histogram distinguishes the 11 state colleges in Massachusetts from the 44 private colleges, which charge much more.

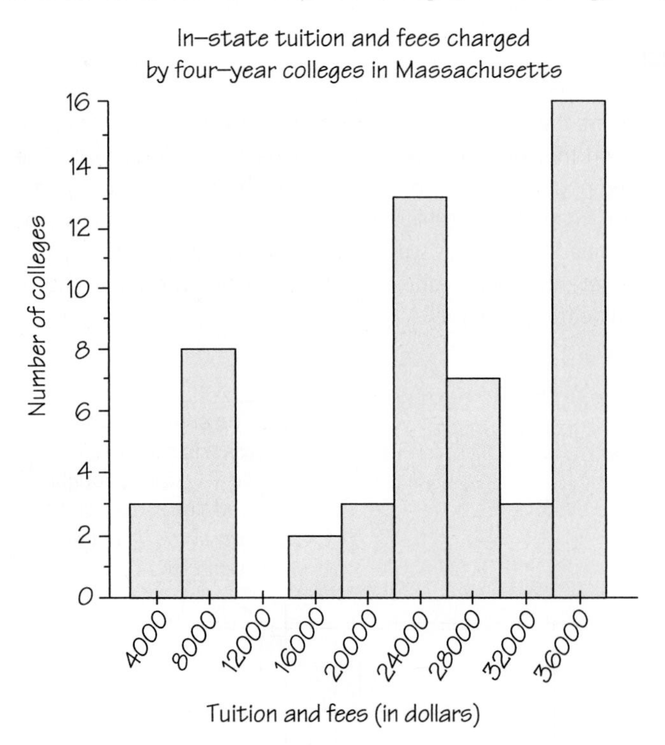

Figure 5.4 Histogram of the tuition and fees charged by four-year colleges in Massachusetts.

5.3 Displaying Distributions: Stemplots

Histograms are not the only way to display distributions graphically. For small datasets, a **stemplot** is quicker to make and presents more detailed information.

Stemplot DEFINITION

A **stemplot** is a display of the distribution of a variable that attaches the final digits of each observation as a *leaf* on a *stem* made up of all but the final digit.

Making a Stemplot

PROCEDURE

To make a **stemplot**:

1. Separate each observation into a *stem* consisting of all but the final (rightmost) digit and a *leaf*, the final digit. (To have meaningful stems, it may be necessary to round or truncate the observed values.) Stems may have as many digits as needed, but each leaf contains only a single digit.
2. Write the stems in a vertical column with the smallest at the top, and draw a vertical line at the right of this column. Include all stems, even if they are not used.
3. Write each leaf in the row to the right of its stem, in increasing order out from the stem.

EXAMPLE 6
Stemplot of "Percent Hispanic"

For the "percent Hispanic" percents in Table 5.1, take the whole-number part of the percent as the stem and the final digit (in this case, the tenths place) as the leaf. The Connecticut entry, 11.6%, has stem 11 and leaf 6. Utah, at 11.3%, places leaf 3 on the same stem. These are the only observations on this stem. We then arrange the leaves in ascending order, as 36, so that 11|36 is one row in the stemplot. Figure 5.5 is the complete stemplot for the data in Table 5.1.

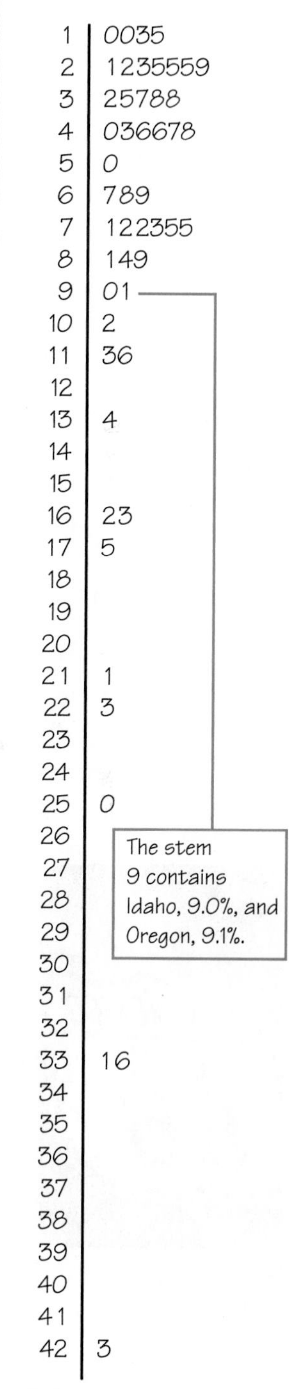

Figure 5.5 Stemplot of the percent of Hispanics among the adult residents of the U.S. states.

If we rotate Figure 5.5 a quarter-turn counterclockwise, the stemplot would look like a histogram (of a distribution skewed to the right). Comparing the stemplot in Figure 5.5 with the histogram in Figure 5.2 reveals the strengths and weaknesses of stemplots. The stemplot, unlike the histogram, preserves the actual value of each observation. But you can choose the classes in a histogram, whereas the classes (the stems) of a stemplot are forced on you. Whether the large number of classes in Figure 5.5 is an improvement over Figure 5.2 is a matter of taste. Stemplots do not work well for large datasets, like the 947 Iowa Test scores in Figure 5.3, because some stems must hold such a large number of leaves.

When the observed values have many digits, it is often best to *round* the numbers to just a few digits before making a stemplot. For example, a stemplot of data containing observations such as

3.468   2.567   2.981   1.095

would have very many stems and most stems would have only 1 or 0 leaves. You can round such values to

3.5   2.6   3.0   1.1

before making a stemplot.

Graphical summaries are good for analyzing the shape of distribution of values. To answer precise questions about features of a dataset, such as its center, however, it helps to have numerical summaries as well. We explore this next to help us obtain a statistical "map" of our data with just the right degree of detail.

5.4 Describing Center: Mean and Median

What kind of gas mileage do you get with the new cars in the Environmental Protection Agency's "midsized cars" category? Table 5.2 gives the city and highway gas mileage for a representative sample of 2011 midsized cars from www.fueleconomy.gov.

Table 5.2

Fuel Economy (Miles per Gallon) for Model Year 2011 Vehicles

Model	City mpg	Highway mpg
Acura RL	17	24
BMW 550i	17	25
Buick Regal	19	30
Cadillac STS	18	27
Chevrolet Malibu	22	33
Ford Fusion AWD	17	24
Infiniti M56	16	25
Kia Optima M6	24	35
Lexus GS350 AWD	18	25
Mitsubishi Galant	21	30
Nissan Maxima	19	26
Toyota Camry	22	32
Toyota Prius	51	48

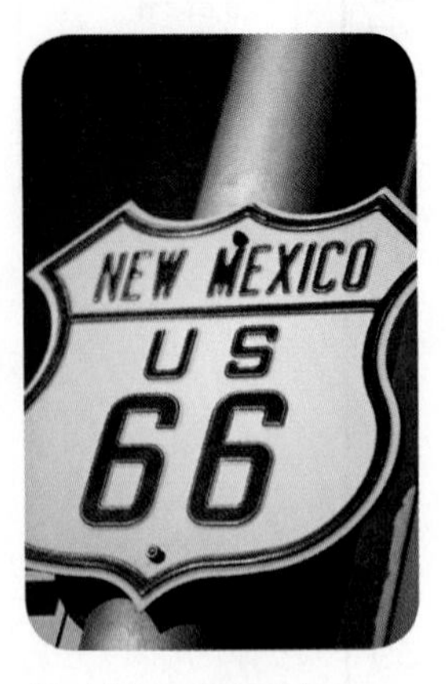

(Corbis/Punchstock.)

We start with graphs. Figure 5.6 is a *dotplot* of the city mileages of the 13 cars in the sample of midsized cars. As is often the case when there are only a few observations, the shape of the distribution is irregular. The most striking feature is a high outlier on the right end of the dotplot. Upon closer examination, to see if the outlier value may be a typographical error, we see that it's the Toyota Prius—the only hybrid gas-electric car in the sample.

Numerical summaries make the comparison that we want more specific. A numerical description of a distribution begins with a measure of its center. The two most common measures of center are the **mean** and the **median**. Basically, the mean is the arithmetic "average value" and the median is the "middle value."

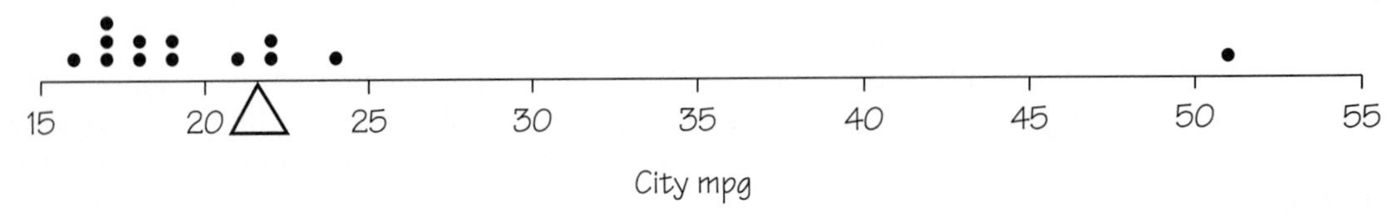

Figure 5.6 Dotplot of the city gas mileages of the sample of midsized cars. The Toyota Prius is an outlier.

We need to explore the precise procedures for calculating these measures and observe how they behave differently.

Finding the Mean $\bar{x}$ PROCEDURE

To find the **mean** of a set of observations:

1. Find the sum of the values.
2. Divide the sum by the number of values.

If the n observations are $x_1, x_2, \ldots, x_n$, the formula for the mean is

$$\bar{x} = \frac{x_1 + x_2 + \cdots + x_n}{n}$$

The common notation for the mean of all the x-values is a bar over the x, and $\bar{x}$ is pronounced "x-bar."

One way to visualize the value of the mean of a dataset is to imagine where the fulcrum would have to be placed for its dotplot to "balance." This analogy tells us the mean is always between the largest and smallest values, and by visual inspection, we can estimate further that the balance point appears to be somewhere between 20 and 25. Let's see what exact value the formula yields.

Using Formulas **Algebra Review**

A formula is a statement that claims that two expressions are equal. When dealing with formulas, it is important to know what each variable represents. Generally, formulas are case-sensitive, which means that an uppercase X may represent a different variable than a lowercase x. If more than one variable from the same type of variables are needed, then subscripts are often used. For an example related to the use of lines in Chapter 6, the slope of a line is determined by two points often expressed as (x_1, y_1) and (x_2, y_2). The formula that relates these two points to the slope of the line that passes between them, commonly denoted as m, is:

$$m = \frac{y_2 - y_1}{x_2 - x_1}$$

In most formulas, the variable on the left is found by substituting for the variable(s) on the right.

The following three steps occur when evaluating a formula to determine the numerical value of one of the variables:

- First, understand what the variables in the formula represent.
- Second, carefully substitute the values that are given into the formula.
- Third, apply order of operations to simplify the expression. The order of operation tells us to do parentheses first, working from the innermost out. Next, we do any exponents. Then we do any multiplication or division, going from left to right, and finally any addition or subtraction, going from left to right. When a fraction is involved, we treat the numerator and denominator as each being enclosed by parentheses.

For example, if we want to calculate the slope of the line that passes through (2, 5) and (−3, 4), we:

- First, assign a meaning to the numbers:

  $(x_1, y_1) = (2, 5)$ and $(x_2, y_2) = (-3, 4)$

- Second, substitute the constants for the variables:

$$m = \frac{y_2 - y_1}{x_2 - x_1} = \frac{4 - 5}{-3 - 2}$$

- Third, simplify by applying order of operations:

$$m = \frac{y_2 - y_1}{x_2 - x_1} = \frac{4 - 5}{-3 - 2} = \frac{-1}{-5} = \frac{1}{5}$$

EXAMPLE 7
Calculating the Mean

The mean city mileage for the 13 midsized cars in Table 5.2 is

$$\begin{aligned}\bar{x} &= \frac{x_1 + x_2 + \cdots + x_n}{n} \\ &= \frac{17 + 17 + 19 + 18 + 22 + 17 + 16 + 24 + 18 + 21 + 19 + 22 + 51}{13} \\ &\approx 21.6 \text{ mpg (miles per gallon)}\end{aligned}$$

We said that the Toyota Prius may not belong with the other cars. If we exclude the Prius, the mean city mileage drops to $\frac{230}{12} \approx 19.2$ mpg. The single outlier adds more than 2 mpg to the mean city mileage. This illustrates an important weakness of the mean as a measure of center: *The mean is sensitive to the influence of extreme observations.* These may be outliers, but a skewed distribution that has no outliers will also pull the mean toward its long tail.

We have used the middle of a distribution as an informal measure of center. The *median* is the formal version of the middle, with a specific rule for calculation. The **median** M is a number where half the observations are smaller and the other half are larger.

Finding the Median M PROCEDURE

To find the **median** of a set of observations:

1. Arrange all observations (including any repeated values) in increasing order (from smallest to largest).
2. If the number of observations is *odd*, the median M is the center observation in the ordered list.

   If the number of observations is *even*, the median M is the mean of the two center observations in the ordered list.

EXAMPLE 8
Calculating the Median

Since we're exploring the gas mileage cars get on the road, you might notice the connection that just as a median divides a road into two halves (with opposite directions of travel), a median divides a dataset into two halves! To find the median city mileage for 2011 midsized cars, arrange the data in increasing order:

16 17 17 17 18 18 **19** 19 21 22 22 24 51

The median is the bold 19, which you can find by eye—there are six observations to the left and six to the right. Or visualize (or construct) a long paper strip divided into 13 equal-sized squares, where the sorted values are written into the squares. When the strip is folded in half, the fold line falls on the median!

What happens if we drop the Toyota Prius? The remaining 12 cars have the following city mileages:

16 17 17 17 18 **18** **19** 19 21 22 22 24

Because the number of observations $n = 12$ is even, there is no single center observation. There is a center *pair* of observations (18 and 19) that has five observations to its left and five to its right. The median M is the mean of the center pair, which is $(18 + 19)/2 = 18.5$.

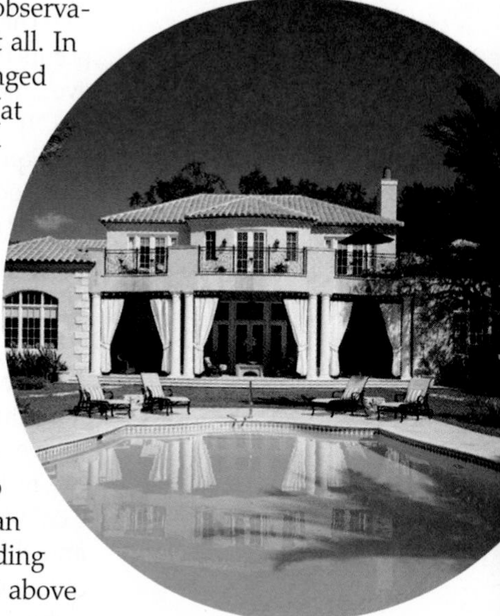
(Dan Forer/Beateworks/Corbis.)

You see that the median resists the influence of extreme observations better than the mean does. A very high value like the Toyota Prius is simply one observation to the right of center, and removing it hardly changed the median at all. In fact, removing an extreme outlier can leave the median completely unchanged while significantly changing the mean. The *Mean and Median* applet (at www.whfreeman.com/fapp9e) is an excellent way to compare the resistance of M and $\bar{x}$. See Applet Exercises 1 and 2.

The median and mean are the most common measures of the center of a distribution. The mean and median are close together in a roughly symmetric distribution and are equal in a perfectly symmetric distribution. In a skewed distribution, the mean is generally farther out in the long tail than is the median (see Figure 5.7). For example, consider the right-skewed distribution of house prices. There are mostly moderately priced houses, with only a few expensive mansions. (Relative to a typical house, there's a lot more room on the number line for extremely high values than extremely low values because a house price could go well into the millions but could never be less than 0.) The mansions increase the mean but do not affect the median. The mean sales price of new homes (including the land) in the United States sold in 2010 was \$271,600, which is well above (i.e., to the right of) the median sales price of \$221,900.

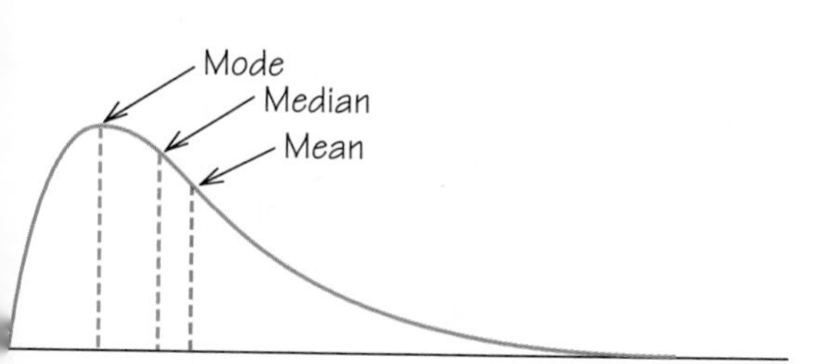

Figure 5.7 The smooth curve describes a right-skewed distribution with one mode. The mean is pulled toward the long tail more than the median is.

Another common numerical summary of a distribution is the **mode**—the most frequently occurring value. Like the mean and median, it is a measure of location for a distribution. However, the mode is not necessarily a good measure of center. Consider a skewed histogram with its maximum height at one end and a long tail on the other end, or a symmetric histogram with a tall peak at each end and a valley in between!

SPOTLIGHT 5.1

Which Mean Do You Mean?

The word "average" can be ambiguous—sometimes used to refer to the mean, sometimes the median, etc. But even if it is clear that we are using the mean, we still need to say what unit is the basis for our averaging. Suppose a tiny school has only 4 classrooms and their class sizes are 10, 3, 4, and 3. Most people would say this school's "average (mean) class size" is $(10 + 3 + 4 + 3)/4 = 5$. But that's not how it seems to the students, half of whom are in a class twice that big! So we could find the mean from the student point of view by taking the mean of what each of the 20 students would report as the size of his or her class. Verify that this yields $134/20 = 6.7$, a number 34 percent larger than 5. It turns out the per-student mean is *always* at least as large as the per-class mean, which is good to know when viewing statistics about schools you are considering attending. (More information on this example can be found in Lawrence Lesser's article "Sizing Up Class Size," in the January 2010 *Mathematics Teacher*.)

Mode DEFINITION

The **mode** is the most frequently occurring value in a set of numerical observations. It is possible for a dataset to have no mode, one mode, or more than one mode.

The mode of the city mileage numbers in Table 5.2 would be 17, because it occurs three times and no other value occurs more than twice. Some datasets, however, may have two or more values that "tie" for the honor of being a mode, such as the "0" and "35" values for COINS in Figure 5.1 (p. 166). (The mode can also work with qualitative data: the mode for sex in Figure 5.1 is female.) We can also identify the mode(s) of a histogram. For example, Figure 5.4 (p. 174) does not give the raw data of individual values, but we can say that the modal class is $34,000–$38,000.

5.5 Describing Variability: Range and Quartiles

The mean and median provide two different measures of the center of a distribution. But a measure of center alone can be limiting. It would not be comfortable to live in a home with a mean temperature of 70 °F if half of the days, it's 40° and the rest of the days, it's 100°! Two neighborhoods with a median house price of $193,000 can still be quite different if one has both mansions and modest homes and the other has little variation among houses. We are interested in the variability of house prices, as well as in their centers. *A useful numerical description of a distribution needs to consist of both a measure of center and a measure of variability.*

The simplest way to measure variability is with the **range,** which is the difference between the smallest and largest observations. For example, the percents of Hispanics in the states are as low as 1.0% (Maine or West Virginia) and as high as 42.3% (New Mexico), so the range would be 42.3% − 1.0% = 41.3%. Likewise, the range of the city mileage numbers in Table 5.2 (p. 176) is 51 − 16 = 35 mpg. The range tells us the full span of the data, but it may be greatly affected by an outlier. Without the Toyota Prius, the preceding answer becomes 24 − 16 = 8 mpg.

Range DEFINITION

The **range** is a measure of variability of a set of observations. It is obtained by subtracting the smallest observation from the largest observation. Equivalently: range = maximum − minimum.

We can improve our description of variability by looking at the variability of the middle half of the data. The first and third **quartiles** delineate the middle half. At the end of the first quarter of a football game, one quarter of the game is complete. Similarly, the first quartile of a distribution or dataset is the point that exceeds one-quarter (or 25 percent) of the values. Q_1 is also the 25th percentile. The third quartile is the point that exceeds three-quarters (or 75 percent) of the values. (You usually won't hear the phrase "second quartile" because it's equivalent to something we already named: the median!) The quartiles break the dataset into four groups with equal numbers of observations. To make the idea of quartiles more exact, we need a procedure to find them.

Finding the Quartiles Q_1 and Q_3 PROCEDURE

To find the **quartiles:**

1. Arrange all observations (including any repeated values) in increasing order.
2. Use the median to split the ordered dataset into two halves—an upper half and a lower half. (If the number of values is odd, don't include the middle position in either half.)
3. The **first quartile, Q_1,** is the median of the lower half.
   The **third quartile, Q_3,** is the median of the upper half.

EXAMPLE 9
Finding Quartiles

The city mileages of the 12 gasoline-powered midsized cars, after sorting, are:

16 17 17 17 18 18 19 19 21 22 22 24

The first quartile is the median of the six observations in the lower half, so $Q_1 = 17$. Similarly, the third quartile is the median of the upper half: $Q_3 = 21.5$.

For an example with an odd number of observations, try the city mileages of all 13 midsized cars in Table 5.2. Below are the mileages in increasing order, with **bold** used to denote the median, which will be excluded to form two equal-sized groups:

16 17 17 17 18 18 **19** 19 21 22 22 24 51

Ignoring the bold **19**, we find the quartiles by finding the median of each half of the dataset: $Q_1 = 17$ and $Q_3 = 22$.

Some software packages or calculators may use a slightly different procedure to find the quartiles, so their results may be a bit different from your own work. Don't worry about this. The differences will be too small to be important.

5.6 The Five-Number Summary and Boxplots

We started by using the smallest and largest observations to indicate the variability of a distribution. These single observations tell us little about the distribution as a whole, but they give information about the tails of the distribution that is missing if we know only Q_1, M, and Q_3. To get a quick summary of both center and variability, combine all five numbers.

Five-Number Summary DEFINITION

The **five-number summary** of a distribution consists of the smallest observation, the first quartile, the median, the third quartile, and the largest observation, written in order from smallest to largest. In symbols, the five-number summary is

Minimum Q_1 M Q_3 Maximum

These five numbers offer a reasonably complete description of center and variability. For the 13 midsized cars (see Table 5.2 on p. 176), you can verify that the five-number summary for city gas mileage is

16 17 19 22 51

and the summary for highway gas mileage is

24 25 27 32.5 48

A **boxplot** can visually represent the variability of the data across these groups from the five-number summary. Figure 5.8 shows boxplots for both city and highway gas mileages for midsized cars.

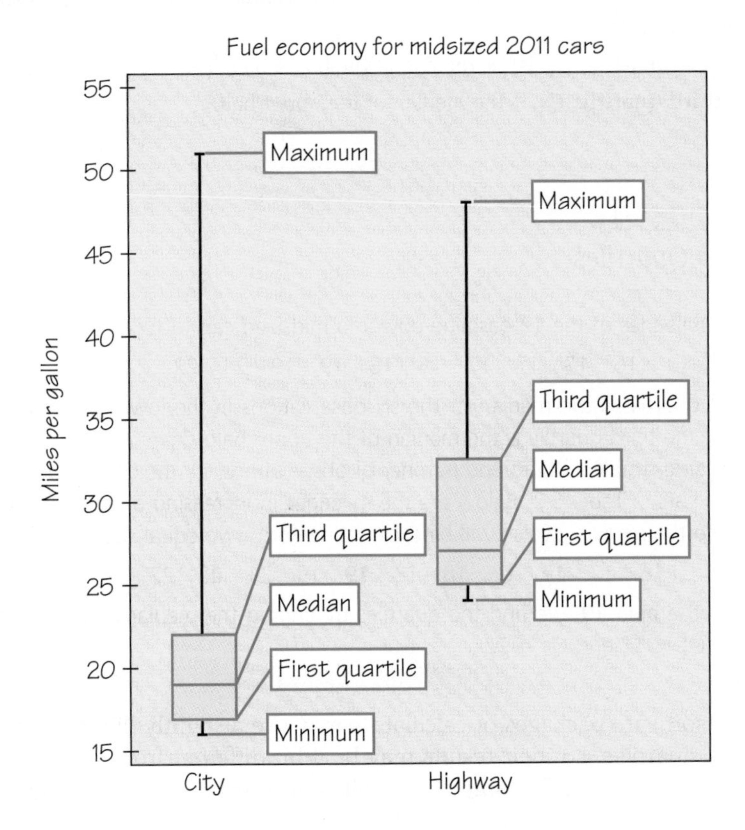

Figure 5.8 Boxplots of the highway and city gas mileages for 13 cars classified as midsized by the Environmental Protection Agency. These boxplots are drawn vertically, but it is equally correct to draw them horizontally.

Boxplot DEFINITION

A **boxplot** is a graph of the five-number summary.

- A central box spans the quartiles Q_1 and Q_3.
- A line somewhere inside the box (often, but not always, near the middle) marks the median M of the dataset.
- Lines extend from the box out to the smallest and largest observations.

Because boxplots show less detail than histograms or stemplots, they are best used for side-by-side comparison of more than one distribution, as in Figure 5.8. When you look at a boxplot, first locate the median, which marks the center of the distribution. Then look at the variability. The quartiles show the variability of the middle half of the data, and the extremes (the smallest and largest observations) show the variability of the entire dataset. So, is there really much of a difference in gas mileages between city and highway? From the boxplots, we see at once that highway mileages are noticeably higher than city mileages: The third quartile city mileage is less than the minimum of highway mileages! We also see that the variability of highway mileages has a somewhat different pattern than the variability of city mileages. Boxplots can also indicate a distribution's skewness, so we have gotten a lot of mileage from this vehicle for exploratory data analysis!

Be aware that some calculators and software packages offer an alternative option for boxplots in which the lines go to the farthest values within 1.5 box-lengths of the quartiles, so they do not automatically go out to the minimum and maximum values. The advantage of this is that any individual values more than 1.5 box-lengths beyond either quartile can be marked as outliers.

5.7 Describing Variability: The Standard Deviation

Although the five-number summary is the most generally useful numerical description of a distribution, it is not the most common. That distinction belongs to the combination of the mean with the **standard deviation**. The mean, like the median, is a measure of center. The **standard deviation**, like the quartiles and extremes in the five-number summary, measures variability. The standard deviation and its close relative, the *variance,* measure variability by looking at how far the observations are from their mean.

EXAMPLE 10
Understanding the Standard Deviation

When you buy stocks or mutual funds, you need to be aware of how to quantify and balance mean gain with the variability or risk of the investment, especially given the volatile years the market has experienced recently. Consider the PIMCO Total Return A (symbol: PTTAX), a fund that invests in intermediate-term fixed-income securities. Here are its annual total returns for a recent 10-year period:

© Justin Guariglia/Corbis

Calendar Year	2000	2001	2002	2003	2004	2005	2006	2007	2008	2009
Return (in percent)	11.56	8.99	9.69	5.07	4.65	2.41	3.51	8.57	4.32	13.33

Figure 5.9 displays the data as points along a number line, with their mean marked by an asterisk (*). The arrows mark two of the deviations from the mean: one is positive and one is negative. We won't get a useful measure of variability by totaling up all the positive and negative deviations because they will always sum to zero! You could prove this using algebra, or at least illustrate it with simple numbers: consider the dataset {2, 3, 7}, which has a mean of 4. Subtract 4 from each of those three numbers to find the three deviations from the mean (−2, −1, and 3, respectively) and note that these add up to 0.

Squaring the deviations makes these numbers all positive, and a reasonable measure of variability is the average of the squared deviations. This average is called the *variance.* The variance is large if the observations are scattered widely around their mean. The variance is small if the observations are fairly close to the mean.

But the variance does not have meaningful units. With the annual return data measured in percent, the variance of the purchase prices has units of "squared percent." Taking the square root of the variance yields the standard deviation, which gets us back to the units of the original variable (in this case, percent). The standard deviation has other uses as well (as discussed in Section 5.9), such as helping us decide if a data value is significantly far from the mean.

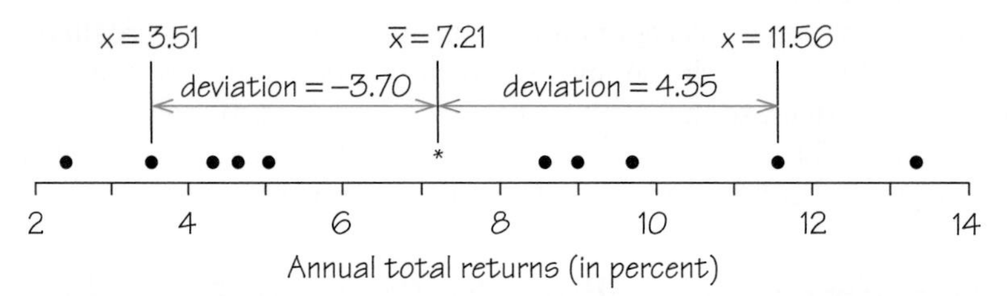

Figure 5.9 The variance and standard deviation measure variability by looking at the deviations of observations from their mean.

Standard Deviation

DEFINITION

The **standard deviation** is a kind of "standard" or average amount that observed data values deviate from their mean. More precisely, it is the square root of what is roughly the mean of the squared deviations. (It turns out that dividing by $n - 1$ instead of the usual n makes this particular formula more accurate, but the justification is beyond the scope of this book.) In symbols, the standard deviation s of n observations $x_1, x_2, \ldots, x_n$ is

$$s = \sqrt{\frac{(x_1 - \bar{x})^2 + (x_2 - \bar{x})^2 + \cdots + (x_n - \bar{x})^2}{n - 1}}$$

Algebra Review — Squares and Square Roots

The expression a^2 is read as "a squared," and means a multiplied by a. When either a positive or negative number is squared, the outcome will always be positive. When zero is squared then the outcome is zero.

If we want to simplify $(-8)^2$, we can easily realize that $(-8)^2 = (-8)(-8) = 64$. However, when the number being squared involves a decimal point, it may be easier to work with a calculator. Make sure that your final answer is consistent with any rounding requirement. For example, if we want to evaluate $(1.32)^2$ and round to two decimal places, our answer would be 1.74.

The expression $\sqrt{a}$ indicates a number that yields a when squared. The expression $\sqrt{a}$ is called the "principal square root of a," and the outcome is never negative. For example, $\sqrt{144} = 12$ because $12^2 = 144$.

Some square root simplifications are easiest to solve with a calculator. Make sure that your final answer is consistent with any rounding requirement. For example, $\sqrt{30} = 5.48$ when rounded to two decimal places.

When a square root involves operations, it may be easier to simplify some of the calculations before taking the square root. For example, if you wish to calculate $\sqrt{\frac{(3.1)(2.7)}{5}}$ and round to one decimal place, you may perform the steps as follows:

$$\sqrt{\frac{(3.1)(2.7)}{5}} = \sqrt{\frac{8.37}{5}} = \sqrt{1.674}$$

This will equal 1.3 when rounded.

For simple datasets, standard deviation often can be estimated mentally by applying the first sentence of the above definition. For example, for the dataset {25, 25, 25, 30, 35, 35, 35}, we can readily see that 30 is the mean and the other numbers are each 5 units away from it. So we might assume that the standard deviation would be a value close to or equal to 5—and it is!

Even for more complex datasets, it is helpful to make a mental estimate first as a way to catch any errors caused by using a calculator. (Another check of reasonableness comes from the fact that the standard deviation is never less than the (positive) difference between the mean and median and is usually close to $\frac{1}{4}$ of the range.) For the 10 return rate values in Example 10, a quick visual inspection might result in an estimate of the mean to be near 7 and a typical amount of deviation from 7 to be something a bit less than 5. Let's keep this estimate in mind as we do the formal calculation in Example 11. For readability, Table 5.3 has the numbers in each step rounded to the nearest tenths place, but you will get more accuracy if you include one or two extra decimal places throughout the process and wait until the end to do any rounding.

EXAMPLE 11
Calculating the Standard Deviation

To find the standard deviation of the 10 return rates, first find the mean:

$$\bar{x} = \frac{11.56 + 8.99 + 9.69 + 5.07 + 4.65 + 2.41 + 3.51 + 8.57 + 4.32 + 13.33}{10}$$

$$= 7.21 \text{ percent} \approx 7.2\%$$

Table 5.3

Step-by-Step Approach to Calculating Standard Deviation

Observations	Deviations (observation minus mean)	Squared Deviations
x_i	$x_i - \bar{x}$	$(x_i - \bar{x})^2$
11.6	$11.6 - 7.2 = 4.4$	$4.4^2 \approx 19.4$
9.0	$9.0 - 7.2 = 1.8$	$1.8^2 \approx 3.2$
9.7	$9.7 - 7.2 = 2.5$	$2.5^2 \approx 6.3$
5.1	$5.1 - 7.2 = -2.1$	$(-2.1)^2 \approx 4.4$
4.7	$4.7 - 7.2 = -2.5$	$(-2.5)^2 \approx 6.3$
2.4	$2.4 - 7.2 = -4.8$	$(-4.8)^2 \approx 23.0$
3.5	$3.5 - 7.2 = -3.7$	$(-3.7)^2 \approx 13.7$
8.6	$8.6 - 7.2 = 1.4$	$1.4^2 \approx 2.0$
4.3	$4.3 - 7.2 = -2.9$	$(-2.9)^2 \approx 8.4$
13.3	$13.3 - 7.2 = 6.1$	$6.1^2 \approx 37.2$
		sum = 123.9

Significant digits, page 658.

The variance is the sum of the squared deviations divided by 1 less than the number of observations, so it would be $\frac{123.9}{10-1} \approx 13.8$. The standard deviation is the square root of the variance, and so we obtain $s \approx \sqrt{13.8} \approx 3.7\%$. This value (3.7%) can be considered small for this context, which suggests that this particular mutual fund happened to have a great deal of stability during a very turbulent decade.

If the 10 observations of the first fund still had a mean of 7.2%, but had less variability, their deviations from 7.2 would be smaller, and the standard deviation would then be even smaller. To explore this dynamics, change 13.33 to 10.33, change 2.41 to 5.41, and then redo the calculation. The mean remains the same, but the resulting standard deviation will be smaller because the numbers have less variability.

Now let's compare the above fund with a different one—the Cohen & Steers Realty Shares (symbol: CSRSX), a mutual fund that invests in real estate investment trusts. Here are its calendar year total returns (in %) for the same 10-year period:

2000	2001	2002	2003	2004	2005	2006	2007	2008	2009
26.63	5.70	2.79	38.09	38.48	14.88	37.13	−19.19	−34.40	32.50

On the one hand, the mean of these 10 numbers is approximately 14.26%, which is almost double the previous fund's mean. However, it comes with a tradeoff—a much higher standard deviation of approximately 25.5%. Scanning the numbers, we see the dramatic lows and highs that make this fund feel like a rollercoaster ride! Knowing how to interpret these numbers is critical when making investment choices to fit your financial goals and tolerance for risk.

SPOTLIGHT 5.2

Calculating Standard Deviation

While the formula in the definition box for standard deviation has conceptual clarity and a straightforward implementation (as in Table 5.3), it can be tedious to apply to large datasets. Even with the most basic calculator, you'll get the same answer faster using this more computationally oriented formula:

$$\sqrt{\frac{(x_1^2 + x_2^2 + \cdots + x_n^2) - n(\bar{x})^2}{n-1}}.$$

If you have a *scientific calculator,* put it into a "STAT MODE" if required, clear out any old data, then enter your data one number at a time. (After each number, press your calculator's data-entry button—it may say DATA or have a symbol such as [Σ+] or [M+].) Once the data is entered, you can find the standard deviation by hitting the key labeled something like [$\sigma n - 1$] or [$\sigma x n - 1$] or [s].

If you have a *graphing calculator* in the TI-83/84+ family (and you already used STAT → EDIT to enter one variable of quantitative data in a column, say, L1), then hit this sequence of buttons: 2ND CATALOG (it's above the "0" key) → stdDev ENTER 2ND L1 (it's above the "1" key) ENTER.

If you use the alternative command sequence STAT → CALC → 1-Var Stats 2ND L1 ENTER, you will get not only the standard deviation *s,* but also other descriptive statistics, including the five-number summary. Keystrokes for other specific models can be found online; for example, in guidebooks downloadable from http://education.ti.com/calculators/downloads/US/#Guidebooks.

More important than the details of hand calculation are the properties that determine the usefulness of the standard deviation:

- s measures variability about the mean $\bar{x}$. Use s to describe the variability of a distribution only when you use $\bar{x}$ to describe the center.
- $s = 0$ only when there is *no variability*. This happens only when all observations have the same value. (If every value is the same, every value equals the mean and thus has zero deviation from the mean!) Otherwise, $s > 0$. As the observations display more variability about their mean, s gets larger.
- s has the same units of measurement as the original observations. For example, if you measure metabolic rates in calories, both the mean $\bar{x}$ and the standard deviation s are also in calories.
- The use of squared deviations makes s even more sensitive than $\bar{x}$ to a few extreme observations. For example, dropping the Toyota Prius from our list of midsized cars drops the standard deviation of city mileages more than 70 percent, from 9.2 mpg with the Prius to 2.5 mpg without it. Distributions with outliers and strongly skewed distributions have large standard deviations. The number s does not give much helpful information about such distributions.

We now have a choice between two descriptions of the center and variability of a distribution: (1) the five-number summary or (2) $\bar{x}$ and s. Because $\bar{x}$ and s are sensitive to extreme observations, they can be misleading when a distribution is strongly skewed or has outliers. In fact, because the two sides of a skewed distribution differ in variability, no single number such as s describes the variability well. The five-number summary, with its two quartiles and two extremes, does a better job.

Choosing a Summary RULE

The five-number summary is usually better than the mean and standard deviation for describing a skewed distribution or a distribution with outliers. Use $\bar{x}$ and s only for reasonably symmetric distributions that are free of outliers.

Although the standard deviation is widely used, it is not a natural or convenient measure of the variability of any possible distribution. The real reason for the popularity of the standard deviation is that it is the natural measure of variability for the special class of distributions called **normal distributions**, which we will discuss next.

Remember that a graph gives the best overall picture of a distribution. Numerical measures of center and variability report specific facts about a distribution, but they do not describe its entire shape; for example, numerical summaries do not disclose the presence of clusters. *Always start with a graph of your data.*

5.8 Normal Distributions

We now have a kit of graphical and numerical tools for describing distributions. What's more, we have a clear strategy for exploring data on a single numerical variable:

1. Always plot your data: make a graph, usually a histogram, a dotplot, or a stemplot.
2. Look for the overall pattern (shape, center, variability) and for striking deviations such as outliers.
3. Calculate a numerical summary to give some description of center and variability.

Here is one more step to add to this strategy:

4. If the overall pattern of a large number of observations is so regular that we can describe it by a smooth curve, then draw that curve superimposed on the histogram.

Figure 5.3 (p. 173) is a histogram of the Iowa Test vocabulary scores of 947 seventh-grade students. Like most histograms from national standardized tests, the histogram is symmetric, is single-peaked, and has a distinctive bell shape. In Figure 5.10, we draw a smooth curve through the tops of the histogram bars to describe the shape. The curve is an idealized description of the distribution. It gives a compact picture of the overall pattern of the data but ignores minor irregularities as well as any outliers. The curve in Figure 5.10 is a *normal curve.* A distribution whose shape is described by a normal curve is a *normal distribution.*

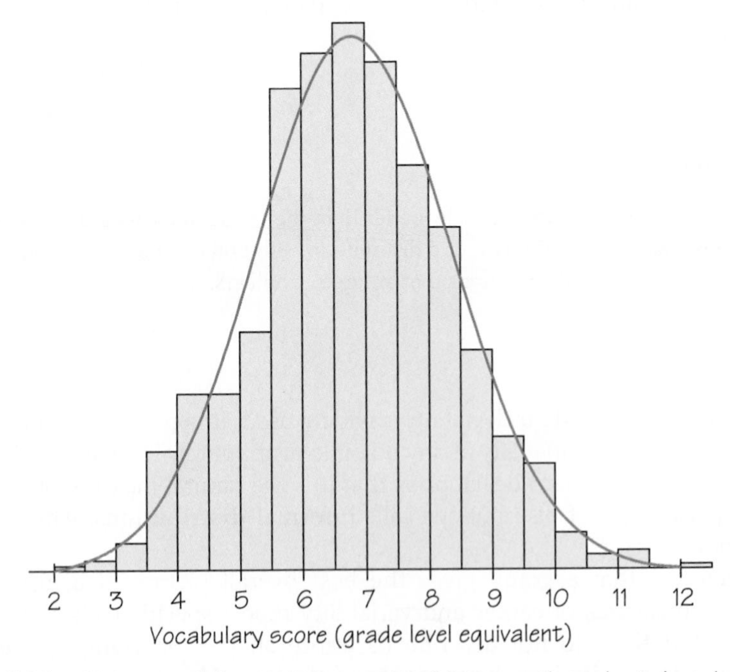

Figure 5.10 A histogram of the vocabulary scores of 947 seventh-grade students in Gary, Indiana. The smooth curve shows the overall shape of the distribution.

Normal Distribution DEFINITION

The *distribution* of a variable tells us what values the variable takes and how often it takes these values. A **normal distribution** is described by a *normal* curve. The area under the curve above any interval of values tells us what proportion of all values of the variable lie in that interval. The total area under the curve is exactly 1.

EXAMPLE 12
From Histogram to Normal Curve

You can think of a normal curve as a smoothed-out histogram when there is symmetry and one mode. Our eyes respond to the *areas* of the bars in a histogram. The bar areas represent proportions of the observations. Figure 5.11a is a copy of Figure 5.10 with the leftmost bars shaded. The area of the shaded bars in Figure 5.11a represents the students with vocabulary scores of 6.0 or lower. This area reflects the proportion $287/947 \approx 0.30$ of Gary, Indiana, seventh graders, so 6.0 is the 30th *percentile*.

Now look at the curve drawn through the bars. In Figure 5.11b, the area under the curve to the left of 6.0 is shaded. We know that the areas of histogram bars represent proportions of all the observations, but we don't worry about the actual total area. Note that all the bars together represent 100 percent of the students, so we treat the total area under the normal curve as $1 = 100\%$. Now, areas under the curve actually *are* proportions of the observations. This curve is a normal curve. The shaded area under the normal curve in Figure 5.11b is the proportion of students with scores of 6.0 or lower. This area turns out to be 0.293, only 0.010 away from the histogram result. You see that areas under the normal curve give quite good approximations of areas given by the histogram.

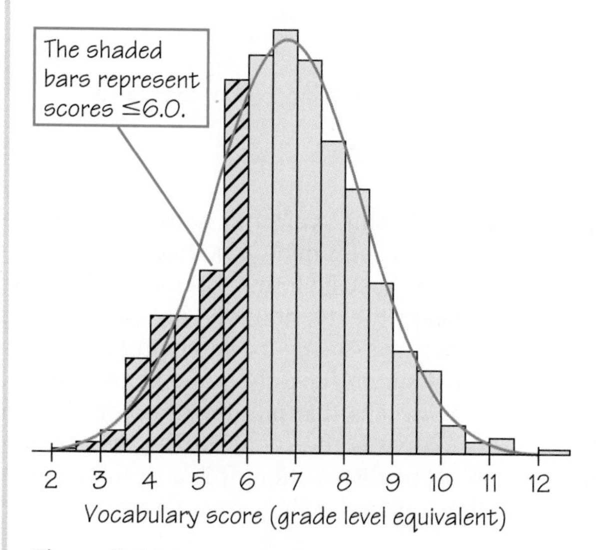

Figure 5.11a The proportion of scores less than or equal to 6.0 from the histogram is 0.303.

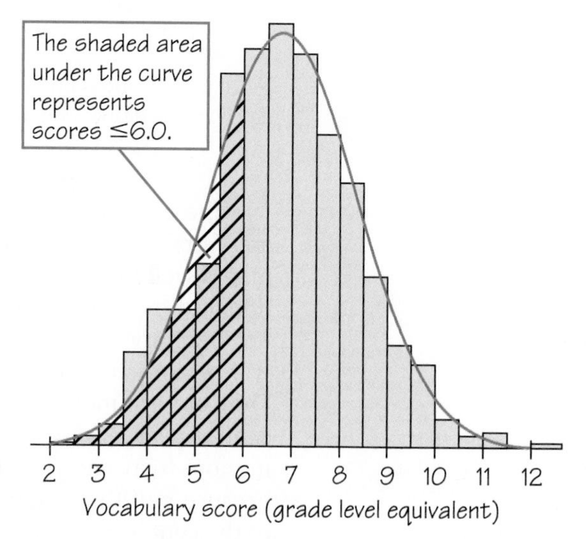

Figure 5.11b The proportion of scores less than or equal to 6.0 from the normal curve is 0.293.

EXAMPLE 13
Heights of American Women

The normal curve is a good approximation of the real-life distribution for a variety of biological measures (height, weight, heart rate, blood pressure, and so on), when examined for a particular species and gender. Figure 5.12 shows the heights of American women between the ages 18–24. The proportion of young women who are between 60 inches (5 feet) and 65 inches tall is given by the area under the curve between 60 and 65. This area is about 0.54, so approximately 54 percent of these women are between 60 and 65 inches tall.

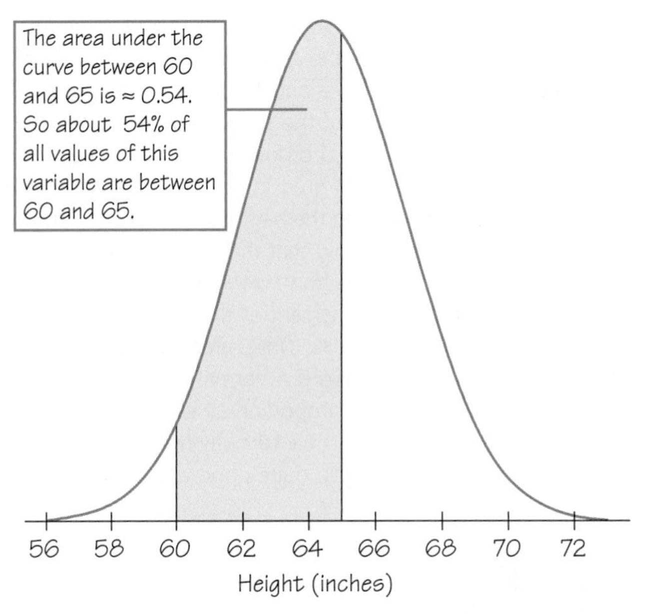

Figure 5.12 Areas under a normal curve describe a normal distribution. This normal curve describes the distribution of heights of American women.

The everyday meaning of "normal" is being typical or natural, and there are certainly some natural phenomena (e.g., Example 13) that are approximated well by the normal distribution. The specific form of a normal distribution and the major role it plays in statistical theory, however, are very special, not ordinary. Normal curves can be specified exactly by an equation, but we will be content with pictures. All normal curves have the same general *shape*. They are symmetric and bell-shaped, with tails that fall down rapidly from a central peak. The *center* of the normal curve is the center of the distribution in more than one way: It is the mean of the distribution. It is also the median because half the observations (half the area under the curve) lie on each side of the center.

What about the *variability* of a normal curve? Normal curves have the special property that their variability is determined completely by a single number, the standard deviation. We have learned how to calculate the standard deviation from a set of observations. For normal distributions, the standard deviation, like the mean, can be found directly from the curve. Here's how. Imagine that you are skiing down a mountain that has the shape of a normal curve. At first, you descend at an ever-steeper angle as you go out from the peak.

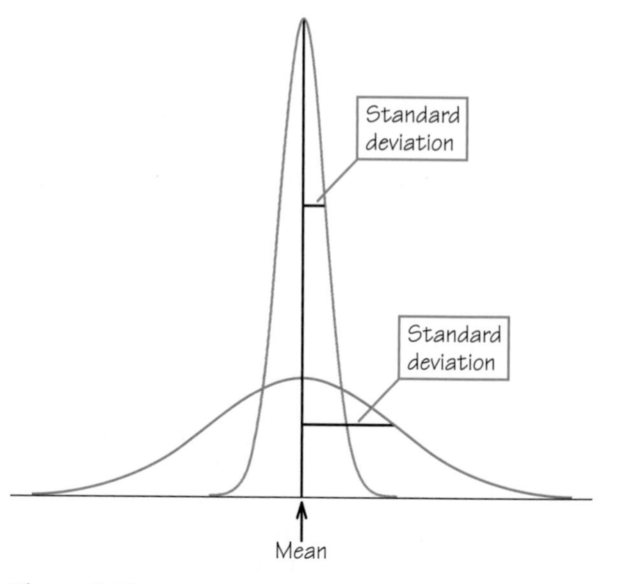

Figure 5.13a Two normal curves with the same mean but different standard deviations. The standard deviation for each curve is the distance from the center (the mean) to the change-of-curvature point on one side of the center.

Fortunately, before you find yourself going straight down, the slope now begins to grow flatter rather than steeper as you continue downhill.

The points at which this change of curvature takes place are located one standard deviation from the mean on either side. You can feel the change as you run your finger along a normal curve, and in that way you can find the standard deviation. Try it. Normal curves with the same standard deviation have exactly the same shape. Changing the mean just moves the center of the curve to a new location, as Figure 5.13b shows. Changing the standard deviation changes the variability of the curve, as Figure 5.13a shows. A normal distribution is completely determined by two numbers: the mean and the standard deviation.

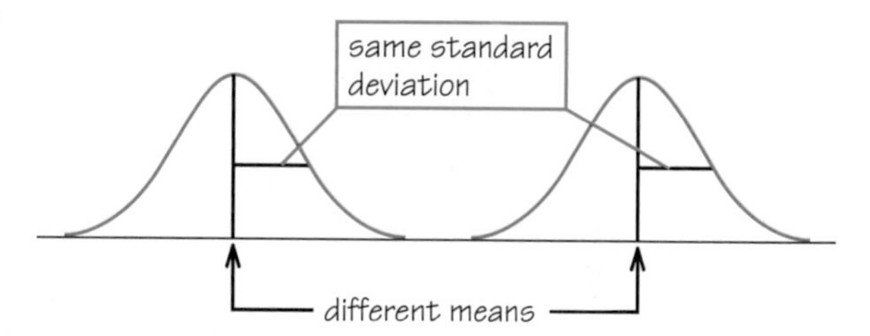

Figure 5.13b Two normal curves with the same standard deviation but different means.

Mean and Standard Deviation of a Normal Distribution — DEFINITION

The **mean of a normal distribution** is at the center of symmetry of the normal curve. The **standard deviation of a normal distribution** is the distance from the center to the change-of-curvature points on either side.

We have often used the quartiles to indicate the variability of a distribution. Because the standard deviation completely describes the variability of any normal distribution, it tells us where the quartiles are. Here are the facts.

Quartiles of a Normal Distribution — DEFINITION

The **quartiles of a normal distribution** are located about 0.67 (which is about 2/3) of a standard deviation away from the mean. In particular, the *first quartile* is located at 0.67 standard deviation below the mean and, by symmetry, the *third quartile* is located at 0.67 standard deviation above the mean.

EXAMPLE 14
Heights of American Women

The distribution of heights of young American women (ages 18–24) is approximately normal, with a mean of 64.5 inches (i.e., 5′ 4.5″) and a standard deviation of 2.5 inches. Figure 5.14 shows this normal curve. The quartiles are 0.67 standard deviation, or

$$(0.67)(2.5 \text{ inches}) \approx 1.7 \text{ inches}$$

away from the mean. The first quartile is 64.5 − 1.7, or 62.8 inches. The third quartile is 64.5 + 1.7, or 66.2 inches. The middle 50 percent of women's heights lie approximately between 62.8 inches and 66.2 inches. These numbers are exact for the normal distribution with a mean of 64.5 inches and a standard deviation of 2.5 inches, but only approximately true for the actual heights of the women because real-life distributions of biological measurements such as heights are only approximately normal.

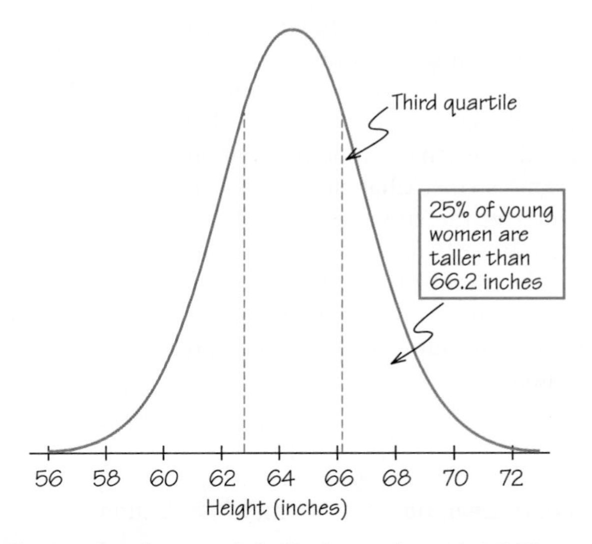

Figure 5.14 The quartiles of a normal distribution are located at 0.67 standard deviation on either side of the mean. For this normal curve, the mean is 64.5 inches and the standard deviation is 2.5 inches.

Why are normal distributions important in statistics? Here are two reasons. First, normal distributions are good models or approximations for some distributions of *real data.* Distributions that are often close to normal include scores on tests taken by many people (such as SAT exams and many psychological tests), repeated careful measurements of the same quantity, and characteristics of biological populations (such as heights of young women and yields of corn). Second, normal distributions are good approximations to the results of many kinds of *chance outcomes,* such as tossing a coin many times. We will return to normal curves when we study the mathematics of chance in Section 8.6. Don't forget that many sets of data do not follow a normal distribution. Most income distributions, for example, are skewed to the right and thus are not normal.

SPOTLIGHT 5.3

Density Estimation

Smooth curves that describe the overall pattern of distributions of data are called *density curves.* Normal curves are one type of density curve. There are many other types used for different purposes. However, you don't have to call for a specific type such as the normal curves. Clever software for "density estimation" will calculate a density curve to describe any set of observations you give it.

The figure shows a strongly skewed distribution, the survival times of 72 guinea pigs in a medical experiment. Two graphs of the distribution are overlaid: a histogram and a density curve produced by software from the data. The histogram and density curve agree on the overall shape and on the "bumps" in the long right tail.

The density curve shows a higher single peak as a main feature of the distribution. The histogram divides the observations near the peak between two bars, thus reducing the height of the peak. Because density estimators don't depend on dividing the data into classes as histograms do, many statisticians prefer them when they need a picture of a distribution.

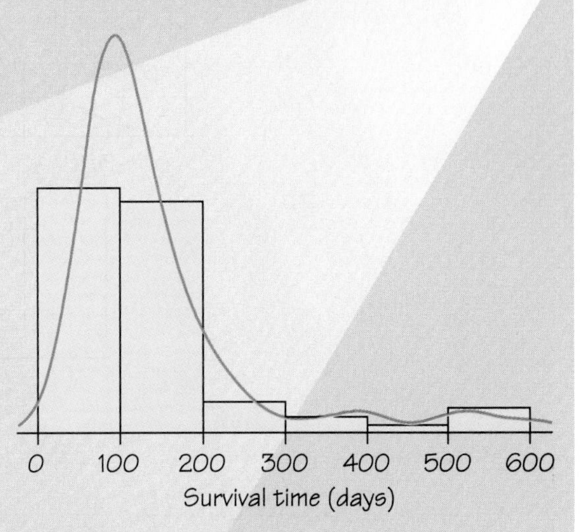

Density estimation software fits this smooth curve to data on the survival time of 72 guinea pigs.

5.9 The 68–95–99.7 Rule for Normal Distributions

Because any particular normal distribution is completely determined by its mean and standard deviation, it is not surprising that all normal distributions are the same in terms of what proportion of observations are within any given number of standard deviations of the mean. Here is an important rule based on this fact.

The 68–95–99.7 Rule for Normal Distributions RULE

According to the **68–95–99.7 rule**, in any normal distribution:

- About 68% of the observations fall within one standard deviation of the mean.
- About 95% of the observations fall within two standard deviations of the mean.
- About 99.7% of the observations fall within three standard deviations of the mean.

Figure 5.15 illustrates the 68–95–99.7 rule. By remembering these three numbers, you can think about normal distributions without making detailed calculations. While the 68–95–99.7 rule is adequate for our purposes here, you can use tables or software that give these and other areas under normal curves. For example, the "68" part of the 68–95–99.7 Rule could be obtained from the TI-84 calculator command 2nd DISTR → normalcdf(−1,1) or the Excel command =NORMSDIST(1)-NORMSDIST(−1).

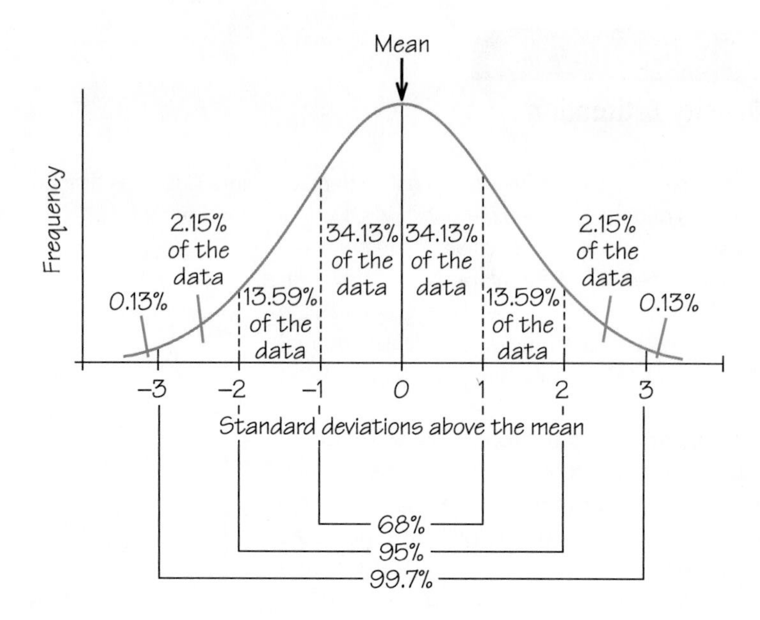

Figure 5.15 The 68–95–99.7 rule for normal distributions.

EXAMPLE 15
Heights of American Women

The heights of women between the ages of 18 and 24 are distributed roughly normally, with a mean of 64.5 inches and a standard deviation of 2.5 inches. Two standard deviations are 5 inches for this distribution. The "95" part of the 68–95–99.7 rule says that the middle 95 percent of young women are between $64.5 - 5$ and $64.5 + 5$ inches tall; that is, between 59.5 inches and 69.5 inches. This fact is exactly true for an exactly normal distribution. It is approximately true for the heights of young women because the distribution of heights is approximately normal.

The other 5 percent of American women have heights outside the range from 59.5 to 69.5 inches. Because the normal distributions are symmetric, half of these women are on the tall side and half on the short side. So the tallest 2.5 percent of young women are taller than 69.5 inches.

EXAMPLE 16
SAT Reasoning Test Scores

The distribution of scores on tests such as the SAT college entrance examination is close to normal. Scores on each of the three sections (math, critical reading, and writing) of the SAT are adjusted so that the mean score is about $\mu = 500$ and the standard deviation is about $\sigma = 100$. This information allows us to answer many questions about SAT scores.

- *How high must a student score to fall in the top 25 percent?*

  The third quartile is $(0.67)(100) = 67$ points above the mean. So scores above 567 are in the top 25 percent (i.e., the top quarter).

- *What percent of scores fall between 200 and 800?*

  Scores of 200 and 800 are three standard deviations on either side of the mean; for example: 500 + 3 × 100 = 800. The "99.7" part of the 68–95–99.7 rule says that 99.7 percent of all scores lie in this interval. (In practice, the SAT makes this 100 percent by reporting as 200 those rare scores below 200 or as 800 those rare scores above 800.)

- *What percent of scores are above 700?*

A score of 700 is two standard deviations above the mean. By the "95" part of the 68–95–99.7 rule, 95 percent of all scores fall between 300 and 700 and 5 percent fall below 300 or above 700. Because normal curves are symmetric, half of this 5 percent are above 700. So, a score above 700 places a student in the top 2.5 percent of test-takers. We can also say that a score of 700 is at the 97.5th percentile of all test takers.

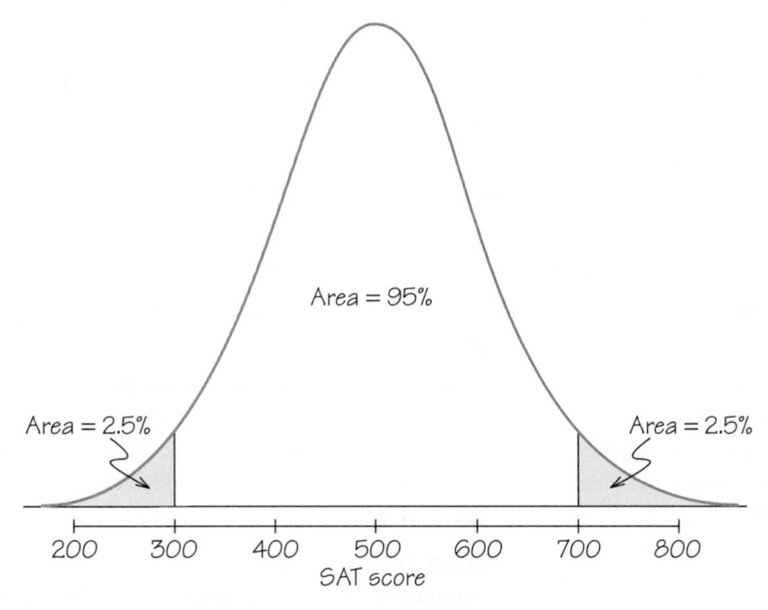

Figure 5.16 Using the 68–95–99.7 rule to find the percentage of SAT section scores that are above 700. This normal curve has a mean of 500 and a standard deviation of 100.

Sketching a normal curve with the points one, two, and three standard deviations from the mean marked can help you use the 68–95–99.7 rule. Figure 5.16 shows the distribution of SAT scores with the areas needed to find the percent of scores above 700. Note that the tails of Figure 5.16, like those of any bell curve, technically stretch out forever in both directions (even as the amount of faraway area becomes vanishingly small). This is another reminder that the bell curve is a very good, but not perfect, model of reality. We know that real-life SAT subtest scores are scaled so that they do not go beyond 200 or 800.

The 68–95–99.7 rule allows you to find selected areas under a normal curve—areas for outcomes bounded by one, two, or three standard deviations away from the mean. You can use software, a graphing calculator, or the *Normal Curve* applet to find *any* area under a normal curve. See Applet Exercises 3 to 5 to practice using the applet.

REVIEW VOCABULARY

Boxplot A graph of the five-number summary. A box spans the quartiles, with an interior line marking the median. Lines extend out from this box to the extreme high and low observations. (p. 182)

Distribution The pattern of how often the variable takes certain values or intervals of values. (p. 167)

Exploratory data analysis The practice of using graphs and numbers to examine data for overall patterns and special features, without necessarily seeking answers to specific questions. (p. 167)

Five-number summary A summary of a distribution that gives the smallest observation, first quartile, median, third quartile, and largest observation, in that order. (p. 181)

Frequency distribution This gives all observed values of the variable and how many times the variable takes on each of these values. (p. 167)

Grouped frequency distribution This gives how many times in each interval of values a variable takes on a particular value. (p. 168)

Histogram A graph of the distribution of outcomes (often divided into classes) for a single numerical variable. The height of each bar is the number of observations in the class of outcomes covered by the base of the bar. All classes should have the same width, and each observation must fall into exactly one class. (p. 168)

Individuals The people, animals, or things described by a dataset. (p. 165)

Left-skewed distribution A distribution in which the longer tail of the histogram is on the left side. (p. 172)

Mean The ordinary arithmetic average of a set of observations. To find the mean, add all the observations and divide the sum by the number of observations summed. (p. 177)

Median The middle of a set of ordered observations. Half the observations fall below the median, and half fall above. (p. 178)

Mode The most frequently occurring value in a set of numerical observations. (p. 180)

Normal distributions A family of distributions that describe how often a variable takes its values by areas under a curve. The normal curves are symmetric and bell-shaped. A specific normal curve is completely described by giving its mean and its standard deviation. (p. 189)

Outlier A data point that falls clearly outside the overall pattern of a set of data. (p. 171)

Quartiles The first quartile (Q_1) of a distribution is the point with one quarter of the observations falling below it; the third quartile (Q_3) is the point with three quarters below it. Q_1 is the median of the lower half of the observations; Q_3 is the median of the upper half. (p. 181)

Range The measure of variability obtained by subtracting the smallest observation from the largest observation. (p. 180)

Relative frequency distribution This gives all observed values of the variable and what fraction (or percentage) of the time each value occurs. (p. 168)

Right-skewed distribution A distribution in which the longer tail of the histogram is on the right side. (p. 172)

68–95–99.7 rule In any normal distribution, 68 percent of the observations lie within one standard deviation on either side of the mean, 95 percent lie within two standard deviations of the mean, and 99.7 percent lie within three standard deviations of the mean. (p. 193)

Standard deviation A measure of the variability of a distribution about its mean as center. It is the square root of the average squared deviation of the observations from their mean. (p. 184)

Standard deviation of a normal curve The standard deviation of a normal curve is the distance from the mean to the change-of-curvature point on either side. (p. 191)

Stemplot A display of the distribution of a variable that attaches the final digits of the observations as leaves on stems made up of all but the final digit. (p. 174)

Symmetric distribution A distribution with a histogram or stemplot in which the part to the left of the median is roughly a mirror image of the part to the right of the median. (p. 173)

Variable A particular characteristic that can take on different values for different individuals. (p. 166)

SKILLS CHECK

1. Here are the first rows of a professor's dataset at the end of a mathematics course:

Name	Major	Points
ADVANI, SURA	COMM	397
BARTON, DAVID	HIST	323
BOAZ, JUDAH	BIOL	446
CHIU, SUN	PSYC	405
DAVIS, LAUREN	PSYC	461

The individuals in these data are

(a) the students
(b) the total points
(c) the majors

2. The number of variables in Exercise 1 is _______.

Figure 5.4 (p. 174) is a histogram of the tuition and fee charges for 55 four-year colleges in Massachusetts. Exercises 3, 4, and 5 are based on this histogram.

3. Which statement can be concluded from the histogram?

(a) There is at least one college with tuition and fees equal to $8,000.
(b) There are no colleges with tuition and fee charges between $8,000 and $16,000.
(c) There are no colleges with tuition and fee charges between $10,000 and $14,000.

4. The number of colleges with tuition and fee charges covered by the leftmost bar in the histogram is _______.

5. The leftmost bar in the histogram covers tuition and fee charges ranging from about

(a) $2000 to $6000.
(b) $3000 to $5000.
(c) $4000 to $8000.

6. The distribution in Figure 5.2 (p. 170) is best described as _______-skewed.

7. Brenner looks at real estate ads for houses in Sarasota, Florida. There are many houses ranging from $200,000 to $400,000 in price. A few houses on the coast, however, have prices of up to $15 million. The distribution of house prices will be

(a) skewed to the left.
(b) roughly symmetric.
(c) skewed to the right.

8. Here are the systolic blood pressures (in mm of mercury) of 10 randomly chosen adults:

147	141	120	124	127
132	98	112	120	128

In a stemplot of these scores, the largest stem is _____.

9. For Figure 5.5 (p. 175), interpret the meaning of 10|2.

(a) 10 states have 2 percent Hispanic population.
(b) 2 states have 10 percent Hispanic population.
(c) 1 state has 10.2 percent Hispanic population.

10. The final (rightmost) digit of an observation would be called a _______ in a stemplot.

11. For Figure 5.5 (p. 175), what is the stem in 10|2?

(a) 10
(b) 2
(c) 1

12. The mean blood pressure of the 10 adults in Skills Check 8 is _______.

13. The median of the blood pressures in Skills Check 8 is

(a) 127
(b) 125.5
(c) 124.9

14. If a single-peaked distribution is skewed to the right, the median is generally to the _______ of the mean.

15. The mode of the 10 blood pressures in Exercise 8 is

(a) 147
(b) 120
(c) 2

16. Between the first quartile and the third quartile lies _________ percent of the observations in a distribution.

17. Which of these is greatest?

(a) The first quartile
(b) The third quartile
(c) The median

18. An outlier's effect on the first quartile is _____ than its effect on the range.

19. In degrees Fahrenheit, a typical January day in Houston has a low of 46 and a high of 63, while in El

Paso has 32 and 57, respectively. Which city has a larger temperature range for January?

(a) Houston, because 63 > 57 and 46 > 32
(b) El Paso, because 57 − 32 > 63 − 46
(c) El Paso, because 46 − 32 > 63 − 57

20. The first quartile of the dataset {1, 2, 3, 4, 5, 6} is ______.

21. Which of these is not in a five-number summary?

(a) Median
(b) Minimum
(c) Mean

22. The five-number summary of the 10 blood pressures in Skills Check 8 is _____. (Remember to list the five numbers in increasing order.)

23. The standard deviation of the 10 blood pressures in Exercise 8 is _______. (Use your calculator.)

(a) 13.23
(b) 13.95
(c) 194.6

24. You have data on the weights (measured in grams) of five crackers. The correct units for the standard deviation of these weights are__________________.

25. What are all the values that a standard deviation s can possibly take?

(a) $0 \leq s$
(b) $0 \leq s \leq 1$
(c) $-1 \leq s \leq 1$

26. To specify the shape of a normal distribution completely, you must give its mean and its _______.

27. If two normal curves have the same mean but different standard deviations, the curve with the larger standard deviation will be______the other curve.

(a) as tall as
(b) taller than
(c) shorter than

28. The steepest part of a normal curve is close to the _______.

29. The scale of scores on an IQ test is approximately normal with a mean of 100 and a standard deviation of 15. The organization Mensa, which calls itself "the high-IQ society," requires an IQ score of 130 or higher for membership. What percent of adults would qualify for membership?

(a) 95 percent
(b) 5 percent
(c) 2.5 percent

30. The length of human pregnancies from conception to birth varies according to a distribution that is approximately normal, with a mean of 266 days and a standard deviation of 16 days. We can expect that about ________ percent of all completed pregnancies are between 234 and 298 days.

CHAPTER 5 EXERCISES

■ *Challenge* ▲ *Discussion*

Make and Model	Vehicle Type	Transmission Type	Number of Cylinders	City mpg	Highway mpg
Mazda MX-5	Two-seater	Manual	4	22	28
Toyota Yaris	Subcompact	Automatic	4	29	35
Honda Accord	Large car	Automatic	6	20	30
Jaguar XF	Midsize car	Automatic	8	16	23

Some exercises require use of a calculator (or software or Internet applet) that will find mean and standard deviation from keyed-in data.

5.1 Displaying Distributions: Histograms

5.2 Interpreting Histograms

1. Above is a small part of a dataset that describes the fuel economy (in miles per gallon) of 2011 model motor vehicles.

(a) What are the individuals in this dataset?
(b) For each individual, what variables are given?
(c) For which of these variables would a histogram be helpful? (That is, which variables do not yield categorical data?)

▲ **2.** Figure 5.17 is a histogram of the lengths of words used in Shakespeare's plays. Because there are so many words in the plays, the vertical axis of the graph is the percent that are of each length, rather than the count.

What is the overall shape of this distribution? What does this shape say about word lengths in Shakespeare?

Do you expect other authors to have word-length distributions of the same general shape? Why?

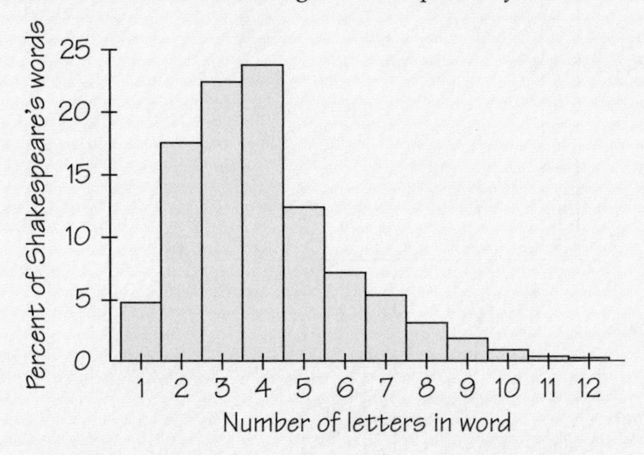

Figure 5.17 Relative frequency histogram of the lengths of words used (the percent rather than the count that are of each length).

▲ **3.** Suppose that you and your friends emptied your pockets of coins and recorded the year marked on each coin. Would you expect the histogram for the distribution of dates to be skewed to the left or right? Explain your answer and make a sketch of this histogram.

4. Make a histogram of the city gas mileages of the midsized cars in Table 5.2 (p. 176). Use classes with widths of 5 mpg. Do you prefer the histogram or the representation in Figure 5.6 (p. 176) of the same data? Why?

5. Burning fuels in power plants or motor vehicles emits carbon dioxide (CO_2), which contributes to global warming. Table 5.4 displays CO_2 emissions per person from countries with populations of at least 20 million.

(a) Why do you think we choose to measure emissions per person rather than total CO_2 emissions for each country?

(b) Display the data of Table 5.4 in a histogram. Describe the shape, center, and variability of the distribution. Which countries appear to be outliers?

■ **6.** A survey of a large college class asked the following questions:

1. Are you female or male? (In the data, male = 0, female = 1.)
2. Are you right-handed or left-handed? (In the data, right = 0, left = 1.)
3. What is your height, in inches?
4. How many minutes do you study on a typical weeknight?

Figure 5.18 shows histograms of the student responses, in scrambled order and without scale markings. Which histogram goes with each variable? Explain your reasoning. Would the 0-1 coding scheme work for someone who is ambidextrous (or transgendered)?

Table 5.4

Carbon Dioxide Emissions, Metric Tons per Person

Country	CO_2	Country	CO_2	Country	CO_2	Country	CO_2
Algeria	2.3	Germany	10.0	Myanmar	0.2	South Korea	8.8
Argentina	3.9	Ghana	0.2	Nepal	0.1	Spain	6.8
Australia	17.0	India	0.9	Nigeria	0.3	Sudan	0.2
Bangladesh	0.2	Indonesia	1.2	North Korea	9.7	Tanzania	0.1
Brazil	1.8	Iran	3.8	Pakistan	0.7	Thailand	2.5
Canada	16.0	Iraq	3.6	Peru	0.8	Turkey	2.8
China	2.5	Italy	7.3	Philippines	0.9	Ukraine	7.6
Colombia	1.4	Japan	9.1	Poland	8.0	United Kingdom	9.0
Congo	0.0	Kenya	0.3	Romania	3.9	United States	19.9
Egypt	1.7	Malaysia	4.6	Russia	10.2	Uzbekistan	4.8
Ethiopia	0.0	Mexico	3.7	Saudi Arabia	11.0	Venezuela	5.1
France	6.1	Morocco	1.0	South Africa	8.1	Vietnam	0.5

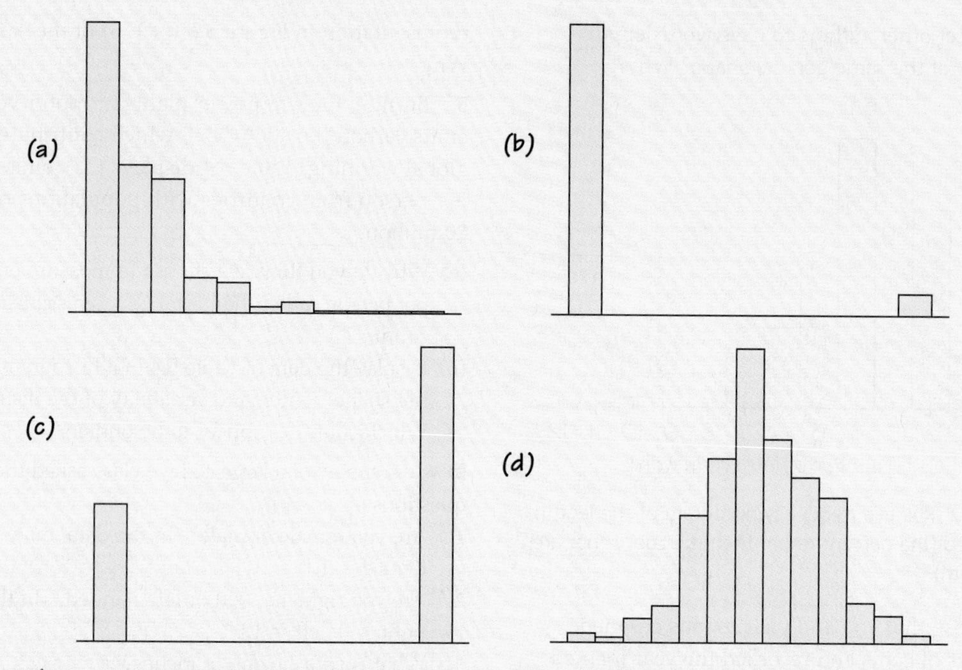

Figure 5.18 Match each histogram with its variable, for Exercise 6.

5.3 Displaying Distributions: Stemplots

7. The population of the United States is aging, though less rapidly than in other developed countries. Figure 5.19 is a stemplot of the percents of residents aged 65 and over in the 50 states, according to the 2010 Census. The stems are whole percents and the leaves are tenths of a percent.

7	8
8	
9	0
10	479
11	49
12	2233345899
13	02334555566778889
14	012344555599
15	69
16	1
17	4

Figure 5.19 Stemplot of the percentages of residents aged 65 and over in the 50 states, for Exercise 7.

(a) Alaska is an outlier with the lowest percentage of older residents (Florida has the highest). What is the percentage for Alaska?

(b) Ignoring Alaska, describe the shape, center, and variability of this distribution.

8. People with diabetes must monitor and control their blood glucose level. The goal is to maintain "fasting plasma glucose" between about 90 and 130 milligrams per deciliter (mg/dl). Here are the fasting plasma glucose levels for 18 diabetics enrolled in a diabetes management class, five months after the end of the class:

78	103	141	148	172	255
95	112	145	153	172	271
96	134	147	158	200	359

(a) Round these values to the nearest 10 and then drop the zero. For example, 141 rounds to 14 and 158 rounds to 16. Make a stemplot of the rounded data.

(b) Describe the main features of the distribution. Are there outliers? How well is the group as a whole achieving the goal for controlling glucose levels?

9. The Survey of Study Habits and Attitudes (SSHA) is a psychological test that evaluates college students' motivation, study habits, and attitudes toward school. A private college gives the SSHA to 18 of its incoming first-year women students. Their scores are (sorted in ascending order):

101	115	129	140	154	165
103	126	137	148	154	178
109	126	137	152	165	200

(a) Make a stemplot of these data. The overall shape of the distribution is irregular, as often happens if only a few observations are available. Are there any outliers?

(b) About where is the center of the distribution (the score with half the scores above it and half below)? What is the variability of the scores (ignoring any outliers)?

10. In 1798, the English scientist Henry Cavendish measured the density of the Earth in a careful experiment with a torsion balance. In sorted order, here are his 29 measurements of the same quantity (the density of the Earth relative to that of water) made with the same instrument. [S. M. Stigler, Do robust estimators work with real data? *Annals of Statistics*, 5 (1977): 1055–1098.

4.88	5.29	5.36	5.47	5.58	5.68
5.07	5.29	5.39	5.50	5.61	5.75
5.10	5.30	5.42	5.53	5.62	5.79
5.26	5.34	5.44	5.55	5.63	5.85
5.27	5.34	5.46	5.57	5.65	

(a) For convenience, round the above numbers to the nearest tenths place (e.g., 4.88 becomes 4.9), and then make a stemplot of the data.

(b) Describe the distribution: Is it approximately symmetric or distinctly skewed? Are there gaps or outliers?

11. Here is a stemplot for percentage of live births to unmarried mothers for each state in the United States in 2007. (*Source:* 2010 report on Centers for Disease Control Web site.)

2	0
2	56
3	13333344
3	5555567777889999
4	0001111112223344
4	5677
5	124

(a) Explain how and why there are repeated stems.

(b) Describe the shape of the distribution.

5.4 Describing Center: Mean and Median

■ **12.** In Malay, the expression for the *mean* is *sama rata*, which roughly translates as "same level." To understand this cultural and conceptual connection, take some poker chips (or other equal-sized, stackable objects) and make stacks with 3, 7, and 8 chips.

(a) Explain how to redistribute chips among the stacks until they are at the same level.

(b) How does this relate to the mean?

13. Refer to the data and the stemplot from Exercise 9:

(a) Find the mean of the 18 values from Exercise 9.

(b) Your stemplot of the scores suggests that the score 200 is an outlier. Find the mean for the 17 observations that remain when you drop the outlier. [*Hint:* can you use the work you did in (a) to avoid calculating this new mean from scratch?]

(c) How does the outlier change the mean?

14. The Major League Baseball career and single-season home run records are held by Barry Bonds of the San Francisco Giants. Here are Bonds's annual home run totals from 1986 (his first year) through 2007 (his last year):

16	25	24	19	33	25	34	46
37	33	42	40	37	34	49	73
46	45	45	5	26	28		

(Lucy Nicholson/Reuters/Corbis.)

(a) Make a stemplot of the data. Are there any outliers?

(b) Find his career mean and median number of home runs. How do these change when you drop his 2001 season total of 73? What general fact about the mean and median does your result illustrate?

▲ **15.** The distribution of income in the United States is skewed to the right. According to the Census Bureau's Current Population Survey report, the mean and median incomes of American households were \$49,777 and \$67,976 in 2009. Explain how you can tell which of these numbers is the mean and which is the median.

16. If a stock gains 50 percent one year and then loses 50 percent of its value the next year, is it accurate to say that its mean growth over the full two-year period was $(50 + -50)/2 = 0\%$? Explain.

17. The basic unit of Census data is the household, not the person. If divorce breaks one household into two, but no individual person's income changes, how (if at all) is mean household income affected?

▲ **18.** Which team is #1? In addition to polls of coaches and journalists, rankings from six computer programs, which have various ways to value factors such as the quality of the opponent played, determine the Bowl Championship Series (BCS) Standings in major college football.

(a) At the end of the 2007 regular season, Hawaii (the only undefeated team) received these computer rankings: 12th, 8th, 14th, 10th, 8th, 13th. The BCS formula throws out the high and low of the six computer rankings and uses the mean of the remaining four ranks. Find this mean.

(b) Why do you think the high and low values are excluded from the mean? Is your reason connected to why the median is sometimes preferred to the mean?

19. Make up an example of a small set of data for which the mean lies in the top 25 percent of the observations.

■ **20.** A sample of five households is selected, and the size of each household is recorded. The median size is 3 and the mode is 5. What is the mean? (*Hint:* Find the only possible dataset.)

5.5 Describing Variability: Range and Quartiles

5.6 The Five-Number Summary and Boxplots

21. The stemplot in Figure 5.19 (p. 200) displays the distribution of the percents of residents aged 65 and over in the 50 states. Stemplots help you find the five-number summary because they arrange the observations in increasing order. Give the five-number summary of this distribution.

22. In chronological order, here are the percents of the popular vote won by each successful candidate in the last 15 presidential elections, starting in 1952:

54.9	57.4	49.7	61.1	43.4	60.7
50.1	50.7	58.8	53.4	43.0	49.2
47.9	50.7	52.9			

(a) Make a stemplot of the winners' percents.

(b) What is the median percent of the vote won by the successful candidate in presidential elections?

(c) Call an election a landslide if the winner's percent falls at or above the third quartile. Find the third quartile. Which elections were landslides?

(d) Find the range.

23. Figure 5.4 (p. 174) is a histogram of the tuition and fees charged by the 55 four-year colleges in the state of Massachusetts. Here are those charges (in dollars), arranged in increasing order. Data for 2007–2008, from the College Board Web site, www.collegeboard.com.

5799	5864	5992	6034	6124
6168	6210	8595	8732	8840
9924	16080	17750	20000	21330
21850	22073	22500	22950	23600
23755	24075	24250	24617	25748
25755	25850	25942	25990	26080
26250	27485	27497	28302	28440
29810	31899	32865	32896	34186
34830	34986	34994	34998	35142
35418	35670	35674	35702	35940
36232	36550	36645	36690	36700

(a) Find the five-number summary and make a boxplot.

(b) What distinctive feature of the Fig. 5.4 histogram do these summaries miss? Remember that numerical summaries are not a substitute for looking at the data.

24. Find the five-number summary of Cavendish's measurements of the density of the Earth in Exercise 10. How is the symmetry of the distribution reflected in the five-number summary?

25. Table 5.4 (p. 199) gives CO_2 emissions per person for countries with populations of at least 20 million. The distribution is strongly skewed to the right. The United States and several other countries appear to be high outliers. Give the five-number summary. Explain why this summary suggests that the distribution is right-skewed.

26. Find the five-number summary of the data from Exercise 8.

▲ **27.** Figure 5.20 shows boxplots of the incomes of a large sample of people who have a high school diploma but no further education and another large group of people with a bachelor's degree but no higher degree. The data come from a Census Bureau survey and represent all people aged 25 to 64 in the United States. Because there are a few extremely high incomes, the boxplot leaves out the highest 5 percent in each group. Based on the plot, compare the distributions of income for these two levels of education. Comment on both center and variability.

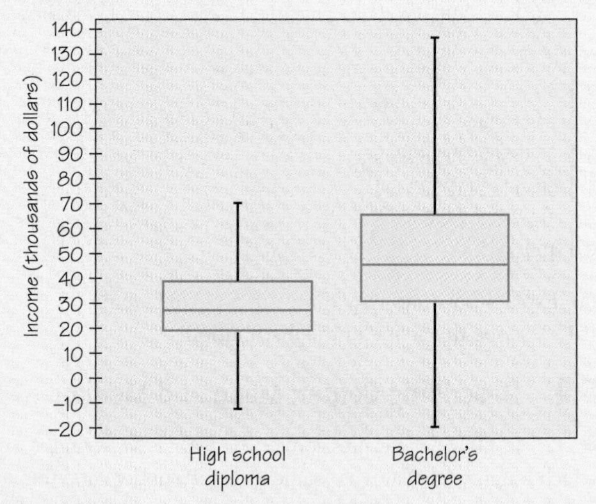

Figure 5.20 Boxplots comparing the incomes of full-time workers aged 25 to 64 years at two levels of education. Because the highest incomes in any large group are very high indeed, the plot omits the top 5 percent of incomes in each group.

28. The data that generates Figure 5.20 include the incomes of 14,959 people whose highest level of education is a bachelor's degree.

(a) What is the position of the median in the ordered list of incomes (1 to 14,959)? From the boxplot, about what is the median income of people with a bachelor's degree?

(b) What is the position of the first and third quartiles in the ordered list of incomes for these people? About what are the numerical values of Q_1 and Q_3?

■ **29.** How much oil that the wells in a given field will ultimately produce is key information in deciding whether to drill more wells. Here are the estimated total amounts of oil recovered from 64 wells in the Devonian Richmond Dolomite area of the Michigan basin, in thousands of barrels. J. Marcus Jobe and Hutch Jobe, A statistical approach for additional infill development, *Energy Exploration and Exploitation*, 18 (2000): 89–103.

2.0	18.5	34.6	47.6	69.5
2.5	20.1	34.6	49.4	69.8
3.0	21.3	35.1	50.4	79.5
7.1	21.7	36.6	51.9	81.1
10.1	24.9	37.0	53.2	82.2
10.3	26.9	37.7	54.2	92.2
12.0	28.3	37.9	56.4	97.7
12.1	29.1	38.6	57.4	103.1
12.9	30.5	42.7	58.8	118.2
14.7	31.4	43.4	61.4	156.5
14.8	32.5	44.5	63.1	196.0
17.6	32.9	44.9	64.9	204.9
18.0	33.7	46.4	65.6	

(a) Make a histogram and describe its main features.

(b) Find the mean and median of the amounts recovered. Explain how the relationship between the mean and the median reflects the shape of the distribution.

(c) Give the five-number summary and explain briefly how it reflects the shape of the distribution.

■ **30.** Look at the histogram of lengths of words in Shakespeare's plays shown in Figure 5.17 (on p. 199). The heights of the bars tell us what percent of words have each length. (Analysis of such tendencies helps determine authorship of newly-discovered manuscripts.)

(a) The median length is the length with half of all words shorter and half longer. What is the median length of words Shakespeare used?

(b) Give the five-number summary for Shakespeare's word lengths.

■ **31.** A common criterion for identifying an outlier in a set of data is if an observation falls more than 1.5 × IQR above the third quartile or below the first quartile. (IQR stands for the interquartile range, which is the difference between the quartiles: $Q_3 - Q_1$.)

So which states are suspected outliers in the distribution of percent of Hispanics among adult residents, as shown in Table 5.1 (p. 169)?

5.7 Describing Variability: The Standard Deviation

32. Do you think the standard deviation of the tuition and fees of the public colleges in Massachusetts is likely to be bigger or smaller than the standard deviation for the private colleges? Why?

33. Many standard statistical methods are intended for use with distributions that are symmetric and have no outliers. These methods start with the mean and standard deviation, $\bar{x}$ and s. An example of scientific data for which standard methods should work well is Cavendish's measurements of the density of the Earth in Exercise 10.

(a) Summarize this dataset by giving $\bar{x}$ and s.

(b) Find the median. Is the median quite close to the mean, as we expect it to be for symmetric distributions?

34. The level of various substances in the blood influences our health. Here are measurements of the level of phosphate in the blood of a patient, in milligrams of phosphate per deciliter of blood, made on six consecutive visits to a clinic.

5.6 5.2 4.6 4.9 5.7 6.4

(a) Find the mean.

(b) Find the standard deviation.

35. The mean $\bar{x}$ and standard deviation s are not generally a complete description. Datasets with different shapes can have the same mean and standard deviation.

(a) To demonstrate this fact, use your calculator to find $\bar{x}$ and s for these two small datasets.

(b) Make a stemplot of each and comment on the shape of each distribution.

Data A:	9.14	8.14	8.74	8.77
	9.26	8.10	6.13	3.10
	9.13	7.26	4.74	
Data B:	7.46	6.77	12.74	7.11
	7.81	8.84	6.08	5.39
	8.15	6.42	5.73	

■ **36.** "Conservationists have despaired over destruction of tropical rainforest by logging, clearing, and burning." These words begin a report on a statistical study of the effects of logging in Borneo. [C. H. Cannon, D. R. Peart, and M. Leighton, Tree species diversity in commercially logged Bornean rainforest, *Science*, 281 (1998): 1366–68.] Researchers compared forest plots that had never been logged (Group 1) with similar plots nearby that had been logged one year earlier (Group 2) and eight years earlier (Group 3). All plots were 0.1 hectare in area. Here are the counts of trees for plots in each group, courtesy of Charles Cannon:

Group 1:	27	22	29	21	19	33
	16	20	24	27	28	19
Group 2:	12	12	15	9	20	18
	17	14	14	2	17	19
Group 3:	18	4	22	15	18	
	19	22	12	12		

Give a complete comparison of the three distributions, using both graphs and numerical summaries. To what extent has logging affected the count of trees? The researchers used an analysis based on $\bar{x}$ and s. Explain why this is reasonably well justified.

(Edward Parker/Alamy.)

■ **37.** This is a standard deviation contest. You must choose four numbers from the whole numbers 0 to 10, with repeats allowed.

(a) Choose four numbers that have the smallest possible standard deviation.

(b) Choose four numbers that have the largest possible standard deviation.

(c) Is more than one choice possible in part (a)? Explain.

(d) Is more than one choice possible in part (b)? Explain.

38. Your data consist of observations on the age of several subjects (measured in years) and the reaction times of these subjects (measured in seconds). In what units are each of the following descriptive statistics measured?

(a) The mean age of the subjects

(b) The standard deviation of the subjects' reaction times

(c) The variance of the subjects' reaction times

(d) The median age of the subjects

5.8 Normal Distributions

5.9 The 68–95–99.7 Rule

39. Some teachers graded "on a (bell) curve" based on the belief that classroom test scores are normally distributed. One way of doing this is to assign a "C" to all scores within one standard deviation of the mean. Then, the teacher would assign a "B" to all scores between one and two standard deviations above the mean, an "A" to all scores more than two standard deviations above the mean, and use symmetry to define the regions for "D" and "F" on the left side of the normal curve. If 200 students take an exam, determine the number of students who would receive a B.

40. The length of human pregnancies from conception to birth varies according to a distribution that is approximately normal, with a mean of 266 days and a standard deviation of 16 days. Draw a normal curve for this distribution on which the mean and standard deviation are correctly located. (*Hint:* First draw the curve and then mark the axis.)

41. Figure 5.21 shows a smooth curve used to describe a distribution that is not symmetric. The mean and median do not coincide. Which of the points marked is the mean of the distribution, and which is the median? Explain your answer.

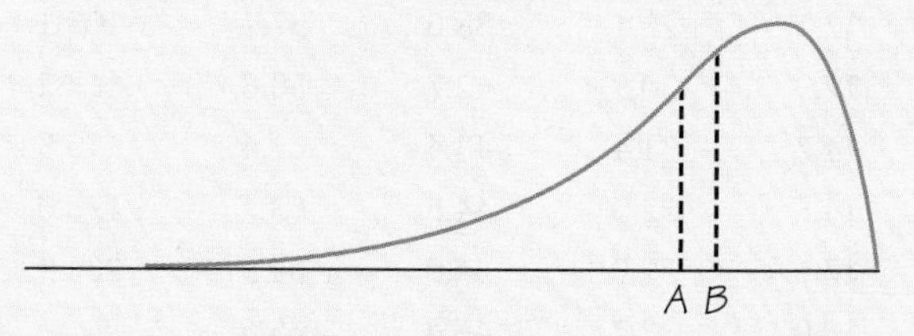

Figure 5.21 A curve describing a left-skewed distribution, for Exercise 41.

42. Sketch a smooth curve that describes a distribution that is symmetric but has two peaks (that is, two strong clusters of observations).

43. Consider the second fund (whose standard deviation is 25.5 percent) discussed in Example 11 (p. 185). Complete these sentences: In about two-thirds of future annual returns, the fund is expected to earn about 14.2 percent each year, plus or minus ______. This means that in two-thirds of future years, the fund may do as well as ______ percent or as poorly as ______ percent.

44. Consider the second fund (whose standard deviation is 25.5 percent) discussed in Example 11 (p. 185).

(a) Complete these sentences, a slight variation of which is commonly used in investment advising: In about 95 percent of future annual returns, the fund is expected to earn about 14.2 percent each year, plus or minus ______. This means that in 95 percent of future years, the fund may do as well as ______ percent or as poorly as _____ percent.

(b) Based on your answers to (a), would this kind of fund be more attractive to an 80-year-old retired person living on a modest fixed pension or to a young working professional? Explain.

45. Bigger animals tend to carry their young longer before birth. The length of horse pregnancies from conception to birth varies according to a roughly normal distribution, with a mean of 336 days and a standard deviation of 3 days. Use the 68–95–99.7 rule to answer the following questions.

(a) Almost all (99.7 percent) of horse pregnancies fall in what interval of lengths?

(b) What percent of horse pregnancies are longer than 339 days?

46. According to the College Board, scores on the math section of the SAT Reasoning college entrance test for the class of 2010 had a mean of 516 and a standard deviation of 116. Assume that they are roughly normal.

(a) What was the interval spanned by the middle 68 percent of scores?

(b) How high must a student score to be in the top 2.5 percent of scores?

47. What are the quartiles of scores from the math section of the SAT Reasoning test, according to the distribution in Exercise 46?

48. The Wechsler Adult Intelligence Scale (WAIS) is the most common "IQ test." The scale of scores is set separately for each age group and is approximately normal, with a mean of 100 and a standard deviation of 15. People with WAIS scores below 70 are generally considered eligible to apply for Social Security disability benefits. By this criterion, what percent of adults are in this IQ category?

49. The yearly rate of return on the Standard & Poor's 500 (an index of 500 large-cap corporations) is approximately normal. From January 1, 1960 through December 31, 2009, the S&P 500 had a mean yearly return of 10.98 percent, with a standard deviation of about 17.46 percent. Take this normal distribution to be the distribution of yearly returns over a long period.

(a) In what interval do the middle 95 percent of all yearly returns lie?

(b) Stocks can go down as well as up. What are the worst 2.5 percent of annual returns?

50. What is the interval of the middle 50 percent of annual returns on stocks, according to the distribution given in the previous exercise? (*Hint:* What two numbers mark off the middle 50 percent of any distribution?)

51. The concentration of the active ingredient in capsules of a prescription painkiller varies according to a normal distribution with $\mu = 10\%$ and $\sigma = 0.2\%$.

(a) What is the median concentration? Explain your answer.

(b) What interval of concentrations covers the middle 95 percent of all the capsules?

(c) What interval covers the middle half of all capsules?

52. Answer the following questions for the painkiller in Exercise 51.

(a) What percent of all capsules have a concentration of the active ingredient higher than 10.4 percent?

(b) What percent have a concentration higher than 10.6 percent?

53. One reason that normal distributions are important is that they describe how the results of an opinion poll would vary if the poll were repeated many times. About 40 percent of adult Americans say they are afraid to go out at night because of crime. Take many randomly chosen samples of 1050 people. The proportions of people in these samples who stay home for fear of crime will follow the normal distribution with a mean of 0.4 and a standard deviation of 0.015. Use this fact and the 68–95–99.7 rule to answer these questions.

(a) In many samples, what percent of samples give results above 0.4? Above 0.43?

(b) In a large number of samples, what interval contains the middle 95 percent of proportions of people who stay home because of crime?

■ **54.** You can compare observations from different normal distributions if you measure in standard deviations away from the mean. Scores expressed in standard deviation units are called *standard scores* (or *z-scores*), and tables and technology commands can convert z-scores into percentiles. A z-score bigger than 3 or less than −3 would definitely be considered an outlier.

(a) Scores on the ACT college entrance exam in a recent year were roughly normal, with a mean of 21.2 and a standard deviation of 4.8. Jermaine scores 27 on the ACT. Express his score in standard deviation units by calculating

$$\text{standard score} = \frac{\text{score} - \text{mean}}{\text{standard deviation}}$$

(b) Scores on the SAT Reasoning college entrance exam in the same year were roughly normal, with mean 1511 and standard deviation 194. Tonya scores 1718 on the SAT. What is her standard score?

(c) Assuming that the ACT and the SAT tests measure the same thing, did Jermaine or Tonya have the better performance?

Chapter Review

Different varieties of the bright tropical flower *Heliconia* are fertilized by different species of hummingbirds. Over time, the lengths of the flowers and the form of the hummingbirds' beaks have evolved to match each other. Here are data on the lengths in millimeters of two varieties of these flowers on the island of Dominica.

Heliconia caribaea Red				
37.40	38.07	38.87	40.66	41.93
37.78	38.10	39.16	41.47	42.01
37.87	38.20	39.63	41.69	42.18
37.97	38.23	39.78	41.90	43.09
38.01	38.79	40.57		
***Heliconia caribaea* Yellow**				
34.57	35.45	36.03	36.66	37.02
34.63	35.68	36.11	36.78	37.10
35.17	36.03	36.52	36.82	38.13

SOURCE: Thanks to Ethan J. Temeles of Amherst College for providing the data. His work is described in Ethan J. Temeles and W. John Kress, Adaptation in a plant-hummingbird association, *Science*, 300 (2003): 630–33.

Exercises 55 to 59 use these data.

55. Make stemplots of the lengths of each of the two varieties (red and yellow). Briefly describe the overall shape of the two distributions.

56. Find the five-number summaries of the two distributions of flower lengths. Make side-by-side boxplots to give a quick picture that compares the two distributions.

57. The biologists who collected the flower length data compared the two *Heliconia* varieties using statistical methods based on the mean and standard deviation.

(a) Find $\bar{x}$ and s for each variety.

(b) Based on Exercise 55, which distribution is more suitable for use of $\bar{x}$ and s as summaries? Why?

58. Your stemplot in Exercise 55 suggests that the distribution of lengths of yellow *Heliconia* flowers is roughly normal. Suppose that the distribution is exactly normal. Use the mean and standard deviation you found in Exercise 57 as the μ and σ of the distribution.

(a) What interval of lengths covers the middle 50 percent of yellow flowers?

(b) What interval of lengths covers the middle 95 percent of yellow flowers?

■ **59.** Continue to work with the normal distribution of lengths of yellow flowers from Exercise 58. The shortest red flower was 37.4 millimeters long. Using the 68–95–99.7 rule and the location of the quartiles in normal distributions, what can you say about what percent of yellow flowers that are longer than 37.4 millimeters?

60. Without a calculator, find the standard deviation of these five numbers: 0, 1, 3, 4, 12. Use the approach in:

(a) the standard deviation definition box, on page 184,

(b) Spotlight 5.2, on page 186.

61. If every number in a dataset is increased by 10, which of these will increase: range, standard deviation, mode, mean, median?

62. Bob is two years older than one brother and five years younger than his other brother. Find the standard deviation of the three brothers' ages.

APPLET EXERCISES

To do these exercises, go to www.whfreeman.com/fapp9e.

1. The *Mean and Median* applet allows you to place observations on a line and see their mean and median visually. Place two observations on the line, by clicking below it. Why does only one arrow appear?

2. In the *Mean and Median* applet, place three observations on the line by clicking below it, two close together near the center of the line and one somewhat to the right of these two. Pull the single rightmost observation out to the right. (Place the cursor on the point, hold down the mouse button, and drag the point.) How does the mean behave? How does the median behave? Explain briefly why each measure acts as it does.

3. In Example 16 (p. 194), we used the fact that SAT section scores are close to normal and are adjusted so that the mean is close to 500 and the standard deviation is close to 100. (Actual scores in a particular year have a slightly different mean and standard deviation.) Use the *Normal Curve* applet with $\mu = 500$ and $\sigma = 100$ to answer these questions:

(a) What proportion of SAT scores is above 640?

(b) What proportion of SAT scores is between 420 and 640? (If you drag one flag across the other, the applet shows the area between the flags.)

4. Because Web browsers have limited resolution, the *Normal Curve* applet can't always get exactly the values you want. Use the applet to come close to exact answers to these questions:

(a) How high must an SAT score be to fall in the top 10 percent of all scores?

(b) How high must an SAT score be to fall in the top 1 percent of all scores?

5. The 68–95–99.7 rule for normal distributions is a useful approximation. You can use the *Normal Curve* applet to see how accurate the rule is. Drag one flag across the other so that the applet shows the area under the curve between the two flags.

(a) Place the flags one standard deviation on either side of the mean. What is the area between these two values? What does the 68–95–99.7 rule say this area is?

(b) Repeat for locations two and three standard deviations on either side of the mean. Again, compare the 68–95–99.7 rule with the area given by the applet.

WRITING PROJECTS

1. Go online and look up information about "statistical quality control" and "six-sigma." Write a paragraph or two about what you learned and how it connects to variability in general and to standard deviation and the normal curve in particular.

2. Many social issues involve data and interpreting data. For example, income inequality (roughly speaking, the gap in income between people toward the top of the income scale and people toward the bottom) has increased in the past few decades. A good place to find data is on the Web site of the Census Bureau (**www.census.gov**). Click on "Income" and look for the latest report on income in the United States. Select a few facts from this detailed collection of income data to describe the extent of income inequality. Write a few paragraphs based on these facts.

3. Let's produce some data and describe them to gain insight into chance behavior. The mathematics of chance is the topic of Chapter 8, but for now, we will concentrate on data rather than math. You need two things: a standard six-sided die and a thumbtack with a rounded back (like a satellite dish). Toss the thumbtack 100 times (to speed things up, you could do 10 tosses of 10 tacks each) and record each outcome (pointing straight up or angled down). Also, toss the die 180 times and record each outcome (1, 2, 3, 4, 5, or 6). Use graphs and numbers to describe each set of results. Is the die roughly balanced, so that all six outcomes come up about equally? What about the thumbtack: Is "point up" or "point down" noticeably more common?

Suggested Readings

CLEVELAND, WILLIAM S. *The Elements of Graphing Data,* rev. ed., Hobart Press, Summit, N.J., 1994. A careful study of the most effective elementary ways to present data graphically, with much sound advice on improving simple graphs.

LESSER, LAWRENCE M. Critical values and transforming data: Teaching statistics with social justice, *Journal of Statistics Education* 15(1) (2007): www.amstat.org/publications/jse/v15n1/lesser.html. Resources for finding social justice data to extend Writing Project 2.

MOORE, DAVID S. *The Basic Practice of Statistics,* 5th ed., W. H. Freeman, New York, 2009. This text is a natural next step to learn more detail on all the material in Part II at about the same mathematical level. The first three chapters provide a more extensive treatment of the material of Chapter 5.

ROSSMAN, ALLAN J., and BETH L. CHANCE. *Workshop Statistics: Discovery with Data,* 4th ed., Wiley, Hoboken, N.J., 2011. A different approach to basic data analysis, using hands-on activities. There are several versions, keyed to graphing calculators and to several different software packages.

Suggested Web Sites

The Web site of the U.S. Census Bureau, **www.census.gov**, is a good source of information on many topics. The latest estimates for the populations of the United States and the world are on the home page, updated regularly. See what data you can find within "American Fact Finder" or the *Statistical Abstract of the United States.* Canadians can find similar help at the Web site of Statistics Canada: **www.statcan.gc.ca**.

Interested in data about schools, colleges, and students? The National Center for Education Statistics, **nces.ed.gov**, is the place to look. Go to the "What's New" section. There are also useful statistics applets at **www.shodor.org/interactivate/activities/**.

Des Jenson/Bloomberg via Getty Images

6

Exploring Data: Relationships

The world is full of relationships between variables. Indeed, it is hard to think of a variable that has no relationship with any other variable. Chapter 5 dealt with one variable at a time, but this chapter looks at the *relationship between two variables,* measured on the *same individuals.* Understanding relationships can have great importance to our lives. It was a major milestone in health awareness to conclude that smoking and life expectancy were connected—that smoking can influence life expectancy. People who smoke more cigarettes per day tend not to live as long as those who smoke fewer. So we call smoking an **explanatory variable** and life expectancy a **response variable**.

Understanding relationships is also crucial in business, such as the music industry. The controversy about file sharing in recent years led people to gather data to see what relationship, if any, there is between free downloading of music and sales of that same music during a given time period. Others have explored what relationship there is between music sales and radio airplay. Music recommendation Web sites help you discover new favorite songs by finding songs that have quantifiable similarities to songs you currently like. Other Web sites allow a songwriter to upload a song and find out how statistically similar it is to the songs in its database that are already known to be hit songs.

This chapter begins by visually exploring relationships between two variables by using a graph called a **scatterplot**. Scatterplots that suggest linear relationships lead us to Section 6.2, where we study a line that fits the data well enough to be able to make predictions. In Section 6.3, we look at correlation, a more precise numerical measure of relationship that tells us just how well the data fits a line. Section 6.4 shows us how to obtain the exact equation of the line we discussed informally in Section 6.2. Finally, Section 6.5 reminds us that these methods have limitations and are not enough to resolve questions about cause and effect.

Response Variable DEFINITION

A **response variable** measures an outcome or result of a study.

Explanatory Variable DEFINITION

An **explanatory variable** is a variable that we think explains or causes changes in the response variables.

6.1 Displaying Relationships: Scatterplot

Even when we know two quantitative variables are related, the relationship is rarely an exact line-shaped pattern, free of any "scatter" or deviation from that pattern. The most useful graph for displaying the relationship between two numerical variables (whether that relationship fits a trend perfectly or not) is a scatterplot.

Scatterplot DEFINITION

A **scatterplot** is a graph of plotted points showing the relationship between two numerical variables measured on the same individuals. In the case of one explanatory and one response variable, the values of the explanatory variable appear on the horizontal (x) axis and the values of the response variable appear on the vertical (y) axis. The values of the explanatory and response variable for one particular individual in the dataset become the x and y coordinates, respectively, of a point representing that individual in the scatterplot.

EXAMPLE 1
Beer and Blood Alcohol

How well does the number of beers a student drinks predict his or her blood alcohol content (BAC)? In a study at The Ohio State University, 16 student volunteers drank a randomly assigned number of cans of beer. Thirty minutes later, a police officer measured their BAC in grams of alcohol per deciliter of blood. Throughout the United States, the legal BAC limit is 0.08. Here are the data:

Student	1	2	3	4	5	6	7	8
Beers	5	2	9	8	3	7	3	5
BAC	0.10	0.03	0.19	0.12	0.04	0.095	0.07	0.06
Student	9	10	11	12	13	14	15	16
Beers	3	5	4	6	5	7	1	4
BAC	0.02	0.05	0.07	0.10	0.085	0.09	0.01	0.05

The students were equally divided between men and women and differed in weight and usual drinking habits. Because of this variation, many students don't believe that the number of drinks ingested predicts BAC well. What do the data say?

Figure 6.1 is a scatterplot of these data. Because we think that the number of beers helps explain BAC, "number of beers" is the explanatory variable. Notice in the

Figure 6.1 caption, the common usage of having the word "against" follow the response variable and precede the explanatory variable. We plot the explanatory variable (number of beers) on the horizontal axis. One student (Student #2) drank 2 beers and had a BAC of 0.03. This student's point on the scatterplot is (2, 0.03), above $x = 2$ and to the right of $y = 0.03$. We have marked this point in Figure 6.1.

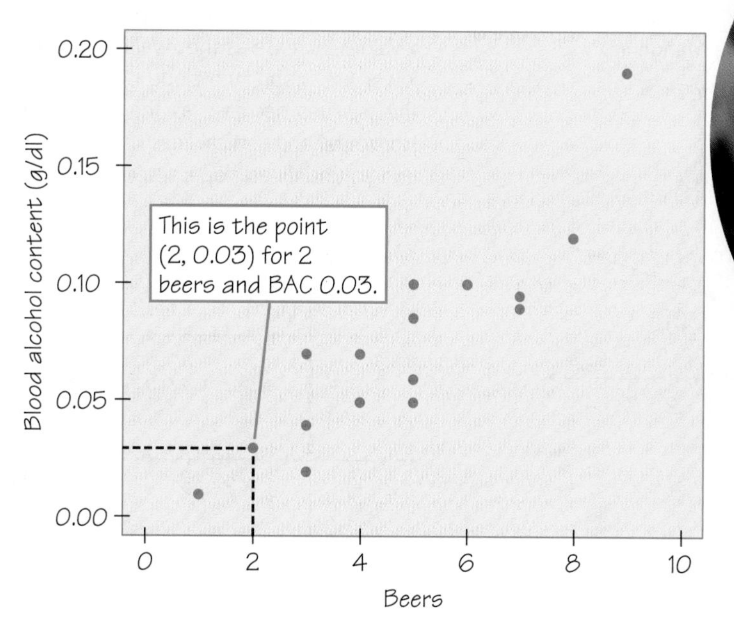

Justin Sullivan/Getty Images

Figure 6.1 A scatterplot of BAC (response variable) against the number of beers a student drinks (explanatory variable).

Examining a Scatterplot

PROCEDURE

1. Describe the *overall pattern* of a scatterplot by the *form, direction,* and *strength* of the relationship. (Be open to the possibility that there may be two or more trends or clusters in the same graph.)
2. Then look for any striking *deviations* from the pattern. Identify each occurrence of an **outlier**—an individual value that falls outside the overall pattern of the relationship, as introduced in Section 5.2.

When one looks at only a single quantitative variable (such as city gasoline mileage from Table 5.2), candidates for outliers are easy to identify numerically because they are usually the minimum or maximum value in the data or they satisfy a numerical criterion such as Exercise 5.31 on p. 203. When the dataset consists of ordered pairs, however, an outlier may or may not include an extreme value in one or both coordinates, so it is even more critical to use a graphical representation (i.e., a scatterplot) to look for deviations from the overall pattern. For example, consider the dataset:

X	0	2	4	5	6	8	10
Y	0	4	8	18	12	16	20

Algebra Review — Slope of a Line

The slope of a line is defined to be the ratio of vertical change to horizontal change. If we think of slope as $m = \frac{\text{change in } y}{\text{change in } x}$, the slope of the line below would be $m = \frac{1}{3}$.

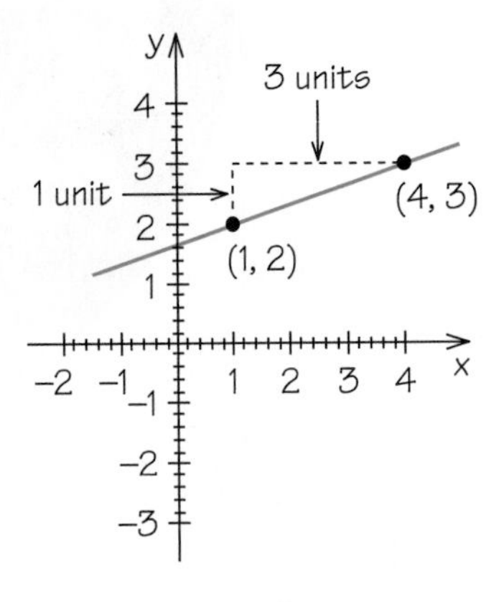

Given two points, the formula to calculate slope is $m = \frac{y_2 - y_1}{x_2 - x_1}$. By letting $(x_1, y_1) = (1, 3)$ and $(x_2, y_2) = (4, 0)$, the slope of the line below would be $m = \frac{y_2 - y_1}{x_2 - x_1} = \frac{0 - 3}{4 - 1} = \frac{-3}{3} = -1$.

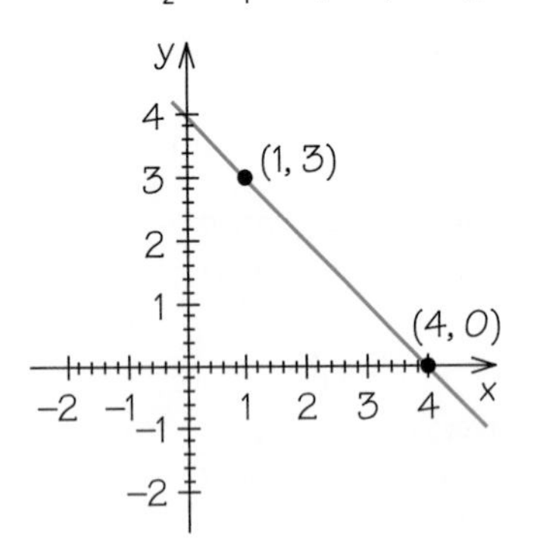

We can determine visually whether a given slope is positive, negative, zero, or undefined. In the first graph, the slope is positive because the y-values increase as the x-values increase. In the lower left graph, the slope is negative because the y-values decrease as the x-values increase. Horizontal and vertical lines have a slope of zero and an undefined slope, respectively.

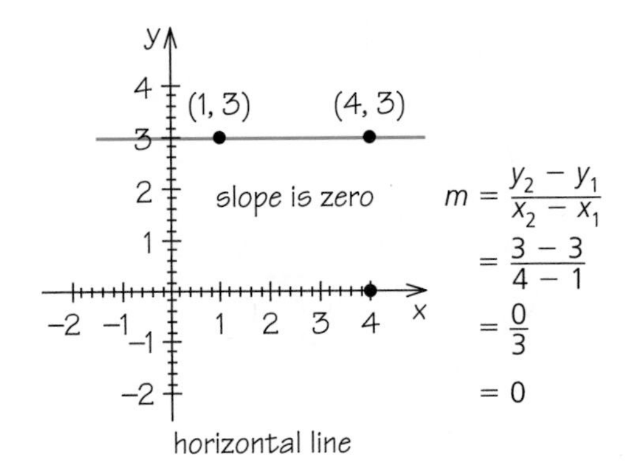

$$m = \frac{y_2 - y_1}{x_2 - x_1} = \frac{3 - 3}{4 - 1} = \frac{0}{3} = 0$$

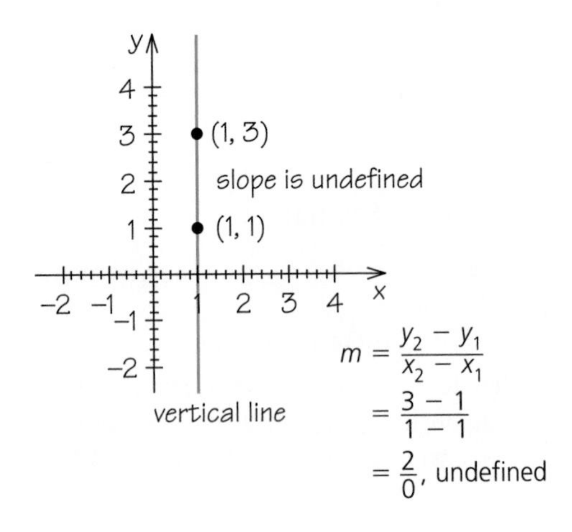

$$m = \frac{y_2 - y_1}{x_2 - x_1} = \frac{3 - 1}{1 - 1} = \frac{2}{0}, \text{ undefined}$$

If you graph the data in the table on the previous page, you will see that the point that deviates from the pattern does not include coordinates that are the minimum or maximum in either variable.

The *form* of the relationship in Figure 6.1 is roughly a straight-line pattern. If you look ahead a bit, Figure 6.3 (p. 215) shows a line drawn through the plot to describe the overall pattern. The *direction* of the relationship is clear: As the number of beers increases, BAC also increases. We call this a **positive association** between the two variables.

Positive Association DEFINITION

Two variables have **positive association** if their changes tend to be in the same direction. This means an increase in one variable tends to accompany an increase in the other variable. Also, a decrease in one variable tends to accompany a decrease in the other variable.

To visualize positive association, think of two neighboring merry-go-round carousel horses whose cycles are "in sync" so that they go up together and down together. If the neighboring horses' cycles are such that when one horse is up, the other is down, and vice versa, then this is the pattern of negative association, which we will see in Example 2.

Negative Association DEFINITION

Two variables have **negative association** if their changes tend to be in opposite directions. This means an increase in one variable tends to accompany a decrease in the other variable.

The *strength* of a relationship describes how closely the points in a scatterplot follow a simple form such as a straight line. Figure 6.1 shows only a small amount of scatter about the straight line, so the relationship is moderately strong. We will soon learn a numerical measure of the strength of a straight-line relationship.

EXAMPLE 2
SAT Mathematics Scores by State

Each year, more than 1 million high school seniors take the SAT Reasoning Test, which has three parts: Mathematics, Critical Reading, and Writing. We sometimes see individual states rated or ranked by the average SAT scores of their seniors. However, this is misleading because the mean SAT score is explained largely by what percent of a state's students take the SAT. For example, the scatterplot in Figure 6.2 shows a negative association between the mean score on the Mathematics section and the percent of test takers for the class of 2010. Each dot represents a particular state.

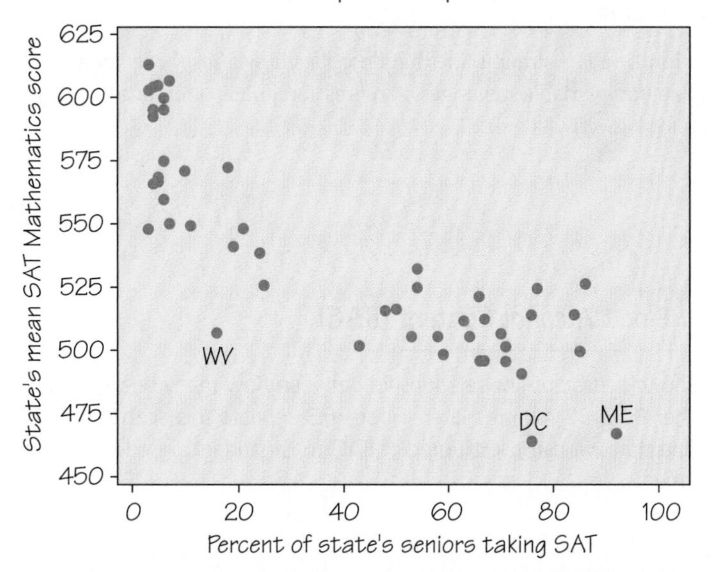

Figure 6.2 A scatterplot of states' mean SAT Mathematics scores (the response variable) against the percent of states' class of 2010 high school seniors who take the SAT (the explanatory variable).

The *form* of Figure 6.2 is a bit irregular, but there are two distinct clusters of states. In each state in the lower-right cluster, a majority or near-majority of high school seniors take the SAT, and the mean scores are low. In the upper-left cluster's states, 25 percent or fewer of seniors take the SAT—and these states have higher mean scores. Clusters in a graph suggest that the data describe several distinct kinds of individuals, and the two clusters in Figure 6.2 indeed describe two distinct sets of states.

There are two common college entrance examinations, the SAT and the ACT, and each state tends to prefer one or the other. In ACT-dominant states (the left cluster in Figure 6.2, where a smaller fraction of those states' seniors take the SAT), most students who do take the SAT are applying to selective, out-of-state colleges. This select group performs well. In SAT-dominant states (the right cluster), a higher percentage of seniors take the SAT, and this broader group has a lower mean score.

The relationship in Figure 6.2 also has a clear *direction*: States in which a higher percent of students take the SAT tend to have lower mean scores. This is true both between the clusters and within each cluster. That is, there is a **negative association** between the two variables.

There are no clear *outliers* in Figure 6.2, but each cluster does include a state whose mean SAT Mathematics score is lower than we would expect from the percent of its students who take the SAT. In the cluster of ACT-dominant states, this occurs with West Virginia (WV). In the cluster of SAT-dominant states, this occurs with the District of Columbia (DC)—which is actually a federal district, not a state—and Maine (ME).

6.2 Making Predictions: Regression Line

If a scatterplot shows a straight-line relationship, we would like to summarize this overall pattern by drawing a line on the scatterplot. A **regression line** summarizes the relationship between two variables, but only in a specific setting: when one variable helps explain or predict the other. That is, regression describes a relationship between an explanatory variable and a response variable.

Regression Line DEFINITION

A **regression line** is a straight line that describes how a response variable y changes as an explanatory variable x changes. A regression line is often used to *predict* the value of y for a given value of x.

EXAMPLE 3
Predicting Blood Alcohol Content (BAC)

Figure 6.1 shows a straight-line relationship between how many beers a student drinks and his or her BAC 30 minutes later. Figure 6.3 repeats this scatterplot and adds a regression line that we can use to predict BAC for a student based on the number of beers consumed.

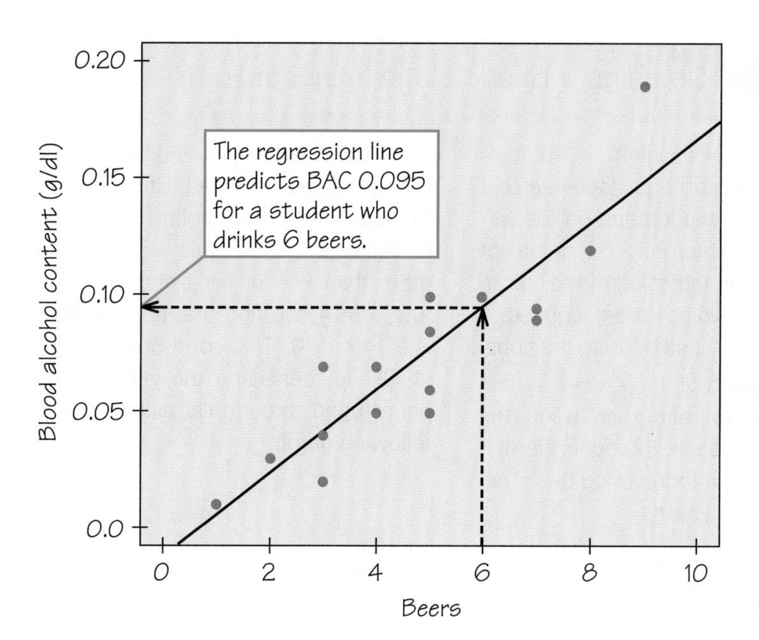

Figure 6.3 A regression line for predicting BAC from the number of beers that a student drinks.

Figure 6.3 shows the prediction in graphical form for a student who drinks 6 beers. Start at $x = 6$, go up to the line, and then head left to the y-axis. We hit the y-axis at BAC $= 0.095$. This is the BAC that corresponds to 6 beers, according to the regression line. (Recall that the legal limit for driving is 0.08.) The line represents only the overall pattern of the data, so the BAC of a randomly chosen student after 6 beers will probably not be exactly 0.095. But because the points for the 16 students in the Ohio State study are not far from the line, we expect the prediction to be reasonably accurate.

It is easier to use the *equation of the line* for prediction. With the application of formulas that will be given in Section 6.4, the equation of the line in Figure 6.3 is

$$\text{predicted BAC} = -0.0127 + 0.01796 \times \text{beers}$$

For a student who drinks 6 beers, we have

$$\text{predicted BAC} = -0.0127 + (0.01796 \times 6) = 0.095$$

Because two points determine a unique line, you could plot a line by using its equation to determine any two particular points that lie on that line, plot those points, and then draw the line through them. For example, from the equation

$$\text{predicted BAC} = -0.0127 + 0.01796 \times \text{beers}$$

we just determined that one point is (6, 0.095). By plugging in $x = 2$, we obtain another point (2, 0.023). Drawing the line through those two points yields the line in Figure 6.3.

Algebra Review Graphing a Line in Slope-Intercept Form

When a linear equation is written in the form $y = mx + b$, it is said to be in slope-intercept form. In this form, b is the location on the vertical axis called the y-intercept. This intercept represents one point on the graph of the line. In order to graph a line, two points are required. A second point can be obtained by using the slope.

Consider the graph of $y = \frac{2}{3}x - 1$ by first plotting the y-intercept and then using the slope to find a second point. By connecting the y-intercept and the point found using the slope, we have the following graph:

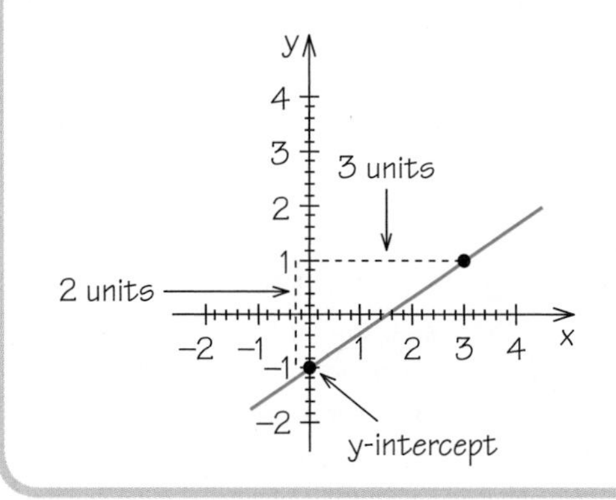

Now consider $y = -0.5x + 2$. First, plot the y-intercept. Instead of using the slope to obtain a second point, it may be easier to evaluate the formula with a value for x other than $x = 0$. For example, if we substitute, $x = 4$, we find that $y = -0.5(4) + 2 = -2 + 2 = 0$. This corresponds to the point (4, 0). By connecting the y-intercept and the point found through calculation, we have the following graph:

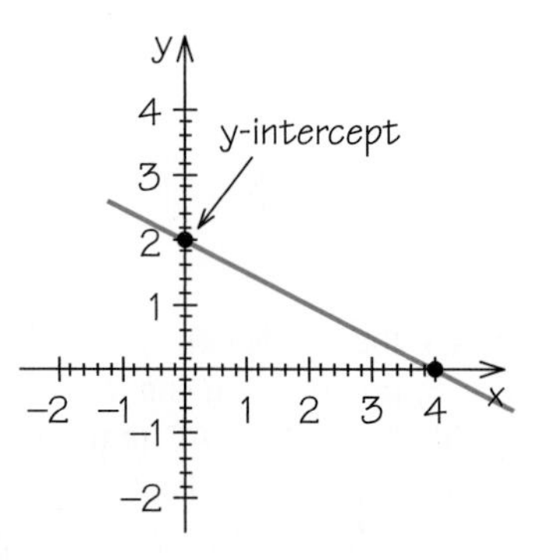

Statistical software and many calculators will give you the equation of a regression line from data that you enter. You should know how to use a regression line even if you don't look into the details needed to calculate the line from data. First, recall some basic facts about the (slope and intercept) coefficients in the equation of a line.

Equation of a Regression Line — DEFINITION

Suppose that y is a response variable (plotted on the vertical axis) and x is an explanatory variable (plotted on the horizontal axis). If we call $\hat{y}$ the predicted value of y, then the resulting regression line for predicting y from x has an equation of the form[1]:

$$\hat{y} = mx + b$$

In this equation, m is the **slope**, which is the amount by which y changes when x increases by 1 unit. The number b is the **y–intercept**, which is the value of y when $x = 0$ (i.e., when the line intercepts the y-axis).

The slope of the line in Example 3 is $m = 0.01796$. This says that as we move to the right along the line, predicted BAC goes up by 0.01796 for each additional beer that a student drinks. So, if a student has 3 additional beers, the BAC would increase by $3 \times 0.01796 = 0.05388$ g/dl. The slope tells us how quickly y changes

[1]The letters m and b are from the slope-intercept form from algebra class, but be aware that some technologies or statistics books use different letters, such as $y = a + bx$. Always play it safe by checking that the letter used for the slope corresponds to the number multiplied by the explanatory variable.

as we change x, so it is important for understanding the data. The slope is positive ($m > 0$) when there is a positive association between the two variables. It is negative when there is a negative association.

You might think that a big slope (either positive or negative) says that there are big changes in y as x changes and that a small slope means that x has little influence on y. Unfortunately, the size of a slope is affected by the units in which we measure the two variables. The slope of the regression in Example 3 is $m = 0.01796$ when we measure BAC y in grams of alcohol per deciliter (g/dl) of blood. That is, when the number of beers consumed increases by 1, BAC increases by 0.01796 grams. There are 1000 milligrams in a gram, so if we measured BAC in milligrams of alcohol per deciliter of blood, the slope would be 1000 times as large: $m = 17.96$. *You can't say how important a relationship is just by looking at how big the slope is.*

The intercept of the regression line in Example 3 is $b = -0.0127$. This is the predicted value of y when $x = 0$. Although we need the value of the intercept to draw the line, it is statistically meaningful only when x can actually take values close to zero. Even then, you should think of the intercept as describing the line rather than taking it seriously as a prediction. If a student drinks no beers, his or her BAC should be exactly zero. The intercept of the estimated regression line in Example 3 is close to zero, but it is not exactly zero.

6.3 Correlation

A scatterplot displays the form, direction, and strength of the relationship ("co-relation") between two numerical variables. Straight-line relations are particularly important because a straight-line pattern is quite common and is easy to interpret. We say a straight-line association is strong if the points lie close to a line and weak if they are widely scattered about a line. But this language is vague. We need to follow our strategy for data analysis by using a numerical measure along with the graph. Correlation is the measure we use. **Correlation** is usually denoted as r, thanks to 19th-century statistician Sir Francis Galton, who was studying related ideas of **r**egression and **r**eversion.

Correlation DEFINITION

The **correlation** measures the direction and strength of the straight-line relationship between two numerical variables.

The correlation r is always a number between -1 and 1, inclusive. It has the same sign as the slope of a regression line for that dataset: $r > 0$ for positive association and $r < 0$ for negative association.

Perfect correlation $r = 1$ or $r = -1$ occurs only when all points lie exactly on a straight line. The correlation moves away from 1 or -1 as the straight-line relationship gets weaker. Correlation $r = 0$ indicates no straight-line relationship.

EXAMPLE 4
Scatterplots and Correlation

In general, it is not always easy to guess the value of r from the appearance of a scatterplot because the appearance of a scatterplot can be altered (without changing the value of r) simply by changing the plotting scales. To make the value of r clearer, the scatterplots in Figure 6.4 all involve the same scale for the horizontal and vertical axes and the same standard deviation value for the x and y variables. From these scatterplots, we are able to see how values of r closer to 1 or −1 correspond to stronger straight-line relationships. You can find online a correlation-guessing applet (try entering into a search engine the words *correlation* and *guessing*) that will give you practice in giving a reasonable visual estimate from a scatterplot with equal axis scales.

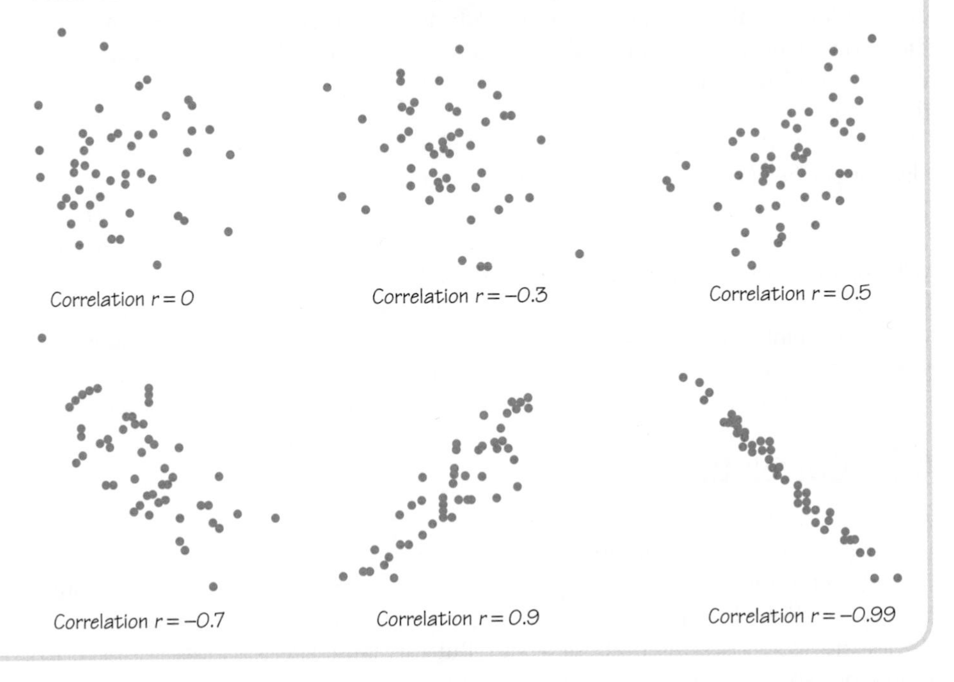

Figure 6.4 How the correlation r measures the direction and strength of straight-line association.

We said that Figure 6.1 (p. 211) shows a moderately strong positive straight-line relationship between how many beers a student drinks and his or her BAC. The correlation between these variables is $r = 0.894$. Figure 6.2, despite the clusters, also shows a very strong straight-line relationship between the percent of a state's high school seniors who take the SAT exam and their mean SAT score. The association is negative: Higher percents taking the SAT go with lower mean scores. The correlation is $r = -0.877$.

Here are more facts about the correlation r:

1. Correlation makes no sense for non-numerical variables (such as ethnicity and occupation).
2. The correlation r measures the strength of only a straight-line relationship. It does *not* measure the strength of a relationship with a curved pattern, no matter how strong that curved relationship is.
3. Unlike regression, correlation makes no distinction between explanatory and response variables. It makes no difference which variable you call x and which you call y in interpreting (or calculating) a correlation.
4. The correlation r does not change when we change the units of measurement of x, y, or both. For example, measuring height in inches rather than centimeters, and weight in pounds rather than kilograms, does not change the correlation between height and weight. (This is different from the case of the slope of a regression line, which changes when the units of x and y change.) The correlation r itself has no unit of measurement; it is just a number.
5. Like the mean and standard deviation, the correlation is affected strongly by a few outlying observations. Use r with caution when outliers appear in the scatterplot, especially for small datasets.

In practice, you will use a calculator or software to find the correlation from keyed-in data, as described in Spotlight 6.1. That's fortunate, because using the PROCEDURE box formula for correlation is quite a bit of work. Nonetheless, students skilled in algebra can see how the formula below helps us appreciate correlation properties such as #3 and #4 on the preceding page.

Formula for Correlation PROCEDURE

Suppose that we have data on variables x and y for n individuals. The means and standard deviations of the two variables are $\bar{x}$ and s_x for the x-values, and $\bar{y}$ and s_y for the y-values. The correlation r between x and y is

$$r = \frac{1}{n-1}\left[\left(\frac{x_1-\bar{x}}{s_x}\right)\left(\frac{y_1-\bar{y}}{s_y}\right) + \left(\frac{x_2-\bar{x}}{s_x}\right)\left(\frac{y_2-\bar{y}}{s_y}\right) + \cdots + \left(\frac{x_n-\bar{x}}{s_x}\right)\left(\frac{y_n-\bar{y}}{s_y}\right)\right]$$

Step 1: Find the mean $\bar{x}$ and standard deviation s_x of the x values. Find the mean $\bar{y}$ and the standard deviation s_y of the y values. (Use your calculator and refer to Spotlight 5.2.)

Step 2: Find the standardized value $(x - \bar{x})/s_x$ for each of the x values.

Find the standardized value $(y - \bar{y})/s_y$ for each of the y values.

Step 3: Insert your numbers from Steps 1 and 2 into the formula for r. Remember that n is the number of (x, y) points or ordered pairs plotted in the scatterplot.

You can solidify your understanding by implementing this procedure in Exercise 28 (p. 233). This formula starts by *standardizing* each observation value (as was done in Exercise 54 (p. 205) in Chapter 5). That is, subtract the mean for that variable from the observation and then divide by the standard deviation. Standardizing turns each original data value into "number of standard deviations above the mean." (Note that a value of, say, –2 indicates 2 standard deviations *below* the mean.) This removes the original units and explains why r has no units and doesn't change when we change the units of x or y. The formula says that the correlation is an average of the products of the standardized x and y values for n individuals.

SPOTLIGHT 6.1

Correlation Calculation

While the PROCEDURE box formula for correlation has conceptual clarity, it can be tedious to apply to large datasets. Even with the most basic calculator, you'll get the same answer faster using this mathematically equivalent but more computationally efficient formula:

$$\frac{(x_1y_1 + x_2y_2 + \cdots + x_ny_n) - n\bar{x}\bar{y}}{(n-1)s_xs_y}.$$

If you have a *scientific calculator*, select the calculator mode to be able to do two-variable regression statistics, clear out any old data, and then enter your new x and y values. Once the data is entered, you can find the correlation (or regression-line slope and y-intercept, for that matter) by hitting the appropriate key(s). On the Internet, you can find Web sites that can help with keystrokes for specific models.

If you have a *graphing calculator* in the TI-83/84+ family (and you have the two variables entered in two columns using STAT → EDIT), then hit this sequence of buttons: STAT → TESTS → LinRegTTest ENTER. Enter what data columns are your independent (x-list) and dependent (y-list) variables. (Note that the labels L1 through L6 are the keys 1 through 6 respectively). Then select Calculate, hit ENTER, and scroll to the end of the output to see the correlation r. (Note that the output also contains the slope and y-intercept coefficients for the least-squares regression line we will see in Section 6.4.)

SPOTLIGHT 6.2

Regression Toward the Mean

Galton studied predicting the heights of men from the heights of their fathers. He found that tall fathers tend to have taller than average sons, but that there is a reversion or regression (i.e., going back) of a son's height toward the average height for sons. If the mean height of men is about 70 inches, verify that this equation demonstrates this effect:

$$\text{son's height} = 0.516 \times (\text{father's height}) + 33.73.$$

It goes the other way, too—a very short father tends to have a son who is also shorter than average, but not quite as short as his father.

More generally, the idea that extreme measurements that include some random variation are likely to be followed by measurements that are not quite as extreme is called "regression toward the mean." This dynamic shows up in many areas involving some combination of skill and luck. Students who are the very top performers on one test will tend to do a bit worse on average on the next test, and the very worst performers will tend to have some improvement. An example of this idea in the world of sports is the "*Sports Illustrated* Cover Jinx," in which athletes appear on the magazine's cover after an "outlier performance," but their subsequent performance is usually less impressive. Singer-songwriter Christine Lavin offers another example in this couplet from her song "Attractive Stupid People":

"but the problem is the kids won't look as good as mom or dad,
and they're always slightly smarter, which drives their pretty parents mad."

The following formula is an algebraic representation of regression to the mean because so long as r is not equal to 1 or −1, the predicted standardized value of y is closer to its mean than the standardized value of x is to its mean. In other words, the $\hat{y}$ value (e.g., the son's predicted height) has a less extreme deviation than the x value (e.g., the height of that son's father).

$$\frac{\hat{y} - \bar{y}}{s_y} = r\frac{x - \bar{x}}{s_x}$$

To make sure that you understand how to obtain a correlation value, it is helpful to practice by taking a particular dataset and trying the formula in the PROCEDURE box as well as trying the method in Spotlight 6.1. For example, let's identify some of the pieces using the data in Example 1. The sample size n is 16, the explanatory variable x is "number of beers" and the response variable y is "blood alcohol content." You would have 16 ordered pairs, such as $(x_1, y_1) = (5, 0.10)$. You can also verify that, to two significant figures, the variable x has a mean $\bar{x} = 4.8$ and standard deviation $s_x = 2.2$, while the variable y has a mean $\bar{y} = 0.074$ and standard deviation $s_y = 0.044$. Now, see if you can plug all the numbers into the formula and obtain the value $r = 0.9$.

6.4 Least-Squares Regression

In Example 3, we used the straight line given by the equation

$$\text{predicted BAC} = -0.0127 + 0.01796 \times \text{beers}$$

to predict BAC from the number of beers consumed. How did we get this equation? We will now see that the equation is the result of saying what we mean by the *best* line for predicting BAC from beers consumed. Once we say exactly what we mean by "the best line," finding that line becomes a mathematical problem. The line in Example 3 is the solution to this problem for the beer and BAC data.

For a given scatterplot, different people might draw different summarizing lines by eye. This is especially true when the points are widely scattered. We need a way to draw a regression line that doesn't depend on our visual guess as to where the line should go. No line will pass exactly through all the points (unless $r = 1$ or -1), but we want one that is as close as possible. We will use the line to predict y from x, so we want a line that is as close as possible to the points in the *vertical* dimension. That's because the prediction errors that we make are errors in the y variable, which is the vertical dimension in the scatterplot.

The table in Example 1 shows that student #12 drank 6 beers and was observed to have a BAC of 0.10. However, the regression-line equation in Example 3 showed that the predicted BAC for a student who drinks 6 beers is 0.095. These values are close, but are not the same. Indeed, from Figure 6.3, we can see the observed data point (6, 0.10) lies a bit off the line, and the vertical deviation of this gap is the prediction error, as follows:

$$\text{prediction error} = \text{observed BAC} - \text{predicted BAC} = 0.10 - 0.095 = 0.005$$

When the observed response lies above the line (e.g., the data point for Student #7, who had only 3 beers but a BAC of 0.07), the error is positive. And when the response lies below the line (e.g., the data point for Student #10, who had 5 beers but a BAC of only .05), the error is negative. The most common way to make the collection of prediction errors for the entire dataset as small as possible is *least-squares regression*. The line in Figure 6.3 is the **least-squares regression line.**

Least-Squares Regression Line DEFINITION

The **least-squares regression line** is the line that makes the sum of the squares of the vertical distances of the data points from the line the least value possible.

SPOTLIGHT 6.3

Regression and Correlation in Action: College Success

Can college success be predicted? Colleges with more applicants than spaces want to admit students who are most likely to succeed. There are many ways (e.g., graduation rate, quality of job obtained upon graduation, etc.) that we might define what successful transition into college means, but most people focus on grades during the first year of college. There are also many choices of what variable might help admissions officers predict this: high school GPA, number of advanced (e.g., AP) classes taken, scores on standardized tests (ACT or SAT), and so on. No one of these variables (or even all of them together) will generate a perfect prediction for an individual person because many other variables (such as work ethic) are not directly taken into account. According to a 2008 report by the College Board, first-year college GPA has a correlation (corrected for having a range restricted by analyzing only admitted and enrolled students) of about 0.5 with any one of the three SAT section tests. This is a measure of (predictive) validity for the SAT, and it turns out that squaring this number yields the interpretation that the SAT alone explains about one-quarter of the variation of first-year college GPA. The (adjusted) correlation between first-year GPA and high school GPA is only slightly higher at 0.54.

Multiple regression extends regression to allow more than one explanatory variable to help explain a response variable. When the SAT and high school GPA are used simultaneously to try to predict first-year college GPA, the adjusted correlation jumps up to 0.62. The dependent variable of the associated regression equation can yield an "index" that admissions officers can use to create a rough ordering of applicants.

"Least squares" refers to our method of handling the errors, not to the function used in the mathematical model we are trying to fit. (Graphing calculators can also use the least-squares method to find the best quadratic model, for example, but our focus here is the more common example of lines.) Now, how can we find this best-fitting line from data? Starting with n observations on variables x and y, finding the line that makes the sum of the squares of the vertical errors as small as possible is a mathematical problem. Here is the solution to this problem.

Finding the Least-Squares Regression Line PROCEDURE

1. From our data on an explanatory variable x and a response variable y for n individuals, calculate the means $\bar{x}$ and $\bar{y}$ and the standard deviations s_x and s_y of the two variables.
2. Calculate the correlation r (recall Spotlight 6.1, p. 219).
3. The regression line's slope m is given by
$$m = r\frac{s_y}{s_x}$$
4. The regression line's y-intercept b is given by
$$b = \bar{y} - m\bar{x}.$$
5. If we call $\hat{y}$ the predicted value of y, then the equation of the least-squares regression line for predicting y from x (we also can say from "regressing y on x") can now be stated:
$$\hat{y} = mx + b.$$

This equation gives insight into the behavior of least-squares regression by showing that it is related to the means and standard deviations of the x and y observations and to the correlation between x and y. For example, it is clear that the slope m always has the same sign as the correlation r. In practice, you don't need to calculate the means, standard deviations, and correlation first. Statistical software or many calculators can give the slope m, intercept b, and equation of the least-squares line from keyed-in values of the variables x and y. (Recall the footnote on the notation in Section 6.2.) Notice that if you confuse whether y is your explanatory or your response variable, you will get a different slope value.

EXAMPLE 5
Least-Squares Regression of BAC on Number of Beers

Go back to the data in Example 1. Use your calculator to verify that the mean and standard deviation of x, number of beers consumed, are

$$\bar{x} = 4.8125 \quad \text{and} \quad s_x = 2.1975$$

The mean and standard deviation of y, BAC, are

$$\bar{y} = 0.07375 \quad \text{and} \quad s_y = 0.04414$$

The correlation between the number of beers and BAC is $r = 0.8943$. The least-squares regression line of BAC y on number of beers x has slope

$$m = r\frac{s_y}{s_x} = 0.8943 \times \frac{0.04414}{2.1975}$$
$$= 0.01796$$

and intercept

$$b = \bar{y} - m\bar{x} = 0.07375 - (0.01796)(4.8125)$$
$$= -0.0127$$

The equation of the least-squares line is therefore

$$\hat{y} = -0.0127 + 0.01796x,$$

just as we claimed earlier.

When doing calculations like this by hand, you may need to carry extra decimal places in the intermediate calculations to get accurate values of the slope and intercept and not round until your final answer. Using software or a calculator with a regression function eliminates this worry.

You now see that correlation and least-squares regression are connected closely. The expression $m = rs_y/s_x$ for the slope says that along the regression line, a change of one standard deviation in x corresponds to a change of r standard deviations in y. When the variables are correlated perfectly ($r = 1$ or $r = -1$), the change in the predicted response is the same (in standard deviation units) as the change in x. Otherwise, because $-1 \leq r \leq 1$, the change in the predicted y is less than the change in x. As the correlation grows less strong, the prediction moves less in response to changes in x.

6.5 Interpreting Correlation and Regression

Correlation and regression are among the most-used statistical methods. Here are a few cautions to keep in mind when you use or see these methods.

Both the correlation r and the least-squares regression line can be influenced strongly by a few outlying points. Always make a scatterplot before doing any calculations. Here is an artificial example that illustrates what can happen.

EXAMPLE 6
Beware the Outlier!

Figure 6.5 shows a scatterplot of data that have a strong positive straight-line relationship. In fact, the correlation is $r = 0.987$, close to the $r = 1$ of a perfect straight line. The line on the plot is the least-squares regression line for predicting y from x. One point is an extreme outlier in both the x and y directions. Let's examine the influence of this outlier.

First, suppose we drop the outlier. The correlation for the five remaining points (the cluster at the lower left) is $r = 0.523$. The outlier extends the straight-line pattern and greatly increases the correlation.

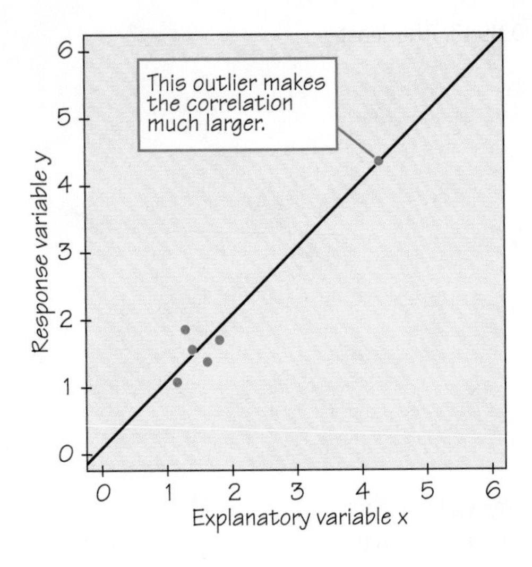

Figure 6.5 The outlier increases the correlation and fixes the location of the least-squares line.

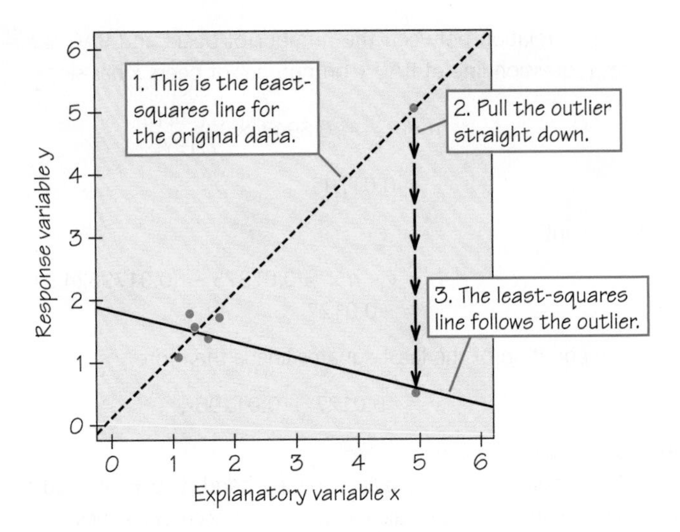

Figure 6.6 Moving the outlier unduly changes the correlation and moves the least-squares line.

Next, grab the outlier and pull it straight down, as in Figure 6.6. The least-squares line chases the outlier down, pivoting until it has a negative slope. This is the least-squares idea at work: The line stays close to all six points. However, its location is determined almost entirely by the one outlier. Of course, the correlation is now also negative, $r = -0.796$. Never trust a correlation or a regression line if you have not plotted the data. You can explore this in an interactive dynamic manner by using an online applet such as http://illuminations.nctm.org/LessonDetail.aspx?ID=L456. (This applet can also be used to verify the comment in Section 6.3 that changing the scales of a graph's axes can affect how a correlation looks.)

Even if the correlation is strong and there are no outliers in the data that we used to find our regression line, we also must not be quick to extrapolate and make predictions well beyond the data collected: Just because the data fits a particular linear trend over a window, there is no guarantee that that trend will continue into the future. For example, the rate of growth of a newborn may fit a line with a steep slope for the first several months, but then the slope (while still positive) starts to decrease.

A good way to see how outlying points can influence the correlation and the regression line is to use the *Correlation and Regression* applet. Applet Exercise 1 (p. 237) asks you to animate Example 6 above, watching *r* change and the regression line move as you pull the outlier down.

Correlation and regression *describe* relationships. *Interpreting* relationships requires more thought. *Often the relationship between two variables is influenced strongly by other variables.* You should always think about the possible effect of other variables before you draw conclusions based on correlation or regression.

EXAMPLE 7
Money Helps SAT Scores?

The College Board, which administers the SAT Reasoning Test, offers this information on its Web site about Class of 2010 seniors who take the test (the 38 percent of test-takers who did not respond to this income question had a mean score of 515):

Family income (in $1000s)	0–20	20–40	40–60	60–80	80–100	100–120	120–140	140–160	160–200	Over 200
Mean Math SAT score	460	479	500	514	529	541	546	554	561	586

This information suggests a strong positive association between test score and a test taker's family income. But there's no direct mechanism—wealthy families are not sending secret bribes to the College Board. It may simply be that children of wealthy parents are more likely to have advantages such as well-educated role models, high expectations, access to extra tutoring or test preparation, smaller class sizes, and schools with more experienced, qualified teachers.

Example 7 brings us to the most important caution about correlation and regression. When we study the relationship between two variables, we often hope to show that changes in the explanatory variable *cause* changes in the response variable. A strong association between two variables is not enough to draw conclusions about cause and effect. Sometimes an observed association really does reflect cause and effect. Drinking more beer does cause an increase in BAC. But in many cases, as in Example 7, a strong association is explained by other variables that influence both x and y. Here is another example.

EXAMPLE 8
Evaluation Correlation?

Grades that students earn in courses are correlated positively with the ratings that students give on anonymous end-of-course surveys administered by the university. One very simple interpretation is that instructors give easy tests with "low standards," which in turn causes students to express appreciation through high instructor ratings. But perhaps there is a third variable that drives the other two variables: A professor who is a skillful teacher and motivator may be more likely both to be rated well and to inspire high performance. Or perhaps a course that includes group projects (rather than only in-class, timed tests) as a significant component of the grade naturally results in higher levels of both performance and satisfaction. Or perhaps courses that have higher grade distributions are more likely to be upper-level courses for majors in that subject, and such students would be more prepared and favorably inclined towards the course.

EXAMPLE 9
Does Running Lead to Winning in Football?

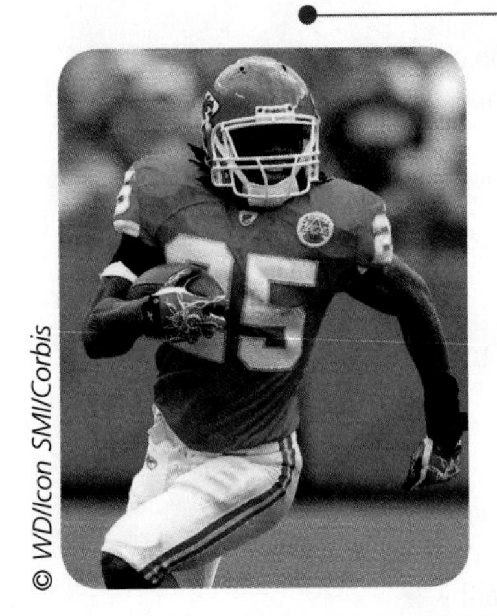

© WD/Icon SMI/Corbis

A football broadcaster discussed how often a team wins when it runs the ball at least 30 times in a game. For the 2010 NFL regular season, the correlation between wins and number of running plays was indeed moderately positive ($r = 0.48$). Could this mean that running causes winning—that all any team has to do to win more games is to run the ball more? No. In the extreme, if a team did only running plays, the other team would simply adjust its defense to focus on and stop the run. Basically, once teams get a good lead in a game (regardless of their mix of special teams, running, and passing), they tend to start running the ball more often as a way to minimize the risk of losing the ball (pass plays are riskier) and to use up the clock faster (an incomplete pass stops the clock). And when teams get far behind late in the game, they begin passing more often as a last chance to catch up before time runs out.

Correlations such as that in Example 9 are sometimes called "nonsense correlations." The correlation is real, but it is nonsense to conclude that increasing the number of running plays will cause an increase in the number of wins that season. So correlations require thoughtful interpretation, not just computation.

Association Does Not Imply Causation RULE

An association between an explanatory variable x and a response variable y, even if it is very strong, is not by itself good evidence that changes in x actually cause changes in y. Causation also requires (1) to demonstrate that you had ruled out the possibility that the change in the response variable was due to any other variable besides the explanatory variable, (2) to show that the association happens under a variety of conditions, and (3) to have a reasonable mechanism or model to explain how x causes changes in y.

Here is a final example in which we use a scatterplot, correlation, and a regression line to understand data.

EXAMPLE 10
What Does Growth Hormone Do in Adults?

In most species, adults stop growing but still release growth hormone from the pituitary gland to regulate metabolism. Physiologists subjected groups of adult rats to various conditions that activated muscle tissue that was either fast-twitch (as sprinters use) or slow-twitch (as distance runners use). They then measured levels of a bioassayable form of growth hormone (BGH) in the blood and in pituitary tissue. Units are 100s of nanograms per milliliter of blood and micrograms per milligram of tissue, respectively.

Here are the data, courtesy of neuroscientist Kristin Gosselink:

blood	15.8	20.0	26.7	25.0	23.0	23.8	24.7	16.3	0.8	0.8
tissue	38.0	36.7	27.8	28.3	34.9	34.1	33.2	32.7	38.1	39.1
blood	0.6	10.8	37.6	41.3	39.0	57.5	84.8	82.8	28.8	16.5
tissue	43.9	42.8	19.3	13.7	11.2	14.2	9.7	9.5	31.7	32.8

SOURCE: G.E. McCall et al., Muscle afferent-pituitary axis: A novel pathway for modulating the secretion of a pituitary growth factor, *Exercise and Sport Science Reviews* 29 (2001): 164-169.

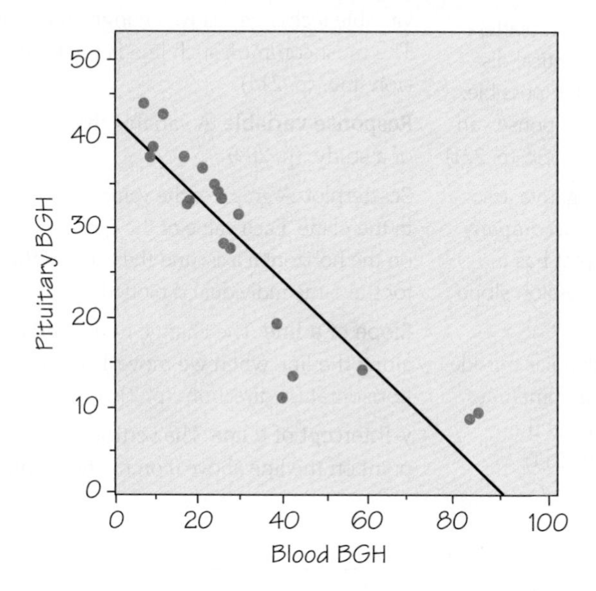

Figure 6.7 This scatterplot of BGH level in pituitary tissue versus BGH level in the blood shows a strong negative association.

Figure 6.7 is a scatterplot of these data. The plot shows a strong negative straight-line association with correlation $r = -0.90$. When there is a higher BGH level in the blood, we can assume that means BGH must have been recently secreted by the pituitary gland so that less BGH now remains in pituitary tissue. The least-squares regression line is

$$\hat{y} = 41.081 - 0.43343x, \text{ or}$$

$$\text{predicted pituitary BGH} = 41.081 + ((-0.43343) \times \text{blood BGH}).$$

The slope $m = -0.43343$ is negative, which reflects how blood and pituitary tissue levels of BGH move in opposite directions. The y-intercept $b = 41.08$ is the estimated amount of BGH that the pituitary gland has if it does not release any into the blood.

Furthermore, the two highest points of the scatterplot represent groups of rats whose slow-twitch muscles were activated, while the five lowest points on the scatterplot involved the activation of fast-twitch muscles in separate groups of rats. The point (37.6, 19.3) comes from a group of rats who were exercised on a treadmill to activate fast and slow-twitch muscles simultaneously. The remaining points represent groups that were untreated. These data come from an *experiment* that assigned rats randomly to treatment (or no treatment) conditions. This makes us reasonably confident that slow-twitch muscle activation *causes* a decrease in BGH secretion and that fast-twitch muscle activation *causes* an increase. We will discuss experiments in detail in the next chapter (in Sections 7.5 and 7.6).

REVIEW VOCABULARY

Correlation A measure of the direction and strength of the straight-line relationship between two numerical variables. Correlations take values between −1 and 1, with the same sign as the regression line slope. (p. 217)

Explanatory variable A variable that attempts to justify the observed outcomes. (p. 210)

Least-squares regression line A line drawn on a scatterplot that makes the sum of the squares of the vertical distances of the data points from the line as small as possible. The regression line can be used to predict the response variable y for a given value of the explanatory variable x. (p. 221)

Negative association Two variables have negative association if an increase in one variable tends to accompany a decrease in the other variable. The scatterplot has a northwest-to-southeast pattern, and the regression slope and correlation are both negative. (p. 213)

Outlier An outlier in a scatterplot is a point that lies outside the overall pattern of the other points. Outliers sometimes strongly influence the value of the correlation and the position of the least-squares regression line. (p. 211)

Positive association Two variables have positive association if an increase in one variable tends to accompany an increase in the other variable. The scatterplot has a southwest-to-northeast pattern, and the regression slope and correlation are both positive. (p. 213)

Regression line Any line that describes how a response variable y changes as we change an explanatory variable x. The most common such line is the least-squares regression line. (p. 214)

Response variable A variable that measures an outcome of a study. (p. 209)

Scatterplot A graph of the values of two variables as points in the plane. Each value of the explanatory variable is plotted on the horizontal axis, and the value of the response variable for the same individual is plotted on the vertical axis. (p. 210)

Slope of a line The change in the vertical (y) direction along the line when we move 1 unit to the right in the horizontal (x) direction. (p. 216)

y-intercept of a line The vertical (y) coordinate of the point on the line above 0 on the horizontal (x) axis. (p. 216)

SKILLS CHECK

1. You have data for many families about the parents' income and the years of education that their eldest child completes. When you make a scatterplot, the explanatory variable on the x-axis

(a) is the parents' income.
(b) is years of education.
(c) doesn't matter.

2. The outcome or result of a study is measured by a/an ______ variable.

3. The data in Example 1 (p. 210) consist of:

(a) 16 ordered pairs of values.
(b) 2 unpaired sets of values, each of which has 16 values.
(c) 32 ordered pairs of values.

4. The explanatory variable is plotted on the ___-axis of a scatterplot.

5. If two variables have a negative association, an increase in one variable tends to accompany ____ in the other variable.

(a) an increase
(b) a decrease
(c) no change

6. You expect to see a _______ association between the parents' income and the years of education that their oldest child completes.

7. Figure 6.8 is a scatterplot of reading test scores against IQ test scores for 15 fifth-grade children. There is one low outlier in the plot. The IQ and reading scores for this child are

(a) IQ = 10, Reading = 124.
(b) IQ = 124, Reading = 72.
(c) IQ = 124, Reading = 10.

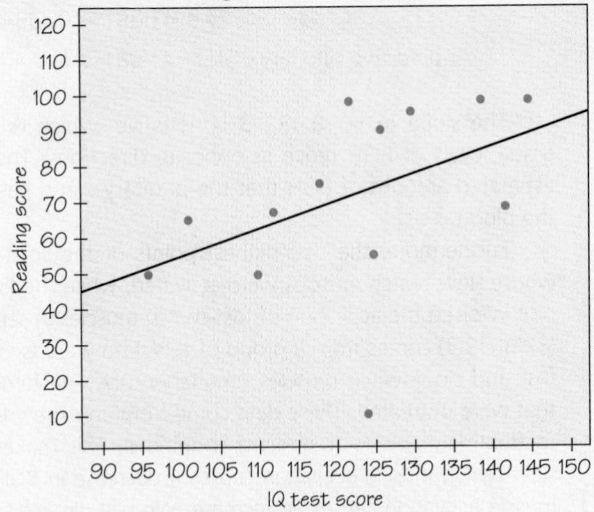

Figure 6.8 A scatterplot of the reading test scores of fifth-grade children (the response variable) against the children's IQ scores (the explanatory variable). For Skills Check 7, 8, 9.

8. The line in Figure 6.8 is a regression line for predicting reading score from IQ score. If another child in this class has an IQ score of 125, then ____ is the multiple of 10 to which that predicted reading score would be closest.

9. The slope of the line in Figure 6.8 is closest to

(a) –1.
(b) 0.
(c) 1.

10. The points on a scatterplot lie close to the line whose equation is $y = 2 - 5x$. The slope of this line is ______ .

11. Starting with a fresh bar of soap, you weigh the bar each day after you take a shower. Then you find the regression line for predicting the weight from the number of days elapsed. The slope of this line will be

(a) positive.
(b) negative.
(c) can't tell without seeing the data.

12. Fred keeps his savings in his mattress. He began with $500 from his mother and adds $100 each year. In the form $y = mx + b$, the equation for his total savings y after x years would be $y =$ __________.

13. The amount of water discharged by the Mississippi River has changed over time in roughly a straight-line pattern. A regression line for predicting water discharged (in cubic kilometers) during a given year is

$$\text{predicted discharge} = -7792 + (4.226 \times \text{year})$$

How much (on average) does the volume of water increase with each passing year?

(a) –7792 cubic kilometers
(b) 4.226 cubic kilometers
(c) 7792 cubic kilometers

14. According to the regression line in the previous exercise, the predicted Mississippi River discharge in the year 2016 is ________ cubic kilometers.

15. You have data on the body weight x and brain weight y for many species of mammals. Body weight is given in kilograms, and brain weight is given in grams. There are 1000 grams in a kilogram. The slope of the regression line for predicting y from x is $m = 1.4$. If brain weight were given in kilograms, the slope would

(a) still be 1.4.
(b) change to 0.0014.
(c) change to 1400.

16. Suppose $y = 2x + 3$, where x and y are measured in meters. If x is re-expressed in centimeters instead, the equation becomes $y =$ ________________ .

17. Given the following set of five ordered pairs, the correlation r equals what?

x	0	1	2	3	4
y	2	3	5	6	14

(a) 0.3
(b) 0.6
(c) 0.9

18. The correlation between brain weight and body weight in Exercise 15 is $r = 0.86$. If brain weight had been measured in kilograms rather than grams, the correlation would have a value of ______________.

19. The points on a scatterplot lie very close to the line whose equation is $y = 5 - 3x$. The correlation between x and y is close to what?

(a) –3
(b) –1
(c) 1

20. High coffee prices give farmers in Indonesia an incentive to cut forest in order to plant more coffee. Here are data on coffee price x (dollars per pound) and percent y of deforestation in a national park for five years:

x	0.29	0.40	0.54	0.55	0.72
y	0.49	1.59	1.69	1.82	2.98

Using a calculator, we can determine that the correlation between x and y has a value of ________, to the nearer hundredths place.

21. Using the table in Example 1 (p. 210) and the prediction equation in Example 3 (pp. 214-215), the prediction error for Student #4 is what?

(a) –0.01
(b) 0.01
(c) 0.13

22. Look again at the coffee data in Skills Check question 20. Using your calculator, you can find that the equation (in $y = mx + b$ form, with m and b taken to the nearest hundredths place) of the least-squares regression line for predicting y from x is $\hat{y} =$ ____________________.

23. In the least-squares method, what is it that is being squared?

(a) each observed y value
(b) the vertical distance of each data point from the line
(c) the sum of the vertical distances of the data points from the line

24. Putting the symbol ^ (i.e., a caret or circumflex) over y says this is the ________ value of y.

25. If the slope and correlation value are equal, that means the standard deviation of y must equal

(a) 1.
(b) the slope.
(c) the standard deviation of x.

26. If the error is positive, the observed response lies ________ the line.

27. There is a strong positive correlation between the number of firefighters at a fire and the amount of damage that the fire does. The reason for this is that

(a) more firefighters cause more damage at the fire scene.
(b) bigger fires require more firefighters and also do more damage.
(c) more damage requires more firefighters to clean it up.

28. Making predictions well beyond the data collected is called ________.

29. Make a scatterplot with the six ordered pairs from the table below. Of the three leftmost ordered pairs in the table, the one that, if deleted, would cause the biggest change in the value of the correlation is

(a) (0, 0)
(b) (8, 3)
(c) (2, 4)

x	0	8	2	0	1	1
y	0	3	4	1	1	0

30. Association (correlation) does not imply that changes in x ________ changes in y.

CHAPTER 6 EXERCISES

■ Challenge ▲ Discussion

Some exercises require use of a calculator (or software or Internet applet) that will find correlation and the slope and intercept of the least-squares regression line from keyed-in data.

1. In each of the following situations, is it more reasonable simply to explore the relationship between the two variables or to view one of the variables as an explanatory variable and the other as a response variable? In the latter case, which is the explanatory variable?

(a) The amount of time spent studying for a statistics exam and the grade on the exam
(b) The weight in kilograms and height in centimeters of a person
(c) The inches of rain in the growing season and the yield of corn in bushels per acre
(d) A student's scores on the SAT math exam and the SAT verbal exam

6.1 Displaying Relationships: Scatterplot

2. Figure 6.9 shows the calories and salt content (in milligrams of sodium) in 17 brands of beef hot dogs. Describe the overall pattern (form, direction, and strength) of these data. In what way is the point marked *A* unusual?

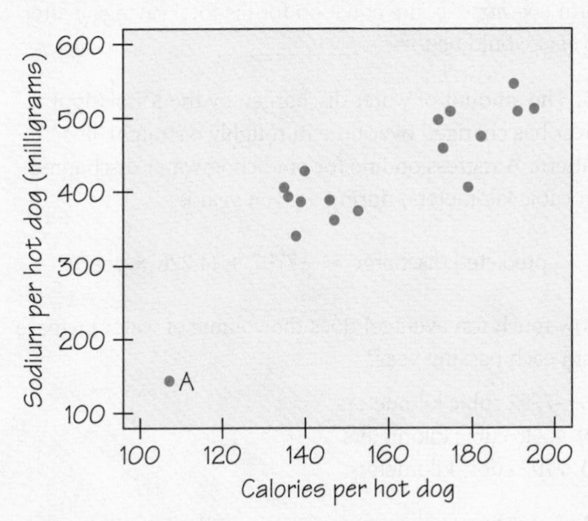

Figure 6.9 A scatterplot of sodium content versus calories in beef hot dogs, for Exercises 2 and 11.

▲ **3.** Figure 6.10 is a scatterplot of data from the World Bank. The individuals are all the world's nations for which data are available. The explanatory variable is a measure of how rich a country is, which is the gross domestic product (GDP) per person. GDP is the total value of the goods and services produced in a country, converted into dollars. The response variable is life expectancy at birth. Three African nations are outliers,

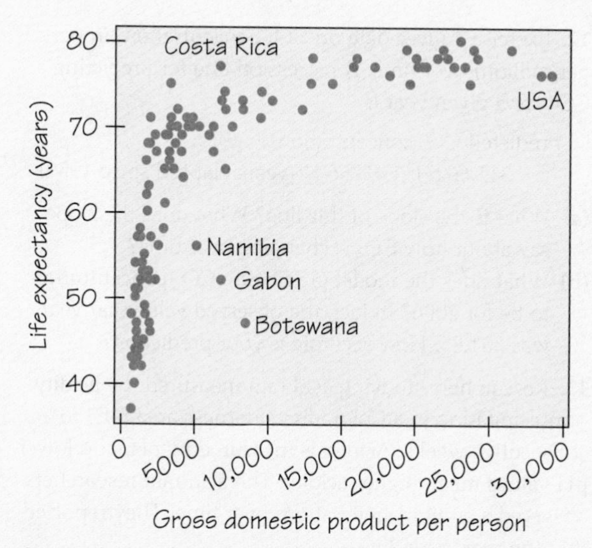

Figure 6.10 A scatterplot of the life expectancy of people in many nations against each nation's GDP per person, for Exercise 3.

with lower life expectancy than usual for their GDP. A full study would ask what special circumstances explain these outliers.

(a) Describe the direction and form of the relationship. Aside from the outliers, it is moderately strong.

(b) Explain why the direction and form of this relationship make sense.

4. Global warming may be due to increased concentrations of greenhouse gases such as carbon dioxide (CO_2). Here are data from the National Oceanic and Atmospheric Administration Web site (**www.noaa.gov**), where atmospheric CO_2 is measured in parts per million by volume and measured at the Mauna Loa Observatory:

CO_2	315.98	324.62	336.78	352.90	368.14	387.35
year	1959	1969	1979	1989	1999	2009

(a) Which is the explanatory variable?

(b) Make a scatterplot. Is the association between these variables positive or negative? Explain why you expect the relationship to have this direction.

(c) Describe the form and strength of the relationship.

5. Table 5.2 (p. 176) gives the city and highway gas mileages for 13 midsized cars. Omit the hybrid car (Toyota Prius) and make a scatterplot, taking city mileage as the explanatory variable. Describe in words the form, direction, and strength of the relationship between highway mileage and city mileage.

6. How fast do icicles grow? Here are data for one set of conditions: no wind, temperature –11°C, and water flowing over the icicle at 12 milligrams per second.

Time (minutes)	10	20	30	40	50
Length (centimeters)	0.6	1.8	2.9	4.0	5.0
Time (minutes)	60	70	80	90	100
Length (centimeters)	6.1	7.9	10.1	10.9	12.7
Time (minutes)	110	120	130	140	150
Length (centimeters)	14.4	16.6	18.1	19.9	21.0

SOURCE: N. Maeno et al., Growth rates of icicles, *Journal of Glaciology,* 40 (1994): 319–26.

Which is the explanatory variable? Make a scatterplot. Describe in words the direction, form, and strength of the relationship.

▲ **7.** How does the fuel consumption of a car change as its speed increases? Here are data for a British Ford Escort. Fuel consumption is measured in miles per gallon of gasoline used and speed is measured in miles per hour.

Speed	6.2	12.4	18.6	24.9	31.1
Fuel	11.2	18.1	23.5	29.4	33.6
Speed	37.3	43.5	49.7	55.9	62.1
Fuel	39.9	37.3	33.8	31.1	28.4
Speed	68.4	74.6	80.8	87.0	93.2
Fuel	26.0	23.8	21.8	20.0	18.3

SOURCE: T. N. Lam, Estimating fuel consumption from engine size, *Journal of Transportation Engineering,* 111 (1985): 339–57.

(a) Which is the explanatory variable?

(b) Make a scatterplot. Describe the form of the relationship. Explain why the form of the relationship makes sense.

(c) How would you describe the direction of this relationship?

(d) Is the relationship reasonably strong or quite weak? Explain your answer.

▲ **8.** Give an example of two variables from everyday life that have a positive association. Give an example of two variables that have a negative association.

9. The following is excerpted and rounded data collected on crickets by Harvard physics professor George W. Pierce in his 1948 book *The Song of Insects.*

Chirps per 15 seconds	44	37	31	25	15
Ground temperature (in °F)	80	68	73	63	55

(a) Which is the explanatory variable?

(b) Make a scatterplot. Is the association between these variables positive or negative?

(c) Describe the form and strength of the relationship.

(d) Do the data seem consistent with a rule of thumb from *The Old Farmer's Almanac* to count the number of chirps in 14 seconds and then add 40 to get the (Fahrenheit) temperature?

10. On the NASA space shuttle, six primary O-rings were used to seal the sections of the two solid-fuel rocket motors and keep hot gases from escaping and catastrophically igniting the liquid hydrogen fuel tank. The number of O-ring erosion problems and the launch temperature (in °F) of that flight are given for 23 successful flights between April 12, 1981 and January 12, 1986:

Mission	O-Ring Incidents	°F	Mission	O-Ring Incidents	°F
1	0	66	13	0	67
2	1	70	14	2	53
3	0	69	15	0	67
4	0	68	16	0	75
5	0	67	17	0	70
6	0	72	18	0	81
7	0	73	19	0	76
8	0	70	20	0	79
9	1	57	21	2	75
10	1	63	22	0	76
11	1	70	23	1	58
12	0	78			

(a) Which is the explanatory variable?
(b) Make a scatterplot. Is the association between these variables positive or negative?
(c) Describe the form and strength of the relationship.
(d) The forecasted temperature the morning of the January 28, 1986 launch of the *Challenger* was between 26–29°F. Should this have been a cause for concern or not? Why?
(e) Would a different conclusion be reached by someone whose scatterplot contained only the seven flights for which there was at least one problem? Explain.

6.2 Making Predictions: Regression Line

11. Figure 6.9 shows the salt content (in milligrams of sodium) and calories in 17 brands of beef hot dogs. If we ignore the outlying point marked *A*, a regression line for predicting sodium from calories passes close to these two observations:

calories = 139, sodium = 386 mg
calories = 191, sodium = 506 mg

Use this fact to estimate the slope of this regression line. [*Hint:* Remember that the slope of a line is the "rise" (vertical change) divided by the "run" (horizontal change) for any two points on the line.]

12. Exercise 4 gives data on CO_2 concentration (in parts per million) over time. A regression line for predicting CO_2 for a given year is

predicted CO_2 concentration
= 311.662 + 1.43866 × (years elapsed since 1959)

(a) What is the slope of this line? What does the slope say about how CO_2 is changing over time?
(b) What does the model predict the CO_2 concentration to be for 2006? In fact, the observed value that year was 381.85. How accurate is your prediction?

13. Researchers studying acid rain measured the acidity of precipitation in a Colorado wilderness area for 150 consecutive weeks. Acidity is measured by pH, and lower pH values mean higher acidity. The acid rain researchers observed a straight-line pattern over time. They reported that the regression line

predicted pH = 5.43 − (0.0053 × weeks)

fit the data well.

Source: W. M. Lewis and M. C. Grant, Acid precipitation in the western United States, *Science*, 207 (1980): 176–77.

(a) Draw a graph of this line. Explain what the line says about how pH was changing over time.
(b) According to the regression line, what was the pH at the beginning of the study (weeks = 1)? At the end (weeks = 150)?
(c) What is the slope of the regression line? Explain what this slope says about the rate of change in pH.

14. A study at the University of Massachusetts, Amherst, published in the May 2007 *Journal of Marriage and Family* found that married women do about one less hour of housework a week for every $7500 they earn as full-time workers outside the home, regardless of the husband's income.

(a) What would be the numerical value of the slope coefficient in the regression model that predicts women's housework from their income? What does the sign of the slope (positive or negative) tell us about the relationship between these variables?
(b) Suppose Lynette's salary is $30,000 greater than Gabrielle's. What would you predict to be the difference in hours of housework they each do?

15. A 21-year-old college student drinks heavily at a party until his BAC is 0.15 g/dl—almost twice the legal driving limit of 0.08. He now stops drinking, and each hour his BAC falls by 0.015 g/dl.

(a) What would be a regression-line equation that would predict BAC from the number of hours after he stopped drinking?
(b) In how many hours would he be able to drive legally?
(c) In how many hours would all the alcohol be out of his body?

16. Suppose that the slope of the regression line of weight on height for a group of young men is $m = 1.1$ when we measure height x in centimeters and weight y in kilograms. That is, when height increases by 1 centimeter, weight increases by 1.1 kilograms. There are 1000 grams in a kilogram. If we measured weight in grams, what would be the slope?

6.3 Correlation

17. Find the correlation between the city and highway gas mileages for the 12 non-hybrid midsized cars in Table 5.2 (p. 176). (That is, omit the Toyota Prius.) Explain why the value of r supports the scatterplot that you made in Exercise 5.

18. Exercise 4 gives data on CO_2 concentration (in parts per million) over time.

(a) Use a calculator to find the correlation r. Explain from looking at the scatterplot why this value of r is reasonable.

(b) Suppose that the concentration had been recorded in parts per billion instead of parts per million. For example, the value 354.16 would become 354,160. How would the value of r change?

19. Find the correlation between city and highway mileage for all 13 midsized cars in Table 5.2 (p. 176), including the Toyota Prius. Compare your r with the value you found in Exercise 17. Explain why adding the Prius changes r in this direction.

20. In Example 7 (p. 225), the positive association between family income and SAT score is clear from the lockstep pattern, but to calculate a specific numerical value for the correlation, we need to make a simplifying assumption. For each income bracket with fixed endpoints, choose its midpoint. Now, use a calculator to calculate a correlation value.

21. Exercise 7 gives data on gas used versus speed for a small car. Make a scatterplot, if you did not do so in Exercise 7. Calculate the correlation. Explain why r is close to 0 despite a strong relationship between speed and gas use.

22. Consider the data in Exercise 6.

(a) Find the correlation between time and icicle length.

(b) If icicle length were measured not in centimeters, but in inches (note: 1 inch = 2.54 cm), how would the correlation change?

23. If heterosexual women always dated men who are three years older than they are, what would be the correlation between the ages of the man and the woman? (*Hint:* Draw a scatterplot for several ages.)

24. We want to find the correlation between:

(a) the heights of fathers and the heights of their adult sons.

(b) the heights of heterosexual married men and the heights of their wives.

(c) the heights of women at age 4 and their heights at age 18.

The answers (in scrambled order) are $r = 0.2$, $r = 0.5$, and $r = 0.8$. Match the r values to the variable pairings and explain your choice.

25. For each of the following pairs of variables, would you expect a substantial negative correlation, a substantial positive correlation, or a small correlation?

(a) The age of used cars and their prices

(b) The weight of new cars and their gas mileages in miles per gallon

(c) The heights and the weights of adult women

(d) The heights and the IQ scores of adult men

▲ **26.** Each of the following statements contains a mistake. Explain what is wrong in each case.

(a) "There is a high correlation ($r = 0.89$) between the hair color of American workers and their income."

(b) "We found a high correlation ($r = 1.09$) between students' ratings of faculty teaching and ratings made by other faculty members."

(c) "The correlation between age and income was found to be $r = 0.53$ years."

■ **27.** Mutual-fund reports often give correlations to describe how the prices of different investments are related. You look at the correlations between three Fidelity funds and the Standard & Poor's 500 Stock Index, which describes stocks of large U.S. companies. The three funds are Dividend Growth (stocks of large U.S. companies), Small Cap Stock (stocks of small U.S. companies), and Emerging Markets (stocks in developing countries). For a recent year, the three correlations are $r = 0.35$, $r = 0.81$, and $r = 0.98$.

(a) Which correlation goes with each fund? Explain your answer.

(b) The correlations of the three funds with the index are all positive. Does this tell you that stocks went up that year? Explain your answer.

■ **28.** *Archaeopteryx* is an extinct beast that had feathers like a bird but teeth and a long bony tail like a reptile. Only six fossil specimens are known. If the specimens belong to the same species and differ in size because some are younger than others, there should be a straight-line relationship between the lengths of a pair of bones from all individuals. An outlier from this relationship would suggest a different species. Here are data on the lengths in centimeters of the femur (a leg bone) and the humerus (a bone in the upper arm) for the five specimens that preserve both bones.

Femur length x	38	56	59	64	74
Humerus length y	41	63	70	72	84

SOURCE: M. A. Houck et al., Allometric scaling in the earliest fossil bird, *Archaeopteryx lithographica, Science*, 247 (1990): 195–98.

(a) Make a scatterplot. Do you think that all five specimens come from the same species?
(b) Find the correlation r step by step, as in the Procedure box in Section 6.3 (p. 219).
(c) Now use the method in Spotlight 6.1 to find r and check that you get the same result as in part (b).

6.4 Least-Squares Regression

29. In Exercise 5, you made a scatterplot of city and highway gas mileage for the 12 non-hybrid midsized cars (omitting the Prius) in Table 5.2.

(a) What is the least-squares regression line for predicting highway mileage from city mileage?

(David R. Frazier Photolibrary, Photo Researchers.)

(b) If a midsized car gets 17 mpg in the city, predict its highway mileage.
(c) Based on the scatterplot you made in Exercise 5, do you expect the prediction in part (b) to be very accurate? Why?

30. In Exercise 6, you made a scatterplot of the length of an icicle and the number of minutes that water has been flowing over the icicle.

(a) What is the equation of the least-squares regression line for predicting icicle length from time?
(b) Use your regression line to predict the length of the icicle after 75 minutes.

31. Redo your scatterplot of highway mileage against city mileage from Exercise 5. Add your regression line from Exercise 29 to the plot. Be sure to show how you were able to plot the line starting with its equation. Finally, use the "up-and-across" method illustrated in Figure 6.3 (p. 215) to show the predicted highway mileage of a car that gets 18 mpg in the city.

32. Redo your scatterplot of icicle length against time from Exercise 6. Add your regression line from Exercise 30 to the plot. Be sure to show how you were able to plot the line starting with its equation. Finally, use the "up-and-across" method illustrated in Figure 6.3 (p. 215) to show the predicted length of the icicle after 75 minutes.

33. Exercise 7 gives data on fuel consumption in miles per gallon (mpg) and speed in miles per hour (mph) for a small car.

(a) From this data, the least-squares regression line to predict gas performance from speed is:

$$\text{predicted fuel} = 27.033 - 0.0125 \times \text{speed}.$$

What are the observed fuel consumption values for speeds of 6.2, 43.5, and 93.2 mph?
(b) What are the predicted fuel consumption values for speeds of 6.2, 43.5, and 93.2 mph?
(c) Draw this line on your scatterplot. Does the plot show a straight-line pattern?
(d) Based on the answer to (c), is this line a useful model? Why or why not?

34. The length of the icicle in Exercise 6 is measured in centimeters. There are 2.54 centimeters in an inch. If length were measured in inches, how would the slope of the regression line you found in Exercise 30 change?

35. The mean height of American women in their early twenties is about 64.5 inches and the standard deviation is about 2.5 inches. The mean height of men the same age is about 68.5 inches, with standard deviation about 2.7 inches.

(a) If the correlation between the heights of married heterosexual men and their wives is about $r = 0.5$, what is the equation of the regression line of the husband's height on the wife's height in young couples?
(b) Predict the height of the husband of a married heterosexual woman who is 67 inches tall.

36. This data, from the National Oceanic and Atmospheric Administration Web site (**www.noaa.gov**), is the mean annual number of named Atlantic storms (hurricanes, tropical storms, and tropical depressions), during five-year windows ending with the year shown in the table.

2007	2002	1997	1992	1987	1982	1977
16.2	13.6	11.0	10.4	8.2	10.0	8.8
1972	**1967**	**1962**	**1957**	**1952**	**1947**	**1942**
11.2	9.2	8.8	10.6	10.4	9.4	7.4

(a) What is the slope of the least-squares regression line of named storms on year? What is the intercept?

(b) Use the regression line to predict the mean annual number of named storms for the five-year window 2013–2017.

■ **37.** Use the equation for the least-squares regression line to show that this line always passes through the point $(\bar{x}, \bar{y})$. That is, set $x = \bar{x}$ and show that the line predicts that $y = \bar{y}$.

■ **38.** Exercise 6 gives data on the growth of an icicle.

(a) Find the mean and standard deviation of the times and icicle lengths. Find the correlation between the two variables. Use these five numbers to find the equation of the regression line for predicting length from time. Verify that your result agrees with what you found in Exercise 30.

(b) Use the same five numbers to find the equation of the regression line for predicting from an icicle's length the time that it has been growing. Use your line to predict the time that an icicle 15 centimeters long has been growing. *There is just one correlation between two variables, but there are two different least-squares lines, depending on which you choose as the response variable.*

■ **39.** Fidelity Investments, like other large mutual fund companies, offers many "sector funds" that concentrate their investments in narrow segments of the stock market. These funds often rise or fall by much more than the market as a whole. Here are the percent returns for 23 Fidelity "Select Portfolios" funds for the years 2002 (when stocks fell) and 2003 (when stocks went up).

2002 return	2003 return	2002 return	2003 return	2002 return	2003 return
−17.1	23.9	−0.7	36.9	−37.8	59.4
−6.7	14.1	−5.6	27.5	−11.5	22.9
−21.1	41.8	−26.9	26.1	−0.7	36.9
−12.8	43.9	−42.0	62.7	64.3	32.1
−18.9	31.1	−47.8	68.1	−9.6	28.7
−7.7	32.3	−50.5	71.9	−11.7	29.5
−17.2	36.5	−49.5	57.0	−2.3	19.1
−11.4	30.6	−23.4	35.0		

Do a careful statistical analysis of these data, using both graphs and whatever numerical measures you think are appropriate. Make a side-by-side comparison of the distributions of returns in 2002 and 2003 and also describe the relationship between the returns of the same funds in these two years. What are your most important findings? (The outlier is Fidelity Gold Fund.)

6.5 Interpreting Correlation and Regression

40. Here is data collected on six individuals:

x	1	2	3	4	10	10
y	1	3	3	5	1	11

(a) Make a scatterplot of the data.

(b) Use your calculator to show that the correlation is about 0.5.

(c) What feature of the data is responsible for reducing the correlation to this value despite a strong straight-line association between x and y in most of the observations?

41. Table 6.1 has four data sets prepared by statistician Frank Anscombe to show the dangers of calculating without first plotting the data.

(a) Without making scatterplots, find the correlation and the least-squares regression line for all four data sets. What do you notice? Use the regression line to predict y for $x = 10$.

(b) Make a scatterplot for each of the data sets and add the regression line to each plot.

(c) In which of the four cases would you be willing to use the regression line to describe the dependence of y on x? Explain your answer in each case.

▲ **42.** Children who watch many hours of TV get lower grades in school on average than those who watch less TV. Explain clearly why this fact does not show that watching TV causes poor grades. In particular, suggest some other characteristics of households where children watch lots of TV that may contribute to poor grades.

▲ **43.** People who use artificial sweeteners in place of sugar tend to be heavier than people who use sugar. Does this mean that artificial sweeteners cause weight gain? Give a more plausible explanation for this association.

▲ **44.** "Based on an examination of 22 companies that announced large layoffs during 1994, Downs found a strong correlation between the size of the layoffs and the compensation of the CEOs." (K. Phillips, *Wealth and Democracy*, Broadway Books, New York, 2002, p. 151.) Discuss why this positive correlation is probably explained by a third variable, the size of the company as measured by its number of employees.

■ **45.** "The positive correlation between health and income per capita is one of the best-known relations in international development. This correlation is commonly thought to reflect a causal link running from income to health. … Recently, however, another intriguing possibility has emerged: that the health-income correlation is partly explained by a causal link running the other way—from health to income." [D. E. Bloom and D. Canning, The health and wealth of nations,

Table 6.1

Four Data Sets for Exploring Correlation and Regression

Data Set A

x	10	8	13	9	11	14	6	4	12	7	5
y	8.04	6.95	7.58	8.81	8.33	9.96	7.24	4.26	10.84	4.82	5.68

Data Set B

x	10	8	13	9	11	14	6	4	12	7	5
y	9.14	8.14	8.74	8.77	9.26	8.10	6.13	3.10	9.13	7.26	4.74

Data Set C

x	10	8	13	9	11	14	6	4	12	7	5
y	7.46	6.77	12.74	7.11	7.81	8.84	6.08	5.39	8.15	6.42	5.73

Data Set D

x	8	8	8	8	8	8	8	8	8	8	19
y	6.58	5.76	7.71	8.84	8.47	7.04	5.25	5.56	7.91	6.89	12.50

SOURCE: Frank J. Anscombe, Graphs in statistical analysis, *The American Statistician*, 27 (1973): 17–21.

Science, 287 (2000): 1207–8.] Explain how higher income in a nation can cause better health. Then explain how better health can cause higher national income. There is no simple way to determine the direction of the link.

■ **46.** The effect of an outside variable can be surprising when individuals are divided into groups. In recent years, the mean SAT score of all high school seniors has increased. But the mean SAT score has decreased for students at each level of high school grades (A, B, C, and so on). Explain how grade inflation in high school can account for this pattern. *A relationship that holds for each group within a population need not hold (and may even be in the opposite direction!) for the population as a whole.*

Chapter Review

47. Recent major recalls of toys with lead paint refocused people on the dangers of lead exposure. Below is data from research exploring the association with student achievement for blood lead levels below the "danger threshold" of 10 mcg/dl set by the Centers for Disease Control [M.L. Miranda et al., The relationship between early childhood blood lead levels and performance on end-of-grade tests, *Environmental Health Perspectives*, 115 (2007): 1242–47].

Blood lead level	1	2	3	4	5
Mean 4th grade reading score	255.9	253.8	252.6	251.0	250.4

Blood lead level	6	7	8	9
Mean 4th grade reading score	249.5	248.5	247.8	249.3

(a) What are the explanatory and response variables?

(b) Do you expect a positive or negative association between these variables? Why? Does the scatterplot support your answer?

48. A study of reading ability in schoolchildren chose 60 fifth-grade children at random from a school. The researchers had the children's scores on an IQ test and on a test of reading ability. Figure 6.11 plots reading test score (response variable) against IQ score (explanatory variable).

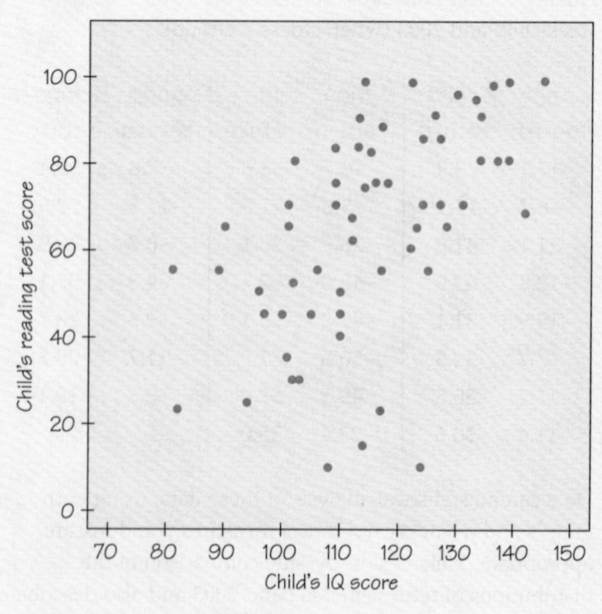

Figure 6.11 IQ and reading test scores for 60 fifth-grade children, for Exercise 48.

(a) Explain why we should expect a positive association between IQ score and reading score for children in the same grade.

(b) Does the scatterplot show a positive association?

(c) A group of four points appear to be outliers. In what way do these children's IQ and reading scores deviate from the overall pattern?

(d) Ignoring the outliers, is the form of the association between IQ score and reading score roughly a straight line? Is it very strong? Explain your answers.

49. A student wonders if tall women tend to date taller people than do short women. She measures herself, her sister, and the women in the adjoining dorm rooms. Then she measures the next person each woman dates and obtains these data (in inches):

Heights of women (x)	66	64	63	65	70	65
Heights of their dates (y)	72	68	70	68	71	64

(a) Based on a scatterplot (with the women's heights as the explanatory variable), do you expect the correlation to be positive or negative? Near ± 1 or not?

(b) Find the correlation r between the heights of the women and their dates.

50. In Exercise 49, you found the correlation r between the heights in inches of several college women and the heights in inches of the next person each woman dates.

(a) How would r change if all the dates were 2 inches shorter than the heights given in the table?

(b) How would r change if heights were measured in centimeters rather than inches? (Note: 1 inch = 2.54 cm)

51. The equation of the least-squares regression line for predicting dates' heights from women's heights for the data in Exercise 49 is

predicted height of date = $41.08 + 0.42 \times$ woman's height

(a) What is the slope of this line?

(b) Explain in simple language what the numerical value of the slope tells us about the heights of the people these women date.

(c) Use the regression line to predict the height of the next person dated by a woman who is 67″ tall.

52. From 2000 to 2005, sales and file sharing (that is, free downloading) intensity were tracked within seven musical genres (rock, alternative, R&B, rap/hip-hop, country, jazz, classical). The correlation between change in sales and file-sharing intensity was –0.648. Is this evidence that file sharing helps or hurts sales? Explain.

53. In issue 49 of *Stats: The Magazine for Students of Statistics*, Schuyler Huck presents a dataset of 100 ordered pairs in which 25 of them are (17, 1), 25 are (18, 2), 25 are (19, 3), and 25 are (20, 4).

(a) Without doing much formal calculation, find the value of r and the slope of the least-squares regression line.

(b) Now, suppose someone adds the 101st point to the dataset: the ordered pair (1, 20). Predict the new value of r and the slope of the regression line, and then do a calculation to see how close your answer is.

APPLET EXERCISES

To do these exercises, go to www.whfreeman.com/fapp9e.

1. In the *Correlation and Regression* applet, imitate Figure 6.5 (p. 224). Click to locate five points at the lower left of the scatterplot, and then click "Show least-squares line."

(a) What is the correlation r for these five points? If necessary, move points with the mouse to get a value near $r = 0.5$, as in Example 6 (p. 223).

(b) Now add an outlier at the upper right that lies exactly on the line. What is the correlation r for the six points?

(c) Use the mouse to drag the outlier down and then to the left. Watch the least-squares line follow this one point. How negative can you make the correlation r?

2. You are going to use the *Correlation and Regression* applet to make different scatterplots with 10 points that have a correlation close to 0.7. *Many patterns can have the same correlation. Always plot your data before you trust a correlation.*

(a) Stop after adding the first two points. What is the value of the correlation? Why does it have this value no matter where the two points are located?

(b) Make a lower-left to upper-right pattern of 10 points with a correlation of about $r = 0.7$. (You can drag points up or down to adjust r after you have 10 points.) Make a rough sketch of your scatterplot.

(c) Make another scatterplot with nine points in a vertical stack at the left of the plot. Add one point far to the right and move it until the correlation is close to 0.7. Make a rough sketch of your scatterplot.

(d) Make yet another scatterplot with 10 points in a curved pattern that starts at the lower left, rises to the right, then falls again at the far right. Adjust the points up or down until you have a smooth curve with a correlation close to 0.7. Make a rough sketch of this scatterplot as well.

3. It isn't easy to guess the position of the least-squares line by eye. Use the *Correlation and Regression* applet to compare a line that you draw with the least-squares line. Click on the scatterplot to create a group of 15 to 20 points from the lower left to the upper right with a clear, positive straight-line pattern (with a correlation of around 0.7). Click the "Draw line" button and use the mouse to draw a line through the middle of the cloud of points from the lower left to the upper right. Note the "thermometer" that appears above the plot. The red portion is the sum of the squared vertical distances from the points in the plot to the least-squares line. The green portion is the "extra" sum of squares for your line; it shows by how much your line misses the smallest possible sum of squares.

(a) You drew a line by eye through the middle of the pattern. Yet the right-hand part of the bar is probably almost entirely green. What does that tell you?

(b) Now click the "Show least-squares line" box. Is the slope of the least-squares line smaller (the new line is less steep) or larger (the line is steeper) than that of your line? If you repeat this exercise several times, you will get the same result consistently. *The least-squares line minimizes the vertical distances of the points from the line. It is not the line through the "middle" of the cloud of points.* This is one reason why it is hard to draw a good regression line by eye.

WRITING PROJECTS

1. Choose two variables that you think have a roughly straight-line relationship. Gather data on these variables and do a statistical analysis: Make a scatterplot, find the correlation, find the regression line (use a statistical calculator or software), and draw the line on your plot. Then write a short report on your work. Some examples of suitable pairs of variables are:

(a) The height and arm span of a group of people

(b) The height and walking stride length of a group of people

(c) The price per ounce and bottle size in ounces for several brands of shampoo and several bottle sizes for each brand

2. Can regression help protect voting rights? This example is adapted from *FAPP* author Lawrence Lesser's work as a statistician for the Texas Legislative Council. To comply with the Voting Rights Act, a state cannot redraw its districts in a way that dilutes the voting strength of a protected group. Because we cannot know how individuals voted, we cannot directly measure if minority and majority persons tend to prefer different candidates. While there are technical details and assumptions we cannot fully discuss here, you can begin to understand how this might be estimated by exploring the following dataset for nine equal-sized districts, where X is the percent of voters who are Hispanic and Y is the percent of voters that voted for the candidate preferred by most Hispanics:

Y	14	7	19	27	37	36	53	48	65
X	12	18	24	36	42	53	68	79	86
District #	1	2	3	4	5	6	7	8	9

(a) Produce a scatterplot, correlation value, and regression equation. Describe the relationship between the concentration of Hispanic population and the proportion of votes that went to the Hispanic-preferred candidate.

(b) Give a practical interpretation of the value of the slope coefficient. Give a practical interpretation of the value of Y that would be predicted when $X = 0$ and when $X = 100$.

Suggested Readings

CLEVELAND, WILLIAM S. *The Elements of Graphing Data*, rev. ed., Hobart Press, Summit, NJ, 1994. A careful study of the most effective elementary ways to present data graphically, with much sound advice on improving graphs such as scatterplots.

LESSER, LAWRENCE M. The "Ys" and "why nots" of line of best fit, *Teaching Statistics*, 21(2) (1999): 54–55.

MOORE, DAVID S. *The Basic Practice of Statistics*, 6th ed., Freeman, New York, 2012. Chapters 4 and 5 of BPS give more extensive treatment of FAPP Chapter 6 material, at about the same mathematical level.

Suggested Web Sites

The Web sites suggested in Chapter 5 as sources for data provide data to investigate relationships as well. How is the number of medical doctors per 100,000 people in each state related to how rich a state is? To infant mortality in the state? To the cost of medical care? You can study these and many other relationships using data from Web sites such as **www.census.gov** or **www.fedstats.gov**.

The *Journal of Statistics Education*, **www.amstat.org/publications/jse/**, has many articles with interesting data and examples. Look, for example, in the Archive for "Exploring Relationships in Body Dimensions" by G. Heinz et al. in the July 2003 issue. Here, you will find information about measuring body dimensions, actual data from 247 men and 260 women, and some examples of both distributions and relationships.

CAUSEweb, **www.causeweb.org**, is a searchable digital library of resources on a wide range of statistics topics offered by the Consortium for the Advancement of Undergraduate Statistics Education.

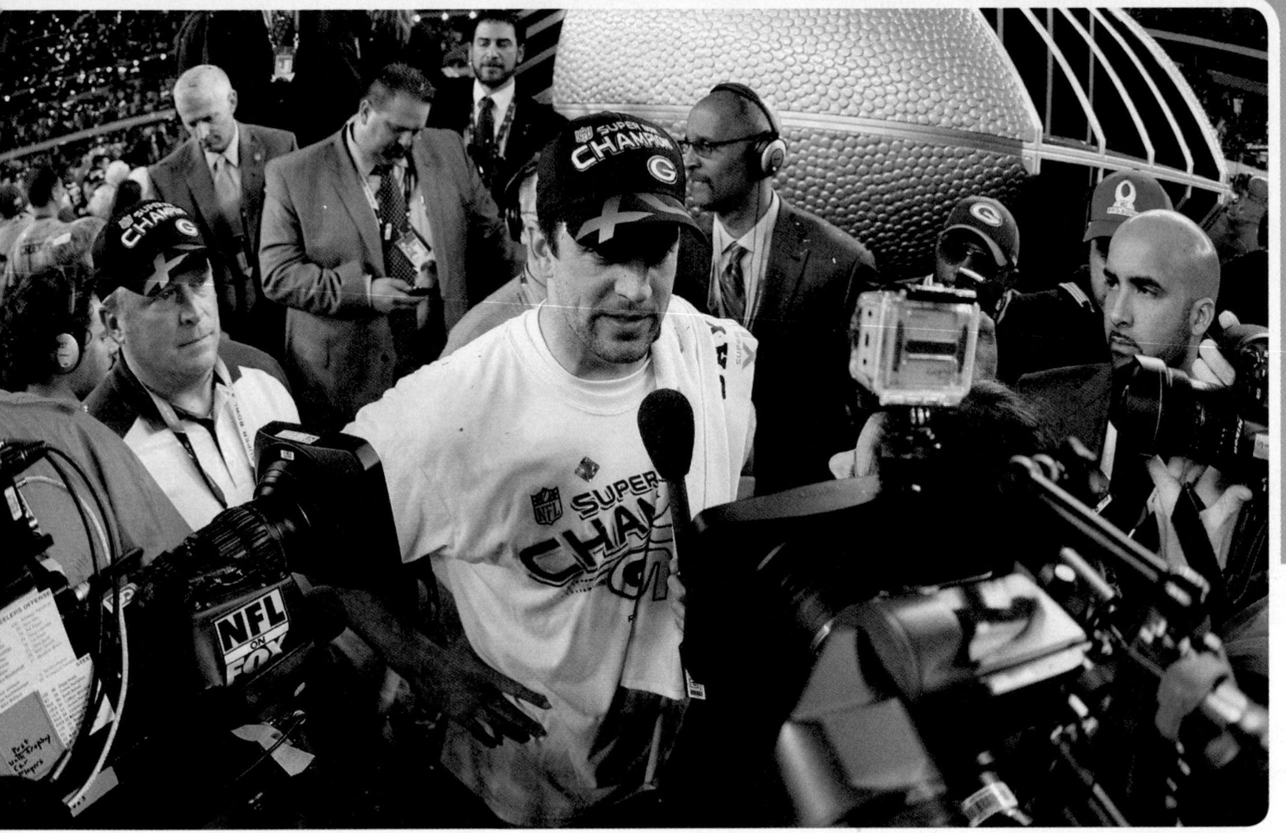

Kevin C. Cox/Getty Images

7

Data for Decisions

The day after Super Bowl XLV in 2011, it was announced that the game was viewed by an estimated 111 million viewers on average, with at least 162 million people watching at least part of the game. Thus, that game became the most-watched show in U.S. history. A politician looks at the latest poll to decide if a new strategy or policy will have enough support. A doctor looks at the latest study in a medical journal to decide if there is strong enough evidence in a new approach to treating a disease for her to recommend it to a patient. Every day that we read a newspaper like *USA Today*, we encounter the results of some kind of poll or study that affects commerce, politics, or the growth of knowledge. Clearly, data can be used to help answer a huge variety of questions in our world.

However, to appreciate fully the graphical and numerical data processing tools that we learned in Chapters 5 and 6, we also have to think about *how data is produced* in the first place. After all, it's straightforward to calculate a mean for two different datasets, but the value of the mean does not tell us whether one of the datasets came from a more reliable process and therefore can be trusted more.

Chapter 7 explores three ways of collecting data: surveys, experiments, and observational studies. In Sections 7.1 through 7.4, we start with the most popular method of data collection, survey sampling, including several pitfalls to avoid when doing it. Section 7.5 covers the collection of data through experiments, progressing through (in increasing design quality) uncontrolled, comparative, and randomized comparative experiments. In Section 7.6, experiments are contrasted with observational studies, which produce weaker evidence but still have their place because experiments are not always possible or ethical to do. The chapter's concluding sections (Sections 7.7 and 7.8) discuss making formal *statistical inference* so that we go beyond describing a sample in front of us (as we did in Chapters 5 and 6) to make an estimate about the entire population from which our sample comes and quantify our uncertainty when doing so.

7.1 Sampling

- A political scientist wants to know what percentage of college students consider themselves to be conservatives.
- An automaker hires a market research firm to learn what percentage of adults aged 18 to 35 recall seeing television advertisements for a new sport utility vehicle.
- Government economists inquire about average household income for Americans.

In all these cases, we want to gather information about a large group of individuals. Time, cost, and inconvenience preclude contacting every person, so we gather information about only part of the group to draw conclusions about the whole. Also, when an observation is destructive, it is necessary to use only a sample. For example, testing a shipment of fuses to see if they are defective would ruin the whole shipment if every single fuse were tested. And if your doctor's appointment includes a blood test, you want only *some* of your blood removed!

Population DEFINITION

The **population** in a statistical study is the entire group of individuals about which we want information.

Sample DEFINITION

A **sample** is a part of the population from which we actually collect information that is used to draw conclusions about the whole. *Sampling* refers to the process of choosing a sample from the population.

For example, let's refer back to the first bulleted item of Section 7.1. The population would be all college students (or perhaps just the millions of college students in the United States) and the sample would be the small subset of students (typically between 500 and 1000) actually selected to participate in the sample. And from the opening scenario of the chapter, the Super Bowl viewer estimate was based on transmissions from "boxes" installed in a representative sample of about 5,000 homes across the United States.

We often draw conclusions about a whole on the basis of a sample. Everyone has sipped a spoonful of soup and judged the entire bowl on the basis of that taste. But a bowl of soup is homogeneous, so the taste of a single spoonful does represent the whole. On the other hand, a spoonful of salad dressing may be misleading because its elements may separate if the bottle has not been shaken recently. Choosing a representative sample from a large and varied population is less easy. The first step in a proper sample survey is to say carefully just what population we want to describe. The second step is to say exactly what we want to measure. These preliminary steps can be complicated, as the following example illustrates.

EXAMPLE 1
How Can a Survey Measure Unemployment?

The monthly unemployment rate comes from the government's Current Population Survey (CPS; www.census.gov/cps/), a sample of about 50,000 households each month conducted by the Census Bureau. To measure unemployment, we must first specify the population that we want to describe. Which age groups will we include? Will we include illegal aliens or people in prisons? The CPS defines its population as all U.S. residents (whether citizens or not), 16 years of age and over, who are civilians and are not in an institution such as a prison. The civilian unemployment rate announced in the news refers to this specific population.

The second question is harder: What does the term *unemployed* mean? Someone who is not looking for work—for example, a full-time student—should not be called unemployed just because she is not working for pay. If you are chosen for the CPS sample, the interviewer first asks whether you are available to work and whether you actually looked for work in the past four weeks. If not, you are neither employed nor unemployed; you are not in the labor force.

If you are in the labor force, the interviewer goes on to ask about employment. Any work for pay, whether for someone else or in your own business, that you performed the week of the survey qualifies you to be counted as employed. So does at least 15 hours of unpaid work in a family business. In addition, you are employed if you have a job but didn't work for reasons such as being on vacation or being on strike. An unemployment rate of 6.7 percent means that 6.7 percent of the sample was unemployed, using the exact CPS definitions of both *labor force* and *unemployed*.

7.2 Bad Sampling Methods

How can we choose a sample that is truly representative of the population? The easiest—but not the best—way to select a sample is to choose individuals close at hand. If we are interested in finding out how many people have jobs, for example, we might be tempted to go to a shopping mall and ask people passing by if they are employed.

Convenience Sample DEFINITION

A **convenience sample** is a sample of individuals who are selected because they are members of a population who are easy (i.e., convenient) to reach. Usually, such a sample cannot be trusted to be representative of the population.

EXAMPLE 2
The Inconvenient Truth About Convenience Samples

Going to the mall, standing by a particular entrance, and surveying as many of the people walking through that entrance as you can seems like a fast, convenient way of finding out Americans' opinions. But people at malls tend to be more prosperous than typical Americans. They are also more likely to be teenagers or retired. The kinds of stores that are near the particular entrance you are standing by could affect the type of people you might more readily encounter. Also, when we decide which people to approach, we may tend (even unconsciously) to avoid poorly dressed or tough-looking individuals. In short, our shopping mall interviews will result in a sample that is not representative of the entire population because we underrepresent certain types of people. For that matter, we also are underrepresenting those Americans who rarely go to malls in the first place.

In your classroom, your professor may try to "sample" the understanding the class has about a topic by simply turning to and calling on the nearest two students on the front row. If students who sit near the front have higher levels of preparation, interest, and engagement, the professor will overestimate how well the class as a whole understands the material.

Bjanka Kadic/Alamy

In both scenarios in Example 2, the inaccuracies obtained cannot simply be explained as a sample's "bad luck." They are likely to happen every time, with the same pattern, because unscientific sampling methods have **bias**. In this context, bias refers matter-of-factly to the built-in systematic error of the procedure itself and not to any political or personal prejudice that the person conducting the poll may have.

Bias DEFINITION

The design of a statistical study has **bias** (i.e., is biased) if it systematically favors certain outcomes.

EXAMPLE 3

Are Online Polls in Line?

The American Family Association (AFA) is a conservative group that claims to stand for "traditional family values." It has often posted online poll questions on its Web site; just click on a response to take part. Because the respondents are people who visit this site, the poll results always support AFA's positions. Well, almost always; a recent AFA online poll asked about allowing same-sex marriage, and before long, email lists and social-network sites favored mostly by young liberals pointed to the AFA poll. Almost 850,000 people responded, and 60 percent of them favored the legalization of same-sex marriage. This example shows that the results of an online poll can be skewed one way or the other by particular characteristics of the people that choose to go to that Web site and participate in the poll.

A related example is the Web site www.ratemyprofessors.com, where students have chosen to evaluate and post comments on more than 1 million college and university instructors worldwide. However, focus group research indicates the ratings may not be representative because the students most motivated to post assessments are those who believe the teacher is extremely bad (or extremely good). A different kind of problem is that this Web site may have no way of keeping out multiple responses from the same student—or even from the instructors themselves! It is usually better to rely on the official end-of-course evaluation data compiled by the university.

Online polls are now everywhere; some sites will even provide help in conducting your own online poll. As Example 3 illustrates, however, you can't trust the results. People who take the trouble to write in, call in, or visit a Web site to respond to an open invitation are not representative of the general population. Polls like these are examples of **voluntary response sampling**.

Voluntary Response Sample DEFINITION

A **voluntary response sample** consists of people who choose themselves by responding to a general appeal. Voluntary response samples are biased because people with strong opinions are more likely to respond and will therefore be overrepresented in the sample.

7.3 Simple Random Samples

In a voluntary response sample, people choose whether to respond. In a convenience sample, the interviewer makes the choice. In both cases, personal choice produces bias. The statistician's remedy (pioneered by such people as George Gallup in the 1930s) is to allow impersonal chance to choose the sample. A sample chosen by chance allows neither favoritism by the sampler nor self-selection by respondents. Choosing a sample by chance avoids bias by giving all individuals an equal chance to be chosen. Any individual, whether rich or poor, young or old, black or white, and so on, has the same chance to be in the sample.

The simplest way to use chance to select a sample is to place slips of paper with the names of all individuals in the population in a hat, shake the hat vigorously, and then draw out only a few (the sample). This is the idea of **simple random sampling**.

Simple Random Sample DEFINITION

A **simple random sample** (SRS) of size n consists of n individuals from the population chosen in such a way that every set of n individuals has an equal chance to be the sample actually selected.

Picturing drawing names from a hat helps us understand what an SRS is. The same picture helps us see that an SRS is a better method of choosing samples than convenience or voluntary response sampling because it doesn't favor any part of the population. But writing names on slips of paper and drawing them from a hat is a slow and inconvenient process, especially if, as in the CPS of Example 1, we must draw a sample of 50,000 participants. We can speed up the process by using a table of random digits. In practice, samplers use computers to do the work, but we can do it by hand for small samples.

Table of Random Digits DEFINITION

A **table of random digits** is a list of the digits 0, 1, 2, 3, 4, 5, 6, 7, 8, 9 with these two properties:

1. Each entry in the table is equally likely to be any of the ten digits 0 through 9.
2. The entries are independent of one another. That is, knowledge of one part of the table gives no information about any other part.

Table 7.1 is a table of random digits. The digits in the table are displayed in groups of 5 to make the table easier to read, and the rows are numbered so we can refer to them, but the groups and row numbers are just for convenience. The entire table is one long string of 1,000 randomly chosen digits.

Here is the procedure to use the random-digit table to choose a simple random sample. If a bigger sample is required, repeat Steps 2 and 3 as needed.

Table 7.1

Random Digits

101	19223	95034	05756	28713	96409	12531	42544	82853
102	73676	47150	99400	01927	27754	42648	82425	36290
103	45467	71709	77558	00095	32863	29485	82226	90056
104	52711	38889	93074	60227	40011	85848	48767	52573
105	95592	94007	69971	91481	60779	53791	17297	59335
106	68417	35013	15529	72765	85089	57067	50211	47487
107	82739	57890	20807	47511	81676	55300	94383	14893
108	60940	72024	17868	24943	61790	90656	87964	18883
109	36009	19365	15412	39638	85453	46816	83485	41979
110	38448	48789	18338	24697	39364	42006	76688	08708
111	81486	69487	60513	09297	00412	71238	27649	39950
112	59636	88804	04634	71197	19352	73089	84898	45785
113	62568	70206	40325	03699	71080	22553	11486	11776
114	45149	32992	75730	66280	03819	56202	02938	70915
115	61041	77684	94322	24709	73698	14526	31893	32592
116	14459	26056	31424	80371	65103	62253	50490	61181
117	38167	98532	62183	70632	23417	26185	41448	75532
118	73190	32533	04470	29669	84407	90785	65956	86382
119	95857	07118	87664	92099	58806	66979	98624	84826
120	35476	55972	39421	65850	04266	35435	43742	11937
121	71487	09984	29077	14863	61683	47052	62224	51025
122	13873	81598	95052	90908	73592	75186	87136	95761
123	54580	81507	27102	56027	55892	33063	41842	81868
124	71035	09001	43367	49497	72719	96758	27611	91596
125	96746	12149	37823	71868	18442	35119	62103	39244

Using a Table of Random Digits

PROCEDURE

Step 1. Label Give each member of the population a numerical label of the same length. Up to 100 items can be labeled with two digits: 01, 02, …, 99, 00. Up to 1000 items can be labeled with three digits, and so on.

Step 2. Pick a Row from the Table Without looking at the table, pick a row (between line numbers 101 and 125, inclusive). From the row you picked, read from Table 7.1 successive groups of digits of the length you used as labels. Your sample contains the individuals whose labels you find in the table. This gives all individuals the same chance because all labels of the same length have the same chance of being found in the table. For example, any pair of digits in the table is equally likely to be any of the 100 possible labels 01, 02, …, 99, 00.

Step 3. Ignore Ineligible Labels From the labels generated by Step 2, ignore the ones that are beyond the size of your population, that duplicate a label already in the sample (unless you want to allow repeated values), or that come after you already have the sample size you need.

EXAMPLE 4
Sampling Songs

Professor Lesser has all 27 songs from the album called *The Beatles One* stored on a digital media player and wants to play 4 randomly chosen songs to accompany his morning commute. Let's follow the three-step procedure for using the random-digit table to choose a simple random sample of size 4 from the 27-song playlist.

Step 1. Give each song a numerical label. Because two digits are needed to label the 27 songs, all the labels will have two digits. In the table here, we have listed the 27 songs with labels from 01 to 27. Always specify how you label the members of the population. If the player had 500 songs, we would label them 001, 002, ..., 499, 500.

01 Love Me Do
02 From Me to You
03 She Loves You
04 I Want to Hold Your Hand
05 Can't Buy Me Love
06 A Hard Day's Night
07 I Feel Fine
08 Eight Days a Week
09 Ticket to Ride
10 Help!
11 Yesterday
12 Day Tripper
13 We Can Work it Out
14 Paperback Writer
15 Yellow Submarine
16 Eleanor Rigby
17 Penny Lane
18 All You Need Is Love
19 Hello, Goodbye
20 Lady Madonna
21 Hey Jude
22 Get Back
23 The Ballad of John and Yoko
24 Something
25 Come Together
26 Let it Be
27 The Long and Winding Road

Step 2. We picked line 125 (you might have picked a different row) from Table 7.1. Because the size of our population is 27, we will grab two digits at a time as we read across that row.

96 74 61 21 49 37 82 37 18 68 18 44 23 51 19 62 10 33 92 44

Step 3. Look at the result of Step 2 and ignore any label larger than 27 (the size of the population), ignore any reoccurrence of a value already selected (in this case, the second 18), and stop once you select the fourth different song.

96 74 61 **21** 49 37 82 37 **18** 68 18 44 **23** 51 **19** 62 10 33 92 44

Corresponding to the selected labels in boldface, our media player will play the song sequence: "Hey Jude," "All You Need Is Love," "The Ballad of John and Yoko," and "Hello, Goodbye."

As an alternative to Table 7.1, you can select an SRS using a spreadsheet (for example, the Microsoft Excel command RANDBETWEEN), statistical software, an online applet (e.g., see Applet Exercise 1), or most types of calculators. For example, you can use (MATH) → PRB → randInt(1, 27, 4) → (ENTER) on the TI-84 calculator. So while digital media players may have a "shuffle" option to put a playlist in a randomized order, you now have a procedure for obtaining a random sample for any situation where you have an ordered list for the population.

Online polls and mall interviews produce samples. We can't trust results from these samples because they are chosen in ways that invite bias. We have more confidence in results from an SRS because it uses impersonal chance to avoid bias. The first question to ask about any sample is whether it was chosen at random. Opinion polls and other sample surveys carried out by people who know what they are doing use random sampling. Most national sample surveys use sampling schemes that are more complex than SRS. For example, they may dial the last four digits of a telephone number at random separately within each exchange (the area code and first three digits). The national sample is pieced together from many smaller samples. The big idea remains the deliberate use of chance to choose the sample. Because SRS is the essential principle behind all random sampling and because it is also the main building block for more complex samples, we will focus our attention on that in this chapter.

EXAMPLE 5

The Plane Truth

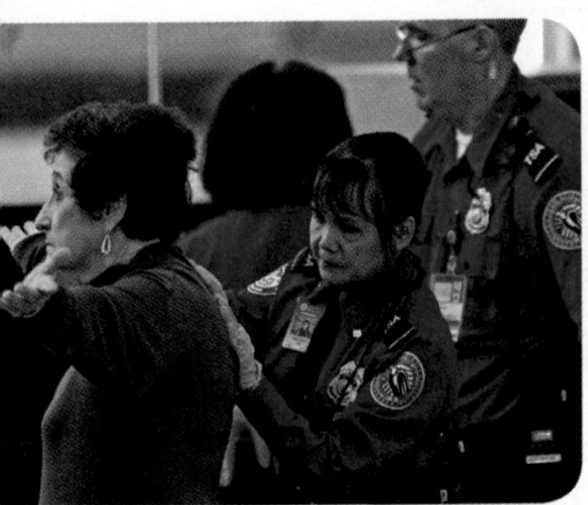

Michael Goulding/ ZUMA Press/Corbis

According to a *USA Today*/Gallup poll, 71 percent of people who have flown at least twice in the past year believe the potential loss of personal privacy from full-body scans or pat-downs is worth it as a method to prevent acts of terrorism. How much do we trust the quality of this survey? Ask first how the sample was selected. On the Gallup Web site, we learn that the results are based on telephone interviews with a randomly selected national sample of 3,018 adults selected using random-digit-dial sampling conducted November 19–21, 2010.

It is a good start toward confidence in the poll to know the intended population, the sample size, the tight window of time (so that there is minimal influence from changes in current events), and—most importantly—random selection. In the next section, we address a few other important considerations.

7.4 Cautions About Sample Surveys

Random sampling eliminates bias in the choice of a sample from a list of the population. Sample surveys of large human populations, however, require more than a good sampling design.

To begin with, we need an accurate and complete list of the population. Because such a list is rarely available, most samples suffer from some degree of **undercoverage**. A sample survey of households, for example, will miss not only homeless people but also prison inmates and students in dormitories. An opinion poll conducted by telephone will miss the 6 percent of U.S. households without residential phones. The results of national sample surveys therefore have some bias if the people not covered—who most often are young or poor people—differ from the rest of the population.

Undercoverage DEFINITION

Undercoverage occurs when some groups in the population are left out of the process of choosing the sample.

A more serious source of bias in most sample surveys is *nonresponse,* which occurs when a selected individual cannot be contacted or refuses to cooperate. Nonresponse to sample surveys often reaches 50 percent or more, even with careful planning and several callbacks. Because nonresponse is higher in urban areas, most sample surveys substitute other people in the same area to avoid favoring rural areas in the final sample. If the people contacted differ from those who are rarely at home or who refuse to answer questions, some bias remains.

Nonresponse DEFINITION

Nonresponse occurs when an individual chosen for the sample can't be contacted or refuses to participate.

EXAMPLE 6
How Bad Is Nonresponse?

The CPS has the lowest nonresponse rate of any poll we know. Only about 4 percent of the households in the CPS sample refuse to take part, and another 3 percent or 4 percent can't be contacted. People are more likely to respond to a government survey such as the CPS, and the CPS contacts its sample in person before doing later interviews by phone. (On a related note, we observe that the national mail participation rate for the 2010 Census was 74 percent.)

What about polls done by the media and by market research and opinion polling firms? We don't know their rates of nonresponse because they won't say. That nondisclosure is a bad sign. The Pew Research Center imitated a careful telephone survey and published the results: Out of 2879 households called, 1658 were never at home, refused the interview, or would not finish the it. That's a nonresponse rate of 58 percent.

Many telephone survey organizations call only landline numbers because, for example, it is against federal law to use automatic dialing devices to call cell phones. Such surveys, however, are excluding an increasing number of people in the United States whose only telephone service is a cell phone. This is not a problem if cell-only adults are very similar to those who can be reached by a landline number. However, the American Association for Public Opinion Research reports that cell-only adults are indeed quite different: "younger, less affluent, less likely to be married, more likely to be renters, more urban, and more liberal." Therefore, researchers of topics likely to be related to these traits need to think carefully how to design and analyze their surveys.

Another danger is that when people do respond, we can't always rely on them to tell the truth. People know that they should take the trouble to vote, for example, so many who didn't vote in the last election will tell a pollster that they did. Fortunately, there are strategies to help improve accuracy.

EXAMPLE 7
Encouraging Honesty

The Centers for Disease Control and Prevention (CDC) previously used face-to-face interviews to ask Americans about their sexual activity and illegal drug use. In a new version of the survey, released in 2007, data were gathered using computer-assisted self-interviews in which participants were alone in a room, heard questions through a headset, and touched a computer screen to give their responses. By not requiring people to give answers to a human being whom they were looking in the face, while they had to look them in the face, this revised method made interviewees more comfortable giving honest answers.

Another strategy for encouraging honesty with sensitive topics is called "randomized response," invented by sociologist S. L. Warner in 1965 and explored in Exercise 16 (p. 272). By introducing randomness into the responses in a structured way, researchers use their knowledge of probability distributions to get reasonably accurate information about the overall group while allowing each potentially embarrassing answer to be "camouflaged." Because the interviewee knows that the researcher has no way, for example, to distinguish which "yes" answers are real and which are simply introduced by the random mechanism, they will feel safe answering honestly a question of the form "Have you ever done [some embarrassing action]?"

Finally, the *wording of questions* strongly influences the answers given to a sample survey. Confusing or leading questions can introduce strong bias, and even minor changes in wording or order can change a survey's outcome. Here are some examples.

EXAMPLE 8
Watch That Wording!

How do Americans feel about government help for the poor? Only 13 percent think we are spending too much on "assistance to the poor," but 44 percent think we are spending too much on "welfare." How do the Scots feel about the movement to become independent from England? Well, 51 percent would vote for "independence for Scotland," but only 34 percent support "an independent Scotland separate from the United Kingdom." It seems that "assistance to the poor" and "independence" are nice, hopeful words, while "welfare" and "separate" are negative words. Other topics that have produced survey results that vary greatly with the wording of the questions include abortion, immigration, gay rights, and affirmative action.

The statistical design of sample surveys is a science, but this science is only part of the art of sampling. Because of nonresponse, false responses, and the difficulty of posing clear and neutral questions, you should analyze critically before fully trusting reports about complicated issues based on surveys of large human populations. Insist on knowing the exact questions asked, the rate of nonresponse, and the date and method of the survey before you trust a poll result.

7.5 Experiments

Sample surveys gather information on part of the population to make conclusions about the whole. When the goal is to describe a population, statistical sampling is the right tool to use.

Suppose, however, that we want to study the response to a stimulus, to see how one variable affects another when we change existing conditions. For example:

- Will a new mathematics curriculum improve the scores of sixth-graders on a standardized test of mathematics achievement?

- Will taking small amounts of aspirin daily reduce the risk of suffering a heart attack?
- Does a woman's smoking during pregnancy reduce the IQ of her child?

Studies that simply *observe and describe* are ineffective tools for answering these questions. **Experiments** give us clearer answers.

Experiment DEFINITION

An **experiment** deliberately imposes a *treatment* on individuals to observe their responses. The purpose of an experiment is to study whether the treatment *causes* a change in the response.

Experiments are the preferred method for examining the effect of one variable on another. By imposing the specific treatment of interest and controlling other influences, we can pin down cause and effect. A sample survey may show that two variables are related, but it cannot demonstrate that one causes the other. Statistics has something to say about how to arrange experiments, just as it suggests methods for sampling.

EXAMPLE 9
An Uncontrolled Experiment

A college regularly offers a review course to prepare candidates for the Graduate Management Admission Test (GMAT) required by most graduate business schools. This year, it offered only an online version of the course. The average GMAT score of students in the online course was 10 percent higher than the longtime average of those who took the classroom review course. Can we conclude that the online course is more effective?

This experiment has a very simple design. A group of subjects (the students) were exposed to a treatment (the online course), and the outcome (GMAT scores) was observed. The design can be represented as

Online course → Observe GMAT scores

or, in general form:

Treatment → Observe response

Most laboratory experiments use a design like that in Example 9: Apply a treatment and measure the response. In the controlled environment of the laboratory, simple designs often work well. But field experiments and experiments with human subjects have more sources of variability that can influence the outcome. With greater variability comes a greater need for statistical design, as we will see in Example 10.

A closer look at the GMAT review course showed that the students in the online review course were indeed quite different from the students who took the classroom course in past years. In particular, they were older and more likely to be employed. An online course appeals to these mature people, but we can't compare their performance with that of the undergraduates who previously dominated the course. The online course might even be less effective than the classroom version. The effect of

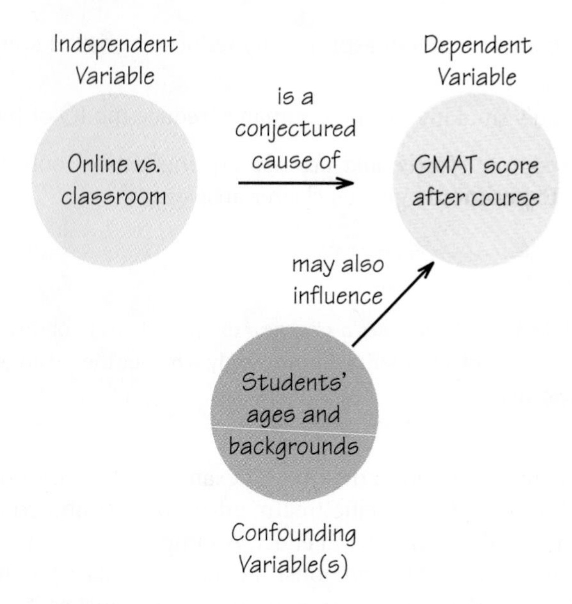

Figure 7.1 Confounding. We can't distinguish the effects of the treatment from the effects of other influences.

online versus in-class instruction is hopelessly mixed up with influences lurking in the background, and this entanglement is displayed in Figure 7.1. We say that student age and background are **confounded** with whatever effect the change to online instruction may have. In everyday usage, someone who is confounded is confused or mixed up. In statistics, confounded variables have their effects mixed together so that it's hard to tell what effect is due to each variable separately.

Confounding DEFINITION

Variables are said to be **confounded** when their effects on the outcome (dependent variable) cannot be distinguished from each other. Such variables may or may not have been intended to be part of the study.

The remedy for confounding is to do a *comparative experiment* (sometimes called a *quasiexperiment*) in which some students are taught in the classroom and other similar students take the course online. The first group is called a **control group**. Most well-designed experiments compare two or more treatments. Of course, comparison alone isn't enough to produce results we can trust. If the treatments are given to groups that differ markedly when the experiment begins, bias will result. For example, if we allow students to choose whether they get online or classroom instruction, older employed students are likely to sign up for the online course. Personal choice will bias our results in the same way that volunteers bias the results of call-in opinion polls. The solution to the problem of bias is the same for experiments and for samples: Use impersonal chance to select the groups.

EXAMPLE 10
A Randomized Comparative Experiment

The college decides to compare the progress of 25 on-campus students taught in the classroom with that of 25 students taught the same material online. Select which students will be taught online by taking a simple random sample of size 25 from the 50 available students. The remaining 25 students form the control group. They will receive classroom instruction. The result is a **randomized comparative experiment** with two groups. Figure 7.2 outlines the design in graphical form.

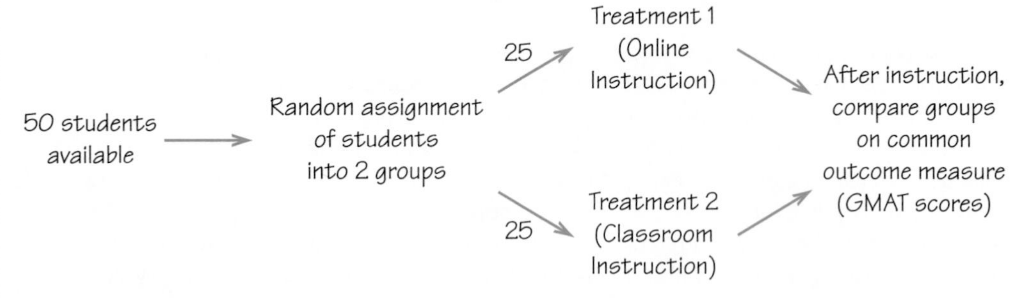

Figure 7.2 The design of a randomized comparative experiment to compare online and classroom instruction.

The selection procedure is exactly the same as it is for sampling:

Step 1: *Label* the 50 students 01 to 50.

Step 2: Go to the *table* of random digits and read successive two-digit groups. The first 25 labels encountered form the online group.

Step 3: Ignore repeated labels and groups of digits not used as labels. For example, if you begin at line 106 in Table 7.1, the first five students chosen are those labeled 41, 50, 13, 15, and 27. Remember that Section 7.3 discusses alternatives to Table 7.1 for selecting a simple random sample that can be used to determine treatment groups at random.

The GMAT experiment is *comparative* because it compares two treatments (the two instructional settings). The experiment could have even had a third treatment if there had been a "hybrid" (part-classroom, part-online) course option. It is *randomized* because the subjects are assigned to the treatments by chance. Randomization creates groups that are similar to each other before we start the experiment. Possible confounding variables act on both groups at once, so their effects tend to balance out and do not greatly affect the results of the study. The *only* difference between the groups is the online versus in-class experience. So if we see a difference in performance, it must be due to the different setting. That is the basic logic of randomized comparative experiments. This logic shows why experiments can give good evidence that the different treatments really *caused* different outcomes.

There is a fine point: The performance of the two groups will differ even if the treatments are identical, just because the individuals assigned at random to the groups differ. It is only differences *larger than would plausibly occur just by chance* that show the effects of the treatments. The laws of probability (in Section 8.1) allow statisticians to say how big of an effect is **statistically significant**. You probably understand this concept intuitively because you would not find it unusual if all the children were girls in a family with two or three children, but you probably would in a family with six or more children. We now present a convention that gives a concrete numerical benchmark to this intuitive gut feeling.

Statistically Significant DEFINITION

An observed effect that is so large that it would rarely occur by chance is called **statistically significant**. ("Rarely" usually means less than 5% of the time.)

SPOTLIGHT 7.1

Ethics in Experiments

The last century saw some unethical experiments; for instance, look up the painful and deadly medical experiments done by the Nazis during World War II or the sexually transmitted disease experiments conducted by the United States in Guatemala (1946–48) and in Tuskegee, Alabama (1932–72). It is understandable that there came to be a great emphasis on having studies approved by an Institutional Review Board (IRB) and on following ethical codes (e.g., the Nuremburg Code, Belmont Report, Declaration of Helsinki, and guidelines adopted by organizations such as the International Statistical Institute, the United Nations Statistics Division, and the American Statistical Association). Key principles of ethics codes include voluntary participation, informed consent, the right to quit at any time, and the avoidance of unnecessary suffering or risk. Also, when evidence in randomized clinical trials accumulates enough to make clear that a treatment is dangerous or is less effective than another, there is a mechanism to stop the data gathering process before the originally scheduled end of the experiment.

For some experiments, however, giving the subject full disclosure about the true purpose of the experiment could prevent getting accurate data. In such cases, an IRB may give the researcher permission to (at least temporarily) withhold a piece of information from participants. Consider this example. Internet field experiments conducted in Sweden by Ali M. Ahmed and Mats Hammarstedt (published in the July 2009 issue of *Economica*) showed that there was discrimination against gay male couples because (fictitious) responses from gay male couples to rental ads got far fewer callbacks and invitations to showings of apartments than did otherwise identical responses from (fictitious) heterosexual couples. Related studies have shown that a fictitious applicant with an Arabic/Muslim male name did far worse than an applicant with a Swedish male name, which in turn did worse than a Swedish female name. Few people are able and willing to acknowledge their prejudices, but this method of research was able to yield an accurate assessment of prejudice in Sweden's rental housing market.

EXAMPLE 11
Cervical Cancer Screening

On October 18, 2007, *The New England Journal of Medicine* reported on a randomized comparative experiment in which 10,154 Canadian women (ages 30–69) were given two different cervical cancer screening tests (the standard Pap test and a test for the DNA of the HPV virus) in a randomly assigned sequence. Although it might have seemed "fair" to assign each woman randomly to one treatment or the other, it was deemed better to give each woman the benefit of both medical tests for ethical reasons. Because of the randomization in the sequence, however, the first test performed on each woman was able to be analyzed as if it had been done alone. The DNA-based test proved to be much more powerful than the Pap test in detecting cancer.

7.6 Experiments versus Observational Studies

The first randomized comparative experiment was published in 1948—a British study of the effectiveness of streptomycin in treating tuberculosis. Randomized comparative experiments quickly became common tools of industrial, academic, and medical research. For example, federal regulations require that the safety and effectiveness of new drugs be demonstrated by randomized comparative experiments. Let's look at a medical experiment as an example.

EXAMPLE 12
St. John's Wort—Treatment for Depression?

Although prescription drugs must pass the test of randomized comparative experiments before being sold, herbs and other "natural remedies" are exempt. Because these treatments are so popular, some are now being studied more carefully. Fans of natural remedies often use extracts of the herb St. John's wort to treat depression. Is the herb safe? Does it work? The *Journal of the American Medical Association* reported a "randomized, double-blind, placebo-controlled clinical trial" in which 200 patients with major depression were assigned at random to take either herb extract or a dummy pill that looked and tasted the same. Results: The herb is safe, but "[i]n this study, St. John's wort was not effective for treatment of major depression."

If you read accounts of medical studies, you will often see language like "randomized, double-blind, placebo-controlled clinical trial." A clinical trial is a medical experiment with actual patients as subjects. "Randomized" and "controlled" tell us that this was a randomized comparative experiment (that's good). A "placebo" looks like a real treatment but is actually a fake pill that has no medication in it and should not actually have an effect in this study. Here we meet a new idea: the importance of the **placebo effect**, a special kind of confounding.

Placebo Effect DEFINITION

The **placebo effect** is the tendency of patients to respond favorably to any apparent treatment (even one that is in reality a "fake treatment") because of their expectations about the treatment.

If depressed patients given St. John's wort are compared with patients who receive no treatment, the first group gets the benefit of both the herb and the placebo effect. Any beneficial effect that St. John's wort may have is confounded with the placebo effect. To prevent confounding, it is important that some treatment be given to all subjects in any medical experiment.

Neither the subjects nor the experimenters who worked with them knew which treatment any specific subject received. Subjects might react differently if they knew they were getting "only a placebo." Knowing that a particular subject was getting "only a placebo" also could influence the health workers who interviewed and examined the subjects. Only the study's statistician knew which treatment each subject received. Because both the subjects and the health workers were "blind" to this information, this study was considered a **double-blind experiment**.

Double-Blind Experiment DEFINITION

A **double-blind experiment** is an experiment in which neither the experimental subjects nor the persons who interact with them know which treatment each subject received.

The difference between the St. John's wort groups and placebo groups was *not statistically significant*; that is, it was no larger than would be expected when we divide 200 depressed patients at random into two groups and do nothing else. Larger numbers of subjects would give more precise results. It's unlikely that there is exactly no difference between St. John's wort and a placebo. If the clinical trial had used 2000 patients rather than 200, it might have picked up a small effect (in either direction). The researchers thought that a group of 200 patients was enough to pick up any effect large enough to be medically important.

The logic of experimentation, the statistical design of experiments, and the laws that govern chance behavior combine to give compelling evidence of cause and effect. Only experimentation can produce the most convincing evidence of causation.

EXAMPLE 13
Smoking and Health

By way of contrast, consider the statistical evidence linking cigarette smoking to lung cancer. We can't ethically assign groups of people to smoke or not, so a direct experiment isn't possible. The most careful studies have selected samples of smokers and nonsmokers, and then followed them for many years, eventually recording the cause of death. These are called **prospective (observational) studies** because they follow the subjects forward in time. (A **retrospective study** looks backward in time.) Prospective studies are comparative, but they are not experiments because the subjects themselves choose whether to smoke. A large prospective study of British doctors found that the death rate from lung cancer among cigarette smokers was 20 times the rate among nonsmokers. Another study of American men aged 40 to 79 found that the lung cancer death rate was 11 times higher among smokers than among nonsmokers. These and many other **observational studies** show a strong connection between smoking and lung cancer.

Observational Study DEFINITION

An **observational study** does not try to manipulate the environment (such as by assigning treatments to people); it simply observes the measurements of variables of interest that result from people's free choices. This kind of study is generally done when assignment of a treatment to a person is unethical (for example, smoking while pregnant) or impossible (such as ethnicity).

The connection between smoking and lung cancer is statistically significant. That is, it is far stronger than would occur by chance. We can be confident that something other than chance links smoking to cancer. But observation of samples cannot tell us *what* factors other than chance might be at work. Perhaps some other factor is involved, such as differences in diet or exposure to air pollution. Perhaps there is something in the genetic makeup of some people that predisposes them both to nicotine addiction and to lung cancer. If so, we would observe a strong link even if smoking itself had no effect on the lungs.

However, the statistical evidence that points to cigarette smoking as a cause of lung cancer is about as strong as nonexperimental evidence can be. First, the connection has been observed in many studies in many countries. This eliminates factors peculiar to one group of people or to one specific study design. Second, there is a *dose-response relationship*: People who smoke more are more likely to get lung cancer than those who smoke less, and quitting cigarettes reduces the cancer risk. Third, specific ways in which smoking could cause cancer have been identified; cigarette smoke contains tars that have been shown by experiments to cause tumors in animals. Finally, no plausible alternative explanation is available. For example, the genetic hypothesis cannot explain the increase in lung cancer among women that occurred as more and more women became smokers. Also, the hypothesis was not supported by studies of identical twins where only one smoked.

It is very difficult and complicated to amass convincing evidence from observational studies and rule out all possible alternative explanations. This is why we have a strong preference for the more conclusive statistical evidence that we get from randomized comparative experiments when it is possible and ethical to conduct them. Despite their status as the "gold standard" of research, however, experiments can also have weaknesses. The most common of these is a contrived condition that makes it hard to say how far results may apply beyond a controlled laboratory setting.

EXAMPLE 14
Is the Experiment Realistic?

Clinical trials give medical treatments to actual patients with the condition that the treatments are supposed to help. But some experiments are less realistic in terms of how well experimental conditions align with the usual circumstances of what is of greatest interest. For example, a researcher studying stages and cycles of sleep observes patients overnight in a special "sleep lab" that can monitor their electroencephalography (EEG) waves and other data. However, individuals may sleep quite differently in a lab setting than in the natural sleeping environment of their bedroom at home.

Another type of example is that some studies on animals may be limited in how reliably their conclusions might apply to humans. Penicillin, for instance, is highly toxic to guinea pigs, but it has been a very helpful medicine for humans.

These are not statistical questions. Researchers must use their understanding of their academic domain to judge how far their results apply. Good statistical design enables us to trust results for the participants in the study at hand, but additional knowledge and judgment is needed to decide the extent to which conclusions might be generalized to other settings.

Figure 7.3 gives a conceptual overview of the types of designs we have covered in these last two sections.

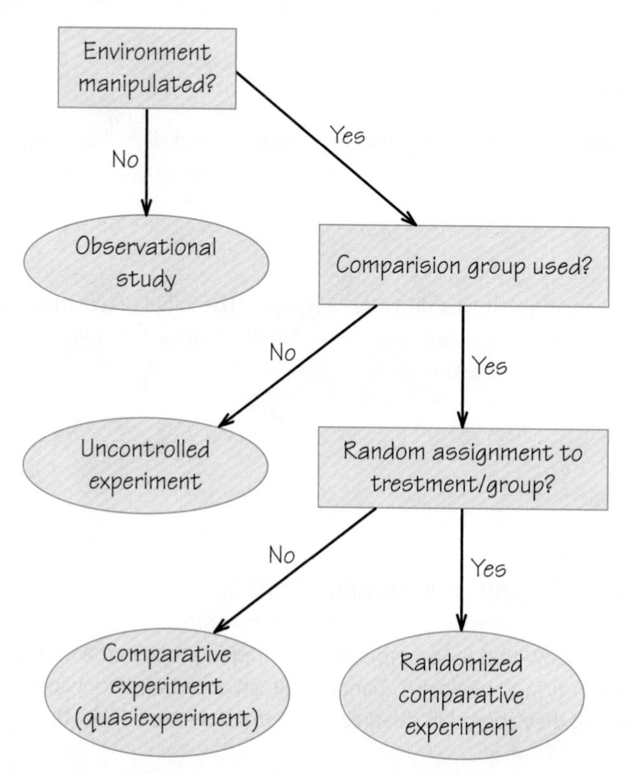

Figure 7.3 This flowchart shows how to distinguish the types of studies in Sections 7.5 and 7.6.

7.7 Inference: From Sample to Population

A market research firm interviews a random sample of 2500 adults. Result: 66 percent find shopping for clothes frustrating and time-consuming. This applies to the 2500 people in the sample. What is the truth about the 230 million American adults who make up the population? Because the sample was chosen at random, it's reasonable to think that these 2500 people represent the entire population fairly well. So the market researchers turn the *fact* that 66 percent of the *sample* find shopping frustrating into an *estimate* that about 66 percent of *all* adults feel this way. That's a fundamental operation in statistics: Use a fact about a sample to estimate the truth about the whole population. We call this *statistical inference.*

Statistical Inference DEFINITION

Statistical inference refers to methods for drawing conclusions about an entire population on the basis of data from a sample.

If the selected individuals were chosen at random, we think that they fairly represent the population and inference makes sense. If we have data from only a convenience sample or a voluntary response sample, the data do not represent the population and we can't use them for inference. *Statistical inference works only if the data come from a random sample or randomized comparative experiment.* That's why this chapter starts with producing reliable data before moving on to inference from the data to a larger population.

To think about inference, we must keep straight whether a number describes a sample or a population (recall Section 7.1). The vocabulary we use is easy to keep straight because *population* and *parameter* start with "p," while *sample, statistic,* and *estimate* all start with the "s" sound (if not "s" itself).

Parameter DEFINITION

A **parameter** is a fixed (usually unknown) number that describes a population.

Statistic DEFINITION

A **statistic** is a number that describes a sample. The value of a statistic is known when we have taken a sample but it can change from sample to sample. We often use a statistic to estimate an unknown parameter.

EXAMPLE 15
Attitudes on Shopping: A Sample Statistic

Sample surveys show that fewer people enjoy shopping than in the past. A survey by the market research firm Yankelovich Clancy Shulman asked a nationwide random sample of 2500 adults if they agreed or disagreed with the statement "I like buying new clothes, but shopping is often frustrating and time-consuming." Of the respondents, 1650 said they agreed. The proportion of the sample who agree is

$$\hat{p} = \frac{1650}{2500} = 0.66 = 66\%$$

The symbol $\hat{p}$ is read "p-hat." The ^ symbol here tells us that a quantity has been estimated, just as the use of $\hat{y}$ in Chapter 6 told us that a value was estimated by using a regression-line model. The number $\hat{p} = 0.66$ is a *statistic*. The corresponding *parameter* is the proportion (call it p) of all adult U.S. residents who would have said "Agree" if asked the same question. We don't know the value of the parameter p, so we use the statistic $\hat{p}$ to estimate it.

(Carol Kohen/Getty Images.)

If Yankelovich took a second random sample of 2500 adults, the new sample would have different people in it. It is almost certain that there would not be exactly 1650 positive responses. That is, the value of the statistic $\hat{p}$ will vary from sample to sample. If the variation when we take repeat samples from the same population is too great, we can't trust the results of any one sample. We are saved by a great advantage of random samples: If we take lots of random samples of the same size from the same population, the variation from sample to sample will follow a predictable pattern.

All of statistical inference is based on one idea: to see how trustworthy a procedure is, ask what would happen if we repeated it many times. So we must ask: "What would happen if we took many samples?" Here's how to answer that question:

- Take a large number of random samples from the same population.
- Calculate the sample proportion $\hat{p}$ for each sample.
- Make a histogram of the values of $\hat{p}$.
- Examine the distribution displayed in the histogram for shape, center, and variability, as well as outliers or other deviations.

In practice, it is too expensive to take many samples from a large population, such as all adult U.S. residents. But we can use a computer to imitate drawing many samples at random from a population that we specify. This is called *simulation*. Here's what happens when we do this.

EXAMPLE 16
What Happens in Many Samples?

Figure 7.4 illustrates a result of choosing many samples and finding the sample proportion $\hat{p}$ for each one. The histogram shows the distribution of the values of $\hat{p}$ from 1000 separate SRSs of size 100 drawn from a population that we suppose has a parameter value $p = 0.6$.

Of course, Yankelovich interviewed 2500 people, not just 100. Figure 7.5 parallels Figure 7.4. It shows a result from choosing 1000 SRSs, each of size 2500, from a population in which the true proportion is $p = 0.6$. The 1000 values of $\hat{p}$ from these samples form the histogram. Figures 7.4 and 7.5 are drawn on the same scale. Comparing them shows what happens when we increase the size of our samples from 100 to 2500. These histograms display the **sampling distribution** of the statistic $\hat{p}$ for two sample sizes. For

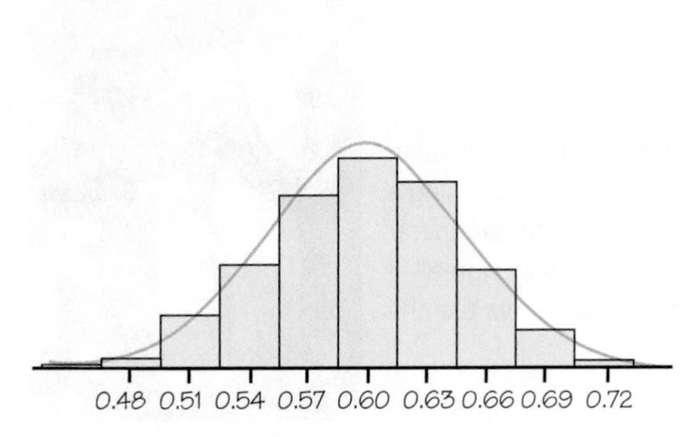

Figure 7.4 Draw 1000 SRSs of size 100 from a population with proportion $p = 0.60$ of successes. The histogram shows the distribution of the 1000 sample proportions $\hat{p}$.

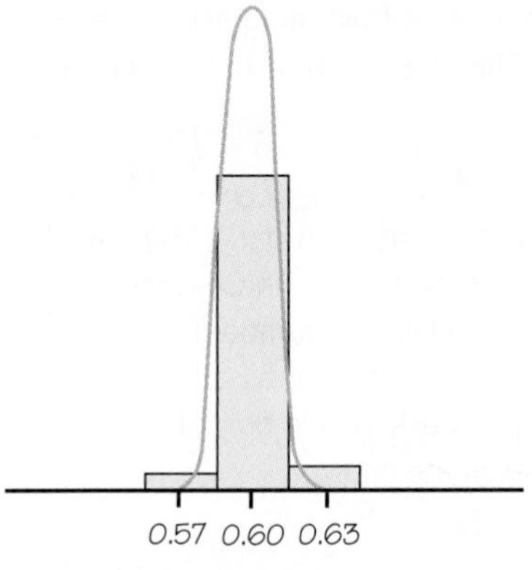

Figure 7.5 Draw 1000 SRSs of size 2500 from the same population as in Figure 7.4. The histogram shows the distribution of the 1000 sample proportions $\hat{p}$, using the same scale as Figure 7.4. The statistic from the larger sample is less variable.

intuition, consider a sample size of only 2. There would be only 3 possible $\hat{p}$ values from a sample of size 2: 0, 0.5, or 1. For a sample of size 100, there would be 101 possible $\hat{p}$ values: 0, 0.01, 0.02, ..., 1.00. Of course, not all values are equally likely—the ones near 0.6 are most common.

Sampling Distribution DEFINITION

The **sampling distribution** of a statistic is the distribution of values taken on by the statistic in all possible samples of the same size from the same population.

Strictly speaking, the sampling distribution is the ideal pattern that would emerge if we looked at all possible samples of the same size from our population. A distribution obtained from a fixed number of trials, like the 1000 trials in these figures, is only an approximation of the sampling distribution. Probability theory, the mathematics of chance behavior, sometimes can describe sampling distributions exactly. Chapter 8 will introduce you to basic probability theory. The interpretation of a sampling distribution is the same, however, whether we obtain it by simulation or by the mathematics of probability.

We can use the tools of data analysis from Chapter 5 to describe any distribution. Let's apply those tools to Figures 7.4 and 7.5.

- **Shape:** The histograms look normal. The normal curves drawn through the histograms describe the overall shape quite well.
- **Center:** In both cases, the values of the sample proportion $\hat{p}$ vary from sample to sample, but the values are centered at 0.6. Recall that we are assuming that $p = 0.6$ is the true population parameter. Some samples have a $\hat{p}$ less than 0.6 and some greater, but there is no tendency to be always low or always high. That is, $\hat{p}$ has *no bias* as an estimator of p. This is true for both large and small samples. (Want the details? The mean of the 1000 values of $\hat{p}$ is 0.598 for samples of size 100 and 0.6002 for samples of size 2500. The median value of $\hat{p}$ is exactly 0.6 for samples of both sizes.)

- **Variability:** The values of $\hat{p}$ from samples of size 2500 have much less variability than the values from samples of size 100. In fact, the standard deviations are 0.051 for Figure 7.4 and 0.0097, or about 0.01, for Figure 7.5.

Although these results describe just two sets of simulations, they reflect facts that are true whenever we use random sampling. We now turn to probability theory to learn the mathematical facts that lie behind the simulations. We'll use the word *success* for whatever we are counting, such as "Agree" responses in the shopping survey. Note that *success* does not necessarily have the positive (or negative) association it does in real life, but is simply a convenient way to identify an outcome.

Sampling Distribution of a Sample Proportion THEOREM

Choose an SRS of size n from a large population that contains population proportion p of successes. Let $\hat{p}$ be the **sample proportion** of successes,

$$\hat{p} = \frac{\text{count of successes in the sample}}{n}$$

Then:

- **Shape:** For large ($n \geq 30$) sample sizes, the sampling distribution of $\hat{p}$ is *approximately normal.*
- **Center:** The *mean* of the sampling distribution of $\hat{p}$ is p.
- **Variability:** The *standard deviation* of the sampling distribution of $\hat{p}$ is

$$\sqrt{\frac{p(1-p)}{n}}$$

(This will be confirmed in Section 8.6, when we apply the Central Limit Theorem.)

Figure 7.6 summarizes these facts in a form that reminds us that a sampling distribution describes the results of lots of samples from the same population.

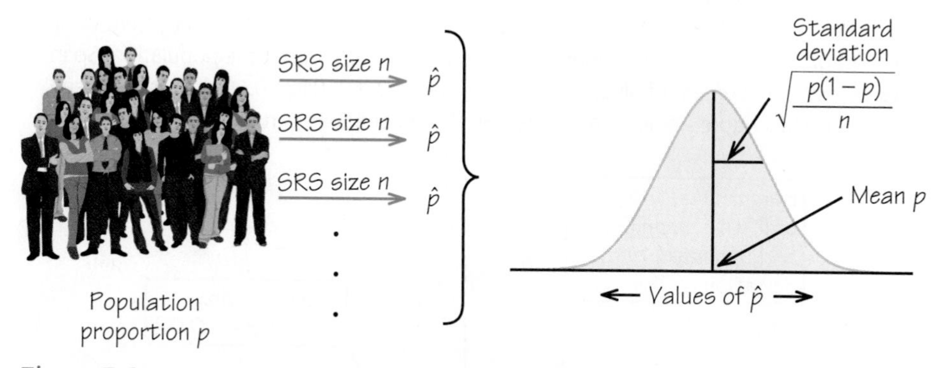

Figure 7.6 Repeat many times the process of selecting an SRS of size n from a population of which the proportion p are successes. The values of the sample proportion of successes $\hat{p}$ have this normal sampling distribution.

EXAMPLE 17
Attitudes on Shopping: A Simulation of Samples

Suppose that 60 percent of all adults in a population find shopping for clothes frustrating and time-consuming. The population proportion is $p = 0.6$. Take a simple random sample of 2500 adults. That's exactly the setting of the second simulation in Example 16 (see Figure 7.5). Now we can apply mathematics to learn how the sample proportion $\hat{p}$ would behave if we took many samples. The distribution of $\hat{p}$ in many samples

- is close to normal;
- has mean 0.6;
- has standard deviation 0.0098.

To show our work for the last number, note that $p(1 - p)/n = (0.6)(0.4)/2500 = 0.000096$, and the square root of 0.000096 is 0.0098.

$$\sqrt{\frac{p(1-p)}{n}} = \sqrt{\frac{(0.6)(0.4)}{2500}} = 0.0098$$

The mean of 0.6 and standard deviation of 0.0098 from the mathematics are very close to the mean of 0.6002 and standard deviation of 0.0097 we observed in our simulation. If the simulation used more than 1000 trials, the results would be still closer to the mathematical truth.

7.8 Confidence Intervals

The sampling distribution shows why we can trust the results of a large random sample: Almost all such samples give results that are close to the truth about the population.

EXAMPLE 18
The 68–95–99.7 Rule Again

In Example 17 the population parameter, the proportion of adults who find shopping frustrating, is $p = 0.6$. If we take SRSs of size 2500, the sample proportions $\hat{p}$ follow the normal distribution with mean of 0.6 and standard deviation of 0.0098, which is about 0.01. The "95" part of the 68–95–99.7 rule from Section 5.9 says that 95 percent of all samples give a $\hat{p}$ within two standard deviations of the truth about the population. So in this example, 95 percent of all samples have $\hat{p}$ within 2×0.01 of 0.60, that is, between 0.58 and 0.62. Figure 7.7 illustrates this use of the 68–95–99.7 rule.

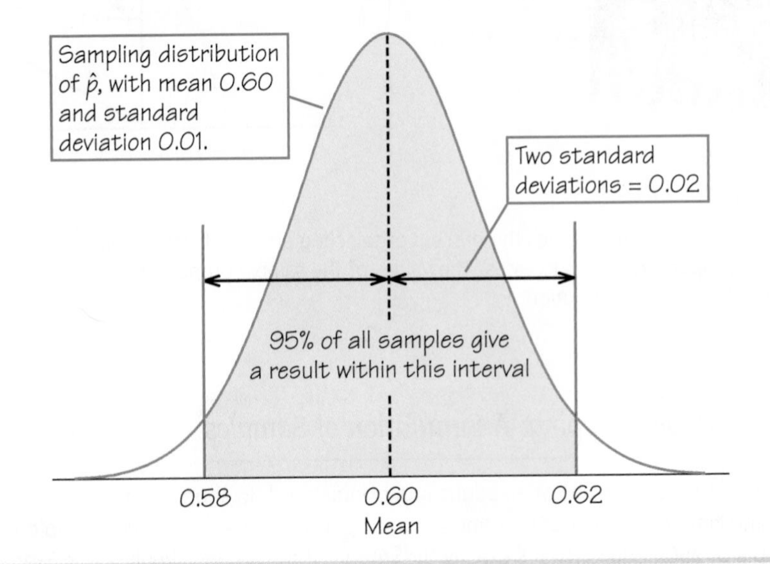

Figure 7.7 The sampling distribution of $\hat{p}$ for Example 17. By the 68–95–99.7 rule, 95 percent of all samples have a sample proportion $\hat{p}$ within 0.02 of the true population proportion $p = 0.6$.

We can repeat this reasoning for any value of the parameter p and the sample size n. It is always true that 95 percent of all samples give a sample proportion $\hat{p}$ within two standard deviations of the population proportion p. That is, 95 percent of all samples catch p in the interval extending two standard deviations on either side of $\hat{p}$. That's the interval

$$\hat{p} \pm 2\sqrt{\frac{p(1-p)}{n}}$$

This formula tells us how close the unknown parameter p lies to the observed statistic $\hat{p}$ in 95 percent of all samples. There is one catch: We can't calculate the interval from the data because the standard deviation involves the population proportion p, and in practice we don't know p. In Examples 17 and 18, we applied the formula for $p = 0.6$, but this may not be the true p for the actual population of all American adults.

What to do? The standard deviation of the statistic $\hat{p}$ does depend on the parameter p, but it doesn't change a lot when p changes. Go back to Example 17 and redo the calculation for other values of p. Here's the result:

Value of p	0.4	0.5	0.6	0.7	0.8
Standard deviation	0.0098	0.01	0.0098	0.0092	0.008

The standard deviations are all 0.01 when rounded to the hundredths place. You see that if we guess a value of p that is reasonably close to the true value, the standard deviation found from the guessed value will be about right. We know that when we take a large random sample, the statistic $\hat{p}$ is almost always close to the parameter p. So we will use $\hat{p}$ as the guessed value of the unknown p. Now we have an interval that we can calculate from the sample data. We call it a *confidence interval.*

Confidence Interval DEFINITION

A 95% **confidence interval** is an interval obtained from the sample data by a method in which 95% of all samples will produce an interval containing the true population parameter.

Choose an SRS of size n from a large population that contains an unknown proportion p of successes. A 95% confidence interval for p is approximately

$$\hat{p} \pm 2\sqrt{\frac{\hat{p}(1-\hat{p})}{n}}$$

The ± sign is read "plus or minus," so 0.5 ± 0.2 yields two numbers: $0.5 - 0.2 = 0.3$ and $0.5 + 0.2 = 0.7$. This can be written as an interval: (0.3, 0.7).

This formula is only approximately correct but is quite accurate when the sample size n is large (≥ 30). Here $\hat{p}$ is the proportion of successes in the sample and $2\sqrt{\hat{p}(1-\hat{p})/n}$, the expression to the right of the ± sign, is the **margin of error**. The margin of error is what is commonly reported in newspapers.

Margin of Error DEFINITION

The **margin of error** is equal to half of the width of a confidence interval. For a 95% confidence interval, it equals about two standard deviations of the sampling distribution of the estimated parameter. If you conducted a very large number of polls, about 95% of the time the difference between a particular poll's result and the true value of the population parameter would be within the margin of error.

This interval is only approximately correct for two reasons. The sampling distribution of the sample proportion $\hat{p}$ isn't exactly normal. And we don't get the standard deviation of $\hat{p}$ exactly right because we used $\hat{p}$ in place of the unknown p. Both of these difficulties go away as the sample size n gets larger. Our method works well enough for many practical uses. More important, it shows how we get a confidence interval from the sampling distribution of a statistic. That's the reasoning behind any confidence interval.

EXAMPLE 19
Risky Behavior in the Age of AIDS

How common is behavior that puts heterosexuals at risk for AIDS? In 1990–91, the National AIDS Behavioral Survey interviewed a random sample of 2673 adult heterosexuals. Of these people, 170 had had more than one sexual partner in the past year. The sample proportion who admit to multiple partners is

$$\hat{p} = \frac{170}{2673} = 0.0636$$

A 95 percent confidence interval for the proportion p of all adult heterosexuals with multiple partners, therefore, is

$$\begin{aligned}\hat{p} \pm 2\sqrt{\frac{\hat{p}(1-\hat{p})}{n}} &= 0.0636 \pm 2\sqrt{\frac{(0.0636)(0.9364)}{2673}} \\ &= 0.0636 \pm 0.0094, \text{ or } 6.36\% \pm 0.94\% \\ &= 0.0542 \text{ to } 0.0730, \text{ or } 5.42\% \text{ to } 7.30\%\end{aligned}$$

A report of these calculations might say, "The study found that 6.36 percent of heterosexuals had more than one sexual partner. The margin of error for this result is 0.94 percent."

Algebra Review
Fractions to Percentages, page 486.

We got the interval in Example 19 by using a formula that catches the true unknown population proportion in 95% of all samples. The shorthand for this is: We are *95% confident* that the true percentage of heterosexuals with multiple partners lies between 5.42 percent and 7.30 percent. The true value lies outside the interval (5.42%, 7.30%) in 5 percent of all samples.

To understand the meaning of "95% confidence," imagine a carnival ring toss game in which each ring tossed represents an estimate from a particular SRS, and the size of the ring and method of tossing is such that (with a large number of tosses) about 95% of the rings will land around the stake (i.e., capture the true value of the population parameter) and 5% won't. In Figure 7.8, the vertical line (the "stake" in our ring toss analogy) is the true value of the population proportion p. The normal curve at the top of the figure is the sampling distribution of the sample statistic $\hat{p}$, which is centered at the true p. The 95% confidence intervals (the "rings") from 25 SRSs appear below, one after the other. The central dots are the values of $\hat{p}$, the centers of the intervals. In the long run, 95% of the intervals will cover or contain the true p and 5% will miss. Of these particular 25 simulated intervals in Figure 7.8, 24 hit and 1 missed. (Now, $\frac{24}{25}$ is 96 percent, not exactly 95%, but remember that the sampling distribution describes what happens in a very large number of samples.) The *Confidence Interval* applet animates Figure 7.8. You can use the applet to watch confidence intervals from one sample after another capture or fail to capture the true parameter.

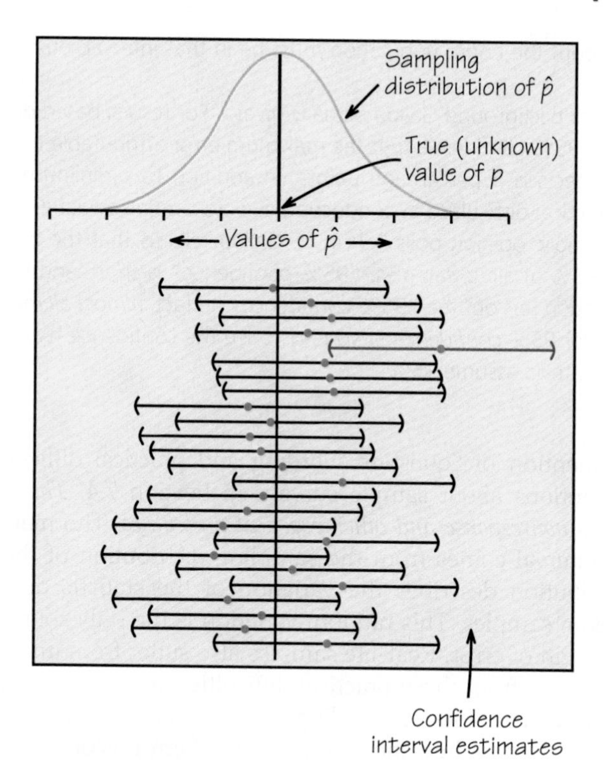

Figure 7.8 A collection of 25 samples from the same population give these 95% confidence intervals. The red dot of each interval is its point estimate of the proportion. Note that 24 of these 25 intervals (i.e., 96%) cover or contain the true population proportion, marked by the vertical line. If this process were done many, many times, 95% of all 95% confidence intervals would cover or contain the true population proportion.

The length of a confidence interval depends on how confident we want to be that the interval does capture the true parameter value. It is common to use 95% confidence, but you can ask for higher or lower confidence if you want. Our 95% confidence interval was based on the middle 95% of a normal distribution. A 99% confidence interval requires the middle 99% of the distribution and therefore is wider (has a larger margin of error). To recall the ring toss analogy of Figure 7.8, you will be a bit more confident that your ring landed around the stake if your ring is a bit wider.

The length of a confidence interval also depends on the size n of the sample. Larger samples give shorter intervals because of the $\sqrt{n}$ in the denominator of the margin of error. But the interval does not depend on the size of the population. This is true so long as the population is much larger than the sample. The confidence interval in Example 19 works for a sample of 2673 from a city with 100,000 adults as well as it does for a sample of 2673 from a nation of 308 million. What matters more is how many people we interview, not what percentage of the population we contact.

EXAMPLE 20
Understanding the News

Here's what the TV news announcer says: "A new Gallup poll on American exercise habits finds that 45% of adults are not engaging in vigorous sports or physical activities. The margin of error for the poll was 3 percentage points." Plus or minus 3%, starting at 45%, is 42% to 48%. People with minimal statistics knowledge may think that the truth about the entire population must be in that interval, but now we know better!

This is the full background Gallup actually gives: "For results based on this sample, one can say with 95% confidence that the maximum error attributable to sampling and other random effects is 3 percentage points. In addition to sampling error, question wording and practical difficulties in conducting surveys can introduce error or bias into the findings of public opinion polls." That is, Gallup tells us that the margin of error works only for 95% of all its samples. "95% confidence" is shorthand for that longer fact. The news report left out the "95% confidence." In fact, *almost all margins of error in the news are for 95% confidence.* If you don't see the confidence level in a scientific poll, it's usually safe to assume 95%.

Gallup's mention of "question wording and practical difficulties" takes us back to our cautions about sample surveys in Section 7.4. *The margin of error does not address nonresponse and other practical difficulties.* The margin of error in a confidence interval comes from the sampling distribution of the statistic. The sampling distribution describes the variation of the statistic due to chance in repeated random samples. This random variation is the only source of error covered by the margin of error. Real-life samples also suffer from undercoverage and nonresponse. Errors from these practical difficulties are usually more serious and harder to quantify than random sampling error. The actual error in sample surveys may be much larger than the announced margin of error. Worse, we can't say how much larger. Statistical conclusions are approximations to a complicated truth, not mathematical results that are simply true.

EXAMPLE 21
Measuring Risky Behavior

What about the National AIDS Behavioral Survey? The interviews were carried out by telephone. This is acceptable for surveys of the general population because about 94 percent of American households have telephones. However, some groups at high risk for AIDS, such as intravenous drug users, often don't live in settled households and are underrepresented in the sample. About 30 percent of the people reached refused to cooperate. A nonresponse rate of 30 percent is not unusual in large sample surveys, but it may cause some bias if those who refuse differ systematically from those who cooperate. The survey used statistical methods that adjust for unequal response rates in different groups. Finally, some respondents may not have told the truth when asked about their sexual behavior. The survey team tried hard to make respondents feel comfortable. For example, Hispanic women were interviewed by Hispanic women, and Spanish speakers were interviewed by Spanish speakers with the same regional accent (such as Cuban, Mexican, or Puerto Rican). Nonetheless, the survey report says that some bias is probably present:

> *It is more likely that the present figures are underestimates; some respondents may underreport their numbers of sexual partners and intravenous drug use because of embarrassment and fear of reprisal, or they may forget or not know details of their own or of their partner's HIV risk and their antibody testing history.* [Joseph H. Catania et al., Prevalence of AIDS-related risk factors and condom use in the United States, *Science,* 258 (1992): 1104.]

SPOTLIGHT 7.2

Truth in Polling

Responsible polling organizations tell the public something about both the precision and limitations of their poll results. Your college student newspaper may not have the resources to conduct polls using random sampling, so it is refreshing when the polls that it publishes from voluntary response samples are accompanied by a disclaimer, such as this one that has been used by *The Prospector,* the student newspaper of The University of Texas at El Paso:

> "This poll is not scientific and reflects the opinions of only those Internet users who have chosen to participate. The results cannot be assumed to represent the opinions of Internet users in general, nor the public as a whole."

Because of this limitation, *The Prospector* simply reports the breakdown of responses given but without any margin of error, since sampling error cannot be quantified from a (voluntary response) sample that is not probability-based.

The Harris Poll accompanies its polls with a more extensive disclaimer:

> "All sample surveys and polls, whether or not they use probability sampling, are subject to multiple sources of error which are most often not possible to quantify or estimate, including sampling error, coverage error, error associated with nonresponse, error associated with question wording and response options, and post-survey weighting and adjustments. Therefore, Harris Interactive avoids the words 'margin of error' as they are misleading. All that can be calculated are different possible sampling errors with different probabilities for pure, unweighted, random samples with 100 percent response rates. These are only theoretical because no published polls come close to this ideal."

REVIEW VOCABULARY

Bias A systematic error that tends to cause the observations to deviate in the same direction from the truth about the population whenever a sample or experiment is repeated. (p. 244)

Confidence interval An interval of values used to estimate a population parameter with a specific level of confidence. A 95 percent confidence interval is an interval computed from a sample by a method that surrounds the unknown parameter 95 percent of the time, so when we calculate the interval for a single sample, we are 95 percent confident that the interval contains the unknown parameter. (p. 263)

Confounding Two variables are confounded when their effects on the outcome of a study cannot be distinguished from each other. (p. 252)

Control group A group of experimental subjects that is given a standard treatment, no treatment, or a fake treatment (such as a placebo). (p. 252)

Convenience sample A sample that consists of the individuals who are easily reachable, such as people passing by in the street. A convenience sample is usually biased. (p. 243)

Double-blind experiment An experiment in which neither the experimental subjects nor the persons who interact with them know which treatment each subject received. (p. 255)

Experiment A study in which treatments are applied to people, animals, or things to observe the effect of the treatments. (p. 251)

Margin of error The number to the right of the ± sign in a 95% confidence interval and equal to half of the width of the full interval. For a 95% confidence interval, it equals about two standard deviations of the sampling distribution of the estimated parameter. If you conducted a very large number of polls, about 95 percent of the time the difference between a particular poll's result and the true value of the population parameter would be within the margin of error. (p. 263)

Nonresponse Some individuals chosen for a sample cannot be contacted or refuse to participate. (p. 249)

Observational study A study (such as a sample survey) that observes individuals and measures variables of interest but does not attempt to influence the responses. (p. 256)

Parameter A number that describes the population. In statistical inference, the goal is often to estimate an unknown parameter or make a decision about its value. (p. 258)

Placebo effect The beneficial effect of a dummy treatment (such as an inert pill in a medical experiment) on the response of subjects. (p. 255)

Population The entire group of people or things about which we want information. (p. 242)

Prospective study An observational study that follows two or more groups of subjects forward in time. (p. 256)

Randomized comparative experiment An experiment to compare two or more treatments in which people, animals, or things are assigned to treatments by chance. (p. 253)

Retrospective study An observational study that uses interviews or records to collect information about past behaviors of subjects in two or more groups. (p. 256)

Sample A part of the population that is actually observed and used to draw conclusions, or inferences, about the entire population. (p. 242)

Sample proportion The proportion $\hat{p}$ of the members of a sample having some characteristic (such as agreeing with an opinion poll question). The sample proportion from a simple random sample is used to estimate the corresponding proportion p in the population from which the sample was drawn. (pp. 259, 261)

Sampling distribution The distribution of values taken by a statistic when all possible random samples of the same size are drawn from the same population. The sampling distributions of sample proportions are approximately normal. (p. 260)

Simple random sample (SRS) A sample chosen by chance, so that every possible sample of the same size has an equal chance to be the one selected. (p. 245)

Statistic A number that describes a sample. A statistic can be calculated from the sample data alone; it does not involve any unknown parameters of the population. (p. 258)

Statistical inference Methods for drawing conclusions about an entire population on the basis of data from a sample. Confidence intervals are one type of inference method. (p. 258)

Statistically significant An observed effect is statistically significant if it is so large that it is unlikely to occur just by chance in the absence of a real effect in the population from which the data were drawn. (p. 253)

Table of random digits A table whose entries are the digits 0, 1, 2, 3, 4, 5, 6, 7, 8, 9 in a completely random order. That is, each entry is equally likely to be any of the 10 digits and no entry gives information about any other entry. (p. 245)

Undercoverage The process of choosing a sample may systematically leave out some groups in the population, such as households without telephones. (p. 248)

Voluntary response sample A sample of people who select themselves by responding to a general invitation to give their opinions. Such a sample is usually strongly biased. (p. 244)

SKILLS CHECK

1. An opinion poll contacts 1021 adults and asks them, "Which political party do you think has better ideas for leading the country in the 21st century?" In all, 723 of the 1021 say "the Democrats." The sample in this setting is

(a) all 220 million adults in the United States.
(b) the 1021 people interviewed.
(c) the 723 people who chose the Democrats.

2. In a part-to-whole relationship, we would say the population is the _______.

3. A committee on community relations in a college town plans to survey local businesses about the importance of students as customers. From the 10,000 businesses listed in the telephone book, the committee chooses 150 businesses at random. Of these, 72 return the questionnaire mailed by the committee. The sample is

(a) all 10,000 businesses in the college town.
(b) the 150 businesses chosen.
(c) the 72 businesses that returned the questionnaire.

4. A call-in poll asks who people are planning to vote for in the next presidential election. People who think major change is needed are likely to be represented in this poll ________ than they should be, if the goal is to get results that are representative of all voters.

5. On January 2, 2008, the *American Idol* Web site (**www.americanidol.com**) had an online poll that asked who respondents liked best among six former contestants. To become part of the sample, you simply clicked on a response. Of the 941,434 responses to this poll, 55 percent went to Clay Aiken. We can conclude that

(a) most Americans prefer Clay Aiken out of those six contestants.
(b) the sample is too small a fraction of the millions of people who watched the TV show to draw any conclusion.
(c) the poll uses voluntary response, so the results tell us little about the population of all adults.

6. A sample consisting of people who chose to have their opinions be part of a poll is a ______ sample.

7. If we use the range of a sample to estimate the range of a population, our estimate would likely be

(a) a bit too high.
(b) a bit too low.
(c) unbiased and right on target.

8. You are using the table of random digits to choose a simple random sample of 6 students from a class of 30 students. You label the students 01 to 30 in alphabetical order. Go to line 113 of Table 7.1 (p. 246). Of the labels corresponding to the six students selected for your sample, the label that is largest is ________.

9. You must choose an SRS of 10 of the 420 retail outlets in New York that sell your company's products. How would you label this population to use Table 7.1 (p. 246)?

(a) 001, 002, 003, ..., 419, 420
(b) 000, 001, 002, ..., 419, 420
(c) 1, 2, 3, ..., 419, 420

10. From an alphabetical list of the 7200 salaried employees of a corporation, you label the employees 0001 to 7200. Using line 111 of Table 7.1 (p. 246), choose an SRS of 5 of the 7200 employees. Of the five employees selected for your sample, the label that is the largest is ________.

11. Which of these is more likely to occur when selecting a sequence of three digits from a very large table of random digits?

(a) 123
(b) 111
(c) The above sequences are equally likely.

12. There are __________ possibilities for each digit drawn from a table of random digits.

13. A sample of households in a community is selected at random from the telephone directory. In this community, 4 percent of households have no telephone and another 35 percent have unlisted telephone numbers. The sample will certainly suffer from

(a) nonresponse.
(b) undercoverage.
(c) false responses.

14. For the survey in Skills Check 3, the nonresponse rate is ________.

15. Nonresponse is a type of

(a) coverage.
(b) sample.
(c) bias.

16. In research about the population of a county containing 100,000 people, a sample of 3000 households was selected to be interviewed. For 1200 of the 3000 households that researchers attempted to contact, there was no one home willing to participate. For this survey, the nonresponse rate was __________ percent.

17. A clinical trial compares an antidepression medicine with a placebo for relief of chronic headaches. There are 36 headache patients available to serve as subjects. To choose 18 patients to receive the medicine, you would

(a) assign labels 01 to 36 and use Table 7.1 to choose 18.
(b) assign labels 01 to 18 because only 18 need to be chosen.
(c) assign the first 18 who signed up to get the medicine.

18. An experiment is designed to see if a treatment is the __________ of the response.

19. A comparative experiment

(a) does not use a treatment.
(b) has two or more groups.
(c) is statistically significant.

20. A study of cell phones and the risk of brain cancer looked at a group of 519 people who have brain cancer. The investigators matched each cancer patient with a person of the same sex, age, and race who did not have brain cancer, then asked about use of cell phones. This kind of study is known as ________.

21. Studies that follow subjects forward in time are called

(a) retrospective.
(b) prospective.
(c) double-blind.

22. A treatment consisting of a "dummy pill" that looks like real medicine (but isn't) is known as a __________.

23. A study of religious practices among college students interviewed a sample of 125 students; 105 of the students said that they prayed at least once in a while. The sample proportion who said they pray is what?

(a) 105
(b) 84
(c) 0.84

24. Suppose that 35 percent of all adults in a population would say "good" and 65 percent would say "bad" if they were asked how they view the state of the economy. An opinion poll asks this question of an SRS of 1000 adults from this population. In repeated samples, the sample proportion $\hat{p}$ who say "good" would follow a normal distribution, with the mean having a value of ________________.

25. Referring to Skills Check 24, the standard deviation of the distribution of the sample proportion of adults who view the economy as "good" is about

(a) 0.00023.
(b) 0.015.
(c) 0.03.

26. Referring to Skills Check 24, the standard deviation of the distribution of the sample proportion of all adults who view the economy as "bad" is ________.

27. The sample survey in Skills Check 23 actually called 150 students, but 25 of the students refused to say whether they pray. This nonresponse could cause the survey result to be in error. The error due to nonresponse

(a) is in addition to the margin of error.
(b) is included in the margin of error.
(c) can be ignored because it isn't random.

28. To the nearer half of a percentage point, the margin of error for a 95% confidence interval is ______ when we use the result of Skills Check 23 to estimate what percentage of all college students pray.

29. A survey of folk music fans yields this 95 percent confidence interval estimate of the proportion of fans who love the music of David Wilcox: 0.74 to 0.86. The estimated mean percentage of fans who love Wilcox's music must be about

(a) 95 percent.
(b) 86 percent.
(c) 80 percent.

30. To the nearer percentage point, the margin of error for the survey in Skills Check 29 is ________.

CHAPTER 7 EXERCISES

■ *Challenge* ▲ *Discussion*

7.1 Sampling

1. A Gallup poll asked, "How would you describe your own personal weight situation right now?" Thirty-eight percent of American adults answered "very/somewhat overweight." Gallup's report said that these results are based on telephone interviews of 1,021 adults conducted November 4–7, 2010.

(a) What is the population for this sample survey?
(b) What is the sample size?

2. Starting with the 2010 Census, the decennial "long form" sample was replaced with the annual American Community Survey (ACS; **http://www.census.gov/acs/**). The main part of the ACS contacts 250,000 households by mail each month, with follow-up by phone and in person if there is no response. Each household answers questions about its housing, economic, and social status. What is the population for the ACS?

7.2 Bad Sampling Methods

▲ **3.** You see a student standing in front of the student center, stopping other students now and then to ask them questions. She says that she is collecting student opinions for a class assignment. Explain why this sampling method is almost certainly biased.

▲ **4.** A member of Congress is interested in whether her constituents favor a proposed gun-control bill. Her staff reports that letters on the bill have been received from 361 constituents, and that 323 of these oppose the bill. What is the population of interest? What is the sample? Is this sample likely to represent the population well? Explain your answer.

▲ **5.** Highway planners made a main street in a college town one-way. Local businesses were against the change. The local newspaper invited readers to call a telephone number to record their comments. The next day, the paper reported:

Readers overwhelmingly prefer two-way traffic flow to one-way streets. By nearly a 7:1 ratio, callers to the newspaper's Express Yourself opinion line on Wednesday complained about the one-way streets that have been in place since May. Of the 98 comments received, all but 14 said no to one-way.

(a) What population do you think the newspaper wants information about?

(b) Is the proportion of this population who favor one-way streets almost certainly larger or smaller than the $\frac{14}{98}$ proportion in the sample? Why?

▲ **6.** Your college wants to gather student opinion about a proposed student fee increase. It isn't practical to contact all students.

(a) Give an example of a way to choose a sample of students that is poor practice because it depends on voluntary response.

(b) Give an example of a bad way to choose a sample that doesn't use voluntary response.

7.3 Simple Random Samples

7. You have just been blessed with quadruplets (all girls). You decide to select their names using an SRS of four names from the following list of the most popular names given to American girls born in the past decade. To do this, use Table 7.1 (p. 246), starting at line 122.

1) Emily	2) Madison	3) Hannah	4) Emma
5) Ashley	6) Alexis	7) Samantha	8) Sarah
9) Abigail	10) Olivia	11) Elizabeth	12) Alyssa
13) Jessica	14) Grace	15) Lauren	16) Taylor
17) Kayla	18) Brianna	19) Isabella	20) Anna

8. **(a)** Would pulling out and lining up several dollar bills to use the eight-digit serial numbers be a reasonable substitute for Table 7.1? Explain.

(b) How about using the telephone numbers on a page of the phone book? Explain.

9. There are approximately 371 active telephone area codes covering Canada, the United States, and some Caribbean areas. (More are created regularly.) You want to choose an SRS of 25 of these area codes for a study of available telephone numbers.

(a) How would you label the area codes to use Table 7.1?

(b) Use Table 7.1 (p. 246), starting at line 125, to choose the first three members of this sample.

10. Each March, the CPS is expanded to gather a wider variety of information. On the Bureau of Labor Statistics (BLS) Web site, you can find data from this survey on 14,959 people aged 25 to 64 whose highest level of education is a bachelor's degree. Think of these people as a population.

(a) To select an SRS of these people, how would you assign labels?

(b) Use Table 7.1 (p. 246), starting at line 107, to choose the first three members of the SRS.

▲ **11.** In using Table 7.1 (p. 246) repeatedly to choose samples, you should not always choose the same row, such as line 101. Why not?

■ **12.** Which of the following statements are true of a table of random digits and which are false? Explain your answers.

(a) There are exactly four 0s in each row of 40 digits.

(b) Each pair of digits has chance 1/100 of being 00.

(c) The digits 0000 can never appear as a group because this pattern is not random.

■ **13.** The last stage of the CPS uses a *systematic sample.* An example will illustrate the idea of a systematic sample. Suppose that we must choose 4 rooms out of the 100 rooms in a dormitory. Because 100/4 = 25, we can think of the list of 100 rooms as 4 lists of 25 rooms each. Choose 1 of the first 25 rooms at random, using Table 7.1 (p. 246). The sample will contain this room and the rooms 25, 50, and 75 places down the list from it. If 13 is chosen, for example, then the systematic random sample consists of the rooms numbered 13, 38, 63, and 88.

(a) Use Table 7.1 (p. 246) to choose a systematic random sample of 5 rooms from a list of 200. Enter the table at line 120.

(b) Your sample gives every room the same chance to be chosen. Explain why.

(c) Despite the answer in part (b), this sample is not an SRS. Explain why.

■ **14.** An ethics institute selected a random sample of 100 U.S. high schools and then gave an in-class survey to all students in each selected school. Of the 29,760 students surveyed, 64 percent have cheated on a test, and 30 percent have stolen from a store. This type of sample is known as a *cluster sample.* Why is this sample not an SRS from the population of all U.S. high school students?

7.4 Cautions About Sample Surveys

▲ **15.** An opinion poll calls 1334 randomly chosen residential telephone numbers, and then the interviewer asks to speak with an adult member of the household to ask, "How many movies have you watched in a movie theater in the past 12 months?"

(a) What population do you think the poll has in mind?

(b) In all, 931 people respond. What is the rate (percent) of nonresponse?

(c) Many responses to this question are likely to be inaccurate. Why?

▲ **16.** Randomized Response: Suppose 30 students in a class participate in a survey in which they each flip a coin and do not tell the result. If the result was "tails," the student is supposed to give an honest answer to the question "Have you ever used a fake ID?" If the result was "heads," the student is supposed to say "Yes" to that question, regardless of what the true answer is.

Suppose the results in the class are 18 "Yes" answers and 12 "No" answers.

(a) If students follow the procedure correctly, is it true that all students who answered "No" have not used a fake ID?

(b) If students follow the procedure correctly, is it true that all students who have not used a fake ID answered "No"?

(c) On average, about half of the students who have not used a fake ID flipped "tails," so what is your best estimate of the true number of students who have not used a fake ID?

(d) Based on the answer to part (c), what is your estimate of the true number and proportion of students who have used a fake ID?

(e) Do we have any way to know which of the 18 "Yes" answers are truthful?

▲ **17.** The wording of questions can strongly influence the results of a sample survey. You are writing an opinion poll question about a proposed amendment to the Constitution. You can ask if people are in favor of "changing the Constitution" or "adding to the Constitution" by approving the amendment. Which of these choices of wording will likely produce a much higher percentage in favor? Why do you think this is true?

7.5 Experiments

▲ **18.** As reported in *College Teaching* in 2006, R. L. Garner randomly assigned 117 undergraduates to "review lecture videos" on statistics research methods. The videos either did or did not have short bits of humor inserted. Students who viewed the humor-added version of the video gave significantly higher ratings in their opinion of the lesson, how well the lesson communicated information, and the quality of the instructor. Even more importantly, that same group of students also recalled and retained significantly more information on the topic.

(a) What is the explanatory variable?

(b) What is the response variable?

(c) Why is this an experiment?

(d) Why were students not initially told that the true purpose of the study was to assess the use of humor?

(e) Why do you think the study was done using a fixed-video format rather than through live teaching?

19. Attitude of Gratitude: As reported in the *Journal of Personality and Social Psychology*, 192 undergraduates were assigned randomly into one of three clusters and asked to keep a regular report on psychological and physical indicators. One cluster was given a prompt to list things in their lives they are grateful for, another cluster's prompt was to list recent hassles, and the third cluster's prompt was to simply list events that recently had an impact on them. The "gratitude group" generally reported higher well-being.

(a) What is the explanatory variable?

(b) What is the response variable?

(c) Why is this an experiment?

(d) Does this experiment address whether it is more reasonable to say that well-being causes gratitude or that gratitude causes well-being?

20. Will owning a video-game system hurt the academic development of young boys? You are interested in tracking time spent playing video games, time spent in academic activities, teacher-reported learning problems, and reading and writing scores four months later. Outline the design of an experiment to study the effect of video-game ownership.

C. Devan/Corbis

21. We want to investigate the question: Will classroom programs explaining the health advantages of drinking water rather than sugary sodas reduce obesity among children aged 7 to 11 years? Because children are already grouped in school classrooms, we must randomize classes rather than individual children. An experiment assigned 15 classes to receive the program and another 14 to form a control group. After 12 months, obesity had increased in the control group and remained steady in the treatment group. Outline the design of the experiment, label the available classes, and use Table 7.1 (p. 246), beginning at line 103, to carry out the random assignment.

▲ **22.** A college allows students to choose either classroom or self-paced instruction in a basic mathematics course. The college wants to compare the effectiveness of self-paced and regular instruction. Someone proposes giving the same final exam to all students in both versions of the course and comparing the average score of those who took the self-paced option with the average score of students in regular sections.

(a) Explain why confounding makes the results of that study worthless.

(b) Given 30 students who are willing to use either regular or self-paced instruction, outline an experimental design to compare the two methods of instruction. Then use Table 7.1 (p. 246), starting at line 108, to carry out the randomization.

23. Will people spend less on healthcare if their health insurance requires them to pay some part of the cost themselves? An experiment on this issue asked if the percentage of medical costs that is paid by health insurance has an effect either on the amount of medical care that people use or on their health. The treatments were four insurance plans, each of which paid all medical costs above a ceiling. Below the ceiling, the plans paid 100 percent, 75 percent, 50 percent, or 0 percent of costs incurred. Outline the design of a randomized comparative experiment suitable for this study.

24. Track down a print or online copy of the Bible. Chapter 1 of the book of Daniel (especially verses 12 through 16) appears to have the first clinical trial in recorded history. Outline the design of the experiment. Discuss how you know whether it is an uncontrolled experiment, a comparative experiment, or a randomized comparative experiment.

25. Stores advertise price reductions to attract customers. What type of price cut is most attractive? Market researchers prepared ads for athletic shoes announcing different levels of discounts (20 percent, 40 percent, or 60 percent). The student subjects who read the ads were also given "inside information" about the fraction of shoes on sale (50 percent or 100 percent). Each subject then rated the attractiveness of the sale on a scale of 1 to 7.

(a) Each treatment in this experiment is a combination of values of two explanatory variables: discount level and fraction on sale. List all the treatments.

(b) Outline a randomized comparative experiment using 60 student subjects. Use Table 7.1 (p. 246) at line 123 to choose the subjects for the first treatment.

26. Healthcare providers are giving more attention to relieving the pain of cancer patients. An article in the journal *Cancer* surveyed a number of studies and concluded that controlled-release (CR) morphine tablets, which release the painkiller gradually over time, are more effective than giving standard morphine when the patient needs it. The "methods" section of the article begins: "Only those published studies that were controlled (i.e., randomized, double-blind, and comparative), repeated-dose studies with CR morphine tablets in cancer pain patients were considered for this review." Explain the terms in parentheses to someone who knows nothing about medical trials.

27. Eye cataracts are responsible for over 40 percent of blindness around the world. Can drinking tea regularly slow the growth of cataracts? We can't experiment on people, so we use rats as subjects. Researchers injected 14 young rats with a substance that causes cataracts. Half the rats also received tea extract; the other half got a placebo. The response variable was the growth of cataracts over the next six weeks. The researchers found that the tea extract did slow cataract growth in the rats.

(a) Outline the design of this experiment.

(b) Use Table 7.1 (p. 246), starting at line 108, to assign rats to treatments.

■ **28.** The rats in the previous exercise were labeled 01 to 14 to use the table of random digits. Unknown to the researchers, the 5 rats labeled 01 to 05 have a genetic defect that favors cataracts. If we simply put rats 01 to 07 in the tea group, the experiment would be biased against tea. We can observe how random selection works to reduce bias by keeping track of how many of these 5 rats get assigned to the tea group. Carry out the random assignment of 7 rats to the tea group 20 times, keeping track of how many of rats 01 to 05 are in the tea group each time. Make a histogram of the count of rats 01 to 05 assigned to tea. What is the average number in your 20 tries?

7.6 Experiments versus Observational Studies

▲ **29.** Could the magnetic fields from power lines cause leukemia in children? Investigators who wanted to explore this question spent five years and $5 million comparing 638 children who had leukemia and 620 who did not. They went into the homes and actually measured the magnetic fields in the children's bedrooms, in other rooms, and at the front door. They recorded facts about nearby power lines for the family home, as well as for the mother's residence when she was pregnant. Result: They found no evidence of more than a chance connection between magnetic fields and childhood leukemia. Explain carefully why this study is not an experiment, and state what kind of study it is.

▲ **30.** A typical hour of prime-time television shows three to five violent acts. Linking family interviews and police records shows a clear association between time spent watching TV as a child and later aggressive behavior.

(a) Explain why this is an observational study rather than an experiment.

(b) Suggest several variables describing a child's home life that may be confounded with how much TV he or she watches.

(c) Explain why confounding makes it difficult to conclude that more TV *causes* more aggressive behavior.

▲ **31.** The Nurses' Health Study has interviewed a sample of more than 100,000 female registered nurses every two years since 1976. Beginning in 1980, the study asked questions about diet, including alcohol consumption. The researchers concluded that "light-to-moderate drinkers had a significantly lower risk of death" than either nondrinkers or heavy drinkers.

(a) Is the Nurses' Health Study an observational study or an experiment? Why?

(b) What does "significant" mean in a statistical report?

(c) Suggest some confounding variables that might explain why moderate drinkers have lower death rates than nondrinkers. (The study adjusted for these variables.)

▲ **32.** The financial aid office of a university asks a sample of students about their employment and earnings. The report says that "for academic year earnings, a statistically significant difference was found between the sexes, with men earning more on the average. No significant difference was found between the earnings of black and white students." Explain both of these conclusions, for the effects of sex and of race on average earnings, in language understandable to someone who knows nothing about statistics. Do not use the words *significant* or *significance* in your answer.

▲ **33.** People who eat lots of fruits and vegetables have lower rates of colon cancer than those who eat little of these foods. Fruits and vegetables are rich in antioxidants such as vitamins A, C, and E. Will taking antioxidant pills help prevent colon cancer? A clinical trial studied this question with 864 people who were at risk for colon cancer. The subjects were divided into four groups: daily beta carotene (related to vitamin A), daily vitamins C and E, all three vitamins every day, and daily placebo. After four years, the researchers were surprised to find no significant difference in colon cancer among the groups.

(a) Outline the design of the experiment. Use your judgment in choosing the group sizes.

(b) Assign labels to the 864 subjects and use Table 7.1 (p. 246), starting at line 118, to choose the first five subjects for the "beta carotene" group.

(c) The study was double-blind. What does this mean?

(d) What does "no significant difference" mean in describing the outcome of the study?

(e) Suggest some characteristics of the kind of people who eat lots of fruits and vegetables that might explain lower rates of colon cancer. The experiment suggests that these variables, rather than the antioxidants, may be responsible for the observed benefits of fruits and vegetables.

▲ **34.** Dr. Megan Moreno sent a cautionary message to a randomly selected half of a sample of MySpace users (ages 18–20) whose public profiles included references to sex and substance abuse. A review of all profiles from the original sample three months later showed that the online profiles of those who had received the email were more likely to have removed the references or to have changed their profile setting to "private." Is this an experiment or observational study, and how do you know?

▲ **35.** In the July 15, 2007 issue of *Cancer*, a study reported on 533,715 women at least 40 years old who were diagnosed with invasive breast cancer and reported to the National Cancer Data Base (NCDB). The study found strong evidence that patients without health insurance were more likely to have a more advanced stage (i.e., III or IV) of cancer. Is this an experiment or observational study, and how do you know?

7.7 Inference: From Sample to Population

36. An opinion poll uses random digit dialing equipment to select 2000 residential telephone numbers. Of these, 631 are unlisted numbers. This isn't surprising, because 35 percent of all residential numbers are unlisted. For each underlined number, state whether it is a parameter or a statistic.

37. In the 1980s, the Tennessee Student Teacher Achievement Ratio experiment randomly assigned more than 7000 children to regular or small classes during their first four years of school. Even though the treatment lasted only for grades K–3, there were differences (in favor of the students who had the smaller classes) that were noticeable even many years later. For example, when these children reached high school, 40.2 percent of Blacks from small classes took the ACT® or SAT® college entrance exam. Only 31.7 percent of Blacks from regular classes took one of these exams. For each underlined number, state whether it is a parameter or a statistic.

38. At a college in Singapore, students were randomly selected and asked to complete a Web-based survey about sexual behavior. Of the 534 students who did, 24 percent reported having had sexual intercourse in the past six months, so $\hat{p} = 0.24$.

(a) What are the mean and standard deviation of the proportion $\hat{p}$ of the sample who have had sexual intercourse in the past six months?

(b) In what interval of values do the proportions $\hat{p}$ from 95 percent of all samples fall?

(c) In what interval of values do the proportions $\hat{p}$ from 99.7 percent of all samples fall?

39. Harley-Davidson motorcycles make up 14 percent of all the motorcycles registered in the United States. You plan to interview an SRS of 500 motorcycle owners.

Tony Harrington/Getty Images

(a) What is the approximate distribution of the proportion of your sample who own Harleys?

(b) In 95 percent of all samples like this one, the proportion of the sample who own Harleys will fall between ____ and ____. What are the missing numbers?

40. Exercise 38 asks what values the sample proportion $\hat{p}$ is likely to take when the population proportion is $p = 0.24$ and the sample size is $n = 534$. What interval covers the middle 95 percent of values of $\hat{p}$ when $p = 0.24$ and $n = 400$? When $n = 1600$? When $n = 6400$? What general fact about the behavior of $\hat{p}$ do your results illustrate?

■ **41.** You can use a table of random digits to *simulate* sampling from a population. Suppose that 60 percent of the population bought a lottery ticket in the last 12 months. We will simulate the behavior of random samples of size 40 from this population.

(a) Let each digit in the table stand for one person in this population. Digits 0 to 5 stand for people who bought a lottery ticket, and 6 to 9 stand for people who did not. Why does looking at one digit from Table 7.1 (p. 246) simulate drawing one person at random from a population with 60 percent "yes"?

(b) Each row in Table 7.1 contains 40 digits. So the first 10 rows represent the results of 10 samples. How many digits between 0 and 5 does the top row contain? What is the percentage of "yes" responses in this sample? How many of your 10 samples overestimated the population proportion of 60 percent? How many underestimated it? You could program a computer to continue this process, say 1000 times, to produce a pattern like that in Figure 7.4 (p. 260).

7.8 Confidence Intervals

42. In a random sample of students who took the SAT Reasoning college entrance exam twice, it was found that 427 of the respondents had paid for coaching courses and that the remaining 2733 had not. Give a 95 percent confidence interval for the proportion of coaching among students who retake the SAT®.

43. A Gallup poll asked each of 1785 randomly selected adults whether he or she happened to attend a house of worship in the previous seven days. Of the respondents, 750 said "yes." Give a 95 percent confidence interval for the proportion of all adults who claim that they attended a house of worship during the week preceding the poll. (The proportion who actually attended may be lower; some people might say "yes" if they often attend, even if they didn't attend that particular week.)

44. A Gallup poll conducted February 1–3, 2010, by telephone interviews of 1025 randomly selected American adults found that 646 Americans say that their sympathies in the Middle East situation lie more with the Israelis than with the Palestinians.

(a) Give a 95 percent confidence interval for the proportion of all American adults whose sympathies in the Middle East situation lie more with the Israelis than with the Palestinians.

(b) In theory, in 19 cases out of 20, the survey results will differ by no more than 3 percentage points in either direction from what would have been obtained by seeking out all American adults. Explain how your results agree with this statement.

▲ **45.** A telephone survey of 880 randomly selected drivers asked, "Recalling the last 10 traffic lights you drove through, how many of them were red when you entered the intersections?" Of the 880 respondents, 171 admitted that at least one light had been red.

(a) Give a 95 percent confidence interval for the proportion of all drivers who ran one or more of the last 10 red lights they met.

(b) A practical problem with this survey is that people may not give truthful answers. What is the likely direction of the bias: Do you think more or fewer than 171 of the 880 respondents really ran a red light? Why?

▲ **46.** A Gallup poll conducted May 3–6, 2010, by telephone interviews of 1029 American adults found that 52 percent of Americans called gay and lesbian relations morally acceptable. This was the first year that this statistic crossed the symbolic 50 percent threshold and is largely due to a change in views among younger men.

(a) How many of the 1029 people interviewed said gay and lesbian relations were morally acceptable?

(b) Gallup says that the margin of error for this poll is plus or minus 4 percentage points. Explain to someone who knows nothing about statistics what "margin of error plus or minus 4 percentage points" means.

(c) Give a 95 percent confidence interval for this survey. Does your margin of error agree with the 4 percentage points announced by Gallup?

▲ **47.** Consider the margin of error formula $2\sqrt{\frac{\hat{p}(1-\hat{p})}{n}}$.

(a) For a fixed value of n, what value of $\hat{p}$ between 0 and 1 causes this formula to attain its largest possible value?

(b) Using the answer to part (a), what would be a simplified (and slightly more conservative) formula for calculating the margin of error?

▲ **48.** A news article reports that in a recent Gallup poll, 78 percent of the sample of 1108 adults said they believe there is a heaven. Only 60 percent said they believe there is a hell. The news article ends, "The poll's margin of sampling error was plus or minus 4 percentage points." Can we be certain that between 56 percent and 64 percent of all adults believe there is a hell? Explain your answer.

▲ **49.** A survey of Internet users found that males outnumbered females by nearly 2 to 1. This was a surprise because earlier surveys had put the ratio of men to women closer to 9 to 1. Later, the article about the research states that surveys were sent to 13,000 organizations and that 1468 of these responded. The survey report claims that "the margin of error is 2.8 percent, with 95 percent confidence."

(a) What was this survey's *response rate*? (The response rate is the percentage of the planned sample that responded.)

(b) Do you think that the small margin of error is a good measure of the accuracy of the survey's results? Explain your answer.

50. A recent Gallup poll found that 68 percent of adult Americans favor teaching creationism along with evolution in public schools. The Gallup press release says:

> *For results based on samples of this size, one can say with 95 percent confidence that the maximum error attributable to sampling and other random effects is plus or minus 3 percentage points.*

Give one example of a source of error in the poll result that is *not* included in this margin of error.

■ **51.** The Internal Revenue Service (IRS) plans to examine an SRS of individual income tax returns from each state that were filed electronically. One variable of interest is the proportion of returns that were filed by a tax practitioner rather than by an individual taxpayer. The total number of e-filed tax returns in a state varies from 4.9 million in California to 97,000 in Vermont.

(a) Will the margin of error for estimating the proportion change from state to state if an SRS of 1000 e-filed returns is selected in each state? Explain your answer.

(b) Will the margin of error change from state to state if an SRS of 1 percent of all e-filed returns is selected in each state? Explain your answer.

■ **52.** Exercise 46 describes a Gallup poll that interviewed 1029 people. Suppose that you want a margin of error half as large as the one you found in that exercise. How many people must you plan to interview?

■ **53.** Though opinion polls usually make 95 percent confidence statements, some sample surveys use other confidence levels. The monthly unemployment rate, for example, is based on the CPS of about 50,000 households. The margin of error in the unemployment rate is announced as about ±0.15 percent with 90 percent confidence. Is the margin of error for 90 percent confidence larger or smaller than the margin of error for 95 percent confidence? Why? (*Hint:* Look again at Figure 7.7, on p. 262.)

Chapter Review

54. The proportion of one's body that is fat is a key indicator of fitness. The many ways to estimate this have different margins of error (given in percentage points):

Method	Calipers pinch	Bioelectrical impedance	Body mass index calculator	Hydrostatic weighing (dunk test)
Margin of error	±3	±4	±10	±1

(a) Which of these tests is the least accurate?
(b) If the pinch test says that you have 21 percent body fat, what is the 95 percent confidence interval for this estimate?

55. Many medical trials randomly assign patients to either an active treatment or a placebo. These trials are always double-blind. Sometimes the patients can tell whether they are getting the active treatment. This defeats the purpose of blinding. Reports of medical research usually ignore this problem. Investigators looked at a random sample of 97 articles reporting on placebo-controlled randomized trials in the top five general medical journals. Only 7 of the 97 discussed the success of blinding. Give a 95 percent confidence interval for the proportion of all such articles that discuss the success of blinding. [Dean Fergusson et al., Turning a blind eye: The success of blinding reported in a random sample of randomised, placebo-controlled trials, *British Medical Journal,* 328 (2004): 432–36.]

56. Tomeka wants to ask a sample of students at her college, "Do you think that Social Security will still be paying benefits when you retire?" She obtains the college email addresses of all 2654 students attending the college.

(a) How would you label the addresses to choose a simple random sample of 100 students?
(b) Use Table 7.1 (p. 246), starting at line 103, to choose the first three labels in the sample.
(c) Tomeka sends her question by email to the 100 addresses in her sample. Although she has chosen an SRS, a serious practical difficulty may make it hard to draw clear conclusions from her sample. What practical difficulty do you expect Tomeka to encounter?

■ **57.** Suppose that exactly 10 percent of all articles in major medical journals that describe placebo-controlled randomized trials discuss the success of blinding. That is, the proportion of "successes" in the population is $p = 0.1$. What is the approximate probability that fewer than 7 percent of an SRS of 97 articles from this population discuss the success of blinding?

58. The ability to grow in shade may help pines found in the dry forests of Arizona resist drought. How well do these pines grow in shade? Investigators planted pine seedlings in a greenhouse in either full light or light reduced to 5 percent of normal by shade cloth. At the end of the study, they dried the young trees and weighed them.

(a) Explain why this study is an experiment.
(b) What are the individuals, the treatments, and the response variable in this experiment?
(c) You have 200 pine seedlings available. Outline the design that you would use for this experiment.

59. The National Children's Study, the largest and most detailed study ever on children's health in the United States, is examining environmental effects on a large sample of children (from roughly 100,000 families) from before birth to age 21 years. Learn more at **http://www.nationalchildrensstudy.gov**.

(a) Explain why this study is an observational study.
(b) Is this observational study prospective or retrospective?
(c) Why couldn't this study be done as an experiment?

APPLET EXERCISES

To do these exercises, go to www.whfreeman.com/fapp9e.

1. Use the *Simple Random Sample* applet to choose the sample of songs in Example 4 (on p. 247). Assign labels 01 to 27 by entering 27 in the "Population 1 to" box and clicking "Reset." Then enter 4 in the "Select a sample of size" box and click "Sample." Which songs from the list in Example 4 (p. 247) make up your sample? Click "Reset" and choose another sample. Which songs did you choose this time? You see that random sampling gives different samples each time—what matters is that all songs have the same chance to be chosen.

2. The *Simple Random Sample* applet is handy when the population or sample is large. (The applet will handle population sizes up to 500.) Skills Check 9 (p. 269) asks you to choose an SRS of 10 out of 420 retail outlets. Use the applet to do this and report your result.

3. You can use the *Simple Random Sample* applet to choose treatment groups at random for a randomized comparative experiment. Exercise 27 (p. 273) asks you to choose the subjects to get the first treatment in an experiment that compares two treatments.

(a) Use the applet to choose an SRS of 7 out of 14 to receive the first treatment. Which subjects make up this group?
(b) The applet allows you to assign subjects randomly to more than two groups. Suppose you had a total of 36 rats and you wanted to assign a different treatment to each of four 9-rat groups. After you choose the

first group, the "Population Hopper" contains the 27 subjects that were not chosen, in scrambled order. Click "Sample" again to choose 9 of these remaining subjects to receive the second treatment. Do this once more to choose the third group. The 9 subjects that remain in the "Population Hopper" form the fourth group. Which of the 36 subjects will receive each of the four treatments?

4. You can use the *Probability* applet to speed up and improve Exercise 41 (p. 275). You have a population in which 60 percent of the individuals play the lottery. You want to take many samples from this population to observe how the sample proportion that plays the lottery varies from sample to sample. Specify the "Probability of Heads" setting in the applet as 0.6 and the number of tosses as 40. This simulates an SRS of size 40 from a large population. Each head in the sample is a person who plays the lottery, and each tail is a person who does not play. By alternating between "Toss" and "Reset," you can take many samples quickly.

(a) Take 50 samples, recording the proportion in each sample that plays the lottery. (The applet gives this proportion at the top left of its display.) Make a histogram of the 50 sample proportions.

(b) Another population contains only 20 percent of people who play the lottery. Take 50 samples of size 40 from this population, record the number in each sample that plays, and make a histogram of the 50 sample proportions. How do the centers of your two histograms reflect the differing truths about the two populations?

5. The idea of an 80 percent confidence interval is that the interval captures the true parameter value in 80 percent of all samples. That's not high enough confidence for practical use, but 80 percent hits and 20 percent misses make it easy to see how a confidence interval behaves in repeated samples from the same population. Go to the *Confidence Interval* applet.

(a) Set the confidence level to 80 percent. Click "Sample" to choose an SRS and calculate the confidence interval. Do this 10 times to simulate 10 SRSs with 10 confidence intervals. How many of the 10 intervals captured the true mean? How many missed?

(b) You see that we can't predict whether the next sample will hit or miss. The confidence level, however, tells us what percentage of responses will hit in the long run. Reset the applet and click "Sample 50" to get the confidence intervals from 50 SRSs. How many hit? Keep clicking "Sample 50" and record the percent of hits among 100, 200, 300, 400, and 500 SRSs. Even 500 samples is not truly "the long run," but we expect the percentage of hits in 500 samples to be fairly close to the confidence level of 80 percent.

WRITING PROJECTS

1. Go to the Web site of the Gallup Organization (**www.gallup.com**). You should be able to find a press release that you can access and read without charge. Newspapers publish short articles based on press releases. Write a news article about two paragraphs long, based on the press release.

2. Recall how Example 8 (p. 250) shows how wording can affect survey results. You can explore this by doing an experiment disguised as a survey. Choose a topic, then design two questions with a key difference in wording.

Use randomization to choose which version of the question you give each person. Don't reveal the design of the experiment to participants until after they have provided their answers. After you have roughly 20 or more responses to each version, compare and interpret your results. If you're interested in reading about an example of such an experiment, see John Rubin's article "Weighing Anchors," in the June 1990 issue of *Omni*.

3. How would you design a double-blind experiment in which participants test which of two brands of tissue they prefer? Conduct this experiment and write up the results. How do the results compare with any claims made in advertisements for the products?

4. Choose an issue of current interest to students at your school. Prepare a short questionnaire (no more than five questions) to determine opinions on this issue. Choose a sample of about 25 students, administer your questionnaire, and write a brief description of your findings. Also write a short discussion of your experiences in designing and carrying out the survey.

Although 25 students are too few for you to be statistically confident of your results, this project centers on the practical work of a survey. You must first identify a population; if it is not possible to reach a wider student population, use students enrolled in this course. Did the subjects find your questions clear? Did you write the questions so that it was easy to tabulate the responses? At the end, did you wish that you had asked different questions?

Suggested Readings

ANDERSON-COOK, C. M., and SUNDAR DORAIRAJ. An active learning in-class demonstration of good experimental design, *Journal of Statistics Education*, 9(1) (2001): **http://www.amstat.org/publications/jse/v9n1/anderson-cook.html**. A good example of issues that arise when designing a randomized experiment. This article includes an applet that students can use to experience the experiment.

BOCK, DAVID E., PAUL F. VELLEMAN, and RICHARD D. DE VEAUX. *Stats: Modeling the World*, 3rd ed., Addison Wesley, Boston, 2010. This is another text on the topic of statistics, aimed at the mathematical level just above this book. The book has a chapter (Chapter 19, "Confidence Intervals for Proportions," pp. 439–58) that uses estimating a population proportion to introduce confidence intervals.

LESSER, LAWRENCE M., and ERIK NORDENHAUG. Ethical statistics and statistical ethics: making an interdisciplinary module, *Journal of Statistical Education*, 12(3) (2004): **http://www.amstat.org/publications/jse/v12n3/lesser.html**. This article's discussion includes ethical issues associated with surveys, experiments, and observational studies.

MOORE, DAVID S. *The Basic Practice of Statistics*, 5th ed., Freeman, New York, 2009. Chapters 8 and 9 discuss samples and experiments, Chapter 14 presents the reasoning of confidence intervals, and Chapter 19 presents confidence intervals for a population proportion.

Suggested Web Sites

The National Council on Public Polls, **www.ncpp.org**, has a statement on "20 Questions a Journalist Should Ask About a Poll" that makes interesting reading. The explanations expand upon our cautions about sample surveys in practice. You can find similar information on the Web site of the American Association for Public Opinion Research, **www.aapor.org**. (Take a look at the "Poll & Survey FAQs.") Also, read the American Statistical Association publication "What Is a Survey? (2nd ed.)" at **www.whatisasurvey.info**.

The single most important sample survey in the United States is probably the government's monthly CPS, carried out by the Census Bureau on behalf of the Bureau of Labor Statistics (BLS). The CPS Web site, **www.bls.census.gov/cps/**, contains a wealth of information.

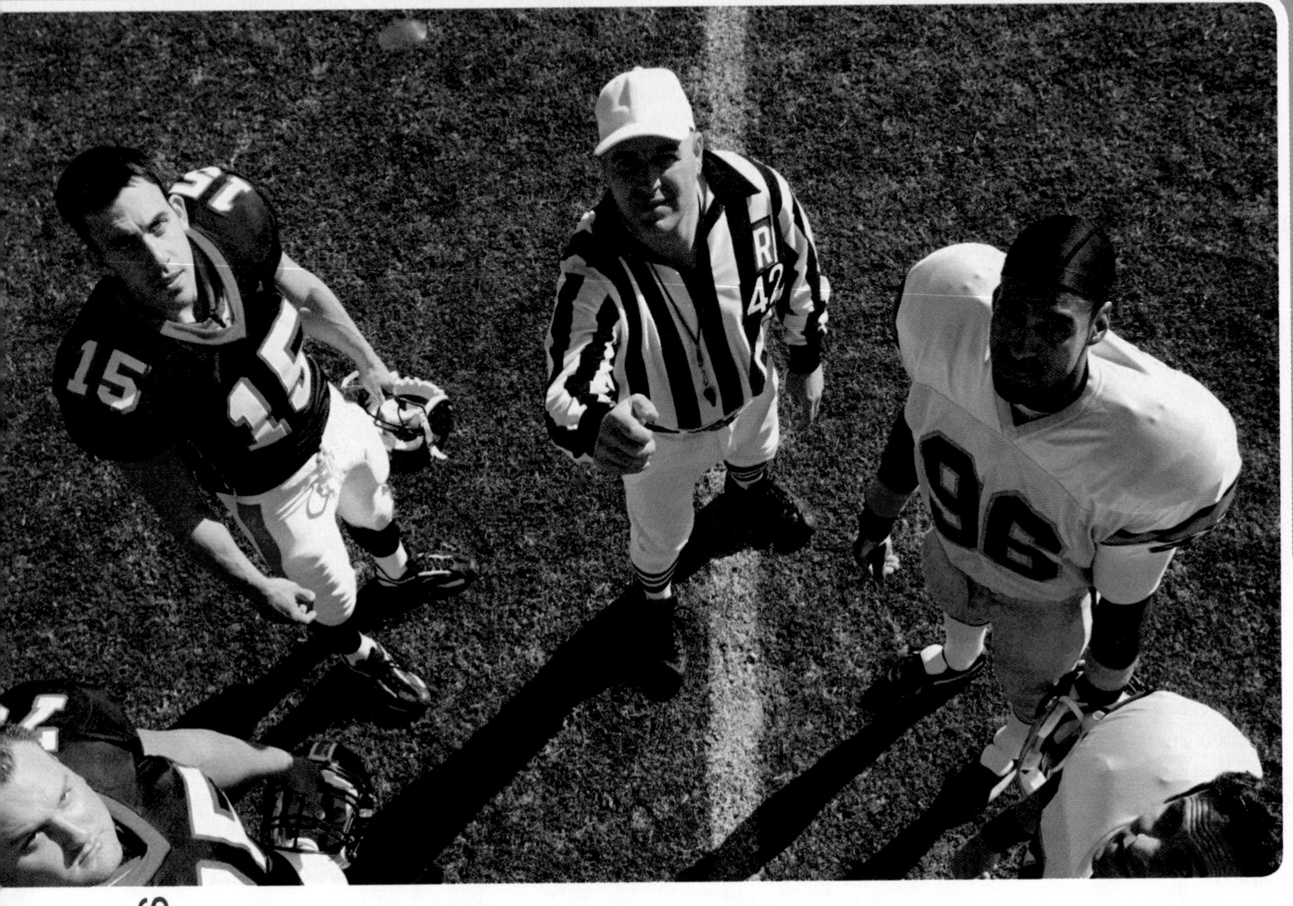

Robert Michael/Corbis

8

Probability: The Mathematics of Chance

Ever wonder how gambling can be a business? A business needs predictable revenue from the service it offers, even when the service is a game of chance. Individual gamblers can never say whether a day at the casino will turn a profit or a loss, but the casino itself takes few chances. Casinos are consistently profitable, and state governments make money both from running lotteries and from selling licenses for other forms of gambling.

It is striking that an individual roll of the dice, spin of the wheel, or flip of a coin is a total unknown, but the aggregate result of thousands of chance outcomes can be known with near certainty. The casino need not load the dice, mark the cards, or alter the roulette wheel. It knows that in the long run, each dollar bet will yield its five cents or so of revenue. It is, therefore, good business to concentrate on free floor shows or inexpensive bus fares to increase the flow of dollars bet. The flow of profit will follow.

A casino is not the only business that relies on the fact that a chance outcome many times repeated is quite predictable. For example, although a life insurance company does not know *which* of its policyholders will die next year, it can predict quite accurately *how many* will die. It sets its premiums according to this knowledge, just as the casino sets its jackpots. These are just some of the many types of companies and people that rely on chance behavior being predictable in the long run.

Section 8.1 introduces the nature of randomness and probability, including how to use sample spaces and rules to calculate numerical values of probabilities. Then Sections 8.2, 8.3, and 8.4 offer further ways to model and calculate probabilities by gaining tools for counting number of possibilities and by modeling real-life phenomena with discrete and continuous models. In Section 8.5, we discuss the mean and standard deviation of such models. Finally, Section 8.6 discusses a central result of inferential statistics known as the Central Limit Theorem, which tells us about how the sample mean behaves regardless of the shape of the population from which it was calculated.

Random DEFINITION

A phenomenon or trial is said to be **random** if individual outcomes are uncertain but the long-term pattern of many individual outcomes is predictable.

In statistics, *random* does not mean "haphazard" or "out of the blue." Randomness is actually a kind of order, an order that emerges in the long run, over many repetitions. Many phenomena, both natural and of human design, are random. The hair colors of children, the spread of epidemics, and the decay of radioactive substances are examples of natural randomness. Indeed, quantum mechanics asserts that at the subatomic level, the natural world is inherently random.

Games of chance are examples of randomness deliberately produced by human effort. Casino dice (see Example 3, p. 285) are carefully machined, and their drilled holes are filled with material equal in density to the plastic body. This guarantees that the side with six spots has the same weight as the opposite side, which has only one spot. Thus, each side is equally likely to land upward. All the odds and payoffs of dice games rest on this carefully planned randomness. Random sampling and randomized comparative experiments are also examples of planned randomness, although they use tables or generators of random digits rather than dice and cards. The reasoning of statistical inference rests on asking, "How often would this method give a correct answer if I used it very many times?" Probability theory, the mathematical description of randomness, is the basis for gambling, insurance, much of modern science, and statistical inference. **Probability** is the topic of this chapter.

8.1 Probability Models and Rules

Toss a coin, or choose a simple random sample (SRS). The result can't be predicted in advance because the result will vary when you toss the coin or choose the sample repeatedly. But there is nonetheless a regular pattern in the results, a pattern that emerges clearly only after many repetitions. This remarkable fact is the basis for the idea of probability.

EXAMPLE 1
Heads Up When Tossing a Coin: Long-Run Frequency Interpretation of Probability

When you toss a coin, there are only two possible outcomes: heads or tails. Figure 8.1 shows the results of tossing a coin 5000 times twice. For each number of tosses from 1 to 5000, we have plotted the proportion of those tosses that gave a head. Trial A (red line) begins tail, head, tail, tail. You can see that the proportion of heads for Trial A starts at 0 on the first toss, rises to 0.5 when the second toss gives a head, then falls to 0.33 and 0.25 as we get two more tails. Trial B (blue line), on the other hand, starts with five straight heads, so the proportion of heads is 1 until the sixth toss.

The proportion of tosses that produce heads is quite variable at first. Trial A starts low and Trial B starts high. As we make more and more tosses, however, the proportions of heads for both trials get close to 0.5 and stay there. If we made yet a third trial at tossing the coin a great many times, the proportion of heads would again settle down to 0.5 in the long run. We say that 0.5 is the *probability* of a head. The probability 0.5 appears as a horizontal line on the graph.

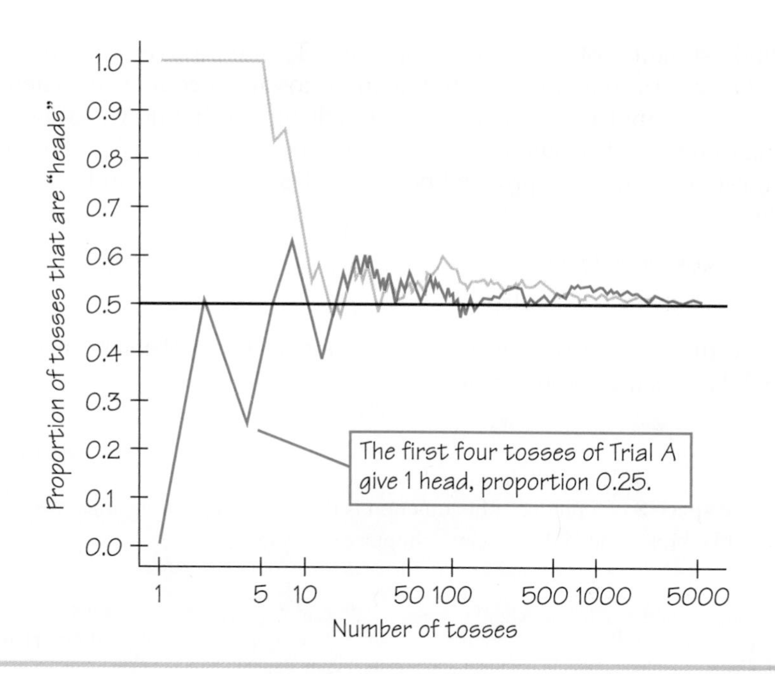

Figure 8.1 The proportion of tosses of a coin that give a head varies as we make more tosses. Eventually, however, the proportion approaches 0.5, the probability of a head. This figure shows the results of two trials of 5000 tosses each. (The horizontal scale is transformed using logarithms to show both short- and long-term behavior.)

Probability DEFINITION

The **probability** of any outcome of a random phenomenon is the proportion of times the outcome would occur in a very long series of repetitions. (We will soon see a concrete expression of this in the Procedure box "Finding Probabilities of Equally Likely Outcomes.") Probabilities can be expressed as decimals, percentages, or fractions.

The *Probability* applet (see Applet Exercise 1, p. 321) animates Figure 8.1. It allows you to choose the probability of a coin landing on "heads" and simulate any number of tosses of a coin with that probability. Try it. You will see that the proportion of heads gradually settles down close to the probability. Equally important, you will also see that the proportion in a small or moderate number of tosses can be far from the probability. *Probability describes only what happens in the long run.* Random phenomena are irregular and unpredictable in the short run.

We might suspect that a coin has probability 0.5 of coming up heads just because the coin has two sides. As Exercise 1 (p. 315) illustrates, such suspicions are not always correct. The idea of probability is *empirical*. That is, it is based on observation rather than theory. Probability describes what happens in a great many trials, and we must actually observe many trials to pin down a probability.

Gamblers have known for centuries that the fall of coins, cards, and dice displays clear patterns in the long run. In fact, a question about a gambling game launched probability as a formal branch of mathematics (see Exercise 9, p. 316). The idea of probability rests on the observed fact that the average result of many thousands of chance outcomes can be known with near certainty. But a definition of probability as "long-run proportion" is vague. Who can say what "the long run" is? We can always toss the coin another 1000 times. Instead, we give a mathematical description of *how probabilities behave*, based on our understanding of long-run proportions. To see how to proceed, think first about a very simple random phenomenon: tossing a coin once. When we toss a coin, we cannot know the outcome in advance. What do we know? We are willing to say that the outcome will be either heads or tails. We believe that each of these outcomes has probability $\frac{1}{2}$. This description of coin tossing has two parts:

- A list of possible outcomes
- The probability for each outcome

We will see that this description is the basis for all the probability models in Section 8.2. Here is the vocabulary we use.

Sample Space DEFINITION

The **sample space S** of a random phenomenon is the set of all possible outcomes that cannot be broken down further into simpler components.

Event DEFINITION

An **event** is any outcome or any set of outcomes of a random phenomenon. That is, an event is a subset of the sample space.

Probability Model DEFINITION

A **probability model** is a mathematical description of a random phenomenon consisting of two parts: a sample space *S* and a way of assigning probabilities to events.

The sample space S can be very simple or very complex. When we toss a coin once, there are only two outcomes, heads and tails. So the sample space is $S = \{H, T\}$. If we draw a random sample of 1000 U.S. residents age 18 and over, as opinion polls often do, the sample space contains all possible choices of 1000 of the 235 million adults in the country. This *S* is extremely large: 2.9×10^{5803}.

EXAMPLE 2
Tossing Two Coins: The Importance of Sample Space

Probabilities can be hard to determine without detailing or diagramming the sample space. For example, E. P. Northrop notes that even the great 18th-century French mathematician Jean le Rond d'Alembert tripped on the question: "In two coin tosses, what is the probability that heads will appear at least once?" Because the number of heads could be 0, 1 or 2, d'Alembert reasoned (incorrectly) that each of those possibilities would have an equal probability of $\frac{1}{3}$, and so he reached the (wrong) answer of $\frac{2}{3}$. What went wrong? Well, {0, 1, 2} could not be the fully detailed sample space because "1 head" can happen in more than one way. For example, if you flip a dime and a penny once apiece, you could display the sample space with a *table*, such as the one on the left.

As you can see, the preceding 2 × 2 table has 2 rows and 2 columns with 2 × 2 = 4 outcomes: {HH, HT, TH, TT}. Another way to generate these 4 outcomes is with a *tree diagram*, in which each possible left-to-right pathway through the branches generates an outcome. For example, going up (to dime "heads") and then down (penny "tails") yields the outcome HT.

Either way, we can see that the sample space has 4, not 3, equally likely outcomes. With the table or tree diagram in view, you may already see that the correct probability of at least 1 head is not $\frac{2}{3}$, but $\frac{3}{4}$. This model could also be applied to find the possible sequences of boys and girls in two-child families: {BB, BG, GB, GG}.

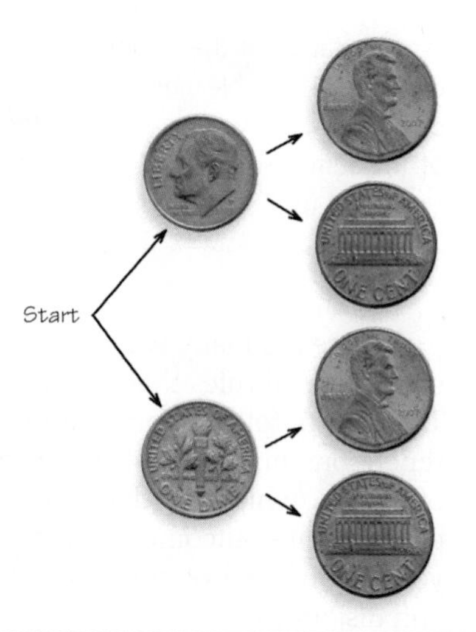

EXAMPLE 3
Pair-a-Dice: Outcomes for Rolling Two Dice

Rolling one six-sided die has an obvious sample space of six equally likely outcomes: {1, 2, 3, 4, 5, 6}. But many board games (and casino games) involve rolling two dice and noting the sum of the spots on the two sides that are facing up. We know from our experience playing games like Monopoly® that the 11 possible sums (2, 3, 4, 5, 6, 7, 8, 9, 10, 11, 12) are *not* equally likely because, for example, there are many ways to get a sum of 7 but only one way to get a sum of 12.

(George Diebold/Stone/Getty Images.)

Just as in Example 2, we can make a sample space diagram that will make it easy to find the exact probabilities and patterns of the various dice sums. Because of the large number of possible outcomes, the table in Figure 8.2 is a more straightforward

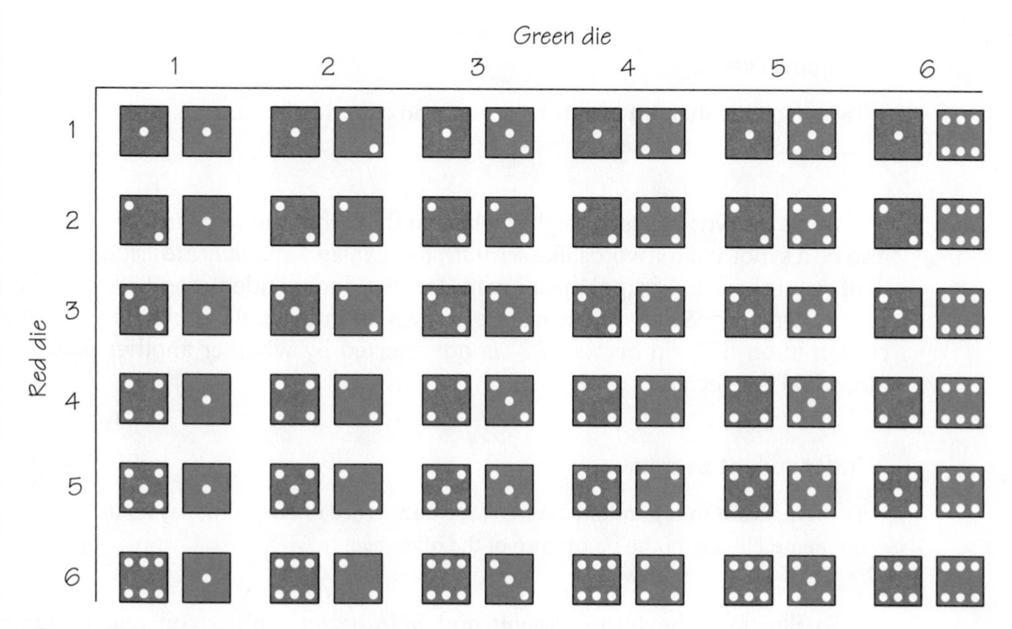

Figure 8.2 Table of the 36 outcomes for rolling two dice.

representation of the sample space S than a tree diagram would be. Figure 8.2 shows $6 \times 6 = 36$ possible (and equally likely) ways to roll two dice. The longest rising diagonal of the table shows the six ways that the sum can equal 7. Therefore, the probability of the sum being 7 is $\frac{6}{36} = \frac{1}{6}$. This table will make it easy to compute probabilities, as Example 4 (p. 289) shows.

There are many ways to assign probabilities, so it is convenient to start with some general rules that any assignment of probabilities to outcomes must obey. These facts follow from the idea of probability as "the long-run proportion of repetitions on which an event occurs." Some rules apply only to special kinds of events, which we define here. Recall that events are really just sets whose elements are outcomes of a random phenomenon, so you will be able to apply your knowledge of set theory terms such as intersection, complement, union, and disjoint.

As Example 4 will show, sometimes it is easier to determine the probability of an event A indirectly by finding the probability of its logical opposite—i.e., the event that A does *not* happen. The special name for this event is consistent with how the word *complement* is used in other contexts: the event that together with its opposite forms a complete whole. We will see in Rules 2 and 3 (p. 289) that the "whole" is the entire sample space.

Complement of an Event

DEFINITION

The **complement of an event** A is the event that A does *not* occur, written as A^C. (The superscript C stands for complement. Some books use the notation $\overline{A}$ or A'.)

Another useful distinction to make is whether or not it is possible for two events to both happen. If it is not possible, then the two events are separate or disjoint.

Disjoint Events

DEFINITION

Two events are **disjoint events** if they have no outcomes in common. Disjoint events are also called *mutually exclusive events.*

While everyday speech might make you think that the word *independent* could also be a synonym for words like *separate, disjoint,* and *mutually exclusive,* independent events have a different meaning in statistics. Independent events do not affect each other's probability of occurrence, just as an individual's probability of being chosen in an SRS (in Section 7.3) is not affected by whether another particular individual is selected.

Independent Events

DEFINITION

Two events are **independent events** if the occurrence of one event has no influence on the probability of the occurrence of the other event.

To illustrate the terms *disjoint* and *independent* with a concrete example, Table 8.1 displays the various possibilities involving the following three events

Table 8.1

Possibilities for Events to Be Disjoint or Independent

	Independent Events	Events Not Independent
Disjoint events	Not possible; if one event happens and is disjoint from the other event, then that *does* influence the other event's probability (because you know that the other event could not have happened!).	Events B and C are disjoint because they cannot both happen from one roll. They are not independent because the probability of B does change (from $\frac{1}{6}$ to 0) if C happens.
Events not disjoint	Events A and B are not disjoint because they could both happen (i.e., red = 1, green = 6). They are independent because the probability of B remains $\frac{1}{6}$ (see Example 3) whether or not A happens.	Events A and C are not disjoint because they both could happen (i.e., red = 1, green = 2). They are not independent because the probability of A does change (from $\frac{1}{6}$ to $\frac{1}{2}$) if C happens.

when the red die and the green die of Example 3 are rolled together:

A = red die shows "1"

B = red and green dice add up to 7

C = red and green dice add up to 3

Now that we have distinguished various terms, we are ready to examine a list of probability rules that are tools for calculating probabilities:

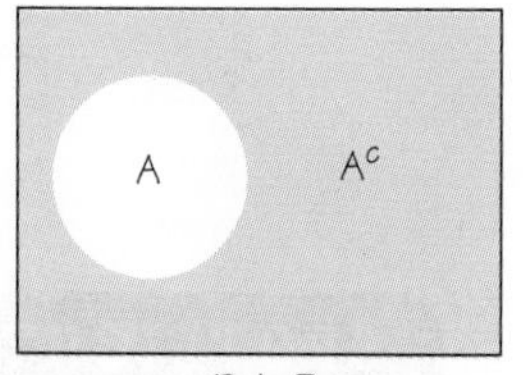

Rule 3

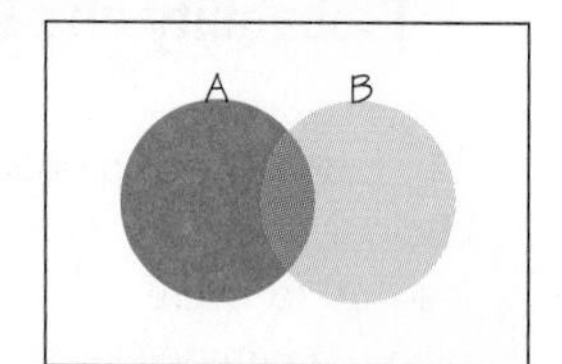

Rule 5

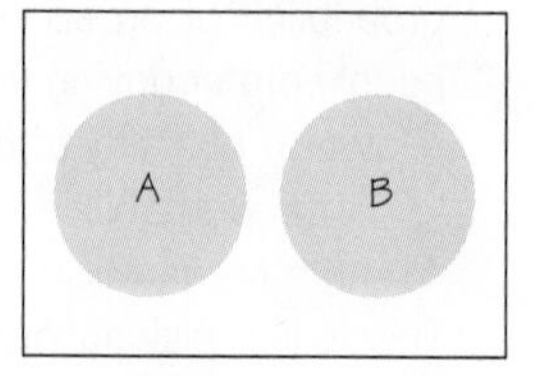

Rule 6

Figure 8.3 Each rectangle represents the whole sample space in these illustrations of Rules 3, 5, and 6.

1. **Any probability is a number between 0 and 1 inclusive.** Any proportion is a number between 0 and 1 inclusive, so any probability is also a number between 0 and 1 inclusive. An event with probability 0 never occurs, an event with probability 1 always occurs, and an event with probability 0.5 occurs in half the trials in the long run.
2. **All possible outcomes together must have a probability of 1.** Because some outcome must occur on every trial, the sum of the probabilities for all possible (simplest) outcomes must be exactly 1.
3. **The probability that an event does not occur is 1 minus the probability that the event does occur.** For example, because the probability the red die lands on "3" is $\frac{1}{6}$, then the probability that the red die's number is *not* 3 is $1 - \frac{1}{6} = \frac{5}{6}$. This is really just another way of saying that the probability that an event occurs and the probability that it does not occur always add to 1, or 100 percent of the sample space. See how the complementary blue and white regions add up to fill the space of the "Rule 3" diagram of Figure 8.3.
4. **If two events are *independent*, then the probability that one event and the other both occur is the product of their individual probabilities.** Table 8.1 explains why events *A* and *B* are independent, so we can find the probability that *A* and *B* both happen as the product of the individual

event probabilities: $\frac{1}{6} \times \frac{1}{6} = \frac{1}{36}$. Note that we can also see from Figure 8.2 that the "overlap" or intersection of events A and B happens in 1 of the 36 outcomes—when the top row and the rising main diagonal intersect (at the roll where red = 1 and green = 6).

5. **The probability that one event or the other occurs is the sum of their individual probabilities minus the probability of their intersection.** [Unlike in usual everyday usage, the mathematical use of "or" is *inclusive*, which means that the event "*A* or *B*" happens so long as at least one of the two events happens. In set theory, this is called "the union of *A* and *B*," and it includes *A*'s and *B*'s "separate property" as well as their "community property."] This general addition rule makes sense if we look at Rule 5 in Figure 8.3. Simply adding the probabilities of the two events would overshoot the answer because we would be incorrectly counting the overlap twice. The way to adjust for this is to subtract the overlap so that it is counted exactly once. Consider events *A* and *B* from Table 8.1. Their intersection corresponds to rolling "red = 1, green = 6", which has a $\frac{1}{36}$ probability. So the probability of the event "*A* or *B*" is $\frac{6}{36} + \frac{6}{36} - \frac{1}{36} = \frac{11}{36}$. Notice that if events *A* and *B* had been disjoint, there would be no overlap to worry about double counting and this rule would simply turn into this next one:

6. **If two events are *disjoint*, the probability that one or the other occurs is the sum of their individual probabilities.** For example, in Table 8.1, events *B* and *C* are disjoint, so the probability of "*B* or *C*" is simply $\frac{6}{36} + \frac{2}{36} = \frac{8}{36}$ or $\frac{2}{9}$.

We can now state Rules 1 to 6 more concisely by using more formal mathematical notation. We use capital letters near the beginning of the alphabet to denote events. If *A* is any event, we write its probability as *P*(*A*). As you apply these rules, remember that they are just another form of intuitively true facts about long-run proportions.

SPOTLIGHT 8.1

Probability and Psychology

Our judgment of probability can be affected by psychological factors. Our desire to get rich quick may lead us to overestimate the tiny probability of winning the lottery. Our feeling that we are "in control" when we are driving may make us underestimate the probability of an accident. (This may be why some people prefer driving to flying even though flying has a lower probability of death per mile traveled.)

The probability of winning (a share of) the 41-state Mega Millions jackpot is 1 in 175,711,536. This is like picking out a particular sheet of typing paper from a stack twice the height of Mt. Everest, or guessing a particular second from a period of about 5.5 years. Without concrete analogies, it is hard to grasp the meaning of very small probabilities, and some players may greatly overestimate their chances of winning even if they buy lots of tickets. For example, suppose someone buys 20 $1 Mega Millions tickets every week for 50 years. She would have spent over $50,000, and yet her probability of winning at least one jackpot in that whole time would still be only 1 in 3368. For comparison, the probability of dying in a car accident during a lifetime of driving is about 50 times greater than this!

Andrew Gelman reports that most people say they would not switch to a situation in which they had a small probability p of dying and a large probability $1-p$ of gaining $1000. And yet, people will not necessarily spend that much for air bags for their cars. Becoming more aware of our inconsistencies and biases can help us make better use of probability when deciding what risks to take.

Probability Rules RULE

Rule 1. The probability $P(A)$ of any event A satisfies $0 \le P(A) \le 1$.

Rule 2. If S is the sample space in a probability model, then $P(S) = 1$.

Rule 3. The **complement rule**: $P(A^C) = 1 - P(A)$.

Rule 4. The **multiplication rule** for *independent* events: $P(A \text{ and } B) = P(A) \times P(B)$.

Rule 5. The *general* **addition rule**: $P(A \text{ or } B) = P(A) + P(B) - P(A \text{ and } B)$.

Rule 6. The *addition rule* for *disjoint* events: $P(A \text{ or } B) = P(A) + P(B)$.

EXAMPLE 4
Probabilities for Rolling Two Dice

Figure 8.2 displays the 36 possible outcomes of rolling two dice. For casino dice, it is reasonable to assign the same probability to each of the 36 outcomes in Figure 8.2. Because all 36 outcomes together must have a probability of 1 (Rule 2), each outcome must have a probability of $\frac{1}{36}$.

What is the probability of rolling a sum of 5? Because the event "roll a sum of 5" contains the four outcomes displayed (along a diagonal) in Figure 8.2, the addition rule for disjoint events (Rule 6) says that its probability is

$$P(\text{roll a sum of 5}) = P(⚀⚃) + P(⚁⚂) + P(⚂⚁) + P(⚃⚀)$$

$$= \frac{1}{36} + \frac{1}{36} + \frac{1}{36} + \frac{1}{36}$$

$$= \frac{4}{36} \approx 0.111$$

Continue using Figure 8.2 in this way to get the full probability model (sample space and assignment of probabilities) for rolling two dice and summing the spots on the up faces. Here it is:

Outcome	2	3	4	5	6	7	8	9	10	11	12
Probability	$\frac{1}{36}$	$\frac{2}{36}$	$\frac{3}{36}$	$\frac{4}{36}$	$\frac{5}{36}$	$\frac{6}{36}$	$\frac{5}{36}$	$\frac{4}{36}$	$\frac{3}{36}$	$\frac{2}{36}$	$\frac{1}{36}$

This model assigns probabilities to individual outcomes. Note that Rule 2 is satisfied because all the probabilities add up to 1. To find the probability of an event, just add the probabilities of the outcomes that make up the event. For example:

$$P(\text{outcome is odd}) = P(3) + P(5) + P(7) + P(9) + P(11)$$

$$= \frac{2}{36} + \frac{4}{36} + \frac{6}{36} + \frac{4}{36} + \frac{2}{36}$$

$$= \frac{18}{36} = \frac{1}{2}$$

What is the probability of rolling any sum other than a 5? The "long way" to find this would be

$$P(2) + P(3) + P(4) + P(6) + P(7) + P(8) + P(9) + P(10) + P(11) + P(12).$$

A much faster way would be to use the complement rule (Rule 3):

$$\begin{aligned} P\text{ (roll sum that is } \textit{not}\text{ 5)} &= 1 - P\text{ (roll sum of 5)} \\ &= 1 - \frac{4}{36} = \frac{32}{36} \approx 0.889 \end{aligned}$$

Another good time to use the complement rule would be to find the probability of getting a sum greater than 3. Compare the calculation of P (sum > 3) with $1 - P$ (sum ≤ 3).

For an example of Rule 5, let event A be "sum is odd" and event B be "sum is a multiple of 3." Earlier in this example, we calculated $P(A) = \frac{1}{2}$. You can verify that $P(B) = \frac{1}{3}$ and $P(A \text{ and } B) = \frac{1}{6}$. And so, $P(A \text{ or } B) = \frac{1}{2} + \frac{1}{3} - \frac{1}{6} = \frac{4}{6} = \frac{2}{3}$.

When the outcomes for a probability model are numbers, we can use a histogram to display the assignment of probabilities to the outcomes. (This is similar to the histogram of a relative frequency distribution you saw in Figure 5.17 on p. 199.) Figure 8.4 is a **probability histogram** of the probability model in Example 4. The height of each bar shows the probability of the outcome at the base of the bar. Because the heights are probabilities, they add to 1. Think of Figure 8.4 as an idealized picture of the results of very many rolls of a die. As an idealized picture, it is perfectly symmetric.

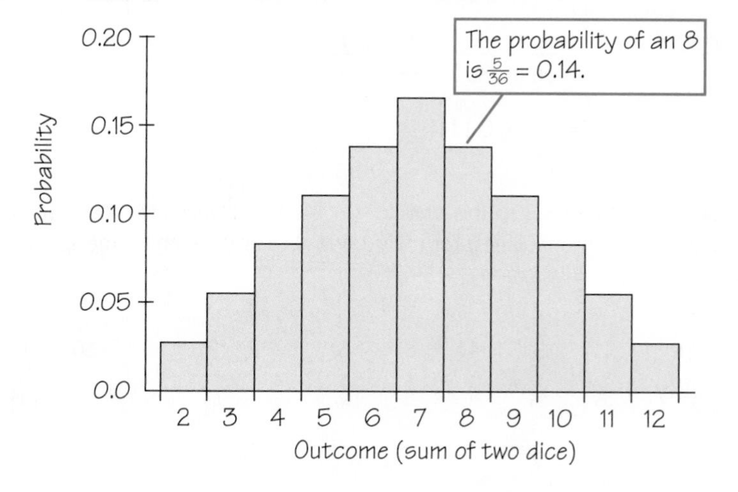

Figure 8.4
A probability histogram showing the probability model for rolling two balanced dice and counting the spots on the up faces.

Example 4 illustrates one way to assign probabilities to events: Assign a probability to every individual outcome, and then add these probabilities to find the probability of any event. This idea works well when there are only a finite (fixed and limited) number of outcomes.

8.2 Discrete Probability Models

We will work with two kinds of probability models. The first kind is illustrated by Example 4 and is called a **discrete probability model**. (The second kind is discussed in Section 8.4.)

Discrete Probability Model DEFINITION

A **discrete probability model** is a probability model with a countable number of outcomes in its sample space.

To assign probabilities in a discrete model, list the probability of all the individual outcomes. By Rules 1 and 2, these probabilities must be numbers between 0 and 1 inclusive and must have a sum of 1. The probability of any event is the sum of the probabilities of the outcomes making up the event.

EXAMPLE 5
Benford's Law: One Is the Likeliest Number that You'll Ever Know

Faked numbers in tax returns, invoices, or expense account claims often display patterns that aren't present in legitimate records. Some patterns, like too many round numbers, are obvious and easily avoided by a clever crook. Others are more subtle. It is a striking fact that the first (leftmost) digits of numbers in legitimate records often follow a model known as Benford's law. Here it is (note that a first digit can't be 0):

First digit	1	2	3	4	5	6	7	8	9
Probability	0.301	0.176	0.125	0.097	0.079	0.067	0.058	0.051	0.046

Check that the probabilities of the outcomes sum exactly to 1. This is, therefore, a legitimate discrete probability model. Investigators can detect fraud by comparing the first digits in records such as invoices paid by a business with these probabilities. For example, consider the events A = "first digit is 1" and B = "first digit is 2." Applying Rule 6 to the table of probabilities yields $P(A \text{ or } B) = 0.301 + 0.176$, which is 0.477 (almost 50 percent). Crooks trying to "make up" the numbers probably would not make up numbers starting with 1 or 2 this often.

Let us use informal reasoning to explore why first digits behave this way. Note that the increase from 1 to 2 is an increase of 100 percent, but from 2 to 3 is only 50 percent, from 3 to 4 is only 33 percent, and so on. So data values that increase at an approximately constant percentage (which a lot of financial data does, for example) will naturally "spend more time" (within any particular power of 10) taking on values whose left digit is 1, and successively less for larger left-digit numbers. To make this more concrete, an index such as the Dow Jones would need more time to get from 1000 to 2000 than to get from 8000 to 9000. We won't make you use the logarithm function, but if you are familiar with it, it may be no surprise that it is in the formula for the probability of the first digit being d: $\log_{10}(1 + \frac{1}{d})$.

8.3 Equally Likely Outcomes

An SRS gives all possible samples an equal chance to be chosen. Rolling two casino dice gives all 36 outcomes the same probability. When randomness is the product of human design, it is often the case that the outcomes in the sample space are all equally likely. Rules 1 and 2 force the assignment of probabilities in this case.

Finding Probabilities of Equally Likely Outcomes PROCEDURE

If a random phenomenon has equally likely outcomes, then the probability of event A is

$$P(A) = \frac{\text{count of outcomes in event } A}{\text{count of outcomes in sample space } S}$$

EXAMPLE 6
Are First Digits Equally Likely?

You might think that first (leftmost) digits are distributed "at random" among the digits 1 to 9. Under such a "discrete uniform distribution," the nine possible outcomes would then be equally likely. The sample space is $S = \{1, 2, 3, 4, 5, 6, 7, 8, 9\}$, and the probability model is:

First digit	1	2	3	4	5	6	7	8	9
Probability	$\frac{1}{9}$	$\frac{1}{9}$	$\frac{1}{9}$	$\frac{1}{9}$	$\frac{1}{9}$	$\frac{1}{9}$	$\frac{1}{9}$	$\frac{1}{9}$	$\frac{1}{9}$

By Rule 6, the probability of the event that a randomly chosen first digit is a 1 or 2 is

$$\begin{aligned} P(1 \text{ or } 2) &= P(1) + P(2) \\ &= \frac{1}{9} + \frac{1}{9} = \frac{2}{9} \approx 0.222 \end{aligned}$$

This answer of 0.222 is less than half of what we found for $P(1 \text{ or } 2)$ using the Benford's law probability model in Example 5—a huge difference that illustrates one way an auditor could easily detect data that was faked—the crook would have too few 1s and 2s. Figure 8.5 displays probability histograms that compare the probability model for random digits with the model given by Benford's law.

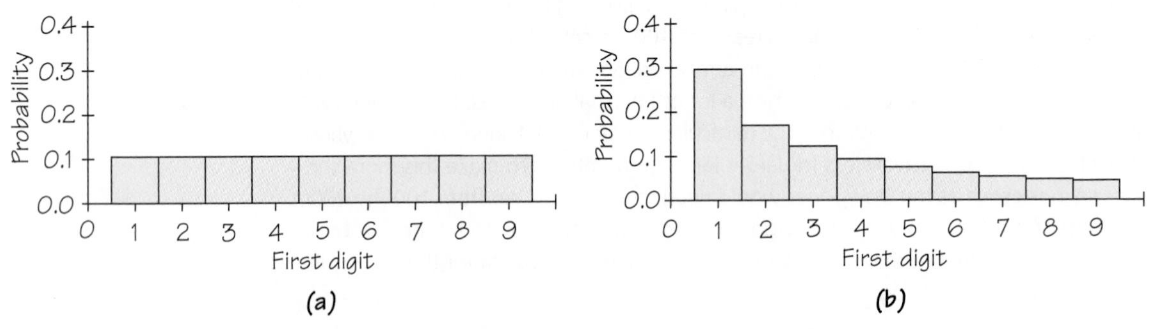

Figure 8.5 Probability histograms of two models for first digits in numerical records. (a) Digits are equally likely. (b) Digits follow Benford's law.

When outcomes are equally likely, we find probabilities by counting outcomes. The study of counting methods is called *combinatorics*, and this is mentioned in the episode "Noisy Edge" (2005) from season 1 of the television crime drama *NUMB3RS*.

Combinatorics DEFINITION

Combinatorics is the study of methods for counting.

One example of a counting method is the **fundamental principle of counting** (from Chapter 2, page 41): If there are a ways of choosing one thing, b ways of choosing a second after the first is chosen, . . ., and z ways of choosing the last item after the earlier choices, then the total number of choice sequences is $a \times b \times \cdots \times z$.

EXAMPLE 7
DNA Sequences

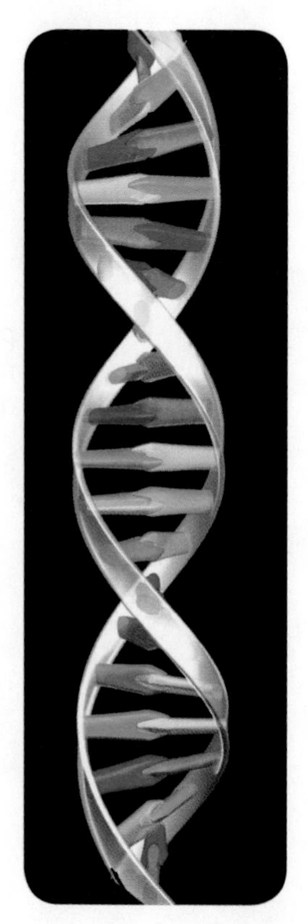
(Pasieka/Photo Researchers, Inc.)

A strand of deoxyribonucleic acid (DNA) is a long sequence of the nucleotides adenine, cytosine, guanine, and thymine (abbreviated A, C, G, T). One helical turn of a DNA strand would contain a sequence of 10 of these acids, such as ACTGCCATGT. How many possible sequences of this length are there?

There are 4 letters that can occur in each position in the 10-letter sequence. Any of the 4 letters can be in the first position. Regardless of what is in the first position, any of the 4 letters can be in the second position, and so on. The order of the letters matters, so a sequence that begins AC will be a different sequence than one that begins CA.

The number of different 10-letter sequences is more than 1 million:

$$4 \times 4 \times 4 \times 4 \times 4 \times 4 \times 4 \times 4 \times 4 \times 4 = 4^{10} = 1{,}048{,}576$$

As big as that number is, consider that it would take a DNA sequence about 3 billion letters long to contain your entire genetic "blueprint"! Knowing the number and frequency of DNA sequences has proven important in criminal justice. When skin or bodily fluids from a crime scene are "DNA fingerprinted," the specific DNA sequences in the recovered material are extremely unlikely to be found in any suspect other than the perpetrator. This result is expressed in Rule A on the next page as well as in Table 8.2 on p. 296.

EXAMPLE 8
Playing Songs

Example 4 (p. 247) of Chapter 7 involved choosing a random sample of 4 different songs from a digital media player with a playlist of 27 different songs. Now we ask: How many possible 4-song samples are possible from a collection of 27 songs? Like DNA sequences Example 7, order matters here. (Performers and DJs know that the same four songs can feel quite different when the songs are played in a different order.) Unlike DNA sequences, listing the same item more than once is not allowed.

(Steve Prezant/Corbis.)

Any of the 27 songs can be chosen to be played first, but only the remaining 26 songs are available to be listed as the second song, so that there are 27×26 choices for the first two songs. Any of these choices leaves 25 songs for the third position, and 24 for the fourth position. Surprisingly, the number of playlists of 4 different songs is almost half a million!

$$27 \times 26 \times 25 \times 24 = 421{,}200$$

This scenario of choosing an ordered subset of k songs from a playlist of n songs is called a **permutation**.

Permutation DEFINITION

A **permutation** is an ordered arrangement of k items that are chosen without replacement from a collection of n items. It can be notated as $P(n, k)$, ${}_nP_k$ or P_k^n and has the formula

$${}_nP_k = n \times (n-1) \times \cdots \times (n-k+1), \text{ which is Rule B below.}$$

Examples 7 and 8 both involve counting the number of arrangements of distinct items. They can each be viewed as specific applications of the fundamental principle of counting, and it is easier to think your way through the counting than to memorize a recipe. Nevertheless, because these two situations occur so often, they deserve to be given their own formal recognition as Rules A and B, respectively.

Counting Ordered Collections of Distinct Items RULE

Rule A. Suppose we have a collection of n distinct items. We want to arrange k of these items in order, and the same item can appear more than once in the arrangement. The number of possible arrangements is

$$\underbrace{n \times n \times \cdots \times n}_{} = n^k.$$

n is multiplied by itself k times.

Rule B. (Permutations) Suppose we have a collection of n distinct items. We want to arrange k of these items in order, and any item can appear no more than once in the arrangement. The number of possible arrangements is

$$n \times (n-1) \times \cdots \times (n-k+1).$$

EXAMPLE 9
Four-Letter Words

Suppose you have 4 Scrabble® tiles that are labeled T, S, O, and P. How many four-letter sequences can be made? Since there are only 4 tiles, the only way to make a 4-letter sequence is to use each letter exactly once, so there are no repeats. So this is a permutation by Rule B, with n and k each equal to 4. To think through the problem, proceed like this: Any of the 4 letters can be chosen first; then any of the 3 that remain can be chosen second; and so on. The number of permutations, therefore, is $4 \times 3 \times 2 \times 1 = 24$.

It turns out that 6 of these 24 sequences of 4 letters are actually words in the English language (STOP and see if you can find them all), so the probability that a permutation (chosen at random from these four tiles) will be an actual word is $\frac{6}{24} = \frac{1}{4}$.

Example 9 shows us that the permutation of all n elements of a collection yields the product of the first n positive integers. This expression of factors is special enough to have its own name—**factorial**—and is also used in Chapters 2 (p. 41) and 11 (p. 384).

Factorial DEFINITION

The **factorial** for a positive integer n equals the product of the first n positive integers. The term "n factorial" is notated $n!$:

$$n \times (n-1) \times (n-2) \times \cdots \times 3 \times 2 \times 1.$$

By convention, we define 0! to equal 1, which can be interpreted as saying there is one way to arrange zero items.

Factorial notation lets us write a long string of multiplied factors very compactly. For example, the expression for permutations in Rule B can now be rewritten as $\frac{n!}{(n-k)!}$. Factorials can be tedious to compute for large values of n, but a scientific calculator should have a key labeled $n!$ or $x!$ and the TI-84+ graphing calculator can find 13! with this sequence: 13 MATH $\rightarrow$ PRB $\rightarrow$! ENTER ENTER. If you have only a basic calculator without a factorial key, the expression $n \times (n-1) \times \cdots \times (n-k+1)$ will involve fewer multiplications than evaluating $\frac{n!}{(n-k)!}$ by first calculating $n!$ and $(n-k)!$ because it has already incorporated all the cancellations between numerator and denominator – namely, cancelling the positive integers from 1 to $n-k$.

EXAMPLE 10
Winning the Texas Lottery?

Most states have lottery games. The Texas Lottery (www.txlottery.org) has a Lotto Texas® game that involves choosing six numbers from the set of whole numbers from 1 to 54. You win (at least a share of) the jackpot so long as the collection of numbers you pick is the same collection that the lottery selects. Repetition is not allowed: The same number can't be picked twice in the same drawing. Unlike permutations, order does not matter here. It doesn't matter in what order the numbered ping pong balls come out of the mixing chamber; all that matters is what numbers are selected to be in that drawing's group of winners.

So while we can't use the permutation approach of Example 8 here, we can use a modification of it. The number of ordered 6-ball sets will be much larger than the number of unordered 6-ball sets because the lottery drawing {2, 14, 15, 21, 30, 33} is the same set of balls as {15, 2, 30, 14, 33, 21}, for example. But from the technique of Example 9, we can see that there would be 6! ways to arrange any particular set of 6 distinct balls. So the number of collections of lottery balls will simply be the number of permutations divided by $k!$, and for Lotto Texas, there are $\frac{_{54}P_6}{6!} = \frac{54 \times 53 \times 52 \times 51 \times 50 \times 49}{6 \times 5 \times 4 \times 3 \times 2 \times 1} = 25{,}827{,}165$ possible sets of numbers. Because only one of these sets of numbers will correspond to the jackpot, the probability of your ticket winning (at least a share of) the jackpot is $\frac{1}{25{,}827{,}165}$.

The scenario of choosing an unordered subset of k balls from a collection of n different balls is called a **combination** (see Rule D, p. 296).

Combination DEFINITION

A **combination** is an unordered arrangement of k items that are chosen without replacement from a collection of n items. It can be notated as $\binom{n}{k}$, $C(n, k)$ or ${}_nC_k$ and is sometimes spoken as "n choose k."

$${}_nC_k = \frac{n \times (n-1) \times \cdots \times (n-k+1)}{k!} \quad \text{or} \quad \frac{n!}{k!(n-k)!}, \text{ which is Rule D.}$$

If it's hard to remember the difference between combinations (Rule D) and permutations (Rule B), consider this: If you order a "combination platter" at a diner, you're asking for a certain set of foods to be on your plate, but you don't care what order they're in. Also, you can use this memory aid: "Permutations Presume Positions; Combinations Concern Collections." For completeness, we also provide a formula (Rule C) for unordered collections in which repetition *is* allowed, but we cannot give a simple explanation in the space we have, and we will not use it again.

Counting Unordered Collections of Distinct Items RULE

Rule C. Suppose that we have a collection of n distinct items. We want to select k of those items with no regard to order, and any item can appear more than once in the collection. The number of possible collections is $\frac{(n+k-1)!}{k!(n-1)!}$.

Rule D. (Combinations) Suppose that we have a collection of n distinct items. We want to select k of these items with no regard to order, and any item can appear no more than once in the collection. The number of possible selections is $\frac{n!}{k!(n-k)!}$.

Rules A, B, C, and D can be summarized in Table 8.2.

Table 8.2

Ways to Choose *k* Items from *n* Distinct Items

	Repetition is allowed	**Repetition is *not* allowed**
Order does matter	Rule A: $\underbrace{n \times n \times \cdots \times n}_{} = n^k$ n is multiplied by itself k times	Rule B (*permutation*): $\frac{n!}{(n-k)!} =$ $n \times (n-1) \times \cdots \times (n-k+1)$
Order does not matter	Rule C: $\frac{(n+k-1)!}{k!(n-1)!}$	Rule D (*combination*): $\frac{n!}{k!(n-k)!}$

We note that the TI-84+ graphing calculator sequence 8 (MATH) → PRB → nPr (ENTER) 3 (ENTER) can find the number of permutations of 3 objects chosen from 8 objects (and for combinations, use nCr instead of nPr).

In Spotlight 8.2, we assume a discrete uniform probability model in which all days of the year are assumed to be equally likely to be a birthday. (While this assumption is not perfectly satisfied, it turns out that any deviations from it only make the probability of a match even higher, such as if all people were born in April!) If we consider the continuous uniform probability model of the day and

SPOTLIGHT 8.2

Birthday Coincidences

If we ignore leap day (February 29), there are 365 possible birthdays a person can have. So if 366 people are gathered, there's a 100 percent chance at least two people share the same birthday. Now, if only 23 people are gathered, what do you think is the probability of any birthday matches? Guess before reading further.

Now imagine these 23 people enter a room one at a time, adding their birthday to a list in the order they enter. Using $n = 365$ and $k = 23$, Rule A gives us the total number of lists of 23 birthdays, and Rule B gives us how many of those lists have birthdays that are all different. (This birthday problem usually disregards leap day, which is why n equals 365, not 366.) Using the rule for Equally Likely Outcomes (assuming each day of the year is equally likely to be a randomly chosen person's birthday), we conclude that the probability of all birthdays being different is the result from Rule B divided by the result from Rule A:

$$\frac{{}_{365}P_{23}}{365^{23}} = \frac{365 \times 364 \times \ldots \times 343}{365^{23}}$$

Alternatively, we could assume independence of birthdays and use Probability Rule 4. The second person who walks in has a 364/365 chance of not matching person #1. The third person who walks in has a 363/365 chance of not matching persons #1 or #2, and so on. Verify that you get the same product by multiplying this string of fractions:

$$\frac{364}{365} \times \frac{363}{365} \times \cdots \times \frac{343}{365}$$

Either way, our final step to find the probability of getting at least one match is to subtract that answer from 1 (using Probability Rule 3), and we obtain the surprisingly high value of 51 percent!

Maybe it is not so surprising if we consider that the combinations formula tells us that there are 253 ways to choose pairs of people from 23 to ask each other if they have the same birthday. (Remember, we are looking for a match between *any* two people, not just between someone and *you*.) A rough alternative way to make the result seem plausible (devised by Manfred Borovcnik) is to say that with 23 people, you would expect to have about two people born in each month. In one month, the chance that one person matches someone else's birthday is about 1/30. Expecting "1/30 of a match" per month adds up to 12/30 of a match for the year, and 12/30 is not much less than 50 percent.

Because we underestimate the number of potential opportunities for "coincidences," we are surprised that they happen as often as they do. But as Jessica Utts notes, if something has a 1 in a million chance of happening to any person on a given day, this rare event will happen to roughly 300 people in the United States each day!

(Michael Rosenfeld/Photographer's Choice/Getty Images.)

time someone is born, then more advanced mathematics shows that only 17 people are required for there to be at least a 50 percent chance that at least two birth times are within 24 hours of each other! In the next two sections, we will examine continuous and discrete probability models in more detail.

8.4 Continuous Probability Models

When we use the table of random digits to select a digit between 0 and 9, the discrete probability model assigns probability $\frac{1}{10}$ to each of the 10 possible outcomes. Suppose that we want to choose a number at random between 0 and 1, allowing *any* number between 0 and 1 as the outcome. You can do this with technology, such

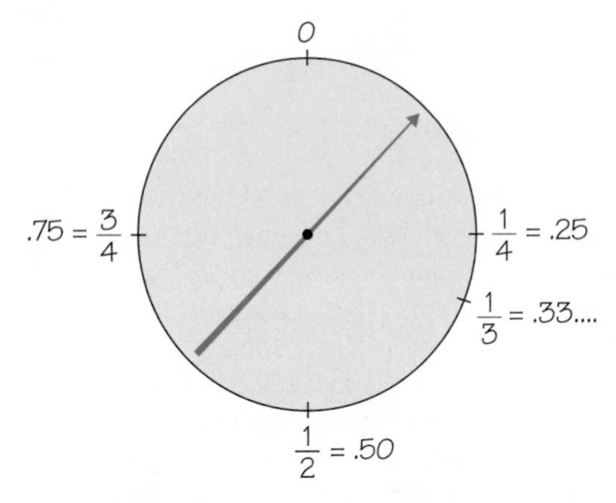

Figure 8.6 This spinner chooses a number between 0 and 1 at random. That is, it is equally likely to stop at any point on the circle.

as the TI-83/84 calculator command sequence (MATH) → PRB → rand or the Microsoft Excel spreadsheet software command =RAND(). You can visualize such a random number by thinking of a spinner needle (Figure 8.6) that turns freely around its center and slowly comes to a stop. The pointer can come to rest anywhere on a circle that is marked from 0 to 1.

The sample space is now an entire interval of numbers:

$$S = \{\text{all numbers } x \text{ such that } x \text{ is between 0 and 1}\}$$

How can we assign probabilities to events such as $\{0.3 \leq x \leq 0.7\}$? As in the case of selecting a random digit, we would like all possible outcomes to be equally likely. But we cannot assign probabilities to each individual value of x and then sum, because there are *infinitely* many possible values—too many to count or list. Instead, we use a second way of assigning probabilities directly to events—*as areas under a curve*. By Probability Rule 2, the curve must have total area 1 underneath it, corresponding to total probability 1. We call such curves **density curves**.

Density Curve — DEFINITION

A **density curve** is a curve that

- is always on or above the horizontal axis and
- has area exactly 1 underneath it.

Continuous Probability Model — DEFINITION

A **continuous probability model** is a probability model that assigns probabilities as areas under a density curve. The area under the curve and above any interval of values is the probability of an outcome in that interval.

The random-number generator will spread its output uniformly across the entire interval from 0 to 1 if we allow it to generate many numbers. The results of many trials are represented by the density curve of a *uniform probability model*. This density curve appears in red in Figure 8.7. It has height 1 over the interval from 0 to 1, and height 0 everywhere else. The area under the density curve is 1, the area of a square with base 1 and height 1. The probability of any event is the area under the density curve and above the event in question.

As Figure 8.7a illustrates, the probability that the random-number generator produces a number X between 0.3 and 0.7 inclusive is

$$P(0.3 \leq X \leq 0.7) = 0.4$$

because the rectangular area under the density curve and above the interval from 0.3 to 0.7 is 0.4. The area of a rectangle is the product of height and length, and the height of this density curve is 1, so the probability of any interval of outcomes will just be the length of the interval: $0.7 - 0.3 = 0.4$.

Also, we can apply Probability Rule 6 to non-overlapping intervals such as:

$$\begin{aligned} P(X < 0.5 \text{ or } X > 0.8) &= P(X < 0.5) + P(X > 0.8) \\ &= 0.5 \qquad + 0.2 \qquad = 0.7 \end{aligned}$$

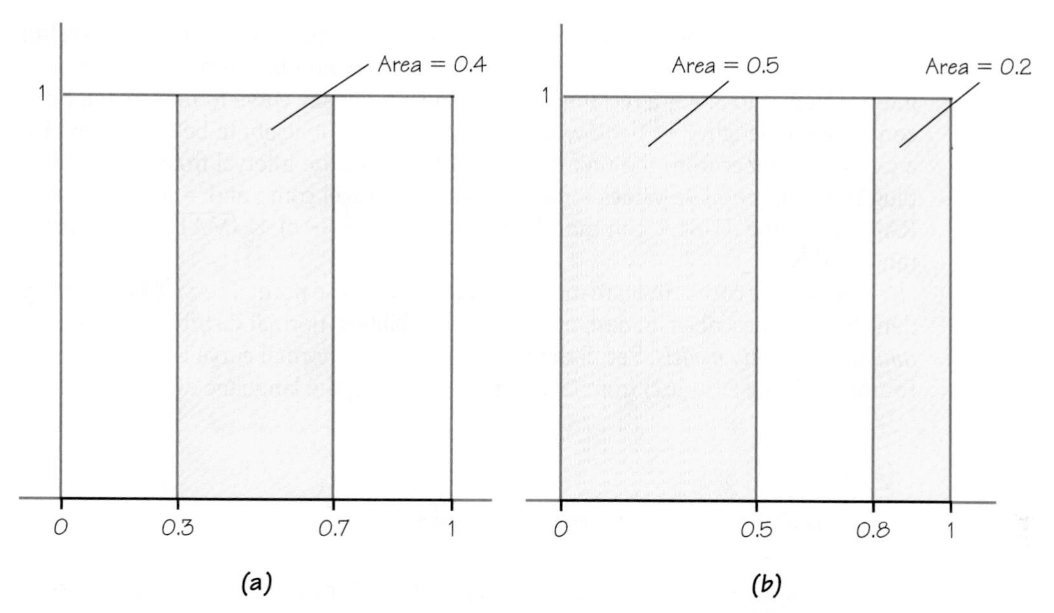

Figure 8.7 Assigning probabilities for generating a random number between 0 and 1 inclusive, for the spinner of Figure 8.6. The probability of any interval of numbers is the area above the interval and under the density curve. (a) The probability of an outcome between 0.3 and 0.7. (b) The probability of an outcome less than 0.5 or greater than 0.8.

The last event consists of two non-overlapping intervals, so the total area above the event is found by adding two areas, as illustrated by Figure 8.7b. This assignment of probabilities obeys all our rules for probability.

The probability model for a continuous random variable assigns probabilities to *intervals* of outcomes rather than to individual point outcomes. In fact, *all continuous probability models assign probability 0 to every individual outcome.* Only intervals of values have positive probability. To see that this is true, consider a specific outcome such as $P(X = 0.6)$ in Figure 8.6. In this example, the probability of any interval is the same as its length. The point 0.6 has no length, so its probability is 0.

EXAMPLE 11
Roundoff Error: Application of the Continuous Uniform Model

Sometimes the reported results of an opinion poll do not add up to 100 percent, but to 99 percent or 101 percent. More generally, before data values are presented, they sometimes get rounded to, say, the nearest whole number for ease of reading. For example, rounding 32.7 to 33 creates a roundoff error of $32.7 - 33 = -0.3$ and rounding 14.17 to 14 yields a roundoff error of $14.17 - 14 = 0.17$. Roundoff error can be critical to keep track of in data analysis and is one of many applications of the continuous uniform probability model. By rounding to the nearest whole number, the absolute value of the roundoff error cannot exceed $\frac{1}{2}$, and it is usually assumed that each roundoff error is equally likely to be any number between -0.5 and 0.5. Note that this example shows that so long as the total area under the density curve is 1, there is no reason the horizontal axis variable has to be between 0 and 1.

Going further, the horizontal axis variable is also not limited to an interval of length 1. For example, consider choosing a random number from the continuous interval from 1 to 3. For a rectangular area under a density curve to remain 1, a horizontal base of length $3 - 1 = 2$ would require the vertical height to be $\frac{1}{2}$. You can pick a random number from the uniform distribution over the interval from a to b if you plug in the appropriate values for a and b into the Excel command $= a + (b - a) \times$ RAND() or the TI-84+ command sequence $a + (b - a) \times$ (MATH) → PRB → rand → (ENTER).

The density curves that are most familiar to us are the normal curves. Because any density curve describes an assignment of probabilities, normal distributions are *continuous probability models*. Recall the total area under a normal curve is 1. Let's revisit Example 17 (pp. 261–262) from Chapter 7 now, using the language of probability.

EXAMPLE 12
Areas Under a Normal Curve Are Probabilities

Suppose that 60 percent of adults find shopping for clothes time-consuming and frustrating. All adults form a population, with population proportion $p = 0.6$. Interview an SRS of 2500 people from this population and find the proportion $\hat{p}$ of the sample who say that shopping is frustrating. We know that if we take many such samples, the statistic $\hat{p}$ will vary from sample to sample according to a normal distribution with

$$\text{mean} = p = 0.6$$

$$\text{standard deviation} = \sqrt{\frac{p(1-p)}{n}} = \sqrt{\frac{(0.6)(0.4)}{2500}} \approx 0.01$$

The 68–95–99.7 rule now gives *probabilities* for the value of $\hat{p}$ from a single SRS. The probability is 0.95 that $\hat{p}$ lies between 0.58 and 0.62. Figure 8.8 shows this probability as an area under the normal density curve.

All that is new is the language of probability. "Probability is 0.95" is shorthand for "95 percent of the time in a very large number of samples."

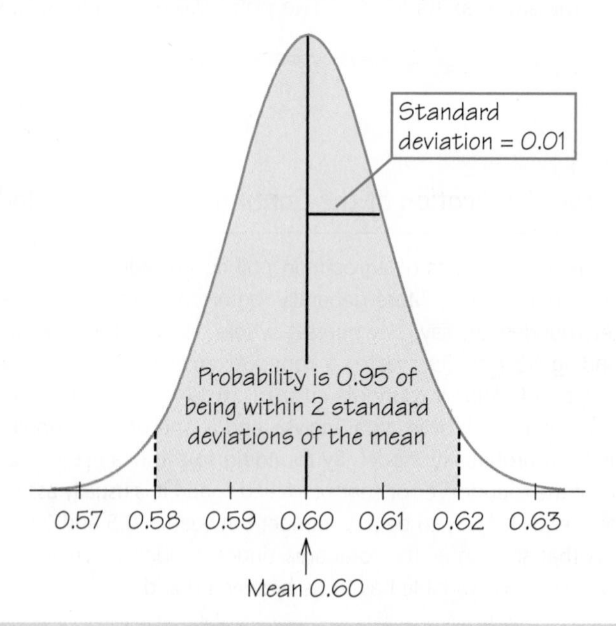

Figure 8.8 Probability shown as the area under a normal curve. The 68–95–99.7 rule gives some probabilities for normal probability models.

8.5 The Mean and Standard Deviation of a Probability Model

Algebra Review
Squares and Square Roots, page 184.

Suppose that you are offered this choice of bets, each costing the same:

Bet *A* pays $10 if you win and you have probability $\frac{1}{2}$ of winning;
bet *B* pays $10,000 if you win and offers probability $\frac{1}{10}$ of winning.

You would very likely choose Bet *B* even though *A* offers a better chance to win. It would be foolish to decide which bet to make just on the basis of the probability of winning. How much you can win is also important. When a random phenomenon has numerical outcomes, we are concerned with their amounts as well as with their probabilities.

What will be the average payoff of our two bets in many plays? Recall that the probabilities are the long-run proportions of plays in which each outcome occurs. Bet *A* produces $10 half the time in the long run and nothing half the time. So the average payoff should be

$$\left(\$10 \times \frac{1}{2}\right) + \left(\$0 \times \frac{1}{2}\right) = \$5$$

Bet *B*, on the other hand, pays out $10,000 on $\frac{1}{10}$ of all bets in the long run. So bet *B*'s average payoff is

$$\left(\$10{,}000 \times \frac{1}{10}\right) + \left(\$0 \times \frac{9}{10}\right) = \$1000$$

If you can place many bets, you should certainly choose B. In general, to take into account values and probabilities at the same time, we can add up the values, each weighted by their respective probability, so that more likely values get more weight. Here is a procedure of the kind of "average outcome" that we used to compare the two bets.

Mean of a Discrete Probability Model PROCEDURE

Step 1: Make a table with two rows. The first row needs to list all the possible numerical outcome values in the sample space.

Step 2: In the second row of the table, list the respective probabilities of each of the outcome values from the first row of the table.

Step 3: Write (or imagine) a third row where each entry is the product of the two items in the same column from the first two rows. Now add up all the values in the third row, and you will get the mean of the discrete probability model.

Now let us express the above procedure with algebraic notation. If there are k possible outcome values, we can use a subscript as an index in labeling each of the k outcomes as follows: $x_1, x_2, \ldots, x_k$. If we write their respective corresponding probabilities $p_1, p_2, \ldots, p_k$, then the mean μ of a discrete probability model can be written as $\mu = x_1p_1 + x_2p_2 + \cdots + x_kp_k$. In Chapter 5, we discussed the mean $\bar{x}$, the average of n observations that we actually have in hand. The mean μ, on the other hand, describes the probability model rather than any one collection of observations. The lowercase Greek letter *mu* (μ) is pronounced "myoo." You can think of μ as a theoretical mean that gives the average outcome that we expect in the long run.

You will sometimes see the mean of a probability model called the *expected value*. This isn't a very helpful name, though, because the mean may not be the value that we "expect," the value that is most likely, or even a value that is possible. For example, in Example 5 (p. 291), the most likely first digit is 1 (which is probably not what we expected), but Example 13 will show us now that the mean is 3.441, which is not even a whole number (and thus, cannot be a first digit)!

EXAMPLE 13
The Mean of the Probability Model for Benford's Law

If first digits in a set of records appear "at random," the probability model for the first digit is as in Example 6:

First digit	1	2	3	4	5	6	7	8	9
Probability	$\frac{1}{9}$	$\frac{1}{9}$	$\frac{1}{9}$	$\frac{1}{9}$	$\frac{1}{9}$	$\frac{1}{9}$	$\frac{1}{9}$	$\frac{1}{9}$	$\frac{1}{9}$

The mean of this model is $\mu =$

$$(1)\left(\frac{1}{9}\right) + (2)\left(\frac{1}{9}\right) + (3)\left(\frac{1}{9}\right) + (4)\left(\frac{1}{9}\right) + (5)\left(\frac{1}{9}\right) + (6)\left(\frac{1}{9}\right) + (7)\left(\frac{1}{9}\right) + (8)\left(\frac{1}{9}\right) + (9)\left(\frac{1}{9}\right) = 5$$

If, on the other hand, the records obey Benford's law, the distribution of the first digit is

First digit	1	2	3	4	5	6	7	8	9
Probability	0.301	0.176	0.125	0.097	0.079	0.067	0.058	0.051	0.046

The mean $\mu =$

$$1 \times 0.301 + 2 \times 0.176 + 3 \times 0.125 + 4 \times 0.097 + 5 \times 0.079 + 6 \times 0.067 +$$
$$7 \times 0.058 + 8 \times 0.051 + 9 \times 0.046 = 3.441.$$

The comparison of means ($3.441 < 5$) reflects the greater probability of smaller first digits under Benford's law. Because the histogram for random digits is symmetric, the mean lies at the center of symmetry. We can't determine the mean of the right-skewed Benford's law model precisely by simply looking at Figure 8.5b (p. 292); calculation is needed.

What about continuous probability models? Think of the area under a density curve as being cut out of solid homogenous material. The mean μ is the point at which the shape would balance. Figure 8.9 illustrates this interpretation of the

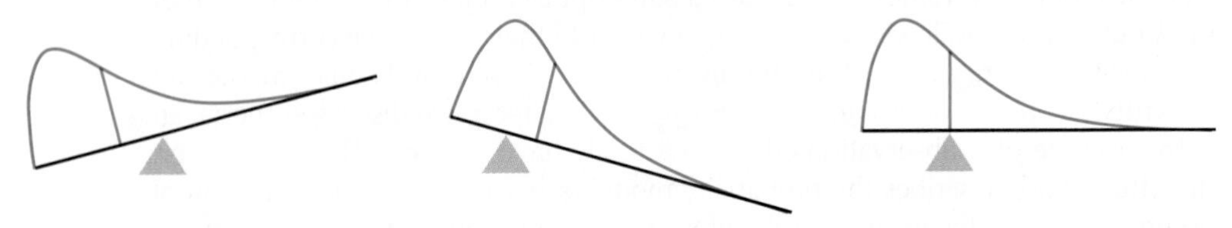

Figure 8.9 The mean of a continuous probability model is the "balance point" for the density curve.

mean. The mean lies at the center of symmetric density curves, such as the uniform density in Figure 8.7 and the normal curve in Figure 8.8. Exact calculation of the mean of a distribution with a skewed density curve requires advanced mathematics. The idea that the mean is the balance point of the probabilities applies to discrete models as well (see Section 5.4), but in the discrete case, we have a formula that gives us this point.

The mean μ is an average outcome in two senses. The definition for discrete models says that it is the average of the possible outcomes not weighted equally but weighted by their probabilities. More likely outcomes get more weight in the average. An important fact of probability, the **law of large numbers**, says that μ is the average outcome in another sense as well.

Law of Large Numbers — THEOREM

Observe any random phenomenon having numerical outcomes with finite mean μ. According to the **law of large numbers,** as the phenomenon is repeated a large number of times,

- the proportion of trials in which an outcome occurs gets closer and closer to the probability of that outcome, and
- the mean $\bar{x}$ of the observed values gets closer and closer to μ.

These facts can be stated more precisely and then proved mathematically. The law of large numbers brings the idea of probability to a natural completion. We first observed that some phenomena are random in the sense of showing long-run regularity. Then we used the idea of long-run proportions to motivate the basic laws of probability. Those laws are mathematical idealizations that can be used without interpreting probability as proportion in many trials. The law of large numbers tells us that in many trials, the proportion of trials on which an outcome occurs will always approach its probability.

The law of large numbers also explains why gambling can be a business. The winnings (or losses) of a gambler on a few plays are highly variable or uncertain; that's why gambling is exciting. It is only *in the long run* that the mean outcome is predictable. The house plays many tens of thousands of times. So the house, unlike individual gamblers, can count on the long-run regularity described by the law of large numbers. The average winnings of the house on tens of thousands of plays will be very close to the mean of the distribution of winnings. Needless to say, gambling games have mean outcomes that guarantee the house a profit, though some games (e.g., blackjack) give the house a smaller advantage than others (e.g., keno).

We know that the simplest description of a distribution of data requires both a measure of center and a measure of variability. The same is true for probability models. The *mean* is the average value for both a set of data and a discrete probability model. All the observations are weighted equally in finding the mean $\bar{x}$ for data, but the values are weighted by their probabilities in finding the mean μ of a probability model. The measure of variability that goes with the mean is the **standard deviation**. In Section 5.7, on page 184, we learned that the standard deviation s of data is the square root of the average squared deviation of the observations from their mean. We apply exactly the same idea to probability models, using probabilities as weights in the average. Here is the definition.

Standard Deviation of a Discrete Probability Model DEFINITION

Suppose that the possible outcomes $x_1, x_2, \ldots, x_k$ in a sample space S are numbers, and that p_j is the probability of outcome x_j. The **standard deviation of a discrete probability model** with mean μ is denoted by the lowercase Greek letter *sigma* (σ) and is given by this formula:

$$\sigma = \sqrt{(x_1 - \mu)^2 p_1 + (x_2 - \mu)^2 p_2 + \cdots + (x_k - \mu)^2 p_k}$$

EXAMPLE 14
Standard Deviation of the Probability Model for Benford's Law

If the first digits in a set of records obey Benford's law, the discrete probability model is

First digit	1	2	3	4	5	6	7	8	9
Probability	0.301	0.176	0.125	0.097	0.079	0.067	0.058	0.051	0.046

We saw in Example 13 that the mean is $\mu = 3.441$. To find the standard deviation,

$$\begin{aligned}\sigma &= \sqrt{(x_1 - \mu)^2 p_1 + (x_2 - \mu)^2 p_2 + \cdots + (x_k - \mu)^2 p_k} \\ &= \sqrt{(1 - 3.441)^2(0.301) + (2 - 3.441)^2(0.176) + \cdots + (9 - 3.441)^2(0.046)} \\ &= \sqrt{1.7935 + 0.3655 + \cdots + 1.4215} \\ &= \sqrt{6.061} \approx 2.46\end{aligned}$$

You can follow the same pattern to find the standard deviation of the equally likely model and show that the Benford's law model, by virtue of clustering near the left side, has less variability than the equally likely model.

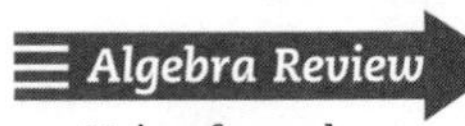

Using formulas, page 177.

Finding the standard deviation of a continuous probability model usually requires advanced mathematics (calculus). Section 5.8 provided the answer in one important case: The standard deviation of a normal curve is the distance from the center (the mean) to the change-of-curvature point on either side.

8.6 The Central Limit Theorem

The key to finding a confidence interval that estimates a population proportion (Chapter 7) was the fact that the sampling distribution of a population proportion is close to normal when the sample is large. This fact is an application of one of the most important results of probability theory, the **Central Limit Theorem**. This theorem says that the distribution of *any* random phenomenon tends to be normal if we average it over a large number of independent repetitions. The Central Limit Theorem allows us to analyze and predict the results of chance phenomena when we average over many observations.

The word limit in Central Limit Theorem reflects that the normal curve is the limit or target shape to which the sampling distribution gets closer and closer as the sample size increases. The theorem also tells us the sampling distribution's mean, and the mean is a measure of *central* tendency.

Central Limit Theorem THEOREM

Draw an SRS of size n from any large population with mean μ and finite standard deviation σ. Then

- The mean of the sampling distribution of $\bar{x}$ is μ.
- The standard deviation of the sampling distribution of $\bar{x}$ is $\frac{\sigma}{\sqrt{n}}$
- The **Central Limit Theorem** says that the sampling distribution of $\bar{x}$ is approximately normal when the sample size n is large ($n \geq 30$).

The first two parts of this statement can be proved from the definitions of the mean and the standard deviation. They are true for any sample size n. The Central Limit Theorem is a much deeper result. Pay attention to the fact that the standard deviation of a mean decreases as the number of observations n increases. Together with the Central Limit Theorem, this supports three general statements that help us understand a wide variety of random phenomena:

Averages are less variable than individual observations.
Averages are more normal than individual observations.
Averages are less variable and more normal than averages of smaller samples.

The *Central Limit Theorem* applet allows you to watch the Central Limit Theorem in action: It starts with a distribution that is strongly skewed, not at all normal. As you increase the size of the sample, the distribution of the mean $\bar{x}$ gets closer and closer to the normal shape.

Let's consider a concrete example with dice. Rolls of a single die have a uniform (flat) probability histogram, which would look like Figure 8.5a but with horizontal axis values going from 1 to 6 and the vertical heights all being 1/6). Now consider the mean when rolling a pair of dice. The mean of two dice is simply half their sum, so, for example, the mean of two dice equaling 4.5 must have the same probability as the sum of the two dice equaling 9. The probability model for the mean of two dice simply divides by 2 each of the numbers in the top row of the table in Example 4 and in the horizontal axis of Figure 8.4, but otherwise has the same probability values and distribution shape of Figure 8.4. The histogram in Figure 8.4 is certainly less variable and closer to looking "normal" than is the flat histogram for rolling a single die. Just as a two-dice sum of 7 (i.e., a mean of 3.5) is more likely than a sum of 12 (i.e., a mean of 6), a mean is more likely to be a value near the center than to be a value near the endpoints. And so the example of two dice allows us to understand how a mean (of two dice) is less variable (i.e., more clustered around 3.5) and more normal in shape than a single observation (rolling one die).

EXAMPLE 15
Heights of Young Women

The distribution of heights of young adult women is approximately normal, with mean 64.5 inches and standard deviation 2.5 inches. This normal distribution describes the population of young women. It is also the probability model for choosing one woman at random from this population and measuring her height. For example, the 68–95–99.7 rule says that the probability is 0.95 that a randomly chosen woman is between 59.5 and 69.5 inches tall.

Now choose an SRS of 25 young women at random and take the mean $\bar{x}$ of their heights. The mean $\bar{x}$ varies in repeated samples; the pattern of variation is the sampling distribution of $\bar{x}$. The sampling distribution has the same center (μ = 64.5 inches) as the population of young women. In statistical terms, the sample mean $\bar{x}$ has no bias as an estimator of the population mean μ. If we take many samples, $\bar{x}$ will sometimes be smaller than μ and sometimes larger, but it has no systematic tendency to be too small or too large.

The standard deviation of the sampling distribution of $\bar{x}$ is

$$\frac{\sigma}{\sqrt{n}} = \frac{2.5}{\sqrt{25}} = \frac{2.5}{5} = 0.5 \text{ inch}$$

The standard deviation σ describes the variation when we measure many individual women. The standard deviation $\sigma/\sqrt{n}$ of the distribution of $\bar{x}$ describes the variation in the average heights of samples of women when we take many samples. The average height is less variable than individual heights.

Figure 8.10 compares the two distributions: Both are normal and both have the same mean, but the average height of 25 randomly chosen women has much less variability. For example, the 68–95–99.7 rule says that 95 percent of all averages $\bar{x}$ lie between 63.5 and 65.5 inches because two standard deviations of $\bar{x}$ make 1 inch. This 2-inch span is just one-fifth as wide as the 10-inch span that catches the middle 95 percent of heights for individual women.

Figure 8.10 The sampling distribution of the average height of an SRS of 25 women has the same center (mean) as the distribution of individual heights but has much less variability because $\frac{2.5}{\sqrt{25}} < 2.5$.

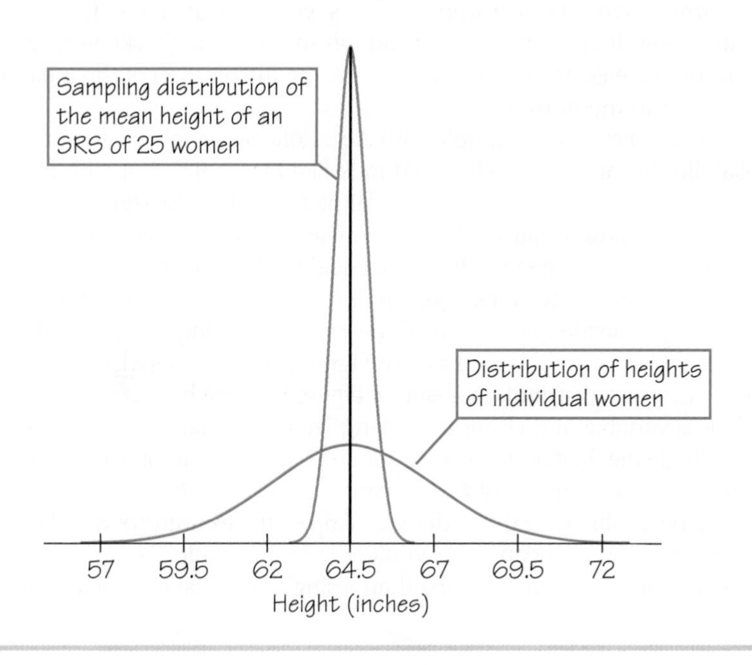

The Central Limit Theorem says that in large samples, the sample mean $\bar{x}$ is approximately normal. In Figure 8.10, we show a normal curve for $\bar{x}$ even though the sample size of 25 is not very large. Is that acceptable? How large a sample is needed for the Central Limit Theorem to work depends on how far from a normal curve the model we start with is. The closer to normality we start, the quicker the distribution of the sample mean becomes normal. In fact, if individual observations follow a normal curve, the sampling distribution of $\bar{x}$ is exactly normal for any sample size. So Figure 8.10 is accurate. The Central Limit Theorem is a striking result because as n gets large, it works for *any* model we may start with, no matter how far it is from normal. Here is an example that starts very far from a normal bell curve shape; its probability histogram would simply be tall spikes at 1 and -1 with nothing in between.

EXAMPLE 16
Roulette Bet: Red or Black

An American roulette wheel has 38 slots, of which 18 are black, 18 are red, and 2 are green. The dealer spins the wheel and whirls a small ball in the opposite direction within the wheel. Gamblers bet on where the ball will come to rest (see Figure 8.11). One of the simplest wagers chooses red (or black). A bet of \$1 on red pays off an additional \$1 if the ball lands in a red slot. Otherwise, the player loses the \$1. The two green slots always belong to the casino, informally referred to as "the house."

Lou bets on red. He wins if the ball stops in one of the 18 red slots. He loses if it lands in one of the 20 slots that are black or green. Because casino roulette wheels are carefully balanced so that all slots are equally likely, the probability model is

Figure 8.11 A gambler may win or lose his night of roulette, but the casino always wins in the long run. *(Ingram Publishing/ PictureQuest.)*

Net Outcome for Gambler

	Win \$1	Lose \$1
Probability	$\frac{18}{38} = 0.474$	$\frac{20}{38} = 0.526$

The mean outcome of a single \$1 bet on red is

$$\mu = (\$1)\left(\frac{18}{38}\right) + (-\$1)\left(\frac{20}{38}\right)$$
$$= -\$\frac{2}{38} = -\$0.053 \text{ (a loss of 5.3 cents)}$$

The law of large numbers says that the mean μ is the average outcome of a very large number of individual bets. In the long run, gamblers will lose (and the casino will win) an average of 5.3 cents per bet. We can similarly find the standard deviation for a single \$1 bet on red:

$$\sigma = \sqrt{(1-(-0.053))^2\frac{18}{38} + (-1-(-0.053))^2\frac{20}{38}}$$
$$= \sqrt{(1.053)^2\frac{18}{38} + (-0.947)^2\frac{20}{38}}$$
$$= \sqrt{0.9972} = \$0.9986$$

Lou certainly starts far from any normal curve. The probability model for each bet is discrete, with just two possible outcomes. Yet the Central Limit Theorem says that the average outcome of many bets follows a normal curve. Lou is a habitual gambler who places 50 \$1 bets on red almost every night. Because we know the probability model for a bet on red, we can simulate Lou's experience over many nights at the roulette wheel. The histogram in Figure 8.12, made from a simulation of 1000 nights, shows Lou's average winnings per bet. As the Central Limit Theorem says, the distribution looks normal.

EXAMPLE 17
Lou Gets Entertainment

The normal curve in Figure 8.12 comes from the Central Limit Theorem and the values of the mean μ and standard deviation σ in Example 16. It has

$$\text{mean } \mu = -\$0.053$$

$$\text{standard deviation} = \frac{\sigma}{\sqrt{n}} = \frac{\$0.9986}{\sqrt{50}} = \$0.141$$

Apply the 99.7 part of the 68–95–99.7 rule from Section 5.9 (p. 193): Almost all average nightly winnings per bet will fall within 3 standard deviations of the mean, that is, between

$$-\$0.053 - (3)(\$0.141) = -\$0.476$$

and

$$-\$0.053 + (3)(\$0.141) = \$0.370$$

What is more interesting to a gambler is not average winnings per bet, but total winnings for the whole 50-bet night. To find this out, Lou can simply multiply both endpoints from the preceding 99.7 percent confidence interval by 50. So Lou's total winnings after 50 bets of \$1 each will almost surely fall between

$$(50)(-\$0.476) = -\$23.80$$

and

$$(50)(\$0.370) = \$18.50$$

Each night, Lou may win as much as \$18.50 or lose as much as \$23.80. Note that he will usually lose more on a bad night than he will win on a good night. Some people find gambling exciting because the outcome, even after an evening of bets, is uncertain. It is possible to beat the odds and walk away a winner. It's all a matter of luck.

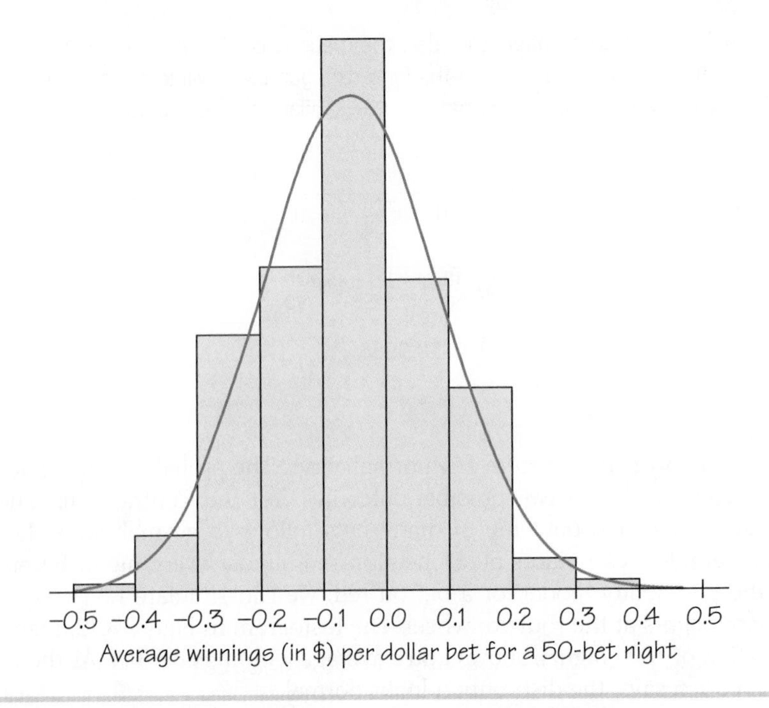

Figure 8.12 (Approximately normal) histogram of a gambler's winnings from a simulation of 1000 nights, where each night had 50 bets on red (or black) in roulette.

The casino, however, is in a different position. It doesn't want rollercoaster excitement, just steady income.

EXAMPLE 18
The Casino Gets Rich

The casino bets with all its customers—perhaps 100,000 individual red/black roulette bets in a week. The Central Limit Theorem guarantees that the distribution of average customer winnings on 100,000 bets is very close to normal. The mean is still the mean outcome for one bet, −$0.053, a loss of 5.3¢ per dollar bet. The key point is that the standard deviation is much smaller when we average over 100,000 bets. It is

$$\frac{\sigma}{\sqrt{n}} = \frac{0.9986}{\sqrt{100{,}000}} = 0.003$$

Here is what the 99.7 percent confidence interval estimate of the average result looks like after 100,000 bets:

$$\begin{aligned} &\text{mean} \pm 3 \text{ standard deviations} \\ &= -\$0.053 \pm (3)(\$0.003) \\ &= -\$0.053 \pm \$0.009, \text{ which generates an interval from } -\$0.062 \text{ to } -\$0.044. \end{aligned}$$

Because the casino covers so many bets, the standard deviation of the average winnings per bet becomes very small. Not only is the mean negative, but the entire 99.7 percent confidence interval is also in the negative region, so the total result is virtually certain to be in the casino's favor. The gamblers' losses and the casino's winnings are almost certain to average between 4.4 and 6.2 cents for every dollar bet.

The gamblers who collectively place those 100,000 bets will lose money. The probable window of their losses is

$$(100{,}000)(-\$0.062) = -\$6200 \quad \text{to} \quad (100{,}000)(-\$0.044) = -\$4400$$

The gamblers are almost certain to lose—and the casino is almost certain to collect—between $4400 and $6200 on those 100,000 bets. What's more, the interval of average outcomes continues to narrow as still more bets are made. That is how a casino can make a business out of gambling. According to *Forbes* magazine, in 2007 the third-richest American (with an estimated worth of $28 billion) was casino mogul Sheldon Adelson. In the 2010 list, Adelson fell to $15 billion—still good for 13th place.

EXAMPLE 19
Diversity: A Stock Response to Risk

For some people, investing in the stock market may feel almost as risky as visiting a casino, especially if all their money is placed in a single stock. Fortunately, the Central Limit Theorem tells us that averages have less risk than individuals. And so we can reduce our risk (and stress!) by taking the money now all in one stock and diversifying the portfolio by spreading that money among many stocks (or bonds,

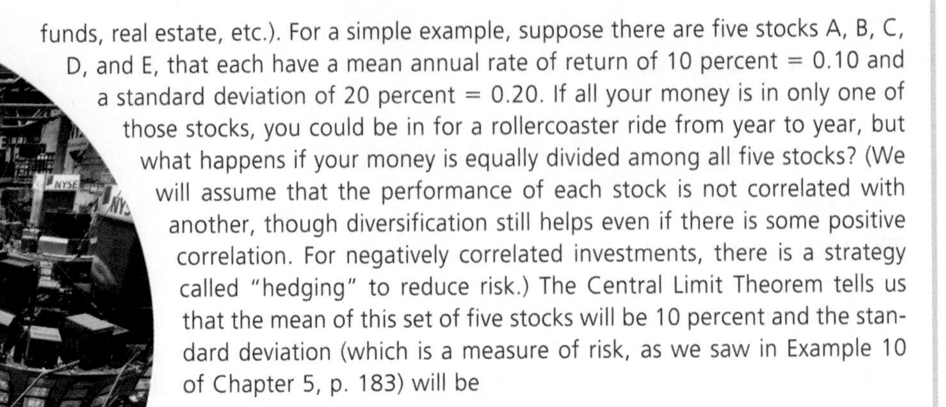

funds, real estate, etc.). For a simple example, suppose there are five stocks A, B, C, D, and E, that each have a mean annual rate of return of 10 percent = 0.10 and a standard deviation of 20 percent = 0.20. If all your money is in only one of those stocks, you could be in for a rollercoaster ride from year to year, but what happens if your money is equally divided among all five stocks? (We will assume that the performance of each stock is not correlated with another, though diversification still helps even if there is some positive correlation. For negatively correlated investments, there is a strategy called "hedging" to reduce risk.) The Central Limit Theorem tells us that the mean of this set of five stocks will be 10 percent and the standard deviation (which is a measure of risk, as we saw in Example 10 of Chapter 5, p. 183) will be

$$\frac{0.20}{\sqrt{5}} = 0.0894$$

which is just under 9 percent, a big drop from the standard deviation of 20 percent for a single stock!

In Chapter 7, we based a confidence interval for a population proportion p on the fact that the **sampling distribution** of a sample proportion $\hat{p}$ is close to normal for large samples. The Central Limit Theorem applies to means. How can we also apply it to proportions? By seeing that *a proportion is really a mean.* Showing this will be our final example of the Central Limit Theorem. While it is more theoretical than our other examples, it gives us an important foundation and brings us full circle to the first figure of the chapter.

EXAMPLE 20
The Sampling Distribution of a Proportion

If we can express the sample proportion of successes as a sample mean, we can apply tools we have learned in Sections 8.5 and 8.6 to derive the formula that was given (in Section 7.7 on page 261) for the standard deviation of the sample proportion.

Consider an SRS of size n from a population that contains proportion p of "having a particular trait." For each of the n individuals, we can define a simple numerical variable x_i to equal one of two values:

1 for a success (i.e., having a particular trait, flipping "heads" with a coin, etc.)

0 for a failure (e.g., not having the trait, flipping "tails" with a coin, etc.).

For example, if the third individual has the trait of interest, then $x_3 = 1$. So the sum of all n of the x_i values is the total number of "successes" (that is, people that had the trait of interest). So the proportion $\hat{p}$ of successes is given by

$$\hat{p} = \frac{\textit{number of successes}}{n} = \frac{x_1 + x_2 + \cdots + x_n}{n} = \bar{x}$$

So $\hat{p}$ is really a mean, and so its sampling distribution (by the Central Limit Theorem) is close to normal when the sample size n is large ($n \geq 30$).

Because $\hat{p}$ is the mean of the x_i, we can find the mean and standard deviation of $\hat{p}$ from the mean and standard deviation of one observation x_i. Each observation has probability p of being a success, so the probability model for one observation x_i is

Outcome	1 (success; heads)	0 (failure; tails)
Probability	p	$1 - p$

Using the tools of Section 8.5, the mean of one observation x_i is therefore

$$\mu = (1)(p) + (0)(1 - p) = p$$

In the same way, after a bit more algebra, the tools of Section 8.5 show that the standard deviation of one observation x_i is

$$\sigma = \sqrt{(1 - p)^2 p + (0 - p)^2(1 - p)} = \sqrt{p(1 - p)}$$

From the Central Limit Theorem (Section 8.6), the standard deviation of the mean of n observations is $\frac{\sigma}{\sqrt{n}}$, so we simply substitute in our expression for σ and obtain:

$$\frac{\sigma}{\sqrt{n}} = \frac{\sqrt{p(1 - p)}}{\sqrt{n}} = \sqrt{\frac{p(1 - p)}{n}}.$$

This last expression is precisely the fact we used in Section 7.7. This formula yields more appreciation of the accumulating proportion of heads graphed in Figure 8.1 (p. 283). The probability of heads (which is what we are calling

SPOTLIGHT 8.3

Probability and Fairness

The fairness of a casino die refers to whether each side is physically designed to have the same probability of landing face up. Probability can also address fairness in life. On June 20, 2011, the Supreme Court ruled that the biggest employment discrimination case in U.S. history (*Wal-Mart Stores v. Dukes*) could not proceed as a class-action lawsuit covering the 1.5 million women who worked at Wal-Mart or Sam's Club stores since December 1998. The highly publicized lawsuit alleged that female employees at those stores were paid less and promoted less often than male employees. Here is a simple way to show how the tools of Example 20 might provide a fairness benchmark. Suppose that 10 people are promoted from a huge pool of qualified workers that is 50 percent women. On average, we would expect 5 of the 10 people promoted to be women, but we know that a fair process may yield a slightly different result simply due to natural variability. Should we be quick to declare "unfairness" if, say, only 3 of the 10 people promoted were women?

Example 20's formula for standard deviation of a proportion gives us an objective, quantifiable answer. By plugging $p = \frac{3}{10} = 0.3$ and $n = 10$ into the formula, we get $\sqrt{\frac{(0.3)(1 - 0.3)}{10}} \approx 0.145$, so two of those standard deviations is 0.290. Note that the 0.5 (the expected proportion under ideal fairness) and 0.3 (the observed proportion of people promoted that were women) differ by 0.2. Since 0.2 is less than two standard deviations, the observed result is "within natural variability" of the fairness benchmark and the departure is not statistically significant (to use the term of Section 7.5). But what if instead there were 100 people promoted and 30 of them were women? Now, the formula for standard deviation for a proportion yields 0.0458. So in this second scenario, $\frac{30}{100}$ is more than three standard deviations below the 0.5 fairness benchmark, and that departure is statistically significant. In other words, it is clear from the 68–95–99.7 rule that less than 5 percent of the data in a normal distribution would be at least this far from the mean (the fairness benchmark).

The probability of obtaining a result with at least this much of a departure (from the fairness benchmark) is called a *p*-value and when it is less than 5 percent, convention says that the result is sufficiently rare or unusual to be considered statistically significant. For some cases involving bias or equity allegations, it suffices to show that one group's outcomes are statistically significantly different from another group's (i.e., "disparate impact"), while other cases require documentation of discriminatory intent.

"success" here) each flip is $p = 0.5$. Plugging this value into the formula for standard deviation of the proportion of heads in n flips yields

$$\sqrt{\frac{\frac{1}{2} \times \left(1 - \frac{1}{2}\right)}{n}} = \frac{1}{2\sqrt{n}}$$

And so, as the number n of flips increases, the amount of variability in the proportion of flips that are heads decreases and the proportion gets closer and closer to the long-run expected value of 0.5. Using the 99.7 part of the 68–95–99.7 rule, the proportion of 10,000 flips that are heads will be within three standard deviations of 0.5. Plugging 10,000 into n in the above formula means the proportion will be within $3 \times 0.005 = 0.015$ of 0.5. This yields a 99.7 percent confidence interval of (0.485, 0.515), so after 10,000 flips the percentage of flips that are heads will almost surely be within only a couple of percentage points of the 50 percent benchmark.

Examples 15 to 20 illustrate the importance of the Central Limit Theorem and the reason for the importance of normal distributions. Thanks to the Central Limit Theorem, we can often replace tricky calculations about a probability model by simpler calculations for a normal distribution.

REVIEW VOCABULARY

Addition rule The probability that one event or the other occurs is the sum of their individual probabilities minus the probability of any overlap they have. (p. 289)

Central Limit Theorem The average of many independent random outcomes is approximately normally distributed. When we average n independent repetitions of the same random phenomenon, the resulting distribution of outcomes has mean equal to the mean outcome of a single trial and standard deviation proportional to $1/\sqrt{n}$. (p. 305)

Combination An unordered collection of k items chosen (without allowing repetition) from a set of n distinct items. (p. 296)

Combinatorics The branch of mathematics that studies how to count and choose elements. (p. 293)

Complement of an event The complement of an event A is the event "A does not occur," which is denoted A^C. (p. 286)

Complement rule The probability that an event does not occur is 1 minus the probability that the event does occur. $P(A^C) = 1 - P(A)$. (p. 289)

Continuous probability model A probability model that assigns probabilities to events as areas under a density curve. (p. 298)

Density curve A curve that is always on or above the horizontal axis and has area exactly 1 underneath it. A density curve describes a continuous probability model. (p. 298)

Discrete probability model A probability model that assigns probabilities to each of a countable number of possible outcomes. (p. 291)

Disjoint events Events that have no outcomes in common. (Also called *mutually exclusive events.*) (p. 286)

Event A collection of possible outcomes of a random phenomenon; a subset of the sample space. (p. 284)

Factorial The product of the first n positive integers, denoted as $n!$ (p. 295)

Fundamental principle of counting A multiplicative method for counting outcomes of multistage processes. (p. 293)

Independent events Events that do not influence each other's probability of occurring. (p. 286)

Law of large numbers As a random phenomenon is repeated many times, the mean $\bar{x}$ of the observed outcomes approaches the mean μ of the probability model. (p. 303)

Mean of a discrete probability model The average outcome of a random phenomenon with numerical values. When possible values $x_1, x_2, \ldots, x_k$ have probabilities $p_1, p_2, \ldots, p_k$, the mean is the average of the outcomes weighted by their probabilities, $\mu = x_1p_1 + x_2p_2 + \cdots + x_kp_k$. (Also called *expected value.*) (p. 301)

Multiplication rule If two events are independent, then the probability that one event and the other both occur is the product of their individual probabilities. $P(A \text{ and } B) = P(A) \times P(B)$, when A and B are independent events. (p. 289)

Permutation An ordered arrangement of k items chosen (without allowing repetition) from a set of n distinct items. (p. 294)

Probability A number between 0 and 1 that gives the long-run proportion of repetitions of a random phenomenon on which an event will occur. (p. 283)

Probability histogram A histogram that displays a discrete probability model when the outcomes are numerical. The height of each bar is the probability of the event at the base of the bar. (p. 290)

Probability model A sample space S together with an assignment of probabilities to events. The two main types of probability models are *discrete* and *continuous*. (p. 284)

Random A phenomenon or trial is random if it is uncertain what the next outcome will be but each outcome nonetheless tends to occur in a fixed proportion of a very long sequence of repetitions. These long-run proportions are the probabilities of the outcomes. (p. 281)

Sample space A list of all possible (simplest) outcomes of a random phenomenon. (p. 284)

Sampling distribution The distribution of values taken by a statistic when many random samples are drawn under the same circumstances. A sampling distribution consists of an assignment of probabilities to the possible values of a statistic. (pp. 260, 310)

Standard deviation of a discrete probability model A measure of the variability of a probability model. When the possible values $x_1, x_2, \ldots, x_k$ have probabilities $p_1, p_2, \ldots, p_k$, the standard deviation is the square root of the average (weighted by probabilities) of the squared deviations from the mean: $\sigma = \sqrt{(x_1 - \mu)^2 p_1 + (x_2 - \mu)^2 p_2 + \cdots + (x_k - \mu)^2 p_k}$. (p. 304)

SKILLS CHECK

1. You read in a book on poker that the probability of being dealt three of a kind in a five-card poker hand is $\frac{1}{50}$. What does this mean?

(a) If you deal thousands of poker hands, the fraction of them that contain three of a kind will be very close to $\frac{1}{50}$.
(b) If you deal 50 poker hands, exactly one of them will contain three of a kind.
(c) If you deal 10,000 poker hands, exactly 200 of them will contain three of a kind.

2. If two coins are flipped and then a die is rolled, the sample space would have ________ different outcomes.

Skills Check questions 3–5 use this model for the blood type of a randomly chosen person in the United States:

Blood type	O	A	B	AB
Probability	0.45	0.40	0.11	?

3. The probability that a randomly chosen American has type AB blood is

(a) 0.044.
(b) 0.04.
(c) 0.4.

4. María has type A blood. She can safely receive blood transfusions from people with blood types O and A. The probability that a randomly chosen American can donate blood to María is ________.

5. What is the probability that a randomly chosen American does not have type O blood?

(a) 0.55
(b) 0.45
(c) 0.04

6. Figure 8.2 (p. 285) shows the 36 possible outcomes for rolling two dice. These outcomes are equally likely. A "soft 4" is a roll of 1 on one die and 3 on the other. The probability of rolling a soft 4 is ________.

7. A discrete probability model has

(a) only two outcomes.
(b) equally likely outcomes.
(c) a countable number of outcomes.

8. The most likely first (leftmost) digit of a number from financial data is ________.

9. In a table of random digits such as Table 7.1 (p. 246), each digit is equally likely to be any of 0, 1, 2, 3, 4, 5, 6, 7, 8, or 9. What is the probability that a digit in the table is a 0?

(a) 1/9
(b) 1/10
(c) 9/10

10. In a table of random digits such as Table 7.1 (p. 246), each digit is equally likely to be any of 0, 1, 2, 3, 4, 5, 6, 7, 8, or 9. The probability that a digit in the table is 7 or greater is ________.

11. Toward the end of a game of Scrabble®, you hold five tiles with the letters A, E, P, R, and S. In how many orders can you arrange these 5 letters (whether or not they form actual words)?

(a) 5
(b) $5 \times 4 \times 3 \times 2 \times 1 = 120$
(c) $5 \times 5 \times 5 \times 5 \times 5 = 3125$

12. Toward the end of a game of Scrabble™, you hold the letters D, O, G, and Q. You can choose 3 of these 4 letters and arrange them in order in ______________ different ways, assuming that you are not trying to form actual words.

13. A 52-card deck contains 13 cards from each of the 4 suits: clubs ♣, diamonds ♦, hearts ♥, and spades ♠. You deal 4 cards without replacement from a well-shuffled deck, so that you are equally likely to deal any 4 cards. What is the probability that *all* 4 cards are clubs?

(a) $\frac{1}{4}$, because $\frac{1}{4}$ of the cards are clubs.
(b) $\frac{13}{52} \times \frac{12}{51} \times \frac{11}{50} \times \frac{10}{49} = 0.0026$
(c) $\frac{13}{52} \times \frac{12}{52} \times \frac{11}{52} \times \frac{10}{52} = 0.0023$

14. You deal 4 cards as in the previous exercise. The probability that you deal *no* clubs is ____________.

15. Figure 5.3 (p. 173) shows that the normal distribution with mean $\mu = 6.8$ and standard deviation $\sigma = 1.6$ is a good description of the Iowa Test vocabulary scores of seventh-grade students in Gary, Indiana. The probability that a randomly chosen student has a score higher than 8.4 is

(a) 0.68.
(b) 0.32.
(c) 0.16.

16. Figure 8.7 (p. 299) shows the density curve of a continuous probability model for choosing a number at random between 0 and 1 inclusive. The probability that the number chosen is less than or equal to 0.4 is ______________.

17. In Figure 8.7 (p. 299), the probability that x is greater than 0.65 is ________.

(a) 0.65.
(b) 0.35.
(c) 1.

18. The total area under a density curve is ________.

19. Annual returns on the more than 5000 common stocks available to investors vary a lot. In a recent year, the mean return was 8.3 percent and the standard deviation of returns was 28.5 percent. The law of large numbers says that

(a) you can get an average return higher than the mean 8.3 percent by investing in a large number of stocks.
(b) as you invest in more and more stocks chosen at random, your average return on these stocks gets ever closer to 8.3 percent.
(c) if you invest in a large number of stocks chosen at random, your average return will have approximately a normal distribution.

20. In Example 14 (p. 304), $k =$ ________.

21. Figure 8.7 (p. 299) shows the density curve of a continuous probability model for choosing a number at random between 0 and 1. The mean of this model is

(a) 0.5 because the curve is symmetric.
(b) 1 because there is area 1 under the curve.
(c) impossible to figure out at this point—this requires advanced mathematics.

22. According to the law of large numbers, as the random phenomenon is repeated a large number of times, $\bar{x}$ gets closer and closer to ________.

23. The expected value is the ________ value.

(a) median
(b) mean
(c) modal

24. The mean payoff of a $\frac{1}{10}$ chance of winning \$500 (with a $\frac{9}{10}$ chance of winning \$0) is ________.

25. The density curve of a continuous probability model would balance on the ________.

(a) mode
(b) median
(c) mean

26. If $\mu = 25$, $\sigma = 16$, and $n = 64$, the standard deviation of the sampling distribution of $\bar{x}$ is ________.

27. Suppose that you are trying to decide between buying many shares of a promising individual stock and spending that same amount of money on a mutual fund consisting of a variety of different stocks. Choosing the mutual fund would result in an investment that is ________ the individual stock.

(a) more variable than
(b) less variable than
(c) as variable as

28. Scores on the SAT® Reasoning Mathematics college entrance test for the class of 2010 were roughly normal, with mean 516 and standard deviation 116. You take an SRS of 100 students and average their SAT® scores. If you do this many times, the mean of the average scores that you get from all those samples would be ________.

29. The number of hours that a light bulb burns before failing varies from bulb to bulb. The distribution of burnout times is strongly skewed to the right. The Central Limit Theorem says that

(a) as we look at more and more bulbs, their average burnout time gets ever closer to the mean μ for all bulbs of this type.

(b) the average burnout time of a large number of bulbs has a distribution of the same shape (strongly skewed) as the distribution for individual bulbs.

(c) the average burnout time of a large number of bulbs has a distribution that is close to normal.

30. Referring to Skills Check 28, the standard deviation of the average scores that you get from all those samples would be ________.

CHAPTER 8 EXERCISES

■ Challenge ▲ Discussion

8.1 Probability Models and Rules

1. Estimating probabilities empirically:

(a) Hold a penny upright on its edge under your forefinger on a hard surface, then snap it with your other forefinger so that it spins for some time before falling. Based on 50 spins, estimate the probability of heads.

(b) Toss a thumbtack with a gently curved back on a hard surface 100 times. (To speed the process up, toss 10 at a time.) How many times did it land with the point up? What is the approximate probability of landing point up? (Try similar experiments with other asymmetrical objects, such as a shoe or a spoon.)

2. While it is a less common way (than probability) of expressing likelihood and does not follow the six rules of probability, statements of "odds" are often encountered in gambling contexts. For example, the odds against an event A happening can be expressed as:

$$\frac{\text{count of outcomes in which A does not happen}}{\text{count of outcomes where A happens}}.$$

This is equivalent to the formula $P(A^C)/P(A)$. If there are 3:2 odds against a particular horse winning a race, what is the probability that the horse wins?

3. The table of random digits (Table 7.1, on p. 246) was produced by a random mechanism that gives each digit probability 0.1 of being a 0. What proportion of the first five lines in the table are 0s? This proportion is an estimate of the true probability, which in this case is known to be 0.1.

4. Probability is a measure of how likely an event is to occur. Match one of the probabilities that follow with each statement about an event. (The probability is usually a much more exact measure of likelihood than is the verbal statement.)

0, 0.01, 0.3, 0.6, 0.99, 1

(a) This event is impossible. It can never occur.

(b) This event is certain. It will occur on every trial of the random phenomenon.

(c) This event is very unlikely, but it will occur once in a while in a long sequence of trials.

(d) This event will occur somewhat more often than not.

In Exercises 5 and 6, describe a reasonable sample space *S* for the random phenomena mentioned. In some cases, you must use judgment to choose a reasonable *S*.

5. A basketball player shoots four free throws.

(a) You record the sequence of hits and misses.

(b) You record the number of shots she makes.

6. A randomly chosen subject (participant) arrives for a study of exercise and fitness.

(a) The subject is either female or male.

(b) After 10 minutes on an exercise bicycle, you ask the subject to rate his or her effort on the Rate of Perceived Exertion (RPE) scale. The RPE goes in whole-number steps from 6 (no exertion at all) to 20 (maximal exertion).

(c) You also measure the subject's maximum heart rate (beats per minute).

7. Consider flipping a dime, nickel, and penny.

(a) Why would a tree diagram be a more convenient way than a table to represent the sample space?

(b) Make a tree diagram.

(c) Use the diagram to find the probability that at least one of the three coins lands on heads.

8. The Punnett square is a diagram that biologists use to determine the probability of offspring having certain genetic makeup. Suppose that *B* represents the gene for brown eyes and *b* represents the gene for blue eyes. In genetics, capital letters refer to dominant traits, so a person receiving both *B* and *b* generally has brown eyes.

This diagram shows the possibilities for the child of two Bb parents. Each parent gives the child one of its two genes with equal probability. What is the probability that this child will receive the genetic makeup for brown eyes? Discuss how this relates to Example 2 (p. 284).

	Mother gives B	Mother gives b
Father gives B	BB	Bb
Father gives b	Bb	bb

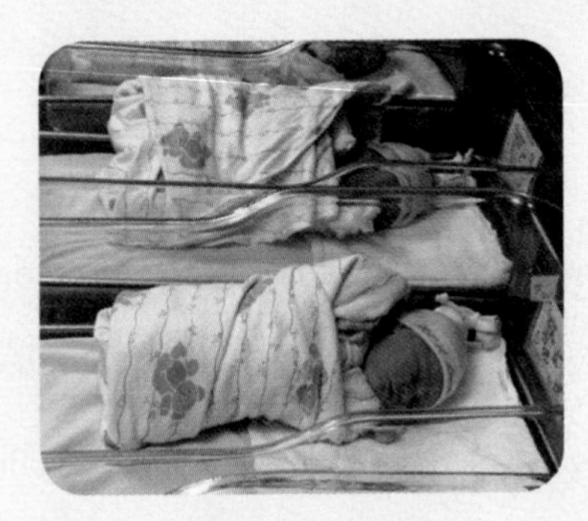

(Lori Adamski Peek/STONE/Getty Images.)

9. We noted that a question posed to Blaise Pascal in 1654 by an experienced gambler launched the formal study of probability. Here's a simplified version of this "Problem of Points." Suppose two players are playing a coin flip game where "heads" earns Player A one point and "tails" earns Player B one point. The winner is the first player to reach a total of four points. The game is interrupted with Player A ahead by a score of 3 to 2. Based on the sample space of possible ways that the game can be finished, what would be a fair division of the jackpot money between players A and B?

10. Choose a young adult (aged 25 to 34 years) at random. The probability is 0.12 that the person chosen did not complete high school, 0.31 that the person has a high school diploma but no further education, and 0.29 that the person has at least a bachelor's degree.

(a) What must be the probability that a randomly chosen young adult has some education beyond high school but does not have a bachelor's degree?

(b) What is the probability that a randomly chosen young adult has at least a high school education?

■ **11.** What is the probability that Laurie rolls doubles (both dice match) each of her first three rolls in the game of Monopoly®? (This matters because rolling three consecutive doubles in that game sends you right to jail!)

12. If you are in the championship round of a tournament and you are better than your opponent, are you better off having the championship determined by playing a single game or by playing a best-two-out-of-three-game series? You might try intuition, simulation, or a tree diagram.

13. Weathering the storm: Suppose there is a forecast of a 70 percent chance of rain. Using the "long run" interpretation of the definition of probability, if it does not rain the next day, is it appropriate to say that the forecaster was "wrong"? Explain.

8.2 Discrete Probability Models

■ **14.** Role-playing games like Dungeons & Dragons® use many different types of dice. One type of die has a tetrahedral (pyramidal) shape with four triangular faces. Each triangular face has a number (1, 2, 3, or 4) next to each of its edges. Because the top of this die is not a face but a point, the way to read it is by the number at the bottom of the face that is visible when the die comes to rest. Suppose that the intelligence of a character is determined by rolling this four-sided die twice and adding 1 to the sum of the results.

(a) Give a probability model for the character's intelligence. (Start with a display in the style of Figure 8.2 on p. 285 adapted for the outcomes of the two rolls of the four-sided die. These outcomes are equally likely.)

(b) What is the probability that the character has intelligence 7 or higher?

15. North Carolina State University posts the grade distributions for its courses online. Students in Statistics 101 in a recent semester earned 21 percent As, 43 percent Bs, 30 percent Cs, 5 percent Ds, and 1 percent Fs. Here is the probability model for the grade of a randomly chosen Statistics 101 student.

Grade	0 (=F)	1 (=D)	2 (=C)	3 (=B)	4 (=A)
Probability	0.01	0.05	0.30	0.43	0.21

(a) Make a probability histogram for this model. Does it have the shape of a normal distribution?

(b) What is the probability that the student got a grade of B or better?

16. How do rented housing units differ from units occupied by their owners? Here are probability models for the number of rooms for owner-occupied units and renter-occupied units, according to the Census Bureau:

# of Rooms	1	2	3	4	5
Owned	0.000	0.001	0.014	0.099	0.238
Rented	0.011	0.027	0.229	0.348	0.224

# of Rooms	6	7	8	9	10
Owned	0.266	0.178	0.107	0.050	0.047
Rented	0.105	0.035	0.012	0.004	0.005

Make probability histograms of these two models, using the same scale. What are the most important differences between the models for owner-occupied and rented housing units?

17. In each of the following situations, state whether or not the given assignment of probabilities to individual outcomes is legitimate; that is, satisfies the rules of probability. If not, give specific reasons for your answer.

(a) Choose a college student at random and record gender and enrollment status: $P(\text{female full-time}) = 0.56$, $P(\text{female part-time}) = 0.24$, $P(\text{male full-time}) = 0.44$, $P(\text{male part-time}) = 0.17$.

(b) Choose a college student at random and record the season of that student's birth: $P(\text{spring}) = 0.39$, $P(\text{summer}) = 0.28$, $P(\text{fall}) = 0$, $P(\text{winter}) = 0.33$.

18. What is the probability that a housing unit has five or more rooms? Use the models in Exercise 16 to answer this question for both owner-occupied and rented units.

■ **19.** Balanced six-sided dice with altered labels can produce interesting distributions of outcomes. Construct the probability model (sample space and assignment of probabilities for each sum) for rolling the dice that is featured in Joseph Gallian's article "Weird Dice" in the February 1995 issue of *Math Horizons*. Instead of using the regular values {1, 2, 3, 4, 5, 6}, one die has the labels 1, 2, 2, 3, 3, 4, and the other die has the labels 1, 3, 4, 5, 6, 8. How does this model compare to the model for regular dice?

8.3 Equally Likely Outcomes

20. If you play the lottery, there are two possibilities—you could either win or not win. Explain whether or not this means that you have a 1 out of 2 chance (i.e., a 50 percent probability) of winning.

21. A party host gives a door prize to one guest chosen at random. There are 42 men and 48 women at the party. What is the probability that the prize goes to a woman?

22. Abby, Boaz, Carmen, Dani, and Eduardo work in a firm's public relations office. Their employer must choose two of them to attend a conference in Paris. To avoid unfairness, the choice will be made by drawing two names from a hat. (This is an SRS of size 2.)

(a) Write down the sample space of all possible choices of two of the five names.

(b) The random drawing makes all choices equally likely. What is the probability of each choice?

(c) What is the probability that Abby is chosen?

(d) What is the probability that neither of the two men (Boaz and Eduardo) is chosen?

23. You toss a balanced coin 10 times and write down the resulting sequence of heads and tails, such as HTTTHHTHHH.

(a) How many possible outcomes are there for the 10 tosses?

(b) What is the probability that your 10-toss sequence is either all heads or all tails?

24. In the Texas Hold 'Em style of poker, play begins with each player being dealt two cards face down. From a standard 52-card deck, how many possible 2-card hands could be dealt to you?

25. A computer assigns three-character log-in IDs that may contain the digits 0 to 9 as well as the letters *a* to *z*, with repeats allowed.

(a) What is the probability that your ID contains no *x*?

(b) What is the probability that your ID contains no digits?

26. Consider a typical combination lock on a locker or briefcase.

(a) If you ask for the three numbers in the combination needed to open the lock, and they are given to you in numerical order as 3–5–8, why is this not enough information to open the lock?

(b) What would be the probability that you could open the lock with one try?

(c) Is such a combination lock accurately named, or is it really a "permutation lock"?

27. You may have heard that a monkey hitting keys at random on a typewriter keyboard for an infinite amount of time could eventually type a particular chosen text, such as the complete works of Shakespeare. Let's focus on a monkey who just types the letters *a*, *p*, and *s* in random order in three-letter sequences.

(a) How many possible three-letter "words" can the monkey type using only these letters?

(b) Which of these are words in an English dictionary?

(c) What is the probability that the word the monkey typed is in an English dictionary?

■ **28.** Mozart composed a 16-bar Viennese minuet ("Musical Dice Game") in which bars 1–7 each have 11 choices, bar 8 has 2, bars 9–15 each have 11, and bar 16 has 1. How many possible versions of this minuet are there?

■ **29.** In poker, a royal flush is a 5-card hand containing (in any order) an ace, king, queen, jack, and 10, all of the same suit.

(a) How many royal flush hands are possible?

(b) What is the number of 5-card hands possible from a 52-card deck?

(c) What is the probability that 5 cards drawn at random from a 52-card deck will yield a royal flush?

30. Biblical permutations: The King James Version of the Old Testament has its 39 books canonized in a different order than the Hebrew Bible does. What mathematical expression would yield the number of possible orders of these 39 books? Is this number larger than you expected?

8.4 Continuous Probability Models

31. Generate two random real numbers between 0 and 1 and take their sum. The sum can take any value between 0 and 2. The density curve is the shaded triangle shown in Figure 8.13.

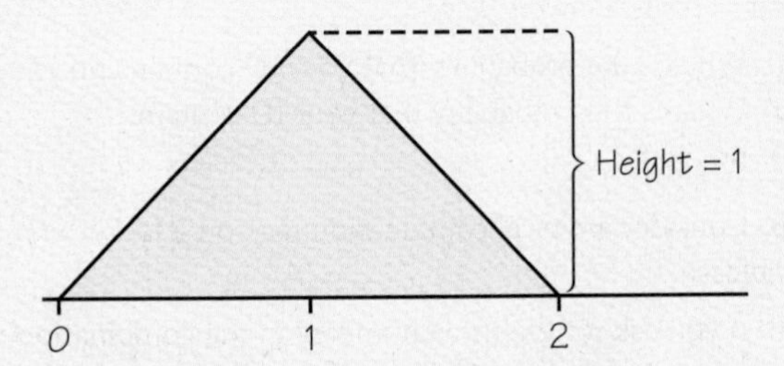

Figure 8.13 The density curve for the sum of two random numbers, for Exercise 31.

(a) Verify by geometry that the area under this curve is 1.
(b) What is the probability that the sum is less than 1? (Sketch the density curve, shade the area that represents the probability, and then find that area. Do this for part (c) as well.)
(c) What is the probability that the sum is less than 0.5?

32. Suppose two data values are each rounded to the nearest whole number. Make a density curve for the sum of the two roundoff errors (assuming each error has a continuous uniform distribution).

33. On the TV show *The Price Is Right*, the "Range Game" involves a contestant being told that the suggested retail price of a prize lies between two numbers that are $600 apart. The contestant has one chance to position a red window with a span of $150 that will contain the price. On one episode, the price of a piano is between $8900 and $9500. If we assume a uniform continuous distribution (i.e., that all prices within the $600 interval are equally likely), what is the probability that the contestant will be successful?

8.5 The Mean and Standard Deviation of a Probability Model

34. You have a campus errand that will take only 15 minutes. The only parking space anywhere nearby is a faculty-only space, which is checked by campus police about once every hour. If you're caught, the fine is $25.

(a) Give the probability model for the money that you may or may not have to pay.
(b) What's the expected value of the money that you will pay for your unauthorized parking?

35. Exercise 15 gives a probability model for the grade of a randomly chosen student in Statistics 101 at North Carolina State University, using the 4-point scale. What is the mean grade in this course? What is the standard deviation of the grades?

36. In Exercise 14, you gave a probability model for the intelligence of a character in a role-playing game. What is the mean intelligence for these characters?

37. Exercise 16 gives probability models for the number of rooms in owner-occupied and rented housing units. Find the mean number of rooms for each type of housing. Make probability histograms for the two models and mark the mean on each histogram. You see that the means describe an important difference between the two models: Owner-occupied units tend to have more rooms.

38. Typographical and spelling errors can be either "nonword errors" or "word errors." A nonword error is not a real word, as when "the" is typed as "teh." A word error is a real word, but not the right word, as when "lose" is typed as "loose." When undergraduates write a 250-word essay (without spell-checking), the number of nonword errors has this probability model:

Errors	0	1	2	3	4
Probability	0.1	0.2	0.3	0.3	0.1

The number of word errors has this model:

Errors	0	1	2	3
Probability	0.4	0.3	0.2	0.1

(a) What is the mean number of nonword errors in an essay?
(b) What is the mean number of word errors in an essay?
(c) How does the difference between the means describe the difference between the two models?

39. Find (and explain how you found) the mean for:
(a) the continuous probability model in Exercise 31.
(b) the probability model in Exercise 32.

40. The idea of insurance is that we all face risks that are unlikely but carry a high cost. Think of a fire destroying your home. Insurance spreads the risk: We all pay a small amount, and the insurance policy pays a large amount to those few of us whose homes burn down. An insurance company looks at the records for millions of homeowners and sees that the mean loss from fire in a year is $\mu = \$250$ per person. (The great majority of us have no loss, but a few lose their homes. The \$250 is the average loss.) The company plans to sell fire insurance for \$250 plus enough to cover its costs and profit. Explain clearly why it would be unwise to sell only 12 policies. Then explain why selling thousands of such policies is a safe business.

▲ **41.** Should you buy the extended warranty on a new washing machine? Suppose there are two outcomes—an 85 percent probability of needing no repairs, and a 15 percent probability of needing a \$200 repair during the warranty period. Based on the mean outcome for this model, what would be a "break-even" price to you for the extended warranty? (The company, of course, will charge more than this so it can make a profit.)

42. An American roulette wheel has 38 slots numbered 0, 00, and 1 to 36. The ball is equally likely to come to rest in any of these slots when the wheel is spun. The slot numbers are laid out on a board on which gamblers place their bets. One column of numbers on the board contains multiples of 3—that is, 3, 6, 9, . . ., 36. Joe places a \$1 "column bet" that pays out \$3 if any of these numbers comes up.

(a) What is the probability model for the outcome of one bet, taking into account the \$1 cost of a bet?

(b) What are the mean and standard deviation for this model?

(c) Joe plays roulette every day for years. What does the law of large numbers tell us about his results?

43. This table shows the prizes and respective probabilities for a lottery:

Net prize	\$1,000,000	\$1000	\$100	\$4
Probability	$\frac{1}{10{,}000{,}000}$	$\frac{1}{10{,}000}$	$\frac{1}{1{,}000}$	$\frac{3}{100}$

On average, how much money from a \$1 ticket comes back to you in prizes?

■ **44.** A friend cuts your cake into two pieces: one is $\frac{1}{3}$ of the cake and the other is $\frac{2}{3}$.

(a) If you flip a coin to decide which piece is yours, what is the expected value of the proportion of the original cake that you will get?

(b) More generally, what is the expected value of your share of the cake if the pieces' proportions are p and $1 - p$?

■ **45.** On five-choice questions on the SAT®, you get 1 point for a correct answer and lose $\frac{1}{4}$ point for a wrong answer.

(a) Find the expected value of a completely random guess on such a question. Does guessing in this situation help you, hurt you, or make no difference?

(b) Suppose you eliminate one of the five choices as definitely not being the correct answer, and then randomly guess among the remaining four choices. Does guessing in this situation help you, hurt you, or make no difference?

■ **46.** In August 2006, El Paso had a storm that was called a "500-year flood."

(a) What is the expected value of the number of 500-year floods in a 1000-year period?

(b) After the city moved to raise money to guard against such a major flood in the future, a city council representative was quoted as protesting, "But we still have 490 more years to deal with this." What false assumption was he making?

Richard McMillin/Shutterstock

8.6 The Central Limit Theorem

47. Newly manufactured automobile radiators may have small leaks. Most have no leaks, but some have one, two, or more. The number of leaks in radiators made by one supplier has mean 0.15 and standard deviation 0.4. The distribution of the number of leaks cannot be normal because only whole-number counts are possible. The supplier ships 400 radiators per day to an auto assembly plant. Take $\bar{x}$ to be the mean number of leaks in these 400 radiators. Over several years of daily shipments, what interval of values will contain the middle 95 percent of the many $\bar{x}$ values?

48. The scores of eighth-grade students on the National Assessment of Educational Progress (NAEP) mathematics test in 2007 have a distribution that is approximately normal, with mean $\mu = 281$ and standard deviation $\sigma = 35$.

(a) Choose one eighth-grader at random. What is the probability that his or her score is higher than 281? Higher than 316?

(b) Now choose an SRS of four eighth-graders. What is the probability that their mean score is higher than 281? Higher than 316?

▲ **49.** Antonio measures the alcohol content of whiskey for his Chemistry 101 lab. He actually measures the mass of 5 milliliters (ml) of whiskey—a chemical calculation—and then finds the percentage of alcohol from the mass. The standard deviation of students' measurements of mass is $\sigma = 10$ milligrams (mg). Antonio repeats the measurement three times and records the mean $\bar{x}$ of his three measurements.

(a) What is the standard deviation of Antonio's mean result?

(b) How many times must Antonio repeat the measurement to reduce the standard deviation of $\bar{x}$ to 5 mg? Explain to someone who knows nothing about statistics the advantage of reporting the average of several measurements rather than the result of a single measurement.

50. In Exercise 42, you found the mean and standard deviation of the outcome of a column bet in roulette. The Central Limit Theorem says that the average outcome of a large number of bets has a distribution that is close to normal.

(a) What is the 99.7 percent confidence interval estimate (mean ± 3 standard deviations) of a gambler's average winnings after 100 bets?

(b) What is the 99.7 percent confidence interval estimate of a gambler's average winnings after 1000 bets?

51. Averages of several measurements are less variable than individual measurements. The true mass of the whiskey sample in Exercise 49 is 4.6 grams, or 4600 mg. Antonio's measurements have the normal distribution with mean 4600 mg and standard deviation 10 mg. In this case, the mean of his three measurements also has a normal distribution.

(a) Sketch on the same graph the two normal curves, for individual measurements and for means of three measurements. Figure 8.10 (p. 306) is an example of this kind of graph.

(b) What interval of values covers the middle 95 percent of Antonio's measurements?

(c) What interval of values covers the middle 95 percent of averages of three measurements?

52. Exercise 14 gives the probability model for the intelligence assigned by chance to a character in a role-playing game. You found the mean intelligence of such characters in Exercise 36. Jermaine plays this character often. What interval covers (approximately) the middle 68 percent of average intelligence scores for 100 of Jermaine's games?

53. The scores of high school seniors on the ACT® college entrance examination in 2010 were roughly normal with mean $\mu = 21.0$ and standard deviation $\sigma = 5.2$.

(a) What is the approximate probability that a single student randomly chosen from all those taking the test scores 26.2 or higher?

(b) Now take an SRS of nine students who took the test. What are the mean and standard deviation of the sample mean score $\bar{x}$ of these nine students?

(c) What is the approximate probability that the mean score $\bar{x}$ of these nine students is 26.2 or higher?

■ **54.** Although cities encourage carpooling to reduce traffic congestion, most vehicles carry only one person. For example, 70 percent of vehicles on the roads in the Minneapolis–St. Paul metropolitan area are occupied by just the driver. You choose 84 vehicles at random.

(a) What are the mean and standard deviation of the proportion of vehicles in your sample that carry only one person?

(b) What is the probability that more than 60 percent of the vehicles in your sample carry only one person? (Use the Central Limit Theorem.)

55. Among high-performing schools (and low-performing schools), there is an unrepresentatively _____ proportion of smaller schools. Explain whether you would complete this sentence with the word "high" or "low" in light of the formula for the standard deviation of the sampling distribution of the mean.

Chapter Review

56. Combination connections:

(a) Give either an intuitive or an algebraic argument to explain why ${}_nC_k = {}_nC_{n-k}$.

(b) Generate several small values of ${}_nC_k$ and explain how they relate to the numbers in Pascal's Triangle (see for example, Chapter 11, page 396).

57. License plates in Florida have the form A12BCD; that is, a letter followed by two digits followed by three more letters.

(a) How many possible different license plates are there?

(b) Jerry would like a plate that ends in AAA. How many such plates are there?

(c) If license plates are issued at random from all possible plates, what is the probability that Jerry will get a plate that ends in AAA?

58. After you tell Jerry the probability that you calculated in the previous exercise, he realizes that he's unlikely to get a plate ending in AAA. So he asks you, "What's the probability I will get a plate in which all four letters are from my name?" These letters are J, E, R, and Y.

(a) Suppose that Jerry insists that the letters appear in order, so that his plate reads J*nm*ERY, where *n* and *m* stand for any number. What is the probability?

(b) Suppose Jerry allows his letters to appear in the plate in any order and also allows repeats. What is now the answer to Jerry's question?

59. Choose a person aged 19–25 years at random and ask, "In the past four days, how many days did you do physical exercise or work out?"

Based on a large sample survey, here is a discrete probability model for the answer you will get:

# of Days	0	1	2	3	4
Probability	0.61	0.17	0.10	0.08	0.04

(a) What is the probability that the person you chose worked out either two or three days in the past four?

(b) What is the probability that the person you chose worked out at least one day in the past four?

60. What is the mean number of days that randomly chosen 19- to 25-year-olds worked out in the past four days? (Use the information in the previous exercise.) If you interview many people in this age group, what does the law of large numbers say about the average number of days that these people work out?

■ **61.** Use the information in Exercise 59 and your result from Exercise 60 to answer these questions.

(a) What is the standard deviation of the number of days in the past four that a randomly chosen 19- to 25-year-old has worked out?

(b) You interview 100 randomly chosen 19- to 25-year olds. You ask each how many days in the past four he or she has worked out and you calculate the average number of days. According to the Central Limit Theorem, there is a probability of 0.95 that your average will fall between what two values?

62. In Example 16 (p. 307), we saw that a $1 bet on red has a mean outcome of –$2/38. It turns out not all $1 bets in American roulette have the same mean outcome. The "5-number bet" {0, 00, 1, 2, 3} pays an additional $6 if one of those 5 numbers comes up—otherwise, the player loses his $1.

(a) Find the expected value for this 5-number bet.

(b) Is this 5-number bet better or worse than a bet on red?

63. Suppose you select 10 people at random. Find the probability of each event below:

(a) At least one match in the day of the week that they were born.

(b) At least one match in the day of the month that they were born (assume 31 days per month).

(c) At least one match in the day of the year that they were born.

APPLET EXERCISES

To do these exercises, go to www.whfreeman.com/fapp9e.

1. When we toss a coin, experience shows that the probability (long-term proportion) of a head is close to $\frac{1}{2}$. Suppose now that we toss the coin repeatedly until we get a head. What is the probability that the first head comes up in an odd number of tosses (1, 3, 5, and so on)? Use the *Probability* applet to estimate this probability. Set the probability of heads to 0.5. Toss coins one at a time until the first head appears. Do this 50 times (click "Reset" after each trial). What is your estimate of the probability that the first head appears on an odd toss?

2. The table of random digits (Table 7.1, p. 246) was produced by a random mechanism that gives each digit probability 0.1 of being a 0.

(a) What proportion of the digits in the first row of Table 7.1 are 0s? This proportion is an estimate, based on 40 repetitions, of the true probability, which in this case is known to be 0.1.

(b) The *Probability* applet can imitate random digits. Set the probability of heads in the applet to 0.1. Check "Show true probability" to show this value on the graph. A head stands for a 0 in the random digit table and a tail stands for any other digit. Simulate 200 digits (40 at a time—don't click "Reset"). If you kept going forever, presumably you would get 10 percent heads. What was the percent of heads in your 200 tosses?

3. One of the few players to have a better field goal percentage than free throw percentage, basketball star Shaquille O'Neal made about half (52.7%) of his free throws in his 21-year NBA career. Use the *Probability* applet to simulate 100 free throws shot independently by a player who has probability 0.527 of making each shot. (Toss 40, 40, and 20 without clicking "Reset.")

(a) What percentage of the 100 shots were made?

(b) Examine the sequence of hits and misses after each click on "Toss" and keep track of the longest run of shots made and the longest run of shots missed. How long were the longest runs in the 100 shots taken? (Sequences of random outcomes often show longer runs than our intuition expects.)

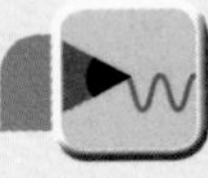

WRITING PROJECTS

1. Psychologists have shown that our intuitive understanding of chance behavior is rather poor. Amos Tversky (1937–96) was a leader in the study of how we make decisions in the face of uncertainty. In its obituary of Tversky, the *New York Times* cited the following example:

> *Tversky asked subjects to choose between two public health programs that affect 600 people. One had a probability of $\frac{1}{2}$ of saving all 600 and a probability of $\frac{1}{2}$ that all 600 will die. The other was guaranteed to save exactly 400 of the 600 people. Most people chose the second program. He then offered a different choice. One program had a probability of $\frac{1}{2}$ of saving all 600 and a probability of $\frac{1}{2}$ of losing all 600, while the other would definitely lose exactly 200 lives. Most people chose the first program.*

Discuss this example. What is the difference between the two choices offered? What is the mean number of people saved by the two options in each choice? What do the reactions of most subjects to these choices show about how people make decisions?

2. There are about 1×10^{44} air molecules in the atmosphere and about 2×10^{22} molecules of air in a single breath taken at rest. What is the probability that the breath that you took just now contained at least one molecule of air that was exhaled by Pythagoras in his last breath? What probability rules did you use to calculate this? What assumptions did you make, and why do you think they were reasonable?

3. Double or Nothing: Gambler's Ruin. We have seen that by betting on "red" in American roulette, you have a $\frac{18}{38}$ chance of winning, therefore doubling the money you bet. Suppose that you have \$5 and you want to bet until either you reach (and stop with) \$10 or you go broke. Is placing individual \$1 bets on red more, less, or equally likely to reach this goal than just placing a single \$5 bet? First, try to give an answer based on intuition, taking into account the casino's advantage.

You could also explore the following formula that gives the probability of going from h dollars to N dollars without going broke by making \$1 bets on red in American roulette:

$$\frac{1 - (20/18)^h}{1 - (20/18)^N}$$

Discuss how the strategy for maximizing the chance of reaching a financial target compares to the strategy for maximizing the length of time that your money lasts (for entertainment value).

Suggested Readings

COMAP. *Principles and Practice of Mathematics*, Springer, New York, 1997. Chapters 4 and 8 present combinatorics and probability at the next level beyond this book.

LESSER, LAWRENCE M. Take a chance by exploring the statistics in lotteries. *Statistics Teacher Network*, 65 (2004): 6–7. This article shows how lotteries can illustrate all major topics of an introductory statistics course, using a graphing calculator. The article is available online at: **www.amstat.org/education/stn/pdfs/STN65.pdf.**

MOSTELLER, FREDERICK, ROBERT E. K. ROURKE, and GEORGE B. THOMAS. *Probability with Statistical Applications*, Addison-Wesley, Reading, MA, 1970. A rich treatment of basic probability that requires only high school algebra but is somewhat sophisticated. Although out of print, this book is a classic that deserves mention.

WAINER, HOWARD. The most dangerous equation. *American Scientist, 95*(3) (2007): 249–56. This article explores many real-world examples of the standard deviation of the sampling distribution of the sample mean.

Suggested Web Sites

"Buffon's needle" is a probability problem first stated in 1777 by Count Buffon: If you drop a needle on a sheet of lined paper, what is the probability that the needle crosses one of the lines? In the simplest case, the length of the needle is the same as the distance between the lines. Some fairly advanced math shows that the answer is $2/\pi$, or about 0.637. A number of Web sites simulate dropping a needle many times to estimate this probability.

A simulation of this activity, as well as many other probability exercises, can be found by clicking on "Probability" on the web site **www.shodor.org/interactivate/activities/**.

You may also be interested in the debate over legalized gambling. For the case against this practice, visit the National Coalition Against Legalized Gambling at **www.ncalg.org**. For the defense by the casino industry, visit the American Gaming Association at **www.americangaming.org**. For an example of a non-casino game involving probability and expected value, try playing the dice game "Skunk" at **illuminations.nctm.org/LessonDetail.aspx?id=L248**.

Photo by Marc Serota/Getty Images

Part III

Voting and Social Choice

The application of mathematics to the study of human beings—their behavior, values, interactions, conflicts, and methods of making decisions—is generally considered to be a recent revolution. Yet the study of voting and social choice, which is very much the root of this revolution, goes back several centuries.

We begin in Chapter 9 with the question of how a group of individuals, each with his or her own set of values, selects one outcome from a list of possibilities. While majority rule is a good system for deciding an election with just two candidates, it turns out that there is no perfect way of deciding an election in which there are three or more candidates.

Group decision-making is often a strategic encounter, and citizens need to be aware of the difficulties that can arise when some participants have an incentive to manipulate the outcome. We turn to this issue in Chapter 10.

In Chapter 11, we consider decision-making bodies in which the individual voters or parties do not have equal power. In particular, we look at weighted voting systems in which a voter's power need not be proportional to the number of votes that he or she is entitled to cast.

In Chapter 12, we analyze not only how the Electoral College influences resource allocation in a campaign but also how polls and the positioning of candidates on a left–right continuum affect the strategies of candidates and the choices of voters.

9

Social Choice: The Impossible Dream

The basic question of social choice, of how groups can best arrive at decisions, has occupied social philosophers and political scientists for centuries. One primary example of a social-choice problem is the selection of a "good" voting system. Indeed, voting is a subject that lies at the very heart of representative government and participatory democracy.

Social-choice theory attempts to address the problem of finding good procedures that will turn individual preferences for different candidates into a single choice by the whole group. An example of such a choice would be the selection of the *winner* of an election. The goal is to find such procedures that will result in an outcome that "reflects the will of the people."

This search for good voting systems, as we shall see, is plagued by a variety of counterintuitive results and disturbing outcomes. In fact, it turns out that one can prove (mathematically) that no one will ever find a completely satisfactory voting system in an election with three or more candidates.

After introducing social choice in Section 9.1, we turn in Section 9.2 to majority rule (for elections in which there are two candidates) and Condorcet's method (for elections in which there are three or more candidates). While majority rule for two candidates has a number of desirable properties, Condorcet's method leads to something known as the "voting paradox." We then move on to Section 9.3 to explore a number of other voting systems and a number of desirable properties. In Section 9.4, we ask if we can "have it all" in terms of the desirable properties that have been discussed. We conclude in Section 9.5 with a voting system known as "approval voting."

9.1 An Introduction to Social Choice

The elections with which we are most familiar often involve only two candidates; however, there are real-world situations in which elections must be held to choose a single winner from among three or more candidates, as in the presidential election of 2008 in which Barack Obama, John McCain, Ralph Nader, Bob Barr, Chuck Baldwin, and Cynthia McKinney were the candidates (although the latter four received only 1,623,550 votes among them, nationwide).

There are several methods that can be used to elect a single candidate from a choice of three or more, and we will investigate some of them in this chapter. Most of these methods use a ballot in which a voter provides a rank ordering of the candidates (without ties) that indicates the order in which he or she prefers them.

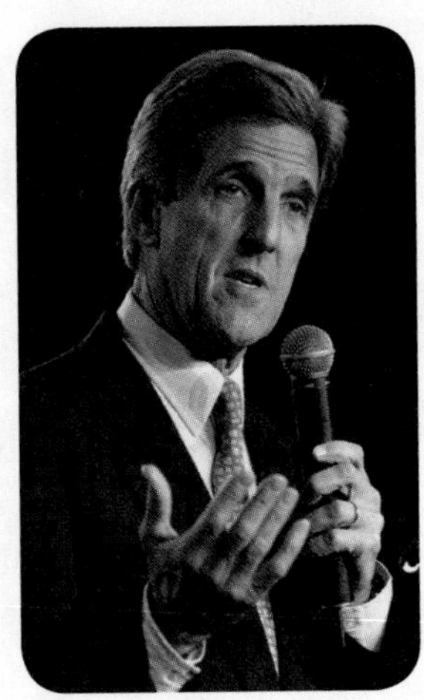

In the 2004 election, President George W. Bush was challenged by Massachusetts senator John Kerry. President Bush received a slim majority of the vote. The election results have been dogged by charges of irregularities in the important swing state of Ohio. *(Left: Reuters/Corbis; right: John Gress/Reuters/Corbis.)*

Preference List Ballot — DEFINITION

A ballot consisting of such a rank ordering of candidates (which we often picture as a vertical list with the most preferred candidate on top and the least preferred on the bottom) is called a **preference list ballot** because it is a statement of the preferences of the individual who is voting.

Preference list ballots allow voters to make a much clearer statement of their preferences than do ballots allowing a single vote. Preference list ballots are already used in a wide range of applications, such as rating football teams and scoring track meets.

Although we do not allow ties in a preference list ballot, most voting rules of interest will, in some elections, result in a tie for the win among two or more of the candidates. In the real world, the number of voters is often so large that ties seldom occur. Nevertheless, to avoid excessive annoyances in the theory that we develop, and to simplify what we do in this chapter, we make the following assumption throughout:

The Number of Voters Assumption — RULE

Throughout this chapter, we consider only elections in which there is an odd number of voters.

With this in hand, we are ready to begin our study of social choice.

9.2 Majority Rule and Condorcet's Method

When a choice is being made between two candidates, the first type of voting system to suggest itself is **majority rule**: Each voter indicates a preference for one of the two candidates, and the one with the most votes wins. With two candidates, there is no real distinction between a ballot that indicates a voter's choice for one of the two candidates and what we have called a preference list ballot. The point is that we can, for example, identify a choice for A (however indicated) with the list that has A over B and a choice B with the list that has B over A.

Majority rule has at least three desirable properties:

1. All voters are treated equally. That is, if any two voters were to exchange (marked) ballots before submitting them, the outcome of the election would be the same.

2. Both candidates are treated equally. That is, if a new election were held and every voter were to reverse his or her vote, then the outcome of the previous election would be reversed as well.
3. It is **monotone**. That is, if some candidate X is a winner of an election, and a new election is held in which the only ballot change made is for some voter to change his or her ballot from not being a vote for X to being a vote for X, then X will remain the winner.

It is easy to devise voting systems for two candidates in which these properties fail, but each such voting system quickly reveals its undesirability. For example, condition 1 is not satisfied by a *dictatorship* (in which all ballots except that of the dictator are ignored); condition 2 is not satisfied by *imposed rule* (in which Candidate X wins regardless of who votes for whom); and condition 3 is not satisfied by *minority rule* (in which the candidate with the fewest votes wins).

But maybe there are voting systems in the two-candidate case that are superior to majority rule in the sense of satisfying the three properties just listed *and* some other properties that we might also wish to have satisfied. This, however, turns out not to be the case. In 1952, Kenneth May proved the following:

May's Theorem THEOREM

Among all two-candidate voting systems that never result in a tie, majority rule is the *only* one that treats all voters equally, treats both candidates equally, and is monotone.

This is an important and elegant result. Thus, mathematical reasoning spares us the trouble of searching for a better voting system for two candidates.

But what if there are three or more candidates? Perhaps we can design a voting system for this situation that, in some way, builds on the success of majority rule in the two-candidate case. In point of fact, there does exist a voting system that arises from precisely this hope, and it is known today as **Condorcet's method.**

Our description of Condorcet's method begins with the observation that if we have a sequence of preference list ballots, then—for each pair of candidates—we can determine who the winner would have been had the election involved only these two in a one-on-one contest using majority rule.

To illustrate this notion of a one-on-one contest, consider the following preference list ballots:

Rank	Number of Voters (3)		
First	A	B	C
Second	B	C	A
Third	C	A	B

In this election, Candidate A would defeat Candidate B in a one-on-one contest (two votes to one), while B would, in turn, defeat C in a one-on-one contest, again by a score of 2 to 1. We'll return to this example in a moment, but we now have at hand all we need to describe Condorcet's voting system for three or more candidates.

Description of Condorcet's Method PROCEDURE

With the voting system known as Condorcet's method, a candidate is a winner precisely when he or she would, on the basis of the ballots cast, defeat every other candidate in a one-on-one contest using majority rule.

Historically, the voting system we are calling Condorcet's method dates back at least to Ramon Llull in the 13th century (see Spotlight 9.1). It was rediscovered and popularized in the 18th century by the Marquis de Condorcet (1743–1794).

EXAMPLE 1
Condorcet's Method

Suppose we have four candidates (*GB*, *AG*, *RN*, and *PB*, with these initials chosen for a soon-to-be-revealed reason) and the following sequence of preference list ballots, where the heading of "6" indicates that 6 of the 15 voters hold the ballot with *GB* over *AG* over *PB* over *RN*, the heading of "5" indicates that 5 of the 15 voters hold the ballot with *AG* over *RN* over *GB* over *PB*, and so on.

	Number of Voters (15)			
Rank	6	5	3	1
First	GB	AG	RN	PB
Second	AG	RN	AG	GB
Third	PB	GB	GB	AG
Fourth	RN	PB	PB	RN

We claim that *AG* is the winner in this election if we use Condorcet's method. Let's check the one-on-one scores for each possible pair of opponents:

AG versus *GB: AG* is over *GB* on 5 + 3 = 8 of the ballots, while the reverse is true on 6 + 1 = 7 of the ballots. Thus, *AG* defeats *GB* by a score of 8 to 7.

AG versus *RN: AG* is over *RN* on 6 + 5 + 1 = 12 of the ballots, while the reverse is true on 3 of the ballots. Thus, *AG* defeats *RN* by a score of 12 to 3.

AG versus *PB: AG* is over *PB* on 6 + 5 + 3 = 14 of the ballots, while the reverse is true on 1 of the ballots. Thus, *AG* defeats *PB* by a score of 14 to 1.

This shows that *AG* is the winner using Condorcet's method.

Like majority rule, Condorcet's method satisfies some very desirable properties, as we'll see later in this section. But it also has a tragic flaw, and this flaw is called **Condorcet's voting paradox**.

Condorcet's Voting Paradox THEOREM

With three or more candidates, there are elections in which Condorcet's method yields *no* winners. In particular, the following ballots (often called the "Condorcet voting paradox ballots") constitute an election in which Condorcet's method yields no winner.

Rank	Number of Voters (3)		
First	A	B	C
Second	B	C	A
Third	C	A	B

SPOTLIGHT 9.1

The Historical Record

The following letter was written by Friedrich Pukelsheim of the University of Augsburg, Germany. He is imagining what Ramon Llull (1232–1316) might say if he were alive today.

Dear Editors:

It is my distinct pleasure to respond "from the beyond" to your kind invitation to set the historical record straight. I was born in 1232 on the Island of Mallorca in the Mediterranean Sea, which in your times is known as a popular tourist place. In my days it was a strong political center of that part of the world, with a population that was a mix of Christians, Jews, and Muslims. It was my dream to persuade people of the virtues of Christian belief by relying, not on force, but on reason.

Unfortunately, people did not find it easy to follow my arguments, so I was more than pleased to discover some down-to-earth applications, including an election system. My idea was to oppose every pair of candidates, one on one, and ask the electors whom of the two they would prefer—very much like a medieval jousting tournament. But how to combine the results from all the duels into a winner of the election? I first proposed electing the candidate who won the most duels, then later suggested a system of successive eliminations.

I wrote three papers on the topic, the second of which I "smuggled" into my novel *Blanquerna* in 1283. More than a century after my death, in 1428, the young German scholar Nicolaus Cusanus (1401–1464) journeyed to Paris to read my works in libraries there. He even copied out the third of my electoral writings, which I had completed on 1 July 1299 in Paris, and his manuscript is the only copy handed down to your days. Reading my papers, Cusanus was inspired to invent his own electoral system. Did he not understand mine, or just find it inadequate? Who knows?

While I had been concerned with electing church officials, Cusanus sought a system to elect the Holy Roman Emperor. In his system, each elector assigns each candidate a rank score, with the lowest candidate getting a score of 1, the second lowest a score of 2, and the best candidate the highest score possible, that is, 10 when there are 10 candidates. The scores are totaled for each candidate and the candidate with the highest score wins. If you are a soccer player or a hockey player, you will have a good sense for one difference between our systems: Whereas I count victories, Cusanus adds up goals. Cusanus applauded himself for having invented an absolutely ingenious and novel electoral system.

Also, I advocated open voting, whereas Cusanus favored a secret ballot. He was concerned that voters might sell their votes, or that the candidates might pressure the voters. Well, that certainly happened all of the time in elections for worldly authorities! But for election to clerical office, I thought it good enough if electors took an oath to vote for the most worthy candidate and submitted themselves to the social control that comes with an open election.

Cusanus was famous in his times, as I was in mine, but fame indeed is transitory. Sure enough, my electoral system was reinvented by the Marquis de Condorcet (1743–1794), and Cusanus's system was proposed afresh by the Chevalier de Borda (1733–1799)—neither of whom, I am sure, wasted a thought on the possibility that "their" systems might already be on record. But, as my works had fallen into oblivion as had those of Cusanus, neither Condorcet nor Borda should be blamed for failing to acknowledge our priority.

My first electoral paper—actually the one that is longest and most detailed, written around 1280—was rediscovered only in 2000, filed away in the Vatican Library. How would you feel if your work attracts fresh attention after more than 700 years? Actually, I am utterly pleased that mine has resurfaced at last! The text was excavated by a mathematician interested in voting systems, Friedrich Pukelsheim of the University of Augsburg, Germany. Since the text is handwritten in Latin, handling it became an interdisciplinary project that brought together experts on medieval manuscripts, Church Latin and theology, and even computer scientists. As a result, my electoral writings are now on the Internet (in the original and in translations into English and German) at www.uni-augsburg.de/llull/.

Looking back on my lack of success in preaching peace among Christians, Jews, and Muslims, and all the writing and copying by hand of my works, I hope you can appreciate how highly I value the printed book (such as this one) and, even more, instant communication worldwide over the Internet. May that ease of communication help facilitate the religious peace that I so dearly sought.

Yours truly,
Ramon Llull (1232–1316)
Left Choir Chapel
San Francisco Cathedral
Palma de Mallorca

The Condorcet voting paradox ballots given above are the same ones we used earlier in illustrating the notion of a one-on-one contest. We pointed out then that *A* defeats *B* one on one and *B* defeats *C* one on one. The additional observation needed is that we also have *C* defeating *A* one on one. Thus, *A* cannot be a winner using Condorcet's method (he or she loses to *C*), *B* cannot be a winner (he or she loses to *A*), and *C* cannot be a winner (he or she loses to *B*). We will revisit Condorcet's voting paradox in Section 9.3.

Notice that because of our assumption that the number of voters is odd, Condorcet's method yields either no winner or a unique winner. (See Exercise 5.)

It is tempting at this point to suggest modifying Condorcet's method as we have presented it by declaring all the candidates to be tied for the win if there is no candidate who defeats each of the others one on one. The drawback to this modification is that a number of the desirable properties possessed by Condorcet's method then evaporate. We'll explore this in the upcoming exercises.

9.3 Other Voting Systems for Three or More Candidates

With three or more candidates, we find no shortage of additional procedures that suggest themselves and that seem to represent perfectly reasonable ways to choose a winner. Closer inspection, however, reveals shortcomings with all of these. We illustrate this with a consideration of several well-known procedures. Additional procedures (and additional shortcomings) can be found in the exercises.

Plurality Voting and the Condorcet Winner Criterion

In **plurality voting**, only first-place votes are considered. Thus, while we will consider plurality voting in the context of preference list ballots, a ballot here might just as well be a single vote for a single candidate. The candidate with the most votes wins, even though this may be considerably fewer than one-half the total votes cast. This is perhaps the most common system in use today. It is how the voters in Florida chose George W. Bush over Al Gore, Ralph Nader, and Patrick J. Buchanan in the presidential election of 2000.

EXAMPLE 2
Plurality Voting and the 2000 Presidential Election

On the evening of December 12, 2000, Al Gore conceded the presidential election of 2000 to George W. Bush, thus bringing to a close one of the most remarkable elections in modern times. The outcome, ultimately decided in the Electoral College, came down to whether Bush or Gore would carry Florida. With more than 6 million votes cast in Florida, the ultimate margin of victory for George W. Bush was only a few hundred votes.

There is little doubt that if the 2000 presidential election had pitted Al Gore solely against any one of the other three candidates, then Gore would have won both the election in Florida and the presidency. The point is that while most of the Buchanan supporters probably would have voted for Bush, the far more numerous Nader supporters probably would have gone largely for Gore. In fact, the illustration of Condorcet's method that we gave in Example 1 is a simplified version of this Florida election (with *GB* standing for George Bush, *AG* for Al Gore, *PB* for Patrick Buchanan, and *RN* for Ralph Nader).

Thus, although plurality voting led to Bush's winning the 2000 election in Florida (and hence the presidency), Gore was, in this example, what is called a **Condorcet winner**: He would have won the election if Condorcet's method had been used.

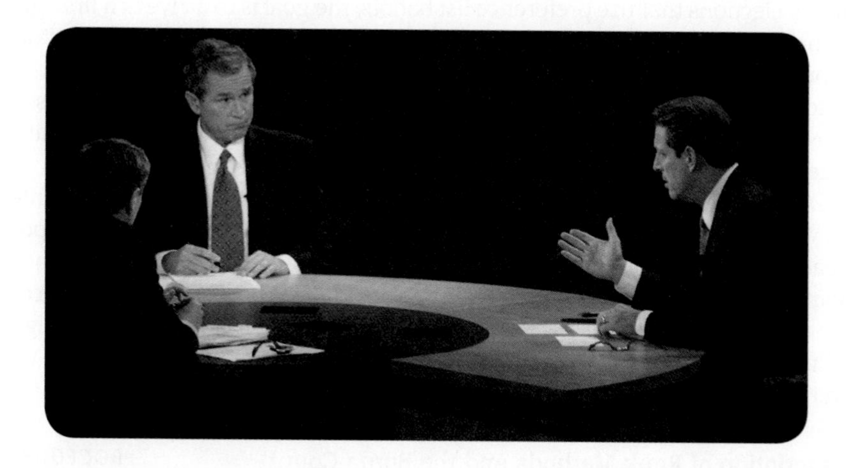

Gov. George W. Bush and Vice President Al Gore debate the issues before the 2000 election, possibly the most controversial election in U.S. history. Gore was the Condorcet winner of the election, but Bush eked out a victory that relied on the rules of the Electoral College. Many voters were suddenly put on notice that the U.S. Constitution makes the election of the president indirect—and not a pure expression of the majority's choice. *(Reuters/Corbis.)*

Condorcet Winner Criterion — DEFINITION

A voting system is said to satisfy the **Condorcet winner criterion** (CWC) provided that, for every possible sequence of preference list ballots, either (1) there is no Condorcet winner (as is often the case) or (2) the voting system produces exactly the same winner for this election as does Condorcet's method.

The CWC is certainly a property that one would like to see satisfied. We record plurality voting's failure in this respect with the following.

The Failure of the CWC with Plurality Voting — THEOREM

The Florida vote in the 2000 presidential election shows that plurality voting fails to satisfy the CWC.

Perhaps a more fundamental drawback of plurality voting is the extent to which the ballots provide no opportunity for a voter to express any preferences except for naming his or her top choice. No use is made, for example, of the fact that a candidate may be no one's first choice but everyone's close second choice.

Finally, there is yet another shortcoming of plurality voting: There are elections in which it is to a voter's advantage to submit a ballot that misrepresents his or her true preferences.

Manipulability — DEFINITION

A voting system is subject to **manipulabity** (or is **manipulable**) if there are elections in which it is to a voter's advantage to submit a ballot that misrepresents his or her true preferences.

For example, in the presidential election of 2000, many voters who ranked Ralph Nader or Patrick J. Buchanan over George W. Bush and Al Gore chose to vote for Bush or Gore rather than to "throw away" their vote on a candidate they felt had no chance. Condorcet's method, it turns out, is not manipulable, and this is one of its most desirable properties. We'll explore this further in the next chapter.

The Borda Count and Independence of Irrelevant Alternatives

In many elections that use preference list ballots, the goal is to arrive at a final group rank ordering of all the contestants that best expresses the desires of the electorate. The purpose is not only to determine the winner—say, the class valedictorian—but also to arrive at who finished second, third, and so on, as in the case of one's rank in the senior class. In other applications, such as an election to a hall of fame, the first few finishers each win, while the remaining nominees are also-rans.

One common mechanism for achieving this objective is to assign points to each voter's rankings and then to sum these for all voters to obtain the total points for each candidate. If there are 10 candidates, for example, then we could assign 9 points to each first-place vote for a given candidate, 8 points for each second-place vote, 7 for each third-place vote, and so forth. The candidate with the highest total number of points is the winner. Subsequent positions are assigned to those with the next-highest tallies.

Description of Rank Methods and the Borda Count — PROCEDURE

A *rank method* of voting assigns points in a nonincreasing manner to the ordered candidates on each voter's preference list ballot and then sums these points to arrive at a group's final ranking. The special case in which there are n candidates with each first-place vote worth $n - 1$ points, each second-place vote worth $n - 2$ points, and so on down to each last-place vote worth 0 points is known as the **Borda count**. The actual point totals are referred to as a candidate's **Borda score**.

The Borda count is named after Jean-Charles de Borda (1733–1799), who was a contemporary of Condorcet.

Rank methods other than the Borda count are common. For example, a track meet can be thought of as an "election" in which each event is a "voter" and each of the schools competing is a "candidate." If the order of finish in the 100-meter dash is school A, school B, school C, school D, then points are often awarded to each school as follows: 5 points for first place, 3 for second place, 2 for third place, and 1 for fourth place.

Sports polls often use point assignments that qualify as rank methods according to our definition. The following example provides an illustration of this.

EXAMPLE 3
Rank Methods and a Basketball Poll

In December of 2010, the Associated Press issued the early-season ranking of the top 25 teams in women's college basketball, shown on the next page.

An interesting question is whether or not this is a ranking system. If it is, who are the candidates and how many are there? In fact, this can be regarded as a ranking system, but the number of candidates is not 25. That is, although 25 teams appeared on each ballot, at least one ballot included each of the teams listed at the bottom in the category "Others receiving votes."

For this to be regarded as a ranking system, the set of candidates must include the entire set of eligible collegiate women's basketball teams. We must also infer that each ballot lists all teams other than that voter's top 25 *below* that voter's top 25, perhaps in alphabetical order. The point assignments are then like those in the newspaper clipping, except that we also assign 0 points for a 26th-place vote, 0 points for a 27th-place vote, and so on. This is why our definition states that a rank method "assigns points in a *nonincreasing* manner" instead of "assigns points in a *decreasing* manner."

We can use this poll to illustrate how total points are arrived at with a ranking method. With the top-ranked team, Connecticut, it's quite easy. Each first-place vote is worth 25, and Connecticut received all 40 first-place votes. This accounts for its total of $25 \times 40 = 1000$ points. But the calculation is more interesting for the second-ranked team, Baylor, and requires some speculation on our part because we don't actually have the ballots to examine. We know that there were 40 ballots (because there were exactly 40 first-place votes altogether), and we know that Baylor had no first-place votes. It stands to reason that Baylor's 943 points must have come from a vast majority of the second-place votes, together with a few lower rankings. (We know that it didn't receive *all* the second place-votes; otherwise, its point total would have been $24 \times 40 = 960$.)

One possibility is that Baylor received:

0 first-place votes (at 25 points each)
33 second-place votes (at 24 points each)
2 third-place votes (at 23 points each)
2 fourth-place votes (at 22 points each)
1 fifth-place vote (at 21 points)
2 sixth-place votes (at 20 points)

The total would then be:

$0 + 792 + 46 + 44 + 21 + 40 = 943$

© Andrew Dieb/ZUMA Press/Corbis

AP POLLS

WOMEN

The top 25 teams in the The Associated Press' women's college basketball poll, with first-place votes in parentheses, records through Sunday, total points based on 25 points for a first- place vote through one point for a 25th-place vote and last week's ranking:

	Record	Pts	Pvs
1. Conn. (40)	9-0	1,000	1
2. Baylor	9-1	943	2
3. Stanford	6-0	930	3
4. Duke	11-0	861	5
5. Xavier	9-0	856	4
6. Tennessee	9-1	762	8
7. West Virginia	10-0	742	9
8. Texas A&M	8-1	716	7
9. UCLA	8-0	666	10
10. N. Carolina	10-0	641	12
11. Ohio St.	7-1	613	6
12. Oklahoma	9-1	567	13
13. Kentucky	7-1	520	14
14. Michigan St.	10-1	460	15
15. Florida St.	8-1	389	17
16. Iowa	9-1	386	19
17. Notre Dame	8-3	341	18
18. St. John's	9-1	304	20
19. Maryland	9-1	225	22
20. Georgetown	8-3	219	11
21. Iowa St.	7-2	201	16
22. DePaul	11-1	144	24
23. Texas	5-3	116	21
24. Syracuse	8-0	84	–
25. Boston Col.	9-0	79	–

Others receiving votes: Georgia 57, Wis.-Green Bay 49, Nebraska 46, Kansas St. 39, Arkansas 12, Bowling Green 8, Kansas 6, Southern Cal 6, Georgia Tech 4, Miami 4, North-western 4.

There is an easy way to calculate the Borda score of a candidate. You can count the number of occurrences of other candidate names that are below this candidate's name. For example, consider the following ballots:

Rank	Number of Voters (5)					Points
First	A	A	A	B	B	2
Second	B	B	B	C	C	1
Third	C	C	C	A	A	0

Because there are three candidates, each first-place vote is $n - 1$, or $3 - 1 = 2$; each second-place vote is $n - 2 = 1$; and each third-place vote is $n - 3 = 0$. If we were to calculate the Borda score of Candidate B algebraically, we would say that B has two first-place votes, worth 2 points each (a total of 4 points), and three second-place votes, worth 1 point each (a total of 3 more points). Thus, the Borda score of Candidate *B* is 4 + 3 = 7.

But instead of calculating this Borda score algebraically, we can mentally replace each occurrence of a letter below *B* by a box, □, and simply count the boxes.

Rank	Number of Voters (5)				
First	A	A	A	B	B
Second	B	B	B	□	□
Third	□	□	□	□	□

Notice that there are seven boxes, giving us the correct value of 7 as the Borda score for Candidate *B*. Of course, you don't actually have to draw any boxes. We are just emphasizing the fact that, in the counting process, it is "spaces" that we are counting, without regard to which letter occurs in the space. A quick glance at the original ballots (without the boxes) reveals that the Borda score of Candidate *A* is 6 and the Borda score of Candidate *C* is 2. When calculating Borda scores this way, be sure that each individual ballot is listed separately, as opposed to using a single list to represent the ballots of several voters (as we often do).

The Borda count certainly seems to be a reasonable way to choose a winner from among several candidates (or to arrive at a group ranking of the candidates). It also has its shortcomings, however, one of which is the failure of a property known as **independence of irrelevant alternatives.**

Independence of Irrelevant Alternatives (IIA) — DEFINITION

A voting system is said to satisfy **independence of irrelevant alternatives (IIA)** if it is impossible for a candidate X to move from nonwinner status to winner status unless at least one voter reverses the order in which he or she had X and the winning candidate ranked.

To describe this property, suppose that an election yields one candidate (call it *A*) as a winner and another candidate (call it *B*) as a nonwinner. Suppose that a new election is now held and that, although some of the voters may have changed their preference list ballots, no one who had previously ranked *A* over *B* changed his or her ballot to rank *B* over *A* now.

If this new election were to yield *B* as a winner, the new outcome would seem strange, especially because none of the relative individual preferences for *A* over *B* had changed in *B*'s favor. The ballot changes responsible for the new outcome involve candidates *other than A* or *B*. One could argue that these other candidates ought to be irrelevant to the question of whether *A* is more desirable than *B* or *B* is more desirable than *A*.

Condorcet's method satisfies IIA. That is, if we have a sequence of preference list ballots that yield *A* as a Condorcet winner and *B* as a nonwinner, then *A* defeats every other candidate, and *B* in particular, in a one-on-one contest according to these ballots. If no voter reverses the order in which he or she ranked *A* and *B*, then *A* will still defeat *B* one on one, and thus *B* remains a nonwinner.

The following illustration shows that the Borda count, unlike Condorcet's method, fails to satisfy IIA. Suppose the initial five ballots are as follows:

Rank	Number of Voters (5)				
First	A	A	A	C	C
Second	B	B	B	B	B
Third	C	C	C	A	A

Our counting procedure shows that the Borda scores are as follows:

Borda score of *A* is 6.
Borda score of *B* is 5.
Borda score of *C* is 4.

The winner is *A* (with 6 points), and *B* is a nonwinner (with 5 points). But now suppose that the two voters on the right change their ballots by moving *C* down between *A* and *B*. The ballots then become as follows:

Rank	Number of Voters (5)				
First	A	A	A	B	B
Second	B	B	B	C	C
Third	C	C	C	A	A

Our counting procedure shows that the Borda scores are as follows:

The Borda score of *A* is 6.
The Borda score of *B* is 7.
The Borda score of *C* is 2.

The Borda count, therefore now yields, *B* as the winner (with 7 points). Thus, *B* has gone from being a nonwinner to being a winner, even though no one changed his or her mind about whether *B* is preferred to *A*, or vice versa.

The above discussion establishes the following theorem.

The Failure of IIA with the Borda Count THEOREM

The Borda count fails to satisfy IIA.

Sequential Pairwise Voting and the Pareto Condition

In our voting-theoretic context, an **agenda** will be understood to be a listing (in some order) of the candidates. This listing is not to be confused with any of the preference list ballots, and, to avoid confusion, we will present agendas as horizontal lists and continue to present preference list ballots vertically.

Description of Sequential Pairwise Voting PROCEDURE

Sequential pairwise voting starts with an agenda and pits the first candidate against the second in a one-on-one contest. The winner then moves on to confront the third candidate in the list, one on one. Losers are deleted. This process continues throughout the entire agenda, and the one remaining at the end wins.

For a given sequence of individual preference list ballots, the particular agenda chosen can greatly affect the outcome of the election, as we'll show in the next chapter. Nevertheless, we will see later in this chapter that sequential pairwise voting arises naturally in the legislative process. Notice also that because of our assumption that the number of voters is odd, there is always a unique winner with sequential pairwise voting.

EXAMPLE 4
Sequential Pairwise Voting

Assume we have four candidates and that the agenda is *A*, *B*, *C*, *D*. Consider the following sequence of three preference list ballots:

Rank	Number of Voters (3)		
First	A	C	B
Second	B	A	D
Third	D	B	C
Fourth	C	D	A

The first one-on-one pits *A* against *B*, and *A* wins by a score of 2 to 1 (meaning that two of the voters—the two on the left—prefer *A* to *B*, and one of the voters prefers *B* to *A*). Thus, *B* is eliminated and *A* moves on to confront *C*. Because *C* wins this one on one (by a score of 2 to 1), *A* is eliminated. Finally, *C* takes on *D*, and *D* wins by a score of 2 to 1. Thus, *D* is the winner.

There is something very troubling about the outcome of the preceding example, especially if you are Candidate *B*. *Everyone* prefers *B* to *D*, but *D* ends up winning! This example shows that sequential pairwise voting fails to satisfy what is called the **Pareto condition.**

Pareto Condition DEFINITION

A voting system is said to satisfy the **Pareto condition** provided that in every election in which every voter prefers one candidate *X* to another candidate *Y*, the latter candidate *Y* is not among the winners.

Again, the Pareto condition (named after Italian economist Vilfredo Pareto, 1848–1923) is a property we would like to see satisfied. But Example 4 (with Candidate *B* in the role of *X* and Candidate *D* in the role of *Y*) establishes the following.

The Failure of the Pareto Condition with Sequential Pairwise Voting THEOREM

Sequential pairwise voting fails to satisfy the Pareto condition.

The following sequence of three preference list ballots illustrates the Pareto condition further:

Rank	Number of Voters (3)		
First	C	C	A
Second	A	A	B
Third	B	B	C

Every one of the three voters prefers *A* to *B*. Hence, if we were using a voting rule that satisfies Pareto, we would conclude that *B* is *not* among the winners. However, we cannot conclude that *A* *is* among the winners. Indeed, there are very reasonable voting rules, like plurality, that satisfy the Pareto condition but would produce *C* as the unique winner using these ballots.

Runoff Systems and Monotonicity

The voting system known as the **Hare system**, which was introduced by Thomas Hare in 1861, is also known by names such as the "single transferable vote system." In 1862, John Stuart Mill described the Hare system as being "among the greatest improvements yet made in the theory and practice of government." Today, the system is used to elect public officials in Australia, Malta, the Republic of Ireland, and Northern Ireland.

Description of the Hare System

PROCEDURE

The Hare system proceeds to arrive at a winner by repeatedly deleting candidates that are "least preferred" in the sense of being at the top of the fewest ballots. If a single candidate remains after all others have been eliminated, he or she alone is the winner. If two or more candidates remain and all of these remaining candidates would be eliminated in the next round (because they all have the same number of first-place votes), then these candidates are declared to be tied for the win.

EXAMPLE 5
The Hare System

Suppose we have the following sequence of preference list ballots:

Rank	Number of Voters (5)				
First	A	A	B	B	C
Second	C	C	C	C	A
Third	B	B	A	A	B

Candidate *C* has only 1 first-place vote (while *B* and *A* have 2 each). Thus, *C* is eliminated in round 1, and the ballots for the second round are as follows:

Rank	Number of Voters (5)				
First	A	A	B	B	A
Second	B	B	A	A	B

In the second round, *B* has only 2 first-place votes (while *A* has 3), so *B* is eliminated in round 2. Because *A* is the only candidate left, he or she is the unique winner of this election.

We now give another example of the Hare system. It is a bit more complicated than Example 5, but as with earlier examples in this section, it will reveal a serious shortcoming of a seemingly very reasonable voting system (the Hare system, in this case).

EXAMPLE 6
The Hare System

Suppose we have the following sequence of preference list ballots, where, as before, the heading of "5" indicates that 5 of the 13 voters hold the ballot with *A* over *B* over *C*, the heading of "4" indicates that 4 of the 13 voters hold the ballot with *C* over *B* over *A*, and so forth.

	Number of Voters (13)			
Rank	5	4	3	1
First	A	C	B	B
Second	B	B	C	A
Third	C	A	A	C

Candidates *B* and *C* have only 4 first-place votes (while *A* has 5). Thus, *B* and *C* are eliminated in the first round, and *A* wins the election.

Now, suppose that the voter in the last column moves Candidate *A* up on his list. Let's look at the new election. Notice that, even though *A* won the last election, the only change we are making in ballots for the new election is one that is favorable to *A*. The ballots for the new election are as follows:

	Number of Voters (13)			
Rank	5	4	3	1
First	A	C	B	A
Second	B	B	C	B
Third	C	A	A	C

If we apply the Hare system again, only *B* is eliminated in round 1, as he or she has 3 first-place votes, as opposed to 4 for *C* and 6 for *A*. Thus, after this round, the ballots are as follows:

	Number of Voters (13)			
Rank	5	4	3	1
First	A	C	C	A
Second	C	A	A	C

We now have *A* on top of 6 lists and *C* on top of 7 lists. Thus, at stage 2, *A* (our previous winner!) is eliminated and *C* is the winner of this new election.

Clearly, this is once again quite counterintuitive. Alternative *A* won the original election, the only change in ballots made was one favorable to *A* (and no one else), and then A lost the next election. This example shows that the Hare system fails to satisfy what is called **monotonicity**.

Monotonicity DEFINITION

A voting system for three or more candidates is said to satisfy **monotonicity** provided that, for every election, if some candidate X is a winner and a new election is held in which the only ballot change made is for some voter to move this winning candidate X higher on his or her ballot (and to make no other changes), then X will remain a winner.

As with the CWC and the Pareto condition, monotonicity is a property we would like to see satisfied. But Example 6 (with Candidate A in the role of X), establishes the following.

The Failure of Monotonicity with the Hare System THEOREM

The Hare system fails to satisfy monotonicity.

The fact that the Hare system does not satisfy monotonicity is considered by many—and with good reason—to be a glaring defect. A 17-voter example in which only a single candidate is eliminated in the first round can also be used to show that the Hare system does not satisfy monotonicity. (See Exercise 32 on p. 353). For an even more glaring version of this defect, one in which alternative A goes from winning to losing because voters move A from last place on their ballots to first place on their ballots, see Exercise 33 (on p. 354).

In spite of these drawbacks, the Hare system is used in important ways today. For example, it is essentially the method that was used to choose Rio de Janeiro as the site of the 2016 Summer Olympics. Chicago was eliminated in the first round on the basis of fewest first-place votes, then Tokyo in the second round, and Madrid in the final round.

There are other runoff systems, some more frequently used than the Hare system. One such example is the following.

Description of the Plurality Runoff Method PROCEDURE

Plurality runoff is the voting system in which there is a runoff (that is, a new election using the same ballots) between the two candidates receiving the most first-place votes. If there are ties, then the runoff is among either those tied for the most first-place votes, or the lone candidate with the most first-place votes along with those tied for the second-most first-place votes (and plurality voting is used).

EXAMPLE 7
Plurality Runoff

The plurality runoff method is somewhat similar in spirit to the Hare system. In fact, you might wonder if they aren't just two different descriptions of the same voting system. That is, you might ask if the plurality runoff method and the Hare system always yield the same winner.

The answer is no, however, as we now demonstrate. Consider the following sequence of preference list ballots:

	Number of Voters (13)			
Rank	4	4	3	2
First	A	B	C	D
Second	B	A	D	C
Third	C	C	A	A
Fourth	D	D	B	B

With the plurality runoff method, *A* and *B* initially tie with 4 first-place votes each, with 3 for *C* and 2 for *D*. In the runoff between *A* and *B*, the ballots are as follows:

	Number of Voters (13)			
Rank	4	4	3	2
First	A	B	A	A
Second	B	A	B	B

With the plurality runoff method, *A* is the winner, defeating *B* in the runoff by a score of 9 to 4.

On the other hand, with the Hare system, we find that the only alternative deleted in the first round is *D*, with only 2 first-place votes. With this deletion of *D*, the ballots are as follows:

	Number of Voters (13)			
Rank	4	4	3	2
First	A	B	C	C
Second	B	A	A	A
Third	C	C	B	B

A and *B* now have only 4 first-place votes compared to the 5 first-place votes that *C* has. Hence, *A* and *B* are now deleted, leaving *C* as the winner with the Hare system.

Alas, the plurality runoff method also does not satisfy monotonicity. Exercise 29 (on p. 353) asks you to verify this.

9.4 Insurmountable Difficulties: Arrow's Impossibility Theorem

All the voting systems for three or more candidates that we have discussed turn out to be flawed in one way or another. You may well ask at this point why we don't simply present *one* voting method for the three-candidate case that has all the desirable properties that we want to satisfy. That is, after all, exactly what we did for the two-candidate case (with majority rule filling the bill, and being the only one to do so by May's theorem).

The answer to this question is extremely important. The difficulties in the three-candidate case are not in any way tied to a few particular systems that we present in a text such as this (or that we choose to use in the real world). The fact is, there are difficulties that will be present *regardless* of what voting system is used, and this applies even to voting systems not yet discovered.

Nothing in the remarkable body of work produced by Nobel laureate Kenneth J. Arrow of Stanford University is as well known or widely acclaimed as the result known as **Arrow's impossibility theorem** (see Spotlight 9.2).

Arrow's Impossibility Theorem THEOREM

With three or more candidates and any number of voters, there does not exist—and there never will exist—a voting system that always produces a winner, satisfies the Pareto condition and IIA, and is not a dictatorship.

Arrow's impossibility theorem isn't obvious, and we won't be saying anything about the proof. But we can state and prove a much weaker result of some interest in its own right. This version is taken from the 2008 text *Mathematics and Politics* (cited in the Suggested Readings on p. 355), and replaces Arrow's assumption of the Pareto condition and non-dictatorship by the CWC.

A Weak Version of Arrow's Impossibility Theorem THEOREM

With three or more candidates and an odd number of voters, there does not exist—and there never will exist—a voting system that satisfies both the CWC and IIA and that always produces at least one winner in every election.

To see why this is true, we'll handle only the case of exactly three voters. Our plan will be to assume that we have some kind of hypothetical voting system that satisfies both the CWC and IIA, and to show that when confronted by the Condorcet voting paradox ballots, it produces *no* winner.

The argument really comes in three separate, but extremely similar, pieces—one for each of the three candidates. Piece 1 argues that *A* can't be among the winners, piece 2 that *B* can't be among the winners, and piece 3 that *C* can't be among the winners. We'll do piece 1 and leave the others for you. The sequence of ballots that we are considering is the following, which we have already seen has no Condorcet winner:

Rank	Number of Voters (3)		
First	A	B	C
Second	B	C	A
Third	C	A	B

Our starting point, however, will be to ask what our hypothetical voting rule must do when confronted by a slightly different sequence of ballots:

Rank	Number of Voters (3)		
First	A	C	C
Second	B	B	A
Third	C	A	B

SPOTLIGHT 9.2

Kenneth J. Arrow

For centuries, mathematicians have been in search of a perfect voting system. Finally, in 1951, economist Kenneth Arrow proved that finding an absolutely fair and decisive voting system is impossible. Arrow is the Joan Kenney Professor of Economics and Professor of Operations Research, Emeritus at Stanford University. In 1972, he received the Nobel Memorial Prize in Economic Science for his outstanding work in the theory of general economic equilibrium. His numerous other honors include the 1986 von Neumann Theory Prize for his fundamental contributions to the decision sciences. He has served as president of the American Economic Association, the Institute of Management Sciences, and other organizations. Dr. Arrow talks about the process by which he developed his famous impossibility theorem and his ideas on the laws that govern voting systems:

> My first interest was in the theory of corporations. In a firm with many owners, how do the owners agree when they have different opinions, for example, about the prospects of the company? I was thinking of stockholders. In the course of this, I realized that there was a paradox involved—that majority voting can lead to cycles. I then dropped that discussion because I was frustrated by it.
>
> I happened to be working with The RAND Corporation one summer about a year or two later. They were very interested in applying concepts of rationality, particularly of game theory, to military and diplomatic affairs. That summer, I felt not like an economist but instead like a general social scientist or a mathematically-oriented social scientist. There was tremendous interest in game theory, which was then new.
>
> Someone there asked me, "What does it mean in terms of national interest?" I said, "That's a very simple matter." He then asked me to write a memorandum on the subject. That memorandum led to a sharper formulation of the social-choice question, and I realized that I had been thinking of it earlier in that other context.
>
> Society must choose among a number of alternative policies. These policies may be thought of as quite comprehensive, covering a number of aspects: foreign policy, budgetary policy, or whatever. Each individual member of the society has a preference, or a set of preferences, over these alternatives. I guess you can say one alternative is better than another. These individual preferences have a property I call rationality or consistency, or more specifically, what is technically known as transitivity: If I prefer *a* to *b*, and *b* to *c*, then I prefer *a* to *c*.
>
> Imagine that society has to make these choices among a set. Each individual has a preference ordering, a ranking of these alternatives. But we really want society, in some sense, to give a ranking of these alternatives. You can always produce a ranking, but you would like it to have some properties. One is that, of course, it be responsive in some sense to the individual rankings. Another is that when you finish, you end up with a real ranking, that is, something that satisfies these consistency, or transitivity, properties. And a third condition is that when choosing between a number of alternatives, all I should take into account are the preferences of the individuals among those alternatives. If certain things are possible and some are impossible, I shouldn't ask individuals whether they care about the impossible alternatives, only the possible ones.
>
> It turns out that if you impose the conditions I just stated, there is no method of putting together the individual preferences that satisfies all of them.
>
> The whole idea of the axiomatic method was very much in the air among anybody who studied mathematics, particularly among those who studied the foundations of mathematics. The idea is that if you want to find out something, to find the properties, you say, "What would I like it to be?" [You do this] instead of trying to investigate special cases. I was really accustomed to this approach. Of course, the actual process did involve trial and error.
>
> But I went in with the idea that there was some method of handling this problem. I started out with some examples. I had already discovered that these led to some problems. The next thing that was reasonable was to write down a condition

Kenneth J. Arrow *(continued)*

that I could outlaw. I constructed another example, another method that seemed to meet that problem, and something else didn't seem very right about it. Then I had to postulate that we have some other property. I found I was having difficulty satisfying all of these properties that I thought were desirable, and it occurred to me that they couldn't be satisfied.

After having formulated three or four conditions of this kind, I kept on experimenting. Lo and behold, no matter what I did, there was nothing that would satisfy these axioms. So after a few days of this, I began to get the idea that maybe there was another kind of theorem here, namely, that there was no voting method that would satisfy all the conditions that I regarded as rational and reasonable. It was at this point that I set out to prove it. It turned out to be a matter of only a few days' work.

It should be made clear that my impossibility theorem is really a theorem [showing that] the contradictions are possible, not that they are necessary. What I claim is that given any voting procedure, there will be some possible set of preference orders for individuals that will lead to a contradiction of one of these axioms.

But you say, "Well, okay, since we can't get perfection, let's at least try to find a method that works well most of the time." Then when you do have a problem, you don't notice it as much. So my theorem is not a completely destructive or negative feature any more than the second law of thermodynamics means that people don't work on improving the efficiency of engines. We're told you'll never get 100% efficient engines. That's a fact—and a law. It doesn't mean you wouldn't like to go from 40% to 50%.

Here, C is clearly a Condorcet winner, and thus it must be the unique winner of the election contested under our hypothetical voting rule. Therefore, C is a winner and A is a nonwinner (for *this* sequence of ballots).

However, because our hypothetical voting rule satisfies IIA, we know that A will remain a nonwinner so long as no voter reverses his or her ordering of A and C. But to arrive at the voting paradox ballots, we can move B (the candidate that is irrelevant to A and C) up one slot in the second voter's list.

Thus, because of IIA, we know that A is a nonwinner when our voting rule is confronted by the voting paradox ballots. This is one-third of the argument. As we mentioned before, similar arguments (see Exercise 39 on p. 354) show that B and C are also nonwinners when our voting rule is confronted by the voting paradox ballots.

We conclude this section with an example that yields a somewhat surprising application of Arrow's impossibility theorem in the context of what are called *social welfare functions.*

EXAMPLE 8
Organ Transplant Policies and Arrow's Impossibility Theorem

Finding an equitable procedure for determining a rank ordering of patients in need of an organ transplant is complicated: there are several criteria that should be considered in arriving at such a "priority ranking." Three such criteria are, for example, (1) the length of time that a patient has been waiting, (2) the probability of success as measured by the numbers of antigens that the patient and donor have matched, and (3) the fraction of the population unsuitable as donors for this potential recipient due to the presence of certain antibodies. A further discussion of these issues occurs in Section 13.3.

Each of the three criteria gives us a ranking (with ties) of the patients according to the more appropriate recipient of the next available organ, according to that particular criterion. Although these rankings are often determined by measurements, the use of different scales for different criteria muddies the water sufficiently so that you might want to work simply with the rankings derived from the measurements, as opposed to working directly with the measurements themselves. This is the context in which we will frame the problem.

So what does the search for a procedure to rank order potential recipients of an organ have to do with voting? In a sense, *everything*, if looked at the right way. We can think of each criterion as a "voter" and each potential recipient as an "alternative." The procedure that we seek is what social choice theorists call a *social welfare function*. It differs from a social choice procedure in that the result of an election is not a single winner or a group tied for the win, but a listing of the alternatives—the priority ranking, in our organ-transplant situation.

For a moment, let's return to the particular task of seeking a priority ranking of the potential recipients of an organ based on how they are ranked according to each of several criteria, like the three we mentioned earlier. What "reasonable" properties might we expect any such procedure to satisfy? Consider the following:

1. If one potential recipient *A* is ranked above another potential recipient *B* with respect to every single criterion, then we should expect *A* to be ranked above *B* in the priority ranking.
2. If potential recipient *A* is ranked above potential recipient *B* in the priority ranking, and there are subsequent changes in how potential recipients are ranked with respect to one or more of the criteria, then potential recipient *B* should not be ranked above potential recipient *A* in the priority ranking unless *B* has moved from being below *A* to being above *A* with respect to at least one criterion.
3. No single criterion should dominate, in the sense that one potential recipient *A*'s ranking above another potential recipient *B*'s ranking, with respect to that criterion, guarantees that *A* will be ranked above *B* in the priority ranking.

Leslie O'Shaughnessy/Visuals Unlimited, Inc.

If we accept these as being required of any "reasonable" procedure, then we have a striking (and highly non-obvious) fact to report: Our task of finding a reasonable procedure is impossible! In fact, this is precisely the statement of Arrow's impossibility theorem in the context of social welfare functions:

There is no social welfare function (for three or more alternatives) that satisfies Pareto (our first condition above), IIA (our second condition above), and non-dictatorship (our third condition above).

9.5 A Better Approach? Approval Voting

Elections in which there are only two candidates present no problem. Majority rule is, as we have seen, an eminently successful voting system in both theory and practice. If there are three or more candidates, however, the situation changes quite dramatically. While several voting systems suggest themselves (plurality, the Borda count, sequential pairwise voting, and the Hare system), each fails to satisfy one or more desired properties (the CWC, IIA, the Pareto condition, and monotonicity). Manipulability is an ever-present problem, as we'll see in the next chapter. Moreover, when all is said and done, Arrow's impossibility theorem says that any search for an ideal voting system of the kind that we have discussed is doomed to failure.

Where does this leave us? More than intellectual issues are at stake here: More than 550,000 elected officials serve in approximately 80,000 governments in the United States. Whether it is a small academic department voting on the best senior thesis or a democratic country electing a new leader, multicandidate elections will be contested in one way or another. If there is no perfect voting system—and perhaps not even a best voting system (whatever that may mean; that is, best in what way?)—what can we do?

Perhaps the answer is that different situations lend themselves to different voting systems, and what is required is a judicious blend of common sense with an awareness of what the mathematical theory has to say. For example, while both the Hare system and the Borda count are subject to manipulability, it seems easier to manipulate the latter. Thus, people may tend to vote more sincerely, rather than strategically, if the Hare system is used instead of the Borda count. This may be a consideration when choosing a voting system for a faculty governance system, for example.

For national political elections, there are also practical considerations. The kind of ballot that we are considering (a preference list ballot) is certainly more complicated than the ballots we now employ, and preference list ballots cannot be used with existing voting machines. There is, however, a voting system that avoids the practical difficulties caused by the type of ballot being used that has much else to commend it. It is called **approval voting**.

Description of Approval Voting PROCEDURE

Under approval voting, each voter is allowed to give one vote to as many of the candidates as he or she finds acceptable. No limit is set on the number of candidates for whom an individual can vote. Voters show disapproval of other candidates simply by not voting for them. The winner under approval voting is the candidate who receives the largest number of approval votes. This approach is also appropriate in situations where more than one candidate can win, for example, in electing new members to an exclusive society such as the National Academy of Sciences or the Baseball Hall of Fame.

EXAMPLE 9
Approval Voting

To illustrate approval voting, suppose that we have nine members of a mathematics department who are trying to choose among five finalists for an open faculty position. They decide to use approval voting—the ballots are indicated in the following table. An X indicates an approval vote. For example, Voter 1, in the first column, approves of Candidates *A*, *C*, and *D*.

	Voter (department member)								
Candidate	1	2	3	4	5	6	7	8	9
A	X	X		X		X	X	X	
B		X	X	X		X	X	X	X
C	X	X			X		X		
D	X		X	X	X			X	
E		X	X		X		X	X	

Counting the Xs in each row shows that Candidate *A* was approved of by six department members (1, 2, 4, 6, 7, and 8), Candidate *B* was approved of by seven, Candidate *C* by four, and *D* and *E* by five. Thus, Candidate *B* wins, with seven approval votes.

Approval voting was proposed independently by several analysts in the 1970s. Probably the best-known official elected by approval voting today is the secretary-general of the United Nations. In the 1980s, several academic and professional societies initiated the use of approval voting. Examples include the Institute of Electrical and Electronics Engineers (IEEE), with about 400,000 members, and the National Academy of Sciences. In Eastern Europe and some former Soviet republics, approval voting has been used in the form wherein one disapproves of (instead of approving of) as many candidates as one wishes.

Is approval voting the perfect voting system? Certainly not. For example, the type of ballot that is used limits the extent to which voter preferences can be expressed. However, it is certainly a voting system with much potential, and the reader wishing to explore it in more detail can start with Brams and Fishburn's 1983 monograph, listed in the Suggested Readings on page 355.

REVIEW VOCABULARY

Agenda An ordering of the candidates to be considered, which is often used in sequential pairwise voting. (p. 337)

Approval voting A method of electing one or more candidates from a field of several in which each voter submits a ballot that indicates of which candidates he or she approves. Winning is determined by the total number of approvals that a candidate obtains. (p. 347)

Arrow's impossibility theorem Kenneth J. Arrow's discovery that any voting system can give undesirable outcomes. (p. 343)

Borda count A voting system for elections with several candidates in which points are assigned to voters' preferences; these points are summed for each candidate to determine a winner. The actual point totals are referred to as a candidate's *Borda score.* (p. 334)

Condorcet's method A voting system for elections with several candidates in which a candidate is a winner precisely when he or she would, on the basis of the ballots cast, defeat every other candidate in a one-on-one contest. (p. 329)

Condorcet winner A Condorcet winner in an election is a candidate who, based on the ballots, would have defeated every other candidate in a one-on-one contest. (p. 333)

Condorcet winner criterion (CWC) A voting system satisfies the Condorcet winner criterion if, for every election in which there is a Condorcet winner, that candidate wins the election when that voting system is used. (p. 333)

Condorcet's voting paradox The observation that there are elections in which Condorcet's method yields no winner. (p. 330)

Hare system A voting system for elections with several candidates in which candidates are successively eliminated in an order based on the number of first-place votes. (p. 339)

Independence of irrelevant alternatives (IIA) A voting system satisfies independence of irrelevant alternatives if the only way a candidate (called *A*) can go from losing one election to being among the winners of a new election (with the same set of candidates and voters) is for at least one voter to reverse his or her ranking of *A* and the previous winner. (p. 336)

Manipulability A voting system is subject to manipulability (or is manipulable) if there are elections in which it is to a voter's advantage to submit a ballot that misrepresents his or her true preferences. (p. 333)

Majority rule A voting system for elections with two candidates (and an odd number of voters) in which the candidate preferred by more than half the voters is the winner. (p. 328)

May's theorem Kenneth May's discovery that, for two alternatives and an odd number of voters, majority rule is the only voting system satisfying three natural properties. (p. 329)

Monotonicity A voting system satisfies monotonicity provided that ballot changes favorable to one candidate (and not favorable to any other candidate) can never hurt that candidate. (p. 341)

Pareto condition A voting system satisfies the Pareto condition provided that every voter's ranking of one candidate higher than another precludes the possibility of this latter candidate winning. (p. 338)

Plurality runoff A voting system for elections with several candidates in which, assuming there are no ties, there is a runoff between the two candidates receiving the most first-place votes. (p. 341)

Plurality voting A voting system for elections with several candidates in which the candidate with the most first-place votes wins. (p. 332)

Preference list ballot A ballot that ranks the candidates from most preferred to least preferred, with no ties. (p. 328)

Sequential pairwise voting A voting system for elections with several candidates in which one starts with an agenda and pits the candidates against each other in one-on-one contests (based on preference list ballots), with losers being eliminated as one moves along the agenda. (p. 337)

SKILLS CHECK

1. A preference list ballot

(a) indicates only a voter's top choice.
(b) is a rank ordering of the candidates, with no ties.
(c) will often have ties.

2. To say that a voting system treats all voters equally means that ______________.

3. To say that a voting system for two candidates treats both candidates equally means that

(a) each wins if he or she receives all the votes.
(b) if all voters reverse their ballots, the election outcome changes.
(c) if any two voters exchange ballots, the election outcome is unchanged.

4. A two-candidate voting system is monotone if ______________.

5. May's theorem says that, with an odd number of voters, among all two-candidate voting systems that never result in a tie, majority rule is the only one that

(a) treats both candidates equally.
(b) treats both candidates equally and all voters equally.
(c) treats both candidates equally and all voters equally and is monotone.

6. When a choice is being made between two candidates, the first type of voting system to suggest itself is ______________.

7. In this chapter, the "number of voters assumption" refers to the assumption that

(a) there is more than one voter.
(b) the number of voters is odd.
(c) the number of voters is even.

8. The winner with Condorcet's method is the candidate who ______________.

9. Which of the following does not satisfy exactly two of the conditions in May's theorem?

(a) A dictatorship
(b) Imposed rule
(c) Minority rule
(d) None of the above

10. The Hare system fails to satisfy ______________.

11. Suppose Condorcet's method is being used in an election in which Candidate *A* is ranked first on more than half of the ballots. Then Candidate *A*

(a) is the unique winner.
(b) is among the winners, but there may be others.
(c) is not necessarily among the winners.

12. The flaw in Condorcet's method is that it ______________.

13. Condorcet's voting paradox refers to the fact that

(a) people vote even though an individual vote virtually never affects the outcome of an election.
(b) the statement "This statement is false" can be neither true nor false.
(c) there are elections in which there is no winner using Condorcet's method.

14. With plurality voting, the winner is the candidate who ______________.

15. George W. Bush's defeat of Al Gore in the state of Florida in the 2000 presidential election shows that

(a) plurality voting does not satisfy the CWC.
(b) majority rule is not monotone.
(c) the Borda count does not satisfy IIA.

16. With the Borda count, the election winner is the candidate who ______________.

17. Instead of by assigning points and doing arithmetic, the Borda score of a candidate can be found by

(a) scanning the ballots and counting the number of occurrences of other candidates below that one.
(b) counting the number of first-place votes and multiplying by 4.
(c) counting the number of candidates that it defeats one on one.

18. Independence of irrelevant alternatives says that a nonwinner can never switch to being a winner unless at least one voter changes his or her ballot in a way that ____________________.

19. The Borda count fails to satisfy
(a) monotonicity.
(b) the Pareto condition.
(c) IIA.

20. The term *single transferrable vote system* is sometimes used to refer to the voting system in this chapter called ____________________.

21. A voting system satisfies the CWC provided that
(a) in every election, there is a Condorcet winner and this candidate is the winner of the election.
(b) in every election, if there is a Condorcet winner, then this candidate is among the winners of the election.
(c) in every election, if there is a Condorcet winner, then this candidate is the unique winner of the election.

22. Sequential pairwise voting is the voting system in which ____________________.

23. Sequential pairwise voting fails to satisfy
(a) monotonicity.
(b) the Pareto condition.
(c) the CWC.

24. The voting system in which a voter can vote for as many candidates as he or she wishes to vote for is called ____________________.

25. Suppose the Borda count is being used in an election in which Candidate *A* is ranked first on more than half of the ballots. Then Candidate *A*
(a) is the unique winner.
(b) is among the winners, but there may be others.
(c) is not necessarily among the winners.

26. A voting system is manipulable if there are elections in which ____.

27. Suppose the Hare system is being used in an election in which Candidate *A* is ranked first on more than half of the ballots. Then Candidate *A*
(a) is the unique winner.
(b) is among the winners, but there may be others.
(c) is not necessarily among the winners.

28. Both the Hare system and the plurality runoff method are defective in that ____________________.

29. Arrow's impossibility theorem says that with three or more candidates and any number of voters, there is no voting system that
(a) is not a dictatorship.
(b) satisfies IIA and is not a dictatorship.
(c) satisfies the Pareto condition and IIA, and is not a dictatorship.
(d) always produces a winner, satisfies the Pareto condition and IIA, and is not a dictatorship.

30. The weak version of Arrow's impossibility theorem asserts that, with three or more candidates and an odd number of voters, there is no voting system that ____________________.

CHAPTER 9 EXERCISES

■ Challenge ▲ Discussion

9.2 Majority Rule and Condorcet's Method

1. In a few sentences, explain why minority rule (the voting procedure for two alternatives that is described on page 329) satisfies conditions (1) and (2) on pages 328–329, but not (3).

2. In a few sentences, explain why imposed rule (the voting procedure for two alternatives that is described on page 329) satisfies conditions (1) and (3) on page 329, but not (2).

3. In a few sentences, explain why a dictatorship (the voting procedure for two alternatives that is described on page 329) satisfies conditions (2) and (3) on page 329, but not (1).

4. Find (or invent) a voting rule for two alternatives that satisfies
(a) condition (1) on page 328, but neither (2) nor (3).
(b) condition (2) on page 329, but neither (1) nor (3).
(c) condition (3) on page 329, but neither (1) nor (2).

5. In a sentence or two, explain why it's impossible, with an odd number of voters, to have two distinct candidates win the same election using Condorcet's method.

6. Construct a real-world example (perhaps involving yourself and two friends) where the individual preference lists for three alternatives are as in the voting paradox of Condorcet.

7. Condorcet's voting paradox shows that with three voters (or three equal-size groups of voters) and the three alternatives *A*, *B*, and *C*, it is possible to have two-thirds prefer *A* to *B*, two-thirds prefer *B* to *C*, and two-thirds prefer *C* to *A*. Find four preference lists that show that

with four voters and the four alternatives A, B, C, and D, it is possible to have three-fourths prefer A to B, three-fourths prefer B to C, three-fourths prefer C to D, and three-fourths prefer D to A.

8. Generalize the result in Exercise 7 from four alternatives to n alternatives: $A_1, \ldots, A_n$.

9. The mathematics department is hiring a new faculty member and the five-person hiring committee has interviewed four candidates: Adam, Beth, Carol, and Dan. They have decided to use Condorcet's method on their five ballots (reproduced in the table below). Who gets the offer?

	Voter 1	Voter 2	Voter 3	Voter 4	Voter 5
1st choice	Adam	Dan	Carol	Dan	Beth
2nd choice	Beth	Beth	Adam	Adam	Adam
3rd choice	Carol	Carol	Beth	Carol	Dan
4th choice	Dan	Adam	Dan	Beth	Carol

10. Suppose that votes on the five mathematics department ballots described in Exercise 9 were distributed according to the table below. Who would get the offer now?

	Voter 1	Voter 2	Voter 3	Voter 4	Voter 5
1st choice	Dan	Beth	Beth	Carol	Carol
2nd choice	Beth	Adam	Adam	Beth	Adam
3rd choice	Adam	Dan	Dan	Dan	Dan
4th choice	Carol	Carol	Carol	Adam	Beth

9.3 Other Voting Systems for Three or More Candidates

11. Plurality voting is illustrated by the 1980 U.S. Senate race in New York among Alfonse D'Amato (D, a conservative), Elizabeth Holtzman (H, a liberal), and Jacob Javits (J, also a liberal). Reasonable estimates (based largely on exit polls) suggest that voters ranked the candidates according to the following table:

22%	23%	15%	29%	7%	4%
D	D	H	H	J	J
H	J	D	J	H	D
J	H	J	D	D	H

(a) Is there a Condorcet winner?
(b) Who won using plurality voting?

12. Condorcet's method can be used to create a new voting system that operates in a manner similar to the Hare system in that it involves repeatedly deleting candidates that are "least preferred." But now we use Condorcet's method to decide what *least preferred* means, and we do this in the following clever kind of way. We tip the ballots *upside-down* and we look for a Condorcet winner from these inverted ballots. Intuitively, a candidate that wins when all the ballots are reversed is "least preferred," according to the original ballots. So this new system works by tipping the ballots upside-down and then repeatedly deleting a Condorcet winner, if there is one. Use this new system to find the winner for the following ballots:

	Voter 1	Voter 2	Voter 3	Voter 4	Voter 5
1st choice	A	D	C	D	B
2nd choice	B	B	A	A	A
3rd choice	C	C	B	C	D
4th choice	D	A	D	B	C

13. Use the voting system introduced in the previous problem to find the winner for the following ballots:

	Voter 1	Voter 2	Voter 3	Voter 4	Voter 5
1st choice	D	B	B	C	C
2nd choice	B	A	A	B	A
3rd choice	A	D	D	D	D
4th choice	C	C	C	A	B

14. (Everyone wins.) Consider the following set of preference lists:

	Number of Voters (9)						
Rank	3	1	1	1	1	1	1
First	A	A	B	B	C	C	D
Second	D	B	C	C	B	D	C
Third	B	C	D	A	D	B	B
Fourth	C	D	A	D	A	A	A

Note that the first list is held by three voters, not just one. Calculate the winner using

(a) plurality voting.
(b) the Borda count.
(c) the Hare system.
(d) sequential pairwise voting with the agenda A, B, C, D.

15. Consider the following set of preference lists:

	Number of Voters (7)				
Rank	2	2	1	1	1
First	C	D	C	B	A
Second	A	A	D	D	D
Third	B	C	A	A	B
Fourth	D	B	B	C	C

Calculate the winner using

(a) plurality voting.
(b) the Borda count.
(c) the Hare system.
(d) sequential pairwise voting with the agenda *B, D, C, A*.

16. Consider the following set of preference lists:

	Number of Voters (8)					
Rank	2	2	1	1	1	1
First	A	E	A	B	C	D
Second	B	B	D	E	E	E
Third	C	D	C	C	D	A
Fourth	D	C	B	D	A	B
Fifth	E	A	E	A	B	C

Calculate the winner using

(a) plurality voting.
(b) the Borda count.
(c) the Hare system.
(d) sequential pairwise voting with the agenda *B, D, C, A, E*.

17. Consider the following set of preference lists:

	Number of Voters (5)				
Rank	1	1	1	1	1
First	A	B	C	D	E
Second	B	C	B	C	D
Third	E	A	E	A	C
Fourth	D	D	D	E	A
Fifth	C	E	A	B	B

Calculate the winner using

(a) plurality voting.
(b) the Borda count.
(c) the Hare system.
(d) sequential pairwise voting with the agenda *A, B, C, D, E*.

18. Consider the following set of preference lists:

	Number of Voters (7)				
Rank	2	2	1	1	1
First	A	B	A	C	D
Second	D	D	B	B	B
Third	C	A	D	D	A
Fourth	B	C	C	A	C

Calculate the winner using

(a) plurality voting.
(b) the Borda count.
(c) the Hare system.
(d) sequential pairwise voting with the agenda *B, D, C, A*.

19. Consider the following set of preference lists:

	Number of Voters (7)				
Rank	2	2	1	1	1
First	C	E	C	D	A
Second	E	B	A	E	E
Third	D	D	D	A	C
Fourth	A	C	E	C	D
Fifth	B	A	B	B	B

Calculate the winner using

(a) plurality voting.
(b) the Borda count.
(c) the Hare system.
(d) sequential pairwise voting with the agenda *A, B, C, D, E*.

20. Consider the following set of preference lists:

	Number of Voters (7)						
Rank	1	1	1	1	1	1	1
First	C	D	C	B	E	D	C
Second	A	A	E	D	D	E	A
Third	E	E	D	A	A	A	E
Fourth	B	C	A	E	C	B	B
Fifth	D	B	B	C	B	C	D

Calculate the winner using

(a) plurality voting.
(b) the Borda count.
(c) sequential pairwise voting with the agenda *A, B, C, D, E*.
(d) the Hare system.

21. An interesting variant of the Hare system was proposed by psychologist Clyde Coombs. It operates exactly as does the Hare system, but instead of deleting alternatives with the fewest first-place votes, it deletes those with the most last-place votes.

(a) Use the Coombs procedure to find the winner if the ballots are as in Exercise 20.

(b) Show that for two voters and three alternatives, it is possible to have ballots that result in one candidate winning if the Coombs procedure is used and a tie between the other two if the Hare system is used.

▲ **22.** In a few sentences, explain why Condorcet's rule satisfies

(a) the Pareto condition.

(b) monotonicity.

▲ **23.** In a few sentences, explain why plurality voting satisfies

(a) the Pareto condition.

(b) monotonicity.

▲ **24.** In a few sentences, explain why the Borda count satisfies

(a) the Pareto condition.

(b) monotonicity.

▲ **25.** In a few sentences, explain why sequential pairwise voting satisfies

(a) the CWC.

(b) monotonicity.

▲ **26.** In a few sentences, explain why the Hare system satisfies the Pareto condition.

▲ **27.** In a few sentences, explain why the plurality runoff method satisfies the Pareto condition.

■ **28.** Use the following ballots to show that the plurality runoff method does not satisfy the CWC:

	Number of Voters (5)		
Rank	2	2	1
First	A	B	C
Second	C	C	B
Third	B	A	A

■ **29.** Use the following ballots to show that the plurality runoff method does not satisfy monotonicity:

	Number of Voters (13)				
Rank	4	3	3	2	1
First	A	B	C	D	E
Second	B	A	A	B	D
Third	C	C	B	C	C
Fourth	D	D	D	A	B
Fifth	E	E	E	E	A

30. Consider the following two elections among Candidates *A*, *B*, and *C*:

	Number of Voters (4)			
Rank	1	1	1	1
First	A	A	B	C
Second	B	B	C	B
Third	C	C	A	A

	Number of Voters (4)			
Rank	1	1	1	1
First	A	A	B	B
Second	B	B	C	C
Third	C	C	A	A

(a) Use these two elections to show that plurality voting does not satisfy IIA.

(b) Use these two elections to show that the Hare system does not satisfy independence of irrelevant alternative.

■ **31.** Construct ballots for the alternatives *A*, *B*, and *C* to show that the Borda count does not satisfy the CWC.

32. Show that the nonmonotonicity of the Hare system can also be demonstrated by the following 17-voter, 4-alternative election. (In a number of recent books, this example is used to show the nonmonotonicity of the Hare system. The 13-voter, 3-alternative example given in the text was pointed out to us by Matt Gendron, an undergraduate at Union College.)

	Number of Voters (17)			
Rank	7	5	4	1
First	A	C	B	D
Second	D	A	C	B
Third	B	B	D	A
Fourth	C	D	A	C

33. The following example illustrates how badly the Hare system can fail to satisfy monotonicity. Consider the following sequence of preference lists:

	Number of Voters (21)			
Rank	7	6	5	3
First	A	B	C	D
Second	B	A	B	C
Third	C	C	A	B
Fourth	D	D	D	A

(a) Show that *A* is the unique winner if the Hare system is used.

(b) Find the winner using the Hare system in the new election, wherein the three voters on the right all move *A* from last place on their preference lists to first place on their preference lists.

▲ **34.** In a few sentences, explain why, with an odd number of voters,

(a) sequential pairwise voting always yields a unique winner.

(b) we can never have exactly two winners with the Hare system.

▲ **35.** In a few sentences, explain why the plurality runoff method can never elect a candidate ranked last on a majority of ballots, assuming there are no ties for first or second place in the voting.

36. Produce ballots showing that plurality voting can, in fact, elect a candidate ranked last on a majority of the ballots.

37. Suppose there are three voters and three alternatives *A*, *B*, and *C*.

(a) If each alternative has exactly one first-place vote, what is the election outcome if the Hare system is used? What if plurality runoff is used?

(b) If an alternative has two or more first-place votes, what is the election outcome if the Hare system is used? What if plurality runoff is used?

(c) Can the Hare system and plurality runoff yield different election outcomes when there are three voters and three alternatives? Explain your answer in one sentence.

38. A voting system is said to satisfy the *majority criterion* if a candidate ranked first by a majority of the voters is always among the winners. For each of the following, either give a sentence or two explaining why the answer is yes, or give a collection of ballots showing that the answer is no.

(a) Does plurality voting satisfy the majority criterion?

(b) Does the Borda count satisfy the majority criterion?

(c) Does the Hare system satisfy the majority criterion?

(d) Does sequential pairwise voting satisfy the majority criterion?

9.4 Insurmountable Difficulties: Arrow's Impossibility Theorem

■ **39.** Complete the proof of the version of Arrow's impossibility theorem from the text by showing that neither *B* nor *C* can be a winner in the situation described. (Your argument will be almost word for word the same as the proofs in the text.)

40. Every voting system *P* can be used to create a new voting system *P** in the manner that we did with Condorcet's method in Exercise 12 on page 351. That is, *P** works as follows: We tip the ballots upside-down and repeatedly delete the "winners" using the voting system *P*. The last candidate (or group of candidates) to be eliminated is the winner. Describe in one sentence the voting system *P* that yields the Hare system as *P**.

9.5 A Better Approach? Approval Voting

41. The 10 members of a board vote by approval voting on eight candidates for new positions on their board, as indicated in the following table. An **X** indicates an approval vote. For example, Voter 1, in the first column, approves of Candidates *A*, *D*, *E*, *F*, and *G*, and disapproves of *B*, *C*, and *H*.

	Voters									
Candidate	1	2	3	4	5	6	7	8	9	10
A	X	X	X			X	X	X		X
B		X	X	X	X	X	X	X	X	
C			X					X		
D	X	X	X	X	X		X	X	X	X
E	X		X		X		X		X	
F	X		X	X	X	X	X	X		X
G	X	X	X	X	X			X		
H		X		X		X		X		X

(a) Which candidate is chosen for the board if just one of them is to be elected?

(b) Which candidates are chosen if the top four are selected?

(c) Which candidates are elected if 80% approval is necessary and at most four are elected?

(d) Which candidates are elected if 60% approval is necessary and at most four are elected?

42. The 45 members of a school's football team vote on three nominees, *A*, *B*, and *C*, by approval voting for the award of "most improved player," as indicated in the following table. An X indicates an approval vote.

	Number of Voters (45)							
Nominee	7	8	9	9	6	3	1	2
A	X			X	X		X	
B		X		X		X	X	
C			X		X	X	X	

(a) Which nominee is selected for the award?

(b) Which nominee gets announced as runner-up for the award?

(c) Note that two of the players "abstained"; that is, approved of none of the nominees. Note also that one person approved of all three of the nominees. What would be the difference in the outcome if one were to "abstain" or "approve of everyone"?

WRITING PROJECTS

1. In the 2000 presidential election in Florida, the final results were as follows:

Candidates	Number of Votes	Percentage of Votes
Bush	2,911,872	49
Gore	2,910,942	49
Nader	97,419	2
Buchanan	17,472	0

Making reasonable assumptions about voters' preference schedules, give a one-page discussion of how the election might have turned out under the different voting methods discussed in this chapter.

2. Frequently in presidential campaigns, the winner of the first few primaries is given front-runner status that can lead to the nomination of his or her party. Moreover, there are often several candidates running in early primaries such as New Hampshire. In one page, consider a recent election and discuss how the nominating process might have proceeded through the campaign if approval voting had been used to decide primary winners.

Suggested Readings

BLACK, DUNCAN. *The Theory of Committees and Elections,* Kluwer, Dordrecht, The Netherlands, 1986. The historical highlights and development of voting methods in the nineteenth and twentieth centuries are traced in this economist's volume.

BRAMS, STEVEN J., and PETER C. FISHBURN. *Approval Voting,* Birkhäuser, Boston, 1983. This volume is a research-level work on development in the recently popular (but rediscovered) method now called *approval voting.* The first chapter, however, is an elementary exposition of this voting method and its uses.

NURMI, HANNU. *Comparing Voting Systems,* Reidel, Dordrecht, The Netherlands, 1987. This monograph provides an excellent treatment, at a somewhat more technical level, of the topics dealt with in this chapter.

SAARI, DONALD G. *Chaotic Elections! A Mathematician Looks at Voting,* American Mathematical Society, Providence, RI, 2001. This expository book begins with the 2000 presidential election and discusses a number of paradoxical results in voting.

TAYLOR, ALAN D., and ALLISON M. PACELLI. *Mathematics and Politics: Strategy, Voting, Power, and Proof,* 2d ed., Springer-Verlag, New York, 2008. Chapters 1 and 7 give an expanded treatment of the topics considered here, with proofs included. This book is also intended for non-math majors.

10

The Manipulability of Voting Systems

People know almost by instinct that you sometimes can achieve the election result you prefer by submitting a ballot that misrepresents your actual preferences. This type of strategic voting is called **manipulation**, and a ballot that misrepresents a voter's true preferences is referred to as an **insincere** or **disingenuous ballot**.

All three of these terms—manipulation, insincere, disingenuous—are widely used in the social-choice literature, but in daily life, we use these terms pejoratively; they aren't exactly warm praise. In fact, your choice to manipulate a voting system typically is no more inherently evil than your submission of a sealed bid for a lamp at an auction at a price considerably below its actual worth. *Strategyproof*—a term with considerably less negative content—is sometimes used in place of *nonmanipulable*, but the latter is more common, so we'll stick with it here.

Charles L. Dodgson was a mathematical lecturer at Oxford University. Dodgson, who used the pen name Lewis Carroll, wrote on mathematical topics and even manipulability. But he achieved greater fame for his satirical works. In the *Alice* books, he refers to the mathematical operations as Ambition, Distraction, Uglification, and Derision, and his characters play nonsensical, easily manipulated games. *(Bettmann/Corbis.)*

Historical references to the manipulability of voting systems include a comment by nineteenth-century mathematician C. L. Dodgson (1832–1898), better known by the pseudonym Lewis Carroll, under which he wrote *Alice's Adventures in Wonderland* (1865). Dodgson commented that voters have a tendency to "adopt a principle of voting which makes it more of a game of skill than a true test of the wishes of the electors" and that it would be "better for elections to be decided according to the wishes of the majority than of those who have the most skill at the game."

But the most famous manipulability quote in the history of social choice is Jean Charles de Borda's reply to a colleague who had pointed out to him how easily the Borda count can be manipulated. "My scheme," Borda replied, "is only intended for honest men!"

After introducing manipulability in Section 10.1, we turn here, as we did in Chapter 9, first to majority rule and Condorcet's method in the case of three or more alternatives. Condorcet's method again shines, leaving the voting paradox as its only blemish. We then revisit the other voting systems introduced in Chapter 9 that apply to elections with three or more candidates, showing that each of them succumbs to some form of manipulation. In fact, there is a striking impossibility result that arises here known as the *Gibbard-Satterthwaite theorem.* It is related to—indeed, some would say equivalent to—Arrow's impossibility theorem. We conclude with a treatment of a striking first cousin of manipulability known as the *chair's paradox.*

10.1 An Introduction to Manipulability

Let's look at an example to illustrate how the Borda count can be manipulated.

EXAMPLE 1
Manipulating the Borda Count with Four Candidates and Two Voters

Suppose there are two voters and four candidates, and suppose the true preferences of the voters are reflected in the following ballots:

Voter 1	Voter 2
A	B
B	C
C	A
D	D

Using the Borda count with point values 3, 2, 1, 0 (or by counting the number of occurrences of other candidates below the one in question, as described in Section 9.3), we see that the Borda scores of the four candidates are as follows:

The Borda score of *A* is 4.
The Borda score of *B* is 5.
The Borda score of *C* is 3.
The Borda score of *D* is 0.

Thus, Candidate *B* wins this election. Voter 1, however, would have preferred to see Candidate *A*—her top choice, according to her true preferences—win this election rather than Candidate *B,* her second choice.

Assume that Voter 1 had known that Voter 2 planned to submit the ballot that he cast above. Could Voter 1 have secured a victory for Candidate *A* by submitting a disingenuous ballot?

The answer here, as we'll show, turns out to be yes. The intuition is fairly transparent: Voter 1 wants to pretend that *B* is not her second choice, but her last choice. Let's see if this is enough to bring about the desired switch in winner from *B* to *A.* The new ballots and Borda scores are as follows:

Voter 1	Voter 2
A	B
C	C
D	A
B	D

The Borda score of *A* is 4.
The Borda score of *B* is 3.
The Borda score of *C* is 4.
The Borda score of *D* is 1.

Close, but not quite what we wanted: Candidates *A* and *C* now tie for the win, and we wanted the winner to be just Candidate *A*. But a moment's inspection reveals that Voter 1 can achieve this if, in addition to plunging Candidate *B* to the bottom of her ballot, she also flip-flops *C* and *D*. That is, the desired ballots (and Borda scores) that yield Candidate *A* as the sole winner are as follows:

Voter 1	Voter 2
A	B
D	C
C	A
B	D

The Borda score of *A* is 4.
The Borda score of *B* is 3.
The Borda score of *C* is 3.
The Borda score of *D* is 2.

In presenting an example of a voting system's susceptibility to manipulation, we will typically present two elections—the original one ("Election 1") in which we assume all ballots are sincere, and the one that contains a disingenuous ballot from a voter ("Election 2"). For example, if we collect the pieces of what we just did, this instance of manipulation of the Borda count could be presented succinctly as follows:

Election 1		
Rank	Number of Voters (2)	
1st choice	A	B
2nd choice	B	C
3rd choice	C	A
4th choice	D	D

Election 2		
Rank	Number of Voters (2)	
1st choice	A	B
2nd choice	D	C
3rd choice	C	A
4th choice	B	D

There are two aspects of manipulation taking place in this example that deserve comment.

First, there is only one voter (the voter on the left, in this example) changing his or her ballot; we call this a **unilateral change** in ballot. An example involving a unilateral change of ballot is sometimes referred to as an instance of "single-voter manipulation" to distinguish it from a situation wherein a group of voters, acting in concert, can change their ballots so that all of them prefer the new winner to the original winner. We'll see examples of group manipulation in Section 10.2.

Second, the original election produced a single winner, as did the new election held after we finished constructing Voter 1's disingenuous ballot. Thus, because we know each voter's sincere preference ranking for the candidates, we also know exactly which of the two election outcomes each voter will prefer. Ties, on the other hand, present a problem. For example, if a voter has sincere preferences that rank *A* over

B over *C* over *D*, then it's not at all obvious whether this voter will prefer an election outcome that ties *A* and *D* to an election outcome that ties *B* and *C* or vice versa.

A voting system is **manipulable** if there is at least one scenario in which some voter can achieve a more preferred election outcome by unilaterally changing his or her ballot. The precise definition follows.

Manipulability DEFINITION

A voting system is said to be **manipulable** if there exist two sequences of preference list ballots and a voter (call the voter Jane) such that

1. Neither election results in a tie.
2. The only ballot change is by Jane.
3. Jane prefers—assuming that her ballot in the first election represents her true preferences—the outcome of the second election to that of the first election.

With this definition at hand, we can now turn to the study of the manipulability of some of the particular voting systems that we introduced in the last chapter.

10.2 Majority Rule and Condorcet's Method

Throughout this section, we assume that the number of voters is odd. In Section 9.2 (on p. 328), we pointed out that with two candidates, majority rule has three very desirable properties: It treats all voters equally, it treats both candidates equally, and it is *monotone*, meaning that a single voter's change in ballot from a vote for the loser to a vote for the winner has no effect on the election outcome. More strikingly, May's theorem told us that among all voting systems in the two-candidate case that never result in a tie, majority rule is the *only* one satisfying these three properties.

But let's consider for a moment what monotonicity is saying in this two-candidate case for voting systems that never yield ties. It says that if you rank *A* over *B* on your ballot, and the election winner is *B*, then the election winner will remain *B* if you switch to a ballot with *B* over *A*. But there are only two possible choices for a ballot in this two-candidate case: *B* over *A* and *A* over *B*. Monotonicity is thus saying that if you rank *A* over *B*, then no unilateral change in your ballot can make the outcome *A*. This is simply the assertion that you can't manipulate the voting system!

Thus, in the two-candidate case, nonmanipulability and monotonicity are exactly the same thing. This allows us to restate **May's theorem** from Section 9.2, with the word *monotonicity* replaced by *nonmanipulability*.

May's Theorem for Manipulability THEOREM

Among all two-candidate voting systems that never result in a tie, majority rule is the only one that treats all voters equally, treats both candidates equally, and is nonmanipulable.

There are examples of two-candidate voting systems that are manipulable, even though they treat all voters equally and both candidates equally. For example, the voting system that declares the winner to be the alternative with the fewest first-place votes is manipulable, as is the one that declares the winner to be whichever alternative has an odd number of first-place votes (even if that's fewer than half). Exercises 1 and 2 ask you to provide an example of voter manipulation for each of these systems.

Turning to the case of three or more candidates, we begin with Condorcet's method, as we did in Chapter 9. Condorcet's method is based on majority rule, and, as we've just seen, majority rule is nonmanipulable. So the following result, as pleasing as it is, comes as no surprise.

The Nonmanipulability of Condorcet's Method THEOREM

Condorcet's method is nonmanipulable in the sense that a voter can never unilaterally change an election result from one candidate to another candidate that he or she prefers.

Paper ballots are still used in elections in many states. A lingering controversy from the 2004 elections is over the use of electronic ballots, which do not leave physical evidence, thus making it extremely difficult to do a recount in potential disputes about the plurality or majority. *(Jonathan Nourok/PhotoEdit.)*

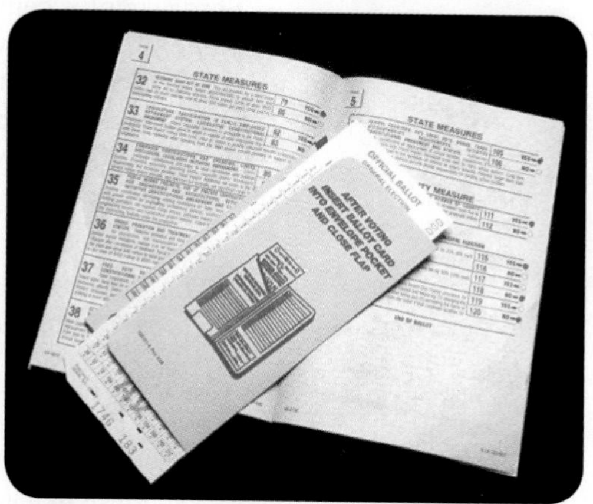

Let's see why Condorcet's method is nonmanipulable, regardless of the number of voters. Suppose that we have an election in which you, as one of the voters, prefer Candidate *A* to Candidate *B*, but *B* wins using Condorcet's method. We'll show that any attempt that you might make to manipulate the election so that *A* becomes the winner is doomed to failure, even if there are more than these two candidates in the election.

Because Candidate *B* is the winner, using Condorcet's method we know that *B* defeats every other candidate in a one-on-one contest based on the ballots cast. In particular, *B* defeats *A* in a one-on-one contest, even with your original ballot that has *A* over *B*. This means that more than half of the other voters ranked *B* over *A*, so, regardless of how you change your ballot, *B* will *still* defeat *A* in a one-on-one contest. While this need not ensure that *B* remains a winner with Condorcet's method, it certainly guarantees that *A* isn't.

EXAMPLE 2
Exploiting the Condorcet Voting Paradox

We had to be careful in stating the theorem that asserts that Concorcet's method is nonmanipulable because, as we've seen, there are elections in which there is no winner using Condorcet's method. With three voters and three candidates, it is possible for a voter (the one on the left in this example) to unilaterally change an election from one that yields his or her second choice as the sole winner (Candidate C in the example), to one in which there is no winner at all, as this example shows:

Election 1

Rank	Number of Voters (3)		
1st choice	A	B	C
2nd choice	C	C	A
3rd choice	B	A	B

Election 2

Rank	Number of Voters (3)		
1st choice	A	B	C
2nd choice	B	C	A
3rd choice	C	A	B

A voter's ability to bring about this kind of change in an election unilaterally, however, is not something that falls within the scope of our formal definition of manipulation. Nevertheless, one could argue that there are situations in which you might well prefer having an election with no outcome at all to having an election in which a candidate other than your top choice emerges as the sole winner.

Hence, you cannot unilaterally cause *A* to be a winner using Condorcet's method, and so your attempt at manipulation will have failed.

We now move on to voting systems with three or more candidates, systems that, unlike Condorcet's method, always produce at least one winner. As you might expect from the results in Chapter 9, in terms of nonmanipulability, these voting systems are not as perfect as one might hope for.

10.3 The Manipulability of Other Voting Systems for Three or More Candidates

Manipulability and the Borda Count

Example 1 showed how a single voter can manipulate an election in which the Borda count is being used. But Example 1 involved four candidates. Is there a simpler example involving only three candidates?

The answer turns out to be no, provided that we continue to interpret the notion of a "more preferred election outcome" to be a switch from a single winner to another single winner (as opposed to a switch creating or breaking a tie). This negative answer is formalized in the following theorem.

The Nonmanipulability of the Borda Count with Exactly Three Candidates THEOREM

With exactly three candidates, the Borda count cannot be manipulated in the sense of a voter unilaterally changing an election outcome from one single winner to another single winner that he or she prefers according to that voter's ballot in the first election, which we take to be sincere preferences.

Let's see why this is true. Suppose the candidates are *A*, *B*, and *C*, and that you prefer *A* to *B*, but *B* is the election winner using the Borda count. We'll show that any attempt you make to manipulate the election by changing your ballot so that *A* emerges as the winner (using the Borda count) is doomed to failure.

Because you prefer *A* to *B*, your sincere ballot can be one of only three possibilities, corresponding to whether *C* is ranked first, second, or third. We'll consider each case in turn.

Case 1. **Your sincere ballot is *A* over *B* over *C*.** No ballot change on your part can increase *A*'s Borda score, and you can decrease *B*'s Borda score by no more than 1. Thus, at best, you can make a unilateral change that results in *A* and *B* having the same Borda score, whereas successful manipulation on your part requires that *A* have a strictly higher Borda score than *B* after your ballot change.

Case 2. **Your sincere ballot is *C* over *A* over *B*.** No ballot change on your part can decrease *B*'s Borda score, and you can increase *A*'s Borda score by no more than 1. Thus, at best, you can make a unilateral change that results in *A* and *B* having the same Borda score, whereas successful manipulation on your part requires that *A* have a strictly higher Borda score than *B* after your ballot change.

Case 3. **Your sincere ballot is *A* over *C* over *B*.** No ballot change on your part can increase *A*'s Borda score or decrease *B*'s Borda score. Thus, after your ballot change, *B* will still have a higher Borda score than *A*, so your attempt at manipulation has failed in this case as well.

So with three candidates, the Borda count is nonmanipulable. With more than three candidates, the Borda count does not fare as well, regardless of how many voters there are.

The Manipulability of the Borda Count with Four or More Candidates THEOREM

With four or more candidates (and two or more voters), the Borda count can be manipulated in the sense that there exists an election in which a voter can change the election outcome unilaterally from one single winner to another single winner that he or she prefers according to that voter's ballot in the first election, which we take to be sincere preferences.

As we've seen in Example 1, the Borda count can be manipulated in the case of four candidates and two voters. This is really half the battle, as we can modify that example to serve in any case in which the number of voters is even, as follows:

1. Any candidates in addition to *A*, *B*, *C*, and *D* can be placed below those four on every ballot.
2. The rest of the voters can be paired off with the members of each pair holding ballots that rank the candidates in exactly opposite orders (thus "canceling each other out" in terms of the Borda scores).

The following example illustrates this method of generalizing our earlier instance of manipulation of the Borda count to the case of five candidates and six voters.

EXAMPLE 3
Manipulating the Borda Count with Five Candidates and Six Voters

Consider the following two elections:

Election 1

A	B	A	E	A	E
B	C	B	D	B	D
C	A	C	C	C	C
D	D	D	B	D	B
E	E	E	A	E	A

Election 2

A	B	A	E	A	E
D	C	B	D	B	D
C	A	C	C	C	C
B	D	D	B	D	B
E	E	E	A	E	A

The ballots of the first two voters (in both elections) are the same as in Example 1 (the manipulation of the Borda count with four candidates and two voters), with the new candidate *E* placed at the bottom of both ballots. The last four voters contribute exactly 8 to the Borda score of each candidate, and so, taken together, they have no effect on who is the winner of the election. This is what we mean by "canceling each other out."

In the first election, as in Example 1, Candidate *B* wins. But if we take these ballots to represent true preferences, the voter on the far left prefers *A* to *B*. Moreover, that voter can achieve this better outcome—Candidate *A*—by submitting the disingenuous ballot that he or she cast in Election 2.

To handle the case where the number of voters is odd, we need to start with a four-candidate, three-voter example of manipulation of the Borda count. Exercise 9 (on p. 372) provides this. We can then modify this example to work for any odd number of voters by again adding pairs of ballots that cancel each other out exactly as we did before. Exercises 10 and 11 (on p. 372) fill in some of the details needed for this part of the argument and ask you to provide the necessary explanations and calculations.

Manipulability of Runoff Systems

EXAMPLE 4
Manipulability of Runoff Systems

Both the plurality runoff rule and the Hare system are manipulable. But rather than give the whole story away, we'll just present the sequences of sincere ballots in each case. Exercises 16 and 17 (on p. 373) ask you to figure out how the leftmost voter in each case can secure a more preferred outcome by a unilateral change of ballot.

Election 1 for the Hare System

A	B	C	C	D
B	A	B	B	B
C	C	A	A	C
D	D	D	D	A

Election 1 for the Plurality Runoff Rule

A	A	C	C	B
B	B	A	A	C
C	C	B	B	A

EXAMPLE 5
Manipulating Sequential Pairwise Voting

Sequential pairwise voting can also be manipulated by a single voter, even in the case of three voters and three candidates. For example, consider the following two elections with the agenda *A*, *B*, and *C*:

Election 1

Rank	Number of Voters (3)		
1st choice	A	B	C
2nd choice	B	C	A
3rd choice	C	A	B

Election 2

Rank	Number of Voters (3)		
1st choice	B	B	C
2nd choice	A	C	A
3rd choice	C	A	B

In Election 1, *A* defeats *B* by a score of 2-to-1, so *A* moves on to meet *C*. But *C* defeats *A* by a score of 2-to-1, so *C* is the winner in Election 1. Election 2 is the result of Voter 1 (on the left) submitting a disingenuous ballot in which he or she has elevated *B* (his or her actual second choice) to first place. It is now clear that *B* first defeats *A* by a score of 2-to-1 and then moves on to defeat *C* by this same score. Hence, *B* is the winner in Election 2. This is an instance of manipulation in which Voter 1 has secured a more preferred outcome by submitting an insincere ballot, because Voter 1 actually prefers *B* to *C* (assuming that his or her ballot in Election 1 represents his or her true preferences). This shows that sequential pairwise voting is manipulable.

Sequential Pairwise Voting and Agenda Manipulability

Thus, sequential pairwise voting can also be manipulated by a single voter, even in the case of three voters and three candidates. But there is another aspect of manipulability that arises with this particular voting system that is of even more interest, and this is something called **agenda manipulation**.

Agenda Manipulation DEFINITION

Agenda manipulation refers to the ability to control who wins an election with sequential pairwise voting by a choice of the agenda.

William H. Riker, in his book *The Art of Political Manipulation,* spoke of the possibility that "those in control of procedures can manipulate the agenda by, for example, restricting alternatives [candidates] or by arranging the order in which they are brought up." The following example provides a striking illustration of this with sequential pairwise voting.

EXAMPLE 6
Agenda Manipulation of Sequential Pairwise Voting

Suppose that we have four candidates and three voters who we know will be submitting the following preference list ballots:

Rank	Number of Voters (3)		
1st choice	A	C	B
2nd choice	B	A	D
3rd choice	D	B	C
4th choice	C	D	A

Now suppose that we have agenda-setting power in the sense that we get to choose the order in which the one-on-one contests will take place. Remarkably, we can arrange for the winner to be whichever of the four candidates we want.

The intuition behind finding an agenda that will yield a certain candidate as the winner arises from the observation that candidates who appear later in the agenda are favored over candidates who appear early in the agenda. For example, if we want *A* to win, we place *A* last and look for which candidates would, in fact, defeat *A* one on one. Here, only *C* defeats *A*, and so we want to arrange for *C* to be eliminated along the way. But *B* defeats *C* one on one, so if we choose the agenda *B*, *C*, *D*, *A*, we have that *C* is eliminated by *B* in the first round, then *D* is eliminated by *B* in the second round, and finally *B* is eliminated by *A* in the third round, leaving *A* as the winner. Exercise 19 (on p. 373) asks you to find the three other agendas that will, in turn, yield *B*, *C*, and *D* as the winner.

Plurality Voting and Group Manipulability

In the real world, all other voting systems pale in comparison to plurality voting in terms of the significance of the role played by disingenuous voting. "Throwing away your vote"—as some accuse Nader voters in Florida of doing in the 2000 presidential election—represents a choice, conscious or otherwise, to forgo obtaining a more desired outcome through strategic considerations.

Ironically, plurality voting, like Condorcet's method, is nonmanipulable according to the formal definition given on page 361. However, a *group* of voters, acting together, can change an election outcome into something they *all* prefer. This observation gives rise to the following definition, theorem, and explanation of why the theorem is true.

Group Manipulability DEFINITION

A voting system is **group-manipulable** if there are elections in which a group of voters can change their ballots so that the new winner is preferred to the old winner by everyone in the group, assuming that the original ballots represent the true preferences of each voter in the group.

The Group Manipulability of Plurality Voting THEOREM

Plurality voting cannot be manipulated by a single individual. However, it is group-manipulable.

First of all, let's see why no individual can manipulate plurality voting. Suppose that you prefer A to B, but B is the winner with plurality voting. Then B has at least one more first-place vote than A. Now, because you prefer A to B, we know that B is not on top of your sincere ballot, so no ballot change that you make can subtract from B's number of first-place votes. Moreover, by moving A to the top of your ballot, you only increase A's number of first-place votes by 1. Thus, the best you can do with a unilateral change in ballots is to move A into a tie with B.

To see that plurality voting is group-manipulable, we only have to look at any real-world election in which a third-party candidate acted as the "spoiler." As we've said, Ralph Nader was exactly that in the state of Florida in the 2000 presidential election. Another example occurs in Exercise 22 (on p. 373).

At this point, we've seen that several of our familiar voting systems for three or more candidates—the Borda count, runoff systems, sequential pairwise voting—can be manipulated. Can't we do better than this in attempting to improve on Condorcet's method? We turn to this question next.

The Green Party holds its convention. Ralph Nader ran for the presidency as a Green in the 2000 election. By doing so, he brought up many questions of social choice—some would say deliberately. Was Nader a spoiler candidate? Were Nader supporters casting sincere votes for him? Would other voters who liked his positions be hedging their bets and voting insincerely if they chose another candidate? *(Mark Leffingwell/AFP/Getty.)*

10.4 Impossibility

Condorcet's method, as we've seen, has a number of very desirable properties, including the following four:

1. Elections (with an odd number of voters) never result in ties.
2. It satisfies the Pareto condition.

3. It is nonmanipulable.
4. It is not a dictatorship.

Unfortunately, Condorcet's voting paradox on page 368 shows that there are elections in which Condorcet's method produces no winner at all.

Can we find a voting system that satisfies all four of these properties and that, unlike Condorcet's method, always yields a winner? Several possibilities suggest themselves. For example, to avoid ties, we could modify any of our usual methods by agreeing to use a fixed ordering of the candidates to break any ties that occur. Or we could extend Condorcet's method by making the winner be the candidate with the best "win-loss record" in one-on-one contests (a method called *Copeland's rule*).

Alas, any such attempt is doomed. In the early 1970s, Allan Gibbard and Mark Satterthwaite independently proved the following remarkable result.

The Gibbard-Satterthwaite Theorem THEOREM

With three or more candidates and any number of voters, there does not exist—and there never will exist—a voting system that always produces a winner, never has ties, satisfies the Pareto condition, is nonmanipulable, and is not a dictatorship.

The Gibbard-Satterthwaite theorem (often called the *GS theorem* for short) is a deep result that is related in important ways to Arrow's impossibility theorem. In particular, you shouldn't find it at all obvious, and we won't be saying anything about the proof. But we can state and prove a much weaker result that is of some interest in its own right.

A Weak Version of the GS Theorem THEOREM

Any voting system for three candidates that agrees with Condorcet's method whenever there is a Condorcet winner—and that additionally produces a unique winner when confronted by the ballots in the Condorcet voting paradox—is manipulable.

Let's see why this is true. With the Condorcet voting paradox, the winner is either *A* or *B* or *C*. For the moment, we'll assume it is *C* (and leave the other two cases to you—see Exercise 29 on p. 374). Consider the following two elections:

Election 1			
Rank	Number of Voters (3)		
1st choice	A	B	C
2nd choice	B	C	A
3rd choice	C	A	B

Election 2			
Rank	Number of Voters (3)		
1st choice	B	B	C
2nd choice	A	C	A
3rd choice	C	A	B

In Election 1, the winner is *C* (our assumption in this case) and in Election 2, the winner is *B* (because we are assuming that our voting system agrees with Condorcet's method when there is a Condorcet winner, as *B* is here). Notice that the voter on the left, by a unilateral change in ballot, has improved the election outcome from being his or her third choice to being his or her second choice. This is what that voter set out to do and this is the desired instance of manipulation.

But the nonintuitive nature of voting and manipulation does not end here. It also turns out that sometimes "more is less" when it comes to "voting power." We illustrate this with the so-called *chair's paradox.*

10.5 The Chair's Paradox

We conclude this chapter with an aspect of manipulability that is so counterintuitive that it is referred to as the **chair's paradox**. To illustrate the situation, we'll consider a hypothetical college in upstate New York that is trying to choose among three academic-year calendars:

- *Terms:* A term system (10 weeks–10 weeks–10 weeks)
- *Semesters:* A semester system (14 weeks–14 weeks)
- *J-Plan:* A January-plan system (12 weeks–4 weeks–12 weeks)

The trustees say that the issue will be decided by majority rule with three voters:

- Someone representing the administration
- Someone representing the student body
- Someone representing the faculty

The administration, however, is given **tie-breaking power**. That is, if each proposed calendar gets one vote, then the one the administration voted for wins. (In this presentation of the paradox, the administration is playing the role of the chair.)

The preferences are given by the following table (and notice that these preference lists exactly mirror the ballots in Condorcet's voting paradox):

	Administration	Students	Faculty
1st choice	J-Plan	Terms	Semesters
2nd choice	Terms	Semesters	J-Plan
3rd choice	Semesters	J-Plan	Terms

We assume that everyone knows everyone else's preferences and that these are real preferences, even after taking into consideration how the other constituencies feel.

The goal now is to analyze the situation and to determine how each of the three will vote if they are all rational in the sense of being willing to vote strategically (that is, to manipulate the system) if it's in their own best interest. This is really a game-theoretic analysis, and it's useful to borrow a couple of pieces of game-theoretic terminology.

First, a choice of which calendar to vote for is called a **strategy**. So each of the voters has three strategies at its disposal: Vote for J-Plan, vote for Terms, and vote for Semesters. The second piece of terminology arises from the observation that if everyone is rational and acting in his or her own self-interest, no one will vote for his or her least-preferred calendar. The point is that voting for either a first or second choice **weakly dominates** the strategies of voting for a third choice, in the sense that the former choices always yield outcomes that are either the same as or better than the latter.

With this, we can see that the administration's strategy of voting for its first choice (the J-Plan) weakly dominates its strategy of voting for its second choice (Terms). That is, if both students and faculty vote for Semesters, the outcome is Semesters regardless of how the administration votes, but otherwise the administration does strictly better by voting for the J-Plan rather than Terms. Hence, assuming that the administration is rational, we know that it will, in fact, vote for its top choice, the J-Plan.

Now, given that we know what the administration will do, the claim is that the faculty's strategy of voting for Semesters weakly dominates its strategy of voting for the J-Plan. That is, if the students vote for Terms, the outcome is the J-Plan regardless of whether the faculty votes for Semesters or the J-Plan. On the other hand, if the students vote for Semesters, then the faculty can secure its best outcome Semesters by voting for Semesters. Assuming that the faculty is rational, we thus know that the faculty will, like the administration, vote for its top choice, which is Semesters.

But let's see where these decisions leave the students. They know that the administration is voting for the J-Plan and that the faculty is voting for Semesters. So if they vote for Terms, then the outcome is the J-Plan—their last choice. However, if they vote for Semesters along with the faculty, then the outcome is Semesters, their second choice. There is no way that the students can secure their top choice, Terms, as the winner. So if they are rational, then they will also vote for Semesters, and thus Semesters will win the election.

So why is this paradoxical? Well, the administration clearly had the most "power," but the eventual winner of the election was its least-preferred calendar! The administration would have been better off handing over the tie-breaking power to either the faculty or the students.

The Chair's Paradox THEOREM

With three voters and three candidates, the voter with tie-breaking power can, if all three voters act rationally in their own self-interest, end up with his or her least-preferred candidate as the election winner.

The chair's paradox represents only one of manipulability's first cousins, some of which involve not only the fields of mathematics and political science but psychology as well. One of the authors relates the following from his early years:

> I recall a third-grade penmanship contest in which each of us had a writing sample taped to the blackboard, and the teacher, Mrs. Levy, announced that we'd get to vote for the one we thought best, with the proviso that the voter couldn't vote for his or her own paper. She also announced that if two or more were tied, we'd have a runoff among those.
>
> I remember being torn as to which of three particular ones to vote for, all of which I thought were very good and considerably better than the rest, including my own. When the votes were counted, these three were, in fact, tied for the win, with my writing sample alone in fourth, and only one vote out of the tie.
>
> After announcing the results, Mrs. Levy went on to say that the runoff would involve not three of us, but four, as she had decided also to vote, and she was voting for me! I don't remember the final tally, or what Mrs. Levy then said to the class, or what my three classmates, all plenty smart enough to realize what had just happened, later said to me. But I do remember sitting back and smiling—absolutely sure of the outcome—as soon as she had announced her intention to vote for me.

A woman casts her ballot on Election Day, the most important day in the American civic ritual of political campaigns and elections. Although she is acting as a responsible citizen, she may also contribute to some remarkable and contradictory results: Condorcet's voting paradox and the Gibbard-Satterthwaite theorem warn us that some elections produce strange results. *(AP Photo/Chris Gardner.)*

REVIEW VOCABULARY

Agenda manipulation The ability to control who wins an election with sequential pairwise voting by a choice of the agenda—that is, a choice of the order in which the one-on-one contests will be held. (p. 365)

Chair's paradox The fact that with three voters and three candidates, the voter with tie-breaking power (the "chair") can end up with his or her least-preferred candidate as the election winner, if all three voters act rationally in their own self-interest. (p. 368)

Disingenuous ballot Any ballot that does not represent a voter's true preferences. Also called an *insincere ballot.* (p. 357)

Gibbard-Satterthwaite (GS) theorem Alan Gibbard and Mark Satterthwaite's independent discovery that every voting system for three or more alternatives and any number of voters that satisfies the Pareto condition, always produces a unique winner, and is not a dictatorship can be manipulated. (p. 367)

Group manipulability A voting system is group-manipulable if there exists at least one election in which a group of voters can change their ballots (with the ballots of voters not in the group left unchanged) in such a way that they all prefer the winner of the new election to the winner of the old election, assuming that the original ballots represent the true preferences of these voters. (p. 366)

Manipulation A voting system is manipulable if there exists at least one election in which a voter can change his or her ballot (with the ballots of all other voters left unchanged) in such a way that he or she prefers the winner of the new election to the winner of the old election, assuming that the original ballots represent the true preferences of the voters. (p. 357)

May's theorem for manipulability Kenneth May's discovery that for two candidates and an odd number of voters, majority rule is the only voting system that treats both candidates equally, treats all voters equally, and is non-manipulable. (p. 360)

Strategy In the chair's paradox, a choice of which candidate (calendar, in our presentation) to vote for is called a *strategy*. This is a special case of the use of the term in general game-theoretic situations. (p. 368)

Tie-breaking power The aspect of the voting rule used in the chair's paradox that says that the winner will be whichever candidate the chair votes for if there is a tie (which happens only if each candidate gets exactly one vote). (p. 368)

Unilateral change A change (in ballot) by one voter, while every other voter keeps his or her ballot exactly as it was. (p. 359)

Weak-dominance One strategy (for example, a choice of whom to vote for) weakly dominates another if it yields an outcome that is at least as good, and sometimes better, than the other. (p. 368)

SKILLS CHECK

1. A "unilateral change in ballot" refers to the fact that

(a) only one candidate's position is being altered.
(b) no communication is taking place.
(c) only one voter is changing his or her ballot.

2. The quote "My scheme is intended only for honest men!" is from ____________________.

3. If a voter has sincere preferences of A over B over C over D, then

(a) she will prefer a tie between A and D to a tie between B and C.
(b) she will prefer a tie between B and C to a tie between A and D.
(c) it's not at all clear which tie—AD or BC—she will prefer.

4. A ballot that misrepresents a voter's true preference is referred to as ____________________.

5. A ballot that does not represent a voter's true preference is often called

(a) an insincere ballot.
(b) a disingenuous ballot.
(c) either (a) or (b).

6. Suppose Voter 1 ranks A over B over C over D and Voter 2 ranks B over C over A over D. Assume the Borda count is being used, so that B wins. If Voter 1 knows that Voter 2 will submit his/her true preferences, then Voter 1 can secure a win for A by submitting the following ballot: ____________.

7. In presenting an example of a voting system's susceptibility to manipulation, we present two elections (Election 1 and Election 2). We assume that

(a) all ballots in Election 1 are sincere.
(b) all ballots in Election 2 are sincere.
(c) both (a) and (b).

8. Nonmanipulability and monotonicity are equivalent if the number of candidates is ________________.

9. The two-candidate voting system in which the winner is the alternative (or alternatives) with the fewest first-place votes

(a) is manipulable.
(b) treats all voters and candidates equally.
(c) both (a) and (b).

10. An example of a two-candidate voting system that is not monotone is ________________.

11. Suppose that two elections show that a voting system is manipulable. Then

(a) neither election results in a tie.
(b) the winners are the same in both elections.
(c) every voter has changed his or her ballot.

12. In the two-candidate case, nonmanipulable is equivalent to ________________.

13. Condorcet's method

(a) can be manipulated but always produces a winner.
(b) is nonmanipulable but sometimes produces no winner.
(c) sometimes results in a tie, so manipulability is hard to assess.

14. May's theorem for manipulability says that, with an odd number of voters, among all voting systems for two candidates that never result in a tie, majority rule is the only one that is nonmanipulable and ________________.

15. With the Borda count, two ballots "cancel each other out" if

(a) they are identical.
(b) each is arrived at by turning the other one upside down.
(c) other voters also hold these same ballots.

16. The Borda count is nonmanipulable in the special case in which ________________.

17. A 6-voter example of manipulation with the Borda count can be modified to yield a 10-voter example by

(a) adding 4 ballots that are identical to each other.
(b) adding 4 ballots that are identical to Voter 1's ballot.
(c) adding 2 pairs of ballots, with the ballots in each pair canceling each other out.

18. With any voting system that satisfies the Pareto condition, an n-voter example of manipulation with k candidates can be modified to yield an n-voter example with $k + j$ candidates by ________________.

19. Of the Hare system and the plurality runoff method,

(a) only the Hare system is manipulable.
(b) only plurality runoff is manipulable.
(c) both are manipulable.

20. Sequential pairwise voting is susceptible to a kind of manipulation called ________________.

21. Plurality voting

(a) cannot be manipulated by a single voter.
(b) can be manipulated by a single voter.
(c) is subject to agenda manipulation.

22. Plurality voting is susceptible to a kind of manipulation called ________________.

23. Group manipulability was discussed in connection with

(a) Condorcet's method.
(b) the Borda count.
(c) sequential pairwise voting.
(d) plurality voting.

24. One strategy weakly dominates another strategy if it yields an outcome that is ________________.

25. Agenda manipulation was discussed in connection with

(a) Condorcet's method.
(b) the Borda count.
(c) sequential pairwise voting.
(d) plurality voting.

26. The deep result in this chapter that is related to Arrow's impossibility theorem is called the ________________.

27. The Gibbard-Satterthwaite theorem says that with three or more candidates and any number of voters, there is no voting system that

(a) is not a dictatorship.
(b) is nonmanipulable and is not a dictatorship.
(c) satisfies the Pareto condition, is nonmanipulable, and is not a dictatorship.
(d) always yields a unique winner, satisfies the Pareto condition, is nonmanipulable, and is not a dictatorship.

28. The weak version of the Gibbard-Satterthwaite theorem asserts that if we have a voting system that agrees with Condorcet's method whenever there is a Condorcet winner and that also produces a unique winner when confronted by the ballots in the Condorcet voting paradox, then the system is ________________.

29. The voters' preferences in the chair's paradox are

(a) precisely the Condorcet voting paradox ballots.
(b) all the same.
(c) dictated by the chair.

30. The chair's paradox is paradoxical because ________________.

CHAPTER 10 EXERCISES

■ Challenge ▲ Discussion

10.2 Majority Rule and Condorcet's Method

1. Consider the voting system for two candidates (*A* and *B*) and three voters in which the candidate with the *fewest* first-place votes wins. Produce two elections that show that this voting system is manipulable.

2. Consider the voting system for two candidates (*A* and *B*) and three voters in which the candidate receiving an odd number of first-place votes wins. Produce two elections that show that this voting system is manipulable.

3. Consider the voting system for two candidates (*A* and *B*) and three voters in which the candidate receiving an even number of first-place votes wins. Produce two elections that show that this voting system is manipulable.

4. There are at least two voting systems for two candidates (*A* and *B*) and three voters that are nonmanipulable and that treat all voters the same (meaning that if two voters were to exchange ballots, then the election outcome would be unchanged).

(a) What does May's theorem tell us about such a voting system?

(b) In one sentence, give an example of such a voting system (that is, produce the rule that determines which of the two candidates, *A* or *B*, wins an election).

(c) In one sentence, give another example that is different from the example you gave in part (b) in that it produces a different winner for at least one election.

5. There are at least three voting systems for two candidates (*A* and *B*) and three voters that are nonmanipulable and that treat both candidates the same (meaning that if all three voters change their ballots, then the election outcome also changes).

(a) What does May's theorem tell us about such a voting system?

(b) In one sentence, give an example of such a voting system; (that is, produce the rule that determines which of the two candidates wins an election).

(c) In one sentence, give two other examples that are different from the example you gave in part (b) in that they produce a different winner for at least one election.

6. Alfonse D'Amato (*D*) won the 1980 U.S. Senate race in New York by defeating Elizabeth Holtzman (*H*) and Jacob Javits (*J*). Reasonable estimates (based largely on exit polls) suggest that voters ranked the candidates according to the following table:

22%	23%	15%	29%	7%	4%
D	*D*	*H*	*H*	*J*	*J*
H	*J*	*D*	*J*	*H*	*D*
J	*H*	*J*	*D*	*D*	*H*

Who would have won if Condorcet's method had been used instead of plurality voting?

10.3 The Manipulability of Other Voting Systems for Three or More Candidates

7. Consider the following election with four candidates and two voters:

B	A
C	D
A	C
D	B

Show that if the Borda count is being used, the voter on the left can manipulate the outcome (assuming the above ballot represents his or her true preferences).

8. Example 3 (on p. 363) showed that the Borda count is manipulable if there are five candidates and six voters. Mimic what was done there to construct an example with seven candidates and eight voters.

9. Use the following election to illustrate the manipulability of the Borda count with three voters and four candidates:

A	B	B
B	A	A
C	C	C
D	D	D

10. Show that the Borda count is manipulable if there are four candidates and five voters. (*Hint:* Start with the ballots in the previous exercise, and then add two ballots that cancel each other out.)

11. Building on the idea in the previous exercise, show that the Borda count is manipulable if there are six candidates and nine voters.

12. Assume the following ballots give the true preferences of the voters and that the Borda count is being used. Show that at least one of the voters can improve the election outcome from her point of view by a unilateral change in her ballot.

B	D	C	B
C	C	A	A
D	A	B	C
A	B	D	D

13. There is a modified version of Condorcet's method called the *weak Condorcet rule:* A candidate is among the winners precisely if he or she would defeat or tie every other candidate in a one-on-one contest. Notice that with an odd number of voters, the weak Condorcet rule is identical to Condorcet's method. Use the following ballots to show that the weak Condorcet rule is manipulable:

A	C	B	D
B	A	D	C
C	B	C	A
D	D	A	B

14. *Copeland's rule* is a voting system that, like Condorcet's method, looks at one-on-one contests. Copeland's rule, however, takes as the election winner the candidate with the best "win-loss record." Use the following ballots to show that Copeland's rule is manipulable:

A	C	A	D
B	E	E	B
C	D	D	E
D	B	C	C
E	A	B	A

15. *Coombs's rule* is the voting system that operates like the Hare system, except that instead of deleting candidates with the *fewest* first-place votes one after another, it deletes candidates with the *most* last-place votes one after another. Use the following ballots to show that Coombs's rule is manipulable:

A	B	B	A	A
B	C	C	C	C
C	A	A	B	B

16. Use the following election to show that the Hare system is manipulable:

A	B	C	C	D
B	A	B	B	B
C	C	A	A	C
D	D	D	D	A

17. Use the following election to show that the plurality runoff rule is manipulable:

A	A	C	C	B
B	B	A	A	C
C	C	B	B	A

18. Use the following election to show that sequential pairwise voting is manipulable. (Assume that the agenda is *A*, *B*, and *C*.)

A	B	C
B	C	A
C	A	B

19. Given the ballots below, mimic what was done in Example 6 (on p. 365) to find an agenda for which

(a) *B* is the winner using sequential pairwise voting.
(b) *C* is the winner using sequential pairwise voting.
(c) *D* is the winner using sequential pairwise voting.

A	C	B
B	A	D
D	B	C
C	D	A

20. Suppose we have an election in which there is a single winner, using plurality voting. In a couple of sentences, explain why we know for sure that there is at least one voter who cannot manipulate this election in the sense of making a unilateral change in his or her ballot that will yield a preferred outcome for that voter, assuming that the original ballot represented his or her true preferences.

21. Suppose we have an election in which there is a single winner, using the Hare system. In a couple of sentences, explain why we know for sure that there is at least one voter who cannot manipulate this election in the sense of making a unilateral change in his or her ballot that will yield a preferred outcome for that voter, assuming the original ballot represented his or her true preferences.

▲ **22.** Suppose that we have a voting system that satisfies unanimity: If every voter ranks the same candidate first, then that candidate is the unique winner. In a few sentences, explain why it is that, if the system fails to satisfy the Pareto condition, it can be manipulated by some group.

23. Assume that the following ballots give the voters' true preferences, and the Borda count is being used. Find a voter who can manipulate this election in the sense of making a unilateral change in his or her ballot that will yield a single winner that is preferred by that voter to the original winner. Explain your answer.

A	E	F	C	B
B	A	D	D	C
F	B	E	E	D
C	F	A	F	F
D	C	B	A	E
E	D	C	B	A

24. Assume that the following ballots give the voters' true preferences, and the Borda count is being used. Find all voters who *cannot* manipulate this election in the sense of making a unilateral change in their individual ballots that will yield a single winner that is preferred by that voter to the original winner. Explain your answer.

A	E	F	C	B
B	A	D	D	C
F	B	E	E	D
C	F	A	F	F
D	C	B	A	E
E	D	C	B	A

25. Use the ballots in Exercise 6 (on p. 372) to show that the plurality rule is group-manipulable.

26. Consider the voting rule in which an alternative is among the winners if it receives at least one first-place vote. In one sentence, explain why this voting system is *not* manipulable.

27. Consider the voting rule in which an alternative is among the winners if it has at least two first-place votes.

(a) In one sentence, explain why this voting system is *not* manipulable.

(b) Explain why the following two elections don't contradict part (a).

Election 1

Rank	Number of Voters (4)			
1st choice	B	A	A	C
2nd choice	C	B	B	B
3rd choice	A	C	C	A

Election 2

Rank	Number of Voters (4)			
1st choice	C	A	A	C
2nd choice	B	B	B	B
3rd choice	A	C	C	A

(c) Intuitively, does it seem to you that Voter 1, on the left in part (b), has secured a better outcome by submitting a disingenuous ballot?

28. Consider the voting system in which the winner is determined by the total number of first- and second-place votes, with ties broken (when possible) according to the number of first-place votes. Thus, a candidate with no first-place votes and three second-place votes would defeat a candidate with two first-place votes and no second-place votes, but a candidate with two first-place votes and three second-place votes would defeat a candidate with one first-place vote and four second-place votes. Given Election 1 below, find a change in Voter 1's ballot that shows that this voting system is manipulable.

Election 1

Rank	Number of Voters (3)		
1st choice	A	C	E
2nd choice	B	D	D
3rd choice	C	A	A
4th choice	D	B	B
5th choice	E	E	C

10.4 Impossibility

■ **29.** Complete the proof of the weak version of the Gibbard-Satterthwaite theorem by handling the case where

(a) the winner with the voting paradox ballots is *A*.
(b) the winner with the voting paradox ballots is *B*.

30. The Gibbard-Satterthwaite theorem says that the following four properties of voting systems cannot be satisfied simultaneously:

(1) Elections always have unique winners.
(2) It satisfies the Pareto condition.
(3) It is nonmanipulable.
(4) It is not a dictatorship.

Which of the four properties are satisfied by a dictatorship?

31. Which of the four properties in Exercise 30 are satisfied by an "antidictatorship," where the election winner is whichever candidate Voter 1 ranks *last* on his or her ballot?

32. Which of the four properties in Exercise 30 are satisfied if we use the plurality rule, with Voter 1's ballot used to break any ties that occur?

10.5 The Chair's Paradox

For Exercises 33 and 34, consider the preference lists from the chair's paradox (reproduced here) and assume that everyone knows that the administration will vote for the

J-Plan, but that no one knows anything about how the faculty will vote.

	Administration	Students	Faculty
1st choice	J-Plan	Terms	Semesters
2nd choice	Terms	Semesters	J-Plan
3rd choice	Semesters	J-Plan	Terms

▲ **33.** In a sentence or two, explain why the students' strategy to vote for Terms does not weakly dominate their strategy to vote for Semesters.

▲ **34.** In a sentence or two, explain why the students' strategy to vote for Semesters does not weakly dominate their strategy to vote for Terms.

WRITING PROJECTS

1. In the chair's paradox, we assume that all voters act rationally. This means that each voter will forgo a strategy that is weakly dominated by another strategy. While this assumption is enough to conclude that the administration will vote for its top choice, it's not actually enough to conclude that the faculty will vote for its top choice. The point is that this latter conclusion required knowing how the administration will vote. Thus, we really need to assume that the faculty *knows* that the administration is rational. But now we can ask what we need to assume about what the students know to conclude that they will vote for Semesters. Answer this (with explanation) and phrase the assumption in terms of the words *knows* and *rational*, as opposed to explicitly speaking of knowing how others will vote.

2. In a paragraph or two, explain why Condorcet's method is not group-manipulable.

Suggested Readings

MOULIN, HERVÉ. *The Strategy of Social Choice,* North Holland, New York, 1983. Manipulability from an economist's point of view.

RIKER, WILLIAM. *The Art of Political Manipulation,* Yale University Press, New Haven and London, 1986. Manipulability from a political scientist's point of view.

TAYLOR, ALAN. *Social Choice and the Mathematics of Manipulation,* Cambridge University Press, Cambridge, U.K., 2005. Manipulability from a mathematician's point of view.

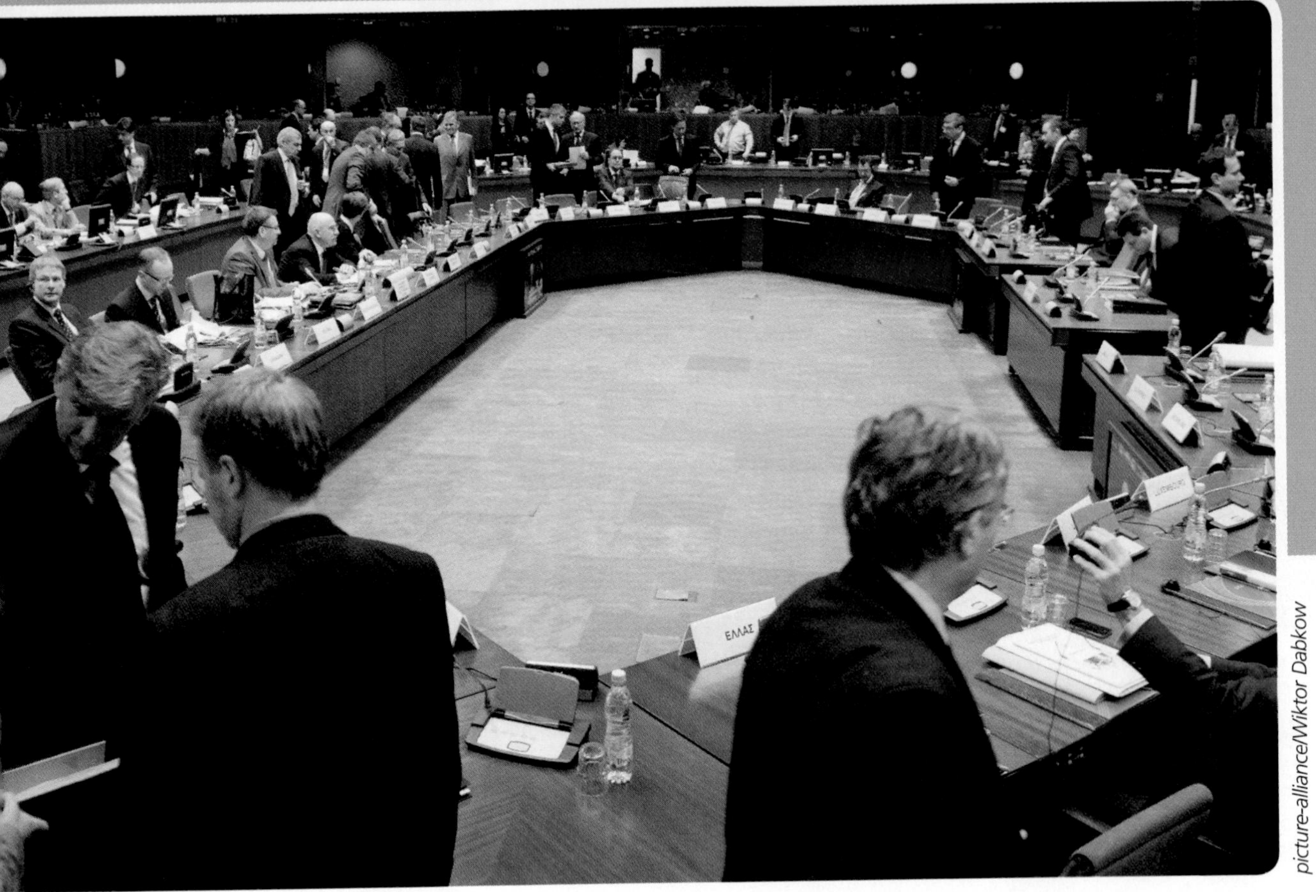

picture-alliance/Wiktor Dabkow

11

Weighted Voting Systems

Voting often is used to decide yes or no questions. Legislatures vote on bills, stockholders vote on resolutions presented by the board of directors of a corporation, and juries vote to acquit or convict a defendant. In this chapter, we shall concentrate on situations where there are just two alternatives, such as "yes" or "no." The theorem of Kenneth May quoted in Chapter 9 says that majority rule is the only system with the following properties:

1. All voters are treated equally.
2. Both alternatives are treated equally.
3. If you vote "no," and "yes" wins, then "yes" would still win if you switched your vote to "yes," provided that no other voters switched their votes.
4. A tie cannot occur unless there is an even number of voters.

There are systems in which the voters appear to be unequal in power, but actually have all the properties required by May's theorem. Any student of politics will attest that not all legislators are equally powerful. (Think of the speaker of the U.S. House of Representatives versus a freshman member, or the prime minister versus a backbencher in Parliament.) Nevertheless, the voting system actually treats the legislators equally: Each has one vote. Our interest is in the voting system itself and not in the influence that some voters might acquire as a result of experience or accomplishment.

In this chapter, we shall consider voting systems that do not treat the voters equally. For example, shareholders of corporations are asked to vote on motions presented by the corporation's board of directors. Each shareholder is allotted one vote per share that he or she owns. If two shareholders own different numbers of shares, they are not treated equally as voters. The voter with the larger number of shares has the greater investment and is given at least the appearance of greater influence. In this case, the stockholders are treated unequally because they are actually unequal. The Council of Ministers of the European Union provides a similar example. Nations with large populations, such as France, have more votes than the smaller nations, such as Austria. Rather than giving the more populous nations more representatives, as the United States does in its House of Representatives, the Council of Ministers gives the ministers from the larger states more votes. Voting systems in which the voters have varying number of votes are called **weighted voting systems**.

Section 11.1 introduces the terminology of weighted voting systems and presents examples to show that the number of votes that a participant is allowed to cast does not always reflect that participant's influence. In the next two sections, we will consider better ways of measuring a voter's power to affect the outcome of a vote. In Section 11.2, we will define and explore the **Shapley-Shubik power index**. The **Banzhaf power index** is another measure of voting power, and we will see it in Section 11.3. You may wonder why there are two measures. The Shapley-Shubik index is preferred if voting is the last step in a negotiation between parties, as sometimes occurs in legislatures. The Banzhaf index is appropriate to use if there is little communication among the voters. Weighted voting is but one type of voting system. Section 11.4 considers more general systems, and we shall see that the two indices developed in sections 11.2 and 11.3 are applicable to all voting systems.

11.1 How Weighted Voting Works

In a **weighted voting system**, each participant has a specified number of votes, called **voting weight**. If the voting weight of Voter A is more than that of Voter B, then A may have more power than B to influence the outcome and certainly won't have less power. (We will see that voters with different numbers of votes may actually have equal power.)

A weighted voting system is provided with a **quota**. If the sum of the voting weights of voters who favor a motion is greater than or equal to the quota, then "yes" wins. If the total voting weight of voters favoring the motion is less than the quota, then "no" wins. The quota must be greater than half of the total weight of all the voters, to avoid situations where contradictory motions can pass, and it cannot be greater than the total weight, or no motion would ever pass.

The European Union's Council of Ministers uses weighted voting, but in the United States, it is unusual for a legislative body to use a weighted voting system. (See Spotlight 11.3 on p. 390.) However, the U.S. Electoral College functions as a weighted voting system when electing the president. The voters are the states. (See Spotlight 11.1.)

SPOTLIGHT 11.1

The Electoral College

In a U.S. presidential election, the voters in each state don't actually cast their votes for the candidates. They vote for electors to represent them in the Electoral College. The number of electors allotted to a state is equal to the size of its congressional delegation, so a state with one congressional district gets 3 electors: 1 for its representative, and 1 for each of its two senators. A state with 25 representatives and 2 senators would get 27 electors. The District of Columbia, while not a state, is entitled by the Twenty-Third Amendment to the U.S. Constitution to send 3 electors to the College.

All states except two select their electors in a statewide contest. Thus, the presidential candidate favored by a plurality of the voters of a state will receive all of the state's electoral votes. For example, in 2008 the Obama-Biden ticket obtained 61 percent of the votes in California and received all 55 of the state's electoral votes. In Maine and Nebraska, there is a different procedure. Two electors (corresponding to the senators) are chosen statewide, and the electors corresponding to the representatives are chosen by congressional district. Nebraska has three congressional districts. In 2008, the Obama-Biden ticket carried one of the Nebraska districts, while the other two districts—and the state as a whole—favored the McCain-Palin ticket.

Effectively, the Electoral College functions as a weighted voting system in which there are 56 participants: the 50 states, the District of Columbia, three Nebraska congressional districts, and two Maine congressional districts. The weights range from 1 for individual congressional districts to 55, and the quota, 270, is a simple majority of the 538 electors.

Notation for Weighted Voting Systems DEFINITION

A weighted voting system is described by specifying the **voting weights** $w_1, w_2, \ldots, w_n$ of the participants, and the **quota**, q. The following notation is a shorthand way of making these specifications:

$$[q : w_1, w_2, \ldots, w_n]$$

The weighted voting system [51 : 40, 60] describes a voting system in which there are two voters, with voting weights 40 and 60, and the quota is 51.

EXAMPLE 1
A Dictator

Suppose there is one voter, D, who has all the power. A motion will pass if and only if D is in favor, and it doesn't matter how the other participants vote. We will call a voter such as D a **dictator**, as in Section 9.1 on page 327. Most weighted voting systems that we will consider do not have a dictator, but if there is one, his or her voting weight must be equal to or more than the quota. The system [51 : 40, 60] has a dictator because the weight-60 voter can pass any motion that he or she wants.

Dummy Voter DEFINITION

A participant in a voting system who never has an opportunity to cast a deciding vote is called a **dummy voter**.

It is easy to identify the dummy voters in some voting systems. For example, if a voting system has a dictator, all the participants except the dictator are dummy voters. There are situations where there are dummy voters, but it is not so obvious who they are.

EXAMPLE 2
Dummy Voters

In the voting system [8 : 5, 3, 1], the weight-1 voter is a dummy because a motion will pass only if it has the support of the weight-5 and weight-3 voters, and then the additional 1 vote is not needed. For a subtler example, consider a committee with members Andy, Beth, Cathy, and Don. They use the weighted voting system [52 : 26, 26, 26, 25]. Any two of Andy, Beth, and Cathy can combine to pass a motion. The combined weight of Don with one other member is just 51, less than the quota. To pass a motion, Don would have to get two of the other members to vote with him. Because the other members have 52 votes, they could pass the motion without Don. Therefore, Don is a dummy voter.

EXAMPLE 3
A Six-Voter System

In 1958, the Board of Supervisors of Nassau County (in Long Island, New York) had six members. Each was to represent a district in the county. Because the districts had significantly different populations, the supervisors were given different voting weights: 9, 9, 7, 3, 1, 1, with 30 votes in all. The quota was set at a simple majority, 16. In our notation, the voting system was [16 : 9, 9, 7, 3, 1, 1]. To pass a measure, two of the three supervisors with voting weight 7 or 9 would be sufficient. The other three supervisors could not cast a decisive vote, even if they all three teamed with one of the higher-weight supervisors, because the combined weight of all four would still be less than the quota. If these three joined two of the higher weight supervisors, a motion would pass, but it would also pass without their votes: In other words, these voters were dummies. This voting system was used by John F. Banzhaf III to call attention to the need for mathematical analysis of weighted voting. See Spotlight 11.3 (on p. 390).

Veto Power DEFINITION

A voter whose vote is necessary to pass any motion is said to have **veto power**.

EXAMPLE 4
Veto Power

In the system [6 : 5, 3, 1], the weight-5 voter has veto power because the other two voters do not have enough combined weight to pass a motion. A dictator always has veto power. It is possible for more than one voter to have veto power as well. In a criminal trial (or any voting situation where a unanimous vote is required), each juror has veto power. Each of the weight-2 voters in the system [5 : 2, 2, 1, 1] has veto power.

The voters in the system [6 : 5, 3, 1] are not equally powerful—the weight-5 voter has veto power and the other two don't—and yet none of the voters are dummies. We can't compare power by comparing the voting weights because the weight-3 voter has the same voting power as the weight-1 voter. Together, they can stop the weight-5 voter from passing a motion, and either one can combine with the weight-5 voter to pass a motion. A **power index** gives a way to measure the share of power that each participant has in a voting system (weighted or otherwise). Spotlight 11.2 gives a brief history of power indices.

Power Index DEFINITION

A **power index** assigns a number to each participant in a decision-making body. A participant's power index indicates his or her ability to influence the decisions of the body. Every power index depends on a mathematical model of the decision-making process.

SPOTLIGHT 11.2

Power Indices

The first widely accepted numerical index for assessing power in voting systems was the **Shapley–Shubik power index**, developed in 1954 by a mathematician, Lloyd S. Shapley, and an economist, Martin Shubik. A particular voter's power as measured by this index is proportional to the number of different permutations (or orderings) of the voters in which he or she has the potential to cast the pivotal vote—the vote in the permutation that first turns the permutation from losing to winning.

The **Banzhaf power index** was introduced in 1965 by John F. Banzhaf III, a law professor who is also well-known as the founder of the anti-smoking organization Action on Smoking and Health (ASH). The Banzhaf index is the one most often cited in court rulings, perhaps because Banzhaf brought several cases to court and established precedent. A voter's Banzhaf index is the number of different possible **voting combinations** in which he or she casts a critical vote—a vote in favor of a motion that is necessary for the motion to pass, or a vote against a motion that is essential for its defeat.

Lloyd S. Shapley

John F. Banzhaf III
(AP Photos.)

Martin Shubik
(Courtesy Yale School of Management Public Affairs.)

11.2 The Shapley-Shubik Power Index

In 1954, Lloyd Shapley and Martin Shubik devised a way to gauge the share of decision-making power of each participant in a voting system. A voter's share of power is called *his or her Shapley-Shubik power index*. The index is defined in terms of *permutations*.

Voting Permutation DEFINITION

A **voting permutation** is an ordered list of all the voters in a voting system. The order in the list can be interpreted as the spectrum of opinion on an issue that the voters are considering.

The spectrum of opinion on the subject of animal welfare, for example, might range from a voter who would outlaw the sale of cow's milk to one who would legalize cockfighting. If an animal welfare bill is being drafted, it must be written so as to receive enough votes to meet the quota.

Pivotal Voter DEFINITION

The first voter in a voting permutation who, when joined by those coming before him or her, would have enough voting weight to win is the **pivotal voter** in the permutation. Each voting permutation has exactly one pivotal voter.

EXAMPLE 5
Pivotal Voters in a Three-Voter System

Alice, Bill, and Cao make decisions by using the voting system [6 : 5, 3, 1]. We observed in Example 4 that the weight-5 voter (Alice) has veto power. Bill and Cao have voting weights 3 and 1, respectively. Let's consider their voting permutations, shown in Table 11.1. (Voters are identified by initials.) Next to each voting permutation, the total weights of the first voter, the first two voters, and all three voters are shown in sequence. The first number in the sequence that equals or exceeds the quota (6) is underlined, and the corresponding pivotal voter's symbol is circled. We see that Alice is pivotal in four permutations, while Bill and Cao are each pivotal in one. Hence Alice's share of the voting permutations is $\frac{4}{6}$, while Bill and Cao each have $\frac{1}{6}$ share. We will notice that although their voting weights differ, Bill and Cao seem to be equally powerful.

Table 11.1

Permutations and Pivotal Voters for the Three-Person Committee

Permutations			Weights			Pivotal Voters		
A	(B)	C	5	<u>8</u>	9		Bill	
A	(C)	B	5	<u>6</u>	9			Cao
B	(A)	C	3	<u>8</u>	9	Alice		
B	C	(A)	3	4	<u>9</u>	Alice		
C	(A)	B	1	<u>6</u>	9	Alice		
C	B	(A)	1	4	<u>9</u>	Alice		

Legislators try to write bills that will be supported by the pivotal voter of the permutation that represents the spectrum of opinion on the topic that the bill addresses. Thus, an animal welfare bill will pass if and only if it is supported by the pivotal voter in the animal welfare permutation; for a tax bill to pass the support of the pivotal voter in the taxation permutation will be needed; a nuclear disarmament treaty will be ratified by the Senate if and only if it is supported by the pivotal member in its corresponding permutation, and so on. Each issue that is considered by a legislature has its own spectrum of opinion, or voting permutation, and its own pivotal voter.

In a presidential election, the campaigns must address multiple issues, each having its own voting permutation. On each issue, the candidates try to address their campaign advertising to the pivotal voter.

EXAMPLE 6
The Permutation from the 2008 Election

In 2008, the Electoral College elected Barack Obama to be president and Joseph Biden to be vice president of the United States. Spotlight 11.1 explains how the Electoral College is effectively a 56-voter weighted voting system.

Each of the voters in the Electoral College is selected by and represents an electorate. Some, such as Nebraska's third congressional district, were heavily in favor of the

McCain-Palin ticket; others, such as Missouri, North Carolina, and Indiana, were almost equally split between the Obama-Biden and McCain-Palin tickets; and still others, such as the District of Columbia, were strongly in the Obama-Biden camp. Table 11.2 lists all the voters in the Electoral College, in decreasing order by their margin in favor of the Obama-Biden ticket. (The margin is the number of popular votes cast for the Obama-Biden ticket divided by the number of votes cast for the McCain-Palin ticket.) A running total of electoral votes gives the total weight of each voter and all who came before it in the table. In listing the states and other voters in this order, we have recorded a permutation of the Electoral College participants. The pivotal voter was Colorado, which brought the running total from 269 to 278, thus exceeding the quota of 270.

Table 11.2

The Permutation Resulting from the General Election for President of the United States in 2008

State	Electors	Obama Margin	Running Total	State	Electors	Obama Margin	Running Total
DC	3	14.15328	3	FL	27	1.05845	338
HI	4	2.70284	7	NE 2nd	1	1.02446	339
VT	3	2.21535	10	IN	11	1.02110	350
RI	4	1.79315	14	NC	15	1.00666	365
NY	31	1.74543	45	MO	11	0.99730	376
MA	12	1.71718	57	MT	3	0.95429	379
MD	10	1.69761	67	GA	15	0.90012	394
IL	21	1.68343	88	SD	3	0.84177	397
DE	3	1.67653	91	AZ	10	0.84115	407
CA	55	1.65100	146	ND	3	0.83794	410
ME 1st	1	1.60538	147	SC	8	0.83337	418
CT	7	1.58520	154	NE 1st	1	0.81974	419
ME	2	1.42893	156	TX	34	0.78776	453
WA	11	1.42436	167	MS	6	0.76548	459
OR	7	1.40464	174	WV	5	0.76449	464
MI	17	1.40219	191	NE	2	0.73584	466
NJ	15	1.37330	206	KS	6	0.73574	472
NM	5	1.36211	211	TN	11	0.73516	483
WI	10	1.32860	221	KY	8	0.71723	491
NV	5	1.29288	226	LA	9	0.68188	500
ME 2nd	1	1.25957	227	AR	6	0.66191	506
PA	21	1.23362	248	AL	9	0.64228	515
MN	10	1.23361	258	AK	3	0.63761	518
NH	4	1.21575	262	ID	4	0.58668	522
IA	7	1.21478	269	UT	5	0.54975	527
CO	9	1.20025	278	OK	7	0.52334	534
VA	13	1.13596	291	WY	3	0.50236	537
OH	20	1.09792	311	NE 3rd	1	0.43162	538

Factorial DEFINITION

If there are n voters, the number of permutations is called the **factorial** of n and is denoted $n!$.

There is a simple formula for $n!$:

Factorial Formula THEOREM

For a positive whole number n,

$$n! = n \times (n - 1) \times (n - 2) \times \cdots \times 2 \times 1$$

and $0! = 1$.

To see why the formula is justified, suppose that $n > 0$ and we are listing all the permutations, as in Table 11.1 (for $n = 3$) or Table 11.3 (for $n = 4$). There are n voters who could be first; when the first voter is selected, there are $n - 1$ remaining voters, any one of whom could be second. Then there are $n - 2$ voters who could be third, $n - 4$ voters who could be fourth, and so on until there is just one voter left who could be last. By the fundamental principle of counting (see Chapter 2), the number of permutations is

$$n! = n \times (n - 1) \times (n - 2) \times \cdots \times 2 \times 1$$

EXAMPLE 7
Calculating n!

Here are the first five, starting with 1!.

$$\begin{aligned} 1! &= 1 \\ 2! &= 2 \times 1 = 2 \\ 3! &= 3 \times 2 \times 1 = 6 \\ 4! &= 4 \times 3 \times 2 \times 1 = 24 \\ 5! &= 5 \times 4 \times 3 \times 2 \times 1 = 120 \end{aligned}$$

It is inefficient to continue like this, because we have repeated the same multiplications over and over. If we notice that $n! = n \times (n - 1)!$, we can speed up the process. Thus the next five factorials would be

$$\begin{aligned} 6! &= 6 \times 5! = 6 \times 120 = 720 \\ 7! &= 7 \times 6! = 7 \times 720 = 5040 \\ 8! &= 8 \times 7! = 8 \times 5040 = 40{,}320 \\ 9! &= 9 \times 8! = 9 \times 40{,}320 = 362{,}880 \\ 10! &= 10 \times 9! = 10 \times 326{,}880 = 3{,}628{,}800. \end{aligned}$$

You can imagine that *n!* continues to increase dramatically—an instance of the combinatorial explosion. Five more steps gets us to $15! = 1{,}307{,}674{,}368{,}000$ (one trillion, three hundred seven billion, six hundred seventy-four million, three hundred sixty-eight thousand), and Google can calculate 100!. It has 158 digits. Here's a puzzle: how many zeros are at the end of 100!? (For information on the Fundamental Theorem of Counting, see Chapter 2, page 41.)

The Shapley-Shubik Power Index DEFINITION

The **Shapley-Shubik power index** of each voter is computed by counting the number of voting permutations in which that voter is pivotal, and dividing that number by $n!$, where n is the number of voters.

If we say that each voter "owns" the permutations in which he or she is pivotal, then each voter's Shapley-Shubik index is his or her share of the voting permutations. A voter's Shapley-Shubik index is equal to the probability that when a voting permutation is selected at random, he or she will be the pivotal voter. In Example 5, we considered the 6 voting permutations of the three-voter system [6 : 5, 3, 1], and we found that Alice, with 5 votes, was the pivotal voter in 4 voting permutations, while Bill and Cao, the other two voters, each were the pivotal voter in one voting permutation. Therefore, the Shapley-Shubik power index of this system is $(\frac{2}{3}, \frac{1}{6}, \frac{1}{6})$.

EXAMPLE 8

The Corporation with Four Shareholders

A corporation has four shareholders, *A*, *B*, *C*, and *D*, with 49, 48, 2, and 1 shares, respectively. It uses the weighted voting system

$$[51 : 49, 48, 2, 1]$$

The $4! = 24$ permutations of the shareholders are shown in Table 11.3. (The arrangement is the same as in Table 11.1, which was to list permutations and find pivotal voters in a three-voter system.) In 10 of the permutations, *A* is the pivotal voter; *B* and *C* are each pivotal voters in 6; and *D* is the pivotal voter in 2 permutations. Therefore, the Shapley-Shubik power index for this weighted voting system is

$$\left(\frac{10}{24}, \frac{6}{24}, \frac{6}{24}, \frac{2}{24}\right),$$

or

$$\left(\frac{5}{12}, \frac{1}{4}, \frac{1}{4}, \frac{1}{12}\right).$$

How to Compute the Shapley-Shubik Power Index

It is practical to calculate the Shapley-Shubik power index of a system with up to four voters by making a list of all the voting permutations and identifying the pivotal voter in each, as we have done in the previous two examples. This is the brute force way of determining the Shapley-Shubik power index. With a computer, brute force can be used to determine the Shapley-Shubik power index of somewhat larger systems, but eventually the combinatorial explosion renders the brute force method impossible to implement.

The Shapley-Shubik power index of the Electoral College is shown in Spotlight 11.4 (on p. 400). The calculations were performed with a Java applet that is available at the Web site listed at the end of this chapter. The applet uses an advanced counting method, *generating functions*, that doesn't rely on tallying individual permutations one at a time.

Table 11.3

Permutations and Pivotal Voters for the Four-Shareholder Corporation

Permutations				Weights				Pivotal Voters			
A	Ⓑ	C	D	49	97	99	100		B		
A	Ⓑ	D	C	49	97	98	100		B		
A	Ⓒ	B	D	49	51	99	100			C	
A	Ⓒ	D	B	49	51	52	100			C	
A	D	Ⓑ	C	49	50	98	100		B		
A	D	Ⓒ	B	49	50	52	100			C	
B	Ⓐ	C	D	48	97	99	100	A			
B	Ⓐ	D	C	48	97	98	100	A			
B	C	Ⓐ	D	48	50	99	100	A			
B	C	Ⓓ	A	48	50	51	100				D
B	D	Ⓐ	C	48	49	98	100	A			
B	D	Ⓒ	A	48	49	51	100			C	
C	Ⓐ	B	D	2	51	99	100	A			
C	Ⓐ	D	B	2	51	52	100	A			
C	B	Ⓐ	D	2	50	99	100	A			
C	B	Ⓓ	A	2	50	51	100				D
C	D	Ⓐ	B	2	3	52	100	A			
C	D	Ⓑ	A	2	3	51	100		B		
D	A	Ⓑ	C	1	50	98	100		B		
D	A	Ⓒ	B	1	50	52	100			C	
D	B	Ⓐ	C	1	49	98	100	A			
D	B	Ⓒ	A	1	49	51	100			C	
D	C	Ⓐ	B	1	3	52	100	A			
D	C	Ⓑ	A	1	3	51	100		B		

In special cases where all (or almost all) of the voters have the same weight, the Shapley-Shubik power index can be calculated by relying on the following two principles:

- Voters with the same voting weight have the same Shapley-Shubik power index.
- The sum of the Shapley-Shubik power indices of all the voters is 1.

EXAMPLE 9
A Nine-Person Committee

Alice is chairperson of a committee. She has 3 votes, and there are eight other members, each with 1 vote. The quota for passing a measure is a simple majority, 6 of the 11 votes. In our notation, this voting system is [6 : 3, 1, 1, 1, 1, 1, 1, 1, 1].

Each weight-1 member has the same power index. Our strategy is to compute Alice's Shapley-Shubik power index first. By subtracting her index from 1, we will get the share of power for the remaining members of the committee. Because there are 8 of them, and they are equally powerful, we can find the index of each weight-1 member by dividing by 8.

There are 9! = 362,880 voting permutations to consider; that's too many. We will take another approach. Alice will be the pivotal voter in any voting permutation where she is in the fourth position, when her vote would bring the total in favor to 6; the fifth position, when she would increase the total weight in favor from 4 to 7; or the sixth position, when the total would increase from 5 to 8 with her vote. If she is in first, second, or third position in the voting permutation, her vote plus those of the member or members before her would not bring the total in favor to 6; and if her position in the voting permutation is after the sixth position, one of the other members of the committee (the sixth one) would be the pivotal voter.

Alice is the pivotal voter in all permutations in which she is the fourth, fifth, or sixth voter, and she is not the pivotal voter when she is in any one of the other 6 positions. Therefore, she is the pivotal voter in $\frac{1}{3}$ of the permutations, and Alice's Shapley-Shubik power index is $\frac{1}{3}$. The remaining $\frac{2}{3}$ of the voting power is shared equally by the 8 other voters. Therefore each has $\frac{2}{3} \div 8 = \frac{1}{12}$ of the power.

The Shapley-Shubik power index of this weighted voting system is, therefore,

$$\left(\frac{1}{3}, \frac{1}{12}, \frac{1}{12}, \frac{1}{12}, \frac{1}{12}, \frac{1}{12}, \frac{1}{12}, \frac{1}{12}, \frac{1}{12}\right)$$

Because $\frac{1}{3} \div \frac{1}{12} = 4$, the Shapley-Shubik model indicates that the Alice is 4 times as powerful as a weight-1 member, although her voting weight is only 3.

The next example is a little more complicated because some of the weight-1 members of Alice's committee have been scheming.

EXAMPLE 10
A Pact

In the committee considered in Example 9, three of the weight-1 members, Bill, Chris, and Dean, make a pact: Chris and Dean will give their votes to Bill. Effectively, the system now has 7 voters: Alice and Bill each have 3 votes, Chris and Dean do not participate, and the five remaining members each have 1 vote. Thus the weighted voting system is [6 : 3, 3, 1, 1, 1, 1, 1]. Let Zoë be a weight-1 member. She is the pivotal voter of a permutation if and only if the voters coming before her in the permutation have a combined weight of exactly 5. There are two kinds of voting permutations that meet this condition (Z is Zoë):

- $Y_1Y_2Y_3ZY_4Y_5Y_6$, where one of Y_1, Y_2, Y_3 is Alice, one of Y_4, Y_5, Y_6 is Bill, and the remaining four Y's are weight-1 voters; and
- $Y_1Y_2Y_3ZY_4Y_5Y_6$, where one of Y_1, Y_2, Y_3 is Bill, and one of Y_4, Y_5, Y_6 is Alice.

Counting these permutations will involve the fundamental principle of counting. Let's count the voting permutations of the first type. There are three places that Alice could occupy before Zoë, and three places that Bill could occupy after Zoë. By the fundamental principle of counting, there are 3 × 3 = 9 ways we could position Alice and Bill in a voting permutation with Alice before Zoë and Bill after Zoë. We can count the number of permutations in each of the nine groups, where Alice, Zoë, and Bill have already been positioned. The four other committee members can be ordered in 4! = 24 ways, and put accordingly into the four open spaces. Using the fundamental principle of counting (again), we see that the number of voting permutations of the first type is 9 × 24 = 216. The number of permutations of the second type, where Bill comes before Zoë and Alice after, is the same. Therefore, there are $2 \times 216 = 432$ voting permutations in which Zoë is pivotal. The Shapley-Shubik index of Zoë is therefore $\frac{432}{7!} = \frac{3}{35}$. The other weight-1 voters have the same Shapley-Shubik index, so the combined share of power of the five weight-1 voters is $5 \times \frac{3}{35} = \frac{3}{7}$. Alice and Bill split the remaining $\frac{4}{7}$ of the power, so each has $(1 - \frac{3}{7}) \div 2 = \frac{2}{7}$ of the power. The Shapley-Shubik index of the system is

$$\left(\frac{2}{7}, \frac{2}{7}, \frac{3}{35}, \frac{3}{35}, \frac{3}{35}, \frac{3}{35}, \frac{3}{35}\right)$$

The pact obviously benefits Bill by making his power equal to Alice's. Alice's power is reduced a bit. Perhaps it is surprising that the Shapley-Shubik indices of the weight-1 members who were not involved in the pact increased slightly.

11.3 The Banzhaf Power Index

Consider the following voting situation: You, along with 100,000 other voters, are deciding if your city should issue bonds to build a convention center. A two-thirds majority is needed to pass the measure. The Shapley-Shubik index would simply say that your power to influence the outcome is the same as that of any other voter, because each voter has an equal chance to occupy the pivotal position (the 66,668th) in a voting permutation. Your index would be 1/100,001. The Banzhaf index would give more information, but it also would be more difficult to compute. Your index would reflect the chance that you would cast the deciding vote. To pass, the measure must be approved by at least two-thirds—or 66,668—of the voters. You would cast the deciding vote (for or against) if *exactly* 66,667 of the other voters were in favor of the measure. Otherwise, neither you nor anyone else will cast a deciding vote.

Your Banzhaf index would be equal to the number of sets consisting of exactly 66,667 voters that could be assembled from the other 100,000 voters, *multiplied by 2*. (The reason for multiplying by 2 is that you could vote "yes," and cast a deciding vote; or you could vote "no," also a deciding vote.) Like the Shapley-Shubik index, the Banzhaf index would recognize that each voter has an equal chance to cast a decisive vote, but it counts the actual number of chances that a voter has to change the outcome.

In contrast to the Shapley-Shubik power index, which is based on counting permutations, the Banzhaf power index is based on counting voting combinations.

Voting Combination DEFINITION

A **voting combination** is a list of the voters indicating how each voted on an issue.

EXAMPLE 11
Voting Combinations in the 2008 Presidential Election

The voting combination for the 2008 Electoral College can be determined from Table 11.2. All voters that had Obama's margin greater than 1.00000 voted for the Obama-Biden ticket; those with Obama's margin less than 1.00000 voted for the McCain-Palin ticket.

In any voting combination, there may be one or more voters who have the power to change the outcome by switching their votes.

Critical Voter DEFINITION

A voter in a given voting combination is a **critical voter** if the outcome would be different if that voter, and no other voter, changed his or her vote.

Although each voting permutation has exactly one pivotal voter, a voting combination may have no critical voters, or it may have many.

EXAMPLE 12
A Criminal Trial

When the jury is unanimous in favor of a motion to convict (or a motion to acquit), then each juror is a critical voter. On the other hand, if all but one juror is in favor of a motion, then the motion fails, and the lone holdout is a critical voter. Voting combinations in which more than one juror opposes a motion have no critical voters.

EXAMPLE 13
The U.S. Presidential Elections

Table 11.2 shows that the Obama-Biden ticket received 365 electoral votes in 2008. The quota for the Electoral College is 270, so the ticket had 95 extra votes. A state that voted for the Obama-Biden ticket was a critical voter if and only if its voting weight was more than 95, and no state has that many electoral votes: thus there were no critical voters. Nevertheless, we have noted that the voting permutation did have a pivotal voter: Colorado.

In the closer election of 2000, which the Bush-Cheney ticket won by only 2 electoral votes, every state that voted for the Bush-Cheney ticket was a critical voter.

Banzhaf Power Index DEFINITION

A voter's **Banzhaf power index** is the number of voting combinations in which he or she casts a critical vote.

We have seen that a juror in a criminal trial casts a critical vote in two voting combinations: one in which the jury is unanimously in favor of a motion, and one in which the juror is the lone holdout, voting against a motion that all other jurors support. Thus, each juror has a Banzhaf index of 2.

SPOTLIGHT 11.3

A Mathematical Quagmire

A county legislature in the United States is usually called a board of supervisors. Unlike state legislators, who represent districts that are carefully drawn to be equal in population, supervisors in some counties represent towns within the county. Because the towns differ in population, some counties use weighted voting to compensate for the resulting inequity.

If each supervisor's voting weight is proportional to the population of the town that he or she represents, there will be situations in which one or more supervisors on a board are dummy voters, even if no supervisor is dictator. In a 1965 law review article, John F. Banzhaf III pointed out that three of the six supervisors of Nassau County, New York, were dummies (see Example 3 on p. 380). The article inspired legal action against several elected bodies that employ weighted voting systems.

The first legal challenge to weighted voting was to invalidate the voting system of the Board of Supervisors of Washington County, New York. In its decision, the New York State Court of Appeals provided a way to fix a weighted voting system: each supervisor's Banzhaf power index, rather than his or her voting weight, should be proportional to the population of the district that he or she represents. The court predicted that its remedy would lead to a "mathematical quagmire."

Five lawsuits filed over a period of 25 years challenged weighted voting in the Nassau County Board of Supervisors. These cases proved to be the mathematical quagmire that the appeals court had feared. The courts attempted to force Nassau County to comply with the Washington County decision. Although the county made a sincere attempt to do so, every voting system that it devised faced a new legal challenge. With conflicting expert testimony, the U.S. District Court finally ruled in 1993 that weighted voting was inherently unfair.

Banzhaf's law review article, which initially drew attention to weighted voting in Nassau County, was aptly titled "Weighted Voting Doesn't Work." (It is included in the "Suggested Readings" section at the end of this chapter.)

Nevertheless, tradition is hard to change. Many boards of supervisors of counties, particularly in the State of New York, still use weighted voting, and legal challenges to the practice, even after the Nassau County decision, have not always succeeded.

EXAMPLE 14
A Three-Member Committee

Let's revisit Alice's committee, which was described in Example 5. The members are Alice, Bill, and Cao, and they use the voting system [6 : 5, 3, 1].

Suppose that the committee votes unanimously in favor of a motion. Let's identify the critical voters. Suppose that Alice switches her vote, as follows:

Alice	Bill	Cao	Votes	Outcome
Yes	Yes	Yes	9	Pass
↓				
No	Yes	Yes	4	Fail

By changing her vote, Alice has changed the outcome. (This is not surprising; we have seen that she has veto power.) In this voting combination, Alice is a critical voter.

Now let's see what happens if Bill changes his vote:

Alice	Bill	Cao	Votes	Outcome
Yes	Yes	Yes	9	Pass
	↓			
Yes	No	Yes	6	Pass

This time, the outcome doesn't change, so Bill is not a critical voter. You can verify for yourself that Cao is also not a critical voter in the combination.

Table 11.4 displays all eight voting combinations for Alice's committee. The entries at the top describe the voting combination with all three members voting "Yes," which was mentioned above. The columns that are marked *X* identify the critical voters in the voting combination to the left. These are the voters whose weight is sufficient to change the outcome if they were to change their votes.

To determine the Banzhaf index, we count the critical votes in each of the eight voting combinations: Alice has critical votes in six voting combinations, while Bill and Cao are each critical in two. Therefore the Banzhaf index of this system is (6, 2, 2).

The Banzhaf index provides a comparison of the voting power of the participants in a voting system. Thus, Alice, with a Banzhaf index of 6, is three times as powerful as Bill or Cao. To determine the way voting power is distributed, we can add the numbers of critical voters for all three voters together to get $6 + 2 + 2 = 10$ critical votes in all. Thus, Alice has 60 percent of the voting power, while Bill and Cao each have 20 percent. The Shapley-Shubik model gives $\frac{2}{3}$ of the power to Alice, while Bill and Cao each have $\frac{1}{6}$, so the models are in close agreement in this case.

Table 11.4

Voting Combinations for the Committee with Voting System [6 : 5, 3, 1]

Voters			Weight of "Yes" votes	Critical Voters		
Alice	Bill	Cao		Alice	Bill	Cao
"Yes"	"Yes"	"Yes"	9	X		
"Yes"	"Yes"	"No"	8	X	X	
"Yes"	"No"	"Yes"	6	X		X
"Yes"	"No"	"No"	5		X	X
"No"	"Yes"	"Yes"	4	X		
"No"	"Yes"	"No"	3	X		
"No"	"No"	"Yes"	1	X		
"No"	"No"	"No"	0			

Counting Combinations

If there are three voters, *A*, *B*, and *C*, and *A* and *C* voted "yes" while *B* voted "no," we might record the voting combination as "Yes, No, Yes." A briefer notation is to visualize voting combinations as **binary numbers.** A whole number *N* is represented in binary form as a sequence of binary digits, or **bits,** which can be 0 or 1. This sequence expresses the way that *N* can be expressed as a sum of powers of 2. For example,

$$5 = 2^2 + 2^0$$

Algebra Review: Counting in Binary

The numeration system that we use every day is a base-10 system, which involves 10 digits, 0 though 9. In order to exceed 9, we add a place and write the number 10. A number such as 307 can be written in expanded form as $3 \times 100 + 0 \times 10 + 7 \times 1$.

Using exponential notation, we could also write the following:

$$307 = 3 \times 10^2 + 0 \times 10^1 + 7 \times 10^0$$

Notice in exponential form that the base is 10. Binary is a base-2 system and has only two digits, namely 0 and 1. If we were to convert a binary number to its base-10 (decimal) equivalent, we would use a base of 2 in the same way we used the base of 10. For example, the binary number 1011 (read as "one zero one one") would be equivalent to $1 \times 2^3 + 0 \times 2^2 + 1 \times 2^1 + 1 \times 2^0$. As a base-10 number, this simplifies to $8 + 0 + 2 + 1 = 11$.

After the number 1, a binary number will have at least two places. Starting with zero, the string of binary numbers in order are

0, 1, 10, 11, 100, 101, 110, 111, 1000, 1001, 1010, 1011, 1100, 1101, 1110, 1111, . . .

When writing these numbers, we do not put commas in, as we do with decimal numbers. For example, the binary number 11011 would not be written as 11,011. Also, when binary numbers are used and the context is not clear whether the number is binary or base-10, a subscript of 2 is often included. Thus 11_2 represents 11 base 2 (which is 3 in base-10) while 11 is the usual number "eleven."

can be represented by a binary number where bits 2 and 0 are equal to 1, and bit 1 is 0. We would say that $101_2 = 5$: This binary number could stand for the voting combination "Yes,No,Yes." The largest number that can be represented with 3 bits is 7, because $111_2 = 7$. The smallest number that can be represented in 3 bits is 0 ($000_2 = 0$). Thus, there are 8 distinct 3-bit binary numbers. Each represents a unique voting combination of three voters.

Number of Voting Combinations THEOREM

The number of voting combinations with n voters is 2^n.

We have seen that the number of voting combinations with 3 voters is equal to the number of 3-bit binary numbers. By the same reasoning, the number of voting combinations with n voters is equal to the number of n-bit binary numbers. The largest n-bit binary number is the sequence of n ones, which represents $2^n - 1$. Because we start counting with 0, there are 2^n n-bit binary numbers.

The Shapley-Shubik index of a voter is the probability that he or she will be pivotal in a randomly selected permutation. We also can interpret the Banzhaf index in terms of probability. A voter's Banzhaf index is simply the number of voting combinations in which he or she is a critical voter. If we divide the voter's Banzhaf index by the number of possible voting combinations, the result will be the probability that the voter will cast a critical vote in a randomly selected voting combination.

EXAMPLE 15
The Probability of Casting a Critical Vote

In Example 14, we found that the Banzhaf power index of a three-member committee with quota and voting weights [6 : 5, 3, 1] was (6, 2, 2). There is a total of $2^3 = 8$ voting combinations, and in 6 of them, Alice (the weight-5 voter) casts a critical vote. Therefore, if a voting combination is selected at random, the probability that Alice will cast a critical vote is $\frac{3}{4}$. Similarly, Bill and Cao each have a probability of $\frac{1}{4}$ of casting a critical vote. In the case of a 12-member jury in a criminal trial, the Banzhaf power index of each juror is 2. There are $2^{12} = 4096$ voting combinations, so the probability that a given juror will cast a critical vote in a randomly selected voting combination is $\frac{2}{4096}$. These two examples illustrate an important difference between the Shapley-Shubik and Banzhaf power indices. Because each voting permutation has exactly one pivotal voter, the sum of the Shapley-Shubik indices of all voters is 1. However, some voting combinations have no critical voters, and other voting combinations have more than one. The sum of the probabilities of the voters casting critical votes is usually not equal to 1.

How to Calculate the Banzhaf Power Index

The brute force approach to calculating the Banzhaf power index is to list all the voting combinations. In each combination, identify the critical voters by circling them on your list. For each voter, count the number of combinations in which he or she is a critical voter. This number is the voter's Banzhaf power index. We followed this procedure in Example 14 for a three-voter system, and as is usual for brute force methods, the combinatorial explosion will make it impractical for larger systems.

To determine the critical voters in a given voting combination, the following principle is useful. For each voting combination, let w be the sum of the weights of the voters who vote "Yes." If $w \geq q$, then we will say the outcome is "Yes" by a margin of $w - q$ **extra votes**. If $w < q$, then we will say that the outcome is "No" by a margin of $q - 1 - w$ extra votes.

Extra Votes Principle THEOREM

The critical voters in any voting combination are the voters whose weights are greater than the number of extra votes in the combination, *and* whose votes agree with the outcome of the combination.

EXAMPLE 16
Voting Combinations in the Five-Vote System

Let's consider some voting combinations in the five-voter system [12 : 7, 4, 4, 3, 1], with voters named Alice, Bill, Carl, Dave, and Ellen, respectively. We'll use binary notation for the combinations, and determine the critical voters in each.

The unanimous combination 11111 has 7 extra votes. No voter has more than 7 votes, so there is no critical voter.

The combination 01111 in which Alice votes "No" and the other voters say "Yes" has outcome "Yes" with a margin of 0 extra votes. Thus all the voters except Alice are critical voters.

The combination 10001 has weight 8, and thus its outcome is "No" with a margin of 3 extra votes. Bill and Carl voted "No," and their weights are more than the margin, so their votes are critical. The "Yes" voters, Alice and Ellen, are not critical, and neither is Dave, because his voting weight is only 3, not more than the margin.

There are 32 voting combinations to consider if we are to determine the Banzhaf power index of this system. If you care to examine the other 29 combinations, you will find that the Banzhaf power index is (18, 10, 10, 6, 6). This means Alice is a critical voter in 18 voting combinations, Bill is critical in 10 combinations, Carl has the same number of critical votes, and Dave and Ellen also are equally powerful, with 6 critical votes each.

EXAMPLE 17
The Corporation with Four Shareholders

The corporation with four shareholders (see Example 8 on p. 385) uses the weighted voting system

$$[51 : 49, 48, 2, 1]$$

Table 11.5 displays a list of all the voting combinations, their outcomes, and their margins. The four columns at the right are marked to indicate the critical voters in each

Table 11.5

Voting Combinations in the Four-Shareholder Corporation

				Critical Voters			
Combination	Weight	Outcome	Extra Votes	A	B	C	D
0000	0	"No"	50				
0001	1	"No"	49				
0010	2	"No"	48	X			
0011	3	"No"	47	X	X		
0100	48	"No"	2	X			
0101	49	"No"	1	X		X	
0110	50	"No"	0	X			X
0111	51	"Yes"	0		X	X	X
1000	49	"No"	1		X	X	
1001	50	"No"	0		X	X	
1010	51	"Yes"	0	X		X	
1011	52	"Yes"	1	X		X	
1100	97	"Yes"	46	X	X		
1101	98	"Yes"	47	X	X		
1110	99	"Yes"	48	X			
1111	100	"Yes"	49				
Critical Votes				10	6	6	2

combination. Counting the critical votes shown in the table, we arrive at the Banzhaf index of the corporation: (10, 6, 6, 2). In this model, A has

$$\frac{10}{24} \text{ or approximately 42 percent}$$

of the voting power, while *B* and *C* each have 25 percent (even though *B* has more shares than *C*). Shareholder *D* has the remaining 8 percent of the voting power, according to the Banzhaf model. In this case, power is distributed exactly as it was by the Shapley-Shubik model.

We have seen that each voting combination for a set of n voters corresponds to a binary number with n bits. Thus, if there are n voters, there will be 2^n voting combinations. Obviously, there is exactly one voting combination where everyone votes "yes" and one voting combination where everyone votes "no." These correspond to the n-bit binary numbers with all bits equal to 1, and all bits equal to 0, respectively.

Combinations were introduced in Chapter 8 for the purpose of determining probabilities. Interpreting a voter's Banzhaf power index as the probability that, in a randomly chosen voting combination, he or she will be a critical voter, we can count voting combinations as combinations were counted in Chapter 8. Recall that the number of voting combinations with n voters and exactly k "yes" votes is denoted ${}_nC_k$ (when speaking, ${}_nC_k$ is pronounced "n choose k"). Thus, the statement that there is exactly one combination of n voters where everyone votes "yes" would be ${}_nC_n = 1$. Similarly, we have ${}_nC_0 = 1$. There are n combinations with exactly one "yes" vote:

$$2^{n-1} = 100\cdots0,\ 2^{n-2} = 010\cdots0,\ \ldots,\ 2^0 = 000\cdots1,$$

where each combination has one 1 and $n - 1$ zeros. Thus, ${}_nC_1 = n$.

Duality Formula for Combinations THEOREM

$${}_nC_k = {}_nC_{n-k}$$

To see that this is true, imagine a combination with k "yes" votes and $n - k$ "no" votes. Suppose that each voter switches his or her vote to the opposite side. The result would be $n - k$ "yes" votes and k "no" votes. Thus, the number of voting combinations of n voters with k "yes" votes is equal to the number of voting combinations of n voters with $n - k$ "yes" votes.

Addition Formula THEOREM

$${}_{n+1}C_k = {}_nC_k + {}_nC_{k-1}$$

This formula is also easy to verify. Suppose that there are $n + 1$ voters, one of whom is Zoë. We would like to determine ${}_{n+1}C_k$: The number of voting combinations with k "yes" votes from a set of $n + 1$ voters. We will separate the voting combinations into two parts, depending on how Zoë votes. If she votes "no," and there are k "yes" votes, then there are ${}_nC_k$ voting combinations where k of the other voters say "yes." If Zoë votes "yes," then a voting combination with k "yes" votes can be assembled by combining Zoë's vote with a combination of the other n voters with $k - 1$ "yes" votes; there are ${}_nC_{k-1}$ of these. Adding, we obtain the addition formula.

The **addition formula** enables us to calculate the numbers ${}_nC_k$. Starting with ${}_0C_0 = {}_1C_0 = {}_1C_1 = 1$, we obtain ${}_2C_1 = {}_1C_1 + {}_1C_0 = 2$. Continuing, it is convenient to display the results in triangular form:

Pascal's Triangle THEOREM

The numbers ${}_nC_k$ can be arranged in the triangle shown below. The number ${}_nC_k$ is located on the nth row (rows are numbered downward; the 1 at the summit is the 0th row) and then counting to the kth entry from the left (again, the 1 at the left end of the row is the 0th entry).

```
            1
          1   1
        1   2   1
      1   3   3   1
    1   4   6   4   1
  1   5  10  10   5   1
1   6  15  20  15   6   1
```

Each entry in **Pascal's triangle** is determined by adding the two entries above its location on the previous row. For example, ${}_6C_3 = 20$ on the last row in the triangle above is obtained by adding ${}_5C_3 + {}_5C_2 = 10 + 10$ on the previous row. Thus, Pascal's triangle is constructed in accordance with the addition formula. The French mathematician and philosopher, Blaise Pascal (1623–1662) is credited with the discovery of his eponymous triangle.

Pascal's triangle is an intriguing pattern, but it is useful to calculate ${}_nC_k$ only when n is relatively small. We saw in Chapter 8 that one can calculate ${}_nC_k$ in more general situations with the following formula:

Combination Formula THEOREM

$${}_nC_k = \frac{n!}{k!(n-k)!}$$

To use the combination formula, cancel before multiplying.

EXAMPLE 18
Calculate ${}_{40}C_4$

From the combination formula, ${}_{40}C_4 = \frac{40!}{4!\,36!}$. Notice that $40! = 40 \times 39 \times 38 \times 37 \times 36!$. Thus we can cancel 36! and obtain

$${}_{40}C_4 = \frac{40 \times 39 \times 38 \times 37}{4 \times 3 \times 2 \times 1} = 91{,}390$$

To verify the combination formula, let ${}_nD_k = \frac{n!}{k!(n-k)!}$. It's our job to show that ${}_nC_k = {}_nD_k$. Recalling that $0! = 1$, we have ${}_nD_0 = \frac{n!}{n!0!} = 1$ and ${}_nD_n = \frac{n!}{0!n!} = 1$. Also, we will see that the numbers ${}_nD_k$ obey the addition formula:

$${}_{n+1}D_k = {}_nD_k + {}_nD_{k-1}$$

or

$$\frac{(n+1)!}{k!(n+1-k)!} = \frac{n!}{k!(n-k)!} + \frac{n!}{((k-1)!(n-k+1)!)}$$

To verify this equation, we have to add the two fractions on the right side. Because $k! = k \times (k-1)!$ and $(n-k+1)! = (n-k+1) \times (n-k)!$, the least common denominator is $k!(n-k+1)!$. Therefore,

$$\frac{n!}{k!(n-k)!} + \frac{n!}{((k-1)!(n-k+1)!)} =$$

$$\frac{n!(n-k+1) + n!k}{k!(n+1-k)!} = \frac{n!((n-k+1)+k)}{k!(n+1-k)!} = {}_{n+1}D_k$$

It follows that if we arrange the numbers ${}_nD_k$ in a triangle, as we did ${}_nC_k$, we will again get Pascal's triangle because the left and right edges are filled with 1s, and each interior entry is equal to the sum of the two entries above it. We thus conclude that ${}_nC_k = {}_nD_k$, and hence the combination formula holds. Efficient counting methods make it possible to compute the Banzhaf power index of large weighted voting systems. The method of counting combinations applies to systems in which most of the voters have the same weight, as in Alice's committee, considered in Example 9.

To use the technique of counting combinations, let's define a new term: a **coalition** is a set of voters who have agreed to vote "Yes" on an issue.

EXAMPLE 19
The Banzhaf Index of the Nine-Person Committee

In Example 9, we calculated the Shapley-Shubik power index of a committee with the voting system [6 : 3, 1, 1, 1, 1, 1, 1, 1, 1]. Alice has voting weight 3; the other 8 members each have voting weight 1.

Let S be a coalition of 3, 4, or 5 members of the committee that does not include Alice. We will suppose that Alice is undecided, and the weight-1 members who do not belong to S are in opposition. If Alice votes with S, the outcome will be "Yes," and if she joins the opposition, the outcome will be "No." In the two possible voting combinations corresponding to the way that Alice votes, Alice will be a critical voter. Alice's Banzhaf power index is $2 \times n$, where n is the number of coalitions with 3, 4, or 5 of the 8 weight-1 members; that is,

$$n = {}_8C_3 + {}_8C_4 + {}_8C_5 = 56 + 70 + 56 = 182$$

Thus, Alice's Banzhaf power index is $2 \times 182 = 364$.

When we calculated the Shapley-Shubik power index, we only had to consider Alice. The other members' indices could then be determined because the Shapley-Shubik indices of all the members add up to 1. There is no fixed sum of the Banzhaf power indices of all the participants, so we have to calculate the indices of the weight-1 voters separately. These voters do have the same voting power, so we have to consider only one of them, Martin. Now we consider a coalition with total weight of 5 and not including Martin. The committee members (other than Martin, he's undecided) who are not in the coalition are in opposition. There will be 2 voting combinations corresponding to each such coalition, depending on how Martin chooses to vote, and he will be a critical voter in each of them. Thus, Martin's Banzhaf index is equal to the number of such coalitions, multiplied by 2. There are two ways to assemble the coalitions:

- If Alice belongs to the coalition, 2 of the 7 weight-1 members other than Martin must be included. There are ${}_7C_2 = 21$ of these coalitions.
- If Alice does not belong to the coalition, then it must include 5 of the 7 weight-1 members (again, not including Martin). There are ${}_7C_5 = 21$ of these coalitions.

Adding the results of these calculations, we find that there are a total of 42 coalitions with total weight 5 and not including Martin. It follows that Martin's Banzhaf power index is $2 \times 42 = 84$.

To summarize, the Banzhaf power index of this voting system is

$$(364, 84, 84, 84, 84, 84, 84, 84, 84).$$

The total number of critical votes is $364 + 8 \times 84 = 1036$. Thus, according to the Banzhaf model, Alice has $\frac{364}{1036}$, or about 35 percent, of the power in the committee, and each weight-1 voter has $\frac{84}{1036}$, or about 8.1 percent, of the power. This is in pretty close agreement with the Shapley-Shubik model, where we found that Alice had $\frac{1}{3}$, or about 33 percent, of the power, while each weight-1 voter had $\frac{1}{12}$, or approximately 8.3 percent, of the power.

In Example 10, we determined the Shapley-Shubik power index of the voting system that results in the nine-person committee if two of the weight-1 voters (Chris and Dean) cede their votes to Bill, another weight-1 member. In the following example, we will determine the Banzhaf power index of the committee with this arrangement.

EXAMPLE 20
The Pact, Revisited

To determine the voting power of each voter in the system [6 : 3, 3, 1, 1, 1, 1, 1] by the Banzhaf model, let's return to the weight-1 voter, Martin. He has the opportunity to cast the deciding vote if there is a coalition with exactly 5 votes in favor, and he's undecided. There are two ways to achieve this total:

- The coalition could consist of just Alice and two of the four voters, other than Martin, who were not involved in the pact. There are ${}_4C_2 = 6$ of these coalitions.
- The coalition could include Bill, and two of the other 4 weight-1 voters. There are also ${}_4C_2 = 6$ such coalitions.

Thus, there are 12 voting combinations in which Martin can cast a deciding "yes" vote and another 12 voting combinations in which he would cast a deciding "no" vote. Therefore, Martin's Banzhaf power index is 24.

Now suppose that Alice is the undecided voter. She will able to cast a decisive vote if the coalition in favor has a total weight of 3, 4, or 5 votes.

- If Bill belongs to the coalition, and the coalition has no more than 2 of the 5 weight-1 members, then Alice has the opportunity to cast a decisive vote, "yes" or "no." The number of coalitions of this sort is

$${}_5C_0 + {}_5C_1 + {}_5C_2 = 1 + 5 + 10 = 16$$

- If Bill is opposed, the coalition could have 3, 4 or all 5 of the weight-1 members. The number of such coalitions is ${}_5C_3 + {}_5C_4 + {}_5C_5 = 16$.

It follows that Alice has the opportunity to cast a decisive "yes" vote in 32 voting combinations, and to cast a decisive "no" vote in another 32 voting combinations. Her Banzhaf power index is 64. Bill has the same Banzhaf power index, so the Banzhaf power index of this committee is (64, 64, 24, 24, 24, 24, 24).

The total number of critical votes in the committee is $64 \times 2 + 24 \times 5 = 248$ when the pact between Bill, Chris, and Dean is in effect. Thus, Alice and Bill each have $64 \div 248 = 25.8$ percent of the power, and each weight-1 member has $24 \div 248 = 9.7$ percent of the power, by the Banzhaf model. In Example 10 (on p. 387), we saw that according to the Shapley-Shubik model, Bill and Alice each had about 28.6 percent of the power and the weight-1 members had about 8.6 percent. The agreement between the two models is, as in the other examples that we have considered, pretty close.

If we compare the probabilistic interpretations of the two power indices, there is disagreement between the model in Example 10 and the one discussed here. Recall that the Shapley-Shubik index of a voter is equal to the probability that the voter will be pivotal in a randomly selected permutation. As noted, these probabilities are 28.6 percent for Alice or Bill and 8.6 percent for a weight-1 member, respectively. The probability that a voter will be critical in a randomly selected voting combination is equal to the voter's Banzhaf index divided by the total number of voting combinations (2^n if there are n voters). Alice and Bill each have a probability of $64 \div 2^7 = 50$ percent of casting a critical vote, while the probability that a weight-1 member will be a critical voter is $24 \div 2^7 = 18.75$ percent.

There are situations in which the differences between the models are significant by any measure, as we will see in the following example.

EXAMPLE 21
The Big Shareholder

Dana holds 100,000 shares of stock in a corporation. There are 1 million shares of stock, and the remaining stock is held by 9000 shareholders, each of whom has 100 shares. A weighted voting system, in which each shareholder's voting weight is equal to the number of shares that he or she owns, is used.

The Shapley-Shubik index of this system is determined by the same strategy that we used in Example 9 (Alice's nine-person committee, discussed on p. 386). Dana is pivotal when he appears in the 4002nd through the 5001st position. If he is 4002nd, then there are $4001 \times 100 = 400{,}100$ shares preceding him, and his 100,000 shares bring the total to a bare majority of 500,100 shares. If there are 5001 or more shares ahead of Dana, the 5001st, a small shareholder, would be the pivotal voter. Thus, Dana is pivotal in 1000 of the 9001 positions, and his Shapley-Shubik power index is $\frac{1000}{9001}$, or about 11.1 percent. The 9000 small shareholders have equal shares of the remaining power: The index of each is

$$\left(1 - \frac{1000}{9001}\right) \div 9000$$

which works out to be 0.0099 percent.

The Banzhaf index can be approximated by referring to its probabilistic interpretation. Imagine a randomly selected coalition involving the 9000 small shareholders. We could obtain such a combination by having each small shareholder toss a coin; those that get heads belong to the coalition.

Dana will have the opportunity to cast a critical vote "yes" or "no" if the number of voters in the coalition is at least 4001 and not more than 5000. Therefore, Dana's probability of casting a critical vote is equal to the probability of getting between 4001 and 5000 heads in 9000 coin tosses. On average, there would be 4500 heads, and by the Central Limit Theorem (see Chapter 8), we can see that the standard deviation is $\sqrt{9000} \times \sigma$, where σ is the standard deviation of the single-coin toss experiment where heads = 1 and tails = 0. For the single toss, $\sigma = \frac{1}{2}$. Therefore, the standard deviation for the number of heads when 9000 coins are tossed is $\sqrt{9000} \times \frac{1}{2}$ (approximately 50). For the 9000-coin toss, the Central Limit Theorem says that the number of heads is normally distributed with mean 4500 and standard deviation, $\frac{\sqrt{9000}}{2}$, which is less than 50. Thus, 68% percent of the voting combinations involving the weight-100 voters will have between 4450 and 4550 "yes" votes, 95 percent will have between 4400 and 4600 "yes" votes, and 99.7 percent will have between 4350 and 4650 "yes" votes. A voting combination with fewer than 4001 or more than 5000 "yes" votes would be more than 10 times the standard deviation away from the mean—a very unlikely event. Therefore, in almost 100 percent of the voting combinations, Dana's vote will be critical.

Daniel Acker/Bloomberg via Getty Images

Let's compare this with the probability that Martin, who has 100 shares, will be a critical voter. If the coalition in favor has exactly 500,000 shares, then Martin can cast a critical vote, "yes" or "no." Such a coalition can be assembled by including Dana and 4000 of the other shareholders, or by having 5000 of the 100-shareholders, but not Dana. Because the average number of heads in our coin-tossing analogy is 4500, the probability of getting a number of heads so far from this average—and Martin's chance of being a critical voter—is very small. Thus Dana is critical in almost all winning coalitions, and the number of winning coalitions in which Martin is critical is negligible by comparison. In the Banzhaf model, Dana has almost 100 percent of the power in this system.

SPOTLIGHT 11.4

The Electoral College: The Presidential Elections of 2012, 2016, and 2020

The following table displays the Shapley-Shubik (SSPI) and Banzhaf (BPI) power indices of the voters in the Electoral College, as compared with the voter's weight as a percent of 538 (PCT), the total weight of all the voters. It shows that for the most part, both measures of power agree closely with the actual share of power that a participant in the college has by virtue of its voting weight. There is an exception, though. California, whose voting weight is slightly more than 10 percent of 538, has more than its share of power by either measure. The power indices shown were calculated with the Power Index applet, which you can find at http://www.math.temple.edu/~conrad/Power/BPIandSSPI.html.

Voter	Weight	PCT (%)	SSPI (%)	BPI (%)
CA	55	10.22	11.10	11.43
TX	38	7.06	7.36	7.25
NY, FL	29	5.39	5.51	5.43
IL, PA	20	3.72	3.74	3.70
OH	18	3.35	3.35	3.32
GA, MI	16	2.97	2.97	2.95
NC	15	2.79	2.77	2.76
NJ	14	2.60	2.59	2.57
VA	13	2.42	2.40	2.39
WA	12	2.23	2.21	2.20
AZ, IN, MA, TN	11	2.04	2.02	2.02
MD, MN, MO, WI	10	1.86	1.83	1.83
AL, CO, SC	9	1.67	1.65	1.65
KY, LA	8	1.49	1.46	1.46
CT, OK, OR	7	1.30	1.28	1.28
AR, IA, KS, MS, NV, UT	6	1.12	1.09	1.10
NM, NV, UT, WV	5	0.93	0.91	0.91
HI, ID, NH, RI	4	0.74	0.73	0.73
AK, DE, DC, MT, ND, SD, VT, WY	3	0.56	0.54	0.55
ME, NE	2	0.37	0.36	0.37
Congressional districts (5 in all)	1	0.19	0.18	0.18

11.4 Comparing Voting Systems

We will say that a coalition is a **winning coalition** if it has sufficient voting power to pass a measure. If a coalition has insufficient voting power to pass a measure, it is a **losing coalition**. (We refer to *voting power* instead of weight to accommodate some unusual voting systems that we will encounter. In the case of weighted voting, a winning coalition is one whose total weight exceeds the quota.)

Different voting systems may have identical sets of winning coalitions. A dictatorship is no different if the dictator's weight is exactly equal to the quota or if it is much more. The dictator will have the same Banzhaf power index (2^n if there are n voters, since the dictator is the critical voter in every voting combination), and the same Shapley-Shubik power index (1, since the dictator is the pivotal voter in every permutation). To compare voting systems—which may be specified with weights or in some other way—we refer to the winning coalitions.

If there are just two voters, A and B, the empty coalition, { }, is surely a losing coalition, and $\{A, B\}$ is a winning coalition. There are only three distinct voting systems with just two voters: In the first, unanimous consent is required to pass a measure, so the only winning coalition is $\{A, B\}$. In the second, A is a dictator, and $\{A\}$ and $\{A, B\}$ are winning coalitions. In the third, B is a dictator, and the winning coalitions are $\{B\}$ and $\{A, B\}$. Although there is an unlimited number of ways to assign weights and a quota to a two-voter system, there are only three ways that the power can be distributed: A as dictator, B as dictator, or consensus rule.

Equivalent Voting Systems DEFINITION

Two voting systems are **equivalent** if there is a way for all the voters of the first system to exchange places with the voters of the second system and preserve all winning coalitions.

The weighted voting systems [50 : 49, 1] and [4 : 3, 3], involving pairs of voters A, B and C, D, respectively, are equivalent because in each system, unanimous support is required to pass a measure. We could have A exchange places with C, and B exchange places with D.

Now consider two voting systems [2 : 2, 1] and [5 : 3, 6] involving the same pair of voters, A and B. In the first, A is a dictator, while in the second, B dictates. By having A and B exchange places with each other, we see that the two systems are equivalent. "Equivalent" does not mean "the same." Voter A would tell you that the system where he is the dictator is not the same as the system where B is the dictator. The systems are equivalent because each has a dictator.

Every two-voter system is equivalent either to a system with a dictator or to one that requires consensus. As the number of voters increases, the number of different types of voting systems increases.

Minimal Winning Coalitions DEFINITION

A **minimal winning coalition** is a winning coalition in which each voter is a critical voter.

In a dictatorship, every coalition that includes the dictator is a winning coalition, but the only *minimal* winning coalition is the one that includes the dictator and no other voters.

EXAMPLE 22
Minimal Winning Coalitions: A Three-Voter System

The three-member committee from Example 5 (on p. 382) uses the voting system [6 : 5, 3, 1]. Let's refer to its members by their initials, *A*, *B*, and *C*, in order of decreasing weight. There are three winning coalitions. One, {*A*, *B*}, has weight 8, more than the quota, but it is minimal because both voters are critical. Another, {*A*, *C*}, with weight 6, is also minimal. The third winning coalition, {*A*, *B*, *C*}, is not minimal because only *A* is a critical voter.

EXAMPLE 23
The Four-Shareholder Corporation

Table 11.5 (on page 394) lists the five winning coalitions in the corporation with the voting system [51 : 49, 48, 2, 1]. In each coalition, the critical voters have been identified. The minimal ones are those in which each voter is marked as critical: {*A*, *B*}, {*A*, *C*}, and {*B*, *C*, *D*}. These minimal winning coalitions are displayed in Figure 11.1.

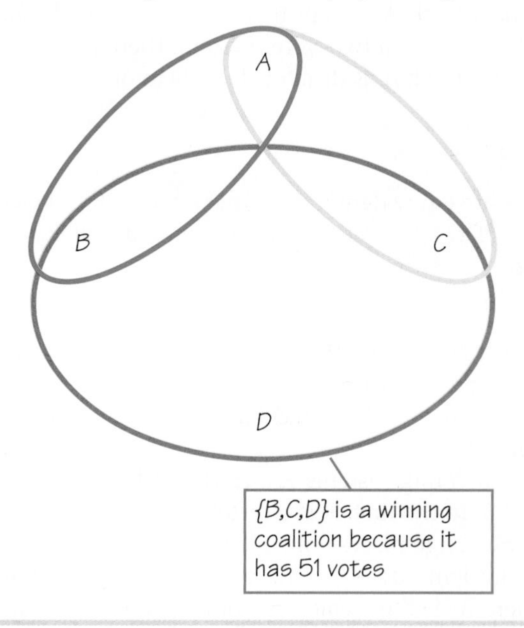

Figure 11.1 Each oval surrounds a minimal winning coalition for the four-shareholder corporation.

A voting system can be described completely by specifying its minimal winning coalitions. If you want to make up a new voting system, instead of specifying weights and a quota, you could make a list of the minimal winning coalitions. You would have to be careful that your list satisfies the following three requirements:

1. Your list can't be empty. You have to name at least one coalition; otherwise, there would be no way to approve a motion.
2. You can't have one minimal winning coalition that contains another one; otherwise, the larger coalition wouldn't be minimal.
3. Every pair of coalitions in the list has to overlap; otherwise, two opposing motions could pass.

In the four-shareholder corporation (see Figure 11.1), you can see that these requirements are satisfied. Now let's construct some voting systems.

EXAMPLE 24
Three-Voter Systems

We would like to make a list of all voting systems that have three participants, *A*, *B*, and *C*. To keep the size of the list manageable, we will insist that no two voting systems on the list be equivalent. To start, suppose that {*A*} is a minimal winning coalition. Requirement 3 tells us that every other minimal winning coalition must overlap with {*A*}, but the only way that could happen would be if *A* also belonged to the other coalition. In this case, requirement 2 would be violated. Thus, {*A*} must be the only minimal winning coalition. This is the voting system where *A* is dictator. Systems where *B* or *C* is dictator are not listed because they are equivalent to this one.

Now suppose that there is no dictator. Every minimal winning coalition must contain either two or all three voters. Let's consider the case in which {*A*, *B*, *C*} is a minimal winning coalition. It is the only winning coalition, because any other winning coalition would have to be entirely contained in this coalition, which requirement 2 doesn't allow. In this voting system, a unanimous vote is required to pass a measure. We will call this system *consensus rule*.

Finally, let's suppose that there is a two-voter minimal winning coalition, {*A*, *B*}. If it is the only minimal winning coalition, then a measure will pass if *A* and *B* both vote "yes" and the vote of *C* does not matter: In other words, *C* is a dummy, and *A* and *B* make all the arrangements. We will call this system the *clique*. Of course, the clique could be {*A*, *C*} or {*B*, *C*}, but these systems are equivalent to the one where {*A*, *B*} is the clique.

There could be two 2-voter minimal winning coalitions, say {*A*, *B*} and {*A*, *C*}. Neither coalition contains the other, and there is an overlap, so all of the requirements are satisfied. In this system, *A* has veto power. We encountered this system in Example 5 (on p. 382)—where Alice had veto power—and we will call it the *chair veto*. There are two other voting systems equivalent to this one, where *B* or *C* is chair.

It is possible that all three two-member coalitions are minimal winning coalitions. Because there are only three voters, any two distinct two-member coalitions will overlap, so the requirements are still satisfied. This system is called *majority rule*.

Table 11.6 lists all five of these three-voter systems. Each system can be presented as a weighted voting system, and suitable weights are given in the table. If we want to make a similar list of all types of four-voter systems, we can start by changing each three-voter system into a four-voter system. This is done by putting a fourth voter, *D*, into the system without including him or her in any of the minimal winning coalitions. This makes *D* a dummy. You may be interested to know that there

Table 11.6

Voting Systems with Three Participants

System	Minimal Winning Coalitions	Weights	Banzhaf Index
Dictator	{A}	[3 : 3, 1, 1]	(8, 0, 0)
Clique	{A, B}	[4 : 2, 2, 1]	(4, 4, 0)
Majority	{A, B}, {A, C}, {B, C}	[2 : 1, 1, 1]	(4, 4, 4)
Chair veto	{A, B}, {A, C}	[3 : 2, 1, 1]	(6, 2, 2)
Consensus	{A, B, C}	[3 : 1, 1, 1]	(2, 2, 2)

are an additional nine 4-voter systems that don't have any dummies. Try to list as many of these systems as you can.

EXAMPLE 25
The U.N. Security Council

The U.N. Security Council has 5 permanent members—China, France, Russia, the United Kingdom, and the United States—and 10 other members that serve two-year terms. To resolve a dispute not involving a member of the Security Council, 9 votes are required, including the votes of each of the permanent members. (Thus, each permanent member has veto power.) The U.N. Security Council voting system is thus specified by describing its minimal winning coalitions as consisting of the five permanent members and four other members. Exercise 38 (on p. 411) is an opportunity to consider this interesting voting system in depth.

EXAMPLE 26
The Scholarship Committee

A university offers scholarships on the basis of either academic excellence or financial need. Each application for a scholarship is reviewed by two professors, who rate the student academically, and two financial aid officers, who rate the applicant's need. If both professors or both financial-aid officers recommend the applicant for a scholarship, the Dean of Admissions decides whether to award a scholarship. Is it possible to assign weights to the professors, the financial-aid officers, and the dean to reflect this decision-making system? The answer is "no."

To see why, let's focus on the minimal winning coalitions. The participants are the two professors, *A* and *B*; the financial aid officers *E* and *F*; and the dean *D*. A scholarship will be offered if approved by the professors and the dean, or by the financial aid officers and the dean. Thus, the minimal winning coalitions (see Figure 11.2) are

$$\{A, B, D\} \quad \text{and} \quad \{D, E, F\}$$

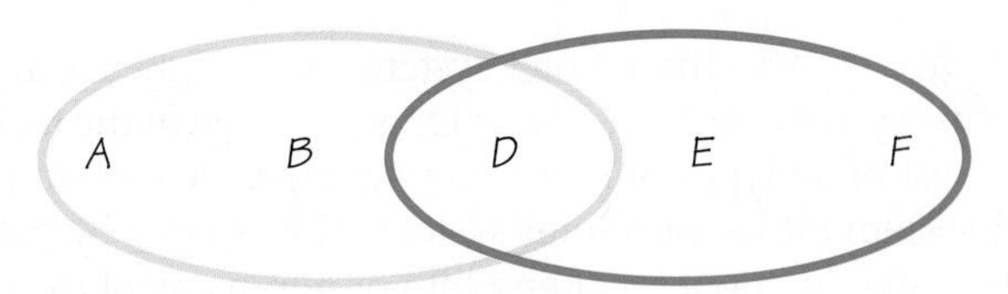

Figure 11.2 The Scholarship Committee: Minimal winning coalitions.

Consider the following two winning coalitions: In the first, all except the financial-aid officer *F*, favors an award; while in the second, only professor *B*, dissents.

$$C_1 = \{A, B, D, E\} \quad \text{and} \quad C_2 = \{A, D, E, F\}$$

In C_1, we notice that *A* is a critical voter and *E* isn't, while in C_2 the tables are turned because *E* is critical while *A* is not. If this were a weighted voting system, then in any winning coalition, the critical voters would all have greater weight than those who are not critical. Thus *A* would have to have both more weight than *E* (because of the situation in C_1) and less weight than *E* (because of C_2), which is impossible.

Although the scholarship committee is not equivalent to any weighted voting system, it is possible to determine the Shapley-Shubik and Banzhaf power indices of each participant.

EXAMPLE 27
Power Indices of the Scholarship Committee

The dean has veto power. Therefore, she will be the pivotal voter in any permutation where she appears last. If she is second-to-last in a permutation, she will still be the pivotal voter, because either both professors or both financial-aid officers must come before her. In the middle position, she will be pivotal if and only if both professors or both aid officers come first. Adding this up, we have $2 \times 4! = 48$ permutations in which the dean is in fourth or fifth position. There are four permutations of the form Prof, Prof, Dean, Aid, Aid because the professors and the aid officers can be in either order, and another four of the form Aid, Aid, Dean, Prof, Prof. The dean is not the pivotal voter when she is first or second because at least three people have to approve a scholarship. We conclude that the dean is pivotal in $48 + 4 + 4 = 56$ permutations in all. Her Shapley-Shubik power index is therefore $\frac{56}{5!} = \frac{7}{15}$. Each of the other participants is equally powerful, and they share the remaining $\frac{8}{15}$ of the power. Thus each professor and each aid officer has a Shapley-Shubik power index of $\frac{2}{15}$.

To compute the Banzhaf index, let's suppose that the dean is undecided, and list all the coalitions that she could cast a decisive vote by either joining or opposing. There are 7 of them:

$$\{A, B\}, \{E, F\}, \{B, E, F\}, \{A, E, F\}, \{A, B, F\}, \{A, B, E\}, \text{and } \{A, B, E, F\}$$

The dean can cast two decisive votes—with or against—each of these coalitions, making her Banzhaf power index 14. Now consider Professor A to be undecided. He can cast a decisive vote when the coalition in favor is one of the following three: $\{B, D\}$, $\{B, D, E\}$, or $\{B, D, F\}$. Because he can vote in two ways in each case, his Banzhaf power index is 6. The remaining participants, B, E, and F, have the same power, so the Banzhaf power index of the scholarship committee is

$$(14, 6, 6, 6, 6).$$

Comparing the two indices reveals that in the Shapley-Shubik model, the dean is $3\frac{1}{2}$ times as powerful as a faculty member or aid officer. In the Banzhaf model, she is only $2\frac{1}{3}$ times as powerful as one of the other members of the Scholarship Committee.

REVIEW VOCABULARY

Addition formula ${}_{n+1}C_k = {}_nC_k + {}_nC_{k-1}$ (p. 395)

Banzhaf power index A count of all voting combinations in which a voter can cast a decisive vote. This is a measure of the actual voting power of that voter. (p. 378)

Bit A binary digit: 0 or 1. (p. 391)

Binary number The expression of a number in base-2 notation. Let b_n denote the nth bit of a whole number N. Bits are numbered starting with the 0th bit on the right. For example, in the binary number 11001101, $b_0 = 1$, $b_1 = 0$, $b_2 = b_3 = 1$, $b_4 = b_5 = 0$, and $b_6 = b_7 = 1$. The decimal form of this number is $2^0 + 2^2 + 2^3 + 2^6 + 2^7 = 205$. (p. 391)

${}_nC_k$ The number of voting combinations in a voting system with n voters, in which k voters say "yes" and $n - k$ voters say "no." This number, referred to as "n choose k," is given by the formula

$${}_nC_k = \frac{n!}{k!(n-k)!}. \text{ (p. 396)}$$

Coalition The set of participants in a voting system who favor a given motion. A coalition may be empty (if, for example, the voting body is unanimously against a motion, the coalition in favor is empty), it may contain some but not all voters, or it may consist of all the voters. (p. 397)

Critical voter A voter who can reverse the outcome by switching his or her vote. (p. 389)

Dictator A participant in a voting system who can pass any issue even if all other voters oppose it and block any issue even if all other voters approve it. (p. 379)

Duality formula ${}_nC_k = {}_nC_{n-k}$ (p. 395)

Dummy voter A participant who has no power in a voting system. A dummy voter is never a critical voter in any voting combination and is never the pivotal voter in any permutation. (p. 379)

Equivalent voting systems Two voting systems are equivalent if there is a way for all the voters of the first system to exchange places with the voters of the second system and preserve all winning coalitions. (p. 401)

Extra votes The number of votes on the dominant side of a given voting combination that could be changed without altering the result. (p. 393)

Extra-votes principle A principle that states that in any voting combination, the critical voters are those on the dominant side whose voting weights exceed the combination's extra votes. (p. 393)

Factorial The number of permutations of n voters (or n distinct objects) is called n-factorial, or, expressed in symbols, $n!$. Because the empty set can be ordered in only one way, $0! = 1$. When n is a positive whole number, $n!$ is equal to the product of all the integers from 1 to n. If n has more than one digit, then $n!$ is a pretty big number: 10! is more than 3 million, and 1000! has 2568 digits. (p. 384)

Losing coalition A coalition that does not have the voting power to pass a motion. (p. 401)

Minimal winning coalition A winning coalition in which each member is a critical voter. (p. 401)

Pascal's triangle A triangular pattern of integers, in which each entry on the left and right edges is 1, and each interior entry is equal to the sum of the two entries above it. The entry that is located k units from the left (starting with $k = 0$), on the row n units below the vertex, is ${}_nC_k$. (p. 396)

Permutation A specific ordering from first to last of the elements of a set; for example, an ordering of the participants in a voting system. (p. 381)

Pivotal voter The first voter in a permutation who, with his or her predecessors in the permutation, will form a winning coalition. Each permutation has one and only one pivotal voter. (p. 381)

Power index A numerical measure of an individual voter's ability to influence a decision, the individual's voting power. (p. 390)

Quota The minimum number of votes necessary to pass a measure in a weighted voting system. (p. 378)

Shapley-Shubik power index The number of permutations of the voters in which a given voter is pivotal, divided by the number of permutations ($n!$ if there are n participants). This is a measure of the actual voting power of that voter. (p. 378)

Veto power A voter has veto power if no issue can pass without his or her vote. (p. 380)

Voting combination A list of voters indicating the vote of each on an issue. There is a total of 2^n combinations in an n-element set, and ${}_nC_k$ combinations with k "yes" votes and $n - k$ "no" votes. (p. 381)

Voting weight The number of votes assigned to a voter in a weighted voting system, or the total number of votes of all voters in a coalition. (p. 378)

Weighted voting system A voting system in which each participant is assigned a voting weight. (Different participants may have different voting weights.) A quota is specified, and if the sum of the voting weights of the voters supporting a motion is at least equal to that quota, the motion is approved. The notation $[q : w_1, w_2, \ldots, w_n]$ is used to denote a system in which there are n voters, with voting weights $w_1, w_2, \ldots, w_n$; and the quota is q. (p. 377)

Winning coalition A set of participants in a voting system who can pass a measure by voting for it. (p. 401)

SKILLS CHECK

1. In the weighted voting system [65 : 60, 30, 10],

(a) the weight-60 voter is a dictator.
(b) the weight-30 voter has veto power.
(c) the weight-10 voter is not a dummy.

2. A voting system has 20 voters, and a simple majority is needed to pass a motion. The quota for this system is ________________.

3. Two daughters and a son administer a trust fund. Each daughter has six votes, and the son has two votes; the quota for passing a measure is 8.

(a) The son is a dummy voter.
(b) The son is not a dummy voter but has less power than a daughter.
(c) The three siblings have equal voting power.

4. Four voters, A, B, C, D use the weighted voting system [6 : 4, 3, 2, 1]. In the permutation $DBCA$, the pivotal voter is ________________.

5. If the last voter in some permutation is the pivotal voter, then

(a) that voter must be the dictator.
(b) that voter has veto power.
(c) all the other voters must be dummies.

6. The Shapley-Shubik index of the weight-3 voter in the voting system [6; 4, 3, 2, 1] is __________.

7. The number 8! (8 factorial) is

(a) more than 1 million.
(b) between 10,000 and 1 million.
(c) less than 10,000.

8. A jury's decision in a criminal trial must be unanimous. In any permutation of the jury's members, the member who is pivotal is ________.

9. In how many ways can six voters respond to a "yes-no" question?

(a) 12
(b) 36
(c) 64

10. If a motion passes in the weighted voting system [6 : 4, 3, 2, 1] with only the weight-2 voter dissenting, then the critical voters are ______________.

11. Four voters, A, B, C, D use the weighted voting system [6 : 4, 3, 2, 1]. The Banzhaf index of B is

(a) 3.
(b) 6.
(c) 14.

12. In the voting system [19 : 8, 7, 6, 5, 4, 3, 2, 1], there is a total of _____ voting combinations. With the combination 10010101, does the motion pass? _______.

13. In the voting system [6 : 4, 1, 1, 1, 1, 1] the Banzhaf index of a weight-1 voter is

(a) $2 \times {}_4C_1$.
(b) $2 \times {}_5C_2$.
(c) $2 \times {}_6C_3$.

14. ${}_6C_3 =$ ________________________.

15. ${}_{12}C_3 =$

(a) ${}_{12}C_9$.
(b) 220.
(c) Both of the above answers are right.

16. ${}_{15}C_7 + {}_{15}C_8 =$ ______________________.

17. ${}_{10}C_5 + {}_{10}C_6 =$

(a) ${}_{11}C_6$.
(b) ${}_{11}C_5$.
(c) ${}_{11}C_7$.

18. The Banzhaf index of the weight-3 voter in the system [7 : 3, 2, 2, 2] is ________________.

19. If voter X is critical in every winning coalition, then

(a) X has veto power.
(b) X is a dictator.
(c) X will be needed to prevent a motion from passing.

20. In a system with n voters, A is a dictator. The Banzhaf index of A is ____________________.

21. If a winning coalition is minimal, the number of extra votes is

(a) zero.
(b) less than the weight of the least powerful member of the coalition.
(c) more than the weight of the least powerful member of the opposing coalition.

22. The minimal winning coalitions of the system [7 : 3, 2, 2, 2], with voters named A, B, C, and D, are ____________________________.

23. The weighted voting system [6 : 4, 3, 2, 1] is equivalent to

(a) [5 : 3, 2, 2, 1].
(b) [8 : 5, 4, 3, 2].
(c) [12 : 8, 5, 5, 2].
(d) All of the above answers are right.

24. If the minimal winning coalitions are {A, B} and {B, C, D}, __________ has veto power.

25. A voting system follows "majority rule." That is, there are $2n + 1$ voters all with weight 1, and the quota is $n + 1$. How many minimal winning coalitions are there?

(a) ${}_{2n+1}C_n$
(b) ${}_{2n+1}C_{n+1}$
(c) Both of the above answers are right.

26. A voter who appears in no minimal winning coalition is a ______________.

27. A voting system has the following winning coalitions: {A, B, C}, {A, B, D}, {B, C, D}, and {A, B, C, D}. Which voter is the most powerful?

(a) A
(b) B
(c) C
(d) D

28. The minimal winning coalitions of a voting system are {A, B, C}, {C, D, E}, and {B, C, D}. In the winning coalition {A, B, C, E}, ______________ are the critical voters.

29. Referring to the voting system in Skills Check Question 28, which voter is *not* critical in the winning coalition {A, C, D, E}?

(a) A
(b) C
(c) D
(d) E

30. In the voting system described in Skills Check Question 28, voter C has ______________ power.

CHAPTER 11 EXERCISES

■ Challenge ▲ Discussion

11.1 How Weighted Voting Works

▲ **1.** How would you explain to the weight-12 voter in the weighted voting system [27 : 14, 14, 13, 12] that he is a dummy?

▲ **2.** Consider a weighted voting system in which there are five voters who have weights 5, 4, 3, 2, and 1, respectively.

(a) What is the least possible quota, and what is the greatest?
(b) If the weight-5 voter has veto power, and no other voter does, what is the quota?
(c) If the weight-1 voter is a dummy, what is the quota?

3. Which voters, if any, have veto power in the weighted voting system [9 : 5, 4, 3]? Is any voter a dummy?

4. Given a voting system [q : 33, 32, 31, 4], such that exactly one of the voters has veto power, answer the following questions:

(a) Which voter has veto power?
(b) Find q.
(c) Is any voter a dummy?
(d) The voter with veto power wields more power than the others. Is there any difference in the power between the other three voters?

5. The various weighted voting systems used by the Board of Supervisors of Nassau County, New York turned out to be the mathematical quagmire described in Spotlight 11.3 (on p. 390). Before the county's weighted voting was declared unconstitutional by a federal district court in 1993, it was changed several times. The weights in use since 1958 were as follows:

		Weights					
Year	Quota	H_1	H_2	N	B	G	L
1958	16	9	9	7	3	1	1
1964	58	31	31	21	28	2	2
1970	63	31	31	21	28	2	2
1976	71	35	35	23	32	2	3
1982	65	30	28	15	22	6	7

Here, H_1 is the presiding supervisor, always from the community of Hempstead; H_2 is the second supervisor from Hempstead; and N, B, G, and L are the supervisors from the remaining districts: North Hempstead, Oyster Bay, Glen Cove, and Long Beach.

(a) In which years were some supervisors dummy voters?
(b) Suppose that the two Hempstead supervisors always vote together. In which years are some of the supervisors dummy voters?

6. Consider the weighted voting system [q : 10, 8, 7, 5, 4, 4, 3].

(a) Find all possible values of q for which no player has veto power.
(b) Find all possible values of q for which one and only one voter has veto power.
(c) Find all possible values of q for which the weight-3 voter is a dummy.

11.2 Shapley-Shubik Power Index

7. For the weighted voting system [51 : 30, 25, 24, 21]:

(a) List all permutations in which the weight-30 voter is pivotal.
(b) List all permutations in which the weight-25 voter is pivotal.
(c) Calculate the Shapley-Shubik index.

8. How would the Shapley-Shubik index in Exercise 7 change if the quota were increased to

(a) 52?
(b) 55?
(c) 58?

9. In the voting system [7 : 3, 2, 2, 2, 2, 2]:

(a) Describe the set of permutations in which the weight-3 voter is pivotal.
(b) How many of these permutations are there?
(c) Use the answer that you have given in part (b) to determine the Shapley-Shubik index of the system.

© Jeff Kowalsky/epa/Corbis

10. Refer to the permutation of the 2008 presidential election. (See Table 11.2 on p. 383.) The Obama-Biden ticket carried Iowa (IA), 828,940 to 682,379. Which state would be the pivotal voter if, at the last minute, the Obama–Biden opponents, McCain–Palin, had broadcast an ad that convinced 5000 voters to switch from the Democratic ticket to the Republican ticket? Assume that no votes are changed outside of Iowa.

11. D was a member of a committee until he discovered that with the voting system in use, he was a dummy. He then resigned from the committee. Did the Shapley-Shubik indices of the other committee members change as a result of *D*'s departure?

▲ **12.** Show that if a state uses the district system to choose its electors in a two-candidate presidential election, as Maine and Nebraska do, then some electoral permutations are impossible. Give an example of an impossible permutation. How would this affect the calculation of the Shapley-Shubik power index?

11.3 Banzhaf Power Index

13. A committee has four members, *A*, *B*, *C*, and *D*. It makes decisions by majority rule.

(a) What is the quota if each member has a voting weight of 1?

(b) List all the voting combinations in which the member named *A* is critical. Use binary notation: 1011 corresponds to the voting combination in which *A*, *C*, and *D* vote "yes," and *B* votes "no."

14. For each of the following weighted voting systems, make a list of all voting combinations in binary notation. Identify the critical voters in each combination, and calculate the Banzhaf power index of each voting system.

(a) [51 : 52, 48]
(b) [3 : 2, 2, 1]
(c) [8 : 5, 4, 3]
(d) [51 : 45, 43, 8, 4]
(e) [51 : 45, 43, 6, 6]

15. Make a table with all voting combinations for the weighted voting system [51 : 30, 25, 24, 21] in the left column (use binary notation). In the next column, put the outcome of the vote ("yes" or "no"), and in a third column, put the number of extra votes in favor of the outcome. This will enable you to identify the critical voters in each combination. Use this table to determine the Banzhaf power index of each participant.

16. If the quota for the voting system in Exercise 15 increases, you can quickly modify the table you made by reducing the extra votes—reverse an outcome when its extra votes become negative. As the number of extra votes decreases, more of a "yes" combination's voters will be critical—until the decision is reversed. Use this method to track changes in the Banzhaf index as the quota increases from the original 51 to 100.

17. Calculate the following terms:[1]

(a) ${}_7C_3$
(b) ${}_{50}C_{100}$
(c) ${}_{15}C_2$
(d) ${}_{15}C_{13}$

18. Calculate the following terms:

(a) ${}_6C_3$
(b) ${}_{100}C_2$
(c) ${}_{100}C_{98}$
(d) ${}_9C_5$

19. A committee has 10 members, and decides measures by weighted voting. The voting weight of the chairperson is 4; each of the 9 other members has weight 1, and the quota is 7. Determine the Shapley-Shubik and Banzhaf power indices of each member.

20. Agnes and Boris, weight-1 voters in the committee described in Exercise 19, cede their votes to a third weight-1 voter, Oxtyl. Recalculate the power indices to reflect this pact. How does the pact affect the voting power of the members who were not involved?

21. Refer to Exercise 5 for a brief history of weighted voting in the Nassau County Board of Supervisors. Assume that the two Hempstead supervisors always agree, so that the board is effectively a five-voter system. Determine the Banzhaf index of this system in each year. You should be able to do this by hand. If you would like to find the index for the full system each year, you may

[1](b) is not an error; how many ways can you (legitimately) get 100 "yes" votes from 50 voters?

use the Power Index calculator applet. (See the suggested Web site at the end of this chapter.)

22. If each member of a 12-person jury in which a unanimous decision is required tosses a coin to determine his or her decision, the probability that a given juror will cast a critical vote is $\frac{1}{2048}$. In some states, civil cases are tried before a 6-person jury, and the quota for a decision is 5 votes. With such a jury, what is the probability that a given juror will cast a critical vote, if each juror uses a coin toss to determine his or her vote?

23. You will need to use the Power Index applet for this. Find the total percentage of power that Nebraska has in the Electoral College by adding the percentages of power of the two electors representing the state and the three individual electors who represent congressional districts, according to the Banzhaf power index. Next, consider what would happen if Nebraska changed its law so that all of its electors would be committed to vote for the ticket that won the statewide contest: there would be 53 participants in the Electoral College, and Nebraska would have a voting weight of 5. Would this change result in an increase in Nebraska's percentage of Banzhaf power, or a decrease?

11.4 Comparing Voting Systems

24. Consider a four-person voting system with voters *A*, *B*, *C*, and *D*. The winning coalitions are {*A*, *B*, *C*, *D*}, {*A*, *B*, *C*}, {*A*, *B*, *D*}, {*A*, *C*, *D*}, and {*A*, *B*}.

(a) List the minimal winning coalitions.

▲ **(b)** A *minimal blocking coalition* is a set of voters that is voting against a measure, has sufficient voting weight to prevent the measure from passing, and in which each participant is a critical voter. Show that *A* has veto power and therefore that {*A*} is a minimal blocking coalition.

(c) Find another minimal blocking coalition.

(d) Determine the Banzhaf power index for this voting system.

■ **(e)** Find an equivalent weighted voting system. *Hint:* If two voters have the same Banzhaf index, give them the same weight.

(f) Calculate the Shapley-Shubik index of this system.

▲ **25.** In Exercise 24, the term *minimal blocking coalition* was defined. Must minimal blocking coalitions overlap, as minimal winning coalitions do?

▲ **26.** A five-member committee has the following voting system. The chairperson can pass or block any motion that she supports or opposes, provided that at least one other member is on her side. Show that this voting system is equivalent to the weighted voting system [4 : 3, 1, 1, 1, 1].

27. Find weighted voting systems that are equivalent to the following:

(a) A committee of three faculty members and the dean. To pass a measure, at least two faculty members and the dean must vote "yes."

(b) A committee of five faculty members, the dean, and the provost. To pass a measure, three faculty, the dean, and the provost must vote "yes."

▲ **28.** A four-member faculty committee and a three-member administration committee vote separately on each issue. The measure passes if it receives the support of a majority of each of the committees. Show that this system is not equivalent to a weighted voting system.

29. Calculate the Banzhaf index of the voting system in Exercise 28. Who is more powerful according to the Banzhaf model, a faculty member or an administrator?

30. Determine the Shapley-Shubik index of the system in Exercise 28. Who is more powerful according to the Shapley-Shubik model, a faculty member or an administrator?

▲ **31.** Explain why a voting system in which no voter has veto power must have at least three minimal winning coalitions.

■ **32.** How many distinct (nonequivalent) voting systems with four voters can you find? Systems that have dummies don't count. The challenge is to find all nine, and to find—if possible—weighted voting systems equivalent to each.

33. A corporation has four shareholders and a total of 100 shares. The quota for passing a measure is the votes of shareholders owning 51 or more shares. The number of shares owned by each shareholder is as follows:

A 48 shares
B 23 shares
C 22 shares
D 7 shares

There is also an investor, *E*, who is interested in buying shares but does not own any shares at present. Sales of fractional shares are not permitted.

(a) List the winning coalitions and compute the number of extra votes for each. Make a separate list of the losing coalitions, and compute the number of votes that would be needed to make the coalition winning.

(b) How many shares can A sell to B without causing any of the winning coalitions listed in part (a) to lose or any of the losing coalitions in part (a) to win?

(c) How many shares can A sell to D without changing the sets of winning or losing coalitions?

(d) How many shares can A sell to E without changing the winning coalitions? Because E is now a dummy, he must remain a dummy after the trade.

34. Which of the following voting systems is equivalent to the voting system in use by the corporation in Exercise 33?

(a) [6 : 2, 2, 2, 2, 1]
(b) [6 : 4, 2, 2, 2, 1]
(c) [8 : 4, 2, 2, 2, 0]
(d) [10 : 6, 4, 2, 2, 1]

35. A nine-member committee has a chairperson and eight ordinary members. A motion can pass if and only if it has the support of the chairperson and at least two other members, or if it has the support of all eight ordinary members.

(a) Find an equivalent weighted voting system.
(b) Determine the Banzhaf power index.
(c) Determine the Shapley-Shubik power index.
(d) Compare the results of parts (b) and (c): Do the power indices agree on how power is shared in this committee?

36. The New York City Board of Estimate consists of the mayor, the comptroller, the city council president, and the presidents of each of the five boroughs. It used to employ a voting system in which the city officials each had 2 votes and the borough presidents each had 1; the quota to pass a measure was 6. This voting system was declared unconstitutional by the U.S. Supreme Court in 1989 (*Morris v. Board of Estimate*).

(a) Describe the minimal winning coalitions.
(b) Determine the Banzhaf power index.

▲ **37.** Here is a proposed weighted voting system for the New York City Board of Estimate that is based on the populations of the boroughs:

[71 : 35, 35, 35, 11.3, 7.3, 9.6, 6.0, 1.8]

Find a simpler system of weights that yields an equivalent voting system.

▲ **38.** The voting system in use by the U.N. Security Council is described in Example 25 (on p. 404).

(a) Show that this voting system is equivalent to the weighted voting system [39 : 7, 7, 7, 7, 7, 1, 1, 1, 1, 1, 1, 1, 1, 1, 1]. (Each permanent member has 7 votes; the other members each have 1 vote.)

(b) Compute the Banzhaf index for the Security Council.

▲ **(c)** The Security Council originally had 5 permanent members and 6 members who served two-year terms. Each permanent member had veto power, and 6 votes were required to resolve an issue. Devise an equivalent weighted voting system and compute its Banzhaf index. Do you think that the addition of 4 more non-permanent members caused the permanent members to lose significant power?

39. Find the minimal winning coalitions of the weighted voting system [7 : 3, 3, 3, 1, 1, 1] and determine the Banzhaf index.

40. A new weight-1 voter joins the system described in Exercise 39. Again, describe the minimal winning coalitions and determine the Banzhaf power index. Does the presence of this new voter increase or decrease the share of power of each weight-1 voter?

41. Compute the Shapley-Shubik power index for the systems in Exercises 39 and 40. How does the addition of the new voter affect the power of the other three weight-1 voters?

▲ **42.** An alumni committee consists of 3 rich alumni and 12 recent graduates. To pass a measure, a majority, including at least 2 of the rich alumni, must approve. Is this equivalent to a weighted voting system? If so, find the weights and a quota; if not, explain why not.

43. List the minimal winning coalitions in the following three-voter systems, and identify each as equivalent to one of the systems listed in Table 11.6 on page 403.

(a) [10 : 9, 5, 1]
(b) [10 : 9, 7, 3]
(c) [10 : 11, 7, 1]
(d) [10 : 6, 4, 2]
(e) [10 : 5, 4, 3]

44. In Skills Check Questions 28–30 (on p. 408), we considered a five-voter system in which the minimal winning coalitions are {A, B, C}, {B, C, D}, and {C, D, E}.

(a) Show that this system is not equivalent to any weighted voting system.
(b) Explain why A and E have equal power, and B and D have equal power.
(c) Make a list of all of the winning coalitions and use it to determine the Banzhaf power index of each voter.

1. The most important weighted voting system in the United States is the Electoral College (see Spotlight 11.1 on p. 378). Three alternate methods to elect the president of the United States have been proposed:

- *Direct election.* The Electoral College would be abolished, and the candidate receiving a plurality of the votes would be elected. Most versions of this system include a runoff election or a vote in the House of Representatives in cases where no candidate receives more than 40 percent of the vote.
- *District system.* This system could be adopted by individual states without amending the U.S. Constitution or passing a federal law. It is now used by two states, Maine and Nebraska. In each congressional district, and in the District of Columbia, the candidate receiving the plurality would select one elector. Furthermore, in each state, including the District of Columbia, the candidate receiving the plurality would receive two electors.
- *Proportional system.* Each state and the District of Columbia would have fractional electoral votes assigned to each candidate in proportion to the number of popular votes that candidate received. With this system, if a candidate received 25 percent of the vote in New Mexico, then that candidate would receive 25 percent, or 1.25, of New Mexico's five electoral votes. Obviously, no actual electors would be involved.

Determine the outcome of a recent election under each of these alternatives. Should the present Electoral College, operating under the unit rule, be replaced by one of these systems? Reference: *The Presidential Election Game*, by Steven Brams, which contains useful references to Senate hearings on Electoral College reform.

2. Write an essay on weighted voting in the Council of Ministers of the European Union. There are 27 member nations, with voting weights ranging from 29 for Germany, France, the United Kingdom, and Italy to 3 for Malta. The sum of the voting weights of all nations in the European Union is 345, and the quota to pass a measure is 255. There is also a requirement that at least 14 of the member nations must support a measure if it is to pass. Thus, the voting system for the European Council of Ministers is not actually presented as a weighted voting system. Is it equivalent to a weighted voting system?

3. California has 10.22 percent of the votes in the Electoral College, but according to Spotlight 11.4 (on p. 400), that state has more than 11 percent of the power in the Electoral College, as measured by either of our power indices. Discuss the appropriateness of each power index as a measure of voting power in the Electoral College. Is the disproportionate power of California in the Electoral College a problem that the United States should address? Assume that California has acquired additional congressional seats as a result of migration. Calculate the Banzhaf index when California has 65, 75, and 100 electors. In each case, the electoral votes that are to be awarded to California are taken from other states. What would happen if all states, except California, adopted the district system for choosing electors? See Writing Project 1 for a discussion of the district system.

4. Spotlight 11.5 demonstrates that an individual voter's probability of being a critical voter in a randomly selected voting combination is inversely proportional to the square root of the number of votes cast. If the voter is electing a representative in a weighted voting system (such as the Electoral College or a minister in the European Union), that representative's chance of being a critical voter is proportional to its Banzhaf index. Thus, the individual voter's probability of casting a vote that would change an outcome (of a U.S. presidential election, for example) is proportional to the voting weight of the representative, divided by the square root of the number of votes cast in the election of the representative. Using the principle that each individual voter should have the same chance of influencing the outcome, it follows that each representative's Banzhaf index should be proportional to the square root of the state's voting population. This is the Banzhaf square root rule. By this rule, what should the weight of each state be in the Electoral College, or in the Council of Ministers of the European Union? *Suggestion*: Make the voting weights roughly proportional to the square roots of the populations, determine the Banzhaf indices, and then make adjustments as necessary.

SPOTLIGHT 11.5

What Are Your Chances of Being a Critical Voter?

An answer to this question was given by Lionel Penrose (1898–1972), a famous geneticist, and father of Sir Roger Penrose, whose tilings you will encounter in Chapter 20. For simplicity, suppose there is an odd number of voters, and let the number be $2n + 1$. Each voter has weight 1 and the quota is a simple majority of $n + 1$ votes. If you belong to a winning coalition and your vote is critical, then the coalition must contain, in addition to you, exactly n other voters. The number of such coalitions is the number of voting combinations of the other $2n$ voters with n "yes" votes; that is,

$$_{2n}C_n = \frac{(2n)!}{(n!)^2}$$

combinations. There is a total of 2^{2n} voting combinations involving the other $2n$ voters, so your probability, which we call P, of being a critical voter—where your vote really makes a difference—is

$$P = \frac{_{2n}C_n}{2^{2n}} = \frac{(2n)!}{2^{2n}(n!)^2}$$

With the number n in the tens of thousands (for a municipal election) up to the tens of millions (for a statewide election in a large state), it seems hopeless to calculate this probability. A formula discovered by Scottish mathematician James Stirling (1692–1770) for factorials tames the expression. It gives an approximation of $n!$, which is quite accurate when n is large. If you would like to see his formula, it can be found by searching the Web for "Stirling's Formula." By using this formula, we find that P is approximately $\frac{1}{\sqrt{\pi n}}$. This approximation improves as n increases, as you can see in the following table.

n	$\frac{_{2n}C_n}{2^{2n}}$	$\frac{1}{\sqrt{\pi n}}$
1	0.5	0.56
5	0.246	0.252
25	0.1123	0.1128
125	0.05041	0.05046

Thus, if 2 million votes are cast in addition to yours ($n = 1{,}000{,}000$) your probability of casting a critical vote, a situation in which your candidate wins by 1 vote, is $\frac{1}{\sqrt{1{,}000{,}000\pi}} = 0.000564$.

This represents a small chance, but your odds are better than they are for winning a big lottery prize! History has recorded very few elections that were decided by a 1-vote margin. Perhaps this is because the mathematical model does not involve politics. In fact, it assumes that voters decide their preferences by tossing coins rather than paying attention to the candidates. The model's purpose is to analyze the voting system itself, and ignore human behavior.

Suggested Readings

BANZHAF, JOHN F. III. Weighted voting doesn't work, *Rutgers Law Review* (1965) Vol. 19, pp. 317–43. The author defines the Banzhaf index and uses it to show that the weighted voting system in use by the Nassau County Board of Supervisors was unfair.

BRAMS, STEVEN J. *Game Theory and Politics*, 2d Ed., Dover Publications, New York, 2004.

Iannucci v. Board of Supervisors of Washington County. This case opened a "mathematical quagmire." You can find the text of the court's opinion with a web search engine.

FELSENTHAL, DAN S., and MOSHÉ MACHOVER, *The Measurement of Voting Power: Theory and Practice, Problems and Paradoxes*, Edward Elgar, Cheltenham, U.K., 1998. This book has a detailed analysis of the Council of Ministers of the European Union, and a thorough treatment of the power indices mentioned in this chapter.

TAYLOR, ALAN D. *Mathematics and Politics: Strategy, Voting Power, and Proof*, Springer-Verlag, New York, 1995. Chapter 4 of this book covers weighted voting systems and their analysis using the Shapley-Shubik and Banzhaf power indices. It has no mathematical prerequisites, but it does include carefully written logical arguments that must be carefully read.

Suggested Web Sites

Temple University's math Web site (**http://www.math.temple.edu/~conrad/Power/BPIandSSPI.html**) provides a Java applet for calculating the Shapley-Shubik and Banzhaf power indices for systems that are too large to be solved with pencil and paper, such as the U. S. Electoral College and the Council of Ministers of the European Union.

AP Photo/Jae C. Hong

12

Electing the President

Electing the president of the United States has been a tricky business since the founding of the republic because of the **Electoral College**—the body that elects the president, currently made up of 538 members, which we analyzed in Chapter 11. As an example, in the 2000 election, Al Gore received 537,000 more popular votes (0.5 percent) than did George W. Bush, but Bush won the electoral-vote tally by 4 votes. The outcome of this election turned on who would win Florida. Thirty-six days after the election, the Supreme Court, in a 5–4 decision, blocked further vote recounts in disputed Florida counties. By winning in Florida by a razor-thin margin of 537 votes (less than 0.01 percent of those cast), George W. Bush won the presidency.

Although analysts thought there might be another divided outcome in 2004, George W. Bush beat John Kerry both in electoral votes (286 to 252) and in popular votes (by 3.3 million, or 2.8 percent). And in 2008, Barack Obama handily defeated John McCain in electoral votes (365 to 173) and in popular votes (by 9.5 million, or 7.4 percent).

In this chapter, we will focus on how presidential elections, in general, can be modeled as games, and what insights mathematics can provide about the strategic aspects of campaigning and elections. (In Chapter 15, we provide a more systematic introduction to game theory and discuss a variety of other applications of the theory.)

The first phase of a presidential election begins in January, when Democratic and Republican candidates seek their party's nomination for president by running in state caucuses and primaries. In the general election in the fall, there are typically only two serious contenders—the nominees of the Democratic and Republican parties—but occasionally there are other significant candidates. One example is Ross Perot, who ran under the banner of the Reform Party in 1992 and garnered 19 percent of the popular vote. Although no minor-party candidate has ever won a presidential election, some have affected which of the two major-party nominees did win, as we will see.

What can mathematics tell us about presidential elections? First, it can clarify what are better and worse campaign strategies in each phase. In addition, it can shed light on the likely effect that different election reforms would have on both campaign strategies and election outcomes. In fact, we will analyze two prominent reform proposals and indicate how they might have changed the outcome in the 2000 election: (1) the use of approval voting and (2) the abolition of the Electoral College and its replacement by direct popular-vote election of the president. We will also briefly discuss a recent third proposal—that states pledge to cast all their electoral votes for the *national* popular-vote winner. Throughout, we will highlight general results that emerge from the mathematical analysis.

We begin by looking at how candidates position themselves in presidential primaries to win their party's nomination. The principal tool of analysis is **spatial models**, which we will describe in the next section and apply to both two-candidate and multicandidate elections.

12.1 Spatial Models for Two-Candidate Elections

While two-candidate contests are most common in the general election, sometimes the nomination race in the Democratic Party or Republican Party also comes down to a contest between just two contenders. As a case in point, Gerald Ford faced one major opponent, Ronald Reagan, in the 1976 Republican race. This race was not decided until the Republican national convention in August, when Ford edged out Reagan in a close vote. The Democratic Party contest between Barack Obama and Hillary Clinton also was close in 2008, though Obama wrapped up his party's nomination well before its national convention.

To model such elections, we assume that voters respond to the positions that candidates take on issues. This is not to say that other factors—such as personality, ethnicity, religion, and race—have no effect on election outcomes, but rather that issues take precedence in a voter's decision.

© RICK WILKING/Reuters/Corbis

How can the position of a candidate on issues be represented? We start by assuming that there is a single overriding issue, or set of issues, on which the candidates must take a definite stand, such as the degree of governmental intervention in the economy. We assume that the attitudes of voters on this issue or dimension can be represented along a left–right continuum, ranging from very liberal on the left (much intervention) to very conservative on the right (little intervention).

To derive conclusions about the behavior of voters from their attitudes and the positions candidates take in a campaign, some assumption is necessary about how voters decide for whom to vote. More important than the attitudes of *individual* voters, however, are the *numbers* of voters who have particular attitudes along the left–right continuum.

Unimodal Distribution

Voter Distribution DEFINITION

A **voter distribution** is a curve that gives, on the vertical axis, the number (or percentage) of voters who have attitudes at that point on the left-right continuum.

The greater the vertical height of the curve, the greater the number of voters who voters have attitudes at that point in the continuum. Figure 12.1a (on page 418) shows one such distribution, which is **unimodal**.

Mode and Unimodal Distribution DEFINITION

A distribution that has one peak or highest point, called the **mode**, is **unimodal**.

For simplicity, we picture the distribution as continuous (see Section 8.4 on p. 297), although in fact, because the number of voters is finite, there cannot be voters at all points along the continuum.

More important than the mode, from the viewpoint of the candidates, is the **median *M*** of a distribution.

Median DEFINITION

The **median *M*** of a voter distribution is the point on the horizontal axis where half the voters have attitudes that lie to the left and half to the right.

The notion of the median of a voter distribution is the same as the notion of a median of a data sample given in Chapter 5 (see Section 5.4 on p. 176).

The Figure 12.1a distribution is *symmetric*: the curve to the left of *M* is a mirror image of the curve to the right. Thus, the same numbers of voters have attitudes that are equal distances to the left and to the right of *M* along the horizontal axis.

Although the *attitudes* of voters are a fixed quantity in the calculations of the candidates, the *decisions* of voters will depend on the positions that the candidates take. Assuming the candidates know the distribution of voter attitudes, what positions are optimal for them?

Assume candidates *A* (red) and *B* (blue) take the positions along the left–right axis shown in Figure 12.1a, where candidate *A* is to the left of *M* and candidate *B* is to the right. Assume that all voters vote for the candidate whose position is closer to their own, and that all voters vote. (We will consider modifications of this simplifying assumption later in the exercises.) Then *A* will certainly attract all the voters to the left of his position, and *B* all the voters to the right of her position. If both candidates are an equal distance from *M*, as shown in Figure 12.1a, they will split any votes in the middle, with those to the left of *M* going to *A* and those to the right going to *B*.

Can either candidate do better by changing his or her position? If *B*'s position remains fixed to the right of *M*, *A* could move alongside *B*, just to her left, and capture all the votes to *B*'s left, as illustrated in Figure 12.1b. Because *A* would have moved to the right of *M*, he would, by changing his position in this manner, receive a majority of the votes and thereby win the election.

By analogous reasoning, there is no reason for *B* to stick to her original position to the right of *M*. By approaching *A*'s original position to the left of *M*, *B* can capture all the votes to *A*'s right (Figure 12.1c). In other words, both candidates, acting rationally, should approach each other and *M*.

If *A* were to move rightward past *M*, but *B* moved leftward only as far as *M*, their positions would be as shown in Figure 12.1d. Now *B* would receive not only the 50 percent of the votes to the left of *M* but also some of the votes that lie between *B*'s position at *M* and *A*'s position (now to the right of *M*).

Clearly, *A* loses by crossing *M* from the left. Hence, there is an incentive for both candidates to move toward *M* but not overstep it. In fact, taking a position at *M* maximizes the minimum number of votes a candidate can guarantee for himself or herself.

Maximin DEFINITION

A position is **maximin** for a candidate if there is no other position that can *guarantee* a better outcome—more votes for that candidate—whatever position the other candidate adopts.

If both candidates choose *M*, voters will be indifferent to the choice between them on the basis of their positions alone and would presumably make their choice on other grounds.

Taking a position at *M*, however, guarantees *A* at least 50 percent of the total vote *no matter what B does*. Moreover, there is no other position that can guarantee *A* more votes, and likewise for *B*.

M is also *stable*, because if one candidate adopts this position, the other candidate has no incentive to choose any position other than *M*. Thus, *M* is both the maximin position for each candidate (it offers a guarantee of a minimum of 50 percent of the votes) and, if *M* is chosen by both candidates, then these choices are in equilibrium (one candidate does worse by departing from it if the other candidate stays at it).

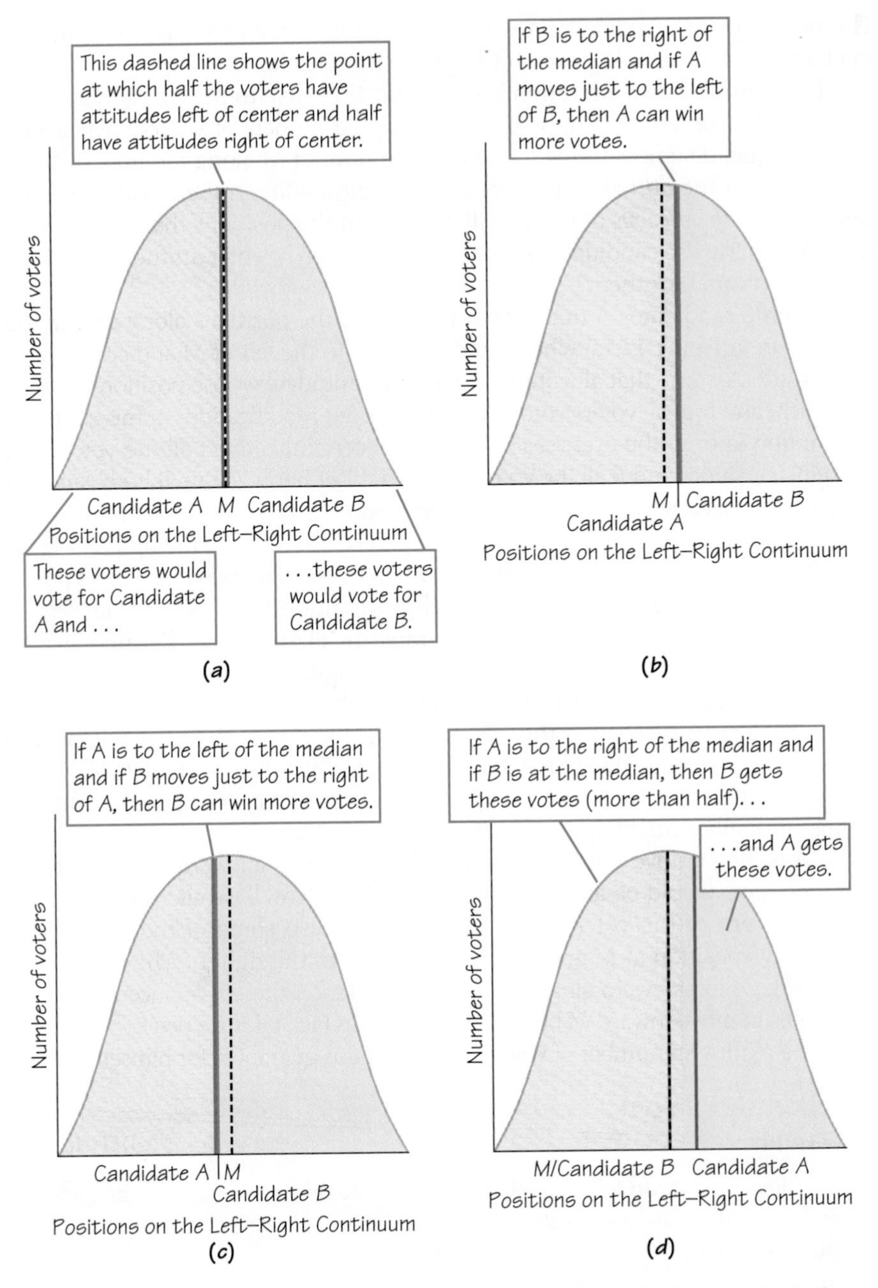

Figure 12.1 Unimodal distribution. The median *M* is the point on the horizontal axis that divides the area under the distribution curve—measuring the number of voters—exactly in half.

Equilibrium DEFINITION

A pair of positions is in **equilibrium** if, once chosen by both candidates, neither candidate has an incentive to depart from it unilaterally (that is, by himself or herself).

More formally, we have the median-voter theorem.

Median Voter Theorem THEOREM

Median-voter theorem: In a two-candidate election with an odd number of voters, M is the unique equilibrium position. (The theorem is applicable if there is an even number of voters, as we will show later.)

We have already shown that if both candidates choose M, these choices are in equilibrium. Is there another equilibrium position or positions? There are two possibilities: (1) It is the same position for both candidates, which we call a *common position,* or (2) it is two distinct positions, one taken by each candidate. If it were a common position, suppose it is to the left of M. (An analogous argument works if it is to the right.) Then one candidate can always do better by moving rightward but staying to the left of M. This contradicts the supposition that the common position is in equilibrium. Now suppose the equilibrium were two distinct positions. Then one candidate can always do better by moving alongside the other candidate but staying closer to M. This contradicts the supposition that these two positions are in equilibrium. Thus, in both cases, one candidate would have an incentive to depart from his or her position—holding the position of the other candidate fixed—so a nonmedian position of one or both candidates cannot be in equilibrium. Therefore, M is the only equilibrium position.

Unimodal Distribution: Median and Mean Different

The median-voter theorem is applicable *whatever* the distribution of the electorate's attitudes. Consider the distribution in Figure 12.2a, which is unimodal but is not symmetric. Applying the logic of the previous analysis, M is once again the maximin and equilibrium position of two candidates, even though the bulk of voters are concentrated at the mode (the higher peak toward the right).

We next compare the median M with the mean, which may be quite different:

Mean of a Voter Distribution DEFINITION

The **Mean** $\bar{l}$ of a voter distribution is

$$\bar{l} = \frac{1}{n}\sum_{i=1}^{k} n_i l_i$$

where

k = number of different positions i that voters take on the continuum

n_i = number of voters at position i

l_i = location of position i on the continuum

$n = \sum_{i=1}^{k} n_i = n_1 + n_2 + \cdots + n_k$ = total number of voters

The symbol Σ (sigma) is the *summation sign*. It signifies that all subscripted terms to its right (e.g., in the definition of n), beginning with the subscript 1 and continuing to the subscript k, are summed and a total obtained.

The notion of a mean is the same as that for data samples in Chapter 5, except that here we are calculating a *weighted* average: The location l_i of each position is weighted by the number of voters, n_i, at that position. Thus, the mean can be thought of as the position of a typical voter—that is, the expected position of a voter drawn randomly from the set of all voters.

Figure 12.2 Unimodal distribution in which the median and mean are different. The distribution is skewed to the left; the median M is to the right of the mean $\bar{l}$.

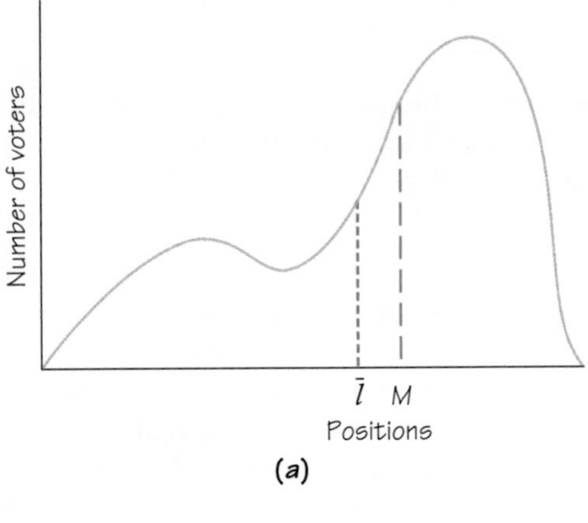

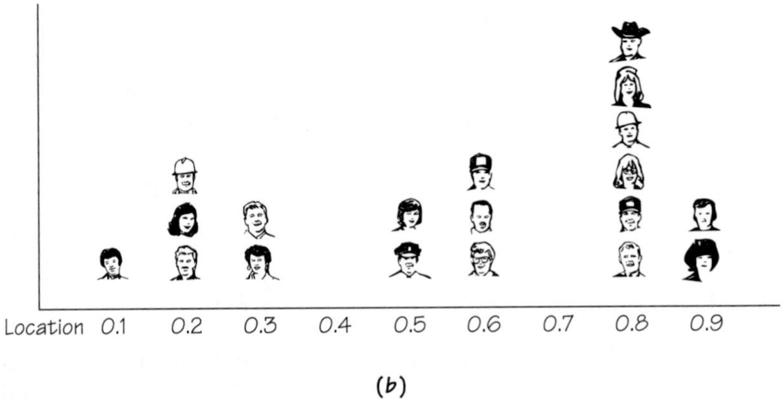

Algebra Review Summation Notation

A *sequence* of numbers is an ordered list of numbers. If there are k numbers in a list, the sequence can be written as $a_1, a_2, \ldots, a_k$. A *summation* is the addition of the numbers in such a sequence. Summation notation has the following parts:

$$\sum_{i=1}^{k} a_i$$

Summation Symbol; Upper Bound of Summation (k); Terms of Sequence (a_i); Index of Summation (i); Lower Bound of Summation (1)

Let's take $\sum_{i=1}^{3} i^2$ as an example. The index (i) indicates that the summation starts at 1 and that successive values of i can be found by adding 1 until the upper bound of 3 is reached. Thus,

$$\sum_{i=1}^{3} i^2 = 1^2 + 2^2 + 3^2 = 14$$

We can find $\sum_{i=1}^{5} 3i$ as follows:

$$\sum_{i=1}^{5} 3i = 3(1) + 3(2) + 3(3) + 3(4) + 3(5)$$
$$= 3 + 6 + 9 + 12 + 15 = 45$$

We also could simplify $\sum_{i=1}^{5} 3i$ as

$$3\sum_{i=1}^{5} i = 3(1 + 2 + 3 + 4 + 5) = 3(15) = 45.$$

If there are two lists, make sure to pair the values from both lists correctly. For example, $\sum_{i=1}^{5} a_i b_i$ can be found for the following table:

i	1	2	3	4	5
a at i^th^ position	2	−5	0	3	−1
b at i^th^ position	3	2	−3	1	4

$$\sum_{i=1}^{5} a_i b_i = 2(3) + (-5)(2) + 0(-3) + 3(1) + (-1)(4)$$
$$= 6 + (-10) + 0 + 3 + (-4) = -5$$

The mean $\bar{l}$ need not coincide with the median M. As an illustration of this point, consider the following **discrete distribution** of $n = 19$ voters at $k = 7$ different positions over the interval between 0 and 1, or [0, 1], which is illustrated in Figure 12.2b.

Discrete Distribution of Voters DEFINITION

A **discrete distribution of voters** is one in which voters are located at only certain positions—not all points—along the left–right continuum.

Position, i	1	2	3	4	5	6	7
Location (l_i) of position i	0.1	0.2	0.3	0.5	0.6	0.8	0.9
Number of voters (n_i) at position i	1	3	2	2	3	6	2

Whereas M is 0.6 because 8 voters lie to the left on the continuum and 8 voters lie to the right, the mean is different:

$$\bar{l} = \left(\frac{1}{19}\right)[1(0.1) + 3(0.2) + 2(0.3) + 2(0.5) + 3(0.6) + 6(0.8) + 2(0.9)] = 0.56$$

Taking a position at 0.56 against an opponent who takes a position at 0.6, a candidate would lose the election by 11(3 + 6 + 2) to 8(1 + 3 + 2 + 2) votes.

The distributions of Figures 12.2a and 12.2b are **skewed** to the left because the area under the curve, or the number of voters, is less concentrated to the left of M than to the right. These more spread-out voters put the mean $\bar{l}$ to the left of the median M. The lesson we derive from these figures is that it may *not* be rational for a candidate to take a position at $\bar{l}$ if the distribution is skewed, either to the right or to the left. The mean $\bar{l}$ describes the position of the average voter, but not the optimal position of a candidate.

Different Unimodal Distributions

A sufficient condition for M and $\bar{l}$ to coincide is that the distribution be **symmetric**, but this condition is not necessary: M and $\bar{l}$ can coincide if a distribution is asymmetric, as illustrated in Figure 12.3a. When M and $\bar{l}$ coincide, a candidate need not take a different position to ensure victory—or at least prevent defeat if his or her opponent adopts the same position. However, as Figure 12.3 demonstrates, the noncoincidence of M and $\bar{l}$ is not necessarily related to the lack of symmetry in a distribution: Half the voters may still lie to the left, and half to the right, of $M/\bar{l}$ if the distribution is asymmetric.

What can we say about equilibrium positions if there is an even number of voters? For example, consider the following discrete distribution of $n = 26$ voters at $k = 8$ different positions over the interval [0, 1], which is illustrated in Figure 12.3b.

Position, i	1	2	3	4	5	6	7	8
Location (l_i) of position i	0	0.2	0.3	0.4	0.5	0.7	0.8	0.9
Number of voters (n_i) at position i	2	3	4	4	2	3	7	1

Figure 12.3 Different unimodal distributions. The median M and mean $\bar{l}$ coincide only in part (a).

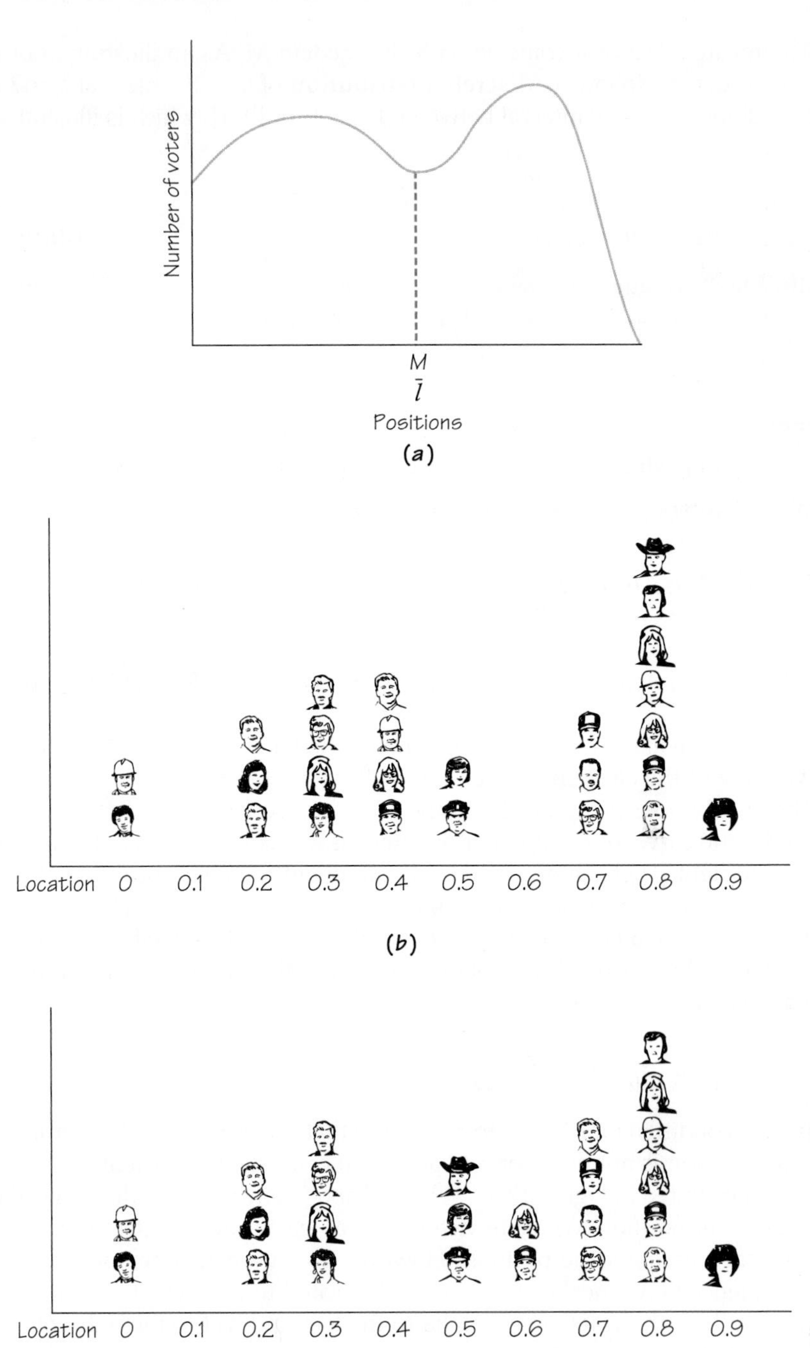

We begin by calculating the mean $\bar{l}$:

$$\bar{l} = \left(\frac{1}{26}\right)[2(0) + 3(0.2) + 4(0.3) + 4(0.4) + 2(0.5) + 3(0.7) + 7(0.8) + 1(0.9)] = 0.5$$

The median M is not 0.5. For an even number of voters, as in this example, M is the average of the two middle positions; at this average, 13 voters lie to the left and 13 to the right. The two middle voters are the 13th and 14th voters when they are lined up in the order of their positions from left to right. The 13th voter is at position 0.4, and the 14th voter is at position 0.5, so M is 0.45. Thus, this discrete distribution does not mimic the continuous distribution in Figure 12.3a, in which the median and mean coincide.

Note that if both candidates position themselves at the median M, their positions will be in equilibrium, as we showed earlier. But it is not the unique equilibrium pair. In this example, there are many other pairs of equilibrium positions. For instance, the same reasoning shows that any pair of positions between 0.4 and 0.5 will be in equilibrium. Moreover, it is easy to show that position 0.4 for one candidate and position 0.5 for the other candidate are in equilibrium. The candidate at 0.4 will get the support of the 13 voters at or to his left, and the candidate at 0.5 will get the support of the 13 voters at or to her right. If either candidate takes a different position, either to the left of 0.4 or to the right of 0.5, he or she will not be assured of 13 votes.

In general, if the number of voters is even, and the two middle voters adopt different positions, then if the candidates adopt those positions or any pair of positions in between, then they will be in equilibrium.

It is possible that, for either an odd or even number of voters, there may not be a median position such that half the voters lie to the left and half to the right of this position. As a case in point, consider the following discrete distribution of $n = 25$ voters at $k = 8$ different positions over the interval [0, 1], which is illustrated in Figure 12.3c.

Position, i	1	2	3	4	5	6	7	8
Location (l_i) of position i	0	0.2	0.3	0.5	0.6	0.7	0.8	0.9
Number of voters (n_i) at position i	2	3	4	3	2	4	6	1

We begin by calculating the mean:

$$\bar{l} = \left(\frac{1}{25}\right)[2(0) + 3(0.2) + 4(0.3) + 3(0.5) + 2(0.6) + 4(0.7) + 6(0.8) + 1(0.9)] = 0.52$$

At 0.6, 12 voters lie to the left of 0.6 and 11 voters lie to the right. At 0.5 and 0.7, the imbalances on the left and the right are even more lopsided. Moreover, there is no position, including the mean $\bar{l} = 0.52$, such that exactly half the voters lie to the left and half lie to the right.

In the absence of such a median position, is there an equilibrium? It is easy to show that 0.6 is indeed the equilibrium for two candidates. It is somewhat more difficult to show that if a distribution is discrete and there is no median position, there is still a unique position for both candidates that is in equilibrium. We call this the **extended median** because it extends the median-voter theorem to the discrete case, in which there may be no median position.

Extended Median DEFINITION

The **extended median** is the equilibrium position of two candiates in the discrete case when there is no median position.

Given the stability of the median or the extended median in a two-candidate, single-issue election, is it any wonder that candidates who want to win try to avoid extreme positions? As shown in Figures 12.2b and 12.3c, even when the greatest concentration of voters does not lie at M but instead at the right mode, a candidate would be foolish to adopt this modal position. For although the right-leaning voters would be very pleased, the candidate's opponent would win the votes of a majority by sidling up to this position but staying just to the left.

Voters on the far left may not be particularly pleased to see both candidates situate themselves at M, which is nearer the right mode in Figure 12.2a. But in a two-candidate race, they would have nobody else to turn to. Of course, if left-leaning voters felt sufficiently alienated by both candidates, they might decide not to vote at all, which has implications we explore further later.

EXAMPLE 1
Location of Department Stores

(Andria Patino/Corbis.)

There is a rather different application of the foregoing analysis to business, which in fact was the first substantive area to which spatial modeling was applied. Consider two competitive retail businesses, such as department stores, that consider locating their stores somewhere along the main street that runs through a city. Assume that because transportation is costly, people will buy at the department store closer to them. Then the analysis says that no matter how the population is distributed along or near the main street, the best location is the median.

Thus, if the city's population is symmetrically distributed—that is, not skewed toward one end or the other of the main street—then this location will, of course, be at the center of the main street. Indeed, clusters of similar stores are frequently bunched together near the center of many main streets, although these stores may not be particularly convenient to people who live far from the city's center. Consequently, their location seems not to be in the public interest. Wouldn't it be better to have some of the same kinds of stores near one end of the main street and some near the other, so no people are discriminated against? In an election, by contrast, not every voter can so easily be satisfied if only one candidate is to be elected, so the median seems the most attractive location in this context.

As an illustration, assume that 13 people have houses equally spaced along a main street, so the median person's house is #7, with six to the left and six to the right. The person in house #1 on the left must walk past six houses to get to one of the two stores at house #7, and so must the person at house #13 on the right. But if one store were at house #4, and the other store were at house #8, no person would have to walk past more than three houses to get to one of the two stores.

12.2 Spatial Models for Multicandidate Elections

Primary elections, in which candidates seek the nomination of one of the major parties, tend to attract more than two candidates. In presidential primaries, in particular, many candidates are likely to jump into the fray, especially in the states that go early in the season, if the incumbent president or vice president is not running (as was the case in 2008).

Under what conditions is entry into a multicandidate race attractive? If no positions offer a potential candidate any possibility of success, then it will not be rational for him or her to enter the primary in the first place. Therefore, the rationality of entering a race, and the rationality of the positions he or she might take once there, are really two aspects of the same decision.

Suppose that two candidates have already entered a primary, and they both take positions at M. Is there any room for a third candidate?

EXAMPLE 2
Entry of a Third Candidate in a Two-Candidate Race

Look at Figure 12.4, where *A* and *B* are both at *M* and therefore split the vote. Now if a third candidate, *C*, enters and takes a position on either side of *M* (say, to the right), the area under the distribution to *C*'s right may encompass less than $\frac{1}{3}$ of the total area and still enable *C* to win a plurality of votes.

To show why this is so, consider the portion of the electorate's vote that *A*/*B* will receive and the portion that *C* will receive. If *C*'s area (yellow) is greater than half of *A*/*B*'s area (blue), *C* will win more votes than *A* or *B*, because *C*'s area includes not only the votes to the right of his or her position but also some votes to the left. More precisely, *C* will attract voters up to the point midway between his or her position on the horizontal axis and that of *A*/*B*; *A* and *B* will split the votes to the left of this midway point. Because *C* picks up some votes to the left of his or her position, less than $\frac{1}{3}$ of the electorate may lie to the right and still enable *C* to win a plurality of more than $\frac{1}{3}$ of the total vote.

By similar reasoning, it is possible to show that a fourth candidate, *D*, could take a position to the left of *A*/*B* and further chip away at the total of the two centrists. Indeed, *D* could beat candidate *C*, as well as *A* and *B*, by moving closer to *A*/*B* from the left than *C* moves from the right.

Clearly, *M* has little appeal, and in fact is quite vulnerable, to a third or fourth candidate contemplating a run against two centrists. Indeed, it is not difficult to show that *whatever* positions two candidates adopt—the same or different—at least one of these candidates will be vulnerable to a third candidate.

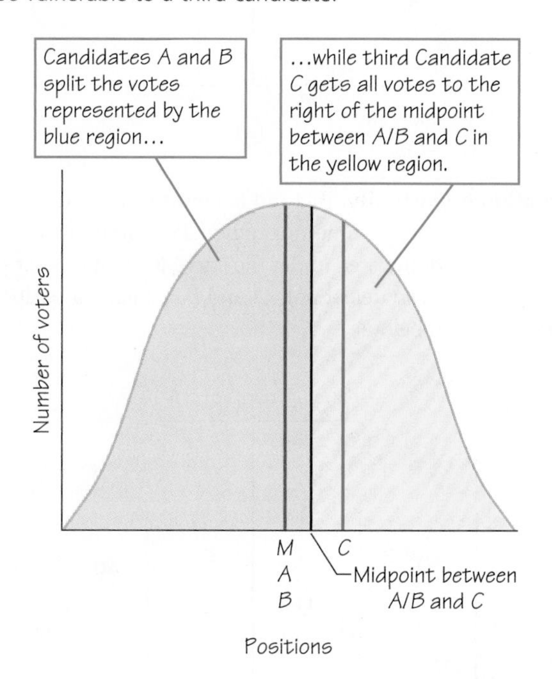

Figure 12.4 Unimodal distribution with three candidates. Candidate *C* can take a position with less than $\frac{1}{3}$ of the voters to his or her right and still win if candidates *A* and *B* at the median *M* and mean $\bar{I}$ split the remainder of the vote.

This is not to say, however, that a third Candidate *C* will necessarily win against *both A* and *B*. There are both obstacles and opportunities for *C*, which are summarized in Figure 12.5. (The reasoning behind these is explored in Exercises 20 and 21 on p. 446.)

1/3-Separation Obstacle

DEFINITION

The **1/3-separation obstacle** occurs when there is little room in the middle, enabling *C* to beat *A* or *B* but, in so doing, causing him or her to lose to the other.

The 1/3-Separation Obstacle and the 2/3-Separation Opportunity

THEOREM

The 1/3-separation obstacle. If *A* and *B* are distinct positions that are equidistant from the median of a symmetric distribution and separated from each other by no more than $\frac{1}{3}$ of the area under the curve (so that no more than $\frac{1}{3}$ of the voters lie between *A* and *B*), *C* can take no position that will displace both *A* and *B* and enable *C* to win (see below).

Figure 12.5 Figures a and b show both an obstacle and an opportunity. The obstacles and opportunities for a third candidate, *C*, to enter a race.

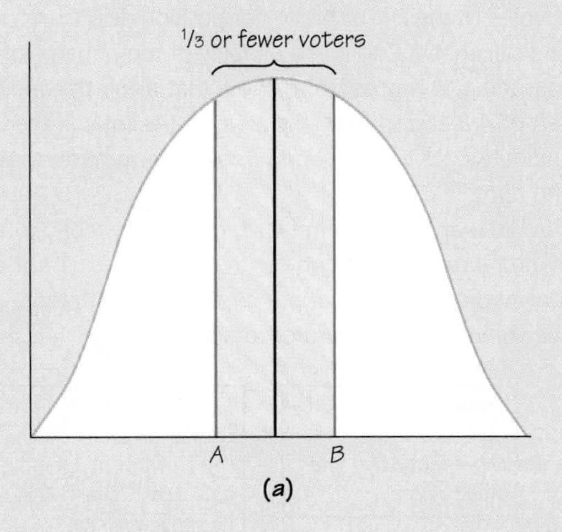

The 2/3-separation opportunity. If *A* and *B* are distinct positions that are equidistant from the median of a symmetric unimodal distribution and separated from each other by at least $\frac{2}{3}$ of the area under the curve (so that at least $\frac{2}{3}$ of the voters lies between *A* and *B*), *C* can defeat both *A* and *B* by taking a position at *M* (exactly between them, as shown below).

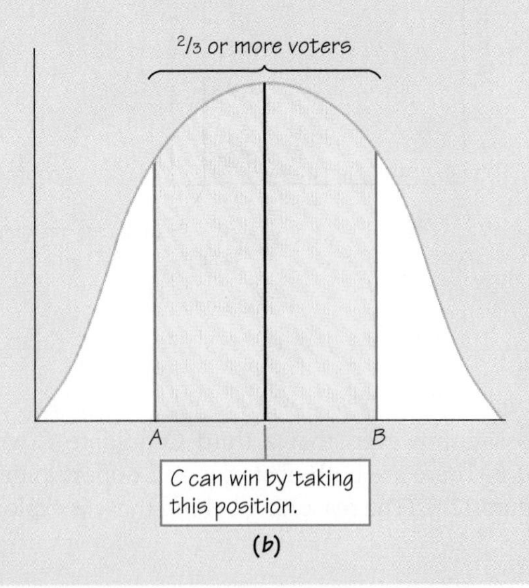

This occurred in the 1912 presidential election, when Theodore Roosevelt ran as the Progressive ("Bull Moose") Party candidate after losing the Republican nomination to William Howard Taft. (Roosevelt had previously been president but had lost favor with his party after sitting out one term.) In the general election, Roosevelt received 27 percent of the popular vote and Taft, 24 percent. Both candidates were handily defeated by the Democratic candidate, Woodrow Wilson, who got 41 percent of the popular vote. There was a fourth candidate in this race, socialist Eugene V. Debs, but he received only 6 percent and was never a serious threat to Wilson on the left. Wilson was also the overwhelming winner in the Electoral College.

Theodore Roosevelt.
(Hulton Getty/Liaison.)

2/3-Separation Opportunity DEFINITION

The **2/3-separation opportunity** occurs when there is a wide separation between A and B, giving enough room in the middle for C to win.

This event has never occurred in a U.S. presidential election. In fact, the 1912 election is the only election in which even one major-party candidate has been defeated by a third-party candidate in the popular vote.

The stability of the two-party system in the United States may be partially explained by the fact that the two major parties, anticipating the possible entry of a third-party candidate, deliberately position themselves far enough away from the median to discourage entry on the left or right—but not so far away as to make entry in the middle advantageous. The following theoretical result gives some insight into how this can be done (and the reasoning behind it is explored in Exercise 22 on p. 446).

Optimal Entry THEOREM

The optimal entry of two candidates, anticipating a third entrant. Assume that A and B are the first candidates to enter an election and anticipate the later entry of C. Assume that the distribution of voters is uniform (rectangular) over [0, 1]. (See Section 8.2 on p. 290.) Then the optimal positions of A and B are to enter at $\frac{1}{4}$ and $\frac{3}{4}$, whereby $\frac{1}{4}$ of the voters lie to the left of one candidate (say, A) and $\frac{1}{4}$ to the right of the other (B). Then C can do no better than win 25 percent of the vote: He or she will be indifferent among entering just to the left of A, just to the right of B, or at any position in between. At none of these positions will C win.

Presidential politics in the United States seems to be a reflection of both the median-voter theorem and optimal entry. For example, the median-voter theorem seems to have been operative in 1968, when the Democratic and Republican nominees, Hubert H. Humphrey and Richard M. Nixon, both presented themselves as centrists. This made them vulnerable to the third-party candidate that year, George Wallace—not in the sense that Wallace could win, but rather that he could throw the election in one direction or the other, or even into the House of Representatives to be decided. In fact, while Wallace won only 14 percent of the popular vote in 1968, he attracted mostly supporters of Richard N. Nixon on the right, who barely defeated Hubert H. Humphrey that year. (Nixon won by less than 1 percent in the popular vote.) Without Wallace in the race, polls show that Nixon's victory would have been far more substantial.

In 1992, Bill Clinton and George Bush were viewed as quite far apart on the left-right spectrum. Ross Perot was generally viewed to be between Clinton on the left and Bush on the right, leaving considerable room in the middle that Perot could better exploit than by trying to displace one of the major-party nominees on the left or right. In winning 19 percent of the popular vote, Perot drew almost equally from each candidate. However, he did not come close to winning, which the optimal entry of the candidates makes difficult. (Clinton won decisively, with 43 percent of the popular vote to Bush's 38 percent.)

12.3 Narrowing the Field

Up to now, we have looked at the spatial game that candidates play as they vie to position themselves optimally in two-candidate and multicandidate races so as to (1) maximize their vote totals or (2) deter new candidates from entering. We will continue to assume that candidates take positions along a left-right spectrum, but now we consider the game from the point of view of the voters. More specifically, we ask the following question in a multicandidate race: When one candidate drops out, perhaps because of performing below expectations in an early primary, to whom will the dropout's supporters shift their votes?

Beating expectations is the name of the game in the early primaries. Thus, when Senator Edmund S. Muskie from Maine ran for the Democratic Party nomination in 1972, he was expected to do well in the neighboring state of New Hampshire. To "win" in this first primary, he had to exceed these expectations, whereas other candidates, who were not expected to do so well, could afford more mediocre performances. As in a horse race, to which primary elections are often compared, the contenders are handicapped; that is, they must beat expectations about their performances if they are to gain momentum.

Momentum, or what George H. W. Bush (the father) called the "Big Mo," can start a **bandwagon**, or a presumption that a candidate will win.

Bandwagon Effect DEFINITION

The **bandwagon effect** induces voters to vote for the presumed winner, independent of his or her merit.

Assume that three candidates take positions, from left to right, as follows: *A*–*B*–*C*. Clearly, if *A* or *C* drops out, their supporters mostly likely will switch to *B*, giving the centrist a boost. But what if *B* is the first to drop out? Then it is unclear whether *A*, *C*, or neither will benefit; it depends on the number of *B*'s supporters who prefer *A* next, *C* next, or neither (and hence may not vote at all). In any event, with *B* out of the race, the winner must be one of the candidates on the extremes.

The possibilities become more interesting when there are four candidates arrayed from left to right as follows: *A*–*B*–*C*–*D*. If one of the extremists, *A* or *D*, drops out, then one of the two centrists, *B* or *C*, will benefit. But what if a centrist, say *C*, drops out? Does this benefit one of the extremists, or does the other centrist (*B*) benefit? At first glance, one might think that, with only one centrist remaining, he or she will surely benefit.

This will not be the case, however, if most *C* supporters prefer *D* to *B*, which is certainly possible. Then the extremist *D* will benefit, which will be most upsetting to *A*'s supporters. Conceivably, *A*'s supporters might encourage *A* to withdraw so they can throw their support to *B*, whom they definitely prefer to *D*.

Does this sound implausible? Think back to the 2000 election, in which our four hypothetical candidates are replaced by the following ordering from left to right: Nader–Gore–Bush–Buchanan. (Ralph Nader was the Green Party candidate on the left and Pat Buchanan the Reform Party candidate on the right.) Just before the election, the polls were showing that Buchanan was not much of a threat to Bush, but Nader—who ended up with 2.7 percent of the popular vote nationwide (Buchanan got only 0.4 percent)—was definitely a threat to Gore. Despite pleas from some of his supporters, Nader refused to withdraw and, consequently, gave Bush a victory in Florida and maybe a few other close states that won him the presidency.

This 2000 scenario is not the same as the previous four-candidate hypothetical scenario, in which we argued that the extremist on the right, *D*, might win if one of the centrists, *C*, dropped out. In the 2000 scenario, the extremist on the left, Nader, could have dropped out to "save" the centrist closer to him, Gore.

Unfortunately for Gore, Nader not only refused to make this sacrifice but contended afterward that Gore's loss was due to Gore's own poor performance, not Nader's presence in the race. We will return to this issue when we discuss the effects of approval voting and the abolition of the Electoral College as possible remedies to the so-called **spoiler** problem.

Spoiler DEFINITION

A **spoiler** is a candidate who cannot win but "spoils" the election for a candidate who otherwise would win.

12.4 What Drives Candidates Out?

So far, we have considered the possibility that candidates drop out, but not why they do so. Presumably, they do so because of their poor performance in the polls or early primaries, where performance depends in part on expectations of how well they will do. But expectations change over time, and this change in turn affects how voters perceive a race—in particular, who is ahead and who is behind. On this basis, voters choose voting strategies that are likely to benefit their favorites.

Polls make public the standing of candidates in a race, as do the returns from presidential primaries. To be sure, the electorate is different in each primary state, whereas a poll is a sample from the entire electorate. While polls and primaries both provide a glimpse of the "state of the electorate," here we will focus on the change of polls over time. But our results are just as applicable to primaries, whose winnowing-out effects are evident in almost every election in which an incumbent is not running for reelection.

Suppose that the election procedure is plurality voting.

Plurality Voting PROCEDURE

Plurality voting is a voting procedure in which each voter votes for one candidate, and the candidate with the most votes wins.

Assume that voters rank candidates. For example, *A B C* indicates that a voter prefers *A* to *B* to *C*. Before a poll, we assume that each voter votes **sincerely**—for his or her favorite candidate—because, in the absence of poll information, there is no reason to do otherwise.

EXAMPLE 3
Poll for Three Candidates

Nine voters, who can be divided into three classes, have the following preferences for candidates *A*, *B*, and *C*:

I. 4: *A C B*
II. 3: *B C A*
III. 2: *C A B*

Because the four voters in Class I prefer *A* to *C* to *B*, the poll would indicate *A* to have 4 votes (44 percent), whereas *B* and *C* would have 3 and 2 votes (33 percent and 22 percent), respectively, if voters vote sincerely.

After the poll, we make the following assumption:

Poll Assumption RULE

Voters adjust, if necessary, their sincere voting strategies to differentiate between the top two candidates revealed in the poll, voting for the one they prefer.

Since Class I and Class II voters chose one of the top two candidates, only the two Class III voters, who voted for *C*, would change their votes. Because they prefer *A* to *B*, it would be in their interest to vote insincerely for *A* (instead of *C*), thereby distinguishing *A* from the other top candidate, *B*, by giving *A* their votes. This would result in *A*'s winning with 6 votes, *B*'s getting 3 votes, and *C*'s getting no votes.

Paradoxically, it is *C* who is the Condorcet winner.

Condorcet Winner DEFINITION

A **Condorcet winner** is a candidate who can defeat each of the other candidates in pairwise contests.

C is preferred to *A* by Class II and III voters (5 to 4) and to *B* by Class I and III voters (6 to 3). Hence, if there were a series of pairwise contests (in any order), *C* would win. Yet the poll not only does not make *C* victorious but instead magnifies *A*'s plurality victory (4 votes) by inducing *C*'s supporters, thinking that their candidate is out of the running, to throw their support to *A*, giving *A* a $\frac{2}{3}$-majority victory (6 out of the 9 votes).

This example can be generalized to yield the following:

Condorcet Winner Unsuccessful THEOREM

Given the poll assumption, a Condorcet winner will always lose if he or she is not one of the top two candidates identified by the poll.

Why this is true is considered in Exercise 32 (on p. 447). Here, we note that even if *C* were given serious consideration in a tight race, *A* would still win with a plurality of votes.

One might think that *C*'s problem is due solely to the fact that the poll assumption puts him or her out of the running by presuming that his or her supporters "jump ship"—desert *C* for the apparently more viable candidates, *A* and *B*. But, surprisingly, even when a Condorcet winner is on top before a poll that distinguishes the top three candidates (instead of the top two), the Condorcet winner may be hurt by the poll after strategic adjustments are made by the voters, as we next illustrate.

EXAMPLE 4
Poll for Four Candidates

Add a fourth candidate, *D*, to the three in the preceding example, and assume that there are 12 voters with the following preferences for the four candidates:

I. 3: *A C B D*
II. 3: *B C A D*
III. 4: *C A B D*
IV. 2: *D A B C*

After the poll establishes that *A*, *B*, and *C* are the top three candidates, the two Class IV voters would be motivated to switch to their second choice, *A*. *A* would thereby increase his or her total from 3 to 5 votes—and win after the poll. Yet *C* is again the Condorcet winner. In staying the same at 4 votes, *C* is hurt, relative to *A*, by the poll. In fact, *C* would lose to *A* after the poll (because *D*'s supporters would continue to vote for *A*), even though *C* was the winner before the poll.

Thus, the poll assumption, which induces strategic adjustments that favor the top two candidates, may hurt the Condorcet winner when a larger number of candidates (possibly including the Condorcet winner) are considered to be contenders. However, when one of the top two contenders distinguished by the poll assumption is the Condorcet winner, we have the following result:

Condorcet Winner Successful THEOREM

Given the poll assumption, a Condorcet winner will always win if he or she is one of the top two candidates identified by the poll.

Why this is true is considered in Exercise 33 (on p. 447). Here, we note that a Condorcet winner need not be the winner in the poll but instead can place second and be successful. If a Condorcet winner places second, and the poll has the effect of turning him or her into a majority winner, then it is proper to say that the poll is instrumental in electing this candidate.

We conclude that a poll may either hurt or help a Condorcet winner. If the poll assumption is modified to distinguish more than two candidates, a Condorcet winner may be hurt *even if he or she is among those distinguished by the poll,* as the previous example showed.

12.5 Election Reform: Approval Voting

The furor caused by the divided outcome in the 2000 presidential election, in which George W. Bush won the electoral vote and Al Gore won the popular vote, has spurred efforts for reform of the election system. But except for calls for abolition of the Electoral College, whose effects we will analyze in the next section, most of the discussion has centered on making balloting more accurate and reliable and eliminating election irregularities, especially those that discriminate among different classes of voters.

Unfortunately, such reforms ignore a fundamental problem that plagues *multicandidate elections*—elections with three or more candidates—namely, that the candidate who wins under plurality voting may not be a Condorcet winner. Indeed, there will be no Condorcet winner if each candidate can be beaten by at least one other candidate. In such a situation, who is the rightful winner?

Chapter 9 presents alternatives to plurality voting. Most of these alternatives allow a voter to rank candidates from best to worst; analysts have investigated their ability to elect a Condorcet winner if one exists. Here, however, we will examine in depth a simple election reform, approval voting, that does not require voters to rank candidates.

Approval Voting DEFINITION

Under **approval voting**, voters can vote for as many candidates as they like or find acceptable. Each candidate approved of receives one vote, and the candidate with the most approval votes wins the election.

What if approval voting had been used in the 2000 presidential election? Nader supporters, knowing that voting for Nader would be only a protest vote because Nader had no chance of winning, might also have voted for Gore. In fact, because polls show that Gore was the second choice of most Nader voters, Gore almost certainly would have won in Florida if there had been approval voting.

To be sure, Bush would have benefited from the approval votes of Buchanan supporters, but the number of these votes would not have come close to matching the number of votes Gore would have received from Nader supporters. There is, therefore, little doubt that Gore was the Condorcet winner in this election because polls show that he could have defeated Bush (with help from Nader voters) as well as each of his less popular opponents in pairwise contests.

Arguments for an election reform like approval voting, however, should not be based on the outcome of only one election. Moreover, even in this one election, we cannot be entirely sure that Gore would have won under approval voting because the nature of the campaign almost surely would have changed if this reform had been in use.

For example, John McCain, the Republican senator from Arizona who defeated George Bush in the New Hampshire primary but ultimately lost the Republican nomination to him, might have run as an independent candidate if there had been approval voting. As a centrist, he would have been attractive to both Democrats and Republicans and, conceivably, could have won under approval voting. But even if McCain had not run, it is likely that the candidates would have pitched their campaign appeals somewhat differently to try to attract as much approval as possible, especially from Nader and Buchanan supporters.

Although we cannot make precise predictions of the effects of approval voting in a presidential election like that of 2000, we can say what voters will do in certain *types* of situations:

Voting Only for a Second Choice THEOREM

In a three-candidate election under approval voting, it is never rational for a voter to vote only for a second choice. If a voter finds a second choice acceptable, he or she should also vote for a first choice.

This is certainly not true under plurality voting. If you were a Nader supporter *and* found Gore also acceptable as a second choice, you should have voted for Gore rather than Nader if (1) you thought Gore could win but Nader could not and (2) electing an acceptable candidate was important to you. (Indeed, some Nader supporters switched to Gore for these reasons.) By comparison, because you lose nothing by voting for *both* Nader and Gore under approval voting—and may gain by doing so (if Nader cannot win, you at least help Gore)—you should never vote for just your second choice (Gore in this example).

Why voting only for a second choice is never rational is considered in Exercise 39 (on p. 447). In effect, a voter whose favorite candidate in a three-candidate race seems to be out of the running can have his or her cake and eat it too, by casting a *sincere* vote for a favorite candidate and a *strategic* vote for a second choice (to try to prevent a worst choice from winning). Roughly speaking, **strategic voting** (for instance, by Nader supporters for Gore in the 2000 presidential election) is voting that is not sincere but nevertheless has a strategic purpose—to elect an acceptable candidate if one's first choice is not viable.

Another general result about approval voting that is helpful to know uses the concept of dichotomous preferences:

Dichotomous Preferences DEFINITION

A voter has **dichotomous preferences** if he or she divides the set of candidates into two subsets—a preferred subset and a nonpreferred subset—and is indifferent among all candidates in each subset.

In other words, a dichotomous voter sees the world in two colors, white and black, and there is nothing in between. True, most of us see grays, but it is useful to analyze the dichotomous case. In this case, voters have a **dominant strategy**, which is a strategy that is at least as good as, and sometimes better than, any other strategy they might choose. With this definition, we can show a general condition under which Condorcet winners will be elected under approval voting:

Effect of Dichotomous Preferences THEOREM

A Condorcet winner will always be elected under approval voting if all voters have dichotomous preferences and choose their dominant strategies.

A dichotomous voter's dominant strategy is to vote for all candidates in his or her preferred subset and no others. This strategy is dominant because the preferred candidates are all assumed to be equally good, so a voter has no reason to distinguish among them. Furthermore, the voter has no reason to vote for a nonpreferred candidate or candidates because they are all equally bad and voting for any one of them could help that candidate win. As an illustration of the effect of dichotomous preferences, consider the following example.

EXAMPLE 5

Dichotomous Preferences

For each class of voters, the preferred subset of candidates is enclosed in the first set of parentheses, and the nonpreferred subset in the second set of parentheses. Thus, the four Class I voters prefer A and B, between whom they are indifferent, to C and D, between whom they are also indifferent:

I. 4: (A B) (C D)
II. 3: (C) (A B D)
III. 2: (B C D) (A)

Assuming that each class of voters chooses its dominant strategy, B wins, with 6 votes to 5 votes for C, 4 votes for A, and 2 votes for D.

In pairwise contests, notice that B is preferred to A by the two Class III voters (Class I and II voters are indifferent between these two candidates), so B would defeat A by 2 to 0 votes. (We assume that indifferent voters express no preference.) Because B is preferred to C by the four Class I voters, and C is preferred to B by the three Class II voters (Class III voters are indifferent between these two candidates), B would defeat C by 4 to 3 votes. Thus, B, the approval-vote winner, is also the Condorcet winner, which must always be the case when voter preferences are dichotomous and voters vote for all their approved candidates.

Insofar as voters in the 2000 presidential election thought equally well (or badly) of Bush and Buchanan on the one hand, and Gore and Nader on the other, they would have preferences like those of Class I voters in the previous example. Of course, most voters probably made finer distinctions, which is allowed by "range voting" (see Suggested Web Sites at the end of this chapter). In this case, there is no guarantee that approval voting will elect a Condorcet winner.

12.6 The Electoral College

As we have noted, the Electoral College had a decisive effect in the 2000 presidential election. In winning the popular vote in Florida by the slimmest of margins, George Bush captured all 25 of Florida's electoral votes, which gave him a majority in the Electoral College. This won him the presidency even though he lost the popular vote.

What is the justification for the Electoral College? Its original purpose was to place the selection of a president in the hands of a body that, while its members would be chosen by the people, would be sufficiently removed from them that it could make more deliberative choices. As for its composition, each state gets 2 electoral votes for its two senators (total for all states: 100). In addition, a state receives 1 electoral vote for each of its representatives in the House of Representatives, whose numbers are based on population (see Chapter 11) and range from 1 representative for the seven smallest states to 53 representatives for the largest state, California. The House has a total of 435 representatives. The District of Columbia, like the smallest states, is given 3 electoral votes. Altogether, there are 538 electoral votes, and a candidate needs 270 to win. In 2000, George W. Bush got 271 electoral votes.

Although there is nothing in the U.S. Constitution mandating that the popular-vote winner in a state receive all its electoral votes, this has been the tradition almost from the founding of the republic. Only in Maine and Nebraska can the electoral votes be split among candidates, depending on who wins each of the two congressional districts in Maine and the three congressional districts in Nebraska. Because the statewide winner receives the two senatorial electoral votes, the closest split possible in these two states is 3–1 in Maine and 3–2 in Nebraska. In the actual election, Gore won all of Maine's 4 electoral votes, and Bush won all of Nebraska's 5 electoral votes, so winner-take-all prevailed in all 50 states and the District of Columbia.

Effectively, then, the presidency is decided by 51 players: members of the Electoral College from each of the 50 states and the District of Columbia, who almost always cast their votes as blocs for a single candidate. The voting weights of states, which depend in part on their populations, are related to their voting power (see Chapter 11).

Although the percentage of voting power of a state closely tracks its percentage of electoral votes, this is not the full story. More important is the power of *individual* voters in each state, based on their ability to be pivotal in their states and their states, in turn, to be pivotal in the Electoral College. Amazingly, individual voters in California are about three times as powerful as individual voters in the smallest states, despite the fact that the smallest states (with only one representative) get a 200 percent (2/1) boost from having 2 "senatorial" electoral votes—besides the 1 electoral vote they are entitled to on the basis of their populations—whereas California receives less than a 4 percent (2/53) boost.

Checking ballots in Florida after the 2000 presidential election. *(Reuters/Corbis.)*

But there is more to the story than just the size of states. Because California was never considered a close state in the 2000 presidential election (polls indicated that it would almost surely go for Al Gore), it received relatively little attention from both candidates. The real battle was fought in the so-called battleground states, or *toss-up states,* where the outcome was expected to be close (as it certainly was in Florida!). These states received the bulk of the candidates' time, money, and other resources.

Instead of viewing the Electoral College as a 105-million-person game in 2000, in which the voters are the players and their power is a function of the size of the states in which they vote, we view it as a game between the two major-party candidates. We will develop two different models: the first in which the candidates seek to maximize their expected popular vote, and the second in which they seek to maximize their expected electoral vote, in the toss-up states that determine the outcome in a close race.

Common to both models is the assumption that the probability, p_i, that a voter in toss-up state i votes for the Democratic candidate is

$$p_i = \frac{d_i}{d_i + r_i}$$

or the proportion of campaign resources that the Democratic candidate (d_i), compared with the Republican (r_i), spends in state i (see Chapter 8). The probability that a voter in state i votes for the Republican candidate will be the probability $1 - p_i = r_i/(d_i + r_i)$. Thus, we ignore the effects of other candidates in the race and assume that either the Democratic or the Republican candidate will win with certainty: $p_i + (1 - p_i) = 1$. This assumption is plausible in light of the fact that no third-party candidate has ever won the presidency and rarely any states.

Expected Popular Vote — DEFINITION

The **expected popular vote (*EPV*)** of the Democratic candidate in toss-up states, EPV_D, is the number of voters, n_i, in toss-up state i, multiplied by the probability, p_i, that a voter in toss-up state i votes for the Democrat, summed across all toss-up states:

$$EPV_D = \sum_{i=1}^{t} n_i p_i$$

where t is the number of toss-up states. EPV_R can be defined similarly.

EPV_D bears some similarity to the expression for the mean ($\bar{l}$) discussed earlier. Whereas $\bar{l}$ is an average weighted by the proportions, n_i/n, EPV_D is an average weighted by the probabilities p_i.

The candidates seek strategies for optimally allocating their resources to each toss-up state. Recall that d_i for the Democrat and r_i for the Republican are the resources that each candidate allocates to state i. For the Democrat, if d_i changes, this affects the value of p_i, the probability that a voter votes for him or her, which in turn affects the value of EPV_D. Thus, the Democrat seeks a strategy d_i that will make EPV_D as large as possible.

Proportional Rule RULE

The strategy of the Democrat that maximizes his or her EPV_D (indicated by the asterisk), given that the Republican also chooses a maximizing strategy, is

$$d_i^* = \left(\frac{n_i}{N}\right)D$$

where $N = \sum_{i=1}^{t} n_i$, the total number of voters in the toss-up states, and $D = \sum_{i=1}^{t} d_i$, the sum of the Democrat's expenditures across all states.

In words, the Democrat should allocate his or her resources in proportion to the size of each state (n_i/N) if the Republican behaves similarly by following a strategy of $r_i^* = (n_i/N)R$, where $R = \sum_{i=1}^{t} r_i$.

To show that d_i^* maximizes EPV_D when the Republican chooses r_i^* requires calculus, but we can readily illustrate why departures from d_i^* by the Democrat will cost him or her popular votes.

EXAMPLE 6
Departures from a Popular-Vote Maximizing Strategy

Suppose that there are three toss-up states with 2, 3, and 4 electoral votes. Assume that the candidates accept public financing for the election, and this limits them to spending the same total of, say, \$63 million (M). If the Republican follows his or her optimal strategy of spending in the proportion 2:3:4 (\$14M:\$21M:\$28M), but the Democrat, ignoring the smallest state, spends in the proportion of 0:3:4 (\$0M:\$27M:\$36M), the Republican will receive, on average,

$$EPV_R = 2[14/(0 + 14)] + 3[21/(21 + 27)] + 4[28/(28 + 36)] = 5.06 \text{ votes}$$

or 56 percent of the 9 votes in the three states.

If the Republican anticipates that the Democrat will spend nothing in the smallest state, the Republican can do even better. By spending only a minuscule amount in the smallest state, the Republican can almost match the Democrat in the other two states and win an average of about 5.5 votes (2 votes from the smallest state and $7/2 = 3.5$ votes from the other two states), or 61 percent.

Now let's assume that the goal of the candidates is not to maximize *EPV* but, instead, their **expected electoral vote (*EEV*)**, which is an entirely different quantity that we define below. To illustrate the difference, a candidate who wants to maximize EEV might think of throwing all of his or her resources into the 11 largest states—and ignoring the 39 other states and the District of Columbia if all states are toss-up states—because the 11 largest states have a majority of electoral votes (271). Moreover, the candidate need not win "big" in these states. Winning them by small margins will work just fine because the candidate will still win *all* the electoral votes of those states and thereby the election.

But this strategy has a problem. An opponent can readily defeat it by spending very small amounts in all the other states, which will defeat the candidate if he or she spends nothing in these states. In addition, by using his or her leftover funds to outspend the candidate in, say, one or two big states, the opponent will end up winning more electoral votes. However, there is a counterstrategy to this strategy, and indeed to every other pure strategy (no randomization—see Chapter 15) one can think of in a winner-take-all system like that of the Electoral College.

To prevent being exploited if an opponent anticipates one's strategy and selects a best counterstrategy against it, each candidate should try to keep secret exactly what he or she intends to do. The best way to keep a secret is to randomize one's choices using mixed strategies. But mixed strategies are difficult to calculate in a system as complicated as that of the Electoral College, so we make simplifying assumptions. First, however, we need a definition.

To determine the probability of winning a state, we need to count the number of ways that a candidate can win a majority of electoral votes in that state and then compute the probabilities that each of these ways occurs. These probabilities, in turn, are used to determine EEV_D. EEV_R can be defined similarly.

Expected Electoral Vote DEFINITION

The **expected electoral vote (*EEV*)** of the Democratic candidate in toss-up states, EEV_D, is

$$EEV_D = \sum_{i=1}^{t} v_i P_i$$

where v_i is the number of electoral votes of toss-up state i, and P_i is the probability that the Democrat wins *more than* 50 percent of the popular votes in this state, which would give the Democrat *all* that state's electoral votes, v_i. EEV_R can be defined similarly.

EXAMPLE 7
Computing the Democrat's Expected Electoral Vote

Consider our earlier example of three states, A, B, and C, with 2, 3, and 4 electoral votes. For simplicity, assume here that the number of electoral votes of each state is equal to the number of voters in that state.

To calculate P_i for each state i, we must determine the probabilities that a majority of voters in States A, B, and C vote Democratic. (We will ignore the possibility of ties in States A and C, which have an even number of voters.) To obtain these probabilities, we multiply the probabilities that individual voters in each state, who are assumed to act independently of each other, vote Democratic or Republican (based on the resources the two candidates allocate to each state).

For State A to vote Democratic, for example, both voters in this state must vote Democratic, so $P_A = (p_A)(p_A) = (p_A)^2$. For State B to vote Democratic, either two of the three voters (in 3 possible ways) or all three voters must vote Democratic, so $P_B = 3[(p_B)^2(1 - p_B)] + (p_B)^3$. For State C to vote Democratic, either three of the four voters (in 4 possible ways) or all four voters must vote Democratic, so $P_C = 4[(p_C)^3(1 - p_C)] + (p_C)^4$. Inserting these probabilities into the formula for EEV_D, we obtain

$$\begin{aligned} EEV_D &= v_A P_A + v_B P_B + v_C P_C \\ &= 2[(p_A)^2] + 3[3(p_B)^2(1 - p_B) + (p_B)^3] + 4[4(p_C)^3(1 - p_C) + (p_C)^4] \end{aligned}$$

We indicated earlier that the strategy of the Democrat that maximizes EEV_D, given that the Republican adopts a similar strategy, is mixed, involving randomizing his or her choice of states in which to allocate resources. Because this randomization is difficult to determine, we simplify the task by assuming that $d_i = r_i$ in each toss-up state i, and necessarily $D = R$ across all these toss-up states.

This assumption is defensible if the candidates have the same total amount to spend in all the states ($D = R$). If they perceive the value of each toss-up state to be the same, which is reasonable, they will allocate equal resources to each. But how much should these amounts be? It is possible to show the following:

The 3/2's Rule RULE

The strategies of the Democratic and the Republican candidates that maximize their *EEV*s are

$$d_i^* = \left(\frac{v_i \sqrt{n_i}}{S}\right) D \qquad r_i^* = \left(\frac{v_i \sqrt{n_i}}{S}\right) R$$

where

$$S = \sum_{i=1}^{t} v_i \sqrt{n_i}$$

In words, the candidates should allocate their resources in proportion to the number of electoral votes of each state (v_i) multiplied by the square root of its size (n_i). The allocations, d_i^* and r_i^*, will be the same if the candidates spend equally in each toss-up state ($d_i = r_i$), as previously assumed.

Algebra Review

Natural and fractional exponents, page 661.

Because the number of electoral votes in each state i (v_i) is approximately proportional to the number of voters in each state (or their populations), n_i, we can substitute for $v_i \sqrt{n_i}$ in the above formulas $v_i \sqrt{v_i} = v_i^{3/2}$. Hence, the maximizing strategies of the candidates can be approximated by

$$d_i^* = \left(\frac{v_i^{3/2}}{T}\right) D \qquad r_i^* = \left(\frac{v_i^{3/2}}{T}\right) R$$

where

$$T = \sum_{i=1}^{t} v_i^{3/2}$$

Thus, if the candidates allocate the same amount of resources to each of the toss-up states, the 3/2's rule is that they should spend approximately in proportion to the 3/2's power of the electoral votes of these states to maximize *EEV*.

EXAMPLE 8
Applying the 3/2's Rule

Assume that States *A*, *B*, and *C* have, respectively, 9, 16, and 25 voters, and these are also their numbers of electoral votes. If each of these states is a toss-up state, the 3/2's rule says that the candidates should allocate their resources in the proportions 27:64:125 because the 3/2's powers of their electoral votes are their numbers of voters multiplied by the square root of these numbers. Thus for State *A*, $9^{3/2} = 9\sqrt{9} = (9)(3) = 27$.

To illustrate the difference between the proportional rule (which maximizes *EPV*) and the 3/2's rule (which maximizes *EEV*), assume that both candidates can spend 100 units of resources. Then the proportional rule says that States *A*, *B*, and *C* should get resources in approximately the amounts 18, 32, and 50, whereas the 3/2's rule says these states should get resources in approximately the amounts 12, 30, and 58.

Clearly, the smallest state gets less and the largest state gets more under the 3/2's rule, whereas the middle state stays about the same. In the actual Electoral College, the voters in California, when it is a toss-up state, are about three times as attractive per capita as voters in the smallest states. Following the 3/2's rule, therefore, the candidates should allocate about three times as much per voter to California as to a small toss-up state with only 3 electoral votes.

In fact, presidential candidates greatly overspend in the largest toss-up states, well out of proportion to their size. This large-state bias is far out of line with the democratic principle of "one person, one vote." For Californians, compared to small-state voters, this principle should read "one person, three votes" when it is a toss-up state.

To be sure, it is not the Electoral College itself that creates this bias but, rather, its winner-take-all feature. If this feature were abolished and the electoral votes of a state were split according to the popular votes received by the candidates—insofar as possible—then the large-state bias would disappear.

There is an even better reform to ensure that the electoral-vote winner is also the popular-vote winner. Let each state pass a law that gives all its electoral votes to the *national* popular-vote winner, which becomes effective when states with a majority of electoral votes (that is, 270) pass such a law. Then the popular vote winner would be guaranteed a victory in the Electoral College. As of February 2011, six states and the District of Columbia had enacted such a law, called the **National Popular Vote law**, which, if passed, would nullify the winner-take-all effect of individual states in the Electoral College.

EXAMPLE 9
Departures from an Electoral-Vote Maximizing Strategy

We illustrated earlier how a candidate's departure from the popular-vote maximizing strategy—the proportional rule—lowers that candidate's expected popular vote, given that the candidate's opponent adheres to this rule. Because of the loss candidates suffer when they depart from the proportional rule, it is an equilibrium strategy for both. This is also true of *some* departures from the 3/2's rule: If a candidate's departure from this rule is "small," he or she will lower his or her expected electoral vote, given that the candidate's opponent sticks to that rule.

Suppose, for example, that the Republican follows the 3/2's rule in allocating resources to States *A*, *B*, and *C* with 9, 16, and 25 voters/electoral votes, respectively. If each candidate has 100 units of resources to spend, we showed in the previous example that this translates into allocating approximately 12, 30, and 58 units to States *A*, *B*, and *C*, respectively.

Now if the Democrat deviates slightly from the 3/2's rule, and allocates 13, 30, and 57 units to these states (more to *A*, less to *C*), he or she increases the chances of winning in *A* and decreases the chances of winning in *C*. It can be shown that this deviation from the 3/2's rule hurts the Democrat because even though his or her chances go up more in *A* than they go down in *C* (because *A* is smaller than *C*), *C* has almost three times as many electoral votes. On balance, this deviation lowers the Democrat's expected electoral vote.

Now suppose that the Democrat makes a "large" deviation from the 3/2's rule, ignoring State *B* entirely and throwing all his or her resources into State *A* (9 electoral votes) and State *C* (25 electoral votes). If he or she wins in these two states, this would give the Democrat 34 of the 50 electoral votes, which is more than enough to win.

If the Republican adheres to the 3/2's rule, he or she will put approximately 12 percent of his or her resources into State *A* and 58 percent into State *C*. By following the 3/2's rule in these two states and ignoring State *B*, the Democrat will put approximately 18 percent into State *A* and approximately 82 percent into State *C*. This translates into *each voter* in *A* and *C* supporting the Democrat with probability 0.58, which means that the Democrat will almost certainly win in both these states, giving him or her an expected electoral vote of almost 34. The Republican will certainly win in State *B* because the Democrat spends nothing in this state, but this is small consolation if the Republican loses in the two other states and receives an expected electoral vote of somewhat more than 16.

This example illustrates why the 3/2's rule is a local maximum but not a global maximum.

Local Maximum and Global Maximum

DEFINITION

A **local maximum** is a maximizing strategy from which small deviations are nonoptimal but large deviations may be optimal. A **global maximum** is a maximizing strategy from which *all* deviations (small or large) are nonoptimal. The proportional rule is a global maximum (and equilibrium) for candidates whose goal is to maximize their expected popular vote, whereas the 3/2's rule is only a local maximum for candidates whose goal is to maximize their expected electoral vote.

If the goal of candidates is to maximize their expected electoral vote, there is no *determinate* maximizing strategy; randomizing one's choices is necessary to prevent exploitation by an opponent. How this randomization is done for some simple games is analyzed in Chapter 15.

In the case of the Electoral College, we illustrated how a radical departure from the 3/2's rule by a candidate, who ignores some state or states entirely, may be the best response to an opponent who follows this rule. But then there is a best response to this best response, and so on, so no determinate strategy is invulnerable.

However, insofar as the candidates view the toss-up states in similar terms, the 3/2's rule offers a good rule of thumb as to how much to spend in each, as a function of its size, to maximize their expected electoral vote. But we must remember that it is only a local maximum. Hence, unlike the proportional rule, which is robust against all other popular-vote maximizing strategies, the 3/2's rule is vulnerable to radically different strategies, such as those in which candidates concentrate their efforts on relatively few states.

Usually those who try such strategies, however, take big risks. In 1964, the Republican presidential candidate, Barry Goldwater, said that he would like to "saw off the Eastern seaboard" (Goldwater was from Arizona), but in the end, in his memorable phrase, he went, "shooting where the ducks [voters] are." In doing so, however, he appeared not so much to want to win as to present voters with "a choice, not an echo," by taking relatively extreme (conservative) positions. Is it any wonder, then, that Goldwater lost in a huge landslide to his Democratic opponent, Lyndon Johnson?

12.7 Is There a Better Way to Elect a President?

The quest for the presidency is the greatest spectacle in American politics. While there is nothing to match its excitement, especially when the race is close, the quieter gamelike features of a presidential campaign are no less consequential.

We have emphasized these features in this chapter, showing how mathematics can be used to analyze optimal positions in two-candidate and multicandidate races, and how polls and presidential primaries may affect who stays in and who drops out of the race. Sometimes, as we have seen, Condorcet candidates—who can beat every other candidate in pairwise contests—may not survive. And, as was dramatically illustrated in the 2000 presidential election, the popular-vote winner may not win in the Electoral College.

Some people think that approval voting would better enable voters to express their preferences, especially in the early presidential primaries, which typically draw many candidates if an incumbent is not running for reelection. However, other election procedures, including those discussed in Chapter 9, possess features that may make them desirable as election reforms.

All these procedures would probably be of most help to centrist candidates, who not only better represent the entire electorate than extremist candidates but who also are more likely to be a party's strongest contender in the general election. In the past 50 years, the biggest losers in presidential elections, Republican Barry Goldwater in 1964 and Democrat George McGovern in 1972, came from the right and left extremes, respectively, of their parties.

Taking the choice of a president out of the hands of voters and putting it into the hands of members of the Electoral College may no longer be justified. The Electoral College, with its winner-take-all feature, creates a large-state bias, as we showed. In the 2000 presidential election, a few hundred voters in one large toss-up state, Florida, determined the outcome.

If one thinks that the votes of *all* voters, wherever they reside, should count equally, then direct popular-vote election of a president would best accomplish this goal. Allocating electoral votes proportionally in each state would approximate this goal. But a better solution would be for states to enact the National Popular Vote law, which would ensure that the electoral-vote winner is the popular-vote winner if states with a majority of electoral votes passed this law.

If approval voting were used, then it would be approval votes rather than the single votes of each voter that would determine the allocation of electoral votes to the candidates. Because the general election in recent years has drawn major third- and fourth-party candidates, approval voting, or one of the other voting procedures discussed in Chapter 9, seems worthy of consideration if one wants to reduce the role of spoilers.

In summary, mathematics illuminates strategic aspects of campaigning and voting in presidential elections not apparent to the naked eye. It also points the way to possible reforms that may ameliorate some of the problems that affect our current system.

REVIEW VOCABULARY

Approval voting This allows voters to vote for as many candidates as they like or find acceptable. Each candidate approved of receives one vote, and the candidate with the most approval votes wins. (p. 432)

Bandwagon effect Voting for a candidate not on the basis of merit but, instead, because of the expectation that he or she will win. (p. 428)

Condorcet winner A candidate who can defeat each of the other candidates in pairwise contests. (p. 430)

Dichotomous preferences Held by voters who divide the set of candidates into two subsets—a preferred subset and a nonpreferred subset—and are indifferent about all candidates in each subset. (p. 433)

Discrete distribution of voters A distribution in which voters are located at only certain positions along the left–right continuum. (p. 421)

Dominant strategy A strategy that is at least as good as, and sometimes better than, any other strategy. (p. 433)

Electoral College A body of 538 electors that selects the U.S. president. (p. 415)

Equilibrium A position is in equilibrium if no candidate has an incentive to depart from it unilaterally. (p. 418)

Expected electoral vote (EEV) The number of electoral votes of each toss-up state, multiplied by the probability that the Democratic (or Republican) candidate wins more than 50 percent of the popular votes in that state, summed across all toss-up states. (p. 437)

Expected popular vote (EPV) The number of voters in each toss-up state, multiplied by the probability that that voter votes for the Democratic (or Republican) candidate, summed across all toss-up states. (p. 435)

Extended median The equilibrium position of two candidates when there is no median. (p. 423)

Global maximum A maximizing strategy from which *all* deviations (small or large) are nonoptimal. (p. 440)

Local maximum A maximizing strategy from which small deviations are nonoptimal but large deviations may be optimal. (p. 440)

Maximin position A position is maximin for a candidate if there is no other position that can guarantee a better outcome—more votes—whatever position another candidate adopts. (p. 417)

Mean ($\bar{I}$) A weighted average, wherein the positions of voters are weighted by the fraction of voters at that position. (p. 419)

Median M The point on the horizontal axis of a voter distribution where half the voters have attitudes that lie to the left and half to the right. (p. 416)

Median-voter theorem In a two-candidate election with an odd number of voters, the median is the unique equilibrium position. (p. 419)

Mode A peak of a distribution. A distribution is **unimodal** if it has one peak. (p. 416)

National Popular Vote law This gives all the electoral votes of a state to the national popular-vote winner if states with a majority of electoral votes enact the law. (p. 439)

1/3-separation obstacle An obstacle for the entry of a third candidate created if two previous entrants are sufficiently close together. (p. 426)

Plurality voting This allows voters to vote for one candidate, and the candidate with the most votes wins. (p. 429)

Poll assumption Voters adjust their sincere voting strategies, if necessary, to differentiate between the top two candidates—as revealed in a poll—by voting for the one they prefer. (p. 430)

Proportional rule A rule by which presidential candidates allocate their resources to states according to their size. This allocation rule maximizes the expected popular vote of a candidate, given that his or her opponent adheres to it. It is a global maximum. (p. 436)

Sincere voting Voting for a favorite candidate, whatever his or her chances are of winning. (p. 429)

Spatial models The representation of candidate positions along a left–right continuum in order to determine the equilibrium or optimal positions of the candidates. (p. 415)

Spoiler A candidate who cannot win but "spoils" the election for a candidate who otherwise would win. (p. 429)

Strategic voting Voting that is not sincere but nevertheless has a strategic purpose—namely, to elect an acceptable candidate if one's first choice is not viable. (p. 433)

3/2's rule Presidential candidates allocate their resources to toss-up states according to the 3/2's power of their electoral votes. This allocation rule maximizes the expected electoral vote of a candidate, given that his or her opponent adheres to it. It is a local maximum. (p. 438)

2/3-separation opportunity An opportunity for the entry of a third candidate created if two previous entrants are sufficiently far apart. (p. 427)

Voter distribution This gives the number (or percentage) of voters who have attitudes at points along the left–right continuum, which can be represented by a curve. The distribution is **symmetric** if the curve to the left of the median is a mirror image of the curve to the right. It is skewed to one side if the area under the curve is less concentrated on that side of the median than the other. (p. 416)

SKILLS CHECK

1. In a two-candidate election, suppose that the attitudes of the voters are distributed symmetrically around the median *M*. Of the two candidates *A* and *B*, *A* is positioned far to the left of *M* and *B* is positioned just to the right of *M*. Which, if either, candidate will receive more votes?

(a) *A* will receive a majority of the votes.
(b) *B* will receive a majority of the votes.
(c) *A* and *B* will both receive exactly one-half of the votes.

2. In a two-candidate election, suppose that the attitudes of the voters are skewed to the left of the median *M*, so they are more spread out to the left than to the right. Assume that Candidates *A* and *B* take positions to the left and right of *M*, respectively, so that there are the same numbers of voters between their positions and *M*. Candidate __________ will receive a majority.

3. In a two-candidate election, which of the following positions is an equilibrium position for both candidates *A* and *B*?

(a) *A* and *B* just to the left and right of *M*
(b) *A* and *B* far to the left and right of *M*
(c) *A* and *B* both at *M*

4. Suppose that voters do not vote for a candidate if he or she is too far away from their ideal positions. If there are two candidates, and the distribution of voters is bimodal around the median *M*, the candidates will be most helped by taking positions at __________.

5. A 1/3-separation obstacle becomes a 2/3-separation opportunity for a third candidate when

(a) less than 2/3 of the voters separate the first two candidates.
(b) exactly 1/3 of the voters separate the first two candidates.
(c) at least 2/3 of the voters separate the first two candidates.

6. In a three-candidate election, suppose the attitudes of the voters are distributed symmetrically around the median *M*. Of the three candidates *A*, *B*, and *C*, *A* is positioned far to the left of *M*, *B* is positioned just to the right of *M*, and *C* is positioned at *M*. Candidate _____ will receive the most votes.

7. Suppose that Candidate *A* is situated so that exactly 1/3 of the voters lie to his left. In a two-candidate race, if Candidate *A* must remain fixed, Candidate *B*—to maximize her vote total—should situate herself

(a) just to the right of *A*.
(b) at the median *M*.
(c) to the right of *M* so that 1/3 of the voters lie to her right.

8. In a three-candidate election, if Candidates *A* and *B* are positioned at *M*, the election-winning position of Candidate *C* is ______________________________.

9. In a three-candidate election, if Candidates *A* and *B* are positioned just to the left and just to the right of *M*, are there election-winning positions for Candidate *C*? What are they?

(a) At *M*
(b) Far to the left or right of *M*
(c) There is no election-winning position for Candidate C.

10. In Exercise 7, suppose that Candidate *A* can move to an optimal position *after* Candidate *B* makes her choice. In this case, Candidate *B*, anticipating Candidate *A*'s optimal move, should move __________.

11. When is a maximin position not an equilibrium position?

(a) When it is not the median
(b) When it is the median
(c) When it is the mean

12. In a four-candidate election, if candidates are aligned in order *A*–*B*–*C*–*D*, candidate _____ benefits if *A* drops out of the race.

(a) *A*
(b) *B*
(c) *C*

13. In a three-candidate plurality election, suppose 12 voters can be divided into three classes according to their preferences: Five voters prefer (in order) *A*, *B*, *C*; 4 voters prefer *C*, *A*, *B*; 3 voters prefer *B*, *C*, *A*. To elect one of their top 2 candidates, which group of voters will not vote sincerely (for their first choice)?

(a) The 5 voters
(b) The 4 voters
(c) The 3 voters

14. In a five-candidate election, if candidates are aligned in order *A*–*B*–*C*–*D*–*E* and each receives about 20 percent of the vote, Candidate ________ benefits most if Candidates *A*, *D*, and *E* drop out.

15. In an election with a large number of candidates, approval voting benefits

(a) candidates at the extreme left and right.
(b) candidates at or near *M*.
(c) only candidates precisely at *M*.

16. It is desirable that two candidates take median positions, but undesirable that two department stores locate themselves at the center of a main street because ________________________.

17. In the election of the president, if the Democrats believe that the Republicans will not allocate resources in a way that maximizes their *EEV*, then

(a) they can successfully counter by allocating their resources according to the method that maximizes *EEV*.
(b) they can successfully counter by not allocating their resources according to the method that maximizes *EEV*.
(c) they cannot successfully counter.

18. Making the poll assumption, the Condorcet winner will always win when ________________.

19. Suppose that polls indicate that an election is very tight between Candidate *A* on the left and Candidate *B* on the right. You support Candidate *C* in the middle and prefer *A* to *B*. It is best for you to approve of

(a) *C* alone.
(b) *A* alone.
(c) both *A* and *C*.

20. In Skills Check 19, suppose that you think Candidates *A* and *B* are equally bad. Your best strategy is to approve of ________.

21. In a three-candidate election, suppose that the candidates are aligned in the order *A*–*B*–*C* and that your top choice is Candidate *B*. Which two-candidate approval-voting strategy would you *never* choose?

(a) Approve of *A* and *B*.
(b) Approve of *B* and *C*.
(c) Approve of *A* and *C*.

22. In a four-candidate approval voting election with 12 voters, if 5 voters approve of *A* and *B*, 4 voters approve of *B* and *C*, and 3 voters approve of *A* and *D*, Candidate ______ will win the election.

23. In Skills Check 22, assume that the 5 voters who approve of *A* and *B* actually prefer *A*. Would they have an incentive to vote strategically?

(a) Yes.
(b) No.
(c) Some would and some wouldn't.

24. Suppose three voting blocs, *A*, *B*, and *C*, control 16, 25, and 36 votes, respectively, and a simple majority of 39 votes are required to win. If each bloc is a toss-up, for every $1 that is allocated to influence the voters in Bloc *A*, about ________ as much should be allocated to influence the voters in Bloc *C*.

25. In Skills Check 24, suppose your opponent spends all his money lobbying Blocs *A* and *B* and ignores Bloc *C*. What would be a good strategy to use in response?

(a) Lobby just *A*.
(b) Lobby just *C*.
(c) Lobby *B* and *C*.

26. Voting for president using the Electoral College would be unbiased if ________________________.

27. How big is the bias factor in Exercise 26 when comparing small and large states?

(a) Very small
(b) About 2:1
(c) About 3:1

28. In the election of the president using the Electoral College, voters in smaller toss-up states have less power than voters in large toss-up states because ____________.

29. The National Popular Vote law is

(a) a constitutional amendment.
(b) a federal law.
(c) a state law.

30. It is impossible for a candidate for president to win the popular vote and yet lose the Electoral College vote under the National Popular Vote law because ____________.

■ Challenge ▲ Discussion

12.1 Spatial Models for Two-Candidate Elections

1. Why do M and $\bar{l}$ not coincide if a distribution is skewed?

2. Assume that M and $\bar{l}$ do coincide. If the distribution is unimodal, will the mode also coincide with M and $\bar{l}$?

3. Show that 0.6 is the equilibrium in Figure 12.3b (on p. 422).

■ 4. Assume that a city is comprised of three equal-sized districts, each of which elects a candidate to the city council. The mayor is elected by the entire city. Show with an example that the median or extended median for the mayor need not be the median or extended median for any of the three city council districts. Does this explain why mayors and city council members often disagree?

■ 5. In Exercise 4, must the median or extended median for the mayor be between the leftmost and rightmost medians, or extended medians, of the three districts? How about the mean $\bar{l}$?

▲ 6. Which is better for consumers: (a) to minimize the maximum distance they must travel to a store; or (b) to foster price competition, which would presumably be encouraged if two stores are located at $M = \frac{1}{2}$?

■ 7. Prove that if a distribution is discrete and there is no median position, there is always an *extended median*. (*Hint:* Show that there is always one position at which a majority of voters lies neither to the left nor to the right, and neither candidate would have an incentive to depart from this position.)

8. Assume that the one voter at 0.1 in Figure 12.2b (on p. 420) decides not to vote because he is "too far away" from the two candidates who take the median position at 0.6. Would either candidate depart from $M = 0.6$ to try to do better if he or she knew that this voter had decided not to vote—but he or she would vote for the closer of two candidates less than a distance of 0.5 away? What if the candidates knew that the three voters at 0.2 had also decided not to vote—but they would vote for the closer of two candidates less than a distance of 0.4 away?

9. If you are a far-left or a far-right voter, are you helping your cause when you announce, like the voters in Exercise 8, that you will not support candidates who are too far away?

10. Consider the two most extreme voters at 0 in Figure 12.3b (those who are farthest from the extended median of 0.6). Would their nonvoting change the extended median? How about, as well, the nonvoting of the somewhat less extreme voter at 0.9? Show when, if at all, M or the extended median will change as fewer and fewer extreme voters decide not to vote in this example?

▲ 11. In Figure 12.2a, $\bar{l}$ is not in equilibrium; one candidate would do better if he or she moved from 0.56 to 0.6. But is 0.6 really a better reflection of the views of the electorate than 0.56?

12. Define an outcome to be in equilibrium if, given that one candidate chooses it, the other candidate cannot do better than take the same position. Show that this definition is equivalent to the text's definition of being in equilibrium.

13. Consider a bimodal distribution, which has two peaks of the same height. Will taking a position at either peak be in equilibrium?

14. Define A's position in a two-candidate race to be *opposition-optimal* if, given that the position of B is fixed, it maximizes A's vote total. Show that A's opposition-optimal position must be adjacent to B's position and closer to M, except when B is at the median. (Roughly speaking, being "adjacent" means being a very small distance away.)

▲ 15. Assume that the population along a main street is uniformly distributed over [0, 1], so there are equal numbers of people located at all equally spaced intervals from $M/\bar{l}$. (This makes the distribution rectangular, or "flat.") It has been argued that the "social optimum" for the location of two stores are at the points $\frac{1}{4}$ and $\frac{3}{4}$, because then no person would have to travel more than $\frac{1}{4}$ of the length of the street to buy at one store. Is this desirable if the population is not uniformly distributed?

▲ 16. What is a social optimum in an election if only one candidate is to be elected? How about five candidates to a city council? Is it better that the city council members' positions all be centered around 0.5, or should they be more spread out?

12.2 Spatial Models for Multicandidate Elections

17. Assume that A and B take the *same* median position. Is there more than one position that C can take to maximize his or her vote total? If so, what are they?

18. Assume that A and B take the *same* nonmedian position. What position should C take to maximize his or her vote total? Is C's position always a winning one?

■ **19.** Assume that A and B take *different* positions, with one possibly being at M. What position should C take to maximize his or her vote total? Is C's position always a winning one?

20. Is there a 1/3-separation obstacle if the distribution is not symmetric but no more than $\frac{1}{6}$ of the area under the curve separates A (on the left) from M, and no more than $\frac{1}{6}$ of the area separates B (on the right) from M? What if these $\frac{1}{6}$-or-less areas on the left and the right are not the same?

21. Is there a 2/3-separation opportunity if the distribution is not unimodal but at least $\frac{1}{3}$ of the area under the curve separates A (on the left) from M, and at least $\frac{1}{3}$ separates B (on the right) from M? (*Hint:* Start by assuming that the distribution is uniform between A and B—and hence not unimodal—and that exactly $\frac{2}{3}$ of the voters lie between A and B. Can C always win by taking a position at M? If not, is there a distribution that affords C this opportunity?)

22. Show that C cannot win under the conditions for the optimal entry of two candidates, anticipating a third entrant (page 427). (*Hint:* Indicate which candidate will win when C enters to the left of A, to the right of B, or in between.)

23. It is known that A, B, and C will enter an election in that order, with A announcing his position first, then B, and finally C. If the distribution is uniform (rectangular) over [0, 1], what position should each candidate take to maximize his or her vote total, anticipating—in the case of A and B—the entry of future candidates? [*Hint:* Start by assuming that A takes a position at $\frac{1}{4}$. Is B's position at $\frac{3}{4}$ optimal, anticipating the entry of C? Or can B do better at some other position (perhaps by influencing C's choice of a maximizing position)?]

■ **24.** If A and B are equidistant from the median of a symmetric distribution and separated from each other by exactly $\frac{1}{2}$ of the area under the curve, under what conditions is this separation an obstacle, and under what conditions is it an opportunity? [*Hint:* Start by constructing examples of symmetric distributions in which C would either win or lose by taking a position at M.]

■ **25.** What are the vote-maximizing positions for four candidates to take if it is known that they will enter in the order A, B, C, D?

12.3 Narrowing the Field

26. Assume that the four candidates in the 2000 presidential election can be arrayed from left to right as follows: Nader–Gore–Bush–Buchanan. Suppose a poll reveals Gore at 48 percent, Bush at 47 percent, Nader at 3 percent, and Buchanan at 2 percent. Would Bush be well advised to offer Buchanan a cabinet position to drop out of the race (as Adams offered Clay the secretary-of-state post after the 1824 election)? What if Bush knew that, after Buchanan dropped out, only half of Buchanan's supporters would switch to him, with most of the remainder not voting, except for a few who would switch to Gore?

27. Assuming the same poll results as in Exercise 26, now suppose that Gore offered the same deal to Nader, knowing that only one-third of Nader supporters would switch to him and the rest would not vote. However, suppose Gore also thought that if Nader dropped out, so would Buchanan, and all Buchanan supporters would vote for Bush. Should Gore set off this train of events?

▲ **28.** Is there any evidence that the four presidential candidates in 2000 might have contemplated "deals" of the kind indicated in Exercises 26 and 27? If you cannot find any evidence, do you think this is because the candidates found such ploys unethical or because they thought they might be found out if they tried to engage in them?

▲ **29.** One tactic that was considered by Nader supporters who thought that their votes for Nader might kill Gore's chances in some states was to swap votes: In close states that Gore might lose if Nader supporters stuck with their candidate, these supporters would switch to Gore if Gore supporters in less contested states, where Gore would almost surely win, would switch to Nader. Thereby the popular-vote totals for the two candidates would not change overall, but Gore would be able to win in the close states he might otherwise lose. Is this a sensible way of dealing with problems created by the Electoral College, which puts a premium on winning in large states?

30. The opposite of a bandwagon effect is an "underdog effect," whereby voters vote for the presumed loser, independent of his or her merit. When would you expect to see such an effect?

12.4 What Drives Candidates Out?

▲ **31.** Would voters be more sincere if polls were banned just before an election? Would such a ban produce better outcomes? According to what criteria?

32. Show why the Condorcet-winner-unsuccessful result (page 431) is true. Is it true that if the poll assumption was modified to differentiate the top three (rather than the top two) candidates from the rest, and the Condorcet winner was not among the top three, that he or she would still lose?

▲ 33. Show why the Condorcet-winner-successful result (page 431) is true. Is it proper that the candidate who comes in second in the poll should win after the results of the poll are announced? Why?

▲ 34. In Example 4 (page 431), after Class IV voters switch from D to A, the vote totals for the top three candidates are A–5, B–3, and C–4. Now assume a second poll is taken, differentiating A and C, the top two contenders, from B. If B supporters switch at this point to their second choice, which candidate will win? Do you consider this a desirable outcome?

▲ 35. Assume there are four classes of voters that rank four candidates as follows:

I. 4: $A\ D\ B\ C$
II. 3: $B\ D\ A\ C$
III. 2: $C\ D\ B\ A$
IV. 1: $D\ C\ B\ A$

Which candidate is the Condorcet winner? Do you find this result strange in the light of what a poll would tell the voters?

36. Assume that there is a poll that differentiates the top two candidates in Exercise 35. Which candidate will win the election after the poll?

▲ 37. Assume that there is a poll that differentiates the top three candidates in Exercise 35. Which candidate will win the election after the poll? Comment on the different outcomes in this exercise and the previous one.

▲ 38. Assume that there are three classes of voters who rank three candidates as follows:

I. 4: $A\ B\ C$
II. 3: $B\ C\ A$
III. 2: $C\ A\ B$

Show that there is no Condorcet winner. Applying the poll assumption to this example, which candidate will win? Is this fair, given the preferences of Class II and III voters, who, together, are a majority?

12.5 Election Reform: Approval Voting

■ 39. Prove the voting-only-for-a-second-choice result (page 433).

40. In a three-candidate election, show that your strategy of voting for your top two choices under approval voting is not always better than voting only for your top choice.

41. Consider a four-candidate election under approval voting. Is there ever a situation in which a voter would vote for a first and a third choice without also voting for a second choice? [*Hint:* Assume that a voter ranks the four candidates $A\ B\ C\ D$ and believes that one of two things can happen: The electorate will favor either liberals (say, A and B) or conservatives (say, C and D) but never favor each side equally.]

42. Is there ever a situation under approval voting in which a voter would vote for a worst choice?

43. Is there ever a situation under approval voting in which a voter would *not* vote for a first choice if he or she finds acceptable one or more lower-ranked candidates? (*Note:* This question asks whether the voting-only-for-a-second-choice result can be generalized to more than three candidates.)

■ 44. Prove the effect-of-dichotomous-preferences result (page 433). [*Hint:* If all voters have dichotomous preferences and vote for all candidates in their preferred subsets, which candidate will get the most approval votes? What does this say about the preferences of voters for the approval-vote winner, compared to their preferences for each of the other candidates?]

45. In the following example, Class I and II voters have dichotomous preferences, but the Class III voter has *trichotomous preferences* (he or she divides the four candidates into three indifference subsets):

I. 2: ($A\ B$) ($C\ D$)
II. 2: (C) ($A\ B\ D$)
III. 1: (D) (C) ($A\ B$)

Is it rational for the Class III voter to vote only for his or her top choice, D? If not, who else should he or she approve of? Which class of voters will be most unhappy if the Class III voter does not vote just for D? Can voters in this class, by voting strategically, do anything about their situation?

▲ 46. Assume the Class III voter's preferences change to a different trichotomous ordering:

III. 1: (D) ($A\ C$) (B)

Suppose, as in Exercise 45, that the Class III voter indicates in an initial poll that he or she intends to vote only for D but then, in response to the poll, switches to voting for the candidates in his or her second-choice subset as well. If there is a new poll, based on these results, what will be the outcome? What if there is a third poll, fourth poll, and so on? Do you regard this result as desirable? Why?

47. In Exercise 35, we saw that under plurality voting the Condorcet winner, D, comes in fourth in a poll and, therefore, cannot be helped by subsequent polling, even when the poll distinguishes the top three candidates and voters differentiate among them:

I. 4: $A\ D\ B\ C$
II. 3: $B\ D\ A\ C$
III. 2: $C\ D\ B\ A$
IV. 1: $D\ C\ B\ A$

What are the outcomes under approval voting—both with and without polling—if voters approve of their (i) top-ranked, (ii) two top-ranked, and (iii) three top-ranked candidates initially? [*Note:* In making adjustments to the poll results, assume that voters approve not only of the preferred of their two top-ranked candidates identified by the poll but also of *all* candidates ranked above their preferred candidate. For example, when the poll based on (i) above identifies A and B as the two top-ranked candidates, with 4 and 3 votes, respectively, the Class III and Class IV voters after the poll will approve not only of their preferred candidate, B, but also of C and D, because they rank the latter two candidates above B.]

▲ **48.** On the basis of your answers to the foregoing problems, do you think approval voting would be beneficial in finding Condorcet winners—either with or without polling—in multicandidate elections?

▲ **49.** If approval voting were adopted, would you recommend that it be followed by a runoff election between the top two candidates if no candidate was approved by more than 50 percent? Why?

12.6 The Electoral College

50. Assume that the National Popular Vote law were adopted. Would the election outcome be the same as if there were direct popular vote (i.e., without the Electoral College), with the winner being the candidate who received the most popular votes?

51. Assume that there are three states with 3, 7, and 9 voters, and that they are all toss-up states. If both the Democratic and Republican candidates choose strategies that maximize their expected popular vote (the proportional rule), and they have the same total resources ($D = R$), what is the expected number of votes that each will receive?

52. Assume that the Republican knows in advance what allocations, d_i, to each state i the Democrat will make in Exercise 51. Then the Republican's optimal response can be shown to be

$$r_i = \frac{\sqrt{n_i d_i}}{\sum_{i=1}^{t} \sqrt{n_i d_i}}(R + D) - d_i$$

instead of that given by the proportional rule (p. 436). Suppose that the Democrat ignores the smallest state and makes proportional allocations to the two largest states. (For concreteness, assume that both candidates have 100 units of resources.) What is the Republican's optimal response? What if the Democrat makes proportional allocations to all three states?

■ **53.** In Exercise 52, show that if the Democrat makes proportional allocations, and the Republican responds optimally according to the formula given there, this formula simplifies to $r_i = (n_i/N)R$, which does not depend on d_i. What does this say about the proportional rule? [*Hint:* If the Republican finds out (say, through a spy) that the Democrat is making proportional allocations, does the Democrat have anything to worry about?]

54. In Exercise 53, if the Democrat has only half the resources of the Republican, would you recommend that he or she behave differently from proportional allocations to maximize his or her expected popular vote? Why? If the Republican allocates his or her resources proportionally to the three states, is there any way the Democrat can allocate his or her resources to win a majority of votes in states with more than half the votes?

55. Instead of maximizing their expected popular vote, assume that the candidates in Exercise 51 want to win in states with more than half the votes. Suppose the candidate who allocates more resources to a state wins that state. Is there any state to which a candidate should not consider allocating resources? Should the states that receive allocations receive equal allocations?

56. In Exercise 51, assume that you can choose specific voters in each state to whom you can allocate resources. Suppose that the candidate who allocates more resources to a voter wins that voter's vote. If your goal is to win the votes of a majority of voters in states that have more than half the votes, which voters would you target, and how much would you spend on each? [*Hint:* First, show which states you would target; then show that these states should receive equal allocations, which in turn should be divided equally among a certain set of voters.]

57. Assume that there are three toss-up states, A, B, and C, with, respectively, 2, 3, and 4 voters, which are also the number of electoral votes of each state. In the text (p. 437), we gave the formulas for the probabilities, P_A, P_B, and P_C, that the Democrat wins a majority of popular votes in each state and, therefore, wins all the electoral votes of that state. Show that the formula for the probability that the Democrat *wins the election* under the Electoral College, PWE_D, is

$$PWE_D = P_A P_B(1 - P_C) + P_A P_C(1 - P_B) + P_B P_C(1 - P_A) + P_A P_B P_C$$

[*Hint:* Winning in any two states is sufficient to win the election.]

■ **58.** Compare the formula for PWE_D with the formula for EEV_D (in the text). Which quantity is it better to maximize? What would be a good resource-allocation strategy for maximizing PWE_D?

59. Is the square root in the formulas for the *EEV* maximizing strategies of the Democratic and Republican candidates related to the square-root rule for the Electoral College discussed in Chapter 11?

WRITING PROJECTS

1. Do you think polling is useful in helping voters choose the "best" candidate? Or would it be better, as in some countries, to ban the publication of polls before an election assuming that such a ban was constitutional? In one to two pages, discuss these questions in light of the theoretical effects polling has when voters react to polls and possibly change their voting strategies. Is there empirical evidence that voters behave in this way?

2. How serious a problem do you think the large-state bias of the Electoral College is? How would you explain the fact that some of the strongest advocates of the Electoral College come from small states? Has the theoretical bias been a reality in the campaign behavior of candidates in recent presidential elections? Discuss in one to two pages.

Suggested Readings

BRAMS, STEVEN J. *The Presidential Election Game,* 2d ed., A K Peters, Wellesley, MA, 2008. Focuses on the strategic aspects of presidential elections—from primaries to conventions to general elections—and also includes an analysis of the "game" played between President Richard Nixon and the Supreme Court over the release of the Watergate tapes that led to Nixon's resignation in 1974. (Nixon has been the only president to resign the presidency.) Approval voting and direct popular-vote election of a president are recommended as election reforms.

BRAMS, STEVEN J., and PETER C. FISHBURN. *Approval Voting,* 2d ed., Springer, New York, 2007. An in-depth analysis of approval voting, which includes several case studies.

BRAMS, STEVEN J. *Mathematics and Democracy: Designing Better Voting and Fair-Division Procedures,* Princeton University Press, Princeton, NJ, 2008. Shows how mathematics can be used to analyze the properties of different democratic procedures and can help to identify those with the most desirable properties.

HINICH, MELVIN J., and MICHAEL C. MUNGER. *Analytical Politics,* Cambridge University Press, Cambridge, U.K., 1997. Extends spatial modeling to more than one dimension, analyzes probabilistic voting, and introduces game-theoretic solution concepts relevant to the study of elections.

SAARI, DONALD G. *Chaotic Elections! A Mathematician Looks at Voting,* AMS [American Mathematical Society], Providence, RI, 2001. Argues that elections—in particular, the 2000 presidential election, but others as well—have chaotic features that can be understood through mathematics. The mathematics used is an unusual kind of geometry that will be accessible to those with some mathematical background.

SHEPSLE, KENNETH A. *Analyzing Politics: Rationality, Behavior, and Institutions,* 2nd ed., Norton, New York, 2010. Rational strategies in voting and elections are a major component of this text, but it also includes sections on collective action and political institutions, such as courts and legislatures. Several case studies illustrate the theory.

Suggested Web Sites

www.fec.gov Federal Election Commission—About federal elections and voting in the United States.

www.ifes.org International Foundation for Election Systems.

www.rangevoting.org The Center for Range Voting.

wiki.electorama.com/wiki/Election-methods_mailing_list Election-methods mailing list.

Ron Sachs/Pool/CNP/Corbis

Fairness and Game Theory

Part IV

Chapter 13
Fair Division

Chapter 14
Apportionment

Chapter 15
Game Theory: The Mathematics of Competition

The central thrust of the first two chapters in Part IV is the fair division of divisible and indivisible objects. Whereas a cake or a parcel of land is divisible, the representatives who are apportioned to the different states are indivisible. Sometimes, however, seemingly indivisible objects, like a car, can be shared, rendering them divisible. By contrast, Chapter 15 focuses on what rational players will choose in different strategic situations, which may be highly unfair to some.

Chapter 13 describes fair-decision schemes in which a group of individuals with different values can be assured of each receiving what he or she views as a fair share when dividing objects like cakes or the goods in an estate.

Chapter 14 discusses the apportionment problem, which is to round a set of fractions to whole numbers while preserving their sum; of course, the sum of the original fractions must be a whole number to start. Apportionment problems occur when resources must be allocated in integer quantities—for instance, when legislators allocated seats in the U.S. House of Representatives to the 50 states.

Chapter 15 introduces the mathematical field called *game theory*, which describes situations involving two or more decision makers having different goals. Game theory provides a collection of models to assist in the analysis of conflict and cooperation as well as strategies for resolution. Interestingly, you will find that the games covered in this chapter provide us with insights into certain social paradoxes that we routinely encounter in our daily lives.

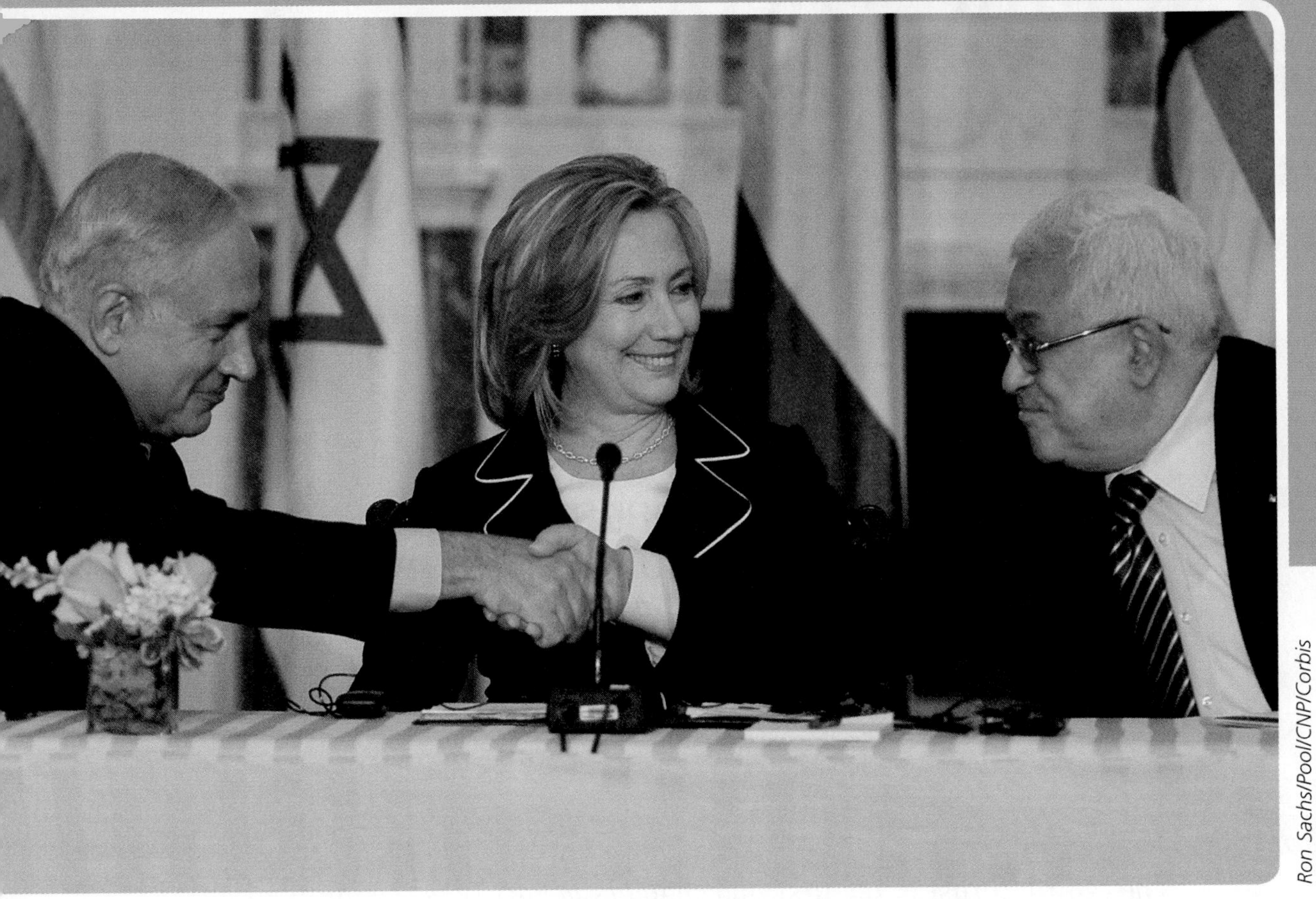

Ron Sachs/Pool/CNP/Corbis

13

Fair Division

When the demands or desires of one party are in conflict with those of another—be it a divorce, a labor-management negotiation, or an international dispute—no one wants to be treated unfairly. And with 1.2 million divorces every year in the United States alone, and crises such as we've seen in the Middle East for decades, it is certainly worth considering how mathematics might help in the search for procedures that can ensure fair and equitable resolutions of such conflicts.

We begin this chapter with one such procedure that was developed in the mid-1990s. The **adjusted winner procedure** allows two parties to settle any dispute involving either issues (as in an international dispute) or objects (as in a divorce or a two-person inheritance) with certain mathematical guarantees of "fairness." Disagreement, it turns out, is both a bad thing and a good thing. On the one hand, disagreement as to how each issue should be resolved typically lies at the heart of a conflict. On the other hand, procedures such as adjusted winner are designed to capitalize on the parties' disagreement as to the importance of each issue, thus allowing each party to end the negotiations thinking that it has been met more than halfway.

But adjusted winner is just one of several so-called fair-division procedures that have been developed over the past 65 years. So following our discussion of adjusted winner, we describe a procedure for handling inheritances that was discovered by the Polish mathematician Bronislaw Knaster during World War II. Staying with real-world applications, we next consider the tricky question of finding a priority ranking for potential recipients of an organ that becomes available for transplantation. This is followed by a discussion of an extremely basic fair-division procedure—taking turns—and the question of what the optimal strategy is when taking turns choosing objects.

Bridging the gap between fair-division procedures with obvious real-world potential, such as divorce and inheritance procedures, and procedures that address fundamental mathematical questions of fairness (as do the procedures treated later in this chapter) is the ancient two-person procedure known as *divide-and-choose.* An application of this procedure to the Convention of the Law of the Sea (also known as the Law of the Sea Treaty) is described.

Divide-and-choose sets the stage for the mathematical investigations of fair division that have gone on for more than half a century. These investigations often have been phrased within the metaphor of "cake cutting." We present three cake-cutting procedures. The first two of these—found by Steinhaus and Banach-Knaster in the 1940s—yield allocations in which each player receives what he or she perceives to be at least his or her fair share of the cake. The last one—found by Selfridge-Conway in 1960—yields allocations in which each player receives what he or she perceives to be a piece at least tied for largest. We conclude with a discussion of Vickrey auctions, a variant of which is used today on eBay.

13.1 The Adjusted Winner Procedure

(Pascal Plessis/AP Photo.)

To illustrate the *adjusted winner procedure,* we will consider an application to the multibillion-dollar world of business mergers. It turns out that one of the most elusive ingredients in the success of a merger is what deal-makers call *social issues*—how power, position, sacrifice, and status are allocated between the merging companies and their executives.

As a case in point, let's revisit the 1998 proposed merger between two giant pharmaceutical companies, Glaxo Wellcome and SmithKline Beecham. While most of the details underlying this aborted deal are still unknown to outsiders, the role of social issues is clearly underscored by reports that the companies "saw nearly 19 billion dollars of stock market value vanish in the clash of two corporate egos."

Exactly what kinds of issues might bring on a "clash of two corporate egos"? While not privy to the details of the Glaxo Wellcome–SmithKline Beecham merger attempt, we can speculate as to their nature. For purposes of illustration, let's assume that the following five social issues were paramount:

1. The name that the combined company would use
2. The location of the headquarters of the combined company
3. The question of who would serve as chair of the combined company
4. The question of who would serve as CEO of the combined company
5. The question of where the necessary layoffs would come from

Each of these five social issues is known to have been a major factor in other recent proposed mergers. For example, when Chrysler merged with Daimler-Benz in 1998, the issue of the choice of a name for the combined company was described as a "standoff" before both sides finally agreed to DaimlerChrysler.

So let's assume that these were the five social issues confronting Glaxo Wellcome and SmithKline Beecham, and let's see how the adjusted winner procedure would have suggested a resolution. The starting point—and something that is quite difficult when dealing with issues (as in a negotiation) as opposed to objects (as in a divorce)—is to have each side quantify the importance it attaches to getting its own way on each of the issues.

With the adjusted winner procedure, quantification is done by having each side—independently and simultaneously—spread 100 points over the issues in a way that reflects the relative worth of each issue to that party. In our present example, let's assume that the companies allocated their 100 points as shown in Table 13.1. The adjusted winner procedure is now used to decide which side gets its way on which issues, but the procedure requires that a compromise of sorts may have to be reached on one of the issues.

Table 13.1

Applying the Adjusted Winner Procedure to a Merger of Two Companies

	Point Allocations	
Issue	**Glaxo Wellcome**	**SmithKline Beecham**
Name	5	10
Headquarters	25	10
Chair	35	20
CEO	15	35
Layoffs	20	25
Total	**100**	**100**

Here's how the procedure works. Suppose that we have two parties and a list of either issues to be resolved in one party's favor or the other's (as in our merger example) or objects to be awarded either to one party or to the other (as in a divorce or a two-person inheritance). To have a single word covering both issues and objects, we will speak of "items." The adjusted winner procedure follows these basic steps:

Basic Steps in the Adjusted Winner Procedure PROCEDURE

Step 0. As described earlier, each party distributes 100 points over the items in a way that reflects each item's relative worth to that party.

Step 1. Each item on which the assigned points differ is initially given to the party that assigned it more points. Add up the total number of points each party feels that he or she has received. The party with the fewest points is now given all the items on which both parties placed the same number of points. Once again, add up the total number of points each party feels that he or she has received. The party with the most points is called the *initial winner;* the other party is called the *initial loser*.

Step 2. For each item given to the initial winner, calculate the "**point ratio.**"

Step 3. Start moving items from the initial winner to the initial loser in ascending order of point ratio. Stop when you get to an item whose move will cause the initial winner to have fewer points than the initial loser. This item will need to be split or shared and is thus called the *shared item.*

Step 4. Let x represent the fractional part of the shared item that will be moved from the initial winner to the initial loser. Write a formula that equates each party's total points after the sharing of this item.

Step 5. Solve the equation and state the final division of items between the two parties.

Let's demonstrate the adjusted winner procedure by continuing with our analysis of the proposed merger between Glaxo Wellcome and SmithKline Beecham. Why the order of transfer given in Step 3 is so important will be explained later.

0. Assume that Glaxo Wellcome and SmithKline Beecham have given us the point assignments shown in Table 13.1.
1. Because Glaxo Wellcome has placed more points on headquarters (25) and chair (35), it is initially "given" these issues, while SmithKline Beecham is initially given name (10), CEO (35), and layoffs (25). Notice that SmithKline Beecham now has 10 + 35 + 25 = 70 of its points, whereas Glaxo Wellcome has only 25 + 35 = 60 of its points.
2. We now calculate the point ratio of the three items held by the initial winner, SmithKline Beecham:

   Layoffs has point ratio 25/20 = 1.25.
   Name has point ratio 10/5 = 2.00.
   CEO has point ratio 35/15 = 2.33.

3. We now start transferring issues from SmithKline Beecham to Glaxo Wellcome until the point totals of the two sides are equal. Because the layoff item has the lowest point ratio, we start to transfer that item first. But we now see that transferring the entire layoff item (worth 25 to SmithKline Beecham and 20 to Glaxo Wellcome) gives Glaxo Wellcome more points (60 + 20 = 80) than SmithKline Beecham has (70 − 25 = 45). Thus, the entire layoff item cannot be transferred. Glaxo Wellcome and SmithKline Beecham will need to compromise on the issue of layoffs (it is the shared item). But compromise may not mean meeting each other halfway. Our goal is to equalize points between the two companies, and a little algebra will tell us exactly the extent to which Glaxo Wellcome and SmithKline Beecham should get their way on the issue of layoffs.
4. Let x be the fractional part of the layoff issue that will be transferred from SmithKline Beecham to Glaxo Wellcome. Because SmithKline Beecham had 25 points on the layoff issue, it loses $25x$ points; because Glaxo Wellcome had 20 points on the layoff issue, it gains $20x$ points.

   Hence, the original point totals of 70 for SmithKline Beecham and 60 for Glaxo Wellcome become, after the transfer, $70 - 25x$ and $60 + 20x$, respectively. Thus, if we want a fraction x that will make SmithKline Beecham's total points equal Glaxo Wellcome's total points, then we need to solve the following equation:

$$70 - 25x = 60 + 20x$$

5. We use algebra to solve this equation:

$$\begin{aligned} 70 - 60 &= 20x + 25x \\ 10 &= 45x \\ 10/45 &= x \end{aligned}$$

   Reducing the fraction, we see that $x = 2/9$. Inserting 2/9 back into the equation, we see that

$$70 - 25(2/9) = 60 + 20(2/9)$$

   or approximately 64 points for each side. Thus, equality of points is achieved when SmithKline Beecham gives up two-ninths of what it wanted on the issue of layoffs and Glaxo Wellcome gets two-ninths of its way.

Solving Linear Equations — Algebra Review

A linear equation is a mathematical statement. In a linear equation with one variable (typically represented by x), that variable has an understood exponent of one, cannot appear in the denominator, and cannot be found under a radical symbol. Here are some examples of linear and non-linear equations.

Linear Equations	Non-linear Equations
$2x = 6$	$x^2 + 1 = 6$
$0.3(x - 5) + 4 = 6.1$	$\frac{3}{x^3} + 4 = 5$
$\frac{1}{2}(x + 2) + 7(2x - 1) = 4$	$\sqrt{x + 1} = 5$

A linear equation can be solved by performing the following steps.

1. Simplify both sides of the equation. If the equation has fractions or decimals, clearing these will generally simplify the remaining steps.
2. Get all variable terms on one side of the equation and combine.
3. Get constant terms on the other side of the equation and combine.
4. Divide both sides by the coefficient of the variable.

Not every linear equation will require all four steps, and typically Step 1 is the longest.

Let's solve the equation $0.4 + 0.2x = 0.6 - 0.2x$. In this equation, we will multiply both sides by 10 to clear decimals.

Step 1: $10(0.4 + 0.2x) = 10(0.6 - 0.2x)$

$4 + 2x = 6 - 2x$

Step 2: $4 + 2x + 2x = 6 - 2x + 2x$

$4 + 4x = 6$

Step 3: $4 + 4x - 4 = 6 - 4$

$4x = 2$

Step 4: $\frac{4x}{4} = \frac{2}{4}$

$x = \frac{1}{2}$ or 0.5

The solution is $x = \frac{1}{2}$ or 0.5.

A linear equation has one of three possible outcomes: one solution, no solution, or infinitely many solutions. (The example above has only one solution.)

Having seen how the adjusted winner procedure works, we must now ask the following question: Exactly what is it about the allocation produced by this scheme that would make someone want to use it? To answer this question, we need three definitions.

Equitable — DEFINITION

A fair-division procedure is said to be **equitable** if each player believes he or she received the same fractional part of the total value.

Envy-Free — DEFINITION

A fair-division procedure is said to be **envy-free** if each player has a strategy that can guarantee him or her a share of whatever is being divided that is, in the eyes of that player, at least as large (or at least as desirable) as that received by any other player, no matter what the other players do.

Pareto-Optimal — DEFINITION

A fair-division procedure is said to be **Pareto-optimal** if it produces an allocation of the property such that no other allocation achieved by any means whatsoever can make any one player better off without making some other player worse off.

The answer to our earlier question is given by the following theorem (whose proof can be found in *Fair Division* by Brams and Taylor, listed in the "Suggested Readings" section at the end of the chapter):

Properties of the Adjusted Winner Allocation THEOREM

For two parties, the adjusted winner procedure produces an allocation based on each player's assignment of 100 points over the items to be divided that has the following properties:

- The allocation is equitable.
- The allocation is envy-free.
- The allocation is Pareto-optimal.

Economists consider Pareto optimality (also named after Vilfredo Pareto) to be an extremely important property, and the order of transfer in Step 3 on page 455 of the adjusted winner procedure is so important because it guarantees that the outcome is Pareto-optimal. The fact that the adjusted winner procedure produces an allocation that is efficient in this sense leads us to hope that it can and will play a future role in real-world dispute resolution.

13.2 The Knaster Inheritance Procedure

The adjusted winner procedure can be applied in the case of an inheritance if there are only two heirs. For *more than two heirs*, there is quite a different scheme, the **Knaster inheritance procedure**, first proposed by Bronislaw Knaster in 1945. It has a drawback, though, in that it requires the heirs to have a large amount of cash at their disposal.

EXAMPLE 1
A Four-Person Inheritance

Suppose (for the moment) that there is just one object—a house—and four heirs—Bob, Carol, Ted, and Alice. Knaster's scheme begins with each heir bidding (simultaneously and independently) on the house. Assume, for example, that the bids are as follows:

Bob	Carol	Ted	Alice
$120,000	$200,000	$140,000	$180,000

Carol, being the high bidder, is awarded the house. Her fair share, however, is only one-fourth of the $200,000 she thinks the house is worth, and so she places $150,000 (which is three-fourths of the $200,000 she bid) into a temporary "kitty."

Each of the other heirs now withdraws from the kitty his or her fair share, that is, one-fourth of his or her bid:

Bob withdraws	$120,000/4 = $30,000
Ted withdraws	$140,000/4 = $35,000
Alice withdraws	$180,000/4 = $45,000

Thus, from the $150,000 kitty, a total of $30,000 + $35,000 + $45,000 = $110,000 is withdrawn, and each of the four heirs now feels that he or she has the equivalent of one-fourth of the estate. Moreover, there is a $40,000 surplus ($150,000 kitty − $110,000 withdrawn), which is now divided equally among the four heirs (so each receives an additional $10,000). The final settlement is

Bob	Carol	Ted	Alice
$40,000	House − $140,000	$45,000	$55,000

This illustrates Knaster's procedure for the simple case in which there is only one object. But what if our same four heirs have to divide an estate consisting of (say) a house (as before), a cabin, and a boat? There are actually two ways to handle this situation, and we'll illustrate both, assuming that our four heirs submit the following bids:

	Bob	Carol	Ted	Alice
House	$120,000	$200,000	$140,000	$180,000
Cabin	60,000	40,000	90,000	50,000
Boat	30,000	24,000	20,000	20,000

The first way is simply to handle the estate one object at a time, proceeding for each object as we just did for the house.

We have already settled the house. Let's handle the cabin the same way. Thus, Ted is awarded the cabin based on his high bid of $90,000. His fair share is one-fourth of this, so he places three-fourths of $90,000 (which is $67,500) into the kitty.

Bob withdraws from the kitty $60,000/4 = $15,000. Carol withdraws $40,000/4 = $10,000, and Alice withdraws $50,000/4 = $12,500. Thus, from the $67,500 kitty, a total of $15,000 + $10,000 + $12,500 = $37,500 is withdrawn.

The surplus left in the kitty is thus $30,000, and this is again split equally ($7500 each) among the four heirs. The final settlement on the cabin is

Bob	Carol	Ted	Alice
$22,500	$17,500	Cabin − $60,000	$20,000

If we were now to do the same for the boat (we leave the details to you), the corresponding final settlement would be

Bob	Carol	Ted	Alice
Boat − $20,875	$7625	$6625	$6625

Putting the three separate analyses (house, cabin, and boat) together, we get a final settlement of

Bob:	Boat + ($40,000 + $22,500 − $20,875 = $41,625)
Carol:	House + (−$140,000 + $17,500 + $7625 = −$114,875)
Ted:	Cabin + ($45,000 − $60,000 + $6625 = −$8375)
Alice:	$55,000 + $20,000 + $6625 = $81,625.

Notice that here, Carol gets the house but must pay $114,875 in cash (and Ted gets the cabin but must put up $8375 in cash). This cash is then disbursed to Bob and Alice. In practice, Carol's having this amount of cash available may be a real problem—the key drawback to Knaster's procedure. Nevertheless, Knaster's procedure shows again that whenever some participants have different evaluations of some objects, there is an allocation in which everyone obtains more than what they would normally consider a fair share.

EXAMPLE 2
Another Way

The second way begins by adding two rows to the chart of bids, one giving the total value of the estate to each heir (arrived at by summing the columns) and the other giving each heir's fair share (which is one-fourth the value of the estate because there are four heirs):

	Bob	Carol	Ted	Alice
House	$120,000	$200,000	$140,000	$180,000
Cabin	60,000	40,000	90,000	50,000
Boat	30,000	24,000	20,000	20,000
Total value	$210,000	$264,000	$250,000	$250,000
Fair share	$52,500	$66,000	$62,500	$62,500

Next, we give each item to the party who values it most. Bob gets the boat, Carol gets the house, and Ted gets the cabin. This is certainly not fair because Carol got the most valuable item and Alice got nothing. We fix this in the following way.

Bob got $30,000 in value but feels slighted since he felt his share was $52,500. The estate gives him the difference in cash: $52,500 − $30,000 = $22,500.

Carol received more than her fair share, so we have her pay the difference to the estate: $200,000 − $66,000 = $134,000.

Ted received $90,000 in value when he believed his fair share was only $62,500, so we have him pay the estate the difference: $90,000 − $62,500 = $27,500.

Alice received nothing when she believed her fair share to be $62,500, so the estate gives her $62,500 in cash.

At this point, every party has his or her fair share. However, the estate has taken in more than it has paid out. This is called the *surplus.*

Surplus = ($134,000 + $27,500) − ($22,500 + $62,500) = ($161,500 − $85,000) = $76,500

We now divide the surplus evenly among the parties.

$76,500 ÷ 4 = $19,125

Finally, we give an additional $19,125 to each party, making the final division (as before):

Bob gets the boat and $41,625 ($22,500 + $19,125).
Carol gets the house and pays the estate $114,875 ($134,000 − $19,125).
Ted gets the cabin and pays the estate $8375 ($27,500 − $19,125).
Alice gets $81,625 cash ($62,500 + $19,125).

We summarize Knaster's inheritance procedure as follows.

Basic Steps in Knaster's Inheritance Procedure with *n* Heirs — PROCEDURE

For each object, the following steps are performed:

Step 1. The heirs—independently and simultaneously—submit monetary bids for the object.

Step 2. The high bidder is awarded the object, and he or she places all but $1/n$ of his or her bid in a kitty. So, if there are four heirs ($n = 4$), then he or she places all but one-fourth—that is, three-fourths—of his or her bid in a kitty.

Step 3. Each of the other heirs withdraws from the kitty $1/n$ of his or her bid.

Step 4. The money remaining in the kitty is divided equally among the n heirs.

13.3 Fair Division and Organ Transplant Policies

In 1984, the U.S. Congress passed the National Organ Transplant Act and established a unified transplant network known as the Organ Procurement and Transplantation Network (OPTN). One of the primary goals of the OPTN was to increase the equity in the national system of organ allocation.

Achieving an equitable system of organ allocation is complicated by factors other than demand exceeding supply. For example, should an available organ go to the patient who needs it the most or to the one for whom the likelihood of a successful transplant is greatest? Should both of these be taken into consideration, and, if so, how? Questions such as these reveal the extent to which an equitable system of organ allocation is a challenging problem in fair division.

To illustrate some of the issues (and paradoxes!) arising in the search for an equitable system for organ allocation, we'll (roughly) follow Peyton Young's synopsis—from his book, listed in the "Suggested Readings" section—of the fair division procedure for kidney allocation adopted by the OPTN in the late 1980s.

There were three (main) criteria used in arriving at a final ranking of those needing a kidney, and each potential recipient was awarded points according to a fixed method that we now describe.

- **Criterion 1: Waiting time.** A list of potential recipients was made according to how long they had been waiting for an organ. For each potential recipient, one calculates the fraction of people at or below the spot on the list he or she occupies and then awards that person a number of points equal to 10 times that fraction. So if there are five people on the list, the first (waiting the longest) gets $10 \times 1 = 10$ points, the second gets $10 \times (4/5) = 8$ points, the third gets $10 \times (3/5) = 6$ points, and so on.
- **Criterion 2: Suitability.** The donor and potential recipient each have six relevant antigens that are either matched or not matched; the likelihood of a successful transplant increases with more matches. Two points are awarded for each match.
- **Criterion 3: Disadvantage.** Each person has antibodies that rule out a certain percentage of the population as being potential donors for that person. For some, only 10 percent are ruled out, whereas for others it may be as high as 90 percent. Those in the latter category are at a serious disadvantage compared to those in the former. Thus, potential recipients are awarded 1 point for each 10 percent of the population they are "sensitized against."

To illustrate this allocation procedure, let's assume we have five potential recipients—*A*, *B*, *C*, *D*, and *E*—with the following characteristics:

Potential Recipient	Months Waiting	Antigens Matched	Percent Sensitized
A	5	2	10
B	4.5	2	20
C	4	0	0
D	2	3	60
E	1	6	90

According to the procedure we described, points would be allocated as follows:

Potential Recipient	Months Waiting	Antigens Matched	Percent Sensitized	Total Points
A	10	4	1	15
B	8	4	2	14
C	6	0	0	6
D	4	6	6	16
E	2	12	9	23

Thus, if one kidney became available, it would go to *E* (with 23 points). Presumably, if two kidneys became available at the same time, *E* would get one and *D* (with 16 points) would get the other.

But now things get interesting. Peyton Young, being well versed in the paradoxes of voting theory, fair division, and apportionment (among other things), observed the following. In the above scenario, what if two kidneys become available, but one is delayed slightly? Presumably, *E* gets the first one, and then we redo the chart with only *A, B, C,* and *D.* This yields the following:

Potential Recipient	Months Waiting	Antigens Matched	Percent Sensitized
A	5	2	10
B	4.5	2	20
C	4	0	0
D	2	3	60

According to the procedure that we described, points would be allocated as follows:

Potential Recipient	Months Waiting	Antigens Matched	Percent Sensitized	Total Points
A	10	4	1	15
B	7.5	4	2	13.5
C	5	0	0	5
D	2.5	6	6	14.5

Thus, *A* (not *D*!) now gets the second kidney, having 15 points to 14.5 for *D.* This is an example of what is called the "priority paradox." For more on this, consult Peyton Young's book, listed in the "Suggested Readings" section.

13.4 Taking Turns

For many of us, an early lesson in fair division happens in elementary school with the choosing of sides for a spelling bee or when picking teams on the playground. In terms of importance, these pale in comparison with the issue of property settlement in a divorce. Remarkably, however, the same fair-division procedure—*taking turns*—is often used in both.

Taking turns is fairly self-explanatory. With two parties (and that's all we'll consider here), one party selects an object, then the other party selects one, then the first party again, and so on. But in this context, there are several interesting questions that suggest themselves:

1. How do we decide who chooses first?
2. Because choosing first is often quite an advantage, shouldn't we compensate the other party in some way, perhaps by giving him or her extra choices at the next turn?
3. Should a player always choose the object he or she most favors from those that remain, or are there strategic considerations that players should take into account?

The answer to Question 1 is often "toss a coin," but there are other possibilities; for example, the two parties could "bid" for the right to go first, as in an auction. The answer to Question 2 is less clear, but in Writing Project 2 (on p. 483), we outline a discussion of the issue it raises.

Question 3, on the other hand, is remarkably interesting, and it is this one that we want to pursue. Let's look at an easy example. Suppose that Bob and Carol are getting a divorce, and their four main possessions, ranked from best to worst by each, are as follows:

	Bob's Ranking	Carol's Ranking
Best	Pension	House
Second best	House	Investments
Third best	Investments	Pension
Worst	Vehicles	Vehicles

If Carol knows nothing of Bob's preferences, then we can assume that she will choose sincerely—selecting at her turn whichever item she most prefers from those not yet chosen. Now, if Bob is also sincere, and if he chooses first, the items will be allocated as follows:

First turn:	Bob takes the pension.
Second turn:	Carol takes the house.
Third turn:	Bob takes the investments.
Fourth turn:	Carol is left with the vehicles.

Hence, Bob gets his first and third favorites (the pension and the investments). However, if Bob opens by choosing the house—and bypassing the pension for the moment—then the allocation will be as follows:

First turn:	Bob takes the house.
Second turn:	Carol takes the investments.
Third turn:	Bob takes the pension.
Fourth turn:	Carol is left with the vehicles.

Thus, by being insincere, Bob does better—getting his first and second favorites (the pension and the house).

In general, then, what is the optimal strategy for rational players to use, assuming that both know the preferences of the other? The answer is something called the **bottom-up strategy**, discovered by the mathematicians D. A. Kohler and R. Chandrasekaran in 1969. We will illustrate it with an example.

Suppose that we have five objects—*A, B, C, D, E*—and Bob is choosing first. Suppose that Bob and Carol have the following rankings of the objects (called **preference lists** in what follows):

Bob	Carol
A	C
B	E
C	D
D	A
E	B

It will turn out that Bob should open with *C* (his third choice) followed by Carol's choice of *D* (skipping over *E,* for the moment). Bob will then take *A,* Carol will follow with *E,* and finally Bob will get *B.* Bob gets his first, second, and third choices without selecting his first choice first! Where does this strategy come from?

The intuition here is quite easy. Let's make two assumptions about rational players: A rational player will never willingly choose his or her least-preferred alternative, and a rational player will avoid wasting a choice on an object that he or she knows will remain available and thus can be chosen later.

With these assumptions as motivation, let's return to the preceding example and think about the mental calculation that Bob will go through in deciding what his first choice will be. Bob knows the eventual sequence of choices will fill in all the following blanks:

Bob: ______ ______ ______
Carol: ______ ______

Now, working mentally from right to left, Bob knows that Carol will not choose *B* because it is at the bottom of her list. Thus, he will get stuck with *B,* and so he will avoid wasting anything but his last choice on alternative *B.* Thus, Bob can pencil in alternative *B* as his last choice:

Bob: ______ ______ *B*
Carol: ______ ______

Bob, placing himself momentarily in Carol's shoes, knows she will reason the same way, and thus he pencils Carol in for the bottom alternative, *E,* on his list:

Bob: ______ ______ *B*
Carol: ______ *E*

Mentally now, Bob reasons as if alternatives *B* and *E* never existed (and the choice sequence had been Bob-Carol-Bob) and continues to pencil in alternatives from right to left, with Bob working from bottom to top on Carol's preference list and Carol working from bottom to top on Bob's preference list. This yields the following sequence of choices mentally penciled in by Bob:

Bob: *C* *A* *B*
Carol: *D* *E*

Remember, this is just a mental calculation that Bob went through to decide upon the actual choice—in this case, *C*—with which he will open. Bob has no guarantee that Carol will, in fact, respond with *D,* so the use of this strategy involves some risk on Bob's part.

This bottom-up strategy can also be viewed as a procedure that a mediator could use to specify a division of several objects between two parties. Given the preference lists of both parties, the mediator could construct a list—exactly as we did for Bob and Carol above—and then offer this to the parties as the suggested allocation. In effect, the mediator is simultaneously playing the role of two rational parties who choose to employ optimal strategies.

13.5 Divide-and-Choose

There are vast mineral resources under the seabed, all of which, one might argue, should be available to both developed and developing countries. In the absence of some kind of agreement, however, what is to prevent the developed countries from mining all of the most promising tracts before the developing countries have reached a technological level where they can begin their own mining operations? Such an agreement, called the **Convention of the Law of the Sea**, went into effect on November 16, 1994, with 161 signatories (including the United States). Also known as the **Law of the Sea Treaty**, it protects the interests of developing countries by means of the following fair-division procedure.

AP Photo/Oddvar Walle Jensen, Scanpix/file

Whenever a developed country wants to mine a portion of the seabed, that country must propose a division of the portion into two tracts. An international mining company called the Enterprise, funded by the developed countries but representing the interests of the developing countries through the International Seabed Authority, then chooses one of the two tracts to be reserved for later use by the developing countries.

Divide-and-Choose PROCEDURE

With **divide-and-choose**, one party divides the object into two parts in any way that he desires, and the other party chooses whichever part she wants.

As a fair-division procedure, the origins of divide-and-choose go back thousands of years. The Hebrew Bible tells the story of Abram (later to be called Abraham) and Lot, who settled a dispute over land via a proposed division by Abram—"If you go north, I will go south; and if you go south, I will go north" (Gen. 13:8–9)—and a choice (of the plain of Jordan) by Lot. Divide-and-choose resurfaced later in Hesiod's book *Theogony*. The Greek gods Prometheus and Zeus had to divide a portion of meat. Prometheus began by placing the meat into two piles, and Zeus selected one.

Actually, a fair-division procedure consists of both rules and strategies, and all we have described so far are the rules of divide-and-choose. But the natural strategies here are quite obvious: The divider makes the two parts equal in his estimation, and the chooser selects whichever piece she feels is more valuable.

Rules and strategies differ from each other in the following sense: A referee could determine whether a rule is being followed, even without knowing the preferences of the players. Strategies represent choices of how players follow the rules, given their individual preferences (and any other knowledge or goals they may have).

The strategies on which we focus in our discussion of fair-division procedures are those that require no knowledge of the preferences of the other players and yet provide some kind of minimal degree of satisfaction even in the face of collusion by the other players. For example, the strategies just given for divide-and-choose guarantee each player a piece that he or she would not wish to trade for that received by the other.

There are, to be sure, other strategic considerations that might be relevant. For example, in divide-and-choose, would you rather be the divider or the chooser? The answer, given our assumptions that nothing is known of the preferences of the others, is to be the chooser. However, if you knew the preferences of your opponent (and how much she may value spite), then you might want to be the divider.

As a final comment on strategic considerations, we need only look to the origins of the well-known expression "the lion's share." It comes from one of Aesop's fables, as reported by Todd Lowry in *Archaeology of Economic Ideas* (1987, p. 130):

> It seems that a lion, a fox, and an ass participated in a joint hunt. On request, the ass divides the kill into three equal shares and invites the others to choose. Enraged, the lion eats the ass, then asks the fox to make the division. The fox piles all the kill into one great heap except for one tiny morsel. Delighted at this division, the lion asks, "Who has taught you, my very excellent fellow, the art of division?" to which the fox replies, "I learnt it from the ass, by witnessing his fate."

13.6 Cake-Division Procedures: Proportionality

The modern era of fair division in mathematics began in Poland during World War II (see Spotlight 13.1). At this time, Hugo Steinhaus asked what is, in retrospect, the obvious question: What is the "natural" generalization of divide-and-choose to three or more people? The metaphor that has been used in this context, going back at least to the English political theorist James Harrington (1611–1677), is a cake. We picture different players valuing different parts of the cake differently because of concentrations of certain flavors or depth of frosting.

Cake-Division Procedure DEFINITION

A **cake-division procedure** for n players is a procedure that the players can use to allocate a cake among themselves (no outside arbitrators) so that each player has a strategy that will guarantee that player a piece with which he or she is "satisfied," even in the face of collusion by the others.

As we have seen, divide-and-choose is a cake-division procedure for two players, if by "satisfied" we mean either "thinks his piece is of size or value at least one-half" or "does not want to trade what she received for what anyone else received." We define the first notion here; envy-free allocations were defined in Section 13.1.

Proportional Procedure DEFINITION

A cake-division procedure (for n players) will be called **proportional** if each player's strategy guarantees that player a piece of size or value at least $1/n$ of the whole in his or her own estimation.

It turns out that for $n = 2$, a procedure is envy-free if and only if it is proportional; that is, for $n = 2$, the two notions of fair division are exactly the same. For $n > 2$, however, all we can say is that an envy-free procedure is automatically proportional. For example, if a three-person allocation is not proportional, then one player (call him Bob) thinks that he received less than one-third. Bob then feels that the other two are sharing more than two-thirds between them, and thus that

SPOTLIGHT 13.1

Sixty Years of Cake Cutting

The modern era of cake cutting began with the investigations of the Polish mathematician Hugo Steinhaus during World War II. His research, and that of dozens of others since, involved dealing with two fundamental difficulties. First, allocation schemes that work in the context of two or three players often do not generalize easily to the context of four or more players. Second, procedures that yield envy-free allocations are considerably harder to obtain than procedures that yield proportional allocations.

The mathematics inspired by these two difficulties constitutes a rather elegant corner of the large and important area of fair division. Steinhaus's investigations in the 1940s led to his observation that there is a rather natural extension of divide-and-choose to the case of three players. This is the "lone-divider procedure" described on page 468. Steinhaus's method was generalized to an arbitrary number of players by Harold W. Kuhn of Princeton University in 1967.

Unable to extend his procedure from three to four players, Steinhaus proposed the problem to some Polish colleagues. Two of them, Stefan Banach and Bronislaw Knaster, solved this problem in the mid-1940s by producing the "last-diminisher procedure" described on page 468.

In addition to the procedures devised by Banach, Knaster, and Kuhn, there are other well-known constructive procedures for obtaining a proportional allocation among four or more players. One of these is by A. M. Fink of Iowa State University and appears in Exercise 33 (on p. 481).

Another constructive procedure of note, although different in flavor from the others, is the 1961 recasting by Lester E. Dubins and Edwin H. Spanier of the University of California at Berkeley of the last-diminisher method as a "moving-knife procedure" (illustrated in Exercise 35 on page 481). The trade-off here involves giving up the "discrete" nature of the last-diminisher method in exchange for the conceptual simplicity of the moving knife.

Although the existence of an envy-free allocation (even for four or more players) was known to Steinhaus in the 1940s, the first constructive procedure for producing an envy-free allocation among three players was not found until around 1960. At that time, John L. Selfridge of Northern Illinois University and, later but independently, John H. Conway of Princeton University found the elegant procedure presented on page 469. Although never published by either, the procedure was quickly and widely disseminated by Richard K. Guy of the University of Calgary and others. Eventually it appeared in several treatments of the problem by different authors.

In 1980, a moving-knife procedure for producing an envy-free allocation among three players was found by Walter R. Stromquist of Daniel Wagner Associates. Then, another procedure, capable of being recast as a moving-knife solution of the three-player case, was found by a law professor at the University of Virginia, Saul X. Levmore, and a former student of his, Elizabeth Early Cook.

In 1992, Steven J. Brams, a political scientist at New York University, and Alan D. Taylor, a mathematician at Union College, succeeded in finding a constructive procedure for producing an envy-free allocation among four or more players. In 1994, Brams, Taylor, and William S. Zwicker (also from Union College) found a moving-knife solution to the four-person envy-free problem. No moving-knife procedure is known that will produce an envy-free allocation among five or more players.

at least one of the two (call her Carol) must have more than one-third. But then Bob will envy Carol, and so the allocation is not envy-free. Because all nonproportional allocations fail to be envy-free, it follows that if an allocation is envy-free, then it must be proportional.

Many procedures that are proportional, however, fail to be envy-free, as we shall soon show. Thus, proportional procedures are fairly easy to come by, but envy-free procedures are fairly hard to come by.

EXAMPLE 3
The Steinhaus Proportional Procedure for Three Players (Lone Divider)

(Brand X Pictures/ Punchstock.)

Given three players—Bob, Carol, and Ted—we have Bob divide the cake into three pieces (call them *X*, *Y*, and *Z*), each of which he thinks is of size or value exactly one-third. Let's speak of Carol as "approving of a piece" if she thinks it is of size or value at least one-third. Similarly, we will speak of Ted as "approving of a piece" if the same criterion applies. Notice that both Carol and Ted must approve of at least one piece.

If there are distinct pieces—say, *X* and *Y*—with Carol approving of *X* and Ted approving of *Y*, then we give the third piece, *Z*, to Bob (and, of course, *X* to Carol and *Y* to Ted), and we are done. The problem case is where both Carol and Ted approve of only one piece and it is the *same* piece.

Let's assume that Carol and Ted approve of only one piece, *X*, and hence (of more importance to us) both *disapprove* of piece *Z*. Let *XY* denote the result of putting piece *X* and piece *Y* back together to form a single piece. Notice that both Carol and Ted think that *XY* is at least two-thirds of the cake because both disapprove of *Z*. Thus, we can give *Z* to Bob and let Carol and Ted use divide-and-choose on *XY*. Because half of two-thirds is one-third, both Carol and Ted are guaranteed a proportional share (as is Bob, who approved of all three pieces).

The procedure just described, which guarantees proportional shares but is not necessarily envy-free and is sometimes called the **lone-divider method**, was discovered by Hugo Steinhaus around 1944. Unfortunately, it does not extend easily to more than three players. It was left to Steinhaus's students, Stefan Banach and Bronislaw Knaster, to devise a method for more than three players. Picking up where Steinhaus left off (and traveling in quite a different direction), they devised the proportional procedure that today is referred to as the **last-diminisher method**. Like the lone-divider method, it is proportional but not envy-free. We illustrate it for the case of four players (Bob, Carol, Ted, and Alice), and we include both the rules and the strategies that guarantee each player his or her fair share.

EXAMPLE 4
The Banach-Knaster Proportional Procedure for Four or More Players (Last-Diminisher)

Bob cuts from the cake a piece that he thinks is of size one-fourth and hands it to Carol. If Carol thinks the piece handed her is larger than one-fourth, she trims it to size one-fourth in her estimation, places the trimmings back on the cake, and passes the diminished piece to Ted. If Carol thinks the piece handed her is of size at most one-fourth, she passes it unaltered to Ted.

Ted now proceeds exactly as did Carol, trimming the piece to size one-fourth if he thinks it is larger than this and passing it (diminished or unaltered) on to Alice. Alice does the same, but, being the last player, simply holds onto the piece momentarily instead of passing it to anyone.

Notice that everyone now thinks the piece is of size at most one-fourth, and the last person to trim it (or Bob, if no one trimmed it) thinks the piece is of size exactly one-fourth. Thus, the procedure now allocates this piece to the last person who trimmed it (and to Bob, if no one trimmed it).

Assume for the moment that it was Ted who trimmed the piece last, so he takes this piece and exits the game. Bob, Carol, and Alice all think that at least three-fourths of the cake is left, so they can start the process over with (say) Bob beginning by cutting a piece from what remains that he thinks is one-fourth of the original cake. Carol and Alice are both given a chance to trim it to size one-fourth in their estimation, and again, the last one to trim it takes that piece and exits the game. The two remaining players both think that at least half the cake is left, so they can use divide-and-choose to divide it between themselves and thus be assured of a piece that is of size at least one-fourth in their estimation.

13.7 Cake-Division Procedures: The Problem of Envy

Divide-and-choose has a property that neither of the last two procedures possesses: It can ensure that each player receives a piece of cake he or she considers the largest or tied for the largest. In the case of only two players, this means that each player can get what he or she perceives to be at least half the cake, no matter what the other player does. Thus, divide-and-choose is an envy-free procedure.

Steinhaus's $n = 3$ proportional procedure (the lone-divider method) is not envy-free. For example, consider the case where Carol and Ted both find one piece unacceptable (and this piece is given to Bob). Carol and Ted will not envy each other when one divides and the other chooses, but Bob may think that this is not a 50-50 split. Indeed, if Bob divided the cake initially into what he thought was three equal pieces, an unequal split of the remaining two-thirds of the cake by Carol and Ted means that Bob will prefer the larger of these two pieces to the one-third he got. Consequently, Bob will envy the person who got this larger piece.

Nor is the last-diminisher method envy-free. For example, if Bob initially cuts a piece of cake of size one-fourth, and no one else trims it, then Bob receives this piece and exits the game. If Carol is the one to make the next initial cut, she may well cut a piece from the cake that she thinks is of size one-fourth but that Bob thinks is of size considerably more than one-fourth. But Bob is out of the game. Thus, if Ted and Alice think this piece is of size less than one-fourth, then Carol receives it, and so Bob will envy Carol.

Nevertheless, there do exist cake-division procedures that are envy-free. We present one of these in what follows.

EXAMPLE 5
The Selfridge-Conway Envy-Free Procedure for Three Players

We start with a cake and three people. The point we wish to arrive at is an envy-free allocation of the entire cake among the three people in a finite number of steps. This task may seem formidable, but quite often in mathematics, an important part of solving a problem involves breaking the problem into identifiable parts. In this case, let's call our starting point *A* and the final point that we want to reach *C*. Now let's identify an appropriate in-between point *B* that makes going from *A* to *C*—via *B*— more manageable. Our in-between point *B* is the following:

> *Point B:* Getting a constructive procedure that gives an envy-free allocation of *part* of the cake.

Can we constructively obtain three pieces of cake, whose union may not be the whole cake, which can be given to the three people so that each thinks he or she received a piece at least tied for largest? This turns out to be quite easy with the solution given by John Selfridge and John Conway:

1. Player 1 cuts the cake into three pieces that he considers to be the same size. He hands the three pieces to Player 2.
2. Player 2 trims at most one of the three pieces to create at least a two-way tie for largest. Setting the trimmings aside, Player 2 hands the three pieces (one of which may have been trimmed) to Player 3.
3. Player 3 now chooses, from among the three pieces, one that he considers to be at least tied for largest.
4. Player 2 next chooses, from the two remaining pieces, one that she considers to be at least tied for largest, with the proviso that if she trimmed a piece in Step 2, and Player 3 did not choose this piece, then she must now choose it.
5. Player 1 receives the remaining piece.

Let's reconsider the five steps of this trimming procedure to assure ourselves that each player experiences no envy. Recall that Player 1 cuts the cake into three pieces, and Player 2 trims one of these three pieces. Now Player 3 chooses, and, as the first to choose, he certainly envies no one. Player 2 created a two-way tie for largest, and at least one of these two pieces is still available after Player 3 selects his piece. Hence, Player 2 can choose one of the tied pieces she created and will envy no one. Finally, Player 1 created a three-way tie for largest and, because of the proviso in Step 4, the trimmed piece is not the one left over. Thus, Player 1 can choose an untrimmed piece and therefore will envy no one.

So far we have gone from point *A* to point *B*: Starting with a cake and three players, we have constructively obtained (in finitely many steps) an envy-free allocation of all the cake, except the part *T* that Player 2 trimmed from one of the pieces. We will now describe how *T* can be allocated among the three players in such a way that the resulting allocation of the whole cake is envy-free. (This is the rest of the **Selfridge-Conway envy-free procedure**.)

The key observation for the $n = 3$ case is that Player 1 will not envy the player who received the trimmed piece, even if that player were to be given all of *T*. Recall that Player 1 created a three-way tie and received an untrimmed piece. The union of the trimmed piece and the trimmings yields a piece that Player 1 considers to be exactly the same size as the one he received. Thus, assume that it is Player 3 who received the trimmed piece. (It could as well be Player 2.) Then Player 1 will not envy Player 3, no matter how *T* is allocated.

The next step ensures that neither Player 2 nor Player 3 will envy another player when it comes time to allocate *T*. Let Player 2 cut *T* into three pieces she considers to be the same size. Let the players choose which of the three pieces they want in the following order: Player 3, Player 1, Player 2.

To see that this yields an envy-free allocation, notice that Player 3 envies no one, because he is choosing first. Player 1 does not envy Player 2, because he is choosing ahead of her; and Player 1 does not envy Player 3 because, as pointed out earlier, Player 1 will not envy the player who received the trimmed piece. Finally, Player 2 envies no one because she made all three pieces of *T* the same size.

Hence, for $n = 3$, the Selfridge-Conway procedure will give an envy-free allocation of all the cake except *T*, followed by an allocation of *T* that gives an envy-free allocation of all the cake.

A naive attempt to generalize to $n = 4$ what we have done for $n = 3$ would proceed as follows: We would begin by having Player 1 cut the cake into four pieces he considers to be the same size. Then we would have Players 2 and 3 trim some pieces (but how many?) to create ties for the largest. Finally, we would have the players choose from among the pieces—some of which would have been trimmed—in the following order: Player 4, Player 3, Player 2, Player 1.

This approach fails because Player 1 could be left in a position of envy. To understand how the approach could fail, consider how many pieces Player 3 might have to trim to create a sufficient supply of pieces tied for largest so that he is guaranteed to have one available when it is his turn to choose. Player 3 might have to trim one piece to create a two-way tie for largest. Player 2 might need to trim two pieces to create a three-way tie for largest (because if there were only a two-way tie for largest, Player 3 might further trim one of these pieces and Player 4 might choose the other). This leaves Player 1 in a possible position of envy because we could have a situation where Player 2 trims two pieces and Player 3 trims a third piece, and Player 4 then chooses the only untrimmed piece. If this happens, Player 1, by being forced to choose a trimmed piece, will definitely envy Player 4.

All is not lost, however, because there are modifications of the Selfridge-Conway procedure that will work for arbitrary n. For more on this, see *Fair Division* (cited in the "Suggested Readings" section).

Although we have used the metaphor of cake cutting throughout our discussion of the problem of envy, the idea of successive trimming is nonetheless applicable to problems of fair division other than parceling out the last crumbs of a cake. The main practical problem in applying the trimming procedure is that many fair-division problems involve goods that cannot be divided up at all, much less trimmed in fine amounts. Such goods are said to be *indivisible*.

It is interesting to recall that when the Allies agreed in 1944 to partition Germany into sectors after World War II (Stage 1), they initially did not reach agreement about what to do with Berlin. Subsequently, they decided to partition Berlin itself into sectors (Stage 2), even though this city fell 110 miles within the Soviet sector. Berlin was simply too valuable a "piece" for the western Allies (Great Britain, France, and the United States) to cede to the Soviets, which suggests how, after a leftover piece is trimmed off, it can be divided subsequently under the trimming procedure.

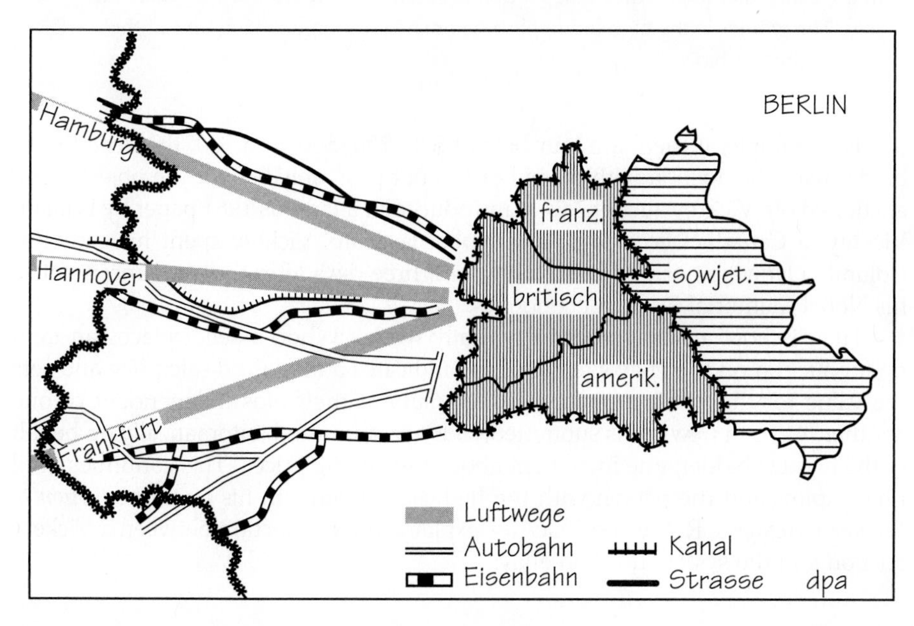

Yet, what if a large piece like Berlin is not divisible? In the settlement of an estate, this might be the house, which may be worth half the estate to the claimants. In this situation, there may be no alternative but to sell this big item and use the proceeds to make the remaining estate more liquid or, in our terms, "trimmable."

13.8 Vickrey Auctions

Among fair-division procedures, auctions are one of the oldest, dating back more than 2500 years. They are used in applications ranging from the determination of who gets to own a promising racehorse to the selection of a contractor to build a new science center at a local college. And, of course, online auction site eBay has become a story unto itself.

But even the two examples above—an auction for a racehorse versus an auction to select a contractor—illustrate some fundamental differences between the kinds of auctions in use today. For example, is the winner the high bidder or the low bidder? With the racehorse, it's the former, whereas with the contractor, it's the latter. Are the bids oral (so everyone knows the last bid) or are they submitted in a sealed envelope (with no one knowing what anyone else bid)? Again, with the racehorse, it's the former whereas with the contractor, it's the latter.

The subject of auctions is both large and important. Indeed, we could well devote an entire chapter (or book) to that topic alone. But we limit ourselves here to one particular kind of auction—known as a **Vickrey auction**—that is reminiscent of what is used today on eBay (and we will describe the latter momentarily as well). To avoid confusion, we will assume the high bidder wins (as opposed to the low bidder winning) in the auctions we are considering, and we will assume that ties in the bidding simply do not occur.

A Vickrey Auction PROCEDURE

In a Vickrey auction, bidders independently submit sealed bids for the object being sold. The winner is the high bidder, but he or she pays only the amount of the second-highest bid.

For example, if there are four bids of $40, $50, $60, and $80, then the fourth bidder wins the auction with his bid of $80 but pays only $60 for the object being auctioned off. Vickrey auctions were introduced in a famous 1961 paper by William Vickrey, a Canadian economist and Nobel laureate. Vickrey spent his career at Columbia University and died in 1996, just three days after the announcement of his Nobel Memorial Prize in Economics.

Here is how an eBay auction typically works: When a seller places an item up for auction on the eBay site, he or she indicates a minimum sale price and sets the value for the bid increments. Bidders submit their bids independent of one another. As each new bid is submitted, bids are submitted automatically on behalf of the highest bidder, one increment above the "going price." This continues until time expires and the person with the highest bid wins. In his text *Introduction to Economic Analysis,* R. Preston MacAfee explains the connection between a Vickery auction and the system used on eBay:

The Vickrey auction underlies the eBay outcome because when a bidder submits a bid in the eBay auction, the current "going" price is not the highest bid, but the second-highest bid, plus a bid increment. Thus, up to the granularity of the bid increment, the basic eBay auction is a Vickrey auction run over time.

Vickrey auctions are interesting because of the answer they provide to the following question: If the object being sold is worth, say, $100 to a bidder, how much less than $100 should he or she bid?

Intuition suggests that in any auction situation, you should bid less than what the object being sold is actually worth to you. In fact, there are mathematical results that suggest that in a standard sealed-bid auction in which the high bidder wins and pays what he or she bid, you should bid only half of what the object is worth to you if there are two bidders, two-thirds of what it is worth to you if there are three bidders, three-fourths if there are four bidders, and so forth.

Remarkably, nothing like this is true with a Vickrey auction. Indeed, with a Vickrey auction, there is a very real sense in which "honesty is the best policy."

Strategy for Bidding in a Vickrey Auction THEOREM

In a Vickrey auction, a bidder can never do better than that achieved by a bid of exactly what the object is worth to that bidder.

EXAMPLE 6
The Vickrey Auction

To see why this is true, let's assume that a lamp is being auctioned off and that it is worth $100 to a bidder named Bob. This means that Bob would prefer winning the lamp and paying less than $100 to losing the auction, but that he would rather lose the auction than wind up paying more than $100 for the lamp.

Let x denote the highest of the bids other than Bob's bid of $100. Then either $x < 100$, in which case Bob wins the auction and pays x dollars, or $x > 100$, in which case Bob loses the auction. We'll consider these two cases separately and show that no bid for Bob can ever do better than his (sincere!) bid of $100.

Case 1: Bob wins the auction (so $x < 100$). Any bid by Bob greater than x yields the same outcome for Bob as does his bid of $100: He wins the auction and gets the lamp for x dollars. So these bids are no better for Bob than his bid of $100. On the other hand, any bid less than x is strictly worse for Bob than that achieved with his bid of $100 because he would lose the auction instead of getting the lamp for less than he actually thought it was worth.

Case 2: Bob loses the auction (so $x > 100$). Any bid by Bob less than x yields the same outcome for Bob as does his bid of $100: He loses the auction. So these bids are no better for Bob than his bid of $100. On the other hand, any bid greater than x is strictly worse for Bob than that achieved with his bid of $100 because he would win the auction and pay more for the lamp than he actually thought it was worth, instead of just losing the auction.

This completes the proof of the theorem. There is something quite satisfying in having a rigorous mathematical proof that establishes—at least in this one context—the fact that honesty is indeed the best policy.

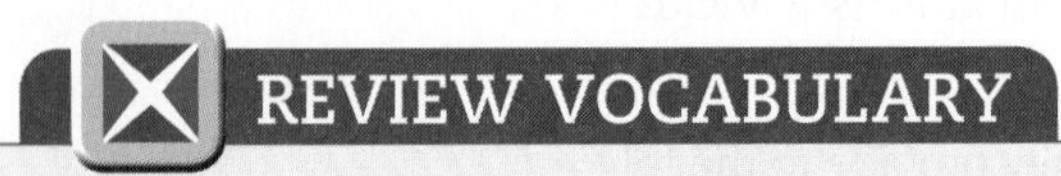

REVIEW VOCABULARY

Adjusted winner procedure A fair-division procedure introduced by Steven Brams and Alan Taylor in 1993. It works only for two players, and begins by having each player independently spread 100 points over the items to be divided so as to reflect the relative worth of each object to that player. The allocation resulting from this procedure is equitable, envy-free, and Pareto-optimal. It requires no cash from either player, but one of the objects may have to be divided or shared by the two players. (p. 453)

Bottom-up strategy A bottom-up strategy is a strategy under an alternating procedure in which sophisticated choices are determined by working backward. (p. 463)

Cake-division procedure A fair-division procedure that uses a cake as a metaphor. Such procedures involve finding allocations of a single object that is finely divisible, as opposed to the situation encountered with either the adjusted winner procedure or the Knaster inheritance procedure. In a cake-division procedure, each player has a strategy that will guarantee that player a piece with which he or she is "satisfied," even in the face of collusion by the others. (p. 466)

Convention of the Law of the Sea An agreement based on divide-and-choose that protects the interests of developing countries in mining operations under the sea. Also referred to as the **Law of the Sea Treaty**. (p. 465)

Divide-and-choose A fair-division procedure for dividing an object or several objects between two players. This method produces an allocation that is both proportional and envy-free (the two being equivalent when there are only two players). (p. 465)

Envy-free A fair-division procedure is said to be envy-free if each player has a strategy that can guarantee him or her a share of whatever is being divided that is, in the eyes of that player, at least as large (or at least as desirable) as that received by any other player, no matter what the other players do. (p. 457)

Equitable A fair-division procedure like adjusted winner is said to be equitable if each player believes he or she received the same fractional part of the total value. (p. 457)

Knaster inheritance procedure A fair-division procedure for any number of parties that begins by having each player (independently) assign a dollar value (a "bid") to the item or items to be divided so as to reflect the absolute worth of each object to that player. The allocation resulting from this procedure leaves each party feeling that he or she received a dollar value at least equal to his or her fair share (and often more so). It never requires the dividing or sharing of an object, but it may require that the players have a large amount of cash on hand. (p. 458)

Last-diminisher method A cake-division procedure introduced by Stefan Banach and Bronislaw Knaster. It works for any number of players and produces an allocation that is proportional but not, in general, envy-free. (p. 468)

Lone-divider method A cake-division procedure introduced by Hugo Steinhaus. It works only for three players and produces an allocation that is proportional but not, in general, envy-free. (p. 468)

Pareto-optimal A fair-division procedure is said to be Pareto-optimal if it produces an allocation with the property that no other allocation achieved by any means whatsoever can make any one player better off without making some other player worse off. (p. 457)

Point ratio The fraction in which the numerator is the number of points one party placed on an object and the denominator is the number of points the other party placed on the object. (p. 455)

Preference lists Rankings of the items to be allocated, from best to worst, by each of the participants. (p. 464)

Proportional A fair-division procedure is said to be proportional if each of n players has a strategy that can guarantee that player a share of whatever is being divided that he or she considers to be at least $1/n$ of the whole in size or value. (p. 466)

Selfridge-Conway procedure A cake-division procedure introduced independently by John Selfridge and John Conway. It works only for three players but produces an allocation that is envy-free (as well as proportional). (p. 470)

Taking turns A fair-division procedure in which two or more parties alternate selecting objects. (p. 463)

Vickrey auction A sealed-bid auction in which the high bidder wins but pays only the amount of the second-highest bid. (p. 472)

1. The adjusted winner procedure

(a) applies only to two-party disputes or disputes that can be recast as two-party disputes.
(b) applies to either two-party or three-party disputes.
(c) applies to n-party disputes for all n.

2. The starting point with adjusted winner is to have each side—independently and simultaneously—spread 100 points over the issues in such a way that it ________________.

3. The *winner* part of the name *adjusted winner* refers to the fact that

(a) each party initially "wins" (that is, is given) each issue on which he or she places more points than the other party.
(b) objects ultimately go to whichever party bid more.
(c) it is impossible for both parties to win with any fair-division scheme.

4. The *adjusted* part of the name *adjusted winner* refers to ________________.

5. In transferring items from one party to the other with the adjusted winner procedure,

(a) the order in which items are transferred is extremely important.
(b) only one item will need to be split or shared.
(c) the order of transfer is obtained by looking at so-called point ratios.
(d) all of the above.

6. With the adjusted winner procedure, suppose A is the party with the highest point total. Then the fraction

$$\frac{A\text{'s point value of the item}}{B\text{'s point value of the item}}$$

is called the item's ________________.

7. With the adjusted winner procedure, the final allocation is

(a) equitable.
(b) envy-free.
(c) Pareto-optimal.
(d) all of the above.

8. With a fair-division procedure, if each player believes that he or she received the same fractional part of the total value, then the procedure is said to be ________________.

9. A fair-division procedure that produces an allocation with the property that no other allocation, achieved by any means whatsoever, can make any one player better off without making some other player worse off is said to be

(a) equitable.
(b) envy-free.
(c) Pareto-optimal.
(d) none of the above.

10. Chris and Terry must make a fair division of three objects. They assign points to the objects (as shown below) and use the adjusted winner procedure. Chris ends up with the ________________.

Object	Chris	Terry
Boat	30	20
Land	50	60
Car	20	20

11. In this chapter, the number of ways to generalize Knaster's inheritance procedure from the case of a single object to the case of several objects is

(a) 0.
(b) 1.
(c) 2.
(d) 3.

12. The inheritance procedure that begins with each heir submitting a monetary bid for each object is known as ________________.

13. Chris and Terry use the Knaster inheritance procedure to divide a coin collection. Chris bids $1000 and Terry bids $800. What is the outcome?

(a) Chris gets the coins and pays Terry $200.
(b) Chris gets the coins and pays Terry $450.
(c) Chris gets the coins and pays Terry $500.

14. Four children bid on two objects (as shown below). Using the Knaster inheritance procedure, Adam ends up with the ________________.

Object	Adam	Beth	Carl	Dietra
House	$80,000	$75,000	$90,000	$60,000
Car	$10,000	$12,000	$13,000	$15,000

15. With the procedure known as *taking turns*, the optimal strategy for players is called

(a) the sincere strategy.
(b) the bottom-up strategy.
(c) the top-down strategy.

16. With taking turns, we assume that a rational player will ______________________.

17. Two people use the divide-and-choose procedure to divide a field. Suppose that Jeff divides and Karen chooses. Which statement is true?

(a) Karen always believes she gets more than her fair share.
(b) Karen can guarantee that she always gets at least her fair share.
(c) Karen can possibly believe she gets less than her fair share.

18. A fair-division procedure is envy-free when each player believes that ______________________.

19. Using the Steinhaus procedure for three players (lone divider), what happens if there is a single portion that is the only one approved of by both nondividers?

(a) One of the other portions is given to the divider.
(b) The two nondividers flip a coin to determine who receives the approved portion.
(c) All portions are returned to the cake, and a different person serves as the new divider.

20. Using the Steinhaus procedure for three players (lone divider), if the two nondividers approve different portions, then ______________________.

21. Using the Banach-Knaster procedure for three or more players (last diminisher), what happens to the first portion after each person has inspected and possibly trimmed it?

(a) The portion goes to the last person to approve the portion, whether or not it was trimmed.
(b) The portion goes to the last person to trim the portion.
(c) The portion goes to the first person to approve and not trim the portion.

22. Using the Banach-Knaster procedure for three or more players (last diminisher), the player who receives the first portion ______________________.

23. Using the Banach-Knaster procedure for three or more players (last diminisher), suppose that Scott initially cuts a piece and passes it among the other people, none of whom trims it. What happens next?

(a) Scott gets this piece.
(b) The last person who is handed the piece keeps it.
(c) The piece is returned to the cake, and someone else cuts a piece.

24. Using the Banach-Knaster procedure for three or more players (last diminisher), when only two people remain, ______________________.

25. For the Selfridge-Conway procedure for three players, which of the following statements is true?

(a) Each of the three players has the opportunity to trim the portions if they appear to be unfair.
(b) Each player receives a portion that he or she believes to be exactly one-third of the total.
(c) The first player may believe that the third player received more than a fair share.

26. For the Selfridge-Conway procedure for three players, the player who will definitely not receive the trimmed piece in Stage 1 is ______________________.

27. An example of a proportional cake-division procedure is

(a) Steinhaus's lone-divider procedure.
(b) the Banach-Knaster last-diminisher procedure.
(c) the Selfridge-Conway procedure.
(d) all of the above.

28. In a Vickrey auction, the winner is the highest bidder, but he pays only ______________________.

29. If the bids in a Vickrey auction are $40, $50, $85, and $90,

(a) the winner is the $90 bidder and he pays $90.
(b) the winner is the $85 bidder and she pays $85.
(c) the winner is the $90 bidder and he pays $85.
(d) the winner is the $85 bidder and she pays $90.

30. The only person mentioned in this chapter in connection with two different fair-division procedures is ______________________.

CHAPTER 13 EXERCISES

■ Challenge ▲ Discussion

13.1 The Adjusted Winner Procedure

1. The 1991 divorce of Ivana and Donald Trump was widely covered in the media. The marital assets included a 45-room mansion in Greenwich, Connecticut; the 118-room Mar-a-Lago mansion in Palm Beach, Florida; an apartment in the Trump Plaza; a 50-room Trump Tower triplex; and just over

$1 million in cash and jewelry. Assume that points are assigned as follows:

	Point Allocations	
Marital Asset	Donald's Points	Ivana's Points
Connecticut estate	10	38
Palm Beach mansion	40	20
Trump Plaza apartment	10	30
Trump Tower triplex	38	10
Cash and jewelry	2	2

Use the adjusted winner procedure to determine a fair allocation of the marital assets. (Exercise 1 courtesy of Catherine Duran.)

2. Suppose that Calvin and Hobbes discover a sunken pirate ship and must divide their loot. They assign points to the items as follows:

Object	Calvin's Points	Hobbes's Points
Cannon	10	5
Anchor	10	20
Unopened chest	15	20
Doubloon	11	14
Figurehead	20	30
Sword	15	6
Cannon ball	5	1
Wooden leg	2	1
Flag	10	2
Crow's nest	2	1

Use the adjusted winner procedure to determine a fair allocation of the loot. (Exercise 2 courtesy of Erica DeCarlo.)

3. This exercise illustrates how the adjusted winner procedure can be used to resolve disputes as well as to achieve fair allocations. Suppose that Mike and Phil are roommates in college, and they encounter serious conflicts during their first week at school. Their resident adviser decides to use the adjusted winner procedure to resolve the dispute. The issues agreed upon, and the (independently assigned) points, turn out to be the following:

Issue	Mike's Points	Phil's Points
Stereo level	4	22
Smoking rights	10	20
Room party policy	50	25
Cleanliness	6	3
Alcohol use	15	15
Phone time	1	8
Lights-out time	10	2
Visitor policy	4	5

Use the adjusted winner procedure to resolve this dispute. (Exercise 3 courtesy of Erica DeCarlo.)

4. Suppose that a labor union and management are trying to resolve a dispute that involves four issues: the base salary of the workers, the annual salary increase that workers can expect, the benefits package the workers will receive, and the amount of vacation time to which each worker will be entitled. Suppose they use adjusted winner to resolve this dispute, with the following point assignments:

Issue	Labor	Management
Base salary	30	50
Salary increases	20	40
Benefits	35	5
Vacation time	15	5

Use adjusted winner to resolve this dispute.

5. Mary and Fred are serving as co-chairs of the mathematics department. There are a number of time-consuming tasks that must be done by one or the other. It really doesn't matter who does which task, except that they disagree on how unpleasant particular tasks are. They decide to use adjusted winner by phrasing items as: "The other co-chair will handle ________________." The items and point assignments are as follows:

The other co-chair will handle:	Mary's Points	Fred's Points
Salary recommendations	11	19
Class schedules	19	9
Hiring	14	20
Department meetings	20	10
Calculus placement	21	11
External review	15	31

(a) Which tasks will Mary do?
(b) Which tasks will Fred do?
(c) Which task do they share, and who takes on more of the burden for this task?

6. Beth and Harvey are co-captains of their intramural softball team. As in Exercise 5, there are a number of time-consuming tasks that must be done by one or the other. It really doesn't matter who does which task, except that they disagree on how unpleasant particular tasks are. Like Fred and Mary, Beth and Harvey decide to use adjusted winner by phrasing items as: "The other co-chair will handle __________." The items and point assignments are as follows:

The other co-chair will handle:	Beth's Points	Harvey's Points
Selection of player positions	9	40
Coordination of game schedule	9	10
Reserving the field	9	11
Scheduling practices	9	12
Checking equipment	9	13
Planning the end-of-season party	55	14

(a) Which tasks will Beth do?
(b) Which tasks will Harvey do?
(c) Which task do they share, and who takes on more of the burden for this task?

7. Make up an example involving two people and several objects for which the adjusted winner procedure can be used, and then use the adjusted winner procedure to determine a fair division.

8. Make up an example involving two people and several issues for which the adjusted winner procedure can be used, and then use the adjusted winner procedure to determine a fair resolution of the dispute.

■ **9.** Suppose we have three items (*X*, *Y*, and *Z*) and three people (Bob, Carol, and Ted). Assume that each person spreads 100 points over the items (as in adjusted winner) to indicate the relative worth of each item to that person:

Item	Bob	Carol	Ted
X	40	30	30
Y	50	40	30
Z	10	30	40

For each of the allocations listed below, indicate

(a) whether or not it is proportional.
(b) whether or not it is envy-free.
(c) whether or not it is equitable.
(d) for the ones that are not Pareto-optimal, another allocation that makes one person better off without making anyone else worse off.

Allocation 1: Bob gets *Z*, Carol gets *Y*, and Ted gets *X*. (This is not Pareto-optimal.)

Allocation 2: Bob gets *Y*, Carol gets *Z*, and Ted gets *X*. (This is not Pareto-optimal.)

Allocation 3: Bob gets *X*, *Y*, and *Z*. (This is Pareto-optimal; explain why.)

Allocation 4: Bob gets *Y*, Carol gets *X*, and Ted gets *Z*. (This is Pareto-optimal.)

Allocation 5: Bob gets *X*, Carol gets *Y*, and Ted gets *Z*. (This is Pareto-optimal.)

13.2 The Knaster Inheritance Procedure

10. If John bids $28,225 and Mary bids $32,100 on their aging parents' old classic car, which they no longer drive, how would you reach a fair division?

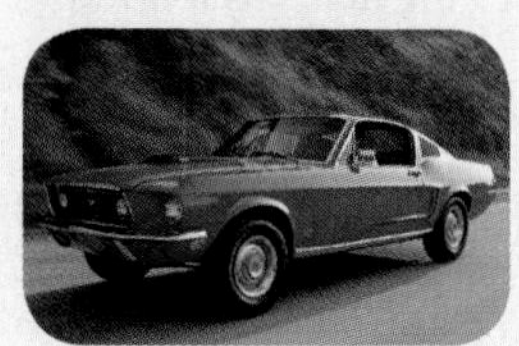

(Car Culture/Corbis.)

11. John and Mary inherit their parents' old house and classic car. John bids $28,225 on the car and $55,900 on the house. Mary bids $32,100 on the car and $59,100 on the house. How should they arrive at a fair division?

12. Can you modify your fair-division procedure in Exercise 11 so that both John and Mary receive one of the two objects while still considering the allocation as fair?

13. Describe a fair division for three heirs, *A*, *B*, and *C*, who inherit a house in the city, a small farm, and a valuable sculpture, and who submit sealed bids (in dollars) on these objects as follows:

	A	B	C
House	$145,000	$149,999	$165,000
Farm	135,000	130,001	128,000
Sculpture	110,000	80,000	127,000

14. Describe a fair division for three children, *E*, *F*, and *G*, who inherit equal shares of their parents' classic car collection and who submit sealed bids (in dollars) on these five cars as follows:

	E	F	G
Duesenberg	$18,000	$15,000	$15,000
Bentley	18,000	24,000	20,000
Ferrari	16,000	12,000	16,500
Pierce-Arrow	14,000	15,000	13,500
Cord	24,000	18,000	22,000

13.3 Fair Division and Organ Transplant Policies

15. Construct the table showing how points would be allocated among the following four potential recipients for a kidney transplant, according to the scheme in Section 13.3.

Potential Recipient	Months Waiting	Antigens Matched	Percent Sensitized
A	9	2	20
B	6	3	0
C	5	4	40
D	2	6	60

16. Does the example in Exercise 15 give rise to the same kind of paradox as in Section 13.3? Explain why or why not.

17. In the scheme for arriving at a priority ranking for organ transplants, how might one change the way points are assigned for "waiting time" so that the kind of paradox that arose in Section 13.3 could not occur?

13.4 Taking Turns

18. Suppose that Bob and Carol rank a series of objects, from most preferred to least preferred, as follows:

Bob	Carol
Car	Boat
Investments	Investments
MP3 player	Car
Boat	Washer-dryer
Television	Television
Washer-dryer	MP3 player

Assume that Bob and Carol use the bottom-up strategy and that Bob gets to choose first. Determine Bob's first choice and the final allocation.

19. Repeat Exercise 18 under the assumption that Carol gets to choose first.

20. Mark and Fred have inherited a number of items from their parents' estate, with no indication of who gets what. They rank the items from most preferred to least preferred as follows:

Mark	Fred
Truck	Boat
Tractor	Tractor
Boat	Car
Car	Truck
Tools	Motorcycle
Motorcycle	Tools

Assume that Mark and Fred use the bottom-up strategy and that Mark gets to choose first. Determine Mark's first choice and the final allocation.

21. Repeat Exercise 20 under the assumption that Fred gets to choose first.

22. Suppose that Donald and Ivana Trump decide to settle their divorce (described in Exercise 1 on p. 476) by taking turns. Assume that they both use the bottom-up strategy, and Donald chooses first. Determine Donald's first choice and the final allocation. (Assume that Donald values the Connecticut estate slightly more than he values the Trump Plaza apartment.)

23. Repeat Exercise 22 under the assumption that Ivana gets to choose first.

24 Suppose that Mary and Fred decide to settle the question of which co-chair performs which tasks (described in Exercise 5 on p. 477) by taking turns. Assume that they both use the bottom-up strategy, and Mary chooses first. Determine the final allocation of tasks.

25. Repeat Exercise 24 under the assumption that Fred chooses first.

13.5 Divide-and-Choose

▲ **26.** If you and another person are using divide-and-choose to divide something between you, would you rather be the divider or the chooser? (Assume that neither of you knows anything about the preferences of the other.)

▲ **27.** Suppose that Bob is entitled to one-fourth of a cake and Carol is entitled to three-fourths. In a few sentences, explain how divide-and-choose can be used to achieve an allocation in which each party is guaranteed to receive at least as much as he or she is entitled to.

28. Suppose that Bob, Carol, and Ted view a cake as having 18 units of value, with each unit of value represented by a small square (as in the accompanying illustration). Suppose, however, that the players value various parts of the cake differently (or that Bob views the cake as being perfectly rectangular, whereas Carol and Ted see it as skewed in opposite ways). We represent this pictorially as follows:

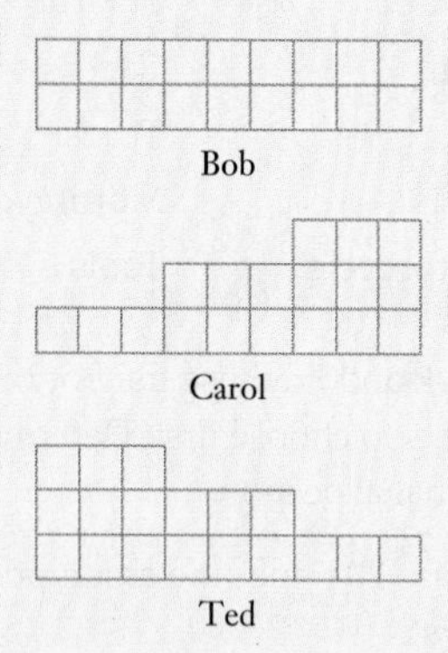

Assume that all cuts that will be made are vertical.

(a) If Bob and Carol use divide-and-choose to divide the cake between them, how large a piece will each receive (assuming they follow the suggested strategies that go with divide-and-choose and that Bob is the divider)?

(b) If Carol and Ted use divide-and-choose to divide the cake between them, how large a piece will each receive (assuming that they follow the suggested strategies that go with divide-and-choose and that Carol is the divider)?

29. Assume that Bob and Carol view the cake as in Exercise 28, but assume also that each knows how the other values the cake, and that neither is spiteful. Suppose they are to divide the cake using the rules, but not necessarily the strategies, of divide-and-choose.

(a) Is Bob better off being the divider or the chooser?

(b) Discuss this in relation to Exercise 26.

13.6 Cake-Division Procedures: Proportionality

30. Suppose that Players 1, 2, and 3 view a cake as in Exercise 28. Notice that each player views the cake as having 18 square units of area (or value). Assume that each player regards a piece as acceptable if and only if it is at least $18/3 = 6$ square units of area (his or her "fair share"). Assume also that all cuts made correspond to vertical lines.

(Angela Wyant/Stone/Getty Images.)

(a) Provide three drawings to show how each player views a division of the cake by Player 1 into three pieces he or she considers to be the same size or value. Label the pieces *A*, *B*, and *C*.

(b) Identify two of these pieces that Player 2 finds acceptable and two that Player 3 finds acceptable.

(c) Show that a feasible assignment of fair pieces can be achieved by letting the players choose in the following order: Player 3, Player 2, Player 1. Indicate how many square units of value each player thinks he or she received. Is there any other order in which players can choose pieces (in this example) that also results in a feasible assignment?

31. Suppose that Players 1, 2, and 3 view a cake as follows:

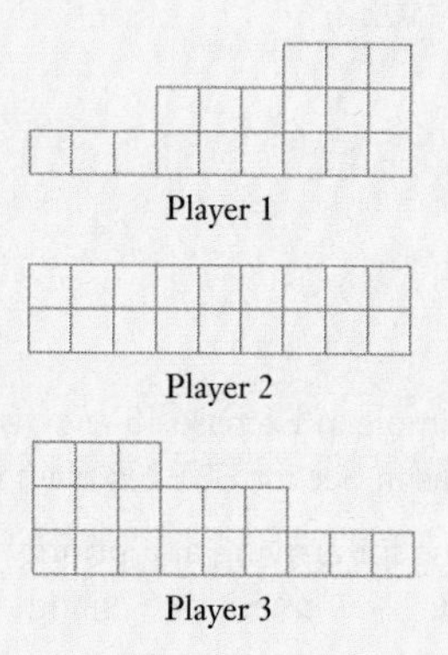

(a) Provide three drawings to show how each player views a division of the cake by Player 1 into three pieces that he or she considers to be the same size or value. Label the pieces *A*, *B*, and *C*. (We are still assuming that all cuts correspond to vertical lines, so this will require a cut along a vertical center line of some of the squares.)

(b) Show that neither Player 2 nor Player 3 finds more than one of the three pieces acceptable (with "acceptable" defined as in Exercise 30).

(c) Identify a single piece that Player 2 and Player 3 agree is not acceptable. (There are actually two such pieces; for definiteness, find the one on the right.)

(d) Assume that Players 2 and 3 give the piece from part (c) to Player 1, and that they reassemble the rest and Players 2 and 3 divide it between themselves using divide-and-choose (with a single vertical cut). Determine what size piece each of the three players will think he or she received (1) if Player 2 divides and Player 3 chooses, and (2) if Player 3 divides and Player 2 chooses.

32. Suppose Players 1, 2, and 3 view a cake as in Exercise 31. Illustrate the last-diminisher method (still restricting attention to vertical cuts and, in addition, assuming that the piece potentially being diminished is a piece off the left side of the cake) by following steps (a) through (f) below:

(a) Draw a picture showing the third of the cake (6 squares) that Player 1 will slice off the cake.

(b) Determine whether Player 2 will pass or further diminish this piece. If he or she would further diminish it, make a new drawing.

(c) Determine whether Player 3 will pass or further diminish this piece. If he or she would further diminish it, make a new drawing.

(d) Determine who receives the piece cut off the cake and what size or value he or she thinks it is. (Actually, we knew what size the person receiving this first piece would think it was, assuming that he or she followed the prescribed strategy. How did we know this?)

(e) Finish the last-diminisher method using divide-and-choose on what remains, with the lowest-numbered player who remains doing the dividing.

(f) Redo step (e) with the other player doing the dividing.

33. The Banach-Knaster last-diminisher method is not the only well-known cake-division procedure that yields a proportional allocation for any number of players. There is also one due to A. M. Fink (sometimes called the *lone-chooser method*). For three players (Bob, Carol, and Ted), it works as follows:

(i) Bob and Carol divide the cake into two pieces using divide-and-choose.

(ii) Bob now divides the piece he has into three parts that he considers to be the same size. Carol does the same with the piece she has.

(iii) Ted now chooses whichever of Bob's three pieces that he (Ted) thinks is largest, and Ted chooses whichever of Carol's three pieces that he thinks is largest.

(iv) Bob keeps his remaining two pieces, as does Carol.

(a) Explain why Ted thinks he is getting at least one-third of the cake.

(b) Explain why Bob and Carol each think they are receiving at least one-third of the cake.

(c) Explain why, in general, this scheme is not envy-free.

■ **34.** In A. M. Fink's procedure (described in Exercise 33), suppose that a fourth person (Alice) comes along after Bob, Carol, and Ted have already divided the cake among themselves so that each of the three thinks that he or she has a piece of size at least one-third. Mimic what was done in the three-person case to obtain an allocation among the four that is proportional. (*Hint:* Begin by having Bob, Carol, and Ted divide the pieces they have into a certain number—how many?—of equal parts.)

35. There is a moving-knife version of the Banach-Knaster procedure that is due to Dubins and Spanier. To describe it, we picture the cake as being rectangular, and the procedure beginning with a referee holding a knife along the left edge, as illustrated below.

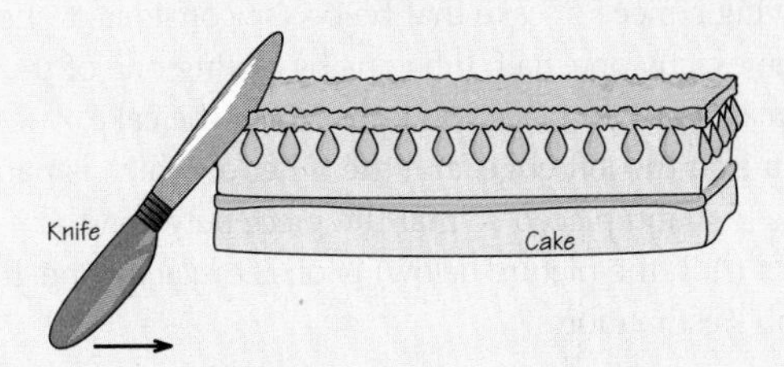

Assume, for the sake of illustration, that there are four players (Bob, Carol, Ted, and Alice). The referee starts moving the knife from left to right over the cake (keeping it parallel to the position in which it started) until one of the players (assume it is Bob) calls "cut." At this time, a cut is made, the piece to the left of the knife is given to Bob, and he exits the game. The knife starts moving again, and the process continues. The strategies are for each player to call "cut" whenever it would yield him or her a piece of size at least one-fourth.

(a) Explain why this procedure produces an allocation that is proportional.

(b) Explain why the resulting allocation is not, in general, envy-free.

(c) Explain why, if you are not the first player to call "cut," there is a strategy different from the one suggested that is never worse for you, and sometimes better.

13.7 Cake-Division Procedures: The Problem with Envy

36. Suppose Players 1, 2, and 3 view the cake as in Exercise 31. Illustrate the envy-free procedure for $n = 3$ (yielding an allocation of part of the cake) by following steps (a) through (c) below. Again, restrict attention to vertical cuts.

(a) Provide a total of three drawings to show how each player views a division of the cake by Player 1 into three pieces he or she considers to be the same size or value. Label the pieces *A*, *B*, and *C*. (This is the same as Exercise 31a.)

(b) Redraw the picture from Player 2's view, and illustrate the trimming of Piece *A* that he or she would do. Label the trimmed piece *A* and the actual trimmings *T*.

(c) Indicate which piece each player would choose (and what he or she thinks its size is) if the players choose in the following order: Player 3, Player 2, Player 1, according to the envy-free procedure.

■ **37.** There is a two-person moving-knife cake-division procedure due to A. K. Austin that leads to each player receiving a piece of cake that he or she considers to be of size exactly one-half. It begins by having one of the two players (Bob) place two knives over the cake, one of which is at the left edge, and the other of which is parallel to the first and placed so that the piece between the knives (*A* in the picture below) is of size exactly one-half in Bob's estimation.

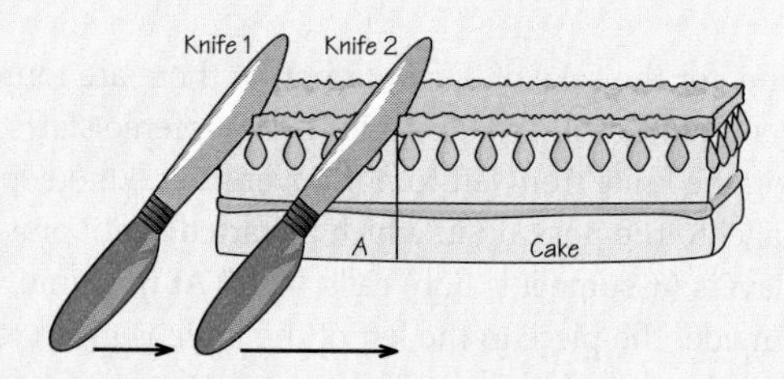

If Carol agrees that this is a 50-50 division, we are done. Otherwise, Bob starts moving both knives to the right—perhaps at different rates—so that the piece between the knives remains of size one-half in his eyes. Carol calls "stop" at the point when she also thinks the piece between the two knives is of size exactly one-half.

(a) If the knife on the right were to reach the right-hand edge, where would the knife on the left be?

(b) Explain why there definitely is a point where Carol thinks the piece between the two knives is of size exactly one-half. (*Hint*: If Carol thinks the piece is too small at the beginning, what will she think of it at the end?)

38. Here are the steps in the Selfridge-Conway procedure for three players:

Stage 1. The initial division

Step 1. Player 1 cuts the cake into, what in his view, is three equal pieces.

Step 2. Player 2, if he thinks one piece is largest, trims from that piece to create what he believes is a two-way tie for largest piece. The trimmings are set aside. If Player 2 thinks that the original split was fair, he does nothing.

Step 3. Player 3 may choose any piece.

Step 4. Player 2 chooses a piece. If the trimmed piece remains, he must choose it. If not, he chooses the one he feels is tied with the trimmed piece for largest.

Step 5. Player 1 gets the remaining piece.

Stage 2. Dividing the trimmings

Assume that Player 3 received the trimmed piece in Stage 1.

Step 6. Player 2 divides the trimmings into what he considers three equal parts.

Step 7. Player 3 chooses one part of the trimmings.

Step 8. Player 1 chooses a piece of the trimmings.

Step 9. Player 2 receives the remaining trimmings.

(a) Explain why Player 1 is envy-free after Stage 1.

(b) Explain why Player 2 is envy-free after Stage 1.

(c) Explain why Player 3 is envy-free after Stage 1.

(d) Explain why Player 1 is envy-free after Stage 2.

(e) Explain why Player 2 is envy-free after Stage 2.

(f) Explain why Player 3 is envy-free after Stage 2.

(Exercise 38 courtesy of Michael Rosenthal.)

13.8 Vickrey Auctions

39. Consider the Vickrey auction for a lamp that is worth \$100 to our bidder Bob (page 473). Suppose someone suggests that Bob would always do as well with a bid of \$80. Show that this is false by playing the role of another bidder who could make Bob regret a choice of bidding \$80 rather than his (sincere) bid of \$100. You are free to make any assumptions that you want about the bids other than Bob's and your own.

40. Consider the Vickrey auction for a lamp that is worth \$100 to our bidder Bob (page 473). Suppose someone suggests that Bob would always do as well with a bid of \$120. Show that this is false by playing the role of another bidder who could make Bob regret a choice of bidding \$120 rather than his (sincere) bid of \$100. You are free to make any assumptions you want about the bids other than Bob's and your own.

WRITING PROJECTS

1. It turns out that there is no way to extend the adjusted winner procedure to three or more players. That is, there are point assignments by three players to three objects so that no allocation satisfies the three desired properties of equability (equal points), envy-freeness, and Pareto optimality. On the other hand, there are separate procedures that will realize any two of the three properties. Thus, tradeoffs must be made, and these may depend on the circumstances. In a few paragraphs, discuss the relative importance of the three properties and circumstances that may affect the choice of which two of the three properties one might wish to have satisfied.

2. If we use taking turns to divvy up a collection of objects between two people (Bob and Carol), then there is an obvious advantage to going first. Assume that we have decided that Bob will, in fact, choose first (say, by the toss of a coin). Let's think about how Carol might be compensated. First of all, if there are only three objects, then the "choice sequence" Bob-Carol-Carol seems to be the only reasonable one. Do you agree? For four objects, however, there are two choice sequences that suggest themselves: Bob-Carol-Carol-Carol and Bob-Carol-Carol-Bob. Do you think that one of these is obviously more fair than the other? What if there are four identical objects? What if both Bob and Carol value object *A* twice as much as *B*, and *B* twice as much as *C*, and *C* twice as much as *D*? What sequences suggest themselves for five objects? For eight objects?

In one page or less, discuss these questions. (For more on this, see *The Win–Win Solution*, in the "Suggested Readings" at the end of the chapter.)

3. One of the most important differences between the three-person and the n-person envy-free procedures is that the latter procedure may take more than two stages. And, of course, the more stages there are, the more cuts and trimmings may be necessary. Do you consider this a serious practical problem, or is it mainly a theoretical problem? In one paragraph, explain your reasons.

4. One often hears of the importance of "process" versus "product," the latter referring to what is achieved and the former referring to *how* it was achieved. In a couple of sentences, comment on the relevance of this to fair division as illustrated by the following rough paraphrasing of an exchange between two old friends, Ralph Kramden (played by Jackie Gleason) and Ed Norton (played by Art Carney) in the 1950s sitcom *The Honeymooners.*

> *Ralph to Ed* (as the two are sitting alone at the dinner table): I can't believe you did that.
>
> *Ed:* Did what, Ralph?
>
> *Ralph:* There were two potatoes there, and you reached right out and took the big one.
>
> *Ed:* What would you have done, Ralph?
>
> *Ralph:* Why, I'd have taken the little one.
>
> *Ed:* You got the little one, Ralph.

Suggested Readings

BRAMS, S. J., and A. D. TAYLOR. *The Win-Win Solution: Guaranteeing Fair Shares to Everybody,* Norton, New York, 1999. Brams and Taylor further discuss adjusted winner, as well as divide-and-choose and taking turns.

BRAMS, S. J., and A. D. TAYLOR. *Fair Division: From Cake-Cutting to Dispute Resolution,* Cambridge University Press, Cambridge, 1996. Brams and Taylor provide a book-length treatment of the kinds of topics introduced in this chapter, as well as divide-and-choose in the political arena, moving-knife procedures for cake cutting, and fairness as it applies to different auction and election procedures.

BRAMS, S. J., and A. D. TAYLOR. An envy-free cake division protocol. *American Mathematical Monthly,* 102 (1995): 9–18. Brams and Taylor describe in detail the finite version of their envy-free procedure for $n = 4$; in addition, they review earlier work on "protocols" (step-by-step procedures) that led up to their constructive solution of the envy-freeness problem for $n > 3$.

ROBERTSON, J., and W. WEBB. *Cake-Cutting Algorithms: Be Fair If You Can,* A. K. Peters, Wellesley, MA, 1998. Robertson and Webb cover a great deal of cake-cutting ground in a text that includes exercises.

YOUNG, P. *Equity in Theory and Practice,* Princeton University Press, Princeton, NJ, 1994. Contains considerably more on fair division in real-world situations, such as the organ transplant example in Section 13.3 on page 461.

10

14

Apportionment

A *dictionary might define the verb to* apportion *to mean to divide some good into shares,* and *apportionment* to be the result of apportioning something. That is what this chapter is about. Many goods cannot be distributed in fractional shares, and the apportionment problem occurs when the shares are rounded. If the rounding of each share is done individually, their sum may be more or less than the whole. Sometimes this does not matter, as when a notation such as "percentages may not add to 100 percent due to rounding" is attached to a table. However, there are instances where the discrepancy matters very much.

In Section 14.1, we will consider some of these instances, starting with a detail-oriented field hockey coach who wants to boost her team's reputation. The apportionment problem actually gets its name from Article 1, Section 2 of the U.S. Constitution, which, as modified by the Fourteenth Amendment, requires that "[r]epresentatives shall be apportioned among the several States according to their respective numbers, counting the whole number of persons in each State ...". There is no evidence that the framers of the Constitution were aware that they were setting a mathematical problem, but representatives must be apportioned in whole numbers, which involves rounding, and those numbers have to add up to a prescribed total. Thus, the apportionment problem is to round a set of numbers, whose sum is a whole number, in such a way that the sum of the rounded numbers is equal to the original sum.

In Section 14.2, we will encounter our first apportionment method. It is relatively easy to apply and seems to be reasonable, but there are surprising problems with it, which are known as *paradoxes.* In 1791, it was proposed and rejected in the first apportionment of Congress, and then it was adopted by Congress in 1850 and used until 1900, when the paradoxes became hard to ignore.

Section 14.3 introduces three apportionment methods, all of which have been used to apportion seats in the House of Representatives. These methods are called divisor methods and each is based on a rule for rounding decimals to whole numbers—such as, for example, "round to the nearest whole number." Divisor methods ensure that the rounded numbers add up to the original sum by dividing each number by a correction factor which is chosen to get the sum right. This correction factor is called the *divisor*, and it is the same for all the numbers in the set. For example, if we are rounding percentages, and the sum of the rounded percentages turns out to be 101 percent, we would choose an appropriate divisor and divide it into each percentage—before rounding—so that when the rounding is done, the sum is 100 percent.

Because there is more than one way to round numbers, there is more than one divisor method. We will explore the differences in Section 14.4.

14.1 The Apportionment Problem

Coach is proud of her field hockey team, and has asked for a poster to display its record for the season. The team played 23 games and won 18, lost 4, and tied 1. Table 14.1 is a draft of the poster. Coach objects because it looks too complicated. "Just express the percents as whole numbers," she says. We round off the percentages: The winning percentage, 78.26 percent, is rounded down to 78 percent; the losing percentage, 17.39 percent, is rounded down to 17 percent; and the tie percentage, 4.35 percent, is rounded down to 4 percent. Because all three percentages were rounded down, the total is only 99 percent. The coach notices and changes the winning percentage to 79 percent. "Now you have 100 percent!"

Table 14.1

The Field Hockey Team's Season

		Percentage
Games won	18	$\frac{18}{23} \times 100\% = 78.26\%$
Games lost	4	$\frac{4}{23} \times 100\% = 17.39\%$
Games tied	1	$\frac{1}{23} \times 100\% = 4.35\%$
Games played	23	100.00%

Algebra Review Fractions to Percentages

Fractions are a way to represent a portion of a whole. When working with fractions, we often are required to convert them to decimals or write them as percentages. The words *percent* and *percentage* are related to each other, but do not have the same meaning. Percent means *per hundred*, and it is accompanied by a number and the percent symbol, %.

Let's consider the fraction $\frac{19}{20}$. To obtain the percentage associated with this fraction, we do the following:

- Convert the fraction to a decimal.

$$\frac{19}{20} = 0.95$$

- Multiply by 100% and simplify:

$$0.95 \times 100\% = 95\%$$

In practice, when we convert a fraction to a percentage, we first convert the fraction to decimal, multiply by 100, and then include the percent symbol. Thus, in the intermediate step, we are changing the value, but it is rectified when the percent symbol is included.

Some fraction-to-percent conversions require rounding, thereby changing the value slightly. For example, let's convert $\frac{3}{13}$ to a percentage and round to a hundredth of a percent. In this process, we must be careful to round at the last step. The process is as follows:

- First, convert the fraction to a decimal. A calculator with 10 places of accuracy will yield 0.2307692308.
- Although we don't need all 10 decimal places, we multiply this number by 100 to obtain 23.07692308.
- We round and then include the percent symbol to get 23.08%.

Apportionment Problem

DEFINITION

An **apportionment problem** is rounding each number in a set of numbers that add up to a whole number. The sum of the rounded numbers must equal the original sum.

Consider a number such as the percentage of games won by the field hockey team, 78.26 percent. It has an integer part, 78, and a fractional part, 0.26. (The fractional part is sometimes called the *decimal part* of the number, but we will stick with *fractional part.*) The principle behind the way that we rounded the above percentages was to round to the nearest whole number. Numbers that have fractional parts less than 0.500 are rounded by dropping the fractional part, and numbers that have fractional parts greater than or equal to 0.500 are rounded to the next-largest whole number. Thus, the field hockey team's winning percentage was rounded to 78 percent. When the same rounding procedure was applied to the losses and ties, each category's fractional part was dropped, and the sum of the rounded numbers was less than 100 percent. Thus the apportionment problem was not solved until Coach imposed an arbitrary solution.

Apportionment Method

DEFINITION

An **apportionment method** is a procedure for solving all apportionment problems without making arbitrary choices.

The U.S. Constitution posed an apportionment *problem* by requiring the seats in the House of Representatives to be apportioned according to the populations of the states. However, the Constitution did not specify an apportionment *method*.

Howard Chandler Christy's "Scene at the Signing of the Constitution of the United States." *(Art Resource, NY.)*

The first census of the United States was conducted in 1790. Delaware's population—55,540—was the least among the 15 states. For the purpose of apportionment, the total population of the nation was 3,615,920, and the House of Representatives was to have 105 members. Thus, the average congressional district should have had a population of $3{,}615{,}920 \div 105 = 34{,}437$. To obtain its fair share of the House seats, we can divide Delaware's population by the average congressional district population. Ideally, Delaware should have had

$$\frac{55{,}540}{34{,}437} = 1.613 \text{ congressional districts}$$

Table 14.2 displays the apportionment proposed in a bill, written by Alexander Hamilton, that Congress passed and sent to President George Washington. Each state's fair share of the 105 congressional districts, or **quota**, was calculated by dividing its population by the average congressional district population of 34,437, as in the case of Delaware, shown above. None of the quotas were whole numbers. The bill gave Delaware two seats in Congress. As you will see in the table, Virginia's quota was 18.310 seats, and Virginia was awarded 18 seats in the House.

Although this apportionment may seem fair enough, on April 5, 1792, President Washington vetoed the bill.[1] Washington came from Virginia, a state that would get less than its quota in the apportionment that Congress proposed. It is impossible to determine if he was just biased for his home state—as the field hockey coach was in favor of her team—because, as we will discover, there were substantial reasons for rejecting the bill.

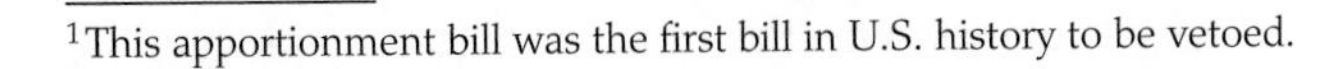

[1] This apportionment bill was the first bill in U.S. history to be vetoed.

Table 14.2

The Congressional Apportionment that George Washington Vetoed

State	Population	Quota	Apportionment
Virginia	630,560	18.310	18
Massachusetts	475,327	13.803	14
Pennsylvania	432,879	12.570	13
North Carolina	353,523	10.266	10
New York	331,589	9.629	10
Maryland	278,514	8.088	8
Connecticut	236,841	6.878	7
South Carolina	206,236	5.989	6
New Jersey	179,570	5.214	5
New Hampshire	141,822	4.118	4
Vermont	85,533	2.484	2
Georgia	70,835	2.057	2
Kentucky	68,705	1.995	2
Rhode Island	68,446	1.988	2
Delaware	55,540	1.613	2
Totals	3,615,920	105	105

Now let's see how to set up an apportionment problem. Although many apportionment problems do not involve the House of Representatives, our terminology refers to *states*, *populations*, and a *house size*. The first step in solving an apportionment problem is to identify the states, the populations, and the house size.

EXAMPLE 1
Rounding Percentages

In the problem of rounding percentages so that their sum is 100, the house size is 100. The categories (such as wins, losses, and ties for a field hockey team) correspond to the states, and the numbers in each category (in our field hockey story, 18 wins, 4 losses, and 1 tie) correspond to the populations of the states.

Standard Divisor DEFINITION

The **standard divisor** is the total population, divided by the house size. If p is the total population, h is the house size, and s is the standard divisor, then

$$s = \frac{p}{h}$$

The second step in addressing an apportionment problem is to determine the standard divisor. Continuing with our example of rounding percentages, the house size is 100 so the standard divisor is the total population, divided by 100. Thus, in our field hockey story, the total population was the number of games played, 23, and the standard divisor was $23 \div 100 = 0.23$.

Quota DEFINITION

In an apportionment problem, the **quota** for a state is the exact share that would be allocated to the state if a *whole number were not required*. To obtain a state's quota, divide its population by the standard divisor.

The third step in solving an apportionment problem is to calculate the quota for each state. In a percentage-rounding problem, the number of individuals in a category is interpreted as the population of a state. The quota for each category is calculated by dividing its population by the standard divisor. If there are many categories, this may entail a bit of work. With a calculator, it is helpful to store the standard divisor in memory to avoid having to enter it repeatedly.

The quotas in a percentage rounding problem are familiar. Let p denote the total population (23 in our field hockey example), and let p_i be the population of the ith category (such as wins, losses, ties in the field hockey example). The standard divisor is $s = p \div 100$, and the quota for the ith category is

$$\begin{aligned} q_i = p_i \div s &= p_i \div (p \div 100) \\ &= 100 \times (p_i \div p) \end{aligned}$$

This is the formula for expressing p_i as a percentage of p. Thus each quota is simply the percentage that we wish to round. In the field hockey scenario, the quotas—$18 \div 0.23 = 78.26\%$ for wins, $4 \div 0.23 = 17.39\%$ for losses, and $1 \div 0.23 = 4.35\%$ for ties—are shown in the right column of Table 14.1.

All apportionment methods start with these three initial steps leading to the determination of the quotas. The next step, rounding the quotas to obtain whole numbers whose sum is the house size, is where the various methods differ.

EXAMPLE 2
The High School Mathematics Teacher

A high school has one mathematics teacher who teaches all geometry, pre-calculus, and calculus classes. She has time to teach a total of five sections, and 100 students are enrolled as follows: 52 for geometry, 33 for pre-calculus, and 15 for calculus. How many sections of each course should be scheduled?

(LWA-Dann Tardif/Corbis.)

This is an apportionment problem because the number of sections is specified (5), and the number of sections allotted to each course must be a whole number. The three courses correspond to the states; the number of students enrolled in each course corresponds to each state's population, and the total number of sections to be taught, 5, is the house size. Thus, the populations of geometry, pre-calculus, and calculus are 52, 33, and 15, respectively. The total population is 100, so the standard divisor is $s = 100 \div 5 = 20$. Table 14.3 displays the calculations of the quotas.

Table 14.3

Calculation of the Quotas for High School Mathematics Courses

Course	Population	Quota	Rounded
Geometry	52	$52 \div 20 = 2.60$	3
Pre-calculus	33	$33 \div 20 = 1.65$	2
Calculus	15	$15 \div 20 = 0.75$	1
Totals	100	5	6

The sum of the quotas is the house size (in this case, 5). It is tempting to round each quota to the nearest whole number, as in the right column of the table, but this makes 6 sections in all—too many! The purpose of an apportionment method is to find an equitable way to round a set of numbers such as these quotas without increasing or decreasing the original sum.

EXAMPLE 3
California's Quota

The Census Bureau recorded the apportionment population of the United States—as of April 1, 2010—to be 309,183,463. There are 435 seats in the House of Representatives; therefore, the standard divisor is

$$309{,}183{,}463 \div 435 = 710{,}767.$$

California's quota was determined by dividing its population—37,341,989—by this standard divisor. Thus,

$$\text{California's quota} = 37{,}341{,}989 \div 710{,}767 = 52.538 \text{ seats.}$$

This quota is slightly more than the quota that was computed with the 2000 census data, 52.447 seats. However, California's apportionment, which must be a whole number, was unchanged at 53 seats.

The apportionment method of largest fractions is named for Alexander Hamilton. *(National Portrait Gallery/Art Resource, NY.)*

Ideally, each state's apportionment should be close to its quota. It is unrealistic to expect that any state will be apportioned its exact quota because each apportionment is required to be a whole number and the quota is unlikely to be a whole number. In choosing an apportionment method, we must decide what we mean by the phrase "each state's apportionment should be close to its quota."

Apportionment always involves rounding, and there are many ways to round. "Rounding down" means discarding the fractional part of a number q to obtain a whole number that we will denote as $\lfloor q \rfloor$. Thus, $\lfloor 7.00001 \rfloor = 7$, $\lfloor 7 \rfloor = 7$, and $\lfloor 6.99999 \rfloor = 6$. "Rounding up" gives the next whole number, denoted as $\lceil q \rceil$. Thus, $\lceil 7.00001 \rceil = 8$, but $\lceil 7 \rceil = 7$.

There are numerous apportionment methods. Each has flaws, and our goal is to understand how to choose a method that is appropriate for a particular apportionment problem.

14.2 The Hamilton Method

The congressional apportionment bill that President Washington vetoed was written by Alexander Hamilton. While it may appear that he just rounded each quota to the nearest whole number, Hamilton was aware that there would be occasions—analogous to the examples of the field hockey team's percentages of wins, losses, and ties not summing to 100 percent, or the high school teacher receiving an extra class to teach—when the total number of seats apportioned in this way would be either more or less than the statutory house size. Hamilton called his method *largest fractions.*

Alexander Hamilton's Method and Upper, Lower Quotas

DEFINITIONS

With the **Hamilton method**, each state receives either its **lower quota** $\lfloor q \rfloor$, its quota rounded down, or its **upper quota**, $\lceil q \rceil$, obtained by rounding the quota up. The states that receive their upper quotas are those whose quotas have the largest fractional parts.

Once the quota for each state has been determined, implementing the Hamilton method is a two-step procedure:

Hamilton Method

PROCEDURE

1. Tentatively assign to each state its lower quota of representatives. Each state whose quota is not a whole number loses a fraction of a seat at this stage, so the total number of seats assigned at this point will be less than the house size. This leaves additional seats to be apportioned.
2. Allot the remaining seats, one each, to the states whose quotas have the largest fractional parts, until the house is filled.

It is possible that a tie will occur, with the quotas of two states having identical fractional parts, but in practice, this rarely happens when large populations are involved.

EXAMPLE 4
The High School Teacher's Dilemma

Let us use the Hamilton method to determine how many sections of geometry, pre-calculus, and calculus the high school teacher in the example above should teach. Recall that in this example, the states correspond to the three subjects to be taught, the populations correspond to the numbers of students who have enrolled in the subjects, and the seats to be apportioned are actually the sections to be taught. We found in Example 2 that the quotas for the three subjects are 2.60, 1.65, and 0.75, respectively (see Table 14.3). Table 14.4 shows how to obtain the apportionment by the Hamilton method.

Table 14.4

Apportioning High School Mathematics Classes by the Hamilton Method

Course	Quota	Lower Quota	Apportionment
Geometry	2.60	2	2
Pre-calculus	1.65	1 ↑	2
Calculus	0.75	0 ↑	1
Totals	5	3	5

The lower quotas are the **tentative apportionments**, and their sum is 3, leaving two sections still to be apportioned. These go to Pre-calculus and Calculus, because the quotas for these courses have the largest fractional parts, 0.65 and 0.75, respectively.

EXAMPLE 5
The Field Hockey Team

In Table 14.1, we saw that the percentages of wins, losses, and ties for the field hockey team were 78.26 percent, 17.39 percent, and 4.35 percent, respectively. These are the quotas that we must round so that they sum to 100 percent. To start, we apportion $\lfloor 78.26 \rfloor = 78\%$, $\lfloor 17.39 \rfloor = 17\%$, and $\lfloor 4.35 \rfloor = 4\%$ to the three categories. These lower quotas add up to 99 percent. The remaining 1 percent to be apportioned goes to the losses because their fraction, 0.39, is the largest. The final apportionment is 78 percent wins, 18 percent losses, and 4 percent ties. Coach would probably veto this apportionment.

EXAMPLE 6
Hamilton's Apportionment

The first congressional apportionment involved 15 states, and the House had 105 seats. According to the 1790 census, the U.S. population was 3,615,920. The standard divisor, 3,615,920 ÷ 105 = 34,437, represents the population of the average congressional district.

Table 14.2 displays Alexander Hamilton's proposed apportionment. Each quota shown in the table was calculated by dividing the state's population by this standard divisor. Adding the lower quotas, we find that their sum, 97, leaves 8 seats to be apportioned. These go to the 8 states whose quotas had the largest fractional parts. Table 14.5 shows the details.

Table 14.5

Alexander Hamilton's Calculation of the Apportionment of Seats in Congress

State	Quota	Lower Quota	Seats Apportioned
Kentucky	1.995	1 ↑	2
South Carolina	5.989	5 ↑	6
Rhode Island	1.988	1 ↑	2
Connecticut	6.878	6 ↑	7
Massachusetts	13.803	13 ↑	14
New York	9.629	9 ↑	10
Delaware	1.613	1 ↑	2
Pennsylvania	12.570	12 ↑	13
Vermont	2.484	2	2
Virginia	18.310	18	18
North Carolina	10.266	10	10
New Jersey	5.214	5	5
New Hampshire	4.118	4	4
Maryland	8.088	8	8
Georgia	2.057	2	2
Totals	105	97	105

In Table 14.2, the states were listed in decreasing order by population; but in Table 14.5, the order is by fractional part of the quota. This makes it easy to determine which eight states are to be awarded their upper quotas, but perhaps thinking of the states in this order may have prompted President Washington's veto. In returning the bill to Congress, he stated that the fractions of seats used in the second step of the Hamilton apportionment were not related to the states' total populations. Actually, the 1792 apportionment bill could have been dismissed because of a technicality. The Constitution requires each congressional district to have a population of at least 30,000, and Delaware, with 55,540 inhabitants, was too small to have two districts. Washington was probably aware of this, making it likely that his objection to the Hamilton method was sincere.

The veto prevented the Hamilton method from being used to apportion the seats in the Third Congress, but in 1850, Congress—forgetting Alexander Hamilton's first attempt at apportionment as well as George Washington's objection to it—reinvented the method, and it remained in use until 1900. The half-century of experience with the Hamilton method revealed a paradox.

Paradoxes of the Hamilton Method

In mathematical terminology, the word *monotone* refers to a change that does not reverse from increase to decrease, or vice versa. A monotone process is either always increasing or always decreasing.

House Monotone DEFINITION

An apportionment method is **house monotone** if it is not possible for any state to receive fewer seats after an increase in the house size than it had before the increase, provided that there is no change in any state's population.

It would be surprising indeed to find an apportionment method that is not house monotone. When President Washington vetoed the Hamilton apportionment of the House of Representatives in 1792, he probably did not know that the Hamilton method is not house monotone—but that was the first paradox of the Hamilton Method to be discovered. In 1881, the Census Bureau provided Congress with a table of congressional apportionments for a range of house sizes from 275 to 350 seats, based on the 1880 Census. In the 19th century, the house size was not fixed as it is now, and Congress routinely increased it as the nation grew. The table revealed a strange phenomenon, when the house size increased from 299 to 300.

EXAMPLE 7
The Apportionment of 1880

In 1880, the total population of the United States was 49,373,329. When the house size was set at 299, the standard divisor was 49,373,329 ÷ 299 = 165,128. Alabama's population was 1,262,505. Thus, Alabama's quota was 1,262,505 ÷165,128 = 7.64561. Alabama received its upper quota, so if the house size of 299 had been chosen, Alabama would have been apportioned 8 seats. All states with quotas having fractional parts less than Alabama's were apportioned their lower quotas. The states following Alabama when comparing fractions were Illinois and Texas, with populations of 3,077,871 and 1,591,729, respectively. Their quotas were 18.63930 and 9.63936, respectively, and they were apportioned their lower quotas.

Moving to the next column of the Census Bureau table, when the house size was 300, the standard divisor was 49,373,329 ÷ 300 = 164,578. Dividing the three states' quotas by this smaller divisor yielded the following quotas: Alabama, 7.67117; Illinois, 18.70159; and Texas, 9.67158. One additional seat was available for apportionment. With this house size, the quota for Illinois had the largest fractional part of the three, followed by Texas; Alabama was in third place. The seat that had been for Alabama, and the 300th seat, went to Illinois and Texas; both states were apportioned their upper quotas. Alabama received its lower quota. Table 14.6 summarizes the result of these calculations.

Table 14.6

The Alabama Paradox. Hypothetical Congressional Apportionments Based on the 1880 Census

	House Size	
State	**299**	**300**
Alabama	8	7
Illinois	18	19
Texas	9	10

The paradox was that Alabama's apportionment decreased as a result of an increase in the number of seats in the House of Representatives. This example shows that Hamilton's method of largest fractions is not house monotone. (Congress avoided controversy by selecting a house size of 325.)

The Alabama Paradox DEFINITION

The **Alabama paradox** occurs when a state loses a seat as the result of an increase in the size of the House of Representatives, with no change in any state's population.

The Alabama paradox validates President Washington's veto message, issued 90 years before its discovery. If the fractional parts of the quotas, which determine the way that the last few seats are apportioned, were related to the populations in any sensible way, the paradox could not have occurred.

EXAMPLE 8
A Mathematics Department Meets the Alabama Paradox

A mathematics department has 30 teaching assistants to cover recitation sections for College Algebra, Calculus I, Calculus II, Calculus III, and Contemporary Mathematics. The enrollments of these courses are given in Table 14.7. The department will use the Hamilton method to apportion the teaching assistants (TAs) to the five subjects. In this problem, the house size is 30 (the number of TAs) and the population is the number of students, 750. The states are the five courses to be offered. The standard divisor is 750 ÷ 30 = 25, which

Table 14.7

Apportioning 30 TAs

Course	Enrollment	Quota	Lower Quota	Apportionment
College Algebra	188	7.52	7	7
Calculus I	142	5.68	5 ↑	6
Calculus II	138	5.52	5	5
Calculus III	64	2.56	2 ↑	3
Contemporary Mathematics	218	8.72	8 ↑	9
Totals	750	30.00	27	30

represents the average number of students per recitation section. Each quota shown in the table was determined by dividing the enrollment of the course by this divisor.

The lower quotas add up to 27, so the three courses whose quotas have the largest fractional parts, Calculus I and III and Contemporary Mathematics, were given their upper quotas.

After the TAs were given their teaching assignments, the graduate school authorized the department to hire an additional TA. To determine which course should get the new TA, the department had to recalculate the apportionment. With 31 TAs, the standard divisor was $750 \div 31 = 24.19355$. The new quotas, determined by dividing each population by this new divisor, are shown in Table 14.8. Now the lower quotas add up to 28, so again three additional TAs go to the subjects whose quotas have the largest fractions. The Calculus III fraction, which had been larger than the College Algebra fraction when there were just 30 teaching assistants, has been surpassed. The new TA was placed in College Algebra, and one of the Calculus III TAs had to be reassigned to Calculus II.

Table 14.8

Apportioning 31 TAs

Course	Enrollment	Quota	Lower Quota	Apportionment
College Algebra	188	7.771	7 ↑	8
Calculus I	142	5.869	5 ↑	6
Calculus II	138	5.704	5 ↑	6
Calculus III	64	2.645	2	2
Contemporary Mathematics	218	9.011	9	9
Totals	750	31.000	28	31

The size of the House of Representatives has been fixed at 435 members by statute since Arizona and New Mexico became states on February 14, 1912. Therefore, the Alabama paradox can no longer occur when apportioning seats in Congress. A second paradox, called the *population paradox,* is associated with a fixed house size. Informally, this paradox occurs when two censuses are compared, and it is found that one state's population increases proportionately more than that of another state, and yet the number of seats apportioned to the latter state increases while the number of seats apportioned to the former state decreases. In extreme cases (we will see one in the next example), state A has an increased population and loses a seat, while state B has a decreased population and gains a seat.

The Population Paradox DEFINITION

Two consecutive apportionments are made by an apportionment method. The house size is the same for both apportionments, but the populations differ. For a state S, let $p_1(S)$ and $p_2(S)$ be the populations measured by the censuses, and $a_1(S)$ and $a_2(S)$ be the numbers of seats awarded to the state. The **population paradox** occurs if there is a pair of states A and B such that

$$p_2(A)/p_1(A) > p_2(B)/p_1(B) \text{ and } a_2(A) < a_1(A), \text{ while } a_2(B) > a_1(B)$$

For example, if the population of A increases between the two censuses, and that of B decreases, then the ratio $p_2(A)/p_1(A)$ is greater than 1, and the ratio $p_2(B)/p_1(B)$ is less than 1. Thus, if the number of seats apportioned to A decreased, and the number of seats apportioned to B increased, the population paradox would have occurred.

Population Monotone DEFINITION

An apportionment method is **population monotone** if it is impossible for the population paradox to occur when the method is used to apportion.

In some countries that have parliamentary systems of government, voters do not elect individual candidates; instead they vote for party lists. Each party nominates a ranked list of candidates. Votes are cast, and each party is apportioned a number of seats, "proportionally" to the number of votes that its list received. If, for example, a party was apportioned 10 seats after an election, then the first 10 candidates on the party's list will be seated in the parliament.

In a parliamentary election, the house size corresponds to the number of seats in the parliament, the states correspond to the political parties, and the population of a "state" is the number of voters that selected that party. The Hamilton method is used in the parliamentary elections of several countries, including Russia. Thus, for example a Russian parliamentary election is subject to paradoxes. (In the context of parliamentary apportionment, the Hamilton method of largest fractions is known as the **Hare method**, after Thomas Hare, whose name also appears in Chapter 9 in connection with the Hare system for deciding multicandidate elections.)

EXAMPLE 9
Apportioning Seats in Parliament

A country has four political parties. Its parliament has 100 members, and seats are apportioned by the Hamilton method after each election so that the number of seats that each party is awarded is as close as possible to being proportional to the number of votes that the party receives.

An election is held, but the parties are unable to form a government, so there is a repeat election. Table 14.9 shows the results of the two elections.

Table 14.9

Election Results

Party	First Election	Repeat Election
Whigs	5,525,381	5,657,564
Tories	3,470,152	3,507,464
Liberals	3,864,226	3,885,693
Centrists	201,203	201,049
Totals	**13,060,962**	**13,251,770**

The three major parties—Whigs, Tories, and Liberals—all received more votes in the second election, but the Centrists received fewer. The quotas for each party, shown in Table 14.10, were determined by dividing each party's votes by the standard divisors

$$13{,}060{,}962 \div 100 = 130{,}609.62$$

for the first election, and 132,517.70 for the second election.

Table 14.10

Quotas for the Parties

Party	First Election	Repeat Election
Whigs	42.3045	42.6929
Tories	26.5689	26.4679
Liberals	29.5861	29.3221
Centrists	1.5405	1.5171

The lower quotas for the results of the first election were 42, 26, 29, and 1, with a sum of 98; thus the Tories and the Liberals, with the largest fractions, get extra seats. The apportionment after the first election was Whigs, 42; Tories, 27; Liberals, 30; and Centrists, 1.

For the repeat election, the lower quotas were the same, but now the largest fractions belong to the Whigs and the Centrists. Therefore, the new apportionment is Whigs, 43; Tories, 26; Liberals, 29; and Centrists, 2.

The Centrists have *gained* a seat, although they received fewer votes in the repeat election, while the Liberals lost a seat even though their vote total increased in the repeat election. This is an instance of the population paradox. This has a disturbing implication: a group of voters might have unintentionally caused the Centrist Party to gain a seat by switching their votes to the Liberal Party.

There is a third paradox of the Hamilton method of largest fractions, the **new states paradox**. This paradox was observed in 1907, when Oklahoma achieved statehood. In 1900, the House of Representatives was apportioned with 386 seats. It was anticipated that Oklahoma's population would warrant 5 seats, so the house size was increased to 391 in 1907, to accommodate the additional seats for Oklahoma. When the new apportionment was calculated, Oklahoma received its 5 seats. The paradox is that the apportionment of two other states changed even though their populations remained the same. Maine's apportionment increased from 3 to 4, and New York's apportionment decreased from 38 to 37. For the details, see Exercise 14 (on p. 522).

Thomas Jefferson favored a method of apportionment biased in favor of states with large populations. *(National Portrait Gallery/Art Resource, NY.)*

14.3 Divisor Methods

The Jefferson Method

The Constitution requires congressional districts to be drawn so that the population of each one is at least 30,000. Recall that President Washington vetoed the Hamilton apportionment bill, perhaps because Delaware's population—only 55,540—was too small for the two congressional districts assigned to it. Thomas Jefferson proposed an apportionment method, now called the **Jefferson method**, to replace the Hamilton method.

In any apportionment, the standard divisor, which we will call s, represents the average district population. In developing his method, Thomas Jefferson specified the population of the *smallest* district in the nation, which we will now call d. Using d, rather than s, as the divisor, each state receives an **adjusted quota** that is *always* rounded *down* to obtain its apportionment. If d is chosen correctly, the total number of seats apportioned will be the statutory house size.

Adjusted quota DEFINITION

The result of dividing a state's population by a divisor that is not the standard divisor is known as the **adjusted quota**.

In effect, the Jefferson method apportions to each state the maximum number of congressional districts with a population no less than d that will be accommodated by the state's population. Any leftover population is divided among these districts.

Once the divisor d is known, the Jefferson apportionment is easy to compute. The apportionment for state X, with population V, is

$$\left\lfloor \frac{V}{d} \right\rfloor$$

The actual divisor that Jefferson used was $d = 33{,}000$. Thus, Virginia's apportionment was

$$\left\lfloor \frac{\text{population of Virginia}}{33{,}000} \right\rfloor = \left\lfloor \frac{630{,}560}{33{,}000} \right\rfloor = \lfloor 19.108 \rfloor = 19$$

Therefore, Virginia received 19 seats, rather than 18, which Hamilton's bill would have allocated. Delaware's apportionment was 1 seat instead of 2, as the following calculation shows.

$$\left\lfloor \frac{\text{population of Delaware}}{33{,}000} \right\rfloor = \left\lfloor \frac{55{,}540}{33{,}000} \right\rfloor = \lfloor 1.683 \rfloor = 1$$

The Jefferson method is one of a class of apportionment methods called *divisor methods*.

Divisor Method DEFINITION

A **divisor method** of apportionment determines each state's apportionment by selecting an appropriate divisor and rounding the resulting adjusted quotas, using a specified rounding rule. Divisor methods differ in the rule used to round the quotient.

The divisor must be chosen carefully to achieve the correct house size. It is possible to select a divisor by trial and error, with the insight that going to a smaller divisor will increase the number of seats apportioned, and a larger divisor will decrease the number of seats.

With the Jefferson method, all the adjusted quotas are rounded down. A divisor greater than or equal to the standard divisor will be too large. The reason is that when the quotas given by the standard divisor are rounded down, the result will be the lower quotas, which sum to less than the house size (except in the very unusual case where all the quotas turn out to be whole numbers). Therefore, start with a divisor that is a bit less than the standard divisor. Let's try this by apportioning wins, losses, and ties for the field hockey team.

EXAMPLE 10
The Field Hockey Team

The field hockey team had 18 wins, 4 losses, and 1 tie last season. The Jefferson method can be used to express this record as percentages. The house size is 100 percent, and the total population is the 23 games played. The standard divisor is 0.23. Table 14.11 shows what happens with a few trial divisors, starting with 0.22, which turns out to be too small, as it apportions a total of 103 percent. For each trial divisor, there are two columns: the adjusted quotas on the left and the adjusted quotas rounded down on the right. The sums of the rounded adjusted quotas are shown, and we see that the divisor 0.227 gets the correct house size, 100. (There is no need to add the adjusted quotas, as their sums play no role in determining the apportionment.)

Table 14.11

Apportioning Wins, Losses, and Ties for the Field Hockey Team

		Divisors					
Category	Population	0.22		0.225		0.227	
Wins	18	81.82	81	80.00	80	79.30	79
Losses	4	18.18	18	17.78	17	17.62	17
Ties	1	4.55	4	4.44	4	4.41	4
Total	23		103		101		100

The apportionment is what Coach wanted: 79 percent wins, 17 percent losses, and 4 percent ties.

Apportioning with the Jefferson Method PROCEDURE

1. Make a table with the "states" in the left column.
2. Choose a divisor that is slightly less than the standard divisor.
3. Divide each state's population by the chosen divisor, to get its adjusted quota, which is rounded down to a whole number, its tentative apportionment. Thus, if the adjusted quota is 9.999, round it to 9.
4. Add the tentative apportionments. If the sum equals the house size, you have completed the apportionment.
5. If the total in Step 4 is less than the house size, choose a smaller divisor and repeat Steps 3 and 4. If the total in Step 4 is larger than the house size, choose a larger divisor—but not larger than the standard divisor—and repeat Steps 3 and 4.

We will implement this procedure for our high school mathematics teacher. The Jefferson divisor d in this context can be interpreted as a minimum class size because the number of classes allocated for each subject is obtained by dividing the number of students enrolled by d, and discarding any fractional section in the quotient. Thus each section will have at least d students.

EXAMPLE 11
The High School Mathematics Teacher (Again)

The teacher can be assigned five classes. There are 52 students enrolled in geometry, 33 in pre-calculus, and 15 in calculus. The calculations to determine her teaching assignment by the Jefferson method are shown in Table 14.12. We have determined previously that the standard divisor is 20, so we'll use 18 as our first trial divisor.

Table 14.12

Apportioning High School Mathematics Classes

		Divisors					
Subject	**Population**	**18**		**15**		**16**	
Geometry	52	2.89	2	3.47	3	3.25	3
Pre-calc	33	1.83	1	2.20	2	2.06	2
Calculus	15	0.83	0	1.00	1	0.94	0
Total	100		3		6		5

This time, our starting divisor was too large. For a second try, we chose $d = 15$, which apportioned 6 sections. Thus the Jefferson divisor has to be more than 15 and less than 18. It turns out that $d = 16$ works as a divisor, resulting in 3 geometry classes and 2 pre-calculus classes. There will be no calculus class because the 15 students enrolled are not enough for a class when the minimum class must have 16 students.

Comparing the results of Example 11 and Example 4, where the Hamilton and Jefferson apportionment methods were applied to the same apportionment problem, we see that the results differ. Examples 10 and 5 also exhibit different apportionments from another apportionment problem. Now let us apportion the TAs in the scenario we encountered in Example 8 on page 494; Will the Alabama paradox occur with the Jefferson method? Recall that with the Hamilton method, Calculus III had 3 TAs when 30 TAs were available, but only 2 TAs when another TA was hired.

EXAMPLE 12

Apportioning TAs with the Jefferson Method

In Example 8, a mathematics department was to apportion its TAs among five courses. The populations were the numbers of students enrolled in each course, the states were the courses, and the house size was the number of TAs that were available. Here are the enrollment data: College Algebra, $p_1 = 188$; Calculus I, $p_2 = 142$; Calculus II, $p_3 = 138$; Calculus III, $p_4 = 64$; and Contemporary Mathematics, $p_5 = 218$. For a house size of 30, the standard divisor was 25.

In Table 14.13, we will follow the procedure used in the previous two examples. The adjusted quotas are not shown, just the tentative apportionments that were obtained by rounding them.

Table 14.13

Apportioning TAs by the Jefferson Method

		Divisors		
Subject	**Population**	23	24	23.5
College Algebra	188	8	7	8
Calculus I	142	6	5	6
Calculus II	138	6	5	5
Calculus III	64	2	2	2
Contemporary Math	218	9	9	9
Totals	**750**	**31**	**28**	**30**

The first trial divisor was 23, less than the standard divisor. The number of sections apportioned was 31, which is what we will need when the additional TA arrives. Attempting to apportion 30 sections, we increase the divisor to 24, but that reduces the seats apportioned to 28. The divisor that we need is therefore between 23 and 24, so we will try 23.5, which indeed produces an apportionment of exactly 30 sections.

There is no evidence of the Alabama paradox in this example. When the house size increases from 30 to 31, Calculus II gets the new TA. In fact, the Jefferson method prevents the Alabama paradox, as the following theorem shows.

Monotonicity of the Jefferson Method THEOREM

The Jefferson method of apportionment is house monotone and population monotone.

If the Jefferson method is used, with a divisor d_1, to apportion a house with h seats and then the house size increases to $h + 1$, a new divisor d_2 must be selected to calculate the new apportionment. Because the new house size is larger, d_2 must be less than d_1. Consider a state with population p. Its apportionment with house size h will be $\lfloor p \div d_1 \rfloor$, and its apportionment with house size $h + 1$ will be $\lfloor p \div d_2 \rfloor$. Since $d_2 < d_1$, it follows that $p \div d_2 > p \div d_1$ and hence $\lfloor p \div d_2 \rfloor$ cannot be less than $\lfloor p \div d_1 \rfloor$. Therefore an increase in the house size, with no change in any state's population, cannot lead to a decrease in any state's apportionment. This shows that the Jefferson method is house monotone. Verifying that the Jefferson method is also population monotone is slightly more complicated.

Here is another distinction between the Hamilton and Jefferson methods—and this time it is in the Hamilton method's favor. The Hamilton method gives each state either its upper quota or its lower quota. With the Jefferson method, no state can receive less than its lower quota as its apportionment—because the lower quota is the initial tentative apportionment—but a state can be apportioned more than its upper quota.

EXAMPLE 13
The 1820 Congressional Apportionment

According to the 1820 census, New York had a population of 1,368,775. The total population of the United States was 8,969,878, and the house size was 213. Therefore, the standard divisor was 8,969,878 ÷ 213 = 42,112, and New York's quota was 1,368,775 ÷ 42,112 = 32.503. The Hamilton method would have apportioned to New York its upper quota, 33 seats. The divisor for the Jefferson method was d = 39,900. Thus, New York's apportionment was $\lfloor 1{,}368{,}775 \div 39{,}900 \rfloor = 34$ seats, 1 more than its upper quota.

Quota Condition DEFINITION

An apportionment method is said to satisfy the **quota condition** if in every situation, each state's apportionment is equal to either its lower quota or its upper quota.

It takes only one example like the 1820 apportionment to show that the Jefferson method does not satisfy the quota condition. In fact, if the house had continued to use the Jefferson method, it would have violated the quota condition in every apportionment since 1850. For example, the Jefferson apportionment of the House according to the 2010 census gives California 55 seats, although its quota is 52.54; and Texas, with quota 35.55, would get 37 seats.

The Hamilton method satisfies the quota condition. This was obvious to Congress in 1850, and it adopted the Hamilton method to apportion the seats in the House of Representatives based on the census of that year.

Congress has never used an apportionment method that satisfies the quota condition and avoids the paradoxes. It would be desirable to use such a method, and in the 1970s, the mathematicians Michel L. Balinski and H. Peyton Young tried to find one. Their research led to a proof that there is no apportionment method that satisfies the quota condition and is population monotone. Thus, while Balinski and Young failed to meet their objective, they did prove that their objective was not attainable. This theorem is like Kenneth Arrow's impossibility theorem, which tells us that there is no completely satisfactory way to decide multicandidate elections based on voter preference schedules (see Section 9.4).

The Jefferson method favors the larger states. It is not an accident that in every example that we have considered, the "state" with the largest population fared better with the Jefferson method than it did with the Hamilton method. Virginia got a greater apportionment and Delaware a smaller apportionment in 1790; the winning percentage for the field hockey team was higher, and the losing percentage lower, when the Jefferson method was used as compared with the Hamilton method, and there were more sections of geometry and no sections of calculus when the Jefferson method was substituted for the Hamilton method.

Let's see why the Jefferson method is biased in favor of larger states. In an apportionment problem, let s be the standard divisor and let d be the divisor used in the Jefferson method. Then, the apportionment given to state X is $\lfloor U \rfloor$, where

$$U = (\text{Population of } X) \div d$$

is the state's *adjusted quota*. The formulas for U and for the quota q for X are similar:

$$q = (\text{Population of } X) \div s$$

In fact, the state's population neatly cancels out of the ratio $U \div q$:

$$U \div q = s \div d$$

Thus, while U and q have different values for each state, the ratio of U to q is always the same number, $M = s \div d$. Multiplying both sides of the identity $U \div q = M$ by q, we obtain $U = M \times q$. Therefore, the Jefferson apportionment for state X is

$$\lfloor M \times (\text{quota for } X) \rfloor$$

Consider the congressional apportionment of 1820. The standard divisor was $s = 42{,}112$, and the Jefferson divisor was $d = 39{,}900$. Thus, the quotient, M, is $42{,}112 \div 39{,}900 = 1.0554$. Now suppose that a state has a quota of q. The state's adjusted quota is $U = (1 + 0.0554) \times q = q + q \times 5.54\%$. In words, this algebraic formula says the adjusted quota is obtained by increasing the state's quota by 5.54 percent. A state with a large quota will get a greater raise under these circumstances than a small state will.

To see how this works with numbers, consider a state X with $q = 18.96$. The adjusted quota is

$$U = 1.0554 \times 18.96 = 20.01$$

The upper quota for X is 19, but X is awarded $\lfloor 20.01 \rfloor = 20$ seats! This violates the quota condition, and in fact, every state whose quota is 18.96 or more will be guaranteed to get at least its upper quota with this value of M. If a state has a quota of $2 \times 18.96 = 37.92$, an identical calculation shows that it will receive at least its upper quota plus 1 seat. On the other hand, consider a small state whose lower quota is 1. To increase its apportionment to 2, its quota must be at least $2 \div M = 1.89502$. Thus, a state with quota 18.96 gets more than its upper quota, and a state with quota 1.89 has to settle for its lower quota.

When parliamentary seats are being apportioned, the Jefferson method's bias favors parties that have attracted the most votes at the expense of less popular parties. The Jefferson method is used by many nations that prefer voters to select party lists rather than to vote for individual candidates for parliament—perhaps to avoid awarding many seats to parties that are perceived by many to be on the fringe. It is believed that this practice favors more stable government because there will be less need for a major party to be forced into a coalition with a small party.

For historical reasons, there are differences in terminology and in the method by which a parliamentary apportionment is calculated. The method is the **d'Hondt method**, after Victor d'Hondt (1841–1901). Countries that use the d'Hondt method to apportion seats in their parliaments include Argentina, Denmark, and Israel.

The procedure for apportioning seats by the d'Hondt method is not the same as the one that we have used for the Jefferson method, *but it leads to the same apportionment*. With the d'Hondt method, seats are apportioned one at a time until the house size is reached.

The d'Hondt Method PROCEDURE

1. Each party is given a priority number, which is reduced as the party accumulates seats. A party's priority number is initially the number of votes that it received.
2. After a party has received n seats, its priority number is equal to the number of votes that it received, divided by $n + 1$.
3. Thus, award the first seat to the party that has the largest number of votes; then divide that party's vote total by 2 to get its new priority number.
4. Award the second seat to the party that now has the highest priority. Continue in this way until all the seats have been distributed.

With the d'Hondt method, it is unnecessary to compute the standard divisor, or the quota for any state. The d'Hondt method provides the order in which members of parliament receive their seats. The first member to be seated is the first on the list of the party with the plurality of votes, and so on. However, it achieves exactly the same result as the Jefferson method. To see why this is so, think of the parties' priority numbers as the divisors. When a party receives the last seat (because its priority is highest), its priority number (before that seat is awarded to it) is the divisor that will produce the Jefferson apportionment. Let us suppose that party P received V votes, that it has currently n seats, and that P has the highest priority to receive the last seat in the house. Then its priority number is $d = V \div (n + 1)$. Thus, $V \div d = n + 1$ (exactly—no rounding needed), so with divisor d, party P will get $n + 1$ seats. The seats awarded to the other parties were awarded with priority numbers larger than d, so the divisor d will award those seats to them as well. They will not get any additional seats, though, because their current priority numbers are less than d.

EXAMPLE 14
Using the d'Hondt Method

We will use the d'Hondt method to apportion the first 30 seats in the parliament of Example 9, based on the data from the first election. The calculations are presented in Table 14.14, which is called a **d'Hondt table**. There is a column in the table for each party. The first entry in each column is the number of votes that the party received—the party's initial priority number. Running down each column, the entries are the number of votes for the parties divided by 2, 3, 4, and so on. When a seat is awarded, the priority number for the party that receives the seat is crossed out because it has been used. The next seat goes to the party with the highest priority that has not been crossed out.

At any stage of the d'Hondt process, we can see how many seats have been assigned to a party: It is the number of entries in its column that have been crossed out. If a party has been assigned n seats so far, the first remaining entry in its column is its population divided by $(n + 1)$.

Table 14.14

A d'Hondt Table

Apportioned	Whigs	Tories	Liberals	Centrists
1	#1 ~~5,525,381~~	#3 ~~3,470,152~~	#2 ~~3,864,226~~	201,203
2	#4 ~~2,762,691~~	#7 ~~1,735,076~~	#5 ~~1,932,113~~	100,602
3	#6 ~~1,841,794~~	#10 ~~1,156,717~~	#9 ~~1,288,075~~	67,068
4	#8 ~~1,381,345~~	#14 ~~867,538~~	#12 ~~966,057~~	50,301
5	#11 ~~1,105,076~~	#17 ~~694,030~~	#16 ~~772,845~~	40,241
6	#13 ~~920,897~~	#21 ~~578,359~~	#19 ~~644,038~~	33,534
7	#15 ~~789,340~~	#25 ~~495,736~~	#23 ~~552,032~~	28,743
8	#18 ~~690,673~~	#28 ~~433,769~~	#26 ~~483,028~~	25,150
9	#20 ~~613,931~~	385,572	#29 ~~429,358~~	22,356
10	#22 ~~552,538~~	347,015	386,423	20,120
11	#24 ~~502,307~~	315,468	351,293	18,291
12	#27 ~~460,448~~	289,179	322,019	16,767
13	#30 ~~425,029~~	266,935	297,248	15,477
14	394,670	247,868	276,016	14,372

As shown in the table, the Whigs get the first seat. The Whigs, first priority number is marked #1 and gets crossed out because it has been used. The greatest remaining priority number is for the Liberals, who get the second seat. As seats are awarded, the priority number are numbered in sequence and crossed out, in inverse order. The number of seats awarded to each party is equal to the number of numbers crossed out in its column. Of the first 30 seats, the Whigs get 13, the Tories get 8, the Liberals get 9, and the Centrists get none.

The Webster Method

A second divisor method was introduced in 1832 by Senator Daniel Webster (1782–1852).

The Webster Method — DEFINITION

The **Webster method** is the divisor method that rounds the quota (adjusted if necessary) to the nearest whole number, rounding up when the fractional part is greater than or equal to $\frac{1}{2}$, and rounding down when the fractional part is less than $\frac{1}{2}$.

Statesman and orator Daniel Webster (1782–1852), who developed a divisor method for apportioning the U.S. House of Representatives. *(National Portrait Gallery/Art Resource, NY.)*

The Webster and Jefferson methods are house monotone and population monotone, but neither satisfies the quota condition. However, the Jefferson method favors the large states, while the Webster method is neutral, favoring neither the large nor the small states. Furthermore, the Webster method rarely violates the quota condition by giving a state more than its upper quota, or fewer seats than its lower quota, and would not have done so in any of the 23 congressional apportionments that have occurred so far.

The Webster Method PROCEDURE

1. Obtain the tentative apportionments by rounding each state's quota q to $\lfloor q \rfloor$ if the fractional part of q is less than 0.5; otherwise, round to $\lceil q \rceil$. (This is the standard way of rounding numbers.)
2. Add the rounded quotas. If their sum is equal to the house size, the job is finished. The tentative apportionments calculated in Step 1 are the final apportionments.
3. When the rounded quotas of the states don't add up to the house size, then calculate adjusted quotas, using a trial divisor as with the Jefferson method. All trial divisors must be larger than the standard divisor if the sum of the rounded quotas is greater than the house size, and smaller than the standard divisor if the sum of the quotas is less than the house size.
4. Round the adjusted quotas from Step 3. If their sum is more than the house size, try a larger divisor; if the sum is less, try a smaller divisor. If the sum is equal to the house size, the rounded adjusted quotas provide the correct apportionment.

EXAMPLE 15
The Field Hockey Team (Again)

As you may remember, the field hockey team has a fine record for the previous season: 18 wins, 4 losses, and 1 tie. We would like to express the record in the form of whole percentages, and Coach is willing to go with the Webster method. The quotas are 78.26 percent for wins, 17.39 percent for losses, and 4.35 percent for ties; these are rounded to get the tentative percentages of 78 percent, 17 percent, and 4 percent, respectively, which add up to 99 percent. The standard divisor was 0.23, and because we need to apportion one more "seat," we will have to try a smaller divisor. Our trials are shown in Table 14.15.

The first divisor that we tried was 0.227. The adjusted quotas are in the left column under that divisor in the table, and to their right are the rounded values. Because the total number of "seats" with that divisor turned out to be 101, we tried the divisor 0.228, with the same result. The divisor 0.229 produced 100 "seats," which was our goal. We never tried any divisor greater than 0.23 because we knew from the start that 0.23 is too large. The apportionment was what Coach preferred: 79 percent wins, 17 percent losses, and 4 percent ties.

Table 14.15

Apportioning Wins, Losses, and Ties by Webster's Method

		Divisors					
Category	Population	0.227		0.228		0.229	
Wins	18	79.2952	79	78.9474	79	78.6026	79
Losses	4	17.6211	18	17.5439	18	17.4672	17
Ties	1	4.4053	4	4.3860	4	4.3668	4
Total	23		101		101		100

EXAMPLE 16
Apportioning Classes (Again)

Let us return to the case of the mathematics teacher who is to teach a total of five classes in geometry, pre-calculus, and calculus. The enrollments are 52 for geometry, 33 for pre-calculus, and 15 for calculus. With a total of 100 students enrolled, and a house size of 5, the standard divisor is 20. The quotas, determined by dividing the enrollments for the three subjects by the standard divisor, are 2.60, 1.65, and 0.75, respectively. The tentative apportionments are 3, 2, and 1; their total, 6, exceeds the house size. We therefore will try a divisor greater than 20. Using 21 as the divisor, we find that the adjusted quotas are 2.48 for Geometry, 1.57 for Pre-calculus, and 0.71 for Calculus. Rounded, these quotients become 2, 2, and 1, respectively, for a total of 5 classes. Thus, Geometry and Pre-calculus are each apportioned 2 classes, and calculus gets 1 class.

To see why the Webster method is not biased in favor of large states or small states, let s be the standard divisor, which is equal to the average district population. Let d be the divisor that is used in the Webster method. Finally, let $M = s \div d$. Each state's adjusted quota $U = M \times q$ where q is the state's quota, as we saw in our discussion of the Jefferson method. When $M > 1$ (this happens when the number of seats tentatively apportioned is less than the house size), the states all receive an across-the-board increase in their quotas—and just as in a company where the workers receive the same percentage raise—the larger states are favored. When $M < 1$, the reverse is true because a large number multiplied by M will decrease more than a small number would. Thus, the Webster method favors neither large nor small states when the tentative apportionment exactly fills the house; it favors small states when the tentative apportionment must be reduced, and it favors large states when the tentative apportionment must be expanded. On balance, the Webster method is neutral because it is equally likely that the tentative apportionment will be less than the house size or that it will be greater than the house size.

The Webster Method Has No Population Bias THEOREM

Among all divisor methods, the Webster method alone shows no bias with regard to state population.

When seats are apportioned to parties after an election, a method that is equivalent to the Webster method is used. This method, called the **Sainte-Laguë method** (after André Sainte-Laguë, 1878–1950), involves a process that is analogous to the d'Hondt method. A Sainte-Laguë table is constructed with a column for each party. The number of votes received is at the top of each party's column, and below that the number of votes is divided by successive odd numbers, 3, 5, 7, ... Seats are apportioned to the parties by treating the entries in the table as a priority list. The party with the highest priority gets a seat, and then that priority number is crossed out. The next seat goes to the party with the largest remaining priority in the Sainte-Laguë table. The countries that use the Sainte-Laguë method to apportion seats to their parliaments include Germany and New Zealand.

If you are curious to know why this works, refer again to the party P that gets the last seat. Let V be the number of votes that it received, and suppose that it has n seats already, before the last seat is awarded. The priority for party P to get the last seat is the highest in the table, and it is $V \div (2n + 1)$. (The $(n + 1)$st odd number is $2n + 1$.) Multiply this priority by 2 to get a number d. Using this d as the divisor, we see that $V \div d = (V/2) \div (V/(2n + 1) = n + ½$, exactly. (The V's cancel each other out, as we divide the quotients by inverting the divisor and multiplying.) This quotient can be rounded up to get $n + 1$ seats. The other parties have lower priority, and so they get no new seats with d as the divisor.

14.4 Which Divisor Method Is Best?

The Jefferson and Webster methods frequently yield different results, although both are divisor methods. We have just seen that one way to compare apportionment methods is to investigate bias based on population, but disputes about apportionment have usually not focused on this type of bias. Instead, they have cited inequities that could be reduced by another choice of divisor method. It is how inequities are measured that forms the basis for these arguments. For an account of two disputes that occurred in 1991, see Spotlight 14.1.

Representative Share DEFINITION

Let A be the apportionment given to a state whose population is V. The quotient $A \div V$ is called the **representative share**. It represents the share of a congressional seat given to each citizen of the state.

In an ideal apportionment, every state would have the same representative share. This is impossible, but we can measure how close a given apportionment is to being ideal by using representative share as the standard of comparison. We are given an apportionment done by some method. The method could be Webster's, Jefferson's, Hamilton's, or some method that we haven't mentioned yet. We would compute the representative share for each state. For each state X, we would compare its representative share with all states that had lesser representative shares. In each such case, we would subtract the representative share of the other state from that of X, to find the absolute difference in representative shares. Then we would find out what would happen if we took one seat from state X and gave it to the other state. If the absolute difference in representative shares is greater than it was before the seat was transferred, in each such comparison, then the apportionment is equitable, from the point of view of representative share.

It would seem to be an impossibly laborious task to determine if an apportionment is equitable. Fortunately, there is a theorem that makes the task rather simple:

SPOTLIGHT 14.1

Legal Challenges to Apportionment

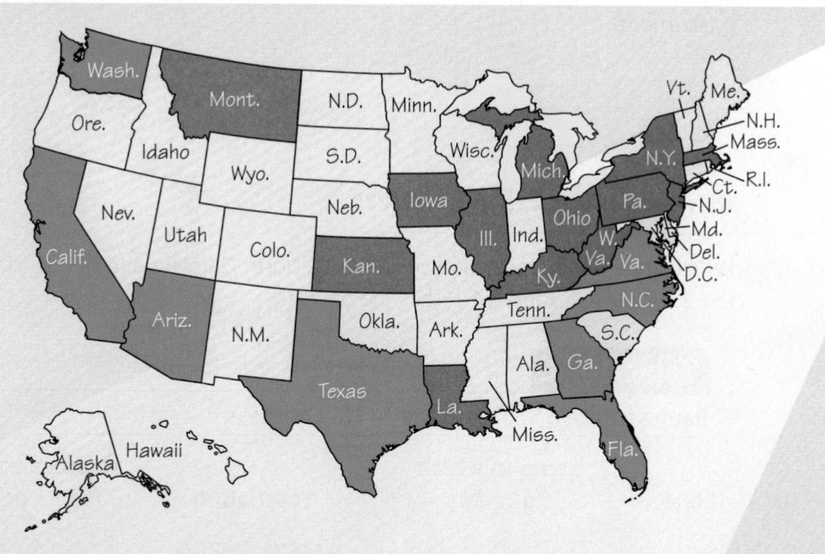

In 1991, the Census Bureau reported the apportionment of congressional seats resulting from the 1990 census. The states colored red on the above map lost representatives, and those in blue gained representatives. New York lost 3 representatives, and Ohio and Pennsylvania lost 2 apiece. Montana, whose apportionment decreased from 2 to 1, sustained the greatest percentage loss, and Montana sued to restore the lost seat. As precedents, Montana referred to the two famous cases, *Baker v. Carr* and *Wesberry v. Sanders*, in which the U.S. Supreme Court required legislative and congressional district boundaries to be drawn so as to make district populations equal.

Montana argued that the correct apportionment would be the one that met the *Baker* and *Wesberry* criterion of having districts as nearly equal in population as possible, and asked the Court to require the Census Bureau to recompute the apportionments using the Dean method, a divisor method that minimizes differences in district populations. This would have resulted in the transfer of a congressional seat from Washington to Montana.

The Montana case coincided with another federal lawsuit, *Massachusetts v. Mosbacher*, which asked the Federal District Court to order the apportionment to be calculated by the Webster method. If Massachusetts had won this suit, it would have gained an additional seat, but Montana would not benefit.

In *U.S. Department of Commerce v. Montana*, the Supreme Court unanimously rejected Montana's claim. The opinion of the Court, written by Justice John Paul Stevens, pointed out that intrastate districts, which were the subject of the *Baker* and *Wesberry* cases, could be equalized in population by drawing district boundaries correctly. Because congressional districts can't cross state lines, some inequity is inevitable in congressional apportionment. The opinion conceded that there were alternative apportionment methods but concluded that the choice of apportionment method was best left to Congress.

Equity and the Webster Method THEOREM

An **apportionment** is **equitable** from the point of view of representative share if and only if it is the same as the apportionment produced by the Webster method.

For example, if seats in a parliament are apportioned by using a Sainte-Laguë table, then the apportionment is equitable from the point of view of representative share. This means that among all apportionments of parliament, this particular apportionment gives all voters as close as possible equal shares of a seat in the parliament.

EXAMPLE 17
Inequity in the 113th Congress

Let's compare the representative shares of North Carolina and Rhode Island based on the 2010 census. Table 14.16 has the data.

Table 14.16

Representative Shares in the 113th Congress

State	Seats in 113th Congress	Population	Representative Share (seats per million)
NC	13	9.565781 million	1.359
RI	2	1.055247 million	1.895

Rhode Island is favored because its representative share is 0.536 seats per million greater than that of North Carolina.

If the apportionment had assigned 14 seats to North Carolina and 1 to Rhode Island, the data would be as in Table 14.17.

Table 14.17

Representative Shares in the 113th Congress If a Rhode Island Seat Should Be Transferred to North Carolina

State	Seats in 113th Congress	Population	Representative Share (seats per million)
NC	14	9.565781 million	1.464
RI	1	1.055247 million	0.948

With this adjustment, North Carolina would have the greater representative share, but the difference, 0.516 seats per million, would be less than the difference in the representative shares before the transfer.

Therefore, in terms of representative share, it would have been more equitable to have given Rhode Island 1 seat and North Carolina 14 seats.

In an ideal apportionment, each congressional district in the nation would have the same population. This is impossible, however, unless congressional districts were permitted to overlap state lines, which is not allowed by the U.S. Constitution.

District Population DEFINITION

The **district population** is the population of each congressional district in the state. The district population of state X is $V \div A$, where V is the state's population and A is its apportionment.

An apportionment can be evaluated by computing differences in district population. Defining what is meant by an equitable apportionment from the point of view of district population is done as in the case of representative share. The equitable apportionment by this standard is the one for which no exchange of seats between any pair of states will lessen the absolute difference in their district populations. It was this basis of comparison that was behind Montana's suit in 1991. (See Spotlight 14.1 on page 509.)

EXAMPLE 18
Comparing District Populations

If we consider differences in district population rather than representative share, it is correct to give North Carolina 13 seats and Rhode Island 2 seats in the 113th Congress. This is made clear in Table 14.18.

Table 14.18
District Populations

State	Population	District Population	District Population after Transfer of Seat so that NC Gets 14 and RI Gets 1
NC	9,565,781	735,829	683,270
RI	1,055,247	527,634	1,055,247
Difference		**208,205**	**371,977**

The transfer of the seat would cause North Carolina's district population to decrease slightly, but it would double Rhode Island's, and the difference in district population would be greater.

For state X, the representative share is $\frac{A}{V}$ and the district population is $\frac{V}{A}$, where A is the apportionment of X and V is its population. Thus:

$$\text{representative share for state } X = \frac{1}{\text{district population for state } X}$$

It may be surprising that these two ways of evaluating the fairness of an apportionment could disagree, but we have just seen that they can.

Edward V. Huntington, a mathematician, pointed out that if relative differences are compared instead of absolute differences, then district population and representative share would give identical comparisons of apportionments—and he suggested a compromise.

Absolute and Relative Difference DEFINITION

Given two positive numbers A and B, with $A > B$, the **absolute difference** is $A - B$ and the **relative difference** is the quotient $\frac{(A - B)}{B} \times 100\%$.

For any two states, it turns out that the relative difference in district populations is equal to the relative difference in representative share (see Exercise 46 on p. 525). Therefore, an apportionment method that minimizes relative difference in representative shares also will minimize the relative difference in district populations.

EXAMPLE 19
Relative Inequity in the 113th Congress

North Carolina will have 14 seats in the 113th Congress, and Rhode Island will have 2 seats. Their representative shares will be 1.359 and 1.895 seats per million, respectively. The relative difference will be

$$\frac{(1.895 - 1.359)}{1.359} = 39.4\%$$

If a seat were transferred from Rhode Island to North Carolina, the representative shares, found in Table 14.17, would be 1.464 seats per million for North Carolina and 0.948 seats per million for Rhode Island. The corresponding relative difference would be

$$\frac{(1.464 - 0.948)}{0.948} = 54.4\%$$

which is greater than the relative difference before the transfer. Therefore, if we measure relative differences instead of absolute differences, the apportionment is fair. To optimize apportionment by the relative difference criterion for equity, Professor Huntington and Joseph Hill, a statistician from the Bureau of the Census, designed a new divisor method. It has been used to apportion seats in the U.S. House of Representatives after each decennial census since 1940.

The Hill-Huntington Method

Like the Jefferson and Webster methods, the **Hill-Huntington method** calculates the apportionment by rounding the quotas, after adjusting them if necessary. The only difference between the three divisor methods is in the rounding procedure.

The Hill-Huntington rounding procedure is related to the geometric mean.

Geometric Mean DEFINITION

The **geometric mean** of two positive numbers A and B is equal to the square root of their product, $\sqrt{A \times B}$.

Consider the rectangle $\mathcal{R}$, displayed in Figure 14.1. The area of $\mathcal{R}$ is the product of the lengths A and B, or $A \times B$. The geometric mean of A and B is equal to the length E of the edge of a square $\mathcal{S}$ with the same area as $\mathcal{R}$, because the area of $\mathcal{R}$ is E^2, and thus $E^2 = A \times B$. Taking square roots, $E = \sqrt{A \times B}$.

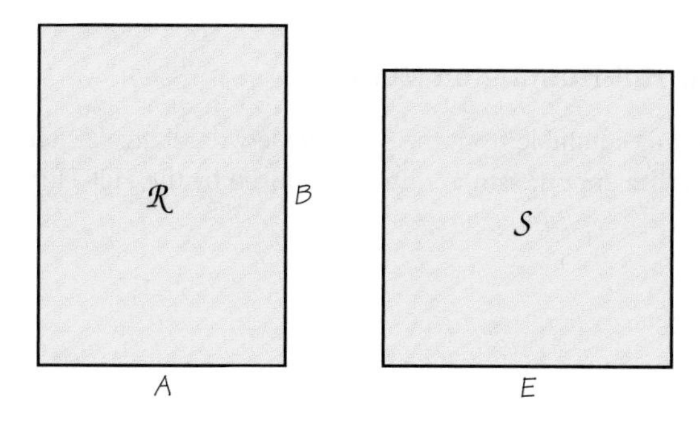

Figure 14.1 The edge of the square S is the geometric mean of the edges of the rectangle $\mathcal{R}$, because the two figures have the same area.

Given a positive number q, let q^* be the geometric mean of $\lfloor q \rfloor$ and $\lceil q \rceil$. This q^* is called the *rounding point* for q. The Hill-Huntington rounding of a number q is equal to $\lfloor q \rfloor$ if $q < q^*$ but is equal to $\lceil q \rceil$ if $q \geq q^*$.

EXAMPLE 20
Hill-Huntington Rounding

Suppose that $q = 7.485$. Jefferson and Webster would round 7.485 down to 7. Because $\lfloor 7.485 \rfloor = 7$ and $\lceil 7.485 \rceil = 8$, the rounding point is $q^* = \sqrt{7 \times 8} = 7.48331 \ldots$, which is less than 7.485. Therefore, the Hill-Huntington rounding of 7.485 is $\lceil 7.485 \rceil = 8$.

Hill-Huntington Apportionment — PROCEDURE

1. Make a table, showing the quota for each state, and the Hill-Huntington rounding point for each quota.
2. Calculate the Hill-Huntington rounding of each quota.
3. Add the rounded quotas from Step 2. If their sum is equal to the house size, they are the apportionment.
4. If the sum of the rounded quotas is not equal to the house size, then compute adjusted quotas, using trial divisors that are greater than the standard divisor if the Hill-Huntington rounded quotas add up to more than the house size, and less than the standard divisor if the sum is less than the house size.

A zero apportionment is impossible with the Hill-Huntington method because the rounding point for quotients between 0 and 1 is $\sqrt{0 \times 1} = 0$. Any quota less than 1 will be rounded to 1.

An apportionment is equitable from the point of view of relative differences (in either representative share or district population) if it is impossible to reduce the relative difference in representative share (or district population) between any pair of states by taking a seat from one and giving it to the other. The reason for considering the Hill-Huntington method is that it provides a way to determine the apportionment that is equitable from the point of view of relative differences.

Equity and the Hill-Huntington Method THEOREM

An apportionment is equitable from the point of view of relative differences if and only if it is the same as the apportionment that is produced by the Hill-Huntington method.

EXAMPLE 21
Percent Effort

Faculty members at a certain university must state the percentage of their time spent in several activities. To comply as accurately as possible, Professor Worktorule has requisitioned five stopwatches to keep track of her activities. Table 14.19 shows, in its left columns, what she recorded over the course of one week.

Table 14.19

Professor Worktorule's Effort by the Hill-Huntington Method

Effort Category	Effort (min.)	Quota	Rounding Point	Tentative Apportionment
Instruction	300	8.33%	8.485	8%
Instructional Support	705	19.58%	19.493	20%
Independent Study	31	0.86%	0.000	1%
Research	2475	68.75%	68.498	69%
Committees	89	2.47%	2.449	3%
Totals	3600	100%	–	101%

The professor is too busy to convert the data into percentages—which the university requires in whole numbers with sum 100 percent—so we'll do it, using the Hill-Huntington method. As with any percentage apportionment problem, the house size is 100, so the standard divisor—one percentage unit—is 36 minutes. Table 14.19 shows in its right columns, the quotas, the rounding points, and the tentative apportionment, obtained by rounding the quotas up or down, depending on whether the quota is above the rounding point or not.

Because too many "seats" were awarded, we'll try divisors larger than the standard divisor, 36. The results are shown in Table 14.20.

The first divisor we tried, 36.5, produced 98 "seats," too few. The left column under that divisor shows the corresponding adjusted quotas. The middle column shows the Hill-Huntington rounding points for the adjusted quotas, and the right column displays the Hill-Huntington tentative apportionment. To increase the number of "seats" apportioned, the next trial divisor must be closer to 36 (but not less than 36). We tried the divisor 36.2, and found that 99 "seats" were apportioned. Our third trial divisor was 36.1 (not shown in the table), which apportioned 101 "seats"—too many. The divisor we needed was therefore between 36.1 and 36.2. We set the divisor equal to 36.15, and the table shows that exactly 100 "seats" were apportioned. The right column of Table 14.20 displays the percentages that Professor Worktorule should put into her effort report.

Table 14.20

Trial Divisors by the Hill-Huntington Method

		Divisors								
		36.5			36.2			36.15		
Category	Pop.	A.Q.	R.P.	T.A.	A.Q.	R.P.	T.A.	A.Q.	R.P.	T.A.
Instruct.	300	8.219	8.485	8	8.287	8.485	8	8.299	8.485	8
Inst. Sup.	705	19.315	19.494	19	19.475	19.494	19	19.502	19.494	20
Ind. Stud.	31	0.849	0.000	1	0.856	0.000	1	0.858	0.000	1
Research	2475	67.808	67.498	68	68.370	68.498	68	68.464	68.498	68
Com.	89	2.438	2.449	2	2.458	2.449	3	2.462	2.449	3
Totals	3600			98			99			100

EXAMPLE 22
The 435th Seat

The Webster and Hill-Huntington methods give almost the same apportionments, based on the 2010 census. The only difference is that the Webster method apportions the last seat to North Carolina, while the Hill-Huntington method gives it to Rhode Island.

The standard divisor s is the apportionment population of the United States, 309,183,463, divided by the number of seats, 435. Thus $s = 710{,}767$. The quotas, obtained by dividing the states' populations by s, are $9{,}565{,}781 \div s = 13.458$ for North Carolina, and $1{,}055{,}247 \div s = 1.485$ for Rhode Island. With the Webster method, both quotas would be rounded down, and only 434 seats would be apportioned. If we reduce the divisor to 708,000, North Carolina's quotient becomes 13.51, which will be rounded up, and Rhode Island's quotient is 1.49, which will be rounded down. North Carolina will get 14 seats, and Rhode Island will get 1.

With the Hill-Huntington method, the rounding point for numbers between 13 and 14 is $\sqrt{(13 \times 14)} = 13.49$, which is above North Carolina's quota. The rounding point for numbers between 1 and 2 is $\sqrt{1 \times 2} = 1.41$, which is below Rhode Island's quota. Therefore North Carolina's quota is rounded down to get its apportionment, 13, and Rhode Island's quota is rounded up to get its apportionment, 2, with the Hill-Huntington method.

The 435th seat was in play between Michigan and Arkansas as a result of the 1940 census, and the resulting dispute led to the permanent adoption of the Hill-Huntington method for the apportionment of seats in the House of Representatives. See Spotlight 14.2.

We have seen that the Jefferson method is biased in favor of populous states and that the Webster method is not biased in regard to state population size. It's natural to ask if the Hill-Huntington method exhibits any bias with respect to state population.

SPOTLIGHT 14.2

Mathematics and Politics: A Strange Mixture

Walter F. Willcox
(Department of Manuscripts and University Archives, Cornell University Libraries.)

Edward V. Huntington
(Courtesy of Harvard University Archives.)

The first American to consider apportionment from a theoretical point of view was Walter Willcox (1861–1964), who strongly advocated the Webster method and had computed the apportionment of the 78th Congress in 1902. His arguments convinced Congress to use the Webster method again in 1912. In 1911, Joseph Hill, a statistician at the Census Bureau, proposed the Hill-Huntington method—with the strong endorsement of Edward V. Huntington, a mathematics professor at Harvard.

In 1920, the two methods were in competition. There were significant differences in the apportionments determined by the two methods, and the result was Washington gridlock: No apportionment bill passed during the decade, and the 1912 apportionment was retained throughout the 1920s. In preparation for the 1930 census results, the National Academy of Sciences formed a committee to study apportionment. In 1929, the committee endorsed the Hill-Huntington method.

The 1930 census was remarkable in that the apportionments calculated by the Webster method were the same as the Hill-Huntington apportionments. Therefore, the House of Representatives was reapportioned, but the method used could be claimed to be either one of the competing methods. The coincidence was almost repeated in the 1940 census, but there was one difference. The Hill-Huntington method gave the last seat to Arkansas, while Webster's method gave it to Michigan. At the time, Michigan was a predominantly Republican state, and Arkansas was in the Democratic column. The vote on the apportionment bill split strictly along party lines, with Democrats supporting the Hill-Huntington method and Republicans voting for the Webster method. Because the Democrats had the majority, the Hill-Huntington method became law.

A divisor method will show bias in favor of large states when the quotas are adjusted by using a divisor that is smaller than the standard divisor. If the quotas must be adjusted downward—that is, a divisor larger than the standard divisor is used—small states are favored. Because the rounding point for the Webster method is halfway between whole numbers, it is just as likely for the divisor to be smaller than the standard divisor as it is for it to be larger.

For any positive number q, the rounding point used by the Hill-Huntington method is closer to $\lfloor q \rfloor$ than to $\lceil q \rceil$ (see Exercise 42 on p. 524). This means that a random number q is more likely to be above the rounding point and thus rounded up to $\lceil q \rceil$ than it is to be less and thus rounded down to $\lfloor q \rfloor$. The difference between the Webster and Hill-Huntington ways of rounding is not significant for relatively large numbers. For example, the Hill-Huntington rounding point between 50 and 51 is 50.498. Therefore, a number q between 50 and 51 will be rounded up to 51 by Hill-Huntington if it is larger than 50.498. The Webster method would round q to 51 if $q > 50.500$.

The differences are more significant when rounding smaller numbers. Hill-Huntington rounds all numbers between 0 and 1 up to 1; Webster rounds only the numbers in the range 0.500–1 up to 1. When the Hill-Huntington method is used for apportionment, the sum of the tentative apportionments is more likely to exceed the house size than it is to be less, especially if there are many states with small populations. Therefore, the Hill-Huntington method is likely to use a divisor larger than the standard divisor. This favors the less populous states.

In conclusion, the Webster method is the best divisor method for general use. It is the only divisor method that is unbiased regarding population size, and it minimizes differences between representative shares. Although the Webster method is capable of violating the quota condition, it is the divisor method least likely to do so.

For apportionment of seats in the U.S. House of Representatives, a slight modification is needed, because no state can receive a zero apportionment. The rounding point for quotas less than 1 is set to zero, rather than 0.5.

There are situations where other apportionment methods could be considered. See Exercise 49 (on p. 525) to explore ways to make teaching assignments.

REVIEW VOCABULARY

$\lfloor q \rfloor$ The result of rounding a number q down; for example, $\lfloor \pi \rfloor = 3$. (p. 491)
$\lceil q \rceil$ The result of rounding a number q up to the next integer; for example, $\lceil \pi \rceil = 4$. (p. 491)

Absolute difference The result of subtracting a smaller number from a larger number. (p. 512)

Adjusted quota The result of dividing a state's quota by a divisor other than the standard divisor. The purpose of adjusting the quotas is to correct a failure of the rounded quotas to sum to the house size. (p. 498)

Alabama paradox A failure of an apportionment method to be house monotone: A state loses a representative solely because the size of the House is increased. This paradox is possible with the Hamilton method but not with divisor methods. (p. 494)

Apportionment method A systematic way of computing solutions of apportionment problems. (p. 487)

Apportionment problem To round a list of fractions to whole numbers in a way that preserves the sum of the original fractions. (p. 487)

d'Hondt table A mechanism for calculating an apportionment that is the same as the one given by the Jefferson method. It typically is used when apportioning seats in parliament to political parties. (p. 504)

District population A state's population divided by its apportionment. (p. 511)

Divisor method One of many apportionment methods in which the apportionments are determined by dividing the population of each state by a common divisor to obtain adjusted quotas. The apportionments are calculated by rounding the adjusted quotas. Divisor methods differ in the way that the rounding of the quotas is carried out. The Jefferson, Webster, and Hill-Huntington methods are divisor methods. (p. 499)

Equitable apportionment An apportionment is equitable from a specified point of view if it cannot be improved, from the specified point of view, by reducing the number of seats apportioned to any one state by one, and increasing the number of seats apportioned to any other state by one. The points of view that we have considered are absolute differences in representative share, absolute differences in district population, and relative differences in either representative share or district population. (Representative share (p. 509), District population (p. 511), Relative differences (p. 512))

Geometric mean For positive numbers A and B, the geometric mean is defined to be $\sqrt{A \times B}$. (p. 512)

Hamilton method An apportionment method that assigns to each state either its lower quota or its upper quota. The states that receive their upper quotas are those whose quotas have the largest fractional parts. (p. 491)

Hare method A method to apportion seats in parliament. It is the same as the Hamilton method. (p. 496)

Hill-Huntington method A divisor method that minimizes relative differences in both representative shares and district populations. This method has been used to apportion seats in the U.S. House of Representatives since 1941. (p. 512)

House monotone An apportionment method is house monotone if no state loses a seat, when an apportionment is recomputed because the house size has increased, with no change in any state's population. All divisor methods are house monotone; the Hamilton method is not. (p. 493)

Jefferson method A divisor method based on rounding all fractions down. Thus, if U is the adjusted quota of state X, the state's apportionment is $\lfloor U \rfloor$. (p. 498)

Lower quota The integer part $\lfloor q \rfloor$ of a state's quota q. (p. 491)

Population monotone An apportionment method is population monotone if the population paradox cannot occur when the method is used. An apportionment method is population monotone if and only if it is a divisor method. (p. 496)

Population paradox A situation in which the apportionment of one state, A, decreases, although its population has increased; while another state, B, loses population (or increases population proportionally less than state A) and gains a seat. (p. 496)

Quota The quota is the quotient $V \div s$ of a state's population divided by the standard divisor s. The quota is the number of seats a state would receive if fractional seats could be awarded. (p. 487)

Quota condition A requirement that an apportionment method should always assign to each state either its lower quota or its upper quota in every situation. The Hamilton method satisfies this condition, but none of the divisor methods do. (p. 502)

Relative difference The relative difference between two positive numbers is obtained by subtracting the smaller number from the larger, and expressing the result as a percentage of the smaller number. Thus, the relative difference between 120 and 100 is 20 percent. (p. 512)

Representative share A state's representative share is the state's apportionment divided by its population. It is intended to represent the amount of influence a citizen of that state would have on his or her representative. (p. 508)

Sainte-Laguë table A mechanism for calculating an apportionment that is the same as the one given by the Webster method. It typically is used when apportioning seats in parliament to political parties. (p. 507)

Standard divisor The ratio $p \div h$ of the total population p to the house size h. In a congressional apportionment problem, the standard divisor represents the average district population. (p. 488)

Tentative apportionment The result of rounding a state's quota or adjusted quota to obtain a whole number. (p. 491)

Upper quota The result of rounding a state's quota *up* to a whole number. A state whose quota is q has an upper quota equal to $\lceil q \rceil$. (p. 491)

Webster method A divisor method of apportionment that is based on rounding fractions the usual way. The Webster method minimizes the *absolute* differences of representative share between states. (p. 505)

SKILLS CHECK

1. A county is divided into three districts with the following populations: Southern, 3600; Western, 3100; Northeastern, 1600. There are 6 seats on the county council to be apportioned. What is the quota for the Southern district?

(a) 2.6
(b) 2.8
(c) 3

2. Two calculus teachers can teach a total of 8 classes. Enrollments are as follows: Calculus I, 200; Calculus II,

100; Calculus III, 52. In this apportionment problem, the population is ______, the standard divisor is ______, and the quotas are ______ for Calculus I, ______ for Calculus II, and ______ for Calculus III.

3. *A*, *B*, and *C* are arguing about fractions of a cent. On a project, they worked exactly 33, 34, and 35 minutes, respectively, and were paid $100. Use the Hamilton method to see who gets his upper quota (in cents!).

(a) *A*
(b) *B*
(c) *C*

4. Round each number in the sum 13.62 + 12.58 + 17.51 + 16.77 + 19.52 = 80 to a whole number.
____ + ____ + ____ + ____ + ____ = 80. (What is it about these numbers that makes this complicated?)

5. The population paradox occurs when

(a) a state's apportionment decreases because the house size increased.
(b) a state's apportionment decreases and its apportionment increases, while another state's apportionment decreases even though its population has increased.
(c) the Jefferson method is used.

6. The Alabama paradox occurred when it was noticed that Alabama would lose a seat, in apportionment by the Hamilton method, if the house size was changed from 299 to ____________.

7. If the Jefferson method is used to do the rounding in Skills Check Question 4, it will be necessary to find a divisor that is

(a) less than 1.
(b) at least 1, but less than 2.
(c) at least 2.

8. Using the Webster method, round the numbers in Skills Check Question 4: ____ + ____ + ____ + ____ + ____ = 80.

9. When rounding the numbers in the sum 20.45 + 30.30 + 49.25 = 100 by the Jefferson method, which number gets its upper quota?

(a) 20.45
(b) 30.30
(c) 49.25

10. Seats in a parliament are apportioned by the Hare (Hamilton) method. Jane had planned to vote for party *B*, but changed her mind and voted for party *A*. If her vote switch caused party *A* to lose a seat, this would be an instance of the ____________ paradox.

11. The parliament in Skills Check Question 10 will be apportioned by the d'Hondt (Jefferson) method. Now is it possible for Jane's vote switch to cause party *A* to lose a seat?

(a) Yes, because the Jefferson method does not satisfy the quota condition.
(b) No, because the Jefferson method is house monotone.
(c) No, because the Jefferson method is population monotone.

12. Use the Jefferson method to apportion the sum 0.8 + 0.9 + 98.3 = 100 as a sum of whole numbers.
____ + ____ + ____ = 100

13. The Jefferson method frequently

(a) gives the smallest state less than its lower quota.
(b) gives the largest state more than its upper quota.
(c) gives a state a lesser apportionment if the house size increases.

14. Use the Webster method to apportion the sum 0.8 + 0.9 + 98.3 = 100 as a sum of whole numbers.
____ + ____ + ____ = 100

15. We want to apportion the sum 1.6 + 3.7 + 5.5 + 89.2 = 100 as a sum of whole numbers. Which method will violate the quota condition?

(a) Hamilton
(b) Jefferson
(c) Webster

16. The sum 1.6 + 2.6 + 3.6 + 4.6 + 5.6 = 18 has to be rounded as a sum of whole numbers. If the ____________ method is used, there will be a tie.

17. When rounding the numbers in the sum 20.45 + 30.30 + 49.25 = 100 by the Webster method, which number gets rounded up?

(a) 20.45
(b) 30.30
(c) 49.25

18. States *A* and *B* have populations of 1 million and 2 million, respectively. If they are apportioned 2 and 3 seats, respectively, then the absolute difference in representative share is ____ per million.

19. If the apportionment in Skills Check Question 18 gave state *A* 1 seat and gave state *B* 4 seats, then

(a) the absolute difference in representative share would increase.
(b) the absolute difference in representative share would decrease.

(c) the absolute difference in representative share would be unchanged.

20. If the criterion is absolute difference in district population, the equitable apportionment of 5 seats to states *A* and *B* in Skills Check Question 18 is ____ for *A* and is ____ for *B*.

21. If the initial calculations leading to the Hill-Huntington apportionment result in a sum that is too large, what happens next?

(a) Adjusted quotas must be calculated, using a divisor slightly larger than the standard divisor.
(b) Adjusted quotas must be calculated, using a divisor slightly less than the standard divisor.
(c) The largest apportionment is reduced.
(d) A different method must be used.

22. The ________ method has been used since 1941 to apportion seats in the U.S. House of Representatives.

23. Which divisor method never apportions to a state fewer seats than its lower quota?

(a) Hill-Huntington
(b) Webster
(c) Jefferson

24. A school principal is apportioning sections of the school's mathematics classes. She wants to set a minimum section size, and to adjust it so that a total of 32 sections are open. She should use the ____ method.

25. The U. S. Constitution says that each state must get at least one representative in the House. Which apportionment method(s) is (are) consistent with this requirement?

(a) Hamilton
(b) Jefferson
(c) Webster
(d) Hill-Huntington

26. Five parties *A, B, C, D, E* participate in a parliamentary election. The parliament has 10 seats. The number of votes received (in thousands) was *A*, 120; *B*, 78; *C*, 50; *D*, 35; and *E*, 20. Here is a d'Hondt table.

A	B	C	D	E
120	78	50	35	20
60	39	25	17.5	10
40	26	16.67	11.67	6.67
30	19.5	12.5	8.75	5

When the seats are apportioned, the seventh seat goes to party ______, the eighth seat goes to party ________, the ninth goes to _______, and the tenth goes to ______.

27. The Hill-Huntington method minimizes relative differences in

(a) district population.
(b) representative share.
(c) both district population and representative share.

28. The divisor method that shows the least bias in favor of either large states or small states is the ______ method.

29. A parliament has 466 seats and there are 13 parties with lists on the ballot. It is proposed that there should be a minimum number *N* of votes to qualify for a seat, and that number should be chosen so that exactly 466 seats are filled. We should point out that this idea is not new; it is the

(a) Hare method.
(b) d'Hondt method.
(c) Sainte-Laguë method.

30. In the election described in Skills Check Question 29, party *A* received exactly 12 million of the votes, and each of the other 12 parties received exactly 1 million votes. Although Party *A* received exactly half of the votes, it will receive __ more than half of the seats. This will violate the ______ condition.

CHAPTER 14 EXERCISES

■ Challenge ▲ Discussion

14.1 The Apportionment Problem

1. Jane has decided to track her daily expenses, and finds them to be as listed in the table at right.

Express these as percentages. If rounded to whole numbers, do the percentages add up to 100 percent?

Rent	$31
Food	16
Transportation	7
Gym	12
Miscellaneous	5

2. A mathematics department uses 20 teaching assistants to aid in its four-semester calculus course. The number of teaching assistants assigned to each level of the course depends on enrollment. Here are the fall enrollments:

Calculus I	500
Calculus II	100
Calculus III	350
Calculus IV	175
Total	**1125**

How many teaching assistants should be assigned to each level of the course?

3. Should the mathematics department in Exercise 2 revise the assignments for its TAs? Grades have been posted for the previous semester, and some students need to repeat the previous level of the course. A total of 45 students move from Calculus II to Calculus I, 41 students move from Calculus III to Calculus II, and 12 students move from Calculus IV to Calculus III.

▲ **4.** Here is a typical apportionment problem. Round the numbers in the following sum to integers:

$$8.37 + 10.33 + 12.38 + 5.47 + 3.45 = 40$$

The rounded numbers must add up to 40. How would you approach this?

▲ **5.** How would you round the numbers in the following sum to integers? The rounded numbers must add up to 60.

$$11.63 + 9.67 + 7.62 + 14.53 + 16.55 = 60$$

14.2 The Hamilton Method

6. Use the Hamilton method to round the numbers in the following sum to whole numbers. The sum of the rounded numbers must be the same as the original sum.

$$2.64 + 1.41 + 2.01 + 0.67 + 0.62 + 0.65 = 8$$

7. Repeat Exercise 6 with the sum

$$0.36 + 1.59 + 0.99 + 2.33 + 2.38 + 2.35 = 10$$

8. The 37th pearl. Three friends have bought a bag guaranteed to contain 36 high-quality pearls for \$14,900 at an auction. Abe contributed \$5900, Beth's contribution was \$7600, and Charles supplied the remaining \$1400. After taking the bag to your house, they pour the 36 pearls from the bag onto the kitchen table.

(a) How many should each friend get if the Hamilton method is used to apportion the pearls according to the size of the contributions?

(b) Charles has noticed something; the bag isn't empty! Another pearl comes out, so recalculate the apportionment.

▲ **(c)** How do you explain the result to Charles?

▲ **9.** A country has three political parties, and it allots seats in its 102-seat parliament by the Hare (Hamilton) method proportionately to the number of votes each receives. In a recent election, the Pro-UFO Party received 254,000 votes, the Anti-UFO Party got 153,000 votes, and the Who Cares Party polled 103,000 votes. Show that two of the parties are tied.

10. A small high school has one mathematics teacher who can teach a total of 5 sections. The subjects that she teaches, and their enrollments, are as follows: geometry, 52; algebra, 33; calculus, 12. Use the Hamilton method to apportion sections to the subjects.

11. Repeat Exercise 10 using the following enrollments: geometry, 77; algebra, 18; calculus, 20.

12. Use the Hamilton method to express the summands of the following expression as whole number percentages of the total:

$$2746 + 1725 + 1921 + 100 = 6492$$

Repeat the calculation for the sum:

$$2814 + 1745 + 1933 + 99 = 6591$$

Do you see a paradox?

13. Abe, Beth, Charles, and David have decided to invest in rare coins. A dealer has offered to sell them a parcel containing 100 identical coins for \$10,000. Each person invests all that he or she can afford, but there is not quite enough money, so Charles asks his Aunt Esther to join the group. The coins will be apportioned by the Hamilton method. Here are the amounts contributed:

Abe	\$3619
Beth	1862
Charles	2258
David	2010
Esther	251
Total	\$10,000

(Alan Carey/Corbis.)

(a) How should the coins be apportioned among the five contributors?

(b) After the coins are distributed, the dealer mentions that there will be $50 in excise tax. Everyone empties his or her wallet: Abe finds $16 more, Beth has $2, Charles has $1, and David finds $32. This adds up to $51, so $1 is returned to Aunt Esther. The apportionment is recalculated, and one of the coins changes hands. Who has to give a coin to whom?

▲ **(c)** Explain what happened.

To see how this situation works out with a different apportionment method, refer to Exercise 31 on page 523.

14. The New States Paradox The census of 1900 recorded the following populations:

Maine	694,466
New York	7,264,183

The house size was 386, and the population of the United States was 74,562,608. The Hamilton method was used to apportion seats in the House of Representatives.

(a) Determine the quotas of New York and Maine. Given that New York was the last state to receive its upper quota, determine the numbers of seats that were apportioned to these two states.

(b) In 1907, Oklahoma became a state. Its population was stipulated to be 1,000,000 for the purpose of apportionment. Using the standard divisor that you found in part (a), determine the apportionment for Oklahoma.

(c) Add the stipulated population of Oklahoma to the 1900 population of the United States, and the number of seats that you found in part (b) to the house size before Oklahoma became a state. With these data, find out if the numbers of seats apportioned to New York and Maine changed.

▲ **(d)** The Hamilton method was never used to apportion seats in the House of Representatives again. Can you see why?

15. A country has five political parties. Here are the numbers of votes each received in a recent election: 5,576,330; 1,387,342; 3,334,241; 7,512,860; and 310,968. Seats in its parliament are apportioned by the Hare (Hamilton) method. Calculate the apportionments for house sizes of 89, 90, and 91. Does the Alabama paradox occur?

14.3 Divisor Methods

▲ **16.** Explain why the tentative Webster apportionment of a state with quota q is $\lfloor q + 0.5 \rfloor$.

17. Reapportion the classes in Exercise 11, using the Jefferson method.

18. Reapportion the classes in Exercise 10, using the Webster method.

▲ **19.** The three friends who bought the pearls (see Exercise 8 on p. 521) ask you to suggest a different apportionment method to distribute their purchase. Before answering, determine the apportionments given by the Jefferson and Webster methods for the 36- and 37-pearl house sizes. Then make your suggestion.

▲ **20.** The three friends have bought a lot of 36 identical diamonds, at a total cost of $36,000; Abe's investment was $15,500, Beth's was $10,500, and Charles's was $10,000. They decided to apportion the diamonds using the Webster method, and they can't make it work out. Can you help?

21. Example 2 (on p. 521) describes the plight of a math teacher who is to teach a total of 5 sections in three different courses. The enrollments were Geometry, 52; Pre-calculus, 33; Calculus, 15.

(a) Make a d'Hondt table and apportion the classes.

(b) Make a Sainte-Laguë table and apportion the classes.

22. A country has a 20-seat parliament. Seats are apportioned to parties by the d'Hondt method. The following table displays the results of a recent election. Make a d'Hondt table and determine the number of seats allocated to each party.

Demopublicans	44,856
Repocrats	34,944
Greenocrats	20,004
Greenicans	19,002
Independents	9,804

23. Referring to the voting data in Exercise 22, make a Sainte-Laguë table to apportion the parliament.

Exercises 24–26 refer to the parliament in Example 9 on page 496. We will use only the data for the first election.

24. Make a Sainte-Laguë table and apportion the first 30 seats.

25. In Example 14 (on p. 504), we used a d'Hondt table to apportion the first 30 seats. The Centrists did not get any of the first 30 seats. Which of seats 31–100 will be the first one that the Centrists receive?

26. Which will be the first seat that the Centrists receive if the Sainte-Laguë table is used to apportion the seats?

▲ **27.** No country uses the Hill-Huntington method to apportion its parliament. Explain why the Hill-Huntington method is not appropriate for this purpose.

▲ **28.** Explain why the new states paradox (see page 497) cannot occur if a divisor method of apportionment is used.

29. Round the following to whole percentages using the Hamilton, Jefferson, and Webster methods:

$$87.85\% + 1.26\% + 1.25\% + 1.24\% + 1.23\% + 1.22\% + 1.21\% + 1.20\% + 1.19\% + 1.18\% + 1.17\% = 100\%$$

Do any of these apportionments show a violation of the quota condition?

30. Round the following percentages to whole numbers, using the methods of Hamilton, Jefferson, and Webster.

$$92.15\% + 1.59\% + 1.58\% + 1.57\% + 1.56\% + 1.55\% = 100\%$$

Do any of these apportionments show a violation of the quota condition?

31. Recalculate the apportionment of the coins in Exercise 13 (on p. 521) by the Webster method. Again, after the excise tax is paid, a coin changes hands. Who gives it to whom?

32. A country has two political parties, the Liberals and the Tories. The seats in its 99-seat parliament are apportioned to the parties according to the number of votes they receive in the election. If the Liberals receive 49 percent of the vote, how many seats do the Liberals get with the Hamilton (Hare) method? With the Webster (Sainte-Laguë table) method? With the Jefferson (d'Hondt table) method?

■ **33.** A country with a parliamentary government has two parties that capture 100 percent of the vote between them. Each party is awarded seats in proportion to the number of votes received.

(a) Explain why the Webster (Sainte-Laguë table) and Hamilton (Hare) methods always will give the same apportionment in this two-party situation.

(b) Explain how to use the result of part (a) to show that the Alabama and population paradoxes cannot occur when the Hamilton method is used to apportion seats between two parties or states.

(c) Explain why the result of part (a) implies that the Webster method satisfies the quota condition when the seats are apportioned between two parties or states.

▲ **(d)** Will the Jefferson and Hill-Huntington methods also yield the same apportionments as the Hamilton method?

14.4 Which Divisor Method Is Best?

34. Determine the relative difference between the numbers 5 and 7.

35. Jim is 72 inches tall and Alice is 65 inches tall. What is the relative difference of their heights?

36. In the 2011 apportionment of Congress, Montana's single congressional district—with the entire state's population of 994,416—is the most populous in the nation. California has 53 seats in the House, each with a constituency of 704,565.

(a) Which state is more favored in this apportionment?

(b) What are the absolute and relative differences in the district populations?

(c) If California and Montana were apportioned 52 seats and 2 seats, respectively, determine the absolute and relative differences in district populations.

▲ **(d)** Does this evidence indicate that the apportionment following the 2010 census is equitable from the point of view of absolute differences in district populations? What about relative differences?

37. The data from Exercise 36 imply that the representative share for California was 1.419315 representatives per million, and the representative share for Montana was 1.005615 representatives per million.

(a) What are the absolute and relative differences in representative share?

(b) If California had been apportioned 52 seats, and Montana had been apportioned 2, determine the absolute and relative differences in representative share.

(c) Which is more equitable, from the point of view of absolute differences in representative share between these two states?

38. Find the Hill-Huntington rounding points for numbers between 0 and 1; between 1 and 2; between 2 and 3; and between 3 and 4.

39. A high school has one math teacher who can teach 5 sections. A total of 56 students have enrolled in the algebra class, 28 have signed up for geometry, and 7 students will take calculus. Use the Hill-Huntington method to decide how many sections of each course to schedule.

40. One year later, the high school described in Exercise 39 still has just one math teacher who teaches 5 sections. The enrollments are algebra, 36; geometry, 61; and calculus, 3. Apportion the classes by the Webster and Hill-Huntington methods. Which apportionment do you think the school principal would prefer?

■ **41.** Suppose that in 2010, the governor of Montana believed that the population of his state was undercounted. What increase in population would be large enough to entitle Montana to take a seat from California if the apportionment is by the Hill-Huntington method? The data needed for this problem are given in Exercise 36.

▲ **42. (a)** Show that for any positive numbers A and B, the geometric mean is less than the arithmetic mean,[2] except when $A = B$; then the two means are equal. (*Hint*: Show that the triangle in Figure 14.2 is a right triangle.)

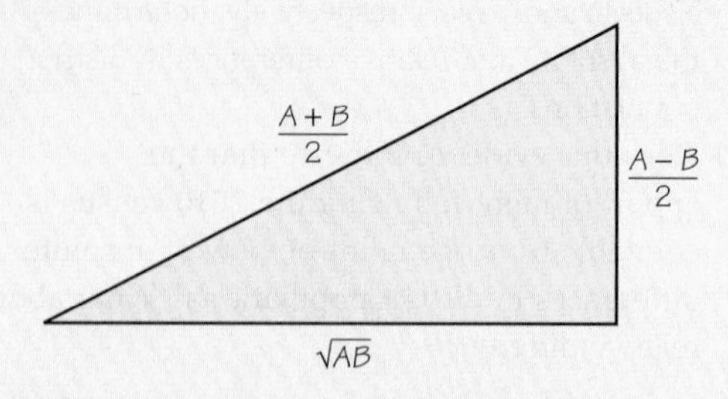

Figure 14.2 Is this a right triangle?

(b) Compare the Webster and Hill-Huntington roundings. Show that for any positive number q, if the two roundings differ, then the Hill-Huntington rounding of q is equal to $\lceil q \rceil$ and the Webster rounding is equal to $\lfloor q \rfloor$.

(c) Explain why the fact established in part (a) implies that the Hill-Huntington method is more favorable to small states than the Webster method.

[2]The arithmetic mean of A and B is equal to $(A + B)/2$.

43. A city has three districts with populations of 100,000; 600,000; and 700,000. Its council has 20 members, and seats on the council are apportioned by the Hill-Huntington method according to the districts. Show that there is a tie. Would a tie occur with any of the other apportionment methods that we have considered?

44. In a 1991 federal lawsuit *Massachusetts* v. *Mosbacher,* Massachusetts claimed that the Hill-Huntington method of apportionment is unconstitutional because it does not reflect the "one person, one vote" principle as well as the Webster method does. Would Massachusetts have gained a seat from Oklahoma if the Webster method had been used to apportion the House of Representatives in 1991? In your calculation, use the following populations and Hill-Huntington apportionments:

State	Population	Apportionment
Massachusetts	6,029,051	10
Oklahoma	3,145,585	6

▲ **45.** The following apportionment method was invented by Congressman William Lowndes of South Carolina in 1822. Lowndes started, as Hamilton did, by giving each state its lower quota. But where Hamilton apportions the remaining seats to the states whose quotas have the largest fractional parts—in other words, the states for which the *absolute difference* between q_i and $\lfloor q_i \rfloor$ is greatest—Lowndes gives the extra seats to the states where the *relative difference* between q_i and $\lfloor q_i \rfloor$ is greatest, raising as many as necessary to their upper quotas to fill the House.

(a) Would this method be more beneficial to states with large populations or small populations than the Hamilton method?

(b) Does the Lowndes method satisfy the quota condition?

(c) Would there be any trouble with paradoxes with the Lowndes method?

(d) Use the method to apportion the 1790 House of Representatives. The populations and quotas resulting from the 1790 Census are in Table 14.2.

▲ **46.** Let the populations of states A and B be p_A and p_B, respectively. The apportionments will be a_A and a_B. Assuming that district populations for state A are larger than district populations for state B, show that the relative difference in district populations is

$$\left(\frac{a_B p_A}{a_A p_B} - 1\right) \times 100\%$$

Also show that this expression is equal to the relative difference in representative share. Hence the relative difference in district populations is equal to the relative difference in representative shares.

▲ **47.** John Quincy Adams, the sixth president of the United States, proposed that the House of Representatives should be apportioned by a divisor method based on the rounding rule that rounds each fraction up to the next whole number.

(a) Is it likely that the initial tentative apportionment will be final?

■ **(b)** Will the divisor that produces the correct apportionment be greater than or less than the standard divisor?

(c) Does the method favor small states or large states?

(d) Is it possible for a state to be apportioned zero seats by using this method?

▲ **48.** The U.S. Constitution requires that each state be apportioned at least one seat in the House of Representatives. The Dean method of apportionment is the divisor method that minimizes absolute differences in district population. Using this information, explain why the Dean method will never give any state zero seats unless the house size is less than the number of states.

▲ **49.** The choice of a divisor method for apportioning classes to subjects according to enrollments, as in Example 2 (on p. 489), depends on what the school principal considers most important.

(a) The principal wants to set a minimum class size. For example, if the minimum class size is 20, and 39 students are signed up for English III, there would be one section, because there are not enough students for two sections with enrollment of at least 20. If there were 40 students, there would be two sections. The minimum class size is adjusted so that as many sections as possible are running. What apportionment method should she use?

■ **(b)** The principal prefers to set a maximum class size. For example, if the maximum class size is 33, and 67 students are taking History I, there will be three sections because there are too many students to fit in two 33-student sections. If there were only 66 students taking History I, there would be two sections. The maximum class size is adjusted so that as many sections as possible are running. What apportionment method should she use, and what will the maximum class size be? (*Hint:* This divisor method is not described in the text but is mentioned in one of the previous exercises.)

(c) The principal wants to cancel any class that has an enrollment of just 1 student. Which apportionment methods should she avoid using?

■ **50.** Let $q_1, q_2, \ldots, q_n$ be the quotas for n states in an apportionment problem, and let the apportionments assigned by some apportionment method be denoted $a_1, a_2, \ldots, a_n$. The *absolute deviation* for state i is defined to be $|q_i - a_i|$; it is a measure of the amount by which the state's apportionment differs from its quota. The *maximum absolute deviation* is the largest of these numbers. Explain why the Hamilton method always gives the least possible maximum absolute deviation.

APPLET EXERCISES

To do these exercises, go to www.whfreeman.com/fapp9e.

A bus company has three lines—*A*, *B*, and *C*—and a total of 48,000 riders. Line *A* has 21,700 riders daily, *B* has 17,200, and *C* has 9100. The company has 40 buses to allocate to the three lines. Use the Apportionment applet to help you find the standard divisor and determine the allocation of the buses according to the methods of apportionment of Hamilton, Jefferson, Webster, and Hill-Huntington.

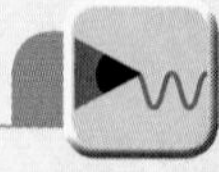

WRITING PROJECTS

1. Does the Hill-Huntington method best reflect the intentions of the Founding Fathers, as these intentions were set down in the Constitution and in the debate during the 1787 Constitutional Convention? Good sources of information here include all the publications listed in the "Suggested Readings" section. This writing project requires that you state your answer to the question and make a case for it.

2. In the apportionment resulting from the 2010 Census, two states, New York and Ohio, lost 2 house seats, and eight other states lost 1 seat. The 12 seats were transferred to states that had experienced dramatic increases in population: Texas got 4 additional seats, Florida received 2 more, and 6 other states got one apiece. If Congress should decide to increase the statutory house size so that no state's delegation in the House would be reduced, how many seats would have been added to the House, and which states would have gotten them? (*Warning:* it's more than 12.) The populations and apportionments for the 50 states are available on the Census Bureau Web site (www.census.gov/population/www/censusdata/apportionment.html).

3. In 2004, the state of Colorado considered an amendment to its constitution regarding the way that electors representing Colorado in the Electoral College would be selected. Although this amendment was not adopted, it provides an interesting apportionment problem. Colorado's electors are selected by the presidential ticket that received a plurality of the votes. The proposed amendment, which would have taken effect in 2004, apportioned Colorado's 9 electoral votes to each ticket in proportion to the number of popular votes received. The apportionment method specified in the amendment was as follows: Determine each ticket's quota, and round to the nearest whole number. Tickets with quotas less than 0.5 receive no electors. If the number of electors thus apportioned is less than Colorado is entitled to have, give the remaining electors to the ticket that received the most votes. If the number of electors is more than Colorado is entitled to have, take electors from that ticket that, among all who received electors, received the smallest number of popular votes. If more electors need to be removed, take from the ticket that received the next smallest number of popular votes, and so on.

Write an essay exploring the implications of this apportionment method, and compare it to others that Colorado could have chosen. Include a discussion of the consequences that would occur if a third-party candidate received some electoral votes as a result of this procedure.

4. The Dean method. The Webster method was proposed in 1832 after New York received an apportionment in excess of its upper quota. Two other apportionment methods were proposed in the same year: the method of John Quincy Adams, which is biased in favor of small states as much as the Jefferson method is biased in favor of large states, and the Dean method. The latter method, invented by James Dean, a professor of mathematics and astronomy at Dartmouth College, gives the equitable apportionment from the point of view of absolute difference in district population. Suppose that state A has population p and its tentative apportionment is n, while state B has population q and tentative apportionment m. If another seat is to be given to one of these states, answer the following questions:

(a) Calculate the absolute difference in district populations if A gets the seat, and repeat the calculation for the situation when B gets the seat.

(b) Show that the difference between the two results in part (a) is equal to $\frac{p}{n^\#} - \frac{q}{m^\#}$, where $n^\#$ denotes the *harmonic mean* (you may have to Google this term) between n and $n + 1$.

(c) Explain the mechanics of the Dean method, including why it is a divisor method that rounds a number r down to $\lfloor r \rfloor$ if r is less than the Dean rounding point, and up to $\lceil r \rceil$ otherwise. It's up to you to figure out the Dean rounding point.

(d) Is the Dean method biased in favor of large states or small states? Is it possible for any state to get an apportionment of zero? Compute the apportionment of the House of Representatives according to the 2010 census by the Dean method. Is the quota condition satisfied? For the source of population data, see the Census Bureau Web site (www.census.gov/population/www/censusdata/apportionment.html).

Suggested Readings

BALINSKI, M. L., and H. P. YOUNG. *Fair Representation: Meeting the Ideal of One Man, One Vote,* Yale University Press, New Haven, CT., 1982. In the 1970s, Balinski and Young analyzed apportionment methods in depth. Their approach was to postulate the desirable properties of an apportionment method as axioms and to deduce from the axioms which method is best. This book combines an account of the history of apportionment of the U.S. House of Representatives with the results of their research.

ERNST, LAWRENCE R. Apportionment methods for the House of Representatives and the court challenges, *Management Science,* 40 (1994): 1207–27. Ernst, who wrote briefs for the government in both the Montana and the Massachusetts cases, reviews the apportionment problem and the arguments in favor of and against each of the divisor methods. The article includes a summary of the arguments used by both sides in the two court cases.

YOUNG, H. PEYTON. *Equity,* Princeton University Press, Princeton, NJ, 1994. Chapter 3 of this book covers apportionment and focuses on which apportionment method is the most equitable.

Suggested Web Sites

The U.S. Census Bureau's Congressional Apportionment site at http://www.census.gov/population/apportionment/ includes a two-page history of apportionment of the Congress.

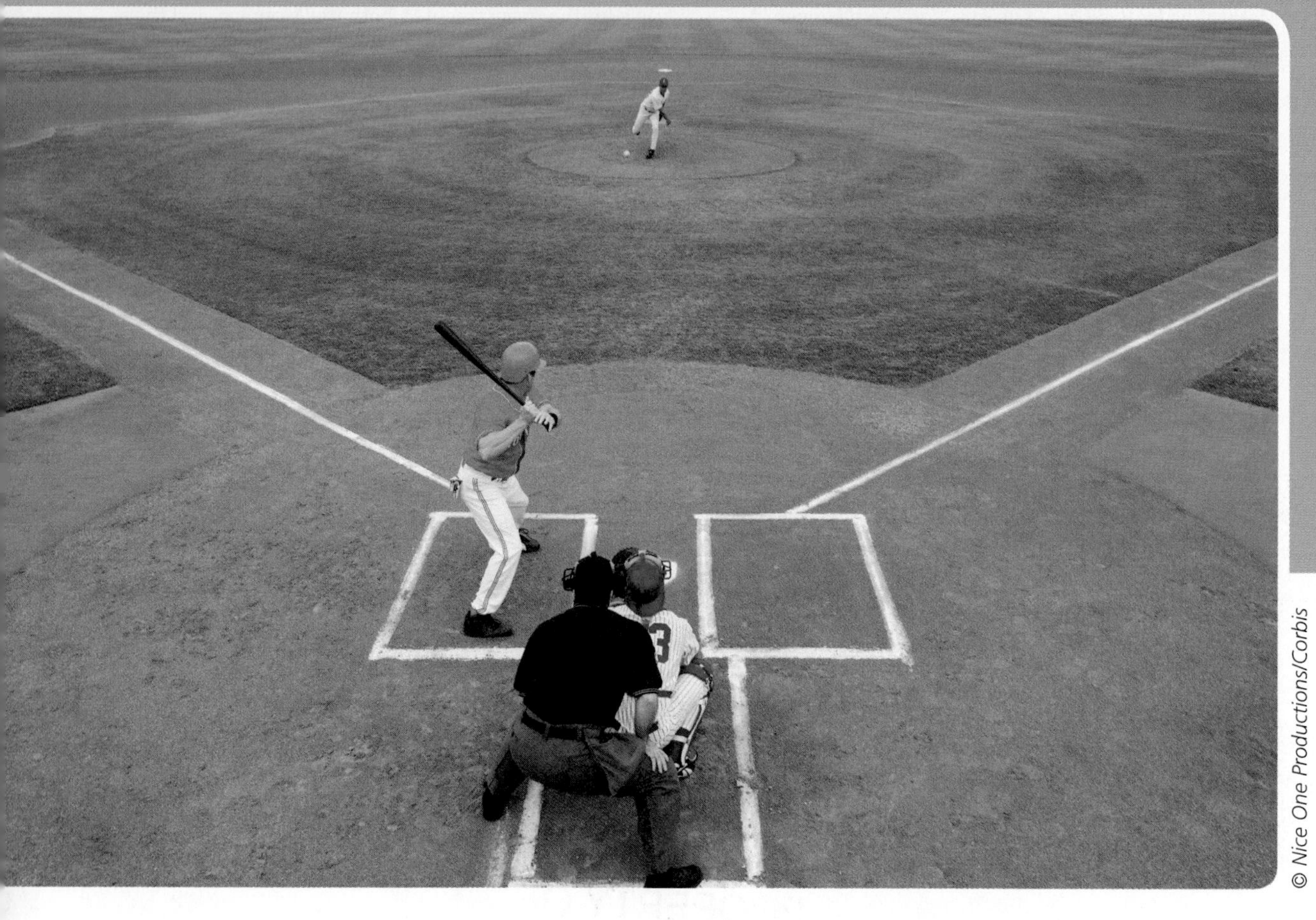

15

Game Theory: The Mathematics of Competition

Conflict has been prevalent throughout human history. It arises whenever two or more individuals with different values or goals compete to try to control the course of events. *Game theory* uses mathematical tools to study situations, called games, involving both conflict and cooperation (see Spotlight 15.1).

The *players* in a game—who may be people, organizations, or even countries—choose from a list of options available to them—that is, courses of action they may take—that are called **strategies**. The strategies chosen by the players lead to *outcomes*, which describe the consequences of their choices. We assume that the players have *preferences* for the outcomes: They like some more than others.

Game theory analyzes the **rational choice** of strategies—that is, how players select strategies to obtain preferred outcomes. Among areas to which game theory has been applied are bargaining tactics in labor–management disputes, resource-allocation decisions in political campaigns, military choices in international crises, and the use of threats by animals in habitat acquisition and protection.

Unlike the subject of *individual* decision making—which researchers in psychology, statistics, and other disciplines study—game theory analyzes situations in which there are at least two players who may find themselves in conflict because of different goals or objectives. The outcome depends on the choices of *all* the players. In this sense, decision making is *collective*, but this is not to stay that the players necessarily cooperate when they choose strategies.

Many interactions involve a delicate mix of cooperative and noncooperative behavior. In business, for example, firms in an industry cooperate to gain tax breaks even as they compete for shares in the marketplace.

Game theory has provided important theoretical foundations in economics, starting with microeconomics but now extending to macroeconomics, industrial organization, and international economics. It also has been increasingly applied in political science, especially in the study of voting, elections, and international relations. In addition, game theory has contributed major insights in biology, particularly in understanding the evolution of species and conditions under which animals—humans included—fight each other for territory or act altruistically.

In the next two sections, we present several simple examples of two-person **total-conflict games**, in which what one player wins the other player loses, so cooperation never benefits the players. We distinguish two different kinds of solutions to such games. Then we analyze two well-known **partial-conflict games**, in which the players can benefit from cooperation but may

SPOTLIGHT 15.1

The Early History of Game Theory

John von Neumann
(Bettmann/UPI/Corbis.)

Oskar Morgenstern
(Courtesy of the Institute for Advanced Study, Princeton University Archives.)

As early as the 17th century, such outstanding scientists as Christiaan Huygens (1629–1695) and Gottfried W. Leibniz (1646–1716) proposed the creation of a discipline that would apply the scientific method to the study of human conflict and interactions. Throughout the 19th century, several leading economists created simple mathematical models to analyze particular examples of competitive encounters. The first general mathematical theorem on this subject was proved for games of perfect information by the distinguished logician Ernst Zermelo (1871–1953) in 1912. A game is said to have *perfect information* if at each stage of the play, every player is aware of all past moves (by itself and others) as well as all future choices that are possible. The theorem stated that any finite game with perfect information, such as checkers or chess, has an optimal solution in *pure* strategies; that is, no randomization or secrecy is necessary. This theorem is an example of an *existence theorem*: It demonstrates that there must exist a best way to play such a game, but it does not provide a detailed plan for playing a complex game, like chess, to achieve victory.

The famous mathematician F. E. Émile Borel (1871–1956) introduced the notion of a *mixed,* or randomized, strategy when he investigated optimal strategies in duels around 1920. The fact that every two-person *zero-sum* game must have a solution in optimal mixed strategies was proved by the Hungarian-American mathematician John von Neumann (1903–1957) in 1928. Von Neumann's result was extended to the existence of equilibrium outcomes in mixed strategies for multiperson games that are either *constant-sum* or *variable-sum* in 1951 by John F. Nash, Jr. (b. 1928), who was portrayed in the movie *A Beautiful Mind* (2001).

Modern game theory dates from the publication in 1944 of *Theory of Games and Economic Behavior* by John von Neumann and the Austrian-American economist Oskar Morgenstern (1902–77). They introduced the first general model and solution concept for multiperson *cooperative games,* which are primarily concerned with coalition formation (by economic cartels, voting blocs, or military alliances) and the resulting distribution of gains or losses. Several other suggestions for a solution to such games have since been proposed. These include the value concept of Lloyd S. Shapley (b. 1923), which relates to fair allocation and serves also as index of voting power (see Chapter 11).

The French artist Georges Mathieu designed a medal for the Musée de la Monnaie in Paris in 1971 to honor game theory. It was the 17th medal to "commemorate 18 stages in the development of Western consciousness." Game theory also has a mascot, the tiger, arising from the Princeton University tiger and the Russian abbreviation of the term *game theory* (ТЕОРИЯ ИГР), where the underlined letters correspond to the sounds of the English *T*, *G*, and *R*, respectively).

have strong incentives not to cooperate. We next turn to the analysis of a larger three-person voting game, in which we show how to eliminate undesirable strategies in stages. Finally, we offer some general comments on solving games and discuss different applications of game theory.

15.1 Two-Person Total-Conflict Games: Pure Strategies

For some games with two players, determining the best strategies for the players is straightforward. We begin with such a case.

EXAMPLE 1
A Location Game

Two young entrepreneurs, Henry and Lisa, plan to locate a new restaurant at a busy intersection in the nearby mountains. They agree on all aspects of the restaurant except one. Lisa likes low elevations, whereas Henry wants greater heights—the higher, the better. In this one regard, their preferences are diametrically opposed. What is better for Henry is worse for Lisa, and likewise what is good for Lisa is bad for Henry.

The layout for their location problem is shown in Figure 15.1. Observe that three routes, Avenue A, Boulevard B, and County Road C (blue lines), run in an east–west direction, and that three highways, numbered 1, 2, and 3 (red lines), run in a north–south direction. Table 15.1 shows the altitudes at the nine corresponding intersections. The same information is shown in three dimensions in Figure 15.2.

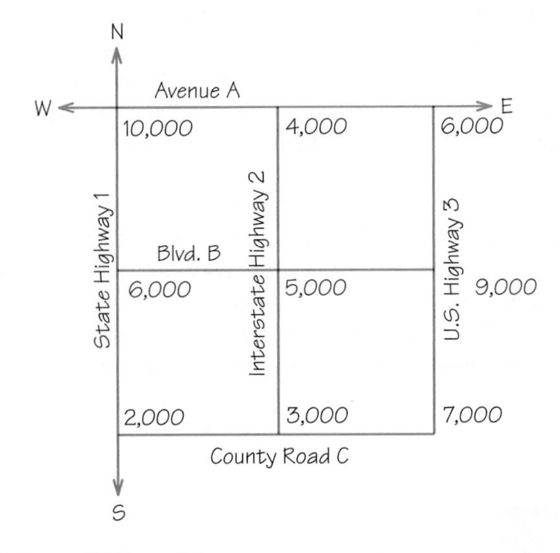

Figure 15.1 The road map for the location of Henry and Lisa's restaurant in Example 1. (The elevations in feet are shown at each intersection.)

Table 15.1

Heights (in thousands of feet) of the Nine Intersections

	Highways		
Routes	1	2	3
A	10	4	6
B	6	5	9
C	2	3	7

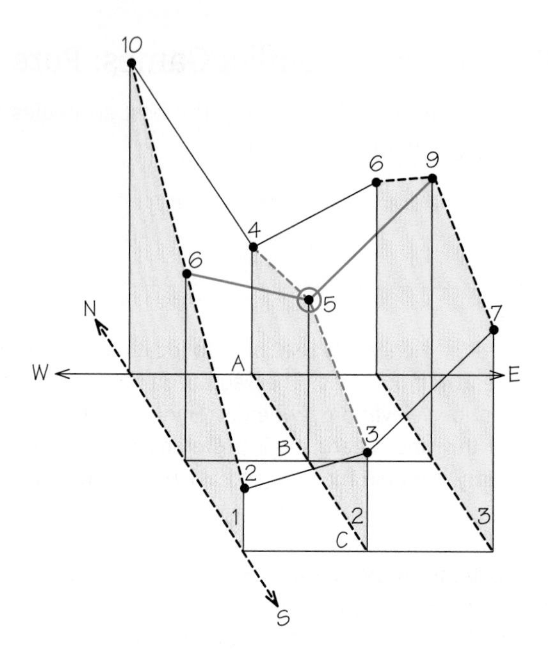

Figure 15.2 Three-dimensional road map showing Henry's and Lisa's possible choices (in thousands of feet).

To maximize the number of customers, Henry and Lisa agree that the restaurant should be at a location where one of the east–west routes intersects one of the three highways. But they cannot agree on which intersection, so they decide to turn their decision into the following competitive game: Henry will select one of the three routes—*A*, *B*, or *C*—and Lisa will simultaneously choose one of the three highways—1, 2, or 3. Because their choices will be made simultaneously, neither one can predict beforehand what the other will do.

Henry, worried that Lisa will choose a low elevation, tries to determine the highest altitude he can guarantee by picking one of the three routes. For each choice of a route, this means considering the worst-case (lowest) elevation on each route. These are the numbers 4, 5, and 2, which are the respective *row minima*, indicated in the right-hand column of Table 15.2. He notes that the highest of these values is 5. By choosing the corresponding route, *B*, Henry can guarantee himself an altitude of at least 5000 feet.

Table 15.2

Heights (in thousands of feet) in Table 15.1, with the Row Minima (maximum circled) and Column Maxima (minimum circled)

		Lisa Highways			
	Routes	1	2	3	Row Minima
	A	10	4	6	4
Henry	B	6	5	9	(5)
	C	2	3	7	2
	Column Maxima	10	(5)	9	

Maximin

DEFINITION

The **maximin** is the maximum value of the minimum numbers in the rows in a table. The strategy of the row player—Henry, in this case—that corresponds to the maximin is called its **maximin strategy**.

The number 5 in the right-hand column of Table 15.2, which is circled, is the **maximin.** Route *B* is Henry's maximin strategy.

Lisa likewise does a worst-case analysis and lists the highest—for her, the worst—elevations for each highway. These numbers, 10, 5, and 9, are the column maxima and are listed in the bottom row of Table 15.2. From Lisa's point of view, the best of these outcomes is 5. If she picks Interstate Highway 2, then she is assured of an elevation of no more than 5000 feet.

Minimax

DEFINITION

The **minimax** is the minimum value of the maximum numbers in the columns. The strategy of the column player—Lisa, in this case—that corresponds to the minimax is called its **minimax strategy**.

The number 5 at the bottom row of Table 15.2, which is circled, is the **minimax**. Highway 2 is Lisa's minimax strategy.

To summarize, Henry has a strategy that will ensure the height is 5000 or higher, and Lisa has a strategy that will ensure the height is 5000 or lower. The height of 5000 at the intersection of Route *B* and Highway 2 is, simultaneously, the lowest value along Boulevard B and the highest along Interstate Highway 2. In other words, the maximin and the minimax are both equal to 5000 for the location game.

Saddlepoint

DEFINITION

When a row minimum and a column maximum are the same, the resulting outcome is called a **saddlepoint**.

The reason for the term **saddlepoint** should be clear from the saddle-shaped payoff surface shown in Figure 15.2. The middle point on a horse saddle (5) is simultaneously the lowest point along the spine of the horse (6 and 9 are higher) and the highest point between the rider's legs (4 and 3 are lower). (In Figure 15.2, the rider would be facing leftward or rightward.) In our example, one might also think of the saddlepoint as a mountain pass: As one drives through the pass, the car is at a high point on a highway (in the north–south direction) and at a low point on a route (in the east–west direction).

The resolution of this contest is for Henry to pick *B* and Lisa to pick 2. This puts them at an elevation of 5000, which is simultaneously the maximin and minimax.

Value

DEFINITION

In total-conflict games, the **value** is the best outcome that both players can guarantee. If a game has a saddlepoint, it gives the value (5 in our example): Players can guarantee this outcome by choosing their maximin and minimax strategies.

Total-conflict games without saddlepoints also have a "value," as we shall see later.

There is no need for secrecy in a game with a saddlepoint. Even if Henry were to reveal his choice of *B* in advance, Lisa would be unable to use this knowledge to beat him. In fact, both players can use the height information in our example to compute the optimal strategy for their opponent as well as for themselves. In games with saddlepoints, players' worst-case analyses lead to the best *guaranteed* outcome—in the sense that each player can ensure that he or she does not do worse than a certain amount (5 in our example) and may do better (if the opponent deviates from a maximin or minimax strategy).

Another well-known game with a saddlepoint is tic-tac-toe. Two players alternately place an X or an O, respectively, in one of the unoccupied spaces in a 3×3 grid with 9 cells. The winner is the first player to have three X's, or three O's, in the same row, in the same column, or along a diagonal; if no player does this when all cells are filled in, the game ends in a tie.

An explicit list of all strategies for either the first- or second-moving player in tic-tac-toe is long and complicated because it specifies a complete plan for all possible contingencies that can arise. For the first-moving player, for example, a strategy might say "put an X in the middle cell, then an X in the corner if your opponent puts an O in a noncorner cell," and so on. While young children initially find this game interesting to play, before long they discover that each player can always prevent the other player from winning by forcing a tie, making the game quite boring. Unlike the game between Henry and Lisa, the list of possible strategies in tic-tac-toe is huge, but only those that force a tie, it turns out, are a saddlepoint.

EXAMPLE 2
The Restricted-Location Game

Assume in our location game that Henry and Lisa are informed by county officials that it is against the law to locate a restaurant on either Boulevard B or Interstate Highway 2. These two choices, which provided our earlier solution, are now forbidden. The resulting location game without these two strategies is given in Table 15.3 (with payoffs again expressed in thousands of feet).

Table 15.3

Heights (in thousands of feet) without Boulevard B and Interstate Highway 2

	Highways	
Routes	1	3
A	10	6
C	2	7

As before, Henry and Lisa can each do a worst-case analysis. Henry is worried about the minimum number in each row, and Lisa is concerned with the maximum number in each column. These are listed in the right column and bottom row, respectively, in Table 15.4.

Henry sees that his maximin is 6, so he can guarantee a height of 6000 feet or more by choosing route *A*. Likewise, Lisa observes that her minimax is 7, so she can keep the elevation of the restaurant down to 7000 feet or less by selecting Highway 3. There is

Table 15.4

Heights (in thousands of feet) in Table 15.3, with Row Minima (maximum circled) and Column Maxima (minimum circled)

		Lisa Highways		
	Routes	1	3	Row Minima
Henry	A	10	6	(6)
	C	2	7	2
	Column Maxima	10	(7)	

a gap of $7 - 6 = 1$ between the minimax and maximin. When the maximin is not the same as the minimax, then a game does *not* have a saddlepoint, but it does have a value (described in the next section).

If Henry plays his maximin strategy, route *A*, and Lisa plays her minimax strategy, Highway 3, then the resulting payoff is 6. However, Henry may be motivated to gamble in this case by playing his other strategy, route *C*. If Lisa sticks to her conservative strategy, Highway 3, then the payoff is 7. Henry will have gained 1 unit (1000 feet), going from 6 to 7.

This is, however, a risky move. If Lisa suspected it, she might counter by selecting Highway 1. The payoff would then be 2, the best for Lisa and the worst for Henry. So Henry's gamble to gain 1 unit (6 to 7) by moving has the risk that he might lose 4 units (6 to 2) if Lisa also moves.

But then there is no incentive for Lisa to play her nonminimax strategy (that is, to play Highway 1) if she believes Henry, in turn, will move back to his maximin strategy (route *A*), leading to a payoff of 10. This is worse than 6 from her viewpoint.

In two-player games that have saddlepoints, like our original 3 × 3 location game and tic-tac-toe, each player can calculate the maximin and minimax strategies for both players before the game is even played. Once the solution has been determined by either mathematical analysis or practical experience (as was probably true of tic-tac-toe), there may be little interest in actually playing the game.

But this is decidedly not the case for much more complex games, like chess, whose solution has not yet been determined—and is unlikely to be in the foreseeable future. Even though computers are able to beat world champions, the computer's winning moves will not necessarily be optimal against those of *all* other opponents. Nevertheless, we know that chess, like tic-tac-toe, has a saddlepoint. (All games of perfect information, in which the players know each other's moves at every step, have a saddlepoint.) What we do not know is whether it yields a win for white, a win for black, or a draw.

Unlike chess, many games, like the 2 × 2 restricted-location game, do not have an outcome that can always be guaranteed. These games, which include poker, involve uncertainty and risk. In such games, one does not want to have one's strategy detected in advance, because this information can be exploited by an opponent. It is no surprise, then, that poker players are told to keep a "poker face," revealing nothing about their likely choices. But this advice is not very helpful in telling the players what actually to do in the game, such as how many cards to ask for in draw poker.

We will show that there are optimal ways to play two-person total-conflict games without a saddlepoint so as not to reveal one's choices. But their solution is by no means as straightforward as that of games with a saddlepoint.

15.2 Two-Person Total-Conflict Games: Mixed Strategies

Probably most competitive games do not have a saddlepoint like the one we found in our first location-game example. Rather, as is illustrated in our restricted-location game—in which the maximin and minimax are not the same—players must try to keep secret their strategy choices, lest their opponent use this information to his or her advantage.

In particular, players must take care to conceal the strategy they will select until the encounter actually takes place, when it is too late for the opponent to alter his or her choice. If the game is repeated, a player will want to *vary* his or her strategy in order to surprise the opponent.

In parlor games like poker, players often use the tactic of *bluffing*. This tactic involves a player's sometimes raising the stakes when he has a low hand so that opponents cannot guess whether or not his hand is high or low—and may, therefore, miscalculate whether to stay in or drop out of the game. (A player would prefer opponents to stay in when he has a high hand and drop out when he has a low hand.) In military engagements, too, secrecy and even deception are often crucial to success.

In many sporting events, a team tries to surprise or mislead the opposition. A pitcher in baseball will not signal the type of pitch he or she intends to throw in advance, varying the type throughout the game to try to keep the batter off balance. In fact, we next consider a confrontation between a pitcher and batter in more detail.

EXAMPLE 3
A Duel Game

The pitcher and the batter use mixed strategies. *(Aflo Sport/Masterfile)*

Assume that a particular baseball pitcher can throw either a blazing fastball or a slow curve into the strike zone and so has two strategies: *fast* (denoted by F) and *curve* (C). The pitcher faces a batter who attempts to guess, before each pitch is thrown, whether it will be a fastball or a curveball, giving the batter two strategies also: guess F and guess C. Assume that the batter has the following batting averages, which are known by both players.

- 0.300 if the batter guesses fast (F) and the pitcher throws fast (F)
- 0.200 if the batter guesses fast (F) and the pitcher throws curve (C)
- 0.100 if the batter guesses curve (C) and the pitcher throws fast (F)
- 0.500 if the batter guesses curve (C) and the pitcher throws curve (C)

A player's batting average is the number of times that he hits safely divided by his number of times at bat. If a batter hit safely 3 times out of 10, for example, his average would be 0.300.

This game is summarized in Table 15.5. We see from the right-hand column in the table that the batter's maximin is 0.200, which is realized when he selects his first strategy, F. Thus, the batter can "play it safe" by always guessing a fastball, which will result in his batting a minimum of 0.200, hardly enough for him to remain on the team.

We see from the bottom row of the table that the pitcher's minimax is 0.300, which is obtained when he throws fast (F). Note that the batter's maximin of 0.200 is less than the pitcher's minimax of 0.300, so this game does not have a saddlepoint. There is a gap of $0.300 - 0.200 = 0.100$ between these two numbers.

Table 15.5

Batting Averages in a Baseball Duel

		Pitcher		Row Minima (maximum circled)
		F	C	
Batter	F	0.300	0.200	(0.200)
	C	0.100	0.500	0.100
	Column Maxima (minimum circled)	(0.300)	0.500	

Each player would like to play so as to win for himself as much of the 0.100 payoff in the gap as possible. That is, the batter would like to average more than 0.200, whereas the pitcher wants to hold the batter down to less than 0.300.

A Flawed Approach

If the batter and pitcher in our example consider how they might outguess each other, they may reason along the following lines:

1. *Pitcher* (to himself): If I choose strategy *F*, I hold the batter down to 0.300 (the minimax) or less. However, the batter is likely to guess *F* because it guarantees him at least 0.200 (his maximin), and it actually provides him with 0.300 against my *F* pitch. In this case, the batter wins all the 0.100 payoff in the gap.
2. *Batter* (to himself): Because the pitcher will try to surprise me with *C* by reasoning as in step 1, I should fool him and guess *C*. I would thus average 0.500, which will show him up for trying to gamble and outguess me!
3. *Pitcher* (to himself): But if the batter is thinking as in step 2—that is, guessing *C*—I, on second thought, should really throw *F*. This will lead to an average of only 0.100 for the batter and teach him to not try to outguess me!

This type of cyclical reasoning can go on forever: "I think that he thinks that I think that he thinks. . . ." It provides no resolution to the players' decision problem.

Clearly, there is no pitch, or guess, that is best under all circumstances. Nevertheless, both the pitcher and the batter *can* do better, but not by trying to anticipate each other's choices. The answer to their problem lies in the notion of a **mixed strategy**.

A Better Idea

The play of many total-conflict games requires an element of surprise, which can be realized in practice by making use of a mixed strategy.

Pure Strategy DEFINITION

Each of the definite courses of action that a player can choose is called a **pure strategy**.

All the choices of players—Henry and Lisa, the batter and the pitcher—that we have considered so far are **pure strategies**, in which each player in the end makes a definite choice.

Mixed Strategy DEFINITION

A **mixed strategy** is a strategy in which the course of action is randomly chosen from one of the pure strategies in the following way: Each pure strategy is assigned some probability, indicating the relative frequency with which that pure strategy will be played. The specific strategy used in any given play of the game can be selected using some appropriate random device.

Note that a pure strategy is a special case of a mixed strategy, with the probability of 1 assigned to just one pure strategy and 0 to all the rest. When a player resorts to a mixed strategy, the resulting outcome of the game is no longer predictable in advance. (For example, if a pitcher throws a curve ball or a fastball with probability 0.5 each, the batter cannot predict which pitch he or she is about to receive.) Rather, the outcome must be described in terms of the probabilistic notion of an **expected value**, which is the average value of the game if it were played many, many times.

Expected Value E DEFINITION

If each of the n payoffs, $s_1, s_2, \ldots, s_n$, will occur with the probabilities $p_1, p_2, \ldots, p_n$, respectively, then the average, or **expected value E**, is given by

$$E = p_1 s_1 + p_2 s_2 + \cdots + p_n s_n$$

We assume that the probabilities sum to 1 and that each probability p_i is never negative. That is, we assume that $p_1 + p_2 + \cdots + p_n = 1$, and $p_i \geq 0$ $(i = 1, 2, \ldots, n)$.

To see how mixed strategies and expected values are used in the analysis of games, we turn to what is perhaps the simplest of all competitive games without a saddlepoint.

EXAMPLE 4
Matching Pennies

In matching pennies, each of two players simultaneously shows either a head *H* or a tail *T*. If the two coins match, with either two heads or two tails, then the first player (Player I) receives both coins (a win of 1 for Player I). If the coins do not match, that is, if one is an *H* and the other is a *T*, then the second player (Player II) receives the two coins (a loss of 1 for Player I). These wins and losses for Player I are shown in Table 15.6.

Table 15.6

Wins and Losses for Player I in Matching Pennies

		Player II	
		H	T
Player I	H	1	−1
	T	−1	1

Payoff Matrix

DEFINITION

A **payoff matrix** (illustrated by Table 15.6) is a table whose rows and columns correspond to the strategies of the two players. The numerical entries give the payoffs to Player I when these strategies are chosen.

Although the entries in our earlier tables for the location game also gave payoffs, they were not monetary, as here. A game represented by a payoff matrix is called *a game in strategic form*.

The two rows in Table 15.6 correspond to Player I's two pure strategies, H and T, and the two columns to Player II's two pure strategies, also H and T. The numbers in the table are the corresponding winnings for Player I and losses for Player II. If two H's or two T's are played, Player I wins 1 from Player II. When one H and one T are played, Player I pays out 1 to Player II.

It is fruitless for one player to attempt to outguess the other in this game. They should instead resort to mixed strategies and use expected values to estimate their likely gains or losses.

To illustrate, assume that Player I randomly selects H half the time and T half the time. This mixed strategy can be expressed as

$$(p, 1 - p) = \left(\frac{1}{2}, \frac{1}{2}\right)$$

Note that the probability p of choosing H, and $(1 - p)$ of choosing T, do indeed sum to 1, as required; in particular, when $p = \frac{1}{2}$, $1 - p = 1 - \frac{1}{2} = \frac{1}{2}$.

This mixture can be realized in practice by the flip of a coin. Player I's resulting expected value is

$$E_H = \frac{1}{2}(1) + \frac{1}{2}(-1) = 0$$

whenever Player II plays H (first column of Table 15.6). Whenever Player II plays T (second column), Player I's resulting expected value is

$$E_T = \frac{1}{2}(-1) + \frac{1}{2}(1) = 0$$

Player I cannot guarantee a better outcome than to choose $p = \frac{1}{2}$, making this strategy optimal (see Exercise 14 on p. 565).

Value

DEFINITION

A player's expected value is the **value** of the game when he or she chooses an optimal strategy. Unlike the use of this notion in games with a saddlepoint, the value here can be realized only by the use of mixed strategies.

The value of 0 in matching pennies is really an expected value and so must be understood in a statistical sense. That is, in a given play of the game, Player I will either win 1 or lose 1. However, his or her expectation over many plays of this game is 0. The optimal mixed strategy for Player II is likewise a 50-50 mix of H and T, which also leads to an expectation of 0, making the game **fair**.

Fair Game

DEFINITION

A **fair game** has a value of 0 and, consequently, it favors neither player when at least one player uses an *optimal* (mixed) strategy—one that guarantees that the resulting expected payoff is the best that this player can obtain against all possible strategy choices (pure or mixed) by an opponent.

Player II gains nothing by knowing that Player I is using the optimal mixed strategy $(\frac{1}{2}, \frac{1}{2})$. However, Player I must not reveal to Player II whether H or T will be displayed *in any given play* of the game before Player II makes his or her own choice of H or T. Even without this information, if Player II knew that Player I was using a particular *nonoptimal* mixed strategy $(p_1, p_2) = (p, 1 - p)$, where $p \neq \frac{1}{2}$ (that is, not choosing a 50-50 mixture between H and T), then Player II could take advantage of this knowledge and increase his or her average winnings over time to something greater than the value of 0. (See Exercise 14 on p. 565)

EXAMPLE 5
Nonsymmetrical Matching

In this game, Players I and II can again show either heads H or tails T. When two H's appear, Player II pays \$5 to player I. When two T's appear, Player II pays \$1 to Player I. When one H and one T are displayed, then Player II collects \$3 from Player I. Note that although the sum of Player I's gains (\$5 + \$1 = \$6) when there are two H's or two T's, and the sum of Player II's gains (\$3 + \$3 = \$6) otherwise, are the same, the game is **nonsymmetrical**.

Nonsymmetrical Game DEFINITION

A two-person total-conflict **nonsymmetrical game** is one in which the row player's gains are different from the column player's gains.

Note that the row player's gains (\$5 and \$1 in our example) are different from the column player's gains (always \$3). In the original matching pennies, on the other hand, the payoff for winning is the same for each player, so that game is *symmetrical*.

The game just described is given by the payoff matrix in Table 15.7, which shows the payoffs that Player I receives from Player II. A worst-case analysis, like that which solved our initial location game, is of little help here. Player I may lose \$3 whether he plays H or T, making his maximin -3. Player II can keep her losses down to \$1 by always playing T (and thus avoiding the loss of \$5 when two H's appear), so Player II's minimax is 1. However, if Player II chooses T and Player I knows this, then Player I will also play T and collect \$1 from Player II. Can Player II do better than lose \$1 in each play of the game?

Table 15.7

Payoffs for Player I in a Nonsymmetrical Matching Game

		Player II	
		H	T
Player I	H	5	−3
	T	−3	1

Consider the situation where Player I uses a mixed strategy $(p, 1 - p)$, which involves playing H with probability p and playing T with probability $1 - p$, where $0 \leq p \leq 1$. Against Player II's pure strategy H, Player I's expected value is

$$E_H = (5)(p) + (-3)(1 - p) = 8p - 3$$

Against Player II's pure strategy T, Player I's expected value is

$$E_T = (-3)(p) + (1 - p) = -4p + 1$$

These two linear equations in the variable p are depicted in Figure 15.3. Note that the four points where these two lines intersect the two vertical lines, $p = 0$ and $p = 1$, are the four payoffs appearing in the payoff matrix.

Algebra Review

Graphing lines in slope-intercept form, page 216.

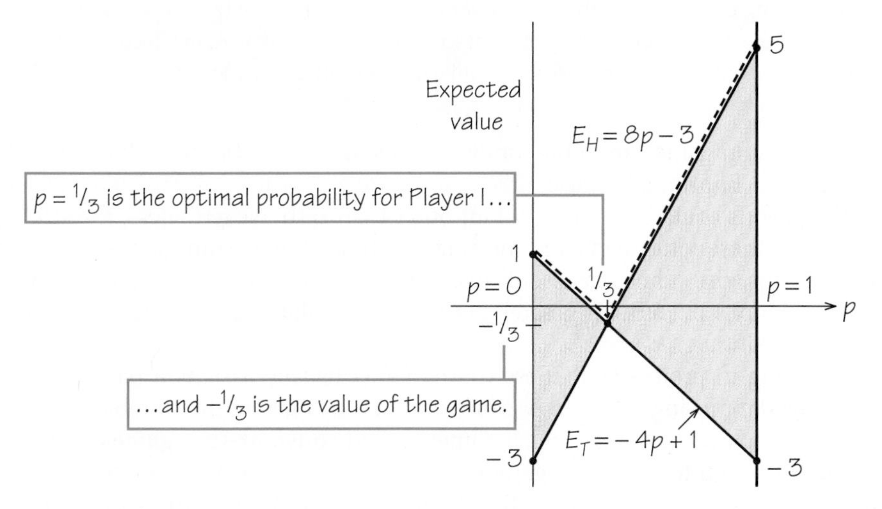

Figure 15.3 Solution to the nonsymmetrical matching pennies game.

The point at which the lines given by E_H and E_T intersect can be found by setting $E_H = E_T$, yielding

$$8p - 3 = -4p + 1$$
$$12p = 4$$

so $p = \frac{1}{3}$. To the left of $p = \frac{1}{3}$, $E_T > E_H$, and to the right, $E_H > E_T$; at $p = \frac{1}{3}$, $E_H = E_T$. If Player I chooses $(p_H, p_T) = (p, 1 - p) = \left(\frac{1}{3}, \frac{2}{3}\right)$, he can ensure

$$E_H = 8\left(\frac{1}{3}\right) - 3 = E_T = -4\left(\frac{1}{3}\right) + 1 = -\frac{1}{3}$$

regardless of what Player II does.

Algebra Review

Solving Linear Equations, page 457.

In other words, Player I's optimal mixed strategy is to pick H and T with probabilities $\frac{1}{3}$ and $\frac{2}{3}$, respectively, which gives Player I an expected value of $-\frac{1}{3}$ and prevents him from having a bigger average loss. As can be seen from Figure 15.3, $-\frac{1}{3}$ is the highest expected value that Player I can guarantee against *both* strategies H and T of Player II. Although T yields Player I a higher expected value for $p < \frac{1}{3}$, and H yields him a higher expected value for $p > \frac{1}{3}$, Player I's choice of $(p_H, p_T) = \left(\frac{1}{3}, \frac{2}{3}\right)$ protects him against an expected loss greater than $-\frac{1}{3}$, which neither of his pure strategies does (each may produce a maximum loss of -3). Put another way, the intersection of E_H and E_T at $p = \frac{1}{3}$ is the minimum of the function given by E_T to the left and E_H to the right (shown by the dashed line in Figure 15.3). If Player II had more than two strategies, this approach to finding a minimum that puts a floor on player I's expected loss can be extended.

A similar calculation for Player II results in the same optimal mixed strategy $\left(\frac{1}{3}, \frac{2}{3}\right)$ and expected value $-\frac{1}{3}$. But because the payoffs for Player II are losses, $-\frac{1}{3}$ means that she gains $\frac{1}{3}$ on the average. It is a coincidence that Player I and II's optimal mixed strategies are identical; in the baseball duel that we will return to in Example 6, this is not the case.

Therefore, this game is unfair, even though the sum of the amounts ($6) that Player I might have to pay Player II when he loses is the same as the sum that Player II might have to pay Player I when she loses. Interestingly, Player II, who will win an average of $33\frac{1}{3}$ cents each time the game is played, is favored, even though she may have to pay more to Player I when she loses (a maximum of $5) than Player I will ever have to pay her (a maximum of $3).

The symmetrical and nonsymmetrical matching games are examples of what are called **zero-sum games**.

Zero-Sum Game DEFINITION

A **zero-sum game** is one in which the payoff to one player is the negative of the corresponding payoff to the other, so the sum of the payoffs to the two players is always zero. These games can be completely described by a payoff matrix, in which the numbers represent the payoffs to Player I, while their negatives are the payoffs to Player II.

Zero-sum games are total-conflict games in which what one player wins the other loses. But not all total-conflict games are zero-sum—in particular, the sum of the payoffs could be some constant other than zero. Nevertheless, the strategic nature of these latter games is the same as that of zero-sum games: What one player wins, the other player still loses. This was true in our location game, in which Henry's payoff was greater the higher the altitude, and Lisa's greater the lower the altitude.

Scoring in professional chess tournaments usually assigns a payoff of 1 for winning, 0 for losing, and $\frac{1}{2}$ to each player for a tie, making the sum of the payoffs to the two players always 1. Such games, called **constant-sum games**, can readily be converted to zero-sum games. Thus, chess could as well be scored -1 for a loss, $+1$ for a win, and 0 for a tie, making the constant 0 in this case. Although constant-sum and zero-sum games have the same strategic nature, constant-sum games are a more general class because the constant need not be zero.

The solution in the symmetrical version of matching pennies illustrated how the mixed strategy of $\left(\frac{1}{2}, \frac{1}{2}\right)$ guarantees each player the value of 0, but we did not give a *solution technique* for finding optimal mixed strategies. In the nonsymmetrical version of matching pennies, we illustrated a procedure that can be applied to *every* payoff matrix in which each player has only two strategies.

We must use more complex methods, which we will not describe here, to find mixed-strategy solutions when one or both players have more than two strategies. However, one should always check first to see whether a game has a saddlepoint before employing any technique for finding optimal mixed strategies.

In our next example, which is the earlier duel between the pitcher and the batter given by the 2×2 payoff matrix in Table 15.5 on page 537, there is no saddlepoint, as we already showed. Thus, the solution will necessarily be in mixed strategies. We now proceed to find what mix is optimal.

EXAMPLE 6
The Duel Game Revisited

In Table 15.8, we add probabilities, which we explain next, to Table 15.5, where F indicates fastball and C indicates curveball. The pitcher should use a mixed strategy $(p_1, p_2) = (p_F, p_C) = (p, 1 - p)$. The probabilities p and $1 - p$ (where $0 \leq p \leq 1$) are indicated below the game matrix and under the corresponding strategies, F and C, for the pitcher. If the pitcher plays a mixed strategy $(p, 1 - p)$ against the two pure strategies, F and C, for the batter, he realizes the respective expected values:

$$E_F = (0.3)p + 0.2(1 - p) = 0.1p + 0.2$$
$$E_C = (0.1)p + 0.5(1 - p) = -0.4p + 0.5$$

Table 15.8

A Baseball Duel with Probabilities

		Pitcher		
		F	C	
Batter	F	0.300	0.200	q
	C	0.100	0.500	$1 - q$
		p	$1 - p$	

As in the nonsymmetrical matching-pennies game, the solution to this game occurs at the intersection of the two lines given by E_F and E_C. Setting the equations of these lines equal to each other yields $p = 0.6$, giving $E_F = E_C = E = 0.260$.

Thus, the pitcher should use his optimal mixed strategy, which selects F with probability $p = \frac{3}{5}$ and C with probability $1 - p = \frac{2}{5}$. This choice will hold the batter down to a batting average of 0.260, which is the value of the game. We stress that 0.260 is an average and must be interpreted in a statistical manner. It says that about one time in four the batter will get a hit, but does not say what will happen on any particular time at bat.

Assume that the batter uses a mixed strategy $(q_1, q_2) = (q_F, q_C) = (q, 1 - q)$, as indicated to the right of the game matrix in Table 15.8. This mixed strategy, when played against the pitcher's pure strategies, F and C, results in the following expected values:

$$E_F = (0.3)q + 0.1(1 - q) = 0.2q + 0.1$$
$$E_C = (0.2)q + 0.5(1 - q) = -0.3q + 0.5$$

The intersection of these two lines occurs at the point $q = 0.8$, giving $E_F = E_C = E = 0.260$. The batter's optimal mixed strategy is therefore $(q_F, q_C) = \left(\frac{4}{5}, \frac{1}{5}\right)$, which gives him the same batting average of 0.260.

We have seen that the outcome of 0.260, which is the value of the game, occurs when either the pitcher selects his optimal mixed pitching strategy $\left(\frac{3}{5}, \frac{2}{5}\right)$ or the batter selects his optimal mixed guessing strategy $\left(\frac{4}{5}, \frac{1}{5}\right)$. This particular result holds true for every two-person zero-sum game; it is the fundamental theorem for such games and is known as the **minimax theorem**.

Minimax Theorem DEFINITION

The **minimax theorem** guarantees that there is a unique game value and an optimal strategy for each player, so that either player alone can realize at least this value by playing this strategy, which may be pure or mixed.

The unique value in our example is 0.260.

15.3 Partial-Conflict Games

The 2×2 matrix games (two players, each with two strategies) presented so far have been total-conflict games: One player's gain was equal to the other player's loss. Although most parlor games, like chess or poker, are games of total conflict, and therefore constant-sum, most real-life games are surely not. (Elections, in which there are usually a clear-cut winner and one or more losers, probably come as close to being games of total conflict as we find in the real world.) We will consider two games of partial conflict, in which the players' preferences are not diametrically opposed, that have often been used to model real-world conflicts.

Partial-Conflict Games DEFINITION

A **partial-conflict game** is one in which both players can benefit by cooperation, but they may have strong incentives not to cooperate.

Variable-Sum Games DEFINITION

Games of partial conflict are **variable-sum games**, in which the sum of payoffs to the players at the different outcomes varies.

There is some mutual gain to be realized by both players if they can cooperate in partial-conflict games, but this may be difficult to do in the absence of either good communication or trust. When these elements are lacking, players are less likely to comply with any agreement that is made. *Noncooperative games* are games in which a binding agreement cannot be enforced. Even if communication is allowed in such games, there is no assurance that a player can trust an opponent to choose a particular strategy that he or she promises to select.

In fact, the players' self-interests may lead them to make strategy choices that yield both lower payoffs than they could have achieved by cooperating. Two partial-conflict games illustrate this problem.

EXAMPLE 7
Prisoners' Dilemma

Prisoners' Dilemma is a two-person variable-sum game. It provides a simple explanation of the forces at work behind arms races, price wars, and the population problem. In these and other similar situations, the players can do better by cooperating. But there may be no compelling reasons for them to do so unless the players have credible threats of retaliation for not cooperating. The name *Prisoners' Dilemma* was first given to this game by Princeton mathematician Albert W. Tucker (1905–1994) in 1950.

Before defining the formal game, we introduce it through a story.

Prisoners' Dilemma STORY

The **Prisoners' Dilemma** involves two persons, accused of a crime, who are held incommunicado. Each has two choices: to maintain his or her innocence or to sign a confession accusing the partner of committing the crime. It is in each suspect's interest to confess and implicate the partner, thereby trying to receive a reduced sentence. Yet if both suspects confess, they ensure a bad outcome—namely, they are both found guilty. What is good for the prisoners as a pair—to deny having committed the crime, leaving the state with insufficient evidence to convict them—is frustrated by their pursuit of their own individual interests.

The game of Prisoners' Dilemma, as we already noted, has many applications, but we will use it here to model a recurrent problem in international relations: arms races between antagonistic countries, which earlier included the superpowers but more recently have included such countries as India and Pakistan and Israel and some of its Arab neighbors. Other countries, such as Iran, may be antagonistic to more than one other country (Israel and the United States).

For simplicity, assume there are two nations, Red and Blue. Each can independently select one of two policies:

A: Arm in preparation for a possible war (noncooperation).
D: Disarm, or at least try to negotiate an arms-control agreement (cooperation).

There are four possible outcomes:

(*D*, *D*): Red and Blue disarm, which is *next best* for both because, while advantageous to each, it also entails certain risks.
(*A*, *A*): Red and Blue arm, which is *next worst* for both because they spend needlessly on arms and are comparatively no better off than at (*D*, *D*).
(*A*, *D*): Red arms and Blue disarms, which is *best for Red* and *worst for Blue*, because Red gains a big edge over Blue.
(*D*, *A*): Red disarms and Blue arms, which is *worst for Red* and *best for Blue*, because Blue gains a big edge over Red.

This situation can be modeled by means of the matrix in Table 15.9, which gives the possible outcomes that can occur. Here, Red's choice involves picking one of the two rows, whereas Blue's choice involves picking one of the two columns.

Table 15.9

The Outcomes in an Arms Race, as Modeled by the Prisoners' Dilemma

		Blue	
		A	D
Red	A	Arms race	Favors Red
	D	Favors Blue	Disarmament

We assume that the players can rank the four outcomes from best to worst, where 4 = best, 3 = next best, 2 = next worst, and 1 = worst. Thus, the higher the number, the greater the payoff, making the resulting game an **ordinal game**: It indicates an ordering of outcomes from best to worst but says nothing about the *degree* to which a player prefers one outcome over another. To illustrate, if a player despises the outcome that he or she ranks 1 but sees little difference among the outcomes ranked 4, 3, and 2, the "payoff distance" between 4 and 2 will be less than that between 2 and 1, even though the numerical difference between 4 and 2 is greater.

The ordinal payoffs to the players for choosing their strategies of *A* and *D* are shown in Table 15.10, where the first number in the pair indicates the payoff to the row player (Red), and the second number the payoff to the column player (Blue). Thus, for example, the pair (1, 4) in the second row and first column signifies a payoff of 1 (worst outcome) to Red and a payoff of 4 (best outcome) to Blue. This outcome occurs when Red unilaterally disarms while Blue continues to arm, making Blue, in a sense, the winner and Red the loser.

Table 15.10

Ordinal Payoffs in an Arms Race, as Modeled by the Prisoners' Dilemma

		Blue	
		A	D
Red	A	(2, 2)	(4, 1)
	D	(1, 4)	(3, 3)

Let's examine this strategic situation more closely. Should Red select strategy *A* or *D*? There are two cases to consider, which depend on what Blue does:

- If Blue selects *A*: Red will receive a payoff of 2 for *A* and 1 for *D*, so it will choose *A*.
- If Blue selects *D*: Red will receive a payoff of 4 for *A* and 3 for *D*, so it will choose *A*.

In both cases, Red's first strategy (*A*) gives it a more desirable outcome than its second strategy (*D*). Consequently, we say that *A* is Red's **dominant strategy** because it is always advantageous for Red to choose *A* over *D*.

In the Prisoners' Dilemma, *A dominates D* for Red, so we presume that a rational Red would choose *A*. A similar argument leads Blue to choose *A* as well—that is, to pursue a policy of arming. Thus, when each nation strives to maximize its own payoffs independently, the pair is driven to the outcome (*A*, *A*), with payoffs of (2, 2). The better outcome for both, (*D*, *D*), with payoffs of (3, 3), appears unobtainable when this game is played noncooperatively.

The outcome (*A*, *A*), which is the product of dominant strategy choices by both players in the Prisoners' Dilemma, is a **Nash equilibrium**.

Nash Equilibrium DEFINITION

When no player can benefit by departing unilaterally (by itself) from its strategy associated with an outcome, the strategies of the players constitute a **Nash equilibrium**. Technically, while it is the set of strategies that define the equilibrium, the choice of these strategies leads to an outcome that we shall also refer to as the *equilibrium*.

Note that in the Prisoners' Dilemma, if either player departs from (*A*, *A*), the payoff for the departing player who switches to *D* drops from 2 to 1 at (*D*, *A*) and (*A*, *D*). Not only is there no benefit from departing, but there is actually a loss, with the *D* player punished with its worst payoff of 1. These losses would presumably deter each nation from moving away from the Nash equilibrium of (*A*, *A*), assuming the other nation sticks to *A*.

Even if both nations agreed in advance jointly to pursue the socially beneficial outcome, (*D*, *D*), (3, 3) is unstable. This is because if either nation alone reneges on the agreement and secretly arms (as North Korea did when it developed nuclear weapons), it will benefit, obtaining its best payoff of 4. Consequently, each nation would be tempted to go back on its word and select *A*. Especially if nations have no great confidence in the trustworthiness of their opponents, they would have good reason to try to protect themselves against the other side's defection from an agreement by arming.

Prisoners' Dilemma

DEFINITION

The **Prisoners' Dilemma** is a two-person variable-sum game in which each player has two strategies, cooperate or defect (not cooperate). Defect dominates cooperate for both players, even though the mutual-defection outcome, which is the unique Nash equilibrium in the game, is worse for both players than the mutual-cooperation outcome.

Note that if 4, 3, 2, and 1 in the Prisoners' Dilemma were not just ranks but numerical payoffs, their sum would be $2 + 2 = 4$ at the mutual-defection outcome and $3 + 3 = 6$ at the mutual-cooperation outcome. At the other two outcomes, the sum, $1 + 4 = 5$, is still different, illustrating why the Prisoners' Dilemma is a variable-sum game.

In real life, of course, people often manage to escape the noncooperative Nash equilibrium in the Prisoners' Dilemma. Either the game is played within a larger context, wherein other incentives are at work, such as cultural norms that prescribe cooperation (though this is just another way of saying that defection from (*D, D*) is not rational, rendering the game not the Prisoners' Dilemma), or the game is played on a repeated basis—it is not a one-short affair—so players can induce cooperation by setting a pattern of rewards for cooperation and penalties for noncooperation.

In a repeated game, factors like reputation and trust may play a role. Realizing the mutual advantages of cooperation in costly arms races, players may inch toward the cooperative outcome by slowly phasing down their acquisition of weapons over time, or even destroying them. (The United States and Russia have been doing exactly this.) They may also initiate other productive measures, such as improving their communication channels, making inspection procedures more reliable, writing agreements that are truly enforceable, or imposing penalties for violators when their violations are detected (as has occurred through reconnaissance or spy satellites).

The Prisoners' Dilemma illustrates the intractable nature of certain competitive situations that blend conflict and cooperation. The standoff that results at the Nash equilibrium of (2, 2) is obviously not as good for the players as that which they could achieve by cooperating—but they risk a good deal if the other player defects.

While saddlepoints are Nash equilibria in total-conflict games, they can never be worse for *both* players than some other outcome (as in partial-conflict games like the Prisoners' Dilemma). The reason is that if one player does worse in a total-conflict or zero-sum game, the other player must do better.

The fact that the players must forsake their dominant strategies to achieve the (3, 3) cooperative outcome (see Table 15.10) makes this outcome a difficult one to sustain in one-shot play. On the other hand, assume that the players can threaten each other with a policy of tit-for-tat in repeated play: "I'll cooperate on each round unless you defect, in which case I will defect until you start cooperating again." If these threats are credible, the players may well shun their defect strategies and try to establish a pattern of cooperation in early rounds, thereby fostering the choice of (3, 3) in the future. Alternatively, they may look ahead, in a manner that will be described at the end of this chapter, to try to stabilize (3, 3).

EXAMPLE 8
Chicken

Let's look at one other two-person game of partial conflict, known as **Chicken**, that can also lead to troublesome outcomes. Two drivers approach each other at high speed. Each must decide at the last minute whether to swerve to the right or not swerve. Here are the possible consequences of their actions:

1. Neither driver swerves, and the cars collide head-on, which is the worst outcome for both because they are killed (payoff of 1).
2. Both drivers swerve—and each is mildly disgraced for "chickening out"—but they do survive, which is the next-best outcome for both (payoff of 3).
3. One of the drivers swerves and badly loses face, which is his next-worst outcome (payoff of 2), whereas the other does not swerve and is perceived as the winner, which is her best outcome (payoff of 4).

These outcomes and their associated strategies are summarized in Table 15.11.

Table 15.11

Payoffs in a Driver Confrontation, as Modeled by Chicken

		Driver 2	
		Swerve	Not Swerve
Driver 1	Swerve	(3, 3)	(2, 4)
	Not Swerve	(4, 2)	(1, 1)

If both drivers persist in their attempts to "win" with a payoff of 4 by not swerving, the resulting outcome will be mutual disaster, giving each driver his or her worst payoff of 1. Clearly, it is better for both drivers to back down and each obtain 3 by swerving, but neither wants to be in the position of being intimidated into swerving (payoff of 2) when the other does not (payoff of 4).

Notice that neither player in Chicken has a dominant strategy. His or her better strategy depends on what the other player does: Swerve if the other does not, don't swerve if the other player swerves, making this game's choices highly interdependent, which is characteristic of many games. The Nash equilibria in Chicken, moreover, are (4, 2) and (2, 4), suggesting that the compromise of (3, 3) will not be easy to achieve because both players will have an incentive to deviate in order to try to be the winner.

Chicken DEFINITION

Chicken is a two-person variable-sum game in which each player has two strategies: to swerve to avoid a collision or not to swerve and possibly cause a collision. Neither player has a dominant strategy. The compromise outcome, in which both players swerve, and the disaster outcome, in which both players do not, are not Nash equilibria. The other two outcomes, in which one player swerves and the other does not, are Nash equilibria.

In fact, there is a third Nash equilibrium in Chicken, but it is in mixed strategies, which can be computed only if the payoffs are not ranks, as we have assumed here, but numerical values. Even if the payoffs were numerical, however, it can be shown that this equilibrium is always worse for both players than the cooperative (3, 3) outcome. Moreover, it is implausible that players would sometimes swerve and sometimes not—randomizing according to particular probabilities—in the actual play of this game, compared with either trying to win outright or reaching a compromise.

The two pure-strategy Nash equilibria in Chicken suggest that, insofar as there is a "solution" to this game, it is that one player will succeed when the other caves in to avoid the mutual-disaster outcome. But there are certainly real-life cases in which a major confrontation was defused and a compromise of sorts was achieved in a Chicken-type game. This fact suggests that the one-sided solutions given by the two pure-strategy Nash equilibria may not be the only pure-strategy solutions, especially if the players are farsighted and think about the possible untoward consequences of their actions.

International crises, labor–management disputes, and other conflicts in which escalating demands may end in wars, strikes, and other catastrophic outcomes have been modeled by the game of Chicken (see Spotlight 15.2 for more on game theorists who have analyzed these and other games). But it can be shown that Chicken, like the Prisoners' Dilemma, is only one of the 78 essentially different 2×2 ordinal games in which each player can rank the four possible outcomes from best to worst.

Chicken and the Prisoners' Dilemma, however, are especially disturbing because the cooperative (3, 3) outcome in each is not a Nash equilibrium. Unlike a constant-sum game, in which the losses of one player are offset by the gains of the other, *both* players can end up doing badly—at (2, 2) in the Prisoners' Dilemma and (1, 1) in Chicken—in these variable-sum games.

15.4 Larger Games

We have shown how to compute optimal pure and mixed strategies, and the values ensured by using them, in 2×2 constant-sum games. In 2×2 variable-sum games, we focused on Nash equilibria as a solution concept in the Prisoners' Dilemma and Chicken, but we found that this notion of a stable outcome did not justify the choice of cooperative strategies in either of these games.

We turn next to a somewhat larger game, in which there are three players, each of whom can choose among three strategies, which is technically a $3 \times 3 \times 3$ game. In this game, we eliminate certain undesirable strategies, but in stages, to arrive at a Nash equilibrium that seems quite plausible.

If one of the three players has a dominant strategy in the $3 \times 3 \times 3$ game, we suppose that this player will choose it, thereby reducing the game to a 3×3 game between the other two players. Of course, if no player has a dominant strategy in a three-person game, it cannot be reduced in this manner to a two-person game.

If this game is not one of total conflict, the minimax theorem, which guarantees players the value in a two-person zero-sum game, is not applicable. Even if the game were zero-sum, the fact that we assume the players in the $3 \times 3 \times 3$ game can only rank outcomes, not assign numerical values to them, means that they cannot calculate optimal mixed strategies in it.

SPOTLIGHT 15.2

The Nobel Prize in Economics

The Nobel Memorial Prize in Economics was awarded to three game theorists in 1994, marking the 50th anniversary of the publication of von Neumann and Morgenstern's *Theory of Games and Economic Behavior* (see Spotlight 15.1).

- *John C. Harsanyi* (1920–2000) of the University of California, Berkeley, a Hungarian-American who emigrated from Hungary to Australia in 1950 and then to the United States in 1956. He is well known for extending game theory to the study of ethics and showing how societal institutions, each of whose members' satisfaction can be measured against that of others, choose among alternatives. His other major contribution was to give a precise definition to "incomplete information" in games in which players may be thought of as different types, and probabilities are assigned to each type.
- *John F. Nash, Jr.* (b. 1928) of Princeton University, an American mathematician who did path-breaking work on both noncooperative game theory (the Nash equilibrium is named after him) and cooperative game theory, especially on bargaining, in which axioms or assumptions are specified and a unique solution that satisfies these axioms is derived. Nash obtained his results in the early 1950s, when he was only in his 20s, after which he became mentally ill and was unable to work. Fortunately, he made a remarkable recovery and has now resumed research.
- *Reinhard Selten* (b. 1930) of the University of Bonn, a German mathematician who proposed significant refinements in the concept of the Nash equilibrium that help to distinguish those that are most plausible in games (often there are many such equilibria, which creates a selection problem). Selten is also noted for pioneering work on developing game-theoretic models in evolutionary biology.

In 2005 mathematician Robert J. Aumann was awarded the prize for a variety of advances in cooperative and noncooperative game theory, and economist Thomas C. Schelling also was awarded the prize for his contributions to the study of conflicts involving promises, threats, and other kinds of commitments. In 2007 three economists, Leonid Hurwicz, Eric S. Maskin, and Roger B. Myerson received the prize for their work on "mechanism design," which analyzes auction, bargaining, voting and other procedures, especially the incentives of players using them to be truthful in their choices.

John C. Harsanyi
(Olivier Laude/Gamma-Liaison.)

John F. Nash, Jr.
(Reuters/Bettmann.)

Reinhard Selten
(Bettmann/Corbis.)

The problem in finding a solution to the reduced 3 × 3 game is not a lack of Nash equilibria. Rather, there are too many! So the question becomes which, if any, are likely to be selected by the players. Specifically, is one more appealing than the others? The answer is "yes," but it requires extending the idea of dominance, discussed in the previous section, to its successive application in different stages of play.

EXAMPLE 9
The Status-Quo Paradox

The $3 \times 3 \times 3$ game that we analyze involves voting, illustrating the applicability of game theory to politics. We will show that the status quo, or existing state of affairs, may be defeated by another alternative, despite its privileged position as the policy in place at the time.

To illustrate this problem, suppose there is a set of three voters, $V = \{X, Y, Z\}$, and a set of three alternatives, $A = \{x, y, z\}$, from which the voters choose. Assume that Voter X prefers x to y to z, indicated by *xyz;* Voter Y's preference is *yzx*; and Voter Z's is *zxy*. These preferences give rise to a *Condorcet voting paradox* (discussed in Chapter 9), because the social ordering, according to majority rule, is *intransitive:* Although a majority (Voters X and Z) prefer x to y, and a majority (Voters X and Y) prefer y to z, a majority (Voters Y and Z) prefer z to x. So there is no **Condorcet winner**—an alternative that would beat all others in separate pairwise contests. Instead, every alternative can be beaten by one other.

Assume that the voting procedure used by the three voters, who choose from among the three alternatives, is the **plurality procedure**, under which the alternative with the most votes wins. If there is a three-way tie (there can never be a two-way tie if there are three voters), we assume that x wins—because it is the status quo—giving X what would appear to be an edge over the other two voters, Y and Z.

To begin, assume that voting is **sincere**.

Sincere Voting DEFINITION

Under **sincere voting**, every voter votes for his or her most-preferred alternative, based on his or her true preferences, without taking into account what the other voters might do (see Chapter 9 for more).

In this case, x will prevail, because there is a three-way tie.

But note that X has a dominant strategy of "vote for x": It is never worse and sometimes better than her other two strategies, whatever the other two voters do. Thus, if the other two voters vote for the same alternative, it wins, and X cannot do better than vote sincerely for x, so voting sincerely is never worse. On the other hand, if the other two voters disagree, X's tie-breaking vote (along with her regular vote) for x will be decisive in x's selection, which is X's best outcome.

Given the dominant choice of x on the part of X, Y and Z face the strategy choices shown in Figure 15.4. Y has one, and Z has two, **dominated strategies**, which are never better and sometimes worse than some other strategy, whatever the other two voters do. For example, observe that "vote for x" by Y always leads to his worst alternative, x, so this strategy is dominated and therefore can be crossed out.

This leaves Y with two *undominated* strategies that are neither dominant nor dominated: "vote for y" and "vote for z." "Vote for y" is better than "vote for z" if Z chooses y (leading to y rather than x), whereas the reverse is the case if Z chooses z (leading to z rather than x). By contrast, Z has a dominant strategy of "vote for z," which leads to outcomes at least as good and sometimes better than his other two strategies, which are therefore dominated and so can be crossed out.

If voters have complete information about each other's preferences, then they can perceive the situation represented by the top matrix of Figure 15.4. Reasoning that no player would ever choose a dominated strategy, they would eliminate such strategies from consideration (these have been crossed out in the first reduction).

Figure 15.4
Sophisticated voting, given *X* chooses "vote for *x*." The dominated strategies of each voter are crossed out in the first reduction, leaving two (undominated) strategies for *Y* and one (dominant) strategy for *Z*. Given these eliminations, *Y* would then eliminate "vote for *y*" in the second reduction, making *z* the sophisticated outcome.

FIRST REDUCTION

		Z (*zxy*)		
		Vote for *x*	Vote for *y*	Vote for *z*
	Vote for *x*	*x*	*x*	*x*
Y (*yzx*)	Vote for *y*	*x*	*y*	*x*
	Vote for *z*	*x*	*x*	*z*

SECOND REDUCTION

		Z (*zxy*)
		Vote for *z*
Y (*yzx*)	Vote for *y*	*x*
	Vote for *z*	*z*

The elimination of these strategies gives the bottom matrix in Figure 15.4. Then *Y*, choosing between "vote for *y*" and "vote for *z*" in this matrix, would cross out "vote for *y*" (second reduction), now dominated because that choice would result in *x*'s winning because it is the status quo. Instead, *Y* would choose "vote for *z*," which is not *Y*'s sincere strategy. This ensures *z*'s election, which is *Z*'s best outcome but only the next-best outcome for *Y*. In this manner, *z*, which is not the first choice of a majority and could in fact be beaten by *y* in a pairwise contest, becomes the sophisticated outcome.

Sophisticated Voting DEFINITION

The *successive* elimination of dominated strategies by voters (insofar as this is possible) is **sophisticated voting**.

Sophisticated voting results in a Nash equilibrium because none of the three players can do better by departing from his or her sophisticated strategy when the other two players choose theirs. This is clearly true for *X* because *x* is her dominant strategy; given *X*'s choice of *x*, *z* is dominant for *Z*; and given these choices by *X* and *Z*, *z* is dominant for *Y*. These "contingent" dominance relations, in general, make sophisticated strategies a Nash equilibrium.

Observe, however, that there are four other Nash equilibria in this game. First, the choice of each of *x*, *y*, or *z* by all three voters are all Nash equilibria because no single voter's departure can change the outcome to a different one, much less a better one, for that player. In addition, the choice of *x* by *X*, *y* by *Y*, and *x* by *Z*—resulting in *x*—is also a Nash equilibrium because no voter's departure would lead to a better outcome for him or her.

In game-theoretic terms, sophisticated voting produces a different and smaller game in which some formerly undominated strategies in the larger game become dominated in the smaller game. The removal of such strategies, sometimes in several successive stages, in effect enables sophisticated voters to determine what outcomes eventually *will* be chosen by eliminating those outcomes that definitely *will not* be chosen. Voters can thereby ensure that their worst outcomes will not be chosen by successively eliminating dominated strategies, given the presumption that other voters do likewise.

How does sophisticated voting affect the status quo? If *X* chooses her dominant strategy of voting for *x*, the status quo and her best outcome, it leads to *X*'s worst outcome (*z*)!

Status-Quo Paradox DEFINITION

This situation, in which supporting the apparently favored outcome hurts, is the **status-quo paradox**.

Clearly, the apparent advantage that the status quo enjoys in the absence of a Condorcet winner disappears if the voters are sophisticated. The strategic situation intervenes and causes them to reassess their sincere strategies in light of the favored position of the status quo. In so doing, they may be led to "gang up" against *X*, handing *X* her worst outcome of *z*.

We stress that *Y* and *Z* do not form a coalition against *X* in the sense of coordinating their strategies and agreeing to act together in a way that can be enforced. Rather, they behave as isolated individuals. At most they could be said to form an "implicit coalition." Such a coalition does not imply communication between its members but simply choices based on their common perceived strategic interests.

So far, we have used payoff matrices to describe games in strategic form. In these games, the row and column players' choices of strategies led to an outcome from which each player received a payoff. These strategy choices were assumed to be simultaneous, though mentally the players might eliminate some sequentially in the manner illustrated.

In the next example of a larger game, we start by assuming simultaneous choices and show what outcome would occur. Then we assume that the choices of the players need not be simultaneous; it is possible for one player to move first. We will use a **game tree** to analyze the *sequential choices* players can then make, as occurs when first you move, then I move, and so on, which are called *games in extensive form.* As we will see, the outcome in such a game may be wholly different from what it is in a game with simultaneous choices, which raises the question of which kind of game is the most realistic model of a situation.

EXAMPLE 10
A Truel

A *truel* is like a duel, except that there are three players. Truels are depicted in several movies, including *The Good, the Bad and the Ugly* (1966), *Reservoir Dogs* (1992), and *Pulp Fiction* (1994).

Each player can either fire, or not fire, his or her gun at either of the other two players. We assume the goal of each player is, *first,* to survive and, *second,* to survive with as few other players as possible. Each player has one bullet and is a perfect shot; no communication (for example, to pick out a common target) leading to a binding agreement with other players is allowed, making the game noncooperative. We will discuss the answers that simultaneous choices, on the one hand, and sequential choices, on the other, give to what is optimal for the players to do in the truel.

If choices are simultaneous, at the start of play, each player will fire at one of the other two players, killing that player.

Why will the players all fire at each other? Because their own survival does not depend an iota on what they do. Since they cannot affect what happens to themselves but can affect how many others survive (the fewer the better, according to the postulated secondary goal), they should all blaze away at each other. Even if the rules of the play permitted shooting oneself, the primary goal of survival would preclude committing suicide. In fact, the players all have dominant strategies to shoot at each other because whether or not a player survives—we will discuss shortly the probabilities of doing so—he or she does at least as well shooting an opponent.

The game, and optimal strategies in it, would change if the players (1) were allowed more options, such as to fire in the air and thereby disarm themselves, or (2) did not have to choose simultaneously but, instead, a particular order of play were specified. Thus, if the order of play were *A,* followed by *B* and *C* choosing simultaneously, followed by any player with a bullet remaining choosing, then *A* would fire in the air, and *B* and *C* would subsequently shoot each other. (*A* is no threat to *B* or *C,* so neither of the latter will fire at *A* and waste a bullet; on the other hand, if one of *B* or *C* did not fire immediately at the other, that player would not survive to get in the last shot, so both *B* and *C* will fire at each other.) Thus, *A* will be the sole survivor.

THE KOBAL COLLECTION/P.E.A

In 1992, a modified version of this scenario was played out in late-night television programming among the three major TV broadcasting networks of the time, with ABC's effectively going first with *Nightline,* its well-established news program, and CBS's and NBC's dueling about which host, David Letterman or Jay Leno, to choose for their entertainment shows. Regardless of their ultimate choices, ABC "won" when CBS and NBC were forced to divide the entertainment audience. In 2002, ABC, presumably to attract a younger audience than that which watched *Nightline,* attempted unsuccessfully to hire Letterman from CBS. Ted Koppel retired in 2005, but *Nightline* continues with other hosts.

More recently, after Jay Leno's program was moved to an earlier time slot on NBC in 2009, and did poorly, it was moved back to its original time slot, displacing Conan O'Brien, who had replaced Leno in his original slot. This led to O'Brien's resignation from NBC and his move to cable TV station TBS, which is a good example of a conflict within an organization.

To return to the original game (all choose simultaneously), the players' strategies of all firing have two possible consequences: Either one player survives (even if two players fire at the same person, the third must fire at one of them, leaving only one survivor), or no player survives (if each player fires at a different person). In either event, there is no guarantee of survival. In fact, if each player has an equal probability of firing at one of the two other players, the probability that any particular player will survive is only 0.25.

The reason is that if the three players are *A, B,* and *C, A* will be killed if *B* fires at him or her, *C* does, or both do. The only circumstance in which *A* will survive is if *B* and *C* fire at each other, which gives *A* one chance in four.

If choices are sequential, no player will fire at any other, so all will survive.

At the start of the truel, all the players are alive, which satisfies their primary goal of survival, though not their secondary goal of surviving with as few others as possible. Now assume that *A* contemplates shooting *B*, thereby reducing the number of survivors, and cannot fire into the air. Looking ahead, however, *A* knows that by firing first and killing *B*, he or she will be defenseless and be immediately shot by *C*, who will then be the sole survivor.

It is in *A*'s interest, therefore, not to shoot anybody at the start, and the same logic applies to each of the other players. Hence, everybody will survive, which is a happier outcome than when choices are simultaneous, in which case everyone's primary goal of survival is not satisfied—or, quantitatively speaking, satisfied only 25 percent of the time.

While sequential choices produce a "happier" outcome, do they provide a plausible model of a strategic situation that mimics what people might actually think and do in such a situation? We believe that the players in the truel, artificial as this kind of shootout may seem, would be motivated to think ahead, given the dire consequences of their actions. Therefore, they would hold their fire, knowing that if one fired first, he or she would be the next target.

In Figure 15.5, we show this logic somewhat more formally with a *game tree,* in which *A* has three strategies, as indicated by the three branches that sprout from *A:*

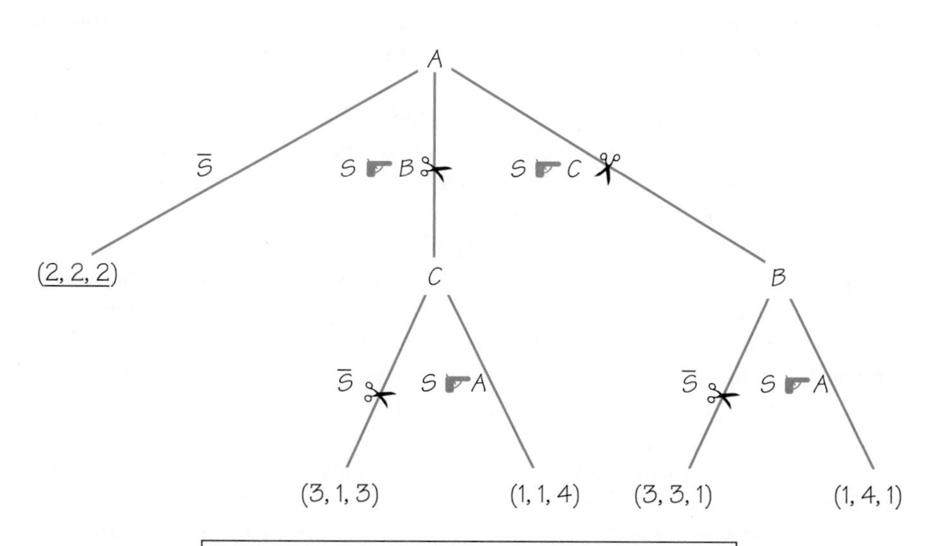

Figure 15.5 A game tree of a truel.

not shoot ($\overline{S}$), shoot *B*, or shoot *C*. The latter two branches, in turn, give survivors *C* and *B*, respectively, two strategies: not shoot ($\overline{S}$) or shoot *A*.

We assume that the players rank the outcomes as follows, which is consistent with their primary and secondary goals: 4 = best (lone survivor), 3 = next best (survivor with one other), 2 = next worst (survivor with two others), and 1 = worst (nonsurvivor). These payoffs are given for ordered triples (*A, B, C*); thus (3, 3, 1) indicates the next-best payoffs for *A* and *B* and the worst payoff for *C*.

Note that play necessarily terminates when there is only one survivor, as is the case at (1, 1, 4) and (1, 4, 1). To keep the tree simple, we assume that play also terminates when either *A* initially chooses $\overline{S}$, or *B* or *C* subsequently chooses $\overline{S}$, giving outcomes of (2, 2, 2), (3, 3, 1), and (3, 1, 3), respectively. Of course, we could allow the two or three surviving players in the latter cases to make subsequent choices in an extended game tree, but this example is meant only to illustrate the analysis of a game tree, not be the definitive statement on truel possibilities. (More will be explored in the exercises.)

In a game in extensive form represented by a game tree, players work backward, starting the analysis at the bottom of the tree. By "bottom" we mean where play terminates; because this is where the tree branches out, the tree looks upside down in Figure 15.5. The players then work up the tree, using backward induction. **Backward induction** is a reasoning process in which players, working backward from the last possible moves in a game, anticipate each other's rational choices.

To illustrate, because *C* prefers (1, 1, 4) to (3, 1, 3), we indicate that *C* would not choose $\overline{S}$ by "cutting" this branch with a scissors. Similarly, *B* would not choose $\overline{S}$. Thus, if play got to the bottom of the tree, *C* would shoot *A* if *C* were the survivor, and *B* would shoot *A* if *B* were the survivor, following *A*'s shooting *C* or *B*, respectively.

Moving up to the next level, *A* would know that if he or she shot *B*, (1, 1, 4) would be the outcome. If he or she shot *C*, (1, 4, 1) would be the outcome, making one or the other the outcome from the bottom level. Choosing between these two outcomes and (2, 2, 2), *A* would prefer the latter, so *A* would cut the two branches, *S* shoots *B* and *S* shoots *C*. Hence, *A* would choose $\overline{S}$, terminating play with nobody shooting anybody else.

This, of course, is the conclusion that we reached earlier, based on the reasoning that if *A* shot either *B* or *C*, he or she would end up dead, too. Because we could allow each player, like *A*, to choose among his or her three initial strategies in a $3 \times 3 \times 3$ game, and subsequently make moves and countermoves from the initial state (if feasible), the foregoing analysis applies to all players.

Underlying the completely different answers given by the simultaneous and sequential choices in a truel is a change in the rules of play. If play is sequential, the players do not have to fire simultaneously at the start, as assumed in a $3 \times 3 \times 3$ strategic-form game. Rather, a player who moves first (*A* in our example)—and then the later players—would not fire, given that play continues until all bullets are expended or nobody chooses to fire.

In the extensive-form analysis, we ask of each player (it need not be *A*): Given your present situation (all alive), and the situation you anticipate will ensue if you fire first, should you do so? Because each player prefers living to the state he or she would induce by being the first to shoot (certain death), no one shoots. This analysis suggests that truels might be more effective than duels, at least when played sequentially, in preventing the outbreak of conflict.

We will not try to develop this argument into a more general model. The main point is that a game tree allows for a look-ahead approach, whereby players compare the present state with possible future states—perhaps several steps ahead—to determine which moves to make. These choices, as we have seen, may lead to radically different outcomes compared with those based on simultaneous choices.

15.5 Using Game Theory

Solving Games

Given any payoff matrix, the first thing we ask is whether it is zero-sum (or constant-sum). If so, we check to see whether it has a saddlepoint by determining the minimum number of each row and the maximum number of each column, as we did in several earlier examples. If the maximum of the row minima (maximin) is equal to the minimum of the column maxima (minimax), then the game has a saddlepoint. The resulting value, and the corresponding pure strategies, provide a solution to the game.

This value will appear in the payoff matrix as the smallest number in its row and the largest in its column. In the 3×3 location game in Example 1, this number was 5 (5000 feet).

As in our voting game, dominated strategies can successively be eliminated in the 3×3 location game. Thus, Route *B* dominates Route *C*, and Highway 2 dominates Highway 3. Having made these eliminations, Highway 2 dominates Highway 1. Having made this elimination, Route *B* dominates Route *A*. Thus, Highway 2 and Route *B* survive the successive eliminations, yielding the saddlepoint of 5. The successive-elimination procedure therefore provides an alternative method for finding the saddlepoint in the 3×3 location game. Unfortunately, it does not work to find the saddlepoint in *all* two-person zero-sum games larger than 2×2.

Recall that instead of eliminating dominated strategies in the 3×3 location game, we eliminated Route *B* and Highway 2, which dominated other strategies, to obtain the 2×2 restricted-location game in Table 15.3. In this game, there were no dominated strategies and, hence, no saddlepoint.

If a two-person zero-sum game does not have a saddlepoint—which was the case not only in the restricted-location game but also for matching pennies, the nonsymmetrical matching game, and the baseball duel—the solution will be in mixed strategies. To find the optimal mix in a 2×2 game, we calculate the expected value to a player from choosing its first strategy with probability p and its second with probability $1 - p$, assuming that the other player chooses its first pure strategy (yielding one expected value) and its second pure strategy (yielding another expected value).

Setting these two expected values equal to each other yields a unique value for p that gives the optimal mix, $(p, 1 - p)$, with which the player should choose its first and second strategies. Inserting the numerical solution of p back into either expected-value equation gives the value of the game, which each player can guarantee for itself, whatever strategy its opponent chooses.

Several general algorithms have been developed to find mixed-strategy solutions to large constant-sum games. This work has mostly been done in the field of linear programming, using such algorithms as the simplex method of G. B. Dantzig and the more recent method of N. K. Karmarkar (see Chapter 4).

In variable-sum games, we also begin by successively eliminating dominated strategies, if there are any. The outcomes that remain do not depend on the numerical values that we attach to them but only on their ranking from best to worst by the players, as illustrated in the three-person voting example.

Care must be taken in interpreting this solution, however. It began with the choice of a dominant strategy by X—and her elimination of her two dominated strategies. Presuming these eliminations, Y and Z were then able to eliminate their own dominated strategies in the first reduction, and Y in turn eliminated a dominated strategy in the second reduction, leading finally to the outcome z, supported by Y and Z.

This solution is a fairly demanding one because it assumes considerable calculational abilities on the part of the players. Less demanding, of course, is that players simply choose their dominant strategies, as is possible in the Prisoners' Dilemma, but of course, games may not have such strategies.

In the game of Chicken, for example, neither player has a dominant (or dominated) strategy, so the game cannot be reduced. In such situations, we ascertain what outcomes are Nash equilibria. There are two (in pure strategies) in Chicken, suggesting that the only stable outcomes in this game occur when one player gives in and the other does not. In the Prisoners' Dilemma, by comparison, the choice by the players of their dominant strategies singles out the mutual-defection outcome as the unique Nash equilibrium, which is worse for both players than the cooperative outcome.

In both Chicken and the Prisoners' Dilemma, there seems no good reason for the choice of the (3, 3) cooperative outcome, at least if each game is played only once, because this outcome is not a Nash equilibrium. However, there is an alternative theory, called the **theory of moves (TOM)**, that assumes different rules of play and renders the cooperative outcomes in both the Prisoners' Dilemma and Chicken stable, given that the players think ahead.

Theory of Moves DEFINITION

Theory of moves (TOM) is a dynamic theory that describes optimal strategic choices in strategic-form games in which the players, thinking ahead, can make moves and countermoves.

TOM is *dynamic* in the sense that it allows players, after choosing strategies that lead to an outcome in a payoff matrix, to make subsequent alternating moves and countermoves, with rows being able to move vertically with changes in his row strategy, and columns being able to move horizontally with changes in her column strategy. The reasoning that the players use in deciding whether to move or not move is backward induction, as illustrated earlier in the truel.

We informally illustrate this reasoning in the Prisoners' Dilemma, starting from each of the four possible outcomes:

- If the play starts at the noncooperative (2, 2) outcome, players are stuck, no matter how far ahead they look, because as soon as one player departs, the other player, enjoying its best outcome at (4, 1) or (1, 4), will not move on. *Result:* The players stay at the noncooperative outcome.
- If play starts at the cooperative (3, 3) outcome, then neither player will defect because if he or she does, the other player also will defect, and both players will end up worse off at (2, 2). Thinking ahead, therefore, neither player will defect. *Result:* The players stay at the cooperative outcome.
- If play starts at one of the (4, 1) or (1, 4) win-lose outcomes, the player doing best (4) will know that if he or she does not move to the cooperative (3, 3) outcome, his or her opponent will move to the noncooperative (2, 2) outcome, inflicting on the best-off player a next-worst (2) outcome. Therefore,

it is in this player's interest—as well as the worst-off player's interest—that the best-off player act cooperatively and move first to (3, 3), anticipating that if he or she does not, the (2, 2) rather than the (3, 3) outcome will be chosen. *Result:* The best-off player will move to the cooperative outcome, where play will stop.

Thus, TOM does not predict unconditional cooperation in the Prisoners' Dilemma but, instead, makes it a function of the starting point of play. As in the truel, a change in rules from simultaneous choices to sequential choices can induce cooperation.

The calculations that we have described for the Prisoners' Dilemma, which are grounded in backward induction and could be formalized by a game tree, are not, we believe, beyond the ability of most players. Farsighted players *can* escape the dilemma in the Prisoners' Dilemma, provided play begins at a state other than the noncooperative one. But we must be careful in interpreting this result: With the change in the rules, the original game changes, so this "solution" to the Prisoners' Dilemma is not for the original dilemma. Similar reasoning in Chicken indicates that if play starts at the cooperative (3, 3) outcome, players will stay at this outcome, but the reasoning in this game is somewhat more complicated than in the Prisoners' Dilemma.

Practical Applications

The element of surprise, as captured by mixed strategies, is essential in many encounters. For example, mixed strategies are used in various inspection procedures and auditing schemes to deter potential violators. When inspection or auditing choices are made random, they are rendered unpredictable.

Police and regulatory agencies monitor certain activities to check for illegal actions. Investigators who conduct surveillance include FBI agents, customs agents, bank auditors, insurance investigators, quality-control experts, and drug testers. The National Bureau of Standards is responsible for monitoring the accuracy of measuring instruments and for maintaining reliable standards. The Nuclear Regulatory Agency demands an accounting of dangerous nuclear material as part of its safeguards program. The Internal Revenue Service attempts to identify those cheating on taxes.

Military or intelligence services may wish to intercept a weapon hidden among many decoys or to plant a secret agent disguised to look like a respectable individual. Because it is prohibitively expensive to check the authenticity of each and every possible item or person, efficient methods must be used to check for violations. Both optimal detection and optimal concealment strategies can be modeled as a game between an inspector trying to increase the probability of detection and a violator trying to evade detection. Since the World Trade Center attack on September 11, 2001, we have seen government agencies take much stronger measures to prevent such evasion.

Some inspection games are constant-sum: The violator "wins" when the evasion is successful and "loses" when it is not. On the other hand, cheating on arms-control agreements may well be variable-sum if both the inspector and the cheater would prefer that no cheating occur to there being cheating and public disclosure of it. The latter could be an embarrassment to both sides, especially if it undermines an arms-control agreement that both sides wanted and the cheating is minor.

We alluded earlier to the strategy of bluffing in poker, which is used to try to keep the other player or players guessing about the true nature of one's hand.

The optimal probability with which one should bluff can be calculated in a particular situation (see Exercise 17 on p. 565). Besides poker, bluffing is common in many bargaining situations, whereby a player raises the stakes (for instance, labor threatens a strike in labor–management negotiations), even if it may ultimately have to back down if its "hand is called."

Perhaps the greatest value of game theory is the framework it provides for understanding the rational underpinnings of conflict in the world today. As a case in point, a confrontation over the budget between the Democratic president Bill Clinton and the Republican Congress resulted in the shutdown of part of the federal government on two occasions between November 1995 and January 1996. Many government workers were frustrated in not being able to do their jobs, even though they knew they would be paid for not working, not to mention the many citizens who were either greatly hurt or substantially inconvenienced by the shutdown. Viewed as a game of Chicken, in which each side wanted not only to get its way for the moment but also to establish a precedent for the future, this conflict was not so foolish as it might seem at first glance.

(Najlah Feanny/Corbis.)

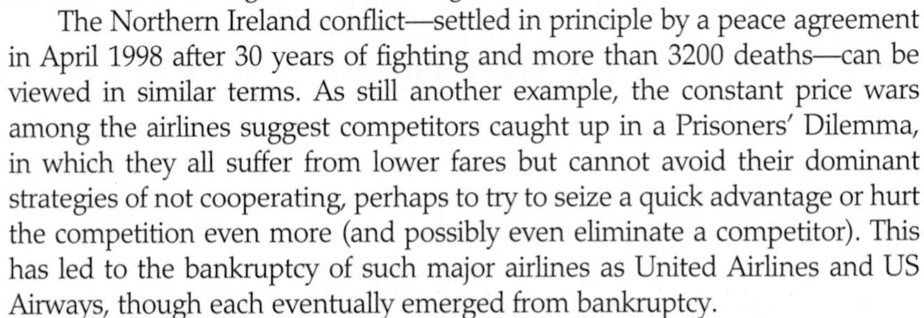
The Northern Ireland conflict—settled in principle by a peace agreement in April 1998 after 30 years of fighting and more than 3200 deaths—can be viewed in similar terms. As still another example, the constant price wars among the airlines suggest competitors caught up in a Prisoners' Dilemma, in which they all suffer from lower fares but cannot avoid their dominant strategies of not cooperating, perhaps to try to seize a quick advantage or hurt the competition even more (and possibly even eliminate a competitor). This has led to the bankruptcy of such major airlines as United Airlines and US Airways, though each eventually emerged from bankruptcy.

To be sure, if the airlines cooperate by colluding on fares, which is definitely not advantageous to consumers, the consumers may be thought of as a collective player whose interests are represented by the government. The government can prosecute the airlines for price fixing, or the consumers themselves can file a class-action suit in a "larger" game. The government has frequently been involved in antitrust suits (for example, against Microsoft) and in setting the rules for auctions of airwaves, in which telecommunication companies—advised by game theorists—have paid billions of dollars for the right to construct cellular phone and other networks.

All in all, game theory offers fundamental insights into conflicts at all levels, especially *seemingly* irrational features which, on second look, are often well conceived and effective.

REVIEW VOCABULARY

Backward induction A reasoning process in which players, working backward from the last possible moves in a game, anticipate each other's rational choices. (p. 556)

Chicken A two-person variable-sum symmetric game in which each player has two strategies: to swerve to avoid a collision, or not to swerve and cause a collision if the opponent has not swerved. Neither player has a dominant strategy; the compromise outcome, in which both players swerve, is not a Nash equilibrium, but the two outcomes in which one player swerves and the other does not are Nash equilibria. (p. 548)

Condorcet winner A candidate that defeats all others in separate pairwise contests. (p. 551)

Constant-sum game A game in which the sum of payoffs to the players at each outcome is a constant, which can be converted to a zero-sum game by an appropriate change in the payoffs to the players that does not alter the strategic nature of the game. (p. 542)

Dominant strategy A strategy that is sometimes better and never worse for a player than every other strategy, whatever strategies the other players choose. (p. 546)

Dominated strategy A strategy that is sometimes worse and never better for a player than some other strategy, whatever strategies the other players choose. (p. 551)

Expected value E If each of the n possible payoffs, $s_1, s_2, \ldots, s_n$, occurs with respective probabilities $p_1, p_2, \ldots, p_n$, then the expected value E is

$$E = p_1 s_1 + p_2 s_2 + \cdots + p_n s_n$$

where $p_1 + p_2 + \cdots + p_n = 1$ and $p_i \geq 0$ $(i = 1, 2, \cdots, n)$. (p. 538)

Fair game A zero-sum game is fair when the (expected) value of the game, obtained by using optimal strategies (pure or mixed), is zero. (p. 539)

Game tree A symbolic tree, based on the rules of play in a game, in which the vertices, or nodes, of the tree represent choice points, and the branches represent alternative courses of action that the players can select. (p. 553)

Maximin In a two-person zero-sum game, the largest of the minimum payoffs in each row of a payoff matrix. (p. 533)

Maximin strategy In a two-person zero-sum game, the pure strategy of the row player corresponding to the maximin in a payoff matrix. (p. 533)

Minimax In a two-person zero-sum game, the smallest of the maximum payoffs in each column of a payoff matrix. (p. 533)

Minimax strategy In a two-person zero-sum game, the pure strategy of the column player corresponding to the minimax in a payoff matrix. (p. 533)

Minimax theorem The fundamental theorem for two-person constant-sum games, stating that there always exist optimal pure or mixed strategies that enable the two players to guarantee the value of the game. (p. 543)

Mixed strategy A strategy that involves the random choice of pure strategies, according to particular probabilities. A mixed strategy of a player is optimal if it guarantees the value of the game. (p. 537)

Nash equilibrium Strategies associated with an outcome such that no player can benefit by choosing a different strategy, given that the other players do not depart from their strategies. (p. 546)

Nonsymmetrical game A two-person constant-sum game in which the row player's gains are different from the column player's gains, except when there is a tie. (p. 540)

Ordinal game A game in which the players rank the outcomes from best to worst. (p. 545)

Partial-conflict game A variable-sum game in which both players can benefit by cooperation but may have strong incentives not to cooperate. (p. 544)

Payoff matrix A rectangular array of numbers. In a two-person game, the rows and columns correspond to the strategies of the two players, and the numerical entries give the payoffs to the players when these strategies are selected. (p. 539)

Plurality procedure A voting procedure in which the alternative with the most votes wins. (p. 551)

Prisoners' Dilemma A two-person variable-sum symmetric game in which each player has two strategies, cooperate or defect. Cooperate dominates defect for both players, even though the mutual-defection outcome, which is the unique Nash equilibrium in the game, is worse for both players than the mutual-cooperation outcome. (p. 544)

Prisoners' Dilemma (Story) The Prisoners' Dilemma involves two persons, accused of a crime, who are held incommunicado. Each has two choices: to maintain his or her innocence or to sign a confession accusing the partner of committing the crime. It is in each suspect's interest to confess and implicate the partner, thereby trying to receive a reduced sentence. Yet if both suspects confess, they ensure a bad outcome—namely, they are both found guilty. What is good for the prisoners as a pair—to deny having committed the crime, leaving the state with insufficient evidence to convict them—is frustrated by their pursuit of their own individual interests. (p. 544)

Pure strategy A course of action a player can choose in a game that does not involve randomized choices. (p. 537)

Rational choice A choice that leads to a preferred outcome. (p. 529)

Saddlepoint In a two-person constant-sum game, the payoff that results when a row minimum and a column maximum are the same, which is the value of the game. The saddlepoint has the shape of a saddle-shaped surface and is also a Nash equilibrium. (p. 533)

Sincere voting Voting for one's most-preferred alternative in a situation. (p. 551)

Sophisticated voting Voting that involves the successive elimination of dominated strategies by voters. (p. 552)

Status-quo paradox The status quo is defeated by another alternative when voters are sophisticated. (p. 553)

Strategy One of the courses of action that a player can choose in a game; strategies are mixed or pure, depending on whether they are selected in a randomized fashion (mixed) or not (pure). (p. 537)

Theory of moves (TOM) A dynamic theory that describes optimal choices in strategic-form games in which players, thinking ahead, can make moves and countermoves. (p. 558)

Total-conflict game A zero-sum or constant-sum game, in which what one player wins the other player loses. (p. 529)

Value The best outcome that both players can guarantee in a two-person zero-sum game. If there is a saddlepoint, that is the value. Otherwise, it is the expected payoff resulting when the players choose their optimal mixed strategies. (p. 533)

Variable-sum game A game in which the sum of the payoffs to the players at the different outcomes varies. (p. 544)

Zero-sum game A constant-sum game in which the payoff to one player is the negative of the payoff to the other player, so the sum of the payoffs to the players at each outcome is zero. (p. 542)

SKILLS CHECK

1. In the following two-person zero-sum game, the payoffs represent gains to row player I and losses to column player II.

$$\begin{bmatrix} 3 & 7 & 3 \\ 8 & 6 & 1 \\ 5 & 9 & 4 \end{bmatrix}$$

What is the maximin strategy for Player I?

(a) Play the first row.
(b) Play the second row.
(c) Play the third row.

2. In Skills Check 1, the minimax strategy of Player II is to play the __________ column.

3. In the following two-person zero-sum game, the payoffs represent gains to row player I and losses to column player II.

$$\begin{bmatrix} 3 & 7 & 2 \\ 8 & 5 & 1 \\ 6 & 9 & 4 \end{bmatrix}$$

What is the minimax strategy for Player II?

(a) Play the first column.
(b) Play the second column.
(c) Play the third column.

4. In Skills Check 3, the maximin strategy for Player I is to play the __________ row.

5. In Skills Check 3, which of the following statements is true?

(a) The game definitely has no saddlepoint.
(b) The game may or may not have a saddlepoint.
(c) The game definitely has a saddlepoint.

For Skills Check Questions 6–11, consider the following three two-person zero-sum games, wherein the payoffs represent gains to the row player I and losses to the column player II:

$$\begin{bmatrix} 3 & 6 \\ 5 & 4 \end{bmatrix}$$

$$\begin{bmatrix} -1 & 3 \\ 2 & 0 \end{bmatrix}$$

$$\begin{bmatrix} 6 & 5 & 6 & 5 \\ 1 & 4 & 2 & -1 \\ 8 & 5 & 7 & 5 \\ 0 & 2 & 6 & 2 \end{bmatrix}$$

6. The __________ games have a saddlepoint?

7. In which two games does neither player have a dominant strategy?

(a) The first two games
(b) The last two games
(c) The first and third games

8. __________ strategies of Player I are dominated in the third game.

9. Which strategy of Player II is dominant in the third game?

(a) Strategy 1
(b) Strategy 2
(c) Strategy 4

10. Strategy __________ of Player I is dominant in the third game.

11. The third game has

(a) no saddlepoint.
(b) one saddlepoint.
(c) more than one saddlepoint.

12. If a game has a saddlepoint, then __________ is the value of the game.

13. In the game of matching pennies, Player I wins a penny if the coins match; Player II wins a penny if the coins do not match. Given this information, it can be concluded that the 2 × 2 matrix that represents this game

(a) has two −1's and two 1's.
(b) has four 1's.
(c) has four −1's.

14. A mixed strategy uses randomization to ______________.

15. Which of these games does not have a saddlepoint?
(a) Tic-tac-toe
(b) Chess
(c) Poker

16. A game is fair if its saddlepoint is equal to ______________.

17. In the following game of batter-versus-pitcher in baseball, the batter's batting averages are given in the game matrix.

		Pitcher	
		Fastball	Curveball
Batter	Fastball	0.400	0.200
	Curve	0.100	0.500

What is the pitcher's optimal strategy?
(a) Throw more curveballs than fastballs.
(b) Throw more fastballs than curveballs.
(c) Throw about the same number of curveballs and fastballs.

18. In Skills Check 17, the batter's optimal strategy is to anticipate ______________.

19. In Skills Check 17, suppose that each player calculates what his or her opponent will do most frequently. Knowing this, the players should
(a) continue to choose their optimal mixed strategies.
(b) always choose their more frequently used mixed strategy.
(c) always choose their less frequently used mixed strategy.

20. In the following game of batter-versus-pitcher in baseball, the batter's batting averages are given in the game matrix.

		Pitcher	
		Fastball	Curveball
Batter	Fastball	0.350	0.250
	Curveball	0.100	0.500

The batter's exact optimal strategy is to anticipate ______________.

21. In Skills Check 20, the pitcher's exact optimal strategy is ______________.

22. Consider the following partial-conflict game, played in a noncooperative manner.

		Player II	
		Choice A	Choice B
Player I	Choice A	(4, 4)	(1, 3)
	Choice B	(3, 1)	(2, 2)

Both players might select *B* because ______________.

23. In Skills Check 22, what strategy constitutes a Nash equilibrium?
(a) Only when both players select *A*
(b) Only when both players select *A* or both select *B*
(c) Only when one player selects *A* and the other selects *B*

24. Consider the game played between the opposing goalie and a soccer player who, after a penalty, is allowed a free kick. The kicker can elect to kick toward one of the two corners of the net or else aim for the center of the goal. The goalie can decide to commit in advance (before the kicker's kick) to either one of the sides, or else remain in the center until he sees the direction of the kick. This two-person zero-sum game can be represented as follows, wherein the payoffs are the probability of scoring a goal:

		Goalie		
		Breaks left	Remains center	Breaks right
Kicker	Kicks left	0.5	0.9	0.9
	Kicks center	0.1	0	0.1
	Kicks right	0.9	0.9	0.5

If we assume that decisions between the left or right side are made symmetrically (with equal probabilities), then this game can be represented by a 2×2 matrix, where $0.7 = \left(\frac{1}{2}\right)(0.5) + \left(\frac{1}{2}\right)(0.9)$:

		Goalie	
		Remains center	Breaks side
Kicker	Kicks center	0	1
	Kicks side	0.9	0.7

The exact optimal strategies of the kicker and goalie are ______________.

25. In the following game, Player I has the preferences of the row player in the Prisoners' Dilemma, and Player II has the preferences of the column player in Chicken.

		Player II	
		Choice A	Choice B
Player I	Choice A	(3, 3)	(1, 4)
	Choice B	(4, 2)	(2, 1)

Does the player with a dominant strategy benefit more than the player without one?

(a) Yes
(b) No
(c) It doesn't make any difference; both players do equally well from choosing their strategies associated with the Nash-equilibrium outcome.

26. In Skills Check 25, the strategies associated with (4, 2) constitute a Nash equilibrium, but those associated with (3, 3) do not, because ________.

27. In Skills Check 25, assume that play starts at outcome (3, 3). According to the TOM, Player II, if it moved to (1, 4), would anticipate that Player I would countermove to (2, 1), after which it could counter-countermove to (4, 2). At outcome (4, 2),

(a) Player I would stop.
(b) Player I would move to (3, 3).
(c) it is impossible to say what Player I would do.

28. The reason for my answer to Skills Check 27 is ________.

29. A game tree is used to

(a) determine the possible strategies of a player.
(b) anticipate each other's choices through backward induction.
(c) plan a deception strategy.

30. TOM is dynamic because ________.

CHAPTER 15 EXERCISES

■ *Challenge* ▲ *Discussion*

15.1 Two-Person Total-Conflict Games: Pure Strategies

Consider the following five two-person total-conflict games, wherein the payoffs represent gains to the row player I and losses to the column player II:

1. $\begin{bmatrix} 6 & 5 \\ 4 & 2 \end{bmatrix}$

2. $\begin{bmatrix} 0 & 3 \\ -5 & 1 \\ 1 & 6 \end{bmatrix}$

3. $\begin{bmatrix} -2 & 3 \\ 1 & -2 \end{bmatrix}$

4. $\begin{bmatrix} 13 & 11 \\ 12 & 14 \\ 10 & 11 \end{bmatrix}$

5. $\begin{bmatrix} -10 & -17 & -30 \\ -15 & -15 & -25 \\ -20 & -20 & -20 \end{bmatrix}$

(a) Which of these games have saddlepoints?
(b) Find the maximin strategy of player I, the minimax strategy of player II, and the value for those games given in part (a).
(c) List strategies in these games that the players should avoid because the resulting payoffs are worse than those for some alternative strategy.

6. If a player has two strategies in a game and one is dominated, must the other strategy be dominant? Why?

7. If a player has more than two strategies in a game and one is dominated, must one of the other strategies be dominant? Why?

15.2 Two-Person Total-Conflict Games: Mixed Strategies

Solve the following three games of batter-versus-pitcher in baseball, wherein the pitcher can throw one of two pitches and the batter can guess either of these two pitches. The batter's batting averages are given in the game matrix.

8.

		Pitcher	
		Fastball	Curveball
Batter	Fastball	0.300	0.200
	Curveball	0.100	0.400

9.

		Pitcher	
		Fastball	Knuckleball
Batter	Fastball	0.500	0.200
	Knuckleball	0.200	0.300

10.

		Pitcher	
		Blooperball	Knuckleball
Batter	Blooperball	0.400	0.200
	Knuckleball	0.250	0.250

11. A businessperson has the choice of either not cheating on his income tax or cheating and making \$1000 if not audited. If caught cheating, he will pay a fine of \$2000 in addition to the \$1000 he owes. He feels good if he does not cheat and is not audited (worth \$100). If he does not cheat and is audited, he evaluates this outcome as −\$100 (for the lost day). Viewing the game as a two-person zero-sum game between the businessperson and the tax agency, what are the optimal mixed strategies for each player and the value of the game?

12. When it is third down and short yardage to go for a first down in American football, the quarterback can decide to run the ball or pass it. Similarly, the other team can commit itself to defend more heavily against a run or a pass. This can be modeled as a 2×2 matrix game, wherein the payoffs are the probabilities of obtaining a first down. Find the solution of this game.

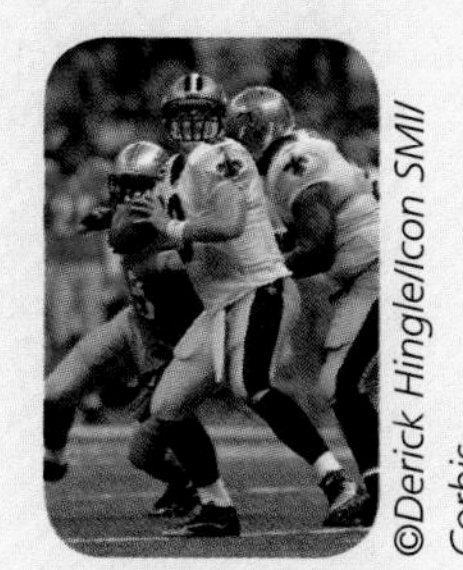

©Derick Hingle/Icon SMI/Corbis

		Defense	
		Run	Pass
Offense	Run	0.5	0.8
	Pass	0.7	0.2

13. You have the choice of either parking illegally on the street or parking in the lot and paying \$16. Parking illegally is free if the police officer is not patrolling, but you receive a \$40 parking ticket if she is. However, you are peeved when you pay to park in the lot on days when the officer does not patrol, and you are willing to assess this outcome as costing \$32 (\$16 for parking plus \$16 for your time, inconvenience, and grief). It seems reasonable to assume that the police officer ranks her preferences in the order (1) giving you a ticket, (2) not patrolling with you parked in the lot, (3) patrolling with you in the lot, and (4) not patrolling with you parked illegally.

(a) Describe this as a matrix game, assuming that you are playing a zero-sum game with the officer.

(b) Solve this matrix game for its optimal mixed strategies and its value.

▲ **(c)** Discuss whether it is reasonable or not to assume that this game is zero-sum.

▲ **(d)** Assuming that you play this parking game each working day of the year, how do you implement an optimal mixed strategy?

14. Describe how a pure strategy for a player in a matrix game can be considered as merely a special case of a mixed strategy.

■ **15. (a)** Describe in detail *one* pure strategy for the player who moves first in the game of tic-tac-toe. This strategy must tell how to respond to all possible moves of the other player. (*Hint:* You may wish to make use of the symmetry in the 3 × 3 grid in this game; that is, there are one "center" box, four "corner" boxes, and four "side" boxes.)

(b) Is your strategy optimal in the sense that it will guarantee the first player a tie (and possibly a win) in the game?

16. In the matching-pennies example, consider the case where player I favors heads H over tails T. For example, assume that player I plays H three-fourths of the time and T only one-fourth of the time—a nonoptimal mixed strategy. What should player II do if he knows this?

17. Assume in the nonsymmetrical matching example that Player II is using the nonoptimal mixed strategy $(p, 1 - p) = \left(\frac{1}{2}, \frac{1}{2}\right)$; that is, he is playing H and T with the same frequency. What should player I do in this case if she knows this?

18. You plan to manufacture a new product for sale next year, and you can decide to make either a small quantity, in anticipation of a poor economy and few sales, or a large quantity, hoping for brisk sales. Your expected profits are indicated in the following table.

		Economy	
		Poor	Good
Quantity	Small	\$500,000	\$300,000
	Large	\$100,000	\$900,000

If you want to avoid risk and believe that the economy is playing an optimal mixed strategy against you in a two-person zero-sum game, then what is your optimal mixed strategy and the resulting expected value? Discuss some alternative ways to go about making your decision.

■ **19.** Consider the following miniature poker game with two players, I and II. Each antes \$1. Each player is dealt either a high card H or a low card L, with probability $\frac{1}{2}$. Player I then folds or bets $1. If Player I bets, then Player II either folds, calls, or raises $1. Finally, if II raises, I either folds or calls.

Most choices by the players are rather obvious, at least to anyone who has played poker: If either player holds H, that player always bets or raises if he or she gets the choice. The question remains of how often one should bluff—that is, continue to play (by calling or raising) while holding a low card in the hope that one's opponent also holds a low card.

This poker game can be represented by the following matrix game, wherein the payoffs are the expected winnings for player I (depending upon the random deal) and the dominated strategies have been eliminated.

		Player II (when holding L)		
		Folds	Calls	Raises
	Folds initially	−0.25	0	0.25
Player I (when holding L)	Bets first, folds later	0	0	−0.25
	Bets first, calls later	−0.25	−0.25	0

(a) Are there any strategies in this matrix game that a player should avoid playing?
(b) Solve this game.
(c) Which player is in the more favored position?

20. Considering the scenario described in Exercise 19, should one ever bluff?

▲ **21.** If Person A threatens Person B but does not intend to carry out his or her threat, we say he or she is bluffing. When is such a bluff rational?

■ **22. (a)** Describe in detail *one* pure strategy for the player who moves second in the game of tic-tac-toe.
(b) Is your strategy in part (a) optimal in the sense that it will guarantee the second player a tie (and possibly a win) in the game?

23. On an overcast morning, deciding whether to carry your umbrella can be viewed as a game between yourself and nature as follows:

		Weather	
		Rain	No rain
You	Carry umbrella	Stay Dry	Lug umbrella
	Leave it home	Get wet	Hands free

Let's assume that you are willing to assign the following numerical payoffs to these outcomes, and that you are also willing to make decisions on the basis of expected values (that is, average payoffs):

(Carry umbrella, rain) = −2
(Carry umbrella, no rain) = −1
(Leave it home, rain) = −5
(Leave it home, no rain) = 3

(a) If the weather forecast says there is a 50 percent chance of rain, should you carry your umbrella or not? What if you believe there is a 75 percent chance of rain?
(b) If you are conservative and wish to protect against the worst case, what pure strategy should you pick?
(c) If you are rather paranoid and believe that nature will pick an optimal strategy in this two-person zero-sum game, then what strategy should you choose?
(d) Another approach to this decision problem is to assign payoffs to represent what your *regret* will be after you know nature's decision. In this case, each such payoff is the best payoff you could have received under that state of nature, minus the corresponding payoff in the previous table:

		Weather	
		Rain	No rain
You	Carry umbrella	0 = (−2) − (−2)	4 = 3 − (−1)
	Leave it home	3 = (−2) − (−5)	0 = 3 − 3

What strategy should you select if you wish to minimize your maximum possible regret?

15.3 Partial-Conflict Games

Consider the following five two-person variable-sum games. Discuss the players' possible behavior when these games are played in a noncooperative manner (with no prior communication or agreements). The first payoff is to the row player; the second, to the column player. Are the Nash equilibria in these games sensible? Why or why not?

24.

	Player II	
Player I	(4, 4)	(1, 3)
	(3, 1)	(2, 2)

25. Battle of the sexes:

		She buys a ticket for:	
		Boxing	**Ballet**
He buys	**Boxing**	(4, 3)	(2, 2)
a ticket for:	**Ballet**	(1, 1)	(3, 4)

26.

	Player II	
Player I	(2, 1)	(4, 2)
	(1, 4)	(3, 3)

27.

	Player II	
Player I	(2, 4)	(4, 3)
	(1, 2)	(3, 1)

28.

	Player II	
Player I	(3, 4)	(2, 3)
	(1, 2)	(4, 1)

▲ **29.** In Exercise 24, players maximize their possible gains by choosing their first strategies, but they minimize their possible losses by choosing their second strategies. Which strategy would you choose, and why?

▲ **30.** Assume that two countries in an arms race assign points to all their own weapons so that the total for each is 1000. Each side can then designate weapons of the *other* side, totaling 100 points, that must be eliminated in the next year, thereby effecting a 10 percent reduction. Would these countries have any reason to lie about how they value their own weapons? Is this procedure practical as an arms-reduction scheme?

15.4 Larger Games

31. For the preferences of the players given in the text—*xyz* for *X*, *yzx* for *Y*, and *zxy* for *Z*—verify that the strategy choices of *x* by *X*, *y* by *Y*, and *x* by *Z* are a Nash equilibrium. Does this equilibrium seem to you defensible as the social choice by the voters? Under what circumstances might the voters choose these strategies rather than their sophisticated strategies?

■ **32.** Under a voting system called *approval voting*, a voter can vote for as many alternatives as he or she wishes. (If there are three alternatives, the only undominated strategies of a voter under approval voting are to vote for his or her best, or two best, choices). If Voters *X*, *Y*, and *Z* have paradox-of-voting preferences of *xyz*, *yzx*, and *zxy*, and *X* is the status quo, show that *x* is the sophisticated outcome, obtained by all voters voting for their two best choices.

■ **33.** Show by example that approval voting is not immune to the status-quo paradox.

34. Odd and Even play Low Person Wins, whose rules are as follows:

(a) Odd announces an odd number between 1 and 5 (inclusive).
(b) Independently, Even announces an even number between 2 and 6 (inclusive).
(c) Whoever announces the lower number gets *twice* this number as its payoff.
(d) Whoever announces the higher number gets the *lower* number as its payoff.

What is the Nash equilibrium of this game, based on the successive elimination of dominated strategies? Is there another Nash equilibrium? What are the similarities and differences between this game and the Prisoners' Dilemma?

■ **35.** Find a two-person zero-sum game with a saddlepoint in which the successive elimination of dominated strategies does *not* lead to the saddlepoint. (Unlike the 3×3 location game, you may restrict yourself to 3×3 games.)

▲ **36.** Why will the first player to act in a truel shoot in the air (if this option is allowed by the rules)? Is this choice optimal if a second player should succeed in firing in the air at the same time?

37. In a sequential truel with no firing in the air allowed, suppose *A*, who hates *B*, goes first; *B*, who hates *C*, goes second; *C*, who hates *A*, goes third. (If a player fires, he will shoot only his *antagonist*—the player he hates.)

(a) Which player is in the best position, and why?
(b) Does the outcome change if *B* hates *A* rather than *C*?

Answer these questions if each player can take only one turn (if alive); and if, after one round, the game continues (if there is more than one player alive) and each player can take more than one turn.

38. If a fourth player is added to the original truel, show that every player will have an incentive to shoot another player (as in a duel).

39. Extend the game tree of the truel in Figure 15.5 to allow the additional possibility that if *A* does not shoot initially, then *B* has the choice of shooting or not shooting *C*. Will *A*, in fact, not shoot initially, and will *B* then shoot *C*?

40. Extend the game tree in Exercise 38 to still another level to allow for the possibility that if *A* does not shoot initially, and *B* shoots *C*, then *A* has the choice of shooting or not shooting *B*. What will happen in this case?

41. Change Exercise 39 to allow for the possibility that if *A* does not shoot initially, and *B* shoots *or does not shoot C*, then *A* has the choice of shooting or not shooting *B*. What will happen in this case?

▲ **42.** What general conclusions would you draw in light of your answers to Exercise 38, 39, and 40?

15.5 Using Game Theory

43. Assume in an auction that the highest bidder wins the item being auctioned, but he or she pays only the second-highest bid. In such a second-price auction, show that all the bidders have a dominant strategy of bidding their "reservation prices," which are the prices that render them indifferent between winning and losing the auction. (*Hint:* Show that bidding more than your reservation price could result in your overpaying, while bidding your reservation price never does; bidding less could result in your not obtaining the item—and paying less than your reservation price—which bidding your reservation price never does.)

APPLET EXERCISES

To do these exercises, go to www.whfreeman.com/fapp9e.

What happens to the value of a game if you or your opponent deviates from the optimal strategy? Can you exploit such deviations by your opponent to your advantage? Explore these possibilities in the *Game Theory* applet.

WRITING PROJECTS

1. Consider a conflict that you, personally, had—with a parent, a boss, a girlfriend or boyfriend, or some other acquaintance—in which each of you had to make a choice without being sure of what the other person would do. What strategies did you seriously consider adopting, and what options do you think the other person considered? What plausible outcomes do you think each set of strategy choices would have led to? How would you rank these outcomes from best to worst, and how do you think the other player would have ranked them? In two to three pages, analyze the resulting game, and state whether you think you and the other person made optimal choices. If not, what interfered with your or the other person's rationality?

2. It is sometimes argued that game theory does not take account of the (irrational?) emotions of people, such as anger, jealousy, or love. What is your opinion about this question? In one to two pages, give an example, real or hypothetical, that supports your position, paying particular attention to whether the players acted consistently with, or contrarily to, their preferences.

3. In tennis, one player often prefers to play from the baseline while her opponent prefers a serve-and-volley game (that is, likes to come to the net). The baseline player attempts to hit passing shots. This player has a choice of hitting "down the line" or "crosscourt." The net player must often correctly guess in which direction the ball will go to cover the shot. In one to two pages, formulate this situation as a duel game and discuss appropriate strategies for the players.

4. Quentin Tarantino's films *Reservoir Dogs* (1992) and *Pulp Fiction* (1994) both have truels, but the choices that the characters make in each are completely different. Does the truel analysis offer any insight into why? Discuss in one to two pages.

5. In one to two paragraphs, discuss the relationship between the status-quo paradox and the chair's paradox (section 10.4).

Suggested Readings

AUMANN, ROBERT J., and SERGIU HART, eds. *Handbook of Game Theory with Economic Applications*, Elsevier, Amsterdam, 1992 (vol. 1), 1994 (vol. 2), 2002 (vol. 3). A comprehensive treatment of game theory and its applications, developed in long chapters written by leading experts.

BINMORE, KEN. *Playing for Real: A Text on Game Theory*. Oxford University Press, Oxford, U.K., 2007. A comprehensive intermediate text.

BRAMS, STEVEN J. *Theory of Moves*, Cambridge University Press, New York, 1994. Describes in detail the theory of moves, and applies it to a wide variety of conflicts.

BRAMS, STEVEN J. *Game Theory and the Humanities: Bridging Two Worlds*, MIT Press, Cambridge, MA, 2011. Unusual applications of game theory to history, literature, the Bible, theology, philosophy, and law.

DIXIT, AVINASH, SUSAN SKEATH and DAVID H. RILEY. *Games of Strategy*, 3d ed., Norton, New York, 2009a. An excellent game-theory text that requires only a minimal mathematical background.

NASAR, SYLVIA. *A Beautiful Mind*, Simon & Shuster, New York, 1998. A biography of John Nash that is also a fascinating account of the early history of game theory. In 2001, a fictionalized version of this biography was made into a movie, which received four Oscars, including Best Picture, in 2002.

OSBORNE, MARTIN J. *An Introduction to Game Theory*. Oxford University Press, New York, 2004. A fine intermediate text with several interesting applications.

Suggested Web Sites

www.economics.utoronto.ca/osborne Martin Osborne's home page (game theory).

kuznets.fas.harvard.edu/~aroth/alroth.html
Alvin Roth's Game Theory and Experimental Economics Page.

www.gametheory.net The most comprehensive general Web site.

FOX NEWS
NBC
YouTube
Google
CATS
arab news
4 WEEKS FREE
PEPSI
Tribune
TNT
EL DIARIO

The Digital Revolution

Chapter 16
Identification Numbers

Chapter 17
Information Science

In the third edition of this book, written in 1993, this part was titled "Coding Information." At that time, mathematical methods to code, store, secure, and transmit information accurately and economically were in place, but computers were expensive, slow, and had little memory; further, there were no broadband or wireless Internet connections. Google, Facebook, Twitter, YouTube, eBay, cell phones, text messages, blogs, iTunes, global position satellites, ebooks, digital cameras, online dating sites, and Amazon.com did not exist. By 2003, when the sixth edition of this book was published, everything had changed: computers were fast, powerful, and inexpensive; broadband and wireless connections were widely available; and online commercial and social networking sites had millions of users.

In recognition of these dramatic changes, the title for this part of the sixth edition of the book was changed to "The Digital Revolution." As of the publication of this edition, there are 4 billion cell phone subscribers, 2 billion Internet users, and 140 million blogs. Little did we know in 2003 that within another decade, the digital revolution would spawn social, political, and commercial revolutions as well. In this part of the text, we examine some of the mathematics that made these revolutions possible.

David Young-Wolff/Photo Edit

16

Identification Numbers

A **code** *is a symbolic way to represent information.* Codes such as hieroglyphics, the Greek alphabet, and Roman numerals existed thousands of years ago. Music scores and the "genetic code" to describe the makeup of deoxyribonucleic acid (DNA) are more recently invented codes. A revolution in information management took place at a grocery store in Ohio in June, 1974, when a 10-pack of Wrigley's Juicy Fruit gum was the first bar-coded retail product purchased. Since then, the use of bar codes and sophisticated product identification numbers for tracking data accurately, efficiently, and inexpensively have become universal.

In Section 16.1, we describe several common coding schemes that are used to assign identification numbers to items such as money orders, airline tickets, bank checks, books, and cars in such a way that certain errors in reading these numbers can be detected. Section 16.2 shows the way the U.S. postal service ZIP codes are assigned. In Section 16.3, we explain how ZIP codes and retail items are assigned bar codes. Section 16.4 tells how social security numbers are assigned and illustrates a method that is used to code personal data such as gender, date of birth, and portions of names.

16.1 Check Digits

Look at the 13-digit **International Standard Book Number (ISBN)** printed on the back cover of this book. The number 978-1-4292-4316-3 (978-1-4292-5482-3 for the paperback version) is a **code**. It distinguishes this book from all others. The last digit, 3, is there solely to detect errors that may occur when the ISBN is entered into a computer. Grocery items, credit cards, overnight mail, magazines, personal checks, travelers cheques, soft-drink cans, automobiles, and many other items that you encounter daily have identification numbers that code data, as well as a digit called a **check digit** for error detection. In this chapter, we examine some of the methods used to assign identification numbers and check digits.

Division by 9 Schemes

Let's begin by considering the U.S. Postal Service money order shown in Figure 16.1. The first 10 digits of the 11-digit number 17620289526 simply identify the money order. The last digit, 6, serves as an **error-detecting** mechanism. Let's see how this mechanism works. The 11th (last) digit of a Postal Service money order number is the remainder obtained when the sum of the

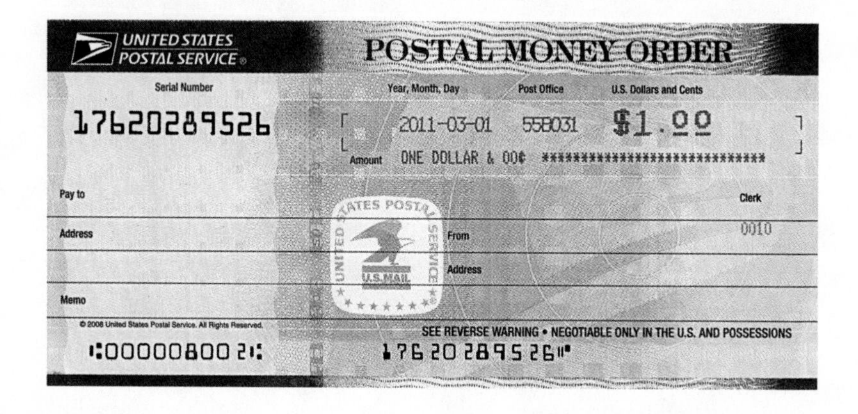

Figure 16.1 Money order with identification number 1762028952 and appended check digit 6. The check digit is the remainder after dividing the sum of the digits by 9.

first 10 digits of the number is divided by 9. In our example, the last digit is 6 because $1 + 7 + 6 + 2 + 0 + 2 + 8 + 9 + 5 + 2 = 42$ and the remainder when 42 is divided by 9 is 6. (Recall that if we divide a positive integer a by a positive integer b, there are unique integers q and r such that $a = bq + r$, where r is nonnegative and less than b. The numbers q and r are called the *quotient* and *remainder*. For example, the remainder when 42 is divided by 9 is 6 because $42 = 9 \times 4 + 6$.)

Now suppose that instead of the correct number, the number 17640289526 (an error in the fourth position) was entered into a computer that had been programmed for error detection in money orders. The machine would divide the sum of the first 10 digits of the entered number, 44, by 9 and obtain a remainder of 8. Because the last digit of the entered number is 6 rather than 8, the entered number cannot be correct. This crude method of error detection will not detect the mistake of replacing a 0 with a 9, or vice versa. Because the value of a sum does not depend on the order in which the numbers are added, this method does not detect the transposition of digits such as 17260289526 instead of 17620289526. (The digits in positions three and four have been transposed.)

American Express travelers cheques, VISA travelers cheques, and Euro banknotes also use a check digit determined by division by 9. In these cases, the check digit is the smallest nonnegative integer such that the sum of the digits, including the check digit, is evenly divisible by 9.

EXAMPLE 1
The American Express Travelers Cheque

The American Express Travelers Cheque with the identification number 387505055 has the check digit 7 because $3 + 8 + 7 + 5 + 0 + 5 + 0 + 5 + 5 = 38$ and $38 + 7$ is evenly divisible by 9.

Division by 7 Schemes

The scheme used on airline tickets and for Avis and National rental cars assigns the remainder after division by 7 of the number itself as the check digit rather than dividing the sum of the digits by 7. For example, the check digit for the number 540047 is 4 because $540047 = 7 \times 77149 + 4$. This method will not detect the substitution of 0 for a 7, 1 for an 8, 2 for a 9, or vice versa. However, unlike the Postal Service method, it will detect transpositions of adjacent digits with the exceptions

Remainders **Algebra Review**

To find the remainder when an integer a is divided by an integer b, we divide a by b using long division. Here is 95 divided by 7:

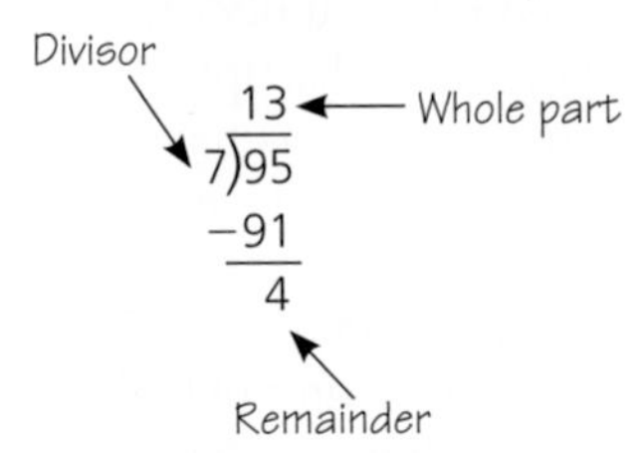

An important feature of a remainder is that it is always less than the divisor.

To use a calculator to find the remainder when a is divided by b, perform the following steps:

Step 1: Enter $a \div b$

Step 2: Enter a − (whole number from Step 1) × b

For example, if we want to find the remainder when 95 is divided by 7, a calculator gives 95 ÷ 7 = 13.571428. If we enter 95 − (7 × 13), we obtain 4. So 4 is the remainder when 95 is divided by 7.

of the pairs 0, 7; 1, 8; and 2, 9. For example, if 5400474 were entered into a computer as 4500474 (the first two digits are transposed), the machine would determine that the check digit should be 3 because 450047 = 7 × 64292 + 3. Because the last digit of the entered number is not 3, the error has been detected.

One can use Google to determine easily the check digits that require division by 7 or 9. To find the Avis check digit for the number 540047, enter "540047 mod 7" in the Search box; for division by 9, enter the number followed by "mod 9." To use a calculator to find the remainder when a is divided by b, first enter $a \div b$. The integer portion of the number is the quotient q. The remainder is $r = a - bq$. Doing this for 123 ÷ 7, we get 17.57, then 123 − 7 × 17 = 4. When a positive integer is divided by 10, the remainder is the first digit after the decimal point, but this is not true when dividing by other integers.

Universal Product Code

The scheme used on grocery products, the **Universal Product Code (UPC)**, is more sophisticated. Consider the number 0 38000 00127 7 found on the bottom of a box of Kellogg's Corn Flakes. The first digit identifies a broad category of goods, the next five digits identify the manufacturer, the next five identify the product, and the last is a check digit. Suppose that this number were entered into a computer as 0 58000 00127 7 (a mistake in the second position). How would the computer recognize the mistake?

For any UPC number $a_1a_2a_3a_4a_5a_6a_7a_8a_9a_{10}a_{11}a_{12}$, the computer is programmed to carry out the following computation: $3a_1 + a_2 + 3a_3 + a_4 + 3a_5 + a_6 + 3a_7 + a_8 + 3a_9 + a_{10} + 3a_{11} + a_{12}$. If the result doesn't end with a 0, the computer knows the entered number is incorrect.

For the incorrect corn flakes number, we have

$$3 \cdot 0 + 5 + 3 \cdot 8 + 0 + 3 \cdot 0 + 0 + 3 \cdot 0 + 0 + 3 \cdot 1 + 2 + 3 \cdot 7 + 7 = 62$$

Because 62 doesn't end with 0, the error is detected. Notice that had we used the correct digit 3 in the second position instead of 5, the sum would have ended in a 0 as it should. This simple scheme detects *all* single-position errors and about 89 percent of all other kinds of errors.

Beginning in January 2005, U.S. retailers were required to have software that could read the 12-digit UPC code used in the United States and the 13-digit

European Article Number (EAN) code used in Europe. This change paves the way for the 13-digit EAN to become the worldwide standard. Existing UPC numbers will be converted to EAN numbers by adding an extra 0 at the beginning. The check digit for a 13-digit EAN number $a_1a_2a_3a_4a_5a_6a_7a_8a_9a_{10}a_{11}a_{12}a_{13}$ is selected so that $a_1 + 3a_2 + a_3 + 3a_4 + a_5 + 3a_6 + a_7 + 3a_8 + a_9 + 3a_{10} + a_{11} + 3a_{12} + a_{13}$ ends with 0. Adding an extra 0 in the front of a UPC number does not affect the check digit. The coefficient 3 for the terms with even subscripts is called a **weight**. Because $1a_i = a_i$, we say that the terms with odd subscripts have weight 1.

Besides error detection, check-digit schemes that use weighted sums can be used to find a digit that has been corrupted in some way. Say, for example, that the packaging for a product with UPC number 1 640002 202034 was damaged or defective in such a way that the second digit was unintelligible. How would we know that it is supposed to be 6? Well, let's call it X. Then we know that the weighted sum

$$1 + 3 \cdot X + 4 + 3 \cdot 0 + 0 + 3 \cdot 0 + 2 + 3 \cdot 2 + 0 + 3 \cdot 2 + 0 + 3 \cdot 3 + 4 = 3X + 32$$

must end with 0. Because 6 is the only digit that makes this true, we know the corrupted digit is 6. This example also illustrates why the weight 3 is superior to the weight 2. While there is only one digit that makes the expression $3X + 32$ end with 0, there are two digits that make $2X + 32$ end with 0: $X = 4$ and $X = 9$.

Bank Identification Numbers

The U.S. banking system uses a variation of the UPC scheme that appends check digits to the numbers assigned to banks. Each bank has an eight-digit routing number $a_1a_2 \cdots a_8$ together with a check digit a_9 so that a_9 is the last digit of $7a_1 + 3a_2 + 9a_3 + 7a_4 + 3a_5 + 9a_6 + 7a_7 + 3a_8$. In this formula, the weights are 7, 3, and 9. The weights were carefully chosen so that all single-digit errors and most transposition errors are detected.

EXAMPLE 2
Bank Routing Number

The First Chicago Bank has the routing number 071000013 on the bottom of all its checks. (See Figure 16.2.) The check digit 3 is the last digit of $7 \cdot 0 + 3 \cdot 7 + 9 \cdot 1 + 7 \cdot 0 + 3 \cdot 0 + 9 \cdot 0 + 7 \cdot 0 + 3 \cdot 1 = 33$. The first four digits of a nine-digit bank routing number identify the bank's Federal Reserve District, office, state, or special collection arrangement; the next four digits are the bank's identification number; the ninth digit is the check digit. The block of numbers 22 63378 shown in Figure 16.2 is the account number. The last block, 0134, is the check number.

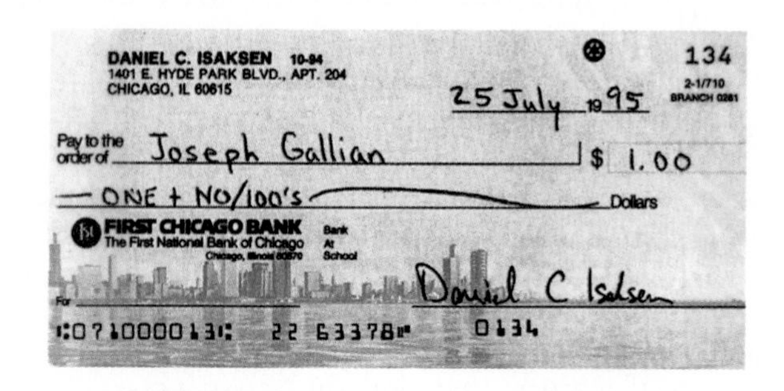

Figure 16.2 A bank check with routing number 071000013. The 3 is the check digit.

You may wonder if there is any advantage to using three weights for the routing number as opposed to using two, as is the case for the UPC error-detection scheme. The answer is yes. While both the UPC scheme and the bank scheme detect 100 percent of single-position errors and some transposition errors involving adjacent digits, the bank scheme will detect most transposition errors of the form $\cdots abc \cdots \to \cdots cba \cdots$, whereas the UPC scheme does not detect such errors.

For example, say that we look at a number that begins with 241. In the UPC scheme, these digits contribute $3 \cdot 2 + 4 + 3 \cdot 1 = 13$ toward the total calculation, while the string 142 (the first and third digits are transposed) also contributes $3 \cdot 1 + 4 + 3 \cdot 2 = 13$ toward the total calculation. So the error is not detected. In contrast, using the bank scheme, 241 contributes $7 \cdot 2 + 3 \cdot 4 + 9 \cdot 1 = 35$ toward the total, while 142 contributes $7 \cdot 1 + 3 \cdot 4 + 9 \cdot 2 = 37$ toward the total. Because the total for the correct number ends with 0, the total for the number that had the transposition error would end with the digit 2. Thus, the error is detected.

Codabar

One of the most efficient error-detection methods is the **codabar scheme** that is used by all major credit-card companies, as well as by many libraries, blood banks, and the South Dakota driver's license department. For a credit card number $a_1a_2a_3a_4a_5a_6a_7a_8a_9a_{10}a_{11}a_{12}a_{13}a_{14}a_{15}$, let T be the number of digits in positions 1, 3, 5, 7, 9, 11, 13, 15 that exceed 4. The check digit a_{16} is chosen so that $2a_1 + a_2 + 2a_3 + a_4 + 2a_5 + a_6 + 2a_7 + a_8 + 2a_9 + a_{10} + 2a_{11} + a_{12} + 2a_{13} + a_{14} + 2a_{15} + T + a_{16}$ is divisible by 10. For example, say a bank intends to issue a credit card with the identification number 312560019643001. We calculate $2 \cdot 3 + 1 + 2 \cdot 2 + 5 + 2 \cdot 6 + 0 + 2 \cdot 0 + 1 + 2 \cdot 9 + 6 + 2 \cdot 4 + 3 + 2 \cdot 0 + 0 + 2 \cdot 1$ to get 66. Then we note that among the digits in the odd-numbered positions, only 6 and 9 exceed 4. So we add 2 to 66 to get 68. The check digit is whatever is needed to bring the final tally to a number that ends with 0. Because $68 + 2 = 70$, the check digit for our example is 2. This digit is appended to the end of the number the bank issues for identification purposes. Errors in input data are detected by applying the same algorithm to the input, including the check digit. If the correct number is entered into a computer, the result will end in 0. If the result doesn't end with 0, a mistake has been detected.

The credit card shown in Figure 16.3 is reproduced from an ad promoting the Citibank VISA card. Notice that the check digit on the card is not valid because the algorithm yields

$$2 \cdot 4 + 1 + 2 \cdot 2 + 8 + 2 \cdot 0 + 0 + 2 \cdot 1 + 2 + 2 \cdot 3 + 4 + 2 \cdot 5 + 6 + 2 \cdot 7 + 8 + 2 \cdot 9 + 3 + 0 = 94,$$

which does not end in 0. This method allows computers to detect 100 percent of single-position errors and about 98 percent of other common errors. The algorithm used to compute the codabar check digit is called the *Luhn algorithm,* after IBM scientist Hans Peter Luhn, who created it in 1953. The Luhn algorithm can be applied to identification numbers of lengths other than 15.

Besides detecting errors, a check digit offers partial protection against fraudulent numbers. A person who wanted to create a phony credit card, bank account number, or driver's license number would have to know the appropriate check-digit scheme for the number to go unchallenged by the computer.

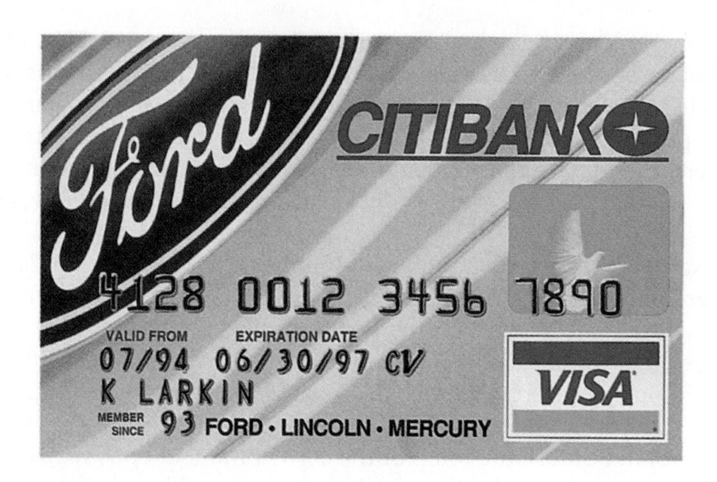

Figure 16.3 VISA card with an invalid number.

International Standard Book Numbers

Thus far, we have not discussed any schemes that detect 100 percent of single errors and 100 percent of transposition errors. As seen on the back of this book and most others published since 2007, there are two identification numbers—a 13-digit number called the *13-digit International Standard Book Number (ISBN-13)* and a 10-digit number, called the *10-digit ISBN (ISBN-10)*. The 10-digit ISBN detects 100 percent of single-digit errors and 100 percent of transposition errors.

A correctly coded 10-digit ISBN $a_1a_2 \cdots a_{10}$ has the property that $10a_1 + 9a_2 + 8a_3 + 7a_4 + 6a_5 + 5a_6 + 4a_7 + 3a_8 + 2a_9 + a_{10}$ is evenly divisible by 11. Consider the 10-digit ISBN of the book that you are now reading: 1-4292-4316-3 (1-4292-5482-3 for paperback version). The 1 at the beginning indicates that the book is published in an English-speaking country, while the next block of digits, 4292, identifies the publisher, W. H. Freeman and Company. The third block for the hardback edition, 4316, is assigned by the publisher and identifies this particular book. The last digit 3, for the hardback version, is the check digit. Let's verify that this number is a legitimate possibility. We must compute $10 \cdot 1 + 9 \cdot 4 + 8 \cdot 2 + 7 \cdot 9 + 6 \cdot 2 + 5 \cdot 4 + 4 \cdot 3 + 3 \cdot 1 + 2 \cdot 6 + 3 = 187$. Because $187 = 11 \cdot 17$, it is evenly divisible by 11, so no error has been detected.

How can we be sure that this method detects 100 percent of the single-position errors? Well, let's say that a correct number is $a_1a_2a_3a_4a_5a_6a_7a_8a_9a_{10}$ and that a mistake is made in the second position. (The same argument applies equally well in every position.) We may write this incorrect number as $a_1a_2'a_3a_4a_5a_6a_7a_8a_9a_{10}$, where $a_2' \neq a_2$. For this error to go undetected, it must be the case that $10 \cdot a_1 + 9 \cdot a_2' + 8 \cdot a_3 + 7 \cdot a_4 + 6 \cdot a_5 + 5 \cdot a_6 + 4 \cdot a_7 + 3 \cdot a_8 + 2 \cdot a_9 + a_{10}$ is evenly divisible by 11. Then, because both $10a_1 + 9a_2 + 8a_3 + 7a_4 + 6a_5 + 5a_6 + 4a_7 + 3a_8 + 2a_9 + a_{10}$ and $10a_1 + 9a_2' + 8a_3 + 7a_4 + 6a_5 + 5a_6 + 4a_7 + 3a_8 + 2a_9 + a_{10}$ are divisible by 11, so is their difference:

$$(10 \cdot a_1 + 9 \cdot a_2 + 8 \cdot a_3 + \cdots + a_{10}) - (10 \cdot a_1 + 9 \cdot a_2' + 8 \cdot a_3 + \cdots + a_{10}) = 9 \cdot (a_2 - a_2')$$

Because a_2 and a_2' are distinct digits between 0 and 9, their difference must be one of $\pm 1, \ldots, \pm 9$. Thus, the only possibilities for the number $9 \cdot (a_2 - a_2')$ are ± 9, ± 18, ± 27, ± 36, ± 45, ± 54, ± 63, ± 72, ± 81—and none of these is divisible by 11. So a single-position error cannot go undetected.

To illustrate with a specific example why the ISBN-10 method detects single errors, let's say the valid ISBN 1-4292-0900-3 is mistaken as 1-2292-0900-3 (an error in position 2). Because the correct second digit in the weighted sum contributes

$9 \cdot 4 = 36$ to the total of 176 and the incorrect second digit contributes $9 \cdot 2 = 18$, we see that the weighted sum of the incorrect number is $176 - 18 = 158$. But 158 is not evenly divisible by 11, so the error is detected.

To verify that the ISBN-10 method detects all adjacent transposition errors, let's suppose that the first two digits are transposed. (The same argument applies to all positions.) Say that the correct number is $a_1a_2a_3 \ldots a_{10}$. As before, for the incorrect number $a_2a_1a_3 \ldots a_{10}$ to go undetected, it must be the case that the difference of the correct number and the incorrect number is evenly divisible by 11. That is,

$$(10a_1 + 9a_2 + 8a_3 + \cdots + a_{10}) - (10a_2 + 9a_1 + 8a_3 + \cdots + a_{10})$$

is evenly divisible by 11. This reduces to the condition that $a_1 - a_2$ is divisible by 11. But the only possible differences of two numbers between 0 and 9 are plus or minus the numbers between 0 and 9, and of these only 0 is divisible by 11. Thus, $a_1 - a_2 = 0$. But then $a_1 = a_2$ and there is no error.

Let's trace through an example to see how the ISBN-10 method detects adjacent transposition errors. Say 1-4292-0900-3 is mistaken as 1-2492-0900-3 (digits in positions 2 and 3 are transposed). Because the second and third digits in the weighted sum of the correct number contribute $9 \cdot 4 + 8 \cdot 2 = 52$ to the total of 176 and the second and third digits of the incorrect number contribute $9 \cdot 2 + 8 \cdot 4 = 50$ to the weighted sum, we know that the weighted sum of the incorrect number is $176 - 2 = 174$, which is not evenly divisible by 11.

With a bit more work, we could prove that every transposition error is detected, not just the transpositions of adjacent digits. This is possible because 11 is prime.

Because the ISBN-10 method, in contrast to the other methods we have described, detects all single-position errors and all transposition errors, why is it not used more? Well, it does have a drawback. Say the next title published by W. H. Freeman is to have 0902 for the third block. (The 10-digit ISBN for all W. H. Freeman books begins with 1-4292.) What check digit should be assigned? Call it a. Then the weighted sum is $10 \cdot 1 + 9 \cdot 4 + 8 \cdot 2 + 7 \cdot 9 + 6 \cdot 2 + 5 \cdot 0 + 4 \cdot 9 + 3 \cdot 0 + 2 \cdot 2 + a = 177 + a$. Because the next integer after 177 that is divisible by 11 is 187, we see that $a = 10$. But appending 10 to the existing 9-digit number would result in an 11-digit number instead of a 10-digit one. This is the only flaw in the 10-digit ISBN scheme. To avoid this flaw, publishers use an X to represent the check digit 10. As a result, not all 10-digit ISBNs consist solely of digits; some end with X. Publishers could avoid this inconsistency by simply refraining from using numbers that require an X.

To expand the inventory of ISBNs and make them compatible with the UPC/EAN numbering scheme for other retail items worldwide, publishers began using a 13-digit ISBN in 2007. The 13-digit ISBN is the same as the 10-digit ISBN number except for a prefix of 978 or 979 and the check digit. The check digit for the 13-digit ISBN is calculated so that the weighted sum using the weights 1, 3, 1, 3, . . . , 1, 3, 1 ends with 0. Thus, the 13-digit ISBN and the 13-digit UPC/EAN numbers used for retail products employ the same check-digit method. During a phase-in period, publishers will use both the 10-digit and 13-digit numbers.

Although all the check-digit schemes that we have described in this chapter that use weighted sums detect 100 percent of single-digit errors, it is impossible to devise a check-digit scheme that employs a single check-digit error that detects 100 percent of multiple-digit errors. In the case of the UPC scheme, for instance, while one error might cause the weighted sum to be say 62 (and thereby detect the error), a second error may result in the sum 70 so that no errors are detected. Consider the single-digit error that we examined earlier for corn flakes. Recall the correct UPC number is 0 38000 00127 7 and its weighted sum is 60. For the incorrect corn flakes number 0 58000 00127 7 (with a single error in position 2), we have

$$3 \cdot 0 + 5 + 3 \cdot 8 + 0 + 3 \cdot 0 + 0 + 3 \cdot 0 + 0 + 3 \cdot 1 + 2 + 3 \cdot 7 + 7 = 62.$$

But if we also make a second error of using 8 in position 4 instead of 0, so that the number becomes 0 58800 00127 7, the errors are not detected because its weighted sum is 70, which does end with 0. Nevertheless, the weight schemes detect most multiple errors because it is rare when a second or third error exactly cancels out previous errors.

A Multiplication by 13 Scheme

After single-digit errors and adjacent transposition errors, the third most common error is one of the form $\cdots abc \cdots \rightarrow \cdots cba \cdots$. In practice, these kinds of errors commonly occur in phone numbers that have matching digits separated by another digit such as 727 5856. A likely mistake when writing or dialing this number is to switch the 8 and the 6, resulting in the number 727 5658. Such an error is called a *jump transposition.* Remarkably, there is a simple way to encode identification numbers so that the three most common errors are detected 100 percent of the time without having to introduce an alphabetic character, as is done for the 10-digit ISBN numbers.

To illustrate the method, suppose that a math instructor wants to post student grades publicly without revealing any information about the students' ID numbers. Assuming the last four digits of each student ID number are different, she could assign each student a six-digit number by multiplying the last four digits of their identification numbers by 13 (adding leading 0s when necessary). For example, a student with an ID number that ends with 8912 is assigned $115856 = 8912 \times 13$. (To preserve confidentiality of the original four digits, students are not informed of the encoding method.) Of course, the instructor can recapture the original four-digit numbers by dividing the encoded numbers by 13. Since all encoded numbers are divisible by 13, the jump transposition error $115856 \rightarrow 115658$ is detected because 115658 is not divisible by 13.

The arguments for verifying that encoding identification numbers as multiples of 13 detect 100 percent of all single-digit errors, all transposition errors involving adjacent digits, and all jump transposition errors are similar to those used to show that the 10-digit ISBN numbers detect errors. In particular, a single-digit error in the number $a_n a_{n-1} \cdots a_i \cdots a_0$ of the form $a_n a_{n-1} \cdots a_i' \cdots a_0$, where $a_i' \neq a_i$ is not detected if and only if $a_n a_{n-1} \cdots a_i' \cdots a_0$ is a multiple of 13. But if both $a_n a_{n-1} \cdots a_i \cdots a_0$ and $a_n a_{n-1} \cdots a_i' \cdots a_0$ are multiples of 13, then so is their difference $(a_n a_{n-1} \cdots a_i \cdots a_0) - (a_n a_{n-1} \cdots a_i' \cdots a_0) = (a_i - a_i')10^i$. But 13 does not divide the term on the right when $a_i \neq a_i'$. Similarly, the transposition of adjacent digits a_i and a_{i-1} is undetected if and only if $9(a_i - a_{i-1})10^{i-1}$ is divisible by 13, which happens only when $a_i = a_{i-1}$. And the jump transposition $\cdots a_i a_{i-1} a_{i-2} \cdots \rightarrow \cdots a_{i-2} a_{i-1} a_i \cdots$ is undetected if and only if $99(a_i - a_{i-2})10^{i-2}$ is divisible by 13, which happens only when $a_i = a_{i-2}$. Incidentally, the arguments just given reveal why we used multiplication by 13 rather than some smaller positive integer. For example, if multiplication by 11 were used to transform the identification numbers instead of 13, then all single-digit errors and all adjacent transposition errors are detected, but not all jump transpositions are, because 11 divides 99.

Euro banknotes, car rental companies, some state driver's license numbers, and the vehicle identification number (VIN) that identifies cars and trucks use alpha-numeric identification numbers. To calculate the check digit, the letters are assigned numerical values. See Spotlight 16.1.

SPOTLIGHT 16.1

The VIN System

Automobiles and trucks are given a VIN by the manufacturer. A typical VIN has 17 alphanumeric characters that code information such as the country where the vehicle was built, manufacturer, make, body style, engine type, plant where the vehicle was built, model year, model, type of restraint, a check digit, and a production sequence number. The check digit is calculated by converting the 26 consecutive letters of the alphabet to the numbers 1, 2, 3, 4, 5, 6, 7, 8, 9, 1, 2, 3, 4, 5, 6, 7, 8, 9, 2, 3, 4, 5, 6, 7, 8, 9 (note the skipped digit after the second 9) to obtain a 16-digit number $a_1a_2 \cdots a_{15}a_{16}$ that is weighted with 8, 7, 6, 5, 4, 3, 2, 10, 9, 8, 7, 6, 5, 4, 3, 2. The check digit is the remainder when the weighted sum $8 \cdot a_1 + 7 \cdot a_2 + \cdots + 3 \cdot a_{15} + 2 \cdot a_{16}$ is divided by 11 unless the remainder is 10, in which case an X is used instead. The check digit is inserted in position 9.

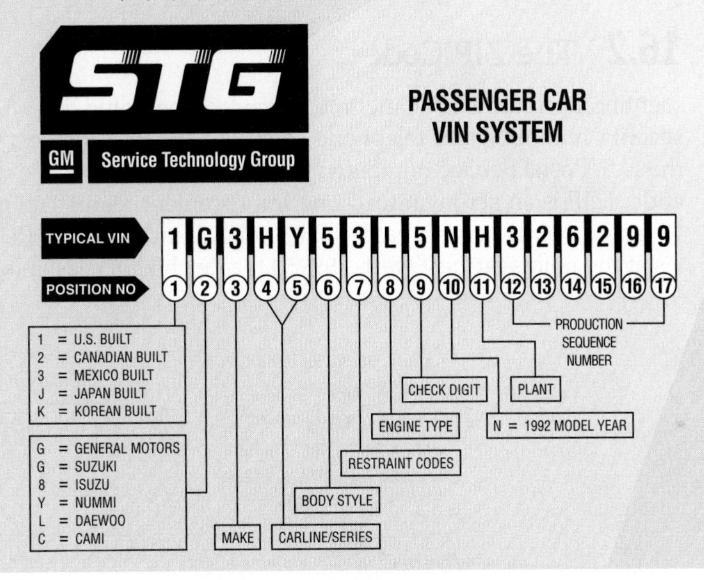

Summary of Error Detection Schemes

Postal money orders. The last digit of a Postal Service money order number is the remainder obtained when the sum of the first 10 digits of the number is divided by 9. This method will detect any single mistake except replacing 0 with a 9 or vice versa.

American Express and VISA travelers cheques; Euro banknotes. The last digit is the smallest nonnegative integer such that the sum of the digits, including the check digit, is divisible by 9. This method will detect any single mistake except replacing 0 with a 9 or vice versa.

Airlines; Avis and National rental cars. The last digit is the remainder after division by 7 of the number itself, excluding the last digit. This method will detect any single error except the substitution of 0 for a 7, 1 for an 8, 2 for a 9, or vice versa. It will detect transpositions of adjacent digits with the exceptions of the pairs 0, 7; 1, 8; and 2, 9.

UPCs. For any UPC number $a_1a_2a_3a_4a_5a_6a_7a_8a_9a_{10}a_{11}a_{12}$, the last digit is chosen so that $3a_1 + a_2 + 3a_3 + a_4 + 3a_5 + a_6 + 3a_7 + a_8 + 3a_9 + a_{10} + 3a_{11} + a_{12}$ is divisible by 10. This method detects 100 percent of all single errors and 89 percent of most others.

Bank routing numbers. For a bank routing number $a_1a_2a_3a_4a_5a_6a_7a_8$, the check digit a_9 is the last digit of $7a_1 + 3a_2 + 9a_3 + 7a_4 + 3a_5 + 9a_6 + 7a_7 + 3a_8$. This method detects 100 percent of all single errors and 89 percent of most others.

Codabars. For a credit card number $a_1a_2a_3a_4a_5a_6a_7a_8a_9a_{10}a_{11}a_{12}a_{13}a_{14}a_{15}$, let T be the number of digits in positions 1, 3, 5, 7, 9, 11, 13, 15 that exceed 4. The last digit a_{16} is chosen so that $2a_1 + a_2 + 2a_3 + a_4 + 2a_5 + a_6 + 2a_7 + a_8 + 2a_9 + a_{10} + 2a_{11} + a_{12} + 2a_{13} + a_{14} + 2a_{15} + T + a_{16}$ is divisible by 10. This method detects 100 percent of all single errors and 98 percent of most others.

ISBN-10s. A correctly coded 10-digit ISBN $a_1a_2 \cdots a_{10}$ has the property that $10a_1 + 9a_2 + 8a_3 + 7a_4 + 6a_5 + 5a_6 + 4a_7 + 3a_8 + 2a_9 + a_{10}$ is evenly divisible by 11. This method detects 100 percent of all single errors and 100 percent all transposition errors.

ISBN-13s. This scheme is the same as the UPC scheme.

16.2 The ZIP Code

Identification numbers sometimes encode geographic data. The ZIP code, social security numbers, and telephone numbers are the foremost examples. In 1963, the U.S. Postal Service numbered every American post office with a five-digit **ZIP code**. (ZIP is an acronym for Zone Improvement Plan.) The numbers begin with 0's at the points farthest east—00601 for Adjuntas, Puerto Rico—and work up to 9's at the points farthest west—99950 for Ketchikan, Alaska (see Figure 16.4).

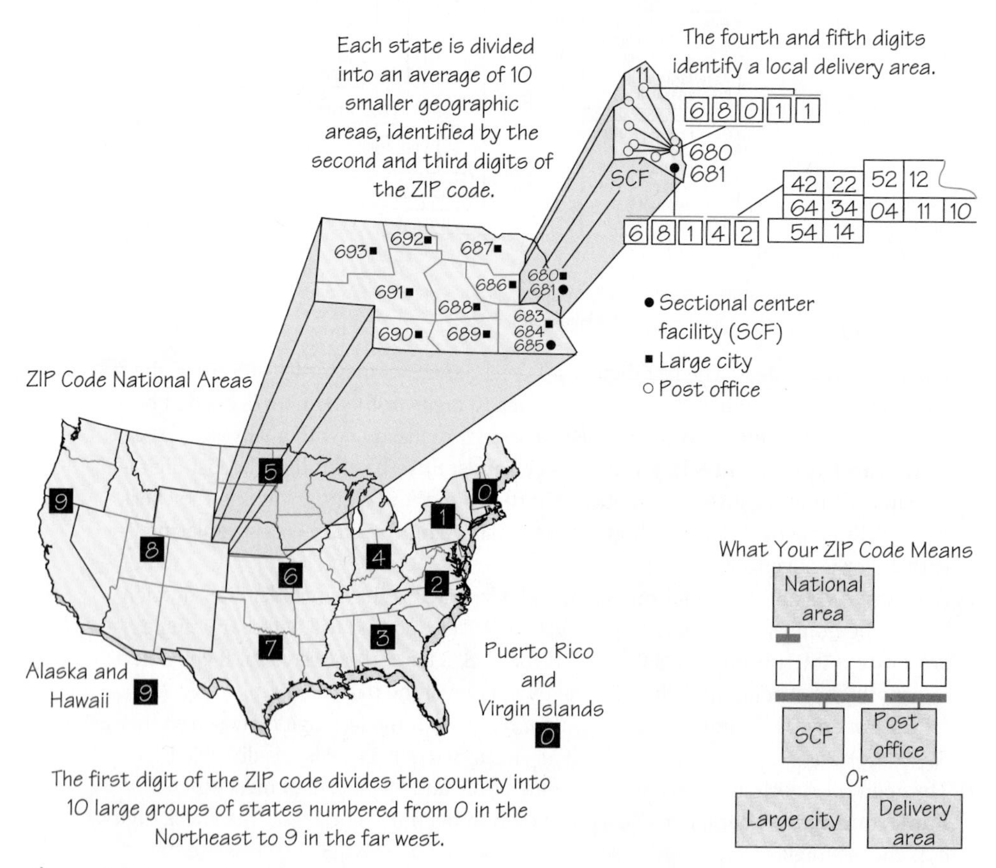

Figure 16.4 ZIP code scheme.

Let's use 55812, one of the ZIP codes for Duluth, Minnesota, as an example.

5 The first digit represents one of 10 geographic areas, usually a group of states. The numbers begin at the points farthest east (0) and end at the points farthest west (9).

58 The second two digits, in combination with the first, identify a central mail-distribution point known as a sectional center. The location of a sectional center is based on geography, transportation facilities, and population density. Although just four centers serve the entire state of Utah, there are six of them to take care of New York City.

12 The last two digits indicate the town or local post office. In many cases, the largest city in a region will be given the digits 01 and surrounding towns assigned succeeding digits alphabetically.

In 1983, the U.S. Postal Service added four digits to the ZIP code. When four digits are added after a dash—for example, 68588-1234—the number is called the **ZIP + 4 code**. Mail with ZIP + 4 coding is eligible for cheaper bulk rates, being easier to sort with automated equipment. It's also helpful for businesses that wish to sort the recipients of their mailings by geographic location. The first two numbers of the four-digit suffix represent a delivery sector, which may be several blocks, a group of streets, several office buildings, or a small geographic area. The last two numbers narrow the area further. They might denote one floor of a large office building, a department in a large firm, or a group of post office boxes.

For businesses that receive an enormous volume of mail, the ZIP + 4 code permits automation of in-house mailroom sorting. For example, the first seven digits of all mail sent to the University of Minnesota Duluth, are 55812-24. The school has designated nine pairs of digits for the last two positions to direct the mail to the appropriate dormitory or apartment complex.

16.3 Bar Codes

In modern applications, bar codes and identification numbers go hand in hand. Bar coding is a method for automated data collection. It is a way to transmit information rapidly, accurately, and efficiently to a computer.

Bar Code DEFINITION

A **bar code** is a series of dark bars and light spaces that represent characters.

To decode the information in a bar code, a beam of light is passed over the bars and spaces via a scanning device, such as a handheld wand or a fixed-beam device. The dark bars reflect very little back to the scanner, whereas the light spaces reflect much light. The differences in reflection intensities are detected by the scanner and converted to strings of 0's and 1's that represent specific numbers and letters. Such strings are called a **binary coding** of the numbers and letters.

Binary Code DEFINITION

Any system for representing data with only two symbols is a **binary code**.

ZIP Code Bar Coding

The simplest bar code is the **Postnet code** used by the U.S. Postal Service and commonly found on business reply forms (see Figure 16.5). For a ZIP + 4 code, there are 52 vertical bars of two possible lengths (long and short). The long bars at the beginning and end are called *guard bars* and together provide a frame for the remaining 50 bars. In blocks of 5, the 50 bars within the guard bars represent the ZIP + 4 code and a 10th digit for error correction. Each block of 5 is composed of exactly 2 long bars and 3 short bars, according to the pattern shown below:

Decimal Digit	Bar Code
1	ııı||
2	ıı|ı|
3	ıı||ı
4	ı|ıı|
5	ı|ı|ı
6	ı||ıı
7	|ııı|
8	|ıı|ı
9	|ı|ıı
0	||ııı

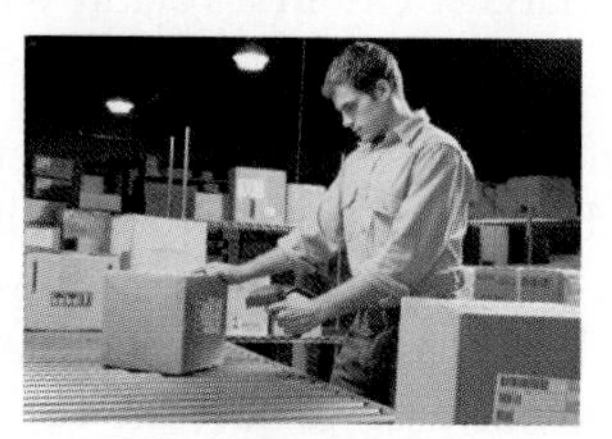

Handheld scanner reading the shipping bar code on a crate. (*Honeywell*)

The 10th digit of a Postnet code number is a check digit chosen so that the sum of the 9 digits of the ZIP + 4 code and the 10th one is evenly divisible by 10. That is, the check digit C for the ZIP + 4 code $a_1a_2 \cdots a_9$ is the digit with the property

Figure 16.5 ZIP + 4 bar code.

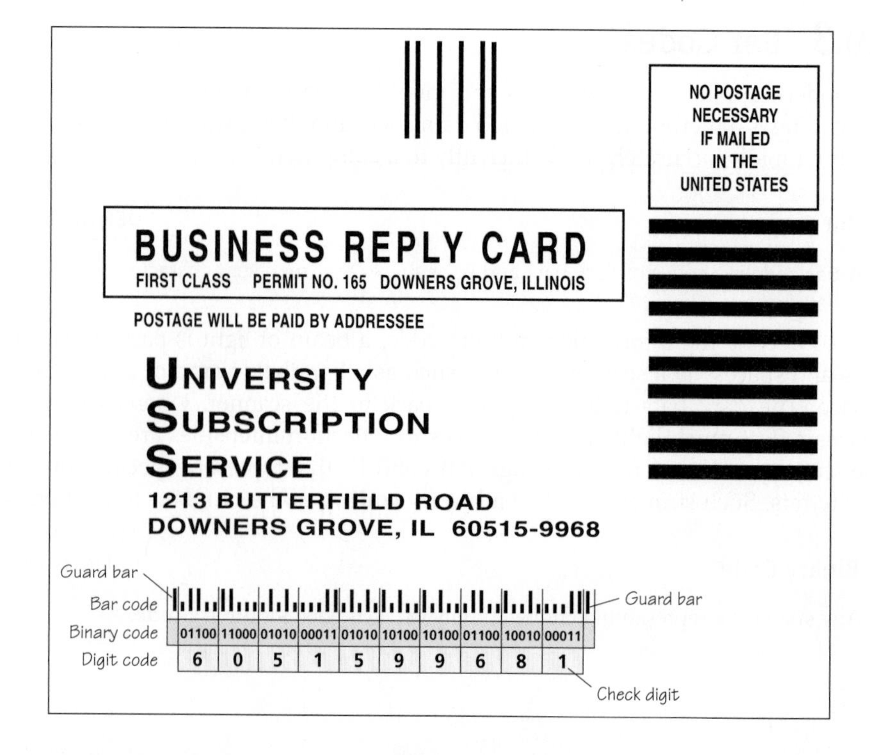

that the sum $a_1 + a_2 + \cdots + a_9 + C$ ends with 0. For example, the ZIP + 4 code 80321-0421 has the check digit 9 because

$$8 + 0 + 3 + 2 + 1 + 0 + 4 + 2 + 1 = 21$$
$$\text{and } 21 + 9 = 30$$

ends with 0.

Because each digit is represented by exactly two long bars and three short ones, any error in reading or printing a single bar would result in a block of five with only one long bar or three long bars. In either case, the error is detected. This is the reason behind the choice of five bars to code each digit rather than four bars. With five bars per digit, there are exactly 10 arrangements composed of two long bars and three short bars. Any misreading of a single bar in such a block is therefore recognizable because it does not match any of the other blocks for the 10 digits. And because the block location of the error is known, the check digit permits the correction of the error. Let's look at an example of an incorrectly printed bar code and see how the error is correctable.

EXAMPLE 3
Detecting and Correcting an Error

The scanner ignores the guard bars at the beginning and the end and reads the remaining bars in blocks of five, as shown below. (We have inserted dashed dividing lines for readability.)

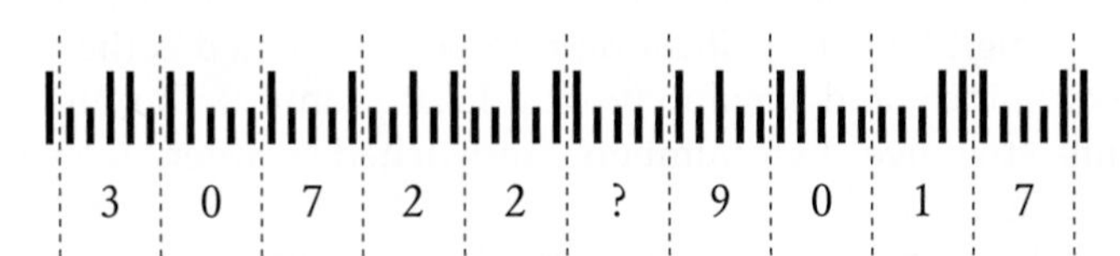

The sixth block is an incorrect one because it has only one long bar. To correct the error, the computer linked with the bar-code scanner sums the remaining 9 digits to obtain 31. Because the sum of all 10 digits ends with 0, the correct value for the sixth digit must be 9.

Beginning in 1993, large organizations and businesses that wanted to receive reduced rates for ZIP + 4 bar-coded mail were required to use a 12-digit bar code called the *delivery-point bar code.* This code permits machines to sort a letter into the order in which it will be delivered by the carrier. Mail for the first location on a mail route occurs first, mail for the second location on a route occurs second, and so on.

The 12-digit bar code uses the Postnet bar scheme to code the 12-digit string composed of the 9-digit ZIP + 4 number, followed by the last two digits of the street address or box number and a check digit chosen so that the sum of all 12 digits is evenly divisible by 10. For example, a letter addressed to 1738 Maple Street with ZIP + 4 code 55811-2742 would have the Postnet bar code for the digits 558112742384 (38 is from the street address and 4 is the check digit).

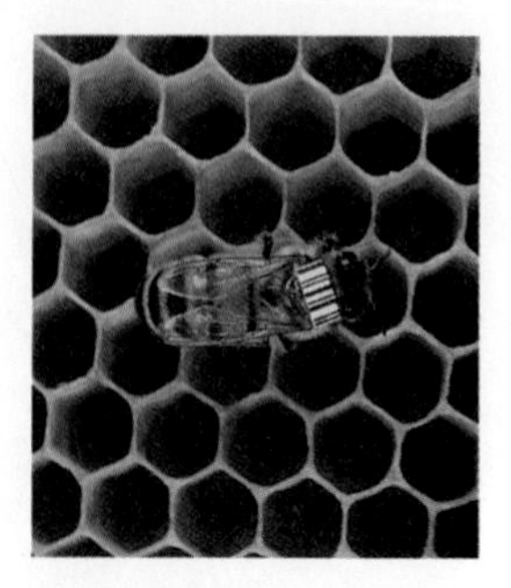

Figure 16.6 Entomologist Stephen Buchmann developed a reliable, inexpensive way to track bees using the same technology that supermarkets use to speed up checkout lines and keep track of inventory. He glued barcode labels onto the backs of 100 bees and placed a laser scanner above the hive. In the past, researchers marked bees with paint or tags, but monitoring activity required the presence of a human observer. *(Scott Camazine/Sue Trainor.)*

In May 2011, the Postnet code was replaced by a new bar code called the *Intelligent Mail Barcode.* The new bar code converts 31 digits of data into 65 vertical bars that encode the type of service, the mail owner, a unique serial number that enables the user to track the letter at every step from arrival at the post office to delivery, and the delivery-point ZIP code. The code uses bars of three lengths and multiple levels to create four states as shown below.

The UPC Bar Code

The bar code that we encounter most often is the UPC, which was first used on grocery items in 1974 and its use has since spread to most retail products. As Figure 16.6 shows, it has other applications as well. The UPC bar code translates 12-digit UPC identification numbers discussed earlier into bars that can be read quickly and accurately by a laser scanner. The number has four components—two five-digit numbers sandwiched between two single digits—as shown in Figure 16.7.

For the UPC 0 38000 00127 7, here is what the four components represent:

0 The first digit identifies the kind of product. For example, a 0 represents general merchandise; a 2 signals random-weight items, such as cheese and meat; a 3 means drug and certain other health-related products; a 4 means products marked for price reduction by the retailer (see Figure 16.7); a 5 signals cents-off coupons.

38000 The next five digits identify the company.

00127 The next five digits, assigned by the manufacturer to identify the product, can include size, color, or other important information (but not price).

7 The final digit is the check digit. This digit is often not printed, but it is always included in the bar code.

Figure 16.7 UPC identification number 4 41120 10640 9. The initial 4 indicates that the number is a savings coupon. The block 41120 identifies the retailer as Kmart. The block 10640 identifies the product. The last digit, 9, is a check digit.

Each digit of the UPC code is represented by a space divided into seven modules of equal width, as illustrated in Figure 16.8. How these seven modules are filled depends on the digit being represented and whether the digit being represented is part of the manufacturer's number or the product number. In every case, there are two light spaces and two dark bars of various thicknesses that alternate. A UPC code has on each end two long bars of one-module thickness separated by a light space of one-module thickness. These two modules are called the *guard*

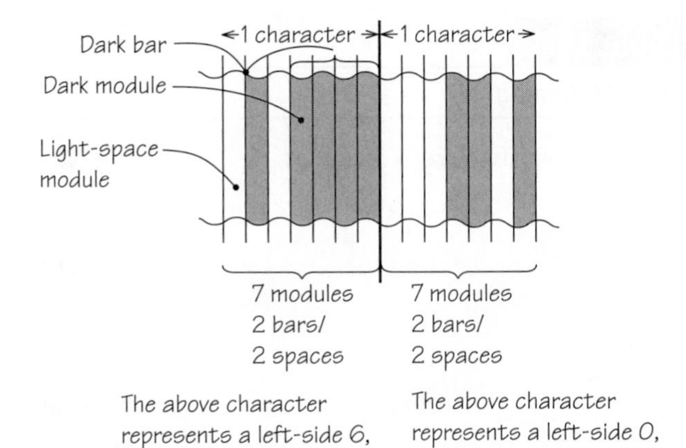

Figure 16.8 UPC bar coding for a left-side 6 and left-side 0, part of the manufacturer's number.

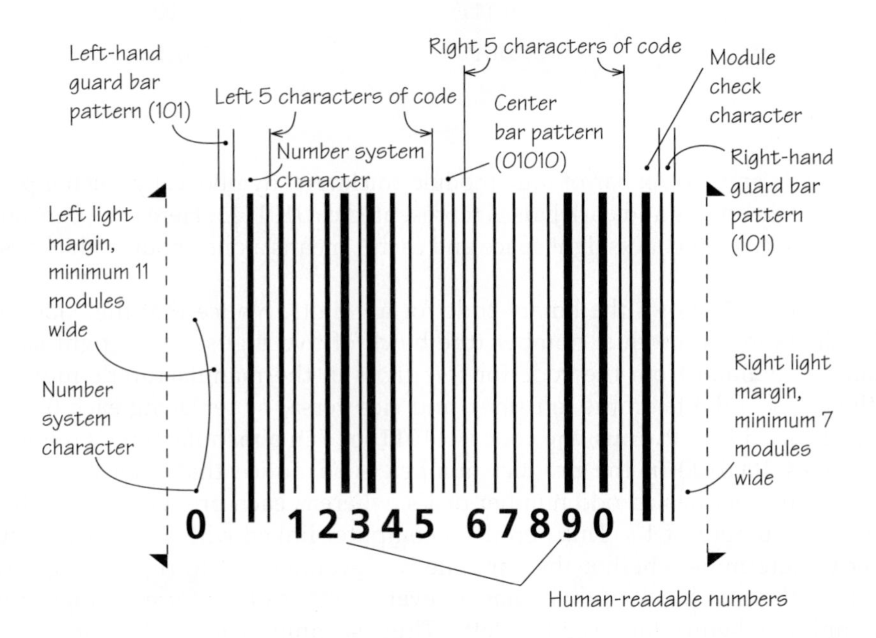

Figure 16.9 UPC bar-code format.

bar patterns (Figure 16.9). The guard bar patterns define the thickness of a single module of each type. They are not part of the identification number. The manufacturer's number and the product number are separated by a center bar pattern consisting of the following five modules: a light space, a dark bar, a light space, a dark bar, and a light space (see Figure 16.9). The center bar pattern is not part of the identification number but merely serves to separate the manufacturer's number and product number. Figure 16.8 shows how the digits 6 and 0 in a manufacturer's number are coded.

Observe the following pattern in Figure 16.8: a light space of one-module thickness, a dark bar of one-module thickness, a light space of one-module

Table 16.1

Binary UPC Coding

Digit	Manufacturer's Number	Product Number
0	0001101	1110010
1	0011001	1100110
2	0010011	1101100
3	0111101	1000010
4	0100011	1011100
5	0110001	1001110
6	0101111	1010000
7	0111011	1000100
8	0110111	1001000
9	0001011	1110100

thickness, and a dark bar of four-module thickness. Symbolically, such a pattern of light spaces and dark bars is represented as 0101111. Here each 0 means a one-module-thickness light space and each 1 means a one-module-thickness dark bar.

Table 16.1 shows the binary code for all digits. Notice that the code for the digits in the product number (the block of five digits on the right side) can be obtained from the code for the digits in the manufacturer's number (the block of digits on the left side), and vice versa, by replacing each 0 by a 1 and each 1 by a 0. Thus, the code 0111011 for 7 in a manufacturer's number becomes 1000100 in the product number. Also notice that each manufacturer's number has an odd number of 1's, whereas each product number has an even number of 1's. This permits a computer linked with an optical scanner to determine whether the bar code was scanned left to right or right to left. (If the first block of digits has an even number of 1's for each digit, the scanning is being done right to left.) Thus, scanning can be done in either direction without ambiguity.

Figure 16.10 Bar code on a building in Japan that can be read by a properly equipped cell phone. *(Ko Sasaki/The New York Times/Redux.)*

New Applications of Bar Coding

New applications of bar coding continue to be found. In 2003, a method of bar coding genetic information about animal species was introduced that provides a convenient, inexpensive way to identify species. (See Spotlight 16.2.) Recently, a new generation of bar codes uses mosaics of black and white rectangles that encode much more information than traditional bar codes (see Figure 16.10). These bar codes can be read by specially equipped cell phones to display video, music, or text on the screen or to link the cell phone to a Web page. A user can point his or her cell phone at the bar code in a magazine, on a billboard, or on the side of a building to receive information about a product or service. A bar code for a movie will allow the viewer to watch a trailer. Scanning the wrapper of a hamburger will provide nutrition information.

SPOTLIGHT 16.2

New Frontier: Bar Coding DNA

In 2003, Paul Hebert from the University of Guelph in Canada proposed the compilation of a public library of DNA bar codes for animal species. Rather than scanning an animal's entire genome, which is expensive and time-consuming, Hebert pinpointed a short piece of a section of a single gene that could be used to distinguish one animal species from another cheaply and quickly. For about $2 per sample, the genetic sequence of this tiny gene section can be converted to a four-color bar code that corresponds to the four nucleotides that make up the genetic code. The bar code identifies the species of its source in the same way that the UPC bar code identifies a retail item. By 2010, more than 65,000 species were bar-coded, and a new field of science was born. The technique has already resulted in improved food safety, disease prevention, and better environmental monitoring. The Consortium for the Barcode of Life has set a goal of bar coding 200,000 species by 2012.

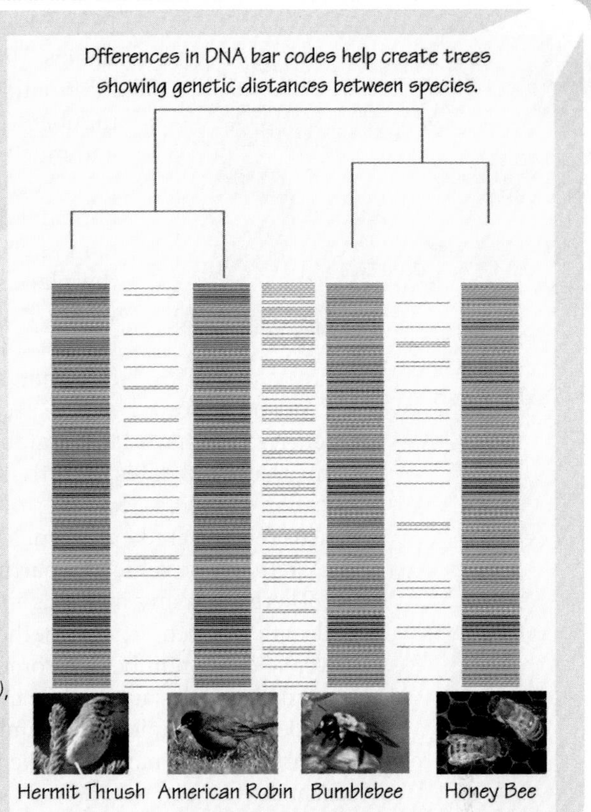

DNA bar codes for the hermit thrush *(George Jameson)*, American robin *(Jeffrey Lepore/Photo Researchers, Inc.)*, bumblebee *(Mark Stoeckle/ The Rockefeller University)*, and Honeybee *(Scott Camazine/Photo Researchers, Inc.)*. *(Mark Stoekle/The Rockefeller University)*

SPOTLIGHT 16.3

History of the Bar Code

1948 Graduate students Norman Joseph Woodland and Bernard Silver at the Drexel Institute of Technology begin working on a bar code.

October 1952 Woodland and Silver receive a U.S. patent.

October 1967 The Association of American Railroad adopts an optical bar code.

December 1971 The Uniform Code Council, originally called the Uniform Grocery Product Code Council, is formed to administer the UPC.

1972 U.S. Supermarket Ad Hoc Committee on a Uniform Grocery Product Code recommends the adoption of the 1972 UPC.

April 1973 An ad-hoc committee composed of grocery executives chooses the linear bar code with 11 digits and a 12th check digit.

June 1974 A 10-pack of Wrigley's Juicy Fruit chewing gum was the first product with a bar code; it is first scanned at a checkout counter in Troy, Ohio. Today, the pack of gum is on display at the Smithsonian Institution's National Museum of American History.

1974 95 percent of the railroad fleet is labeled with a bar code.

February 1977 European Article Numbering Association is formed in Belgium.

September 1981 U.S. Department of Defense adopts the use of bar codes for marking all products sold to the U.S. military.

1992 Norman Joseph Woodland is awarded the 1992 National Medal of Technology by President George H.W. Bush.

16.4 Encoding Personal Data

Consider this social security number: 189-31-9431. What information about the holder can be deduced from the number? Only that the holder obtained it in Pennsylvania (see Spotlight 16.4). Figure 16.11 shows an Illinois driver's license number: M200-7858-1644. What information about the holder can be deduced from this number? This time, we can determine the date of birth, sex, and much about the person's name.

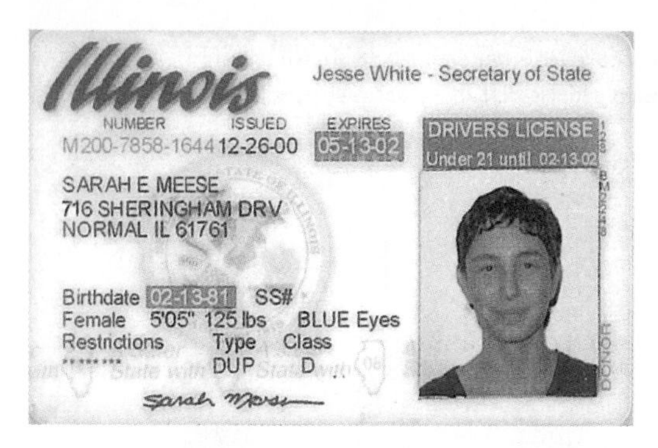

Figure 16.11 Illinois driver's license.

These two examples illustrate the extremes in coding personal data. The social security number has no personal data encoded in the number. It is entirely determined by the place and time that it is issued, not the individual to whom it is assigned. In contrast, in some states, driver's license numbers are determined entirely by personal information about the holders. It is no coincidence that the unsophisticated social security numbering scheme predates computers. Agencies that have large databases that include personal information such as names, sex, and dates of birth find it convenient to encode these data into identification numbers. Examples of such agencies are the National Archives (where census records are kept), genealogical research centers, the Library of Congress, and state motor vehicle departments.

There are many methods in use to encode personal data such as name, sex, and date of birth. These methods are perhaps most widely used in assigning driver's license numbers in some states. Coding license numbers solely from personal data enables automobile insurers, government entities, and law enforcement agencies to determine the number from the personal data. Many states encode the surname, first name, middle initial, date of birth, and sex by very sophisticated schemes.

In one scheme that is based on sound, the first four characters of the license number are obtained by applying the **Soundex Coding System** to the surname as follows:

1. Delete all occurrences of h and w. (For example, *Schworer* becomes *Scorer* and *Hughgill* becomes *uggill.*)
2. Assign numbers to the remaining letters as follows:

$$a, e, i, o, u, y \to 0 \qquad l \to 4$$
$$b, f, p, v \to 1 \qquad m, n \to 5$$
$$c, g, j, k, q, s, x, z \to 2 \qquad r \to 6$$
$$d, t \to 3$$

3. If two or more letters with the same numeric value are adjacent, omit all but the first. (For example, *Scorer* becomes *Sorer* and *uggill* becomes *ugil.*)
4. Delete the first character of the original name if still present (*Sorer* becomes *orer*).
5. Delete all occurrences of a, e, i, o, u, and y.
6. Retain only the first three digits corresponding to the remaining letters; append trailing 0's if fewer than three letters remain; precede the three digits obtained in step 6 with the first letter of the surname.

Figure 16.12 shows three examples.

SPOTLIGHT 16.4

Social Security Numbers

Numbers	State
001–003	New Hampshire
004–007	Maine
008–009	Vermont
010–034	Massachusetts
035–039	Rhode Island
040–049	Connecticut
050–134	New York
135–158	New Jersey
159–211	Pennsylvania
212–220	Maryland
221–222	Delaware
223–231 and 669–699	Virginia
232–236	West Virginia
232, 237–246, and 681–690	North Carolina
247–251 and 654–658	South Carolina
252–260 and 667–675	Georgia
261–267, 589–595, and 766–772	Florida
268–302	Ohio
303–317	Indiana
318–361	Illinois
362–386	Michigan
387–399	Wisconsin
400–407	Kentucky
408–415 and 756–763	Tennessee
416–424	Alabama
425–428, 587–588, and 752–755	Mississippi
429–432 and 676–679	Arkansas
433–439 and 659–665	Louisiana
440–448	Oklahoma
449–467 and 627–645	Texas
468–477	Minnesota
478–485	Iowa
486–500	Missouri
501–502	North Dakota
503–504	South Dakota
505–508	Nebraska
509–515	Kansas
516–517	Montana
518–519	Idaho
520	Wyoming
521–524 and 650–653	Colorado
525, 585, and 648–649	New Mexico
526–527, 600–601, and 764–765	Arizona
528–529 and 646–647	Utah
530, 680	Nevada
531–539	Washington
540–544	Oregon
545–573 and 602–626	California
574	Alaska
575–576 and 750–751	Hawaii
577–579	District of Columbia
580	Virgin Islands
580–584 and 596–599	Puerto Rico
586	Guam
586	American Samoa
586	Philippine Islands
700–728	*through July 1, 1963, reserved for railroad employees* (Railroad employees had their own retirement plan before social security began, so that plan was integrated into the social security system, and they received extra benefits as well.)

Source: Social Security Administration.

What is the advantage of this method? It is an error-correcting scheme. Indeed, it is designed so that likely misspellings of a name nevertheless result in the correct coding of the name. For example, frequent misspellings of the name *Erickson* are *Ericksen, Eriksen, Ericson,* and *Ericsen*. Observe that all of these yield the same coding as *Erickson.* If a law enforcement official, a genealogical researcher, or a librarian wanted to pull up the file from a data bank for someone whose name was pronounced "Erickson," the correct spelling isn't essential because the computer searches for records that are coded as E-625 for

Figure 16.12 The Soundex Coding System.

Step 1: Schworer → Scorer; Step 2: → Scorer 220606
Step 3: → Sorer 20606; Step 4: → orer 0606; Step 5: → rr 66; Step 6: → S-660

Step 1: Hughgill → uggill; Step 2: → uggill 022044
Step 3: → ugil 0204; Step 4: → ugil 0204; Step 5: → gl 24; Step 6: → H-240

Step 1: Schmidlapper → Scmidlapper; Step 2: → Scmidlapper 22503401106
Step 3: → Smidlaper 250340106; Step 4: → midlaper 50340106; Step 5: → mdlpr 53416; Step 6: → S-534

all spelling variations. The search feature of a Web site where many mathematicians post their research papers uses the Soundex Coding System. This system was designed for the U.S. Census Bureau when much census information was obtained orally.

There are many schemes for encoding the date of birth and the sex in driver's license numbers. For example, the last five digits of Illinois and Florida driver's license numbers capture the year and date of birth as well as the sex. In Illinois, each day of the year is assigned a three-digit number in sequence beginning with

SPOTLIGHT 16.5

Census Records at the National Archives

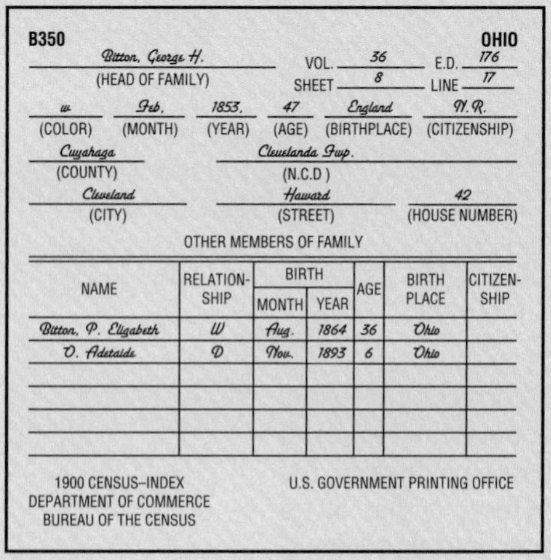

B350 OHIO
Bitton, George H. (HEAD OF FAMILY) VOL. 36 E.D. 176
SHEET 8 LINE 17
w (COLOR) Feb. (MONTH) 1853. (YEAR) 47 (AGE) England (BIRTHPLACE) N.R. (CITIZENSHIP)
Cuyahaga (COUNTY) Clevelanda Twp. (N.C.D)
Cleveland (CITY) Haward (STREET) 42 (HOUSE NUMBER)
OTHER MEMBERS OF FAMILY

NAME	RELATIONSHIP	BIRTH MONTH	BIRTH YEAR	AGE	BIRTH PLACE	CITIZENSHIP
Bitton, P. Eligabeth	W	Aug.	1864	36	Ohio	
O. Adetaide	D	Nov.	1893	6	Ohio	

1900 CENSUS–INDEX
DEPARTMENT OF COMMERCE
BUREAU OF THE CENSUS
U.S. GOVERNMENT PRINTING OFFICE

One of the best places to look for information pertaining to family history is the old censuses that are kept up by the National Archives in Washington, D.C. By law, census records are open to the public 72 years after the census was taken. The data from 1880, 1900, 1910, and 1920 censuses (records from 1890 were destroyed by fire) were put on cards during the 1930s as a Works Progress Administration (WPA) project. This information was coded using the Soundex system so that names that sound alike regardless of how they are spelled are grouped together. On old documents, family names were so often misspelled—especially those that were not of British origin—that genealogists say several variations of a name may apply to one set of ancestors. To look for a surname on the index, the researchers must work out the Soundex code. This code, together with the state record, identifies a page number on microfilm where the data are located. A typical census Soundex card is shown. Note the Soundex code in the upper-left corner (B350).

001 for January 1. However, each month is assumed to have 31 days. Thus, March 1 is given the number 063 because both January and February are assumed to have 31 days. These numbers are then used to identify the month and day of birth of male drivers. For females, the scheme is identical except that 600 is added to the number. The last two digits of the year of birth, separated by a dash (probably to obscure the fact that they represent the year of birth), are listed in the fifth and fourth positions from the end of the driver's license number. Thus, a male born on October 13, 1940, would have the last five digits 4-0292 ($292 = 9 \cdot 31 + 13$), whereas a female born on the same day would have 4-0892.

The scheme to identify birth date and sex in Florida is the same as in Illinois except that each month is assumed to have 40 days and 500 is added for women. Moreover, a dash occurs between the two digits for the year and the three digits for the day. For example, the five digits 49-585 belong to a woman born on March 5, 1949.

In this chapter, we have investigated how mathematics is used to append a check digit to an identification number for error detection. In the next chapter, we will show how codes consisting of 0's and 1's can be devised so that errors can be corrected.

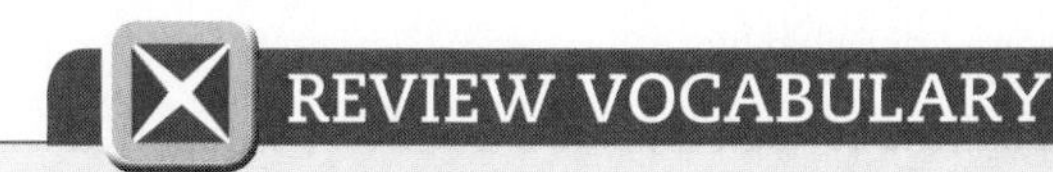

REVIEW VOCABULARY

Bar code A code that employs bars and spaces to represent information. (p. 583)

Binary code A coding scheme that uses two symbols, usually 0 and 1. (p. 583)

Check digit A digit included in an identification number for the purpose of error detection. (p. 573)

Code A group of symbols that represent information, together with a set of rules for interpreting the symbols. (p. 573)

Codabar An error-detecting method used by credit cards, libraries, blood banks, the Canadian Social Insurance program, and others. (p. 577)

Error-detecting code A code in which certain types of errors can be detected. (p. 573)

International Standard Book Number (ISBN) An identification number used on books throughout the world that contains a check digit for error detection. (p. 573)

Postnet code The bar code used by the U.S. Postal Service for ZIP codes. (p. 584)

Soundex Coding System An encoding scheme for surnames based on sound. (p. 590)

Universal Product Code (UPC) A bar code and identification number that is used on most retail items. The UPC code detects 100 percent of all single-digit errors and most other types of errors. (p. 575)

Weights Numbers used in the calculation of check digits. (p. 576)

ZIP code A five-digit code used by the U.S. Postal Service to divide the country into geographic units to speed sorting of the mail. ZIP stands for Zone Improvement Plan. (p. 582)

ZIP + 4 code The nine-digit code used by the U.S. Postal Service to refine ZIP codes into smaller units. (p. 583)

SKILLS CHECK

1. When a single incorrect digit is entered, an error-detecting code

(a) sometimes will detect the error.

(b) always will detect the error but may not be able to correct it.

(c) always will detect and correct the error.

2. If a U.S. Postal Service money order is numbered 1012065994*X*, where *X* indicates that the last digit is obliterated, *X* is ________________.

3. If the first five digits of a valid U.S. Postal Service money order are rearranged, the resulting number

________________ will have the same check digit as the original number.

(a) always

(b) sometimes

(c) never

4. If the number $19a_3a_4 \cdots a_{10}7$ is a valid Postal Service money order number and the number $7Xa_3a_4 \cdots a_{10}7$ is also a valid Postal Service money order number, the value of X is ________.

5. If $a_1a_2 \cdots a_{11}$ is a valid U.S. Postal Service money order number, then $a_1a_2 \cdots a_{10}-a_{11}$ is evenly divisible by

(a) 9.

(b) 10.

(c) 11.

6. The sum of the digits of a correctly coded American Express Travelers Cheque identification number is evenly divisible by ________________.

7. Is the number 105408970012 a legitimate airline ticket number?

(a) Yes.

(b) No, but if the final digit is changed to a 5, the resulting number, 105408970015, is legitimate.

(c) No, but if the final digit is changed to a 3, the resulting number, 105408970013, is legitimate.

8. If the digits of a VISA travelers cheque, excluding the check digit, add up to 36, the check digit is ________.

9. If 8103955 is a valid Avis identification number, which of the three following numbers is *not* detected as invalid by the Avis check-digit scheme?

(a) 1103955

(b) 8173955

(c) 8703955

10. If an American Express Travelers Cheque is numbered X425036791, where X indicates that the first digit is obliterated, X is ____.

11. If the first two digits of a valid airline ticket identification number are transposed, the resulting number will ____________ be valid.

(a) always

(b) sometimes

(c) never

12. A correctly coded UPC number has a weighted sum that is evenly divisible by ____________.

13. Identify each binary UPC code below as the manufacturer's part of the number or as the product part of the number.

(a) 100111

(b) 0011001

(c) 0001011

14. The check digit that should be appended to the UPC code 0-14300-25433 is ____________.

15. The bank routing number error detection scheme detects

(a) all transportation and most single-digit errors.

(b) all single-digit errors and most transpositions.

(c) all single-digit errors and all transpositions.

16. The check digit that should be appended to the bank routing number 01500085 is ____________.

17. The bank routing number error detection scheme

(a) detects the same errors as the UPC scheme.

(b) detects fewer errors than the UPC scheme.

(c) detects more errors than the UPC scheme.

18. A correctly coded 10-digit ISBN has a weighted sum that is evenly divisible by ____________.

19. Suppose the 10-digit ISBN 0-1750-3549-0 is reported incorrectly as 0-1750-3540-1. Which of the following statements is true?

(a) This error will not be detected by the check digit.

(b) While this particular error will be detected, the check digit does not detect all 2-digit errors in ISBNs.

(c) All 2-digit errors in a 10-digit ISBN are detectable by the check digit.

20. If a 10-digit ISBN number has an X in the check-digit position, the remainder of the weighted sum divided by 11 is________.

21. A valid 13-digit ISBN has a weighted sum evenly divisible by

(a) 9.

(b) 10.

(c) 11.

22. The ISBN-10 error detection scheme detects ________ percent of single-digit errors and ________ percent of transposition errors.

23. If an error in an identification number is made by transposing the first and third digits, the error

(a) usually is detected by the UPC scheme.

(b) always is detected by the bank scheme.

(c) always is detected by the ISBN-10 scheme.

24. If the weighted sum of the first 12 digits of a 13-digit ISBN is 56, the 13th character is ________.

25. If the ZIP code for a home begins with a 9, in which part of the United States is it located?

(a) the East

(b) the Midwest
(c) the West

26. The sum of the 10 digits of a ZIP code is evenly divisible by _______.

27. If a scanner misreads exactly one bar of a Postnet code, the computer will
(a) not always detect the error.
(b) always detect the error but will not always be able to correct it.
(c) always detect the error and correct it.

28. If the sixth digit of the Postnet code 20001-5800-7 is incorrect, the correct Postnet code is _______.

29. Which of the codes listed below does not use a check digit?
(a) Soundex
(b) Social security
(c) VIN

30. If the identification 48945 has been encoded by multiplying the original number by 13, the original number is _______________.

CHAPTER 16 EXERCISES

■ Challenge ▲ Discussion

16.1 Check Digits

1. Determine the check digit for a money order with identification number 3953981640.

2. Determine the value of X so that 7X345417803 is a valid money order identification number.

3. Determine the check digit for the Avis identification number 873345672.

4. Determine two values for X that will make 723459X0161 a valid Postal Service money order number.

5. Determine the check digit for the airline ticket number 30860422052.

6. Suppose a money order with the identification number and check digit 21720421168 is erroneously copied as 27750421168. Will the error be detected? Explain your reasoning.

7. Determine the check digit for the travelers cheque with identification number 661340874.

8. Determine the check digit for an Avis rental car with identification number 540047.

9. If a Postal Service money order with identification number $19a_3a_4 \ldots a_{10}$ has the check digit 5, what is the check digit for the Postal Service money order number $33a_3a_4 \ldots a_{10}$?

10. Which of the error detection schemes below will detect some errors of the form *abc* → *cba* (jump transposition)?
(a) UPC code
(b) Bank scheme
(c) 13-digit ISBN

11. If an Avis identification number $a_1a_2 \ldots a_{10}a_{11}8$ has the check digit 5, what is the check digit for the Avis identification number $a_1a_2 \ldots a_{10}a_{11}6$?

12. Find the check digit for the UPC number 03608072089.

13. Suppose that the packaging of a retail item were damaged in such a way that the first digit of a 12-digit UPC code was scratched off, but the remaining 11 digits were 88072303584; determine the first digit.

14. Determine the ISBN-10 check digit for the number 0-547-16509.

15. Determine the ISBN-13 check digit for the number 978-0-547-16509.

16. When the eighth edition of this textbook was in preparation, the publisher sent the author of this section the following ISBNs for the book: ISBN-10: 1-4292-0900-3; ISBN-13: 978-1-4292-0890-0. How did the author know the ISBN-13 was wrong, and how did he know how to correct it? (This really happened!)

17. When calculating the check digit for a 13-digit ISBN, why can you disregard the first two digits? (Try it for Exercise 16.)

18. Determine the check digit for the bank routing number 09100001.

19. Determine the check digit for an American Express Travelers Cheque with identification number 461212023.

■ 20. Suppose that a check digit is assigned to a four-digit number by appending the remainder after division by 7. If the number 36806 has a single-digit error in the first position, determine the possibilities for the correct number.

21. Determine whether the MasterCard number 3541 0232 0033 2270 is valid.

22. Suppose that the digit indicated by a question mark in the MasterCard number 426452002177?337 is unreadable. What is the unreadable number?

23. Suppose that a valid credit card number of the form $11a_3a_4 \ldots a_{16}$ is changed to $55a_3a_4 \ldots a_{16}$. Is the result a valid credit number?

24. Suppose that a valid credit card number of the form $22a_3a_4 \ldots a_{16}$ is changed to $55a_3a_4 \ldots a_{16}$. Is the result a valid credit number?

25. Suppose a correctly coded credit card number $a_1a_2 \ldots a_{16}$ is modified in such a way that the sum produced by the Luhn algorithm is increased by 10. Explain why the credit card error detection method does not detect the error. What happens if the sum is increased by 12?

26. If a number of the form $a_1a_2 \ldots a_{13}a_{14}34$ is a valid credit card number, what is the check digit for a credit number of the form $a_1a_2 \ldots a_{13}a_{14}5$?

27. If a credit card number of the form $83a_3 \ldots a_{15}4$ is a valid credit card number, what is the check digit for a credit number of the form $19a_3 \ldots a_{15}$?

28. Replace the question mark in the number JM1GD222?J1581570 with a digit that will result in a valid VIN. (See Spotlight 16.1 on page 581 for a description of the method to be used.)

29. Create a check digit for the UPC number 38137009213 using the weights 7, 1, 7, 1, 7, 1, . . ., 7, 1, instead of 3, 1, 3, 1, 3, 1, . . ., 3, 1. Test to see whether this check digit will detect single-digit errors by trying several examples.

30. Create a check digit for the UPC number 38137009213 using the weights 2, 1, 2, 1, 2, 1, . . ., 2, 1, instead of 3, 1, 3, 1, 3, 1, . . ., 3, 1. Is the error caused by replacing the 3 in the first position with an 8 detected? What about the error caused by replacing the 1 in the third position with a 6? Explain why or why not.

31. If the weights 5, 1, 5, 1, 5, 1, . . ., 5, 1 were used for the UPC code, which single-digit errors would go undetected?

32. Exercises 29, 30, and 31 reveal that using the weights 1, 3, or 7 for a particular position detects all errors in that position, whereas using weights 2 or 5 in a position does not detect all errors. Using this observation, make a guess about error-detection capability using weights 9, 4, 6, or 8.

33. Enter the number 036000260809 in a Google search box. Is it a valid UPC number? Now change any one digit. Is the new number a valid UPC number? Try a different change.

34. Instead of calculating the codabar check digit as described in this chapter, use the following method: To assign a check digit to a 15-digit identification number $a_1a_2a_3a_4a_5a_6a_7a_8a_9a_{10}a_{11}a_{12}a_{13}a_{14}a_{15}$, first write the expression $2a_1 + a_2 + 2a_3 + a_4 + 2a_5 + a_6 + 2a_7 + a_8 + 2a_9 + a_{10} + 2a_{11} + a_{12} + 2a_{13} + a_{14} + 2a_{15}$. Then, instead of adding the individual terms, add the digits of each term so that the term 10 is counted as $1 + 0$; 12 is counted as $1 + 2$; 14 is counted as $1 + 4$; and so on. (For example, the number 312560019643001 yields $2 \cdot 3 + 1 + 2 \cdot 2 + 5 + 2 \cdot 6 + 0 + 2 \cdot 0 + 1 + 2 \cdot 9 + 6 + 2 \cdot 4 + 3 + 2 \cdot 0 + 0 + 2 \cdot 1 = 6 + 1 + 4 + 5 + 12 + 0 + 1 + 18 + 6 + 8 + 3 + 0 + 0 + 2$, which we change to $6 + 1 + 4 + 5 + 1 + 2 + 0 + 1 + 1 + 8 + 6 + 8 + 3 + 0 + 0 + 2 = 48$.)

The check digit is whatever is needed to bring the final tally to a number that ends with 0. So, in our example, the check digit is 2. Use this method to determine the check digit for the number given by the first 15 digits in Exercise 21. Is it the same as the check digit for the number in Exercise 21 on page 596?

35. Explain why the check-digit method described in Exercise 34 gives the same result as the Codabar method described in this chapter.

■ **36.** State a general criterion for the detection of an error of the form $\cdots abc \cdots \rightarrow \cdots cba \cdots$ for the routing number of a checking account.

■ **37.** The 10-digit ISBN 0-669-03925-4 is the result of a transposition of two adjacent digits not involving the first or last digit. Determine the correct ISBN.

38. Explain why the bank scheme will detect the error $751 \cdots \rightarrow 157 \cdots$ but the UPC scheme will not.

39. Suppose that the check digit a_9 for the bank routing number was chosen to be the last digit of $3a_1 + 7a_2 + a_3 + 3a_4 + 7a_5 + a_6 + 3a_7 + 7a_8$ instead of using the method described in this chapter. How would this compare with the actual check digit?

40. Explain why an error caused by transposing the first two digits of a Postal Service money order is not detected by the check-digit scheme. Explain why the same is true for the second and third digits. What about the last two digits?

41. Suppose that a company assigns an extra digit to every employee social security number by appending a 0 if the sum of the digits is even and a 1 if the sum is odd. If a 2 were mistakenly read as a 7, would the error be detected? What if a 2 were mistakenly read as an 8? Try a few other experiments with single-digit errors. (For experiments, you can use three-digit numbers instead of nine-digit numbers.) Determine which errors are detected by this method. Explain your reasoning. Approximately what percentage of errors is detected by appending the extra digit?

42. Explain why an error caused by transposing any two digits of an American Express Travelers Cheque is not detected by the check-digit scheme.

43. When using the travelers cheque, credit card, or UPC number algorithms for detecting errors, does the computer have to know which digit is the check digit?

44. Explain why the Postal Service money order check-digit scheme does not detect the mistake of substituting a 0 for a 9, or vice versa.

45. Which digit never appears as a check digit on a Postal Service money order?

46. Which digit never appears as a check digit on an American Express Travelers Cheque?

47. Which digits never appear as a check digit for an airline identification number?

48. Suppose that four-digit numbers $a_1a_2a_3a_4$ are assigned a check digit a_5 so that $a_1 + 2a_2 + a_3 + 2a_4 + a_5$ is evenly divisible by 10. Test the number 43216 created in this way to see whether the method detects adjacent-digit transposition errors.

49. Starting with the 10-digit ISBN 0-7167-4782-0, create three new numbers by transposing any two different digits. (They need not be adjacent.) Are these errors detected by the scheme?

50. Suppose that in an Avis identification number, an 8 is mistaken for a 5. Is the error detected? What if a 9 is mistaken for a 2?

51. Give an argument to show that the 10-digit ISBN error-detection method will detect a transposition error involving the first and third digits. Does the same argument work for the fourth and sixth digits?

■ **52.** Suppose the check digit a_{10} of 10-digit ISBNs were chosen so that $a_1 + 2a_2 + 3a_3 + 4a_4 + 5a_5 + 6a_6 + 7a_7 + 8a_8 + 9a_9 + 10a_{10}$ is divisible by 11 instead of using the method described in the chapter. How would this compare with the actual check digit?

■ **53.** Consider a UPC number in which the digits 7 and 2 appear consecutively (that is, the number has the form $\cdots 72 \cdots$). Will the error caused by transposing these digits (that is, the number is taken as $\cdots 27 \cdots$) be detected? What if the digits 6 and 2 were transposed instead? State the general criterion for the detection of an error of the form $\cdots ab \cdots \rightarrow \cdots ba \cdots$ by the UPC scheme.

54. If the first three digits of a routing number for a checking account are 537 and the 5 and 3 are transposed, will the error be detected? If the first three numbers are 237 and the 2 and 7 are transposed, will the error be detected?

■ **55.** The Canadian province of Quebec assigns a check digit a_{12} to an 11-digit driver's license number $a_1a_2 \cdots a_{11}$ so that $12a_1 + 11a_2 + 10a_3 + 9a_4 + 8a_5 + 7a_6 + 6a_7 + 5a_8 + 4a_9 + 3a_{10} + 2a_{11} + a_{12}$ is divisible by 10. Criticize this method. Describe all single-digit errors that are undetected by this scheme.

56. Speculate on the reason why telephone numbers, social security numbers, and serial numbers on most currency do not have check digits.

57. Suppose that a company uses a check-digit scheme similar to the UPC scheme, except that instead of using the UPC weights 3, 1, 3, 1, . . ., it uses w, 1, w, 1, ... If two of the ID numbers used by the company are 73215674 and 73215661, determine w.

■ **58.** If a publishing company has headquarters in both the United States and Germany and publishes the same book in both countries, it is likely that the 10-digit ISBN for the book will be identical except for the first and last digits (because the first digit for U.S. publications is 0 and the first digit for German publications is 3). If the last digit of the U.S. edition is 1, what is the last digit for the German publication?

16.3 Bar Codes

59. Determine the ZIP + 4 code and check digit for each of the following Postnet bar codes:

(a)

(b)

(c)

60. Determine the ZIP + 4 code and check digit for each of the following Postnet bar codes:

(a)

(b)

(c)

61. In each of the following Postnet bar codes, exactly one mistake occurs (that is, a long bar appears instead of a short one, or vice versa). Determine the correct ZIP code.

(a)

(b)

(c)

62. Below is a 12-digit delivery-point bar code. Determine the ZIP + 4 number, the last two digits of the street address, and the check digit.

63. If the check digit for the ZIP + 4 code for a house on 1738 Maple Street is 3, what would the check digit for the delivery-point code be?

64. Find the check digit for the ZIP + 4 code 50037-2452.

65. Explain why any two errors in a particular block of five bars in a Postnet code are always detectable. Explain why not all such errors can be corrected.

66. Change 173 into a Postnet code.

67. Form all possible strings consisting of exactly three *a*'s and two *b*'s and arrange the strings in alphabetical order. (For example, the first two possibilities are *aaabb* and *aabab*.) Do you see any relationship between your list and the Postnet code?

68. The back cover of recently published books includes a bar code that has the 10-digit ISBN above the bars and a 13-digit identification number below the bars. (See below.) Examine the bar code on several books. How does the number below the bar code differ from the UPC code? How is the number below the bar code related to the ISBN?

69. Suppose that the first block of a UPC bar code following the guard bar pattern that a scanner reads is 1000100. Is the scanner reading left to right or right to left?

70. The following is an actual identification number and bar code from a roll of wallpaper. What appears to be wrong with them? Speculate on the reason for the apparent violation of the UPC format.

Building Regulations: 1985 Class 0
FINE ART WALLCOVERINGS LTD.
HOLMES CHAPEL, CHESHIRE
MADE IN ENGLAND
FABRIQUE EN ANGLETERRE

16.4 Encoding Personal Data

71. Judging from the information in Spotlight 16.4 (on p. 591), which three states were most likely to have had the smallest populations when social security numbers were allocated to the states?

72. What geographical information was used in allotting the first three digits of social security numbers to the states?

73. If a social security number begins with 0, the number was issued in a state in the

(a) Northeast.

(b) Midwest.

(c) West.

74. What demographic information was used in allotting the first three digits of social security numbers to the states?

75. As of 2010, there were seven states with a population under 1 million. Use the data in Spotlight 16.4 to identify those seven states.

■ **76.** The Canadian postal system has assigned each geographic region a six-character code composed of alternating letters (not including D, F, I, O, Q, and U) and digits, such as P7B5E1 and K7L3N6. Discuss the advantages that this scheme has over the five-digit ZIP code used in the United States.

77. Given that the letters D, F, I, O, Q, and U are not used in the Canadian postal code (see the previous exercise), determine the maximum possible number of postal codes.

78. Determine the Soundex code for the names *Hu*, *Lee*, and *Shaw*.

79. Determine the Soundex code for *Skow*, *Sachs*, *Lennon*, *Lloyd*, *Ehrheart*, and *Ollenburger*.

80. Determine the Soundex code for *Smith*, *Schmid*, *Smyth*, and *Schmidt*.

81. Determine the total number of codes possible using the Soundex Coding System.

82. In Florida, the last three digits of the driver's license number of a female with birth month m and birth date b are $40(m - 1) + b + 500$. For both males and females, the fourth and fifth digits from the end give the year of birth. Determine the last five digits of a Florida driver's license number for a female born on July 18, 1942.

83. Explain why an Illinois driver's license number that ends with the last five digits 03217 is suspicious.

84. Determine the last five digits of an Illinois driver's license number for a male born on June 18, 1942.

85. In Illinois, one obtains the last three digits of the driver's license number for a female by adding 600 to the number for a male with the same birthday. In Florida, 500 is added to the number for a male. Why can't Florida use 600?

86. Explain why an Illinois driver's license number that ends with 77061 cannot be valid.

87. Determine the birth date of a person whose Illinois driver's license number ends with 58818.

88. Provide three names that share the same Soundex code as *Gallihan*.

89. Another math book describes the Soundex algorithm for the surname code as follows:

(a) Leave the first letter alone, then cross off all occurrences of the letters *a*, *e*, *i*, *o*, *u*, *y*, *h* and *w*.

(b) Cross off the second of any double letters.
(c) Leave the first letter alone, and replace each of the other letters with the appropriate number (using the same assignment as given on page 590).
(d) The code is the first letter of the surname followed by the first three numbers.

Compare the codes for *Jackson*, *Mnack*, and *Shaw* using this method and the method given on page 590. (The method on page 590 is the correct one.)

90. The state of Washington encodes the last two digits of the year of birth into driver's license numbers (in positions 8 and 9) by subtracting the two-digit number from 100. For example, a person born in 1942 has 58 in positions 8 and 9, whereas a person born in 1971 has 29 in positions 8 and 9. Speculate on the reason for subtracting the birth year from 100.

91. Driver's license number-assignment schemes that use personal data sometimes produce the same number for different people. Speculate about circumstances under which this is more likely to occur.

92. Apply the Soundex code to common ways to misspell your name. Do they give the same code as your name does?

93. Why would the Soundex system of coding be a poor method for encoding names in China?

WRITING PROJECTS

1. Prepare a two-page report on coded information in your location. Possibilities for investigation include driver's license numbers in your state, student ID numbers and bar codes at your school, and bar codes used by your school library and city library. Identify the coding schemes and, when possible, determine whether a check digit is employed. Include samples. The "Suggested Readings" section at the end of this chapter lists sources of information that will assist you.

2. Prepare a two-page report on the driver's license coding schemes used by Michigan, Maryland, and Washington (Michigan and Maryland use the same method). J. Gallian's "Assigning Driver's License Numbers" has the information you will need (see the "Suggested Readings" section at the end of this chapter).

3. Use the Web to find material for a two-page report on the history of the bar code.

4. Use the Web to find material for a two-page report on the Barcode of Life project.

5. Use the Web to find material for a two-page report on smart card technology.

6. Use the Web to find material for a two-page report on the Soundex system.

Suggested Readings

GALLIAN, J. The mathematics of identification numbers, *College Mathematics Journal*, 22 (1991): 194–202. A survey of check-digit schemes associated with identification numbers. Available at www.d.umn.edu/~jgallian/ident.pdf.

GALLIAN, J. Assigning driver's license numbers, *Mathematics Magazine*, 64 (1992): 13–22. Discusses various methods used by the states to assign driver's license numbers. Several of these methods include check digits for error detection. Available at www.d.umn.edu/~jgallian/license.pdf.

GALLIAN, J. Error detection methods, *ACM Computing Surveys*, 28 (1996): 504–17. A detailed description of many error-detection methods. Available at www.d.umn.edu/~jgallian/detection.pdf.

GALLIAN, J., and S. WINTERS. Modular arithmetic in the marketplace, *American Mathematical Monthly*, 95 (1988): 548–51. A detailed analysis of the check-digit schemes presented in this chapter. In particular, the error-detection rates for the various schemes are given. Available at www.d.umn.edu/~jgallian/marketplace.pdf.

KIRTLAND, J. *Identification Numbers and Check Digit Schemes*, Mathematical Association of America, Washington, DC, 2001. This book provides more examples and exercises for the check-digit schemes discussed in this chapter.

Suggested Web Sites

www.d.umn.edu/~jgallian/fapp7 This Web site enables users to calculate check digits using the various methods discussed in this chapter.

http://osama-oransa.blogspot.com/2010_08_01_archive.html This blog article explains several kinds of cryptographic methods.

© Michael Kai/Corbis

17

Information Science

With the widespread use of the Internet, cell phones, satellite TV and radio, Facebook, Twitter, downloads, YouTube, iTunes, and text messages, the digital revolution has brought about a number of mathematical challenges, such as how to transmit data accurately, how to represent information compactly, how to keep information secret, and how to search the Web efficiently.

In the first two sections of this chapter, we illustrate how messages consisting of 0's and 1's can be created so that errors in transmission can be corrected automatically, as well as how data can be stored efficiently. Section 17.3 explains various ways to protect the security of messages. In the last section, we describe how mathematical logic relates to the Internet.

17.1 Binary Codes

Coded data made up of two states (or symbols) is called a **binary code**. Binary codes are the hidden language of computers. The Postnet code (short and long bars) and the Universal Product Code (UPC) bar code (white spaces and dark bars) explained in Chapter 16 are two examples of binary codes. Morse code (dots and dashes) and Braille (bumps and flat markings) are two more. The opinions of two critics rating movies with U for "thumbs up" and D for "thumbs down" could be conveyed by a binary code with the four messages UU, UD, DU, and DD. Fax machines, CDs, DVDs, high-definition television signals, cell phones, and space probes represent data as strings of 0's and 1's rather than the usual digits 0 through 9 and letters *A* through *Z*. In this section, we will illustrate one way binary codes can be devised so that errors in the transmission of the code can be corrected.

The idea behind error-correction schemes is simple and one you often use. To illustrate, suppose that you are reading the employment section of a newspaper and you see the phrase "must have a minimum of bive years' experience." Instantly you detect an error because *bive* is not a word in the English language. Moreover, you are fairly confident that the intended word is *five*. Why is that? Because *five* is a word derived from *bive* by changing a single letter, and it makes the phrase understandable. In other phrases, words such as *bike* or *give* might be sensible alternatives to *bive.* Using the extra information provided by the context, we often are able to infer the intended meaning when errors occur.

To demonstrate the way binary error-correcting schemes work, suppose that the National Aeronautics Space Administration (NASA) sends a spacecraft to land at 1 of 16 possible landing sites on Mars. The spacecraft orbits Mars while surveying the sites for the most favorable landing

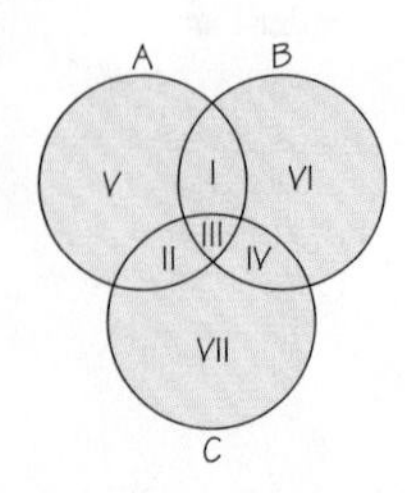

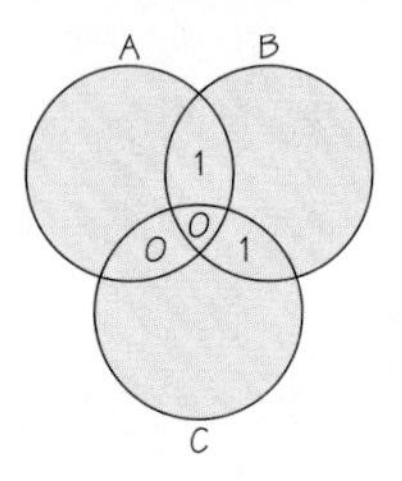

Figure 17.1 Diagram for message 1001.

conditions. NASA officials have coded the 16 landing sites with four-digit strings of 0's and 1's such as 0000, 0001, 0010, and 0100. (Recall when counting that if each of k events can occur in $n_1, n_2, \ldots, n_k$ ways, then the total number of ways all k events can occur is the product $n_1 n_2 \cdots n_k$. In this case, there are $2 \cdot 2 \cdot 2 \cdot 2 = 16$ strings; see Table 17.1 on page 605 for the complete list.)

Once the best site has been selected, NASA will inform the spacecraft where to land by sending the code for the site. However, signals sent through space are subject to interference called *noise.* The noise might cause the spacecraft to interpret the signal as 0001 when the signal actually sent was 1001. Fortunately, over the past 60 years, mathematicians and engineers have devised highly sophisticated schemes to build extra information into messages composed of 0's and 1's that often permits the correct message to be inferred even though the message may have been received incorrectly (see Spotlights 17.1 and 17.2).

As a simple example, let's assume that our message is 1001. We will build extra information into this message with the aid of the diagram in Figure 17.1.

SPOTLIGHT 17.1

The Ubiquitous Reed-Solomon Codes

One of the mathematical ideas underlying current error-correcting techniques for everything from computer hard-disk drives to CD and DVD players was introduced in 1960 by Irving Reed and Gustave Solomon. Reed-Solomon codes made possible the stunning pictures of the outer planets sent back by the space probes *Voyager 1* and *2*. They make it possible to scratch a CD or a DVD and still enjoy the content.

"When you talk about CD players, digital television, and various other digital systems—all of those Reed-Solomon [codes] are an integral part of the system," says Robert McEliece, a coding theorist at Caltech.

Why? Because digital information consists of 0's and 1's, and a physical device may confuse the two occasionally. *Voyager 2*, for example, was transmitting data at incredibly low power over billions of miles. Error-correcting codes are a kind of safety net, mathematical insurance against the vagaries of an imperfect material world.

In 1960, the theory of error-correcting codes was only about a decade old. Through the 1950s, a number of researchers began experimenting with a variety of error-correcting codes. But the Reed-Solomon paper, McEliece says, "hit the jackpot." "In hindsight, it seems obvious," Reed later said. However, he added, "Coding theory was not a subject when we published the paper." The two authors knew they had a nice result; they didn't know what impact the paper would have.

Irving Reed (left) and Gustave Solomon
At the Jet Propulsion Laboratory in 1989 to monitor the encounter of *Voyager 2* with Neptune. *(Rex Ridenhouse.)*

Five decades later, the impact is clear. The vast array of applications has settled the questions of the practicality and significance of Reed-Solomon codes. Billions of dollars in modern technology depend on ideas that stem from Reed and Solomon's original work.

Source: Adapted from the article "The Ubiquitous Reed_ Solomon codes" by Barry Cipra, with permission from *SIAM News*, January 1993, p. 1.

SPOTLIGHT 17.2

Vera Pless

Vera Pless
A leader in the field of coding theory.
(Courtesy of Vera Pless.)

Vera Pless was born on March 5, 1931, to Russian immigrants on the West Side of Chicago. The neighborhood was intellectually stimulating, and there was a tradition of teaching each other things. At age 12, Pless was taught some calculus by a mathematics graduate student. She accepted a scholarship to attend the University of Chicago at age 15. The program at Chicago emphasized great literature but paid little attention to physics and mathematics. At age 18, with no more than one pre-calculus course in mathematics, she entered the prestigious graduate program in mathematics at the University of Chicago, where, at that time, there were no women on the mathematics faculty, or even women colloquium speakers. After receiving her master's degree, Pless took a job as a research associate at Northwestern University while pursuing a Ph.D. there. In the midst of writing her thesis, she moved to Boston with her husband and continued to work on her thesis at home. She defended her thesis two weeks before her daughter was born.

Over the next several years, Pless stayed at home to raise her children and taught part time at Boston University. When she decided to work full time, she found that women were not welcome at most colleges and universities. Some people told her outright, "I would never hire a woman." Fortunately, there was an Air Force lab in the area that had a group working on error-correcting codes. Although she had never even heard of coding theory, she was hired because of her background in algebra. In 1975, she went to the University of Illinois–Chicago, where she remained until her retirement. Having written more than 100 research papers and a widely used book on coding theory, Pless is a leader in the field.

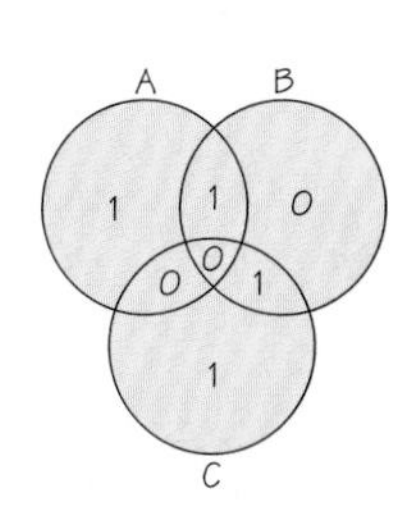

Figure 17.2 Diagram for encoded message 1001101.

Begin by placing the four message digits in the four overlapping regions I, II, III, and IV, with the digit in the first position (starting at the left of the sequence) in region I, the digit in the second position in region II, and so on. For regions V, VI, and VII, assign 0 or 1 so that the total number of 1's in each circle is even (see Figure 17.2).

Using the diagram, we have now encoded our message 1001 as 1001101. Now suppose that this encoded message is received as 0001101 (an error in the first position). How would we know an error was made? We place each digit from the received message in its appropriate region, as in Figure 17.3.

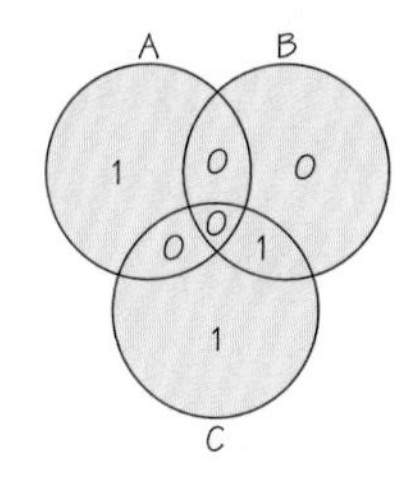

Figure 17.3 Diagram for received message 0001101.

Noting that there is an odd number of 1's in both circles A and B, we instantly realize that something is wrong because the intended message had an even number of 1's in each circle. How do we correct the error? The answer involves parity. (Parity refers to the oddness or evenness of a number; even integers have **even parity** and odd integers have **odd parity**.) Because circles A and B have the wrong parity and C does not, the error is located in the portion of the diagram in circles A and B, but not in circle C, that is, region I (shaded in Figure 17.4). Here we also see the advantage of using only 0's and 1's to encode data. If you have only two possibilities and one of them is incorrect, then the other one must be correct. Because the 0 in region I is incorrect, we know that 1 is correct.

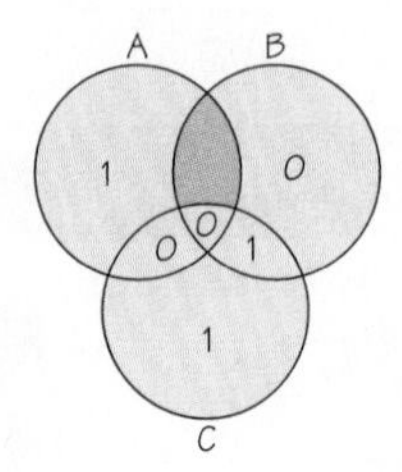

Figure 17.4 Circles A and B, but not C, have the wrong parity.

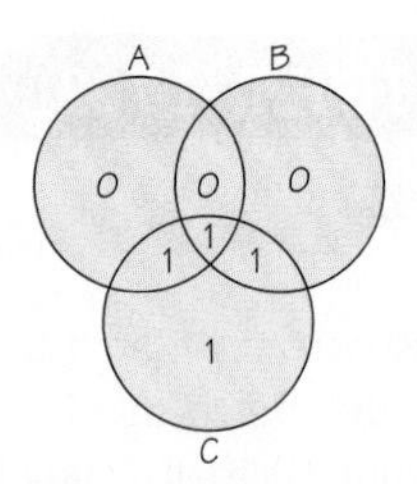

Figure 17.5 Diagram for encoded message 0111001.

For practice, we will do another example. Consider the message 0111. Proceeding as before, we place 0 in region I and 1's in regions II, III, and IV. For regions V, VI, VII, we assign 0 or 1 so that the total number of 1's in each circle is even (see Figure 17.5). Then the message 0111 is encoded as 0111001. If this code word is received as, say, 0111011 (error in the sixth position), the diagram for the received word is shown in Figure 17.6. Then circle B has an odd number of 1's, and circles A and C have an even number. If the received message has only one error, then the error must be in circle B but not circles A and C. This tells us that the entry is region VI is incorrect. So, the error can be corrected.

Because the circle diagram method was designed to correct a received message with exactly one error, if a received message has two or more errors, the diagram method will never yield the correct message. For example, if the encoded message 0111001 is received as 1111011 (errors in positions 1 and 6), then only circle A has an odd number of 1's (see Figure 17.7), so our decoding method assumes that there is an error in a region in circle A but not in regions in circles B and C. Thus, our method incorrectly assumes that a single error occurred in region V (shaded in Figure 17.8).

The diagram technique can be used to encode all 16 possible binary messages of length 4, as shown in Table 17.1. The encoded messages are called **code words**. The three digits appended to each string of length 4 provide the "extra information" that is sufficient to infer the intended four-digit message, so long as the received seven-digit message has at most one error.

Because the diagram method can be used only to encode four-digit binary messages, it is not practical. We used it merely to illustrate a simple way to encode and decode messages. In the next section, we explain a method that can be used in more general settings.

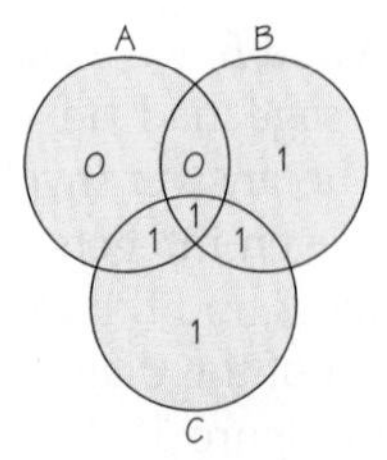

Figure 17.6 Diagram for encoded message 0111011.

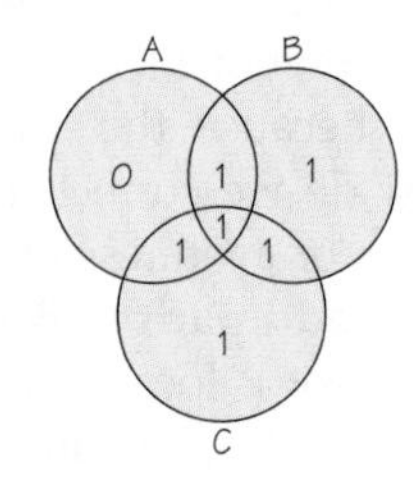

Figure 17.7 Diagram for encoded message 1111011.

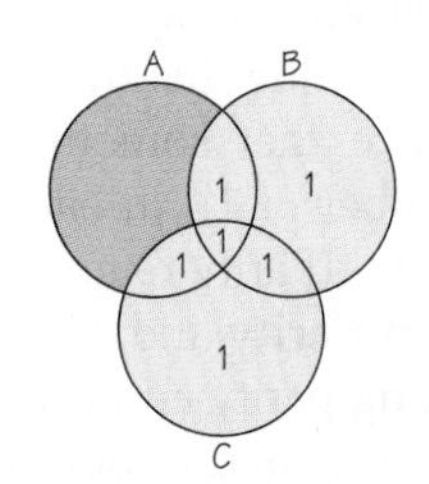

Figure 17.8 Circle A, but not B and C, has the wrong parity.

Table 17.1

Code Words

Message	→	Code Word	Message	→	Code Word
0000	→	0000000	0110	→	0110010
0001	→	0001011	0101	→	0101110
0010	→	0010111	0011	→	0011100
0100	→	0100101	1110	→	1110100
1000	→	1000110	1101	→	1101000
1100	→	1100011	1011	→	1011010
1010	→	1010001	0111	→	0111001
1001	→	1001101	1111	→	1111111

17.2 Encoding with Parity-Check Sums

Strings of 0's and 1's with extra digits for error correction can be used to send full-text messages. A simple way to do this is to assign an empty space the string 00000, the letter *a* the string 00001, *b* the string 00010, *c* the string 00100, and so on. Because there are 32 possible binary strings of length 5, the five unassigned strings can be used for special purposes, such as indicating uppercase letters or numerals.

For example, we might use the string 11111 to indicate a "shift" from lowercase to uppercase when it precedes the code for a letter (i.e., 1111100010 represents *B*). This is analogous to the shift key on a keyboard. Similarly, we could use 11110 to indicate a "shift" from letters to numerals. Here, 11110 followed by the code for *a* represents the numeral 0, 11110 followed by the code for *b* represents the numeral 1, and so on up to 9. Punctuation marks could be handled in the same fashion. Rather than using the circle diagram method, messages are encoded by appending extra digits determined by the parity of various sums of certain portions of the messages. We illustrate this method for the 16 messages shown in the left-hand column of Table 17.1. (See also Spotlight 17.3.)

Our goal is to take any binary string $a_1a_2a_3a_4$ and append three check digits $c_1c_2c_3$ so that any single error in any of the seven positions can be corrected. This is done as follows: Choose

$$\begin{aligned}
c_1 &= 0 \text{ if } a_1 + a_2 + a_3 \text{ is even}\\
c_1 &= 1 \text{ if } a_1 + a_2 + a_3 \text{ is odd}\\
c_2 &= 0 \text{ if } a_1 + a_3 + a_4 \text{ is even}\\
c_2 &= 1 \text{ if } a_1 + a_3 + a_4 \text{ is odd}\\
c_3 &= 0 \text{ if } a_2 + a_3 + a_4 \text{ is even}\\
c_3 &= 1 \text{ if } a_2 + a_3 + a_4 \text{ is odd}
\end{aligned}$$

The sums $a_1 + a_2 + a_3$, $a_1 + a_3 + a_4$, and $a_2 + a_3 + a_4$ are called **parity-check sums**. They are so named because their function is to guarantee that the sum of various components of the encoded message is even. Indeed, c_1 is defined so that $a_1 + a_2 + a_3 + c_1$ is even. (Recall that this is precisely how the value in region V in Figure 17.2 was defined.) Similarly, c_2 is defined so that $a_1 + a_3 + a_4 + c_2$ is even, and c_3 is defined so that $a_2 + a_3 + a_4 + c_3$ is even.

SPOTLIGHT 17.3

Neil Sloane

In the middle of Neil Sloane's office, which is in the center of AT&T Bell Laboratories, which in turn is at the heart of the information age, there sits a tidy little pyramid of shiny steel balls stacked up like oranges at a neighborhood grocery. Sloane has been pondering different ways to pile up balls of one kind or another for most of his professional life. Along the way, he has become one of the world's leading researchers in the field of sphere packing, a field that has become indispensable to modern communications. Without it, we might not have CDs, DVDs or satellite photos of Neptune. "Computers would still exist," says Sloane. "But they wouldn't be able to talk to one another."

To exchange information rapidly and correctly, machines must code it. As it turns out, designing a code is a lot like packing spheres: Both activities involve cramming things together into the tightest possible arrangement. Sloane, fittingly, is also one of the world's leading coding theorists, not least because he has studied the shiny steel balls on his desk so intently.

Here's how a code might work. Imagine, for example, that you want to transmit a child's drawing that uses every one of the 64 colors found in a jumbo box of Crayola crayons. For transmission, you could code each of those colors as a number—say, the integers from 1 to 64. Then you could divide the image into many small units, or pixels, and assign a code to each one based on the color that it contains. The transmission then would be a steady stream of those numbers, one for each pixel.

In digital systems, however, all those numbers would have to be represented as strings of 0's and 1's. Because there are 64 possible combinations of 0's and 1's in a six-digit string, you could handle the entire Crayola palette with 64 different six-digit "code words." For example, 000000 could represent the first color, 000001 the next color, 000010 the next, and so on.

Neil Sloane
At work, wearing his famous "Codemart" T-shirt (952 points in a sphere). *(Courtesy of Neil Sloane/AT&T Labs.)*

But in a noisy signal, two different code words might look practically the same. A bit of noise, for example, might shift a spike of current to the wrong place, so that 001000 looks like 000100. The receiver then might color someone's eyes wrongly. An efficient way to keep the colors straight in spite of noise is to add four extra digits to the six-digit code words. The receiver, programmed to know the 64 permissible combinations, now could spot any other combination as an error introduced by noise; then it would correct the error automatically to the "nearest" permissible color.

In fact, says Sloane, "If any of those 10 digits were wrong, you could still figure out what the right crayon was."

Source: Adapted from the article "Math in a Million Dimensions" by David Berreby, *Discover*, October 1990.

Let's revisit the message 1001 that we considered in Figure 17.1. Then, $a_1a_2a_3a_4 = 1001$ and

$$c_1 = 1 \text{ because } 1 + 0 + 0 \text{ is odd}$$
$$c_2 = 0 \text{ because } 1 + 0 + 1 \text{ is even}$$

and

$$c_3 = 1 \text{ because } 0 + 0 + 1 \text{ is odd.}$$

So, because $c_1c_2c_3 = 101$, we have $1001 \rightarrow 1001101$.

Now how is the intended message determined from a received encoded message? This process is called **decoding**. Say, for instance, that the message 1000, which has been encoded using parity-check sums as u = 1000110, is received as v = 1010110 (an error in the third position). We simply compare v with each of the 16 code words (that is, the possible correct messages) in Table 17.1 and decode it as the one that differs from v in the fewest positions. (Put another way, we decode v as the code word that agrees with v in the most positions.) This method works even if the error in the message is one of the check digits rather than one of the digits of the original message string. When there is more than one code word that differs from v in the fewest positions, we do not decode. To carry out this comparison, it is convenient to define the distance between two strings of equal length.

Distance Between Two Strings DEFINITION

The **distance between two strings** of equal length is the number of positions in which the strings differ.

For example, the distance between v = 1010110 and u = 1000110 is 1 because they differ in only one position (the third). In contrast, the distance between 1000110 and 0111001 is 7 because they differ in all seven positions. Thus, our decoding procedure is simply to decode any received message w as the code word w' that is "nearest" to it. More specifically, for any received word w, we determine the distance between w and all code words and decode w as the one for which the distance is a minimum. Table 17.2 shows the distance between v = 1010110 and all 16 code words. From this table, we see that v will be decoded as 1000110 = u because it differs from u in only one position, whereas it differs from all others in the table in at least two positions. This method is called **nearest-neighbor decoding**.

Table 17.2

Distances from 1010110 to Code Words

v	1010110	1010110	1010110	1010110	1010110	1010110	1010110	1010110
Code word	0000000	0001011	0010111	0100101	1000110	1100011	1010001	1001101
Distance	4	5	2	5	1	4	3	4
v	1010110	1010110	1010110	1010110	1010110	1010110	1010110	1010110
Code word	0110010	0101110	0011100	1110100	1101000	1011010	0111001	1111111
Distance	3	4	3	2	5	2	6	3

Assuming that errors occur independently, the nearest-neighbor method decodes each received message as the one it most likely represents.

Nearest-Neighbor Decoding DEFINITION

The **nearest-neighbor decoding** method decodes a received message as the code word that agrees with the message in the most positions, provided that there is only one such code word.

The scheme we have just described was first proposed in 1948 by Richard Hamming, a mathematician at Bell Laboratories. It is one of a family of codes that are called the *Hamming codes*. Strings obtained from all possible messages of a given length of 0's and 1's by appending extra 0's and 1's using parity-check sums, as illustrated earlier, are called **binary linear codes**. The strings with the appended digits are called **code words**.

Binary Linear Code/Code Words

DEFINITION

A **binary linear code** is a set of words composed of 0's and 1's obtained from all possible messages of a given length by using parity-check sums to append check digits to the messages. The resulting strings are called **code words**.

You should think of a binary linear code as a set of n-digit strings in which each string is composed of two parts: the message part, consisting of the original messages, and the remaining check-digit part.

The longer the messages are, the more check digits are required to correct errors. For example, binary messages consisting of six digits require four check digits to ensure that all messages with one error can be decoded correctly.

Given a binary linear code, how can we tell whether it will correct errors and how many errors it will detect? It is remarkably easy. We examine all the code words to find one that has the fewest number of 1's, excluding the *zero code word* consisting entirely of 0's. Call this minimum number of 1's in any nonzero code word the *weight* of the code.

Weight of a Binary Code

DEFINITION

The **weight of a binary code** is the minimum number of 1's that occur among all nonzero code words of that code.

The test for the error-detecting and error-correcting capacities of a code is the following.

Test for Error Detection and Correction Capacity

PROCEDURE

1. Calculate the weight t of the code.
2. The code will detect any $t - 1$ or fewer errors.
3. If t is odd, the code will correct any $(t - 1)/2$ or fewer errors.
4. If t is even, the code will correct any $(t - 2)/2$ or fewer errors.

Applying this test to the code in Table 17.1, we see that the weight is 3, so it will correct any $(3 - 1)/2 = 1$ error or it will detect any $3 - 1 = 2$ errors. Be careful here. We must decide *in advance* whether we want our code to correct single errors or detect any two errors. It can do whichever we choose, but not both. If we decide to detect errors, then we will not decode any message that was not among our original list of encoded messages (just as *bive* is not a word in the English language). Instead, we simply note that an error was made and, in most applications, request a retransmission. An example of this occurs when a bar-code reader at the supermarket detects an error and therefore does not emit a sound (in effect, requesting a rescanning). On the other hand, if we decide to correct errors, we will decode any received message as its nearest neighbor.

Here is an example of another binary linear code. Let the set of messages be {000, 001, 010, 100, 110, 101, 011, 111} and append three check digits c_1, c_2, and c_3 using

$$c_1 = 0 \text{ if } a_1 + a_2 + a_3 \text{ is even}$$
$$c_1 = 1 \text{ if } a_1 + a_2 + a_3 \text{ is odd}$$
$$c_2 = 0 \text{ if } a_1 + a_3 \text{ is even}$$
$$c_2 = 1 \text{ if } a_1 + a_3 \text{ is odd}$$
$$c_3 = 0 \text{ if } a_2 + a_3 \text{ is even}$$
$$c_3 = 1 \text{ if } a_2 + a_3 \text{ is odd}$$

For example, if we take $a_1a_2a_3$ as 101, we have

$$c_1 = 0 \text{ because } 1 + 0 + 1 \text{ is even}$$
$$c_2 = 0 \text{ because } 1 + 1 \text{ is even}$$
$$c_3 = 1 \text{ because } 0 + 1 \text{ is odd.}$$

So we encode 101 by appending 001, that is, 101 → 101001. The entire code is shown in Table 17.3.

Table 17.3

Code Words

Message	→	Code Word	Message	→	Code Word
000	→	000000	110	→	110011
001	→	001111	101	→	101001
010	→	010101	011	→	011010
100	→	100110	111	→	111100

Because the minimum number of 1's of any nonzero code word is three, this code will either correct any single error or detect any two errors, whichever we choose.

It is natural for you to ask how the method of appending extra digits with parity-check sums enables us to detect or even correct errors. Error detection is obvious. Think of how a computer spell-checker works. If you type *bive* instead of *five*, the spell-checker detects the error because the string *bive* is not on its list of valid words. On the other hand, if you type *give* instead of *five*, the spell-checker will not detect the error because *give* is on its list of valid words.

Our error-detection scheme works the same way, except that if we add extra digits to ensure that our code words differ in many positions—say, t positions—then even as many as $t - 1$ mistakes will not convert one code word into another code word. And if every pair of code words differs from each other in at least three positions, we can correct any single error because the incorrect received word will differ from the correct code word in exactly one position, but it will differ from all others in two or more positions.

Thus, in this case, the correct word is the unique "nearest neighbor." So the role of the parity-check sums is to ensure that code words differ in many positions. For example, consider the code in Table 17.1. The messages 1000 and 1100 differ in only the second position. But the two parity-check sums $a_1 + a_2 + a_3$ and $a_2 + a_3 + a_4$ will guarantee that encoded words for these messages will have different values in positions 5 and 7 as well as in position 2. It is the job of mathematicians to discover the appropriate parity-check sums to correct several errors in long codes.

17.3 Data Compression

Binary linear codes are fixed-length codes. In a fixed-length code, each code word is represented by the same number of digits (or symbols). In contrast, the Morse code (see Spotlight 17.4), designed for the telegraph, is a **variable-length code**; that is, a code in which the number of symbols for each code word may vary.

Notice that in Morse code, the letters that occur most frequently are represented by the fewest number of symbols, whereas letters that occur less frequently are coded with more symbols. By assigning the code in this manner, telegrams could convey more information per line than would be the case for fixed-length codes or a randomly assigned variable-length coding of the letters. Morse code is an example of data compression.

Data Compression DEFINITION

Data compression is the process of encoding data so that the most frequently occurring data are represented by the fewest symbols.

Figure 17.9 shows a typical frequency distribution for letters in English-language text material.

Data compression provides a means to reduce the costs of data storage and transmission. A **compression algorithm** converts data from an easy-to-use format to one optimized for compactness. Conversely, an uncompression algorithm converts the compressed information back to its original (or approximately original) form. Downloaded files in the ZIP format are an example of a particular kind of data compression. When you "unzip" the file, you return the compressed data to its original state. In some applications, such as datasets that represent images, the original data need only be recaptured in approximate form. In these cases, there are algorithms that result in a great saving of space. Graphics Interchange Format (GIF) encoding returns compressed data to its exact original form, while

SPOTLIGHT 17.4

Morse Code

The Morse code is a ternary code consisting of short marks, long marks and spaces (see figure on the right). It was invented in the early 1840s by Samuel Morse as an efficient way to transmit messages using electronic pulses through telegraph wires. The code enabled operators to send strings of short pulses, long pulses, and pauses representing characters transformed into indentations on paper tape that could be easily converted back to characters. Although Morse code was widely used up until the mid-twentieth century, it has gradually been supplanted by more machine-friendly codes. Because the Morse code uses data compression, sending messages using Morse code is faster than text messaging. Many Nokia cell phones can convert text messages to Morse code.

A	·—	N	—·
B	—···	O	———
C	—·—·	P	·——·
D	—··	Q	——·—
E	·	R	·—·
F	··—·	S	···
G	——·	T	—
H	····	U	··—
I	··	V	···—
J	·———	W	·——
K	—·—	X	—··—
L	·—··	Y	—·——
M	——	Z	——··

Morse code

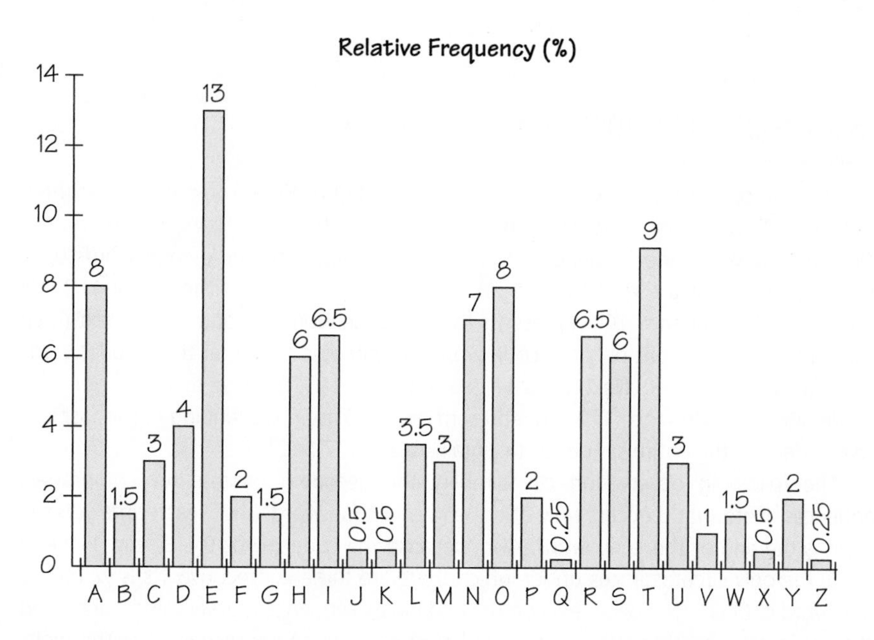

Figure 17.9 A widely used frequency table for letters in normal English usage.

Joint Photographic Experts Group (JPEG) encoding and Motion Picture Expert Group (MPEG) encoding return data only approximately to its original state.

EXAMPLE 1
Data Compression

Let's illustrate the principles of data compression with a simple example. Biologists are able to describe genes by specifying sequences composed of the four letters *A, T, G,* and *C*, which represent the four nucleotides adenine, thymine, guanine, and cytosine, respectively. One way to encode a sequence such as *AAACAGTAAC* in fixed-length binary form would be to encode the letters as

$$A \rightarrow 00 \qquad C \rightarrow 01 \qquad T \rightarrow 10 \qquad G \rightarrow 11$$

The corresponding binary code for the sequence *AAACAGTAAC* is then

00000001001110000001

On the other hand, if we knew from experience that *A* occurs most frequently, *C* second most frequently, and so on, and that *A* occurs much more frequently than *T* and *G* together, the most efficient binary encoding would be

$$A \rightarrow 0 \qquad C \rightarrow 10 \qquad T \rightarrow 110 \qquad G \rightarrow 111$$

For this encoding scheme, the sequence *AAACAGTAAC* is encoded as

0001001111100010

Notice that this binary sequence has 20 percent fewer digits than our previous sequence, in which each letter was assigned a fixed length of 2 (16 digits versus 20 digits). However, to realize this savings, we have made decoding more difficult. For the binary sequence using the fixed length of two symbols per character, we decode the sequence by taking the digits two at a time in succession and converting them to the corresponding letters. For the compressed coding, we can decode by examining the digits in groups of three.

EXAMPLE 2
Decode 0001001111100010

Consider the compressed binary sequence 0001001111100010. Look at the first three digits: 000. Because our code words have one, two, or three digits and neither 00 nor 000 is a code word, the sequence 000 can represent only the three code words 0, 0, and 0. Now look at the next three digits: 100. Again, because neither 1 nor 100 is a code word, the sequence 100 represents the two code words 10 and 0. The next three digits, 111, can represent only the code word 111 because the other three code words all contain at least one 0. Next, consider the sequence 110. Because neither 1 nor 11 is a code word, the sequence 110 can represent only 110 itself. Continuing in this fashion, we can decode the entire sequence to obtain *AAACAGTAAC*.

The following observation can simplify the decoding process for compressed sequences. Note that 0 occurs only at the end of a code word. Thus, each time you see a 0, it is the end of the code word. Also, because the code words 0, 10, and 110 end in a 0, the only circumstances under which there are three consecutive 1's is when the code word is 111. So, to decode a compressed binary sequence quickly using our coding scheme, insert a comma after every 0 and after every three consecutive 1's. The digits between the commas are code words.

EXAMPLE 3
Code AGAACTAATTGACA and Decode the Result

Recall: $A \rightarrow 0$, $C \rightarrow 10$, $T \rightarrow 110$, and $G \rightarrow 111$. So

$$AGAACTAATTGACA \rightarrow 01110010110001101101110100$$

To decode the encoded sequence, we insert commas after every 0 and after every occurrence of 111 and convert to letters:

0,	111,	0,	0,	10,	110,	0,	0,	110,	110,	111,	0,	10,	0
A,	*G*,	*A*,	*A*,	*C*,	*T*,	*A*,	*A*,	*T*,	*T*,	*G*,	*A*,	*C*,	*A*

Delta Encoding

For datasets of numbers that fluctuate little from one number to the next, the method of compression called *delta encoding* works well.

EXAMPLE 4
Delta Encoding

Consider the following closing prices (rounded to the nearest integer) of the Standard & Poor's (S&P) index of the stock prices of 500 companies in July 2010:

1027 1023 1028 1060 1070 1078 1079 1095 1095 1096
1065 1071 1083 1070 1094 1103 1106 1102 1102

These numbers use 94 characters in all (counting spaces). To compress this dataset using the delta method, we start with the first number and continue by listing only the change from each entry to the next. So, our list becomes

1027 −4 5 32 10 8 1 16 0 1 −31 6 12 −13 24 9 3 −4 0

This time we have used only 51 characters, counting the minus signs, to represent the same data, a savings of almost 46 percent.

For comparison, we note that the S&P 500 stock prices for July 2000 were as follows:

1470 1446 1457 1479 1476 1481 1493 1496 1510 1510
1494 1482 1496 1480 1464 1474 1452 1450 1420 1431

Huffman Coding

The methods we have shown previously are too simple for general use, but in 1951, a graduate student named David Huffman (see Spotlight 17.5) devised a scheme for data compression that became widely used. As was the case for the first scheme we discussed, Huffman coding assigns short code words to those characters with high probabilities of occurring and long code words to those with low probabilities of occurring. A *Huffman code* is made using a so-called *code tree,* by arranging the characters from top to bottom according to increasing probability; it proceeds by combining, at each stage, the two least probable combinations and repeating this process until there is only one combination remaining. To illustrate the method, say we have a dataset of six letters that occur with the following probabilities:

A	0.125
B	0.051
C	0.215
D	0.173
E	0.210
F	0.226

Rearranging them in increasing order, we have:

B	0.051
A	0.125
D	0.173
E	0.210
C	0.215
F	0.226

Because *B* and *A* are the two least likely to occur, we begin our tree by merging them with the one with the smallest probability on the left (that is, *BA* rather than *AB*), adding their probabilities, and rearranging the resulting items in increasing order:

D	0.173
BA	0.176
E	0.210
C	0.215
F	0.226

SPOTLIGHT 17.5

David Huffman

Large networks of IBM computers use it. So do high-definition televisions and an electronic device that takes the brainwork out of programming a video recorder. All these digital wonders rely on the results of a 60-year-old term paper by an MIT graduate student—a data-compression scheme known as *Huffman encoding*.

In 1951, David Huffman and his classmates in an electrical engineering graduate course on information theory were given the choice of writing a term paper or taking a final exam. For the term paper, Huffman's professor had assigned what at first appeared to be a simple problem: Students were asked to find the most efficient method of representing numbers, letters, or other symbols using binary code. Huffman worked on the problem for months, developing a number of approaches, but he couldn't prove that any of them were the most efficient. Finally, he despaired of ever reaching a solution and decided to start studying for the final. Just as he was throwing his notes in the garbage, the solution came to him. "It was the most singular moment of my life," Huffman says. "There was the absolute lightning of sudden realization. It was my luck to be there at the right time and also not have my professor discourage me by telling me that other good people had struggled with the problem," he says. When presented with his student's discovery, Huffman recalls, his professor exclaimed: "Is that all there is to it?!"

David Huffman *(Matthew Mulbry.)*

"The Huffman code is one of the fundamental ideas that people in computer science and data communications are using all the time," says Donald Knuth of Stanford University. Although others have used Huffman's encoding to help make millions of dollars, Huffman's main compensation was dispensation from the final exam. He never tried to patent an invention from his work and experienced only a twinge of regret at not having used his creation to make himself rich. "If I had the best of both worlds, I would have had recognition as a scientist, and I would have gotten monetary rewards," he says. "I guess I got one and not the other."

But Huffman received other compensation. A few years ago, an acquaintance told him that he had noticed that a reference to the code was spelled with a lowercase *h*. Remarked his friend to Huffman, "David, I guess your name has finally entered the language."

David Huffman died October 7, 1999.

Source: Adapted from the article "Profile: David Huffman" by Gary Stix, *Scientific American*, September 1991, pp. 54, 58.

This time, D and BA are the two least likely remaining entries, so we merge them with D on the left because it has smallest probability, add their probabilities, and re-sort from smallest to largest. This gives:

E	0.210
C	0.215
F	0.226
DBA	0.349

Next, we combine E and C with E on the left and re-sort to get:

F	0.226
DBA	0.349
EC	0.425

Then we combine F and DBA with F on the left and re-sort again:

EC	0.425
$FDBA$	0.575

And finally we combine EC and $FDBA$ with EC on the left to obtain:

$$ECFDBA \quad 1.000$$

To create the *Huffman tree*, start with a vertex at the top labeled with the final string $ECFDBA$ 1.000, then work your way backwards by creating two branches corresponding to each merger of the two least likely entries at each stage; assign 0 to the branch with the lower probability and 1 to the other branch, as shown in Figure 17.10. To assign a binary code word to each letter, we follow the path from the end of the tree to each letter by assigning, at each merging juncture, 0 or 1 according to the label on the branch taken. Thus, we have:

A	1111
B	1110
C	01
D	110
E	00
F	10

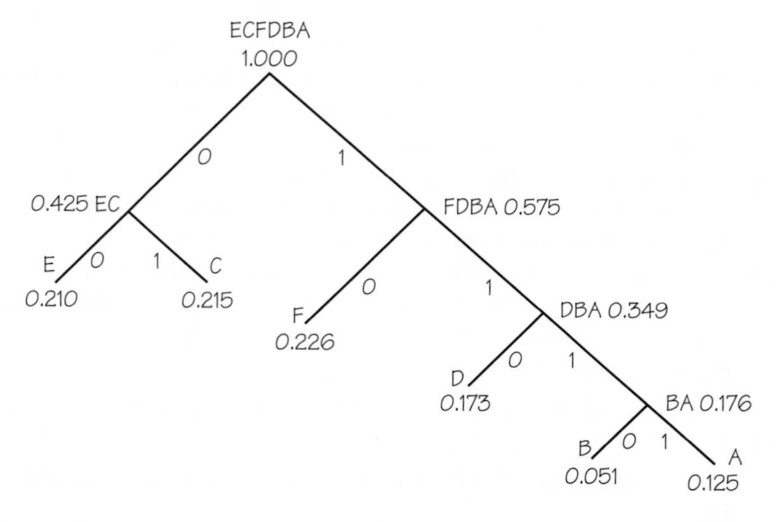

Figure 17.10
A Huffman tree.

Notice that the letters that occur least often have the longest codes and the letters that occur most often have the shortest codes. Decoding codes created from a Huffman tree is possible because at each stage, there is only one way that a particular string could have occurred. Here is an example. Consider the Huffman code created using the code words given in the previous display:

$$111010000110100111111010010$$

How can we determine the corresponding string of letters that has this Huffman code? The method is quite simple. Starting at the beginning of the string, every time we find a string that corresponds to a code word, we replace that string with its code word and continue. So, because all of our code words have at least two digits, we look at the first two digits. If they correspond to a code word, then decode them as that letter and continue in the same fashion. If not, then look at the next digit. If the three digits correspond to a code word, then decode it as the corresponding letter. If not, then these three digits and the next one are a four-digit code word.

Looking at our example 111010000110100111111010010, we see that neither 11 nor 111 is a code word but 1110 is the code word for B. So, replacing 1110 with B, we have *B*1000011010011111010010. The next possibility is 10, which is the code word for *F*, so we have *BF*00011010011111010010. Next, we have 00, which is the code word for *E*, giving us *BFE*011010011111010010. Continuing in this way, we obtain *BFECFFCACEF*. Of course, in practice, coding and decoding are done by computers.

17.4 Cryptography

Thus far, we have discussed ways in which data can be encoded to detect errors or correct errors in transmission. In many situations, there is also a desire for security against unauthorized interpretation of coded data (that is, a desire for secrecy). The process of disguising data is called *encryption*.

Cryptography DEFINITION

Cryptography is the study of methods to make and break secret codes. The process of coding information to prevent unauthorized use is called **encryption**.

Historically, encryption was used primarily for military and diplomatic transmissions. Today, encryption is essential for securing electronic transactions of all kinds. Cryptography is what allows you to have a Web site safely receive your credit card number. Cryptographic schemes prevent hackers from charging calls to your cell phone. Cryptography is also used for authenticating electronic transactions. In 1998, history was made when President Bill Clinton and Ireland's Prime Minister Bertie Ahern used digital signatures to sign an intergovernmental document. Each leader had a unique signing code and a digital certificate that served as a "digital ID," thereby ensuring that the document truly was approved by them. Although modern encryption schemes are extremely complex, we will illustrate the fundamental concepts involved with a few simple examples.

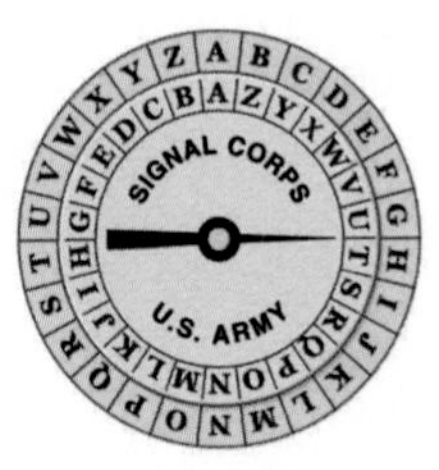

Among the first known cryptosystems is the **Caesar cipher**, used by Julius Caesar to send messages to his troops. To encrypt a message with the method employed by Caesar, we use the following table to replace each letter in the top row with the letter below it:

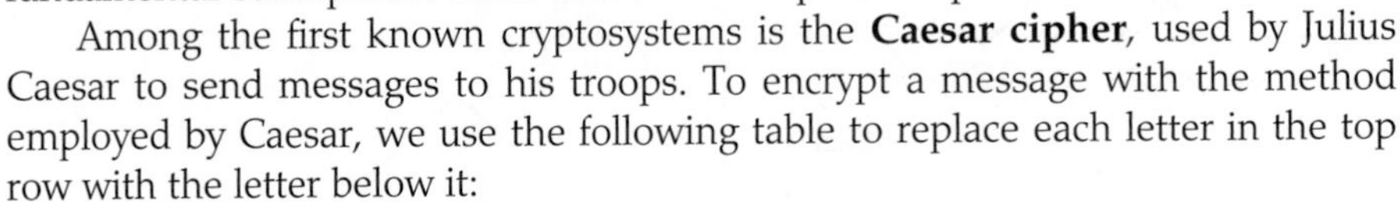

A B C D E F G H I J K L M N O P Q R S T U V W X Y Z
D E F G H I J K L M N O P Q R S T U V W X Y Z A B C

For example, the message ATTACK AT DAWN is encrypted as DWWDFN DW GDZQ.

To decrypt the message, replace each letter with the letter above it in the table. Obviously, it would not require much effort for someone to "crack" this code. To describe more sophisticated schemes for transmitting messages secretly, it is convenient to introduce a special kind of arithmetic used in cryptography. Recall that if we divide a positive integer a by a positive integer n, there are unique integers q and r such that $a = nq + r$, where r is nonnegative and less than n. The number q is called the *quotient* and r is called the *remainder*. For any positive integers a and n, we define a mod n (read: "a modulo n" or just "a mod n") to be the remainder when a is divided by n. Thus,

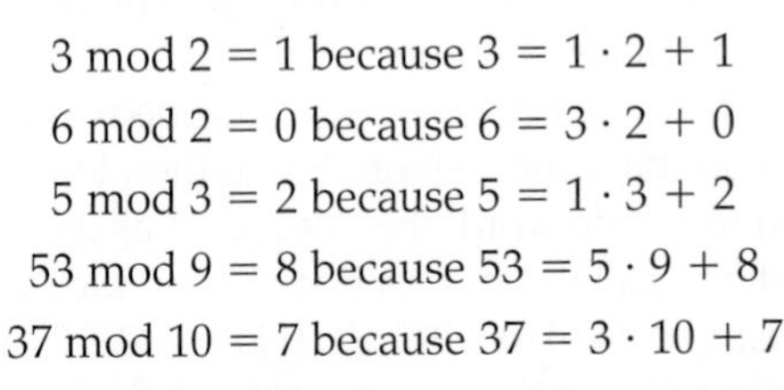

$$3 \bmod 2 = 1 \text{ because } 3 = 1 \cdot 2 + 1$$
$$6 \bmod 2 = 0 \text{ because } 6 = 3 \cdot 2 + 0$$
$$5 \bmod 3 = 2 \text{ because } 5 = 1 \cdot 3 + 2$$
$$53 \bmod 9 = 8 \text{ because } 53 = 5 \cdot 9 + 8$$
$$37 \bmod 10 = 7 \text{ because } 37 = 3 \cdot 10 + 7$$

$$66 \bmod 11 = 0 \text{ because } 66 = 11 \cdot 6 + 0$$
$$105 \bmod 26 = 1 \text{ because } 105 = 4 \cdot 26 + 1$$
$$342 \bmod 85 = 2 \text{ because } 342 = 4 \cdot 85 + 2$$
$$62 \bmod 85 = 62 \text{ because } 62 = 0 \cdot 85 + 62$$

To find the quotient and remainder for small integers, simply use long division. For larger numbers such as 2751 mod 13, use a calculator to divide 2751 by 13 to obtain 211.615385. Then the integer part 211 is the quotient q. To find the remainder r when a is divided by n, note that $r = a - nq$. So, we have $r = 2751 - 13 \cdot 211 = 2751 - 2743 = 8$. The easiest way to find 2751 mod 13 is to enter "2751 mod 13" into a Google search box.

Arithmetic involving mod n is called **modular arithmetic**. Although this arithmetic may appear unfamiliar, you often unconsciously use it. For example, if it is now September, what month will it be 25 months from now? Of course, you answer "October," but the interesting fact is that you didn't arrive at the answer by starting with September and counting off 25 months. Instead, without even thinking about it, you simply observed that $25 = 2 \cdot 12 + 1$ so that $25 \bmod 12 = 1$, and you added one month to September.

Similarly, if it is now Wednesday, you know that in 23 days, it will be Friday. This time, you arrived at your answer by noting that $23 = 3 \cdot 7 + 2$ (that is, $23 \bmod 7 = 2$), so you added 2 days to Wednesday instead of counting off 23 days. Because $51 = 2 \cdot 24 + 3$, if it is now 1 P.M., in 51 hours, it will be 4 P.M. ($51 \bmod 24 = 3$). If you travel east 390 degrees, you end up 30 degrees east of where you began ($390 \bmod 360 = 30$). Applications of modular arithmetic include the check-digit schemes described in Chapter 16, where arithmetic mod 7, 9, 10, and 11 were used. The parity-check sums (discussed in Section 17.2) use addition mod 2. An application of modular arithmetic to genetics is described in Spotlight 17.6.

SPOTLIGHT 17.6

Modeling the Genetic Code

The way that genetic material is composed can be conveniently modeled using modulo 4 arithmetic. A DNA molecule is made up of two long strands in the form of a double helix. Each strand is made up of strings of the four nitrogen bases adenine (A), thymine (T), guanine (G), and cytosine (C). Each base on one strand binds to a complementary base on the other strand. Adenine is always bound to thymine, and guanine is always bound to cytosine. To model this situation, we identify A with 0, T with 2, G with 1, and C with 3. Thus, the DNA segment ACGTAACAGGA and its complementary segment TGCATTGTCCT are identified by 03120030110 and 21302212332.

Using modulo 4 arithmetic, $0 + 2 = 2$, $2 + 2 = 0$, $1 + 2 = 3$, and $3 + 2 = 1$, we see that adding 2 to any of the integers 0, 1, 2, or 3 interchanges 0 and 2 and 1 and 3. So, for any DNA segment $a_1a_2 \ldots a_n$ represented by strings of 0's, 1's, 2's, and 3's, we see that its complementary segment is represented by $a_1a_2 \ldots a_n + 22 \ldots 2$, where we add the integers in each component using modulo 4. In particular, $03120030110 + 22222222222 = 21302212332$.

Source: Adapted from *Discrete Mathematics* by S. Washburn, T. Marlow, and C. Ryan, Addison-Wesley, 1999.

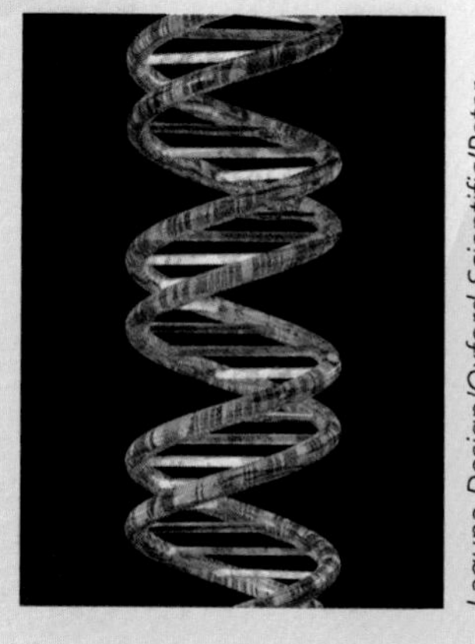

Laguna Design/Oxford Scientific/Peter Arnold

With modular arithmetic, we can describe the Caesar cipher easily as follows. Begin by saying that the letter A is in position 0, B is in position 1, C is in position 2, and so on. Then the Caesar cipher replaces the letter in position i with the letter in position $(i + 3) \bmod 26$. This formula expresses the fact that the Caesar cipher shifts each letter from A through W three positions to the right, while X, Y, and Z are replaced with A, B, and C, respectively. (Think of X, Y, and Z as "wrapping" around to the beginning of the alphabet.) It is customary to use the term **Caesar cipher** for any encryption scheme that shifts each letter a fixed number of positions. When the shift is anything other than three positions, we will specify the amount of the shift.

Decimation Cipher

Rather than encrypting a message by adding a fixed number to the position of every letter and using modular arithmetic, as is done in the Caesar cipher, another simple way to encrypt a message is to multiply each position by a fixed number and use modular arithmetic. This method of encryption is called the **decimation cipher**.

To begin, we assign the 26 letters the numbers 0 through 25, in order, and select any odd integer k between 3 and 25 except 13. (For the method to work, k and 26 must have no prime divisors in common.) Then, in the message, a letter with numerical value i is replaced with the letter with numerical value ki modulo 26. For example, if k is 5, then D is replaced with P because D has value 3 and $5 \times 3 = 15$, which is assigned to P. When ki exceeds 26, we use modulo 26 arithmetic to determine the replacement for the letter. Thus, J is replaced by T because J has the value 9 and $5 \times 9 = 45$ and $45 \bmod 26 = 19$ and T has value 19. The value of k is called the **key**.

To decode an encrypted message, we use the same method except that we multiply the numerical value for each encrypted letter by 21. The number 21 is used to decrypt the message because it has the property that $5 \times 21 = 105$ and $105 \bmod 26 = 1$. As a consequence, for any integer x, we have $(x \times 5 \times 21) \bmod 26$ equals $(x \times 1) \bmod 26 = x \bmod 26$. Thus, multiplying an integer x by 5 and then the result by 21 gets us back to x when we use modulo 26. In general, if a number k is used to encrypt a message modulo 26, the value j used to decrypt the message has to be chosen so that $kj \bmod 26 = 1$. Given a particular value for k, we can find the corresponding j by trial and error. Table 17.4 shows the values corresponding to each choice of k.

Table 17.4

Decimation Cipher Decryption Values

Encryption Value	Decryption Value
3	9
5	21
7	15
9	3
11	19
15	7
17	23
19	11
21	5
23	17
25	25

EXAMPLE 5
Decimation Cipher

To illustrate the decimation cipher, let's encrypt the message ATTACK AT DAWN using the key 3.

Message	A	T	T	A	C	K	A	T	D	A	W	N
Position	0	19	19	0	2	10	0	19	3	0	22	13
Position × 3	0	57	57	0	6	30	0	57	9	0	66	39
New position	0	5	5	0	6	4	0	5	9	0	14	13
Encrypted message	A	F	F	A	G	E	A	F	J	A	O	N

Linear Cipher

We can combine the methods used in the Caesar cipher and the decimation cipher easily by replacing the letter represented by the integer x by the letter corresponding to the integer $(kx + s) \bmod 26$, where k is any allowable decimation key and s is the shift that we want. For example, choosing $k = 11$ and $s = 6$ the letter D, which corresponds to 3, is replaced by M, which corresponds to 13, because $(11 \cdot 3 + 6) \bmod 26 = 13$. This method is called a **linear cipher**.

To decrypt a message that has been created using the linear cipher formula $(kx + s) \bmod 26$, we must undo shifting by s by shifting by $-s$, as well as undo the step of multiplying by the encryption value k by multiplying by the decryption value corresponding to k given in Table 17.4. In our example of encoding the letter with the numerical value x with the formula $(11x + 6) \bmod 26$, we can determine the starting value of x that results in 13 by solving the equation $(11x + 6) \bmod 26 = 13 \bmod 26$. This is done by subtracting 6 from both sides to get $11x \bmod 26 = 7 \bmod 26$, which reverses the shift of 6, then multiplying by 19, which we see from Table 17.4 reverses multiplying by 11, to get $19 \cdot 11x \bmod 26 = 19 \cdot 7 \bmod 26$. Because $19 \cdot 11 \bmod 26 = 1$ and $19 \cdot 7 \bmod 26 = 3$, this simplifies to $x = 3$, which corresponds to the letter D. In cases where working backwards results in a negative value after the reverse shift such as $19 \cdot (-5) \bmod 26$, Google still gives the correct answer.

Vigenère Cipher

Modular arithmetic also provides the basis for a more sophisticated cryptosystem called the **Vigenère cipher**. For this method, we first select a **key word**, which can be any word. The letters of the key word are then used to determine the amount of shifting for each letter of our message.

EXAMPLE 6
Vigenère Cipher

We will use the Vigenère cipher to encrypt the message ATTACK AT DAWN. Choosing the key word MATH, we shift the first letter of the message by 12 because *M* is in position 12; the second letter of the message is shifted by 0 (unchanged) because *A* is in position 0; the third letter of the message is shifted by 19 because *T* is in position 19, and so on. A shift of j means that the letter in position i is replaced by the letter in position $(i + j) \bmod 26$. When we have used all the letters of the key word, we start over at the beginning. To encrypt ATTACK AT DAWN using the key word MATH, we first note that the letters in the key word MATH are in positions 12, 0, 19, and 7, respectively. So the *A* in ATTACK is converted to *M* (0 + 12 = 12), the first *T* in ATTACK is converted to *T* (19 + 0 = 19), the second *T* in ATTACK is converted to *M* ((19 + 19) mod 26 = 12), and so on. The first two lines of Table 17.5 show the position numbers for the letters of the message and the key word. The third line of the table is obtained from the first two by adding the values in the columns using mod 26 and converting the results back to letters.

Table 17.5

Encryption Using a Vigenère Cipher

ATTACK AT DAWN	0	19	19	0	2	10	0	19	3	0	22	13
MATHMA TH MATH	12	0	19	7	12	0	19	7	12	0	19	7
MTMHOK TA PAPU	12	19	12	7	14	10	19	0	15	0	15	20

The **Vigenère square** in Table 17.6 provides a quick way to use the Vigenère cipher to encrypt messages with any key word. Use the left column for each letter of the key word and the top row for the corresponding letter of the message to be encrypted. The letter at the intersection of that row and column is the shifted letter. For example, if the message letter is *Q* and the corresponding letter from the key word is *M*, then the letter *C* at the intersection of the row that begins with *M* and the column that begins with *Q* is the encrypted letter for *Q*.

TABLE 17.6

Vigenère Square

	A	B	C	D	E	F	G	H	I	J	K	L	M	N	O	P	Q	R	S	T	U	V	W	X	Y	Z
A	A	B	C	D	E	F	G	H	I	J	K	L	M	N	O	P	Q	R	S	T	U	V	W	X	Y	Z
B	B	C	D	E	F	G	H	I	J	K	L	M	N	O	P	Q	R	S	T	U	V	W	X	Y	Z	A
C	C	D	E	F	G	H	I	J	K	L	M	N	O	P	Q	R	S	T	U	V	W	X	Y	Z	A	B
D	D	E	F	G	H	I	J	K	L	M	N	O	P	Q	R	S	T	U	V	W	X	Y	Z	A	B	C
E	E	F	G	H	I	J	K	L	M	N	O	P	Q	R	S	T	U	V	W	X	Y	Z	A	B	C	D
F	F	G	H	I	J	K	L	M	N	O	P	Q	R	S	T	U	V	W	X	Y	Z	A	B	C	D	E
G	G	H	I	J	K	L	M	N	O	P	Q	R	S	T	U	V	W	X	Y	Z	A	B	C	D	E	F
H	H	I	J	K	L	M	N	O	P	Q	R	S	T	U	V	W	X	Y	Z	A	B	C	D	E	F	G
I	I	J	K	L	M	N	O	P	Q	R	S	T	U	V	W	X	Y	Z	A	B	C	D	E	F	G	H
J	J	K	L	M	N	O	P	Q	R	S	T	U	V	W	X	Y	Z	A	B	C	D	E	F	G	H	I
K	K	L	M	N	O	P	Q	R	S	T	U	V	W	X	Y	Z	A	B	C	D	E	F	G	H	I	J
L	L	M	N	O	P	Q	R	S	T	U	V	W	X	Y	Z	A	B	C	D	E	F	G	H	I	J	K
M	M	N	O	P	Q	R	S	T	U	V	W	X	Y	Z	A	B	C	D	E	F	G	H	I	J	K	L
N	N	O	P	Q	R	S	T	U	V	W	X	Y	Z	A	B	C	D	E	F	G	H	I	J	K	L	M
O	O	P	Q	R	S	T	U	V	W	X	Y	Z	A	B	C	D	E	F	G	H	I	J	K	L	M	N
P	P	Q	R	S	T	U	V	W	X	Y	Z	A	B	C	D	E	F	G	H	I	J	K	L	M	N	O
Q	Q	R	S	T	U	V	W	X	Y	Z	A	B	C	D	E	F	G	H	I	J	K	L	M	N	O	P
R	R	S	T	U	V	W	X	Y	Z	A	B	C	D	E	F	G	H	I	J	K	L	M	N	O	P	Q
S	S	T	U	V	W	X	Y	Z	A	B	C	D	E	F	G	H	I	J	K	L	M	N	O	P	Q	R

	A	B	C	D	E	F	G	H	I	J	K	L	M	N	O	P	Q	R	S	T	U	V	W	X	Y	Z
T	T	U	V	W	X	Y	Z	A	B	C	D	E	F	G	H	I	J	K	L	M	N	O	P	Q	R	S
U	U	V	W	X	Y	Z	A	B	C	D	E	F	G	H	I	J	K	L	M	N	O	P	Q	R	S	T
V	V	W	X	Y	Z	A	B	C	D	E	F	G	H	I	J	K	L	M	N	O	P	Q	R	S	T	U
W	W	X	Y	Z	A	B	C	D	E	F	G	H	I	J	K	L	M	N	O	P	Q	R	S	T	U	V
X	X	Y	Z	A	B	C	D	E	F	G	H	I	J	K	L	M	N	O	P	Q	R	S	T	U	V	W
Y	Y	Z	A	B	C	D	E	F	G	H	I	J	K	L	M	N	O	P	Q	R	S	T	U	V	W	X
Z	Z	A	B	C	D	E	F	G	H	I	J	K	L	M	N	O	P	Q	R	S	T	U	V	W	X	Y

Encrypting Credit Card Data on the Web

Suppose that you want to purchase a compact disc from Amazon.com. Should you be concerned that a hacker will intercept your credit card number during the transaction? As you might expect, your credit card number is sent to Amazon in encrypted form to protect the data.

To describe one way that this encryption can be done, we need to perform addition of binary strings. We add two binary strings $a_1a_2 \ldots a_n$ and $b_1b_2 \ldots b_n$ as follows:

$$\begin{array}{r} a_1a_2 \cdots a_n \\ +\ b_1b_2 \cdots b_n \\ \hline c_1c_2 \cdots c_n \end{array}$$

where $c_i = 0$ if $a_i = b_i$ and $c_i = 1$ if $a_i \neq b_i$. Equivalently, $c_i = (a_i + b_i) \bmod 2$. (Add a_i and b_i in the ordinary way, but replace 2 by 0.)

EXAMPLE 7
Sum of Binary Strings

$$\begin{array}{r} 11000111 \\ +01110110 \\ \hline 10110001 \end{array} \qquad \begin{array}{r} 00111011 \\ +\ 01100101 \\ \hline 01011110 \end{array} \qquad \begin{array}{r} 10011100 \\ +\ 10011100 \\ \hline 00000000 \end{array}$$

We can now explain one way to send credit card numbers over the Web securely. When you place an order with Amazon, the company sends your computer a randomly generated string of 0's and 1's called a **key**. This key has the same length as the binary string corresponding to your credit card number, and the two strings are added (think of this process as "locking" the data). The resulting sum is then transmitted to Amazon. Amazon in turn adds the same key to the received string, which then produces the original string corresponding to your credit card number. (Adding the key a second time "unlocks" the data.)

To illustrate the idea, say you want to send an eight-digit binary string such as $s =$ 10101100 to Amazon (actual credit card numbers have very long strings), and Amazon sends your computer the key $k = 00111101$. Your computer returns the string $s + k =$ $10101100 + 00111101 = 10010001$ to Amazon, and Amazon adds k to this string to get $10010001 + 00111101 = 10101100$, which is the string representing your credit card number. If someone intercepts the number $s + k = 10010001$ during transmission, it is of no value without knowing k. This method works because of the property of binary addition that $a_1a_2 \cdots a_n + b_1b_2 \cdots b_n = 00 \cdots 0$ if and only if the two strings are identical. Thus, $(s + k) + k = s + (k + k) = s + 00 \cdots 0 = s$. The method is secure because the key sent by Amazon is randomly generated and used only one time.

You can tell when you are using an encryption scheme on a Web transaction by looking to see if the Web address begins with "https" rather than the customary "http." You will also see a small padlock in the status bar at the bottom of the browser window.

Public Key Cryptography

The decimation cipher and the Vigenère cipher have the drawback that anyone who knows the key for encryption will also know how to decrypt the encrypted messages. In the mid-1970s, Ronald Rivest, Adi Shamir, and Leonard Adleman devised an ingenious method that permits each person who is to receive a secret message to tell publicly how to scramble messages sent to him or her. And even though the method used to scramble the message is known publicly, only the person for whom it is intended will be able to unscramble the message.

To illustrate their method for transmitting messages secretly, we need the following property of modular arithmetic:

$$(ab) \bmod n = ((a \bmod n)(b \bmod n)) \bmod n$$

This property allows you to replace integers that are greater than or equal to n with integers that are less than n to simplify calculations. You should think of it as saying, "mod before you multiply."

EXAMPLE 8
Multiplication Property for Modular Arithmetic

$(17 \cdot 23) \bmod 10 = ((17 \bmod 10)(23 \bmod 10)) \bmod 10$
$= (7 \cdot 3) \bmod 10 = 21 \bmod 10 = 1$ (Note also that $(17 \cdot 23) \bmod 10$
$= 391 \bmod 10 = 1$)

$(22 \cdot 19) \bmod 8 = ((22 \bmod 8)(19 \bmod 8)) \bmod 8$
$= (6 \cdot 3) \bmod 8 = 18 \bmod 8 = 2$

$(100 \cdot 8) \bmod 85 = ((100 \bmod 85)(8 \bmod 85)) \bmod 85$
$= (15 \cdot 8) \bmod 85 = 120 \bmod 85 = 35$

Algebra Review: Prime and Composite Numbers

An integer greater than 1 is prime if the only positive integers that divide it with the remainder 0 are the number itself and 1. There are infinitely many prime numbers. The first ten are 2, 3, 5, 7, 11, 13, 17, 19, 23, 29.

A composite number is a positive integer that is greater than 1 and is not prime. A list of the first 1000 prime numbers can be found by searching Google for "1000 primes." One can use a list of primes and a calculator to determine whether any positive integer N is prime by dividing N one by one by all the primes less than or equal to $\sqrt{N}$. If one of the quotients is an integer, N is not prime. For example, to test whether 147 is prime you need only divide 147 by 2, 3, 5, 7, and 11 (because these are all primes less than or equal to $\sqrt{147}$). In this case $147 \div 7 = 21$ so 147 is not prime. In contrast, when 151 is divided by 2, 3, 5, 7 and 11 the quotient is never an integer, so 151 is prime.

We now describe the Rivest, Shamir, and Adleman method by way of a simple example. Say that we wish to send the message "IBM." We convert the message to digits by replacing A by 1, B by 2, ..., and Z by 26. So the message IBM becomes 9213. The person to whom the message is to be sent has picked two distinct primes p and q, say, $p = 5$ and $q = 17$. The receiver also has picked a number r, such as 3, that has no divisors in common with the least common multiple m of $(p - 1) = 4$ and $(q - 1) = 16$ other than 1, and published $n = pq = 85$ and $r = 3$ in a public directory. To decode our message, the receiver must find a number s so that $(r \cdot s) \bmod m = 1$ (this is where knowledge of p and q is necessary). That is, $(3 \cdot s) \bmod 16 = 1$. This number is 11. (The number s can be found by calculating successive powers of r mod m. When 1 is reached, the previous power of r is s. In our example, we have $3 \bmod 16 = 3$, $3^2 \bmod 16 = 9$, $3^3 \bmod 16 = 11$, $3^4 \bmod 16 = 1$, so $s = 3^3 \bmod 16 = 11$.)

To send our message to this person, we consult the public directory to find $n = 85$ and $r = 3$, then send the "scrambled" numbers $9^3 \bmod 85$, $2^3 \bmod 85$, and $13^3 \bmod 85$ rather than 9, 2, and 13, and the receiver will unscramble them. Thus, we send:

$$9^3 \bmod 85 = 49$$
$$2^3 \bmod 85 = 8$$
$$13^3 \bmod 85 = 72$$

Now, the receiver must take the numbers that he or she receives—49, 8, and 72—and convert them back to 9, 2, and 13 by calculating $49^{11} \bmod 85$, $8^{11} \bmod 85$, and $72^{11} \bmod 85$.

Provided that the numbers are not too large, the search engine at www.google.com will do modular arithmetic by simply entering the number and the mod value in the format a^n mod m. For numbers such as 49^{11} that are too large, entering an expression such as "49^11 mod 85" in a Google search box does not return a value. Instead, we use the "mod before you multiply" method by entering smaller powers and "moding" as we go, such as (49^5 mod 85)(49^6 mod 85) mod 85. Doing so gives us the value 9. Thus, the receiver has determined the code for I correctly. The Modular Arithmetic Calculator (at http://www.math.uga.edu/~bjones/calc/) can do modular arithmetic for extremely large numbers.

Similarly, we may calculate $8^{11} \bmod 85$ and $72^{11} \bmod 85$. Notice that without knowing how $n = pq$ factors, we cannot find the least common multiple of $p - 1$ and $q - 1$ (in our case, 16), and therefore the s that is needed to determine the intended message.

The procedure just described is called the **RSA public key encryption scheme** in honor of Rivest, Shamir, and Adleman, who discovered it. The method is practical and secure because efficient methods exist for finding very large prime numbers (say, about 100 digits long) and for multiplying large numbers, but no one knows an efficient algorithm for factoring large integers (say, about 200 digits long).

The algorithm is summarized here. In practice, the messages are not sent one letter at a time. Rather, the entire message is converted to decimal form, with A represented by 01, B by 02, . . . , and a space by 00. The message is then broken up into blocks of uniform size and the blocks are sent. See step 2 under Sender.

Receiver

1. Pick distinct very large primes p and q and compute $n = pq$. (The number n is called the *key*.)
2. Compute the least common multiple of $p - 1$ and $q - 1$; let's call it m.
3. Pick r so that it has no divisors in common with m other than 1 (any such r will do).

4. Find s so that $(rs) \bmod m = 1$. (To find s, simply compute $r^2 \bmod m$, $r^3 \bmod m$, $r^4 \bmod m$, . . . until you reach $r^t \bmod m = 1$. Then $s = r^{t-1} \bmod m$.)
5. Publicly announce n and r, but keep p, q, and s secret.

Sender

1. Convert the message to a string of digits.
2. Break up the message into uniformly sized blocks of digits, appending 0's in the last block if necessary. Call them $M_1, M_2, \ldots, M_k$. For example, for a string such as 2105092315, we could use $M_1 = 2105$, $M_2 = 0923$, and $M_3 = 1500$.
3. Check to see that the greatest common divisor of each M_i and n is 1. If not, n can be factored and the code is broken. (In practice, the primes p and q are so large that they exceed all M_i, so this step may be omitted.)
4. Calculate and send $R_i = M_i^r \bmod n$.

Receiver

1. For each received message R_i, calculate $R_i^s \bmod n$.
2. Convert the string of digits back to a string of characters.

Let's do another example step by step with $p = 7$, $q = 11$, and the message "HI."

Receiver

1. $n = 77$.
2. The least common multiple m of $7 - 1 = 6$ and $11 - 1 = 10$ is 30.
3. We pick $r = 7$.
4. Because $7^4 = 1 \bmod 30$, we have $s = 7^3 \bmod 30 = 13$.
5. Make public $n = 77$ and $r = 7$.

Sender

1. HI converts to 89.
2. We will send 8 and 9 individually (that is, our blocks have size 1).
3. The greatest common divisor of 8 and 77 is 1, and the greatest common divisor of 9 and 77 is 1, so we can proceed.
4. Send $8^7 \bmod 77 = 57$ and $9^7 \bmod 77 = 37$.

Receiver

1. $57^{13} \bmod 77 = (57^7 \bmod 77)(57^6 \bmod 77) \bmod 77 = 8$;
   $37^{13} \bmod 77 = (37^7 \bmod 77)(37^6 \bmod 77) \bmod 77 = 9$.
2. 89 converts to HI.

This method works because of a basic property of modular arithmetic and the choice of r. As a result of choosing the number m as we described, m has the property that for each positive integer x having no common divisors with n except 1, we have $x^m \bmod n = 1$. So, in the case of our first example with $n = 85$, $m = 16$, and $r = 3$, for the original message 9 and the received message 49, we have

$$\begin{aligned} 49^{11} \bmod 85 &= (9^3)^{11} \bmod 85 = 9^{33} \bmod 85 = 9 \cdot 9^{32} \bmod 85 \\ &= 9 \cdot (9^{16} \bmod 85)^2 = 9 \cdot 1^2 = 9. \end{aligned}$$

(We have written $49^{11} \bmod 85$ in the form $9 \cdot (9^{16} \bmod 85)^2$ because Google can calculate the latter but not the former.)

In 2002, Rivest, Shamir, and Adleman received the Association for Computing Machinery A. M. Turing Award, which is considered to be the "Nobel Prize of Computing," for their seminal contribution to public key cryptography.

17.5 Web Searches and Mathematical Logic

With the number of Web pages indexed by large Internet search engines such as Google numbering in billions, computer scientists and mathematicians attempt to manage massive datasets by taking advantage of the associated network structure, which represents the interrelations of the data. The algorithm used by the Google search engine, for instance, ranks all pages on the Web using these interrelations to determine their relevance to the user's search. Factors such as the frequency, location near the top of the page of key words, font size, and number of links are taken into account. Spotlight 17.7 discusses the "Six Degrees of Kevin Bacon" game that illustrates the so-called small-world phenomenon of the interconnectedness of

SPOTLIGHT 17.7
Six Degrees of Kevin Bacon

The actor Kevin Bacon is probably better known for a game named after him than for any of the movies he has been in. The game works like this. Every actor who has been in a film with Kevin Bacon has Bacon number 1. Every actor who does not have Bacon number 1 but has been in a film with someone with Bacon number 1 has Bacon number 2. Any actor who does not have Bacon number 1 or 2 but has been in a film with someone who has Bacon number 2 has Bacon number 3, and so on. For example, Nicole Kidman has not been in a film with Bacon, but she appeared in *The Interpreter* with Sean Penn, and Sean Penn was in *Mystic River* with Kevin Bacon. So Kidman's Bacon number is 2.

This game was conceived by three college students who first explained it to the public on an MTV show hosted by Jon Stewart. It goes by the name "Six Degrees of Kevin Bacon" because nearly every actor has a Bacon number that is no more than 6. In fact, because Bacon has made so many movies, it is a challenge to think of an actor with a Bacon number exceeding 3. The game is an example of what scientists call the "small-world phenomenon," by which they mean that every person is connected to every other person by a surprisingly short number of links. You can play the Bacon game at www.cs.virginia.edu/oracle/.

The first experiment involving the small-world phenomenon occurred in 1967, when social psychologist Stanley Milgram mailed a series of traceable letters from points in Kansas and Nebraska to "targets" in Boston. The letters could be sent only to someone whom the holders knew on a first-name basis and whom they thought was more likely to know the target than they were themselves. The data revealed a median chain length of about 6.

Mathematicians have their own version of the Bacon game called the "Erdös number," where coauthors of research papers with Paul Erdös have Erdös number 1. The author of this spotlight has Erdös number 3. Erdös's Bacon number is 3. In April 2004, a person with Erdös number 4 auctioned on eBay the opportunity for someone to get an Erdös number of 5 by offering to write a joint paper with the highest bidder. The auction was halted when someone bid $1 million as a protest to the idea of selling coauthorships. Before the auction was halted, the highest bid was $1031. That bidder also refused to pay as a protest. Interestingly, the actress Danica McKellar, who played Winnie Cooper on the TV series *The Wonder Years*, has a Bacon number of 2 and an Erdös number of 4.

data. In this section, we will show how a branch of mathematics called **Boolean logic**, named after the 19th-century mathematician George Boole (1815–1864), can be used to make search engine queries more efficient.

An *expression* in Boolean logic is simply a statement that is either true or false. For example, the expression, "There are infinitely many prime numbers," is either true or false. For purposes of mathematical reasoning, it is not necessary that we know whether this statement is true or false, but simply that it is one or the other. (It was proved to be true by Euclid more than 2000 years ago.) The expression "The integer 51 is prime" is an example of an expression that is false because $51 = 3 \cdot 17$. When combining expressions in logic, we avoid statements that are subject to opinion or various interpretations, such as "Math is cool." When we enter a phrase such as "college football" as a query to a search engine, the search engine automatically interprets it as the expression "This Web page contains the phrase 'college football.'" The search engine then returns the list of Web pages for which this expression is true.

In this section, we discuss how complex expressions can be constructed by connecting individual expressions with the *connectives* AND, OR, and NOT. For example, to obtain Web pages containing the phrase "football" but not pages that also contain "NFL" or "college," we could formulate the query using the expression "football AND (NOT NFL) AND (NOT college)." The search engine interprets this as, "This Web page contains the phrase 'football' AND it is NOT the case that this Web page contains the word 'NFL' AND it is NOT the case that this Web page contains the word 'college.'" The parentheses are not necessary but are sometimes useful, as we will see shortly.

Each search engine has slightly different conventions for formulating queries. Virtually every search engine has a hyperlink on its Web page that explains how to formulate queries. Although we use traditional terminology and notation from logic for our connectives, most popular search engines employ a more user-friendly format for advanced searches. For Google, "Find results with all of the words" is a substitute for our AND connective; "with at least one of the words" plays the role of our OR connective; and "without the words" is the same as our NOT connective.

Our interest is finding out how we can use Boolean logic to decide whether two different expressions have the same meaning. For example, is the expression, "football AND (NOT NFL) AND (NOT college)," equivalent to "football AND NOT (NFL AND college)"? Or is it equivalent to the expression "football AND NOT (NFL OR college)"? To answer these questions, we will now take a closer look at the connectives AND, OR, and NOT.

The NOT connective allows us to take an expression P and create a new expression NOT P, called the *negation* of P. If P is true, then NOT P is false. If P is false, then NOT P is true. Rather than writing NOT P, we will use the more standard mathematical notation $\sim P$ (many people use $\neg P$ instead). The negation relationship can be summarized in the following format, known as a **truth table**:

P	$\sim P$
T	F
F	T

Notice that T and F are used here as shorthand for *true* and *false*, respectively. The left column of the truth table shows the two possible values of P: T and F. The right column shows the values of $\sim P$ for each of the corresponding values of P.

The AND connective allows us to combine two expressions, P and Q, into a new expression P AND Q called the *conjunction* of P and Q. The new expression is true when both statements P and Q are true and is otherwise false. The mathematical notation for P AND Q is $P \wedge Q$. (You can remember this by noting that the symbol $\wedge$ has a shape like the first letter of AND.) This relationship can also be summarized in a truth table as follows:

P	Q	$P \wedge Q$
T	T	T
T	F	F
F	T	F
F	F	F

Here, the first two columns are used to show all possible values of P and Q, and the right column shows the value of $P \wedge Q$.

Finally, the OR connective allows us to combine two expressions P and Q into a new expression P OR Q, called the *disjunction* of P and Q, which is true if either P or Q, or both, are true and is otherwise false. The mathematical notation for P OR Q is $P \vee Q$. This relationship is summarized by the following truth table:

P	Q	$P \vee Q$
T	T	T
T	F	T
F	T	T
F	F	F

In everyday circumstances, the word *or* is used in two distinct ways. In some situations, "P or Q" means either P is valid, or Q is valid, or both are valid; whereas in other situations, "P or Q" means one of P or Q is valid and the other is not. A typical example of the former is the criterion for admission to an entertainment event that states: "Must be at least 18 years old or accompanied by an adult." On the other hand, a menu entry that says, "Price includes soup or salad" is an example of the latter. To distinguish between these two usages, mathematicians call the first the *inclusive or* and the second the *exclusive or.* When you encounter a mathematical statement of the form P OR Q, the inclusive or is meant. In nonmathematical ambiguous situations, some people use the term *and/or* to mean the inclusive or.

The connectives $\sim$, $\wedge$, and $\vee$ are more universal than they appear. Consider the statement "Pam won't go unless Quintin goes." Can we use these connectives to formulate a symbolic equivalent statement (that is, a symbolic statement with the same truth table)? Ignoring truth tables and just thinking about it, we realize that the only way this statement can be false is if Quintin does not go and Pam does. For convenience, let P stand for "Pam goes" and Q stand for "Quintin goes." Then the only way the statement can be false is for Q to be false and P to be true. That is, we need $\sim Q \wedge P$ true only when Q is false and P is true. So let's look at a truth table for $\sim Q \wedge P$:

P	Q	$\sim Q$	$\sim Q \wedge P$
T	T	F	F
T	F	T	T
F	T	F	F
F	F	T	F

Indeed, we have used the connectives to formulate a symbolic statement that is equivalent to the original one.

By mixing the three operations defined thus far, complicated expressions such as $(P \vee \sim Q) \wedge (\sim P \wedge Q)$ can be formed. To determine the truth table for such an expression, it is best to break it up into component pieces, as shown in the following table:

P	Q	~P	~Q	P ∨ ~Q	~P ∧ Q	(P ∨ ~Q) ∧ (~P ∧ Q)
T	T	F	F	T	F	F
T	F	F	T	T	F	F
F	T	T	F	F	T	F
F	F	T	T	T	F	F

An expression such as $(P \vee \sim Q) \wedge (\sim P \wedge Q)$ for which all the values are false is called a *logical contradiction.*

A statement involving three expressions P, Q, and R such as $P \wedge Q \wedge R$ appears to be ambiguous. Does this expression mean that P and Q are first combined into a new expression $P \wedge Q$ and this new expression is then combined with R? In other words, should we interpret this expression as $(P \wedge Q) \wedge R$? Perhaps the intention was to connect P with the single expression $Q \wedge R$. In this case, the expression is interpreted as $P \wedge (Q \wedge R)$. Just as with the case for arithmetic statements such as $5 + 3 + 6$, which we can interpret to mean $(5 + 3) + 6$ or $5 + (3 + 6)$, it turns out that both interpretations are the same.

For example, consider the three constitutional requirements for the office of president of the United States. A candidate for president must be at least 35 years old, must be a natural-born U.S. citizen, and must have lived in the United States for at least 14 years. Let P be the statement "A candidate must be at least 35 years old," let Q be the statement "A candidate must be a natural-born U.S. citizen," and let R be the statement "A candidate must have lived in the United States for at least 14 years."

The expression $(P \wedge Q) \wedge R$ can then be interpreted as "A candidate must be at least 35 years old and a natural-born U.S. citizen and also must have lived in the United States for at least 14 years." The expression $P \wedge (Q \wedge R)$ can be interpreted as "A candidate must be at least 35 years old and also a natural-born U.S. citizen who has lived in the United States for at least 14 years." Both of these descriptions are effectively the same. Both expressions are true only if each of P, Q, and R is true.

One way to verify this is to construct the truth table for expression $(P \wedge Q) \wedge R$ and the truth table for $P \wedge (Q \wedge R)$ and show that they give the same values for every possible value of P, Q, and R. Thus, because the order of operations does not matter in this case, we can simply write $P \wedge Q \wedge R$ without worrying about any possible ambiguity. The same thing is true for $P \vee Q \vee R$. We refer to this by saying that the connectives $\wedge$ and $\vee$ have the *associative property*. Notice this terminology is consistent with the "associative property" of real-number addition and multiplication: $(a + b) + c = a + (b + c)$ and $(ab)c = a(bc)$.

In some cases, however, the ambiguity is not easy to resolve. For example, consider the expression $P \wedge Q \vee R$, which can be interpreted as either $(P \wedge Q) \vee R$ or as $P \wedge (Q \vee R)$. Using the statements P, Q, and R as before, $(P \wedge Q) \vee R$ can be interpreted as "A candidate must be at least 35 years old and a natural-born U.S. citizen, or must have lived in the United States at least 14 years." On the other hand, $P \wedge (Q \vee R)$ can be interpreted as "A candidate must be at least 35 years old, and be a natural-born citizen or have lived in the United States at least 14 years." These two descriptions are certainly not the same! For example, because Arnold Schwarzenegger is not a natural-born U.S. citizen, the first expression excludes him as a candidate for president, whereas the second one includes him. One way to see exactly how the two statements differ is to compare the truth table for $(P \wedge Q) \vee R$ to the truth table for $P \wedge (Q \vee R)$. The truth table for $(P \wedge Q) \vee R$ is as follows:

P	Q	R	$(P \wedge Q)$	$(P \wedge Q) \vee R$
T	T	T	T	T
T	T	F	T	T
T	F	T	F	T
T	F	F	F	F
F	T	T	F	T
F	T	F	F	F
F	F	T	F	T
F	F	F	F	F

The truth table for $P \wedge (Q \vee R)$ is as follows:

P	Q	R	$(Q \vee R)$	$P \wedge (Q \vee R)$
T	T	T	T	T
T	T	F	T	T
T	F	T	T	T
T	F	F	F	F
F	T	T	T	F
F	T	F	T	F
F	F	T	T	F
F	F	F	F	F

Because the last columns of these two truth tables differ for some values of P, Q, and R, the two expressions are not equivalent. For example, notice that if P is false and Q and R are both true, then $(P \wedge Q)$ is false. Because R is true, however, $(P \wedge Q) \vee R$ is true. On the other hand, when P is false, $P \wedge (Q \vee R)$ is false regardless of whether $(Q \vee R)$ is true or false. To avoid ambiguity, it is often best to use parentheses.

A way to avoid ambiguity without using parentheses is to adopt a convention on the order of operations. For example, in arithmetic, the convention is that multiplication takes precedence over addition. Therefore, $3 + 4 \times 5$ is determined by first evaluating 4×5 and then adding 3. Of course, we could have written $3 + (4 \times 5)$ to avoid the ambiguity altogether. Similarly, in Boolean logic we adopt the convention that $\wedge$ (AND) takes precedence over $\vee$ (OR). Therefore, the expression $P \wedge Q \vee R$, by convention, is to be interpreted as $(P \wedge Q) \vee R$. Furthermore, the convention states that ~ (NOT) takes the highest precedence of all. Thus, $\sim P \wedge \sim Q \wedge R$ is interpreted as $((\sim P) \wedge (\sim Q)) \vee R$.

EXAMPLE 9
Applying Boolean Logic to a Web Search

Let's revisit the Web queries mentioned before. Let P represent the query "football," which corresponds to the expression, "This Web page contains the phrase 'football'." Let Q represent the expression, "This Web page contains the word 'NFL'." Let R represent the expression, "This Web page contains the word 'college'." We now translate the query "football AND (NOT NFL) AND (NOT college)" as $P \wedge (\sim Q) \wedge (\sim R)$ and write its truth table thus:

P	Q	R	~Q	~R	$P \wedge (\sim Q) \wedge (\sim R)$
T	T	T	F	F	F
T	T	F	F	T	F
T	F	T	T	F	F
T	F	F	T	T	T
F	T	T	F	F	F
F	T	F	F	T	F
F	F	T	T	F	F
F	F	F	T	T	F

For every possible value of *P, Q,* and *R,* the truth table gives us the value of our expression. The fourth and fifth columns of the table are not strictly necessary, but they are helpful in determining the values in the last column. As expected, this table tells us that the expression $P \wedge (\sim Q) \wedge (\sim R)$ is true precisely when *P* is true, *Q* is false, and *R* is false.

Two expressions are said to be **logically equivalent** if they have the same value, true or false, for each possible assignment of the Boolean variables. To decide whether two expressions are logically equivalent, we construct the truth tables for each one and then check if they have the same values for each of the possible assignments of the Boolean variables. If they do, the expressions are logically equivalent. If they differ for even one case, however, then the expressions are not equivalent. So, to determine whether the expression, "football AND (NOT NFL) AND (NOT college)," is logically equivalent to the expression, "football AND NOT (NFL AND college)," we need only compare their corresponding truth tables.

EXAMPLE 10
Logically Equivalent Expressions

We first determine the truth table for the expression, "football AND NOT (NFL AND college)," which is represented as $P \wedge \sim(Q \wedge R)$. Its truth table is as follows:

P	Q	R	$Q \wedge R$	$\sim(Q \wedge R)$	$P \wedge \sim(Q \wedge R)$
T	T	T	T	F	F
T	T	F	F	T	T
T	F	T	F	T	T
T	F	F	F	T	T
F	T	T	T	F	F
F	T	F	F	T	F
F	F	T	F	T	F
F	F	F	F	T	F

This truth table differs in the last column from the truth table for $P \wedge (\sim Q) \wedge (\sim R)$ in the previous example. For instance, when *P* is true, *Q* is true, and *R* is false, we see that $P \wedge (\sim Q) \wedge (\sim R)$ is false, but $P \wedge \sim(Q \wedge R)$ is true. Therefore, we must conclude that $P \wedge (\sim Q) \wedge (\sim R)$ is not logically equivalent to $P \wedge \sim(Q \wedge R)$.

On the other hand, the expression, "football AND NOT (NFL OR college)," is represented by $P \wedge \sim(Q \vee R)$. Its truth table is as follows:

P	Q	R	$Q \vee R$	$\sim(Q \vee R)$	$P \wedge \sim(Q \vee R)$
T	T	T	T	F	F
T	T	F	T	F	F
T	F	T	T	F	F
T	F	F	F	T	T
F	T	T	T	F	F
F	T	F	T	F	F
F	F	T	T	F	F
F	F	F	F	T	F

Because the last column of this table agrees with the last column for the expression $P \wedge (\sim Q) \wedge (\sim R)$, we know that the expression "football AND (NOT NFL) AND (NOT college)" is equivalent to the expression "football AND NOT (NFL OR college)."

Applying Logic to Message Routing

The AND operator in the truth table on page 626 is also used by computers to deliver messages over the Internet with a device called a *router.* Recall that if P and Q are expressions, then the statement $P \wedge Q$ is true when both P and Q are true and is false otherwise. Computer scientists use an analogous operation on 0 and 1 by allowing P and Q to represent 0 or 1 and defining that $P \wedge Q = 1$ when P and Q are 1 and $P \wedge Q = 0$ otherwise. When $\wedge$ is used in this way, it is called the *bitwise* AND. Notice that we can obtain an operation table for the bitwise AND from the table for the logical AND on page 627 by substituting 1 for T and 0 for F. In particular, we have

P	Q	$P \wedge Q$
1	1	1
1	0	0
0	1	0
0	0	0

The bitwise AND operation can be extended to binary strings of equal length by applying it individually to corresponding entries. Thus, $11001001 \wedge 01101101 = 01001001$ because both strings have a 1 only in positions 2, 5, and 8. In general, for any binary string s, we can use the bitwise AND to copy whichever entries of s we desire while converting all the other entries of s to 0. For example, if we have a list of binary strings of length 8 and we wish to modify these strings by copying the entries in positions 2, 7, and 8 and changing all other entries to 0, we simply take each string in the list and combine it with 01000011 using the bitwise AND operator. Thus, we have, $11101001 \wedge 01000011 = 01000001$; and $00011110 \wedge 01000011 = 00000010$. When doing the bitwise AND operation on binary strings, it is convenient to put one string directly below the other. In those positions where both entries are 1, the result is 1; otherwise, the result is 0.

s	11101001	s	00011110
t	<u>01000011</u>	t	<u>01000011</u>
$s \wedge t$	01000001	$s \wedge t$	00000010

The bitwise AND operation is used by computers to determine when certain entries of two binary strings match. Say, for example, that a computer would take a particular action if two binary strings s and t of equal length match in the first three positions. This happens precisely when $s \wedge 11100000 = t \wedge 11100000$. The reason why this works is that $x \wedge 1 = 1$ only when $x = 1$. Thus,

- if s and t both begin with 1, then both $s \wedge 11100000$ and $t \wedge 11100000$ will begin with 1;
- if s and t both begin with 0, then both $s \wedge 11100000$ and $t \wedge 11100000$ will begin with 0;
- if s and t begin with different digits, then $s \wedge 11100000$ and $t \wedge 11100000$ will begin with different digits.

The same reasoning applies to positions 2 and 3. The string of five 0's at the end of 11100000 ensures that $s \wedge 11100000$ and $t \wedge 11100000$ will both end with five 0's. So, checking that $s \wedge 11100000 = t \wedge 11100000$ checks whether s and t agree in the first three positions while disregarding the other positions.

For devices on a network to communicate with each other, each one must be given a unique identifier. This is done with an **Internet Protocol (IP) address**.

Internet Protocol Address DEFINITION

An **Internet Protocol (IP) address** is a sequence of four numbers between 0 and 255 separated by dots assigned to routers, computers, printers, and fax machines that allows those linked electronically to identify and communicate with each other uniquely.

Each device on the Internet is assigned an IP address. From the IP address, we can determine the **network address** of the device, which specifies the network or subnet that the device is a part of. While each device on the network or subnet will have a different IP address, they will all have the same network address. To demonstrate how computers determine the network address from an IP address, we must first explain how to convert the decimal form of each component of an IP address such as 131.212.66.17, which is convenient for humans, to their binary forms of length 8, which are convenient for computers. To convert an IP address in decimal to binary, we express each decimal number as a sum of distinct powers of 2 ranging from $128 = 2^7$ to $1 = 2^0$. For example, $213 = 128 + 64 + 16 + 4 + 1 = 2^7 + 2^6 + 2^4 + 2^2 + 2^0$. Next, make a row of the powers of 2 from 128 to 1 and beneath each one place a 1 if that power of 2 appears in the sum and a 0 if it does not. For the number 213, we have

128	64	32	16	8	4	2	1
1	1	0	1	0	1	0	1

Reading off the sequence of 0's and 1's, we have that the binary form of 213 is 11010101. Because 00000000 is the binary form of length 8 of 0, and 11111111 is the binary form of $255 = 128 + 64 + 32 + 16 + 8 + 4 + 2 + 1$, all the integers between 0 and 255 can be written as binary strings of length 8 using leading 0's as needed.

Each IP address is assigned a companion number called a **subnet mask** that also consists of four numbers, each of which ranges from 0 to 255. To determine the network address from the IP address, one performs the bitwise AND operation to the binary forms of the IP address and its companion subnet mask. The resulting number is the network address.

EXAMPLE 11
Network Address for IP Address 131.212.66.17 With Subnet Mask 255.255.255.0

We determine the network address corresponding to the IP address 131.212.66.17 with subnet mask 255.255.255.0. Because the binary form of 131.212.66.17 is 10000011.11010100.01000010.00010001 and the binary form of the subnet mask is 11111111.11111111.11111111.00000000, combining them using the bitwise AND gives

IP address	10000011.11010100.01000010.00010001
Subnet mask	11111111.11111111.11111111.00000000
Bitwise AND	10000011.11010100.01000010.00000000.

Thus, the network address is 10000011.11010100.01000010.00000000 = 131.212.66.0 (Some authors omit one or more 0's when they appear at the end of a network address.)

From Example 11, it appears that the network address corresponding to an IP address with subnet mask 255.255.255.0 is simply the same as the IP address with the last number 17 changed to 0. This is correct in this case, but for other subnet masks, the network address is not readily apparent from the IP address.

EXAMPLE 12
Network Address for IP Address 131.212.66.56 with Subnet Mask 255.255.255.240

We determine the network address corresponding to the IP address 131.212.66.56 with subnet mask 255.255.255.240. Because the binary form of 131.212.66.56 is 10000011.11010100.01000010.00111000 and the binary form of the subnet mask is 11111111.11111111.11111111.11110000, combining them using the bitwise AND gives

IP address	10000011.11010100.01000010.00111000
Subnet mask	11111111.11111111.11111111.11110000
Bitwise AND	10000011.11010100.01000010.00110000.

In this example, the network address is 10000011.11010100.01000010.00110000 = 131.212.66.48. Because the subnet mask for this IP address ends with four 0's, even if we changed the last four bits in the IP address, we still would get the same network address. Any decimal number whose binary form starts with 0011 is a number between 48 and 63, so any device with an IP address that begins with 131.212.66 and ends with any number between 48 and 63 will be in the network whose address is 131.212.66.48. We note that a device with an IP address that begins with 131.212.66 and ends with a number that is not between 48 and 63 is not on the same network as any device that starts with 131.212.66 and ends with a number between 48 and 63 because the devices will have different network addresses.

The network addresses allow routers to deliver messages to IP addresses in much the same way the U.S.Postal Service delivers mail to home addresses. That is, the Postal Service first checks to see if the mail is to be sent to a local address, and if not, it is relayed to a larger mail center. Likewise, if a message is to be sent from one computer to another, a local router checks to see if both computers are on the same network. If so, the message is sent over that network; otherwise, it is routed to a larger network. In this case, the local router searches its memory for the network address that most closely matches that of the destination network address and relays the message to a router in a larger network that has that address in its memory. A postal analogy to this situation might be a letter from Duluth, Minnesota, addressed to Rochester, Minnesota, being routed through Minneapolis.

REVIEW VOCABULARY

Binary linear code A code consisting of words composed of 0's and 1's obtained by using parity-check sums to append check digits to messages. (p. 608)

Boolean logic Logic attributed to George Boole that uses operations such as $\wedge$, $\vee$, and ~ to connect statements. (p. 626)

Caesar cipher A cryptosystem used by Julius Caesar whereby each letter is shifted the same amount. (p. 616)

Code word A string of digits composed of a message and check digits. (p. 608)

Compression algorithm A procedure for converting data from one format to another one optimized for compactness. (p. 610)

Cryptography The study of how to make and break secret codes. (p. 616)

Data compression The process of encoding data so that the most frequently occurring data are represented by the fewest symbols. (p. 610)

Decimation cipher A cryptosystem that uses multiplication by a fixed value to shift each letter. (p. 618)

Decoding The process of translating received data into code words. (p. 607)

Distance between two strings The distance between two strings of equal length is the number of positions in which they differ. (p. 607)

Encryption The process of encoding data to make it unreadable by unauthorized people. (p. 616)

Even parity Even integers are said to have even parity. (p. 603)

IP address A sequence of four numbers that uniquely identifies a device on a network. (p. 632)

Key A string used to encode and decode data. (p. 618)

Key word A word used to determine the amount of shifting for each letter while encoding a message. (p. 619)

Linear cipher A cryptosystem that replaces a letter with numerical value x by the letter with numerical value $(kx + s)$ mod 26. (p. 619)

Logically equivalent Two expressions are said to be logically equivalent if they have the same values for all possible values of their Boolean variables. (p. 630)

Modular arithmetic Addition and multiplication involving modulo n. (p. 617)

Nearest-neighbor decoding A method that decodes a received message as the code word that agrees with the message in the most positions. (p. 607)

Network address The portion of an IP address that identifies a local network. (p. 632)

Odd parity Odd integers are said to have odd parity. (p. 603)

Parity-check sums Sums of digits whose parities determine the check digits. (p. 605)

RSA public key encryption scheme A method of encoding that permits each person to announce publicly the means by which secret messages are to be sent to him or her. (p. 623)

Subnet mask A companion number to an IP address that allows a router to determine the network portion of an IP address. (p. 632)

Truth table A tabular representation of an expression in which the variables and the intermediate expressions appear in columns and the last column contains the expression being evaluated. (p. 626)

Variable-length code A code in which the number of symbols for each code word may change. (p. 610)

Vigenère cipher A cryptosystem that uses a key word to determine how much each letter is shifted. (p. 619)

Vigenère square A table that can be used to encrypt any message quickly using any key word with the Vigenère cipher. (p. 620)

Weight of a binary code The minimum number of 1's that occur among all nonzero code words of a code. (p. 608)

SKILLS CHECK

1. I[illegible]ilm critics listed their ratings of a film with [illegible]p (U) or thumbs down (D)—for example, one thu[illegible] is UDU—and critic number 1's rating is first, [illegible]umber 2's rating is second, critic number 3's rating [illegible]d—the number of possible outcomes is

[illegible] [illegible].
[illegible] 8.
[illegible]) 9.

2. A four-digit binary message was encoded using Table 17.1 and the message 1010010 was received. Using the nearest neighbor method, the decoded four-digit message is ______________________.

3. Using the circle diagram method to encode the message 1011, the encoded message is

(a) 1011001.
(b) 1011010.
(c) 1010001.

4. The distance between received words 1011001 and 1000101 is ______________________.

5. Using the nearest-neighbor method and the code in Table 17.2, the word 1110011 decodes as

(a) 0110010.
(b) 1100011.
(c) 1010001.

6. The weight of the binary linear code {0000000, 0011111, 0101011, 0110100} is ______.

7. If the two messages 0 and 1 are encoded as 000 and 111, respectively, the number of errors that the code can correct is

(a) 0.
(b) 1.
(c) 2.

8. If a binary linear code has weight 4, the maximum number of errors that it will detect is ____________.

9. If every pair of code words differs in at least five positions, then nearest-neighbor decoding can decode words accurately that have

(a) two errors.
(b) three errors.
(c) four errors.

10. The sum of the binary string 1011001 and 1001101 is ______________.

11. Using the encoding scheme A → 0, B → 10, C → 11, the string 010110 decodes as

(a) ABCB.
(b) ABCA.
(c) ABACA.

12. 3^5 mod 20 is equal to ______________.

13. The Caesar cipher encrypts GO HOME NOW as

(a) JR KRPH QRZ.
(b) DL ELJB KLT.
(c) Neither of these.

14. The permissible values for the key of a decimation cipher are ________.

15. Using the Vigenère cipher and the key word ADAM to decrypt EIEIO, we obtain

(a) ELELR.
(b) EFEFL.
(c) EFEWO.

16. The name for the cipher obtained when the value $s = 0$ is used in the linear cipher formula $kx + s$ is ______________________.

17. The name for the cipher obtained when the value $k = 1$ is used in the linear cipher formula $kx + s$ is

(a) Vigenère.
(b) Caesar.
(c) decimation.

18. Describe the cipher obtained when the value $k = 1$ and $s = 0$ are used in the linear cipher formula $kx + s$: ______________.

19. If a message was encrypted using the decimation cipher with the key 9, what value would you use to decrypt it? (Answer this question without looking at Table 17.4.)

(a) 3
(b) 5
(c) 7

20. 13^7 mod 33 would be entered in a Google search box as ______________.

21. The value of 17^{15} mod 33 cannot be obtained directly by entering it in a Google search box. What is another expression that you could enter in a Google search box to find the value?

(a) (17^5 mod 33)^3 mod 33
(b) (17^8 mod 33)(17^7 mod 33) mod 33
(c) Either (a) or (b)

22. If we use $p = 7$, $q = 17$, and $r = 5$ in the RSA scheme, the value of s is ____________.

23. Using the RSA scheme with $n = 91$ and $s = 5$, the message 4 decodes as

(a) 11.
(b) 20.
(c) 23.

24. As it is described in this chapter, the RSA scheme cannot be used to send the message GO with the key $n = 121$ because ________.

25. Which messages are the hardest to break?

(a) Messages encrypted with the Vigenère cipher
(b) Messages encrypted with the decimation cipher
(c) Messages encrypted with the RSA scheme

26. The statement "*P* OR *Q*" is true if and only if ________________.

27. In the statement, "John was born in 1940 or '41," the *or* being used is

(a) inclusive.
(b) exclusive.
(c) Not enough information is given here to decide.

28. Two Boolean expressions are logically equivalent if the last column of their truth tables ________.

29. When the statement "*P* AND NOT *Q*" is true, it must be the case that

(a) *P* is true.
(b) *P* is false.
(c) either *P* or *Q* is false.

30. $00111001 \wedge 11111001 =$ ________________.

CHAPTER 17 EXERCISES

■ Challenge ▲ Discussion

17.1 Binary Codes

1. Use the circle diagram method shown in Figures 17.1 and 17.2 (on pp. 602, 603) to verify the code words in Table 17.1 (on p. 605) for the messages 0101, 1011, and 1111.

2. Use the circle diagram method to decode the received messages 0111011 and 1000101.

3. Find the distance between each of the following pairs of words:

(a) 11011011 and 10100110
(b) 01110100 and 11101100

4. Referring to Table 17.1, use the nearest-neighbor method to decode the received words 0000110 and 1110100.

5. If the code word 0110010 is received as 1001101, how is it decoded using the circle diagram method?

6. Suppose that a received word has the circle diagram arrangement shown here:

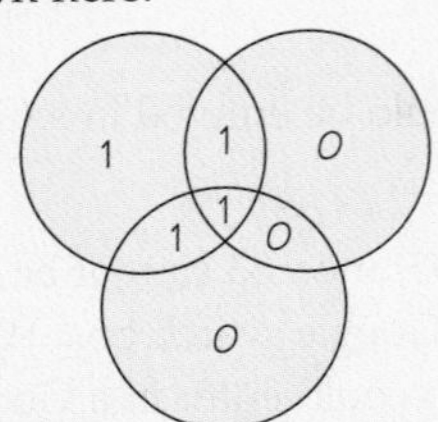

What can we conclude about the received word?

17.2 Encoding with Parity-Check Sums

7. Determine the binary linear code that consists of all possible three-digit messages with three check digits appended using the parity-check sums $a_2 + a_3$, $a_1 + a_3$, and $a_1 + a_2$. (That is, $c_1 = 0$ if $a_2 + a_3$ is even, $c_1 = 1$ if $a_2 + a_3$ is odd, and similarly for c_2 and c_3.)

8. Let *C* be the code

$$\{0000000, 1110100, 0111010, 0011101, 1001110, 0100111, 1010011, 1101001\}$$

What is the error-correcting capability of *C*? What is the error-detecting capability of *C*?

9. Find all code words for binary messages of length 4 by appending three check digits using the parity-check sums $a_2 + a_3 + a_4$, $a_2 + a_4$, and $a_1 + a_2 + a_3$. Will this code correct any single error?

10. Consider the binary linear code

$$C = \{00000, 10011, 01010, 11001, 00101, 10110, 01111, 11100\}$$

Use nearest-neighbor decoding to decode 11101 and 01100. If the received word 11101 has exactly one error, can you determine the intended code word? Explain your reasoning.

11. Construct a binary linear code using all eight possible binary messages of length 3 and appending three check digits using the parity-check sums $a_1 + a_2$, $a_2 + a_3$, and $a_1 + a_3$. Decode each of the received words below by the nearest-neighbor method:

001001, 011000, 000110, 100001

12. Extend the code words listed in Table 17.1 to eight digits by appending a 0 to words of even weight and a 1 to words of odd weight. What are the error-detecting and error-correcting capabilities of the new code?

13. Extend the code words listed in Table 17.2 (on p. 607) to eight digits by appending a 0 to words of even weight and a 1 to words of odd weight. What are the error-detecting and error-correcting capabilities of the new code?

14. Suppose that the weight of a binary linear code is 6. How many errors can the code correct? How many errors can the code detect?

15. How many code words are there in a binary linear code that has all possible messages of length 5 with three check digits appended? How many possible received words are there with this code?

17.3 Data Compression

16. Suppose we code a four-symbol genetic set $\{A, C, T, G\}$ into binary form as follows:

$A \to 0 \quad C \to 10 \quad T \to 110 \quad G \to 111$

Convert the sequence *ACAAGTAAC* into binary code.

17. Use the code in Exercise 16 to determine the sequence of symbols represented by the binary code 001100001111000.

18. Suppose that we code a five-symbol set $\{A, B, C, D, E\}$ into binary form as follows:

$A \to 0 \quad B \to 10 \quad C \to 110$
$D \to 1110 \quad E \to 1111$

Convert the sequence of *AEAADBAABCB* into binary code. Determine the sequence of symbols represented by the binary code 01000110100011111110.

19. Use the code in Exercise 18 to convert the sequence *EABAADABB* into binary code. Determine the sequence of letters represented by the binary code 00100011001110111010.

■ **20.** Devise a variable-length binary coding scheme for a six-symbol set $\{A, B, C, D, E, F\}$. Assume that *A* is the most frequently occurring symbol, *B* is the second most frequently occurring symbol, and so on.

21. Judging from the frequency table in Figure 17.9, what are the three most frequently occurring consonants in English text material? What is the most frequently occurring vowel?

22. Explain why Morse code must include a space after each letter, but fixed-length codes do not.

23. Following are the closing values (rounded to the nearest integer) of the Dow Jones Industrial Average stock market values for the period from September 17, 2007 to September 28, 2007. Use the delta function method to compress these values. What percentage reduction in characters is there?

13403 13739 13816 13767 13820
13759 13779 13878 13913 13896

24. The following numbers were encoded using delta function encoding. Determine the original numbers.

1207 373 −57 −97 −234 −105 178 −73 275
79 −183 −146 −94 129

25. Assume that a dataset of four letters occurs with the following probabilities:

A	0.425
B	0.210
C	0.215
D	0.150

Use a Huffman tree to create a Huffman code for *A*, *B*, *C*, and *D*.

26. Given that the delta function was used to create values in the following list

13403 336 77 −49 53 −61 20 99 35 −17,

recreate the original list.

27. Decode the following binary string,

1110010000100111011001101 0,

which has been encoded using the Huffman code given on page 613.

28. Use a Huffman tree to assign a binary code to the letters that occur with the following probabilities:

A	0.025
B	0.150
C	0.015
D	0.170
E	0.200
F	0.225
G	0.215

29. Suppose that a Huffman tree has been used to create a binary code for the letters *A* through *J*, and the results include $B = 111110$, $J = 111111$, and $G = 11110$. If the code has only two code words of length 6 and one of length 5, what can you say about the probability of the occurrence of the letters *B*, *J*, and *G*?

17.4 Cryptography

30. For each part below, explain how modular arithmetic can be used to answer the question.

(a) If today is Wednesday, what day of the week will it be in 16 days?

(b) If a clock's hands indicate that it is now four o'clock, what time will the clock indicate in 37 hours?

(c) If a military person says it is now 0400, what time would it be in 37 hours? (Instead of A.M. and P.M., military people use 1300 for 1:00 P.M., 1400 for 2:00 P.M., and so on.)
(d) If it is now July 20, what day will it be in 65 days?

31. Use the Caesar cipher to encrypt the message RETREAT. Decrypt the message DGYDQFH, which was encrypted using the Caesar cipher.

32. The message ADDAOS was encrypted using the decimation cipher with the key 7. Decrypt it.

33. Using 0, 1, 2, . . ., 25 to label the positions of the letters *A, B, C*, . . ., *Z*, suppose that we create a cipher by replacing the letter in position i with the letter in position $(i + 8) \bmod 26$. How many iterations of this cipher must be done before a message will return to its original state?

34. If you attempted to use the decimation cipher with the key 13, how would the word MESSAGE be encrypted?

35. Explain why 2 cannot be used as the key in a decimation cipher.

36. Use the decimation cipher with the key 5 to encrypt RETREAT.

37. Given that BEATLES was used as the key word for the Vigenère cipher to encrypt SSLETRY TXOGPW, decrypt the message.

38. Use the Vigenère cipher with the key word CLUE to encrypt the message THE WALRUS WAS PAUL.

39. Use the Vigenère cipher with the key word HELP to encrypt the message PHONE HOME.

40. Use the linear cipher method with $k = 5$ and $s = 4$ to encrypt GOOD.

41. Given that the received message ZVW was encrypted using the linear cipher method with $k = 11$ and $s = 9$, decrypt the message.

42. Add the following pairs of binary strings:

(a) 10111011 and 01111011
(b) 11101000 and 01110001

43. For any two binary strings u and v of the same length, define $u + v$ to be the string obtained by summing the strings as explained in Example 7 (on p. 636). Find $u + v$ for each of the following cases:
(a) $u = 1100001$ and $v = 0011100$
(b) $u = 1011010$ and $v = 0111001$

44. Suppose that u and v are two binary words of length 7 whose distance apart is 7. What string is $u + v$?

45. Given the binary word $u = 1010110$, find all binary words of length 7 that are a distance of 1 from u.

46. Suppose that u and v are binary code words of the same length. Using the method of summing binary strings as shown in Example 7, explain why the distance from u to v is the same as the weight of $u + v$.

47. Suppose that u, v, and w are binary code words of the same length. Using the method of summing binary strings as shown in Example 7, explain why the distance from u to v is the same as the distance from $u + w$ to $v + w$.

48. All binary linear codes have the property that the sum of two code words is another code word. Use this fact to determine which of the following sets cannot be a binary linear code:

(a) {0000, 0011, 0111, 0110, 1001, 1010, 1100, 1111}
(b) {0000, 0010, 0111, 0001, 1000, 1010, 1101, 1111}
(c) {0000, 0110, 1011, 1101}

49. Use the RSA scheme with $p = 5$, $q = 17$, and $r = 3$ to determine the numbers sent for the message VIP.

50. Use the RSA scheme with $p = 5$, $q = 17$, and $r = 3$ to decode the received numbers 52 and 72.

51. In the RSA scheme with $p = 5$, $q = 17$, and $r = 5$, determine the value of s.

52. Why can't we use the RSA scheme with $p = 7$, $q = 11$, and $r = 3$?

53. Explain why we can't employ the RSA scheme to send the message "NO" with $p = 7$ and $q = 11$ using blocks of length 2, but we can send it if we use blocks of length 4.

54. Use the search box at www.google.com to compute $8^{11} \bmod 85$ and $72^{11} \bmod 85$. (To compute $8^{11} \bmod 85$, enter "8^11 mod 85.")

17.5 Web Searches and Mathematical Logic

55. Show that $P \vee (P \wedge Q)$ is logically equivalent to P.

56. Show that $\sim(P \vee Q)$ is logically equivalent to $\sim P \wedge \sim Q$.

57. Show that $\sim(P \wedge Q)$ is logically equivalent to $\sim P \vee \sim Q$. (This relationship and the one in Exercise 56 are known collectively as De Morgan's Laws.)

58. Show that $P \vee (Q \wedge R)$ is logically equivalent to $(P \vee Q) \wedge (P \vee R)$.

59. Show that $P \wedge (Q \vee R)$ is logically equivalent to $(P \wedge Q) \vee (P \wedge R)$.

60. A patron at a restaurant tells the waiter to bring her the chef's recommendation, so long as it has "lots of anchovies or is not spicy, and in addition, the portion must be large." The waiter goes to the kitchen and tells the chef to prepare a dish that has "lots of anchovies and is also large or is spicy and is also large." Did the waiter communicate the patron's wishes correctly to the chef? Use truth tables to support your answer.

61. The implication connective $P \to Q$ is defined by the following truth table:

P	Q	$P \to Q$
T	T	T
T	F	F
F	T	T
F	F	T

Use truth tables to show that $P \to Q$ is logically equivalent to $\sim P \vee Q$.

62. The Minnesota Vikings football coach tells his team before the last game of the regular season that if the team wins, they will be in the playoffs. Use the truth table given in Exercise 61 to verify that if the Vikings lose and are still in the playoffs, the coach made a truthful statement to the team.

63. Using the implication connective and other connectives, variables, and truth tables, determine whether the statement "If it snows, there will be no school" is logically equivalent to the statement "It is not the case that it snows and there is school."

64. Suppose that s and t are binary strings of length 8. How would you use the bitwise operator $\wedge$ to determine whether the last three digits of s and t match? How would you determine whether s and t match in positions 2, 4, 6, and 8?

65. Using $\wedge$ to denote the bitwise AND operator, compute the following items:

(a) $11110001 \wedge 00101110$

(b) $01110001 \wedge 10111110$

66. Explain why $01100000 \wedge 1011111$ is undefined.

67. In practice, a computer checks to see if $s \wedge 11100000 = t \wedge 11100000$ by checking if $s \wedge 11100000 + t \wedge 11100000 = 00000000$, where addition is done using mod 2 in each component. Explain why this works.

68. If s is an 8-digit binary string, determine $s \wedge 11111111$ and $s \wedge 00000000$.

69. If s is a binary string of length 8 and $s \wedge 01010101 = 00000001$, what is the most that you can say about s?

70. Given a binary string s of length 8, how could you use the bitwise AND operator to determine whether the digits in positions 1, 3, and 5 are 0's?

71. Find four binary strings s that satisfy $s \wedge 11100111 = 01100010$.

72. How many binary strings s of length 8 are there that satisfy $s \wedge 11100011 = 10000001$?

73. Determine the network address for the IP address 8.20.15.1 with subnet mask 255.000.000.000.

74. Determine the network address for the IP address 8.20.15.1 with subnet mask 255.255.000.000.

75. Determine whether the IP address 172.16.17.30 with subnet mask 255.255.255.240 has the same network address as the IP address 172.16.17.15 with subnet mask 255.255.255.240.

WRITING PROJECTS

1. Prepare a two-page report on applications of modular arithmetic. Explain the calculation of the check digits described in Exercises 7, 9, and 11 with modular arithmetic. Use modular arithmetic to describe the error-detection schemes used in Chapter 16.

2. Use the Web to find information for a two-page report on the Braille system of coding.

3. Use the Web to find information for a two-page report on Morse code.

4. Use the Web to find information about smart card technology and write a two-page report on your findings.

Suggested Web Sites

www.d.umn.edu/~jgallian/fapp7 This site implements the nearest-neighbor decoding method for seven-digit binary strings using the code given in Table 17.1 on page 605.

http://www.math.csusb.edu/notes/quizzes/tablequiz/tablepractice.html This interactive page generates random truth table problems involving two simple statements P and Q, indicates if your solutions are correct, and provides the correct truth tables upon request.

http://turner.faculty.swau.edu/mathematics/materialslibrary/truth/ This site provides the truth table for any Boolean expression that you want.

Getty Images/Flickr RF

On Size and Growth

Part VI

Mathematics is the study of patterns and relationships. Mathematicians can explain why there are no King Kongs, analyze designs on ancient pottery, and suggest new and beautiful artistic designs. They search for and classify numerical, geometric, and even abstract patterns. In these chapters, we follow some of those searches. We concentrate on geometric patterns, but they lead to numerical considerations as well, so we explore those, too.

In Chapter 18, "Growth and Form," we look at how the sizes of objects influence their forms. We investigate some big things, such as King Kong, tall trees, mile-high buildings, and mountains. Seeing the underlying principles of scaling will help you to appreciate why objects in the world have the shapes and sizes that they do.

We start with a simple numerical pattern in Chapter 19, "Symmetry and Patterns," which leads to questions about esthetically pleasing proportions and the importance of bilateral symmetry. We expand our notion of symmetry and discover surprising limitations that even broader notions of symmetry face. We examine the beauty of fractal patterns, ones that resemble themselves at finer and finer scales, in nature and in traditional art from Africa and elsewhere.

Chapter 20, "Tilings," answers the question of how to arrange objects symmetrically on a surface. What shapes can we use? What patterns can arise if the objects themselves are symmetrical, or if we allow irregular shapes but demand that they all face the same way? Most curious of all, you can arrange shapes in a pattern that does not repeat but is nevertheless systematic.

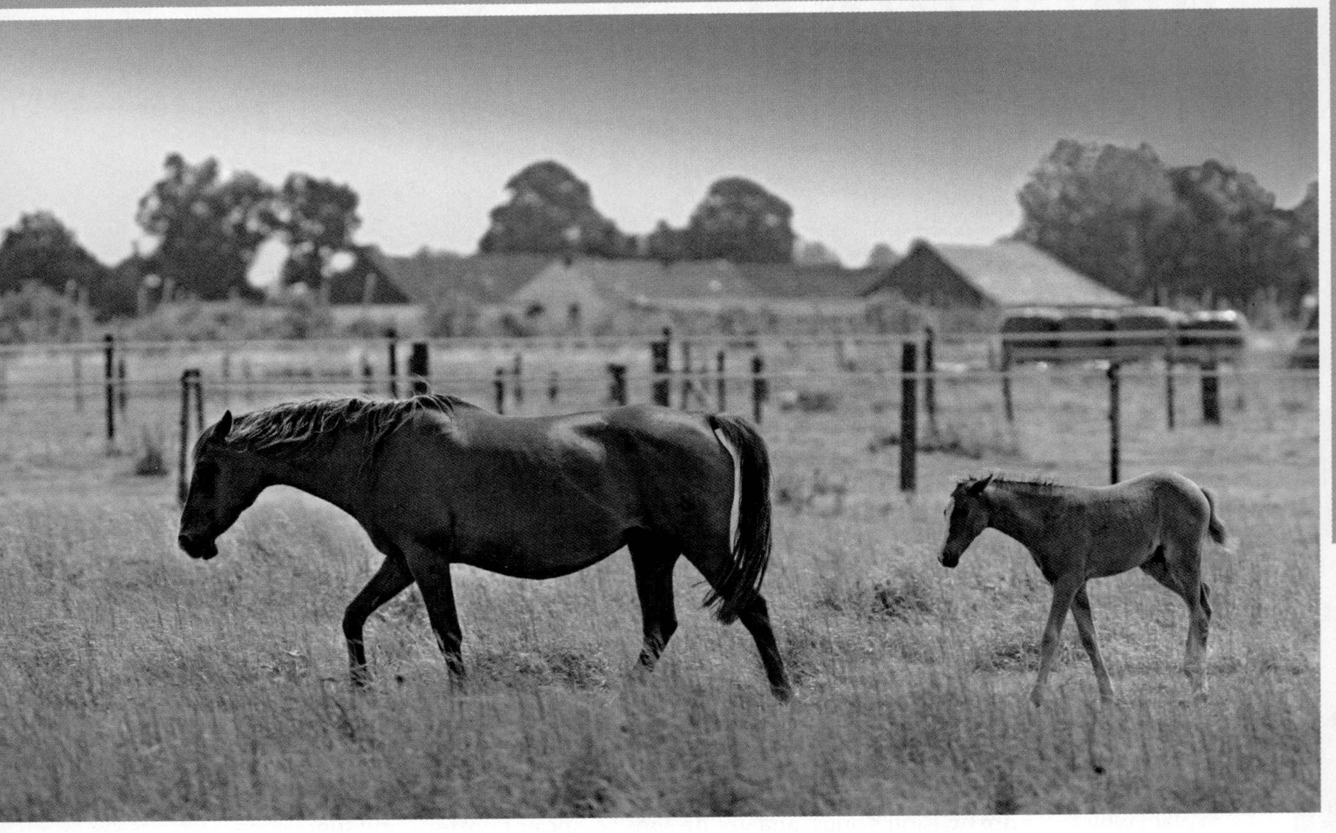

18

Growth and Form

Films show us giant creatures, including King Kong, Godzilla, and the auliphants in *The Lord of the Rings*. Literature has the giant of "Jack and the Beanstalk," and the Brobdingnagians of *Gulliver's Travels,* and the giants in the Battle of Hogwarts in *Harry Potter and the Deathly Hallows*. "Fakelore" gives us Paul Bunyan and his blue ox, Babe.

Could such beings ever exist? (See Figure 18.1.)

Every species has to adapt and survive at the different sizes from babyhood to mature adult. For example, the giant panda ranges from 1 pound (lb) at birth to 275 lb in adulthood. A baby panda could be crushed by its mother; an adult panda needs to eat a lot.

For contrast, consider the horse. A newborn foal that weighed as little as a newborn panda would be too small to keep up with the herd and could not survive. An adult horse weighs much more than a panda and has to consume far more food, but the horse can move much more quickly and cover greater distances to find sustenance.

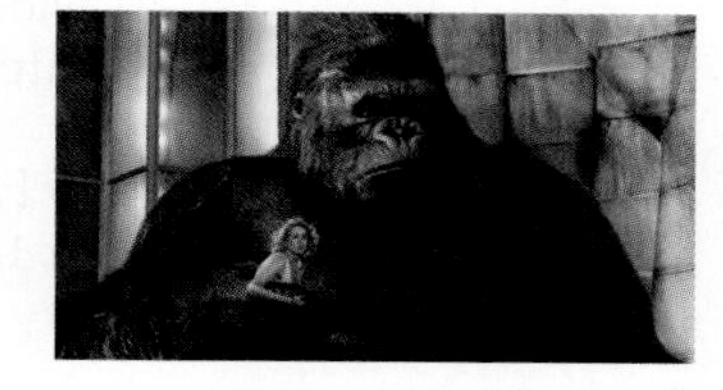

Figure 18.1 Could King Kong actually exist? *(Universal/Wing Nut Films/The Kobal Collection)*

There have been large land mammals (mammoths) and huge sea mammals (the blue whale)—not to mention the dinosaurs. But the tallest humans have been only 9 to 10 feet (ft) tall, and even the tallest dinosaur, *Supersaurus,* stood only 40 ft high.

What about supergiants and utterly huge monsters? The fact that they have never existed suggests physical limits to size. Using a few simple principles of geometry, we show that no objects or living beings could exist, unchanged in shape, on a vastly different scale, larger or smaller.

In Section 18.1, we determine what it means for two objects to be similar in shape. We see how to scale up an object while maintaining its shape, and then investigate what the change does to surface area and to volume. We show how to compare two objects measured in different units, such as feet and meters, in Section 18.2. In Section 18.3, we enlarge a mountain as a case study; we then wonder how high it could be without collapsing on itself, and how tall a building can be. In Section 18.4, we bid farewell to King Kong, note the tallest living beings, and consider consequences for a life form to exist at a different scale. Section 18.5 examines the effect of the disparity between surface area and volume as a being is scaled up. Finally, Section 18.6 looks at how humans grow from birth to maturity.

18.1 Geometric Similarity

The powerful mathematical idea that we use is **geometric similarity**, which you encountered in geometry in terms of similar triangles. We apply the concept to real objects here.

> **Geometric Similarity** DEFINITION
>
> Two objects are **geometrically similar** if they have the same shape, regardless of the materials of which they are made; they do not have to be the same size.

Although similar objects need not be the same size, corresponding distances must be proportional. For example, when a photo is enlarged, it is enlarged by the same factor in both the horizontal and vertical directions—in fact, in any direction (such as a diagonal). We call this factor the **linear scaling factor** (or **length scaling factor**).

> **Linear (Length) Scaling Factor** DEFINITION
>
> The **linear (length) scaling factor** of two geometrically similar objects is the ratio of a length of any part of the second to the corresponding part of the first.

In Figure 18.2, the linear scaling factor is 3; the enlargement is three times as wide and three times as high as the original. In fact, every pair of points becomes three times as far apart.

Objects can be scaled down as well as up; for example, the smaller photograph in Figure 18.2 is geometrically similar to the larger one, with a linear scaling factor of 1/3.

Figure 18.2 Two geometrically similar photographs. *(© Design Pics Inc./Alamy)*

How Area and Volume Scale

The enlargement can be divided into $3 \times 3 = 9$ rectangles, each the size of the original. Hence, the enlargement has $3 \times 3 = 3^2 = 9$ times the area of the original. More generally, if the linear scaling factor is some general number L

(not necessarily 3), the resulting enlargement has an area $L \times L = L^2$ ("L squared") times the area of the original.

How Area Scales RULE

The *area* of a scaled-up object goes up in proportion to the square of the linear scaling factor.

We symbolize the relationship between the area A and the linear scaling factor L by

$$A \propto L^2$$

where the symbol $\propto$ is read as "is proportional to" or "scales as."

What about enlarging three-dimensional objects? If we take a cube and enlarge it by a linear scaling factor of 3, it becomes three times as long, three times as high, and three times as deep as the original (see Figure 18.3).

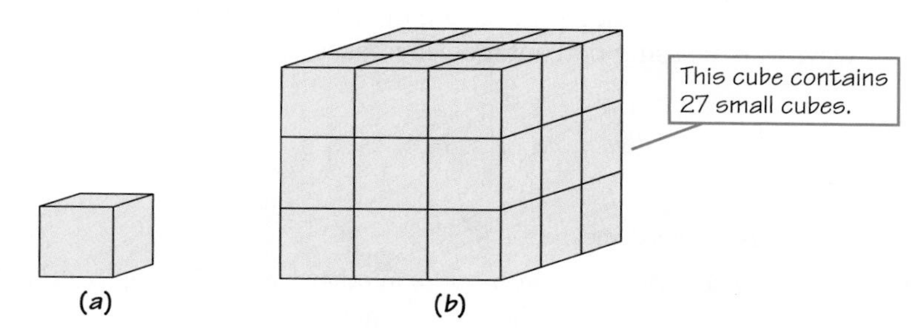

Figure 18.3 Cube (b) is made by enlarging cube (a) by a factor of 3.

What about volume? How much bigger is the volume of the enlarged cube? The enlarged cube has three layers, each with $3 \times 3 = 9$ little cubes, each the same size as the original. Thus, the total volume is $3 \times 3 \times 3 = 3^3 = 27$ times as much as the original cube. Denoting the volume by V, we can write

$$V \propto L^3$$

Thus, for an object enlarged by a linear scaling factor of L, the enlargement has $L \times L \times L = L^3$ ("L cubed") times the volume of the original. Like the relationship between area and L^2, this relationship holds even for irregularly shaped objects, such as science-fiction monsters.

How Volume Scales RULE

The *volume* of a scaled-up object goes up with the cube of the linear scaling factor.

You can see, however, that the area of each face (side) of the enlarged cube is $3^2 = 9$ times as large as that of a face of the original cube, just as the area of the photo enlarged by a factor of 3 has 9 times the area of the original. The total surface area of the enlarged cube is 9 times the total surface area of the original cube.

More generally, for objects of any shape, the total *surface area* of a scaled-up object goes up with the *square* of the linear scaling factor. Thus, the surface area of an object scaled up by a factor of L is L^2 times the surface area of the original. This feature holds true even for irregular shapes.

Before we discuss scaling real three-dimensional objects, you should understand some language pitfalls in describing increases and decreases.

The Language of Growth, Enlargement, and Decrease

Streaming Movies Costs Netflix **10 Times Less Than** *Mailing Them*

—http://gizmodo.com/5622167

Commonwealth Challenge is a 17½-month co-educational program for at-risk teens ages 16 to 19. . . . Cost per cadet is $15,000. On a per capita basis, the cost . . . is:

- *85% less than high school*
- *320% less than job corps*
- *433% less than juvenile corrections*
- ***600% less than** adult corrections*
- *660% less than private military-style programs*

— Virginia National Guard Association 2010 State Legislative Action Plan, http://www.vnga.org/legislation.shtml

Let's explore why the bolded phrases are incorrect and confusing. First, we set out correct ways of discussing percentages, increases, and decreases:

Meanings of Percentage RULE

"x% of A" or "x% as large as" means $\frac{x}{100} \times A$.

"x% more than A" means A plus x percent of A, in other words, $\left(1 + \frac{x}{100}\right) \times A$. Saying that A has "increased by x%" means the same thing.

"x% less than A" means A minus x% of A, in other words, $\left(1 - \frac{x}{100}\right) \times A$. Saying that A has "decreased by x%" means the same thing.

You no doubt surmise that "10 times less than" really means that the cost of streaming the movies, for example, is "one-tenth the cost of" mailing them. But the latter doesn't sound as grand. Writing "10 times" helps inflate the savings in the mind of the reader, and "less" clues the reader about which is cheaper.

What about the cadet program? Following the same logic, you would guess (correctly) that adult corrections (prison) must cost 6 times as much as the cadet program, or $6 \times \$15{,}000 = \$90{,}000$ per teen per year. Similarly, we could interpret "433 percent less than juvenile corrections" as meaning that the juvenile corrections costs $4.33 \times \$15{,}000 \approx \$65{,}000$ per teen per year.

So far, so good. How then should we understand the claim that the program cost is "85% less than high school"? Following the same logic would lead us to calculate that high school costs $0.85 \times \$15{,}000 \approx \$13{,}000$ per teen per year—and that is right on the mark for Fairfax County, Virginia.

But wait a minute—then the cadet program ($15,000) costs *more*—not less—than high school ($13,000)! That can't be what was meant.

If something costs $100 and the cost is reduced 85%, you would pay $15. So maybe the claim means that the cadet program costs just 15% of the cost of high school? That would mean that high school costs $100,000 (so that at 85% less, the cadet program costs $15,000). You aren't paying that much a year even for college, so that can't be right, either.

Maybe what was meant instead was that the program costs 85% as much as high school? Call the cost of high school C; then we would have $0.85C = \$15{,}000$. Solve for C by dividing both sides by 0.85, and you'll find that C is a little under \$18,000. Indeed, there are Virginia school districts (e.g., Alexandria) whose average cost per student is that high.

What was really meant? It certainly would be useful to avoid all the confusion by using consistent and precise language.

The terms *of, times,* and *as much as* refer to *multiplication* of the original amount, while the terms *more, larger,* and *greater* refer to *adding* to the original amount. For instance, "five times as much" means the same as "four times more than" (the original plus four times as much). Similarly, the relationship of the original amount to the larger amount can be expressed either in multiplicative terms ("one-fifth as much" or "20% as much") or in subtractive terms ("four-fifths less than" or "80% less than")—but don't mix the two.

Using *times* together with *more* is, to use a football analogy, "piling on." It is designed to impress you by attempted exaggeration.

People often say "five times more than" when they mean "five times as much." About the streamed movies, the author wrote "10 times less" to mean "one-tenth as much." All you can do is be aware of the potential confusion, try to figure out what was meant, and be careful in your own expression. In particular:

Correct Comparisons RULE

Don't use *times* with *more* or with *less.*

Finally, we need to distinguish *percent* from *percentage points:* If support for the president decreased from 60% to 30%, it dropped 30 *percentage points* but decreased 50% (because the drop of 30 percentage points is 50% of the original 60 percentage points).

EXAMPLE 1
What About That Cadet Program?

Returning to the cadet program, how could you state correctly what the author was trying to say and still manage to use the word *less*?

SOLUTION As we noted above, the cadet program costs "85% as much as high school," or its cost is "85% of the cost of high school." If you want to be sure to use the word *less*, then you can say that the program costs "15% less than high school." But that doesn't sound as impressive as "85% less"—and it shouldn't, because it isn't true that the program costs 85% less.

18.2 How Much Is That in . . . ?

We are interested in the limits of size and want to compare objects of different sizes, such as a gorilla with King Kong. However, it is not easy to compare two objects measured in different units—say, the gorilla is measured in inches and pounds, but King Kong is measured in centimeters and kilograms. Consequently, we explore how to convert units from one measurement system to another.

We introduce two systems of units in which physical quantities are commonly measured and give a table of *conversion factors* and examples of how to convert from one system to the other.

U.S. Customary System

Table 18.1 lists units of the U.S. *customary system* of measurement and their abbreviations. Please note in the table the systematic way of converting from one unit to another and the use of scientific notation. The symbol ≈ means "is approximately equal to."

Table 18.1

Units of the U.S. Customary System

Distance:	
1 mile (mi)	= 1760 yards (yd) = 5280 feet (ft) = 63,360 inches (in.)
1 yard (yd)	= 3 feet (ft) = 36 inches (in.)
1 foot (ft)	= 12 inches (in.)
Area:	
1 square mile	$= 1 \text{ mi} \times 1 \text{ mi} = 5280 \text{ ft} \times 5280 \text{ ft}$
	$= 27{,}878{,}400 \text{ ft}^2 \approx 2.8 \times 10^7 \text{ ft}^2$
	$= 63{,}360 \text{ in.} \times 63{,}360 \text{ in.}$
	$= 4{,}014{,}489{,}600 \text{ in.}^2 \approx 4 \times 10^9 \text{ in.}^2$
	= 640 acres
1 acre	$= 43{,}560 \text{ ft}^2$
Volume:	
1 cubic mile	$= 1 \text{ mi} \times 1 \text{ mi} \times 1 \text{ mi}$
	$= 5280 \text{ ft} \times 5280 \text{ ft} \times 5280 \text{ ft}$
	$= 147{,}197{,}952{,}000 \text{ ft}^3$
	$\approx 1.5 \times 10^{11} \text{ ft}^3$
	$= 63{,}360 \text{ in.} \times 63{,}360 \text{ in.} \times 63{,}360 \text{ in.}$
	$\approx 2.5 \times 10^{14} \text{ in.}^3$
1 U.S. gallon (gal)	$= 4$ U.S. quarts (qt) $= 231 \text{ in.}^3$
Mass:	
1 ton (t)	= 2000 pounds (lb)

Metric System

The world generally uses the metric system in science, industry, and commerce. It was proposed by Gabriel Mouton, a vicar in Lyons, in 1670 and was adopted in France in 1795. The fundamental unit of length, the *meter* (m), was originally 1/10,000,000 of the distance from the North Pole to the equator, along the meridian through Paris. The length is now defined as the distance that light travels in a vacuum in $\frac{1}{299,792,458}$ second (s). The second, in turn, is defined as the time that it takes an atom of the metal cesium to vibrate 9,192,631,770 times.

All other units of length, area, and volume are *defined* in terms of the meter. For example, a centimeter (cm) is a hundredth of a meter.

Mass is the quantity of matter. The metric unit of mass, the *kilogram* (kg), is defined as the mass of a platinum-iridium standard kept in Paris. Since you can't determine the mass of a sack of potatoes by comparing it to that, we measure the mass indirectly by seeing how much force gravity exerts on it—that is, we weigh it on a scale calibrated in pounds or kilograms. However, a mass of 1 kg would "weigh" (register on the scale) only one-sixth as much on the Moon.

Table 18.2 lists the units of the metric system.

Table 18.2

Units of the Metric System

Distance:	
1 meter (m)	= 100 centimeters (cm)
1 kilometer (km)	= 1000 meters (m)
	= 100,000 centimeters (cm) = 1×10^5 cm
	= 1,000,000 millimeters (mm) = 1×10^6 mm
Area:	
1 square meter (m^2)	= 1 m × 1 m
	= 100 cm × 100 cm = 10,000 (cm^2) = 1×10^4 cm^2
1 hectare (ha)	= 10,000 m^2
Volume:	
1 liter (L)	= 1000 cm^3 = 0.001 m^3
1 cubic meter (m^3)	= 1 m × 1 m × 1 m
	= 100 cm × 100 cm × 100 cm
	= 1,000,000 cm^3 = 1×10^6 cm^3 (or cc)
Mass:	
1 kilogram (kg)	= 1000 grams (g) = 1×10^3 g
1 metric ton (tonne)	= 1000 kg

Converting Between Systems

What are the conversion factors between the U.S. customary system and the metric system? Since 1959, the fundamental units of the U.S. customary system, the yard (for length) and the pound (for mass), have been *defined* in terms of metric units, so that we have *exactly*

$$1 \text{ yd} = 0.9144 \text{ m}$$
$$1 \text{ lb} = 0.45359237 \text{ kg}$$

Table 18.3 illustrates the conversion factors. Most Internet search engines offer conversions; try entering, for example, "180 cm in ft." So, although you do not need to memorize conversion factors, you will find it useful in life to memorize a few rough approximations:

1 cm ≈ 0.4 in.	1 m ≈ 3 ft	1 km ≈ 0.6 mi	1 mi ≈ 1.6 km
1 m^2 ≈ 10 ft^2	1 L ≈ 1 qt	1 gal ≈ 4 L	1 kg ≈ 0.5 lb

Table 18.3

Conversions Between the U.S. Customary System and the Metric System

Distance:	
1 in.	= 2.54 cm
1 ft	= 12 in. = 12 × 2.54 cm = 30.48 cm = 0.3048 m ≈ 0.3 m
1 yd	= 0.9144 m ≈ 1 m
1 mi	= 5280 ft = 5280 × 30.48 cm
	= 160,934.4 cm ≈ 1.609 km ≈ 1.6 km
1 cm	≈ 0.393701 in. ≈ 0.4 in.
1 m	≈ 39.37 in. ≈ 3.281 ft ≈ 3 ft
1 km	≈ 0.621 mi ≈ 0.6 mi
Area:	
1 ft^2	≈ 0.09290 m^2 = 929.0 cm^2 ≈ 1000 cm^2
1 m^2	≈ 10.76 ft^2 ≈ 10 ft^2
1 hectare (ha)	≈ 2.5 acres
Volume:	
1 ft^3	≈ 28.32 liters (L)
1 gallon	≈ 3.785 liters (L) ≈ 4 L
1 cubic meter (m^3)	= 1000 liters ≈ 264.2 U.S. gallons ≈ 35 ft^3
1 liter (L) = 1000 cm^3	≈ 1.057 U.S. quarts (qt) ≈ 0.2642 U.S. gallons ≈ 1 qt
Mass:	
1 lb	= 0.45359237 kg ≈ 0.5 kg
1 kg	≈ 2.205 lb ≈ 2 lb

In the following examples, we explain how to convert measurements between systems.

EXAMPLE 2
What's That in Feet?

An international student tells his American student friends that he is 180 cm tall. They ask how much that is in feet and inches.

We approach this conversion by using the scaling factor 1 cm = $\frac{1}{2.54}$ in.:

$$180 \text{ cm} = 180 \text{ cm} \times 1$$
$$= 180 \text{ cm} \times \frac{1 \text{ in.}}{2.54 \text{ cm}}$$
$$\approx 70.9 \text{ in.} = 70.9 \text{ in.} \times \frac{1 \text{ ft}}{12 \text{ in.}} = \frac{70.9}{12} \text{ ft} \approx 5.9 \text{ ft}$$

However, because we normally give height in feet and a whole number of inches, the height is

$$70.9 \text{ in.} = 5 \times (12 \text{ in.}) + 10.9 \text{ in.} \approx 5 \text{ ft, } 11 \text{ in.}$$

Another way to approach the problem is by means of a proportion:

$$\frac{\text{height in in.}}{\text{height in cm}} = \frac{\text{length of 1 in. in inches}}{\text{length of 1 in. in cm}} = \frac{1 \text{ in.}}{2.54 \text{ cm}}$$

so that

$$\text{height in in.} = \text{height in cm} \times \frac{1 \text{ in.}}{2.54 \text{ cm}}$$
$$= 180 \text{ cm} \times \frac{1 \text{ in.}}{2.54 \text{ cm}} \approx 70.9 \text{ in.}$$

EXAMPLE 3
Got Gas?

Although in the United States we have traditionally measured the efficiency of cars in miles per gallon (mpg), the rest of the world measures it in liters per 100 kilometers. The conversion between these two measures is more complicated than other conversions because the U.S. measure has distance (mi) in the numerator and quantity of fuel (gal) in the denominator, while the other measure has quantity of fuel (L) in the numerator and distance (km) in the denominator. We need to take this difference into account when doing the conversion.

For example, according to the Environmental Protection Agency (EPA), the most efficient gasoline vehicle of all time was the two-passenger 2000 Honda Insight, at 61 mpg on the highway. (This model was discontinued in 2006 due to poor sales!) What is the equivalent in liters per 100 km?

SOLUTION

$$61 \text{ mpg} = 61 \times \frac{1 \text{ mi}}{1 \text{ gal}}$$
$$\approx 61 \times \frac{1.609 \text{ km}}{3.785 \text{ L}} = 61 \times \frac{1.609}{3.785} \times \frac{\text{km}}{\text{L}} \approx 25.93 \frac{\text{km}}{\text{L}}$$
$$= \frac{25.93}{1} \times \frac{100 \text{ km}}{100 \text{ L}} = \frac{1}{\frac{1}{25.93}} \times \frac{100 \text{ km}}{100 \text{ L}}$$
$$\approx \frac{100 \text{ km}}{3.9 \text{ L}}$$

or about 4 L per 100 km. (Europeans would call such a car a "4-liter car.") The key steps in the solution are to multiply both units by 100, and then divide both numerator and denominator of the fraction by 25.93, so as to get exactly 100 km in the numerator of the result.

Proposed U.S. new-car labels, as shown in Figure 18.4, would indicate fuel economy in terms of gallons per 100 mi.

Figure 18.4 New fuel-economy label for 2013-model U.S. cars. *(Courtesy of Environmental Protection Agency.)*

18.3 Scaling a Mountain

Gravity exerts an enormous effect on the size and shape that objects and beings can assume. **Weight** (force under the Earth's gravity) is the reading at sea level on a scale (such as your bathroom scale) *calibrated in pounds or kilograms of mass.*

Suppose that the two cubes in Figure 18.3 on page 645 are made of steel and that the first is 1 ft on a side and the second is 3 ft on a side. A cubic foot of steel weighs about 500 lb; we say that the **density** of steel is 500 lb per cubic foot, or 500 lb/ft^3. The cube that is 1 ft on a side weighs 1 ft^3 $\times$ 500 lb/ft^3 = 500 lb. The weight W of an object of volume V and uniform density D is

$$W = DV$$

Each cube's bottom face supports the weight of the entire cube. **Pressure** is the force per unit area, so the pressure exerted on the bottom face by the weight of the cube is equal to the weight of the cube divided by the area of the bottom face, or

$$P = \frac{W}{A}$$

The first cube weighs 500 lb and has a bottom face with area of 1 ft^2, so the pressure exerted on this face is 500 lb/ft^2.

The second cube is 3 ft on a side. The area of the bottom face increases with the square of the linear scaling factor, so it is $3^2 \times 1$ ft^2 = 9 ft^2. As we saw earlier, volume goes up with the cube of the linear scaling factor. So this larger cube has a volume of $3^3 \times 1$ ft^3 = 27 ft^3. Because both cubes are made of the same steel, the larger cube has 27 times as much steel as the smaller one. Hence, it weighs 27 times as much as the smaller cube, or 27 $\times$ 500 lb = 13,500 lb.

When we divide this weight by the area of the bottom face (9 ft^2), we find that the pressure exerted on the bottom face is 1500 lb/ft^2, or three times the pressure on the bottom face of the original cube. This makes sense because over each 1-ft^2 area stands 3 ft^3 of steel. In general, if the linear scaling factor for the cube is L, the pressure on the bottom face is L times as much. Using the notation of proportionality, where $\propto$ stands for "is proportional to," we have $A \propto L^2$ and $W \propto V \propto L^3$, so

$$P = \frac{W}{A} \propto \frac{L^3}{L^2} \propto L$$

EXAMPLE 4
What About a 10-Foot Cube?

SOLUTION If we scale the original cube of steel up to a cube that is 10 ft on a side, then the dimensions are

$$10 \text{ ft} \times 10 \text{ ft} \times 10 \text{ ft}$$

The total volume is

$$V = \text{length} \times \text{width} \times \text{height} = 10 \text{ ft} \times 10 \text{ ft} \times 10 \text{ ft} = 1000 \text{ ft}^3$$

The weight of the cube is

$$W = D \times V = \frac{500 \text{ lb}}{\text{ft}^3} \times 1000 \text{ ft}^3 = 500{,}000 \text{ lb}$$

The area of the bottom face is

$$A = \text{length} \times \text{width} = 10 \text{ ft} \times 10 \text{ ft} = 100 \text{ ft}^2$$

The pressure on the bottom face is

$$P = \frac{W}{A} = \frac{500{,}000 \text{ lb}}{100 \text{ ft}^2} = \frac{5000 \text{ lb}}{\text{ft}^2}$$

This is 10 *times*—not "10 times *more* than"—the pressure on the bottom face of the original 1-ft cube.

At some scale factor, the pressure on the bottom face will exceed the steel's ability to withstand that pressure—and the steel will deform under its own weight. That point for steel is reached for a cube that is about 3 miles (mi) on a side; the pressure exerted by the cube's weight exceeds the **crushing strength** of steel, which is about 7.5 million lb/ft^2. Because 3 mi $= 3 \times 5280$ ft $= 15{,}840$ ft, a 3-mi-long cube of steel would be more than 15,000 times as high as the original 1-ft cube; that is, the linear scaling factor is more than 15,000. The pressure on the bottom face of the cube would, therefore, be more than 15,000 times as much as for the 1-ft cube, or more than $15{,}000 \times 500$ lb/ft^2 $= 7.5$ million lb/ft^2.

EXAMPLE 5
What About Burj Khalifa?

Burj Khalifa in Dubai, completed in 2010 at a cost of \$1.5 billion, is the world's tallest skyscraper at 2684 ft (how much is that in meters?), not counting radio and television antennas (see Spotlight 18.1). What is the pressure at the bottom of its walls?

SOLUTION The building is made of reinforced concrete, which weighs 160 lb/ft^3. Although the building tapers toward the top, we are not far off if we model it as straight up and down; that is, as a rectangular solid. Consider one of its supporting walls. The volume of the wall is its height H times the area A of its base, or $V = HA$. The weight of the wall is $W = DV = DHA$. The pressure at the bottom is

$$P = \frac{W}{A} = \frac{DV}{A} = \frac{DHA}{A} = DH = \frac{160 \text{ lb}}{\text{ft}^3} \times 2684 \text{ ft} = \frac{429{,}440 \text{ lb}}{\text{ft}^2}$$

or approximately 430,000 lb/ft^2.

So the pressure at the bottom of the wall from the wall's weight alone is about 430,000 lb/ft^2. That's not counting the contents of the tower, which also must be supported.

Could we have a Super Burj Khalifa 10 times as high? The bottom of its walls would have to support about $10 \times 430{,}000\ \mathrm{lb/ft^2} = 4.3$ million $\mathrm{lb/ft^2}$. The crushing strength of reinforced concrete under the Earth's gravity is 8.5 million $\mathrm{lb/ft^2}$, which would leave some safety margin.

One World Trade Center in New York City, whose cornerstone was laid July 4, 2004, and which is scheduled to be completed in 2014, would be the tallest building in the United States, at 1776 ft.

SPOTLIGHT 18.1

A Mile-High Building?

In 1956, the famous American architect Frank Lloyd Wright (1867–1959) proposed a mile-high tower for the Chicago lakefront. Burj Khalifa, at 2684 ft, is just over half that high. In the text, we focus on the problem of holding up the weight of such a structure.

But there are other limits to the height of a building, for example, the bending of the building in the wind, which can go up dramatically with height. Bending can be controlled by making the building stiffer.

The terrorist destruction of the World Trade Center towers in 2001 resulted not directly from the aircraft impacts but from the subsequent fires and collapse of the towers' structure.

Even if designed to better resist fires and impacts, however, a mile-high building might not be practical. For example, the enormous number of people (perhaps 100,000) living, working, or visiting in such a building would create enormous traffic problems (pedestrian, parking, deliveries) for blocks around.

Cost per square foot of usable area is an important consideration. Even if the building did not taper, the space in the upper floors might not justify their additional expense. With increasing height, an increasingly larger proportion of the cross-sectional area of all floors must be devoted to services, such as elevators; everyone entering the building and going to any floor needs to start in an elevator on the ground floor. In an emergency evacuation, the people must walk downstairs!

Some architects, however, maintain that the main limit on the height of a building is human physiology. Differences in air pressure between the top and bottom of a building limit how fast elevators can rise or drop without discomfort to passengers, thereby enforcing long travel times for "vertical commuters." Human psychology also might present some limits.

Are skyscrapers now dinosaurs, except as exercises in one-upmanship? Telecommuting and outsourcing of work to decentralized locations may make it obsolete to bring office workers together at a single site in a dense, expensive downtown.

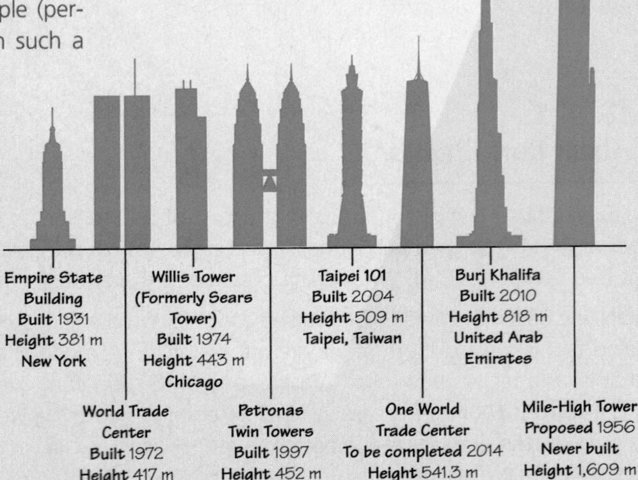

EXAMPLE 6
How High Can a Mountain Be?

The height of mountains also is limited, by gravity, their composition, and their shape. How tall can a mountain be?

SOLUTION We build a simple mathematical model of a mountain. Suppose that it is made of granite, a common material, with uniform density. Granite weighs 165 lb/ft^3 and has a crushing strength of about 4 million lb/ft^2.

In the interests of both realism and simplicity, we assume that the mountain is a solid cone whose width at the base is the same as its height. Let's model Mount Everest, the tallest mountain on Earth, at about 6 mi high. The base, then, is a circle with diameter (distance across) 6 mi. The radius (half the diameter) is 3 mi (Figure 18.5). Because we took round numbers (6 mi) for the height and width, we record as significant only the first digit or two of the results of the calculations.

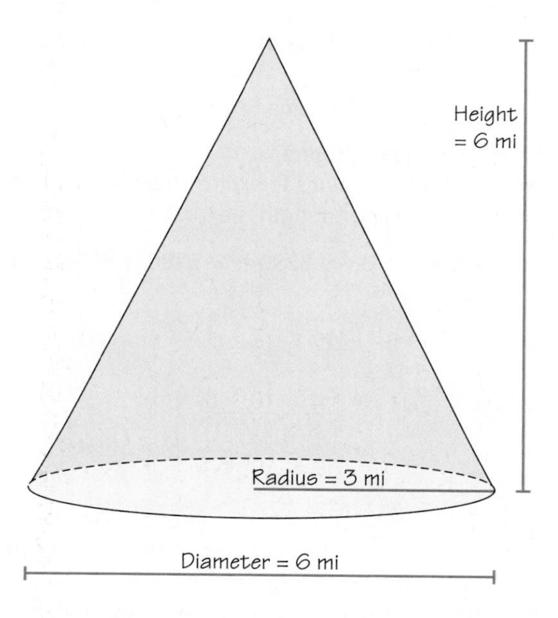

Figure 18.5 Model of Mount Everest as a cone of granite.

What does the model Everest weigh? The relevant formula is $W = DV$, or

$$\text{weight} = \text{density} \times \text{volume}$$

We already know the density of granite (165 lb/ft^3), so to find the weight, we need the formula for the volume of a cone of radius r and height h:

$$V = \text{volume} = \frac{1}{3}\pi r^2 h$$

Using a radius of 3 mi and a height of 6 mi and π (pi), which is approximately 3.14, we find that the model Everest has a volume of about 57 mi^3.

To find the weight of 57 mi^3 of granite, we need to convert units because the density is given in pounds per cubic foot (lb/ft^3). Let's convert to units of feet as follows:

$$\begin{aligned} 1\ \text{mi}^3 &= 1\ \text{mi} \times 1\ \text{mi} \times 1\ \text{mi} \\ &= 5280\ \text{ft} \times 5280\ \text{ft} \times 5280\ \text{ft} \\ &\approx 1.5 \times 10^{11}\ \text{ft}^3 \end{aligned}$$

Thus,

$$57 \text{ mi}^3 \approx 57 \times 1.5 \times 10^{11} \text{ ft}^3 \approx 8.6 \times 10^{12} \text{ ft}^3$$

So we have

$$\begin{aligned} W = \text{weight of mountain} &= \frac{165 \text{ lb}}{\text{ft}^3} \times 8.6 \times 10^{12} \text{ ft}^3 \\ &\approx 1.4 \times 10^{15} \text{ lb} \\ &\approx 1.4 \text{ quadrillion lb} \end{aligned}$$

Now that we know the weight of the mountain, we want to find the pressure on the base of the cone and compare it with the crushing strength of granite. (Everest is standing, so if our model is any good, that pressure will be below the crushing strength.) Physics tells us that the weight of the mountain is spread evenly over the base of the cone (though we are oversimplifying the geology underlying mountains). Because

$$P = \frac{W}{A} \qquad \left(\text{pressure} = \frac{\text{weight}}{\text{area}}\right)$$

we need to calculate the area of the base of the cone. The shape is a circle, and the familiar formula

$$A = \text{area} = \pi r^2$$

gives an area of 28 mi^2 for a radius of 3 mi.

Once again, we need to convert units to express the pressure in pounds per square foot, the units in which the crushing strength is expressed. We get

$$A = \text{area} = 28 \text{ mi}^2 = 28 \times 5280 \text{ ft} \times 5280 \text{ ft} \approx 7.8 \times 10^8 \text{ ft}^2$$

Then

$$\begin{aligned} P &= \frac{W}{A} \\ &= \frac{1.4 \times 10^{15} \text{ lb}}{7.8 \times 10^8 \text{ ft}^2} \\ &= 1.8 \times 10^6 \text{ lb/ft}^2 \end{aligned}$$

This number is about half the crushing strength of granite, 4×10^6 lb/ft^2.

For a mountain to come close to the limitation of the crushing strength of granite, it would have to be only about 10 mi high, not quite twice as high as Everest. Other physical considerations suggest a maximum height of at most 15 mi. The fact that no current mountains are that high may be a consequence of the Earth's high amount of volcanic activity and the structural deformation of the Earth's crust.

What about mountains made of other materials—glass, ice, wood, or old cars? They couldn't be nearly as high. The pressure would cause glass to flow, ice to melt, and old cars to compact. What about mountains on another planet, or on an asteroid? Their potential height depends on the gravity there.

18.4 Sorry, No King Kongs

Unfortunately, the resistance of bone to crushing is not nearly as great as that of steel or granite. This fact helps to explain why there couldn't be any King Kongs (unless they were made of steel or granite!). A King Kong scaled up by a factor of, say, 20 would weigh $20^3 = 8000$ times as much. Although the weight increases with the cube of the linear scaling factor, the ability to support the weight—as measured by the cross-sectional area of the bones, like the area of the bottom face of the cube in Figure 18.3—increases only with the square of the linear scaling factor.

These simple consequences of the geometry of scaling apply not only to supermonsters but also to other objects, such as trees.

EXAMPLE 7
How Tall Can a Tree Be?

Galileo suggested that no tree could grow taller than 300 ft. The world's tallest trees are giant sequoias (Figure 18.6), which grow only on the West Coast of the United States and hence were unknown to Galileo. The tallest known sequoia today is 379 ft in Redwood National Park in Northern California. (To protect it, the National Park Service refuses to disclose its exact location.)

What limits the height of a tree? If the roots do not adequately anchor it, a tall tree can blow over. (This happened in 1990 to the world's then-tallest tree, the Dyerville Giant in Humboldt Redwoods State Park in California.) The tree could buckle or snap under its own weight and the force of a strong wind. The wood at the bottom will crush if there is too much weight pressing upon it. Finally, there is a limit to how far the tree can lift water and minerals from the roots to the leaves.

Figure 18.6 Even giant sequoias can grow no taller than their form and materials allow. *(Michael Rothman.)*

SOLUTION Could a tree be a mile high? To make a rough estimate of the pressure at the base of the tree due to gravity, let's model the tree as a perfectly vertical cylinder. Over each square foot at the bottom, there is 5280 ft^3 of cells of wood, which weighs about half as much as water. A convenient scientific fact to know involving the metric system is that water weighs just about 1 gram (g) per cubic centimeter. So, to calculate the weight, we first translate 1 ft^3 into metric measurement:

$$1 \text{ ft}^3 = (12 \text{ in.})^3 = (12 \times 2.54 \text{ cm})^3 \approx 28{,}317 \text{ cm}^3$$

So, 1 ft^3 of water weighs about

$$28{,}317 \text{ g} = 28.317 \text{ kg} = 28.317 \times 2.205 \text{ lb} \approx 62 \text{ lb}$$

Consequently, 5280 ft^3 of water weighs about 5280×62 lb $\approx$ 330,000 lb. The weight of the same volume of wood is about half as much, or about 165,000 lb. Therefore, the pressure at the bottom of the tree would be about 165,000 lb/ft^2.

This is an overestimate because we assumed that the tree does not taper. A tree that tapers steadily looks like an elongated cone; using a more realistic cone model (as we did in the last section for a mountain), you would find that the pressure at the bottom of the tree would be one-third of 164,000 lb/ft^2, or about 55,000 lb/ft^2.

A biological organism needs a safety factor of at least two to four times the absolute minimum physical limits, so a mile-high tree would need from 110,000 to 220,000 lb/ft^2 of upward pressure for water and minerals. Tension in the string of water molecules from root to leaf ranges from 80,000 to 3.2 million lb/ft^2, for different kinds and heights of trees, so this consideration does not rule out mile-high trees.

At more than about 500 lb/in.$^2 \approx$ 70,000 lb/ft^2, though, the bottom of the tree would begin to crush under this weight. On this basis, a mile-high tree is barely feasible, with little margin of safety. However, researchers who hauled themselves up to the top of the tallest trees in 2004 found a much lower limit, at least for giant sequoias. With increasing height, leaves are smaller, dryer, and less efficient at photosynthesis. The researchers estimated that trees can't top out higher than 400 to 427 ft. The tallest reliably measured tree was a North American Douglas fir, measured in 1902 at 413 ft.

There are other considerations. The taller the tree, the greater the area from which it must draw water and minerals, for which nearby trees also compete. Moreover, for a tree to grow very tall, it would have to live for a very long time. Evolution and time may select against extremely tall trees, or maybe, for no reason at all, they have just never evolved.

Algebra Review Significant digits

All measurements introduce some degree of uncertainty due to the deficiencies of the measuring device or the person doing the measuring. It is this uncertainty that we take into consideration when dealing with significant digits. The rules for determining the number of significant digits are as follows:

- Any non-zero number is significant.
- Any zero which lies to the right of the decimal point and to the right of a non-zero digit is significant.
- Any zero between significant digits is significant.

For example, 10,134 has five significant digits, while 2300.00 has six. Because the leading zero is not considered a significant digit, 0.200 has three significant digits, while 0.000034 has two.

Suppose that you had a box and measured to find a height of 4.8 inches, length of 5.4 inches, and width of 9.4 inches.

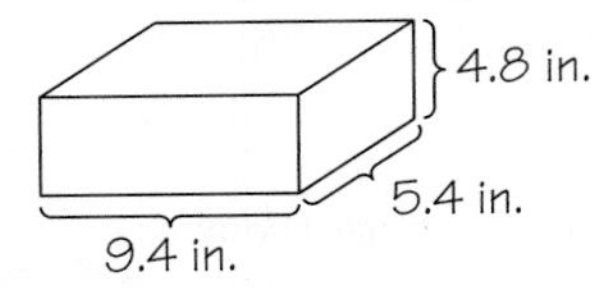

Each of these measurements has two *significant digits*. The volume of the box is found by multiplying length, width, and height. When this is done, we get a result of 243.648 cubic in. However, because we had at most only two significant digits, we must round this answer to 240 cubic in.

In the context of significant digits, when multiplying (or dividing) numbers, we determine which of the original numbers has the smallest number of significant digits and round our product (or quotient) to match this smallest accuracy.

When adding or subtracting, first perform the calculation with all digits, then round the sum or difference to have the same decimal places as the number with the least number of digits following the decimal point.

For example, consider the numbers 0.012, 2.30, and 0.001. The product of these numbers is 0.0000276. Because 0.001 has only one significant digit, the rounded answer would be 0.00003. The sum of the numbers is 2.313. Because 2.30 has the smallest accuracy in terms of decimal places, the rounded sum would be 2.31.

Some calculations involve numbers that are considered *exact*. Numbers such as exact conversion factors do not figure into determining the number of significant digits in a calculation.

SPOTLIGHT 18.2

Is There Any Advantage to Being Gigantic?

Gigantism has been a common feature of land animals since the beginning of the Jurassic period, over 200 million years ago. Toward the end of the Jurassic era, many sauropods reached 10 to 20 metric tons, some weighed as much as 50 metric tons, and a few may have exceeded 100 metric tons and 150 feet in length, rivaling the largest modern whales.

Why do animals become gigantic? Some reasons are simple. The bigger an animal is, the safer it is from predators, and the better it is able to kill prey. Antelope are easy prey for lions, hyenas, and hunting dogs, but adult elephants and rhinos are nearly immune—and their young benefit from the protection of their huge parents. For herbivores, being gigantic means being taller and therefore able to access higher foliage. Giraffes and elephants can reach over 18 feet high, and elephants can use their great bulk to push over even taller trees.

Other reasons for being gargantuan are less obvious, although important. The cost of locomotion decreases with increasing size; thus, it is much cheaper for a 5-ton elephant to walk a mile than it is for a 5-ton herd of gazelle to move the same distance. Metabolic rate also decreases with increasing size. A shrew must, therefore, frantically eat more than its own weight each day. The elephant, on the other hand, needs to take in only 5% of its own weight. And whereas big herbivores have long digestive systems that allow them to process and digest tougher plants, small herbivores can survive only on higher-quality foods. Also, as size increases, great bulk acts as a form of mass insulation. Large animals are, thereby, less affected by temperature extremes.

© David Noton Photography/Alamy

But there are disadvantages to being big. Because big animals eat more, there cannot be as many of them. Before human hunting, the population of elephants and rhinos in Africa was in the low millions. Rodents, in contrast, number in the countless billions. Nor can giants do a lot of things that smaller creatures can do, such as burrow into the ground, climb trees, or fly.

On land, only dinosaurs and mammals have become gigantic; reptiles have never done so. (The biggest tortoises and lizards have only weighed 1 metric ton.) One reason may be the rate of growth. Land reptiles cannot grow rapidly. It takes many years for an alligator to reach 100 pounds, whereas an ostrich does so in less than a year.

Adapted from Gregory S. Paul, in *The Scientific American Book of Dinosaurs, 2000*

[also published in *Scientific American*, April 2001, p. 116]

18.5 Dimension Tension

A large change in scale forces a change in either materials or form. A major manifestation of the **problem of scale** is the tension between weight and the need to support it. For example, a real building or machine must differ from a scale model: The balsa wood or plastic of the model would never be strong enough for the real thing, which would need aluminum, steel, or reinforced concrete.

Another way to compensate is to redesign the object to distribute its weight better. Let's go back to the original cube. It supports all its weight on its bottom face. In the version scaled up by a factor of 3, each small cube of the bottom layer has a bottom face supporting that cube's weight plus the weight of the two cubes piled on top of it.

Let's redesign the scaled-up cube, concentrating for simplicity only on the front face, with its nine small cubes. We take the three cubes on top and move them to the bottom, alongside the three that are already there. We take the three cubes on the second level, cut each in half, and put a half-cube over each of the six ground-level cubes (see Figure 18.7). We have the same volume and weight that we started with, but now there is less pressure on the bottom face of each small cube. Of course, the new design is not geometrically similar to the object that we started with; it's no longer a cube. We have solved the scaling problem by changing the proportions.

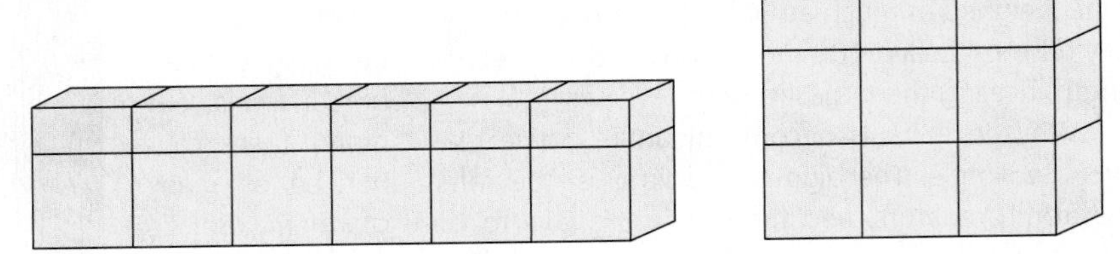

Figure 18.7 Nine small cubes rearranged to support greater weight.

We observe in nature both strategies for scaling: change of materials and change of form. Small animals (such as insects) do not have bony internal skeletons. Larger animals generally do. Animals made of similar materials but differing greatly in size, such as a mouse and an elephant, must differ in shape. If a mouse were scaled up to the size of an elephant, it would need the disproportionately thicker legs of the elephant to support its weight and the elephant's thick hide to contain its tissue.

Some dinosaurs, like *Supersaurus* (which weighed 30 tons), had special adaptations to lighten their weight, such as hollow bones, just as some birds have. Hollow bones are stronger: Of two bones of the same weight and length, the hollow one is wider across at its midpoint because of the air it contains, and the greater the width, the greater the resistance to fracture.

Falls, Jumps, and Flight

The need to support weight can be thought of as a tension between volume and area. As an object is scaled up, its volume and weight go up together, so long as the density remains constant (for example, no air bubbles introduced into the steel to make it into a Swiss cheese!). At the same time, the ability to support the weight goes up with the cross-sectional area, like the bottom face of the steel cube.

Area-Volume Tension

DEFINITION

Area-volume tension is a result of the fact that as an object is scaled up, the volume increases faster than the surface area and faster than areas of cross sections.

Because volume V is proportional to the cube of the linear scaling factor L, we have $V \propto L^3$; taking each side to the one-third power, $L \propto V^{1/3}$. The fact that surface area A is proportional to the square of the linear scaling factor becomes

$$A \propto L^2 \propto (V^{1/3})^2 = V^{2/3}$$

so that surface area scales as the two-thirds power of volume.

In any crowded city, you can observe tension between length, area, and volume. Consider an apartment building that spans a city block. The area of parking spaces on the adjacent streets is proportional to the perimeter of (length around) the building. But the number of cars belonging to people in the building is proportional to the number of apartments, which is proportional to the volume of the building. So the higher the building, the greater the parking tension.

Natural and Fractional Exponents — Algebra Review

Consider a positive real number a. When n is a natural number, the exponential expression a^n means

$$\underbrace{a \cdot a \cdot a \cdots a \cdot a}_{n \text{ of these}}.$$

An exponential expression such as 4^3 simplifies to $4 \cdot 4 \cdot 4$, or 64. In a^n, a is called the base and n is called the exponent or power. Thus, a^n can be read as "a to the nth power."

Two special rules are used in exponential notation:

- The multiplication rule: $a^m \cdot a^n = a^{m+n}$.
  For example, $d^2 \cdot d^3 = d^{2+3} = d^5$.
- The power rule: $(a^m)^n = a^{mn}$
  For example, $(b^2)^3 = b^{2 \cdot 3} = b^6$.

Exponents can come in fractional form. When n is a natural number, $a^{1/n}$ means $\sqrt[n]{a}$. For example,

- $36^{\frac{1}{2}} = \sqrt{36} = 6$
- $8^{\frac{1}{3}} = \sqrt[3]{8} = 2$

Fractional exponents follow the multiplication and power rules. For example,

- $3^{\frac{1}{2}} \cdot 3^{\frac{1}{2}} = 3^{\frac{1}{2}+\frac{1}{2}} = 3^1 = 3$
- $8^{\frac{2}{3}} = \left(8^{\frac{1}{3}}\right)^2 = \left(\sqrt[3]{8}\right)^2 = 2^2 = 4$

In general, expressions in the form of $a^{\frac{m}{n}}$ can be simplified by rewriting as $(\sqrt[n]{a})^m$ or $\sqrt[n]{a^m}$.
For example, $c^{\frac{2}{3}} = \sqrt[3]{c^2}$ and $4^{\frac{5}{2}} = (\sqrt{4})^5 = 2^5 = 32$.

In some cities, zoning tries to help the situation by putting shops on the ground floor, which cuts out one floor of apartments. If the residents' cars are away during the day, customers and employees of the shops can park where the apartment dwellers do at night. A more common solution is an underground garage, usually with several levels (with an area for cars proportional to the volume of the building). However, garages that were designed for one car per apartment have proven inadequate now that families tend to have more than one car.

Other examples of dimensional tension solutions include the old-fashioned diner, with its serving counter in the form of S-shapes to expand its effective length, and your small intestine, which coils its 20-ft length to fit into your abdomen.

Area-volume tension has many other practical consequences, some of them related to our childhood fantasies. We can forget about humans "leaping tall buildings in a single bound," "soaring like an eagle," or "diving miles below the sea." Consider the following examples.

EXAMPLE 8
Falls

Area-volume tension affects how animals respond to falling, another of gravity's effects. A mouse may be unharmed by a 10-story fall, and a cat by a 2-story fall, but many humans are injured by falling while running, walking, or even just standing.

What is the explanation? The energy acquired in falling is proportional to the weight of the falling object, and hence to its volume. This energy must be absorbed either by the object or by what it hits, or must be otherwise dissipated at impact—for example, as sound. The fall is absorbed over part of the surface area of the object, just as the weight of the cube was distributed over its base. With scaling up, volume—hence weight, hence falling energy—goes up much faster than area. As size increases, the hazards of falling from the same height increase.

EXAMPLE 9
Jumps

A flea can jump as high as 50 cm (20 in.) vertically, many times its own height. Some people believe that if a flea were as large as a person, it could jump 1000 ft into the air. Imagining—against our earlier arguments—that there could be so large a flea, we can deduce its limits: A scaled-up flea could jump about the same height as a small flea. The strength of a muscle is proportional to its cross-sectional *area* (see Spotlight 18.3). A jump involves suddenly contracting the muscle through its length, so it turns out that the ability to jump is proportional to the *volume* of muscle. But the volume of the flea and the volume of its leg muscles go up in proportion.

Let's say that a real flea's leg muscles account for 1% of its body. If we scale the flea up to the size of a person (without any change in its form), the enlarged flea's leg muscles will still make up 1% of its body. For either flea, each bit of muscle has the same power: In a jump, it propels 100 times its own weight, and it can do so to the same height. Both the weight of the flea and the power of its legs go up proportionately. In fact, the maximum heights that people, fleas, grasshoppers, and kangaroos can jump from standing are all within a factor of 3 of each other.

EXAMPLE 10
Flight

Wouldn't it be nice to be able to fly? Well, you have to be able to stay up. The power necessary for sustained flight is proportional to the **wing loading**, which is the weight supported divided by the area of the wings. We know that in scaling up, weight grows with the cube of the length of the bird or plane, and wing area with the square of the length. So the wing loading is proportional to the length of the flying object.

For example, if a bird or plane is scaled up proportionally by a linear scaling factor of 4, it will weigh $4^3 = 64$ times as much but will have only $4^2 = 16$ times as much wing area. So each square foot of wing must support 4 times as much weight.

Once you're up, you have to keep moving. To stay level, an airborne object must fly fast enough to maintain the lift on the wings. The minimum necessary speed is proportional to the square root of the wing loading. Combining this fact with the first consideration, we conclude that the minimum speed goes up with the square root of the length. A bird scaled up by a factor of 4 must fly $\sqrt{4} = 2$ times as fast. (Hovering helicopters, hummingbirds, and insects maintain lift by moving their wings directly rather than through forward motion.)

Take, for instance, a sparrow, whose minimum speed is about 20 miles per hour (mph). An ostrich is 25 times as long as a sparrow, so the minimum speed for an ostrich would be $\sqrt{25} \times 20 = 100$ mph. Have you seen any flying ostriches lately? Heavy birds have to fly fast or not at all!

Of course, ostriches are not just scaled-up sparrows, nor are eagles (nor are airplanes!). Larger flying birds have disproportionately larger wings than a sparrow to keep the wing loading down. The largest animal ever to take to the air was *Quetzalcoatlus northropi*, a flying reptile that lived 65 million years ago. It had a wingspan of 36 ft, weighed about 100 lb, and was as tall as a giraffe. Recent research suggests that it might have been able to "fly," taking advantage of air currents, 10,000 miles or more, nonstop.

You have to stay up, you have to keep moving—and you have to get up there. Here, basic aerodynamics imposes further limits. Paleontologists originally thought that *Q. northropi* weighed 200 lb and had a 50-ft wingspan. Even though that works out to about the same wing loading as for 100 lb and a wingspan of 36 ft, other considerations from aerodynamics show that at the larger size, the reptile couldn't have gotten off the ground.

SPOTLIGHT 18.3

Scaled to Fit

Big isn't always beautiful when it comes to the U.S. military's physical fitness tests.

Paul Vanderburgh, of the Department of Health and Sport Science at the University of Dayton, has spent more than a dozen years researching how a person's body mass affects performance on such tests, which consist of distance runs, push-ups, sit-ups, and abdominal crunches. The Arnold Schwarzeneggers of the world actually tend to score lower.

Vanderburgh emphasizes that some larger people (like Schwarzenegger) have more muscle, not more fat. Nevertheless, he and fellow researcher Todd Crowder found that scores for larger and heavier (though muscular) men and women are 15% to 20% lower than for their smaller and lighter counterparts. "A person's strength doesn't increase as fast as their size," explains his student Liz Trouten. "The extra muscle that big people have doesn't make up for their size."

Vanderburgh noticed at the U.S. Military Academy that, even at similar fitness levels, smaller cadets tend to score higher on physical fitness tests than larger cadets. "Fitness testing is a big part of cadets' grade point averages, and the stakes are pretty high. The test results affect class rank and even a cadet's first assignment, so it matters a lot how well a cadet does."

For example, a larger cadet with a fitness test score of 256, which Vanderburgh compares to a grade of C+, may not be eligible for certain awards and assignments. However, a smaller cadet with a perfect score of 300 "would get lots of attention."

But that doesn't mean the C+ cadet isn't worthy. "In fact, if these two cadets were scale models of each other, these two performances would be biologically the same, and they should receive the same score."

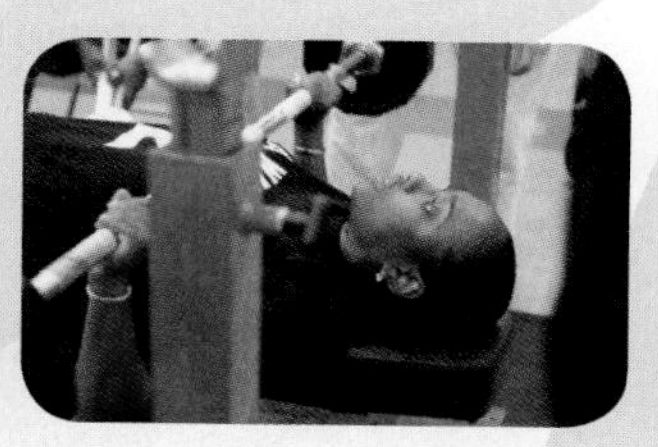

Photo by Spc. Hannah Frenchick, 20th Public Affairs Detachment

Vanderburgh gives another example using the scale-model approach. Take a woman who is 5 feet 5 inches tall, weighs 130 pounds, and scores a perfect 300 on the fitness test. If she were 5 feet 8 inches tall and 30% heavier, she would score only 250.

To compensate for this body mass "penalty," Vanderburgh and Crowder developed a correction factor, which multiplies the score by a number based on weight, "to place everybody on an even playing field."

This formula is similar to the Flyer Handicap, developed by Vanderburgh and colleague Lloyd Laubach. The handicap adjusts a runner's race time based on age and body weight. "A higher body weight is definitely a handicap for performance, whether it be running a marathon or military physical fitness tests," Laubach said. (A Web calculator for the Flyer Handicap is at http://academic.udayton.edu/PaulVanderburgh/weight_age_grading_calculator.htm.)

Source: Adapted from the article "Scaled to Fit," by Kristen Wicker in the *University of Dayton Quarterly* (Winter 2006–2007) 21–22.

Keeping Cool (and Warm)

Area-volume tension is also crucial to an animal's thermal equilibrium. Both warm-blooded and cold-blooded animals gain or lose heat from the environment in proportion to body surface area.

Warm-Blooded Animals

A warm-blooded animal's basal metabolism, or rate of food intake needed to maintain body heat, depends primarily on its surface area, the temperature of its environment, and the insulation provided by its coat or skin. Other factors being equal, a scaled-up mammal scales up its food consumption with *surface area* (proportional to the square of the linear scaling factor), *not with volume* (proportional to its cube). For example, a mouse eats about half of its weight in food every day, while a human consumes only about one-fiftieth of its own weight, because the mouse has more surface area per unit volume.

Thus, the metabolic rate should be proportional to the surface area. Using proportionality notation, we can find how the metabolic rate changes with the mass of the animal. We know that mass is proportional to volume, which in turn is proportional to the cube of length, or

$$M \propto V \propto L^3$$

Taking each side to the one-third power, we have

$$M^{1/3} \propto V^{1/3} \text{ or } L \propto M^{1/3}$$

Meanwhile, the metabolic rate (call it R) is proportional to surface area, so

$$R \propto A \propto L^2 \propto (M^{1/3})^2 = M^{2/3}$$

So, based on area-volume tension, we would expect metabolic rate to scale as the two-thirds power of body mass. But it doesn't—instead, it scales as the *three-quarters* power of body mass; that is, $R \propto M^{3/4}$. The least-squares line (see Section 6.4 on p. 220) through the points in the "mouse-to-elephant" curve of Figure 18.8 has a slope of 0.74, very close to three-quarters. (The logarithmic coordinates used in this graph are explained in Section 18.6.)

Why the difference from the two-thirds that area-volume tension would predict? And does the small difference between two-thirds and three-quarters matter? The answers lie in further considerations from geometry, physiology, and physics. A plant or animal needs a network of vessels (like the blood system) to

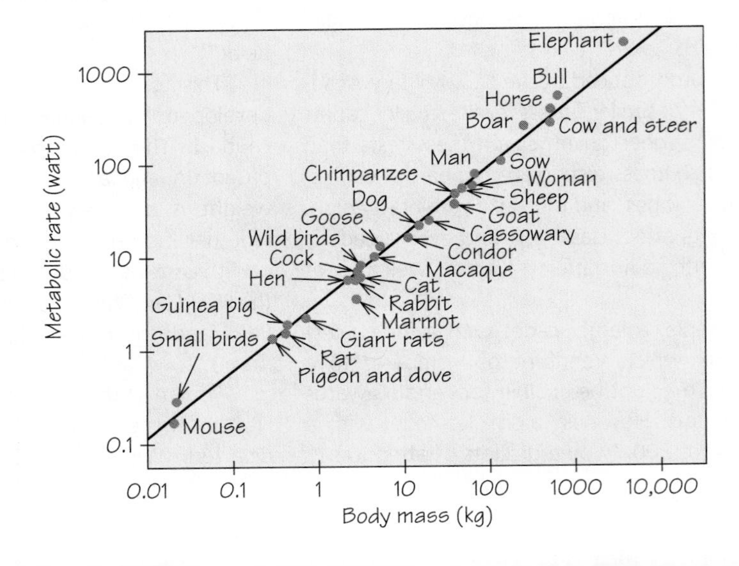

Figure 18.8 Metabolic rates for mammals and birds, when plotted against body mass on logarithmic coordinates, tend to fall along a single straight line. *(Adapted from* F. G. Benedict, *Vital Energetics: A Study in Comparative Basal Metabolism,* Carnegie Institute of Washington, Washington, D.C. Publication No. 503, 1938.*)*

transport resources to, and wastes away from, every part of the animal's tissues. The terminal branches (capillaries in the blood system) tend to be just about the same size in all species, for reasons of the physics involved.

To minimize the energy involved in transport, the network of vessels needs to be organized as a fractal-like tree, with smaller and smaller vessels branching off. (See Section 19.5 to learn about fractal patterns.) With same-size smallest branches at the ends, minimization of energy demands that the metabolic rate scale as the three-quarters power of body mass. Fractal branching makes it possible for the circulatory system of a whale, with 10^7 times the mass of a mouse, to have only 70% more branches than the mouse has.

EXAMPLE 11
Dives

Sperm whales (and some other species) regularly hold their breath and stay underwater for an hour. Why can't we? In part, because we aren't as large as whales. A mammal's breath-holding ability depends on how much air it can hold in its lungs, which is proportional to its mass. It also depends on how fast it uses up air—in other words, on its metabolic rate, which is proportional to the three-quarters power of its mass. Hence, the limit of duration of a dive should be proportional to

$$\frac{M}{M^{3/4}} = M^{1/4}$$

For a 90,000-lb sperm whale, this limit is proportional to $90{,}000^{1/4} = 17.3$, while the corresponding figure for a 150-lb human is $150^{1/4} = 3.5$. So the sperm whale should be able to hold its breath for about $17.3/3.5 \approx 5$ times as long. However, humans cannot hold their breath for one-fifth of an hour (12 minutes)! This fact tells us that the whale has special adaptations to make long dives possible. The stars of the 2005 film *March of the Penguins*, emperor penguins, weigh 80–90 lb but can dive for as long as 20 minutes. Their special adaptations are more blood per pound of body weight, an abundance of myoglobin (which can store oxygen) in their tissues, and slowing their heart rate during dives.

Cold-Blooded Animals

Mammals and birds regulate their metabolism and maintain a constant internal body temperature. Cold-blooded animals, such as alligators and lizards, have a somewhat different issue. They absorb heat from the environment for energy, but they must also dissipate any excess heat to keep their temperatures below unsafe levels. The amount of heat that must be gained or lost is proportional to total volume because the entire animal must be warmed or cooled. But the heat is exchanged through the skin, so the rate is proportional to surface area.

Dimetrodon was a large, mammal-like reptile that roamed present-day Texas and Oklahoma 280 million years ago. (See Figure 18.9a.) *Dimetrodon* had a great "sail," or fan, on its back. As an individual grew, and as the species evolved, the sail grew. But it did not grow according to geometric similarity, the kind of growth we refer to as **proportional growth**.

Proportional Growth DEFINITION

Proportional growth is growth according to geometric similarity, where the length of every part of the organism enlarges by the same linear scaling factor.

Figure 18.9
(a) *Dimetrodon* may have evolved a sail to absorb and dissipate heat efficiently. *(Robert F. Walters.)*
(b) A toucan's bill helps it lose heat. *(© Picture Press/Alamy)*

Instead, the area of *Dimetrodon's* sail grew in proportion to the volume of the animal, a fact that strongly suggests to paleontologists that the sail was a temperature-regulating organ. Larger specimens of *Dimetrodon* didn't look like scaled-up smaller ones. We would say that the sail grew disproportionately compared to the rest of the animal. An individual twice as long would have eight (2^3) times as much weight and volume and a sail with eight times as much area. If it had grown according to geometric similarity, the sail would have been twice as high and twice as wide, and hence would have had only four times as much area.

Dimetrodon was a large animal, but heat regulation is even more important for small animals; like human babies, they can lose heat quickly because of their high ratio of surface area to volume. Paleontologists believe that birds evolved from dinosaurs and that feathers are modified reptilian scales. The wings of birds and insects may have evolved not for flight but as temperature-control devices. Birds that live in hot climates, such as the toco toucan (*Ramphastos toco*), tend to have large bills and use them to lose heat by increasing bloodflow to the bill.

Some scientists have speculated that African pygmies are small in part because a small body can better lose heat in the hot, humid climate of the Ituri Forest in the Congo, where pygmies live. The discovery announced in late 2004 of "hobbit-sized" people (1 m tall) who lived on the island of Flores in Indonesia 13,000 years ago, suggests another explanation. Being marooned on the island with a self-limiting food supply (they hunted pygmy elephants!) made large size—and a corresponding need for more calories—a disadvantage.

Other scientists have suggested that ancestors of human beings began walking on two legs in part to keep cool in a hot climate. Walking upright exposes less body area to the rays of the sun than walking on all fours and also reduces the amount of water needed by about one-half.

18.6 How to Grow

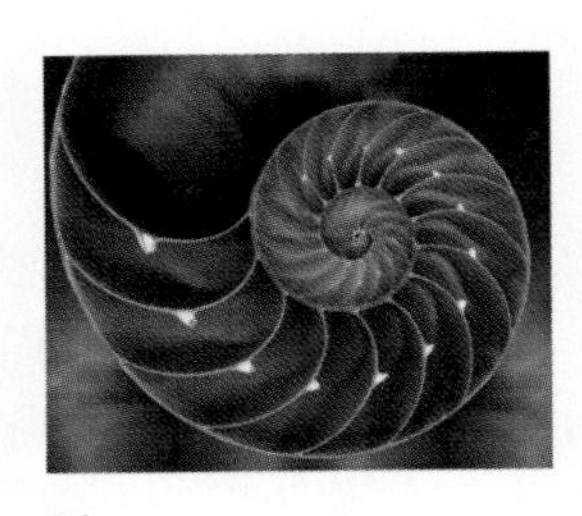

Figure 18.10
A chambered nautilus shell. *(Photodisc/Punchstock.)*

A large change of scale forces adaptive changes in materials or form. However, within narrow limits—in most cases, up to a factor of 2—creatures can grow according to geometric similarity. That is, they can grow proportionally, so that their shape is preserved. A striking example of such growth by a far greater factor is the chambered nautilus (*Nautilus pompilius*). Each new chamber that it adds to its shell is larger than, but geometrically similar to, the previous chamber and also similar to the shape of the shell as a whole—an *equiangular,* or *logarithmic,* spiral (see Figure 18.10).

Most living things grow over the course of their lives by a factor greater than 2. We've seen with *Dimetrodon* that a big specimen was not just a scaled-up small one. Nor is a human adult simply a scaled-up baby: A baby's head is relatively much larger than an adult's, and its arms are disproportionately shorter. In growth from baby to adult, the body does not scale up as a whole. Different parts of the body scale geometrically, each with a different linear scale factor. That is, a baby's eyes grow to perhaps twice their original size, while the arms grow by a factor of about 4.

Although the laws for growth can be much more complicated than for proportional growth (or even for the allometric growth that we discuss later), more sophisticated mathematics—for example, differential geometry, the geometry of curves and surfaces—permits analysis of complex and interlocking scalings. For a model of the process in which a baby's head changes shape to grow into an adult head, we can use graph paper: First, we put a picture of the baby's skull on graph paper. Then we determine how to deform the grid until the pattern matches an adult skull. (See Figure 18.11 and Spotlight 18.4.) The same idea lies at the heart of computerized "morphing," the process in which the face of one person can be changed smoothly into the face of another, with different scalings for different parts of the face.

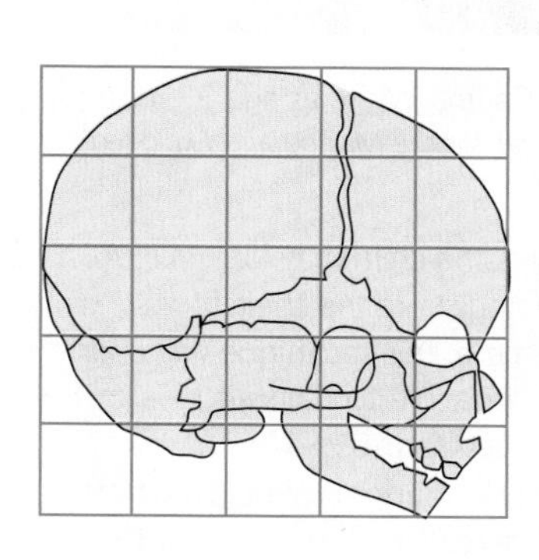

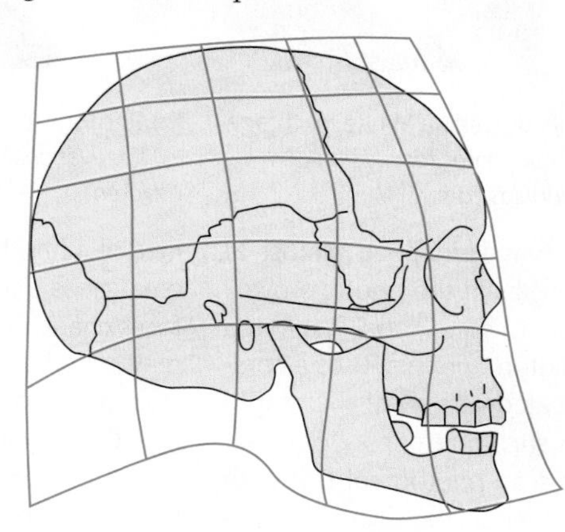

Figure 18.11 Modeling the changes in the shape of a human head from infancy to adulthood.

Allometric Growth

If we measure the arm length or head size for humans of different ages and compare these measurements with body height, we observe that humans do not grow proportionally; that is, in a way that maintains geometric similarity. The head of a newborn baby may be one-third of the baby's length, but an adult's head is usually close to one-seventh of the individual's height. The arm, which at birth is one-third as long as the body, is by adulthood closer to two-fifths as long (see Figure 18.12a).

Graphing provides a way to test for differential growth. We plot body height on the horizontal axis and arm length on the vertical axis (see Figure 18.12b). A straight line would indicate proportional growth; that is, according to geometric similarity. We do get a straight line from 9 months (0.75 years) on up. But up to 9 months, we get a curve, which indicates that the ratio of arm length to height does not remain constant over the first year.

Is there an orderly law by which we can relate arm length to height? Let's plot again, this time using a different scale. For this **base-10 logarithmic scale**, we mark off equal units, as usual. But instead of labeling the marked points with 0, 1, 2, 3, and so on, we label them with the corresponding powers of 10: $10^0 = 1$, $10^1 = 10$, $10^2 = 100$, $10^3 = 1000$, and so on, which are also called **orders of magnitude**.

Plotting a point on such a scale is not easy because the point midway between 1 and 10 is not 5.5; rather, it is closer to 3. Special graph paper (available in most college bookstores) marks smaller divisions and makes points easier to plot; paper marked with log scales on both axes is called **log-log paper**, while **semilog paper** has a logarithmic scale on just one axis. Also, many computer plotting packages can produce logarithmic scales.

SPOTLIGHT 18.4

Helping to Find Missing Children

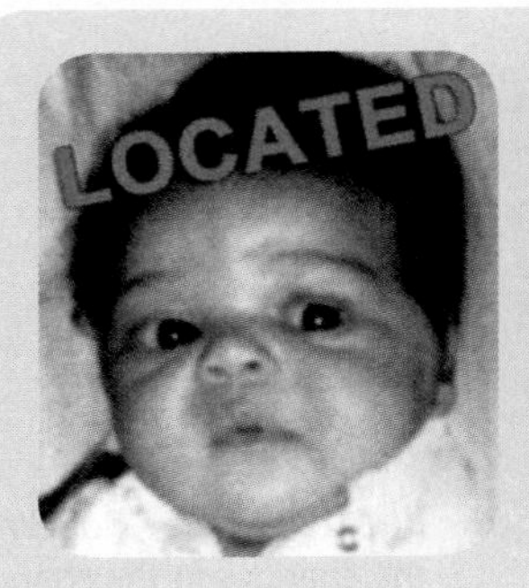

(a) Photograph at age 19 days of Carlina White, with age progression to age 19. *(National Center for Missing and Exploited Children, www.missingkids.com.)*

(b) Photograph of Carlina White at age 23 in January 2011. *(National Center for the Missing and Exploited Children)*

What might a child who was kidnapped almost at birth look like 19 years later (as in the photo above)? The National Center for Missing and Exploited Children (NCMEC), in Arlington, Virginia, uses a computer and a more sophisticated version of the graph-paper technique to answer such questions. Computer age-progression specialists scan photographs of both the missing child and an older sibling or a biological parent at age 19. Then the face of the missing child is stretched, depending on age, to reflect craniofacial growth, and then merged with the image of the sibling or parent at 19 years old. The result is a rough idea of what the missing child may look like. As mathematicians and biologists refine their models of how faces change over time, this technique will improve. It may even become possible to gain an idea of how a child may look at age 40 or 65.

Carlina White, kidnapped almost at birth, as a teenager began to realize that her "mother" was not her birth mother. Eventually, she contacted the NCMEC, which found three possible matches. In January 2011, at age 23, she was at last reunited with her birth mother, a happy resolution to the longest-known stranger-abduction case.

Algebra Review Base-10 Logarithms

A logarithm represents the power that a number must be raised to in order to get another number. For example, 3 must be raised to the second power in order to get 9. Using a logarithm, this would be written as

$$\log_3 9 = 2.$$

This equation would be read as "the base-3 logarithm of 9 is 2" or "log base-3 of 9 is 2." The corresponding exponential equation is

$$3^2 = 9.$$

Notice that the subscript (or *base*) of the logarithmic equation is also the base of the exponential equation.

Because our numeration system is base 10, we often use a base-10 logarithm, which also is called the *common logarithm*. For example, because $10^3 = 1000$, we know that $\log_{10} 1000 = 3$. We can express common logarithms without writing the base of 10. For example, $\log 1000 = 3$.

Some important properties of base-10 logarithms are as follows:

- $\log_{10} 10 = 1$, because $10^1 = 10$
- $\log_{10} 1 = 0$, because $10^0 = 1$

The following two properties are true when $x > 0$.

- $10^{\log_{10} x} = x$,
- $\log_{10} x^r = r \log_{10} x$

For example, let's simplify $10^{2+3\log_{10} x}$ for $x > 0$.

$$\begin{aligned} 10^{2+3\log_{10} x} &= 10^{2+\log_{10} x^3} \\ &= 10^2 \cdot 10^{\log_{10} x^3} \\ &= 100x^3 \end{aligned}$$

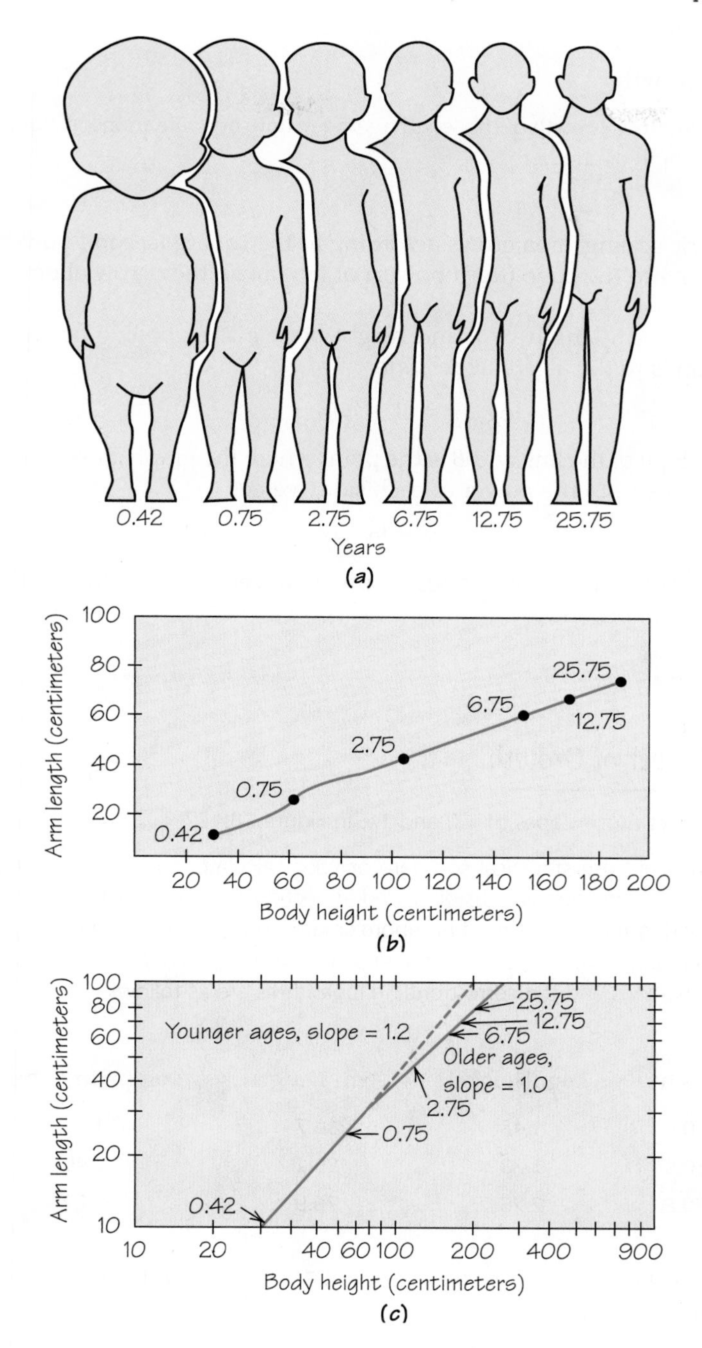

Figure 18.12
(a) The proportions of the human body change with age.
(b) A graph of human body growth on ordinary graph paper. The numbers shown beside the points indicate the age in years; they correspond to the stage of human development shown in part (a).
(c) A graph of human body growth on log-log paper.

We could use a logarithmic scale for either height or arm length, or for both. When logarithmic scales are used for both, as in Figure 18.12c, the data plot closely to a straight line. Looking carefully, we can discern two different straight lines: a steeper one that fits early development (we will see shortly that it has slope 1.2), and a less steep one (with slope 1.0) that fits development after 9 months of age.

The change from one line to another after 9 months indicates a change in pattern of growth. The pattern after 9 months, characterized by the straight line with slope 1, is indeed proportional growth (sometimes called **isometric growth**). For the pattern before 9 months, the slope is 1.2. The fact that it is greater than 1 means that arm length is increasing relatively faster than height. This early growth also follows a definite pattern, called **allometric growth**.

Allometric Growth DEFINITION

Allometric growth is growth of the length of one feature at a rate proportional to a power of the length of another.

In geometric scaling, area grows according to the square (second power), and volume according to the cube (third power) of length, so they grow allometrically with length.

If we denote arm length by y and height by x, a straight-line fit on log-log paper corresponds to the algebraic relation

$$\log_{10} y = B + a \log_{10} x$$

where a is the slope of the line and B is the point where the graph crosses the vertical axis. If we raise 10 to the power of each side, we get

$$y = bx^a$$

where $b = 10^B$. This equation describes a **power curve:** y is a constant multiple of x raised to a certain power.

EXAMPLE 12
Finding the Power of Growth

How do we arrive at those slopes of 1.0 and 1.2 in Figure 18.12?

SOLUTION You could use statistical software or your calculator to find the equation of the least-squares regression line (as discussed in Section 6.4 on p. 220) through the points on the log-log plots. Here we find approximate values for the slope a for each line from the coordinates of the points at the ends of the lines, for ages 0.42, 0.75, and 25.75. The observations and the corresponding logarithms are as follows:

Age	Height	Log (Height)	Arm Length	Log (Arm Length)
0.42	30.0	1.48	10.7	1.03
0.75	60.4	1.78	25.1	1.40
25.75	180.8	2.26	76.9	1.89

The slope for the line from age 0.42 to age 0.75 is the vertical change over the horizontal change in terms of log units:

$$\frac{\log 25.1 - \log 10.7}{\log 60.4 - \log 30.0} = \frac{1.40 - 1.03}{1.78 - 1.48} = \frac{0.37}{0.30} \approx 1.2$$

The slope for the line from age 0.75 to age 25.75 is

$$\frac{\log 76.9 - \log 25.1}{\log 180.8 - \log 60.4} = \frac{1.89 - 1.40}{2.26 - 1.78} = \frac{0.49}{0.48} \approx 1.0$$

So $a = 1.2$ up to 9 months, and $a = 1.0$ after 9 months. Up to 9 months, arm length grows according to $(\text{height})^{1.2}$. After 9 months, arm length grows according to (height)$^{1.0}$, and we get $y = bx^{1.0}$, which is a linear relationship describing proportional growth; that is, growth according to geometric similarity. On ordinary graph paper, proportional growth appears as a straight line, allometric growth as a curve. On log-log paper, both patterns appear as straight lines.

Allometry was used by paleontologists to determine that all specimens (just six!) of the earliest bird, *Archaeopteryx*, are indeed of the same species, and that the puzzling minute fossil fish *Palaeospondylus* (found only in Scotland) is probably just the larval stage of a better-known fish.

In this chapter, we have explored the limitations on life imposed by dwelling in three dimensions. In Chapters 19 and 20, we will see that dimensionality also imposes surprising limits on artistic creativity in devising patterns.

REVIEW VOCABULARY

Allometric growth A pattern of growth in which the length of one feature grows at a rate proportional to a power of the length of another feature. (p. 669)

Area-volume tension A result of the fact that as an object is scaled up, the volume increases faster than the surface area and faster than areas of cross sections. (p. 660)

Base-10 logarithmic scale A scale on which equal divisions correspond to powers of 10. (p. 667)

Crushing strength The maximum ability of a substance to withstand pressure without crushing or deforming. (p. 653)

Density Mass per unit volume. (p. 652)

Geometrically similar Two objects are geometrically similar if they have the same shape, regardless of the materials of which they are made. They need not be the same size. Corresponding linear dimensions must have the same factor of proportionality. (p. 644)

Isometric growth Proportional growth. (p. 669)

Linear (length) scaling factor The number by which each linear dimension of an object is multiplied when it is scaled up or down; that is, the ratio of the length of any part of one of two geometrically similar objects to the length of the corresponding part of the second. (p. 644)

Log-log paper Graph paper on which both the vertical and the horizontal scales are logarithmic scales; that is, the scales are marked in orders of magnitude 1, 10, 100, 1000, . . . , instead of 1, 2, 3, 4, ... (p. 667)

Orders of magnitude Powers of 10. (p. 667)

Power curve A curve described by an equation $y = bx^a$, so that y is proportional to a power of x. (p. 670)

Pressure Force per unit area. (p. 652)

Problem of scale As an object or being is scaled up, its surface and cross-sectional areas increase at a rate different from its volume, forcing adaptations of materials or shape. (p. 659)

Proportional growth Growth according to geometric similarity, where the length of every part of the organism enlarges by the same linear scaling factor. (p. 665)

Semilog paper Graph paper on which only one of the scales is a logarithmic scale. (p. 667)

Weight Force under gravity. (p. 652)

Wing loading Weight supported divided by wing area. (p. 662)

SKILLS CHECK

1. A penny and a nickel are

(a) not geometrically similar because they are made of different materials.

(b) not geometrically similar because they are of different sizes.

(c) geometrically similar because they have the same shape and proportional dimensions.

2. A scale model of a carillon stands 10 in. tall, and the actual carillon stands 100 ft tall. The linear scaling factor of the carillon compared to its model is ____________ .

3. If a model car is built to a scale of 1 to 10, and the actual car has a length of 15 ft, what should be the length of the model?

(a) 1.5 ft
(b) 15 ft
(c) 150 ft

4. You want to enlarge a 2½-in. by 3½-in. photograph to an 5-in. by 7-in. copy. Assuming that the cost of photographic paper is proportional to its area and that 2½-in. by 3½-in. reprints cost 40 cents each, you would expect to pay ________________ for the large copy.

5. If a medium 10-in. pizza costs $10 and a similar 14-in. pizza costs $14, which costs less per square inch?

(a) The 10-in. pizza
(b) The 14-in. pizza
(c) They are about the same price per square inch.

6. An artist plans to melt 1000 pennies and re-form a larger penny proportional in all dimensions to an ordinary penny. The linear scaling factor of the large penny compared with the ordinary penny is ________.

7. The actor Elijah Wood, who plays Frodo Baggins in the movie version of J. R. R. Tolkien's *Lord of the Rings*, is 5 ft, 6 in. tall, but his character is barely 4 ft tall. Put correctly, how much shorter is Frodo than Wood?

(a) 138% shorter
(b) 73% shorter
(c) 27% shorter

8. The volume of a scaled-up object goes up with the _________ of the linear scaling factor.

9. Which of the following is precise language to express the comparison that Ahmed has $100 and Burke has $50?

(a) Ahmed has 100% more than Burke.
(b) Burke has 200% less than Ahmed.
(c) Burke has twice as less as Ahmed.

10. The Apple iPhone 4 features a screen with 960 pixels in one direction and 640 in the other. For that screen to be geometrically similar to a 1080 p high-definition screen, which has 1920 pixels horizontally, the high-definition screen would have to have _______ pixels vertically (which it doesn't).

11. A kilometer is approximately equal in length to

(a) 5 mi.
(b) 3 mi.
(c) $\frac{3}{5}$ mi.

12. The distance for the marathon race was established in 1921 as 42.195 km. Converted to the U.S. customary system, the distance is ________________.

13. A 2-liter bottle contains approximately

(a) 2 quarts.
(b) 1 gallon.
(c) 10 pints.

14. A weight of 130 lb is approximately the same as ________ kg.

15. Coffee costs about $8 per pound in the United States. If a Canadian dollar (Cdn$) exchanges for U.S.$0.98, what is the approximate cost in Canadian dollars of 500 g of coffee?

(a) Cdn$2
(b) Cdn$4
(c) Cdn$8

16. A common speed limit in European neighborhoods is 30 km/h, which is about __________ mph.

17. Which of the following is a unit in the metric system?

(a) minimeter
(b) miniliter
(c) millimeter

18. A mile is approximately ________ km.

19. A speed of 60 mph is approximately

(a) 3.2×10^3 mm/min.
(b) 88 ft/s.
(c) 50 m/s.

20. A sculpture weighs 140 lb and is supported by three legs, each of which is 0.5 in. by 0.5 in. by 2 in. high. The legs exert a pressure of _______ lb/in.^{2} on the floor.

21. The mass of a scaled-up object goes up with the

(a) square root of the volume.
(b) square of the linear scaling factor.
(c) density.

22. In comparing flight speeds of birds, an analysis of wing loading leads to the conclusion that __________ birds fly faster than __________ birds.

23. Scaling analysis leads to the general conclusion that

(a) smaller animals can jump much higher than larger animals.
(b) larger animals can jump much higher than smaller animals.
(c) all animals can jump to about the same height.

24. If an object is scaled linearly so that its volume grows to 8 times its original volume, its surface area is scaled to ______ times its original surface area.

25. A large change in scale forces a change in

(a) form.
(b) form and materials.
(c) form or materials.

26. Assuming that a catfish maintains the same shape and proportions as its grows, and that a catfish 8 in. long weighs about 1 lb, a 2-lb catfish is about _____ in. long.

27. Metabolic rate scales as the _______ power of mass.

(a) one-third
(b) two-thirds
(c) three-quarters

28. The base-10 logarithm of 1000 is ____ .

29. The population of Mexico is expected to grow 0.7% per year from 2012 through 2015. The population will be growing

(a) proportionally.
(b) allometrically.
(c) by a constant amount each year.

30. Allometric growth is growth of the length of one feature at a rate ________ the length of another.

CHAPTER 18 EXERCISES

■ Challenge ▲ Discussion

Most of these exercises require a calculator; one with square roots will suffice.

18.1 Geometric Similarity

1. Your digital camera probably takes pictures with an aspect ratio of 4 to 3, meaning that the longer side is 4/3 times as long (in pixels) as the shorter side. For example, you probably can take a "small" picture with 640 pixels by 480 pixels, or perhaps a "large" picture with 2592 pixels by 1944 pixels (for a total of 2592 × 1944 pixels, or just a little more than 5 megapixels). Photographic prints from your digital camera are available in various sizes of paper, quoted in inches: 4 × 6, 5 × 7, and 8 × 10.

(a) Which of the paper sizes, if any, is geometrically similar to the original digital image?
(b) If a 4 × 6 print is made by scaling the shorter side of the digital image to be exactly 4 in., how long should the longer side of the image be on the print?
(c) If a 4 × 6 print is made by scaling the longer side of the digital image to be exactly 6 in., how long should the shorter side of the image be on the print? (*Hint:* The paper isn't wide enough!)

For Exercises 2–3, refer to the following: The area of a circle of radius r is πr^2. Expressed in terms of the diameter, $d = 2r$, the area is $\frac{1}{4}\pi d^2$. If we apply a linear scaling factor L to the diameter, then the area of the scaled circle—as in the case of the square that we considered in the text—changes with L^2, the square of the linear scaling factor.

2. A natural application of this idea is to pizza. The prices at Vince's pizza restaurant in Beloit, Wisconsin, are $7.47, $8.61, $9.39, $11.03, and $12.53, respectively, for small (10-in.), medium (12-in.), large (14-in.), extra-large (16-in.), and XX-large (18-in.) cheese pizzas.

(a) What is the linear scaling factor for an XX-large pizza compared with a small one?
(b) How many times as large in area is the extra-large pizza compared with the small one?
(c) How much pizza does each size give per dollar? What "hidden" assumptions are you making about how the pizzas are scaled up?
(d) The corresponding prices for a pizza with six toppings are $12.25, $15.39, $18.32, $21.50, and $25.64. Is there any size of these for which you get more pizza per dollar than some size of the cheese pizzas?

(Curiously, all the prices are, to the nearest cent, exactly $0.50 higher than two years earlier! That is an example of arithmetic scaling.)

3. The *NBC Nightly News* on June 21, 2010 featured a story about food portions, stating that dinner plates were 9 in. in diameter in the 1960s but now they are 12 in. in diameter, "making room for one-third more food."

(a) What is the linear scaling factor for a 12-in. plate compared to a 9-in. plate?
(b) How many times as large in area is a 12-in. plate compared to a 9-in. plate?
(c) What percentage greater is the area of a 12-in. plate compared to a 9-in. plate?

4. Dollhouses and their furnishings are usually built to a scale of exactly 1 in. to 1 ft, meaning that an item 1 ft long in a real house is 1 in. long in a dollhouse.

(a) What is the linear scaling factor for a dollhouse?
(b) If a dollhouse were made of the same materials as a real house, how would their weights compare?

5. According to *Time* (March 7, 2005), men's brains on average are 10% larger than women's, even though men on average are only 8% taller. (The article mainly discusses the many differences in brain structure that likely outweigh any size differences.) If the brain scales linearly with height, and men are 8% taller, what percentage larger would you expect their brains to be?

6. At our house, we have some 10-in. frying pans and a 12-in. one; the 12-in. one weighs a lot more, cooks food more slowly, and never gets as hot. Suppose that a 10-in. frying pan weighs 1 lb, apart from its handle. How much would a geometrically similar 12-in. frying pan weigh? How much would it weigh if it had the same thickness of metal as the 10-in. pan?

7. The human figures in Lego® sets are 4 cm tall (without hats or helmets).

(a) What is the linear scaling factor of a Lego figure if it represents a human who is 160 cm tall?

(b) How does the volume of a real human compare with the volume of a Lego figure?

(c) The car in one Lego set is 10 cm long. Using the linear scaling factor in part (a), how long would a real car be?

8. Recent dollar coins (Presidential, Susan B. Anthony, Sacagawea) have been largely rejected by the public, which finds them too small and too light. Suppose that you are commissioned to design a new $5 coin. (Whom should it depict?) The sole requirement is that it be made of the same material as the quarter but weigh four times as much. A quarter can be described geometrically as a circular cylinder approximately 24.26 mm in diameter and 1.75 mm thick. Because your new dollar should weigh four times as much, it needs to have four times the volume of a quarter. [The formula for the volume of a cylinder is $\pi \times (\text{diameter}/2)^2 \times$ height.]

Sacagawea dollar
(United States Mint)

Presidential series dollar. Coins shown actual size (26.50 mm in diameter, 2.00 mm thick).
(United States Mint)

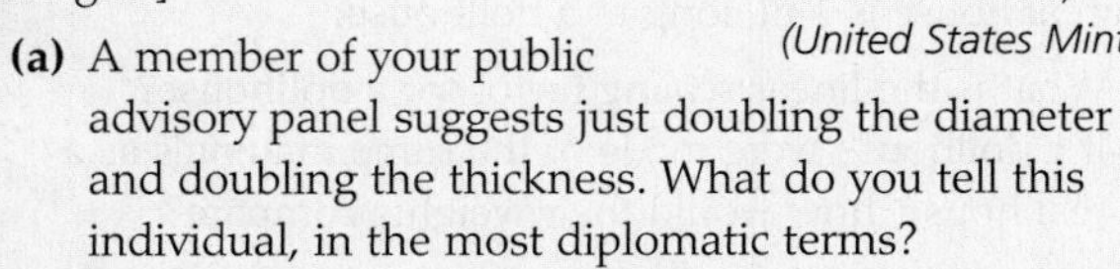

(a) A member of your public advisory panel suggests just doubling the diameter and doubling the thickness. What do you tell this individual, in the most diplomatic terms?

(b) If you double the diameter, how thick does the coin need to be?

(c) Another member feels that the result of part (b) would be inconveniently large and proposes instead to scale up the quarter proportionality. (She studied a previous edition of this book.) What would the dimensions be for this coin?

9. Criticize the following statement and write a correct version.

"Two Tylenol four times a day, let's see. . . . That would be eight tablets a day, but . . . I could take two Aleve . . . and that would be . . . *four times fewer pills!*"

—TV commercial for Aleve, aired April 2, 2010

10. Criticize the following statement and write a correct version.

"The war funding bill . . . passed this week, but with three times fewer Democrats voting for it than did the last year."

—Christiane Amanpour, *ABC World News with Diane Sawyer*, July 30, 2010

Hint: Actual count: 2010: 148 D, 160 R in favor; 2009: 221 D in favor, 5 R in favor (originally 368 to 60, with 200 D, 168 R in favor.)

11. Criticize the following statement.

"An Intel Atom–based [ultrathin computer notebook] system that costs 30 to 35 percent less than a given Yukon platform is going to have 100 to 150 percent less performance."

—Joanna Stern, "AMD Q+A: Yukon to Outperform Atom, Not Destined for Netbook," **http://blog.laptopmag.com/amd's-netbook-counterstrike-ultrathin-notebooks-with-new-yukon-platform**.

12. Criticize the following statement.

"The total weight after drying should be between 40 and 150 percent less than before drying."

—Steve Smith, "How to Dry Wood Chips," **http://www.ehow.com/how_6300829_dry-wood-chips.html**

13. Abuses of the language of comparison aren't hard to find. For example, the phrase "times less than" occurs in more than 11 million Internet documents. Search on the Internet and find an abuse of "times" and "less than" together. Figure out what the author meant to say, and write it in correct language.

14. The phrase "times more than" occurs in more than 68 million Internet documents. Search on the Internet and find either an abuse of "times" and "more than" together. Figure out what the author meant to say, and write it in correct language.

18.2 How Much Is That In . . . ?

15. In 2011, the cost of mailing a lightweight airmail letter from the United States to most of western Europe was \$0.98. How much was that in euros (€), the currency of the European Union (EU) in June 2011, when the exchange rate was €1 = \$1.43? (For comparison, the cost then of an airmail letter to the United States varied from country to country in the EU, ranging from €0.51 to €1.70.)

16. The cost of mailing a lightweight letter from the United States to Canada in 2011 was US\$0.75. How much was that in Canadian dollars when the exchange rate was US\$1 = Cdn\$1.02? (The postage cost from Canada to the United States was Cdn\$1.00.)

17. In Germany, the fuel efficiency of cars is measured in liters of fuel per 100 km (L/100 km). A typical average in a compact station wagon is 7.3 L/100 km. What is that in miles per gallon (mpg)?

18. According to EPA ratings in 2010, the highest-mileage 2010 car was the gasoline/electric hybrid Toyota Prius, at 51 mpg in the city. How many liters of gasoline does such a Prius use to travel 100 km in the city?

19. Consider a real locomotive that weighs 88 tons and an HO-gauge scale model of it, for which the linear scaling factor is 1/87.

(a) How much would an exact scale model weigh in tons?
(b) What assumptions are involved in your answer to part (a)?
(c) How much would an exact scale model weigh in pounds?
(d) In kilograms?
(e) In tonnes (1 tonne = 1 metric ton = 1000 kg)?

20. What's wrong in the following quotations?

(a) "President Bush visited California, where 12 forest fires have charred more than 700,000 square miles."
—Steve Stadelman, WTVO television news, Channel 17, Rockford, Illinois, October 2007. (Curiously, exactly the same number, 700,000 also appeared in news reports for California fires in 2003, 2000, and 1987.)

(b) "The population of the USA has topped 300 million. . . . If current trends continue, it is expected to reach 400 billion by 2043. This makes it an acceleration of growth. . . ."
—*Significance* 3 (4) (December 2006) p. 146

21. Gasoline is sold in the United States by the U.S. gallon and in Europe by the liter (1 U.S. gal = 231 in.^{3}; 1 L = 1000 cm^3). What was the equivalent cost, in U.S. dollars per U.S. gallon, for gasoline in Germany priced in euros at €1.423 per liter, when €1 = \$1.43, in June 2011?

22. In 1991, Edward N. Lorenz, a meteorologist who was an early researcher into chaos and dynamical systems (discussed in Chapter 23), received the Kyoto Prize in Basic Sciences, consisting of a gold medal and 45 million Japanese yen (¥). If US\$1 = ¥125 at the time, what was the value of the cash award in 1991 U.S. dollars? (In Chapter 21, we show how to convert such an amount to its value in today's dollars.)

23. The year 2008 marked the 50th anniversary of the installation of length markers on the sidewalk of the Harvard Bridge across the Charles River between Boston and Cambridge, Massachusetts (where in 1908 Harry Houdini performed one of his "escapes"). The bridge is marked at 10-smoot intervals, where 1 smoot was the height of an MIT fraternity pledge named Oliver Smoot. The length of the bridge is 620 m or 364.4 smoots and one ear. How long is a smoot in feet and inches? (Oliver Smoot later became the head of the International Standards Organization.)

18.3 Scaling a Mountain

24. Burj Khalifa has the fastest elevators in the world, at about 18 m/s going up. (They are limited to 8 m/s going down so that passengers' middle ears can adjust to the change in pressure.) Refer to Example 5 (p. 653).

(a) With no stops, how long would it take to get to the top of Burj Khalifa?
(b) With no stops, how long would it take to get to the top of the Petronas Towers, which are 1483 ft high and whose elevators run at 41 ft/s?
(c) The answers to parts (a) and (b) suggest that building designers find less than a minute a reasonable standard for getting to the top of a building (though the Burj Khalif elevators could be set to run faster). How fast would the elevators in a mile-high building have to be to achieve that standard?

25. Calculate the speed, in miles per hour, of the elevators in the three buildings in parts (a), (b), and (c) of Exercise 24.

For Exercises 26−31, refer to the following.
The Canadian dollar for many years fell steadily in value in terms of the U.S. dollar, but that trend has been reversed in the past few years. In January 2002, a U.S. dollar was worth Cdn\$1.60. In June 2011, US\$1 = Cdn\$1.02. How should you measure how much one currency has depreciated (lost value) against another? Let the *home*

currency be the one whose change in value you are interested in, and let the *target currency* be the one in terms of which you will measure the change. In our example, we want to track the change in value of the U.S. dollar (home currency) in terms of the Canadian dollar (target currency).

There are two competing practices. Both begin by calculating the change as measured in the target currency, the new value minus the old value—here, Cdn\$1.02 − Cdn\$1.60 = Cdn\$−0.58.

- Option A, used by the International Monetary Fund and the British periodical *The Economist,* divides this difference by the new trading value (Cdn\$1.02) and multiplies by 100 to get a result in percent—here, −57%.
- Option B, sometimes called the "popular method," divides instead *by the old trading value,* Cdn\$1.60, here getting −36% as the percentage change in value of the U.S. dollar in terms of the Canadian dollar.

So, depending on the method used, we can say that the U.S. dollar lost either 57% or 36% in value against the Canadian dollar.

26. To get a feeling for how the results of these two options come out, we consider an artificial example where the numbers are simple. Suppose that in January 2011 the imaginary currency of the imaginary country Middle Earth, the Middie (*M*), traded at US\$1; but in January 2012, its value was \$2 (thanks to increased exports from Middle Earth of power rings and Frodo bobbleheads and also to the Great Recession in the United States.).

(a) Using Option A, calculate how much percentage value the Middie gained against the dollar (the target currency).

(b) Calculate how much percentage value the Middie gained against the dollar using Option B.

27. We use the same data as in Exercise 26, but look at matters from the perspective of an American entrepreneur importing rings of power from Middle Earth. In January 2011, \$1 = *M*1; in January 2012, \$1 = *M*0.50.

(a) Using Option A, calculate how much percentage value the dollar lost against the Middie (the target currency).

(b) Using Option B, calculate how much percentage value the dollar lost against the Middie.

28. In January 2002, when euro (€) currency coins and bills were introduced, the conversion to U.S. dollars was US\$1 = € 1.160. Nine years later, in January 2011, the dollar had declined severely, so that \$1 = €0.708.

(a) Using Option A, calculate how much percentage value the dollar lost against the euro.

(b) Using Option B, calculate how much percentage value the dollar lost against the euro.

29. We use the same data as in Exercise 28, but look at matters from the perspective of a European considering a vacation in the United States. In January 2002, €1 = \$0.862; in January 2011, €1 = \$1.411.

(a) Using Option A, calculate how much percentage value the euro gained against the dollar.

(b) Using Option B, calculate how much percentage value the euro gained against the dollar.

▲ **30.** Using your results from either Exercises 26−27 or 28−29, answer the following questions:

(a) Why don't the numbers agree? If the dollar loses a certain percentage against another currency, shouldn't that currency gain the same percentage against the dollar? Why or why not?

(b) Which option, A or B, seems to give a better sense of the effect of the change in relative values of currencies?

▲ **31.** For both Options A and B, answer the following questions:

(a) Can either give a percentage loss that is more than 100%? Would it make sense to speak of a currency declining more than 100%?

(b) Is the percentage from Option A—loss or gain—always higher than the Option B percentage? Always lower? If neither, do you see any pattern?

(c) Which option would you expect a person to use who wants to make a decline seem large? To make a gain seem large?

18.4 Sorry, No King Kongs

32. The weight of a 1-ft cube of steel is 500 lb. What is the pressure on the bottom face in

(a) pounds per square inch?

(b) atmospheres (1 atm = 14.7 lb/sq in.)?

33. In the photo at right of a statue of Paul Bunyan in Akeley, Minn., he would appear to stand about 33 ft tall and be geometrically similar to an ordinary muscular man.

(a) What is the linear scaling factor of the Paul Bunyan statue compared to a man 6 ft tall?

(b) If a man 6 ft tall weighs 200 lb, how much would Paul Bunyan weigh?

© Phil Schermeister/CORBIS

For Exercises 34 and 35, refer to the following.
A mature gorilla weighs 400 lb and stands 5 ft tall; its two feet combined have an area of about 1 ft^2.

34. (a) Give an estimate of the gorilla's weight when it was half as tall.
(b) What assumptions are involved in your estimate?
(c) When the gorilla is standing, what is the pressure on its feet in pounds per square inch?

35. Suppose that King Kong is a gorilla scaled up with a linear scaling factor of 10.
(a) How much does King weigh?
(b) What is the pressure on King's feet in pounds per square inch?

36. You may want a waterbed, but waterbeds are not allowed in your building. Apart from the danger of flood if the bed should puncture or leak, the weight is an issue.
(a) Suppose that a queen-size waterbed mattress is 80 in. long by 60 in. wide by 12 in. high, and water weighs 1 kg/L. How much does the water in the mattress weigh in pounds?
(b) If the weight of the mattress and frame is carried by four legs, each 2 in. by 2 in., what is the pressure, in pounds per square inch, on each leg?
(c) How does the pressure on the legs of the waterbed compare with the pressure that a person exerts on his or her feet—for example, a 130-lb person with a total foot area of about one-quarter of a square foot in contact with the ground?

37. If you aren't allowed to have a waterbed, how about a spa (hot tub)? Find the weight of the water in a spa that is in the shape of a cylinder 6 ft in diameter and 3.5 ft deep. (Hint: The volume of a cylinder is $\pi r^2 h$, where r is the radius and h is the height.)

38. What does the largest giant sequoia tree (named "Hyperion") weigh? Model the tree as a (very elongated) cone. Assume that the tree is 379 ft high, has a circumference of 40 ft at the base, and that the density of the wood is 31 lb/ft^3. (The volume of a cone of height h and radius r is $\frac{1}{3}\pi r^2 h$.)

39. A 6-ft-tall indoor holiday tree needs four strings of lights to decorate it. How many strings of lights are needed for an outdoor tree that is 30 ft high? (Contributed by Charlotte Chell of Carthage College, Kenosha, Wisconsin.)

For Exercises 40 and 41, refer to the following.

An ancient measure of length, the cubit, was the distance from the elbow to the tip of the middle finger of a person's outstretched arm. So the length of a cubit depended on the person, though there was some attempt at standardization. Most estimates place the cubit between 17 and 22 in.

40. According to classical Greek sources, Pythagoras (sixth century B.C.) used geometric scaling to model the height of Hercules, the heroic figure of classical mythology. Pythagoras compared the lengths of two racecourses, one (according to tradition) paced off by Hercules and the other by a man of average height. Both were 600 "paces" long, but the one by Hercules was longer because of his longer stride. A normal man in the time of Pythagoras would have been about 5 ft tall.
(a) If the distance paced off by Hercules was 30% longer than the other racecourse, how tall was Hercules? What does your calculation assume?
(b) The sources give two conflicting answers, that Hercules was 4 cubits tall and 4 cubits 1 "foot" tall. What range does this give for his height in feet and inches? In centimeters? (Assume that a Greek "foot" was the length of a modern foot.)

41. Goliath [of David and Goliath, as related in the Bible (I Samuel 17:4)] was "six cubits and a span." A "span" was originally the distance from the tip of the thumb to the tip of the little finger when the hand is fully extended, about 9 in. What range of heights would this indicate for Goliath in feet and inches? In centimeters?

For Exercises 42–45, refer to the following.

The body mass index (BMI) is the basis for the National Heart, Lung, and Blood Institute's weight guidelines. BMI is body weight (in kilograms) divided by the square of height (in meters). A BMI of 25 through 29 is considered "overweight"; a BMI of 30 or over is considered "obese." Some 55% of American adults have a BMI of 25 or above. (Note: BMI is not a useful measure for young children, pregnant or breastfeeding women, the frail elderly, or very muscular people.) For practice with this concept, calculate your own BMI.

42. Calculate the BMI for a woman 160 cm tall who weighs 65 kg. Is she overweight according to the institute's guidelines?

43. How much in kilograms must a man weigh who is 190 cm tall if he is not to be considered overweight according to the institute's guidelines?

44. Suppose that weight and height are measured instead in U.S. customary units of pounds and inches. We can still calculate body weight divided by the square of height using these units. What conversion factor is necessary to convert this number to the BMI?

▲ **45.** Calculate the BMI of a man who is 6′2″ tall and weighs 217 lb.

18.5 Dimension Tension

▲ **46.** Jonathan Swift's Gulliver traveled to Lilliput, where the Lilliputians were human-shaped but only about 6 in. tall. In other words, they were geometrically similar in shape to ordinary human beings but only one-twelfth as tall. What would a Lilliputian weigh? Are Lilliputians ruled out by the size-shape and area-volume considerations in this chapter? If you think they are, what considerations do you find convincing? If not, why not?

47. **(a)** What would you expect an individual *Q. northropi* to weigh if it had half the wingspan of an adult?
(b) If an individual weighed half as much as an adult, what would you expect its wingspan to be?

48. In the children's story *Peter Pan,* Peter and Wendy can fly. We may suppose that they are 4 ft tall, so they are about 12 times as tall as a sparrow is long. What should their minimum flying speed be?

49. Icarus of Greek legend escaped from Crete with his father, Daedalus, on wings made by Daedalus and attached with wax. Against his father's advice, Icarus flew too close to the sun; as a result, the wax melted, the wings fell off, and he plunged into the sea and drowned. What must have been his minimum cruising speed? What assumptions does your answer involve?

▲ **50.** Recent years have seen the beginnings of human-powered controlled flight, in the Gossamer Condor and other superlightweight planes, which have disproportionately large wings compared with geometric scaling up of birds. The Gossamer Condor is far longer than an ostrich, but it flies at only 12 mph. How can it do that?

51. Justify the claim in Example 10 on page 662 that a *Q. northropi* weighing 200 lb with a wingspan of 50 ft would have had the same wing loading as one weighing 100 lb with a wingspan of 36 ft.

52. The largest and heaviest aircraft in service today is the An-225—and we indeed mean "the" because there is only one! It was used to bring humanitarian equipment from Japan for Haiti, as well as—in a single flight—216,000 meals for U.S. military personnel. The plane has a wing area of 905 m^2 and a maximum takeoff weight of 1.3 million lb. What is its wing loading, in kg/m^2?

53. The cult movie *Them* (1954) features enormous ants (8 m long by 3 m wide). We can investigate the feasibility of such a scaled-up insect by considering its oxygen consumption. A common ant, 1 cm long, needs 24 milliliters (mL) of oxygen per second for each cubic centimeter of its volume. Because an ant has no lungs, it absorbs oxygen through its "skin" at a rate of 6.2 mL per second per square centimeter. Suppose that the tissues of a scaled-up ant would have the same need for oxygen for each cubic centimeter, and that its skin could absorb oxygen at the same rate as a common ant.

(a) Compared with a common ant, how many times as large is an enormous ant's
 (i) length?
 (ii) surface area?
 (iii) volume?
(b) What proportion of such an ant's oxygen need could its skin supply?
(c) What can you conclude about the existence of such insects? (Adapted from George Knill and George Fawcett, Animal form or keeping your cool, *Mathematics Teacher,* May 1982, 395–397.)

For Exercises 54–57, refer to the following.

Maybe some trees could grow to a mile high, but they just don't live long enough to have the chance. In this problem, we try to determine how fast the height of a tree increases. We can measure indirectly how much mass the tree adds in a year by the area of the annual tree ring added. Here are two relevant facts:

- As you may have noticed from stumps, as a tree grows older, its annual rings get less wide. Although the width of the ring varies somewhat from year to year with the amount of rainfall and other factors, the total *area* of each annual ring is roughly the same over the years, meaning that *the tree adds roughly the same amount of mass each year.* Call that amount a; then the mass M of the tree is $M = at$, where t is its age in years.

- Over a large range of tree sizes and tree species, the diameter d of a tree of a species is approximately proportional to the three-halves power of the height h of the tree. (Different species have different constants of proportionality.) Thus, $d \propto h^{3/2}$ (this is shown in Exercise 54).

Now, if we assume that the bulk of the mass of the tree is in the trunk, and if we model the trunk either as a long cylinder or as a thin cone, the mass is proportional to the volume, so $M \propto d^2h$. Then

$$at = M \propto d^2h \propto (h^{3/2})^2h = h^4$$

so $h \propto t^{1/4}$. In other words, *the tree grows in height as the fourth root of its age.*

54. Suppose that a tree grows to 10 m in 15 years (as the red oak in our front yard did). How tall will it be (if it lives long enough) when it is 60 years old?

55. How long would it take the tree in Exercise 54 to grow to be 40 m tall?

56. Giant sequoias can reach 100 m after about 1000 years. If it could keep on growing at the same rate of its addition of mass, how long would it take a giant sequoia 100 m tall to grow to 200 m?

■ **57.** The branching of trees is similar to the branching of circulatory systems in the bodies of animals. For similar reasons, the area of the cross section of the tree at its base scales as the three-fourths power of the tree's mass; that is, $A \propto M^{3/4}$. Assume that most of the mass is in the trunk and model the tree either as a tall cylinder ($V = \pi r^2h$) or as a cone ($V = \pi r^2h/3$). Show that the diameter d of a tree is approximately proportional to the three-halves power of the height; that is, $d \propto h^{3/2}$.

▲ **58.** Some humans, such as the Bushmen of the Kalahari Desert in Africa, live in desert environments, where it is important to be able to do without water for periods of time. Would you expect such an environment to favor short people or tall ones? (Adapted from A. Zherdev, Horseflies and flying horses, *Quantum,* May–June 1994, 32–37, 59–60.)

▲ **59.** Smaller birds and mammals generally maintain higher body temperatures than do larger ones. Explain why you would expect this to be so. (Adapted from A. Zherdev, Horseflies and flying horses, *Quantum,* May–June 1994, 32–37, 59–60.)

18.6 How to Grow

60. Listed below are the numbers of species of reptiles and amphibians on some Caribbean islands, together with the approximate areas of the islands. (Suggested by Florence Gordon of the New York Institute of Technology, with contributions from Kevin Mitchell and James Ryan of Hobart and William Smith Colleges, Geneva, N.Y. This table is adapted from Tables 15 and 16 in P. J. Darlington, *Zoogeography: The Geographic Distribution of Animals,* Wiley, New York, 1957, pp. 483–484.)

Island	Area (mi^2)	Species
Redonda	1	3
Saba	4.9	5
Montserrat	40	9
Trinidad	2000	80
Puerto Rico	3400	40
Jamaica	4500	39
Hispaniola	30,000	84
Cuba	40,000	76

(a) Plot number of species versus area on ordinary graph paper and then on log-log graph paper. If you don't have log-log paper available, use a calculator or spreadsheet to take the logarithms ($\log_{10}$) of all the numbers and graph logarithm of number of species versus logarithm of area on ordinary graph paper. (Note: Trinidad is an outlier from the general pattern; see Chapter 6.)

(b) Is the relationship that you graphed in part (a) proportional? allometric?

(c) What would be the expected number of species on an island of 400 mi^2?

(d) For each 10-fold increase in the island's size, what happens to the number of species, approximately?

61. Listed in the next page are the weights and wingspans of some birds and of some fully loaded airplanes. (Idea and most data contributed by Florence Gordon of the New York Institute of Technology.)

Bird	Weight (lb)	Wingspan (ft)
Crow	1	2.9
Harris hawk	2.6	3.2
Blue-footed booby	4	3
Red-tailed hawk	4	4
Horned owl	5	5
Turkey vulture	6.5	6
Eagle	12	7.5
Golden eagle	13	7.3
Whooping crane	16.1	7.5
Vulture	18.7	9.3
Condor	22	9.9
Quetzalcoatlus northropi	100	36
Plane		
Boeing 737	117,000	93
DC9	121,000	93.5
Boeing 727	209,500	108
Boeing 757	300,000	156.1
Boeing 707	330,000	145.7
DC8	350,000	148.5
Howard Hughes's "Spruce Goose"	400,000	320.9
DC10	572,000	165.4
Boeing 747	805,000	195.7
Boeing 747-400	895,000	212.6
Anton An-225	1,323,000	290.2

(a) Use a calculator or spreadsheet to take the logarithms ($\log_{10}$) of all the numbers and then graph logarithm of weight versus logarithm of wingspan on ordinary graph paper.

(b) For the birds, is the relationship that you graphed in part (a) proportional? Allometric? How about for the planes?

(c) Does the same relationship of wingspan to weight seem to hold for birds and planes?

62. Each December, the Friends of the Beloit-Janesville (Wisconsin) Symphony Orchestra raises funds with a holiday sale of poinsettia plants. The choices available, with cost in 2010 and average number of flowers per plant, are shown below.

Diameter of Flowerpot	Average Number of Flowers	Cost
6 in.	6	\$10
7 in.	9	\$18
8 in.	18	\$28

(a) Plot on ordinary graph paper cost versus diameter of flowerpot. Do you observe linear scaling?

(b) Use a calculator or spreadsheet to take the base-10 logarithms ($\log_{10}$) of the cost and then graph the logarithm of cost versus diameter of flowerpot.

(c) Repeat parts (a) and (b) for cost versus average number of flowers.

(d) What would you estimate for the number of flowers and the cost for a 9-in.-diameter flowerpot?

WRITING PROJECTS

1. A human infant at birth usually weighs between 5 and 10 lb and has a height (length) between 1 and 2 ft, with the shorter babies having the lesser weight. Considering the weight and height of an adult human, write a paragraph arguing that human growth must not be just proportional growth.

2. The principle that area scales with the square of length, and volume with the cube, has important consequences for the depiction and interpretation of data in graphic form. Suppose that we wish to indicate in an artistic way that the weekly income of a U.S. carpenter is twice that of a carpenter in (mythical) Rotundia. We draw one moneybag for the Rotundian and another one "twice as large" for the American. (Illustration from Darrell Huff, *How to Lie with Statistics,* Norton, New York, 1954, p. 69.)

What's the problem? Well, first, people tend to respond to graphics by comparing areas. Because the larger moneybag is twice as high and twice as wide as the smaller one, its image has four times the area. Second, we are used to interpreting depth and perspective in drawings in terms of three-dimensional objects. Because the larger bag is also twice as thick as the smaller, it has eight times the volume. The graphic leaves the subconscious impression that the U.S. carpenter earns eight times as much instead of twice as much. With these ideas in mind, evaluate—in a paragraph each—the following data depictions.

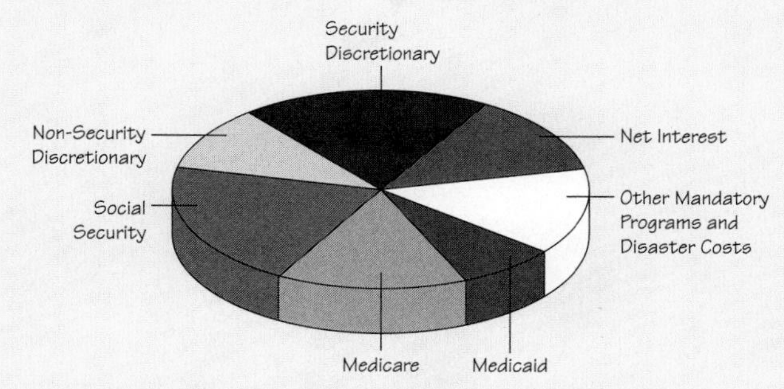

(a) U.S. Federal Budget Policy Outlays by Category for 2015. (U.S. Federal Budget for 2011.) Source: **http://www.whitehouse.gov/sites/default/files/omb/budget/fy2011/assets/tables.pdf**, p. 153.

This display shows expected expenditures of the federal government in 2015, totaling $4.4 trillion (including an anticipated deficit of almost $1 trillion). The actual percentages of expenditure are:

Security Discretionary	21%
Net Interest	13%
Other	12%
Medicaid	8%
Medicare	15%
Social Security	20%
Non-Security Discretionary	11%.

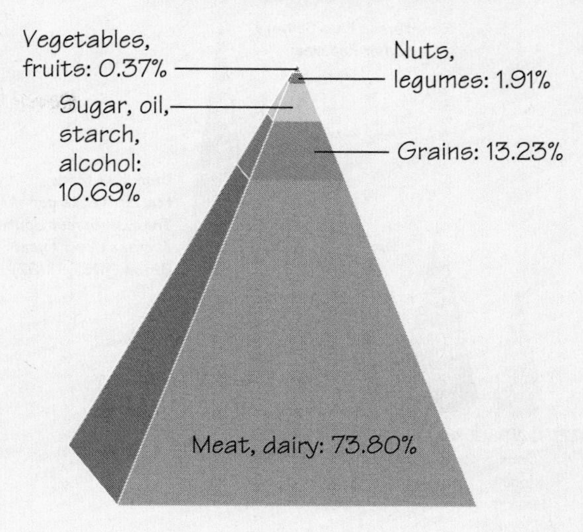

(b) Federal subsidies for food production, 1995–2005. ("Why That Salad Costs More Than a Big Mac," *Readers Digest*, October 2010, p. 72.)

(c) U.S. colleges as classified by enrollment. (From David S. Moore, *Statistics: Concepts and Controversies*, 4th ed., W. H. Freeman, New York, 1997, p. 217.)

3. Evaluate in a paragraph each of the following depictions (a–c).

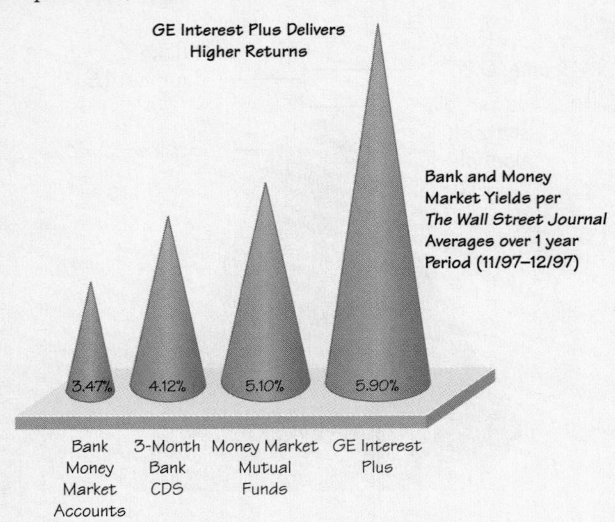

(a) Average bank and money market yields, per the *Wall Street Journal,* for 1/97–12/97.
(Source: Flyer from GE Interest Plus.)

(b) Marijuana arrest rates in some California cities, 2006–2008. (*New York Times,* October 23, 2010, p. A19.) (See figure at right.)

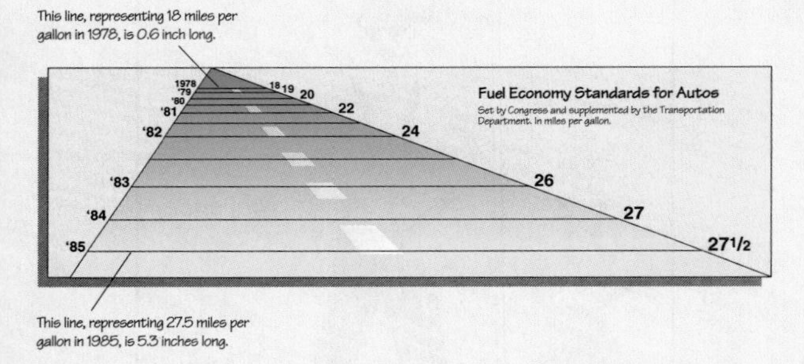

(c) Fuel economy standards for autos. (From Edward R. Tufte, *The Visual Display of Quantitative Information,* Graphics Press, 1983, p. 57, as adapted from *The New York Times,* August 9, 1978, p. D2.)

4. With the ideas of Writing Projects 2 and 3 in mind, collect and evaluate similar depictions of data from magazines and newspapers.

Imbalance in Arrests
Marijuana possession arrest rates in some of California's largest cities, 2006–08

Los Angeles
Arrest rate per 100,000 blacks
523
73
Arrest rate per 100.000 whites

San Diego
835
145

San Jose
619
121

Fresno
500
98

Long Beach
1,461
246

Bakersfield
502
82

Riverside
383
80

Source: "Arresting Blacks for Marijuana in California: Possession Arrests in 25 Cities, 2006–08"

5. Dolls and human figures are usually scaled to be geometrically similar to actual humans. But are dolls designed to represent babies or adult humans? Go to a toy store and measure the height, the vertical height of the head, and the arm length of some dolls and other figures. Scale your measurements to compare them with Figure 18.12; from that comparison, try to estimate the ages of the humans that the figures resemble. Write up your procedure, data, calculations, and conclusions in a page or two.

6. (Refer to Exercises 42–45.) Because body weight is average density times body volume, BMI is average density times a quantity that has units of length. Discuss whether BMI makes sense as a measure of being overweight. Would dividing by a different power of height make for a better measure? (See Keith Devlin, "Top 10 Reasons Why the BMI Is Bogus," NPR, July 4, 2009, **http://www.npr.org/templates/story/story.php?storyId5106268439**.)

Suggested Readings

ADAM, JOHN A. *Mathematics in Nature: Modeling Patterns in the Natural World,* Princeton University Press, Princeton, NJ, 2003.

BONNER, JOHN TYLER. *Why Size Matters: From Bacteria to Blue Whales,* Princeton University Press, Princeton, NJ, 2006.

DUDLEY, BRIAN A. C. *Mathematical and Biological Interrelations,* Wiley, New York, 1977. An excellent and extended introduction to graphing, scale factors, and logarithmic plots.

GOULD, STEPHEN JAY. "Size and shape." In *Ever Since Darwin*, Norton, New York, 1977, Chap. 21.

HALDANE, J. B. S. "On being the right size." In *Possible Worlds and Other Papers,* Harper, New York, 1928. Reprinted in James R. Newman (ed.), *The World of Mathematics,* vol. 2, Simon & Schuster, New York, 1956, pp. 952–957. Also reprinted in John Maynard Smith (ed.), *On Being the Right Size and Other Essays* by J. B. S. Haldane, Oxford University Press, Oxford, 1985, pp. 1–8. Succinctly surveys area-volume tension, flying, the size of eyes, and even the best size for human institutions.

LARRICK, RICHARD P., and JACK B. SOLL. The MPG illusion. *Science* 320 (20 June 2008) 1593–1594.

McMAHON, T. A., and J. T. BONNER. *On Size and Life,* Scientific American Library, New York, 1983. Astonishingly beautiful and informative book on the effects of size and shape on living things.

SCHMIDT-NIELSEN, KNUT. *Scaling: Why Is Animal Size So Important?* Cambridge University Press, New York, 1984.

SILLETT, STEVE. The tallest trees. *National Geographic,* 216 (4) (October 2009).

WEIBEL, EWALD R. *Symmorphosis: On Form and Function in Shaping Life,* Harvard University Press, Cambridge, MA, 2000.

Suggested Web Sites

physics.nist.gov/cuu/Units/index.html In-depth information on SI, the modern metric system.

www.missingkids.com National Center for Missing and Exploited Children.

www.usmint.gov U.S. Mint.

www.thusness.com/bmi.t.html BMI calculator and further links.

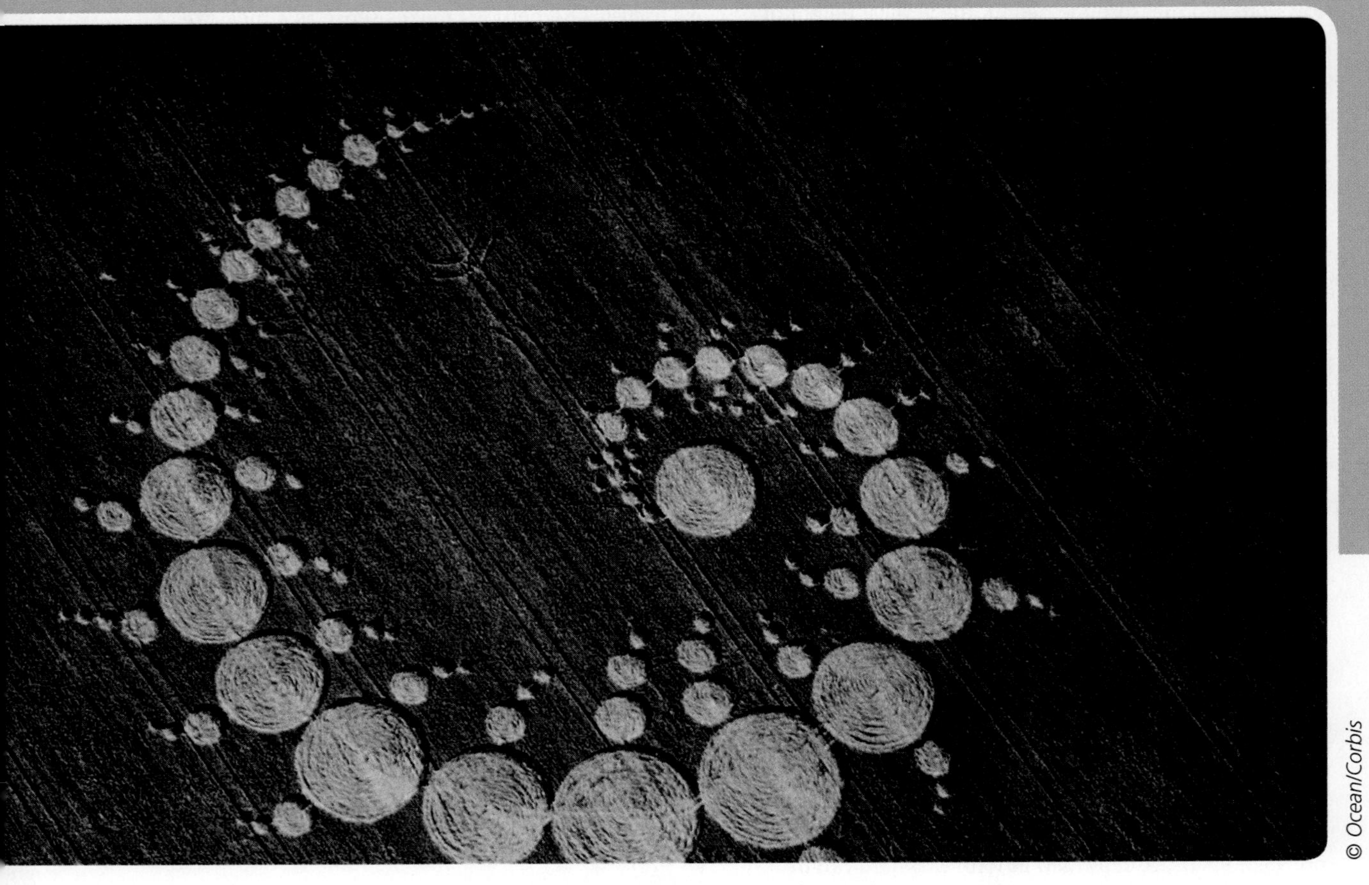

19

Symmetry and Patterns

In the narrowest sense, symmetry refers to a mirror-image reflection, such as between the sides of the human body. We also see symmetry in rotations, as in the two-dimensional spirals of sunflowers and daisies (Figure 19.1) and even in a three-dimensional form on pineapples, pinecones, sunflowers, and cacti (Figure 19.2a). The chambered nautilus (Figure 19.2b) may stretch your notion of symmetry (as will other examples in this chapter). It has neither reflection nor rotational symmetry, but its successive sections are geometrically similar (in the sense of Chapter 18: same shape but different size), and the resulting spiral has the same shape at any size: A photographic enlargement superimposed on it would fit exactly.

Taken in a wide sense, symmetry includes notions of *balance, proportionality* (geometric similarity), and *repetition* according to a pattern.

What does mathematics have to do with symmetry? *Mathematics* is the *science of patterns*; it allows us to identify, classify, and appreciate more deeply different kinds of symmetry.

In Section 19.1, we show how a simple pattern in numbers is related both to spirals in plants and to a numerical proportion that provided an ancient standard of beauty. That proportion, the golden ratio, has found expression in architecture through the ages, while the formula for calculating it is also used in the Consumer Price Index (CPI) and was crucial in formulating the standard for high-definition television (HDTV). Temporarily leaving nature for art, in Section 19.2 we explore what it means for a pattern to reproduce copies of a design element ("motif") symmetrically. What may be utterly surprising is that though there is an infinite variety of motifs, there are very few symmetrical patterns for arranging copies of one. We illustrate these patterns and provide (in Section 19.3) notation, decision trees, and examples.

In Section 19.4, we step back, examine common elements of symmetry patterns, and come to a unified sense of what symmetry is by thinking of symmetries as transformations that preserve a pattern, interacting among themselves to generate further symmetries. Finally, in Section 19.5 we ask what happens if we allow scaling the design element down (or up) as part of the notion of symmetry; along with the chambered nautilus, we get fractals, with their own kind of symmetry and beauty.

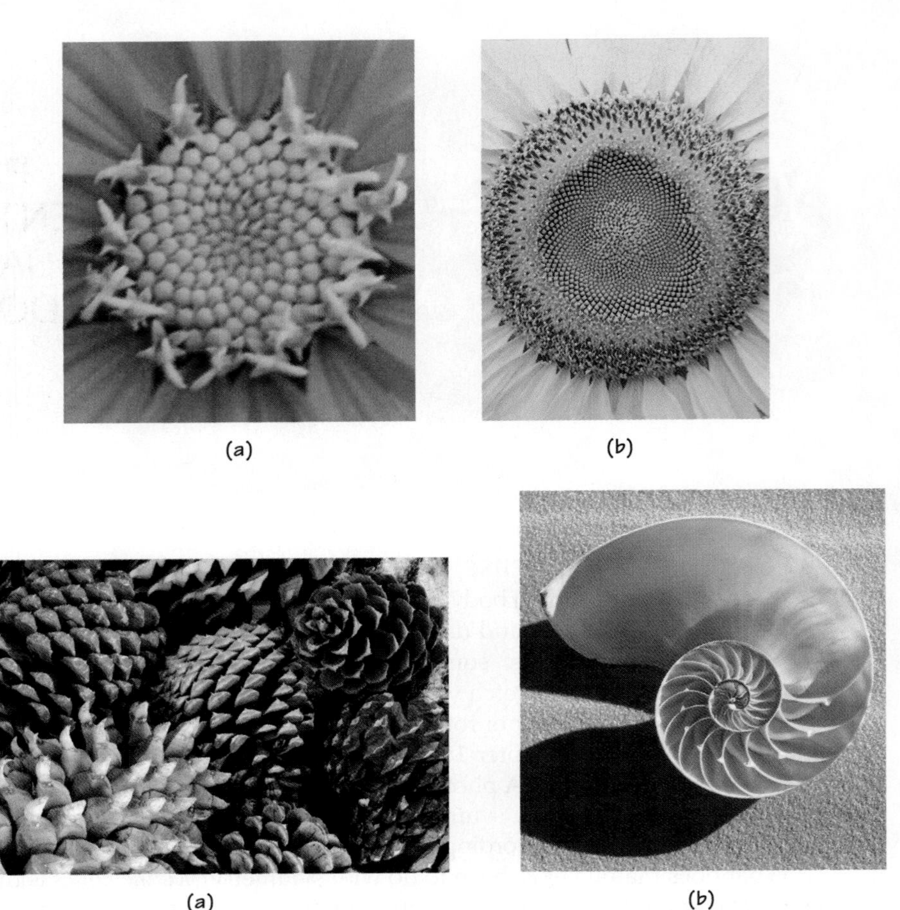

Figure 19.1 (a) This daisy has in its center 13 spirals to the right and 21 to the left. *(© Kenneth Peterson)* (b) This sunflower has 55 spirals in one direction and 89 spirals in the other direction. *(Harvey Lloyd/The Stock Market.)*

Figure 19.2 (a) Spirals of scales on a pinecone: 8 right, 13 left. *(From Verner E. Hoggatt, Jr., Fibonacci and Lucas Numbers, Houghton Mifflin, New York, 1969, p. 81.)* (b) A chambered nautilus shell. *(James Randkler/Tony Stone Images.)*

19.1 Fibonacci Numbers and the Golden Ratio

Fibonacci Numbers

Associated with the geometric symmetry in flowers and pinecones is a kind of *numeric symmetry,* with a "proportion" in the sense of a ratio of numbers. Strangely, the number of spirals in plants is not just any whole number but always comes from a particular sequence of numbers called **Fibonacci numbers** (see Spotlight 19.1).

Fibonacci Numbers (Fibonacci Sequence) DEFINITION

Fibonacci numbers occur in the sequence

$$1, 1, 2, 3, 5, 8, 13, 21, 34, 55, 89, 144, 233, 377, \ldots$$

This sequence begins with the numbers 1 and 1 again, and each next number is obtained by adding the two preceding numbers together.

Sometimes a sequence of numbers is specified by stating the value of the first term or first several terms and then giving an equation to calculate succeeding

SPOTLIGHT 19.1

Leonardo of Pisa ("Fibonacci")

Born in 1170, Leonardo of Pisa has been known as "Fibonacci" for the past century and a half. This nickname, which refers to his descent from an ancestor named Bonaccio, is modern, and there is no evidence that he went by it.

Leonardo was the greatest mathematician of the Middle Ages. He introduced calculation with Hindu-Arabic numerals into Italy to replace the Roman numerals and abacus then in use, and he also wrote about geometry, algebra, and number theory.

We know little of Leonardo's life apart from a short autobiographical sketch:

> I joined my father after his assignment by his homeland Pisa as an officer in the customhouse located at Bugia [Algeria] for the Pisan merchants who were often there. He had me marvelously instructed in the Arabic-Hindu numerals and calculation. I enjoyed so much the instruction that I later continued to study mathematics while on business trips to Egypt, Syria, Greece, Sicily, and Provence and there enjoyed discussions and disputations with the scholars of those places.

Leonardo of Pisa ("Fibonacci")
A portrait of unlikely authenticity.
(From Columbia University, D. E. Smith Collection.)

(Translated by L. E. Sigler in Leonardo Pisano, *The Book of Squares: An Annotated Translation into Modern English,* Academic Press, New York, 1987, p. xvi.)

The *Liber abbaci* contains a famous problem about rabbits, whose solution is now called the Fibonacci sequence. Leonardo did not write further about it.

terms from preceding ones. This is called a *recursive rule,* and the sequence is said to be defined by **recursion**. Let's denote the nth Fibonacci number by F_n; then the Fibonacci sequence can be defined by

Recursion for the Fibonacci Sequence PROCEDURE

$$F_1 = 1, F_2 = 1, \quad \text{and} \quad F_{n+1} = F_n + F_{n-1} \quad \text{for} \quad n \geq 2$$

The recursive rule just expresses in algebraic form that the next Fibonacci number is the sum of the previous two.

Look at the daisy in Figure 19.1a. There are 13 spirals streaming out in the clockwise direction and 21 harder-to-count ones in the counterclockwise direction—two consecutive Fibonacci numbers. The common grocery pineapple (*Ananas comosus*) has three sets of spirals, one each along the three directions through each hexagonally shaped scale: 8 spirals to the right, 13 to the left, and 21 vertically—again, consecutive Fibonacci numbers.

Why are the numbers of spirals in plants the same numbers that appear next to each other in a purely mathematical sequence? There is no easy answer for this phenomenon, which is called **phyllotaxis.** There are several intricate theories about the dynamics of the plant's growth.

Algebra Review: Sequences

A sequence is a listing of objects, generally numbers, called terms. Terms in a sequence are listed in a specific order and are separated by commas. Here is the sequence of positive multiples of three:

$$3, 6, 9, 12, \ldots$$

The above sequence goes on forever, and is therefore called an *infinite sequence*. Sequences that have a definite end are called *finite*.

Subscripts are used in sequences to describe where in the ordered list a term is located. For example, t_1 could be the first term and for the above sequence, $t_1 = 3$. Because the pattern continues in the sequence, we know that $t_5 = 15$. In general, t_n represents the nth term, t_{n-1} is the term that precedes t_n, and t_{n+1} is the term that follows.

There are two types of rules that can govern how terms in a sequence are calculated, *explicit* and *recursive*. An explicit rule dictates how to find the nth term by direct calculation. A recursive rule defines the first or first few terms and then makes a general statement about the sequence.

If we examine the sequence of positive multiples of three, we have the following.

Explicit

Term number	Term	Comment
1	3	$t_1 = 3 \cdot 1$
2	6	$t_2 = 3 \cdot 2$
3	9	$t_3 = 3 \cdot 3$
4	12	$t_4 = 3 \cdot 4$

An explicit rule about this sequence is $t_n = 3n$.

Recursive

Term number	Term	Comment
1	3	$t_1 = 3$
2	6	$t_2 = t_1 + 3$
3	9	$t_3 = t_2 + 3$
4	12	$t_4 = t_3 + 3$

A recursive rule about this sequence is $t_1 = 3$ and $t_n = t_{n-1} + 3$ for $n \geqslant 2$ (or $t_{n+1} = t_n + 3$ for $n \geqslant 1$).

The Golden Ratio

During the last several centuries, an attractive myth has arisen that the ancient Greeks considered a specific numerical proportion essential to beauty and symmetry. Known in modern times as the **golden ratio, golden mean,** or even **divine proportion,** this proportion was investigated by Euclid in Book II of his *Elements*. There is little evidence connecting this proportion to Greek esthetics, but let's pursue the golden ratio briefly because of its intimate connection to the Fibonacci sequence and because it does have some appeal as a standard for beautiful proportion.

Golden Ratio DEFINITION

The value of the **golden ratio,** which is usually denoted by the Greek letter phi (ϕ), is

$$\phi = \frac{1 + \sqrt{5}}{2} = 1.618033\ldots$$

The basic esthetic claim is that a **golden rectangle**—one whose height and width are in the ratio of 1 to ϕ—is the most pleasing of all rectangles. The Greeks treated lengths geometrically, so for them it was important to construct lengths

SPOTLIGHT 19.2

How the Greeks Constructed a Golden Rectangle

In constructing a golden rectangle, the Greeks started from a 1-by-1 square (outlined in black in the figure), which they made by constructing perpendiculars at the two ends of a horizontal segment of unit length. To extend the square to a golden rectangle, they bisected the original segment, getting a new point that divides it into two pieces of length one-half each. Using this new point and a compass opening equal to the distance from it to a far corner of the square (shown by the diagonal line in the figure), they could add the blue length to the length one-half to get an interval (in red at bottom) with total length ϕ.

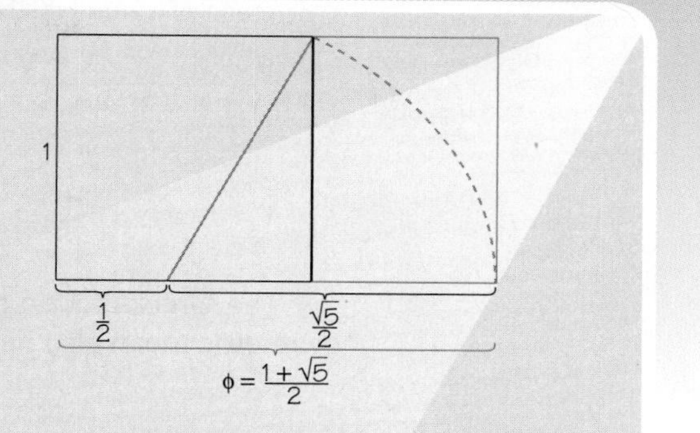

A golden rectangle has the pleasing property that if you cut a square-shaped piece off one end of it, the rectangle that remains is again a golden rectangle.

using straightedge and compass. In Spotlight 19.2, we show how to construct a golden rectangle that is 1 unit by ϕ units.

Why would anyone think that this is an attractive ratio? And where did it come from? The answer lies not in Fibonacci numbers but in the Greeks' pursuit of balance in their study of geometry.

Given two line lengths, one way to find a length that "strikes a balance" between the two is to average them. For lengths l (the larger) and w (the smaller), their average, or *arithmetic mean,* is $m = (l + w)/2$, and it satisfies

$$l - m = m - w$$

The length m strikes a balance between l and w, in terms of a common *difference* from the two original lengths. More generally, the arithmetic mean of n numbers or lengths is their sum divided by n. (See Chapter 5 for its use in statistics.)

The Greeks, however, preferred a balance in terms of *ratios* rather than differences. They sought a length s, the **geometric mean**, that gives a common ratio

$$l \div s = s \div w \qquad \text{or} \qquad \frac{l}{s} = \frac{s}{w}$$

Hence $lw = s^2$, which expresses the geometric fact that s is the side of a square whose area equals the area of an l by w rectangle (the Greeks thought in terms of geometric objects). In geometry, the geometric mean s is called the *mean proportional* between l and w (see Figure 19.3).

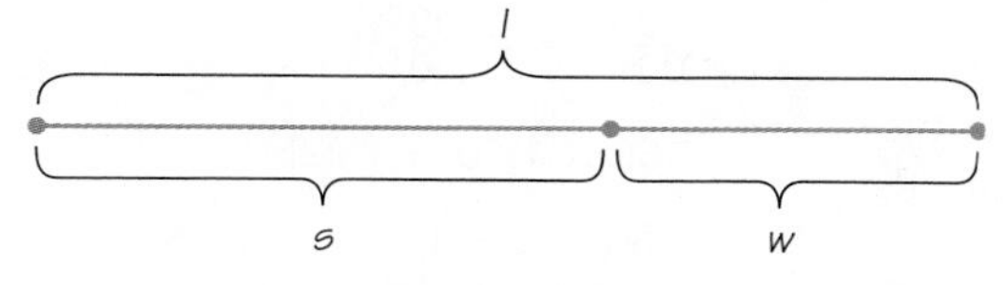

Figure 19.3 The line segment of length l is divided so that the length of s is the geometric mean between l and $w = l - s$. The dividing point divides the length l in the golden ratio.

Geometric Mean DEFINITION

The quantity $s = \sqrt{lw}$ is the **geometric mean** of l and w. More generally, the geometric mean of n numbers is the nth root of the product of all n factors: The geometric mean of $x_1, \ldots, x_n$ is $\sqrt[n]{x_1 \times \cdots \times x_n}$. For example, the geometric mean of 1, 2, 3, and 4 is $\sqrt[4]{1 \times 2 \times 3 \times 4} = \sqrt[4]{24} = 24^{1/4} \approx 2.213$.

The Greeks found symmetry and proportion in the geometric mean, but the geometric mean also has important practical applications (see Spotlight 19.3).

SPOTLIGHT 19.3

The Consumer Price Index: An Application of the Geometric Mean

The Bureau of Labor Statistics (BLS) uses the geometric mean—not the arithmetic mean—to calculate the Consumer Price Index (CPI), which tracks changes in the cost of the goods and services that people buy.

The geometric mean takes into account substitutions that consumers make when prices change. For example, if the price of beef goes up but the price of chicken doesn't, then consumers may buy less beef and substitute the cheaper chicken for some beef.

Suppose that, overall, U.S. families consume equal dollar values of beef and chicken. A typical family might consume weekly 5 lb of beef at \$4/lb and 10 lb of chicken at \$2/lb, for \$20 each and a total cost of \$40. We say that beef and chicken each have a *relative market share* of 0.5 (50% beef, 50% chicken, by dollar value).

What if beef goes up to \$6/lb but chicken stays at \$2/lb? The *relative price change* in beef is \$6/\$4 = 1.5 and the relative price change in chicken is \$2/\$2 = 1.00 (no change). If the average family continues to eat just as much beef and chicken as before, the cost is now \$50, an increase of 25%. Because \$30 goes for beef and \$20 for chicken, the relative market shares (0.6 and 0.4) have changed. The relative price change for the family's meat is \$50/\$40 = 1.25, which is just the arithmetic mean of the two relative price changes (1.50 and 1.00). A more general formulation is:

$$\begin{aligned}\text{relative price change} &= (\text{old market share of beef}) \frac{\text{new cost of beef}}{\text{old cost of beef}} \\ &\quad + (\text{old market share of chicken}) \times \frac{\text{new cost of chicken}}{\text{old cost of chicken}} \\ &= 0.5 \times \frac{6.00}{4.00} + 0.5 \times \frac{2.00}{2.00} \\ &= \frac{1.50 + 1.00}{2} = 1.25\end{aligned}$$

A family that eats no beef sees no increase. A family that eats only beef sees an increase of 50%. The CPI is an average over *all* families, weighted by the dollar value that each consumes.

If instead we use the geometric mean, we get a relative price change of $\sqrt{1.50 \times 1.00} \approx 1.225$. The more general formulation is

$$\begin{aligned}\text{relative price change} &= \left(\frac{\text{new cost of beef}}{\text{old cost of beef}}\right)^{(\text{old market share of beef})} \\ &\quad \times \left(\frac{\text{new cost of chicken}}{\text{old cost of chicken}}\right)^{(\text{old market share of chicken})} \\ &= \left(\frac{6.00}{4.00}\right)^{0.5} \times \left(\frac{2.00}{2.00}\right)^{0.5} \\ &= \sqrt{1.50 \times 1.00} \approx 1.225\end{aligned}$$

This relative price change, a 22.5% increase, is less than the 25% using the arithmetic mean.

The intention of the CPI is to measure the change in the cost of goods and services that still

The Consumer Price Index: An Application of the Geometric Mean *(continued)*

yield the same level of satisfaction to consumers. Use of the arithmetic mean presumes that a family buys the same amount of beef and chicken (5 lb beef, 10 lb chicken) as before. Use of the geometric mean presumes that a family buys the same *relative dollar value* of each meat as before, hence \$24.50 (12.25 lb) of chicken and \$24.50 (4.08 lb) of beef, for a total of $\$49 = 1.225 \times \$40$. Buying 2.25 lb more chicken and 0.92 lb less beef is supposed to yield the "same satisfaction" as before.

Because the geometric mean is always less than or equal to the arithmetic mean (see Exercise 19), the geometric mean gives a lower figure for inflation than using the arithmetic mean would produce.

Social Security payments, some wage increases, and income tax rates are all automatically geared to the CPI, which we treat in detail in Chapter 21.

Justin Guariglia/Corbis

The Greeks were interested in cutting a single line segment of length l into lengths s and w, where $l = w + s$, so that s would be the mean proportional between w and l. Surprisingly, the ratio ϕ arises, as we show. Denote by x the common ratio

$$\frac{l}{s} = \frac{s}{w} = x$$

Substituting $l = s + w$, we get

$$x = \frac{l}{s} = \frac{s + w}{s} = \frac{s}{s} + \frac{w}{s} = 1 + \frac{w}{s}$$

But w/s is just $1/x$, so we have

$$x = 1 + \frac{1}{x}$$

Multiplying through by x gives

$$x^2 = x + 1 \qquad \text{or} \qquad x^2 - x - 1 = 0$$

This is a quadratic equation of the form

$$ax^2 + bx + c = 0$$

with $a = 1$, $b = -1$, and $c = -1$. We apply the famous quadratic formula,

$$x = \frac{-b \pm \sqrt{b^2 - 4ac}}{2a}$$

to get the two solutions

$$x = \frac{1 + \sqrt{5}}{2} = 1.618033\ldots \qquad \text{and} \qquad \frac{1 - \sqrt{5}}{2} = -0.618033\ldots.$$

The negative second solution does not correspond to a length. The first solution is the golden ratio ϕ. It occurs often in other contexts in geometry; for example, ϕ is the ratio of a diagonal to a side of a regular pentagon (see Figure 19.4).

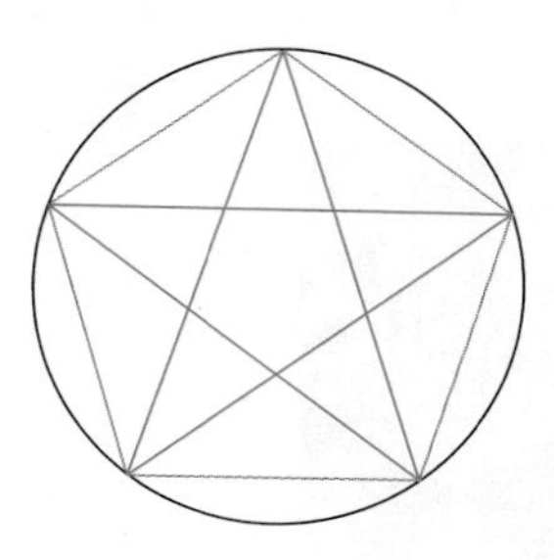

Figure 19.4 In a pentagon with equal sides, ϕ is the ratio of a diagonal to a side. The five-pointed star formed by the diagonals was the symbol of the followers of the ancient Greek mathematician Pythagoras.

The term *golden ratio* was not used in antiquity and there is no evidence that the Great Pyramid was designed to conform to ϕ, nor the Greeks used ϕ in the proportions of the Parthenon, nor that Leonardo da Vinci used ϕ in proportions for the human figure (Figure 19.5a). The area from the top of the head of the "Mona Lisa" to the top of her bodice may form a golden rectangle, but da Vinci left no documents saying that was his intention or design principle.

Some have claimed that the impressionists Gustave Caillebotte (1848–1894) and Georges Seurat (1859–1891) used the golden ratio to design some of their paintings, but the painters themselves left no word about it. Wolfgang Amadeus Mozart (1756–1791), who was fascinated by mathematics as a student, may have constructed the lengths of parts of some of his piano sonatas with an eye to the golden ratio, but we do not have evidence that this was his intention.

Moreover, experiments show that people's preferences for dimensions of rectangles cover a wide range, with golden rectangles not holding any special place. A ratio near 1.8 seems the most popular; this ratio is close to the aspect ration of high-definition TV (see Spotlight 19.4)

It is true that human bodies exhibit ratios close to the golden ratio, as you can see by comparing your overall height to the height of your navel. The Swiss-born architect Le Corbusier (Charles-Edouard Jeanneret [1887–1965]) used the golden ratio (including a navel-height feature) as the basis for his "Modulor" scale of proportions (Figure 19.5b).

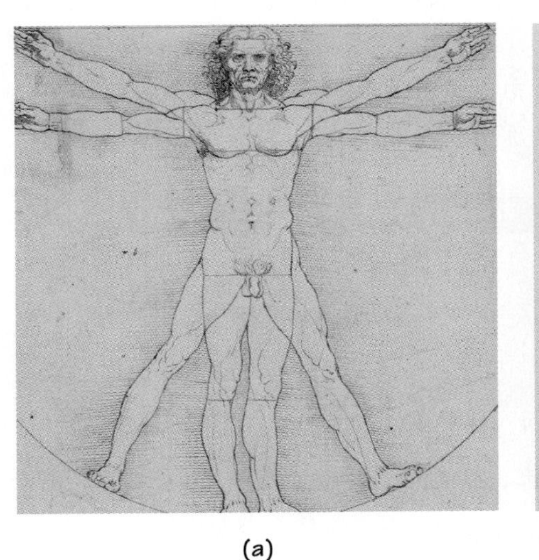

(a)

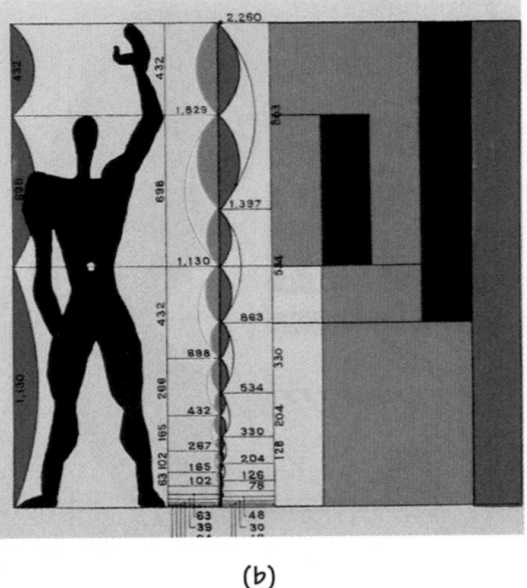

(b)

Figure 19.5

(a) Leonardo da Vinci's "Vitruvian Man" (ca. 1490), based on body proportions by Vitruvius (architect and engineer, first century B.C.). Despite claims on the Web and in the thriller *The Da Vinci Code,* neither Vitruvius nor Leonardo suggested using ϕ for human proportions or anything else. (*Accademia, Venice, Italy/Scala/Art Resource, New York.*)

(b) Le Corbusier, however, did use ϕ in his "Modulor" scale of proportions.

(*Le Corbusier, "Le Modulor," 1945. © 2000 Artists Rights Society [ARS], New York/ADAGP, Paris/FLC.*)

SPOTLIGHT 19.4

Are We Trying to Reclaim the "Glory That Was Greece"*?

The *aspect ratio* of an object, such as a computer screen, is the ratio of its width to its height. For a golden rectangle, this is phi, approximately 1.61. An aspect ratio is usually expressed as two numbers separated by a colon; for example, most digital cameras take pictures that have an aspect ratio of 4:3 (pronounced "4 to 3").

Through the ages, artists have produced canvases of varying sizes and aspect ratios. The introduction of photography on a mass-consumer scale involved standardizing sizes for photographic roll film and for photographic paper prints. Popular sizes were designated 110 ("pocket Instamatic"), at 17 mm × 13 mm, whose aspect ratio is 17:13 ≈ 1.31, and 135 ("35 millimeter film"), at 36 mm × 24 mm, for an aspect ratio of 1.5. Meanwhile, motion picture engineers settled in 1932 on a standard frame size for movies of 22 mm × 16 mm, with an aspect ratio of 1.375.

Later, television was developed to use an aspect ratio of 4:3, or 1.33. Still, movies could be shown on TV with minimal cropping. Until 1968, TV was in black and white only, while movies in theaters offered color. Filmmakers sought a further advantage in providing the "panoramic vision" of widescreen. Through either optical compression and the use of anamorphic lenses, or else cropping the top and bottom of the standard frame, movies in the United States evolved to an aspect ratio of 1.85, while those in some European countries evolved to 1.66, with some "widescreen" films at 2.35.

For HDTV, a 16:9 ≈ 1.78 ratio was chosen as a compromise. Dr. Kerns H. Powers had drawn equal-area rectangles of each aspect ratio, superimposed them with a common center, and observed that they all fit inside an outer rectangle with an aspect ratio of 1.77. They also all contained a smaller common rectangle with the same aspect ratio. The idea is that video shot at other aspect ratios could be shot so as to keep the main action within the inner rectangle, to allow for potential broadcast in HDTV, while video shot at 16:9 could be shot to keep the main action within the 4:3 rectangle (which extends to the top and bottom of the outer rectangle).

When video shot at 16:9 is presented on a TV set with a 4:3 aspect ratio but with the original aspect ratio preserved, either black bars appear above and below the image ("letterboxing") or the image is cropped at the sides. When 4:3 video is presented on HDTV, the original material is often placed in the

*From Edgar Allen Poe's poem "To Helen."

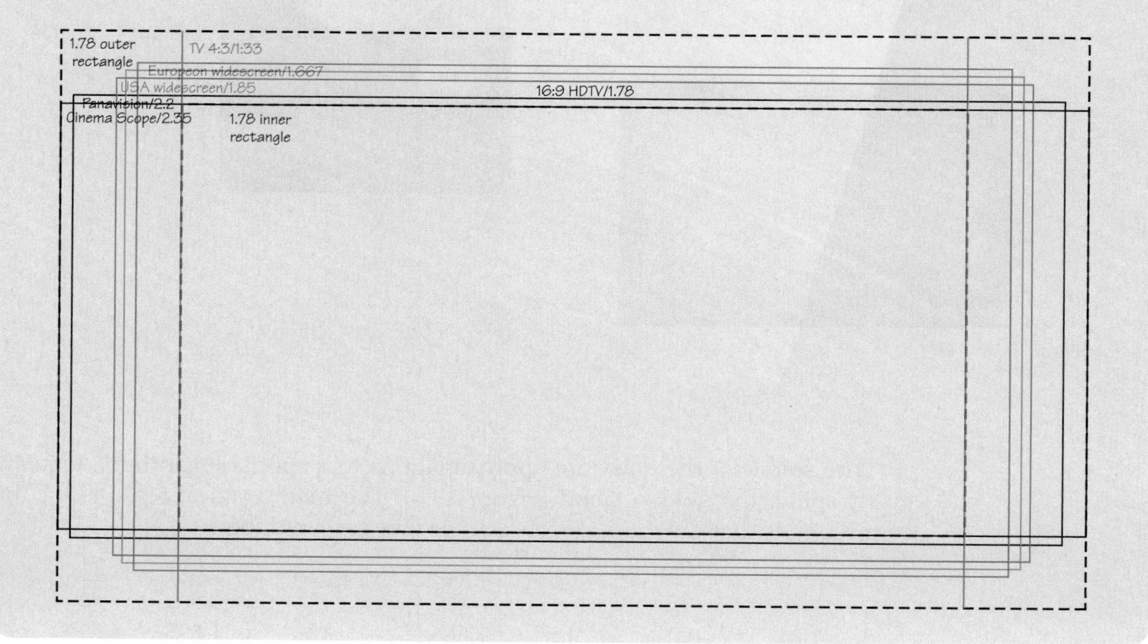

Are We Trying to Reclaim the "Glory That Was Greece"? *(continued)*

middle of the screen and black bars appear on sides of the image ("pillarbox" effect); the alternative is to enlarge the image but crop at top and bottom.

The 1.77 value is—not coincidentally!—the geometric mean of 1.3333 . . . and 2.35. One reason for the final choice of 16:9 is that that ratio is a very close approximation to 1.77 that can be achieved by relatively small integers.

Standard HDTV resolutions of "720 p" (1280 × 720 pixels) and "1080 p" (1920 × 1080 pixels) both have exactly a 16:9 aspect ratio (and pixels with a 1 : 1 aspect ratio).

But there are other video fronts. With a computer, the user can usually choose the display resolution, such as 640 × 480 pixels (4 : 3) or 2560 × 1600 pixels (16:10). Those aspect ratios are not as high as 1.78, but with a "stretching" option available to avoid letterboxing or pillarboxing. However, for the sharpest clarity, the display resolution of the computer video card should be matched to the native resolution of the monitor used, so that the pixels of an image are in 1-to-1 correspondence with screen pixels.

Yet another common display unit is found on smart phones. The iPhone, for example, has an aspect ratio of 1.5. When you are watching a movie on your smart phone, either the image must be compressed horizontally or the edges of each frame are trimmed.

So over time, there has been an evolution of aspect ratios, from 1.31 for roll film and 1.33 for TV to an eventual 1.78 for HDTV—a compromise closer to the Greeks' golden ratio.

© Imageroller/Alamy

© Chuck Eckert/Alamy

© Nathan Griffith/Corbis

The spirals of the daisy are approximations to a special logarithmic (equiangular) spiral, the golden spiral (Figure 19.6). The mathematical reasons for this connection is that the ratios of consecutive Fibonacci numbers

$$\frac{1}{1} \quad \frac{2}{1} \quad \frac{3}{2} \quad \frac{5}{3} \quad \frac{8}{5} \quad \frac{13}{8} \quad \frac{21}{13} \ldots$$

$$1.0 \quad 2.0 \quad 1.5 \quad 1.666\ldots \quad 1.6 \quad 1.625 \quad 1.615\ldots$$

provide alternately under- and overapproximations that narrow in on $\phi = 1.618033\ldots$, and a golden spiral gets wider by a factor of ϕ for each quarter turn.

(The spiral of the chambered nautilus shell of Figure 19.2b [p. 686] is also an equiangular spiral, but the widening factor is approximately 1.3 rather than the golden ratio.)

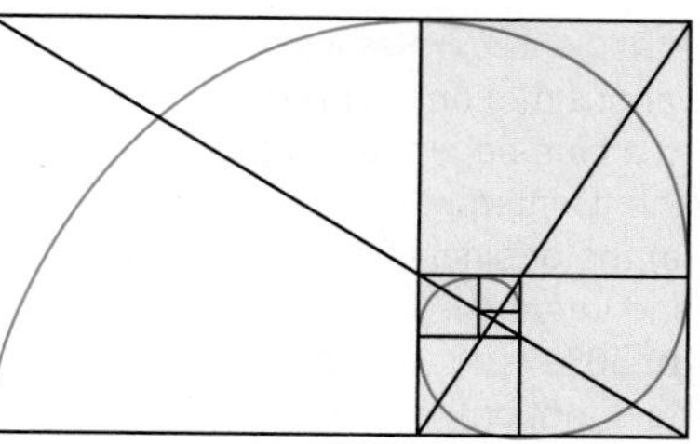

Figure 19.6 A logarithmic spiral determined by a sequence of golden rectangles and corresponding squares.

For reasons that we do not understand, some ratios in the DNA molecule are close to the golden ratio; for example, the length of one full cycle of a strand in the double helix is about 1.62 times its width. Perhaps the most surprising appearance of the golden ratio is in connection with black holes, regions of space in which the gravitational field is so strong that nothing can escape (even light). A rotating black hole loses energy and, up to a point, heats up as it does so; after that point—when the mass of the hole equals its angular momentum times the square root of ϕ—the hole starts to cool down instead.

19.2 Rosette, Strip, and Wallpaper Patterns

The spiral distribution of the seeds in a sunflower head and the spiraling of leaves around a plant stem are instances of *similarity* and *repetition,* two key aspects of symmetry. They also illustrate *balance,* which refers to regularity in *how* the repetitions are arranged. In considering patterns with repetition, we distinguish the individual element or figure of the design (the *motif*) from the *pattern* of the design—*how the copies of the motif are arranged.*

We focus on exploring and classifying the fundamentally different ways in which a flat design can be symmetrical. The ideas that we discuss were used by scientists to discover what crystalline forms are possible. Although there is a limitless number of chemical structures, and of motifs that people can make, what is quite surprising is that there is only a limited number of ways to arrange atoms in a structure or motifs in a design in a symmetrical way.

How can we enumerate all the ways that designs can be put together without counting the designs themselves? The key mathematical idea is to look at what you can do to the pattern without changing its appearance.

Rigid Motions

Mathematicians describe various kinds of symmetry by using the geometric notion of a **rigid motion,** also known as an **isometry** (which means "same size"). A transformation or "motion" of the plane is a mapping of the entire plane into itself. A rigid motion preserves the distances between every pair of points. You can imagine picking up the plane and moving it as a whole, perhaps rotating it, possibly flipping it over—but *without changing its size or shape.* For example, moving every point 1 unit in the same direction constitutes a rigid motion.

Rigid Motion DEFINITION

A **rigid motion** is a transformation of the entire plane that preserves the size and shape of figures. In particular, any pair of points is the same distance apart after the transformation as before.

Figure 19.7 shows the results of how various motions affect the rectangle and its interior in Figure 19.7a. In Figure 19.7b, each side is shrunk by 50%—which is not a rigid motion because the size of the rectangle changes. For

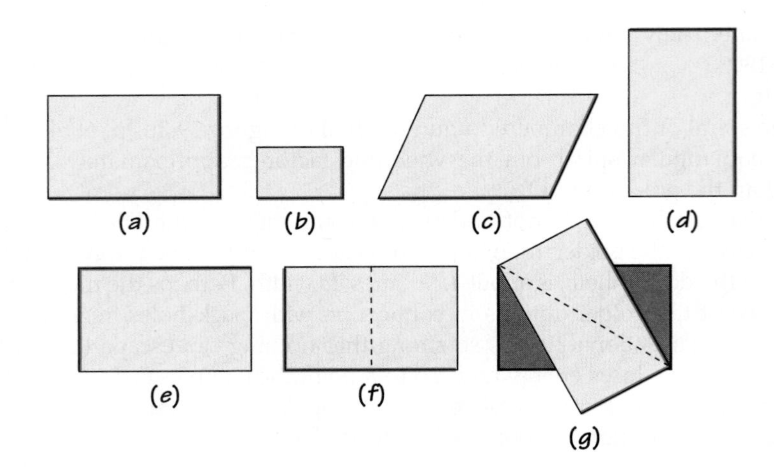

Figure 19.7 Results of various motions applied to a blue-edged rectangle and its interior: (a) the original rectangle and interior; (b) 50% reduction (not a rigid motion); (c) shearing (not a rigid motion); (d) quarter-turn; (e) half-turn; (f) reflection along the vertical line down the middle; (g) reflection along a diagonal line.

Figure 19.7c, we shear ("squash") the rectangle; again, this is not a rigid motion because the shape of the rectangle changes. In Figure 19.7d, we rotate the plane 90° (a quarter-turn) counterclockwise around the center of the rectangle: This is a rigid motion. Similarly, in Figure 19.7e, rotating the plane by 180° (a half-turn) is a rigid motion.

In Figure 19.7f, we reflect the rectangle along a vertical mirror down the middle. The right and left halves exchange places.

Figure 19.7g shows the result of reflecting across a diagonal of the rectangle. All reflections and all rotations are rigid motions. So are all **translations,** which move every point in the plane a certain distance in the same direction.

The only remaining kind of rigid motion of the plane is a combination of reflection and translation. **Glide reflection** is the kind of pattern that your footprints make as you walk: Each successive element of the design (footprint) is a reflection of the previous one (Figure 19.8). The motion combines translation ("glide") with a reflection across a line parallel to the direction of the translation.

Glide Reflection DEFINITION

A **glide reflection** is a translation (= glide) combined with alternating reflection in a line parallel to the translation direction.

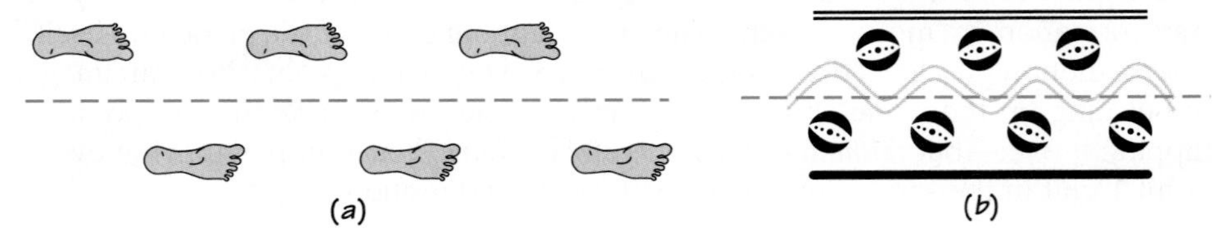

Figure 19.8 Glide reflection of (a) footprints; (b) design elements on a pot from San Ildefonso Pueblo, in New Mexico.

Any rigid motion of the plane must be one of the following:

- Reflection (across a line)
- Rotation (around a point)
- Translation (in a particular direction)
- Glide reflection (across a line)

Performing one rigid motion after another results in a rigid motion that (surprisingly) must be one of the four types that we have just explored.

Preserving the Pattern

In terms of symmetry, we are especially interested in rigid motions like those of Figures 19.7e and 19.7f that **preserve the pattern**—that is, ones for which the pattern looks exactly the same, *with all the parts appearing in the same relative places,* after the motion is applied.

You might enjoy thinking of applying these motions as "The Pattern Game": You turn your back, I apply a transformation, and then you turn back and see if you can tell whether anything is changed.

The 90-degree rotation of Figure 19.7a into Figure 19.7d does not preserve the pattern. The moved rectangle doesn't fit exactly over the original rectangle. On the other hand, the 180-degree rotation in Figure 19.7e does preserve the pattern. It's true that the top of the original rectangle is now on the bottom of the transformed version, but you can't tell. A rotation by any multiple of 180° also would preserve the pattern.

Similarly, the reflection across the vertical line in Figure 19.7f preserves the pattern, while the one in Figure 19.7g, along a diagonal, does not. Spotlight 19.5 discusses possible biological consequences of **reflection symmetry** or imperfections in it.

The pattern of footsteps in Figure 19.8a is not preserved under reflection along the direction of walking; there is not a left footprint directly across from a right footprint. The pattern is preserved under a glide reflection along the direction of

SPOTLIGHT 19.5

"Strive Then to Be Perfect"

Is there no such thing as objective and universal beauty, as claimed by American feminist Naomi Wolf in her book *The Beauty Myth*?

Stand in front of a mirror and look at yourself. Do your left and right sides look exactly symmetrical? What about the part in your hair, freckles on your face, evenness of your shoulders, bending of your ears?

Symmetry may be a proxy for fitness. Symmetrical racehorses tend to run faster; male lions with lopsided facial whisker-spot patterns die younger. The more symmetrical a flower is, the more nectar it produces, making it a better food source for pollinating insects; and correspondingly, insects prefer symmetrical flowers, giving such flowers a better chance of being pollinated.

Perhaps because of association with fitness, symmetry may affect mate selection among animals. Female zebra finches prefer males with symmetrical leg bands. Fruit flies and female barn swallows prefer males with symmetrical tails; a particular parasite can lead to an uneven tail.

What about people? Both male and female Britons, as well as Tanzanian hunter-gatherers, find facial symmetry more attractive than asymmetry.

Michelangelo's *David*
(*Roger Antrobus/Getty Images.*)

Perfectly symmetrical female faces that are computer- generated from composites of individual photos appear more attractive to men than photos of actual women's faces.

walking, as well as by a translation of two steps, or of four steps, and so on—but not by a translation of one step.

We classify a pattern by determining which rigid motions preserve it; they are the **symmetries of the pattern.**

You may think of a pattern as a recipe for repeating a figure (motif) indefinitely. Of course, any pattern in nature or art has only a finite number of copies of the figure. If the recipe for repetition is clear, we can imagine that we are looking at just a part of a pattern that extends indefinitely.

Patterns in the plane can be divided into those that have indefinitely many repetitions in

- no direction—the **rosette patterns** (sometimes called *band patterns* or *frieze patterns*),
- exactly one direction (and its reverse)—the **strip patterns,** or
- more than one direction (and their reverses)—the **wallpaper patterns**

Rosette Patterns

A rosette is an ornament or badge made to resemble a flower, and the term *rosette pattern* is used to describe the possible symmetries of such an ornament or a single flower. The repetition aspect of symmetry consists of the repetition of the petals around the stem. Translations and glide reflections do not come into play. The pattern is preserved under a rotation by certain angles corresponding to the number of petals. There may or may not be reflections that preserve the pattern, depending on whether the petal itself has reflection symmetry. Most flowers do (Figure 19.9a), but some do not.

An everyday example of the rosette pattern—a human-made one—that does not have reflection symmetry is a pinwheel (Figure 19.9b). If there is no reflection symmetry, the motif of the pattern (the element that is repeated) is an entire petal. If there is reflection symmetry, the motif is just half a petal because the entire pattern can be generated by rotation and reflection of a half petal. The fact that these are the only possibilities is sometimes called *Leonardo's theorem,* after Leonardo da Vinci, who, in the course of planning the design of churches, needed to decide if chapels and niches could be added without destroying the symmetry of the central design.

(a)

(b)

Figure 19.9 (a) A flower; each petal has reflection symmetry. *(Photodisc)* (b) A pinwheel with seven symmetrical "leaves," each asymmetric, hence pattern *c7* *(Ryan McVay/Lifesize/Getty Images)*

Leonardo realized that there are two different classes of rosettes, the ones without reflection symmetry (*cyclic rosettes*) and those with it (*dihedral rosettes*) (see Figure 19.9). The respective notations for the patterns are *cn* and *dn*, where *n* is the number of times that the rosette coincides with its original position in one complete turn around the center. A cyclic pattern has no lines of reflection symmetry, while the dihedral pattern *dn* has *n* different lines of reflection symmetry. The flower in Figure 19.9a with its 34 petals (a Fibonacci number!) has dihedral pattern *d34* (if we idealize it) because each petal has reflection symmetry, while the pinwheel in Figure 19.9b has pattern *c7*.

Strip Patterns

We illustrate the different kinds of strip patterns, and their "ingredient" symmetries, with patterns in the art of the Bakuba people of the Democratic Republic of the Congo, who are noted for their fascination with pattern and symmetry (see Spotlight 19.6).

SPOTLIGHT 19.6

Patterns Created by the Bakuba People

Among the Bakuba people of the Democratic Republic of the Congo (see the shaded area of the map), it is considered an achievement to invent a new pattern, and every Bakuba king had to create a new pattern at the outset of his reign. The pattern was displayed on the king's drum throughout his reign and, for some kings, on his dynastic statue.

When missionaries first showed a motorcycle to a Bakuba king in the 1920s, he showed little interest in it. But the king was so enthralled by the novel pattern the tire tracks made in the sand that he had it copied and gave it his name.

Source: Adapted from Jan Vansina, *The Children of Woot*, University of Wisconsin Press, Madison, 1978, p. 221.

Two women with raffia cloths from the Bakuba village of Mbelo, July 1985. *Left:* Mpidi Muya with embroidered raffia (a kind of fiber) cloth. *Right:* Muema Kenye with plush and embroidered raffia cloth. *(Dorothy K. Washburn.)*

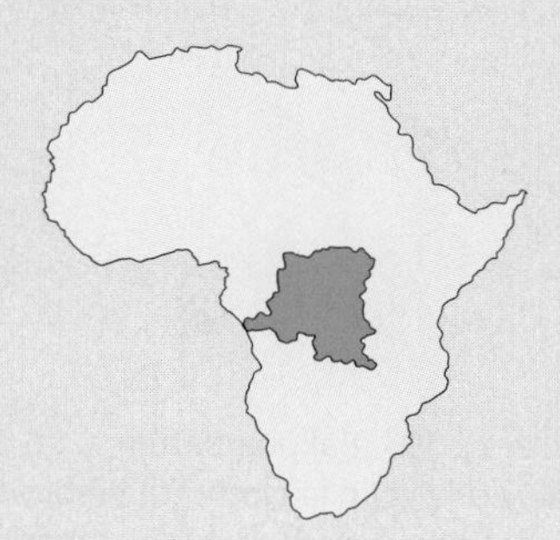

The pattern made by tire tracks fascinated the Bakuba people. *(Travis Amos.)*

All the strip patterns offer repetition and **translation symmetry** along the direction of the strip. For simplicity, we position patterns horizontally. For example, in Figure 19.10b, what you can imagine as eyes and a nose (if you rotate the page clockwise) is constantly shifted (translated) along the pattern by the same amount each time.

The pattern may have no other rigid motions that preserve it apart from translation, as in Figure 19.10a.

The other simplest rigid motion to check is reflection across a line. For a strip pattern, the horizontal center line of the strip may be a reflection line, as in Figure 19.10b; we say that the pattern has symmetry across a horizontal line. There may instead be reflection across a *vertical* axis, such as the vertical lines through or between the *V*'s in Figure 19.10c.

What kind of **rotational symmetry** can a strip pattern have? The only possibility for a strip pattern is a rotation by 180° (a half-turn), because any other angle won't even bring the strip back into itself. (We don't count rotations of 360° or integer multiples [full turns], because any pattern is preserved under these.) Figure 19.10d shows a strip pattern that is unchanged by a 180-degree rotation about any point at the center of the small crosshatched regions.

What about glide reflections? A row of alternating p's and b's has glide reflection, as shown in the illustration below, in which the dotted line through the middle of the pattern is the line in which reflection takes place:

Glide: p p p p p p p p p

Glide reflection: p---------b---------p---------b---------p---------b---------p---------b---------p

For glide reflection, a p is translated as far as the next b and is then reflected upside down. Figure 19.10e shows a Bakuba pattern whose only symmetry (except for translation) is glide reflection.

Having examined symmetries on strip patterns, we can ask: What *combinations* of the four are possible? It turns out that apart from the five kinds of patterns we have already seen, there are only two other possibilities: We can have vertical line reflection, half-turns, and glide reflection, either with horizontal line reflection (Figure 19.10g) or without (Figure 19.10f).

Figure 19.10 Bakuba patterns. (a) Carved stool; (b) pile cloth; (c) pile cloth; (d) embroidered cloth; (e) embroidered cloth; (f) carved back of wooden mask; (g) carved box.

Mathematical analysis reveals the following.

There Are Only Seven Ways to Strip RULE

There are only seven ways to repeat a pattern along a strip

That this number is so small is quite surprising because there are myriad different design elements (motifs). Two designs may look entirely different yet share the same pattern of reproducing their design elements.

Wallpaper Patterns and Crystal Structures

So far, we have classified the patterns that have no translation repetition (the rosette patterns) and those with repetition in one direction (the strip patterns). What about repetitions in more than one direction—say, in two different directions across a plane? It turns out that there are exactly 17 ways to do so, called *wallpaper patterns*. We give illustrations, notation, and a flowchart in Spotlight 19.7.

SPOTLIGHT 19.7

The 17 Wallpaper Patterns

There are exactly 17 wallpaper patterns. Here we give an example of each, together with a flowchart for identifying them. Crystallographers have standard notations and abbreviations for the patterns. The full notation consists of four symbols:

1. The first symbol is *c* (for "centered") if all rotation centers lie on the reflection lines, or *p* (for "primitive") otherwise.
2. The second symbol indicates rotational symmetry. It is either *1, 2, 3, 4,* or *6,* corresponding to rotational symmetry of, respectively, 360°, 180°, 120°, 90°, and 60°. The symbol is the largest applicable number. For example, if symmetries of 360°, 120°, and 60° are present, the symbol is *6*.
3. The third symbol is either *m, g,* or *1,* corresponding to the presence of "mirror," "glide," or no reflection symmetry.
4. The fourth symbol (*m, g,* or *1*) is for describing symmetry relative to an axis at an angle to the symmetry axis of the third symbol.

(*Note:* The patterns *p31m* and *p3m1* are exceptions to this scheme.)

Below each pattern illustration, we give both the standard abbreviation (on top) and the full notation (below).

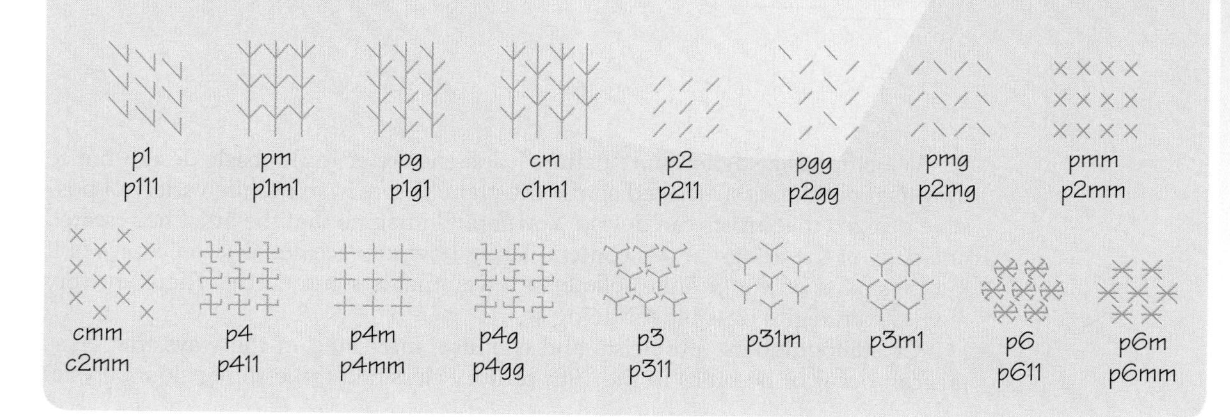

The 17 Wallpaper Patterns *(continued)*

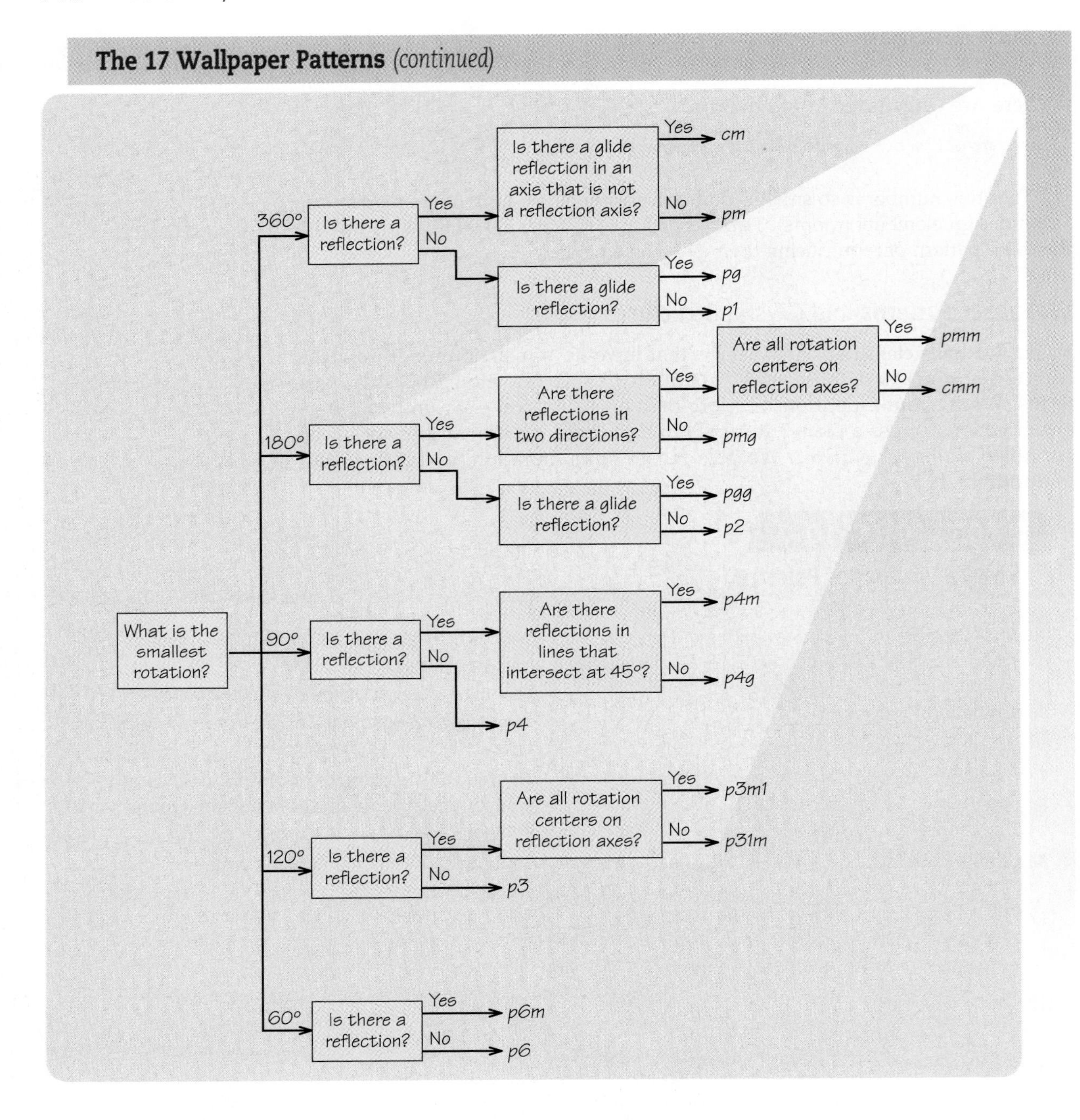

We emphasize again that "pattern" does not refer to the basic design but to how its repetition is structured across the plane. There is an infinite variety of possible designs that artists can devise. You should imagine that the artist has created one copy of the design and is contemplating how to place equal-sized copies of it in other parts of the (infinite) plane, in a way that is symmetrical. There are very few (17) strategies possible for doing so.

Crystallographers (physicists and chemists interested in the ways that crystals can occur or be built) in the 19th century classified three-dimensional crystal structures in terms of combinations of symmetry elements. They proved—after several years of coming up with different totals!—that there are exactly 230 patterns for crystals. Mathematicians have refined the classification of patterns further to take into account colors that are repeated in a symmetrical way.

19.3 Notation for Patterns

It's useful to have a standard notation for patterns. **Crystallographic notation** is commonly used. For the strip patterns, it consists of four symbols: (an example is *pma2*):

1. The first symbol is always a *p*, which indicates that the pattern repeats (is "periodic") in the horizontal direction.
2. The second symbol is *m* if there is a vertical line of reflection. Otherwise, it is *1*.
3. The third symbol is
   - *m* (for "mirror"), if there is a horizontal line of reflection (in which case there is also glide reflection)
   - *a* (for "alternating"), if there is a glide reflection but no horizontal reflection
   - *1*, if there is no horizontal reflection or glide reflection
4. The fourth symbol is *2*, if there is half-turn rotational symmetry; otherwise, it is *1*.

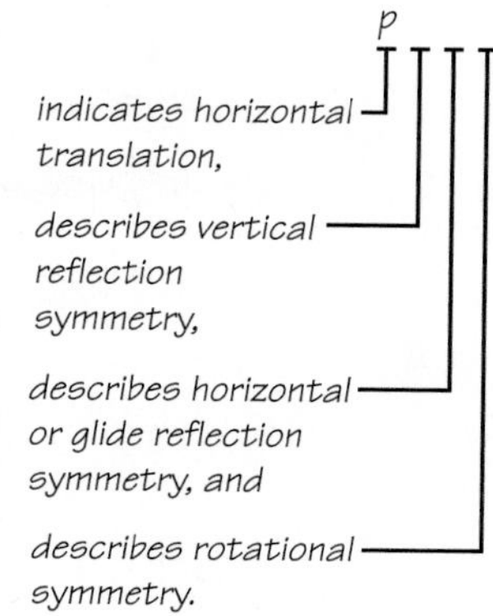

Figure 19.11 Scheme for strip pattern notation.

A *1* always means that the pattern does *not* have the symmetry corresponding to that position. In the notation:

EXAMPLE 1
Bakuba Patterns

We use the flowchart of Figure 19.12 on page 704 to analyze some of the Bakuba patterns of Figure 19.10.

SOLUTIONS Figure 19.10a does not have a vertical reflection, so we branch right, and the pattern notation begins to take shape as *p1_ _*. The figure has neither a horizontal reflection nor a glide reflection, so we branch right again, filling in the third position in the notation to get *p11_*. A half-turn preserves part of but not all the pattern, so we conclude that we have a *p111* pattern.

Figure 19.10b does not have vertical reflection, so we branch right to *p1_ _*. The figure does have horizontal reflection, so we branch left and left, concluding that the pattern is *p1m1*.

Figure 19.10f has vertical reflection, so we branch left to *pm_ _*. The figure does not have horizontal reflection, so we branch right but cannot yet fill in the third symbol. The figure does have a half-turn symmetry (and glide symmetry), with center on the middle of the three lines between any pair of closest triangles. So the pattern is *pma2*.

Remember the Bakuba people's fascination with tire treadmarks? Apart from esthetic value, certain symmetries are important for practical purposes. The Museum

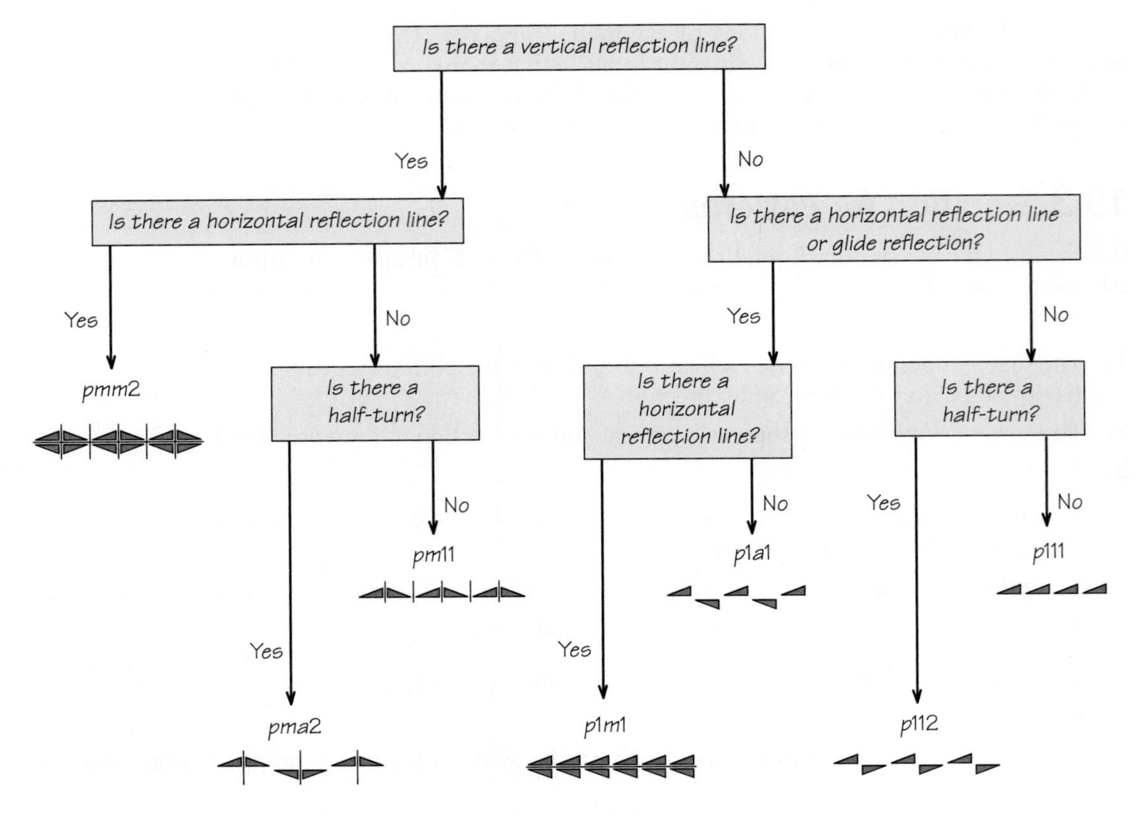

Figure 19.12 A flowchart for identifying the seven strip patterns and classifying them according to crystallographic notation.

of Transport in Glasgow, Scotland, includes all kinds of vehicle tires. However, only five of the seven strip patterns appear among treads of all the tires there. Examining Figure 19.12, can you guess which two patterns do not appear, what they have in common, and the practical reason why they are not used in tire treads?

Imperfect Patterns

In applying these classification schemes to patterns on real objects, we need to take into account that the pattern may not be perfectly rendered. Also, patterns not on flat surfaces—for example, the pattern around the rim of a bowl or around the body of a jar—require some latitude in interpretation.

EXAMPLE 2
Patterns on Pueblo Pottery

The pitchers in Figure 19.13 are from a thousand-year-old Pueblo site at Starkweather Ruin near Reserve, New Mexico. Consider the patterns on the bodies of the pitchers, which continue on the back sides. Let's suppose that they could be unwrapped and continued as strip patterns, but we'll disregard the patterns on the spouts and handles.

Figure 19.13 Reserve black-on-white pitchers from the Pueblo II "horizon" (culture) (A.D. 900–1100), excavated 1935–1936 from Starkweather Ruin by Professor Paul H. Nesbitt and students from Beloit College. These pots are no longer available for study. In 2005, they were returned to representatives of Native American tribes who claim descent from the Pueblo culture and who requested return of the artifacts for reburial. (*Courtesy of Logan Museum of Anthropology, Beloit College, photos by Paul J. Campbell.*)

We immediately come up against the question of the perfectness of the patterns. In Figure 19.13a, the "teeth" (represented by the zigzagging of lightning bolts) in the left design element on the main body are "sharper" than those on the right. Is this lack of pattern, or just lack of perfection in executing one? For our analysis, we opt for the latter.

Similarly, what are we to make of the diagonal lines on the pitcher in Figure 19.13b? In the narrowest interpretation, these lines are part of the pattern, and rigid motion that is to qualify as a symmetry of the pattern must preserve them. More liberally, we may consider the lines as a kind of shading, a way to make the region appear gray. Indeed, to an observer at a distance, that is the effect of the lines.

For the pattern on the body of the pitcher in Figure 19.13c, we notice that the jagged white line in the design element on the left has three "steps," while the one on the right has four. If we were really strict, we would decide that the two are different design elements. But there is a similarity that we do not want to deny. We attribute the variations to artistic license and, for our purposes, consider the two jagged lines to be the same.

SOLUTIONS We follow the flowchart in Figure 19.12 and get the following for each of these pitchers:

- Figure 19.13a: Is there a vertical reflection? *No*. Is there a horizontal reflection or glide reflection? *No*. Is there a half-turn? *No*. Hence the pattern is *p111*.
- Figure 19.13b (narrow interpretation of the diagonal lines): Is there a vertical reflection? *No*. Is there a horizontal reflection or glide reflection? *No*. Is there a half-turn? *Yes* (around the center of each cross). The pattern is *p112*.
- Figure 19.13b (liberal interpretation—diagonal lines as shading, their direction doesn't have to be preserved): Is there a vertical reflection? *Yes* (on a vertical line through the center of a cross). Is there a horizontal reflection? *Yes* (through the center of a cross). The pattern is *pmm2*.
- Figure 19.13c: Is there a vertical reflection? *No*. Is there a horizontal reflection or glide reflection? *No*. Is there a half-turn? *Yes* (around the center of each jagged white line). The pattern is *p112*. (This pitcher has the interesting feature that the patterns on the neck and the body are mirror images of each other.)

Women made the pots at Starkweather. They strongly preferred the symmetry of half-turns. Very few of the pots have any reflection symmetry, either reflection or glide. The avoidance of reflection symmetry was a consistent feature of pottery of the indigenous peoples of the Western Hemisphere.

19.4 Symmetry Groups

We mentioned earlier that the key mathematical idea about detecting and analyzing symmetry is to look not at the motifs of a pattern but at its symmetries, the transformations that preserve the pattern.

The symmetries of a pattern have some notable properties:

- If we combine two symmetries by applying first one and then the other, we get another symmetry.
- There is an identity, or "null," symmetry that doesn't move anything but leaves every point of the pattern exactly where it is.
- Each symmetry has an inverse, or "opposite," that undoes it and also preserves the pattern. A rotation is undone by an equal rotation in the opposite direction, a reflection is its own inverse, and a translation or glide reflection is undone by another of the same distance in the opposite direction.
- In applying a number of symmetries one after the other, we may combine consecutive ones without affecting the result ("associativity"). For example, if we have symmetries A followed by B followed by C, we can do either of the following: first combine A with B, apply that symmetry, and then apply C; or first apply A and then follow that by applying the combination of B with C. That is, we can "associate" adjacent symmetries, but we must observe the overall order (A, B, C) in which they occur.

EXAMPLE 3
The Symmetries of a Rectangle

Consider the rectangle of Figure 19.14. Its symmetries, the rigid motions that bring it back to coincide with itself (even as they interchange the labeled corners), are as follows:

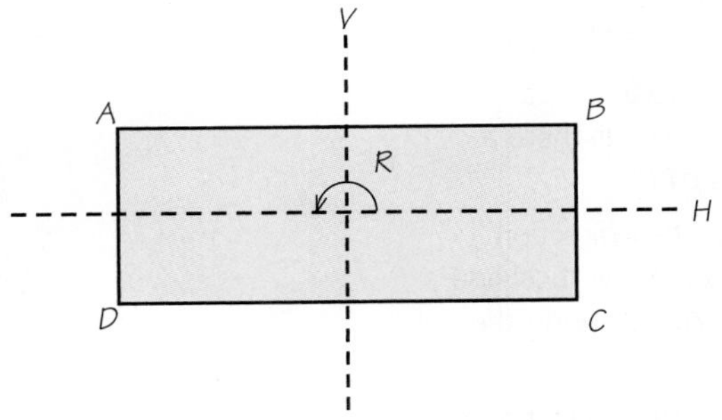

- The identity symmetry I, which leaves every point where it is;
- A 180-degree (half-turn) rotation R around its center;
- A reflection V in the vertical line through its center;
- A reflection H in the horizontal line through its center.

You should convince yourself that the symmetries fulfill the four properties above.

V
A
B
R
H
D
C

Figure 19.14 A rectangle, with reflection symmetries and 180-degree rotation symmetry marked.

SOLUTION

- Combining any pair by applying first one and then the other is equivalent to one of the others. It's handy to have a notation for this combining; if we apply V first and then H, we will write the result as $V \circ H$—in other words, *we apply the sequence of actions from left to right.* You can check that the result is the same as applying R; that is, $V \circ H = R$. Check this by following where the corner A goes to under the symmetries. Practice combining symmetries by making yourself a "multiplication table" of them.

- The element I is an identity element.
- Each element is its own inverse. For example, rotating by 180°, then doing it again, gets the rectangle back to coincide with itself; in our notation, $R \circ R = I$.
- Try some examples to verify that associativity holds. For instance, check that $R \circ H \circ V = (R \circ H) \circ V = R \circ (H \circ V)$. In other words, applying R then H then V, we get the same result if we combine the first two and then apply the third, or if we apply the first one and then apply the combination of the second two.

The four properties of symmetries of an object are common to many kinds of mathematical objects. The properties characterize what mathematicians call a *group*. Various familiar collections of numbers, together with operations on them, form groups.

EXAMPLE 4
A Group of Numbers

The positive real numbers form a group under multiplication:

SOLUTION

- Multiplying two positive real numbers yields another positive real number.
- The positive real number 1 is an identity element.
- Any positive real number x has an inverse $(1/x)$ in the collection.
- In multiplying several numbers, it doesn't matter if we first multiply some adjacent pairs of numbers; that is, it doesn't matter how we group or parenthesize the multiplication. For instance, $2 \times 3 \times 4$ is equal to $2 \times (3 \times 4) = 2 \times 12$ and also to $(2 \times 3) \times 4 = 6 \times 4$.

Group — DEFINITION

A **group** is a collection of elements $\{A, B, \ldots\}$ and an operation $\circ$ between pairs of them such that the following properties hold:

- *Closure:* The result of one element operating on another is itself an element of the collection ($A \circ B$ is in the collection).
- *Identity element:* There is a special element I, called the *identity element*, such that the result of an operation involving the identity and any element is that same element ($I \circ A = A$ and $A \circ I = A$).
- *Inverses:* For any element A, there is another element, called its *inverse* and denoted A^{-1}, such that the result of an operation involving an element and its inverse is the identity element ($A \circ A^{-1} = I$ and $A^{-1} \circ A = I$).
- *Associativity:* The result of several consecutive operations is the same regardless of grouping or parenthesizing, provided that the consecutive order of operations is maintained: $A \circ B \circ C = A \circ (B \circ C) = (A \circ B) \circ C$.

EXAMPLE 5
A Group of Non-Numbers

With all your experience with arithmetic, numbers are concrete to you, even if thinking of them in terms of a group is not. Here, we look at a very simple "abstract" group. The group is a collection of just three elements {A, B, C}, and it is convenient to show how the operation ∘ behaves by giving a table of its results as follows:

•	A	B	C
A	A	B	C
B	B	C	A
C	C	A	B

The table is organized so that, for example, we find the result of A ∘ B by looking in the row for A and the column for B, finding B. So A ∘ B = B. Similarly, C ∘ B = A. We confirm that indeed this set is a group under the operation.

SOLUTION Since all the entries in the table are from {A, B, C}, the set is closed under the operation. You should identify which element serves as an identity element. What is the inverse to A? To B? To C? To check associativity would require checking the results of all possible products X ∘ Y ∘ Z, where each of X, Y, and Z can be any of A, B, or C. We won't go to that (tedious) length, but you should check just one example. For instance, (A ∘ B) ∘ C = B ∘ C = A while A ∘ (B ∘ C) = A ∘ A = A. Also, can you see why there are $3^3 = 27$ products to check?

This particular abstract group can be interpreted concretely in several ways. One interpretation is in terms of an equilateral triangle.

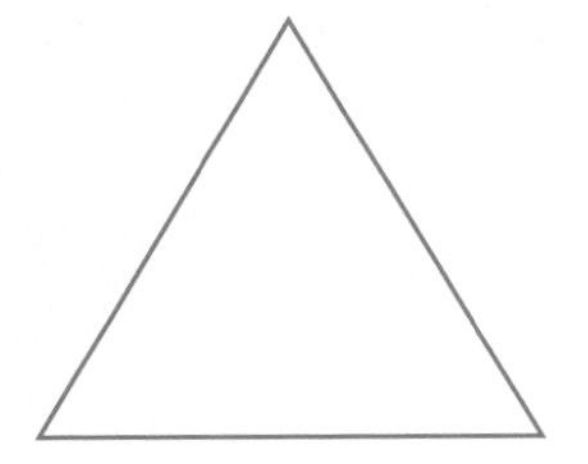
An equilateral triangle.

Each of A, B, and C is a rotation of the triangle about its center. A is a rotation by 0°, B is a rotation clockwise by 120° (one-third of a complete turn), and C is a rotation counterclockwise by 120°. The operation ∘ just gives the result of doing one rotation followed by another. For example, B ∘ C is first to rotate the triangle 120° clockwise, then rotate it 120° counterclockwise—which leaves it as if it had not rotated at all; that is, as rotated by 0°. Hence, B ∘ C = A.

We have in fact explored here some of the symmetries of an equilateral triangle. (What other symmetries does an equilateral triangle have? Think of it as a rosette pattern.)

Symmetry Group of a Pattern DEFINITION

The symmetries that preserve a pattern form the **symmetry group of the pattern**.

EXAMPLE 6
Group Theory in Your Room, or More Than Once Upon a Mattress

Have you ever turned a mattress? The purpose of this usually seasonal practice is to even out lumps and sags and make the mattress last longer. For this example, make yourself a "mattress" from a (non-square) rectangular sheet of paper. (It's a lot easier to flip a piece of paper than a real mattress!) Label the initial position of the "mattress": write "UP" in the middle of the top surface of the paper and label the edges in clockwise order "TOP," "RIGHT," "BOTTOM," and "LEFT." Then turn the sheet over so that the top edge is again at the top. Label this surface of the paper "BOTTOM" in the middle and label the edges in clockwise order "TOP," "RIGHT," "BOTTOM," and "LEFT." Now turn the sheet over again to the initial position.

There are various ways to "turn" a mattress, but they all require that the mattress fit back on the bed. What are those ways? We'll designate the turning over of the paper that you just did as follows:

- *flip* (turns it over, left and right switch) and denote it by *F*. A flip amounts to a 180-degree rotation around an axis running from bottom to top of the mattress. Satisfy yourself that there are only two other ways to turn the mattress:
- *rotation* (up side stays up, top and bottom switch), denoted by *R*, which amounts to a 180-degree rotation of the mattress as seen from above, i.e., around a vertical axis through the center of the mattress; and
- *toss* (turn it over and switch top and bottom), denoted by *T*, which corresponds to a 180-degree rotation of the mattress around a horizontal axis across the middle of the short side of the mattress.

How many different positions are there for the mattress? Satisfy yourself that there are really only four positions for the mattress that fit it back on the bed: the results of the turns *F, R,* and *P*—together with the "un-turn," which we denote by *I*. These four turns together form a group {*I, F, R, P*}. The group operation is to perform one turn followed by another.

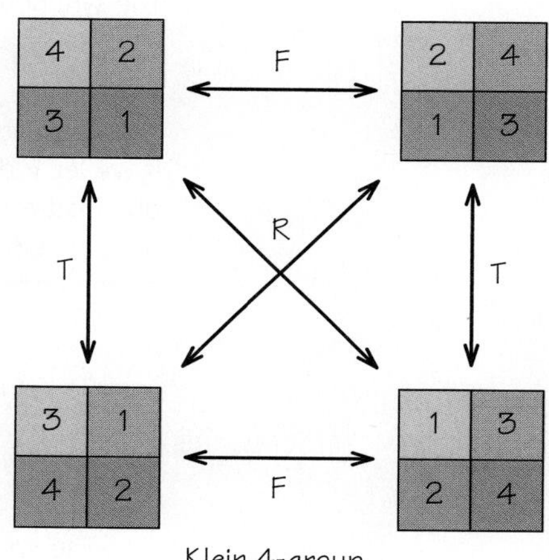

Klein 4-group

Performing one of the turns, followed by a subsequent one in the next season, actually gives the same result as a single turn; for example, doing *R* followed by *T* puts the mattress in the same position as doing just *F*. This fact is the *closure* property of the group.

The *identity element* of the group is the "un-turn" *I*. Each turn is its own *inverse*: performing it twice in a row puts the mattress back in the initial position *I*.

Associativity of the turns is true, though it is tedious to check, but this necessary property of a group indeed holds.

The same structure of operations holds for turning your pillow. Any group with four elements, each of which is its own inverse, has basically the same abstract structure, called the *Klein 4-group* (after Felix Klein [1849–1925], who classified geometries by their symmetry groups).

EXAMPLE 7
Symmetry Groups of Strip Patterns

Each of the strip patterns of Figure 19.10 has a different group of symmetries. What do they have in common, and how do they differ?

SOLUTIONS The pattern of Figure 19.10a is preserved only by translations. If we let *T* denote the smallest translation to the right that preserves the pattern, then the pattern is also preserved by $T \circ T$ (which we write as T^2), by $T \circ T \circ T = T^3$, and so forth. Although the pattern looks the same after each of these translations by different distances, we can tell them apart if we number each copy of the motif and observe which other motif it is carried into under the symmetry.

For instance, T^2 takes each motif into the motif two to the right. The symmetry T has an inverse T^{-1} among the symmetries of the pattern: the smallest translation to the left that preserves the pattern. Moreover, $T^{-1} \circ T^{-1}$ (which we write as T^{-2}), $T^{-1} \circ T^{-1} \circ T^{-1} = T^{-3}$, and so forth are also symmetries. The entire collection of symmetries of the pattern is

$$\{\ldots, T^{-3}, T^{-2}, T^{-1}, I, T, T^2, T^3, \ldots\}$$

From this listing, it is natural to think of the identity I as being T^0. All the strip patterns are preserved by translations, so the symmetry group of each includes the subgroup consisting of all translations in this list. We say that the group is **generated by** T, and we write the group as $\langle T \rangle$, where between the angle brackets, we list symmetries that, in combination, produce all the group elements.

The symmetry group of Figure 19.10e includes, in addition, a glide reflection G and all combinations of the glide reflection with the translations. Doing two glide reflections is equivalent to doing a translation, which we express as $G^2 = T$. The glide is only "half as far" as the shortest translation that preserves the pattern. Check that $G \circ T = T \circ G$. The symmetry group of the pattern is

$$\{\ldots, G^{-3}, G^{-2} = T^{-1}, G^{-1}, I, G, G^2 = T, G^3, \ldots\} = \langle G \rangle$$

The pattern of Figure 19.10c is preserved by vertical reflections at regular intervals. If we let V denote reflections at a fixed particular location, the other reflections can be obtained as combinations of V and T. To get a handle on what each of the symmetries does, it helps to make a "simplified" copy of the strip (we use V's), number fixed positions on the page, and identify individual copies of the V's with letters, as in the following:

1	2	3	4	5
V_a	V_b	V_c	V_d	V_e

The symmetries move the V's among the numbered positions. Let V be the reflection across the vertical line through the middle of position 3, and let T be the translation that moves each V one square to the right. To familiarize yourself with the symmetries, write out the result for each of V, T, and VT (V followed by T). Where does V_e end up? (For convenience, we can omit the operation sign between the two symmetries.)

The symmetry group of the pattern, the list of all of the symmetries, is

$$\{\ldots, T^{-3}, T^{-2}, T^{-1}, I, T, T^2, T^3, \ldots;$$
$$\ldots, T^{-3}V, T^{-2}V, T^{-1}V, V, TV, T^2V, T^3V, \ldots\}$$

This group is notable because not all its elements satisfy the commutative property that $A \circ B = B \circ A$, which you are used to for numerical operations ($a + b = b + a$; $a \times b = b \times a$). In fact, we do not have $VT = TV$, but instead $VT = T^{-1}V$. Verify this fact by working out the effect of $T^{-1}V$, using your simplified strip from above, and compare with what you got for VT earlier.

We can express this group compactly as $\langle T, V \mid VT = T^{-1}V \rangle$, where we list two symmetries that generate the group and indicate relations that hold between them.

We have made a transition from thinking about patterns in geometrical terms to reasoning about them in algebraic notation—in effect, applying one branch of mathematics to another. This kind of cross-fertilization is characteristic of contemporary mathematics.

The concept of a group is a fundamental one in the mathematical field of abstract algebra. The generality ("abstractness") is exactly why groups and other algebraic structures arise in so many applications, in areas ranging from crystallography, quantum physics, and cryptography, to error-correcting codes (see Chapters 16 and 17) and anthropology (describing kinship systems).

19.5 Fractal Patterns and Chaos

We noted earlier that similarity and repetition are key aspects of symmetry, as are balance and proportionality. In most of our examples, the repetitions of a motif have been at the same size. Exceptions were the chambers of the nautilus in Figure 19.2b (on p. 686) and the varying sizes of leaves and seeds in plants that feature the spiral pattern of phyllotaxis (Figures 19.1 and 19.2a, on p. 686). These exhibit a kind of "proportion," or numeric symmetry— symmetry with changes of scale.

Another example of similarity with changes of scale are the nested dolls (known as "matrioshka") shown in Figure 19.15. They feature a linear scaling factor (see Section 18.1 on p. 644) between one doll and another. Each part of one doll (face, arm, and so forth) has the same proportion (scaling factor) to the corresponding part of a second doll.

Figure 19.15 Nested matrioshka dolls from Russia exhibit symmetry at different scales. *(Photodisc Green/Getty Images.)*

Fractals

Fractals are another example of symmetry in which linear scaling is used. The word **fractal** was invented in 1975 by Benoit Mandelbrot from the Latin word *fractus* meaning "broken into fragments" (of varied sizes), from which we get *fragment* and also *fracture* and *fraction*. Mandelbrot defined a fractal in strict mathematical terms that we formulate more informally as follows:

Fractal DEFINITION

A **fractal** is a pattern that exhibits similarity at ever finer scales.

What this means is that no matter how closely we zoom in, we still see the same pattern.

We show various fractals in Figure 19.16. Figure 19.16a looks to us like an orchid with pronounced "bee guides" to the pollen. With its vertical mirror line, the overall pattern has *d1* rosette symmetry. However, the basic motif of the lacy wings is repeated at an infinite number of scales. In Figure 19.16b, the "suckers" on the "tentacles" appear in smaller and smaller sizes as the "tentacles" wind their way toward the point at the center. The pattern in Figure 19.16c has overall rosette symmetry of type *c2*, but the "seahorse" motif, with two large "seahorses" foot-to-foot in the center, is repeated in diminishing sizes throughout. Figure 19.16d features (to our imagination) "spiky snowmen," with smaller ones growing out of the sides of larger ones. What do they look like to you? And does the overall pattern as a rosette have symmetry *c1, c2, d1,* or *d2*?

(a) "Paradise." (b) "Purgatory."

(c) "r-crest." (d) "Scarab 2."

Figure 19.16 Various fractals. *(Courtesy of Noel Giffin and the Spanky Fractal Database.)*

A famous example of a fractal pattern is M. C. Escher's print "Circle Limit IV" (Figure 19.17). As you examine the angels and devils closer and closer to the boundary of the circle, you notice that they are not necessarily geometrically similar to the ones at the center. However, if you imagine that the print is the image of a hemisphere, figures farther away from your viewpoint should indeed appear smaller.

Figure 19.17 M. C. Escher's fractal pattern "Circle Limit IV." *(© 2005 M. C. Escher Company—Holland. All rights reserved. www.mcescher.com.)*

Apart from their beauty and the opportunity that they offer as an art form (there is even "fractal music"!), fractals have two major applications:

- Fractals with simple rules can mimic very well the structure of a leaf, a shell, a tree, or a mountain (see Figure 19.18). This fact not only allows us to model natural phenomena using fractals but also suggests that they are produced by corresponding simple "rules of nature." Moreover, computer special effects in films can use fractals to mimic nature very closely, as in *Star Trek II: The Wrath of Khan,* for landscapes on the Genesis planet, and in *Return of the Jedi,* for the moons of Endor and the Death Star. In *The Perfect Storm,* fractals were used to add surface textures and even to scatter light inside water drops. Figure 19.19 shows how a recursive rule, like the one for forming the Fibonacci numbers, can form a fractal. (This is an example of an **iterated function system [IFS]**, a topic that is discussed in detail in Chapter 23 in connection with mathematical chaos.)
- Fractals form the basis for an important method of image compression, similar in efficiency to the better-known Joint Photographic Experts Group

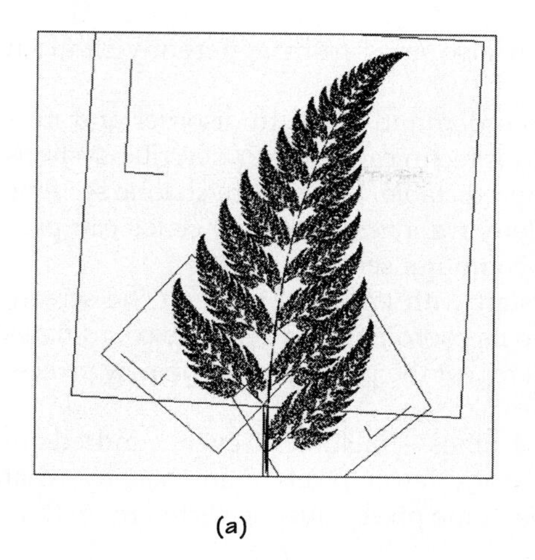

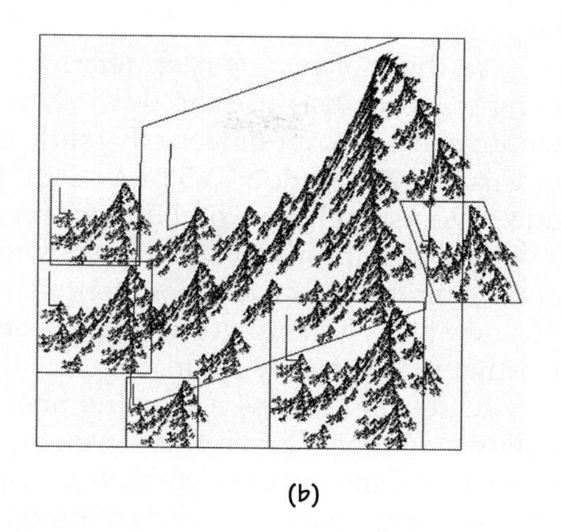

Figure 19.18 (a) Barnsley's fractal fern and (b) a snowy mountain landscape, both with templates showing how they were formed using reflections and linear "distortions" in addition to linear scaling. Each leaf of the fern and each mountain peak is in fact just a smaller copy of the entire image. *(Fred Solomon.)*

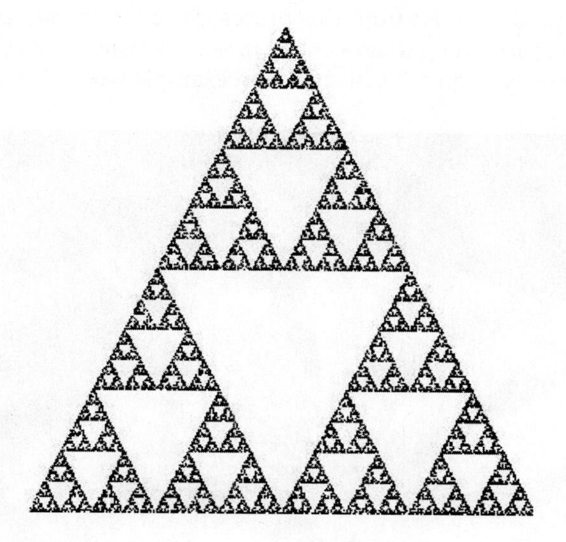

Figure 19.19 A Sierpiński "triangle." Start with the big triangle and remove its middle triangle. Then do the same for the three remaining smaller triangles. Recursively, do the same for each subsequent smaller triangle. Can you guess the area of the resulting figure? (*Hint:* It may be less than you think.) *(Annalisa Crannell, Franklin and Marshall College.)*

(JPEG) algorithm. The key idea is to store not the millions of bits that make up an image but instead a much smaller number (maybe thousands) of rules for generating patterns that can be found in the image. A simple example (which doesn't use fractals) is that you can compress a checkerboard of a million pixels that alternate between black and white to just two simple rules: If the current pixel is white, the next one is black, and if the current one is black, the next one is white. A more realistic example is Microsoft's original 1992 Encarta Encyclopedia. It contained thousands of articles and photographs, plus color animations and hundreds of maps—all on one CD-ROM, thanks to fractal data compression.

Symmetry in Chaos

While the patterns in rosettes, strips, wallpaper patterns, and some fractals can be produced from very simple rules for the symmetries involved, you may be surprised to learn that symmetry can also arise from apparently random behavior.

We think of symmetry as referring to order, and chaos to disorder and randomness. Scientists use the word *chaos* in a technical sense to describe systems whose behavior over time is inherently unpredictable. We explore chaotic systems in Chapter 23 (Section 23.5 on p. 848). Here, we investigate how chaos can produce astonishingly beautiful designs on a computer screen.

One way to produce a graphic is to start with an initial pixel on the screen, apply a mathematical function (formula) to its coordinates to generate coordinates of a new pixel, light up the new pixel, then repeat the process with the new pixel—in other words, iterate recursively.

Iterate the process a large number of times—millions or even hundreds of millions of times. Since the screen has many fewer pixels than that, by what mathematicians call the *pigeonhole principle,* some pixels must be visited more than once— maybe even thousands of times.

The clue to producing art from this process is "color by number": Choose the color for each pixel according to how many times it is visited, and choose the colors with an eye to beauty. Figure 19.20 shows an example with *d5* symmetry that was

Figure 19.20
"Emperor's Cloak," with *d5* symmetry. This work of art was produced by iterating a chaos-producing function, starting at one point and successively generating new points according to a fixed rule. The color bar shows the coloring of pixels according to how often they are visited by the iterations. *(Figure 1.13, p. 20, of Michael Field and Martin Golubitsky,* Symmetry in Chaos: A Search for Pattern in Mathematics, Art, and Nature, *Oxford University Press, New York, 1992.)*

SPOTLIGHT 19.8

The Father of Fractals

Benoit Mandelbrot (1924–2010) was born in Poland, spent part of his time growing up in France, and came to the United States to work for IBM. He found that many phenomena feature both repeating patterns and power curves (see Section 18.6 on p. 666). The patterns are repeated at a change of scale, as Barnsley's fern in Figure 19.18 and the Sierpiński triangle in Figure 19.19 show, and can be described by very simple rules. In 1975, he coined the term *fractals*, and the systems of rules became known as iterated function systems. The Web site for the course that he pioneered at Yale University (see the "Suggested Web Sites" section at the end of this chapter) lists 100 or so examples of fractal phenomena in nature and society. Mandelbrot's book, *The (mis)Behavior of Markets: A Fractal View of Risk, Ruin, and Reward* (see the "Suggested Readings" section at the end of this chapter) applies fractals to the movement of stock prices

Benoit Mandelbrot
(Roger Ressmeyer/Corbis.)

produced by 30 million iterations. The scale on the right in Figure 19.20 shows the colors for the number of times that pixels were hit; unhit pixels stay black. The order in which pixels are visited appears to be completely chaotic and is irrelevant to the final image.

- If you ignore the first thousand or so pixels visited, *it doesn't matter what pixel you start from—you get the same image!* But the pixels are visited in different orders.
- The formulas are variations on the *logistic map,* which we discuss in Chapter 23 in connection with biological populations.

REVIEW VOCABULARY

Crystallographic notation A four-symbol notation used by crystallographers (and mathematicians) to classify strip patterns and wallpaper patterns. (p. 703)

Divine proportion Another glorifying term for the golden ratio. (p. 688)

Fibonacci numbers The numbers in the sequence 1, 1, 2, 3, 5, 8, 13, 21, 34, ... Each number after the second one is obtained by adding the two preceding numbers. (p. 686)

Fractal A pattern that exhibits similarity at ever-finer scales. (p. 711)

Generated A group is generated by a particular set of elements if composing them and their inverses in combinations can produce all elements of the group. (p. 710)

Geometric mean The geometric mean of two numbers a and b is $\sqrt{ab}$. (p. 689)

Glide reflection A glide reflection is a translation (= glide) combined with alternating reflection in a line parallel to the translation direction. Example: pbpbpb. (p. 696)

Golden ratio, golden mean Inflated names for the number $\phi = \frac{1 + \sqrt{5}}{2} = 1.618033\ldots$ (p. 688)

Golden rectangle A rectangle the lengths of whose sides are in the golden ratio. (p. 688)

Group A group is a collection of elements with an operation on pairs of them such that the collection is closed under the operation, there is an identity for the operation, each element has an inverse, and the operation is associative. (p. 707)

Isometry Another word for rigid motion. Angles and distances, and consequently shape and size, remain unchanged by a rigid motion. For plane figures, there are only four possible isometries: reflection, rotation, translation, and glide reflection. (p. 695)

Phyllotaxis The spiral pattern of shoots, leaves, or seeds around the stem of a plant. (p. 687)

Preserves the pattern A transformation preserves a pattern if all parts of the pattern look exactly the same after the transformation has been performed. (p. 697)

Recursion A method of defining a sequence of numbers, in which the next number is given in terms of previous ones. (p. 687)

Reflection symmetry Mirror-image symmetry. (p. 697)

Rigid motion A transformation of the plane that preserves the size and shape of figures. In particular, any pair of points is the same distance apart after the transformation as before. (Also called *isometry*.) (p. 695)

Rosette pattern A pattern whose only symmetries are rotations about a single point and reflections through that point. (p. 698)

Rotational symmetry A figure has rotational symmetry if a rotation about its "center" leaves it looking the same, like the letter *S*. (p. 700)

Strip pattern A pattern that has indefinitely many repetitions in one direction. (p. 698)

Symmetry of the pattern A transformation of a pattern is a symmetry of the pattern if it preserves the pattern. (p. 698)

Symmetry group of the pattern The group of symmetries that preserve the pattern. (p. 708)

Translation A rigid motion that moves everything a certain distance in one direction. (p. 696)

Translation symmetry An infinite figure has translation symmetry if it can be translated (slid, without turning) along itself without appearing to have changed. Example: *AAA* (p. 700)

Wallpaper pattern A pattern in the plane that has indefinitely many repetitions in more than one direction. (p. 698)

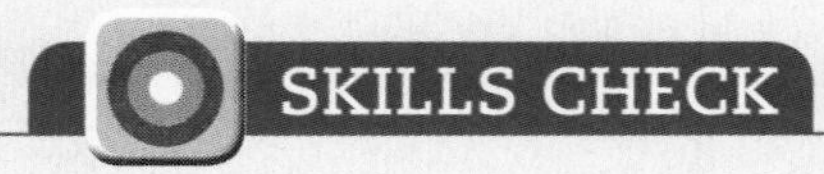

SKILLS CHECK

1. Symmetry includes notions of
(a) balance.
(b) similarity.
(c) repetition.
(d) all of the above.

2. Many people think that mathematics is just about numbers, but in fact mathematics is the study of ______________.

3. Which of the following rectangles is an approximate golden rectangle?
(a) 10 by 16
(b) 6 by 13
(c) 8 by 11

4. The geometric mean of 4 and 36 is ______________.

5. Which artist claimed to use the golden ratio in his work?
(a) Leonardo da Vinci
(b) Wolfgang Amadeus Mozart
(c) Neither

6. In the Fibonacci sequence, ______ follows 13 and 21.

7. Recursion means
(a) cursing over and over.
(b) giving a next number in terms of previous ones.
(c) giving previous numbers in terms of the current one.

8. Of pineapples, pine beetles, and pine bark, only ____________ exhibit Fibonacci numbers.

9. Phyllotaxis is
(a) the town in Pennsylvania where fractals were invented, near Fillodoughphia.
(b) more taxis than you need.
(c) a spiral pattern in plants.

10. The geometric mean of 4, 16, and 125 is _______.

11. A rigid motion always moves any pair of points
(a) in the same direction.
(b) to another pair of points the same distance apart.
(c) to their mirror images.

12. The capital letters ___, ___, ___, ___, ___, ___, and ___ each have a rotation isometry.

13. Assume that the following two patterns continue in both directions. Which of these patterns has a reflection isometry?

ZZZZZZZZZ
UUUUUUUUU

(a) ZZZZZZZZZ only
(b) UUUUUUUUU only
(c) Neither

14. This strip pattern

╜╒ ╜╒ ╜╒ ╜╒

has ____________ and ____________ isometries.

15. What isometries does this wallpaper pattern have?

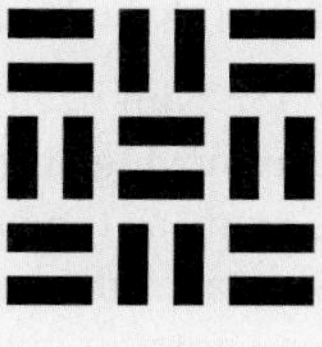

(a) Translation and reflection only
(b) Translation and rotation only
(c) Translation, rotation, and reflection

16. This wallpaper pattern

has ____________ and ____________ isometries.

17. If a horizontal strip pattern has a glide reflection isometry, then
(a) it always has a horizontal reflection isometry.
(b) it may also have a horizontal reflection isometry.
(c) it cannot have a horizontal reflection isometry.

18. If a strip pattern has both vertical and horizontal reflection isometries, then it always has a ____________ isometry.

19. Consider the strip pattern in the raffia cloth held by the woman in the photo on the right. What isometries does it have?
(a) Vertical reflection
(b) Horizontal reflection
(c) Glide reflection

(Dorothy K. Washburn.)

20. There are ____ ways to repeat a pattern along a strip.

21. The key mathematical idea for detecting and analyzing patterns is
(a) cataloging possible motifs.
(b) figuring out the symmetries.
(c) looking for the golden ratio.

22. The symbol *p* indicates that a strip pattern has ______________ symmetry.

23. The symbol *2* indicates that a pattern has
(a) rotational symmetry.
(b) reflection symmetry.
(c) too much symmetry.

24. The symbol *m* indicates that a wallpaper pattern has ____________ symmetry.

25. The symmetry group of a non-square rectangle has how many elements?
(a) 4
(b) 6
(c) 8

26. The symmetry group of the strip pattern *pmm2* has __________ elements.

27. The symmetry group of a square has how many elements?
(a) 2
(b) 4
(c) 8

28. In the mattress group, the result of a flip followed by a rotation is a ______________.

29. In mathematics, chaos refers to
(a) randomness.
(b) disorder.
(c) unpredictability.

30. A fractal is a pattern that exhibits ____________ at ever finer scales.

CHAPTER 19 EXERCISES

■ *Challenge* ▲ *Discussion*

19.1 Fibonacci Numbers and the Golden Ratio

1. Examine the "scales" on the surface of a pineapple, which are arranged in spirals around the fruit in three distinct directions. For each direction, how many spirals are there?

2. Repeat Exercise 1, but for a pinecone from your area.

3. Repeat Exercise 1, but for a sunflower.

4. Here are two primitive models of natural increase of biological populations, similar to those that Fibonacci hypothesized around the year 1200, based on the following situation: A pair of newborn male and female rabbits is placed in an enclosure to breed.

(a) Suppose that the rabbits start to bear young one month after their own birth. At the end of each month, they have another male-female pair, which in turn mature and start to bear young one month later. Assuming that none of the rabbits die, how many pairs of rabbits will there be at the end of six months from the start (just before any births for that month)? (*Hint:* Draw a month-by-month chart of the situation at the end of the month, just before any births.)

(b) Repeat part (a), but assume instead that the rabbits start to bear young exactly two months after their own birth.

5. New houses are to be built along one side of a street ("Leonardo's Lane"), divided into equal-sized lots. Each house is either a single-family detached house, taking up one lot, or a duplex, taking up two lots. Suppose that there are n lots on the street. How many different arrangements (orderings) of houses are there, for $n = 1$, 2, 3, 4, 5, and in general? (This exercise was inspired by a puzzle by Paul Dixon at the Web site by Ron Knott, cited in the "Suggested Web Sites" section at the end of this chapter.)

6. When it snows in the winter, the local school district superintendent must decide by 5 a.m. whether to declare a "snow day" and cancel school. The 900 faculty and staff are notified by a robocall broadcast, but formerly a binary "telephone tree" was used, in which the superintendent called two people and each person who received a call called two others. Suppose that each call takes exactly 1 minute.

(a) Draw the telephone tree of calls for, and determine how many calls take place in, the first 1 minute, 2 minutes, 3 minutes, 4 minutes, and 5 minutes.

(b) How many calls does it take to notify all the faculty and staff? How long does that take?

(This exercise was inspired by a puzzle at the Web site by Ron Knott.)

7. Here is a trick to "prove" that you can calculate faster than a person with a calculator. Turn your back and ask a friend to write down any two positive integers, then add them to get a third, then add the second and third to get a fourth, and so on, adding each time the last two integers until there are 10 numbers. Have your friend show you the list, whereupon you write down right away the total of all 10, while your friend begins to add them up on the calculator (to prove that you're right). The secret: The total is always 11 times the seventh number, and multiplying by 11 is pretty easy to do in your head—just add each pair of neighboring digits, carrying if necessary. Suppose that your friend writes down m and n as the first two numbers. Show that indeed the total of all 10 numbers is 11 times the seventh number. (Adapted from Martin Gardner, *Mathematical Circus,* Knopf, New York, 1979.)

8. The game of Fibonacci Nim begins with n counters. Two players take turns removing at least one counter, but no more than twice as many as the opponent just did. The winner is the player who takes the last counter. One other rule: The first player may not win immediately by taking all the counters on the first turn. (Adapted from Martin Gardner, *Mathematical Circus,* Knopf, New York, 1979.)

(a) Play this game taking turns with an opponent and starting with different numbers n of counters and try to come up with a strategy for one player or the other to win. (*Hint*: The key is that any positive integer can be represented uniquely as a sum of Fibonacci numbers.)

(b) Proceed as in part (a), but with the rule changes that the player who takes the last counter loses and the first player may not take all but one counter.

9. Put the golden ratio $\phi = (1 + \sqrt{5})/2$ into the memory of your calculator.

(a) Look at the value of ϕ. Now square it (either use the $\boxed{x^2}$ button or multiply it by itself). What do you observe?

(b) Back to ϕ. Now take its reciprocal (either use the $\boxed{1/x}$ button or divide it into 1). What do you observe?

(c) What formula explains what you saw in part (a)?

(d) What formula explains what you saw in part (b)?

10. The golden ratio satisfies the equation $x^2 = x + 1$. Show that $(1 - \phi)$ also satisfies the equation, so that $(1 - \phi) = (1 + \sqrt{5})/2$ is the other solution to $x^2 - x - 1 = 0$.

11. The geometric mean has interpretations in both arithmetic and geometry.

(a) Find the geometric mean of 3 and 27.

(b) Find the length of a side of a square that has the same area as a rectangle that is 4 by 64.

12. Here's further practice on arithmetic and geometric interpretations of the geometric mean.

(a) Find the geometric mean of 4 and 9.

(b) You are to make a golden rectangle with 6 inches of string. How wide should it be, and how high?

13. What is the geometric mean of 3, 6, and 12?

14. What is the geometric mean of 2, 4, 8, 16, and 32? (Such a sequence, in which each successive number is the same constant times the previous one, is called a *geometric sequence*.)

15. Another sequence closely related to the Fibonacci sequence is the *Lucas sequence*, which is formed using the same recursive rule but different starting numbers. The nth Lucas number L_n is given by

$$L_1 = 1,\ L_2 = 3,\ \text{and } L_{n+1} = L_n + L_{n-1} \text{ for } n \geq 2$$

(a) Calculate L_3 through L_{10}.

(b) Calculate the ratio of successive terms of the Lucas sequence:

$$\frac{L_2}{L_1}, \frac{L_3}{L_2}, \ldots, \frac{L_{10}}{L_9}$$

What do you notice?

16. As in Exercise 15, but start with the pair of numbers 1 and 4.

17. As in Exercise 15, but start with a pair of numbers of your choice. Based on your result and those in Exercises 15 and 16, what is your hunch?

18. For a sequence specified by a recursive rule, finding an explicit expression for the nth term is not easy, nor is the form necessarily simple. An exact expression for the nth term of the Fibonacci sequence is given by the Binet formula:

$$F_n = \frac{1}{\sqrt{5}}\left(\frac{1+\sqrt{5}}{2}\right)^n - \frac{1}{\sqrt{5}}\left(\frac{1-\sqrt{5}}{2}\right)^n$$

(a) Verify the formula for $n = 1$ and $n = 2$ by multiplying out, not by using a calculator.

(b) Use the Binet formula and your calculator to find F_5.

(c) In fact, the second term on the right of the equation gets closer and closer to 0 as n gets large. Because we know that the Fibonacci numbers are integers, we can just round off the result of calculating the first term. Find F_{13} by calculating the first term with your calculator and rounding.

19. For two positive numbers x and y, show that the arithmetic mean $(x + y)/2$ is always greater than or equal to the geometric mean $x^{1/2} y^{1/2} = \sqrt{xy}$. Try some values for x and y and convince yourself, then demonstrate algebraically that it is true in general. When does equality hold? [*Hint*: Suppose that the claim is false, so that $(x + y)/2 < \sqrt{xy}$.) Square both sides of the inequality, bring all terms to one side, factor, and observe a contradiction.]

■ **20.** You may remember having to work problems like, "If Joe can dig a ditch in 3 days, and Sam can dig it in 4, how long will it take the two of them working together?" The answer is related to the harmonic mean of 3 and 4. The formula for the harmonic mean of two numbers x and y is

$$\frac{2}{1/x + 1/y}$$

(a) Calculate the answer for Joe and Sam, which is one-half of the harmonic mean of 3 and 4. Explain why this is the correct answer.

(b) Show that the harmonic mean of two positive numbers is always less than or equal to the geometric mean. (Thus, in light of Exercise 19, we have the general conclusion that $H \leq G \leq A$, where H stands for the harmonic mean, G for the geometric mean, and A for the arithmetic mean.) (*Hint:* Suppose that the claim is false. Simplify the fraction that is the harmonic mean, square both sides of the inequality, and proceed as in Exercise 19.)

(c) Show once more that the harmonic mean of two positive numbers is always less than the geometric mean, but this time do it with less work: let $A = 1/x$ and $B = 1/y$, and discover one connection (equation) between the harmonic mean of x and y and the arithmetic mean of A and B, and a second connection between the geometric mean of x and y and the geometric mean of A and B. Then use Exercise 19 on A and B.

(d) What should be the formula for the harmonic mean of three numbers? Of n numbers?

21. Shari Lynn Levine, a high school student, published an article in *The Fibonacci Quarterly* that investigated the "Beta-nacci" sequence that results if, instead of bearing one pair of baby rabbits per month, mature rabbits bear two pairs every month, starting when they reach two months of age. Here, we ask you to rediscover some of Shari's results.

(a) How many rabbits will there be each month for the first 12 months?

(b) What is the recursive rule for the nth Beta-nacci number B_n?

(c) For the terms of the sequence in part (a), calculate the ratios B_{n+1}/B_n of successive terms. (*Motivating hint:* It's not the golden ratio this time.)

(d) Suppose that the ratio of successive terms approaches a number x. We show how to find x exactly. For very large n, we have $B_{n+1} \approx xB_n \approx x^2B_{n-1}$. Inserting these values into the recursive rule for the sequence and dividing by B_{n-1} gives the equation $x^2 = x + 2$. Solve this equation for x (you can use the quadratic formula). Make a table of values of $3B_n$ versus 2^n. From the evidence, can you suggest a formula for B_n?

22. Generalize Exercise 21, parts (a) through (d), as follows:

(a) to the case of each pair of rabbits having three pairs of rabbits (the "Gamma-nacci" sequence).

(b) to the case of each pair of rabbits having q pairs of rabbits.

For Exercises 23 and 24, refer to the following. We have seen that the golden ratio is a positive root of the quadratic polynomial $x^2 - x - 1$. We can generalize this polynomial to $x^2 - mx - 1$ for $m = 1, 2, 3, \ldots$ and consider the positive roots of those polynomials as generalized means—the "metallic means family," as they are sometimes known. In particular, for $m = 2, 3, 4$, and 5, we have respectively the silver, bronze, copper, and nickel means. It is surely surprising that these numbers arise both in connection with quasicrystals (investigated in Chapter 20) and in analyzing the behavior of some dynamical systems (investigated in Chapter 23) as the systems evolve into chaotic behavior.

23. Use the quadratic formula to find expressions in terms of square roots for the silver, bronze, copper, and nickel means, and approximate these to three decimal places. Find a general expression in terms of a square root for the mth metallic mean.

24. Just as the golden mean arises as the limiting ratio of consecutive terms of the Fibonacci sequence, each of the metallic means arises as the limiting ratio of consecutive terms of generalized Fibonacci sequences. A generalized Fibonacci sequence G can be defined by

$$G_1 = 1, G_2 = 1, \quad \text{and} \quad G_{n+1} = pG_n + qG_{n-1},$$

where p and q are positive integers. The Fibonacci sequence itself is the case $p = q = 1$.

(a) Try various small values of p and q and determine which mean they lead to.

(b) Divide the equation for G_{n+1} by G_n. Assume that G_{n+1}/G_n and G_n/G_{n-1} both tend toward the same number x as n gets large, replace those quantities by x, and simplify the resulting equation. What must be the value of x?

(c) What happens to the sequence and to the mean if we allow one or both of p and q to be negative integers?

19.2 Rosette, Strip, and Wallpaper Patterns

25. Determine whether each of the following statements is always or sometimes true. Drawing some sketches may be helpful.

(a) A line reflection preserves collinearity of points. That is, if the points A, B, and C are in a straight line (collinear), then their images reflected in some other line also lie in a straight line.

(b) A line reflection preserves betweenness. That is, if the collinear points A, B, and C (with B between A and C) are reflected about a line, then the image of B is between the images of A and C.

(c) The image of a line segment under a line reflection is a line segment of the same length.

(d) The image of an angle under a line reflection is an angle of the same measure.

(e) The image of a pair of parallel lines under a line reflection is a pair of parallel lines.

26. Determine whether each of the following statements is always or sometimes true. Drawing some sketches may be helpful.

(a) The image of a pair of perpendicular lines under a line reflection is a pair of perpendicular lines.

(b) The image of a square under a line reflection is a square.

(c) Label the vertices of a square A, B, C, and D in a clockwise direction. Then their images A', B', C', and D' under a line reflection also follow a clockwise direction.

(d) The length of the perimeter of a geometric figure is equal to the length of the perimeter of its image under a line reflection.

(e) The image of a vertical line under a line reflection is always a vertical line.

27. Which of the capital letters of the alphabet, when drawn in the most symmetrical way, has the following symmetries? For example, assume that the upper and lower loops of B are the same size.
(a) A horizontal line of reflection symmetry
(b) A vertical line of reflection symmetry
(c) A rotational symmetry

28. Repeat Exercise 27 for the lowercase letters.

29. In *The Complete Walker III* (3d ed., Knopf, New York, 1984, p. 505), Colin Fletcher's answer to "What games should I take on a backpacking trip?" is the game he calls "Colinvert": "You strive to find words with meaningful mirror (or half-turn) images." Some of the words he found are

MOM WOW pod Mud bUM

(a) Which of his words reflect into themselves?
(b) Which of his words rotate into themselves?
(c) Find some more words or phrases of these various types—the longer, the better.

30. Repeat Exercise 29, but for words written vertically instead of horizontally.

31. For each of the following patterns, identify the rigid motions that preserve the pattern:
(a) CCCCCCCCCC
(b) GGGGGGGGGG
(c) HHHHHHHHHH
(d) MMMMMMMMMM

32. Repeat Exercise 31, but for the following patterns:
(a) SSSSSSSSSS
(b) bdbdbdbdbd
(c) dbpqdbpqdbpq

19.3 Notation for Patterns

33. What is the notation (such as *d4* or *c5*) for the symmetry pattern of a regular pentagon (which has all five sides equal)?

34. What is the notation for the symmetry pattern of a snowflake?

35. Give the notation (such as *d4* or *c5*) for the symmetry patterns of the rosettes in hubcaps (a) through (c) below, disregarding the logos in the centers. (Can you identify the make of car for each hubcap?)

36. Repeat Exercise 35 for hubcaps (d) through (f).

(*a*) (*b*) (*c*)

(*d*) (*e*) (*f*)

(All hubcap photos courtesy of Joe Gallian, University of Minnesota, Duluth.)

37. Repeat Exercise 35 for corporate logos (a) through (c) below. (Bonus: Can you identify the corporations?)

(a)

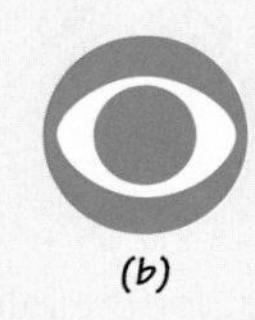
(b)

(c)

38. Repeat Exercise 35 for automobile logos (d) through (f) below.

(d)

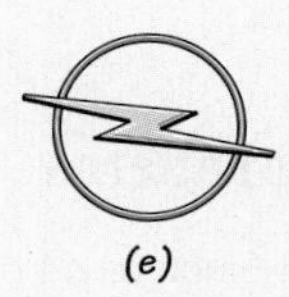
(e)

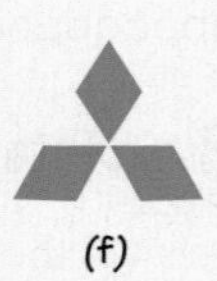
(f)

For Exercises 39–40, refer to the following.
Step patterns are found in Celtic illuminated manuscripts, metalwork, and stone crosses. Square ones were constructed by first designing on a square lattice one quarter of the pattern (say, the top right), using horizontal and diagonal lines to produce a prototype such as the following:

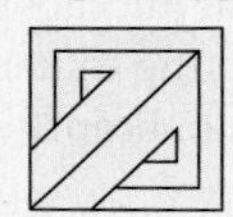

Then three copies were added, either by (1) rotating the original successively by 90° [as in accompanying illustration (a)], or else by (2) reflecting it across its right and bottom edges [as in illustration (b)]. (Based on research by Mark A. M. Lynch of Glasgow Caledonian University, Scotland.)

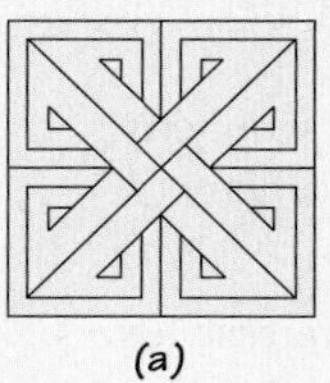
(a)

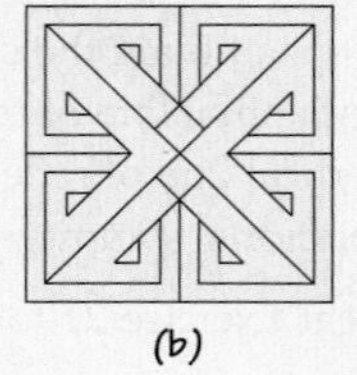
(b)

39. Identify the rosette pattern for:
(a) step pattern (a).
(b) step pattern (b).

40. Which rosette pattern would result if the prototype, unlike the one above, has reflection symmetry across its diagonal from the top left to the lower right and
(a) strategy (1) is used.
(b) strategy (2) is used.

41. Use the flowchart in Figure 19.12 (on p. 704) to identify the notation for the types of strip patterns from the pottery and basketry shown in the illustrations below.

(a) (d) (g)

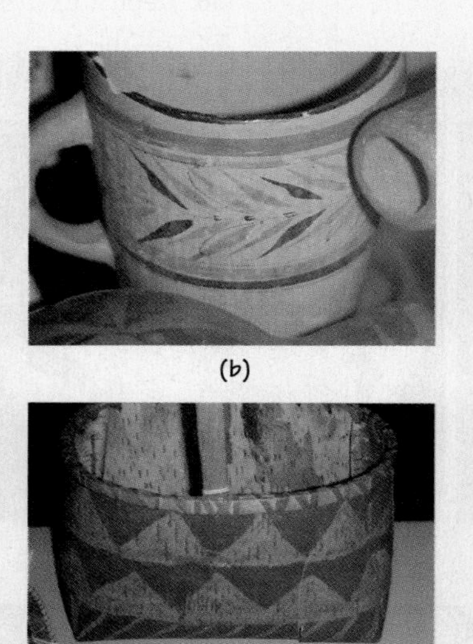
(b) (e)

(c) (f)

Pottery and basketry from the Americas.
(a), (b) Mexico, modern.
(c) Lower Central America, pre-Columbian.
(d) Pomo people, California, early 20th century.
(e) Woodland Indians, central North America, early 20th century.
(f) Pomo people, California, mid-20th century; originally from the collection of Dr. Herbert Zim, editor of the "Golden Guides" series of nature books.
(g) Woodland Indians, central North America, early 20th century. *(Courtesy of Logan Museum of Anthropology, Beloit College, photos by Darrah Chavey.)*

42. In each of the four accompanying examples, two adjacent triangles of an infinite strip are shown. (Contributed by Margaret A. Owens, California State University, Chico.)

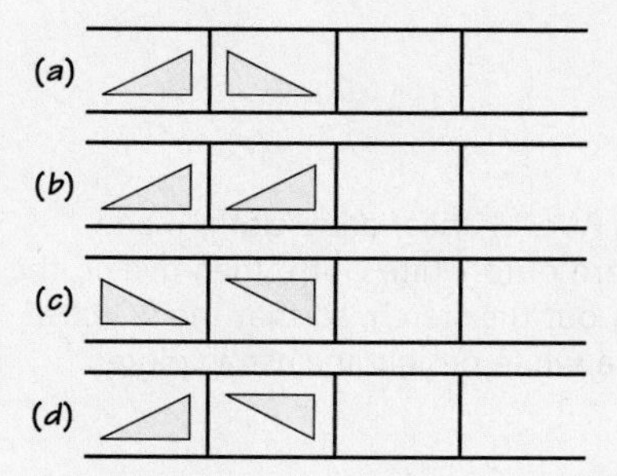

For each example:

(a) Determine a motion (translation, reflection, rotation, or glide reflection) that takes the first (left) triangle to the second (right) one.

(b) Draw the next four triangles of the infinite strip that would result if the second triangle is moved to the next space by another motion of the same kind, and so on.

(c) Identify (by notation) the resulting strip as one of the seven possible strip patterns.

43. Repeat Exercise 41 for the accompanying eight strip patterns, all of which appear on the brass straps for a single lamp from 19th-century Benin in West Africa. (From H. Ling Roth, in Great Benin.)

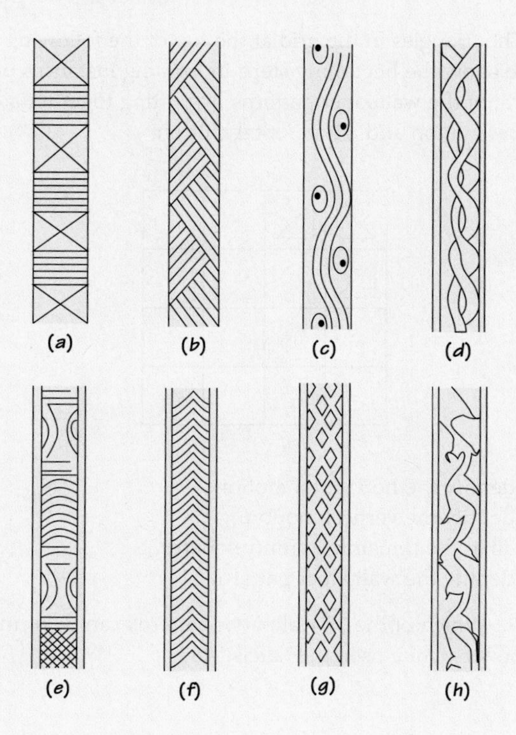

Note that the patterns are roughly carved, so you will need to discern the intent of the artist.

44. Repeat Exercise 41 for the accompanying patterns from San Ildefonso Pueblo, New Mexico.

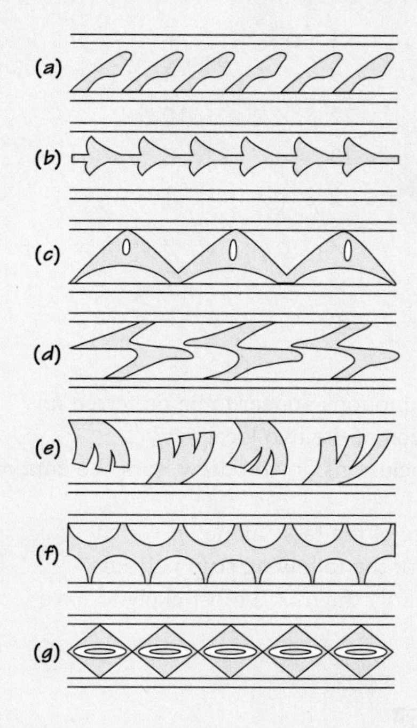

45. The following table shows comparative data about the frequency of occurrence of strip designs of various types on Chinese porcelain and smoking pipes (Begho, in what is now Ghana) from two continents.

Frequency of Strip Designs on Porcelains from the Chinese Yuan (1280–1368) and Ming (1368–1644) Dynasties Porcelains and on African Begho Smoking Pipes

	Yuan (1280–1368) and Ming (1368–1644) Dynasties	
Strip Type	Number of Examples	Percentage of Total
*p*111	29	18
*p*1*m*1	1	1
*pm*11	66	42
*p*112	20	13
*p*1*a*1	21	13
*pma*2	13	8
*pmm*2	9	6
TOTAL	159	

	Begho	
Strip Type	Number of Examples	Percentage of Total
*p*111	4	2
*p*1*m*1	9	4
*pm*11	22	10
*p*112	19	8
*p*1*a*1	2	1
*pma*2	9	4
*pmm*2	165	72
TOTAL	230	

(a) Which types of motions appear to be preferred for designs from each of the two localities?
(b) What other conclusions do you draw from the data of this table?
(c) On the evidence of the table alone, in which locality is each of the following strip patterns (but not necessarily the motif) most likely to have been found?

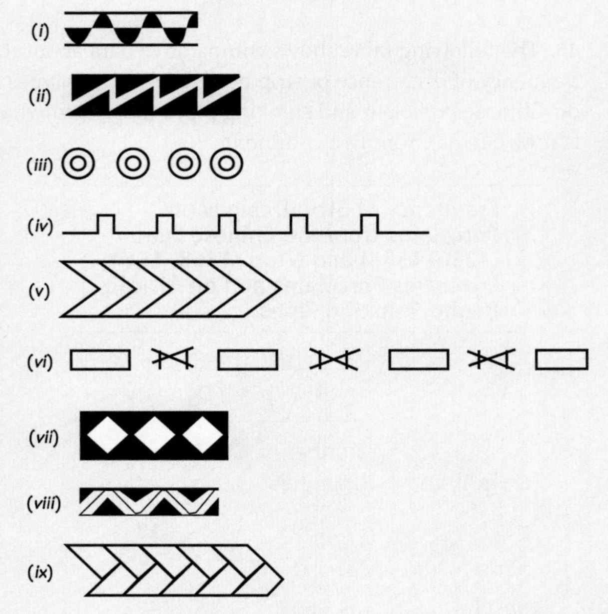

46. For the Nigerian Yoruba cloths (a) and (b) in the following illustration, use the flowchart in Spotlight 19.7 (on p. 701) to identify (by notation) the type of wallpaper pattern.

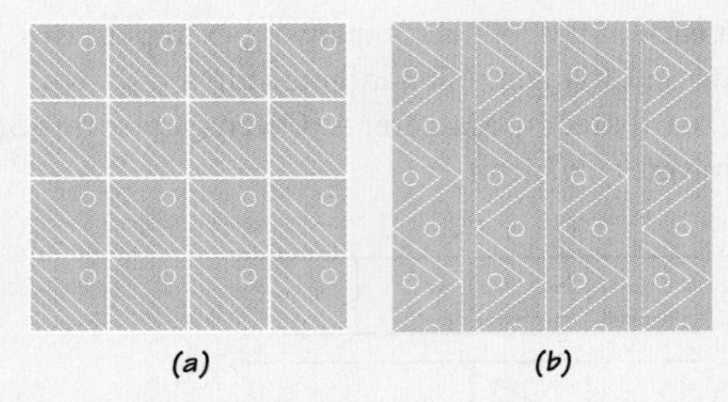

(a) (b)

Patterns on Yoruba (West Africa) adire cloth, made by starching a pattern onto white cloth, then dyeing the cloth before rinsing out the starch, so that the starched portion remains as a white design against a colored background.

47. For the Nigerian Yoruba cloths (c) and (d) in the accompanying illustration, use the flowchart in Spotlight 19.7 to identify (by notation) the type of wallpaper pattern.

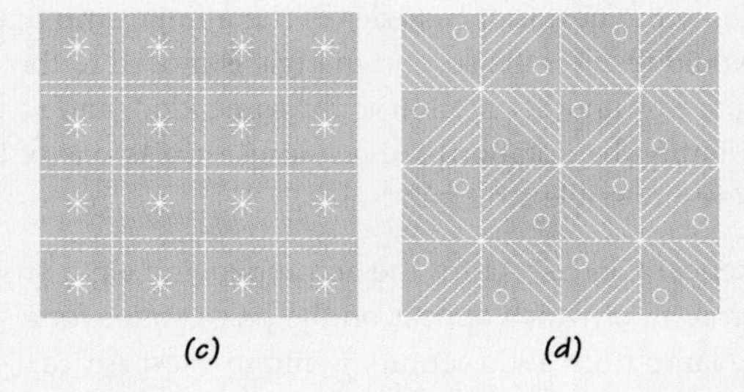

(c) (d)

48. The triangles in the grid at the top of the following figure show the beginning steps in forming instances of several of the wallpaper patterns by putting together a vertical motion and a horizontal motion.

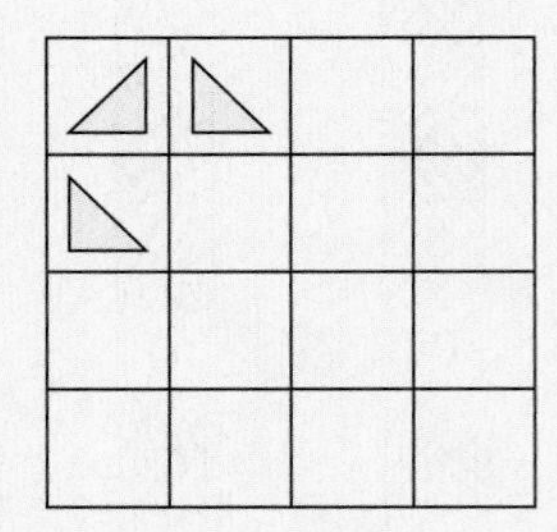

(a) Identify the horizontal motion.
(b) Identify the vertical motion.
(c) Fill in the remaining empty squares.
(d) Identify the wallpaper pattern.

■ **49.** Which of the 17 wallpaper patterns can be formed by the technique used in Exercise 48?

For Exercises 50–55, refer to the following:
In Chapter 20, we study both repeating and nonrepeating plane patterns, from the point of view of their basic building blocks (tiles). Here we ask you to analyze the repeating patterns from figures in that chapter according to wallpaper type, using the flowchart of Spotlight 19.7. Identify all the symmetries and give the notational type for the wallpaper pattern of:

50. Figure 20.10 (on p. 741).

51. Figure 20.11 (on p. 741).

52. Figure 20.15 (on p. 744).

53. The figure in Spotlight 20.2 (on p. 740).

54. The hexagonal regular tiling at upper right in Figure 20.5 (on p. 736).

55. The convex hexagon tiling of type 3 in Figure 20.9 (on p. 738).

56. Visit the Web site **escher.epfl.ch/escher/**, which features an interactive Java program called Escher Web Sketch. Experiment with choosing wallpaper patterns using crystallographic notation. For each one, draw on the screen a colored design for the motif; the program will reproduce the motif using the pattern.

19.4 Symmetry Groups

57. For positive integers a and n, the expression a mod n means the remainder when a is divided by n. Thus, $23 \bmod 4 = 3$, because $23 = 5 \cdot 4 + 3$, and we say that "23 is equivalent to 3 modulo 4." (See Chapter 17 for further details about this modular arithmetic.) Every positive integer is equivalent to 0, 1, 2, or 3 modulo 4. Consider the collection of elements {0, 1, 2, 3} and the operation $\oplus$ on them defined by $a \oplus b = (a + b) \bmod 4$. Show that under this operation, the collection forms a group.

58. Explain, by referring to the properties of a group, whether the collection of all real numbers is a group under the operation of (a) addition; (b) multiplication.

59. Explain why the table for the operation * below shows that the elements indicated do not form a group under *.

*	A	B	C
A	B	B	B
B	B	C	A
C	C	A	B

60. Consider the table for the operation # below.

#	A	B	C	D	E	F
A	A	B	C	D	E	F
B	B	A	D	C	F	E
C	C	E	A	F	B	D
D	D	F	B	E	A	C
E	E	C	F	A	D	B
F	F	D	E	B	C	A

(a) Explain why the elements form a group under #. (Don't bother to check associativity.)
(b) What do you notice about $F \# E$ vs. $E \# F$? (This is the smallest example of a group that is noncommutative.)

61. Construct the table for the mattress group of Example 6, putting down the side of the table the first turn made and, across the top of the table, the subsequent turn.

62. The mattress group is a commutative group, meaning that the order of the turns doesn't make any difference. How can you tell that from the table of Exercise 61? (A commutative group is sometimes called an *abelian* group, after Niels Henrik Abel [1801–1829], a Norwegian mathematician who died young of turberculosis.)

63. A problem about mattress turning is that people usually don't remember the immediately previous position or immediately previous turn that they made months ago. So the next time, they could wind up just turning the mattress back to the position it was in two seasons ago. Suppose that the mattress has a lengthwise stripe on the UP side and a sidewise stripe on the DOWN side. Show that following the rule "Turn the mattress with the stripe showing as the axis" in fact cycles the mattress through its four positions.

64. A king mattress is officially 76 × 80 in., which is almost square. We investigate a square mattress. Make yourself a "square mattress" from a sheet of paper and investigate its group.
(a) What are the possible turns for a square mattress?
(b) Make a table for this mattress group.

65. Show that the collection of numbers {1, 3, 5, 7} under multiplication modulo 8 (see Exercise 57 on p. 725) has the structure of the Klein 4-group.

66. Show that the dihedral rosette group $d4$ has the structure of the Klein 4-group.

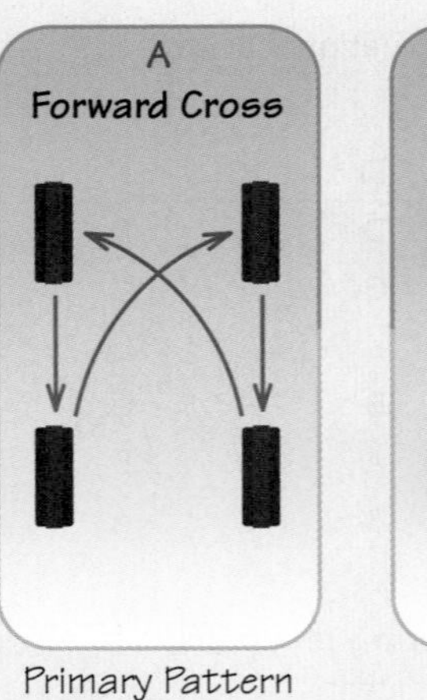

Primary Pattern for Front-Wheel Drive Vehicles

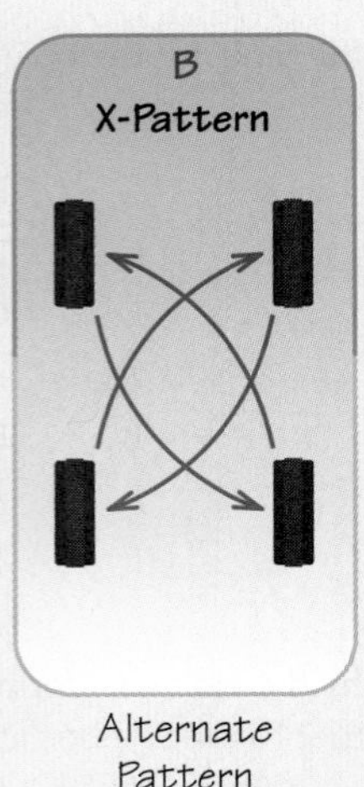

Alternate Pattern

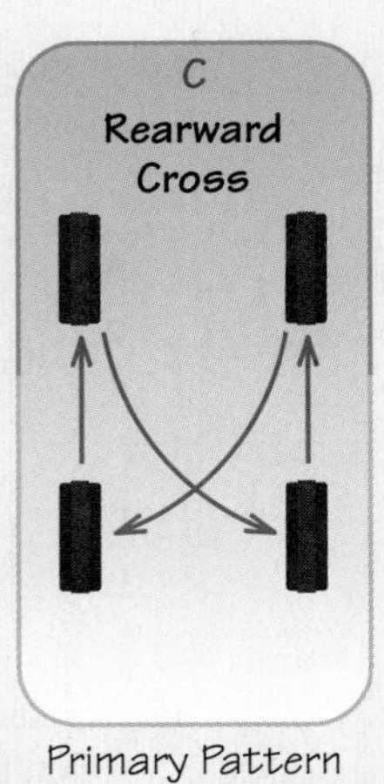

Primary Pattern for Rear- and Four-Wheel Drive Vehicles

For Exercises 67–70, refer to the following:

Like mattresses, car tires need to be rotated so as to promote even wear; wear on a tire varies with wheel position. The three main rotation schemes are shown above.

67. Show that successive tire rotations using scheme A form a group. Is it the Klein 4-group?

68. Observe that repeating scheme B will never take the front right tire to the back right wheel. Is there any combination of schemes A and C that will produce the result of scheme B?

69. A tire rotation scheme is designed so that no tire will remain where it was. Such a rearrangement (permutation) of objects is called a *derangement*. Two more tire rotation schemes are shown below. Are there still more schemes that are derangements? One way to record a tire rotation is to label the original tire positions clockwise from the left front: 1 for left front, 2 for right front, 3 for right rear, and 4 for left rear. Record the results of a scheme by writing in turn where each tire goes; for example, scheme D below produces the derangement 4321 because tire 1 goes to position 4, tire 2 goes to position 3, and so forth.

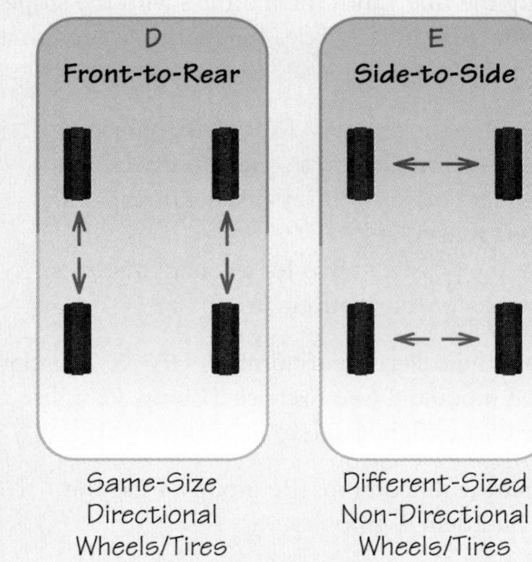

Same-Size Directional Wheels/Tires

Different-Sized Non-Directional Wheels/Tires

70. Do the five tire rotation schemes A, B, C, D, and E, plus the identity rotation and any others that you found in Exercise 69, form a group? Why or why not?

71. For the traditional North American beadwork shown below, answer the following questions.

(Courtesy of Dr. Ron Eglash, RPI. See http://csdt.rpi.edu/na/loom/index.html.)

(a) Which rosette pattern does it have?
(b) Specify two rigid motions that together generate the group of the pattern.
(c) List the elements of the group.

72. Repeat Exercise 71 for the Plains Indian embroidery shown below.

(Courtesy of Dr. Ron Eglash, RPI.)

73. What is the group of symmetries of a square?

74. What is the group of symmetries of
(a) an equilateral triangle (all three sides equal)?
(b) an isosceles triangle (two equal sides) that is not equilateral?
(c) a scalene triangle (no pair of sides equal)?

75. (a) Give a numerical example to show that the operation of subtraction on the integers is not associative.

(b) Repeat part (a), but for division on the positive real numbers.

76. What are the elements of the group of symmetries of (a) Figure 19.10b? (b) Figure 19.10f? (See p. 700 for these images.)

77. What are the elements of the group of symmetries of (a) Figure 19.10d? (b) Figure 19.10g?

78. What are the elements of the group of symmetries of the dihedral pattern *d8*? (See the flower in Figure 19.9a on p. 698.)

79. What is the group of symmetries of the cyclic pattern *c8*?

■ **80.** What is the group of symmetries of a cube?

■ **81.** What is the group of symmetries of a general rectangular solid (its length, width, and height are all unequal)?

19.5 Fractal Patterns and Chaos

▲ **82.** Explore Sprott's Fractal Gallery at **sprott.physics.wisc.edu/fractals.htm**, which features a "Fractal of the Day" and accompanying fractal music. There are various "rooms" in the gallery—including "Iterated Function Systems," "Natural Fractals" (We particularly like "Broccoli" and "Trees"), and "Publication Quality Attractors" ("SMKBNZQA" is our favorite)—together with PC programs for generating such fractals. What are your favorites, and why?

83. Explain how the pattern of the following illustration is fractal.

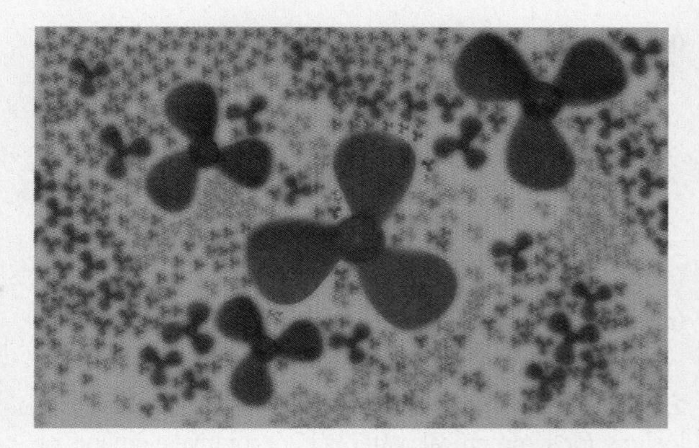

Exercises 84–86 use applets that require a computer with a Web browser equipped with Java and Flash plug-ins. These plug-ins are available at links from the Web site **http://csdt.rpi.edu/**.

84. The Web site **http://csdt.rpi.edu/african/MANG_DESIGN/culture/mang_homepage.html** has information about a fractal-patterned ivory hatpin from the Mangbetu culture in Africa. The site includes a tutorial on producing similar designs using reflection, rotation, translation, and scaling. Work your way through the tutorial and then create a Mangbetu-style artifact.

85. Cornrow hairstyles are fractal in nature. At the Web site **http://csdt.rpi.edu/african/CORNROW_CURVES/**, you can see how and why, including a tutorial on designing cornrow hairstyles using reflection, rotation, translation, and scaling. Work your way through the tutorial and then create a hairstyle. The Web site also includes instructions for actual braiding, with a short video.

86. Download fractal-creation software and accompanying documentation and use the software to create your own fractal. Recommended software:
For Windows: Fractint, from **spanky.triumf.ca/www/fractint/fractint.html**
For Macintosh: FractaSketch, from **www.info.ucl.ac.be/~pvr/fracta.html** with draft of manual (shows how to make fractal trees and leaves).

WRITING PROJECTS

1. Generations of children have enjoyed the popular toy Spirograph®, which allows the user to trace out symmetric patterns. A pencil or pen is placed in a hole in one of several plastic circular disks with teeth on the outside rim. The disk is then meshed in the teeth of another plastic circle and rotated around its inside or outside. Each plastic piece is labeled with the number of teeth that it has on its circumference.

Either obtain a copy of Spirograph or a closely related toy, or else visit the Web site **www.wordsmith.org/~anu/java/spirograph.html**, which offers an interactive Java application (which you can download) that mimics what the Spirograph toy does.

(a) Experiment to determine, from the numbers of teeth on the rotating circular disk and the fixed circle, what symmetry pattern the result will have.

(b) Choose a rotating circular disk and a fixed circle for which the ratio of the number of teeth reduces to a whole number. For each of several "offsets" (holes to choose for the pencil or pen), trace overlapping designs. What symmetry pattern do you get for the design taken as a whole? Repeat this experiment for other pairs of pieces and try to reach a general conclusion.

(c) Write up, in a page or so, a description of your experiments and what conclusions you reached.

2. (Project for a team of two or three) Explore your campus looking for symmetrical patterns in decorative elements of walls, floor, carpets, and ceilings. Find one example each of a rosette pattern, a strip pattern, and a wallpaper pattern. Take a digital photo of each and incorporate your photos into a document of three pages or so that explains to the reader where the pattern can be found, what symmetries (translation, rotation, reflection, or glide reflection) it has, and how you identified the notation for it.

3. (Project for a team of two or three) Visit a store that sells wallpaper and ask for a few old samples. Identify three that have different patterns according to the flowchart in Spotlight 19.7 (on p. 701). Write in a page or two your explanations of how you identified the patterns, and attach the wallpaper samples to your report.

Suggested Readings

BELCASTRO, SARAH-MARIE, and THOMAS C. HULL. Classifying frieze patterns without using groups, *College Mathematics Journal,* 33 (March 2002): 93–98. Elementary analysis of why there are only seven ways to repeat a pattern along a strip.

CROWE, DONALD W. *Symmetry, Rigid Motions and Patterns, High School Mathematics and Its Applications (HiMAP)* Module 4, COMAP, Lexington, MA, 1987. Reprinted in smaller format in *The UMAP Journal,* 8(3) (1987): 207–236. Instructional module on rigid motions of the plane, strip patterns, and wallpaper patterns, with worksheets.

HAYES, BRIAN. Group theory in the bedroom, *American* Scientist, 93 (5) (September–October 2005) 393ff, at **http://www.americanscientist.org/issues/id.3465,y.2005,no.9,content.true,page.1,css.print/issue.aspx**. Reprinted, with Afterthoughts, in *Group Theory in the Bedroom, and Other Mathematical Diversions,* Hill and Wang, New York, 2008, pp. 219–237, 252–253.

LEE, KEVIN D. KaleidoMania!: Interactive Symmetry, Windows/Macintosh program, Key Curriculum Press, 1999. Lets the user construct rosette, strip, and wallpaper patterns.

LIVIO, MARIO. *The Golden Ratio: The Story of Phi, the World's Most Astonishing Number.* Broadway Books, New York, 2002. Reviewed in *Notices of the American Mathematical Society,* 52 (3) (March 2005): 344–347, at **http://www.ams.org/notices/200503/rev-markowsky.pdf**.

MANDELBROT, BENOIT B., and RICHARD L. HUDSON. *The (mis)Behavior of Markets: A Fractal View of Risk, Ruin, and Reward.* New York: Basic Books, 2004.

MARKOWSKY, GEORGE. Misconceptions about the golden ratio, *College Mathematics Journal,* 23 (1) (January 1992): 2–19.

POSAMENTIER, ALFRED S., and INGMAR LEHMANN. *The (Fabulous) Fibonacci Numbers,* Prometheus Books, Amherst, NY, 2007.

POSAMENTIER, ALFRED S., and INGMAR LEHMANN. *The Glorious Golden Ratio,* Prometheus Books, Amherst, NY, 2011.

WALSER, HANS. *The Golden Section,* Mathematical Association of America, Washington, DC, 2001.

WASHBURN, DOROTHY K., and DONALD W. CROWE. *Symmetries of Culture: Theory and Practice of Plane Pattern Analysis,* University of Washington Press, Seattle, 1988. An introduction to the mathematics of symmetry, splendidly illustrated with photographs of patterns from cultures all over the world. Includes a complete analysis of patterns with two colors and proofs that there are only four rigid motions in the plane and exactly seven strip patterns.

Suggested Web Sites

http://www.geom.uiuc.edu/software/tilings/Tessellation resources. Lists programs for various platforms that allow the user to create designs featuring the rosette, strip, and wallpaper patterns.

escher.epfl.ch/escher/ Interactive Escher Web Sketch program that allows a user to design repeating patterns. Choose a wallpaper pattern using crystallographic notation and draw on the screen a colored design for the motif; the program then reproduces the motif using the pattern. The software (for Windows, Macintosh, and Unix) can also be downloaded.

www.geom.uiuc.edu/java/Kali/ Interactive Java Kali Web program that lets the user draw pictures under the action of rosette, strip, or wallpaper groups. Versions for various platforms can be downloaded.

www.wordsmith.org/~anu/java/spirograph.html Interactive Spirograph Java application (which you can download) that lets you do electronically what the Spirograph toy does.

http://csdt.rpi.edu/african/African_Fractals/ An African fractals site.

classes.yale.edu/fractals/index.html Web site for Benoit Mandelbrot's course in fractals at Yale. Features many applets for different kinds of IFS (for example, incorporating randomness), including a fractal music composer.

http://www.math.smith.edu/phyllo//index.html Phyllotaxis: An interactive site for the study of plant pattern formation, created by Pau Atela and Christophe Golë.

http://www.maths.surrey.ac.uk/hosted-sites/R.Knott/Fibonacci/fib.html Fibonacci numbers and the golden section. Splendidly illustrated extensive Web pages by Ron Knott about Fibonacci numbers and the golden ratio: their occurrences in nature, their applications, puzzles, and much more.

Walter Bibikow/eStock Photo

20

Tilings

Our ancestors had an artistic impulse to cover floors and walls with patterns and mosaics, in buildings ranging from Roman dwellings to Muslim mosques (see Figure 20.1). They expressed the same desire for patterns in other decorative arts as well—carpets, fabrics, baskets, and even linoleum.

They use repeated shapes ("tiles") to cover a flat surface, without gaps or overlaps. Such patterns, apart from their esthetic appeals, can also have practical applications. In manufacturing, for example, stamping components from a sheet of metal is most economical if the shapes of the components fit together without gaps—in other words, if the shapes form a **tiling**.

We build tiling patterns "from the ground up." We start with one or more types of tile and ask if they can fit together in a pattern, and if so, how. Surprisingly, there is no way to decide this question in all cases. For some sets of tiles, we can exhibit tilings; for others, we can prove that there can't be any tilings. But mathematicians have proved that there is no algorithm (mechanical step-by-step process) that can tell for every conceivable set of tile shapes which of the two situations holds. (See Chapter 9, pp. 324–355, for other examples of "unattainable ideals" in regard to voting.)

We begin in Section 20.1 with just one kind of tiling, in which all the tiles are the same size. We first analyze just which regular polygons can serve as tiles. There aren't many that work, so in Section 20.2, we consider polygons that don't have all equal sides. Along the way, we encounter the contributions of an amateur mathematician who contributed solutions to a simple-sounding problem that is still not completely resolved.

In Section 20.3, we let our imagination loose and consider tiles in the shapes of horsemen, fish, or any other artistic shape that you might like. Which of them can tile? In Section 20.4, we let the tiles appear both upside down as well as right side up.

Finally, in Section 20.5, we ask what would happen if we had two kinds of tiles. Some pairs of tiles can tile the plane in a pattern that never repeats. Such tilings have recently been recognized in medieval Persian architecture, and a three-dimensional version is the structure of atoms in new ultra-strong alloys.

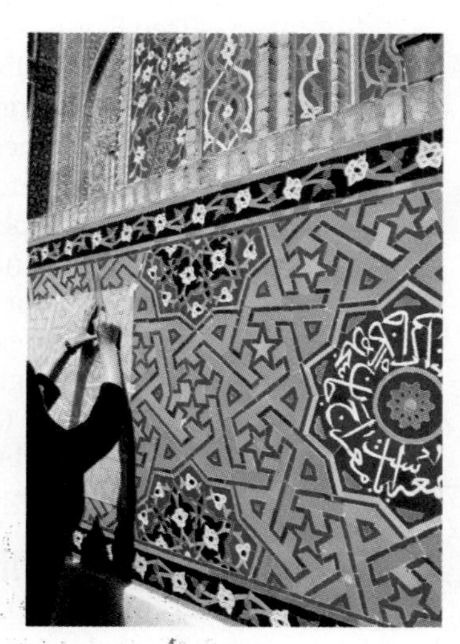

Figure 20.1 Arab mosaic. *(Jose Fuste Raga/Corbis.)*

20.1 Tilings with Regular Polygons

Tiling (Tessellation) DEFINITION

A **tiling (tessellation)** is a covering of the entire infinite plane by nonoverlapping regions called *tiles.*

The simplest tilings use only one size and shape of tile. They are known as *monohedral tilings.*

Monohedral Tiling DEFINITION

A **monohedral tiling** is a tiling that uses only one size and shape of tile.

The simplest tiles are **regular polygons**, which have all sides the same length and all angles equal. A square is a regular polygon with four equal sides and four equal interior angles; a triangle with all sides equal (an **equilateral triangle**) is also a regular polygon. A polygon with five sides is a pentagon, one with six sides is a hexagon, and one with n sides is an ***n*-gon**. Regular polygons have a high degree of symmetry. Each has the reflection and rotation symmetries of a dihedral rosette pattern (see Section 19.2 on p. 695). In three dimensions, the corresponding highly symmetrical figures are called *regular polyhedra* (see Spotlight 20.1).

An **exterior angle** of a polygon is an angle formed by one side and the extension of an adjacent side (see Figure 20.2). At each vertex of the polygon, there are two exterior angles, depending on which side we extend; but we will consider only one of them. Let us agree to extend the sides consistently in turn as we proceed counterclockwise around the polygon, as in Figure 20.2, producing the set of exterior angles A through E, one at each vertex.

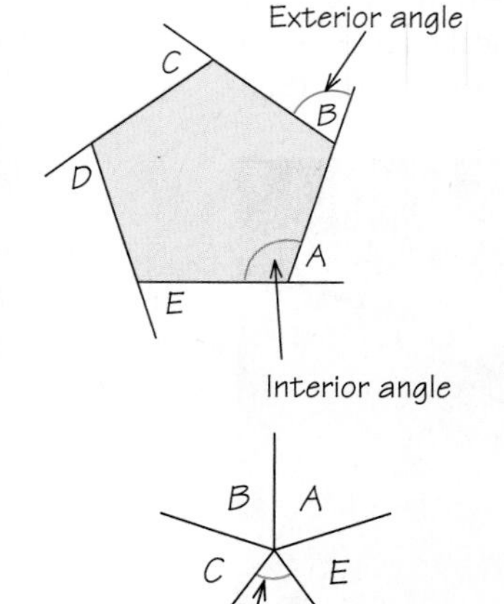

Figure 20.2 The exterior angles of a regular pentagon, like those of any regular polygon, add up to 360 degrees. Each exterior angle measures 72 degrees.

By a convention dating back to the ancient Babylonians, angles are measured in degrees, with the total of angles around a point being 360 degrees. If we bring a set of exterior angles together at a point, we can see that they add up to 360 degrees (see Figure 20.2). Hence, for a regular polygon with n sides, each exterior angle must measure $360°/n$. For example, a square, with $n = 4$ sides, has 4 exterior angles in a set, each measuring 90 degrees; a regular pentagon, with $n = 5$ sides, has 5, each measuring 72 degrees; a regular hexagon, with 6 sides, has 6, each measuring 60 degrees.

Each exterior angle is paired with a corresponding **interior angle** (the angle inside the polygon formed by the two adjacent sides), and the pair add up to a straight line, or 180 degrees. For a regular polygon with more than six sides, each interior angle is between 120 degrees and 180 degrees. That's because if $n > 6$, the exterior angle is $360°/n$, which has to be less than 60 degrees; since (interior angle) $= 180° -$ (exterior angle), the interior angle has to be at least 120 degrees (and can't be any larger than 180 degrees if the polygon is not to bend back upon itself). This consideration will shortly prove crucial in determining how regular polygons can fit together to form tilings.

Regular Tilings

Edge-to-Edge DEFINITION

A tiling is **edge-to-edge** if all the tiles are polygons and for every tile, each edge coincides with the entire edge of a bordering tile.

SPOTLIGHT 20.1

Regular Polyhedra and Buckyballs

The three-dimensional analogue of a regular polygon is a regular polyhedron, a convex solid whose faces are regular polygons that are all alike (same number of sides, same size), with each vertex surrounded by the same number of polygons. Although there are infinitely many regular polygons, there are only five regular polyhedra, a fact proved by Theaetetus (414–368 B.C.). They were called the *Platonic solids* by the ancient Greeks.

If the restriction that the same number of polygons meet at each vertex is relaxed, five additional convex polyhedra are obtained, all of whose faces are equilateral triangles. If we allow more than one kind of regular polygon, 13 further convex polyhedra are obtained, known as the *semiregular polyhedra* or *Archimedean solids* (although there is no documented evidence that Archimedes studied them—but in the early 1600s, Johannes Kepler catalogued them all). Once inflated, the truncated icosahedron—whose faces are pentagons and hexagons—is known throughout the world as a regulation soccer ball. Drawings of it appear in the work of Leonardo da Vinci.

The truncated icosahedron is also the structure of C_{60}, a form of carbon known as buckminsterfullerene and, more familiarly, the "buckyball." A total of 60 carbon atoms lie at the 60 vertices of this molecule, which was discovered in 1985. It is named after R. Buckminster Fuller (1895–1983), inventor and promoter of the geodesic dome. The molecule buckminsterfullerene resembles a dome.

The buckyball is part of a family of carbon molecules, the *fullerenes*, in which each carbon atom is joined to three others. Then, 30 years before the discovery of fullerenes, mathematicians had shown that a convex polyhedron in which every vertex has three edges must have 12 pentagon faces and may have any number of hexagon faces, from 0 on up, except for 1.

That there must be 12 pentagons follows from a famous equation due to Leonhard Euler (1707–1783). For any convex polyhedron, it must be true that $v - e + f = 2$, where v is the number of vertices, e is the number of edges, and f is the number of faces of the polyhedron.

In 2003, astronomers and mathematicians advanced a remarkable new theory about why the universe does not show as much historic fluctuation in temperature as other models predict. This lack of fluctuation could be explained by the universe being in the shape of a dodecahedron (the figure shown here with 12 pentagonal sides), with opposite faces coinciding. This theory harks back to Kepler, who had conceived of the universe in terms of the five regular polyhedra nested within one another.

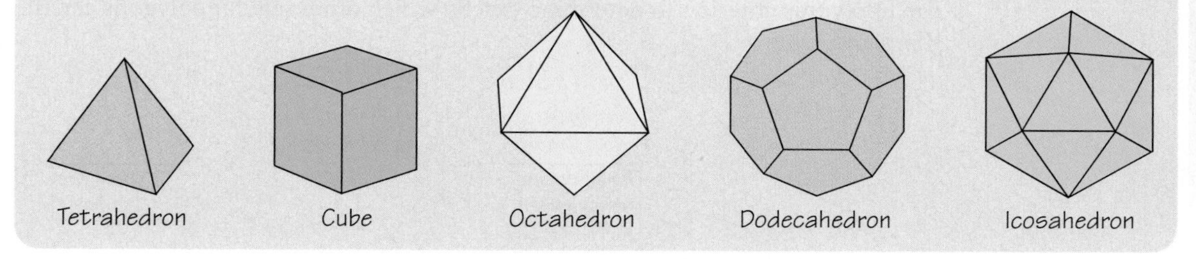

Figure 20.3 shows one tiling that is not edge-to-edge and another that is.

Regular Tiling DEFINITION

A **regular tiling** is an edge-to-edge tiling that uses only one kind of regular polygon.

A square tile is the simplest case. Apart from varying the size of the square, which would change the scale but not the pattern of the tiling, we can get

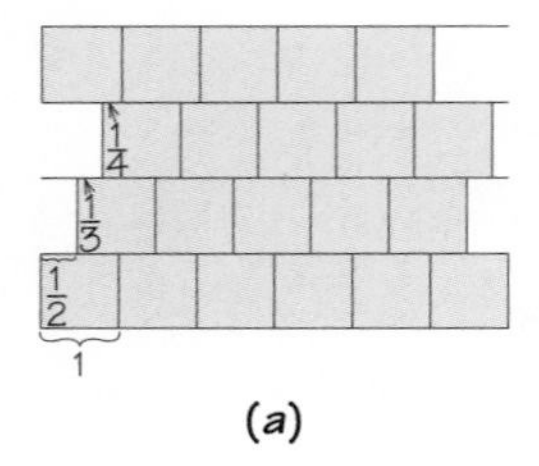

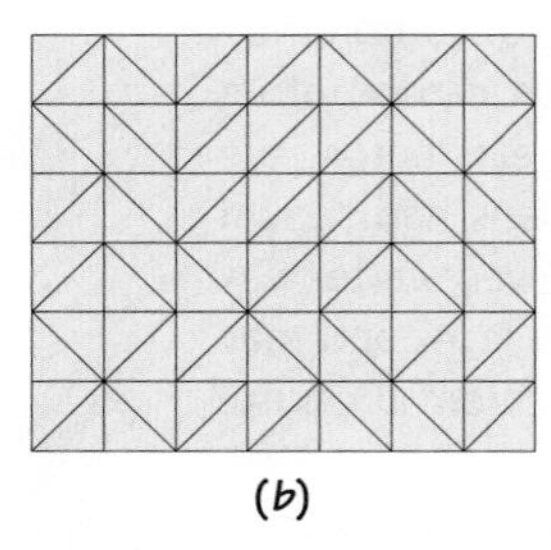

Figure 20.3 (a) A tiling that is not edge-to-edge. The horizontal edges of two adjoining squares do not exactly coincide. (b) A tiling by right triangles that is edge-to-edge.

different tilings by offsetting one row of squares some distance from the next (see Figure 20.3a). However, there is only one edge-to-edge tiling pattern using a square.

What about tilings with regular triangles? We can get a regular tiling with equilateral triangles by arranging them in rows and alternately inverting triangles. (See the upper-left pattern in Figure 20.5 on page 736.) As with squares, there is only one pattern of equilateral triangles that forms an edge-to-edge tiling. (Any tiling by squares can be refined to one by triangles by drawing a diagonal of each square [as in Figure 20.3b], but the resulting right triangles are not regular [equilateral].)

What about tiles with more than four sides? An edge-to-edge tiling with regular hexagons is easy to construct. (See the upper-right pattern in Figure 20.5 on page 736.)

However, if we look for a tiling with regular pentagons, we won't find one. How do we know whether we're just not being clever enough or there really isn't one to be found? This is the kind of question that mathematics is uniquely equipped to answer. In the other sciences, phenomena may exist even though we have not observed them; such was the case for bacteria before the invention of the microscope. However, we can conclude with certainty that there is no edge-to-edge tiling with regular pentagons.

The proof is easy and numerical. As we calculated earlier, the exterior angles of a pentagon are each 72 degrees; each corresponding interior angle is thus 108 degrees (see Figure 20.4). How many pentagons can meet at a point? The total of all of the angles around a point must be 360 degrees. As you can see in Figure 20.4, four pentagons at a point would be too many (their angles would add to $4 \times 108° = 432°$, so they'd have to overlap), and three would be too few (their angles would add to $3 \times 108° = 324°$, so some of the area wouldn't be covered). Because 108 does not evenly divide into 360, *regular pentagons can't tile the plane*.

With this argument, we can do something that is characteristic of mathematics—we can *generalize* it to a criterion for when a regular polygon can tile the plane; namely, when the total of its interior angles divides 360 evenly. We can apply this criterion to determine exactly which other regular polygons can tile the plane.

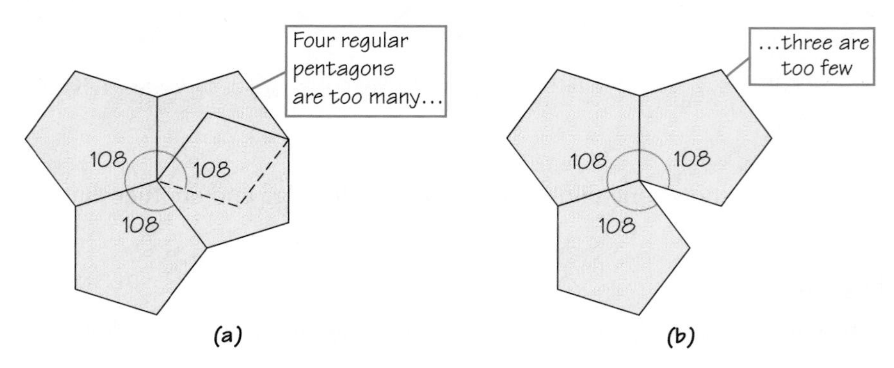

Figure 20.4 Polygons that come together at a vertex in a tiling must have interior angles that add up to 360 degrees—no more, no less.

EXAMPLE 1
Identifying the Regular Tilings

SOLUTION A regular hexagon has interior angles of 120 degrees; 120 divides into 360 evenly, and three regular hexagons fit together exactly around a point. A regular 7-gon (heptagon)—or any regular polygon with more than 6 sides—has interior angles that are larger than 120 degrees but smaller than 180 degrees. Now 360 divided by 120 gives 3, and 360 divided by 180 gives 2—and there aren't any other possibilities in between. Angles between 180 degrees and 120 degrees divided into 360 degrees will give a result between 2 and 3, and which consequently is not an integer. So there are no regular tilings of the plane with polygons of more than six sides.

Only Three Regular Tilings THEOREM

The only regular tilings are the ones with equilateral triangles, with squares, and with regular hexagons.

The follow-up question, of course, is which *combinations* of regular polygons of different numbers of sides can tile the plane edge-to-edge.

Vertex Type DEFINITION

In an edge-to-edge tiling by regular polygons, the **vertex type** of a vertex is the arrangement of the polygons around the vertex.

To describe a vertex type, we list the sizes of polygons, separated by periods, in either clockwise or counterclockwise order starting from the smallest number of sides. For example, $4 \cdot 4 \cdot 4 \cdot 4$ (or 4^4 for short) denotes 4 squares meeting at a vertex, with 4 angles of 90 degrees each. Similarly, $4 \cdot 6 \cdot 12$ denotes a square followed by a hexagon then by a dodecagon (12-gon); see the tiling in the middle of the bottom row of Figure 20.5. Two vertices have the same type even if one has the polygons in clockwise order and the other has them in counterclockwise order; both clockwise and counterclockwise versions of the 4.6.12 type occur in that tiling in Figure 20.5. In both cases, the sum of the interior angles at the vertex is 360 degrees = 90 degrees (the square) + 120 degrees (the hexagon) + 150 degrees (the dodecahedron—do you see why it is 150 degrees?).

Semiregular Tiling DEFINITION

A systematic tiling that uses a mix of regular polygons with different numbers of sides but in which all vertex types are alike—the same polygons in the same order, clockwise or counterclockwise—is called a **semiregular tiling** (see Figure 20.5).

As before, the technique of adding up angles at a vertex (to be 360 degrees) can eliminate some impossible combinations, such as "pentagon, pentagon, pentagon"

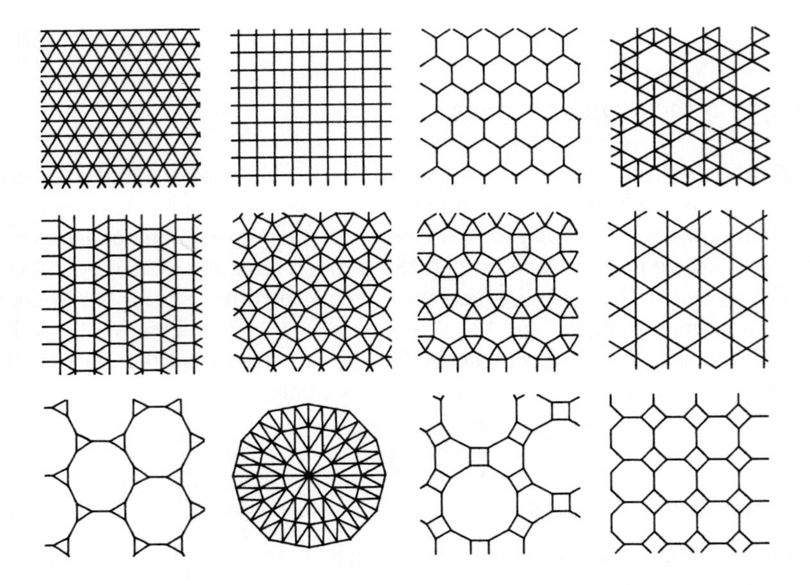

Figure 20.5 The three regular tilings . . . and the eight semiregular tilings, plus a "mystery" tiling that does not belong to either group. Can you identify it? (*Hint*: It uses just one tile, which isn't regular.)

(Figure 20.4). Once we have found an arrangement that is numerically possible, we must confirm the actual existence of each tiling by constructing it (showing that it is geometrically possible). For example, even though a possible arrangement of regular polygons around a point is "triangle, square, square, hexagon," it is not possible to construct a tiling with that vertex figure at every vertex.

The result of such an investigation is that in a semiregular tiling, no polygon can have more than 12 sides. In fact, polygons with 5, 7, 9, 10, or 11 sides do not occur either. Figure 20.5 exhibits all the semiregular tilings.

If we abandon the restriction about the vertex types being the same at every vertex, then there are *infinitely many* systematic edge-to-edge tilings with regular polygons, even if we continue to insist that all polygons with the same number of sides have the same size.

20.2 Tilings with Irregular Polygons

What about edge-to-edge tilings with irregular polygons, which may have some sides longer than others or some interior angles larger than others? We will look just at monohedral tilings (in which all tiles have the same size and shape) and investigate in turn which triangles, **quadrilaterals** (four-sided polygons), hexagons, and so forth, can tile the plane.

The most general shape of a triangle has all sides of different lengths and all interior angles of different sizes. Such a triangle is called a **scalene triangle**, from the Greek word for "uneven." We can always take two copies of any triangle and fit them together to form a **parallelogram**, a quadrilateral whose opposite sides are parallel (Figure 20.6a). It's easy to see that we then can use such parallelograms to tile the plane by making strips and then fitting layers of strips together edge-to-edge (Figure 20.6b).

Tiling with Triangles THEOREM

Any triangle whatever—whether it has all three sides equal, two sides equal, or no sides equal—can tile the plane.

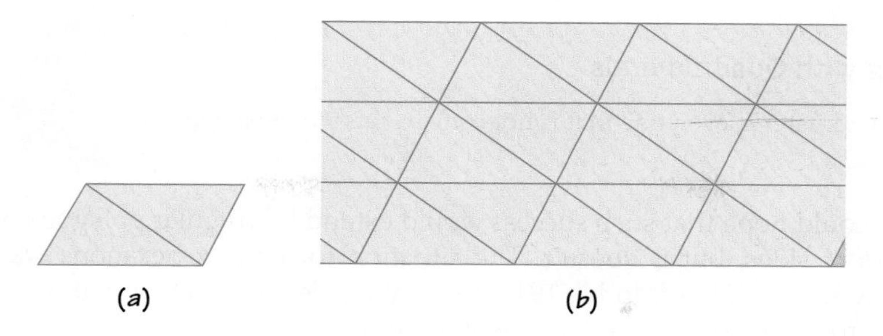

Figure 20.6 (a) Two triangles form a parallelogram. (b) Any triangle (even a scalene one) can tile the plane.

What about quadrilaterals? We have seen that squares tile the plane, and rectangles certainly will, too. We have just noted that any parallelogram will tile. What about a quadrilateral (four-sided polygon) with its opposite sides not parallel, as in Figure 20.7a? The same technique as for triangles will work. We fit together two copies of the quadrilateral, forming a hexagon whose opposite sides are parallel. Such hexagons fit next to each other to form a tiling, as in Figure 20.7b.

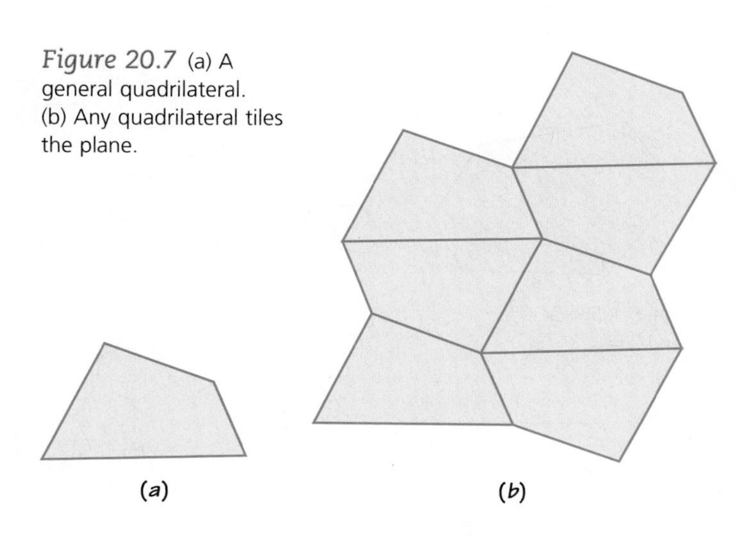

Figure 20.7 (a) A general quadrilateral. (b) Any quadrilateral tiles the plane.

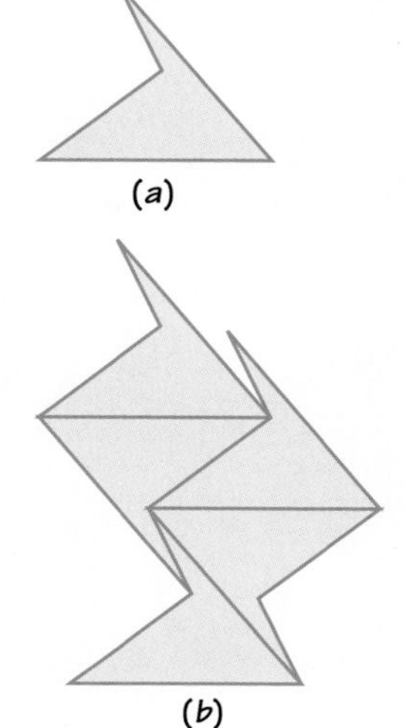

Figure 20.8 (a) A general nonconvex quadrilateral. (b) Any quadrilateral, convex or not, tiles the plane.

The quadrilaterals shown in Figure 20.7 are all **convex**, meaning that if you take any two points on the tile (including the boundary), the line segment joining them lies entirely within the tile (again, including the boundary). The quadrilateral of Figure 20.8a is not convex, but the same approach works for using it to form a tiling (Figure 20.8b).

Convex DEFINITION

A tile is **convex** if for any two points on it (including its boundary), all the points on the line segment joining them also belong to the tile (including its boundary).

Tiling with Quadrilaterals THEOREM

Any quadrilateral, even one that is not convex, can tile the plane.

We could hope that such success would extend to irregular polygons with any numbers of sides, but it doesn't. The situation for convex hexagons was determined by Karl Reinhardt in his 1918 doctoral thesis. He showed that for a convex hexagon to tile, it must belong to one of three classes. Examples of the three classes are shown in Figure 20.9. Tilings with a hexagon of type 2 use both ordinary and mirror-image versions of the hexagon.

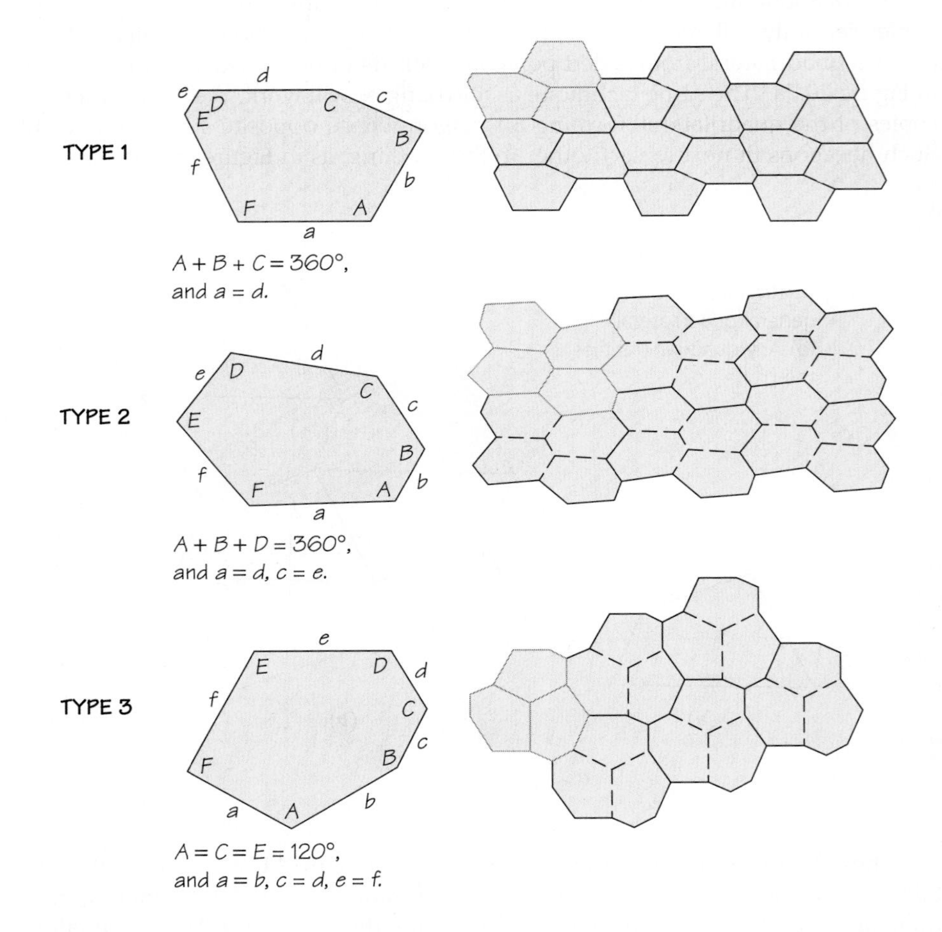

Figure 20.9 The three types of convex hexagon tiles.

Tiling with Hexagons THEOREM

Exactly three classes of convex hexagons can tile the plane.

Reinhardt also explored convex pentagons and found five classes that tile. For example, any pentagon with two parallel sides will tile. Reinhardt did not complete the solution, as he did for hexagons, by proving conclusively that no other pentagons could tile. He claimed that it would be very tedious to finish the analysis. Still, he felt that he had found them all. In 1968, after 35 years of working on the problem on and off, R. B. Kershner, a physicist at Johns Hopkins University, discovered three more classes of pentagons that will tile. Kershner was sure that he had found all pentagons that tile, but, like Reinhardt, he did not offer a complete proof, which "would require a rather large book."

When an account of the "complete" classification into 8 types appeared in the July 1975 issue of *Scientific American*, the article provoked an amateur mathematician to discover a 9th type! Marjorie Rice, a housewife with no formal education in mathematics beyond high school "general mathematics" taken 36 years earlier, devised her own mathematical notation and found four more types (see Spotlight 20.2). A 14th type was found by a mathematics graduate student in 1985. Since then—26 years and counting—no new types have been discovered, yet no one knows if the classification is complete.

With the situation so intricate for convex pentagons, you might think that it must be still worse for polygons with seven or even more sides. In fact, however, the situation is remarkably simple, as Reinhardt proved in 1927.

Tiling with Polygons with More Sides THEOREM

A convex polygon with seven or more sides cannot tile.

M. C. Escher and Tilings

The Dutch artist M. C. Escher (1898–1972) was inspired by the great variety of decoration in tilings in the Alhambra, a 14th-century palace built during the last years of Islamic dominance in Spain. He devoted much of his career of making prints to creating tilings with tiles in the shapes of living beings (a practice forbidden to Muslims). Those prints of interlocking animals and people have inspired awe and wonder among people all over the world. Figures 20.10 through 20.13 illustrate a few of his drawings and finished works. Like Marjorie Rice, he, too, developed his own mathematical notation for the different kinds of patterns for the tilings.

20.3 Using Only Translations

Following Escher's artistic genius, we investigate monohedral tilings in which the tile used may be nonconvex and have boundary any curve whatever (rather than one consisting of only straight edges). You may wonder just how much liberty can be taken in shaping a tile and how you could design an Escher-like tiling yourself.

[Technical note: We restrict ourselves to *isohedral tilings,* in which any tile can be moved to any other tile by one of the wallpaper groups of symmetries of Spotlight 19.7, pages 701–702.] In practical terms, if you copied the tiling onto a sheet of clear plastic, chose any tile on the paper original and any tile on the plastic, and made them coincide (perhaps by turning the plastic over), then all the tiles on the plastic would coincide exactly with tiles on the original.]

SPOTLIGHT 20.2

In Praise of Amateurs

Marjorie Rice
(Courtesy of Kathy Rice.)

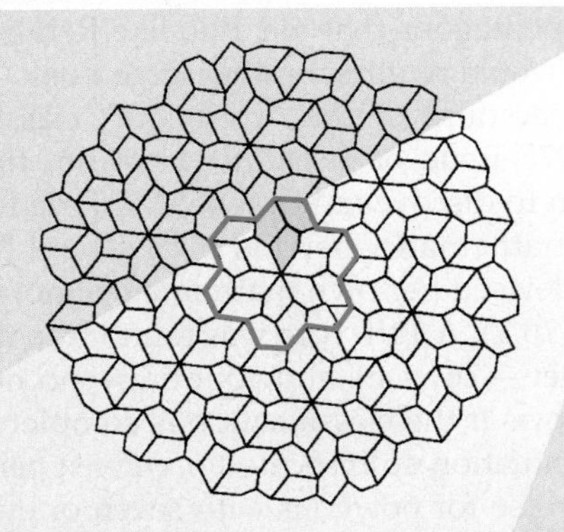

This tiling in the headquarters of the Mathematical Association of America in Washington, D.C., was discovered in 1995 by Marjorie Rice. The angles of each pentagon tile are 60 degrees, 90 degrees, 120 degrees, and 150 degrees— all multiples of 30 degrees. The tiling is periodic, although not every pentagon is surrounded in the same way. Three pentagons form a fundamental block, and the outlined group of 18 pentagons tiles by translations.

R. B. Kershner's claim to have found all convex pentagons that tile was read by many puzzle enthusiasts, including Richard James III and Marjorie Rice. James found a tiling that Kershner had missed.

Rice, a San Diego housewife and mother of five, read about James's new tile. "I thought I would see if I could find still another type. It was a delightful new puzzle to me."

With no formal education in mathematics beyond a general mathematics course in high school, she not only worked out her own method of attack, but she invented her own notation as well.

"I began drawing little diagrams on my kitchen counter when no one was there, covering them up quickly if someone came by, for I didn't wish to have to explain what I was doing. I was searching for a new type, and a few weeks later, I found it." Over the next two years, she found three additional new tilings.

Rice was born in 1923 in St. Petersburg, Florida, and went to a one-room country school.

"When I was in the sixth or seventh grade, our teacher pointed out to us the Golden Section in the proportions of a picture frame. This immediately caught my imagination and I never forgot it. I've . . . been especially interested in architecture and the ideas of architects and planners such as Buckminster Fuller. I've come across the Golden Section again in my reading and considered its use in painting and design."

After high school, Rice worked until her marriage in 1945. She was drawn back into mathematics by her children, finding solutions to their homework problems "by unorthodox means, since I did not know the correct procedures." She became especially interested in textile design and the works of M. C. Escher. As she pursued the pentagonal tilings, she produced some imaginative Escher-like patterns. (See Figure 20.19 and the figure here.)

What makes a person pursue a problem so patiently and persistently? Marjorie Rice was not trained for it, nor was she paid, but she gained great personal satisfaction from the pursuit.

Intense spirit of inquiry and keen perception are the forte of all such amateurs. No formal education provides these gifts. Lack of a mathematical degree separates these "amateurs" from the "professionals," yet their curiosity and ingenious methods make them true mathematicians.

Source: Adapted from Doris Schattschneider, "In Praise of Amateurs," in David A. Klarner (ed.), *The Mathematical Gardner*, pp. 140–166, plus Plates I–III, Wadsworth, Belmont, CA, 1981.

The simplest case is when the tile is just *translated* in two directions; that is, when copies are laid edge-to-edge in rows, as in Figure 20.10. Each tile must fit exactly into the ones next to it, including its neighbors above and below. We say that each tile is a **translation** of each other one because we can move one to coincide with another without doing any rotation or reflection.

Figure 20.10 Escher No. 128 (Bird), from Escher's 1941–1942 notebook. *(© 1967 M. C. Escher Foundation, Baarn, Holland, all rights reserved.)*

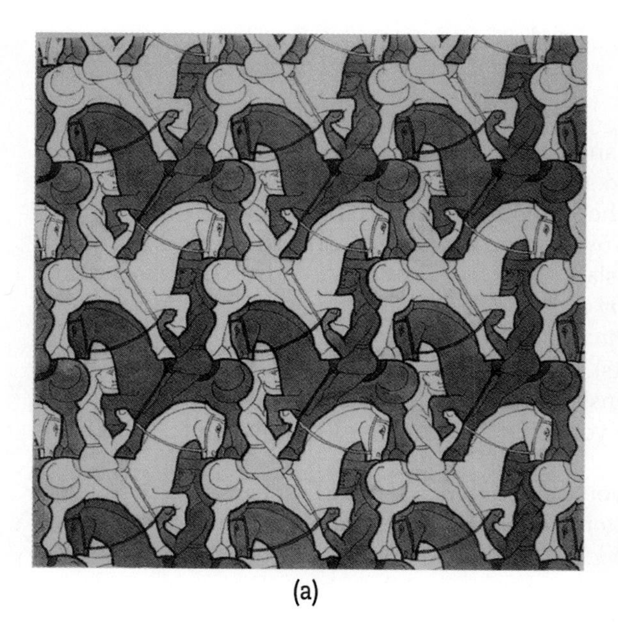

(a)

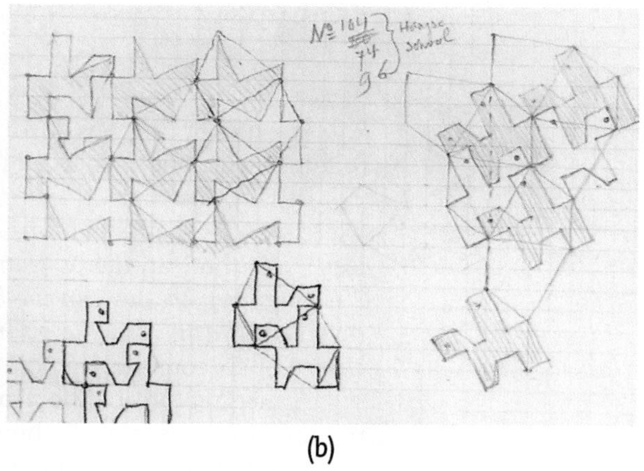

(b)

Figure 20.11 (a) Escher No. 67 (Horseman), from Escher's 1941–1942 notebook. *(© 1947 M. C. Escher Foundation, Baarn, Holland, all rights reserved.)* (b) Sketch showing the tile design for the Horseman print. *(© 1947 M. C. Escher Foundation, Baarn, Holland, all rights reserved. From the collection of Michael S. Sachs.)*

When is it possible for a tile to cover the plane in this manner? The boundary of the tile must be divisible into matching pairs of opposing parts that will fit together. Figures 20.10 and 20.11 illustrate two basic ways that this can happen. In the first, two opposite pairs of sides match; in the second, three opposite pairs of sides match.

Translation Criterion RULE

A tile can tile the plane by translations alone if either

1. There are four consecutive points A, B, C, and D on the boundary such that
   (a) the boundary part from A to B is congruent by translation to the boundary part from D to C, and
   (b) the boundary part from B to C is congruent by translation to the boundary part from A to D (see Figure 20.12a) or
2. There are six consecutive points A, B, C, D, E, and F on the boundary such that the boundary parts AB, BC, and CD are congruent by translation, respectively, to the boundary parts ED, FE, and AF (see Figure 20.12b).

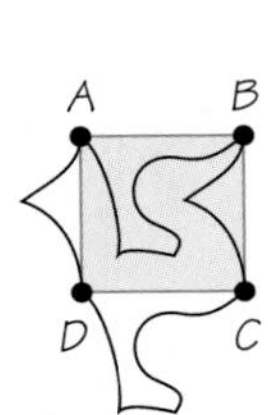

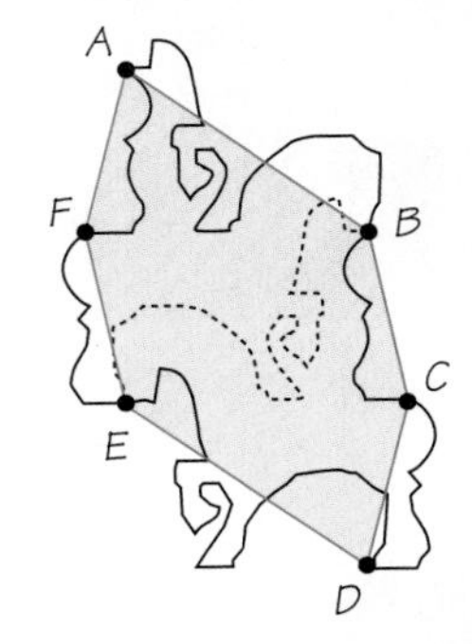

Figure 20.12 Individual tiles traced from the Escher prints of Figures 20.10 and 20.11, with points marked to show they fulfill the criteria for tiling by translations. The two horsemen form a block that tiles by translations, although a single horseman can tile by itself if we allow mirror-image reflections, too.

The tiles for Figures 20.10 and 20.11 are shown in outline form in Figure 20.12, together with points marked to show how the tiles fulfill the criterion.

In fact, alternative 1 of the criterion is a special case of alternative 2 (see Exercises 21–24 and 26 on p. 759). Moreover, alternative 2 completely "characterizes" tiles that can tile by translations. That is, not only is it true that *if alternative 2 is true, then the tile can tile by translations,* but also that the criterion works "in reverse": *If a tile can tile by translations, then alternative 2 must be true* (for some choice of six consecutive points).

A nice feature of the Translation Criterion is that if you can find points as required for alternative 2, then you can join them in order, as in Figures 20.12a and b, to see how to do the tiling.

To create tilings, though, you can proceed exactly as Escher did. His notebooks show that he designed his patterns in just the way that we now describe.

EXAMPLE 2
Tiling Starting from a Parallelogram

SOLUTION For the first alternative of the criterion, start from a parallelogram, make a change to the boundary on one side, and then copy that change to the opposite side. Similarly, change one of the other two sides and copy that change on the side opposite it (see Figure 20.13). Revise as necessary, always making the same change to opposite sides. You might find it useful (as Escher did) to make your designs on graph paper, or you can work by cutting and taping together pieces of heavy paper.

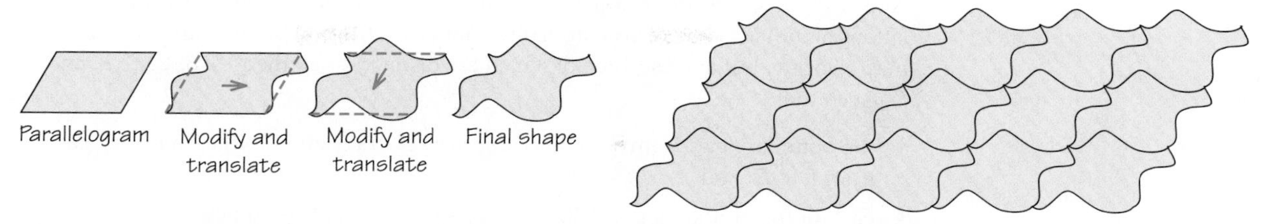

Figure 20.13 How to make an Escher-like tiling by translations, from a parallelogram base.

EXAMPLE 3

Tiling Starting from a Hexagon

SOLUTION For the second alternative, start from a **par-hexagon**, a hexagon whose opposite sides are equal and parallel. This is one of the kinds of hexagons that tile the plane. Again, make a change on one boundary and copy the change to the opposite side, and do this for all three pairs of opposite sides (see Figure 20.14).

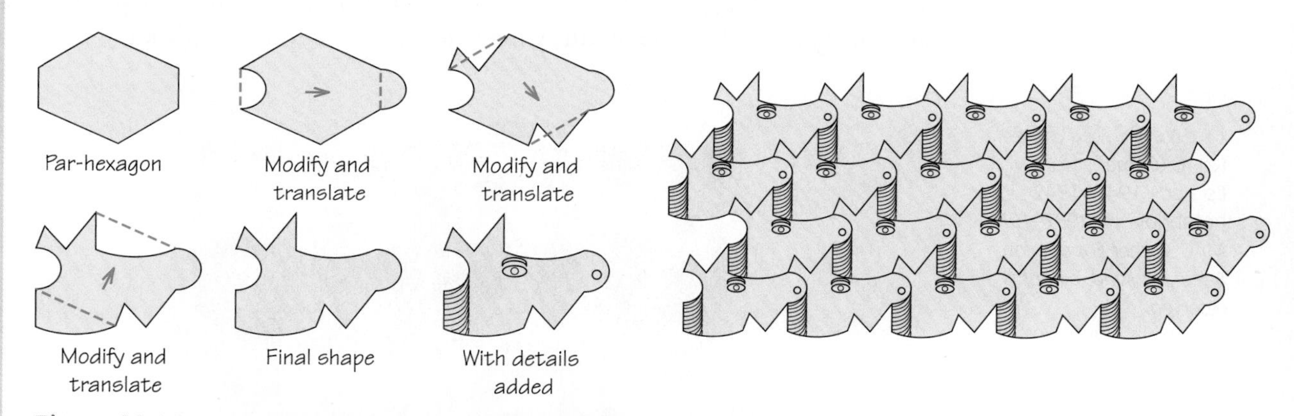

Figure 20.14 How to make an Escher-like tiling by translations, from a par-hexagon base.

20.4 Using Translations Plus Half-Turns

If the tiling is to allow half-turns, so that some of the figures are "upside down," the part of the boundary of a right-side-up figure has to match the corresponding part of itself in an upside-down position. For that to happen, that part of the boundary must be **centrosymmetric**; that is, symmetric about (unaltered by) a 180-degree rotation around its midpoint. The key to some of Escher's more sophisticated monohedral designs, and the fundamental principle behind some further easy recipes for making Escher-like tilings, is the **Conway criterion**, formulated by John H. Conway (b. 1937) of Princeton University.

Conway Criterion RULE

A tile can tile the plane by translations plus half-turns (including possibly by translations alone) if there are six consecutive points on the boundary (some of which may coincide, but at least three of which are distinct)—call them *A, B, C, D, E,* and *F*—such that

- the boundary part from *A* to *B* is congruent by translation to the boundary part from *E* to *D,* and
- each of the boundary parts *BC, CD, EF,* and *FA* is centrosymmetric.

The first condition means that we can match up the two boundary parts exactly, curve for curve, angle for angle. The second condition means that each of the remaining boundary parts is brought back into itself by a half-turn around its center. Either condition is automatically fulfilled if the boundary part in question is a straight-line segment.

The tiles for Figures 20.15 and 20.16 are shown in outline form in Figure 20.17, together with points marked to show how the tiles fulfill the Conway criterion. Figure 20.15 shows that Escher sketched tiny circles exactly where we have red dots in Figure 20.17a.

Mathematicians do not know whether the Conway criterion completely characterizes tiles that can tile by translations and half-turns. Tiles that fulfill the Conway criterion can tile by translations and half-turns, but there might be tiles that tile that way but do not satisfy the criterion—however, nobody knows of any.

Figure 20.15 Escher No. 6 (*Camel*), from Escher's 1941–1942 notebook. *(© 1937–1938 M. C. Escher Foundation, Baarn, Holland, all rights reserved.)*

Figure 20.16 Escher No. 88 (*Sea Horse*). *(© 1947 M. C. Escher Foundation, Baarn, Holland, all rights reserved.)*

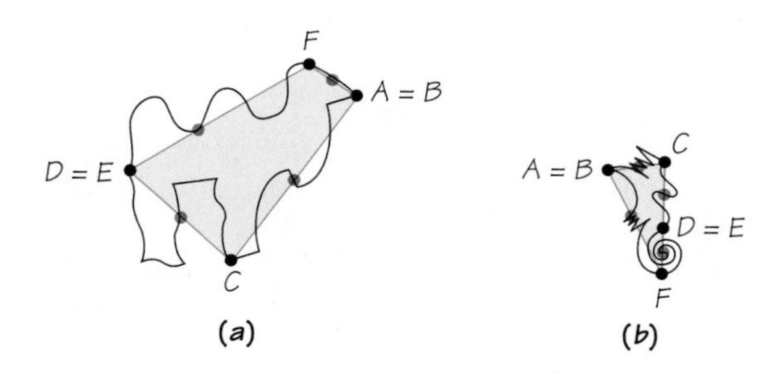

Figure 20.17 Individual tiles traced from the Escher prints of Figures 20.15 and 20.16, with points marked to show that they fulfill the Conway criterion for tiling by translations and half-turns *(around the red dots).*

(However, the Conway criterion does completely characterize tiles that produce the wallpaper pattern *p2* of Spotlight 19.7 on page 701: Any tile that satisfies the criterion can be used to make a *p2* pattern, and any tile that can produce that pattern must satisfy the criterion.)

We have considered tilings that tile by translations alone or by translations plus half-turns. You may have noticed differences among the figures, in the following ways:

- In Figures 20.10 and 20.14, all the birds and the horseheads face the same way, and none are upside down. What if we wanted to allow some of them to be upside down?
- In Figures 20.15 and 20.16, although there are upside-down camels and seahorses, all the right-side-up ones face the same way and all the upside-down ones face the opposite way. What if we wanted to allow both kinds to face both ways?
- In Figure 20.11, there are no upside-down horsemen, but the dark horsemen face left while the light horsemen face right. The two kinds of horsemen are mirror images of each other; but we avoided letting the tile occur in mirror-image reflections of itself by building the mirroring into the tile itself. However, we noted in Figure 20.12 that a single horseman could be a tile by itself if we allow mirror-image reflections.

In addition to the Translation Criterion and the Conway Criterion, there are seven other sufficient criteria for a tile to produce a tiling, depending on what operations we allow on the tile: mirror reflections, or rotations by various angles (60 degrees, 90 degrees, 120 degrees, or 180 degrees). (We have already considered 180-degree rotations; they are half-turns in the Conway Criterion.) We can't consider them all here, or show the extent to which Escher used them; but in the "Suggested Readings" and "Suggested Web Sites" sections at the end of the chapter, we point to relevant sources and explanations.

There is a simple way to implement the Conway Criterion to make Escher-like tilings by starting from any triangle or any quadrilateral. We illustrate this creative process in the following examples.

EXAMPLE 4
Tiling Using a Triangle

SOLUTION For a triangle, modify half of one side, then rotate that side around its center point to extend the modification to the rest of the side, thereby making the new side centrosymmetric. Then you can do the same to the second and third sides (Figure 20.18).

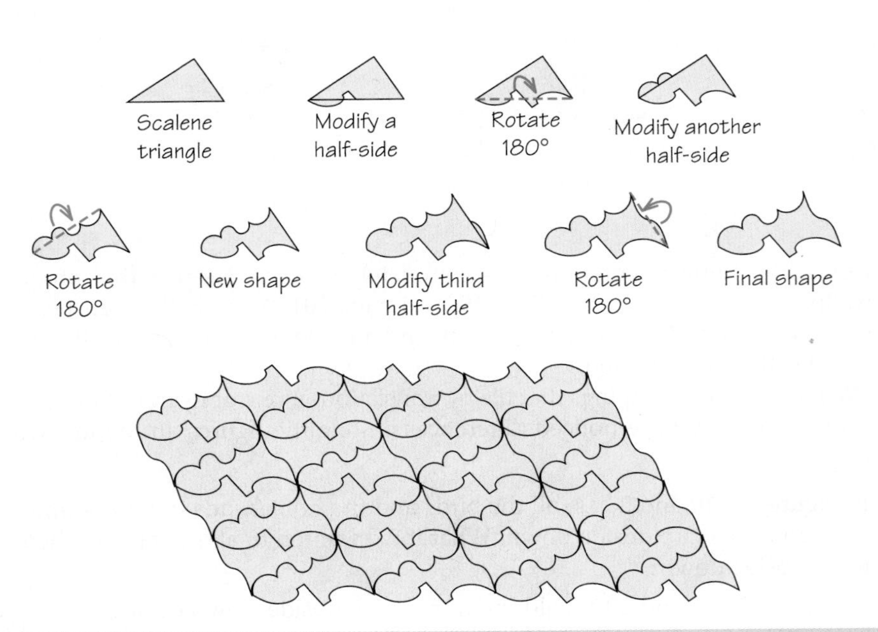

Figure 20.18 How to make an Escher-like tiling by translations and half-turns, using a scalene triangle base.

EXAMPLE 5
Tiling Using a Quadrilateral

SOLUTION For the quadrilateral, do the same process as in Example 4, modifying each of the four sides, or as many as you wish (Figure 20.19).

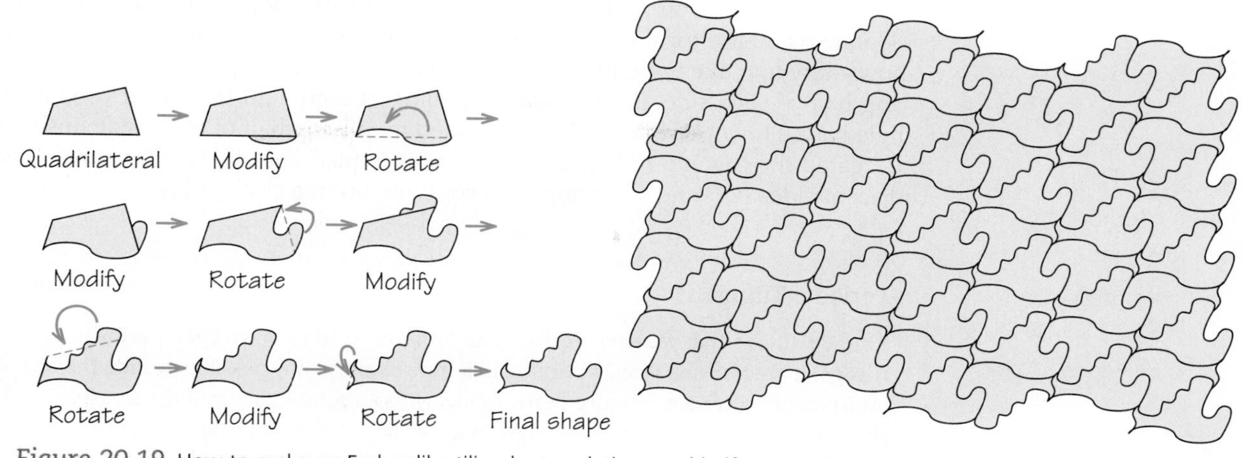

Figure 20.19 How to make an Escher-like tiling by translations and half-turns, using a quadrilateral base.

The same approach will work with some of the sides of some pentagons and hexagons that tile. Because not all sides can be modified, there is less freedom for designing tiles, so it is more difficult to make the resulting tiles resemble intended figures. Figure 20.20 shows the beautiful results achieved by Marjorie Rice using one of the unusual tilings by pentagons that she discovered.

Figure 20.20 *Fish*, by Marjorie Rice, based on one of her unusual tilings by pentagons.

Sketches in Escher's notebook indicate how he designed many of his prints. For the bird tiling of Figure 20.10, the single bird below the tiling shows that he modified the sides of a square. For the horsemen tiling of Figure 20.11a, the sketches in Figure 20.11b show that he modified the pairs of sides of a par-hexagon. We redraw the two fundamental figures more clearly in Figure 20.12. (The horsemen tiling also has a reflection symmetry, taking a leftward-facing light horseman to a rightward-facing dark horseman; but we have not discussed criteria for producing a tiling with such a symmetry.)

As can be seen in faint lines in Figure 20.15 (p. 744), Escher used a parallelogram as a base for the camel tiling. In Figure 20.17a (p. 745), the blue overlay shows how to make the tiling starting from a more general quadrilateral by modifying half of each side. For the seahorse tiling of Figure 20.16, Escher used a triangle base. However, once more he avoided modifying half of every side; instead, he treated the triangle *ACF* (Figure 20.17b) as a quadrilateral *ACDF* in which two adjacent sides (*CD* and *DF*) happen to continue on in a straight line.

Periodic Tilings PROCEDURE

All the patterns that we have exhibited and discussed so far have been **periodic tilings**. If we transfer a periodic tiling to a transparency, it is possible to slide the transparency a certain distance horizontally, without rotating it, until the transparency exactly matches the tiling everywhere. We can also achieve the same result by moving the transparency in some second direction (possibly vertically) by a certain (possibly different) distance.

In a periodic tiling, you can identify a **fundamental region**—a tile, or a block of tiles—with which you can cover the plane by translations at regular intervals. For example, in Figure 20.10, a single bird forms a fundamental region. In Figure 20.15, two adjacent camels, one right side up and one upside down, form a fundamental region. In the terminology of Chapter 19, the periodic tilings are ones that are preserved under translations in more than two directions.

20.5 Nonperiodic Tilings

In Figure 20.3a (on p. 734), the second row from the bottom is offset one-half of a unit to the right from the bottom row, the third row from the bottom is offset one-third of a unit further, and so forth. Because the sum $\frac{1}{2} + \frac{1}{3} + \frac{1}{4} + \cdots + \frac{1}{n}$ never adds up to exactly a whole number, there is no direction (horizontal, vertical, or diagonal) in which we can move the entire tiling and have it coincide exactly with itself.

Nonperiodic Tiling DEFINITION

A **nonperiodic tiling** is a tiling in which there is no regular repetition of the pattern by translation.

EXAMPLE 6
A Nonperiodic Tiling Through Randomness

SOLUTION Consider the usual edge-to-edge square tiling. For each square, flip a coin. Depending on the result, divide the square into two right triangles by adding either a rising or a falling diagonal (see Figure 20.3b on p. 734). Because what happens in each individual square is unconnected to what happens in the rest of the tiling, this random tiling by right triangles has no chance of being periodic.

Penrose Tiles and Quasicrystals

For all known cases, if a single tile can be used to make a nonperiodic tiling of the plane, then it also can be used to make a periodic tiling. It is still an open question whether this property is true for every possible shape in two dimensions. In 1993, Conway discovered an example in three dimensions of a single convex polyhedron that tiles space non-periodically but cannot be used to make a periodic tiling.

For a long time, mathematicians also tended to believe the more general assertion that if you can construct a nonperiodic tiling with a set of one or *more* tiles, you can construct a periodic tiling from the same tiles. In 1964, however, a set of tiles was found that permits only nonperiodic tiling. It contains 20,000 different shapes! Over the next several years, smaller sets were discovered with the same property, with as few as 100 shapes. But it was still amazing when Sir Roger Penrose (b. 1931), a mathematical physicist at Oxford, announced in 1975 that he had found a set that tiles only nonperiodically—consisting of just two tiles! (See Figure 20.21 and Spotlight 20.3.)

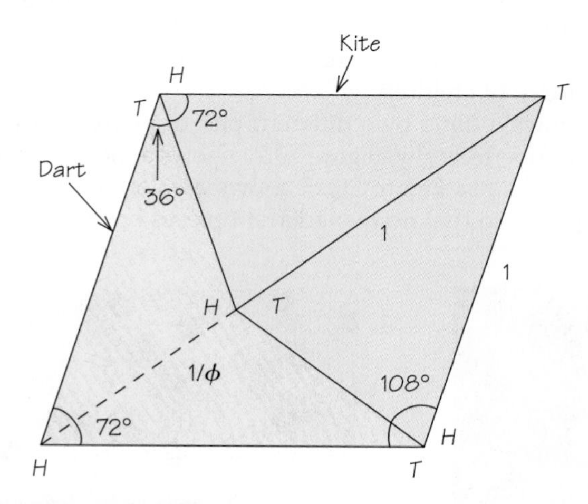

Figure 20.21 Construction of Penrose's "dart" (beige area) and "kite" (blue area). The length $1/\phi \approx 0.618$ is the reciprocal of the golden ratio ϕ.

SPOTLIGHT 20.3

Sir Roger Penrose

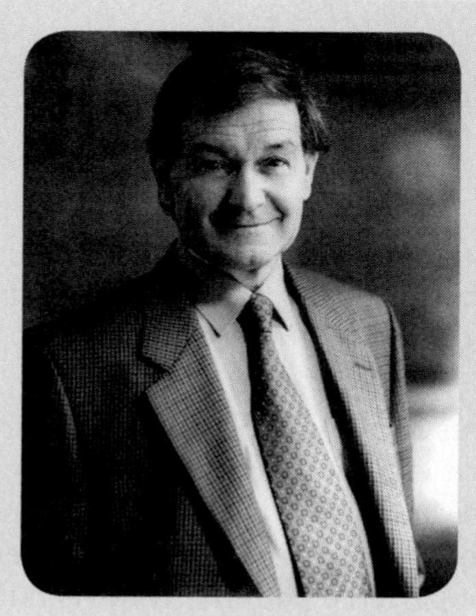

Sir Roger Penrose
(Courtesy Sir Roger Penrose)

Sir Roger Penrose, a professor at the University of Oxford, received a doctorate in mathematics but has been seriously interested in physics for many years; he was one of the first to conjecture the existence of black holes. He discovered what are now called the Penrose pieces in 1973. His latest endeavor has been to try to establish that the mind is not a machine; that is, that the ideas and concepts of artificial intelligence cannot explain human consciousness.

A chance meeting with M. C. Escher resulted in Penrose sending Escher some art by Penrose's grandfather, which inspired some of Escher's prints.

Penrose called his tiles "darts" and "kites," and both of these *Penrose tiles* can be obtained from a single rhombus. A **rhombus** is a quadrilateral with four equal sides and equal opposite interior angles. The particular rhombus from which the Penrose tiles are constructed has interior angles of 72 degrees and 108 degrees. If we cut the longer diagonal in two pieces so that the longer piece is the golden ratio, or $(1 + \sqrt{5})/2 \approx 1.618$ times as long as the shorter (see Chapter 19, page 688), and connect the dividing point to the remaining corners, we split the rhombus into a dart and a kite (Figure 20.21).

Label the front and back vertices of the dart with H (for head) and its two wing tips with T (for tail), and do the reverse for the kite. Then the rule for fitting the pieces together is that only vertices with the same letter may meet: Heads must go to heads, tails must go to tails. Thus, the rules don't allow the pieces to fit together as a rhombus (which would allow them to tile periodically).

A prettier method of enforcing the rules, proposed by Conway, is to draw circular arcs of different colors on the pieces and require that adjacent edges must join arcs of the same color. The result is the pretty patterns of Figure 20.22. In fact, Conway thinks of the darts as children, each with two hands. The rule for fitting the pieces together is that children are required to hold hands. Penrose patterns become dancing circles of children.

Figure 20.23 shows a tiling by a different pair of pieces, both rhombuses, that tile the plane only nonperiodically. Figure 20.24 shows a modification of the Penrose pieces into two bird shapes. Figure 20.25 shows a coloring of one particular tiling with the Penrose pieces so that no two adjacent pieces have the same color.

Figure 20.22 A Penrose tiling with specially marked tiles, forming what is known as the *cartwheel tiling*. *(From Sir Roger Penrose.)*

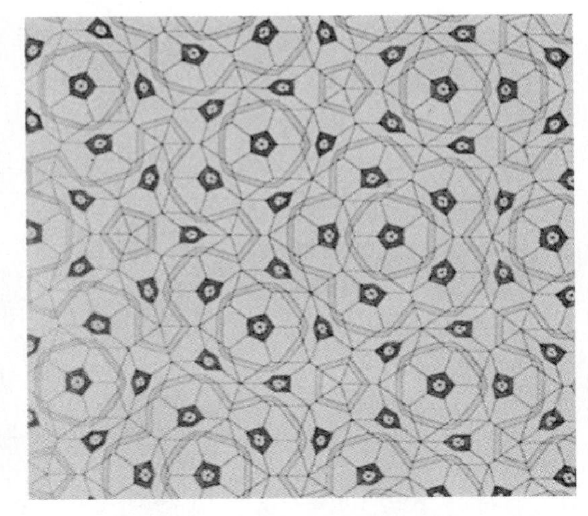

Figure 20.23 A Penrose nonperiodic tiling made with two rhombus shapes, one thin and one fatter. The fatter one has a yellow stripe across one end. *(Tiling by Sir Roger Penrose.)*

Although tilings with Penrose's pieces cannot be periodic, the tilings possess unexpected symmetry. In Chapter 19 we explore our intuitions of symmetry in terms of *balance, similarity,* and *repetition.* Patterns made with the Penrose pieces certainly involve repetition, but it is the balance in the arrangement that we seek. What balance can there be in a nonperiodic pattern? It turns out that some Penrose patterns have a single line of reflection. But most surprising of all is that every Penrose pattern has a kind of fivefold rotational symmetry.

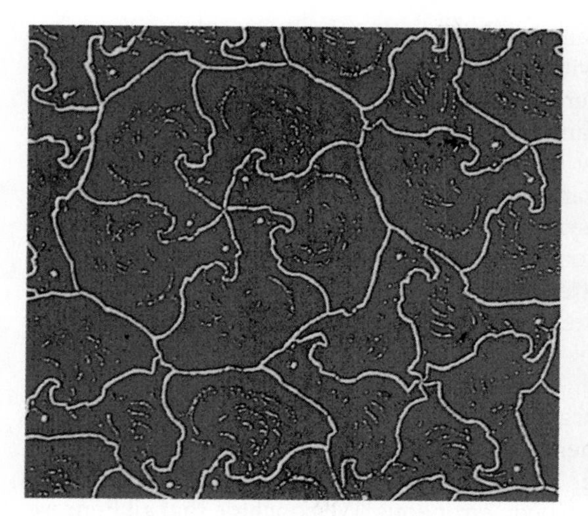

Figure 20.24 A modification of a Penrose tiling by refashioning the kites and darts into bird shapes. *(Tiling by Sir Roger Penrose.)*

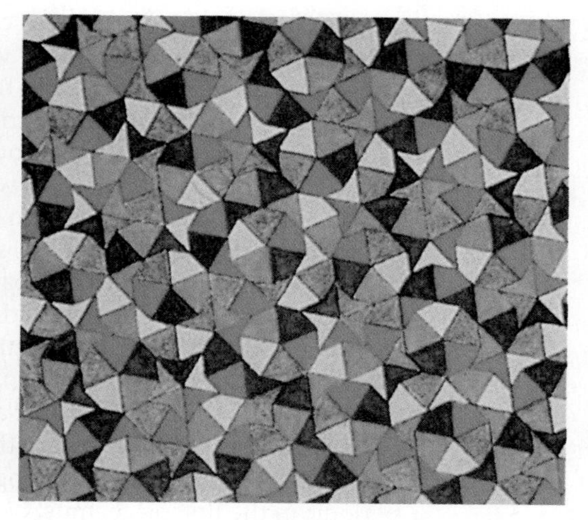

Figure 20.25 A Penrose tiling by kites and darts, colored with five colors. A Penrose tiling can always be colored using just three colors in such a way that two tiles that share an edge have different colors. *(Tiling by Sir Roger Penrose.)*

EXAMPLE 7
How Does a Penrose Pattern Have Fivefold Symmetry?

SOLUTION Look again at Figure 20.21 on page 749, which shows how to split a rhombus into the Penrose dart and kite pieces. Except in the recess of the dart and the matching part of the kite, each of the internal angles of the kite and of the dart is either 72 degrees or 36 degrees.

Now, 72 degrees goes into 360 degrees 5 times, and 36 degrees goes into 360 degrees 10 times. If we recall that it is the interior angles that matter in arranging polygons around a point, we see why it might be possible for a Penrose pattern to have fivefold or tenfold rotational symmetry.

A Penrose pattern with tenfold rotational symmetry is impossible, but there are exactly two Penrose patterns that tile the entire plane with fivefold rotational symmetry about one particular point. We show finite parts of these patterns in Figure 20.26. For each pattern, the center of rotational symmetry is at the center of the figure, surrounded by either five darts or five kites.

Figure 20.26 Successful deflation (that is, the systematic cutting up of large tiles into smaller ones) of patches of tiles of a Penrose nonperiodic tiling.

For any other Penrose pattern, the pattern as a whole does not have fivefold rotational symmetry. However, what is surprising is that the pattern must have arbitrarily large finite regions with fivefold rotational symmetry. You can see this feature in the regions of Figure 20.23 that are enclosed by yellow lines. In Conway's metaphor, whenever a chain of children (darts) closes, the region inside has fivefold symmetry.

Conway invented a process called *inflation,* which turns any Penrose pattern into a different Penrose pattern with larger darts and kites. The inflation operation—we don't give the details here—systematically cuts up the darts and kites into triangles and regroups the triangles into larger darts and kites.

We can use inflation to show that a Penrose pattern must be nonperiodic. Suppose (contrary to what we want to establish) that some Penrose pattern is periodic; that is, it has translation symmetry. Let d be the distance along the translation direction to the first repetition. Performing inflation does the same thing to each repetition, so the inflated pattern must still have translation symmetry and a distance d along the translation direction to the first repetition. Keep on performing inflation, time after time, until the darts and kites are so large that they are more than d across.

The pattern, as we have just argued, must still have translation symmetry at a distance d—but it can't because there's no repetition inside a single tile! We reach a contradiction. So what's wrong? Our initial supposition, that the pattern was periodic in the first place, must have been erroneous. We conclude that all Penrose tilings are nonperiodic.

Despite their being nonperiodic, all Penrose patterns are somewhat alike, in the following remarkable sense.

Figure 20.27 Penrose toilet paper. *(Mario Ruiz/ Time Magazine.)*

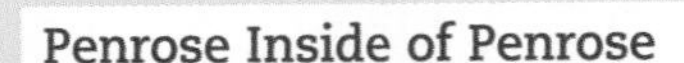

Penrose Inside of Penrose THEOREM

The subpattern of any finite region in one Penrose pattern is contained somewhere inside every other Penrose pattern. In fact, any subpattern occurs infinitely many times in every Penrose pattern.

The nonperiodicity of Penrose filings found a surprising application in 1997—to bathroom tissue. Quilted bathroom tissue is embossed with a pattern to keep the layers together (Figure 20.27). If the pattern is regular, then the multiple layers on the roll can produce lumpy ridges and grooves. Using a nonrepeating Penrose pattern averts the lumpiness. However, the company used Penrose's pattern without his permission, and Penrose sued successfully.

Penrose tilings have another feature that allows us to characterize them as *quasiperiodic,* or somewhere between periodic and random. (Noting the precise definition of *random* would take us too far afield.) Robert Ammann introduced lines onto the two rhombic Penrose pieces (used in Figure 20.23 on p. 750) that are now known as *Ammann bars.* In any Penrose tiling, these bars line up into five sets of parallel lines, each set rotated 72 degrees from the next, forming a pentagonal grid (see Figure 20.28). The distance between two adjacent parallel bars is one of only two values, either A or B. Do you want to guess what the ratio of the longer A is to the shorter B? You don't think it could possibly be anything but the golden ratio of Chapter 19, do you? And so it is.

EXAMPLE 8
Musical Sequences

What about the order in which the A's and B's occur, as we move from left to right in Figure 20.28? Is there any pattern to that?

SOLUTION From the limited part of the pattern that we can observe, we see the sequence as

$$ABAABABAABABA$$

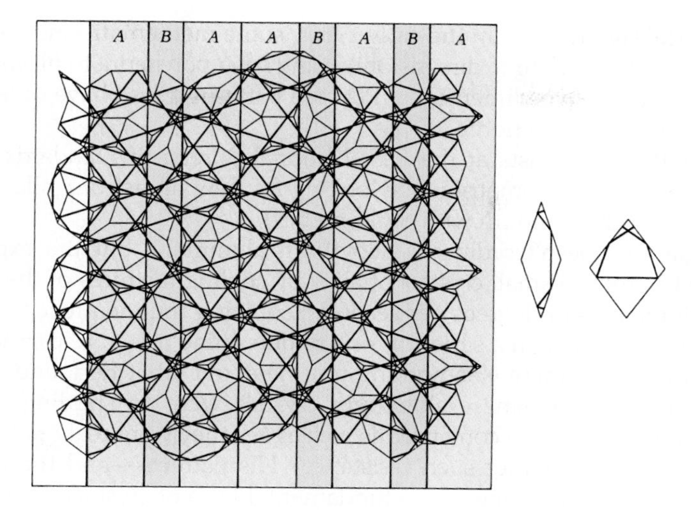

Figure 20.28 Penrose tilings with Ammann bars. Specially placed lines on the tiles produce five sets of parallel bars in different directions.

You might think from the figure that the pattern continues repeating the group

ABAAB

indefinitely; after all, there are five symbols in this group. But such is not the case. The sequence of intervals between Ammann bars is nonperiodic: it cannot be produced by repeating any finite group of symbols. We can think of it as a one-dimensional analogue of a Penrose tiling. The notation is reminiscent of the melody pattern of songs: Many popular songs follow the pattern *ABA*, with the first and the last sections having the same melody but the middle section being different. Consequently, a sequence of intervals between Ammann bars is known as a *musical sequence*.

There is some regularity in musical sequences. Two *B*'s can never be next to each other, nor can we have three *A*'s in a row. Just as any finite part of any Penrose tiling occurs infinitely often in any other Penrose tiling, any finite part of any musical sequence appears infinitely often in any other one. The order of the symbols is neither periodic nor random, but between the two—quasiperiodic.

The ratio of darts to kites in an infinite Penrose tiling, or of *A*'s to *B*'s in a musical sequence, is exactly the golden ratio, approximately 1.618. So if you are going to play with sets of Penrose pieces to see what kinds of patterns you can create, you will need about 1.6 times as many darts as kites.

As pointed out by geometers Marjorie Senechal (Smith College) and Jean Taylor (Rutgers University), Penrose tilings have three important properties:

- They are constructed according to rules that force nonperiodicity.
- They can be obtained from a substitution process (inflation and deflation) that features self-similarity at different scales (like the fractals in Chapter 19).
- They are quasiperiodic.

These properties are somewhat independent, meaning that one or two may be true of a tiling without all three being true.

Quasicrystals and Barlow's Law

Although Penrose's discovery was a big hit among geometers and in recreational mathematics circles in the mid-1970s, few people thought that his work might have practical significance. In the early 1980s, some mathematicians even generalized Penrose tilings to three dimensions, using solid polyhedra to fill space nonperiodically. Like the two-dimensional Penrose patterns, these have orderly fivefold symmetry but are nonperiodic.

Yet in 1982, scientists at the U.S. National Bureau of Standards discovered unexpected fivefold symmetry while looking for new ultrastrong alloys of aluminum (mixtures of aluminum with other metals).

Manganese doesn't ordinarily alloy with aluminum, but the experimenters were able to produce small crystals of alloy by cooling mixtures of the two metals at a rate of millions of degrees per second. Following routine procedures, chemist Daniel Shechtman began a series of tests to determine the atomic structure of the special crystals. But there was nothing routine about what he found: The atomic structures of the manganese-aluminum crystals were so startling that it took Shechtman three years to convince his colleagues they were real.

Why did he encounter such resistance? His patterns—and the crystals that produced them—defied one of the fundamental laws of crystallography. Like our discovery that the plane cannot be tiled by regular pentagons, **Barlow's law**, also called the **crystallographic restriction**, says that a crystal must be periodic and hence can have only rotational symmetries that are twofold, threefold, fourfold, or sixfold. If there were a center of fivefold symmetry, there would have to be many such centers. Barlow proved this impossible.

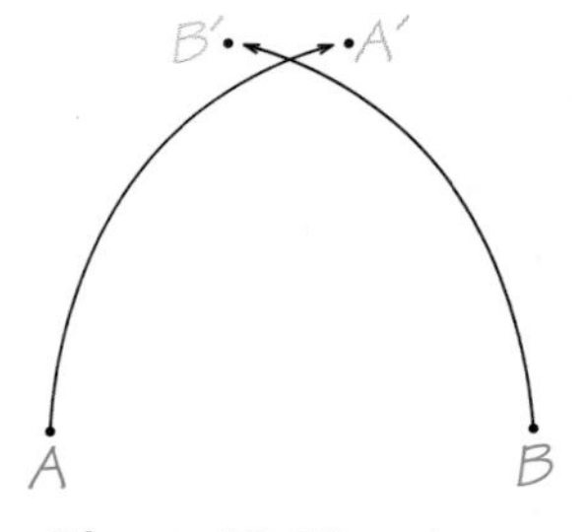

Figure 20.29 Barlow's proof that no pattern can have two centers of fivefold symmetry.

Peter Barlow (1776–1862) argued by contradiction, similar to Conway's proof in which we saw earlier that Penrose patterns are not periodic. Suppose (contrary to what we intend to show) that there is more than one fivefold rotation center. Let A and B be two of these that are closest together (see Figure 20.29). Rotate the pattern of Figure 20.28 by one-fifth of a turn clockwise around B, which carries A to some point A'. Because the pattern has fivefold symmetry around B, the point A', which is the image of the fivefold center A, must itself be a fivefold center.

Now use A as a center and rotate the pattern by one-fifth of a turn clockwise, which carries B to some point B'. As we just argued in the case of A', B' must also be a fivefold center. But A' and B' are closer together than A and B, which is a contradiction. Hence our original supposition must be false, and a pattern can have at most one fivefold rotation center (as the patterns in Figure 20.26 in fact do) and so cannot be periodic.

For chemists, crystals are modeled well by periodic three-dimensional tilings; an array of atoms with no symmetry whatever would not be considered a crystal. Since Barlow's law shows that fivefold symmetry is impossible in a *periodic* tiling, no one suspected until Penrose's discovery that there could be symmetric *non*periodic tilings, nor, until Schechtman's alloys, that real atoms could arrange themselves in such a way.

Schechtman's alloys, since they are not periodic, are not crystals, though in other respects they do resemble crystals. It is scientifically more fruitful to extend the concept of crystals to include them than to rule them out. They are now known as *quasicrystals* (see Spotlight 20.4).

Once again, as so often happens in history, pure mathematical research anticipated scientific applications. Penrose's discovery, once just a delightful piece of recreational mathematics, has prompted a major reexamination of the theory of crystals. Barlow's law is not refuted since it applies only to periodic crystals, not to quasicrystals.

SPOTLIGHT 20.4

Quasicrystals

In 1984, working at the University of Pennsylvania, Paul Steinhardt and Don Levine did a computer simulation of what a three-dimensional Penrose pattern would be like. They decided to call such structures *quasicrystals*. Later that fall, their chemist colleague Daniel Shechtman showed that quasicrystals really exist. He produced images of an alloy of aluminum and manganese that were amazingly similar to images from the computer simulations. In short order, sevenfold, ninefold, and other symmetries were also shown to occur in real materials.

In 1991, Sergei Burkov showed that quasiperiodic tilings can be made using only a single kind of 10-sided tile, *provided the tiles are allowed to overlap*. With overlaps, the resulting patterns are no longer tilings. They are called *coverages*. In late 1998, scientists presented electron microscope photos that demonstrated that atoms really do form such coverages.

The current theory is that quasicrystals are packings of copies of a single type of atom cluster, with each cluster sharing atoms with its neighbors; that is, overlapping nearby clusters. The clusters form a quasiperiodic pattern that maximizes their density, thereby minimizing the energy of the atoms involved.

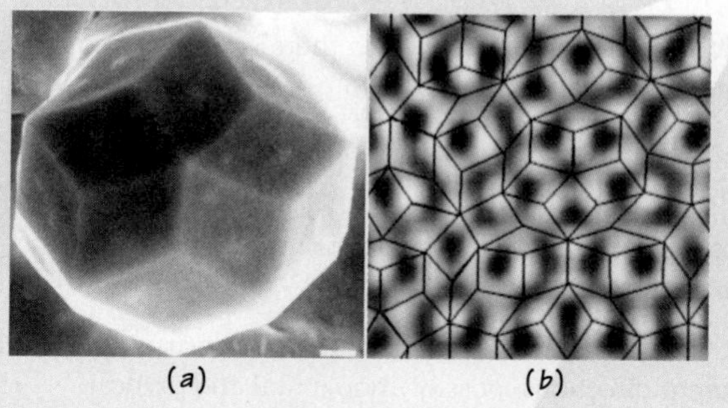

(*a*) (*b*)

(a) A scanning electron microscope image of the quasicrystal alloy $Al_{5???}Li_3Cu$ (the question marks indicate uncertainty about how many aluminum atoms are involved). The fivefold symmetry can be seen in the five rhombic faces that meet at a single point in the center of the photograph, forming a starlike shape. (b) This image of the quasicrystal material $Al_{65}Co_{20}Cu_{15}$ was obtained with a scanning tunneling microscope. The resulting image has been overlaid with a nonperiodic tiling to display the local fivefold symmetry. *[Both adapted from Hans C. von Baeyer, "Impossible Crystals,"* Discover *11(2) (February 1990): 69–78, 84.]*

In 2007, Steinhardt and Peter J. Lu announced the discovery of decagonal and Penrose tilings in medieval Islamic architecture in Iran.

REVIEW VOCABULARY

Barlow's law, or the **crystallographic restriction** A law of crystallography that states that a crystal may have only rotational symmetries that are twofold, threefold, fourfold, or sixfold. (p. 754)

Centrosymmetric Symmetric by 180 degrees of rotation around its center. (p. 743)

Convex A tile is convex if for any two points on it (including its boundary), all the points on the line segment joining them also belong to the tile (including its boundary). (p. 737)

Conway criterion A criterion for determining whether a shape can tile by means of translations and half-turns. (p. 743)

crystallographic restriction See Barlow's Law. (p. 754)

Edge-to-edge tiling A tiling is edge-to-edge if all the tiles are polygons and for every tile, each edge coincides with the entire edge of a bordering tile. (p. 732)

Equilateral triangle A triangle with all three sides equal. (p. 732)

Exterior angle The angle outside a polygon formed by one side and the extension of an adjacent side. (p. 732)

Fundamental region A tile or group of adjacent tiles that can tile by translations. (p. 748)

Interior angle The angle inside a polygon formed by two adjacent sides. (p. 732)

Monohedral tiling A tiling with only one size and shape of tile. (The tile is allowed to occur also in "turned-over," or mirror-image, form.) (p. 732)

***n*-gon** A polygon with *n* sides. (p. 732)

Nonperiodic tiling A tiling in which there is no repetition of the pattern by translation. (p. 748)

Parallelogram A convex quadrilateral whose opposite sides are equal and parallel. (p. 736)

Par-hexagon A hexagon whose opposite sides are equal and parallel. (p. 743)

Periodic tiling A tiling that repeats at fixed intervals in two different directions, possibly horizontal and vertical. (p. 748)

Quadrilateral A polygon with four sides. (p. 736)

Regular polygon A polygon whose sides and angles are all equal. (p. 732)

Regular tiling An edge-to-edge tiling that uses only one kind of regular polygon. (p. 733)

Rhombus A parallelogram whose sides are all equal—four equal sides and equal opposite interior angles. (p. 750)

Scalene triangle A triangle with no sides equal. (p. 736)

Semiregular tiling A systematic tiling that uses a mix of regular polygons with different numbers of sides but in which all vertex types are alike—the same polygons in the same order, clockwise or counterclockwise. (p. 735)

Tessellation A tiling. (p. 732)

Tiling A covering of the entire infinite plane by nonoverlapping regions, called *tiles*. (p. 731)

Translation A rigid motion that moves everything a certain distance in one direction. (p. 740)

Vertex type The pattern of polygons surrounding a vertex in a tiling. (p. 735)

SKILLS CHECK

1. In a tiling of the plane, the tiles

(a) are allowed to overlap, so long as no area is left uncovered.
(b) are not allowed to overlap.
(c) must meet edge to edge.

2. The exterior angle of a regular octagon is ________.

3. In a tiling of the plane, the tiles

(a) have to be regular polygons.
(b) have to be polygons but need not be regular.
(c) can be any shape at all.

4. A regular tiling can be constructed using polygons with _____, _____, or _____ sides.

5. Regular octagons and squares can form a semiregular tiling of the plane with

(a) two octagons and one square at each vertex.
(b) two octagons and two squares at each vertex.
(c) a varying configuration at the vertices.

6. A semiregular tiling has a square, a regular dodecagon (12-gon), and another regular polygon at each vertex. This other polygon has ________ sides.

7. A tessellation

(a) allows overlapping pieces.
(b) is not the same as a tiling.
(c) covers the entire infinite plane.

8. There are _____ regular polyhedra.

9. How many semiregular tilings are there?

(a) 5
(b) 8
(c) Infinitely many

10. The smallest number of sides that a polygon can have and not be able to tile the plane is ________.

11. A tiling of the plane can be formed using as a tile

(a) some but not all convex quadrilaterals.
(b) any convex quadrilateral, but no nonconvex quadrilaterals.
(c) any quadrilateral.

12. A regular polygon with ______ or more sides cannot tile the plane.

13. A tiling of the plane can be formed using which of the following as a tile?

(a) Some but not all pentagons
(b) Any pentagon with at least two right angles
(c) Any pentagon with at least three right angles

14. Any quadrilateral can tile the plane by ________.

15. Regular pentagons

(a) can't tile the plane.
(b) can tile the plane, but only if you are very careful.
(c) don't occur in any tilings.

16. An artist famous for works based on tilings is __________.

17. A convex irregular polygon

(a) can never tile the plane.
(b) can always tile the plane.
(c) cannot tile the plane if it has more than six sides.

18. The tile below can be used to tile the plane using __________.

19. Which of the following is true?

(a) If a polygon fulfills the Conway criterion, it can tile the plane by translations.
(b) If a polygon fulfills the Conway criterion, it can tile the plane by translations and half-turns.
(c) If a polygon doesn't fulfill the Conway criterion, it can't tile the plane at all.

20. The __________ criterion says that the tile below can be used to create a tiling of the plane using __________ and __________.

21. In a nonperiodic tiling of the plane,

(a) the pattern never repeats.
(b) the pattern is not repeated by any translation.
(c) there must be at least three kinds of tiles.

22. Penrose tilings are __________-periodic.

23. A Penrose dart has the property that

(a) opposite angles are congruent.
(b) it is nonconvex.
(c) the edges are all of different lengths.

24. A rhombus always has the property that __________.

25. Ammann bars are

(a) an Arab delicacy that comes from Jordan.
(b) jazz venues where musical sequences are played.
(c) sets of parallel bars in a Penrose pattern.

26. Barlow's Law prohibits the existence of crystals with __________ symmetry.

27. Quasicrystals

(a) do not exist in nature.
(b) are not regular enough to be used in quasi–New Age ceremonies.
(c) are symmetric nonperiodic tilings.

28. In a Penrose tiling, the proportion of darts to kites is __________.

29. A nonperiodic tiling

(a) is an impossibility.
(b) requires at least two kinds of tiles.
(c) does not have translation symmetry.

30. The process that takes a Penrose pattern into a different Penrose pattern with larger darts and kites is called __________.

CHAPTER 20 EXERCISES

■ *Challenge* ▲ *Discussion*

Hint: For the exercises about determining whether a shape will tile the plane, you should make a number of copies of the shape and experiment with placing them. One easy way to make copies is to trace the shape onto a piece of paper, staple half a dozen other blank sheets behind that sheet, and use scissors to cut through all the sheets along the edges of the traced shape on the top sheet.

20.1 Tilings with Regular Polygons

1. Determine the measure of an exterior angle and of an interior angle of a regular octagon (8 sides).

2. Determine the measure of an exterior angle and of an interior angle of a regular decagon (10 sides).

3. Discover a formula for the measure of an interior angle of a regular n-gon.

4. Using the formula from Exercise 3 and either your calculator or a short computer program, make a chart of the interior-angle measures of regular polygons with 3, 4, . . ., 12 sides.

■ **5.** Use the chart of interior-angle measures from Exercise 4 to determine all the possible vertex types of regular polygons (with at most 12 sides) surrounding a point.

■ **6.** Which of the vertex types of Exercise 5 do not occur in a semiregular tiling?

■ **7.** In addition to the vertex types of Exercise 5, exactly five others are possible, each involving one polygon with more than 12 sides. None of these vertex types leads to a semiregular tiling. The five many-sided polygons involved in these five vertex types have 15, 18, 20, 24, and 42 sides. Determine the other polygons in these vertex types.

For Exercises 8 and 9, refer to the lower-left corner of Figure 20.5 (on p. 736), which shows a tiling by isosceles triangles.

8. Use the center vertex to determine the measures of the angles of the isosceles triangle tile.

9. Every vertex except the center vertex has the same vertex type, in terms of the measures of the angles surrounding the vertex. What is that vertex type?

10. Give a numerical reason why a semiregular tiling could not include both polygons with 12 sides and polygons with 8 sides (with or without any polygons with other numbers of sides).

20.2 Tilings with Irregular Polygons

11. You know that a regular pentagon cannot tile the plane. Suppose that you cut one in half. Can this new shape tile the plane? (See Figure 19.4 on p. 692 for a regular pentagon that you can trace.)

12. For each tile below, show how it can be used to tile the plane. (Adapted from *Tilings and Patterns,* by Branko Grünbaum and G. C. Shephard, Freeman, New York, 1987, p. 25.)

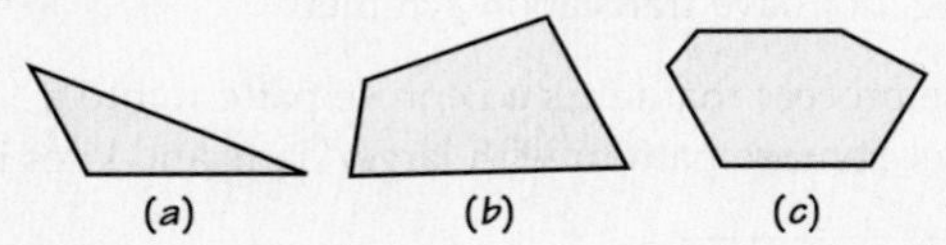

13. If you "unbulge" an ordinary soccer ball so that each of its sewn pieces is flat, you get a polyhedron, but it is not a regular polyhedron. It is a truncated icosahedron, one of the semiregular polyhedra. Some of the faces are regular hexagons, and some are regular pentagons. Explain why not all of its faces can be regular hexagons. All the vertices have the same vertex type; what is it? How many pentagons are there and how many hexagons? How many vertices are there?

14. A soccer ball should be as round as possible, which suggests using a greater number of polygons but each of which is shorter across. A ball with 92 faces was used for European soccer competitions: 12 pentagons, 20 hexagons, and 60 triangles (see the figure at right), all regular polygons. What are the possible vertex types for a polyhedron made up of pentagons, hexagons, and triangles? Explain why the polyhedron underlying this soccer ball is not semiregular.

20.3 Using Only Translations

Refer to tiles (a) through (g) below in doing Exercises 15 and 16.

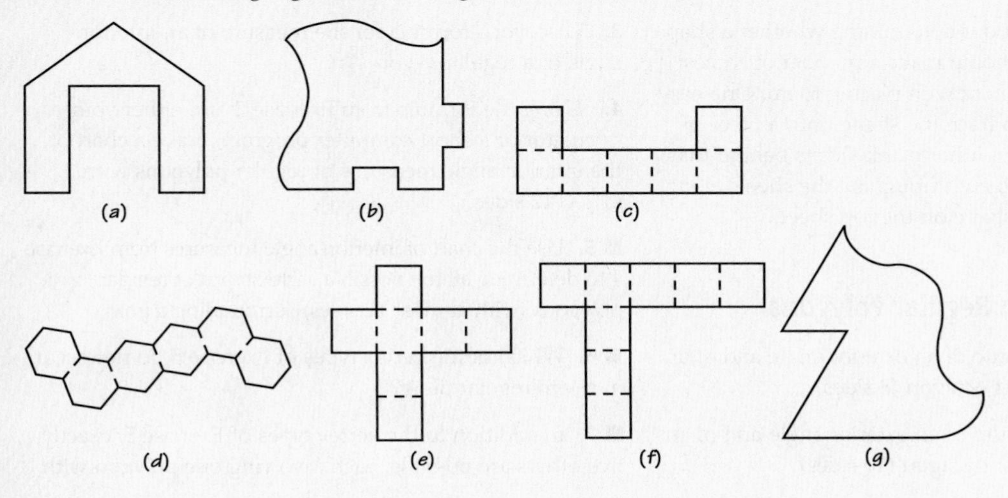

15. For each tile (a) through (c), determine whether it can be used to tile the plane by translations. (From *Tiling the Plane,* by Frederick Barber et al., COMAP, Lexington, MA, 1989, pp. 1, 8, 9.)

16. Repeat Exercise 15, but for tiles (d) through (g).

17. Start from a par-hexagon of your choice and modify it to tile the plane by translations. (You will probably find it useful to do your work on graph paper. If you choose a regular hexagon, there is special graph paper, ruled into regular hexagons, that would be particularly good to use.) Can you draw a design on the tile so as to make an Escher-like pattern?

18. Start from a parallelogram of your choice and modify it to tile the plane by translations. (You will probably find it useful to do your work on graph paper.) Can you draw a design on the tile so as to make an Escher-like pattern?

Refer to the following information in doing Exercises 19–22.

A particularly simple kind of polygon, called a *polyomino,* is one made of squares joined edge-to-edge. The name is a generalization of "domino"; indeed, there is only one kind of domino (two squares joined at an edge to form a rectangle). There are just two "trominos" (short for "triominos"), the straight tromino and the L-tromino.

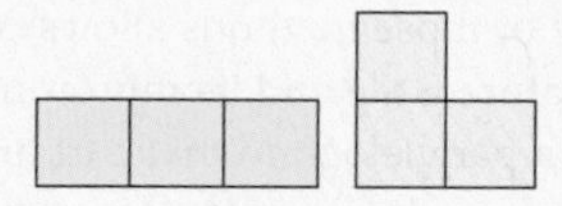

The straight tromino has the shape of a rectangle, so it can tile the plane by translations; and the L-tromino has the shape of a hexagon.

19. Is the L-tromino convex? Does the result about what hexagons can tile the plane (on page 738) give any information about whether the L-tromino can tile the plane or not?

20. Find a tiling of the plane using just the L-tromino and translations of it. Is there more than one way to do the tiling?

21. Show how alternative 2 of the Translation Criterion can be applied to the L-tromino.

■ **22.** Give an argument why alternative 1 of the Translation Criterion cannot be applied to the L-tromino. (*Hint:* Label each of the eight corners of the component squares of the tromino with the letters S, T, . . ., Z. Let these be our candidates for the points *A, B, C,* and *D* of the criterion.) Each of the sides of the tromino that is two units long has nowhere to go under a translation. Any application of the criterion must divide each side into two pieces, so their midpoints must be two of the points *A, B, C,* and *D.* Make a similar argument about two corners of the tromino. Thus, we have four points, which can be labeled consecutively *A, B, C,* and *D,* starting at any one of them. Show that none of the four possibilities "works." (This argument can be generalized to show that trying *A, B, C,* and *D* at points other than the corners of the squares won't work either.)

Refer to the following information in doing Exercises 23–28.

Demonstrate to your own satisfaction that there are exactly five shapes of tetrominos (each made of four squares joined at edges)—plus differing mirror images of two of them—as shown below. In the order shown, they are called the *square, straight, T-, L-,* and *skew* tetrominos. The straight and the square tetromino certainly can tile the plane by translations.

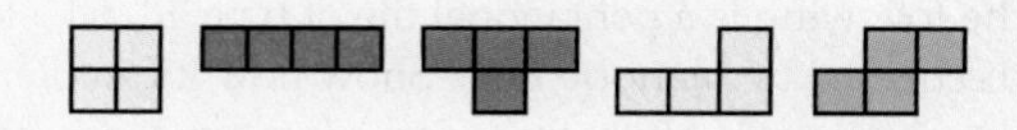

You will definitely find it useful to make yourself several copies of each of the polyominos mentioned below, by cutting them out of graph paper.

23. Apply alternative 1 of the Translation Criterion to the straight-tetromino and show how it can tile.

24. Show how alternative 2 of the Translation Criterion applies to the T-tetromino.

25. Apply alternative 1 of the Translation Criterion to the skew-tetromino and show how it can tile.

26. Show how alternative 2 of the Translation Criterion applies to the L-tetromino.

27. Show how alternative 2 of the Translation Criterion can be applied to the skew-tetromino, and show how it can tile.

28. In Exercises 25 and 27, we indulged in what appears to be "overkill," proving the same fact in two different ways. But those exercises should give you the idea that alternative 2 of the Translation Criterion can reduce to (and hence is more general than) alternative 1 if some points are allowed to coincide. For such a reduction, which pairs of points must coincide? (We are allowed to relabel the remaining four distinct points.)

20.4 Using Translations Plus Half-Turns

29. Explain how the final shape of Figure 20.18 (on p. 746) satisfies the Conway Criterion by identifying relevant points *A, B, C, D, E,* and *F.*

30. Do the same as in Exercise 29, but for the final shape of Figure 20.19 (on p. 747).

For Exercises 30 and 31, refer to tiles (a) through (g) on page 760.

31. For each tile (a) through (c), determine whether it can be used to tile the plane by translations and half-turns.

32. Repeat Exercise 31, but for tiles (d) through (g).

33. Show how an arbitrary pentagon with two parallel sides can tile the plane.

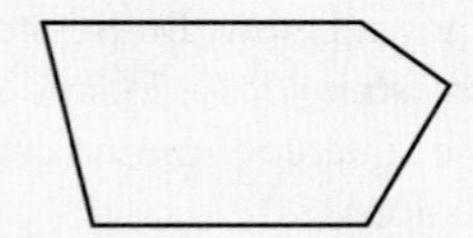

34. The following is a pentagonal tile of type 13, which was discovered by Marjorie Rice. Show how it can tile the plane. (*Hint:* Carefully trace and cut out a dozen or so copies and try fitting them together.)

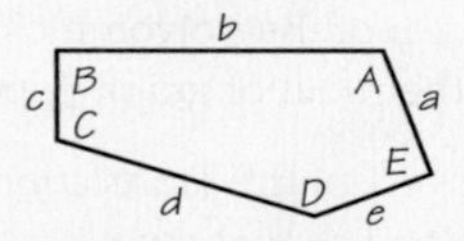

The parts of this pentagon satisfy the following relations: $A = C = D = 120°$, $B = E = 90°$, $a = e$, and $a + e = d$. [Adapted from "In Praise of Amateurs," by Doris Schattschneider, in David A. Klarner (ed.), *The Mathematical Gardner,* Wadsworth, Belmont, CA, 1981, p. 162.]

35. Start from a triangle of your choice and modify it to tile the plane by translations and half-turns. (You will probably find it useful to do your work on graph paper.) Can you draw a design on the tile so as to make an Escher-like pattern?

36. Start from a quadrilateral of your choice and modify it to tile the plane by translations and half-turns. (You will probably find it useful to do your work on graph paper.) Can you draw a design on the tile so as to make an Escher-like pattern?

Refer to the information about polyominos preceding Exercise 19, and to the following, in doing Exercises 37–40.

We saw earlier that all the dominos, trominos, and tetrominos tile the plane by translations. Here, we investigate the 12 pentominos, shown as follows with a letter notation for each. (If you allow mirror images to count as different pentominos, there are 18.) It will be useful for you to make several copies of each of the pentominos discussed below.

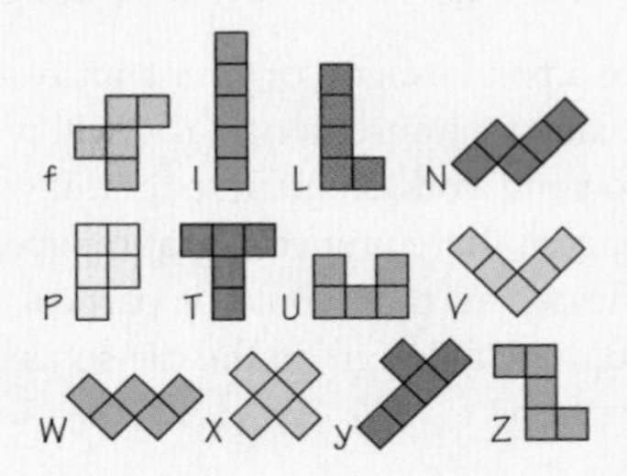

37. Just by experimenting, determine which of the pentominos can tile the plane by translations. (*Hint:* There are nine.)

38. Apply the Conway criterion to the f-pentomino, and show how it can tile by translations and half-turns.

39. Apply the Conway criterion to the U-pentomino, and show how it can tile by translations and half-turns.

40. Apply the Conway criterion to the T-pentomino, and show how it can tile by translations and half-turns.

41. In the text, we discuss criteria and methods for generating Escher-like patterns that involve just translations or translations and half-turns. A slight variation on one of those methods allows construction of tilings that feature a tile and its mirror image.

Begin with a parallelogram made from two congruent isosceles triangles, as shown in the figure below. Each of these triangles has two sides equal. Be sure that the two triangles are arranged so that they have one of the equal sides in common, forming a diagonal of the parallelogram.

Make any modification to half of the third side of one of the triangles. Mirror-reflect that modification across the side, then translate the reflection to become the modification of the other half of the side. Take the complete modification of this side, and translate it to become the modification of the opposite side of the parallelogram.

Modify in any way one of the two remaining sides of the parallelogram, and make the same modification to the opposite side (that is, translate the modification, without rotation or reflection). Then reflect this modification across the diagonal of the parallelogram.

The result is a modified parallelogram that tiles by translation and splits into two pieces that are mirror images of each other. Escher used a similar technique, but starting from a par-hexagon made from two quadrilaterals, in his Horseman print, as shown in his sketch in Figure 20.11b.

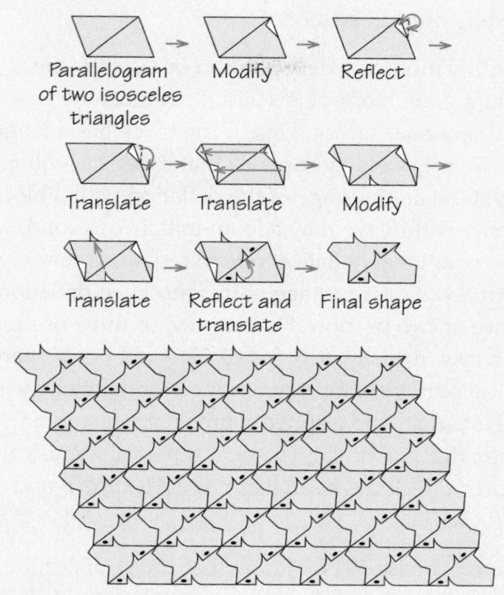

Use this technique to produce a tiling of your own design. Can you draw a design on the tile so as to make an Escher-like pattern?

42. Show that the modified parallelogram in Exercise 41 fulfills the Conway criterion by identifying the six points of the criterion.

20.5 Nonperiodic Tilings

43. In this chapter, we have been concerned mostly with tiling the plane, with some attention to using crystals to fill space. We can also consider a simpler case—tiling the line. For the line, a tile is a line segment of a particular length.

(a) What are the monohedral tilings of the line?

(b) What are the periodic tilings of the line that use two tiles of different lengths?

Exercises 44–46 connect nonperiodic tilings to the Fibonacci numbers and golden ratio of Chapter 19 but do not require any information from that chapter. (Thanks for this idea to David J. Wright, Oklahoma State University.) Just as the plane can be tiled quasiperiodically with Penrose tiles, the line can be tiled quasiperiodically with a pair of tiles, provided that their lengths are in the right proportion. Let the lengths of the tiles be a and b, with $b < a$ and a exactly c times as long as b, so that $cb = a$. Thus, scaling up a tile of length b by a factor of $c > 1$ produces a tile of length a. Similarly, we determine the scale factor c so that scaling up a tile of length a produces a tile of length a followed by a tile of length b, that is, $ca = a + b$.

44. (a) Using substitution, eliminate a and reduce the two equations to a single equation in just c and b alone.

(b) Use the quadratic formula to solve for the two possible values of c. Since we want $c > 1$, we choose the larger value.

45. We can define an inflation process for a line segment consisting of a's and b's: First, replace each original b by an a and each original a by two adjacent segments a and b. For example, we would replace aba by $(ab)(a)(ab)$, where we have inserted parentheses for clarity.

(a) Start with just a single b and repeat the inflation process, showing the stages, until you reach a stage with 21 segments.

(b) How many segments are there at each stage? (If we continue this process forever, we tile a half-line to the right; we tile the entire line by reflecting this right half-line over to cover the left half-line. The result is called a *Fibonacci tiling* of the line.)

(c) If a line segment contains m copies of the a tile and n copies of the b tile, how many tiles will the inflation of the segment contain?

46. We can similarly define a deflation process: Replace each original adjacent pair ab by an a and each remaining original a by a b.

(a) Apply this deflation process repeatedly to the stage in your answer to Exercise 45, part (a), that has 21 segments. What do you end up with?

(b) Apply this deflation process repeatedly to the periodic sequence of tiles *ababababab*. . . . What do you end up with?

(c) Devise your own periodic sequence of a and b tiles. Apply the deflation process to it repeatedly. What do you end up with? What do you conjecture?

(d) In what sense is the Fibonacci tiling quasiperiodic?

For Exercises 47–50, refer to the following: The rabbit problem in Chapter 19 (Exercise 4, on p. 718) leads us directly into nonperiodic patterns and musical sequences. Let A denote an adult pair of rabbits and B denote a baby pair. We record the population at the end of each month, just before any births, in a particular systematic way—as a string of A's and B's. At the end of their second month of life, a rabbit pair will be considered to be adult and give birth to a baby pair. At the end of the first month, the sequence is just A, and the same is true at the end of the second month. When an adult pair A has a baby pair B, we write the new B immediately to the right of the A. So at the end of the third month, the sequence is AB; at the end of the fourth, it is ABA because the first baby pair is now adult; at the end of the fifth month, we have $ABAAB$.

Mathematicians and computer scientists call this manner of generating a sequence a *replacement system* or *substitution system*. At each stage, we replace each A by AB and each B by A.

47. What is the sequence at the end of the sixth month?

48. Why can't we ever have two B's next to each other?

49. Why can't we ever have three A's in a row?

50. Show that from the fourth month on, the sequence for the current month consists of the sequence for last month followed by the sequence for two months ago.

For Exercises 51–58, refer to the following:
We can define inflation and deflation of a sequence of A's and B's, and musical sequences themselves, without reference to Penrose patterns, and thereby arrive at an example of a nonperiodic pattern in one dimension. Inflation consists of replacing each A by AB and each B by A, and deflation consists of replacing each AB by A and each A by B; inflation and deflation undo each other on musical sequences. Call a sequence musical if it results from applying inflation to the sequence consisting of a single B. Then inflation and deflation preserve musicality: If we inflate or deflate a musical sequence, we get another musical sequence. Another way to think of this relationship is that a musical sequence is self-similar under inflation and deflation.

■ **51.** Let the lone B be considered the first stage of inflation. Show that at the nth stage of inflation, for $n \geq 3$, there are F_n (the nth Fibonacci number—see Section 19.1 on p. 686) symbols in the sequence, of which F_{n-1} are A's and F_{n-2} are B's. (*Hint:* Check it for $n = 1, 2, 3,$ and 4.)

52. Show that no musical sequence contains AAA or BB.

53. Show that no musical sequence ends in AA or in $ABAB$.

54. Show that apart from the lone sequence B, every musical sequence is an initial subsequence of all the succeeding musical sequences.

55. Slightly modified, deflation can be used to check whether a finite block of A's and B's can belong to a musical sequence or not. First, if the block has a length greater than 1, we may suppose that it begins with an A (why?). So at any stage of the deflation with a block beginning with B, we may add an initial A. Second, we add the additional deflation rule to replace an ending AA with BA. If, at any stage of this modified deflation, we arrive at two or more B's in a row, or three or more A's in a row, then the original block could not be part of a musical sequence. Otherwise, the original block will deflate eventually to a single symbol, at which point we conclude that the original block is a part of a musical sequence. Check the two blocks $ABAABABAAB$ and $ABAABABABA$.

56. From Exercise 54, we know that each application of inflation to a musical sequence simply extends it. By successive inflation, then, we build an infinite sequence. Show that as we approach this limiting sequence, the ratio of B's to A's tends toward the golden ratio ϕ.

57. Conclude from Exercise 56 that the sequence cannot be periodic, nor settle into a period after a finite "burn-in" period. Thus, the sequence is nonperiodic. (*Hint:* ϕ is not a rational number; that is, it cannot be represented as a ratio m/n of whole numbers m and n.)

58. Show that any finite block of A's and B's that occurs in the infinite sequence must occur over and over again (just as any patch of tiles in a Penrose pattern occurs infinitely often in the pattern). Thus, the infinite sequence is self-similar.

WRITING PROJECTS

1. Get computer software to make some tilings of your own. Check for software at **www.geom.uiuc.edu/ software/tilings/TilingSoftware.html**. Print out your tilings and describe, in a sentence or two each, how you made them.

2. You can build a model of a buckyball by weaving strips of paper in a hexagonal pattern, much as peoples in Africa and elsewhere weave baskets and balls. Background material and instructions on doing this are at **http://www.ccd.rpi.edu/Eglash/csdt/african/hex/intro.html.** Make such a buckyball, preferably with colored strips. Explain why the result is a buckyball. The construction depends on the fact that when strand is wrapped around a rim of the same width, a 60-degree angle results. Explain why this is so, and experiment to try to determine what angles are formed when the strand and rim are not the same width (e.g., when the strand is twice as wide as the rim or half as wide).

Suggested Readings

LEE, KEVIN. TesselMania!, TesselMania! Deluxe. A computer program for Macintosh or Windows that produces Escher-like tilings. Free demo downloadable at **www.worldofescher.com/down/tessdemo.exe** or **britton.disted.camosun.bc.ca/jbtessinst.htm**. A review of pros and cons can be found at **http://www.tessellations.org/tesselmania0.htm**. One major con is that the full-capability program seems to have disappeared from the market.

LU, PETER J., and PAUL J. STEINHARDT. Decagonal and quasi-crystalline tilings in medieval Islamic architecture, *Science* 315 (23 February 2007) 1106–1110, **www.sciencemag.org/cgi/content/full/315/5815/1106**.

RANUCCI, ERNEST, and JOSEPH TEETERS. *Creating Escher-Type Patterns,* Creative Publications, Oak Lawn, IL., 1977.

SCHATTSCHNEIDER, DORIS. M. C. *Escher: Visions of Symmetry,* 2d ed., Harry N. Abrams, New York, 2004.

SCHATTSCHNEIDER, DORIS. Will it tile? Try the Conway criterion! *Mathematics Magazine* 53 (4) (September 1980) 224–233.

SEYMOUR, DALE, and JILL BRITTON. *Introduction to Tessellations,* Dale Seymour Publications, Palo Alto, CA, 1989. An excellent introduction to tessellations, including how to make Escher-like tessellations.

STEPHENS, PAM, and JIM McNEILL. *Tessellations: The History and Making of Symmetrical Designs,* Crystal Productions, Glenview, IL, 2001.

TEETERS, JOSEPH L. How to draw tessellations of the Escher type, *Mathematics Teacher,* 67 (1974): 307–310.

WILLSON, JOHN. *Mosaic and Tessellated Patterns: How to Create Them,* Dover Publications, Mineola, NY, 1983.

Suggested Web Sites

www.geom.uiuc.edu/software/tilings Lists programs for various platforms that allow the user to design tilings.

www.reocities.com/SiliconValley/Pines/1684/Penrose.html A Java applet to play with Penrose tiles.

www.geom.uiuc.edu/apps/quasitiler An interactive Web program QuasiTiler 3.0 that draws Penrose tilings and their generalizations in higher dimensions.

www.geometrygames.org/ Interactive Windows and Macintosh program that lets the user design tilings on the plane, the sphere, and the hyperbolic plane. Spherical tilings can be realized as polyhedra.

www.geometrygames.org/KaleidoTile/index.html An interactive Windows and Macintosh Classic (PPC) program to create tilings on the plane, a sphere, and the hyperbolic plane.

demonstrations.wolfram.com/ComplementTiling/ An interactive Web program to make Escher-like tilings, with links to other interactive demonstrations.

www.eschertiles.com/links.html Links to the official M. C. Escher site and to sites with Escher-inspired tilings and art.

www.angelfire.com/mn3/anisohedral/isohedral.html Gives the nine known sufficient criteria for a tile to tile the plane, taking into account translations, reflections, and rotations.

Your Money and Resources

Part VII

This part concentrates on numerical patterns of growth and decline in the realms of finance, resources, and biology. The unifying concept is a population, whether of dollars, barrels of oil, or tons of fish. How much interest will your savings account earn in the next year? How much will the monthly payment be on your credit card, your car loan, or a home mortgage? How much would you need to save to pay for a child's college education or for your retirement? What will inflation do to your savings? How much should you pay for a stock? These are problems of daily life for which mathematics provides custom-tailored models. In Chapter 21, "Savings Models," and Chapter 22, "Borrowing Models," you become familiar with the mathematics and terminology of situations that you will face repeatedly in everyday life.

Good mathematical models are often versatile and flexible, and the financial models of these chapters apply broadly to important problems in other areas of life. Growth of money at interest is like growth of biological populations. Inflation of a currency or depreciation of an asset is like the decay of a radioactive substance. Finding out how long a retirement "nest egg" will last is similar to determining how long it will be before a nonrenewable resource, such as oil or coal, may be exhausted. Managing a trust fund, such as the endowment of a college, presents problems similar to the management of a renewable biological resource, such as a forest or a fishery. In Chapter 23, "The Economics of Resources," we explore these similarities, together with the profound effect that economic conditions can have on natural resources.

Finally, you will see the surprisingly large and puzzling consequences that very small changes can produce in a physical system or biological population as a result of behavior that mathematicians call chaos.

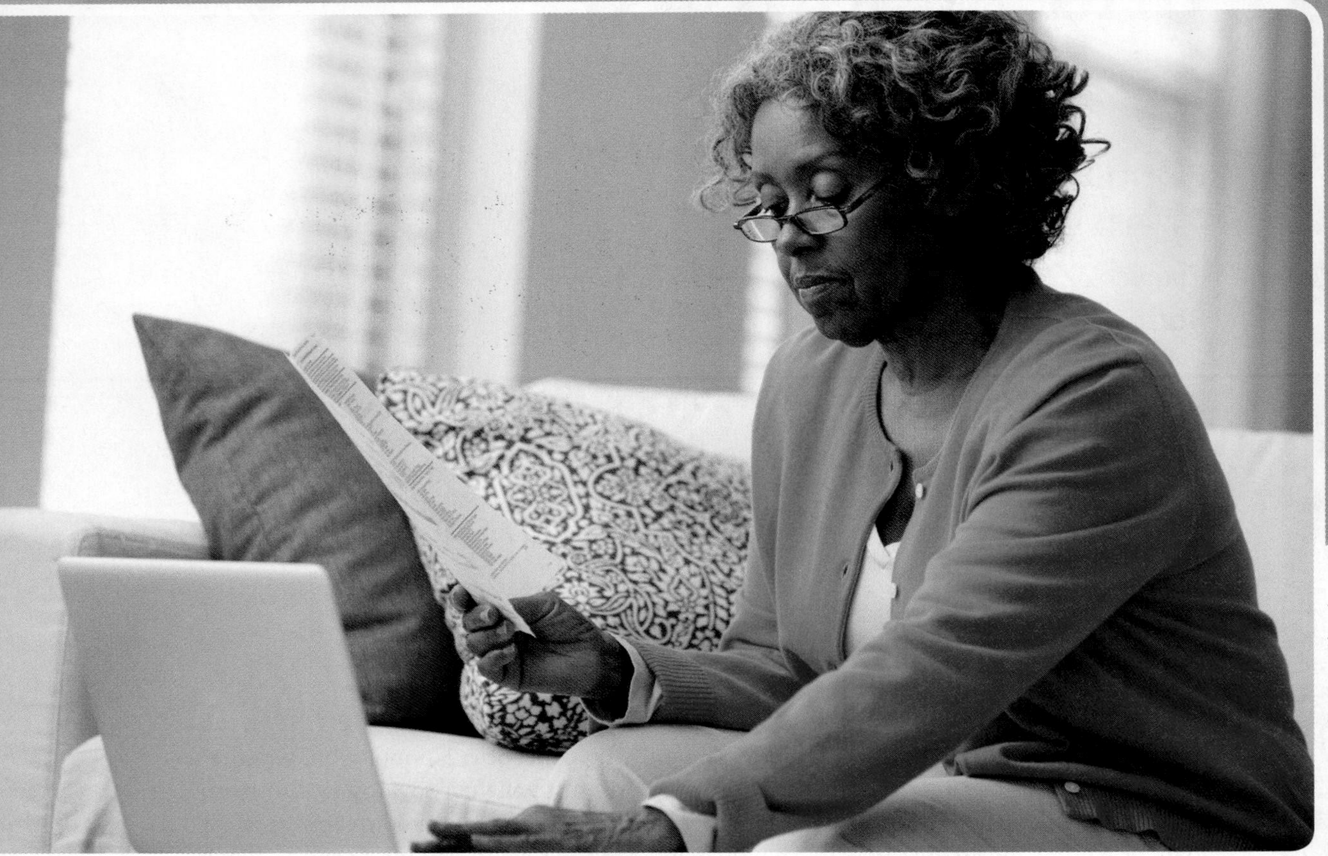

21

Savings Models

How much interest will your savings account earn in the next year? What difference does it make how the interest is calculated? How much should you save for a comfortable retirement in the face of inflation? In this chapter, we consider such questions and show how the underlying mathematical models also explain the recent mortgage and bank crisis and the effect of interest-rate changes on stock prices.

In Section 21.1, we look at one way that money can grow at interest, called *simple interest* or *arithmetic* (or *linear*) *growth,* followed in Section 21.2 by the more usual compound interest, also known as *geometric* (or *exponential*) *growth.* The compound interest rate over an interval of time can be converted to an equivalent simple interest rate, called the *effective rate,* which for a period of a year is known as the annual percentage yield (APY). That is the rate that must be advertised by banks that try to attract your savings. Interest can be compounded any number of times in a year; in Section 21.3, we find a limit to how much can be realized, even with compounding infinitely often.

For regular deposits to a savings account—to save up for a car, the down payment on a house, or other purchase—there is a formula that relates the size of the regular deposit, the interest rate, the amount to be accumulated, and the time to realize it. We derive this savings formula in Section 21.4 and show you how to use it to find any of the quantities if you know the others.

However, dollars saved may lose value due to inflation, so in Section 21.5, we show how to take inflation into account in your savings plans. Inflation, a "decay" of dollars, is usually expressed in terms of the Consumer Price Index (CPI). We show how to convert from the CPI to the rate of inflation and how to determine the real rate of growth of an investment under inflation.

This chapter and the next encompass the world of savings and loans that you will experience in your lifetime, ranging from your parents' (and your) saving for your college education, to making time payments on a car or home loan, to saving—and withdrawing savings—for retirement.

21.1 Arithmetic Growth and Simple Interest

When you open a savings account, your primary concerns are the safety and the growth of the "population" of your savings. Suppose that you deposit \$1000 in an account that "pays interest at a rate of 10% annually." (This is an unrealistic rate, particularly in this era of very low interest rates. We use it solely because it makes the calculations transparent.) Assuming that you make no other deposits or withdrawals, how much is in the account after 1, 2, and 5 years?

The $1000 is the **principal**, the initial balance of the account. At the end of one year, **interest** is added. The amount of interest is 10% of the principal, or

$$10\% \times \$1000 = 0.10 \times \$1000 = \$100$$

So the balance at the beginning of the second year is $1100.

Algebra Review
Fractions to Percentages, page 486.

We express an interest rate either as a percentage or as a decimal fraction. "Percent" means "per 100," so you can think of the symbol "%" as standing for "1 per 100" or $\frac{1}{100} = 0.01$. So to convert from a percentage to a decimal fraction, divide the percentage by 100 by moving the decimal point two places to the left. An interest rate of 10% is 10/100 or 0.10; an interest rate r as a decimal number is $100r\%$. (*Caution:* A common error in using the formulas in this chapter is to forget to express the percentage as a decimal; for example, for $r = 5\%$, don't substitute 5 for the r in the formula, but instead substitute 0.05.)

With **simple interest**, interest is paid only on the original balance, no matter how much interest has accumulated. With simple interest, for a principal of $1000 and a 10% interest rate, you receive $100 interest at the end of the first year; so at the beginning of the second year, the account contains $1100. But at the end of the second year, you again receive only $100; so at the beginning of the third year, the account contains $1200. In fact, at the end of each year, you receive just $100 in interest.

Simple Interest DEFINITION

Simple interest is interest that is paid on the original principal only, not on any accumulated interest.

The formulas for simple interest are themselves simple.

Simple Interest RULE

For a principal P and an annual rate of interest r, the interest earned in t years is

$$I = Prt$$

and the total amount A accumulated in the account is

$$A = P + I = P + Prt = P(1 + rt)$$

You may find this method for interest rather strange if you are accustomed from your savings account to a different system of awarding interest—**compound interest**, which we will consider shortly. However, simple interest is often used for the following transactions:

- private loans between individuals, because it is easy to calculate;
- commercial loans for less than one year—not just because it is easy to calculate, but also because for low interest rates, simple interest differs negligibly from compound interest; and
- financing of corporations and the government through bonds. A bond is a loan with repayment at the end of a fixed term and simple interest in the meantime, paid usually annually or semiannually.

EXAMPLE 1
Simple Interest on a Student Loan

Let's suppose that you have exhausted the amount that you can borrow under federal loan programs and need a private direct student loan for \$10,000.

PNC Financial Services, headquartered in Pittsburgh, PA, quoted an example for its "Lowest Tier Pricing" rate in October 2011 of 3.21% ("Highest Tier" was 11.21%). PNC offers an interest-only repayment option, under which you make monthly interest payments while you are in school and pay on the principal only after graduation. Under this plan, PNC earns simple interest from you while you are in school.

How much monthly interest would you pay for such a \$10,000 loan?

SOLUTION The principal is $P = \$10{,}000$, the annual interest rate is $r = 3.21\% = 0.0321$ per year, and the number of years is $t = \frac{1}{12}$ year. The interest for one month would be $I = Prt = \$10{,}000 \times 0.0321 \times \frac{1}{12} = \$26.75$. (Actually, PNC would adjust the interest rate quarterly, so the payment could increase. Also, the initial interest rate might not be 3.21% but could be more than 11% since it would depend on the creditworthiness of you and your cosigner.)

We frequently observe the kind of growth corresponding to simple interest, called **arithmetic growth** or **linear growth**, in other contexts.

Arithmetic Growth DEFINITION

Arithmetic (pronounced with accent on the "met" syllable) **growth** (also called *linear growth*) is growth by a constant amount in each time period.

For example, the population of medical doctors in the United States grows arithmetically because the medical schools graduate the same number of doctors each year (and the number of doctors dying is also fairly constant). The concept of linear growth has appeared already in this book in the discussions of linear programming (Chapter 4) and linear regression (Chapter 6).

21.2 Geometric Growth and Compound Interest

What you probably expected to happen to the savings account discussed in the last section is that during the second year, the account would earn interest of 10% not on the *initial* balance of \$1000 (as with simple interest) but on the *new* balance of \$1100. Then, at the end of the second year, 10% of \$1100, or \$110, would be added to the account.

Thus, during the second year, you would earn interest on both the principal of \$1000 and on the \$100 interest earned during the first year. With this method, you receive more interest during the second year than during the first; that is, the account grows by a greater amount during the second year. At the beginning of the third year, the account contains \$1210, so at the end of the third year, you receive \$121 in interest. Again, this is larger than the amount that you received at the end of the preceding year. Moreover, the increase during the third year,

$$\text{third-year interest} - \text{second-year interest} = \$121 - \$110 = \$11$$

is larger than the increase during the second year,

$$\text{second-year interest} - \text{first-year interest} = \$110 - \$100 = \$10$$

Thus, not only is the account balance increasing each year, but the amount added also increases each year.

Compound Interest DEFINITION

Compound interest is interest that is paid on both the original principal and accumulated interest.

Savings institutions usually compound interest and credit it to accounts more often than once a year—for example, quarterly (four times per year). With an interest rate of 10% per year and quarterly compounding, you get one-fourth of the rate, or 2.5%, paid in interest each quarter year. The "quarter" (three months) is the **compounding period**, or the time elapsing before interest is paid.

Compounding Period DEFINITION

The **compounding period** is the fundamental interval on which compounding is based, within which no compounding is done.

Consider again a principal of $1000. At the end of the first quarter, you have the original balance plus $25 interest, so the balance at the beginning of the second quarter is $1025. During the second quarter, you receive interest equal to 2.5% of $1025, or $25.625, which is rounded up in posting to your account (since the fraction is half a cent or more) to $25.63. Continuing in this manner, the balance at the end of the first year is $1103.82. (You should "read" all calculations in this chapter by confirming them on your calculator.)

Even though the account was advertised as paying 10% interest, the interest for the year is $103.82, which is 10.382% of the principal of $1000.

Practical note: Without rounding the interest for each quarter, the interest for the year would have been $1103.81 (as shown in Table 21.2 on page 771.) Table 21.1 shows the results of calculation with rounding done only at the end of the year, while savings institutions must round at each posting and credit the rounded amount to your account. A spreadsheet program could duplicate the results of their computer programs; but in this table and in later calculations, we take the simpler route of rounding only at the final answer. Any differences will be very small; and if your answers differ by just a few cents, that will be OK.

Table 21.1

Compound Interest on $1000, at an Interest Rate of 10% Compounded Quarterly

Date	Beginning Balance	Interest on Principal	Interest on Interest	Total Interest Added	Ending Balance
January 1	1000.00				
March 31	1000.00	25.00	0.00	25.00	1025.00
June 30	1025.00	25.00	0.63	25.63	1050.63
September 30	1050.63	25.00	1.27	26.27	1076.90
December 31	1076.90	25.00	1.92	26.92	1103.82

If interest is compounded monthly (12 times per year) or daily (365 or 366 times per year), the resulting balance is even larger, as shown in Table 21.2 (together with the results of continuous compounding, which we discuss later). We will show you shortly the compound interest formula for these calculations.

Table 21.2

Comparing Compound Interest: The Value of $1000, at 10% Annual Interest, for Different Compounding Periods*

Years	Compounded Yearly	Compounded Quarterly	Compounded Monthly	Compounded Daily	Compounded Continuously
1	1100.00	1103.81	1104.71	1105.16	1105.17
5	1610.51	1638.62	1645.31	1648.61	1648.72
10	2593.74	2685.06	2707.04	2717.91	2718.28

*Without rounding at posting of interest and neglecting leap years; the difference in most cases is no more than 1 cent.

Terminology for Interest Rates

We have seen that an account at a particular annual rate of interest can produce different amounts of interest, depending on how the compounding is done. To help prevent confusion on the part of consumers, the Truth in Savings Act establishes specific terminology and calculation methods for interest.

A **nominal rate** is any stated rate of interest for a specified length of time, such as a 3% annual interest rate on a savings account or a 1.5% monthly rate on a credit-card balance. By itself, a nominal rate *does not indicate or take into account whether or how often interest is compounded.*

Table 21.2 shows that at an annual interest rate of 10% (a nominal rate) compounded daily for one year, $1000 yields $105.16 in interest, which is 10.516% of the principal. If instead you had $1000 at *simple* interest of 10.516% for one year, you would earn exactly the same amount of interest.

Effective Rate and APY DEFINITION

The **effective rate** is the rate of simple interest that would realize exactly the same amount of interest over the same length of time. For a year, the effective rate is called the **annual percentage yield (APY)**.

So, for the $1000 at 10% compounded daily, the effective rate is 10.516%. Later in this section, we will give a formula for calculating the effective rate for any interest rate and any compounding period.

To keep the different interest rates straight, we use:

- r for a nominal annual rate, which may or may not be compounded; and
- i for the rate during a compounding period, which can be a day, a month, or a year. Since there is no compounding done inside a compounding period, the effective rate for a compounding period is the nominal rate. We call i the *periodic rate.*

Rate Per Compounding Period RULE

For a nominal annual rate r compounded m times per year, the rate per compounding period is

$$\text{periodic rate} = \boxed{i = \frac{r}{m}} = \frac{\text{nominal annual interest rate}}{\text{number of compounding periods per year}}$$

For that \$1000 in savings at 10% compounded quarterly, we have $r = 10\%$ and $m = 4$, so $i = 2.5\%$ per quarter.

We denote the number of compounding periods per year by m. We use r only for an annual rate and t for the number of years.

To avoid confusion, we don't use the terminology *annual percentage rate* because that term has a special legal meaning just for loans (see Section 22.2, page 806).

Geometric Growth

Here, we look for the underlying mathematical pattern of compounding. We continue to use the values from our previous example; namely, an initial balance of \$1000, an annual interest rate $r = 10\%$, quarterly compounding (so $m = 4$), and hence quarterly interest rate $i = 2.5\%$. For quarterly compounding, you have at the end of the first quarter,

$$\text{initial balance} + \text{interest} = \$1000 + \$1000(0.025) = \$1000(1 + 0.025)$$

and at the end of the second quarter,

$$\begin{aligned}\text{initial balance} + \text{interest} &= \$1000(1 + 0.025) + [\$1000(1 + 0.025)](0.025)\\ &= [\$1000(1 + 0.025)] \times (1 + 0.025)\\ &= \$1000(1 + 0.025)^2\end{aligned}$$

The pattern continues in this way, so that you have $\$1000(1 + 0.025)^4$ at the end of the fourth quarter. You use the calculator button marked $\boxed{y^x}$ to evaluate expressions like $(1.025)^4$; on a spreadsheet, use the caret key ^ (Shift-6), as in 1.025^4.

More generally, with initial principal P and interest rate i (= $100i$%) per compounding period, you have at the end of the first compounding period,

$$P + Pi = P(1 + i)$$

This amount can be viewed as a new starting balance. Hence, in the next compounding period, the amount $P(1 + i)$ grows to

$$P(1 + i) + P(1 + i)i = P(1 + i)(1 + i) = P(1 + i)^2$$

The pattern continues, and we reach the following conclusion.

Compound Interest Formula RULE

An initial principal P in an account that pays interest at a periodic interest rate i per compounding period grows after n compounding periods to

$$A = P(1 + i)^n$$

For convenience, we convert the general interest formula into one that is specific for years and annual rates. An annual rate of interest r with m compounding periods per year gives a rate $i = r/m$ per compounding period, and t years contain $n = mt$ compounding periods.

Compound Interest Formula for an Annual Rate — RULE

An initial principal P in an account that pays interest at a nominal annual rate r, compounded m times per year, grows after t years to

$$A = P\left(1 + \frac{r}{m}\right)^{mt}$$

Notation for Savings — DEFINITION

A	amount accumulated, sometimes denoted *FV* for "future value"
P	initial principal, sometimes denoted *PV* for "present value"
r	nominal annual rate of interest
t	number of years
m	number of compounding periods per year
$n = mt$	total number of compounding periods
$i = r/m$	periodic rate, the interest rate per compounding period

The amount added each compounding period is proportional to the amount present at the time of compounding; we are adding Pi to the amount P. This type of growth is called **geometric growth.**

Geometric Growth (Exponential Growth) — DEFINITION

Geometric growth (also called **exponential growth**) is growth proportional to the amount present.

EXAMPLE 2
Compound Interest

Suppose that you have a principal of $P = \$1000$ invested at 10% nominal interest per year. Using the compound interest formula $A = P(1 + i)^n$, you can determine the amount in the account after 10 years—once you know the compounding period.

SOLUTION

- *Annual compounding.* The annual rate of 10% gives $i = 0.10$, and after 10 years, the account has

$$\$1000(1 + 0.10)^{1\times 10} = \$1000(1.10)^{10} = \$2593.74$$

- *Quarterly compounding.* Then $i = r/m = 0.10/4 = 0.025$, and after 10 years ($mt = 4 \times 10 = 40$ quarters) the account contains

$$\$1000\left(1 + \frac{0.10}{4}\right)^{4\times 10} = \$1000(1.025)^{40} = \$2685.06$$

- *Monthly compounding.* Then $i = r/m = 0.10/12 = 0.008333$. The amount in the account after 10 years ($mt = 12 \times 10 = 120$ months) is

$$\$1000\left(1 + \frac{0.10}{12}\right)^{12\times 10} = \$2707.04$$

These entries are found in the last row of Table 21.2 on page 771.

In doing the calculations, use as many decimal places as your calculator or spreadsheet carries and don't round off until the final result. We show intermediate results with enough decimal places to give the final result to the nearest cent.

Natural and fractional exponents, page 661.

Effective Rate

For an interest rate i per compounding period, a principal of \$1 grows to $(1 + i)^n$ in n periods, so the interest earned on that \$1—which is the effective rate of interest for the n periods—is the final principal $\$(1 + i)^n$ minus the original principal of \$1, or $\$(1 + i)^n - \$1$. Hence, we have the following formula for the effective rate:

Formula for Effective Rate RULE

$$\text{effective rate} = (1 + i)^n - 1$$

Mostly, we will be interested in the effective rate on an annual basis. For a nominal *annual* interest rate r compounded m times, the interest rate per compounding period is $i = r/m$, and an amount of \$1 grows in one year to

$$\left(1 + \frac{r}{m}\right)^m$$

The effective *annual* rate of interest (the APY) is the amount of interest earned

$$\left(1 + \frac{r}{m}\right)^m - 1$$

divided by the original principal. Since that principal is \$1, we have:

Formula for APY RULE

$$\text{APY} = \left(1 + \frac{r}{m}\right)^m - 1$$

where APY = annual percentage yield (effective annual rate)
r = nominal interest rate
m = number of compounding periods per year

EXAMPLE 3
Finding the APY

With a nominal annual rate of 6% compounded monthly, what is the APY?

SOLUTION

$$\text{APY} = \left(1 + \frac{0.06}{12}\right)^{12} - 1 = 0.0617 = 6.17\%$$

In some cases, you know the principal, the current balance, and the interval of time, and you want to learn the interest rate. For example, money market funds typically report earnings to investors each month, based on interest rates that vary from day to day, but often do not report the average interest rate. We can find the equivalent average effective *daily* rate, from which we can calculate the APY.

The compound interest formula gives the end-of-month balance as $A = P(1 + i)^n$, where P is the balance at the beginning of the month, i is the average daily interest rate, and n is the number of days that the statement covers. So we have

$$\frac{A}{P} = (1 + i)^n$$

Taking the nth root gives

$$1 + i = \left(\frac{A}{P}\right)^{1/n} \qquad i = \left(\frac{A}{P}\right)^{1/n} - 1$$

EXAMPLE 4
Money Market Account

Suppose that the monthly statement from the fund reports a beginning balance (P) of \$7373.93 and a closing balance (A) of \$7382.59 for 28 days (n). What is the effective daily rate?

SOLUTION We thus have

$$i = \left(\frac{7382.59}{7373.93}\right)^{1/28} - 1 = 0.0000419194$$

Thus, the average effective daily rate is 0.00419194%. Compounding daily for a (non-leap) year, we would have $(1 + 0.0000419194)^{365} = 1.01542$, for an APY of 1.54%.

Simple Interest Versus Compound Interest

The amounts in accounts paying interest at 10% per year with compound and simple interest are shown in Table 21.3 and in the graph in Figure 21.1, which dramatically illustrate the exponential growth of money at compound interest compared with the linear growth at simple interest.

Table 21.3

The Growth of \$1000: Compound Interest Versus Simple Interest

Years	Amount in Account from Compounded Interest	Amount from Simple Interest
1	1100.00	1100.00
2	1210.00	1200.00
3	1331.00	1300.00
4	1464.10	1400.00
5	1610.51	1500.00
10	2593.74	2000.00
20	6727.50	3000.00
50	117,390.85	6000.00
100	13,780,612.34	11,000.00

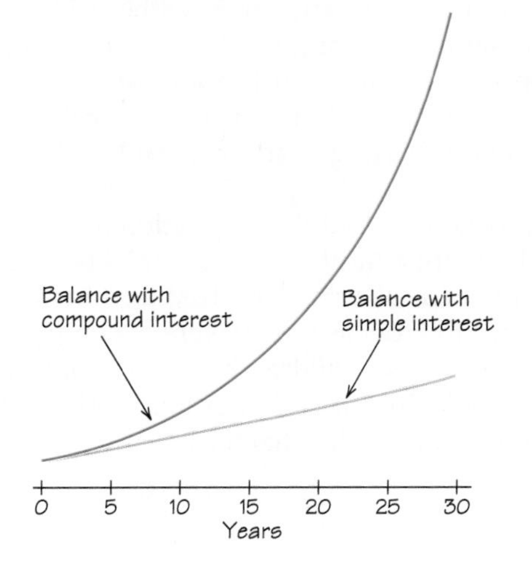

Figure 21.1 The growth of \$1000: compound interest and simple interest. The straight line explains why growth at simple interest is also known as *linear growth*.

Algebra Review: Integer Exponents and Exponential Relations

When multiplying two exponential expressions with the same base, the rule is to add the exponents:

$$a^m \cdot a^n = a^{m+n}$$

When dividing, subtract the exponents:

$$\frac{a^m}{a^n} = a^{m-n}, \ a \neq 0$$

This last rule is easily understood when simplifying a problem such as $\frac{2^5}{2^3}$ by using cancelling, as follows:

$$\frac{2^5}{2^3} = \frac{2 \cdot 2 \cdot \not{2} \cdot \not{2} \cdot \not{2}}{\not{2} \cdot \not{2} \cdot \not{2}} = 2^2 \text{ and } \frac{2^5}{2^3} = 2^{5-3} = 2^2$$

When working with negative integer exponents, in general, $a^{-n} = \frac{1}{a^n}, a \neq 0$. We can apply the same cancellation technique used above to see why this is the case:

$$\frac{2^3}{2^5} = \frac{\not{2} \cdot \not{2} \cdot \not{2}}{2 \cdot 2 \cdot \not{2} \cdot \not{2} \cdot \not{2}} = \frac{1}{2^2} \text{ and } \frac{2^3}{2^5} = 2^{3-5}$$
$$= 2^{-2} = \frac{1}{2^2}$$

When zero is an exponent, $a^0 = 1, a \neq 0$. For example,

$$\frac{2^3}{2^3} = \frac{\not{2} \cdot \not{2} \cdot \not{2}}{\not{2} \cdot \not{2} \cdot \not{2}} = 1 \text{ and } \frac{2^3}{2^3} = 2^{3-3} = 2^0 = 1$$

Equations such as $y = 10^x$ are called *exponential equations*. In such equations, the independent variable, x, is an exponent. To understand the behavior of such a relation, we can find its graph by plotting points:

x	$y = 10^x$
2	$10^2 = 100$
1	$10^1 = 10$
0	$10^0 = 1$
−1	$10^{-1} = \frac{1}{10}$
−2	$10^{-2} = \frac{1}{100}$

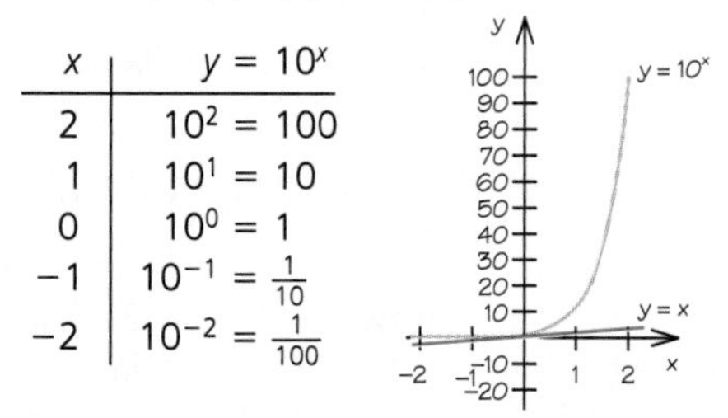

By comparing the graphs of $y = 10^x$ and $y = x$, we see that as values of x increase, the y-values of the exponential equation increase faster than that of the linear equation. This will be true for exponential equations of the form $y = a^x$, where $a > 1$.

In some situations, the contrast is not so immediately dramatic at first glance. The amount of carbon dioxide in the atmosphere, which contributes to global warming, has been growing since 1750 as a result of burning fuels. The amount is growing *superexponentially;* that is, faster than exponentially: The growth rate, or "interest rate," increases every year. The current growth rate is about 0.5% per year. That would seem to be a very low rate of "interest," but it is "interest" on what is already a large "principal" of carbon dioxide. The international Kyoto Protocol that went into effect in early 2005 (without U.S. participation) aims to lower worldwide emissions. Even at a fixed lower constant increase, though, the "principal" of carbon dioxide atoms in the atmosphere would still go up—and global warming would intensify—but just arithmetically, instead of superexponentially. We are in effect "saving" carbon dioxide into the atmosphere, at an unknown future cost.

We noted earlier (page 769) that the population of U.S. medical doctors grows as if it were at simple interest (arithmetic growth) because the same number of doctors graduate from medical school each year. On the other hand, general human populations tend to grow as if they were at compound interest (geometric growth), because the number of children born—the "interest"—increases as the population—the "balance"—increases. We examine models for population growth and for consumption of resources in Chapter 23.

The distinction between arithmetic growth and geometric growth is fundamental to the major theory of demographer and economist Thomas Robert Malthus (1766–1834). He claimed that human populations grow geometrically but food supplies grow arithmetically, so that populations tend to outstrip their ability to feed themselves (see Spotlight 21.1).

The situation of nuclear waste generated by a nuclear power plant is more complicated. The absolute volume of waste added each year depends on the size and output of the power plant, not on the growing amount of waste in storage. Hence, the volume of waste grows arithmetically. What about the total amount of radioactive material in the storage dump? The waste is a mixture of radioactive and nonradioactive substances. Over time, the radioactive ingredients decay very slowly into nonradioactive ones. While the radioactivity of waste already in storage is decreasing, new amounts of radioactive material are added each year. The situation requires a hybrid model that incorporates positive arithmetic growth (adding to the dump) accompanied by (much smaller) negative geometric growth (radioactive decay). The situation is like turning on the faucet in the bathtub while leaving the drain hole open a little. The faucet determines how fast water runs in, the height of the water determines how fast it runs out, and those two rates determine what happens to the volume of water in the tub.

SPOTLIGHT 21.1

Thomas Robert Malthus

Thomas Robert Malthus (1766–1834), a 19th-century English demographer and economist, based a well-known prediction on his perception of the different patterns of growth of the human population and growth of the "population" of food supplies.

He believed that human populations increase geometrically, but food supplies increase arithmetically—so that increases in food supplies will eventually be unable to match increases in population. He concluded, however, that over the long run, there also would be restrictions on the natural growth of human populations, including war, disease, and starvation—hardly an optimistic forecast and doubtless responsible for the dreary image associated with his views.

Some observers suggest that the genocide in Rwanda in 1994 was indirectly a result of overpopulation compared to available food resources.

Thomas Robert Malthus
(*The Granger Collection, New York.*)

21.3 A Limit to Compounding

The rows in Table 21.2 (page 771) show a trend: More frequent compounding yields more interest. But as the frequency of compounding increases, the interest tends to a limiting amount, shown in the far-right column.

Why is this so? Basically, because the extra interest from more frequent compounding is *interest on interest.* For example, in the first row of Table 21.2, the $3.81 extra interest from compounding quarterly is interest on the $100 yearly interest. The $3.81 is less than 10% of the $100 because the $100 interest is not on deposit for the whole year. Just part of it is credited to the account (and begins earning interest) at the end of each quarter. As compounding is done more and more often, smaller and smaller amounts of interest on interest are added.

Let's see what happens with the crazy interest rate of 100% per year compounded m times per year. This crazy rate makes the numbers simple (but it is nowhere close to the crazy 66,000% rate in Zimbabwe in December 2007, or the incredible 313 million% rate in Yugoslavia in 1994). Later, we examine interest rates closer to those in stable economies. For an initial balance of $1, the amount at the end of one year—from the compound interest formula, with $P = \$1$ and $i = 100\%$—is

$$A = \$1 \times \left(1 + \frac{100\%}{m}\right)^m = \$\left(1 + \frac{1.00}{m}\right)^m$$

As m increases, this amount, which is just $(1 + 1/m)^m$, gets closer and closer to a special number called $\boldsymbol{e} \approx 2.71828$ (see Spotlight 21.2). This is illustrated in Table 21.4, where the dots (ellipses) indicate that more decimal places follow.

Table 21.4

Yield of $1 at 100% Interest, Compounded *m* Times Per Year

m	$\left(1 + \frac{1}{m}\right)^m$
1	2.0000000 . . .
10	2.5937424 . . .
100	2.7048138 . . .
1,000	2.7169239 . . .
1,000,000	2.7182804 . . .

SPOTLIGHT 21.2

The Number *e*

The number e is similar to the number π in several respects. Both arise naturally, π in finding the area and circumference of circles, and e in compounding interest continuously (e is also the base for the system of "natural" logarithms). In addition, neither is rational (expressible as the ratio of two integers, such as 7/2) or even algebraic (the solution of a polynomial equation with integer coefficients, such as $x^2 = 2$); we say that they are *transcendental* numbers. Finally, no pattern has ever been found in the digits of the decimal expansion of either number.

For a general interest rate r, as m becomes larger and larger, the limiting amount is e^r, and the interest method is called **continuous compounding**. The APY is $(e^r - 1)$. (You can calculate powers of e using the [e^x] button on your calculator. On some calculators, this button is the [2nd] function of the button marked [LN] or [ln x]. For example, to calculate $e^{0.10}$, press [2nd], press [ln x] and enter 0.10. You get 1.105170918.)

Continuous Compounding DEFINITION

Continuous compounding is the method of calculating interest that yields what compound interest tends toward with more and more frequent compounding per period.

EXAMPLE 5
Continuous Compounding

For \$1000 at an annual rate of 10%, compounded m times in the course of a single year, what is the balance at the end of the year?

SOLUTION It is

$$\$1000\left(1 + \frac{0.10}{m}\right)^m$$

This quantity gets closer and closer to $\$1000e^{0.1} = \$1105.17\ldots$ as the number of compoundings m is increased. No matter how frequently interest is compounded— daily, hourly, every second, infinitely often ("continuously")—the original \$1000 at the end of one year cannot grow beyond \$1105.17. The values after 5 and 10 years are shown in the lower rows of Table 21.2 on page 771.

Continuous Interest Formula RULE

For a principal P deposited in an account at a nominal annual rate r, compounded continuously, the balance after t years is

$$A = Pe^{rt}$$

We illustrate with \$1000 at 10%. For one year, we have $t = 1$ and

$$A = \$1000e^{0.10} = \$1105.17$$

To find the amount in the account after 5 years, we have $t = 5$:

$$A = \$1000e^{(0.10)(5)} = \$1000e^{0.5} = \$1648.72$$

exactly as shown in the rightmost column of Table 21.2 (page 771).

It makes virtually no difference whether compounding is done daily or continuously over the course of a year. Most banks apply a daily periodic rate (based on compounding continuously) to the balance in the account each day and post interest daily (rounded to the nearest cent). The daily nominal rate is $r/365$, so each day the balance of the account is multiplied by $e^{r/365}$, the daily effective rate. Except for the rounding in posting interest, the effect is the same as continuous compounding throughout the year, because the compound interest formula gives $A = P\,(e^{r/365})^{365}$, which is the same as the formula $A = Pe^r$ from the continuous interest formula.

For example, for a principal of \$1000 and an interest rate of 5%, interest compounded daily over a year yields an amount

$$\$1000\left(1 + \frac{0.05}{365}\right)^{365} = \$1051.2675,$$

while continuous compounding yields $\$1000e^{0.05} = \$1051.2711$.

21.4 A Model for Saving

The compound interest formula tells the fate over time of a single deposited amount, but another common question that arises in finance is: What size deposit do you need to make regularly in an account with a fixed rate of interest, to have a specified amount at a particular time in the future?

This question is important in planning for a major purchase in the future or accumulating a retirement nest egg. In Chapter 22, we apply the same concepts and formula to paying off a mortgage and making installment payments on a car.

PhotoAlto/Eric Audras/Getty Images

EXAMPLE 6
A Savings Plan

A graduate at her first job saves \$100 per month, deposited directly into her credit union account on payday, the last day of the month. The account earns 1.8% per year, compounded monthly. How much will she have at the end of five years, assuming that the credit union continues to pay the same interest rate?

SOLUTION Note that she makes the first deposit at the end of the first month and the last deposit at the end of the 60th month. The monthly interest rate is $i = r/12 = 0.018/12 = 0.0015$.

It's easier to look at the deposits in reverse time order. The last deposit is made on the last day of the five years, so it earns no interest and contributes just \$100 to the total.

The second-last deposit earns interest for one month, contributing $\$100(1 + i)$. Similarly, the third-last contribution is on deposit for two months, contributing $\$100(1 + i)^2$.

Continuing in the same way, we find that the first deposit earns interest for 59 months and contributes $\$100(1 + i)^{59}$. The total of all of the contributions is

$$\$100 + \$100(1 + i)^1 + \$100(1 + i)^2 + \cdots + \$100(1 + i)^{59}$$
$$= \$100[1 + (1 + i)^1 + (1 + i)^2 + \cdots + (1 + i)^{59}]$$

This expression is known as a **geometric series** because the successive terms have geometric growth: Each succeeding term is a constant—in this case, $(1 + i)$—times the preceding term. For the sum of such a series with general ratio x, we have the formula in the following box.

Formula for Sum of a Geometric Series RULE

$$S = a_0 + a_0x + a_0x^2 + a_0x^3 + \cdots + a_0x^{n-1} = a_0\,[1 + x + x^2 + x^3 + \cdots + x^n]$$
$$= a_0\left[\frac{x^n - 1}{x - 1}\right], x \neq 1$$

The formula is easy to derive. Multiply S by x, getting

$$xS = a_0x + a_0x^2 + a_0x^3 + \cdots + a_0x^{n-1} + a_0x^n$$

and then subtract that quantity from S itself:

$$S - xS = a_0 + a_0x + a_0x^2 + a_0x^3 + \cdots + a_0x^{n-1}$$
$$- a_0x - a_0x^2 - a_0x^3 - \cdots - a_0x^{n-1} - a_0x^n$$

With most terms canceling, we are left with

$$S - xS = a_0 - a_0x^n,$$
$$S(1 - x) = a_0\,(1 - x^n),$$
$$S = a_0\left[\frac{1 - x^n}{1 - x}\right] = a_0\left[\frac{x^n - 1}{x - 1}\right], x \neq 1$$

That this formula works for all x (except $x = 1$) can be confirmed by multiplying both sides by $(x - 1)$ and watching terms on the left cancel. (You should do this confirmation for $n = 4$.)

In our example, we have $x = 1 + i$, and the formula becomes

$$1 + (1 + i)^1 + (1 + i)^2 + \cdots + (1 + i)^{n-1} = \frac{(1 + i)^n - 1}{i}$$

We have $n - 1 = 59$, or $n = 60$ months, and $i = 0.0015$, the interest rate per month. The total accumulation after five years is

$$A = \$100\left[\frac{(1 + 0.0015)^{60} - 1}{0.0015}\right] = \$6273.37$$

For a uniform deposit of d per compounding period (deposited at the end of the period) and an interest rate i per period, the amount A accumulated after n compounding periods is given by the **savings formula**:

Savings Formula RULE

$$A = d\left[\frac{(1 + i)^n - 1}{i}\right] = d\left[\frac{(1 + \frac{r}{m})^{mt} - 1}{\frac{r}{m}}\right].$$

where

$A =$	amount accumulated
$d =$	regular deposit of payment at the end of each period
$n = mt$	number of periods
$r =$	nominal annual interest rate
$m =$	number of compounding periods per year
$t =$	number of years
$i = r/m$	periodic rate, the interest rate per compounding period

The expression on the right gives the amount accumulated in terms of the nominal annual interest rate r, the number m of compounding periods per year, and the number t of years, using the relations $i = r/m$ and $n = mt$.

Algebra Review: Sums of Arithmetic and Geometric Sequences

An arithmetic sequence occurs when you *add* the same amount to a term to obtain the next term. This relationship is linear. A geometric sequence occurs when you *multiply* the same amount to a term to obtain the next term. This relationship is exponential.

Suppose that each sequence has a first term a. For the arithmetic sequence, we add d to obtain the next term, and for the geometric sequence, we multiply by r. We can find the nth term of each sequence by looking for a pattern:

Term Number	Arithmetic	Geometric
1	a	a
2	$a + d$	ar
3	$a + 2d$	ar^2
4	$a + 3d$	ar^3
⋮	⋮	⋮
n	$a + (n - 1)d$	ar^{n-1}

The nth term of an arithmetic sequence is $a + (n - 1)d$, and the nth term of a geometric sequence is ar^{n-1}.

Let's look at two specific sequences where the first term is 3. The arithmetic has a common difference (d) value of 2, and the geometric has a common ratio (r) value of 2. The first eight terms of each sequence are as follows:

Arithmetic: 3, 5, 7, 9, 11, 13, 15, 17
Geometric: 3, 6, 12, 24, 48, 96, 192, 384

If we want to find the sum of the terms for the arithmetic sequence, we notice a special relation. The sum of the first and last term is 20. That sum appears three more times:

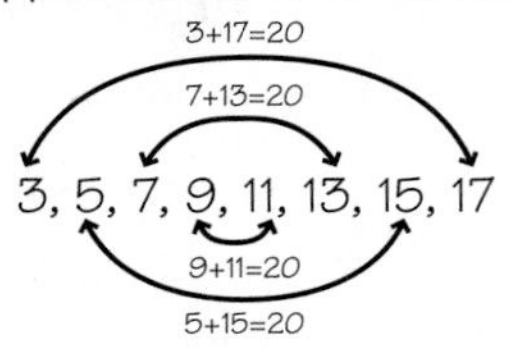

The sum of the 8 terms will be $4 \cdot 20 = 80$. In general, to find the sum of the first n terms of an arithmetic sequence, the pattern will be

$$\frac{\text{number of terms, } n}{2}(\text{first term} + \text{last term})$$

A series is made from a sequence of terms that are added up. For the geometric series, we will write the sum of the eight terms as

$$3 + 6 + 12 + 24 + 48 + 96 + 192 + 384$$

By adding these eight terms, we see that the sum is equal to 765.

Because the sum is the same as

$$3(1 + 2 + 2^2 + 2^3 + 2^4 + 2^5 + 2^6 + 2^7)$$

we can use the sum of a geometric series rule, which is

$$S = a_0\left[\frac{x^n - 1}{x - 1}\right], x \neq 1$$

We have $a_0 = 3$, $x = 2$, $x^n = 2^8$. Substituting into the sum rule, we have

$$3\left(\frac{2^8 - 1}{2 - 1}\right) = 3\left(\frac{256 - 1}{1}\right) = 3(255) = 765$$

The savings formula involves four quantities: A, d, i, and n. If any three are known, the fourth can be found. A common situation is for A, i, and n to be known, with d (the regular payment) to be found, because the practical concern for most people is how much their monthly payment will be.

Since we often want to solve for d, we solve the savings formula algebraically once and for all for d to get the *payment formula:*

Payment Formula RULE

$$d = A\left[\frac{i}{(1 + i)^n - 1}\right] = A\left[\frac{\frac{r}{m}}{(1 + \frac{r}{m})^{mt} - 1}\right]$$

Sometimes the purpose of saving is to accumulate a fixed sum by a particular date. Such a savings plan is called a **sinking fund** because you sink money into it. If the same amount is deposited regularly, the sinking fund is called an **annuity**.

Sinking Fund DEFINITION

A **sinking fund** is a savings plan to accumulate a fixed sum by a particular date, usually through equal periodic deposits.

Annuity DEFINITION

An **annuity** is a specified number of equal periodic payments.

EXAMPLE 7
Sinking Funds

Suppose that your parents had started saving for your college education when you were born. How much would they have had to save each month to accumulate $15,000 (to pay for just your first year of college!) over 18 years with an account earning a steady 5% per year, compounded monthly?

SOLUTION Applying the payment formula with $A = \$15{,}000$, monthly rate $i = r/m = 0.05/12$, and $n = mt = 12 \times 18 = 216$, we get

$$d = \$15{,}000\left[\frac{\frac{0.05}{12}}{\left(1 + \frac{0.05}{12}\right)^{216} - 1}\right] = \$42.96$$

This sounds like a manageable amount to contribute, but it doesn't take into account inflation, costs beyond the first year, or the higher cost of a private college. In the next section, we investigate how to take inflation into account.

Saving for Retirement (It's Never Too Early to Start)

Financial advisers stress the importance of beginning early to save for retirement. Many firms offer a *401(k) plan* (named after a section of law regulating pensions), which allows an employee to make monthly contributions to a retirement account. The plan has the advantage that income tax on the contributions is deferred until the employee withdraws the money during retirement. That means, for example, that an employee making a $100 monthly contribution may see a reduction in take-home pay of only $75 or less, since taxes are not withheld on the contribution.

Sometimes a company's pension plan consists of just contributing company stock to the employee's individual 401(k) account. In 2002, the bankruptcy of Enron Corporation resulted in thousands of its employees losing almost their entire retirement savings. Those savings consisted largely of Enron stock contributed by Enron, which fell from $90 per share to under $1 per share in just a couple of months. The Enron bankruptcy illustrated how unwise it is for most of an employee's retirement fund to consist of stock in just one company, particularly if—as was the case for Enron—the employee is not free to sell the stock. Even more people lost retirement savings and jobs when the stock of WorldCom declined more than 99% in 2002, after news of financial fraud by its management.

EXAMPLE 8
Retirement Fund Annuity Savings

Suppose that you start a 401(k) plan when you turn 23 and contribute \$50 at the end of each month until you turn 65 and retire. Suppose that you put your contributions into a very safe long-term investment that returns a steady 5% annual interest compounded monthly. How much will be in your fund at retirement?

SOLUTION Apply the savings formula with d = \$50, $i = 0.05/12$, and $n = mt = 12 \times (65 - 23) = 504$. We get

$$A = \$50\left[\frac{\left(1 + \frac{0.05}{12}\right)^{504} - 1}{\frac{0.05}{12}}\right] = \$85{,}567.43$$

At first glance, that may seem like a lot of money, but it is not so much if that's all you have to live off for the rest of your life. (Of course, there is also Social Security.) In the exercises later in this chapter, we explore the effects of saving more each month, getting a higher interest rate, saving on taxes, and—especially—having inflation erode the value of your savings.

Annuities are a common way for lotteries to pay out grand prizes and for retirees to receive funds saved up for retirement. We examine an example of each in Section 22.4 (pages 817–820), where we turn the savings formula around to get a formula (the amortization formula) to relate your savings to a regular payout. In Chapter 22, you find out how much monthly income for a fixed period—or for life—\$86,000 could buy.

21.5 Present Value and Inflation

Suppose that you want to make a one-time deposit of amount P that will grow to a specific amount A in n compounding periods from now by earning interest at a rate i per period. The quantities A, P, i, and n are related through the compound interest formula, $A = P(1 + i)^n$. The quantity P is called the **present value** of the amount A to be paid n compounding periods in the future.

Present Value DEFINITION

The **present value** of an amount to be paid or received at a specific time in the future is what that future payment would be worth today, as determined from a given interest rate and compounding period.

Present Value RULE

The present value P of an amount A to be paid t years in the future, earning a nominal annual rate of interest r compounded m times per year—that is, after $n = mt$ compounding periods at a rate $i = r/m$ per compounding period—is

$$P = \frac{A}{(1 + i)^n} = \frac{A}{\left(1 + \frac{r}{m}\right)^{mt}}$$

EXAMPLE 9
Certificates of Deposit

A certificate of deposit (CD) pays a fixed rate of interest for a term specified in advance, from a month to 10 years. The best local rate in December 2010 was for a 45-month CD at 2.00% compounded daily. How much would we need to set aside now in such a CD to have $12,000 in 45 months?

SOLUTION We find the present value of $12,000 45 months from now, with $r = 0.02$, $m = 365$, and $t = 45/12 = 3.75$, so that $mt \approx 1369$. (Whether the 45-month span includes a leap day, or even four Februarys instead of four 31-day Augusts, will make negligible difference.) The present value formula gives

$$P = \frac{A}{\left(1 + \frac{r}{m}\right)^{mt}} = \frac{\$12{,}000}{\left(1 + \frac{0.02}{365}\right)^{365 \times 3.75}} \approx \frac{\$12{,}000}{(1 + 0.0000547945)^{1369}} = 11{,}132.79$$

In times of economic inflation, prices increase. When the rate of inflation is constant, the compound interest formula can be used to project prices.

Inflation DEFINITION

Inflation is a rise in prices from a set base year.

Annual Rate of Inflation DEFINITION

The **annual rate of inflation,** a (= $100a$%), is the additional proportionate cost of goods one year later. Goods that cost $1 in the base year will then cost $\$(1 + a)$.

EXAMPLE 10
Inflation

Suppose that there is constant 3% annual inflation from mid-2009 through mid-2013. What will be the projected price in mid-2013 of an item that costs $100 in mid-2009?

SOLUTION The compound interest formula applies with $P = \$100$, $a = r = 3\%$, $m = 1$, and $t = 4$. The projected price is $A = P(1 + r)^t = \$100(1 + 0.03)^4 = \$112.55$.

Inflation and Depreciation as Exponential Decay

During constant-rate inflation, prices grow geometrically (exponentially) and the value of the dollar goes down geometrically: one is growing exponentially; the other is decaying exponentially.

Exponential Decay DEFINITION

Exponential decay is geometric growth with a negative rate of growth.

Let a (for "additional") represent the annual rate of inflation; what costs $1 now will cost $\$(1 + a)$ this time next year. For example, if the inflation rate were

$a = 25\%$, then what costs \$1 now would cost \$1.25 this time next year. A dollar next year would buy only 0.8 (= 1/1.25) times as much as a dollar buys today. In other words, a dollar next year would be worth only \$0.80 in today's dollars; by next year, a dollar would have lost 20% of its purchasing power. Notice that although the inflation rate is 25%, the loss in purchasing power is 20%. This may seem peculiar; why aren't they the same percentage?

The situation may be clearer if you consider inflation at a rate of 100%. Then a dollar next year is worth only 50% of a dollar today; what costs \$1 today will cost \$2 next year. The reason for the difference in the percentages is that the percent inflation (100%) uses today's price level (\$1) as a base (prices rise from \$1 to \$2, hence they rise 100%), while the percentage loss in purchasing power uses as a base the larger price level next year (\$2) (so loss in purchasing power is 50% of \$2).

For an annual inflation rate a, a year from now a dollar will buy only a fraction of what a dollar today can buy.

Relative Purchasing Power of a Dollar a Year from Now with Inflation Rate a RULE

$$\frac{\$1}{1+a} = \$1 - \frac{\$a}{1+a}$$

The loss in purchasing power is the fraction $a/(1 + a)$. (You should calculate what these expressions become for $a = 25\%$.) The quantity $i = -a/(1 + a)$ behaves like a negative interest rate. We can use the compound interest formula to find the relative purchasing power of P dollars t years from now as

$$A = P(1 + i)^t = P\left[1 - \frac{a}{1+a}\right]^t.$$

The actual posted price of an item, at any time, is said to be in **current dollars**. That price can be compared with prices at other times by converting all prices to **constant dollars**, dollars of a particular year.

EXAMPLE 11
Deflated Dollars

Suppose that there is 25% annual inflation from mid-2012 through mid-2016. What would be the value of a dollar in mid-2016 in constant mid-2012 dollars? The inflation figure is unrealistic (we hope!), but it makes the calculations easy so that you can focus on the ideas.

We have $a = 0.25$, so $i = -a/(1 + a) = -0.25/1.25 = -0.20$. This, not 25%, is the negative interest rate, the rate at which the dollar is losing purchasing power. We have $t \approx 4$ years, so the purchasing power of \$1 four years from mid-2012, relative to (or in terms of) 2012 dollars, would be

$$\$1(1 + i)^4 = \$(1 - 0.20)^4 = (0.80)^4 = \$0.41 \text{ in 2012 dollars}$$

For a more realistic rate of 3% annual inflation from mid-2012 through mid-2016, we would have $a = 0.03$ and negative interest rate $i = -1/(1 + a) = -0.03/1.03 \approx -0.0291262$. In 2016, the purchasing power of \$1 would be $\$1(1 + i)^4 = \$(1 - 0.0291262)^4 = (0.970874)^4 \approx \$0.8885$ in 2012 dollars. Notice that "losing" 3% to inflation each year for 4 years amounts to losing almost—but not quite as much as—a total of $4 \times 3\% = 12\%$. (Just as with the previous inflation rate, where losing 25% per year for 4 years doesn't completely reduce the value to 0.)

In Example 11, we may think of the value of the dollar as "depreciating" each year. Depreciation of the value of equipment is similar.

EXAMPLE 12
Depreciation

If you bought a car at the beginning of 2012 for $12,000 and its value in current dollars depreciates at a rate of 15% per year, what will be its value at the beginning of 2015 in current dollars?

SOLUTION We have $P = \$12{,}000$, $i = -0.15$, and $n = 3$. The compound interest formula gives

$$A = P(1 + i)^n = \$12{,}000(1 - 0.15)^3 = \$7369.50 \text{ in 2015 dollars}$$

The Consumer Price Index

In our preceding model, we supposed that inflation stayed constant over a period of time. That is not generally the case. However, based on regular measures of inflation, we can determine the equivalent today of a price in an earlier year or how much a dollar in that year would be worth today in purchasing power.

The official measure of inflation is the Consumer Price Index (CPI), determined by the Bureau of Labor Statistics (BLS). Here, we describe and use the CPI-U, the index for all urban consumers, which covers about 80% of the U.S. population and is the index of inflation that is usually referred to in newspaper and magazine articles.

Each month, the BLS determines the average cost of a "market basket" of goods, including food, housing, transportation, clothing, and other items. It compares this cost with the cost of the same (or comparable) goods in a base period. The base period used to construct the CPI-U is 1982–1984. The index for 1982–1984 is set to 100, and the CPI-U for other years is calculated by using the proportion

$$\frac{\text{CPI for other year}}{100} = \frac{\text{cost of market basket in other year}}{\text{cost of market basket in base period}}$$

For example, the cost of the market basket in 1976 (in 1976 dollars) was 0.569 times the cost in 1982–1984 (in 1982–1984 dollars), so the CPI for 1976 is 100×0.569, or 56.9.

Table 21.5 shows the average CPI for each year from 1913 through 2010, with estimates for 2011 and 2012. This table can be used to convert the cost of an item in dollars for one year to what it would cost in dollars in a different year, using the proportion cost in CPI for year A:

$$\frac{\text{cost in year A}}{\text{cost in year B}} = \frac{\text{CPI for year A}}{\text{CPI for year B}}$$

EXAMPLE 13
The Price of Our House and the Value of a Dollar

Where my family and I live, housing is relatively inexpensive. We bought our house in mid-1992 for $133,000 (close to the median price of U.S. housing at that time). What would be the equivalent cost in mid-2012 dollars?

Patti McConville/Getty Images

SOLUTION We see from Table 21.5 that the CPI for 1992 is 140.3 and the CPI for 2012 is estimated to be 232.0. The table gives the average value for each year, which is very close to the value at midyear. Month-by-month values are available at the Bureau of Labor Statistics Web site (http://stats.bls.gov/cpi/).

Using the proportion, we have

$$\frac{\text{cost in 2012}}{\text{cost in 1992}} = \frac{\text{CPI in 2012}}{\text{CPI in 1992}}$$

or

$$\frac{\text{cost in 2012}}{\$133{,}000} = \frac{232.0}{140.3}$$

so that

$$\text{cost in 2012} = \$133{,}000 \times \frac{232.0}{140.3} \approx \$220{,}000$$

That's what our house would sell for if its price exactly matched inflation.

The ratio 232.0/140.3 ≈ 1.654 is the *scaling factor* for converting 1992 dollars to 2012 dollars. What we are observing is a proportion, or *numerical similarity*, between 1992 dollars and 2012 dollars, analogous to the geometric similarity discussed in Chapter 18 (page 644). To convert from 2012 dollars to 1992 dollars, we would multiply by 1/1.654 ≈ 0.605.

Spotlight 19.3 (on page 690) describes how the CPI is calculated, using the geometric mean (defined and introduced on page 512).

Table 21.5

U.S. Consumer Price Index (1982–1984 = 100)

—	—	1931	15.2	1951	26.0	1971	40.5	1991	136.2
—	—	1932	13.7	1952	26.6	1972	41.8	1992	140.3
1913	9.9	1933	13.0	1953	26.7	1973	44.4	1993	144.5
1914	10.0	1934	13.4	1954	26.9	1974	49.3	1994	148.2
1915	10.1	1935	13.7	1955	26.8	1975	53.8	1995	152.4
1916	10.9	1936	13.9	1956	27.2	1976	56.9	1996	156.9
1917	12.8	1937	14.4	1957	28.1	1977	60.6	1997	160.5
1918	15.1	1938	14.1	1958	28.9	1978	65.2	1998	163.0
1919	17.3	1939	13.9	1959	29.1	1979	72.6	1999	166.6
1920	20.0	1940	14.0	1960	29.6	1980	82.4	2000	172.2
1921	17.9	1941	14.7	1961	29.9	1981	90.9	2001	177.1
1922	16.8	1942	16.3	1962	30.2	1982	96.5	2002	179.9
1923	17.1	1943	17.3	1963	30.6	1983	99.6	2003	184.0
1924	17.1	1944	17.6	1964	31.0	1984	103.9	2004	188.9
1925	17.5	1945	18.0	1965	31.5	1985	107.6	2005	195.3
1926	17.7	1946	19.5	1966	32.4	1986	109.6	2006	201.6
1927	17.4	1947	22.3	1967	33.4	1987	113.6	2007	207.3
1928	17.1	1948	24.1	1968	34.8	1988	118.3	2008	215.3
1929	17.1	1949	23.8	1969	36.7	1989	124.0	2009	214.5
1930	16.7	1950	24.1	1970	38.8	1990	130.7	2010	218.1
								2011 (est.)	226.0
								2012 (est.)	232.0

Note: This the CPI-U index, which covers all urban consumers, about 80% of the U.S. population. Each index is an average for all cities for the year. The basis for the index is the period 1982–1984, for which the index was set to be equal to 100. For each year, the figure is the average during the year, hence is usually close to the value at midyear.

SOURCE: ftp://ftp.bls.gov/pub/special.requests/cpi/cpiai.txt

SPOTLIGHT 21.3

What Is a Derivative?

There are many ways to save and invest:

- Some offer guaranteed rates of interest (savings accounts, CDs, bonds, annuities);
- others offer rates that vary with interest rates in the general economy (money market accounts, special inflation-protected U.S. Treasury savings bonds, life insurance); and
- still others are not savings but investments (stocks, mutual funds, "derivatives").

Thanks to the Great Depression, when many banks failed and depositors lost everything, deposits in most banks and credit unions are now insured by the federal government up to $250,000 per account holder. If the bank or credit union fails, the government takes it over and depositors get their money back.

A bond is a loan at simple interest with repayment at the end of a fixed term. With bonds, investors loan money for civic construction projects, company expansions, and other purposes. Bonds, which are issued by states, municipalities, and corporations, do not offer any guarantee. The bond issuer (the borrower) could default, meaning that the bond holder (the investor) would no longer receive interest payments and might lose the original investment.

Bonds usually can be bought and sold after being issued. The price can vary with sentiment about credit-worthiness of the issuer and with changes in interest rates: If rates go above the rate that a bond is paying, the bond becomes less valuable and its price will fall, so that its "yield" (effective interest rate relative to the current price) rises. If interest rates instead fall, the bond becomes more valuable because it is paying a higher rate; its price will rise and its yield rate decline toward the current interest rate.

Some bonds can be "called," meaning that the issuer can end the contract by paying back the bond holder before the end of the bond term. Bonds tend to be called when interest rates fall: The issuer calls bonds previously issued at a high interest rate, then offers new bonds at a lower rate, thus saving on the cost of borrowing.

Stocks (also called *equities*) are another way for a company to raise capital, and an opportunity for investors to share in the financial future of the company by owning shares in it. Usually, a part of the company's profits is distributed each year to shareholders as dividends. In addition, the shares themselves, which are generally traded on an exchange, may rise in market value. On the other hand, a company that reinvests all profit, or makes little profit, or incurs losses, may pay no dividends, and its price may fall. The value of a company's stock depends on "fundamentals" (measures of the "health" of the company), on the general state of the economy as a whole, and on current events.

Derivative DEFINITION

A **derivative** is a financial instrument whose value "derives" from the value of an underlying asset such as a stock, bond, commodity, mortgage, option, etc.

An example of a "first-order" derivative is shares in a *mutual fund*. A mutual fund holds an array of other companies' stocks and bonds, perhaps concentrating on stocks of a particular kind ("green" companies, foreign companies, companies focused on growth, etc.). Investors in the mutual fund do not individually own any shares of the stocks that it invests in; the value of the mutual fund shares derives from the investments that it holds. *Hedge funds* are basically less-regulated mutual funds.

A popular kind of mutual fund is an *index fund*. An index fund selects investments to try to mirror movements of the index of a particular market and thus capture proportionate gains (but also suffers proportionate losses). Common indices include the Dow Jones Industrial Average and the Standard and Poor's (S&P) 500 Index. At another level removed, there are index funds of index funds.

Traditional derivatives include agricultural commodity *futures* (corn, soybeans, pork bellies), used by farmers and food processors to hedge against the risk of poor crops; and currency futures, used by companies to lock in a fixed exchange rate for future purchases of foreign goods and raw materials. There are even weather derivatives, whose price depends

What Is a Derivative? *(continued)*

on the number of sunny days or amount of rainfall in a particular region!

Derivatives arise also in the context of loans. A *credit derivative* bases its value not on the underlying loan, such as a mortgage, but on the risk that the loan will not be repaid. The buyer of the credit derivative does not own the mortgage loan itself, hence is not entitled to be paid by the homeowner.

A common form of credit derivative is a *credit default swap*; credit default swaps currently in force amount to $62 trillion worldwide (almost $10,000 for every person on the globe). One party buys protection from a second against default on a debt owed by a third. The "protection buyer" pays a fee to the "protection seller" in return for the seller making good on the debt if the debtor defaults. The buyer thus "swaps" risk to the seller. But you aren't going to want to believe this: The debtor can owe the debt to someone else entirely unrelated to the protection buyer or seller! In such a case, the protection buyer and seller are gambling together on whether the debtor will default. This has happened in connection with the "sovereign debt" of countries such as Greece and Ireland.

Where does the ordinary person come into all this? A handful of investors made billions in the past few years through credit default swaps on "bundles" of regular home mortgages. Well, not all were so "regular"; too many turned out to be subprime mortgages, loans to people with poor credit ratings, who were consequently saddled with higher interest rates and who then defaulted in large numbers as the economy forced them out of their jobs. (Spotlight 22.3, on p. 816, looks in more detail at the mortgage crisis.)

Those credit default swaps involved not the original mortgages, nor the bundles of them—which themselves were derivatives of the underlying assets of the homes involved. The owners of the bundles had issued "second-order" securities, backed by the bundles—not by the homes—and the credit default swaps were on those securities. But the decline in housing prices reduced the value of the underlying asset of homes, hence of the bundles, hence of the securities based on the bundles—derivatives of derivatives. The securities became "toxic assets": No one was willing to buy them at anywhere near the original price, and banks have been unwilling to sell them at big losses.

Attempts by holders of the securities to foreclose on the homes have been to some degree thwarted by the fact that the securities holders are not the direct owners of the mortgage loans. In some cases, there has been so much trading of loans and of derivatives on them, without sufficient documentation, that it is impossible to establish to whom a loan is owed.

Warren Buffett, chairman of Berkshire Hathaway, the country's largest holding company, and one of the world's richest men, was prescient in 2002 when he called financial derivatives "time bombs" and "financial weapons of mass destruction." Financial derivatives, combined with greed and exploitation, indeed destroyed the prosperity that the world enjoyed in the early years of the 21st century.

Real Growth Under Inflation

It's natural to think that if your investment is growing at 6% per year and inflation is at 3% per year, then the real growth in the value (purchasing power) of your investment is $6\% - 3\% = 3\%$. That is a handy approximation, but it is not exactly right. Let's see why.

Suppose that you invest $500 for a year at 6% and inflation is 3%. At the beginning of the year, you have $500, which at $5 per pound could buy 100 pounds of steak. At the end of the year, you have $\$500(1 + 0.06) = \$530$, but steak now costs $\$5(1 + 0.03) = \$5.15$ per pound. How much steak would that buy? $\$530/\$5.15 \text{ lb} = 102.91 \text{ lb}$. In other words, in terms of purchasing power, or real gain, your investment has grown only 2.91%. This is not a great deal different from 3%, but it is different, and the difference is greater for higher rates of interest and inflation.

Consider an investment principal P and a market basket of goods of value m. Let the annual yield (rate of interest) of the investment be r and the rate of inflation be a. We calculate the rate of real growth g of the investment as follows:

At the beginning of the year, the investment would buy quantity $q_{\text{old}} = P/m$ of the market basket. At the end of the year, the investment would buy

$$q_{\text{new}} = \frac{P(1 + r)}{m(1 + a)}$$

quantity of the market basket. Notice that the gain of r in the investment multiplies the principal by $(1 + r)$, while the erosion due to inflation divides the principal by $(1 + a)$. Here, you see that the two influences on the investment have directly opposite effects.

SPOTLIGHT 21.4

Using a Spreadsheet for Financial Calculations

Both commercial (e.g., Microsoft Excel) and open-source (e.g., Open Office) spreadsheets have commands that use the formulas developed in this chapter in situations of compound interest:

PMT (*rate, nper, pv, fv, type*)
PV (*rate, nper, pmt, fv, type*)
FV (*rate, nper, pmt, pv, type*)
RATE (*nper, pmt, pv, fv, type, guess*)

where

rate = interest rate per payment (compounding) period
nper = total number of payments
pv = principal, or present value
pmt = payment made each (compounding) period
fv = future value, or cash balance after last payment
type = 0 if payment made at end of period, 1 if at beginning
guess = guess at interest rate per payment period

Both *pv* and *pmt* will be negative if they correspond to payment by you. The guess can be omitted. Also, for simplicity, all problems in this chapter have payments at the end of the period, so *type* = 0; since that is the default for the spreadsheet, it too can be omitted. (Note: OpenOffice uses semicolons for separators instead of commas.)

We show how some of our previous examples can be solved by using a spreadsheet:

Example 2 (compound interest), with monthly compounding: We want the amount accumulated in an account of \$1000 after 10 years at 10% interest compounded monthly. We have *rate* = 10%/12, *nper* = 10 × 12 = 120, *pmt* = 0 (since we make no payments after depositing the principal), and *pv* = −1000. Put into a cell in the spreadsheet =FV(10%/12, 10*12, 0, −1000) and see \$2707.04 emerge.

Example 6 (savings plan): We want to determine how much a regular \$100 per month deposit will amount to after 5 years at 1.8% per year compounded monthly. We have *rate* = 1.8%/12, *nper* = 5 × 10 = 60, *pmt* = −100 (remember, payments are outlays), and *pv* = 0. Put into a cell =FV(1.8%/12, 5*12, −100, 0) and see \$6273.37 emerge.

Example 7 (sinking fund): Your parents want to save \$15,000 over 18 years at 5% compounded monthly. Put into a cell =PMT(5%/12, 18*12, 0, 15000) and see (\$42.96)—it's in red because it is \$−42.96, meaning that that is how much they must deposit monthly.

Example 9 (certificate of deposit): We want to deposit an amount that will accumulate \$12,000 in 45 months ≈ 1369 days at 2.0% interest compounded daily. Put in = PV(2.0%/365, 1369, 0, 12000) and see (\$11,132.79), which is in red because it is a deposit.

The growth of the investment, relative to how many market baskets it could have bought originally, is

$$g = \frac{q_{new} - q_{old}}{q_{old}} = \frac{\frac{P(1+r)}{m(1+a)} - \frac{P}{m}}{\frac{P}{m}} = \frac{1+r}{1+a} - 1 = \frac{r-a}{1+a}$$

In the last expression, the numerator is the difference of the two rates (6% − 3% in our example), which is divided by a quantity greater than 1 if there is inflation. You should confirm that this formula gives 2.91% for $r = 6\%$ and $a = 3\%$.

One way to understand why this is the correct formula is to realize that the gain itself is not in original dollars but in deflated dollars.

The relationship between interest rate, inflation rate, and rate of real growth is called *Fisher's effect,* after the American economist Irving Fisher (1867–1947).

Real Rate of Growth RULE

The real (effective) annual rate of growth of an investment at annual interest rate r with annual inflation rate a is

$$g = \frac{r-a}{1+a}$$

REVIEW VOCABULARY

Annual percentage yield (APY) The effective interest rate per year. (p. 771)

Annual rate of inflation The additional proportionate cost of goods one year later. (p. 785)

Annuity A specified number of equal periodic payments. (p. 783)

Arithmetic growth Growth by a constant amount in each time period. (p. 769)

Compound interest Interest that is paid on both the original principal and the accumulated interest. (p. 768)

Compound interest formula The formula for the amount in an account that pays compound interest periodically. For an initial principal P and an effective rate i per compounding period, the amount after n compounding periods is $A = P(1 + i)^n$. (p. 772)

Compounding period The fundamental interval on which compounding is based, within which no compounding is done. Also called simply *period*. (p. 770)

Constant dollars Costs are expressed in constant dollars if inflation or deflation has been taken into account by converting the costs to their equivalent in dollars of a particular year. (p. 786)

Continuous compounding Payment of interest in an amount toward which compound interest tends with more and more frequent compounding. (p. 779)

Current dollars The actual cost of an item at a point in time; inflation or deflation before or since then has not been taken into account. (p. 786)

Derivative A financial instrument whose value "derives" from the value of an underlying asset. (p. 789)

e The base for continuous compounding, geometric (exponential) growth, and natural logarithms; $e = 2.71828\ldots$. (p. 778)

Effective rate The rate of simple interest that would realize exactly as much interest over the same period of time. (p. 771)

Exponential decay Geometric growth at a negative rate. (p. 785)

Exponential growth Geometric growth. (p. 773)

Geometric growth Growth proportional to the amount present. (p. 773)

Geometric series A sum of terms, each of which is the same constant times the previous term; that is, the terms undergo geometric growth. (p. 780)

Inflation A rise in prices from a set base year. (p. 785)

Interest Money earned on a savings account or a loan. (p. 768)

Linear growth Arithmetic growth. (p. 769)

Nominal rate A stated rate of interest for a specified length of time; a nominal rate does not take into account any compounding. (p. 771)

Present value The value today of an amount to be paid or received at a specific time in the future, as determined from a given interest rate and compounding period. (p. 784)

Principal Initial balance. (p. 768)

Savings formula The formula for the amount A accumulated after $n = mt$ periods, with a uniform deposit of d at the end of each compounding period and interest rate $i = r/m$ per period:

$$A = d\left[\frac{(1+i)^n - 1}{i}\right] = d\left[\frac{(1+\frac{r}{m})^{mt} - 1}{\frac{r}{m}}\right]. \text{ (p. 781)}$$

Simple interest Interest that is paid on the original principal only, not on any accumulated interest. (p. 768)

Sinking fund A savings plan to accumulate a fixed sum by a particular date, usually through equal periodic deposits. (p. 783)

SKILLS CHECK

1. Simple interest is an example of

(a) linear growth.
(b) arithmetic growth.
(c) constant growth.

2. If a savings account pays 3% simple annual interest, a deposit of $250 will earn ____________ in 2 years.

3. An 18% annual rate on a credit-card balance is an example of

(a) an effective rate.
(b) a nominal rate.
(c) an adjusted rate.

4. If you deposit $1000 at 6.2% simple interest, the balance after three years is ______________.

5. Suppose that you deposit $180 at the beginning of each year into a savings account that pays 2% simple interest per year at the end of the year. After two years, the amount in the account is

(a) $360.00.
(b) $367.20.
(c) $374.40.

6. If $800 is invested for one year at 6% compounded quarterly, the amount of interest earned is _____.

7. If a single deposit is made into a compound interest CD, the account earns

(a) interest only for the first period.
(b) the same amount of interest each period.
(c) more interest in each subsequent period.

8. If you deposit $1000 at 6.2% interest compounded quarterly, the balance after three years is

______________.

9. Compound interest is an example of

(a) geometric growth.
(b) exponential growth.
(c) humongous growth.

10. Suppose that you deposit $15 at the end of each month into a savings account that pays 2% interest compounded monthly. After a year, ____________ is in the account.

11. Compound interest is paid on

(a) just the initial principal.
(b) the current balance.
(c) just the accumulated interest.

12. Suppose that you deposit $10 at the end of each day into a savings account that pays 2% interest compounded daily. After a year, ____________ is in the account.

13. Suppose that you open an account that pays 1% interest compounded monthly with a deposit of $1000. At the end of one year, you will have approximately

(a) $1010.
(b) $1120.
(c) $1200.

14. The APY for 3% compounded monthly is ____.

15. Suppose that you invest $250 in an account that pays 4.5% interest compounded quarterly. After 30 months, how much is in your account?

(a) $279.08
(b) $279.59
(c) $279.71

16. The APY for 3% compounded daily is ________.

17. Which of the following pays more interest?

(a) 6% compounded annually
(b) 6% compounded monthly
(c) 6% compounded continuously

18. If you deposit $1000 at 6.2% interest compounded continuously, the balance after three years is ____________.

19. If an account with $100 pays 6% interest compounded continuously, at the end of one year, you will

(a) be fabulously wealthy.
(b) be getting 6% APY.
(c) have earned only pennies more interest than 6% simple interest.

20. The APY for 3% compounded continuously is ________.

21. Which of the following is the most generous interest rate for a one-year CD?

(a) 6% simple interest
(b) 5.9% compounded annually
(c) 5.9% compounded continuously

22. The value of *e* is approximately ________.

23. The number *e* is

(a) irrational.
(b) irrelevant.
(c) irrotational.

24. Depositing $100 on a child's annual birth date at ages 1 through 18 years is an example of a(n) ____________.

25. The sum of the geometric infinite series $1 + \frac{1}{2} + \frac{1}{4} + \frac{1}{8} + \cdots$ is

(a) infinite.
(b) 2.
(c) *e*.

26. If your investments earned 4.5% last year but inflation was 1.0%, the real rate of growth of your investments was ________%.

27. An example of exponential decay is

(a) the depreciation of factory equipment.
(b) a retirement annuity.
(c) the CPI.

28. If a new car costs $18,000 and loses value at a rate of 20% per year, its value after 3 years is _____.

29. If your investment is growing at a rate less than the rate of inflation,

(a) you have a positive real growth in your investment.
(b) you do not have a positive real growth in your investment.
(c) you do not have enough information to determine whether real growth is positive or negative.

30. According to the CPI, a market basket of goods that cost $100 in mid-2000 would cost ____________ in mid-2012.

CHAPTER 21 EXERCISES

■ *Challenge* ▲ *Discussion*

The exercises below require a scientific calculator with buttons for powers $\boxed{y^x}$, exponential $\boxed{e^x}$, and natural logarithm $\boxed{\ln x}$, or else a computer spreadsheet.

21.1 Arithmetic Growth and Simple Interest

1. In Example 1 on page 769, we considered a student loan from PNC Financial Services at that institution's "Lowest Tier Pricing" interest rate of 3.21% in October 2011. Its "Highest Tier Pricing" then had an interest rate of 11.21%. Under the same circumstances as Example 1 ($10,000 principal, interest-only repayment), how much interest would you pay over the 54 months of deferred payments on the principal?

2. Suppose that you need $30,000 for your last year of college. You could go to a private lending institution and apply for a signature student loan; rates range from 7% to 14%. However, your Aunt Sally is willing to lend you the money from her retirement savings, with no repayment until after graduation. All she asks is that in the meantime, you pay her each month the amount of interest that she would otherwise get on her savings (since she needs that to live on), which is 4%. What is your monthly payment to her, and how much interest will you pay her over the academic year (9 months)? (Aunt Sally will be glad to hear from you every month, anyway!)

3. On November 15, 2010, you could buy a 10-year U.S. Treasury note ("T-note," a kind of bond) for \$10,000 that pays 2.65% simple interest every year through November 15, 2020. How much total interest would it earn by then?

4. On November 15, 2010, you could buy a 30-year U.S. Treasury bond for \$10,000 that pays 4.25% simple interest every year through November 15, 2040. How much total interest would it earn by then?

21.2 Geometric Growth and Compound Interest

5. An often-heard claim is that "the amount of information in the world doubles every three days." Presumably the claim refers to the amount of data, which can be quantified in terms of number of bits. (A bit is the smallest unit of storage in a computer.) Show that the claim is absolutely preposterous by doing a little arithmetic and comparing your result with the estimated number of particles in the universe (10^{70}). In particular:

(a) Start with one bit of data and double the number of bits every third day. How long does it take to get past 10^{70}? (*Hint:* Don't just keep multiplying by 2 over and over. Convince yourself that since the amount of data increases by a factor of 2 every 3 days, then it increases by a factor of $2^2 = 4$ every 6 days, a factor of $4^2 = 16$ every 12 days, a factor of $16^2 = 256$ every 24 days, and so forth.)

(b) Part (a) involves a lot of multiplying by 2, even if you do it efficiently. Another approach is to use the fact that $2^{10} = 1024 \approx 1000$. Thus, the amount of data increases by a factor of more than 1000 every $3 \times 10 = 30$ days, or every month (except February, but the 31-day months make up for it). By when will the total surely be past 10^{70}?

6. In a "FoxTrot" cartoon by Bill Amend (9/10/2006), shown below, the girl Paige confronts a math problem in which "a math teacher assigns one second of homework the first week of school, two seconds the second week, four seconds the third, and so on." She is asked whether she would agree to this weekly homework doubling for the duration of the 36-week school year. How much homework (in hours) would this plan require in week 36?

7. (Requires spreadsheet.) Write a spreadsheet program that reproduces Table 21.1 (on page 770).

8. (Requires spreadsheet.) Write a spreadsheet program that reproduces Table 21.1 but rounds the interest added to the nearest cent.

9. (Requires spreadsheet.) Write a spreadsheet program that reproduces Table 21.2 (on page 771).

10. (Requires spreadsheet.) Write a spreadsheet program that reproduces Table 21.2 but rounds to the nearest cent the interest added for yearly and quarterly compounding.

11. You deposit \$1000 at 2% per year in a money market account. What is the balance at the end of one year, and what is the annual yield, if the interest paid is

(a) simple interest?
(b) compounded annually?
(c) compounded quarterly?
(d) compounded daily?

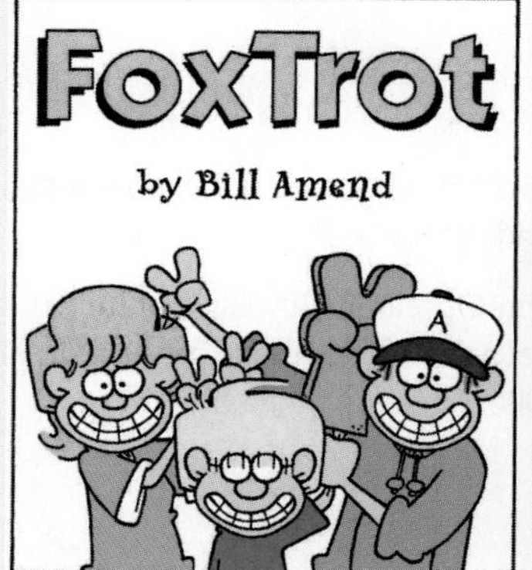

12. Repeat Exercise 11, but for $1000 at 1.2% per year in a savings account.

13. I had a CD with National City Bank through 2010 that paid 4.69% interest compounded daily. What was the APY for this rate?

14. I have an independent retirement account (IRA) with First Community Credit Union of Beloit, Wisconsin, which currently pays dividends at 0.75% per year, compounded monthly, for accounts with balances up to $2000. What is the APY for such a rate?

15. U.S. savings bonds are a common form of present or award that also helps to underwrite the debt of the United States. The interest is exempt from state and local income taxes and may also be exempt from federal income tax if used to pay for college tuition and fees. In December 2010, the Rodel Exemplary Teacher Initiative, which addresses the shortage of effective teachers in Arizona's neediest schools and encourages excellent teachers to stay in the profession, chose 13 teachers each to receive a $10,000 U.S. savings bond. Paper Series EE Savings Bonds issued between November 2010 and April 2011 cost 50% of face value (so $5000 for a $10,000 bond) and earn 0.60% interest, compounded semiannually, for the 20 years until their maturity. What is the APY? Will such a bond double in value in 20 years? (If not, the U.S. Treasury will make a one-time adjustment at the end of 20 years to ensure that the bond doubles in value to the full $10,000 face value.)

16. A Paper Series EE Savings Bond is sold at half face value, and the U.S. Treasury guarantees that it will double in value to the full face value by 20 years from the issue date. What is the minimum APY for such a bond?

17. By 2015, the concentration of the greenhouse gas carbon dioxide in the Earth's atmosphere will be 400 parts per million (ppm) by volume.

(a) The concentration in recent years has been increasing at about 2 ppm per year. If that trend continues, what will the concentration be at about the time that you retire, in, let us say, 2055?

(b) Actually, the concentration in recent years has been increasing not by a steady amount per year but by gradually increasing amounts, corresponding to an increase of about 0.5% per year (caused by the increase in burning of fuels on Earth). If that trend continues, what will the concentration be at about the time that you retire, in, let us say, 2055?

18. Suppose that on the statement for a money market account this month, the initial balance was $7722.54, the statement was for 27 days, and the final balance was $7744.70. Calculate the APY.

19. Repeat Exercise 18, but for the following month, which had an initial balance of $7744.70, a period of 34 days, and a final balance of $7770.84.

20. The atmospheric concentration of carbon dioxide was 315 ppm in 1958. What steady rate of compound-interest growth would explain the concentration growing to 400 ppm in 2015? [Your answer will be less than the 0.5% estimate for the current rate of increase mentioned in Exercise 17(b), which suggests that the rate of increase is itself slowly increasing.]

21. *The rule of 72* is a rule of thumb for finding how long it takes money at interest to double: If r is the annual interest rate, then the doubling time is approximately $72/(100r)$ years.

(a) Calculate the balance at the end of the predicted doubling time for each $1000, with annual compounding, for the small growth rates of 3%, 4%, and 6%.

(b) Repeat part (a) for the intermediate interest rates of 8% and 9%.

(c) Repeat part (a) for the larger interest rates of 12%, 24%, and 36%.

(d) What do you conclude about the rule of 72?

22. If carbon dioxide in the atmosphere continues to increase at 0.5% per year, how long will it take for the concentration to double from 390 ppm in 2010?

21.3 A Limit to Compounding

23. Use your calculator to evaluate for n = 1, 10, 100, 1000, and 1,000,000:

(a) $\left(1 + \frac{1}{m}\right)^m$

(b) $\left(1 + \frac{2}{m}\right)^m$

(c) As m gets larger, what numbers are the expressions in parts (a) and (b) tending toward?

24. (Contributed by John Oprea of Cleveland State University.) Use your calculator to evaluate for m = 1, 10, 100, 1000, and 1,000,000:

(a) $\left(1 - \frac{1}{m}\right)^m$

(b) $\left(1 - \frac{2}{m}\right)^m$

(c) As m gets larger, what numbers are the expressions in parts (a) and (b) tending toward?

25. You have $1000 on deposit at your bank at an annual rate of 1.2%. How much interest do you receive after one year if the bank compounds

(a) continuously?
(b) daily, using 365 days in a year?

26. Suppose that you have a bank account with a balance of $4532.10 at the beginning of the year and $4632.10 at the end of the year. Your bank advertises "continuous compounding," but in fact, it compounds continuously over each 24-hour day and posts interest to accounts daily.

(a) What effective rate did you receive?
(b) What nominal rate is the calculation based on?
(c) What difference is there between what the bank is doing and true continuous compounding?

27. Suppose that you have an investment that earns 0% in the first year, but 10% in the second year.

(a) What rate of interest, compounded annually, would yield the same return after two years?
(b) What rate of interest, compounded continuously, would yield the same return after two years?

28. Suppose that you have an investment that earns 10% in the first year, 20% in the second year, and 30% in the third year.

(a) What rate of interest, compounded annually, would yield the same return after three years? (The answer here is related to the geometric mean discussed in Chapters 14 and 19, but you do not need to use that to solve the problem.)
(b) What rate of interest, compounded continuously, would yield the same return after three years?

(Thanks for the idea to Yi Cheng, Indiana University South Bend.)

29. We saw on p. 780 that a nominal rate of 5% compounded continuously yields an effective annual rate of 5.12711%.

(a) What effective annual rate does a nominal rate of 4% yield with continuous compounding?
(b) The difference between the effective rate under continuous compounding, $e^r - 1$, and the nominal rate r is $e^r - 1 - r$. You can't calculate this formula in your head, but you can approximate it closely with one that you can: $e^r - 1 - r \approx \frac{1}{2} r^2$. Thus, for $r = 4\% = 0.04$, the difference is $\frac{1}{2}(0.04)(0.04) = \frac{1}{2}(0.0016) = 0.0008 = 0.08\%$.

So the effective rate is about 4.08%. Apply this formula to approximate the difference for a nominal rate of 5%, and compare the result with the 5.12711%.

30. Repeat Exercise 29 but for a nominal rate of 2%.

31. **(a)** Suppose that your house went down in value a total of 25% over the last three years but then will go up a total of 25% over the next three years. Will you end up with more than, less than, or the same as the original value?
(b) Suppose that your friend's house went up in value a total of 25% over three years but then went down a total of 25% over the succeeding three years. Did your friend's house end up at more than, less than, or the same as the original value?

32. More frequent compounding yields greater interest, but with diminishing returns as the frequency of compounding is increased. For small interest rates, there is little difference in yield for compounding annually, quarterly, monthly, daily, or continuously. Investigating doubling times with continuous compounding leads to understanding why the rule of 72 of Exercise 21 (on p. 796) works. Recall that for continuous compounding at annual rate r, the balance A at the end of t years is Pe^{rt} for an initial principal of P. For the initial principal to double, we have $2P = A = Pe^{rt}$, so $e^{rt} = 2$. Taking the natural logarithm of both sides yields $rt = \ln 2$, where ln stands for the natural logarithm, represented on a calculator by a button marked either [ln] or [LN] (not [log] or [$\log_{10}$], which stands for a different kind of logarithm). Using the button gives $\ln 2 = 0.693$. So we have $rt = 0.693$, from which we can determine t if we know r.

Calculate the doubling times for continuous compounding at 3%, 6%, and 9%, and compare them with those predicted by the rule of 72. What do you conclude? Why do you think people prefer the rule of 72 over a rule of 69.3?

21.4 A Model for Saving

33. Suppose that you want to save up $2000 for a semester abroad two years from now. How much do you have to put away each month in a savings account that earns 2% interest compounded monthly?

34. Repeat Exercise 33, except that you have found a better deal: 3% interest compounded monthly.

35. Parents struggle for the first few years after their child is born but are finally able to start saving toward the child's college education when the child goes to school at age 6 (because the parents stop paying for daycare). If they save $400 per month in a credit union account paying 2.5% interest compounded monthly, how much will they have for college expenses 12 years later?

36. Suppose that you save for retirement by contributing the same amount each month from your 23rd birthday until your 65th birthday, in an account that pays a steady 4% annual interest compounded monthly.

(a) How much will be in your fund at age 65 if you save $100 a month?

(b) How much will be in your fund if you get a steady return of 7.5% compounded monthly?

(c) How much will be in your fund if you get a steady return of 10% compounded monthly? (This is comparable to the average annual return on the New York Stock Exchange from 1950 to 2000.)

37. A colleague feels that he will need $1 million in savings to afford to retire at age 65 and still maintain his current standard of living. A younger colleague, age 30, decides to begin saving for retirement based on that advice. How much does the younger colleague need to save per month to have $1 million at retirement if the fund earns a steady 3% annual interest compounded monthly?

38. The younger colleague of Exercise 37 is not satisfied with a 3% return, which he could get with long-term CDs. Instead, he wants to take the riskier route of investing in the stock market, which has over its history returned an APY of about 10% per year (although over 2000–2009, the return was negative). Assuming that over the 35 years until his retirement the stock market behaves just that way (a big assumption!), how much would he need to invest each month to achieve his goal of $2 million by age 65?

39. Many young people do not start saving right away for retirement, although by the time that they do, they may be earning more and thus be able to afford to save more each month. How much will be in your fund at age 65 if you don't start saving until age 35 (by which time you hope interest rates will have risen) and at that age, start saving $100 per month in an account paying a steady 6% annual interest compounded monthly?

40. Suppose instead that you have children young, pay for their college expenses, and finally start saving for retirement at age 45. How much do you have to save per month, with a steady return of 6% compounded monthly, to accumulate $250,000 by age 65?

41. Suppose that you just turned 25, are single, and are in a 25% bracket for federal income tax and a 7% bracket for state and local income taxes. (In 2010 this corresponded to an income, beyond exemptions and deductions, of $34,650–$83,900 for a single person.) This means that you pay an income tax rate ("marginal rate") of 32% on part of your income (but a lower rate on the rest). Suppose that you commit to saving $100 per month toward retirement.

(a) How much will be in your fund at age 65 if you can get a steady return of 6% compounded monthly?

(b) How much will you pay in income tax on that $100?

■ **42.** Instead of saving $100 per month—money on which you have already paid taxes, "after-tax" dollars—you have the alternative option offered in the tax code of participating in a tax-deferred retirement account (TDA), either through payroll deduction at work [e.g., as part of a 401(k) plan] or through an independent retirement account (traditional IRA). The money that goes into such a fund consists of "pre-tax" dollars: You do not pay tax on the money until you withdraw it (usually at retirement). Since you don't pay income tax on the money that you put into the retirement plan, so you can actually put in more than $100 per month while reducing your take-home pay by only $100.

(a) How much can you put into the retirement fund each month?

(b) How much will be in your fund at age 65 if you can get a steady return of 6% compounded monthly?

(c) Suppose that when you turn 65, you withdraw the entire amount in your account and pay the deferred taxes that are owed on it, say a total of 32% (federal, state, and local combined). How much do you net?

■ **43.** There is yet another alternative to the two options for saving toward retirement in Exercises 41 and 42. Instead of saving after-tax dollars or contributing to a tax-deferred plan, you can take the money as income, pay income tax on it, and make a deposit into a Roth IRA. For this special kind of retirement account, the interest earned over the years is not taxed. (There are further advantages and disadvantages.)

Suppose that you put $100 per month in after-tax dollars into a Roth IRA account. Assuming the same savings account or safe investment as in Exercises 41 and 42 that pays a steady return of 6% compounded monthly, how much will be in your account, tax-free, at age 65? How does that compare with the answers to Exercises 41(a) and 42(c)?

■ **44.** We continue the theme of Exercises 41–43 by comparing in algebraic terms three kinds of investments for retirement: an ordinary after-tax investment, a tax-deferred investment (such as a tax-deferred annuity or an IRA), and a Roth IRA. Let an investment earn interest at a steady annual yield r and let your income (in whatever year you receive it) be taxed at rate τ.

(a) Ordinary after-tax investment: Explain why if you earn $\$E$, pay taxes on it, let what remains earn interest, and pay tax each year on that year's interest, the $E grows after n years to $\$E(1 - \tau) \times [1 + r(1 - \tau)]^n$.

(b) Ordinary IRA: Explain why if you earn $E, defer taxes on it, let it earn interest, and defer taxes on all the interest, then the $E grows after n years to $\$E(1 + r)^n (1 - \tau)$.

(c) Roth IRA: Explain why if you earn $E, pay taxes on it, let what remains earn interest, and pay no taxes on all the interest, the $E grows after n years to $\$E(1 - \tau)(1 + r)^n$.

(d) Which investment gives the best return after n years?

45. Apart from CDs, returns on investments are rarely the same from year to year, as they vary with prevailing interest rates. How should you calculate an "average" rate of return over several years? Consider a mutual fund that delivers 100% return one year and loses 50% the next year. Calculate just the ordinary average (the arithmetic mean) to get [100% + (−50%)]/2 = 25%. That sounds good, but check what happens to a $1000 investment: It grows to $2000, then halves back to $1000—for a 0% gain. The customary way used in finance to calculate the "average" return is to use the geometric mean. If the initial value of the portfolio was P, and its value after n years is A, then the average annual rate of return is the value of r that solves $(1 + r)^n = A/P$, or $r = (A/P)^{1/n} - 1$.

(a) Use this formula to determine the average annual rate of return for a portfolio with returns of 10%, −25%, and 25% in three consecutive years.

(b) Is the average rate that the formula finds a nominal rate or an effective rate?

46. As in Exercise 45(a), but for consecutive returns over four years of 10%, –20%, –10%, and 7%.

21.5 Present Value and Inflation

47. Classify the following growth and decay scenarios as linear (arithmetic), exponential (geometric), or neither:

(a) The amount of caffeine in the bloodstream decreases by 10% every hour.

(b) The amount of trash in a landfill increases by 350 tons per week.

(c) The amount of alcohol in the bloodstream decreases by 10 grams (the amount in a standard drink) per hour.

(d) Your age increases every day.

(Adapted from Terence Blows, Northern Arizona University.)

48. Classify as in Exercise 47, but for

(a) The mean concentration of carbon dioxide in the atmosphere increases by 2 ppm (parts per million) per year.

(b) The mean concentration of carbon dioxide in the atmosphere increases 0.5% per year.

(c) Your knowledge of mathematics and its applications increases with each section of this book that you study.

(d) The number of people in the world increases by 1.3% per year.

(Adapted from Terence Blows, Northern Arizona University.)

49. What is the present value of $10,000, 4 years from now, at an APY of 5%?

50. What is the present value of $150,000, 10 years from now, at an APY of 3%?

51. As you will see in Chapter 22, if you had a 30-year $200,000 mortgage at 8% on a house or apartment, three-quarters of the way through the mortgage—after 22.5 years of payments—you will still owe half the amount, or about $100,000! You also would have paid about $300,000 in interest. (Current interest rates are much lower, but even at 4.75%, you would still owe $80,000 after 22.5 years.) What is the present value of $100,000, 22.5 years from now, at an interest rate of 8%? (If you put this much into a down payment, but made the same-size payments as for the 30-year mortgage on $200,000, you would own the house free and clear after 22.5 years.)

52. If you have a 30-year $200,000 mortgage at 6.48% on a house or apartment, after 10 years of payments you will still owe about $170,000. What is the present value of $170,000, 10 years from now, at an interest rate of 6.48%?

53. Suppose that inflation proceeds at a constant rate of 2% per year from mid-2012 through mid-2015.

(a) Find the cost in mid-2015 of a basket of goods that cost $1 in mid-2012.

(b) What will be the value of a dollar in mid-2015 in constant mid-2012 dollars?

54. A 1963 Chevy Bel Air, a classic car today, cost $2400 new in mid-1963. How much would that be in 2012 dollars?

Joseph H. Dennis

55. My first-semester college mathematics book cost \$10.75 in 1962. What would the equivalent price be in 2012 dollars? How does that compare with what you paid for this book? (My book had black-and-white text and figures and no photographs, color or otherwise.

56. In 1970, before an oil embargo by the Organization of Petroleum Exporting Countries (OPEC), gasoline cost about 25 cents per gallon. In 1974, after the embargo, it cost about 70 cents per gallon. What would the equivalent prices be in 2012 dollars? How do they compare with the price of gasoline today?

Refer to the following in doing Exercises 57 and 58. From Table 21.5 (p. 788), you can determine the average rate of inflation from one year to another. For example, you find the inflation from 1990 to 2000 by subtracting the two index numbers and dividing by the earlier one: $(172.2 - 130.7)/130.7 = 31.752\%$. However, the average rate of inflation is not this number divided by the number of years (10). We must take into account compounding of the rate of inflation. We set $(1 + a)^{10} = 1.31752$ and find $a = (1.31752)^{1/10} - 1 = 2.80\%$.

57. Find the average rate of inflation from 2002 to 2012. Is 3% a good approximation?

58. If inflation had been 3% each year from 2002 to 2012, what would the CPI have been in 2012?

59. (Suggested by Ed Barbeau's column "Fallacies, Flaws, and Flimflam" in *The College Mathematics Journal.*) Suppose that you get a year-end pay raise of 5%, but over the year, there has been inflation of 10%—so in effect you have suffered a pay decrease in terms of what your salary will buy. What is the percentage decrease?

60. Suppose that you get a year-end pay raise of 5% but over the year there has been inflation of 2%. What is the percentage increase in your purchasing power?

Refer to the following in doing Exercises 61 and 62. A typical new assistant professor at a liberal arts college starts at age 30 with a salary of \$45,000, while colleagues retiring at age 65 make about twice that. One college gives annual pay raises of inflation plus 1 percentage point, plus a promotion raise (to associate professor) of \$1500 after (usually) 6 years and another promotion raise (to full professor) of \$1500 after (usually) another 6 years.

■ **61.** (Spreadsheet helpful.) Can a new assistant professor who starts now expect to be making the equivalent of \$90,000 in today's dollars when she retires 35 years from now if inflation holds steady at

(a) 1.5%?

(b) 2.5%?

■ ▲ **62.** (Spreadsheet helpful.) Repeat Exercise 61, but suppose that you are the vice president for academic affairs at the college. Suggest a salary policy that would result in the new assistant professor, when she retires in 35 years, making the equivalent of

(a) \$90,000 in today's dollars.

(b) \$135,000 in today's dollars. (She would prefer that!—and hence she would be more likely to accept an offer to come work at your college.)

■ **63.** Surprise! Just for fun, one of your friends wrote your name on an Illinois State Lottery ticket, and you are the sole winner of \$40 million! You discover, however, that you don't get the \$40 million all at once. In fact, it is paid in 20 equal annual installments of \$2 million each. All you get right away is the first installment of \$2 million. So, what is the prize really worth to you? That depends on the rate of inflation over the years and the interest rate that you use to discount future income (e.g., if you had the money now, what rate could you get if you invested it yourself?). Assume a constant rate of $a = 2\%$ inflation over the 19 years until your last payment, and calculate the present value of your prize winnings in today's dollars by using the formula for present value combined with the formula for the sum of a geometric series. Do the calculation for a rate of interest of $r = 4.5\%$. *Hint:* Each year into the future, the present value is multiplied by yet another factor of $1/(1 + a) \times 1/(1 + r)$: $1/(1 + a)$ for inflation, and $1/(1 + r)$ for not being paid interest.

Actually, the checks will come from an insurance company from which Illinois purchases an annuity, whose price depends on long-term interest rates.

■ **64.** Repeat Exercise 63, but for an interest rate of $r = 6.5\%$.

■ **65.** (Spreadsheet helpful.) Your roommate (a business major) has already planned her retirement and started funding it in 2012. She plans to retire in 2047 at age 57 on \$100,000 per year in 2047 dollars, living on just the interest on her investments. Assume that she realizes a steady 7.2% and assume a steady 3% annual inflation.

(a) What must the size of her nest egg be, and what should her monthly investment be over the 35 years, to achieve this goal?

(b) What will be the value in 2012 dollars of her 2047 income of \$100,000?

(c) What will be the value in 2012 dollars of her income of \$100,000 in 2075 (when she is 85)? (Suggested by Terence Blows, Northern Arizona University, Flagstaff, Ariz.)

■ **66.** (Spreadsheet helpful.) You think what your roommate means in Exercise 65 is that she wants to retire in with a steady income of $100,000 a year in 2012 dollars. You also feel that she should plan to receive that same value of income for 43 years in case she lives to 100 (2% of your classmates will). What is the present value in 2012 of the planned stream of 43 years of retirement income?

In the savings formula, the interest rate i appears twice. The particular ways in which i is involved make it impossible to solve it algebraically to get an explicit formula for i. However, with the help of a spreadsheet, you can find i approximately when the other quantities are given. Exercises 67–70 deal with such situations.

67. (Spreadsheet helpful.) Suppose that you decide to lease a new car. (Leasing is cheaper than buying, because over the lease period you pay only about half the cost of buying the car.) At the end of the 48-month lease period, you either return the car or else need to make a lump-sum payment of $5000 if you want to keep the car. You decide to save up, just in case you decide to keep the car; if you don't keep this car, you will still have saved a good down payment on a new car. You feel comfortable with saving $70/month (over and above your lease payments). How high an annual nominal interest rate on savings do you need to accumulate $5000 in 48 months, with interest compounded monthly?

68. (Spreadsheet helpful.) A 1990 advertisement reads, "If you had put $100 per month into this fund starting in 1980, you'd have $37,747 today." Assume that deposits were made on the last day of the month, starting in January 1980, through December 1989, and that interest was paid monthly on the last day of the month (120 months).

(a) How much money was deposited during this period?

(b) What annual rate of interest, compounded monthly, would lead to the result described in the advertisement? What is the APY?

69. (Spreadsheet helpful.) The Powerball lottery drawing on December 25, 2002, resulted in the largest prize for a single winner anywhere ever, with a jackpot of $314.9 million. The winner had the option of an annuity in 30 equal annual installments of $10.5 million (the first payment being right away) or else an instant lump sum of $170 million. Taking the lump sum (most lottery winners do) makes sense, particularly if the winner needs a large sum of money now or can earn a higher rate of interest than the annuity is based on. The present value of the payment t years from now is given by the compound interest formula $A = P(1 + i)^t$, where $P = 10.5$ million and i is the annual rate of interest built into the annuity. Hence, we have $P_k = A/(1 + i)^t$ for the tth payment. The complete stream of 30 payments has present value (in millions of dollars)

$$170 = 10.5\left[1 + \frac{1}{1+i} + \frac{1}{(1+i)^2} + \cdots + \frac{1}{(1+i)^{29}}\right]$$

We use the geometric series formula with $x = 1/(1 + i)$, finding that the right-hand side is also equal to

$$10.5\left[\frac{1 - \dfrac{1}{(1+i)^{30}}}{1 - \dfrac{1}{1+i}}\right]$$

Enter $i = 0.04$ (annual rate) in the preceding expression; the result is larger than 170. For $i = 0.05$, the result is smaller than 170. Make changes in the value of i until you determine to two decimal places the rate i that gives the closest value to 170. This is the rate on which the annuity is based. Is it a nominal rate or the effective rate?

70. (Spreadsheet helpful.) Suppose that your parents are willing to lend you $20,000 for part of the cost of your college education and living expenses. They want you to repay them the $20,000, without any interest, in a lump sum 15 years after you graduate, when they will be about to retire and move. Meanwhile, you will be busy repaying federally guaranteed loans for the first 10 years after graduation. But you realize that you can't repay the lump sum without saving up. So you decide that you will put aside money in an interest-bearing account every month for the five years before the payment is due. You feel comfortable with the idea of putting aside $275 a month (the amount of the payment on your government loans).

How high an annual nominal interest rate on savings do you need to accumulate the $20,000 in 60 months, if interest is compounded monthly? Enter into a spreadsheet the values $d = 275$, $r = 0.05$ (annual rate), and $n = 60$, and the savings formula with r replaced by $r/12$ (the monthly interest rate). You will find that the amount accumulated is not enough. Change r to 0.09; it's more than enough. Try other values until you determine r to two decimal places.

APPLET EXERCISES

To do these exercises, go to www.whfreeman.com/fapp9e.

How important is it to begin a retirement fund at an early stage of one's career? In the Saving for Retirement applet, you will discover that early funding of a retirement plan can make a huge difference in the ultimate amount that will be available when a person retires.

WRITING PROJECTS

1. Plan your retirement. Decide on a retirement age and desired income (in today's dollars). Estimate yield on investments, inflation rate, and Social Security benefits. (A few notes: By the time of your retirement, a woman retiring in her mid-60s will likely live an average of 20 years more, a man 18 years more. Social Security income goes up with inflation. Annuities are available whose income grows to keep up with inflation. Various other financial products, such as a life annuity, can make sure that you don't outlive your retirement income. We discuss life annuities in Chapter 22. Also, do not neglect consideration of taxes and tax deferral of income, as considered in Exercises 41–44.) Write up your assumptions, justifications for them, calculations, and conclusions in three to four pages. Be sure to note any additional factors that you think should be taken into account but which your analysis does not include; don't be afraid to consider possibilities whose financial impact you can't calculate exactly.

2. Exercises 41–44 (p. 798) ask you to look at various forms of tax-deferred and ordinary savings and compare them on the basis of amount of tax-free income accumulation at age 65.

Ordinary savings have the important advantage that, at any time, you can do anything you want with the money accumulated so far (buy a car, put down money on a house, and so on). A second advantage is that the money is free and clear, in that taxes have already been paid on it.

A tax-deferred (e.g., IRA or 401(k)) retirement fund has the advantage of postponing taxation of the funds, but the disadvantage that if you withdraw funds before age $59\frac{1}{2}$, you must pay income tax in the year of withdrawal and in addition a 10% penalty for "early withdrawal." (These plans were given the advantage of tax deferral to encourage individuals to save for retirement—hence the penalty for withdrawing money earlier.)

A third option, the Roth IRA, has some of the advantages and disadvantages of each of the other plans.

Look into the details of the rules for 401(k) plans and Roth IRAs, compare your answers in Exercises 41–44, and devise and describe your own plan for how you will save for retirement. Your report should run two to three pages.

3. Should you stock up on "forever" stamps? In 2007 the U.S. Postal Service began selling such stamps, which will suffice for first-class postage at any time in the future, even when rates rise. These stamps cost $0.44 in early 2011. Would it be a good investment to buy a "lifetime supply" of them? To render your judgment, compare the historic increases postal rates (see **www.vaughns-1-pagers.com/economics/postal-rates.htm**) with the CPI (page 788). For example, you could convert postal rates at each of the dates of change to the cost in today's dollars, and then see if there appears to be a recent trend that you can project into the future.

4. Based on ticket sales and current dollars, the top-grossing domestic film of all time was *Avatar,* which earned $760,507,625 since release in 2009. But adjusted for inflation of ticket prices, *Gone with the Wind* (1939, plus re-releases) comes out on top at $1,606,254,800 in 2011 dollars (see **boxofficemojo.com/alltime/adjusted.htm**; the comparison, since it considers proceeds rather than audience share, does not take into account the larger population today compared to 1939). Use a spreadsheet to analyze how movie ticket prices have risen since 1913 compared to the CPI. The CPI data from Table 21.5 can be downloaded from **data.bls.gov/cgi-bin/surveymost?cu**. (Check the first box, click "Retrieve data," then in the next screen adjust the year "From" to 1913.) Find the data on movie ticket prices at **boxofficemojo.com/about/adjuster.htm**. Prepare a short report, with a graph. (Thanks to Martin Campbell for the idea.)

5. The federal minimum wage was set at $7.25 per hour starting July 24, 2009. Has the federal minimum wage kept pace with inflation? Use a spreadsheet to analyze how it has risen since 1938 compared to the CPI. The CPI data from Table 21.5 can be downloaded from **data.bls.gov/cgi-bin/surveymost?cu**. (Check the first box, click "Retrieve data," then, in the next screen, adjust the year "From" to 1938). Find the data on federal minimum wage rates at **www.dol.gov/esa/minwage/chart.htm**. Prepare a short report, with a graph.

6. The assumptions in many of the exercises of this chapter of a constant interest rate, rate of return on an investment, or tax rates (as in Exercises 41–44) holding over a long period of time are simplifications that simply won't be true. Interest rates fluctuate (though you can lock in a long-term constant rate by buying a long-term bond or certificate of deposit), and the tax rate may change (with your income, your state of residence, and changes in tax laws). If your marginal tax rate (the rate you pay on one additional dollar of income) is lower in one year than the tax rate you expect to pay in retirement, what kind of retirement investment is better for you that year? If you have a windfall one year and your marginal tax rate is higher that year than the tax rate that you expect to pay in retirement, what kind of retirement investment is better for you that year? How should you take the various factors into account as you make investment and savings decisions early in life versus later in life near retirement?

Suggested Readings

KASTING, MARTHA. *Concepts of Math for Business: The Mathematics of Finance.* UMAP Modules in Undergraduate Mathematics and Its Applications: Module 370–372. COMAP, Inc., Arlington, MA, 1980.

LINDSTROM, PETER A. *Nominal vs. Effective Rates of Interest.* UMAP Modules in Undergraduate Mathematics and Its Applications: Module 474. COMAP, Inc., Arlington, MA, 1988. Reprinted in Paul J. Campbell (ed.), UMAP Modules: Tools for Teaching 1988, COMAP, Inc., Arlington, MA, 1989, pp. 21–53. A learning module, requiring no more background than this chapter, that teaches about nominal and effective rates of interest and how to calculate them. Gives real examples of banks using different options for calculating interest.

MILLER, CHARLES D., VERN E. HEEREN, and JOHN HORNSBY. Consumer mathematics. In *Mathematical Ideas,* 11th ed., Pearson Education/Addison Wesley Longman, Boston, 2007.

VEST, FLOYD, and REYNOLDS GRIFFITH. The mathematics of bond pricing and interest rate risk. *Consortium (COMAP),* 59 (Fall 1996): HiMAP Pullout Section 1–6.

Suggested Web Sites

www.bls.gov/cpi/ Home page for the inflation tables prepared by the BLS.

www.bls.gov/data/inflation_calculator.htm CPI Inflation Calculator. Converts dollar value from any year to its equivalent buying power in any other year.

www.westegg.com/inflation/ Inflation calculators for the United States (1800–2010), Canada, and Italy, by S. Morgan Friedman, with links to sites about the current purchasing power of amounts of currencies of other countries in the past.

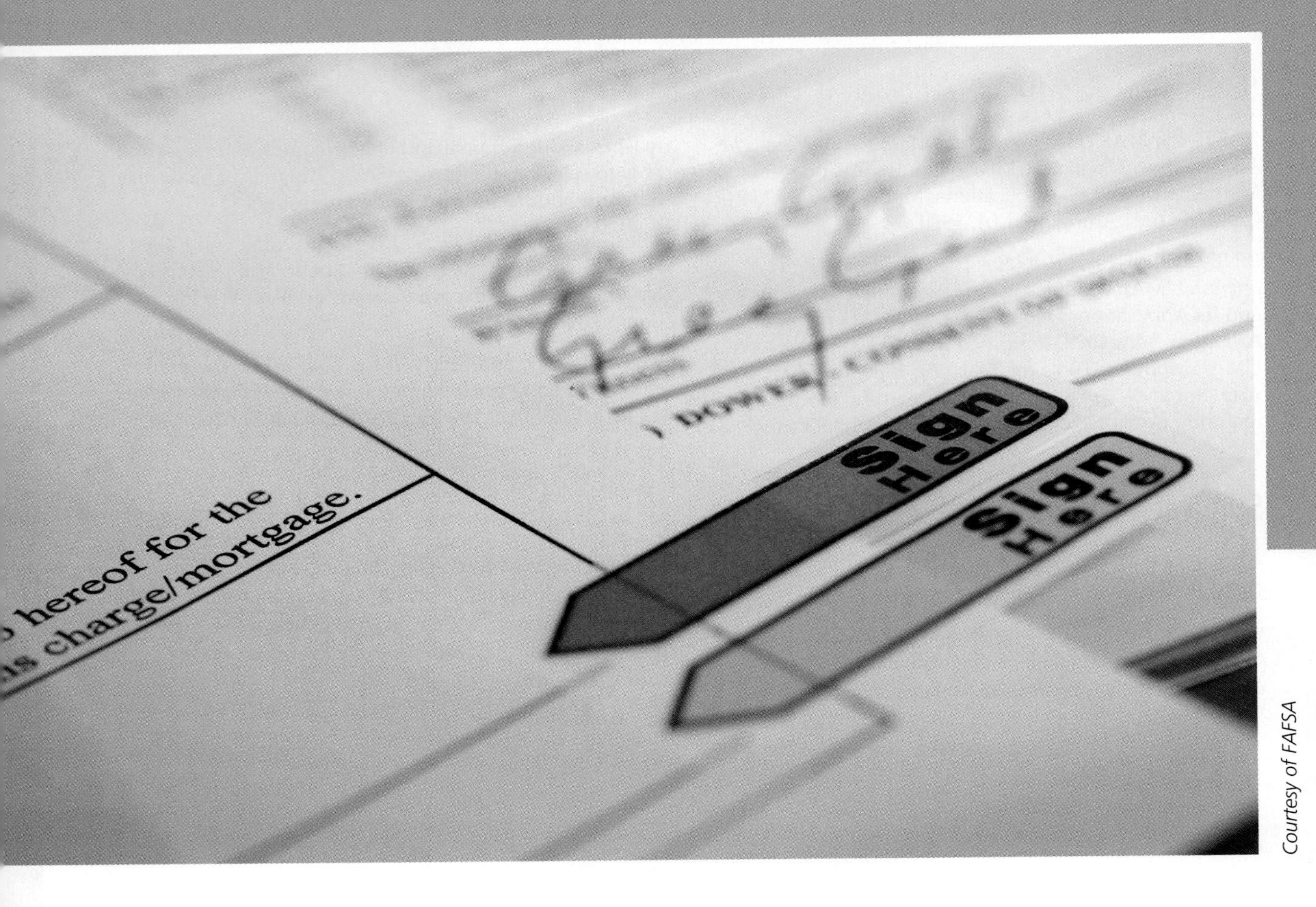

Courtesy of FAFSA

22

Borrowing Models

In the previous chapter, we looked at consumer financial models for saving and formulas for calculating the amount accumulated. Savings or investments would not earn interest unless they could be loaned to someone to make productive use of the money.

In this chapter, we examine the other side of consumer finance—borrowing. You may have a student loan, you are likely going to need to borrow (or have already borrowed) to buy a car, you will almost certainly borrow if you buy a house or apartment, and you are borrowing if you use a credit card. For any such loan, you pay "finance charges," which include interest and perhaps other "fees" as well. We investigate and compare some common kinds of loans.

Section 22.1 (re)acquaints you briefly with simple interest, while Section 22.2 does the same with compound interest, in the contexts of student loans and credit cards. Section 22.3 considers "conventional" loans (such as the mortgage on a house), reviews the savings formula from Chapter 21, and derives from it the formula for the payment on a conventional loan. Finally, Section 22.4 investigates annuities, one of the ways that you can receive the grand prize in a lottery and a way of providing for retirement and old age. If you have a grasp of the ideas behind compound interest and can use the compound interest formula (page 772) and the savings formula (page 781), which we repeat shortly for your convenience, you can proceed with this chapter without first reading Chapter 21.

22.1 Simple Interest

The amount of **interest** charged on a loan is determined by the **principal**, by the amount borrowed, and by the method used to calculate the interest. With **simple interest**, the borrower pays a fixed amount of interest for each period of the loan. The interest rate is usually quoted as an annual rate.

For a principal P and an annual rate of interest r, the interest owed after t years is

$$I = Prt$$

and the total amount A due on the loan is

$$A = P(1 + rt)$$

EXAMPLE 1
Simple Interest on a Federal Direct Student Loan

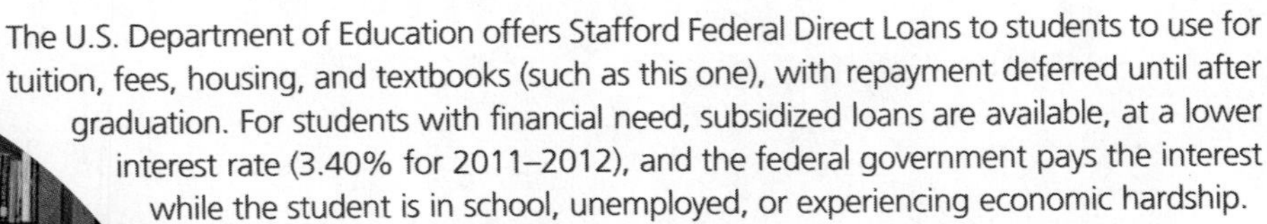

The U.S. Department of Education offers Stafford Federal Direct Loans to students to use for tuition, fees, housing, and textbooks (such as this one), with repayment deferred until after graduation. For students with financial need, subsidized loans are available, at a lower interest rate (3.40% for 2011–2012), and the federal government pays the interest while the student is in school, unemployed, or experiencing economic hardship.

Dennis MacDonald/Alamy

However, any eligible student can take out an unsubsidized Stafford loan, at 6.8% interest, in which interest is charged from when you receive the loan until it is repaid in full. One option is to pay each quarter the interest due, and defer paying back the principal until six months (the "grace period") after you leave school.

Suppose that you took out an unsubsidized Stafford loan for \$5500 (the maximum amount for a first-year student) on September 1, 2011, before your freshman year and you plan to pay the interest as you go and begin paying back the principal on December 1, 2015, after graduating on June 1, 2015 (so you will have had the loan for 4 years + 3 months = 51 months).

How much is the monthly interest, how much total interest will you have paid over the 51 months, and how much will you owe when you start to pay back the loan?

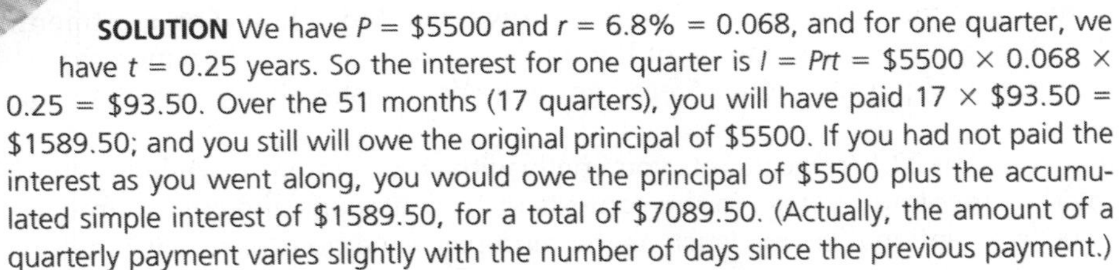

SOLUTION We have P = \$5500 and $r = 6.8\% = 0.068$, and for one quarter, we have $t = 0.25$ years. So the interest for one quarter is $I = Prt$ = \$5500 × 0.068 × 0.25 = \$93.50. Over the 51 months (17 quarters), you will have paid 17 × \$93.50 = \$1589.50; and you still will owe the original principal of \$5500. If you had not paid the interest as you went along, you would owe the principal of \$5500 plus the accumulated simple interest of \$1589.50, for a total of \$7089.50. (Actually, the amount of a quarterly payment varies slightly with the number of days since the previous payment.)

22.2 Compound Interest

Compounding is the calculation of interest on interest. A common example is the balance on a credit card. So long as there is an outstanding balance, the interest owed is calculated on the entire balance, including any part of it that was previously calculated as interest and added to the balance in earlier months.

EXAMPLE 2
Credit Card Interest

Suppose that you owe \$1000 on your credit card, the company charges 1.5% interest per month, and you just let the balance ride. How much interest do you pay in the first year?

SOLUTION Your interest the first month is 1.5% of \$1000, or 0.015 × \$1000 = \$15. The new balance owed is (1 + 0.015) × \$1000 = \$1015. Your interest the second month is not 1.5% of \$1000, or \$15 (as would be the case for simple interest), but 1.5% of \$1015, or 0.015 × \$1015 = \$15.23. Therefore, the new balance for the second month is

$$(1 + 0.015) \times \$1015 = \$(1 + 0.015) \times (1 + 0.015) \times \$1000 = \$1030.23$$

rounded (up) to the nearest cent. (We are not factoring in the extra charges for your failure to make minimum payments, which would add up pretty fast!) After 12 months of letting the balance ride, it becomes

$$(1.015)^{12} \times \$1000 = \$1195.62$$

(where we neglect the rounding of interest to the nearest cent that occurs at each billing). In other words, the actual interest for the year comes to \$195.62, which is 19.562% of \$1000. So, although the quoted rate of interest is 1.5% per month, which seems as if it should amount to $12 \times 1.5 = 18\%$ per year, the effective interest rate is actually greater. (Actually, credit cards charge interest by the day, so the interest rate for a month will vary with the number of days in the month.)

We apply two formulas from Chapter 21: the **compound interest formula** (page 772) and the **savings formula** (page 781), phrasing them for loans. Here is the compound interest formula, followed by an example:

Compound Interest Formula RULE

If a principal P is loaned at interest rate i per compounding period, then after n compounding periods (with no repayment), the amount owed is

$$A = P(1 + i)^n$$

This formula just generalizes what we saw happen with the credit card balance. We give the formula in a slightly more elaborate version below, to make the connection to multiple compoundings per year.

General Compound Interest Formula RULE

For a principal P loaned

- at a nominal annual rate of interest rate r
- with m compounding periods per year (so the interest rate $i = r/m$ per compounding period), the amount owed
- after t years (hence $n = mt$ compounding periods) with no payment of interest or principal is

$$A = P(1 + i)^n = P\left(1 + \frac{r}{m}\right)^{mt}$$

EXAMPLE 3
Not Repaying a Guaranteed Student Loan

Before July 1, 2010, another kind of student loan was available (and you may have one). That was a "federal guaranteed student loan," under which you got your loan not from the government but from a bank. If you failed to pay back the loan, the bank was paid by the federal government instead. (In other words, the bank charged an origination fee for the loan and then earned interest on it with no risk to the bank; because this arrangement did not seem fair, the Obama administration abolished this kind of loan.)

With a guaranteed student loan, if you do not make the payments due, the interest continues to accumulate and is usually *capitalized* every quarter. That means that the interest during each quarter is simple interest on the amount due at the beginning of the quarter, and at the end of the quarter, that interest is added to the amount due. In other words, the compounding period is one quarter.

Suppose that six months after you graduate, when you begin to have to make payments on your student loans, you owe \$5000 on your guaranteed student loan at 6.8% interest per year. However, you aren't able to find a job, so you fail to make any payments for the next nine months. (This would be very foolish, since after 270 days of nonpayment, the loan would be in default and all kinds of bad things would happen!) How much would you owe then?

SOLUTION The principal P is \$5000. The quarterly interest rate is $i = 6.8\%/4 = 1.7\%$ and there are $n = 3$ compounding periods. The compound interest formula gives the amount owed as $A = \$5000(1 + 0.017)^3 = \$5259.36$.

Terminology for Loan Rates

The interest on a loan depends on whether compounding is done and how the interest is calculated. Just like the Truth in Savings Act mentioned in Chapter 21 (on page 771), the Truth in Lending Act establishes terminology and calculation methods for interest.

A **nominal rate** is any stated rate of interest for a specified length of time. For instance, a nominal rate could be a 1.5% monthly rate on a credit card balance. By itself, such a rate does not indicate or take into account whether or how often interest is compounded.

The **effective rate** *takes into account compounding.* It is the rate of simple interest that would realize exactly as much interest over the same period of time.

We saw that \$1000 at a yearly interest rate of 18% (a nominal rate), calculated as 1.5% per month compounded monthly, yields \$195.62 in interest owed at the end of the year, which is 19.562% of the original principal. Hence, the effective annual rate is 19.562%. In other words, a \$1000 loan at simple interest of 19.562% for one year would owe exactly the same interest.

Finally, when stated per year ("annualized"), the effective rate is called the **effective annual rate**. (In connection with savings, the effective annual rate is the annual percentage yield of Chapter 21, discussed on page 771.)

To keep the rates straight, we use i for a nominal rate for the specified **compounding period**—such as a day, month, or year—*within which no compounding is done;* this rate is the effective rate for that length of time. For a nominal rate compounded m times per year, we have $i = r/m$. For that \$1000 credit card balance at 18% compounded monthly, we have $r = 18\%$ and $m = 12$, so $i = 1.5\%$ per month.

The Truth in Lending Act introduced the term **annual percentage rate (APR)**.

Annual Percentage Rate (APR) DEFINITION

The **annual percentage rate (APR)** is the number of compounding periods per year times the rate of interest per compounding period:

$$\text{APR} = m \times i$$

In the example of the credit card balance, the interest is compounded monthly, or $m = 12$ times per year, and the interest rate for the compounding period is $i = 1.5\%$, so the APR is $12 \times 1.5\% = 18\%$. The APR is the rate that the Truth in Lending Act requires the lender to disclose to the borrower. *The APR is not the effective annual rate* (as we have already seen in the credit card example), and Spotlight 22.1 explains further.

SPOTLIGHT 22.1

What's the Real Rate?

Financial experts agree that the real, "true" rate of interest for savings or loans is the effective annual rate.

The 1991 Federal Truth in *Savings* Act requires that savers be told the annual percentage yield (APY) (discussed on page 774 in Chapter 21), which is just the effective annual rate.

The 1968 Federal Truth in *Lending* Act, however, requires that borrowers be told the APR, which is *not* the same as the effective annual rate. The APR is the rate of interest per compounding period times the number of compounding periods per year. Thus, a credit card rate of 1.5% per month translates to an APR of 18%. The APR does not take into account compounding. Hence, it is not equivalent to—indeed, it understates—the true cost of borrowing; that is, the effective annual rate. For the credit card loan, with monthly compounding, the effective annual rate is

$$(1 + 0.015)^{12} - 1 \approx 19.562\%$$

The APR also ignores costs that are sometimes involved in borrowing, such as a flat charge for making the loan in the first place (called a "loan-processing fee"), charges for late payments, and charges for failing to make a minimum payment.

Stafford Federal Direct Loans have a 1% origination fee, and one-half of the loan is disbursed at the start of each semester; both of these factors raise the effective interest rate. For Federal Direct PLUS Loans to parents of students, the origination fee is 4%. The fees are intended to cover in part the cost of loans that default.

For home mortgage loans, however, the Truth in Lending Act requires that lenders include in the APR some of the upfront costs referred to as "closing costs": any "loan origination" fee, "loan-processing" fee, and "points" (additional charges to get a reduced interest rate). The APR does not include title insurance, appraisal, credit-report fees, or transaction taxes.

Closing costs are paid at the closing of the sale, while interest is paid over the life of the loan. However, the APR treats the closing costs included in it as if they will be amortized over the term of the mortgage, despite the fact that they were paid beforehand. Here, too, the APR understates the true costs.

However, very few people hold a mortgage to its maturity. The median life of a 30-year mortgage is only about 5 years; that is, half of all mortgage holders pay off their mortgage before 5 years are up, usually because they sell their homes and move elsewhere. Thus, for almost all home loans, the APR also includes interest that will never be paid.

Also, we must take into account inflation. One advantage of buying a home with a fixed-rate mortgage is that your payment stays the same, but your earnings and the value of your home are likely to go up with inflation: You are thus paying back the loan with dollars of lesser value. For any loan in a time of inflation, *Fisher's effect* comes into play: If your loan has an effective annual rate of 7% but inflation is running at 3.5% per year, the true cost to you of the loan is not exactly 7% − 3.5% = 3.5%. Instead, for an effective annual rate of r and an inflation rate of a, the cost of the loan at the beginning of the first year is indeed $r - a$ (3.5% in our example), but at the end of the first year, it is

$$g = \frac{r - a}{1 + a}$$

For $r = 7\%$ and $a = 3.5\%$, we get $g = 3.38\%$. The reason that this is less than the expected 3.5% is that at the end of the first year, you are paying back the loan with dollars that have been inflated for a year. As inflation mounts over the term of a mortgage, the cost g goes down steadily each year. For example, at the end of 5 years of steady inflation at 3.5%, the total inflation has been $a = (1 + 0.035)^5 - 1 \approx 18.8\%$, and we have $g = 2.95\%$.

A final—and major—consideration is that interest paid on your home mortgage is deductible from taxable income on federal, state, and some local income tax returns. Thus, your home ownership is subsidized by other taxpayers (just as you help subsidize other home buyers), and the cost to you of the loan is reduced further.

22.3 Conventional Loans

A common situation that you are likely to encounter is a loan—for a house, a car, or college expenses—to be paid back in equal periodic installments. Your payments are said to **amortize** (pay back) the loan. In these so-called **conventional**

loans, each payment pays the current interest and also repays part of the principal. *As the principal is reduced, there is less interest owed, so less of each payment goes to the interest and more toward paying off the principal.*

We remind you of the savings formula from Chapter 21 (on page 781):

Savings Formula RULE

The amount A that is accumulated

- at a nominal annual rate of interest rate r
- with m compounding periods per year (so interest rate $i = r/m$ per compounding period)
- after t years (hence $n = mt$ compounding periods)
- by a uniform deposit d at the end of each compounding period is

$$A = d\left[\frac{(1+i)^n - 1}{i}\right] = d\left[\frac{(1+\frac{r}{m})^{mt} - 1}{\frac{r}{m}}\right]$$

For the loan situation, the "uniform deposit" becomes the monthly payment:

d = the monthly payment, made at the end of each month

EXAMPLE 4
Buying a House

Let's suppose that you buy a house with a $100,000 loan to be paid off over 30 years in equal monthly installments. Suppose that the interest rate for the loan is 6.00%. How much is your monthly payment?

SOLUTION Imagine changing the setup slightly so that instead of making monthly payments, you are supposed to pay off the entire principal and interest at the end. Meanwhile, you make payments to a savings fund that you're building up to pay off the loan, and the savings fund earns the same rate of interest as the loan costs. The interest rate of 6.00% on the loan is compounded monthly, so the monthly rate is 0.5%. At the end of 30 years, the principal and interest on the loan will (by the compound interest formula) amount to

$$\$100{,}000 \times (1 + 0.005)^{12\times30} \approx \$602{,}257.52$$

Integer Exponents, page 776.

On the other hand, saving d each month for 30 years at 6.00% interest compounded monthly, we know from the savings formula that you will accumulate

$$d\left[\frac{(1 + 0.005)^{360} - 1}{0.005}\right]$$

To make d just the right amount to pay off the loan exactly, we need to solve the equation

$$d\left[\frac{(1 + 0.005)^{360} - 1}{0.005}\right] = \$100{,}000 \times (1 + 0.005)^{12\times30} \approx \$602{,}257.52$$

for the value of d, getting d = $599.55 as your monthly payment. The total of the payments is "only" 360 × $599.55 = $215,838.00—on a loan of just $100,000. (Usually, the bank will round up your regular monthly payment to the next nearest cent, with the consequence that your very last payment will be slightly less than the usual monthly payment.)

We put this idea into a more general setting: *Paying off a conventional loan is like saving.* You can think of paying off the loan as making payments to a savings account that earns interest at the same rate as the loan. At the end of the loan term, the savings balance will exactly equal the principal and interest on the loan. Let the loan amount be P, the effective interest rate per compounding period be i, the number of compounding periods be n, and the loan payment be d. We equate the principal and interest on the loan (from the compound interest formula) with the savings balance (from the savings formula):

$$P(1 + i)^n = d\left[\frac{(1 + i)^n - 1}{i}\right]$$

The quantity P is sometimes called the *present value of an annuity* of n payments of d, each at the end of a compounding period with interest i per period. This terminology is used in the financial mode of some calculators, such as the TI-83.

Solving the above equation for d requires a little algebra. To make things simpler, let $b = (1 + i)^n$, so

$$Pb = d\left[\frac{b - 1}{i}\right]$$

Then

$$d = P\left[\frac{b}{\frac{b-1}{i}}\right] = P\left[\frac{bi}{b - 1}\right]$$

Now divide numerator and denominator by b, getting

$$d = P\left[\frac{i}{1 - b^{-1}}\right]$$

Substituting $(1 + i)^n$ back for b, we get the usual form of the **amortization payment formula:**

Amortization Payment Formula RULE

A conventional loan amount P

- at a nominal annual rate of interest rate r
- with m compounding periods per year (so interest rate $i = r/m$ per compounding period)
- for t years (hence $n = mt$ compounding periods) can be paid off by uniform payments at the end of each compounding period in the amount

$$d = P\left[\frac{i}{1 - (1 + i)^{-n}}\right] = P\left[\frac{\frac{r}{m}}{1 - (1 + \frac{r}{m})^{-mt}}\right]$$

Algebra Review Solving Linear Equations, page 457.

EXAMPLE 5
Repaying Your Student Loan

The standard repayment option for federal direct student loans is repayment over 10 years with a minimum monthly payment of \$50 at the end of each month, beginning six months after you graduate. For the student loan of Example 1 (on p. 806), if you didn't pay the interest over the 51 months, what will your monthly payments be on the \$7089.50 that you will owe at the start of repayment?

SOLUTION With the amortization payment formula, it's easy to figure out your monthly payment. We have $P = \$7089.50$, monthly interest rate $i = r/m = \frac{0.068}{12} = 0.00566667$, and $n = mt = 12 \times 10 = 120$ months for the payback. We find the payment d as

$$d = P\left[\frac{\frac{r}{m}}{1-\left(1+\frac{r}{m}\right)^{-mt}}\right] = \$7089.50\left[\frac{\frac{0.068}{12}}{1-\left(1+\frac{0.068}{12}\right)^{-12\times 10}}\right] \approx \$81.59$$

So your monthly payment will be \$81.59. (That is for this loan; you may owe more for loans for your other years in college.) Hence, over the lifetime of the loan, you will pay $119 \times \$81.59 = \$9709.21$ plus (because you really owed a fraction of a cent less per month) a smaller last payment of \$80.94, for a total of \$9790.15. Of the total, \$9790.15 – \$7089.50 = \$2700.65 is interest during the 10-year payback period, plus the \$1589.50 interest accrued during the deferment and grace periods, for a total of \$4290.15 in interest.

You also paid a \$55 origination fee when you first got the loan; so you will pay about 80% as much more in interest and fees as the original principal. If you were to stretch your payments over more years (which is permissible in some circumstances), an even greater proportion would be interest.

You can check these amounts, and those for your own loans, at www2.ed.gov/offices/OSFAP/DirectLoan/RepayCalc/dlentry1.html and (in greater detail, with more options) at www.finaid.org/calculators/loandiscountanalyzer.phtml.

EXAMPLE 6
Buying a Car

You decide to buy a new Wheelmobile car. After a down payment, you need to finance (borrow) an additional \$12,000. After you compare interest rates offered by the car dealership, local banks, and your credit union, the best deal you can find is 4.9% compounded monthly over 48 months. What is your monthly payment?

We have $P = \$12{,}000$, monthly interest rate $i = \frac{0.049}{12}$, and $n = 48$. Using the amortization formula, we have

$$d = P\left[\frac{\frac{r}{m}}{1-\left(1+\frac{r}{m}\right)^{-mt}}\right] = \$12{,}000.00\left[\frac{\frac{0.049}{12}}{1-\left(1+\frac{0.049}{12}\right)^{-48}}\right] \approx 275.81$$

How much interest do you pay? You make payments totaling $47 \times \$275.81 + \$275.72$ (last payment) = \$13,238.79, so the interest is \$13,238.79 − \$12,000 = \$1,238.79.

What if you could get a 60-month loan at the same rate? Then your monthly payments would be \$225.91; over 60 months, you would pay \$13,554.29, of which \$1554.29 would be interest.

If you had bought a Plushmobile instead, with \$24,000 to finance, you would have borrowed twice as much, and your monthly payment would have been twice as much.

A car loan is usually for 48 or 60 months, but when you buy a home, you usually borrow a great deal more money and pay it off over a much longer period. The usual term for a home mortgage is 30 years.

EXAMPLE 7
A 30-Year Mortgage on a Median-Priced Home

Let's suppose that you are a family with the U.S. median household income of about \$50,000, that you want to buy a median-priced home (\$171,900 in September 2011) with a 30-year fixed-rate mortgage at 4.25%. Recall that the median (discussed in Section 5.4, on pages 178–179) means that half are below and half are above. Suppose that you can make a down payment of only $10,000 (plus closing costs of about $4000). Can you afford such a home?

SOLUTION Lenders have "affordability" guidelines that suggest that a family can afford to spend about 28% of its monthly income on housing. Thus, by these guidelines, you can afford $0.28 \times \$50{,}000/12 = \$1166.67$ per month.

What is the monthly payment on the loan? The principal is $P = \$161{,}900$, the monthly interest rate is $i = 0.0425/12 \approx 0.00354167$, and $n = 360$ months. The amortization formula gives a monthly payment of

$$d = P\left[\frac{\frac{r}{m}}{1-\left(1+\frac{r}{m}\right)^{-mt}}\right] = \$161{,}900\left[\frac{\frac{0.0425}{12}}{1-\left(1+\frac{0.0425}{12}\right)^{-12\times 30}}\right] \approx \$796.45$$

Well, that sounds good. Unfortunately, there is more to the mortgage than just the amount needed to amortize the loan. Your payment will also have to cover real estate taxes, mortgage insurance, and homeowner's insurance on the property. On a $171,900 home, these may add $500 or more to the monthly payment, which will then total about $1300.

So, the median household cannot afford the median-priced home, even while housing prices stay depressed and interest rates stay low.

The spreadsheet formula PMT can be used to determine the payment amount; see Spotlight 21.4 on page 791 for details and examples.

A payment on an amortized loan includes both the current interest and a portion toward repaying the principal. You are "building **equity**" in a house as you pay off the mortgage.

Equity DEFINITION

Equity is the amount of principal of a loan that has been repaid.

EXAMPLE 8
Home Equity

My wife's parents sold their house in rural Minnesota to move to the town where we live. They had bought their house in 1980 with a 30-year mortgage for $100,000 at an 8% interest rate. After 22 years, how much *equity* did they have in the house—that is, how much of the principal had been repaid? And how much did they still owe on the house?

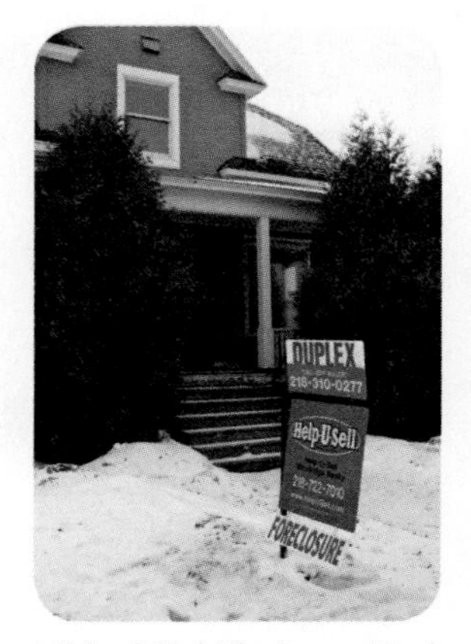

Michael Siluk/The Image Works

SOLUTION What may shock you is that when they sold their house in May 2002—after 269 months of payments, almost exactly three-quarters of the 30 years of the mortgage—they had only $50,000 in equity (hence, they still owed $50,000 on the house) but had already paid $147,000 in interest. *Three-quarters of their payments had gone to interest.*

We can use the amortization formula to determine just how much equity they had after 269 months of payments, but first we need to determine their monthly payment. We see $P = \$100{,}000$, $n = 360$ months, and $i = \frac{0.08}{12}$ monthly interest, getting $d = \$733.76$.

Now we use the formula again, this time "in reverse." Knowing $i = \frac{0.08}{12}$ and $d = \$733.76$, we find out how much of the loan would have been paid off by the remaining $n = 360 - 269 = 91$ payments of $733.76:

$$d\left[\frac{1-(1+i)^{-n}}{i}\right] = \$733.76\left[\frac{1-\left(1+\frac{0.08}{12}\right)^{-91}}{\frac{0.08}{12}}\right] \approx \$49{,}940.03$$

This is how much my parents-in-law had yet to pay, so their equity was $100,000 − $49,940.03 = $50,059.97. (The above formula would not apply if they had made larger or additional payments.)

Figure 22.1 and Table 22.1 show that equity builds up very slowly at first but rapidly later. (The values shown do not take into account any possible increase or decrease in the value of the house itself, the effect of inflation, nor the effect of making higher monthly payments or other extra payments.) In fact, the amount of principal in a payment grows by a factor of $1 + i$ from one payment to the next, so the equity at any point is the sum of a geometric series (discussed in Section 21.4, on pages 780–781) whose common ratio is $1 + i$.

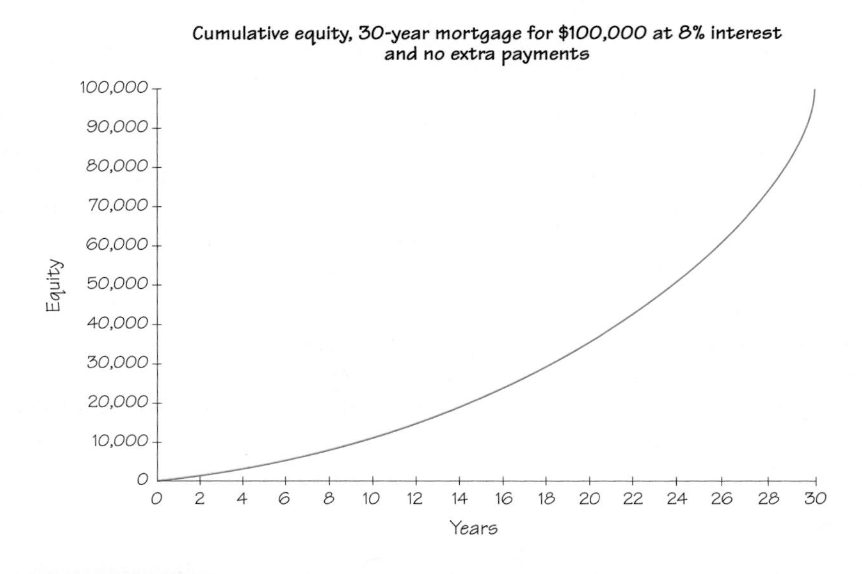

Figure 22.1 Equity grows almost exponentially, especially in the later years of a mortgage.

Table 22.1

Equity in a 30-Year Mortgage for $200,000 at 5% Interest

End of Year	1	2	3	4	5	10	15	20	25	30
Equity ($ × 10^3)	3	6	9	13	16	37	64	99	143	200

When you buy a home, you have several options for the mortgage: a conventional 30-year mortgage, a conventional 15-year mortgage, or an adjustable-rate mortgage (ARM) for either length of time but with an interest rate that can vary.

You might expect the payment on a 15-year mortgage to be double that of a 30-year mortgage. On the contrary, the payment is only 55% more (for a 4% mortgage) to 26% more (for a 9% mortgage). This range includes the prevailing mortgage rates over the past 20 years. Moreover, over the course of a $200,000 mortgage at 5%, you would pay $187,000 in interest over 30 years but only $85,000 over 15 years. At 9%, the interest totals are $380,000 versus $166,000. (Some financial counselors advise taking a 30-year mortgage and making extra payments when you can afford them, rather than incurring the higher payment obligation of a 15-year loan, on which, if you encounter tight personal financial circumstances, you might not be able to make the payments.) In Spotlight 22.2, we discuss what we did in our own circumstances and mention other options.

SPOTLIGHT 22.2

What We Did with Our House, and What Else You Could Do

We bought our house in 1992. We were offered a choice between a 30-year fixed-rate mortgage at 8.375% and a 30-year ARM at 6.875% whose rate could be raised (or lowered) by up to 2% every year. When we asked, we were also quoted slightly lower rates for corresponding 15-year mortgages.

We were planning to stay in the house much longer than the median of 5 years, and we were concerned that inflation might force the ARM considerably higher. Also, we did not want the obligation of the higher payments of a 15-year mortgage, in case our circumstances changed (such as through job loss or death). Some loans provide for penalties for paying off the loan early, but in our case (thanks to state law), there was no penalty for making extra payments (if we could afford them).

We chose the 8.375% fixed-rate 30-year mortgage (and made some extra payments). People in other circumstances, or with a different tolerance for risk, would no doubt have decided otherwise. Had we been sure then that interest rates would not go higher in the 1990s, we would have gone for the ARM. But hindsight is always better than foresight. Homeowners with mortgage interest rates such as ours later refinanced at much lower prevailing rates, near 5% for a fixed-rate 30-year mortgage.

Currently, about one-third of borrowers take ARMs rather than fixed-rate. Newer mortgage "products" include interest-only mortgages and shared appreciation mortgages (SAMs). With an interest-only ARM, payments are (just slightly) lower than for a conventional 30-year mortgage, but you accumulate no equity (at least, not by paying off the loan; the market value of the house may rise). After five to seven years, you start also paying off the principal—which means that your payments go up then. In some such loans, the interest rate—and your payments—fluctuate as frequently as every month.

In a SAM, interest payments are lower or nonexistent, but the lender receives a portion of any appreciation (rise in value) when the house is sold. In a nationally reported instance in 2003, a single mother received a no-interest SAM loan to finance the $30,000 down payment on a $223,000 house in Pleasanton, California, through the city's affordable housing program. Four years later, she sold the house for $385,000, and the "affordable housing" lenders got 60% of the $162,000 appreciation, or $97,000. She herself realized $65,000 (minus the cost of the sale) but complained bitterly, saying that she would have been better off to have put the loan on her credit card! Critics have termed SAMs an urban form of sharecropping.

Because very few mortgages are held for the full term, it is useful to compare the status of mortgages after 5 years, the median length of time that Americans remain in a home. Table 22.2 shows the equity after 5 years for a variety of interest rates. For a 30-year mortgage, the equity after 5 years may be less than the cost of selling the home through a realtor. Of course, the resale value of the home also may be higher after 5 years.

A mortgage with an interest rate that can vary is called an **adjustable-rate mortgage (ARM)**. Often such mortgages have a substantially lower interest rate (hence, a lower payment) than a fixed-rate mortgage. The ARM's interest rate may

Table 22.2

Equity (in thousands of dollars) on a $200,000 Mortgage After 5 Years

Term (years)	Interest Rate					
	4%	5%	6%	7%	8%	9%
15	54	51	48	45	42	40
30	19	16	14	12	10	8

SPOTLIGHT 22.3

The Mortgage Crisis

Late 2007 saw the development and widening consequences of what has become known as the "mortgage crisis." To understand what that was, why it took place, and how it will have widespread effects for some time, you need to know what happens when you get a mortgage to buy a house compared to what used to happen.

In the "good" old days, you would go to a local bank (or savings and loan, or credit union). If you proved "creditworthy"—meaning that after careful consideration, the personnel felt that you could repay the loan—the bank would lend you its money, raised from its depositors. The interest rate depended on your credit rating and down payment. You paid back the loan at a fixed rate of interest, usually over 30 years. If interest rates went down, you could refinance the loan at a lower rate by taking out a new loan to pay off the old one; if rates went up, your payments would not rise.

Meanwhile, the value of your house usually went up 5% to 10% per year, your income went up, and your payments stayed the same. (In fact, if you have only 10% equity in your house and it goes up in value 10% in one year, you have made 100% on your investment! This kind of "leveraging" can make real estate investments very profitable—so long as prices keep rising.)

What changed? Efforts to extend home ownership to a wider proportion of the population resulted in banks making more "subprime" loans (loans to people with poorer credit histories who are less likely or able to repay them), with lower down payments but higher rates of interest (because of the greater risk). Some of those loans were "predatory lending," at high rates of interest to people who could not possibly make the payments.

Certainly, real estate speculation played a role, as did greed. House prices doubled from 2000 to 2006 (the "housing bubble"), but median income adjusted for inflation remained stagnant—and then fell with increasing unemployment, making it increasingly hard to afford houses. Banks countered with ARMs and interest-only mortgages. They also maximized their income with upfront charges ("origination fee" plus "points" paid by the buyer to lower the interest rate); and at the same time, they minimized their risk by immediately "flipping" the mortgages (selling them to bigger banks or other investors, to whom buyers would then make their payments). All these factors led to banks making more loans that were riskier.

What went wrong? Interest rates rose, and people with ARMs saw their payments rise beyond their ability to pay. At the same time, the "pyramid" of housing prices could not continue with the higher mortgage rates, so housing prices fell. When your house becomes "underwater"—worth less than the balance remaining on the mortgage—you might be better off just walking away from it (especially because you build up almost no equity in the first few years of the mortgage). Mortgage defaults lowered the value of investments in bundles of mortgages.

Houses are worth perhaps hundreds of billions of dollars less than a few years ago, and big investment banks still have mortgages on their hands that are worth hundreds of billions of dollars less than they paid for them. That means that for some time to come, banks have less money to lend and can (indeed, must) demand more creditworthy clients and higher rates of interest.

How does all this affect you? Financial institutions have less money to lend for any purpose—buying a home, buying a car, starting a business, etc.—hence the "credit crunch" following the mortgage crisis. We can only hope that the worst will be over before you are in the market for a house, car, or business loan.

Where you won't see a direct effect is in the Consumer Price Index (CPI) (discussed in Section 21.5, on pages 787–788), which is based on the rental value of houses, not their prices. If the doubling of housing prices in 2000–2006 had been taken into account, the CPI would have risen by 5% per year rather than 3%.

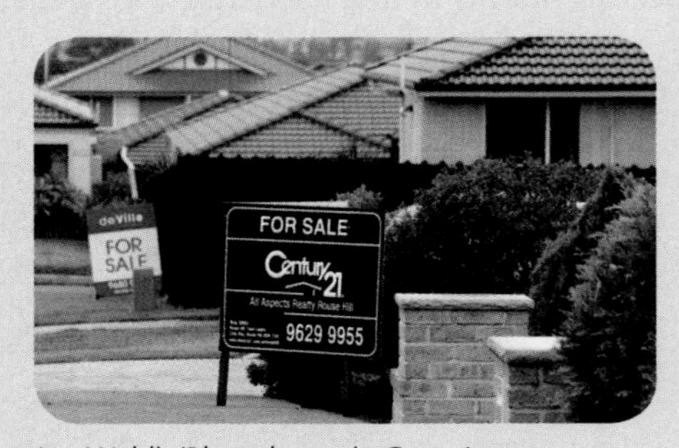

Ian Waldie/Bloomberg via Getty Images

go up or down with interest rates in the economy. Usually, the rate can be raised or lowered only every year or two, and then by a limited percentage. An ARM may be attractive if you plan to pay off the mortgage after only a few years or because it allows lower payments, it facilitates buying a more expensive home, or you do not plan to keep the home long (hence, you would be selling before the interest rate could rise substantially).

Does it pay to buy a house or apartment? Apart from the joys of ownership, you need to take into account the upfront expenses of closing costs (perhaps \$3000), plus the back-end expense of selling the house (usually 6%–7% if through a realtor, so say \$12,000). Consulting the table, you might think that you would finally be in the black on your house as an investment after 6 years (and you have had the pleasure of living in it rent-free!).

However, we have not yet taken into account the ongoing expenses of maintenance, repairs, insurance, and real estate taxes (perhaps \$5000–\$10,000 per year). Of course, if your house is rising in value 5% (\$10,000) per year or more, it's a different story. The growth of home ownership, which rose to 73% before the bursting of the housing bubble in 2007, depended on such a steady rise in value. Renting may be more attractive if you anticipate moving in just a few years; each year, one-sixth of Americans move.

22.4 Annuities

We recall from Section 21.4 (on page 783) the concept of an **annuity**:

Annuity DEFINITION

An **annuity** is a specified number of equal periodic payments.

We restrict our discussion to annuities for which payments are made at the end of each period and that period is also the compounding period.

An annuity can be interpreted as involving borrowing. For example, winners of lotteries are often offered the choice of receiving either the jackpot amount paid as an annuity over a number of years or else a smaller lump sum to be paid immediately. The cost to the lottery administration is the same. If the winner wants an annuity, the administration buys one from an insurance company for the lump sum. You can think of the insurance company as borrowing the lump sum in exchange for making the payments of the annuity. In effect, the insurance company is amortizing the lump sum over the duration of the annuity.

EXAMPLE 9
Winning the Lottery

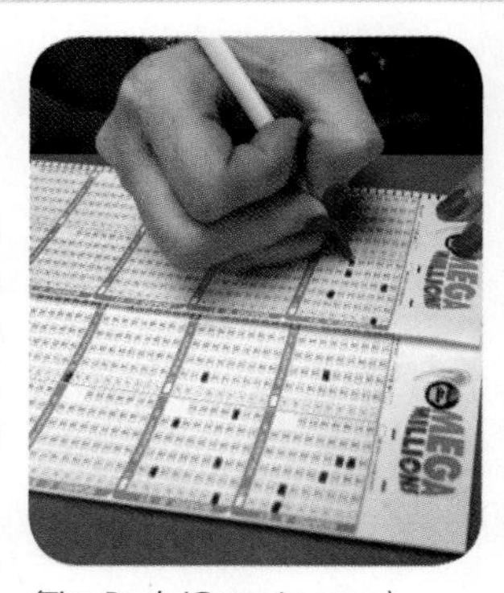

(Tim Boyle/Getty Images.)

On March 6, 2007, two winning tickets (one from New Jersey, one from Georgia) shared the world record for a lottery prize, a jackpot of \$390 million. Each ticket's share was half the total, or \$195 million. One option was to receive the \$195 million as an annuity in 26 equal annual installments of \$7.5 million each, the first payment being right away. However, each winner chose instead an instant lump sum of \$116,557,083. What was the interest rate of the annuity?

SOLUTION The insurance company offering the annuity regarded \$116,557,083 as the present value P (page 784) of the annuity. To consider payments at the end of each period, we must subtract the first payment, leaving $195 − $7.5 million = $187.5 million to be paid in 25 equal installments at the end of each year, with $116,557,083 − $7,500,000 = $109,057,083 as the present value of the stream of payments.

Solving for P in the amortization payment formula gives

$$P = d\left[\frac{1-(1+i)^{-n}}{i}\right]$$

Converting P to millions, we have P = $109.057083, d = $7.5, and $n = 25$. It is impossible to solve algebraically for i because i is involved in the equation both by itself and as part of a power; so we must use either a calculator with a financial mode or a spreadsheet. Either way, we find that $i = 4.69\%$.

As you save for retirement, it is probably wise to save part of your funds in the form of a tax-deferred annuity. If you do not, you can still sell holdings in other forms and purchase an annuity at retirement.

EXAMPLE 10
How Much Do You Need to Retire?

Suppose that your father wants to retire at age 65 with an annuity that pays $1000 per month for 25 years and is willing to assume that at retirement, the long-range steady interest rate will be 4% per year compounded monthly. What amount should he expect to pay for such a stream of income?

SOLUTION We apply the amortization formula with d = $1000, $r = 0.04$, $m = 12$, and $t = 25$, to find the amount P:

$$P = d\left[\frac{1-\left(1+\frac{r}{m}\right)^{-mt}}{\frac{r}{m}}\right] = \$1000\left[\frac{1-\left(1+\frac{0.04}{12}\right)^{-12\times 25}}{\frac{0.04}{12}}\right] \approx \$189{,}452.48$$

So purchasing such an annuity would cost $189,452.48.

Such annuities differ in a crucial way from the lottery annuity in Example 9. If you retire at 65 and purchase an annuity, you would be in trouble if you live longer than the term of the annuity (past 90), because the payments would stop and you would have no further income from the annuity. (About 2% of U.S. children born since 2000 can expect to live to age 100.) Similarly, if you die sooner, your estate would still get the payments due after your death, but they wouldn't have helped you meet your living expenses while you were alive.

An approach that avoids these two disadvantages is the **life income annuity**: You receive a fixed amount of income per month for as long as you live. How much you receive per month is based on the life expectancy of people your age, as determined from population data. There are many variations on life annuities, such as payments that increase with anticipated cost-of-living increases, or payments that last until both you and your life partner die (see Spotlight 22.4). But we focus on a simple one-life annuity.

The insurance company that sells you the annuity makes money on your policy if you die younger than average and loses money if you die older than

SPOTLIGHT 22.4
What Is an Actuary?

The Truth in Savings Act and the Truth in Lending Act specify that the APY for savings and the APR for loans must be calculated "according to the actuarial method."

Actuaries are financial experts who assess the costs of risks and investigate the probability of various contingencies—for example, death, default, or cancellation—that might occur. Actuaries are crucially involved in setting premiums. Their calculations take into account historical rates—such as the percentage of female 85-year-olds who live to be 86, or the percentage of unmarried male drivers under age 25 who have auto accidents—and project those rates and the accompanying costs into the future.

Other actuaries concentrate on setting up and evaluating pension and fringe benefit plans. For example, the city of Beloit, Wisconsin, hired a consulting actuary to estimate the current and future costs of free lifetime medical benefits to families of police and firefighters.

Another major activity of actuaries is managing return on investment. Contrary to popular belief, insurance companies (particularly life insurance companies) do not earn all their money from premiums paid. In fact, a substantial portion of their income comes from return on investment of financial *reserves*, funds that they are required to have to meet current and future insurance obligations.

Becoming an actuary requires training in mathematics, statistics, economics, and finance and includes a sequence of professional exams taken over several years.

© Olix Wirtinger/Corbis

average. As in any kind of insurance, over a large number of people, the company expects gains to balance losses. This is a manifestation of the law of large numbers (discussed in Section 8.5 on page 303). However, the company will reinvest the annuity funds, and hence its profits will depend on the rate that those investments earn compared to the rate that it pays on the annuity.

How much the annuitant (the purchaser of the annuity) receives per month depends on gender. Because women on average live longer than men, the monthly payment to a woman can be lower.

EXAMPLE 11
Life Income Annuity

Suppose that your 65-year-old father retiring with \$250,000 in a life income annuity. According to the table from one particular insurance company, he would receive \$6.3448 per month for every \$1000, so his monthly income would be \$1586. According to the Social Security Administration actuarial life table, his life expectancy at age 65 is about 17 years = 204 months. If he lived exactly that long, he would receive a total of 204 × \$1586 = \$323,544.

However, simple algebra cannot be used to find the rate of interest that the annuity would need to earn to last that long. We use the RATE function in a spreadsheet (for more details, see Spotlight 21.4 on page 791); entering =RATE(204, 1586, –250000) gives a monthly rate of 0.2636%, for an effective annual rate of (1+0.002636)^(12) – 1 ≈ 3.21%.

Now let's consider instead the case of your mother retiring now, also at age 65 and also with $250,000 savings in a life income annuity; she would receive $5.9010 per month for every $1000, or $1475 per month. Her life expectancy would be about 19.72 years = 237 months. If she lived exactly that long, she would receive a total of 237 × $1475 = $349,575. The rate of interest that her annuity would need to earn to last that long can be calculated from the amortization formula; using =RATE(237, 1475, −250000) gives a monthly rate of 0.2997%, for an effective annual rate of 3.66%. The difference of this rate from the one for your father probably reflects the company's use of life expectancies (which vary by region of the country) different from those for the nation as a whole.

Notice that a man and a woman who save the same amount receive different monthly incomes at retirement: The woman receives less per month but for longer—93% as much for 16% longer. Yet their living expenses are likely to be the same. That consideration has resulted in some companies offering "merged gender" rate schedules for annuity payments, so that the individual receives the same monthly payment regardless of gender.

REVIEW VOCABULARY

Adjustable-rate mortgage (ARM) A loan whose interest rate can vary during the course of the loan. (p. 815)

Amortization payment formula The formula for installment loans that relates the principal P, the interest rate i per compounding period, the payment d at the end of each period, and the number of compounding periods n needed to pay off the loan:

$$d = P\left[\frac{i}{1 - (1+i)^{-n}}\right],\ P = d\left[\frac{1 - (1+i)^{-n}}{i}\right].\quad \text{(p. 811)}$$

Amortize To repay in regular installments. (p. 809)

Annual percentage rate (APR) The number of compounding periods per year times the rate of interest per compounding period. (p. 808)

Annuity A specified number of equal periodic payments. (p. 817)

Compound interest formula The formula for the amount in an account that pays compound interest periodically. For an initial principal A and effective rate i per compounding period, the amount after n compounding periods is $A = P(1 + i)^n$. (p. 807)

Compounding period The fundamental interval for compounding, within which no compounding is done. Also called simply *period.* (p. 808)

Conventional loan A loan in which each payment pays all the current interest and also repays part of the principal. (pp. 809–810)

Effective annual rate (EAR) The effective rate per year. (p. 808)

Effective rate The actual percentage rate, taking into account compounding. (p. 808)

Equity The amount of principal of a loan that has been repaid. (p. 813)

Interest Money charged on a loan. (p. 805)

Life income annuity An annuity with regular payments for as long as you live. (p. 818)

Nominal rate A stated rate of interest for a specified length of time; a nominal rate does not take into account any compounding. (p. 808)

Principal Initial balance. (p. 805)

Savings formula The formula for the amount in an account to which a regular deposit is made (equal for each period) and interest is credited, both at the end of each period. For a regular deposit of d and an interest rate i per compounding period, the amount A accumulated is

$$A = d\left[\frac{(1+i)^n - 1}{i}\right].\quad \text{(p. 807)}$$

Simple interest The method of paying interest on only the initial balance in an account and not on any accrued interest. For a principal P, an interest rate r per year, and t years, the interest I is $I = Prt$. (p. 805)

SKILLS CHECK

1. The interest charged on Federal Direct Stafford Loans is

(a) simple interest.
(b) compounded daily.
(c) compounded quarterly.

2. If you borrow $1000 at 5% interest per year, compounded quarterly, and pay back the principal and interest after four years, the amount that you pay back is ________.

3. A nominal rate of interest

(a) takes into account any compounding involved.
(b) is always stated as an annual rate.
(c) neither of the above.

4. If you put $100 at the end of each month for two years in an account that pays 6% annual interest compounded monthly, at the end of the two years, you have ____________.

5. An effective interest rate

(a) always takes inflation into account.
(b) is the same as the nominal rate.
(c) takes compounding into account.

6. APR stands for ________________.

7. The nominal rate of interest for a loan is

(a) the same as the effective rate.
(b) less than the effective rate.
(c) never greater than the effective rate.

8. If a store credit account charges 1.5% interest each month, the effective annual rate is ________.

9. Credit card interest is

(a) computed using compound interest.
(b) computed using simple interest.
(c) included in the late fees.

10. If a store credit account charges 1.5% interest each month, the APR is ________.

11. The APR calculated for a loan takes into account the amount of

(a) the loan.
(b) the loan plus any loan-processing fee.
(c) the loan, any loan-origination fee, and any other closing costs.

12. The median length of time that Americans hold a mortgage is ________.

13. The Truth in Lending Act requires that borrowers be told the

(a) APY of the loan.
(b) APR of the loan.
(c) effective annual rate of the loan.

14. Your credit union offers to finance a $6000 conventional loan at 4% to be repaid in four years of monthly payments. Your monthly payment is ______.

15. After 15 years of minimum payments on a 30-year mortgage, the balance remaining is about

(a) one-third of the original balance.
(b) one-half of the original balance.
(c) two-thirds of the original balance.

16. If you finance $15,000 for 3 years at 6% compounded monthly, the monthly payments will be ____.

17. Equity in a 30-year conventional mortgage grows

(a) linearly.
(b) logarithmically.
(c) exponentially, but slowly.

18. Monthly payments for a 15-year 6% mortgage are about ______ times the payments for a 30-year mortgage of the same amount and the same interest rate.

19. An ARM

(a) has variable interest rates but maintains a fixed payment amount.
(b) has variable payment amounts.
(c) is always a better alternative to fixed-rate mortgages.

20. In a 30-year mortgage, most of the amount of the first few payments goes toward______.

21. With a conventional loan,

(a) each payment pays current interest and repays part of the principal.
(b) you pay exactly the same current interest rate as everyone else.
(c) you have to buy a conventional house—no condos, triple-deckers, or yurts allowed.

22. A convenient rule of thumb is that for a 30-year mortgage at 6%, the monthly payment is about 0.6% of the loan. So, on a $100,000 mortgage, the monthly payment is about $600. About ______ of the first payment goes toward interest.

23. Most people

(a) never pay off their home mortgage loan.
(b) pay off their home mortgage loan early.
(c) make late payments on their home mortgage loan.

24. A high rate of inflation is likely to mean ________ payments on your ARM.

25. Equity refers to

(a) fair lending practices.
(b) principal paid off on a loan.
(c) a union that represents actors and stage managers for live theatrical performances.

26. ARM stands for ____________.

27. Paying off a conventional loan is like

(a) saving.
(b) slaving.
(c) caving.

28. If you just won a lottery jackpot paid in 25 equal annual installments of $1 million each at 6% annual effective interest, the present value of the jackpot is ____________.

29. A life income annuity is designed to pay a fixed amount each period until

(a) the annuity runs out of money.
(b) you die.
(c) you reach your life expectancy.

30. A professional who assesses the costs of risks for life insurance, auto insurance, health insurance, or pensions is called a(n) ___________.

CHAPTER 22 EXERCISES

■ *Challenge* ▲ *Discussion*

22.1 Simple Interest

1. Suppose that you take out an unsubsidized Stafford loan on September 1 before your senior year for $7500 (the maximum allowed) and plan to begin paying it back on December 1 after graduation (so you will have had the loan for 15 months, including the six-month grace period after leaving school). The interest rate is 6.8%. How much will you owe then, and how much of that will be interest?

2. Assume the same situation as in Exercise 1, but you borrow $7500 on September 1 before your junior year and plan to begin paying it back on December 1 after graduation and grace period 27 months later. How much will you owe then, and how much of that will be interest?

3. Suppose that you borrow $5500 for your first year and $6500 for your second year (the maximum amounts), as unsubsidized Stafford loans. Suppose that each loan begins on September 1 of its year, that you finish college in four years, and that you begin repayment on December 1 after graduation. What is your total debt then, and how much of that is interest?

4. Assume the same situation in Exercise 3, but you also borrow $7500 for each of your third and fourth years (again, the maximum amounts), again on September 1. You finish college in four years, and you begin repayment on December 1 after graduation. What is your total debt then, and how much of that is interest?

22.2 Compound Interest

5. If you borrowed $15,000 to buy a new car at 4.9% interest per year, compounded annually, and paid back all the principal and interest at the end of 5 years, how much would you pay back?

6. Assume the same situation as in Exercise 5, but the interest is compounded monthly.

7. If you borrowed $200,000 to buy a house at 6% interest per year, compounded annually, and paid back the principal and interest at the end of 30 years, how much would you pay back?

8. Assume the same situation as in Exercise 7, but the interest is compounded monthly (this is the usual case).

9. A recent credit card bill of mine showed an APR of 17.24%.

(a) What is the corresponding daily interest rate (the bank uses a 365-day year for this purpose)?
(b) What is the effective annual rate?

10. I received an offer for a credit card with 0% fixed APR for the first 12 months, followed by one of several rates depending on credit history. The highest was a 22.74% APR (and the company reserves the right to change the APR "at any time for any reason").

(a) What is the corresponding daily interest rate for the 22.74% APR?
(b) What is the effective annual rate?

22.3 Conventional Loans

11. Suppose that your unsubsidized Stafford loans plus accumulated interest total $20,000 at the time that you start repayment, the interest rate is 6.8% APR, and you elect the standard repayment plan of a fixed amount each month for 10 years.

(a) What is your monthly payment?
(b) How much will you pay in interest?

12. Suppose that your unsubsidized Stafford loans plus accumulated interest total $31,811 at the time that you start repayment and the interest rate is 6.8% APR.

(a) If you elect the standard repayment plan of a fixed amount each month for 10 years, what is your monthly payment?
(b) How much will you pay in interest?
However, because your accumulated outstanding federal loans total more than $30,000, you can elect to repay over 25 years instead. If you do that:
(c) What is your monthly payment?
(d) How much will you pay in interest?

Refer to the following for Exercises 13 and 14.
Your parents (if their credit rating qualifies) can take out a federal Direct PLUS loan to pay for the total remaining cost of your undergraduate education, after any other financial aid (such as a Stafford loan). The simple interest rate is 7.9% after July 1, 2010. (There are also fees, which we disregard here.) The standard repayment plan is for fixed monthly payments to be made over 10 years, and your parents can elect to defer the start of repayment until six months after your graduation.

13. Suppose that your parents take out a PLUS loan on your behalf on September 1 before your senior year for $10,000 and begin paying it back six months after graduation. How much is their monthly payment?

14. If your parents instead take out a PLUS loan for $10,000 on September 1 before each of your four years of college, how much is their monthly payment if they begin paying it back six months after your graduation?

15. In January 2011, a dealership in southern Wisconsin was offering a new 2011 Toyota Corolla sedan for $19,859. One option for financing was 2.9% over 36 months. What was the monthly payment?

Courtesy of Toyota

16. Assume the same situation as in Exercise 15, but for the second financing option of 2.9% over 48 months.

17. Assume the same situation as in Exercise 15, but for the third financing option of 2.9% over 60 months.

18. In December 2010, Kevin Lauterbach, 29, of Coral Springs, Florida, who had "mildly damaged credit," bought a 2008 Jeep Liberty with no money down and a 72-month loan for $19,000 with a 4.75% rate. (*New York Times,* February 28, 2011, p. A3).

(a) What would the monthly payment have been?
(b) How much interest would he have paid over the course of the 72 months?
■ **(c)** In fact, he was instead required to make payments every two weeks. How much was that payment?

19. In January 2011, a 2010 Corolla sedan listed at $19,084 with a choice of either 0% interest or else $1250 cash back (deducted from purchase price) and 2.9% interest. Which alternative gives lower monthly payments for which lengths of time (36, 48, 60 months)?

20. As in Exercise 19, but for a Dodge Journey minivan listed at $29,715 (after $1000 trade-in), for which the choice was either 0% interest over 60 months plus $1000 cash back or else $3000 cash back and 6% interest?

21. Suppose that you have good credit and can get a 30-year mortgage for $100,000 at 5%. What is your monthly payment?

22. Assume the same situation as in Exercise 21, except that your credit is not as good and the rate that you are offered is 7.125%.

23. Assume the same situation as in Exercise 21, but you inquire about a 15-year loan instead. You are offered 3.75%. What is your monthly payment?

24. Assume the same situation as in Exercise 23, but your credit is not as good, and you are offered 6.75%. What is your monthly payment?

25. For the mortgage in Exercise 21, how much equity would you have after 5 years?

26. For the mortgage in Exercise 22, how much equity would you have after 5 years?

27. For the mortgage in Exercise 23, how much equity would you have after 5 years?

28. For the mortgage in Exercise 24, how much equity would you have after 5 years?

Refer to the following for Exercises 29 and 30.
When interest rates drop, it may become attractive to refinance your home. Refinancing means that you acquire a new mortgage to borrow the current principal due

on your home and use the proceeds to pay off your old mortgage. You then begin a new 15- or 30-year mortgage at the new, lower interest rate. A second factor that reduces your monthly payment is that the equity you accumulated under the old mortgage reduces the amount that you have to borrow under the new mortgage. Suppose that you have been paying for 5 years on a 30-year mortgage for $200,000 with a fixed rate of 6%. Your monthly payment is $1199.10, you have $13,890.81 in equity, so $200,000 − $13,891.20 = $186,108.80 remains to be paid. We consider two refinancing offers current in October, 2011.

29. The first offer is from a local bank for a 30-year fixed-rate mortgage at 4.25% with closing costs of $2639.07.

(a) What is the new monthly payment?
(b) How much less is that per month than the old payment?
(c) How many months will it take for the savings in payments to make up for the closing costs?

30. The second offer is on the Internet (from a company you have never heard of) for a 30-year fixed-rate mortgage at 3.99% with closing costs of $1800.

(a) What is the new monthly payment?
(b) How much less is that per month than the old payment?
(c) How many months will it take for the savings in payments to make up for the closing costs?

31. In a 2/28 "hybrid" adjustable-rate mortgage (ARM), the initial interest rate is fixed for 2 years, then is adjusted every 6 months. (You usually pay "points" up front at closing in exchange for the "rate lock" for the first 2 years.) Suppose that you buy a house with a $200,000 mortgage with a 2/28 ARM with initial rate of 3%; and suppose that 2 years later, the interest rate goes up to 5%.

(a) What was your payment originally, at 3%?
(b) What is your new payment? (Careful: The amount of the loan is no longer $200,000, and you have only 28 years to pay it off.)

32. In a 5/1 "hybrid" adjustable-rate mortgage (ARM), the initial interest rate is fixed for 5 years, then is adjusted annually. (You usually pay "points" at up front at closing in exchange for the "rate lock" for the first 5 years.) Suppose that you buy a house with a $200,000 mortgage with a 5/1 ARM with initial rate of 4%; and suppose that 5 years later, the interest rate goes up to 6%.

(a) What was your payment originally, at 4%?
(b) What is your new payment? (Careful: The amount of the loan is no longer $200,000, and you have only 25 years to pay it off.)

Refer to the following for Exercises 33–37, about credit card payments.

Many credit cards use a similar formula for the minimum payment, which is the new balance (if less than $25), or else the greatest of $25 or

> 1% of the new balance (excluding interest and late fees), plus the interest billed, rounded down to the nearest dollar.

Any late fees are then added on to this calculated amount. Moreover, when any interest is due, there is a minimum charge of $1.50.

■ **33.** (Requires spreadsheet.) Suppose that your credit card has an APR of 18% interest rate, corresponding to approximately 1.5% per month. (The actual interest applied is daily interest, at a daily rate of 18%/365; but for simplicity we use a uniform approximate monthly rate. Also, the amount of interest owed depends on exactly when in the month your payment is received.)

(a) Why do we say "approximately" 1.5% per month for an APR of 18%?
(b) How many months will it take to pay off a new balance of $3117.83 by making the minimum payment each month?
(c) How much will you have paid altogether? How much of that is interest?

■ **34.** (Requires spreadsheet.) As in Exercise 33, but you screw up and miss the first payment, incurring a $35 late fee and an increase to a penalty APR of 30%, corresponding to approximately 2.5% per month.

(a) How many months—of paying your bill on time!—will it take to pay off the balance of $2500 by making the minimum payment each month (on time!)? (You must pay the $35 late fee the first month, over and above the minimum payment on the $2500 and one month's interest.)
(b) How much will you pay altogether?

■ **35.** The purpose of such a complicated formula for the minimum payment on a credit card is to avoid the situation of a customer who makes just the minimum payment nevertheless falling farther and farther behind. For example, formerly some banks set the minimum payment at balance due (if < $10) or else the larger of $10 or 2% of the total new balance (including interest). However, for a high enough interest rate, paying 2% of the balance due will not cover the interest, so the balance actually would increase (this is called negative amortization). How high would the APR have to be to make this happen? (Careful: It's not just 12 × 2%.)

■ **36.** (Requires spreadsheet.) A well-known national credit card calculates minimum payment due as the new balance (if less than $35), or else the greatest of: (1) $35; (2) 2% of the new balance (excluding new late fees); (3) interest charged on the statement plus 1% of the new balance (excluding late fees and new interest charged on the statement), not to exceed 4% of the new balance. Then any late fees are added and the total is rounded to the nearest whole dollar.

(a) With a monthly interest rate of 1.5%, how many months will it take to pay off a new balance of $5000 by making the minimum payment each month?

(b) How much will you pay altogether?

■ **37.** (Requires spreadsheet.) As in Exercise 36, but for a monthly interest rate of 2.5%.

■ **38.** (Requires spreadsheet.) A bank or credit union may offer to let you agree in advance to skip a payment (e.g., on a car loan but usually not on a mortgage or a credit card)—in exchange for a processing fee (such as $35) to be added to the principal. If you skip the payment, interest continues to accrue for that month on the remaining principal plus the added processing fee. You continue regular payments as usual in the same amount as before, except that the last payment is a larger "balloon" payment to pay off the loan. Suppose that you borrowed $11,158.05 from your credit union for a 60-month home improvement loan at 9%. Verify that your monthly payment is $231.62 and that after 12 months of payments you still owe $9307.74. You receive an offer to skip the 13th payment, for $35 added to the principal, and you do so. How much will the balloon payment be?

39. Despite filters, lots of spam gets into my email. For a while, I was getting mortgage offers, such as "$160,000 for less than $735 per month" (for a 30-year mortgage). What would be the corresponding interest rate?

40. Suppose that you and two friends decide to live off-campus in your senior year. One of them (who has wealthy parents) suggests that instead of renting an apartment, you could buy a house together, live in it for your senior year, then rent it out or sell it. Assuming that (with the help of her parents and their good credit rating) you could get a mortgage for $180,000 to buy a house near the campus, what would be the monthly mortgage payment on a 30-year mortgage at 5.75%?

© *Tetra Images/Corbis*

For Exercises 41 and 42, refer to the following: Payday lenders provide small loans until the borrower's next payday. The borrower receives the desired cash in exchange for a postdated check in the amount of the loan plus a fee, which is usually a percentage of the loan amount. In many states, there are now more payday loan offices than McDonald's fast food outlets. The average loan amount is $300.

41. For one payday lender, the fee for a $100 loan for up to two weeks is $15. What is the APR if the loan is for the full two weeks?

42. Another payday lender charges $18.62 for a $100 loan for 7 to 14 days. What is the APR if the loan is for 7 days?

For Exercises 43 and 44, refer to the following: Many income tax preparation services, including the large national chains, offer refund anticipation loans (RALs), or "rapid refunds." These are similar to payday loans in providing an advance on anticipated income—in this case, a tax refund. The loan is repaid when the Internal Revenue Service (IRS) pays the refund, usually about 7 to 17 days after the loan is made. The cost of the loan is deducted from the proceeds to the client, so this is a discounted loan. The RAL business takes in about $2 billion each year (including tax preparation and check-cashing fees) to arrange payment of the earned-income tax credit to working parents, about 7% of the total of this aid to poor families. A RAL for an anticipated refund usually is issued for a flat fee, often $88, and the average loan is $1500.

43. Suppose that the RAL speeds the refund by 7 days. What is the APR for the average RAL?

44. Suppose that the RAL speeds the refund by 17 days. What is the APR for the average RAL?

22.4 Annuities

45. The largest amount won by an individual in a U.S. lottery was $314.9 million, by Jack Whittaker of West Virginia on Christmas Day, 2002. (His subsequent life has been far from a fairy tale, as a Google search will reveal.) Instead of receiving $314.9 million in 30 equal annual payments, including one immediately, he chose a lump sum, which came to $170 million. What was the corresponding interest rate of the annuity?

■ **46.** Winners of the Powerball lottery can elect either an immediate lump sum (almost all do) or an annuity. In the latter case, the advertised jackpot amount is paid in 30 annual payments, including 1 immediate payment. To keep up with inflation, each payment is 4% more than the previous year's; such an annuity is called a *graduated annuity*. On October 10, 2007, Eugene and Stanislawa

Markiewicz took their prize of $20 million in the form of a graduated annuity.

(a) What was the amount of their first payment, and how much will they receive in their last payment in October 2036?

(b) The winners could have chosen a lump sum of $9,402,914.90 instead. What was the corresponding interest rate of the annuity?

47. Suppose that a man retires at age 65, and in addition to Social Security, he needs $2000 per month in income. Based on an expected lifetime of 204 more months, how much would he have to invest in a life income annuity earning 3% APR to pay that much per year?

48. Assume the same situation as in Exercise 45, but for a woman at age 65, whose expected lifetime is 237 more months.

49. Put off by the high monthly car payments of Exercises 15–17, you might be attracted instead to leasing. You could lease a 10% cheaper 2010 Toyota Corolla LE for $169 per month for 36 months. Because this is a lease, not a loan, the dealer does not have to disclose anything about interest rates. However, the "capitalized cost" (after down payment of $1180 and "acquisition fee" of $650) is $16,197. You will have the option at the end of the lease to purchase the car for $10,123. So in effect, your monthly payments are paying for the difference of $16,197 − $10,123 = $6074. However, your payments (including that first one in advance) total only $169 × 36 = $6084.

(a) The purchase price, if you bought instead of leased, would be $18,300, with 0% interest. What would be the monthly payment over 36 months? Over 60 months?

▲ **(b)** Why are the payments for leasing a car so much lower than for purchasing?

▲ **(c)** How do the manufacturer and the dealer make money on such a deal?

APPLET EXERCISES

To do these exercises, go to www.whfreeman.com/fapp9e.

There are two ways to buy a car: save up and pay cash or borrow the money. In the Buying a Car: Cash vs. Loan applet, you can explore just how much more expensive it is to borrow the money.

WRITING PROJECTS

1. Locate current advertised incentives for a car that you would consider buying and compare them in an essay of two to three pages. For each option, give the price, the interest rate, the term, and how much interest you would pay over the course of the loan.

2. A substantial proportion of new cars today are not sold but leased. Contact a local car dealer about a car that you are interested in and find out the details on leasing. Compare the cost of the lease and associated expenses with the cost of purchasing and owning the car. Include estimated maintenance, repair, and insurance costs for each option. Which seems like a better deal, and why? Consult the "Suggested Web Sites" section at the end of this chapter. Write two to three pages describing and comparing the two options.

3. Banks often offer choices of mortgages with various combinations of interest rates and "points." A point is 1% of the mortgage amount. Points are paid to "buy down" the interest rate for the mortgage; they are paid upfront to the bank at the closing of the house sale. For example, you may have a choice between a mortgage at 6% with 2 points (2%) and a mortgage at 8% and no points. Which would you choose, and why? Does it make a difference how long you are planning to own the home? Or how expensive the home is? Write a page justifying your decision.

4. One of the advantages of buying a home with a fixed-rate mortgage is that your payment stays the same but your earnings and the value of your home are likely to go up with inflation. You are paying back the loan with dollars of lesser value.

Consider the following scenario. Suppose that you buy a "starter" two-bedroom home for $105,000 under a special program for first-time home-buyers that requires a down payment of only $5000. You have a 30-year fixed-rate mortgage for $100,000 at 7%, on which the monthly payment is $665.30. You also have a $2000 one-time expense in closing costs and annual costs of $200 for insurance and $2000 for property taxes.

You live in the home for five years and spend $10,000 on maintenance, upkeep, and improvements. You then sell the home for $125,000, pay a realtor $9000 to sell it, and pay closing costs of $500 (for title insurance and other costs). Finally, it costs $3000 to move.

(a) Make out a balance sheet of revenue and expenses. How did you make out on owning the home?

(b) Remember that you also got to live in the home without paying rent. Translate the cost of owning the home into an equivalent monthly rent.

5. Explore actual costs of homes in your area, mortgages with local banks (including closing costs), and property taxes and insurance. Come up with data on a particular mortgage, and the costs and benefits of refinancing, and make out a corresponding balance sheet for five-year ownership.

Suggested Readings

KASTING, MARTHA. *Concepts of Math for Business: The Mathematics of Finance* (UMAP Modules in Undergraduate Mathematics and Its Applications: Module 370–372), COMAP, Inc., Arlington, MA, 1980.

MILLER, CHARLES D., VERN E. HEEREN, and JOHN HORNSBY. Consumer mathematics. In *Mathematical Ideas,* 11th ed., Pearson Education/Addison Wesley, Reading, MA, 2008.

YAREMA, CONNIE H., and JOHN H. SAMPSON. Just say "Charge it!" *Mathematics Teacher 94* (7) (October 2001), 558–564. Shows how to apply the savings formula and the amortization formula and graph the results on the TI-83 calculator. Notes that the 78% of undergraduates in the United States who have credit cards carry an average debt of more than $2700, with 10% owing more than $7000.

Suggested Web Sites

www.lendingtree.com/stmrc/calculators1.asp Java applet calculators (for any platform) to calculate payments and amortization schedules for conventional loans, adjustable-rate mortgages, auto loan vs. home-equity loan, and credit-card payoff. (Note: Lending Tree, Inc., is a loan broker; the mention here of calculators at its Web site does not imply endorsement of its other services by this book's authors, editors, or publisher.)

www.leaseguide.com/ A guide to how car leasing works, including what "money factor" means, and how leasing cost is determined.

www.edmunds.com/calculators/ Commercial site offering a calculator to compare rebate vs. interest-rate offers for car purchase. (*Note:* Edmunds is a loan broker; mention here of calculators at its Web site does not imply endorsement of its other services by this book's authors, editors, or publisher.)

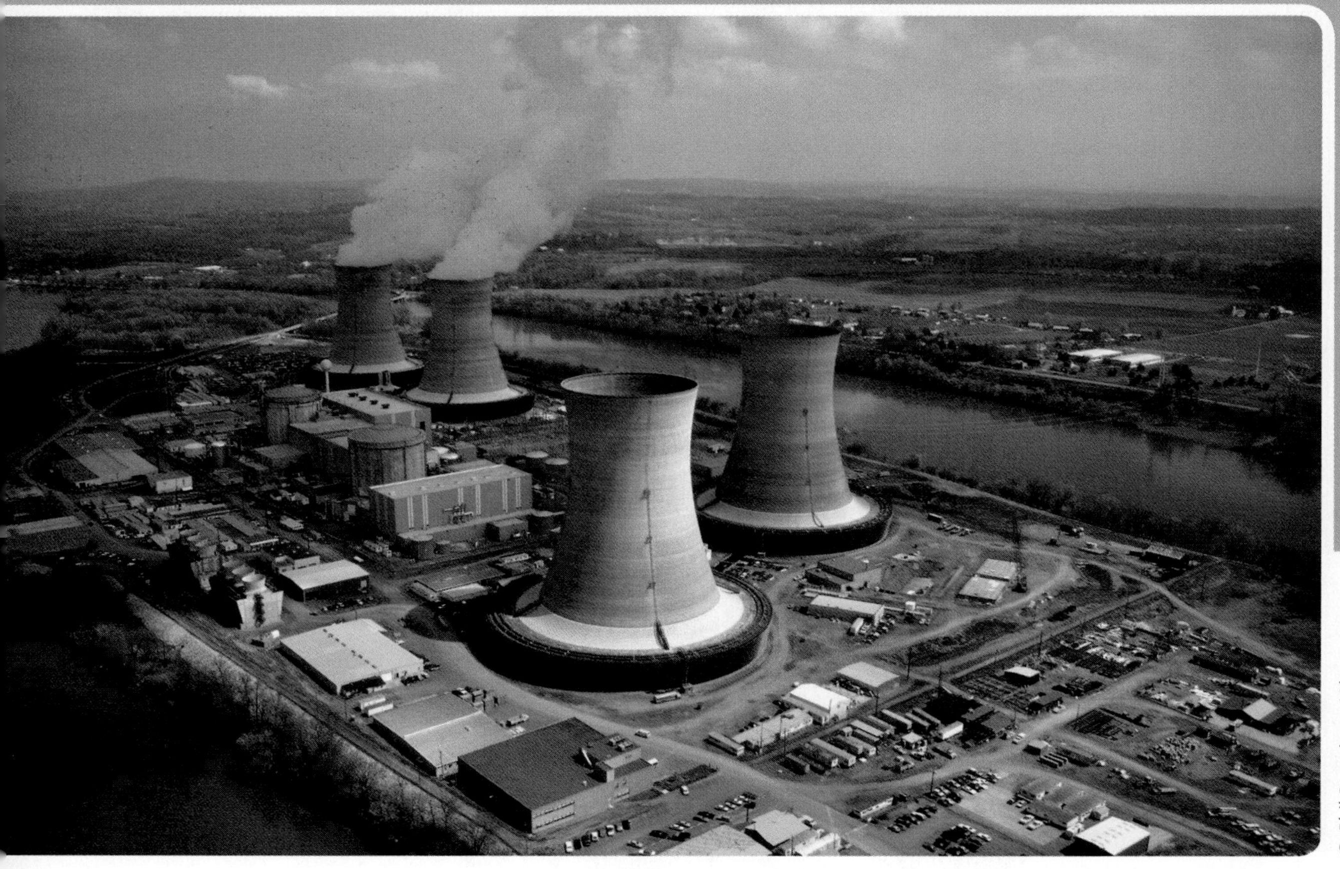

23

The Economics of Resources

In Chapters 21 and 22, we explored mathematical models for saving, accumulating, and borrowing—the building up of resources. In this chapter, we model processes in the other direction—the use, decay, depletion, or spending down of resources, including resources that tend to replenish themselves regularly.

We use here only three formulas from Chapter 21, "Savings Models," specialized to an annual interest rate r and number of years n:

Compound interest formula (page 772): If a principal P is deposited into an account that pays interest at rate r per year, then after n years the account contains the amount

$$A = P(1 + r)^n$$

Savings formula (page 781): For a uniform deposit of d per year (deposited at the end of the year) and an interest rate r per year, the amount A accumulated after n years is

$$A = d\left[\frac{(1 + r)^n - 1}{r}\right]$$

Continuous interest formula (page 779): For a principal P in an account at a nominal annual rate r, compounded continuously, the balance after n years is

$$A = Pe^{rn}$$

A reader who can use these formulas can proceed in this chapter without first reading Chapter 21 or 22.

We begin in Section 23.1 with a geometric growth model for human populations, applying it to the current U.S. population. Limits to growth of any population lead us to the logistic model, which takes into account the carrying capacity of the environment. Surprisingly, such a model also applies to the spread of a technology or new product. In Section 23.2, we calculate how long a nonrenewable resource, such as oil, can last at a current rate of use and also at a constantly increasing rate of use; and we show how the logistic model fits well the history of oil consumption. We turn in Section 23.3 to renewable resources, such as forests and fisheries. We examine how a population's growth curve determines equilibrium population sizes and which harvesting policies can produce yields that are sustainable year after year.

In Section 23.4, we tackle resources, such as nuclear waste, that decay and that we *want* to become exhausted. In Section 23.5, we consider economic factors, which influence harvesting efforts and consequently the population of the resource. But even biologically and economically sound policies are subject to the unpredictability of weather and other factors; what's worse, as we will see in Section 23.6, even in the absence of such chance effects, the dynamics of a population can vary in an apparently chaotic fashion—even to extinction.

23.1 Growth Models for Biological Populations

We encountered geometric growth models for savings accounts in Chapter 21. Growth is proportional to the amount present, and such growth is expressed in terms of compound interest and its formula. We now use a geometric growth model to make rough estimates about sizes of human populations. In addition to the **rate of natural increase**—the annual birth rate minus the annual death rate—we must take into account net migration. The sum of the two, in the terminology of financial models, is the effective rate.

Birth, death, and migration rates rarely remain constant for long, so projections must be made with care. In the short run, however, predictions based on the model may be useful. Let's apply this model to two questions about the population of the United States.

EXAMPLE 1
Predicting the U.S. Population

(Blaine Harrington III/Corbis.)

The U.S. population increased at an average effective growth rate of 0.95% per year (including immigration) to 310.5 million at the beginning of 2011. What is the anticipated population at the beginning of 2015? What is it if the effective rate of growth changes to 1.2% per year or to 0.7% per year?

SOLUTION We apply the compound interest formula with an initial population size ("principal") of 310.5 million. Using a year as the compounding period and the formula $A = P(1 + r)^n$, where $n = 4$, the projected population size in 2015 for a rate $r = 0.0095$ is

$$\begin{aligned}\text{population in 2015} &= (\text{population in 2011}) \times (1 + \text{growth rate})^4\\ &= 310{,}500{,}000\,(1 + 0.0095)^4\\ &= 310{,}500{,}000\,(1.0385)\\ &\approx 322{,}000{,}000\end{aligned}$$

Because of the limited accuracy of the estimates of population and growth rate, we round off the final answer. The result of a calculation can't be more precise than the ingredients.

In the same way, with a growth rate of 1.2% per year, we predict a population of 326 million, while a growth rate of 0.7% per year yields 319 million.

So an uncertainty of one-fourth of 1 percentage point in the growth rate has major implications, even over fairly short time horizons. The presence or absence of 7 million people would have a significant impact on our social and economic systems! Indeed, much of the concern over long-range funding of social security programs results from uncertainties over birth and immigration rates. Figure 23.1a gives a graph of the U.S. population in 2007, structured by age and sex. Figure 23.1b shows possible futures for India.

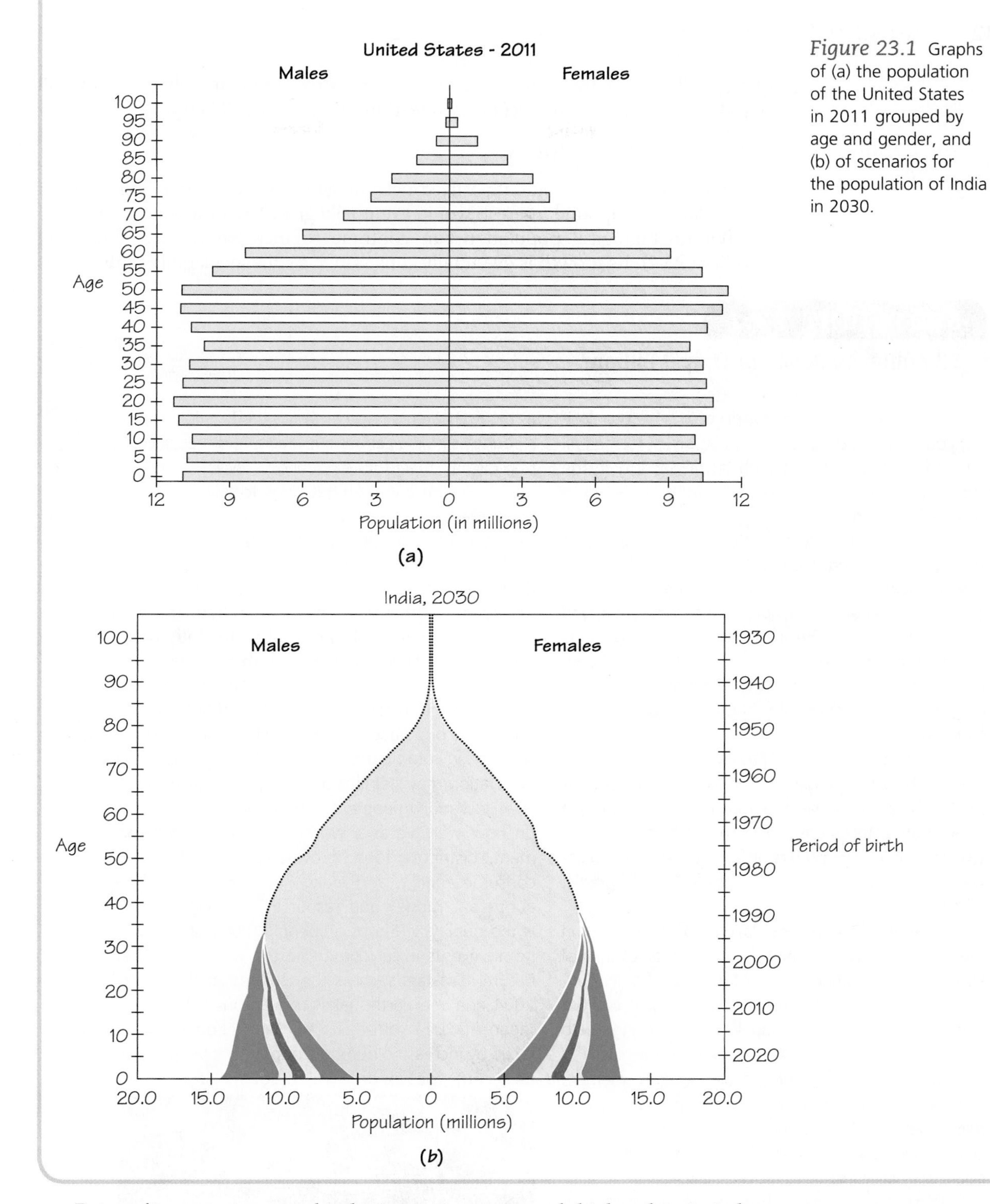

Figure 23.1 Graphs of (a) the population of the United States in 2011 grouped by age and gender, and (b) of scenarios for the population of India in 2030.

Rates of increase in most developing nations are much higher than in industrialized nations. With a growth rate of a little less than 2.4%, Nigeria, Africa's most populous country, will grow from 158 million in mid-2010 to 217 million in mid-2025, an increase of more than one-third in only 15 years. Such projections raise concern over providing sufficient food and resources for all people.

It is not just the number of people that is crucial, but also the **population structure**. In poorer countries, the proportion of the population over 60 years of age will be 20% by 2050, compared with 8% now; in Japan, where the overall population is expected to decline by one-third by then, it will be 40%.

Limitations on Growth

A population that keeps adding a fixed percentage each year, like a bank account accumulating compound interest, would eventually grow to astronomical numbers.

But no biological population can continue to increase without limit (see Spotlight 23.1). Its growth is eventually constrained by the availability of resources

SPOTLIGHT 23.1

12 Billion by 2050—or Only 9 Billion?

How many people can the world hold? Are developing countries heading for a population disaster? Will falling fertility play havoc with Social Security in the United States? Will aging result in 50% of Japanese being over 60 in 2100?

Answers come from mathematical modeling of the future from predicted trends. The best analyses suggest a probability distribution over a range of estimates. They project separately by age, gender, education, and other characteristics. They try to factor in improvements in agriculture, spread of diseases (especially HIV), changes in urbanization, increases in economic aspirations, and the potential for climate change (for example, from global warming).

A basic concept is *total fertility rate* (TFR), the average number of births per woman. Absent catastrophes (such as war or disease), a rate of 2.1 continues a population at the same size. A model that assumes a value above 2.1 will predict an ever-increasing population; one that assumes a value below 2.1 will predict an ever-dwindling one.

Most of the world's population growth will occur in the lesser-developed countries, whose TFR values are well above 2.1 (for example, it is 4.7 for Africa). Many countries in Europe have TFR values less than 2.1; without immigration, they will lose population and struggle with fewer workers to provide social benefits to the elderly.

China's situation illustrates *demographic momentum*. Even though its fertility rate has been below replacement level for 20 years, the number of women in the childbearing years was (and still is) so large that China's population will continue to grow until 2040.

The most sophisticated models try to assess how the TFR will vary with changing social circumstances. The most important single factor affecting fertility is the education of women. There is a strong negative association between level of education and TFR, and the effect can be very large: In India and China, women with some college education have on average only half as many children as women with no education. Hence, a country's policies about education, and their success, may directly affect future population levels.

As we have revised this book for successive editions, we have seen population projections change. The estimates have decreased because fertilities have declined. The key questions are how to model such declines, whether they will continue, and how they will adapt to other world changes. In Spotlight 21.1 on page 777, we saw Thomas Robert Malthus's oversimplified prediction that mere arithmetic growth in food supplies would limit the geometric increase of human populations. Some demographers now think that population growth will remain a serious problem in some parts of the world, but that global population may stabilize or even decline after 2050.

How many people the world will have depends on how well we as a world conserve the environment, distribute food, provide jobs, produce and consume energy, and make other critical decisions about our money and resources. The key concern is the quality of life of all people. Political and economic events in far corners of the world, and even natural disasters such as the tsunami at the end of 2004 and the earthquakes in Haiti in 2010 and in Japan in 2011, affect us all. Neglecting problems faced by increasing numbers of poor people provides no security, peace, or moral refuge for anyone.

such as food, shelter, and psychological and social "space." There may be a maximum population size that can be supported by the available resources, the **carrying capacity** of the environment.

Carrying Capacity DEFINITION

The **carrying capacity** of an environment is the maximum population size that it can support with the available resources.

As the population increases toward the carrying capacity—which we will denote by M—the growth rate decreases. For a population at carrying capacity ($P = M$), the growth rate is zero.

If the population ever exceeds the carrying capacity, the growth rate becomes negative (because $P > M$) and the population decreases. The carrying capacity is the long-range capacity to support the population, so the population could exceed it for short periods of time. This could happen either because the population grows very rapidly and surges above the carrying capacity or because of a sudden decrease in the food supply, thus temporarily lowering the carrying capacity, as happens to deer and other animals in winter. The **logistic model** is a simple model, but it provides excellent predictions for some populations.

Logistic Model DEFINITION

The **logistic model** for population growth takes carrying capacity into account by reducing the natural increase rP by a factor $(1 - P/M)$ of how close the population size P is to the carrying capacity M:

$$\text{growth rate } P' = rP\left(1 - \frac{P}{M}\right)$$

EXAMPLE 2
Logistic Model for the U.S. Population

How well does the historical U.S. population fit a logistic model?

SOLUTION The U.S. population from 1790 to 1950 closely followed a logistic model with $r = 0.031$, P = population in 1790 = 3,900,000, and M = 201 million. In the first decades after 1790, the population was a small fraction of this carrying capacity, and it grew at close to the rate r of 3.1% per year (a rate higher than in many developing nations today). By 1920, the U.S. population had reached 106 million, and the growth rate had slowed by about one-half, to about 1.5% per year (see Figure 23.2).

The 2011 U.S. population of 310.5 million far exceeds the carrying capacity of 201 million that the model suggested. The structure of the U.S. population changed, from a large proportion of people making their living on family farms to a highly urbanized society. The average number of children per family shrank. As the structure changed, the model based on the prior structure gradually became invalid. A logistic model to the data through 2010 suggests a carrying capacity in excess of 400 million, which leaves room for growth toward the 420 million projected by the Census Bureau for 2050.

Figure 23.2 U.S. population by census year, showing actual growth, exponential (geometric) growth, and logistic growth.

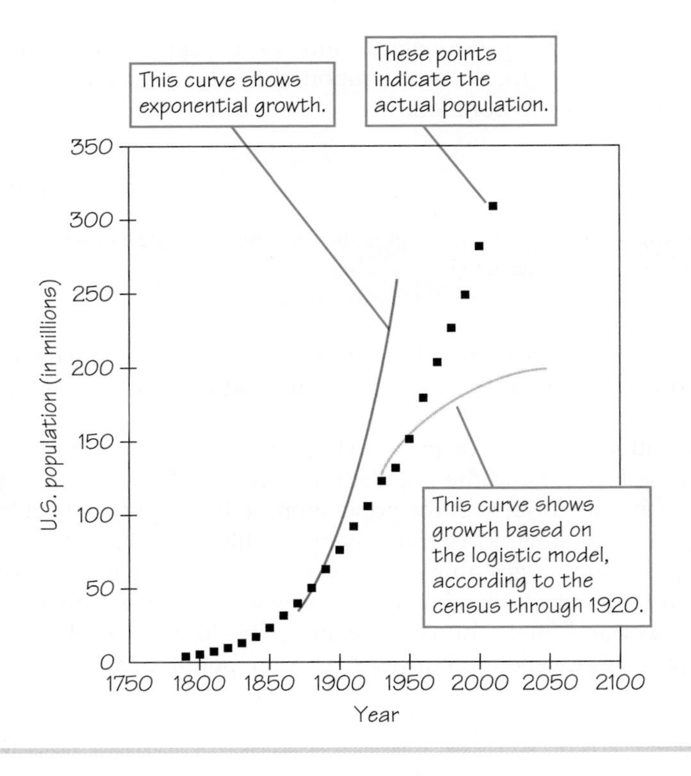

What about the world population? Figure 23.3 shows the historical trend—which does not show any sign of tapering off yet toward a carrying capacity—and three projections for the future. By the time you may retire, say in the 2050s, you will know which came true.

Figure 23.3 World population from 1820, with high, medium, and low projections from the United Nations (UN). *(Courtesy Loren Cobb, GFDL and Creative Commons license)*

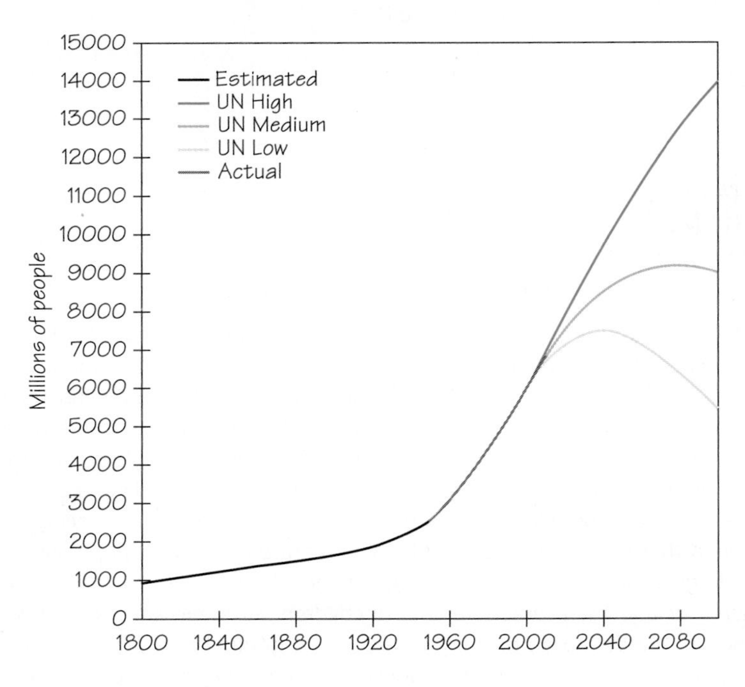

The logistic model applies not only to population growth limited by carrying capacity but also to modeling the spread of a technology or product, such as smart phones or flat-screen plasma TVs. Initially, sales are slow, and then they begin to climb rapidly.

Finally, as the market gets saturated, new sales slow. We will see in the next section that the logistic model also can apply to exhaustion of nonrenewable resources.

23.2 How Long Can a Nonrenewable Resource Last?

People use resources, some of which are renewable, but others are not. In this section, we model depletion of **nonrenewable resources**. In the next, we treat renewable resources.

Nonrenewable Resource DEFINITION

A **nonrenewable resource** is one that does not tend to replenish itself.

Radius Images/Corbis

Gasoline, coal, and natural gas are important examples, while lottery winnings and inheritances could be examples from personal affairs. There is no practical way to recover or reconstitute these resources after use. Some substances, such as aluminum or the sand used to make glass, are potentially recyclable, but to the extent that we do not recycle them, they, too, are nonrenewable.

For a nonrenewable resource, only a fixed supply S is available. Even without human population increases, we face dwindling nonrenewable resources. We are interested in the question: How long will the supply of a resource last?

So long as the rate of the resource use remains constant, the answer is easy. If we are using U units per year and continue using U units per year, then the supply will last S/U years. This kind of calculation is the basis for statements such as those claiming that at the current rate of consumption, U.S. recoverable coal reserves will last 250 years or that the U.S. strategic reserve of gasoline (stored in underground salt domes in the South) will last 60 days.

However, the rate of resource use tends to increase with population and with a higher standard of living. For example, projections for use of electric power often assume that use will increase by a fixed percentage each year. This is the simplest situation (apart from constant usage) and one that we can easily model.

Suppose that $U_1 = U$ is the rate of use of the resource in the first year (this year), and that usage increases $r = 0.05 = 5\%$ each year. Then the usage in the second year is

$$U_2 = U_1 + 0.05U_1 = 1.05U$$

and usage in the third year is

$$U_3 = U_2 + 0.05U_2 = 1.05U_2 = 1.05(1.05U) = (1.05)^2U$$

Generalizing, we see that usage in year i will be $(1.05)^{i-1}U$. Total usage over the next five years, for example, will be

$$U + (1.05)^1U + (1.05)^2U + (1.05)^3U + (1.05)^4U$$

This situation should remind you of the accumulation of regular deposits plus interest (see Chapter 21). Here the usage U corresponds to a deposit, and the increasing rate of use r corresponds to the annual interest rate. We may think of the situation as making regular withdrawals (with interest) from a fixed supply of the nonrenewable resource. The savings formula gives

$$A = d\left[\frac{(1 + r)^n - 1}{r}\right]$$

In the resource situation, A is the accumulated amount of the resource that has been used up at the end of n years, and U is the initial rate of use. We have

$$A = U\left[\frac{(1 + r)^n - 1}{r}\right]$$

To find out how long the supply S will last, we set the supply S equal to the cumulative use A over n years and then determine what n has to be. We have

$$S = U\left[\frac{(1 + r)^n - 1}{r}\right]$$

We perform some algebra to isolate the term involving n, and we get

$$(1 + r)^n = 1 + \frac{S}{U}r$$

At this point, to isolate n, we have to take the natural logarithm of both sides:

$$\ln[(1 + r)^n] = n \ln(1 + r) = \ln\left(1 + \frac{S}{U}r\right)$$

Doing this gives the final expression

$$n = \frac{\ln[1 + (S/U)r]}{\ln(1 + r)}$$

which may look complicated but is quite easy to evaluate on a calculator. The expression S/U is called the **static reserve**, and n is called the **exponential reserve**.

Static Reserve DEFINITION

The **static reserve** is how long the supply S will last at a particular constant annual rate of use U, namely, S/U years.

Exponential Reserve DEFINITION

The **exponential reserve** is how long the supply S will last at an initial rate of use U that is increasing by a proportion r each year, namely

$$n = \frac{\ln[1 + (S/U)r]}{\ln(1 + r)} \text{ years.}$$

Algebra Review: Using Natural Logarithms

In the equation $A = P(1 + i)^n$, when the variables P, i, and n are stated, A can be found by direct calculation. Solving for P is also fairly straightforward. We find that

$$P = \frac{A}{(1 + i)^n} \text{ or } P = A(1 + i)^{-n}$$

However, if we want to solve our equation, $A = P(1 + i)^n$, for the exponent n, we would use the natural logarithm (which is the logarithm with a base of e).

We start by dividing both sides by P to obtain $\frac{A}{P} = (1 + i)^n$. Next, we take the natural logarithm (written as *ln*) of both sides. In doing so, we will use the following property of nature logarithms:

$$\ln a^n = n \ln a$$

Taking the natural logarithm of both sides of $\frac{A}{P} = (1 + i)^n$ gives the following:

$$\ln\left(\frac{A}{P}\right) = \ln(1 + i)^n \text{ or } \ln\left(\frac{A}{P}\right) = n \ln(1 + i)$$

By dividing both sides by $\ln(1 + i)$, we have a final answer of $n = \frac{\ln\left(\frac{A}{P}\right)}{\ln(1 + i)}$.

Let's solve $2^x = 10$. Because $2^3 = 8$ and $2^4 = 16$, we can estimate that the value of x is between 3 and 4. Although we know the value of x is closer to 3, we can't state its exact value without logarithms. By taking the natural logarithm of both sides to solve, we have the following:

$$\begin{aligned} 2^x &= 10 \\ \ln 2^x &= \ln 10 \\ x \ln 2 &= \ln 10 \\ x &= \frac{\ln 10}{\ln 2} \end{aligned}$$

Using a calculator and rounding to two decimal places, we find that the approximate power is 3.32.

EXAMPLE 3
U.S. Coal Reserves

Coal accounts for 30% of U.S. energy use, including 50% of electricity. Recoverable reserves of U.S. coal would last about 250 years at the current rate of use, so the static reserve is 250 years. How long would the supply last if the rate of use increases 1% per year, about the rate of growth of the U.S. population?

SOLUTION The corresponding exponential reserve is

$$n = \frac{\ln[1 + (250)(0.01)]}{\ln 1.01} = \frac{\ln 3.5}{\ln 1.01} \approx 126 \text{ years}$$

That's quite a difference!

We must not take such projections as exact predictions. Estimates of supplies of a resource may underestimate how much is available, and previously unknown sources may be discovered or the technology may be improved to extract previously unavailable supplies. In addition, as supplies dwindle, the economic considerations of supply, demand, and price come into play. We will never completely run out of any resource. It will always be available "at a price."

We must not take such projections lightly, either, because we are discussing resources that, once used, are gone forever. In any projection, it is very important to examine the assumptions because small differences in the rate of increase of use can make big differences in the exponential reserve.

EXAMPLE 4
Using Up Retirement Savings

Suppose that you begin retirement with $1 million in savings, and you don't trust banks or the stock market, so you keep it all under your mattress. Suppose that it costs you $50,000 per year to live at your accustomed standard of living, and there is no inflation. How long will your retirement nest egg last? How long will it last if inflation is constant at 2.8%, the average annual inflation from 2000 to 2008?

SOLUTION The static reserve is $1,000,000/$50,000 per year = 20 years. With inflation, it will cost you increasingly more per year to live, so you should realize that your savings will last only for the length of the exponential reserve, which is

$$n = \frac{\ln(1 + 20(0.028))}{\ln(1 + 0.028)} \approx 16 \text{ years}$$

You have a fine strategy if you expect to live just 16 more years and want to die broke!

In our examples so far, we have assumed that the resource is just sitting there, waiting to be used up. For many natural resources, however, we have to find and develop new sources. As doing that becomes more difficult and more costly, the exponentially increasing demand outstrips the ability to meet that demand at some point.

Such a situation is modeled well by the logistic model famously applied to oil by M. King Hubbert, director of Shell Oil Company's research laboratory.

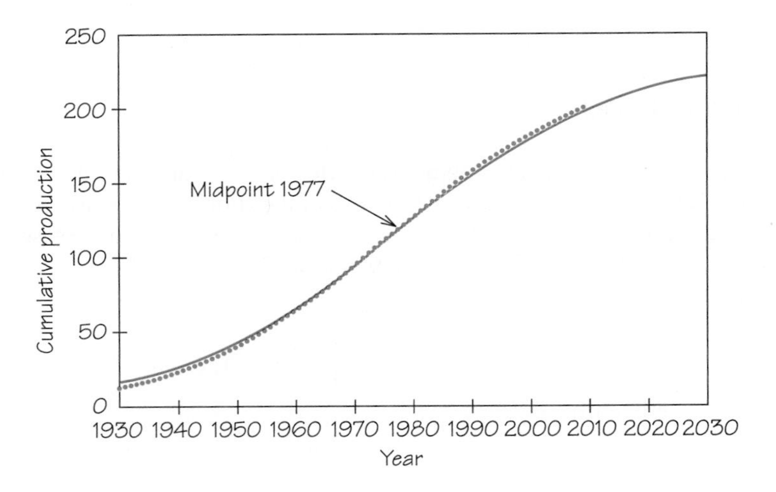

Figure 23.4 Logistic model (solid curve), assuming ultimate production of 240 billion barrels, and actual data (points) for cumulative U.S. crude oil production, in billions of barrels, through 2009.

Source: Adapted from Seppo A. Korpela, Oil depletion in the United States and the world, www.mecheng.osu.edu/files/u57/opmatalk.pdf, with revised and further data from the Energy Information Administration through 2009.

Figure 23.4 shows data for cumulative U.S. oil production through 2001 compared with the logistic curve for the ultimate production of M = 240 gigabarrels (240 billion barrels). In 1956, Hubbert predicted that U.S. production would peak in the early 1970s (it did) and decline steadily thereafter (it did that, too, except for a blip from Alaska in the 1980s). (See Figure 23.5, whose curve is similar to but "heavier in the tails" than the normal distribution curve of Chapter 5.)

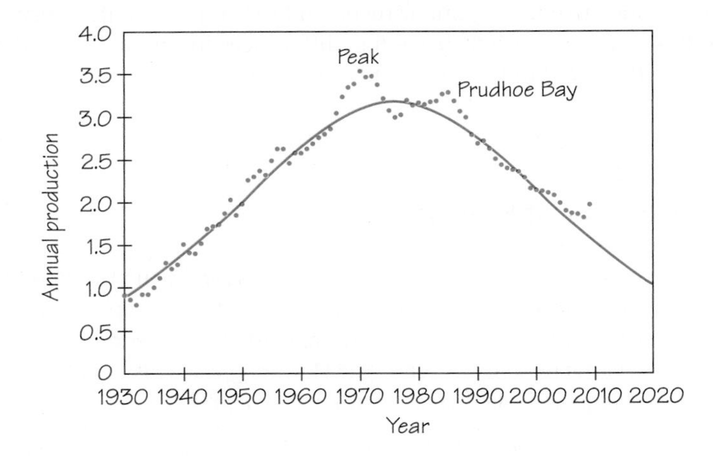

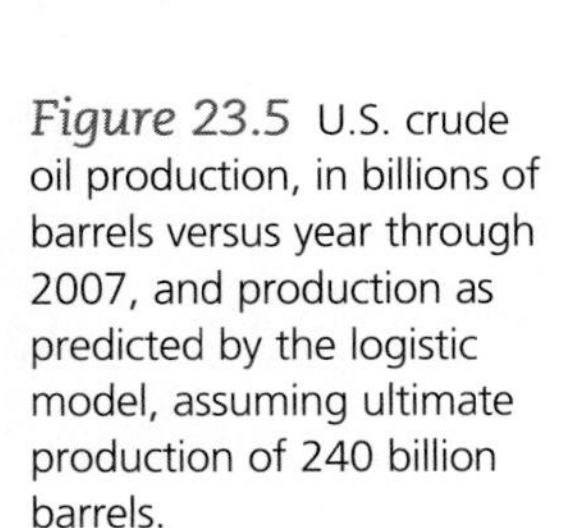

Figure 23.5 U.S. crude oil production, in billions of barrels versus year through 2007, and production as predicted by the logistic model, assuming ultimate production of 240 billion barrels.

Source: Adapted from Seppo A. Korpela, Oil depletion in the United States and the world, www.mecheng.osu.edu/files/u57/opmatalk.pdf, with revised and further data from the Energy Information Administration through 2009.

If we rearrange the logistic equation on page 833 by dividing both sides by P and doing a little algebra, we get

$$\frac{P'}{P} = r - \frac{r}{M}P$$

You can recognize this as the equation of a straight line $y = a - bx$, where P'/P takes the role of y and P takes the role of x. In other words, for a logistic model, if we graph P'/P against P, we get a straight line. Figure 23.6 shows that the data fit the Hubbert model well.

The world's oil and gas are running out far faster than most people realize or than their governments are willing to acknowledge (see Exercises 11 and 12 on p. 862). The need for "affordable" fuels will likely soon dominate the political agendas of the entire world, particularly because experts have long predicted that world oil output would peak around 2006. In fact, production leveled off in 2005–2009 at about 85 million barrels per day, or 31 billion barrels per year.

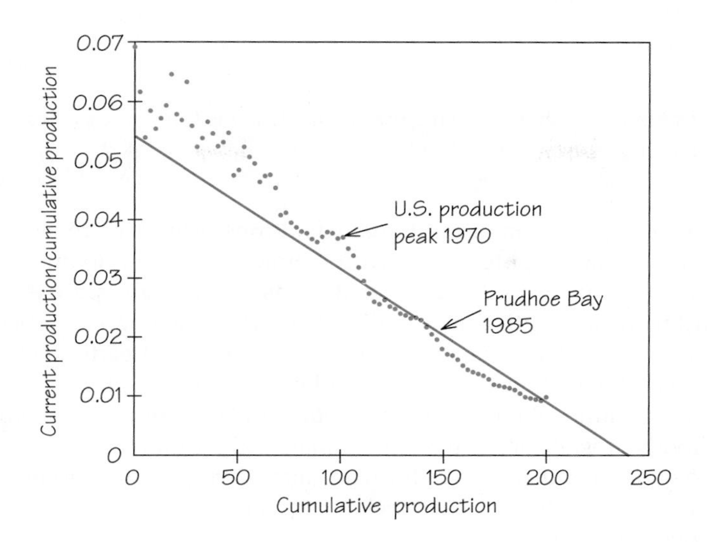

Figure 23.6 Display of the fit of data to the logistic model (solid line) for U.S. oil production. The line follows the equation $y = 0.054 - 0.000225x$.

Source: Adapted from Seppo A. Korpela, Oil depletion in the United States and the world, www.mecheng.osu.edu/files/u57/opmatalk.pdf, with revised and further data from the Energy Information Administration through 2009.

23.3 Radioactive Decay

Some populations grow (Section 23.1), others are eaten away through use (Section 23.2), and some just decay on their own. In fact, there are some "resources" that we *want* to become exhausted, such as nuclear waste.

"Decaying" resources can have enormous economic consequences, whether they involve

- unintended release of radioactive materials,
- safe disposal of nuclear waste,
- building and securing nuclear power plants, or even
- assuring reliable supplies of radioactive materials for medical purposes.

Radioactive materials are characterized by exponential decay, which we encountered in Section 21.5 (on p. 785) as geometric growth with a negative rate of growth, in connection with depreciation and inflation. Here, we see it in a different light, with a different way of measuring it.

Unlike the purchasing power of the dollar, which declines at an irregular rate as inflation varies, a radioactive substance emits particles and decreases in quantity at a predictable continuous rate. The negative growth rate r is usually written instead as $-\lambda$, where the (positive) quantity λ is called the *decay constant*.

Decay Constant DEFINITION

The **decay constant** for a substance decaying exponentially is the proportion of the substance that decays per unit time.

The amount of radioactive substance remaining is given by the continuous interest formula of Section 21.3 (p. 779): For an original amount P, the amount A remaining after t time units is

$$A = Pe^{-\lambda t}$$

However, the rate of decay of the substance is often described instead in terms of the *half-life*. A substance decaying geometrically never completely vanishes, even after millions of years. Because there is no time until it is all gone, we settle for measuring how long it takes until half of it is gone.

Half-life DEFINITION

The **half-life** of a substance decaying exponentially is the time that it takes for one-half of a quantity to decay.

For instance, iodine can occur in many **isotopes**—different versions of iodine with different atomic weights. Iodine with atomic weight 131, iodine-131, is produced in nuclear reactors and nuclear explosions; it is a **radioisotope** (also called **radionuclide**), meaning that it is radioactive. There was great concern about release of iodine-131 across Japan and the Pacific after the earthquake, tsunami, and subsequent nuclear reactor disasters in Japan in early 2011.

Iodine in the atmosphere falls to the ground and gets into water supplies, milk, and other foods. It is absorbed into the human body from food and concentrated in the thyroid gland; radioactive iodine can cause thyroid cancer. Iodine-131 from the earlier reactor disaster at Chernobyl in Ukraine in 1986 has caused thousands of cancers over the years.

Iodine-131 was also released from 1945 to 1963 during atmospheric testing of nuclear weapons by the United States, the Soviet Union, and Great Britain; by France until 1974; and by China until 1980. (Taking potassium iodide pills when exposure to radioactive iodine is anticipated floods the body with "good" iodine-127 and inhibits the uptake of iodine-131; potassium iodide is also used in "iodized" table salt to prevent iodine deficiency in people who eat little seafood.)

The half-life of iodine-131 is 8 days. This means that of 1 gram (g) of iodine-31 now, only 0.5 g will remain after 8 days (the rest will decay into nonradioactive xenon gas). Furthermore, in $2 \times 8 = 16$ days, only 0.25 g will remain; and in $3 \times 8 = 24$ days, only 0.125 g (one-eighth of the original amount) will remain.

EXAMPLE 5
Radioactivity in Your Home

The major radioactivity exposure for most Americans is from radon, which is formed from decay of uranium. Radon-222 is a gas that enters many homes in the United States from the underlying soil and rock. It causes 10–15% of lung cancers in the United States. (The rest are due to smoking.)

The half-life of radon-222 is 3.82 days. This means that of 1 g of radon-222 now, only 0.5 g will remain in 3.82 days; the rest will decay into other elements. Furthermore, in $2 \times 3.82 = 7.64$ days, only 0.25 g will remain; and in $4 \times 3.82 = 15.28$ days, only 0.0625 g (one-sixteenth of the original amount) will remain.

How long will it take for a quantity of radon-222 to decay to one-thousandth (0.1%) of the original amount?

SOLUTION Let's write H for the half-life. Then the amount remaining after time t (measured in the same time units as the half-life) is

$$A = P\left(\frac{1}{2}\right)^{t/H}$$

This and the earlier expression $A = Pe^{-\lambda t}$ are two equivalent ways of expressing the decay. To convert between decay constant and half-life, set the two expressions equal:

$$Pe^{-\lambda t} = P\left(\frac{1}{2}\right)^{t/H}, \quad \text{so} \quad e^{-\lambda t} = \left(\frac{1}{2}\right)^{t/H}$$

At this point, we take the natural logarithm of both sides, getting

$$\frac{t}{H}\ln\left(\frac{1}{2}\right) = -\lambda t \ln e = -\lambda t \times 1 = -\lambda t$$

Observe that $\ln\frac{1}{2} = -\ln 2 \approx -0.693$ (check this on your calculator) and divide by $(-t)$, getting

$$\frac{1}{H}\ln 2 = \lambda;$$

multiplying through by H gives the fundamental relationship:

Decay Constant and Half-Life Relationship RULE

$$\lambda H = \ln 2 \approx 0.693$$

For radon-222, we have $H = 3.82$ days and want to determine the t when $A = 0.001P$:

$$0.001P = P\left(\frac{1}{2}\right)^{t/3.82}$$

Dividing both sides by P gives

$$0.001 = \left(\frac{1}{2}\right)^{t/3.82}$$

and taking the natural logarithm of each side gives

$$\ln 0.001 = -6.91 = \frac{t}{3.82}\ln\frac{1}{2} = \frac{t}{3.82}(-0.693)$$

where the values $\ln 0.001 = -6.91$ and $\ln\frac{1}{2} = -0.693$ you get from your calculator. Solving for t, we find

$$t = \frac{6.91 \times 3.82}{0.693} \approx 38 \text{ days}$$

EXAMPLE 6
Carbon-14 Dating

Carbon-14 dating is a method of determining the age of organic materials, including mummies, charcoal from ancient fires, parchment, and cloth.

The element carbon, which is present in the food that we eat and in all living things, always has small traces of a radioactive form called *carbon-14*. Plants and animals continually absorb carbon-14 during their lives, from the air (for plants) and from food (for animals), so that the concentration in their bodies stays the same while they are alive. Once they die, however, no new carbon-14 gets absorbed, and the carbon-14 that is already present decays.

Because we know the concentration of carbon-14 in living things, and we know how long it takes carbon-14 to decay, we can calculate from a sample how long ago a plant or animal was living.

The half-life of carbon-14 is 5730 years; the value of its decay constant is $\lambda = \ln 2/H \approx 0.693/(5730 \text{ yr}) \approx 1.209 \times 10^{-4}/\text{yr} = 0.0001209/\text{yr}$. In other words, about 12 in 100,000 carbon-14 atoms decay each year. In each gram of carbon, approximately 814 carbon-14 atoms decay each hour.

An approximate age of a sample can be determined by working backward by half-lives. Suppose that a sample is decaying at 26 atoms per hour per gram of carbon. Table 23.1 shows that the 814 atoms per hour per gram of carbon would decrease to approximately 26 atoms per hour per gram of carbon in approximately 29,000 years; so that is the approximate age of the sample. (An age of 0 for the sample denotes the time of death of the living body.)

Table 23.1

Estimating the Age of a Carbon Sample

Age in Half-Lives	Age in Years	Decays per Hour per Gram of Carbon
0	0	$\left(\frac{1}{2}\right)^0 (814) = 814$
1	5,730	$\left(\frac{1}{2}\right)^1 (814) = 407$
2	11,460	$\left(\frac{1}{2}\right)^2 (814) = 203.5$
3	17,190	$\left(\frac{1}{2}\right)^3 (814) = 101.8$
4	22,920	$\left(\frac{1}{2}\right)^4 (814) = 50.9$
5	28,650	$\left(\frac{1}{2}\right)^5 (814) = 25.5$

How much of the original carbon-14 would be left after 50,000 years? (This is roughly the practical age limit for carbon-14 dating of the typically small samples available.) How many atoms would be decaying per hour per gram of carbon?

SOLUTION We have

$$A = Pe^{-\lambda t} = Pe^{-0.0001209 \times 50000} \approx Pe^{-6.05} \approx 0.0024P,$$

so only about 0.24% of the original amount remains. This remaining amount would be decaying at a rate of $0.0024 \times 814 \approx 2$ atoms per hour per gram of carbon. An accurate estimate for the age of a sample much less than a gram might take weeks or months of tallying counts on a Geiger counter.

Now we turn the question around to determine the age from the observed number of decays: How old is a sample that is decaying at a rate of 105 grams per hour per gram of carbon?

SOLUTION The formula that relates t, the age of the sample in years, and N, the number of carbon-14 atoms disintegrating per gram per hour, is

$$\left(\frac{1}{2}\right)^{t/5730} = \frac{N}{814}$$

Solving for N gives

$$N = 814 \times \left(\frac{1}{2}\right)^{t/5730}$$

Using natural logarithms, we can solve for t as

$$\ln N = \ln 814 + \frac{t}{5730} \ln \frac{1}{2}, \qquad t = 55{,}403 - 8{,}267 \ln N$$

Thus, a sample decaying at a rate of 105 atoms per hour per gram of carbon would be $t = 55{,}403 - 8{,}267 \times \ln 105 \approx 17{,}000$ years old.

Although we are interested in the exhaustion and decay of nuclear waste, we are also interested in maintaining supplies of some radioactive materials:

- Even before the Japanese disaster, there was concern about nuclear power—not just in terms of safety, but also whether enough uranium could be mined for a "nuclear future": a great many more reactors to meet the world's demand for electricity without adding to its output of carbon dioxide.
- Some radioactive isotopes are used for medical imaging and as tracers in the body, such as technetium-99m (the "m" is for "metastable"), which is used in 20 million diagnostic procedures each year. In 2010, two of the five

reactors that produce molybdenum-99 (which decays into technetium-99m) closed for repairs, reducing capacity by two-thirds. The short half-life of molybdenum-99 (66 hours) and the even shorter half-life of technetium-99m (6 hours) mean that weekly supplies of "fresh" molybdenum-99 are essential.

23.4 Sustaining Renewable Resources

A **renewable natural resource** is a resource that tends to replenish itself, such as fish, wildlife, and forests. How much can we harvest and still allow the resource to replenish itself?

Renewable Resource DEFINITION

A **renewable resource** is one that tends to replenish itself.

Other renewable resources are biological populations. We concentrate on the subpopulation with commercial value. For a forest, this might be trees of a commercially useful species and appropriate size. We measure the population size in terms of its **biomass**, the physical mass of the population. For example, we measure a fish population in pounds rather than in number of fish, and a forest not by counting the trees but by estimating the number of board feet of usable timber.

Reproduction Curves

Our models for growth include many simplifications. Complicated factors that can affect populations, such as climatic or economic change, may mean that the only way to understand a population is to plot a graph of its size over time. Either from data, a model, or both, we can construct a **reproduction curve** to predict next year's likely population size (biomass) based on this year's size. Although the precise shape of the curve varies from one population to another, the shape in Figure 23.7

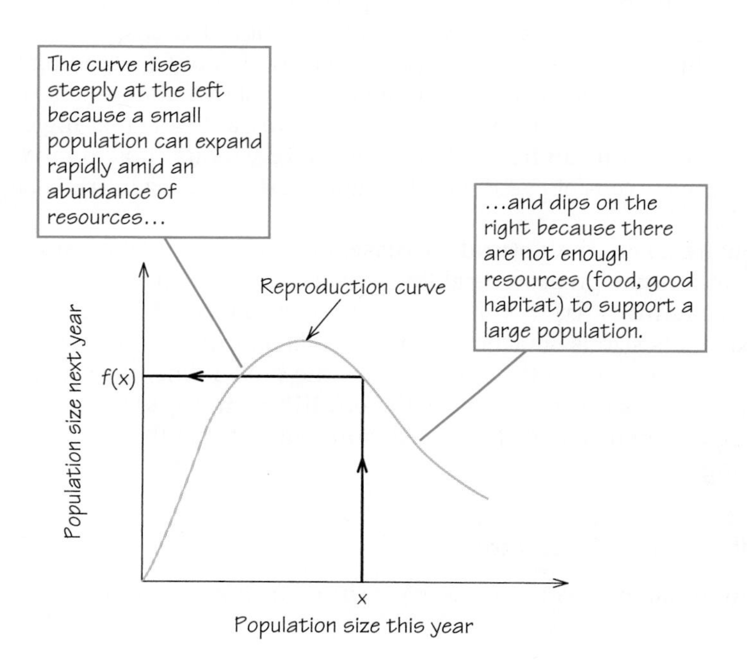

Figure 23.7 A typical reproduction curve.

Algebra Review: Function Notation

A function is a relationship between two sets. Those two sets are called the *domain* and *range*. The domain can be thought of as the set of allowable *input* values and is generally associated with x. The range can be thought of as the set of *output* values and is generally associated with y.

The relationship between the two sets in a function is that every value of x is assigned to only one value of y. The most common notation used when identifying a function is

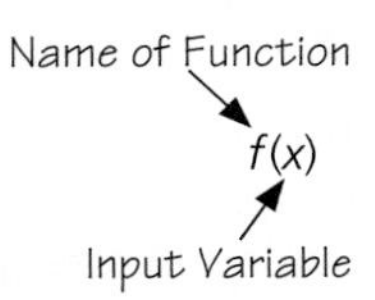

This notation is read as "f evaluated at x" or "f of x." Note: In function notation, multiplication is not implied by $f(x)$, as it would be in $(a)(b)$ or ab.

As an example, let's look at the squaring function, which takes a number and squares it. The squaring function can be written as $f(x) = x^2$. In this function, x would be the input and the square of x would be the output, so if 2 is in our domain, then 4 would be in our range.

When we graph a function, f, our points are of the form $(a, f(a))$. For the squaring function, the points are of the form (a, a^2).

Another important function is the identity function, $f(x) = x$. In the identity function, a number is assigned to itself. So, with an input of a, the output is also a. Some points on the graph of the identity function are (0, 0), (1, 1), and (2, 2).

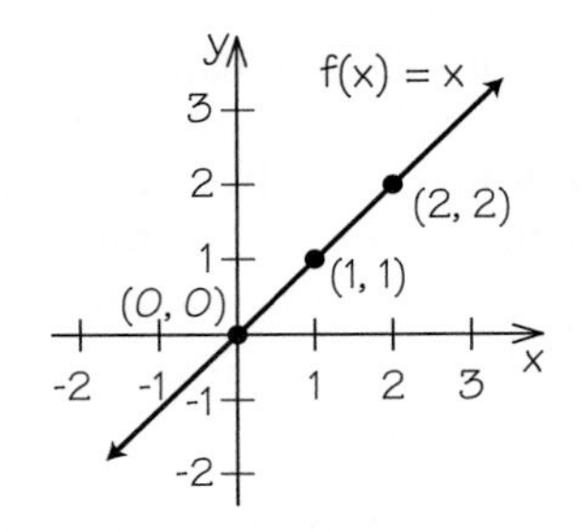

is typical. For all possible sizes, it shows the size this year (on the horizontal axis) and the size next year (on the vertical axis).

Let x on the horizontal axis be a typical size of the population in the current year. The size next year is given by the height of the curve above the point marked x. This value is denoted by $f(x)$. (You can think of f as standing for "function of," or even as "forthcoming.") Figure 23.8a shows the same reproduction curve, plus the broken line $y = x$ (which makes a 45-degree angle with the horizontal axis). You can trace what happens for various choices for x. For an x for which the curve is above the broken line, next year's size, $f(x)$, is larger than this year's, x.

In Figure 23.8b, the **natural increase**, or gain in population size, is shown as the length of the green vertical line from the broken line to the curve, which in algebraic terms is $f(x) - x$. For an x for which the curve is below the broken line, next year's size is smaller than this year's and $f(x) - x$ is negative. For the size labeled x_e, for which the curve crosses the broken line, the size is the same next year as it was this year. This is the **equilibrium population size**. (Zero is also an equilibrium size for the population, but not one that we are interested in attaining!)

Equilibrium Population Size — DEFINITION

An **equilibrium population** size does not change from year to year.

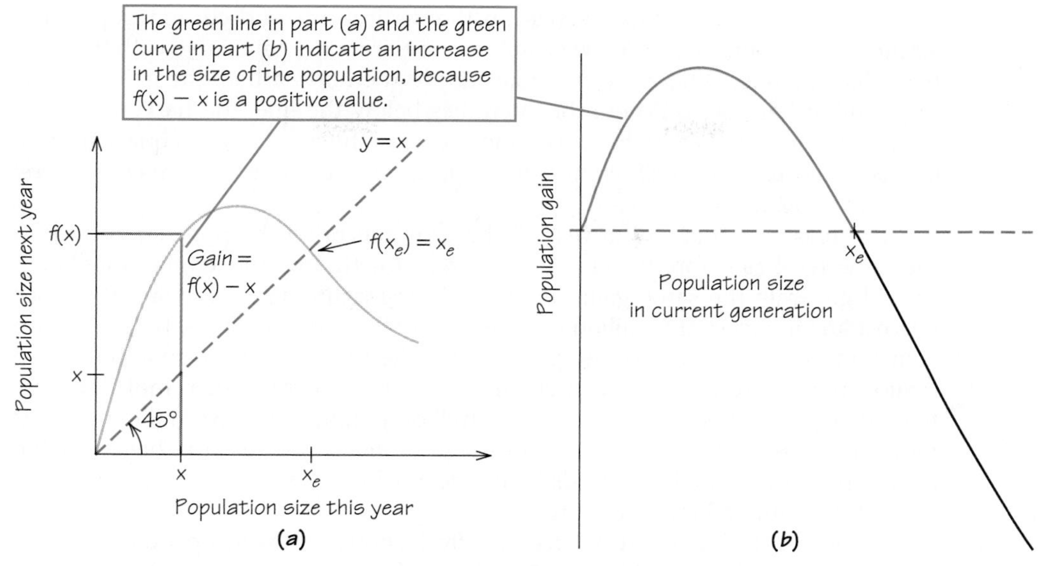

Figure 23.8 Depiction of the natural increase (gain) in population from one year to the next. The population size x_e is the equilibrium population size, for which the population one year later is the same, or $f(x_e) = x_e$.

Sustained-Yield Harvesting

In the case of fishing, for example, the **yield** (harvest) h depends on both the population x of fish and the amount of fishing effort (number of boats, hours of fishing).

If the fleet fishes the same banks with the same total effort in a year when there are only half as many fish as usual, it can expect to catch only half as many fish. A simple model is that for any particular level of fishing effort, the harvest is proportional to the fish population, so that a graph of harvest versus population would be a straight line with a steeper slope corresponding to greater effort.

For each level of fishing effort, there is a level of **sustainable yield**, one that could be sustained year after year because the fish population would recover to the same level after each harvest. Figure 23.9 shows the gain curve and two possible

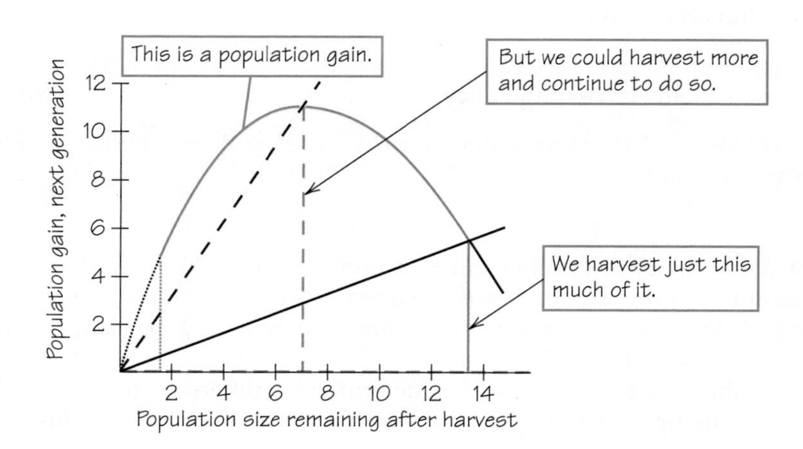

Figure 23.9 This population curve shows the effects of harvesting on a fish population. The scales for population and gain sizes are in millions of pounds.

harvest lines, with gain and harvest on the vertical axis and current population along the horizontal axis. We focus first on comparing the gain curve with the solid black harvest line. Where the curve lies above the harvest line, the gain exceeds the number harvested. Where the curve lies below the line, the harvest exceeds the annual growth. Where the curve and the line intersect, harvest equals growth; *next year the population will return to the same initial size and we can expect to harvest the same amount.*

The reason is that we harvest all the gain and bring the population back to the same level as before the current season's growth, from which the population should generate the same gain next year. In the figure, the curve and the black harvest line intersect at 5 million pounds, where the harvest equals the population gain (indicated by height of the green line). The post-harvest population is 13.5 million pounds (on the horizontal axis at the foot of the green line); the pre-harvest population was 13.5 + 5 = 18.5 million pounds. The 13.5 million pounds remaining after harvest will again grow by 5 million pounds (the height of the green line) to 18.5 million pounds by the next fishing season, when we can again harvest the gain of 5 million pounds.

However, at a higher level of fishing effort, shown by the steeper dashed black harvest line, the fleet could harvest 11 million pounds, leaving a post-harvest population of 7 million pounds (at the foot of the dashed green line). The pre-harvest population was 11 + 7 = 18 million pounds, about the same as in the previous scenario. The population gain curve indicates that the 7 million pounds left after the harvest can be expected to grow by 11 million pounds (the height of the dashed green line) by the start of the next fishing season—the same total pre-harvest level of 18 million pounds as the previous year. As before, the cycle can repeat, but this time with more than double the harvest each year.

Finally, we consider the dotted black harvest line at the far left. It corresponds to harvesting 5 million pounds (the height of the dotted green line) and leaving just 1.5 million pounds (at the foot of the dotted green line). The main difference between this and the first scenario is that here, *if the fleet takes 6.5 million pounds instead of 5 million, it wipes out the fish altogether.* Even if it takes only 5.5 million pounds, the fish grow back the next year by only 3 million, to a total of 4 million pre-harvest; and trying to harvest 5 million then extinguishes the fish population.

You can see that how much can be harvested "safely" and on a sustainable basis depends on knowing where we are on the population gain curve. But that curve is difficult to determine, as is estimating how many fish are out there. Moreover, the curve is an ideal that varies with changes in the weather, the environment, and other fish populations.

Sustained-Yield Harvesting Policy

DEFINITION

A **sustained-yield harvesting policy** is a policy that, if continued indefinitely, will maintain the same yield.

For a sustainable yield, the same amount is harvested every year and the population remaining after each year's harvest is the same. To achieve this stability, the harvest must exactly equal the natural increase each year, the length of the green vertical line in Figure 23.8a.

Each value of x between 0 and x_e determines a different green vertical line and corresponding sustained-yield harvest (Figure 23.10 a and b). This harvest

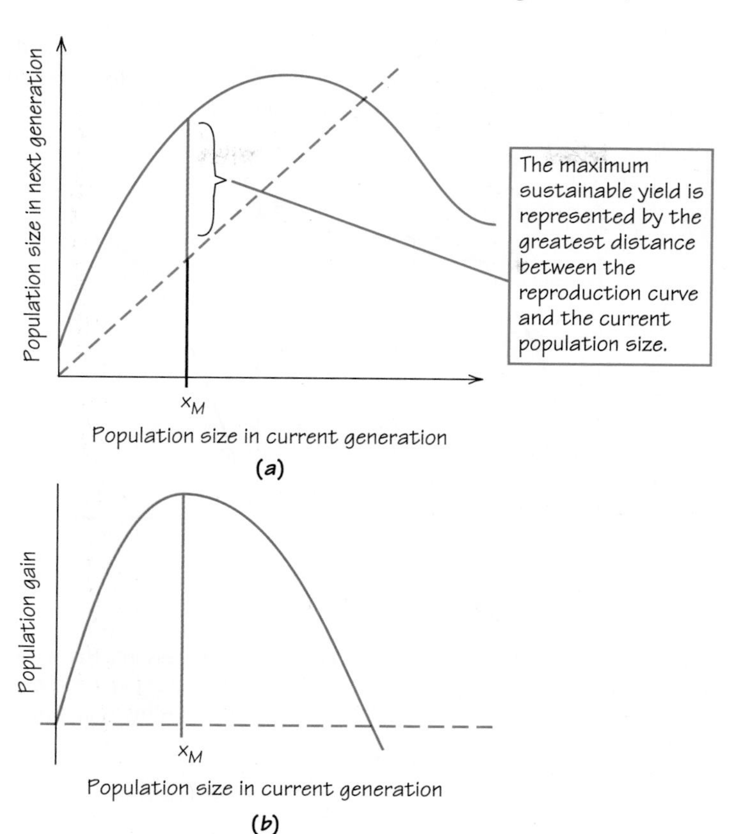

Figure 23.10 Two formulations of the reproduction curve, with the population size x_M corresponding to the maximum sustainable yield. (a) Population size in next generation. (b) Population gain.

can vary from 0 (for $x = 0$ or $x = x_e$) to some maximum value (for some x between 0 and x_e). A goal for a timber company or a fishery could be to harvest the **maximum sustainable yield**: to select an x whose colored vertical line is as long as possible, marked as x_M in Figure 23.10b. For example, in Figure 23.9, leaving $x_M = 7$ million pounds (horizontal axis) of fish in the sea after the harvest gives a maximum sustainable yield of 11 million pounds (height of the dashed green line). Such a yield level is difficult to determine, and we will see that economic and environmental factors—plus the inevitable involvement of mathematical chaos—also come into play.

EXAMPLE 7
Decline of a Fishery

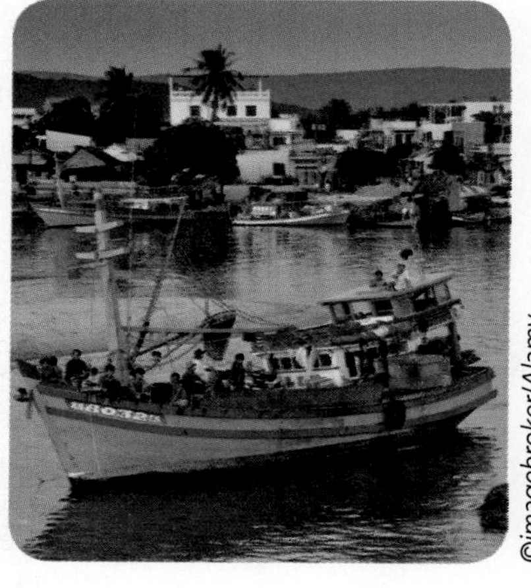
©imagebroker/Alamy

A successful fishing effort attracts more fishers and more boats. A lower price for fish means that fishers have to fish more hours to maintain their income. Whichever is the case, the long-run effect can be catastrophic. How so?

SOLUTION Figure 23.11 shows the sudden decline of a fishery with only a 20% increase beyond the original fishing effort. Extinction of the resource could result from additional fishing effort. The yield, shown as a vertical green line, is much greater for fishing effort geared to maximize sustainable yield (Figure 23.11a) than for effort 20% greater (Figure 23.11b). The fishing effort can be measured by the slope of the dashed red harvest line; the slope is steeper in Figure 23.11b. With the greater fishing effort, the yield declines from the maximum sustainable yield of 2.4 million pounds to less than 1 million pounds, and the population of fish pre-harvest shrinks from 9 million pounds to 7.5 million pounds.

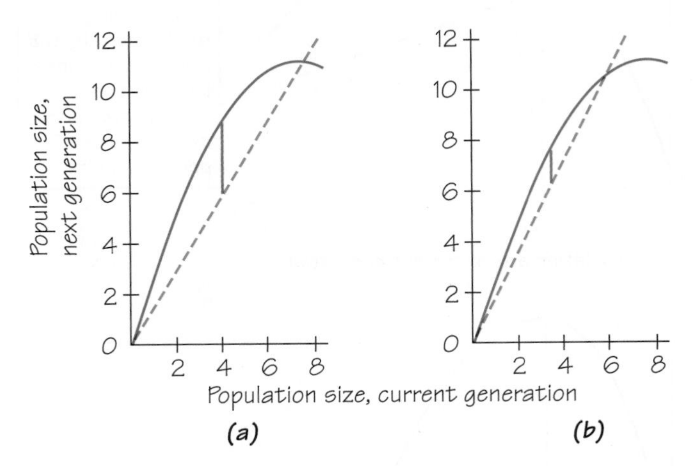

Figure 23.11 Deterioration of a fishery with increasing fishing pressure. The yield, shown as a vertical green line, is much greater in (a), which shows fishing effort geared to maximum sustainable yield, while (b) shows a 20% greater effort (as measured by the slope of the dashed red line).

Dynamics of a Population over Time

The line $y = x$ provides a convenient way to trace the evolution of the population over several years (see Figure 23.12a), by alternating steps vertically to the curve and horizontally to the line $y = x$. Begin with the first year's population on the horizontal axis and go up vertically to the curve. The height is the population in the second year. Proceed horizontally from the curve over to the line $y = x$. Proceeding vertically from there to the curve yields a height that is the population in the third year. The result is a **cobweb diagram**, so called because it resembles a spiderweb. A cobweb diagram gives a convenient visual representation of the evolution of the population over time.

Figure 23.12 shows several traces for the same reproduction curve, each starting from a different initial population on the horizontal axis. The resulting variation is surprising and can be "chaotic" in a very specific mathematical sense, showing how apparently random behavior can result from strict rules.

23.5 The Economics of Harvesting Resources

We consider two models: one for a cattle ranch and one that can apply to either a fishing boat or to a tree farm.

We assume that the price p received is the same for each harvested unit and does not depend on the size of our harvest. In effect, we assume that our operation is a small part of the total market, not substantially affecting overall supply and hence price.

We want to stay in business, so we do not extinguish the resource for quick profits. For any given population size, we harvest just the natural increase.

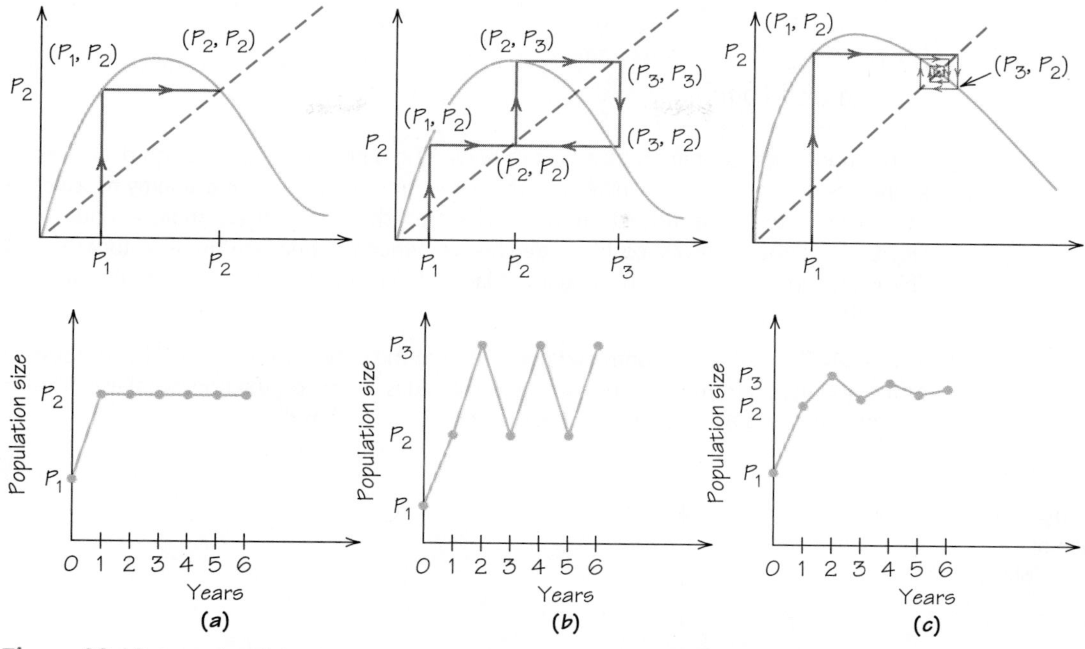

Figure 23.12 Examples of the dynamics, over time, for the same reproduction curve but different starting populations. (a) The population goes in one year to the equilibrium population and stays there year after year. (b) After initial adjustment, the population cycles between values over and under the equilibrium population. (c) The population spirals in toward the equilibrium population.

EXAMPLE 8
Cattle Ranching

If we assume that the cost of raising and bringing a steer to market is the same for every steer and does not depend on how many steers we bring to market, what should our sustainable-yield harvesting policy be?

SOLUTION Because the cost does not depend on the population size, the cost curve is a horizontal line (Figure 23.13).

So long as the selling price per unit is higher than the harvest cost per unit, we make a profit. The points of view of economics and biology agree because the maximum profit occurs for the maximum sustainable yield.

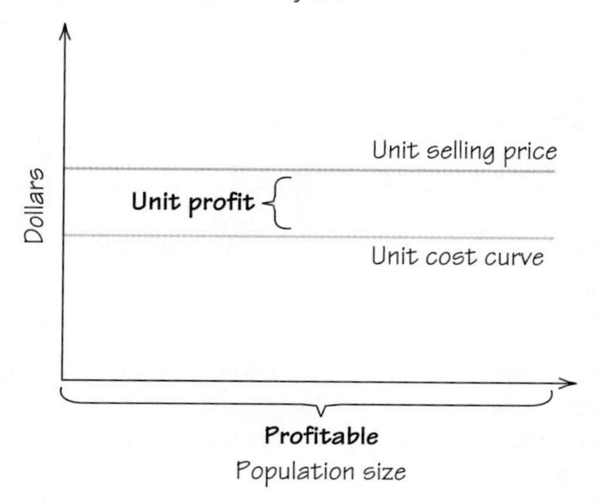

Figure 23.13 The unit cost, unit revenue, and unit profit of harvesting one unit, as a function of population size, for the cattle ranch.

EXAMPLE 9
Fishing and Logging

In this model, we assume that the cost of harvesting a unit of the population decreases as the size of the population increases. This is the familiar principle of **economy of scale**. For example, the same fishing effort yields more fish when fish are more abundant. Similarly, a logger's harvest costs per tree are less when the trees are clumped together. This is the logger's motivation to clear-cut large stands. What would a sustainable-yield harvesting policy be?

SOLUTION The cost curve slopes downward and to the right, as in Figure 23.14. The size of the population from which one unit is harvested is shown on the horizontal axis. The cost of harvesting a single unit is measured on the vertical axis.

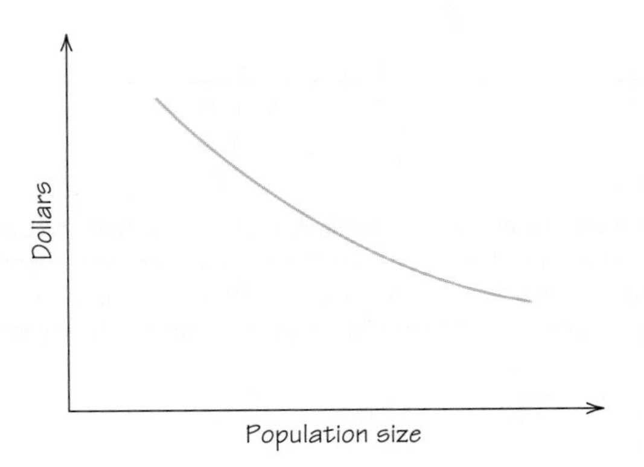

Figure 23.14 The unit cost, as a function of population size, for fishing or logging.

An optimal sustainable-yield harvesting policy depends on the relation between price and costs. There are two cases, as shown in Figure 23.15.

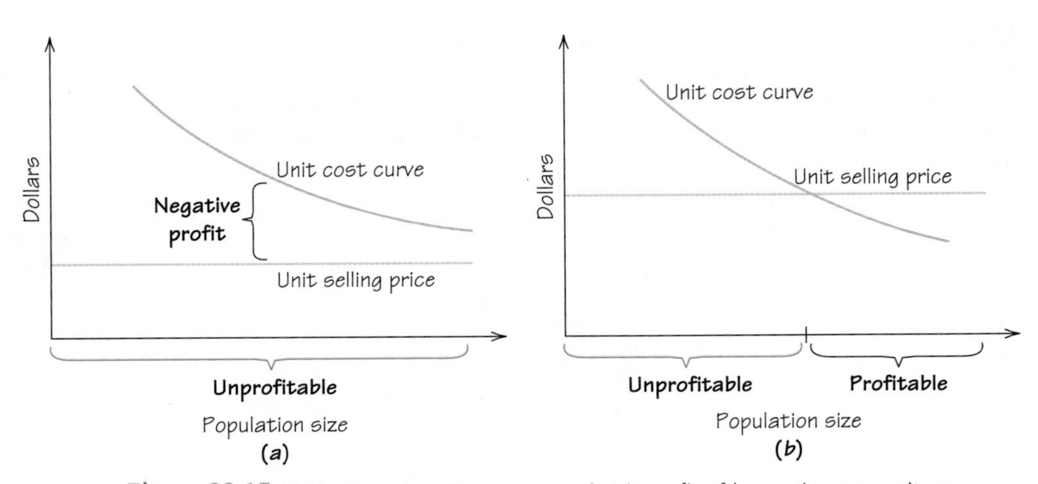

Figure 23.15 The unit cost, unit revenue, and unit profit of harvesting one unit, as a function of population size, for fishing or logging. (a) The market price is below the harvesting cost for all population sizes. (b) The operation is profitable for populations above a certain minimum size.

- *The unit cost curve lies entirely above the unit price line* (Figure 23.15a). The price for a harvested unit is less than the cost of harvesting it, no matter how large the population. It is impossible to make a profit.
- *The unit cost curve intersects the unit price line* (Figure 23.15b). Above a certain population size, the price for a harvested unit is more than the cost of harvesting it, so profit is possible. Some population size, call it x_Q, gives a maximum net profit.

Using calculus, it can be shown that x_Q is actually larger than x_M, the population that gives the maximum sustainable yield (see Figure 23.16). Compared to harvesting the maximum sustainable yield, economic considerations result in harvesting less, but also in maintaining a larger stock of the population.

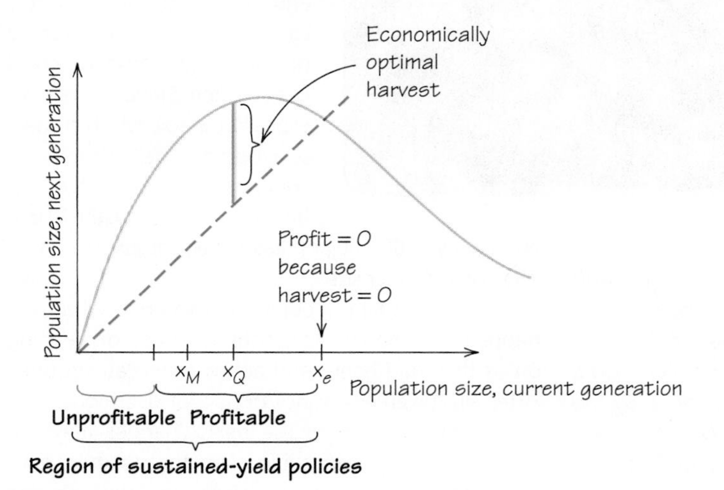

Figure 23.16 A reproduction curve showing regions of profitability for sustained-yield policy, with the economically optimal population size x_Q marked.

Our simple models fail to take into account a very critical feature of a modern economy: *the time value of money,* as measured by the interest that capital can earn. We now investigate how the time value of money makes biological populations susceptible to overexploitation and even extinction.

Why Eliminate a Renewable Resource?

Some species, such as the passenger pigeon, were harvested to extinction. In other cases, an entire ecosystem has been destroyed (see Spotlight 23.2). Why would anyone eliminate a renewable resource? Our approach helps explain why.

Sustained-yield policies involve revenues that will be received, year after year, in the future. The value of these revenues should be discounted to reflect the lost investment income that we could earn if instead we had the revenues today. For funds invested at a return of $100r\%$ per year, compounded annually, the present value P of an amount A to be received in n years in the future is related to A by the compound interest formula $A = P(1 + r)^n$.

The economic goal is to maximize the sum of the present values of all future receipts from harvesting. The optimal harvesting policy thus must depend on the expected rate of return r. We don't delve into the details of the calculations here but instead just give the results of the analysis.

SPOTLIGHT 23.2

The Tragedy of Easter Island

© Anders Haglund/Naturbild/Corbis

Easter Island is famous for its isolation—it is 1400 miles from the nearest island—and for its hundreds of huge stone statues. For 30,000 years before the arrival of people in about 400, Easter Island maintained a lush forest, with several species of land birds. By the time of the first visit by Europeans in 1722, the island was barren, denuded of all trees and bushes over 10 feet high, and with no native animals larger than an insect. The 2000 or so islanders had only three or four leaky canoes made of small pieces of wood.

What happened? Careful analysis of pollen in soil samples tells the sad story. The settlers and their descendants cut wood to plant gardens, build canoes, make sledges and rollers to move the huge statues, and burn for cooking and warmth in the winter. In addition to crops that they raised and chickens that they brought to the island and cultivated, they ate palm fruit, fish, shellfish, the meat and eggs of birds, and the meat of porpoises that they hunted from seagoing canoes. The population of the island grew to 7000 (or perhaps even 20,000).

By 1500, the forest was gone. Most tree species, all land birds, half the seabirds, and all large and medium-sized shellfish had been extinguished. There was no firewood, no wood for sledges and rollers to transport hundreds of statues at various stages of completion, and no wood for seaworthy canoes. Without canoes, fishing declined and porpoises could not be taken. Stripping the trees exposed the soil, which eroded, so crop yields fell. The people continued raising chickens, but warfare and cannibalism ensued. By 1700, the population had shrunk to 10% to 25% of its former size.

Why didn't the people notice earlier what was happening, imagine the consequences of keeping on as they had been, and act to avert catastrophe? After all, the trees did not disappear overnight.

From one year to the next, changes may not have been very noticeable. The forests may have been regarded as communal property, with no one charged with limiting exploitation or ensuring new growth. There was no quantitative assessment of the resources available and the need for conservation versus the long-term needs of the "public works" program of erecting statues. Moreover, the religion of the people, the prestige of the chiefs, and the livelihood of hundreds depended on the statue industry. There was no perceived need to limit the population and no technology for birth control.

Once the large trees were gone, there was no means for excess population to emigrate.

(Adapted from Jared Diamond, "Easter's end," *Discover*, 16(8) (August 1995): 63–69.)

Again, there are several cases to consider:

1. The unit cost of harvesting exceeds the unit price received, for all population sizes. Then it is impossible to make a profit.
2. For small r: for some population size x, the unit cost of harvesting equals the unit price received. Then there is a size between x and x_e (the equilibrium

population size) for which the present value of the total return is maximized and the population and its yield are sustained.

3. For larger r, *the economically optimal policy may be to harvest the entire population, immediately extinguish the resource, and invest the proceeds.* The unit price exceeds the unit cost for *all* population sizes.

Let's put this in the simplest and starkest terms. Suppose that you own a resource, such as a forest, whose cost of harvesting is small relative to its value. If the rate at which the forest population grows is greater than what you can earn on other investments, it pays to let the forest keep on growing.

On the other hand, if the forest grows more slowly than the rate of return on other investments, the economically optimal harvesting policy is to cut down all the trees now and invest the money. You could then start raising cattle on the land—and right there, you have the scenario that is resulting in deforestation all over the world.

The sobering fact is that *very few economically significant renewable resources can sustain annual growth rates over 10%.* Many of them, like whales and most forests, have growth rates in the 4% to 5% range. These values—even a growth rate of 10%—are far below the return that many investors expect on their investment. For example, over the long run, the U.S. stock market has yielded an average 10% return, but venture capital firms expect to exceed a 25% profit.

The concept of maximum sustainable yield is an attractive ideal if the expectations of investors are low enough. However, there are still difficult problems:

- One problem is "the tragedy of the commons," discussed by ecologist Garrett Hardin. Several hundred years ago, English shepherds would graze their flocks together on common land. The grass of the commons could support only a fixed number of sheep. Each shepherd could reasonably think that adding just one or two more sheep to his flock would not overtax the commons. Yet if each did so, there could be disaster, with all the sheep starving. Many natural-products industries, such as fisheries, are a form of commons. Small overexploitation by each harvester can produce disastrous results for all. Global warming may be a tragedy of a worldwide commons.
- How, in the presence of human needs or greed, can we anticipate and prevent overexploitation and possible extinction of a resource? By and large, it has been politically impossible to force a harvesting industry to reduce current harvests to ensure stability in the future.
- In some industries, such as a fishery, growth of the population may be abundant one year but meager another. Moreover, varied harvesting pressure and selectivity can magnify fluctuations in the age structure and the abundance of the resource. So a steady yield cannot be guaranteed nor sustained without some risk of damaging the resource. For example, a few good years in a row may provoke increased investment in fishing capacity; then attempting to harvest at the same levels in succeeding normal or below-normal years results in overfishing. This exact scenario destroyed the California sardine fishery in the 1930s, the Peruvian anchovy fishery in 1972, and much of the North Atlantic fishery in the 1980s. The part of the Atlantic Ocean off northwest Africa was nearly picked clean in the past decade.

Were the fishing fleets and regulators mentioned above at fault for extinguishing these fisheries by overexploiting a dependable resource? Or were the extinctions due to chance variations of the fish stocks? In the next section, we examine a third possibility—that the fish stocks followed simple rules that nevertheless produced "chaotic" behavior of stocks; that is, wide variation from one year to the next. When we do not see the pattern, we interpret such behavior as randomness, much as the moves in a chess game may appear random and inexplicable to someone who does not know the rules of the game.

23.6 Dynamical Systems and Chaos

In this and the two previous chapters, we have considered systems that change over time: bank accounts, the amount due on a loan, and the size of a population. Other examples are a dripping faucet, a playground swing, a pinball play, the solar system, the business cycle, epidemics, the passage of a drug through the human body, and the weather. Some of these are very predictable (interest on a bank account), while others are notoriously unpredictable (the weather). Some involve no outside influences (the amount due on a loan, assuming that you don't get behind on payments!), while others are the result of many contributing factors (the business cycle).

In some systems (such as the population of a country), the state of the system may depend largely on its states at previous times (e.g., last year's population), while in other systems (such as an epidemic) chance may play a large role (e.g., in who and how many become infected).

We are interested in modeling systems, such as a fishery, as they operate without influence from outside or from chance. The applicable mathematical tool is a **dynamical system**.

Dynamical System DEFINITION

A **dynamical system** is a mathematical model for a system whose state evolves with time and whose future states depend deterministically on its present and past states.

To make this definition meaningful, we need to be explicit about what we mean by **deterministically**.

Deterministic DEFINITION

A system is **deterministic** if its changes through time depend only on natural and mathematical laws and are not substantially affected by what we consider to be chance or free will.

An example of a deterministic system is the path of a golf putt, which is governed by gravity, terrain, wind, and the force imparted by the golfer. A non-example is the outcome of a vigorous toss of a coin or a random number generator; although the result, like the golf putt, is determined by physical laws, we consider the result to be random. Another non-example is the outcome of an election, which involves choices by humans.

Mathematical Chaos

We think of **chaos** as referring to general confusion, unpredictability, and apparent randomness. Mathematicians and other scientists use the word to describe systems whose behavior over time is inherently unpredictable.

Chaos DEFINITION

A dynamical system exhibits **chaos** if it is:

1. *Orbitally dense*—any state is near one that eventually will recur;
2. *Transitive*—from any state, you can eventually get close to any other; and
3. *Sensitive*—a small change in the initial state can produce widely diverging results later.

EXAMPLE 10
Chaos in Manhattan

You may already know from experience that getting around Manhattan can be a chaotic experience, in the ordinary sense of the word. Manhattan's subway system is also chaotic in the mathematical sense:

1. *Orbitally dense:* Subway trains "orbit," periodically visiting subway stops, and everybody lives near a subway stop.
2. *Transitive:* You can get close to anywhere else in Manhattan by taking the subway.
3. *Sensitive:* If you get on the wrong train, you could wind up miles from where you wanted to go.

Because the system covers the island of Manhattan, (1) is actually a consequence of (2). Also, anyone who has gotten on the wrong subway train or bus realizes that (3) is an inevitable consequence of (1) and (2). So in fact, (1) and (3) both follow from (2)—a conclusion that is true not just of this Manhattan example but of a large class of dynamical systems.

The most noticeable property of a chaotic system is sensitivity—that a small change now can make a big difference later.

This feature is sometimes known as the **butterfly effect**, from the title of a 1979 talk by meteorologist E. N. Lorenz: "Predictability: Does the Flap of a Butterfly's Wings in Brazil Set Off a Tornado in Texas?" (The phrase probably traces to a 1953 science-fiction story by Ray Bradbury, "A Sound of Thunder," in which history is changed by a time-traveler who steps on and kills a prehistoric butterfly.)

We can get a feel for chaotic systems by playing with some **iterated function systems (IFS)**. The fancy name just means that we take an initial value, apply a function to it, then repeat over and over. This is exactly what we did earlier, geometrically, with reproduction curves for populations. (See Section 19.5, pages 759–759, for more about IFS and their connection to fractals.)

Iterated function system (IFS) DEFINITION

An **iterated function systems (IFS)** is a sequence of elements (numbers or geometric objects) in which each element is produced by applying a consistent function (rule) to the previous element.

EXAMPLE 11
Doubling on a "Stone Age" Calculator

Imagine that you have a calculator that keeps only the last two digits of a number.

It has a special key marked DBL that doubles the number in the display and keeps only the last two digits. For example, DBL applied to 52 gives 04 (not 104).

Let's start with two numbers that are as close together as can be on this calculator, such as 37 and 38. As we push the DBL key over and over again, will the result stay close?

SOLUTION

37, 74, 48, 96, 92, 84, 68, 36, 72, 44, 88, 76, 52, 04, . . .

38, 76, 52, 04, 08, 16, 32, 64, 28, 56, 12, 24, 48, 96, . . .

Already, by the fourth iteration, the two sequences are far apart.

The function used in this iterated function system is

$$f(x) = 2x \bmod 100$$

where the mod notation of modular arithmetic (introduced in Chapter 17, p. 617) means to take the remainder when $2x$ is divided by 100.

EXAMPLE 12
The Solar System

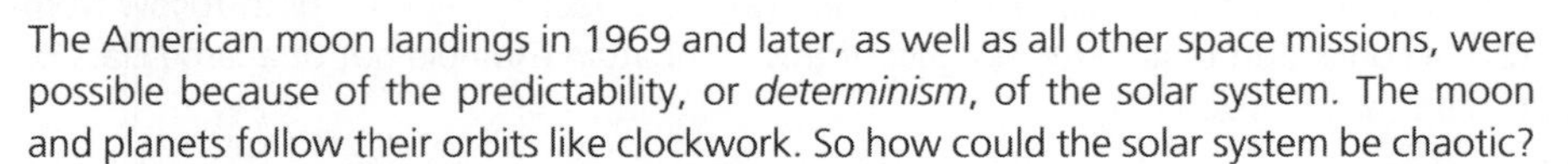

The American moon landings in 1969 and later, as well as all other space missions, were possible because of the predictability, or *determinism*, of the solar system. The moon and planets follow their orbits like clockwork. So how could the solar system be chaotic?

SOLUTION Over tens of millions of years, the orbit of each planet is chaotic, meaning that the slightest change in its position or velocity—due to, say, a comet passing nearby—could produce a huge difference later.

More down-to-earth examples of physical systems that can exhibit chaos include the fluttering of a falling autumn leaf, heart arrhythmias, and the Tilt-a-Whirl amusement park ride.

Chaos in Biological Populations

If we measure this year's population as a fraction x of the carrying capacity, and do the same for next year's population as a fraction $f(x)$, the "logistic map" model—which, despite the similar name, is not the same as the logistic model!—takes the form

$$f(x) = \lambda x(1 - x)$$

where $\lambda = 1 + r$ is the amount by which the population is multiplied each year. When expanded, the equation has the familiar form of a quadratic in x:

$$f(x) = -\lambda x^2 + \lambda x$$

EXAMPLE 13
The Logistic Map Population Model

What behaviors can occur in the logistic map model?

SOLUTION For different values of the parameter λ and different starting values for the population fraction, each of the behaviors of Figure 23.12 on page 849 can occur:

- $\lambda = 2.8$ and the starting population fraction $x = 0.357$ produces the values 0.357, 0.643, 0.643, . . ., the pattern shown in the lower graph in Figure 23.12a.
- $\lambda = 3.1$ and the starting population fraction $x = 0.235$ produces the values 0.235, 0.557, 0.765, 0.558, 0.765, 0.558, . . ., the pattern shown in the lower graph in Figure 23.12b.
- $\lambda = 2.5$ and the starting population fraction $x = 0.550$ produces the values 0.550, 0.619, 0.590, 0.605, 0.598, 0.601, 0.599, . . ., the pattern shown in the lower graph in Figure 23.12c.

In other words, for population growth rates (values of λ) that are fairly close together (2.5, 2.8, 3.1), the population evolves very differently. This is a surprising and non-intuitive conclusion.

But there is more. For $\lambda = 4$ and any starting population fraction, the population does not settle down into any of the patterns of Figure 23.12; year after year, it wanders "unpredictably" all over the place (Figure 23.17). This is *chaotic behavior*. It is deterministic, complex, and—in the long run—unpredictable. In the short run, the behavior is completely predictable. For example, from this year's population fraction, the equation tells us exactly what next year's will be. Repeating the use of the equation, we can determine what it will be the following year. But as the years pass, any sense of pattern gets lost in the complexity.

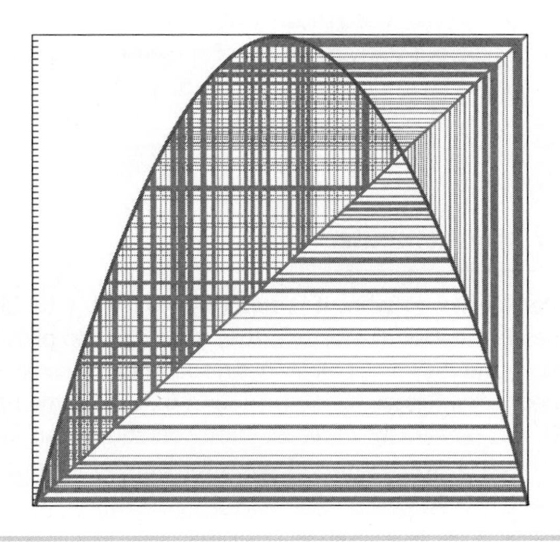

Figure 23.17 The chaotic behavior of a population. Start at any population value at a point on the diagonal 45-degree line, move vertically to the curve (moving up or down as needed), and move horizontally over to the 45-degree line (moving left or right as needed). Repeating this process can take the population through a wide variety of values, which signals chaos in the underlying process.

This potentially chaotic behavior of a biological population is bad news for those who manage a ranch or any biological population, whether in the wild or in captivity. In recent years, the lobster catch in Maine has been much higher than in previous years, reaching record levels, for no discernible reason.

On the other hand, in the late 1950s, the annual harvest of Dungeness crabs off the central California coast declined from 12 million pounds to less than 1 million pounds without any evidence of disease, heightened predation, or increased crabbing effort. Researchers who modeled that population in 1994 found that booms and busts are the rule. The population can remain nearly level for generations before suddenly exploding or crashing without warning.

Searching for an environmental cause for these fluctuations could be futile because there may not be one. Moreover, observing the population over a few generations provides no help in predicting future behavior.

EXAMPLE 14
Childhood Disease Epidemics

The incidence of childhood diseases such as chickenpox and measles varies greatly from year to year. Why?

SOLUTION There are three plausible explanations for the fluctuations:

- There is an underlying regular cycle that is perturbed and occasionally overwhelmed by random events ("noise").
- There is no pattern because the fluctuations are due solely to chance.
- There is no discernible pattern because such fluctuations are inherent in the epidemiology of the disease, a chaotic system.

The first explanation, a perturbed cycle, fits chickenpox, with a cycle of one year. For measles, either the second or the third explanation may be correct, depending on the size of the community. For small communities, chance is an adequate explanation. For large communities, historical data from before the era of mass immunization suggest that measles cases were chaotic. That doesn't mean that they occurred at random, but rather that they were unpredictable. Research also shows that there is a critical community size above which a disease will not die out solely by chance. For measles, this size is about 250,000.

EXAMPLE 15
Chaos in Your Home Computer

Does your computer always do what it is supposed to? The Intel i5 chip common in many computers has 731 million transistors. They are subject to physical laws and environmental conditions, and they don't all always respond in the same amount of time to do their commanded tasks. Researchers have found that the same program (of several billion instructions), run repeatedly under identical conditions, can take greatly varying amounts of time. However, the pattern of times for some programs displays clear evidence of deterministic chaos.

What you need to understand about chaos is that behavior that appears to be random can be produced even with very simple systems that are completely governed by deterministic rules. Just because the behavior appears chaotic does not mean a lack of underlying order and structure, though discovering that structure may be difficult.

Even if we discover the structure, prediction may elude us because of chaos. If we had an absolutely correct model of how weather behaves and measurements at every location on and above the Earth, we might still not be able to forecast the weather accurately a week ahead of time.

What about the fishery extinctions? Perhaps the fishers and the fish were victims not of greed or chance, but of the chaotic nature of the reproduction curve for the fish.

REVIEW VOCABULARY

Biomass A measure of a population in common units of equal value. (p. 843)

Butterfly effect A small change in initial conditions of a system can make an enormous difference later on. (p. 855)

Carrying capacity The maximum population size that can be supported by the available resources. (p. 833)

Chaos Complex but deterministic behavior that is unpredictable in the long run. (p. 854)

Cobweb diagram A kind of graphical portrayal of the evolution of a dynamical system, such as a population. (p. 848)

Compound interest formula The formula for the amount in an account that pays compound interest periodically. For an initial principal P and effective rate r per year, the amount after n years is $A = P(1 + r)^n$. (p. 829)

Decay constant For a substance decaying exponentially, the proportion that decays per unit time. (p. 839)

Deterministic A system is deterministic if its future behavior is completely determined by its present state, past history, and known laws. (p. 854)

Dynamical system A system whose state depends only on its states at previous times. (p. 854)

Economy of scale Costs per unit decrease with increasing volume. (p. 850)

Equilibrium population size A population size that does not change from year to year. (p. 844)

Exponential reserve The amount of time that a fixed amount of a resource will last at a constantly increasing rate of use. A supply S, at an initial rate of use U that is increasing by a proportion r each year, will last

$$\frac{\ln\left(1 + \frac{S}{U}r\right)}{\ln(1 + r)} \text{ years. (p. 836)}$$

Half-life For a substance decaying exponentially, the time that it takes for one-half of a quantity to decay. (p. 840)

Isotope A form of a chemical element whose atomic nucleus contains the same number of protons as other forms (the atomic number of the element) but a different number of neutrons (giving it a different atomic weight). (p. 840)

Iterated function system (IFS) A sequence of elements (numbers or geometric objects) in which each next element is produced from the previous one by applying a consistent function (rule) to the previous element. (p. 855)

Logistic model A particular population model that begins with near-geometric growth but then tapers off toward a limiting population (the carrying capacity). (p. 833)

Maximum sustainable yield The largest harvest that can be repeated indefinitely. (p. 847)

Natural increase The growth of a population that is not harvested. (p. 844)

Nonrenewable resource A resource that does not tend to replenish itself. (p. 835)

Population structure The division of a population into subgroups. (p. 831)

Rate of natural increase Birth rate minus death rate; the annual rate of population growth without taking into account net migration. (p. 830)

Renewable natural resource A resource that tends to replenish itself; examples are fish, forests, and wildlife. (p. 843)

Reproduction curve A curve that shows population size in the next year plotted against population size in the current year. (p. 843)

Savings formula The formula for the amount in an account to which a regular deposit is made (equal for each period) and interest is credited, both at the end of each period. For a regular deposit of d and an effective interest rate r per year, the amount A accumulated after n years is

$$A = d\left[\frac{(1 + r)^n - 1}{r}\right]. \text{ (p. 829)}$$

Static reserve The amount of time that a fixed amount of a resource will last at a constant rate of use; a supply S used at an annual rate U will last S/U years. (p. 836)

Sustainable yield A harvest that can be continued at the same level indefinitely. (p. 845)

Sustained-yield harvesting policy A harvesting policy that can be continued indefinitely while maintaining the same yield. (p. 846)

Yield The amount harvested at each harvest. (p. 845)

SKILLS CHECK

1. The carrying capacity of a population is

(a) the largest recorded population.
(b) the largest supportable population.
(c) the change in population.

2. The logistic population model can model not only human or animal populations, but also ________.

3. A population with a total fertility rate of 2.1 will

(a) double in each generation.
(b) remain exactly the same size from generation to generation.
(c) decrease.

4. China's population exhibits demographic ________.

5. The logistic curve is a model for a population that is growing

(a) linearly.
(b) exponentially.
(c) with a ceiling.

6. U.S. oil use has grown according to a(n) ________ model.

7. Management of a nonrenewable resource can be modeled by

(a) logistic growth.
(b) making withdrawals from a savings account.
(c) paying back a loan.

8. If we have enough reserves of a product to last 200 years at the current rate of use, but the rate of use increases by 10% per year, the supply will last ________ years.

9. The static reserve is always ________ the exponential reserve.

(a) greater than
(b) equal to
(c) less than

10. If we have enough reserves of a product to last 1000 years at the current rate of use, but the rate of use increases by 1% per year, the supply will last ________ years.

11. If a sample contains 20 g of carbon-14, whose half-life is 5730 years, in how many years will there be only 5 g of carbon-14 remaining in the sample?

(a) 4 years
(b) 11,460 years
(c) 22,920 years

12. Plutonium-239 (produced in breeder reactors and used in atomic bombs) has a half-life of approximately 24,200 years. Of 4,000 grams of plutonium-239 now (the approximate content of one atomic bomb), in ________ years there will be only 500 grams of it remaining.

13. The equilibrium population size is the same as the

(a) carrying capacity.
(b) intersection point of the reproduction curve and the diagonal.
(c) natural increase of the population.

14. The shape of the reproduction curve reflects that a small population has abundant resources and can grow quickly by the curve ________ steeply at the left.

15. A sustained-yield harvest

(a) is possible at exactly one size of a population.
(b) is not always possible.
(c) is possible at various sizes of a population.

16. A cobweb diagram shows the ________ of a population.

17. If the starting population for a reproduction curve is changed, the subsequent population pattern will

(a) eventually return to the same pattern.
(b) always change to a different pattern.
(c) sometimes change to a different pattern.

18. Harvesting at an economically optimal level maintains a ________ stock of the population than harvesting the maximum sustainable yield.

19. Economic considerations

(a) always work against the conservation of a resource.
(b) never interfere with conservation of a resource.
(c) always affect the conservation of a resource in some way.

20. The harvest yield that leads to the maximum net profit under sustainable-yield harvesting is always ________ than the maximum sustainable yield.

21. A system whose current state depends solely on its previous states is called

(a) a dynamical system.
(b) a stable system.
(c) an optimal system.

22. For the logistic model $f(x) = 4x(1 - x)$, a starting population fraction that will immediately lead to zero population growth is ________.

23. Chaotic behavior appears to be random and is

(a) actually not random.
(b) actually always random.
(c) sometimes random.

24. For the logistic model $f(x) = 3x(1 - x)$, if the starting population fraction is 0.5, the next population fraction is ________.

25. Pressing a digit and then repeatedly pressing the SIN key is a model of

(a) an iterated function system.
(b) chaos.
(c) randomness.

26. The "butterfly effect" refers to a feature of chaos called ________________.

27. Mathematical chaos occurs in

(a) heart arrhythmias.
(b) the solar system.
(c) amusement park rides.

28. Biological populations can be at risk of extinction from ________, ________, and/or ________.

29. Mathematical chaos

(a) happens only in mathematics.
(b) is just another word for random events.
(c) can happen in deterministic systems.

30. The three characteristics of mathematical chaos in a dynamical system are ________, ________, and ________.

CHAPTER 23 EXERCISES

■ Challenge ▲ Discussion

23.1 Growth Models for Biological Populations

1. (Spreadsheet helpful.) For many years, China has been the world's most populous country. However, India has been catching up, with 1180 million in mid-2010 and growing at 1.5% per year, versus China, with 1338 million and growing at only 0.5% per year. If these rates continue, when will India have more people than China?

2. The population of the less-developed countries (excluding China) in mid-2010 of 4.318 billion is expected to grow at 1.7% per year. (This is an annual yield, so you may think of it as compounded annually.) If this growth rate continues until mid-2025, what will be the size of the population then?

3. If the growth rate of the less-developed countries of Exercise 2 in mid-2020 were actually 1.6% rather than 1.7%, what would be the size of the population in mid-2025?

4. (Spreadsheet helpful.) If the growth rate of the less-developed countries of Exercise 2 decreased by $\frac{1}{25}$ of a percentage point (0.04%) per year from 2010 through 2015, beginning in mid-2010, what would be the size of the population in mid-2015?

5. An advertisement for Paul Kennedy's book *Preparing for the Twenty-First Century* (Random House, 1993) asked: "By 2025, Africa's population will be: 50%, 150% or 300% greater than Europe's?" The population of Europe in mid-2010 of 739 million is expected to increase 1% (not per year, just 1%) by 2025. The population of Africa in mid-2010 of 1030 million is expected to increase at about 2.4% per year. What answer would you give to that question?

6. Similar to the rule of 72 used in banking and explained in Exercises 13 and 14 in Chapter 22 (on p. 823), the rule of 69 says that if a country's population continues to grow at a constant rate of r% per year, then it will double in size every $69/r$ years. (As noted in Chapter 22, a rule of 69.3 would be even more accurate, but the difference between that and the rule of 69 is negligible.) Apply the rule of 69 to estimate the doubling times for the following populations (figures are for mid-2010):

(a) Africa, 1030 million, 2.4%
(b) United States, 309 million, 0.95%

7. Do the calculations as in Exercise 6, but for:

(a) China, 1.338 billion, 0.5%
(b) The world as a whole, 6.892 billion, 1.2%

8. Wisconsin's electricity demand increased 3.03% per year for the 35 years from 1970 to 2005. If that trend continues, when will Wisconsin need to have twice as much generating capacity as it did in 2005?

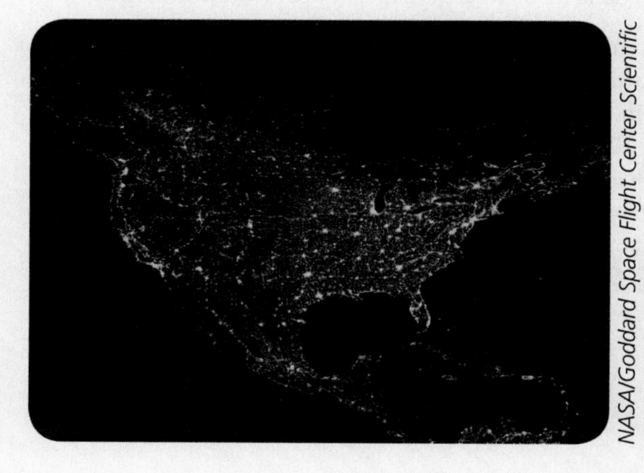

NASA/Goddard Space Flight Center Scientific Visualization Studio

For Exercises 9 and 10, refer to the following:

Is Warren Sanderson right about world population growth slowing down (see Spotlight 23.1 on p. 832)? How much difference does it make in projections if we look at the world as a whole or break it down by countries or regions? In Exercises 9 and 10, we investigate this question, first projecting as a whole and then projecting by regions and adding.

9. The population of the world in mid-2010 of 6.892 billion is expected to increase 1.2% per year.

(a) Project the population to mid-2025 (by then you may have finished having children, if you have any) and to mid-2050 (by then you may be thinking about retiring).
(b) What are the assumptions involved in your projections?

10. Divide the countries of the world into three groups with differing rates of increase (see the table). (Why is this useful?)

Group	Population Mid-2010 (billions)	Rate of Growth (%)
More-developed countries	1.237	0.2
Less-developed countries (excluding China)	4.318	1.7
China	1.338	0.5

(a) Redo the projections in Exercise 9a for the years 2025 and 2050 by projecting each group separately and adding the totals. Is there a major difference from the results in Exercise 9a?
(b) Will the world be able to support the numbers of people that you project? What problems will these greater numbers of people cause? What could be done to avert those problems? Do you think that anything will be done before there is some kind of worldwide crisis?

(In mid-1995, the world population was 5.7 billion and growing at 1.5% per year. Those figures led to projections as in Exercise 9 of 8.3 billion for 2020 and 11.1 billion for 2040. The corresponding growth rates for the groups of Exercise 10 were 0.2%, 2.2%, and 1.1%, which led to projections of 8.9 billion for 2020 and 12.7 billion for 2040.)

23.2 How Long Can a Nonrenewable Resource Last?

11. At the start of 2010, world oil reserves (including oil sands) totaled 1476 billion barrels, while daily consumption was 84.4 million barrels in 2009 (down from 2008, due to the global recession). The U.S. Energy Information Administration (EIA) projected in 2010 that world consumption would increase 1.0% per year through 2035.

(a) What was the static reserve for oil at the start of 2010?
(b) What was the exponential reserve for oil at the start of 2010?
(c) What considerations may affect the answers to parts (a) and (b) over time?

12. At the start of 2010, world natural gas proven reserves totaled 6609 trillion cubic feet, while annual consumption was 107 trillion cubic feet in 2009. The EIA projected in 2010 that world consumption would increase 1.0% per year through 2035.

(a) What was the static reserve for natural gas at the start of 2010?
(b) What was the exponential reserve for natural gas at the start of 2010?
(c) What considerations may affect the answers to parts (a) and (b) over time?

13. Can our energy problems be solved by increasing the supply? [Thanks for the idea to Evar D. Nering of Arizona State University, in "The mirage of a growing fuel supply," *New York Times* (June 4, 2001) Op-Ed page.]

(a) Suppose that we have a 100-year supply of a resource (such as oil, for which known world reserves will last less than 100 years at the current world rate of use). That is, the resource would last 100 years at the current rate of consumption. Suppose that the resource is consumed at a rate that increases 2.5% per year. (This is the average increase in consumption for oil in the United States since 1973.) How long will the resource last?
(b) Suppose that we underestimated the supply and actually have a 1000-year supply at the current rate of use. How long will that last if consumption increases 2.5% per year?
(c) Let's think big and suppose that there is 100 times as much of the resource as we thought—a 10,000-year supply. How long will that last if consumption increases 2.5% per year?

14. In this problem, we explore the consequences of reducing the rate of growth of oil use. Suppose that we halve the growth rate from the 2.5% per year given in Exercise 13 to 1.25% per year. [Thanks for the idea to Evar D. Nering of Arizona State University, in "The mirage of a growing fuel supply," *New York Times* (June 4, 2001) Op-Ed page.]

(a) How long will the 100-year supply last?
(b) How long will the 1000-year supply last?
(c) How long will the 10,000-year supply last?

15. We continue the ideas of Exercises 13 and 14, but with a more radical hypothesis.

(a) How long would the 100-year supply last if we reduced our consumption by just $\frac{1}{2}$% per year—that is, if we used $\frac{1}{2}$% less each year instead of 2% more?
(b) What if we used 1% less each year?

16. By the time there is concern about using up a nonrenewable resource, it may be too late. Suppose that a resource has a static reserve of 10,000 years, but consumption is growing at 3.5% per year.

(a) How long will the resource last?
(b) How long will it be before half the resource is gone?
(c) How much longer will the resource last if, after half of it is gone, consumption is stabilized at the then-current level?
(d) What implications do you see to your answers?

17. Do a calculation to criticize the claim in the following quotation: "The United States holds 437 billion tons of known (coal) reserves, enough energy to keep 100 million large electric generating plants going for the next 800 years or so." [*Forbes* (December 15, 1975), p. 28; thanks to Albert A. Bartlett.]

For Exercises 18–20, refer to the following:

The formula for the average growth rate over a period of time is even simpler than the one for the exponential reserve of a resource. If usage at the beginning is N_0 and it is N after an interval of t years, then the average annual rate of growth is

$$\frac{1}{t}\ln\left(\frac{N}{N_0}\right)$$

18. The U.S. population was 62.95 million in 1890 and 309 million in 2010 (at roughly the same times of the year). What was the annual rate of growth, to two decimal places?

19. The U.S. population was 3.93 million in 1790 and 62.95 million in 1890. What was the average annual rate of growth, to two decimal places?

20. The average increase in oil consumption in the United States during 1993–2004 was nearly 2%. In fact, consumption in 1993 was 6.291 billion barrels and consumption in 2004 was 7.588 billion barrels. What is a more accurate (two-decimal-place) estimate of the average annual percentage increase in oil consumption? (From 2004 through 2007, growth was zero, due to much higher prices for oil; and growth was again zero from 2008 to 2011, due to the world recession.)

23.3 Radioactive Decay

21. The carbon-14 in a carbon sample is decaying at approximately 6.5 atoms per hour per gram of carbon. Determine the approximate age of the sample.

22. The "Ice Man" is the popular name for the body of a man that was found in 1991, preserved in a glacier in the Tyrolean Alps. Carbon-14 dating revealed that he died about 5,000 years ago. About how many atoms of carbon-14 are breaking down today per hour per gram of carbon of his tissue?

23. The Nuclear Test-Ban Treaty of 1963 brought an end to the atmospheric testing of nuclear weapons by all nations but France and China. This testing released the radioactive isotope strontium-90 into the atmosphere. Just as with iodine-131, it settled out of the air onto grass in fields, was eaten by cows, and wound up in milk that children drank. In the body, strontium-90 mimics calcium and is absorbed into the bones, where its radiation can cause cancer; its half-life is 28.8 years. Assuming that all of the strontium-90 absorbed into the bones of children in 1962 still remains, approximately what proportion remains 50 years later, in 2012? [Contributed by John Oprea of Cleveland State University.]

24. After one week, what percentage of a supply of molybdenum-99 (with a half-life of 66 hours) remains?

For Exercises 25–32, refer to the following.

The physical half-life of a radioactive substance is unaffected by the biochemistry inside the human body. However, such a substance may not remain in the body for good but instead is excreted at some rate, thereby reducing the radiation exposure. The rate of excretion can be expressed in the form of a *biological* half-life. Let the physical half-life be H_p and the biological half-life be H_b. A combined *effective half-life* H_e can be calculated from

$$1/H_e = 1/H_p + 1/H_b$$

The two paths for eliminating the radioisotope act in parallel. In mathematical terms, H_e is the *harmonic mean* of the two half-lives. Sometimes physical decay is the dominant influence (Exercises 25–27), and sometimes biological clearing is (Exercise 28).

25. Iodine-131 actually has benign uses; it is used in treating thyroid cancer, hyperthyroidism, and non-Hodgkins lymphoma. Its physical half-life is 8 days; its biological half-life is 138 days. What is its effective half-life in the body?

26. Apart from uranium itself, the four major radioactive isotopes present in reactors and their waste are iodine-131, strontium-90, strontium-89, and cesium-137.

Of the four, the isotope with the longest physical half-life is cesium-137, at 30 years. Found in high levels in water near the ruined Japanese reactors in 2011, it is absorbed by ocean plants and moves up the food chain with increasing concentrations. Its biological half-life is 70 days. What is its effective half-life in the body?

27. Technetium-99m has a physical half-life of 0.25 days and a biological half-life of 1 day. What is its effective half-life in the body?

28. Phosphorus-32 is used to treat excess production of red blood cells in bone marrow. It has a physical half-life of 14.3 days and a biological half-life of 1155 days. What is its effective half-life in the body?

29. The physical half-life of carbon-14 is 5568 years, while its biological half-life is 40 days. The relatively short biological half-life explains why over a lifetime, the proportion of carbon-14 in a body equilibrates to the level in the atmosphere. What is the effective half-life of carbon-14 in the body?

30. Exercise 23 asked how much strontium-90 remains in the body after 50 years, applying the physical half-life of 28.8 years. However, strontium-90 is also cleared from the body with a biological half-life of about 50 years. Based on the effective half-life, give a different answer to what proportion of strontium-90 remains in the body after 50 years.

31. Reinterpret the formula involving H_e in terms of decay constants and explain why the formula makes sense.

32. Joe can split a cord of wood in 1 hour, while Sam can split a cord in 2 hours. Explain why the amount of time that it would take both of them working together to split a cord is the harmonic mean of 1 and 2.

23.4 Sustaining Renewable Resources

23.5 The Economics of Harvesting Resources

For Exercises 33–37, refer to the following:
We suppose that a population has the reproduction curve shown in the following figure, with units of thousands of tons of biomass. The mathematical description is that the population in the following year, x_{n+1}, depends on the population x_n in the current year (after any harvest) according to

$$x_{n+1} = f(x_n) = \tfrac{1}{5}x_n(20 - x_n) = 4x_n(1 - 0.05x_n)$$

for x_n between 0 and 10, in units of millions of pounds. We start with a population this year of x_1 whose value we vary. (For these exercises, a spreadsheet or a programmable calculator is useful.)

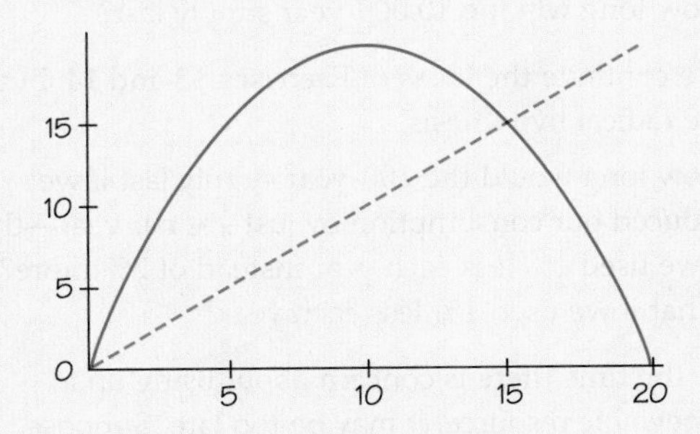

33. Start with $x_1 = 5$. Calculate numerically the population in the first few years, draw a cobweb diagram, and briefly describe the qualitative behavior of the population.

34. Repeat Exercise 21 with the starting value $x_1 = 10$.

35. Repeat Exercise 21 with the starting value $x_1 = 7$, going at least as far as x_{10}.

36. Try to find a starting value (besides 0, 5, 10, and 15) that leads to a stable population over time.

37. What is the equilibrium population size?

For Exercises 38–48, refer to the following:

We suppose that the population has the reproduction curve shown in the following figure, with units of thousands of tons of biomass, whose mathematical description is

$$x_{n+1} = f(x_n) = 3x_n(1 - 0.05x_n)$$

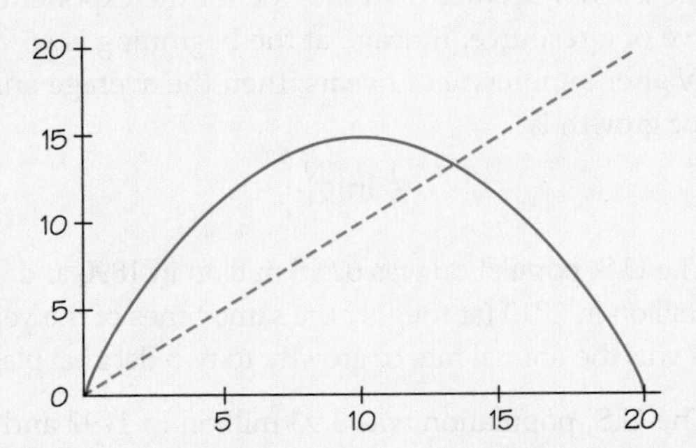

38. Start with $x_1 = 5$. Calculate numerically a population in the first 10 years, draw a cobweb diagram, and briefly describe the qualitative behavior of the population.

39. Repeat Exercise 26 with the starting value $x_1 = 10$.

40. Try to find a starting value (besides 0 and 20) that leads to extinction of the population.

41. What is the equilibrium population size?

▲ **42.** Which of the reproduction curves—the one for Exercises 33–36 or the one for Exercises 38–41—seems to you more realistic as a model of a biological population, and why?

43. What is the significance of the red dashed line in the preceding figures and its intersection with the blue curve?

44. The following figure shows the annual population gain in the absence of any harvesting. Determine the maximum sustainable yield to one decimal point.

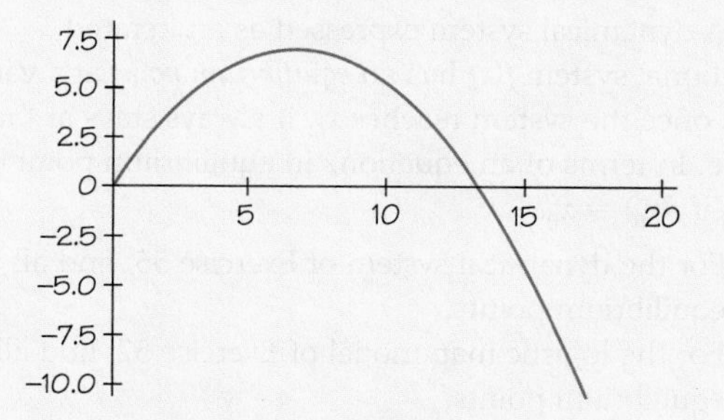

45. Suppose that the population of Exercises 38–44 starts with $x_1 = 11$ and that each year, we harvest half of the population. For example, in year 1, we harvest 5.5 million pounds, leaving the remaining 5.5 million pounds to reproduce (according to the reproduction curve) for the next year. Calculate numerically the population in the first 10 years, draw a cobweb diagram, and briefly describe the qualitative behavior of the population.

46. Repeat Exercise 45 but for a starting population $x_1 = 5$.

47. Harvesting a set proportion of a population is unrealistic for some situations, such as fishing, in which we can't know the size of the population or when we have harvested half of it. A more realistic situation for fishing is that increasing harvests attract increasing fishing effort (e.g., more boats). Repeat Exercise 33 with $x_1 = 11$ and a harvesting strategy that harvests 1 million tons the first year and every year harvests an extra 1 million tons (over the harvest of the previous year).

48. Suppose that the population of Exercises 26–35 has been overharvested to the point that only 1 million tons remain at the end of a particular year. If there is no harvesting at all until a year after a year with a population of 11 million tons, when can harvesting resume?

49. A reproduction curve for a population is shown in the following figure. Estimate the equilibrium population size and the maximum sustainable yield. (The units are in millions of pounds.)

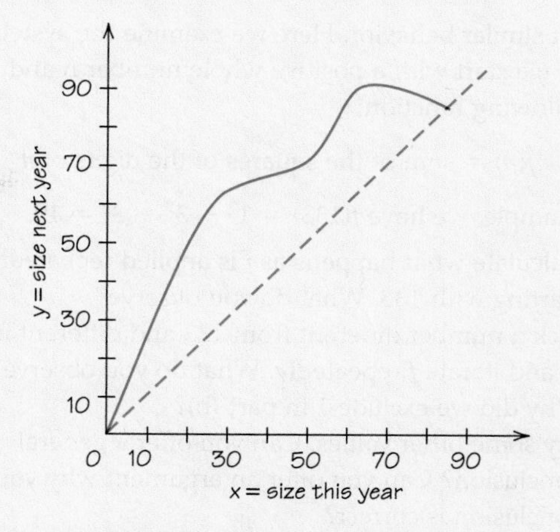

50. Suppose that a reproduction curve for a certain population is as in the following figure, where the units are in millions of pounds.

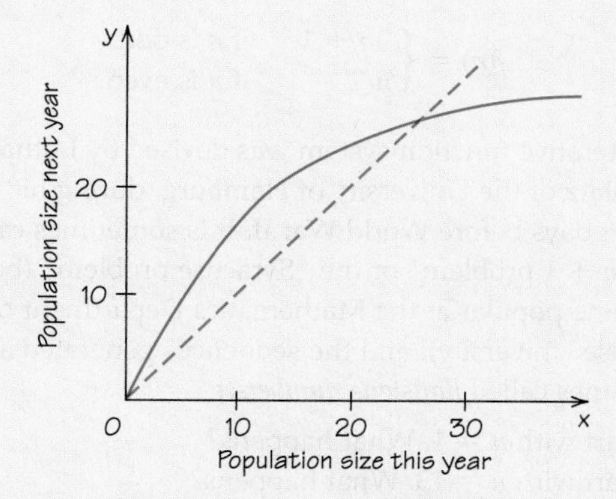

(a) Estimate the sustainable yields corresponding to a population of size 10 remaining after the harvest.

(b) Estimate the maximum sustainable yield.

23.6 Dynamical Systems and Chaos

51. In doubling on a "Stone Age" calculator (see Example 11 on p. 856):

(a) What do you notice about the two sequences that were produced?

(b) Suppose that we had started with a different seed, say, 39. What would happen as we iterate the doubling?

(c) Explain what you observe in parts (a) and (b) and give a general argument about why it is true.

▲ **52.** Explain why the logistic model on page 833 is not the same as the logistic map model on page 857, where we have $f(x) = \lambda x(1 - x)$.

53. You saw in Figure 23.12 that a logistic model can result in a stable value, produce cycling between several values, or result in chaos. Other dynamical systems can exhibit similar behavior. Here we examine the system in which we start with a positive whole number n and iterate the following function:

$$f(n) = \text{sum of the squares of the digits of } n$$

For example, we have $f(133) = 1^2 + 3^2 + 3^2 = 19$.

(a) Calculate what happens as f is applied repeatedly, starting with 133. What do you observe?
(b) Pick a number different from 133 and different from 1, and iterate f repeatedly. What do you observe?
(c) Why did we exclude 1 in part (b)?
(d) Try some other values. Can you offer a general conclusion? Can you offer an argument why your conclusion is correct?

54. The behavior of some very simple dynamical systems is still not completely known. Consider the system that starts with a positive whole number n and gives as the next number:

$$f(n) = \begin{cases} 3n + 1 & \text{if } n \text{ is odd} \\ n/2 & \text{if } n \text{ is even} \end{cases}$$

[This iterative function system was devised by Lothar O. Collatz of the University of Hamburg, during his student days before World War II. It is sometimes called the "$3n + 1$ problem" or the "Syracuse problem" (because it became popular at the Mathematics Department of Syracuse University), and the sequences generated are sometimes called *hailstone numbers*.]

(a) Start with $n = 1$. What happens?
(b) Start with $n = 13$. What happens?
(c) Start with $n = 12$. What happens?

What you observe is known to happen for all $n < 10^{40}$, but after more than 60 years, mathematicians have been unable to show that it happens for every n whatsoever.

55. (Requires programmable calculator, spreadsheet, or BASIC programming.) A population model slightly different from the logistic model is given by the iterative function system

$$g(x) = x + rx(1 - x)$$

where x is a fraction of the limiting population and r is a growth rate.

(a) Set $r = 3$, start with $x = x_1 = 0.01$, and calculate the first 20 values $x_1, \ldots, x_{20}$.
(b) In part (a), you should have found $x_{10} = 0.722914$. Replace this value with the rounded-up value $x_{10} = 0.723$ and continue on to calculate x_{20}.
(c) Now replace x_{10} with the rounded-down value $x_{10} = 0.722$ and continue on to calculate x_{20}.

56. A dynamical system expressed as an iterated functional system $f(x)$ has an *equilibrium point* at a value x_0 if, once the system reaches x_0, it always stays at that value. In terms of an equation, an equilibrium point exists at x_0 if $f(x_0) = x_0$.

(a) For the dynamical system of Exercise 55, find all equilibrium points.
(b) For the logistic map model of Exercise 52, find all equilibrium points.

■ **57.** (Requires a programmable calculator, a spreadsheet, BASIC programming, or preferably the software listed in the "Suggested Web Sites" section at the end of this chapter.) The behavior of the logistic population model $f(x) = \lambda x(1 - x)$ depends on the value of the positive parameter λ. As λ increases from 0 to 4, the system changes from one behavior to another through the following possible states:

- The population simply dies out.
- The population tends toward a nonzero equilibrium point.
- The population oscillates between 2 points.
- The population oscillates between 4 points, then 8 points, then 16 points, and so on.
- The population oscillates between numbers of points that are not powers of 2, until at last . . . the population oscillates between 3 values.
- The population behaves chaotically.

Explore what happens for various values of λ between 0 and 4, trying to identify where the shifts in the system's behavior take place.

WRITING PROJECTS

1. Based on the calculations you did in Exercises 9 and 10, write a one- to two-page guest editorial for a newspaper. Describe your projections and how you arrived at them, how serious a problem you think population growth is, what problems it is likely to cause, what you think needs to be done, and what the implications are for your own life.

2. Identify a particular regional, national, or world nonrenewable primary resource (such as coal) or a secondary resource (one, such as electric power, that is produced from primary resources). Research how much of it is available now and what the current rate of consumption is. Determine the static reserve. Estimate the growth rate in consumption, taking into account human population increase, and determine the exponential reserve. What social and technological factors contribute to the increasing rate of consumption? Brainstorm how those factors could be changed. Write an essay of three to five pages.

3. Identify a particular regional, national, or world renewable resources (such as timber or clean drinking water). Research how much of it is produced now, how much is harvested now, and what the current rate of consumption is. Estimate the growth rate in consumption, taking into account human population increase. For how long can this resource continue to meet the demand? What social and technological factors contribute to the increasing rate of consumption? Brainstorm how those factors could be changed. Write an essay of three to five pages.

Suggested Readings

BARTLETT, ALBERT A. *The Essential Exponential! For the Future of Our Planet.* Center for Science, Mathematics, & Computer Education, University of Nebraska–Lincoln (126 Morrill Hall, Lincoln, NE 68588-0350), 2004.

CLOVER, CHARLES. *The End of the Line: How Overfishing Is Changing the World and What We Eat,* The New Press, New York, 2006.

COHEN, JOEL. *How Many People Can the Earth Support?* Norton, New York, 1995.

GLEICK, JAMES. *Chaos: Making a New Science,* Viking, New York, 1987.

KALMAN, DAN. Chapter 13, "Logistic Growth," and Chapter 14, "Chaos in Logistic Models," in *Elementary Mathematical Models: Order Aplenty and a Glimpse of Chaos,* Mathematical Association of America, Washington, DC, 1997.

MEADOWS, DONELLA H., JORGEN RANDERS, and DENNIS L. MEADOWS. *Limits to Growth: The 30-Year Update,* Chelsea Green, White River Junction, VT, 2004.

MEYER, MICHAEL. How much is left? *Scientific American 303* (3) (September 2010) 74–81.

PETERSON, IVARS. *Newton's Clock: Chaos in the Solar System,* W. H. Freeman, New York, 1993.

SCHWARTZ, RICHARD H. *Mathematics and Global Survival,* 4th ed., Ginn Press, Needham Heights, MA, 1998.

Suggested Web Sites

www.census.gov/ipc/www/idb/ U.S Census Bureau population statistics and projections for all countries, with option to display population pyramids.

www.prb.org/ Population Reference Bureau (PRB) population statistics and rates of growth by regions.

www.techkiva.com/teachcensus/PopPyr.htm Dynamic population pyramid for the U.S. 1950–2050.

www.maths.anu.edu.au/~briand/chaos Downloadable Java applets to illustrate chaos.

math.bu.edu/DYSYS/applets/ Downloadable Java applets designed to accompany Robert L. Devaney's series of booklets *A Toolkit of Dynamics Activities,* but can be used independently.

staff.science.uva.nl/~alejan/dynamicstour.html A Java applet for the logistic population model of Figures 23.12 (on p. 849) and 23.17 (on p. 889).

www.census.gov/ipc/www/idb An international database of demographic data for 227 countries, with projections and capability to produce population pyramids for various years.

http://forio.com/simulate/simulation/pontifexconsult/how-much-oil-is-left/ A computer "lab" that allows the user to make various assumptions about reserves, consumption, productivity, and cost of production and then simulate the future reserves and consumption of oil.

ANSWERS TO SKILLS CHECK EXERCISES

Chapter 1

1. c
2. 7; 8
3. b
4. 3
5. a
6. 20
7. a
8. *B*; *E*
9. c
10. 7; 7
11. b
12. 1; 2; 4; 5; 10
13. c
14. 6
15. a
16. 9; 22
17. c
18. 6
19. b
20. 4
21. a
22. 12
23. a
24. 3
25. b
26. digraph; graph; digraph
27. c
28. 8; 13
29. a
30. 13

Chapter 2

1. c
2. 90
3. a
4. 0
5. b
6. 27
7. c
8. 26
9. c
10. 33
11. b
12. *V*
13. b
14. 94
15. c
16. 54
17. b
18. 2600
19. a
20. 234
21. b
22. 13; 3
23. c
24. 7; 8; 9
25. b
26. 14
27. a
28. 18
29. c
30. 16

Chapter 3

1. c
2. 20
3. b
4. 14 min
5. b
6. 2 min
7. a
8. 4; 2; 3
9. c
10. T_6; T_4
11. c
12. 4; 2; 3
13. b
14. 10 min
15. c
16. 14
17. c
18. 17
19. b
20. 3
21. b
22. 1; 4
23. a
24. 1
25. c
26. 2
27. c
28. 2
29. a
30. 3

Chapter 4

1. a
2. −2; −3
3. a
4. 6; 2
5. c
6. 14; 4
7. a
8. $x + 2y = 8$
9. b
10. 92
11. b
12. 2; 5
13. a
14. 25; 6
15. c
16. $12
17. b
18. convex
19. c
20. 4; 4
21. c
22. 3; 2
23. c
24. 3
25. a
26. 3; 1
27. c
28. 57
29. c
30. −6

Chapter 5

1. a
2. 2 (Major and Points)
3. c
4. 3
5. a
6. right
7. c
8. 14 (which corresponds to any values between 140 and 149)
9. c
10. leaf
11. a
12. 124.9
13. b
14. left
15. b
16. 50
17. b
18. less
19. b
20. 2
21. c
22. 98, 120, 125.5, 132, 147
23. b
24. grams
25. a
26. standard deviation
27. c
28. points at which the curvature changes
29. c
30. 95

Chapter 6

1. a
2. response
3. a
4. x
5. b
6. positive
7. c
8. 70
9. c
10. −5
11. b
12. $100x + 500$
13. b
14. 727.6
15. b
16. $0.02x + 3$
17. c
18. the same: $r = 0.86$
19. b
20. $r = 0.96$
21. a
22. $5.18x - 0.88$
23. b
24. predicted; estimated; fitted
25. c
26. above
27. b
28. extrapolation
29. c
30. cause

Chapter 7

1. b
2. whole
3. c
4. more
5. c
6. voluntary response
7. b
8. 25
9. a
10. 6694
11. c
12. ten
13. b
14. 52%
15. c
16. 40
17. a
18. cause
19. b
20. an observational study
21. b
22. placebo
23. c
24. 0.35

25. b
26. 0.015
27. a
28. 6.5%
29. c
30. 6%

Chapter 8

1. a
2. 24
3. b
4. 0.85
5. a
6. 1/18
7. c
8. 1
9. b
10. 0.3
11. b
12. 24
13. b
14. 0.30
15. c
16. 0.4
17. b
18. 1
19. b
20. 9
21. a
22. the population mean μ
23. b
24. $50
25. c
26. 2
27. b
28. 516
29. c
30. 11.6

Chapter 9

1. b
2. if any two voters exchange ballots, the election outcome is unchanged
3. b
4. a switch in a ballot from being a vote for the loser to being a vote for the winner doesn't change the election outcome
5. c
6. majority rule
7. b
8. defeats every other candidate in a one-on-one contest
9. d
10. monotonicity
11. a
12. sometimes produces no winner at all
13. c
14. receives the most first-place votes
15. a
16. has the highest Borda score
17. a
18. reverses the order in which this nonwinner and the winner were ranked
19. c
20. the Hare system
21. c
22. one-on-one contests take place according to an ordering of the candidates called an "agenda"
23. b
24. approval voting
25. c
26. it is to a voter's advantage to submit a ballot that misrepresents his or her preferences
27. a
28. they fail to satisfy monotonicity
29. d
30. satisfies the Condorcet winner criterion and independence of irrelevant alternatives, and always produces at least one winner in every election

Chapter 10

1. c
2. Borda
3. c
4. either an insincere ballot or a disingenuous ballot
5. c
6. A over D over C over B
7. a
8. two
9. c
10. one in which the winner has the fewest first-place votes
11. a
12. monotonicity
13. b
14. treats both candidates equally and all voters equally
15. b
16. there are only three candidates
17. c
18. placing the additional *j* candidates at the bottom of each ballot (in any order whatsoever)
19. c
20. agenda manipulation
21. a
22. group manipulation
23. d
24. at least as good, and sometimes better, than the other
25. c
26. the Gibbard-Satterthwaite theorem
27. d
28. manipulable
29. a
30. the chair has the most power but fares the worst

Chapter 11

1. c
2. 11
3. c
4. *C*
5. b
6. 1/4
7. b
8. the last juror in the permutation
9. c
10. the voters with weights 4 and 3
11. b
12. 256; no
13. a
14. 20
15. c
16. ${}_{16}C_8$
17. a
18. 8
19. a
20. 2^n
21. b
22. {A, B, C}, {A, B, D}, {A, C, D}
23. d
24. *B*
25. c
26. dummy
27. b
28. *A*, *B*, and *C*
29. a
30. veto

Chapter 12

1. b
2. *B*
3. c
4. one of the modes
5. c
6. *B*
7. a
8. just to the left or just to the right of *M*
9. c
10. to the median *M*
11. a
12. *B*
13. c
14. *C*
15. b
16. because only one candidate can win, and a median choice is not too far away from any voter, whereas two department stores near the ends of a main street are closer to more consumers than one at the center
17. b
18. he or she is one of the top two candidates identified by the poll
19. c
20. *C* alone
21. c
22. *B*
23. a
24. three times
25. c
26. it did not favor voters who live in the largest states

27. c
28. of the large-state bias of the Electoral College
29. c
30. the law mandates that the popular-vote winner wins if states with a majority of electoral votes pass it

Chapter 13

1. a
2. reflects the relative worth of each issue to that party
3. a
4. the transfer of items (or parts thereof) from one party to the other until points are equalized
5. d
6. point ratio
7. d
8. equitable
9. c
10. the boat, the car, and part of the land
11. c
12. the Knaster inheritance procedure
13. b
14. cash only
15. b
16. never willingly choose his or her least-preferred item and avoid wasting a choice on an item that he or she knows will remain available and can be chosen later
17. b
18. no other player received more than he or she did
19. a
20. each nondivider receives a portion of which he or she has approved
21. b
22. leaves the game
23. a
24. they use the divide-and-choose procedure
25. c
26. the first player
27. d
28. the amount of the second-highest bid
29. c
30. Bronislaw Knaster

Chapter 14

1. a
2. 352; 44; 4.545; 2.273; 1.182
3. c
4. 14; 13; 17; 17; 19
5. b
6. 300
7. a
8. 14; 13; 17; 17; 19
9. c
10. population
11. c
12. 0; 0; 100
13. b
14. 1; 1; 98
15. b
16. Hamilton
17. a
18. 0.5 seat
19. a
20. 2; 3
21. a
22. Hill-Huntington
23. c
24. Jefferson
25. d
26. *D*; *A*; *B*; *C*
27. c
28. Webster
29. b
30. 5; quota

Chapter 15

1. c
2. third
3. c
4. third
5. c
6. third
7. a
8. Three
9. c
10. 3
11. c
12. the value of the saddlepoint
13. a
14. prevent a player from being exploited by always choosing a pure strategy
15. c
16. zero
17. c
18. that the pitcher will throw more fastballs
19. a
20. a fastball with probability 0.8 and a curveball with probability 0.2
21. to throw a fastball and a curveball with equal probabilities of 0.5 each
22. it precludes their worst outcomes from being chosen
23. b
24. (kick center, kick side) = (1/6, 5/6) for the kicker, and (remain center, break side) = (1/4, 3/4) for the goalie
25. b
26. each player can do better if it moves by itself to Choice *B*
27. a
28. Player I does not benefit by returning to (3, 3).
29. b
30. players must think ahead about what moves are optimal in order to make optimal choices in the present

Chapter 16

1. a
2. 1
3. a
4. 3
5. a
6. 9
7. b
8. 0
9. a; b
10. 8
11. b
12. 10
13. product; manufacturer; manufacturer
14. 9
15. b
16. 9
17. c
18. 11
19. b
20. 10
21. b
22. 100; 100
23. c
24. 4
25. c
26. 10
27. c
28. 20001-2800-7
29. a; b
30. 3765

Chapter 17

1. b
2. 1011
3. b
4. 3
5. b
6. 3
7. b
8. 3
9. a
10. 0010100
11. b
12. 3
13. a
14. 3, 5, 7, 9, 11, 15, 17, 19, 21, 23, 25
15. c
16. decimation
17. b
18. Every letter is unchanged
19. a
20. 13^7 mod 33
21. c
22. 29
23. c
24. *n* must be the product of two distinct primes
25. c
26. either *P* or *Q* is true, or both are true
27. b
28. are identical
29. a
30. 00111001

Chapter 18

1. c
2. 120
3. a
4. $1.60
5. b
6. 10
7. c
8. cube
9. a
10. 1280
11. c
12. 26.2187575 mi; officially, 26 miles and 385 yards
13. a
14. 60
15. c
16. 18 (more exactly, 19)
17. c
18. 1.6
19. b
20. 187
21. c
22. heavy; light (or big; small)
23. a
24. 4
25. c
26. 10
27. c
28. 3
29. a (or b)
30. proportional to a power of

Chapter 19

1. d
2. patterns
3. a
4. 12
5. c
6. 34
7. b
8. pineapples
9. c
10. 20
11. b
12. H; I; N; O; S; X; Z
13. b
14. translation; rotation
15. c
16. translation; reflection
17. b
18. half-turn rotation
19. c
20. 7
21. b
22. translation
23. a
24. reflection
25. a
26. infinitely many
27. c
28. toss
29. c
30. self-similarity

Chapter 20

1. b (or c)
2. 45°
3. c
4. 3; 4; 6
5. a
6. 6
7. c
8. 5
9. b
10. 5
11. c
12. 7
13. a
14. translations
15. a
16. M. C. Escher
17. c
18. translations and half-turns
19. b
20. Conway; translations; half-turns
21. b
22. quasi
23. b
24. opposite sides are equal in length (or opposite sides are parallel, or opposite angles are congruent)
25. c
26. fivefold
27. c
28. the golden ratio
29. c
30. inflation

Chapter 21

1. a, b, or c
2. $15
3. b
4. $1186.00
5. b
6. $49.09
7. c
8. $1202.71
9. a (or b)
10. $181.66
11. b
12. $3686.64 ($3696.74 in a leap year)
13. a
14. 3.0416%
15. b
16. 3.04533%
17. c
18. $1204.42
19. c
20. 3.04545%
21. c
22. 2.718
23. a
24. annuity (or sinking fund)
25. b
26. 3.46535%
27. c (or a)
28. $9216
29. b
30. $134.73

Chapter 22

1. a
2. $1219.89
3. c
4. $2543.20
5. c
6. annual percentage rate
7. c
8. 19.56%
9. a
10. 18%
11. a
12. 5 years
13. b
14. $135.47
15. c
16. $456.33
17. c
18. 1.25 to 1.5
19. b
20. interest
21. a
22. $500
23. b
24. higher
25. b (or a, or c)
26. adjustable rate mortgage
27. a
28. $13.55 million if you get the first million right away, $12.78 million if you have to wait a year for the first million
29. b
30. actuary

Chapter 23

1. b
2. the spread of a technology or of a product
3. b
4. momentum
5. c
6. exponential
7. b
8. 32
9. a
10. 241
11. b
12. 72,600
13. b
14. rising
15. b
16. dynamics (or growth, or evolution, or variation)
17. c
18. larger
19. c
20. smaller
21. a
22. 0 (or 1)
23. a
24. 0.75
25. a
26. sensitivity to initial conditions
27. a, b, and c
28. greed; chance; chaotic variation
29. c
30. near-periodic; transitive; sensitive (to initial conditions)

Chapter 1

1. (a)

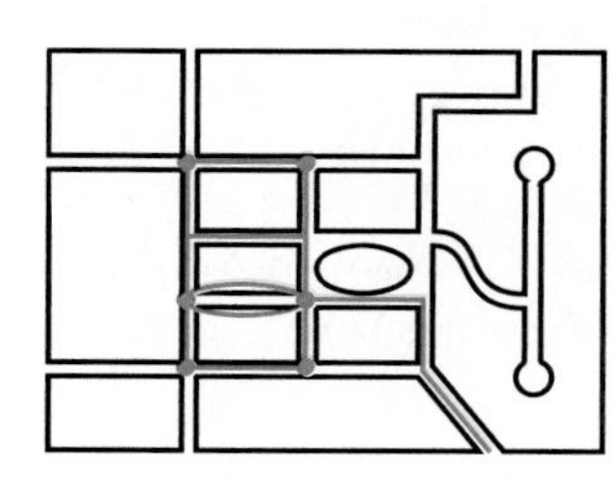

(b) *ADGIJHKLMLKHGHEDEBA* (other answers possible)

3. (a) eight vertices
(b) 13 edges
(c) *A*: 4; *B*: 2; *C*: 3; *D*: 3; *E*: 4; *F*: 4; *G*: 3; *H*: 3
(d) *A*, *D*, and *F*
(e) *E*, *G*, and *H*

5. (a) No
(b) *EC*, *AD*, *BD*, and *AC*
(c) 5; 6

7. *E*: 0; *A*: 1; *H*, *D*, and *G*: 2; *B* and *F*: 3; C: 5

9. (a) *CGDBC* (Answers can vary.)
(b) **(i)** *BD*; *BFD*; **(ii)** *CBF*; *CGDF*; *CGDBF*; **(iii)** *GDBCG* (Answers can vary.)

11. Drawings can vary. Possible renderings for **(a)** and **(b)** include the following:

(a)

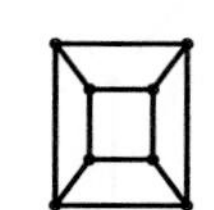

(b)

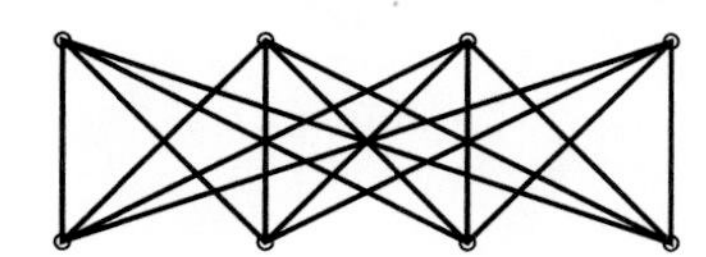

(c) Yes

13. (a) 4; 4
(b) 7; 6
(c) 10; 14

15. 2

17. (a)

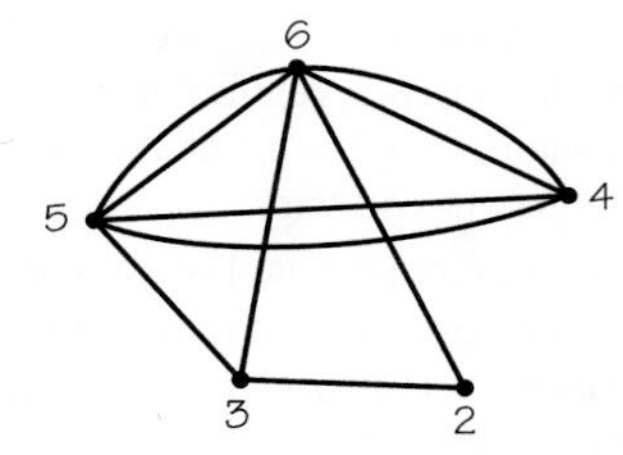

(b) No. To have an Euler circuit a graph can not have odd-valent vertices.

19. Drawings can vary. Possible renderings include the following:

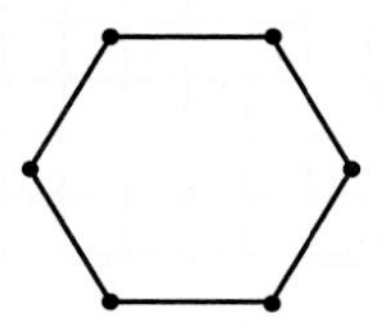

21. Drawings can vary. Possible renderings include the following:

(a) **(b)**

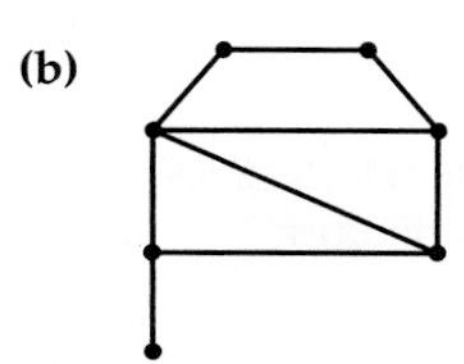

23. (a) Not all edges are traveled by worker;
(b) end of route not the same as beginning of route; not realistic; no Euler circuit in graph

25. Since this graph is connected and even-valent, it has an Euler circuit. Any Euler circuit will solve the problem efficiently.

27.

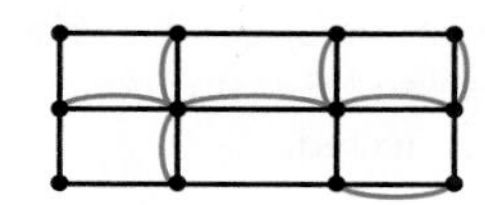

29. (a) 3; **(b)** and **(c)** Answers will vary.

31. Do not choose edge 2, but edges 1 or 10 could be chosen.

33. Answers will vary.

35. (a) *A*, *C*, *E*, and *H* are odd-valent.
(b) 2

37. (a) No.
(b) 4
(c) Answers will vary. One solution would involve duplicating edges *CD*, *DE*, *FG*, and *AB* in the accompanying graph.

39. Answers will vary; no

41. Answers will vary. Possible answers include *AECDABDCBEA*.

43. Drawings can vary. Possible renderings for **(a)**, **(b)**, and **(c)** include the following:

(a)

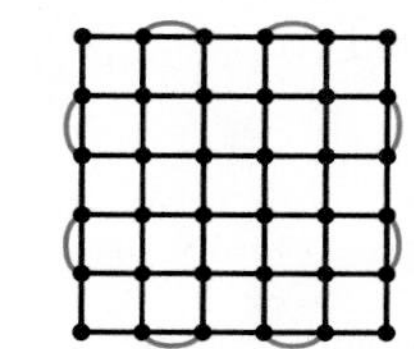

(b)

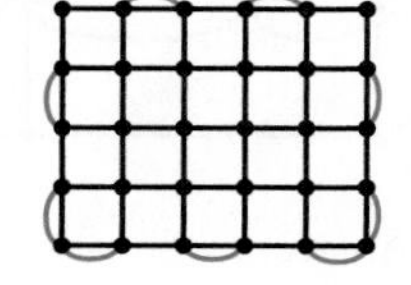

(c)

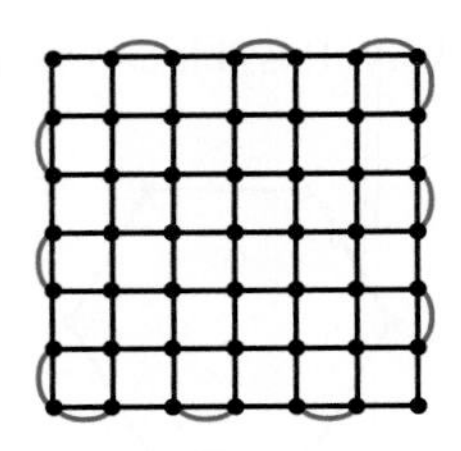

(b) Yes; no

45. (a) 2
(b) Yes
(c) 2
(d) No

47. (a) *ADEADCAFCFBA*
(b) 95 minutes

49.

51. There are many circuits that achieve a minimum length of 44,000 feet.

53. (b) and **(c)**; Additional answers will vary.

55. For each edge e of G, add another edge joining the vertices that are endpoints of e to obtain H. If G is connected, H does have an Euler circuit because whatever the valences of G, graph H has valences that are doubles of these and remains connected. Hence, H is even-valent and connected.

57. (a) Drawings can vary. Possible renderings include the following:

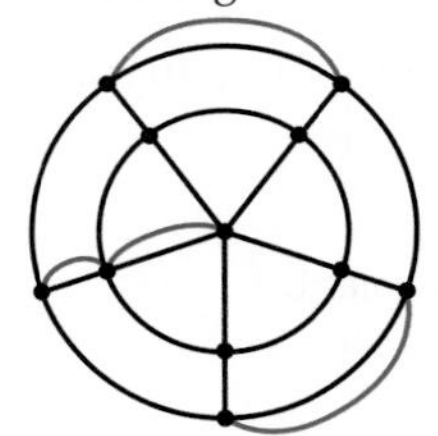

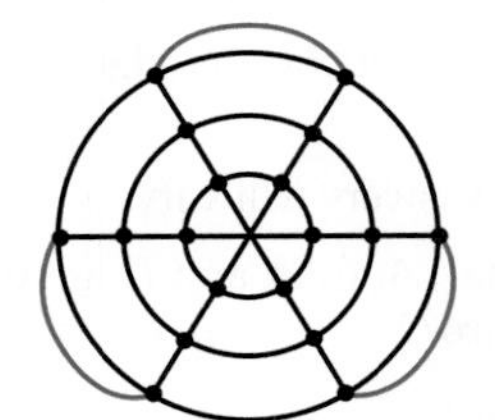

(b) The best eulerization for the four-circle, four-ray case adds two edges.
(c) Answers will vary.

59. (a) Yes.
(b) 15

61. Answers will vary.

63.

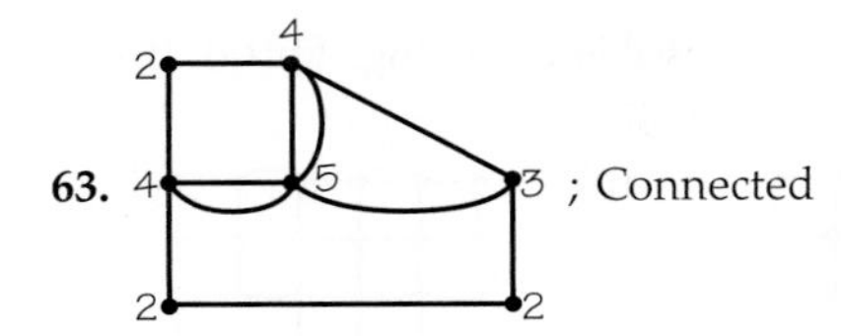

65. Answers will vary.

67. Answers will vary. Possible answers include: *ABDEFBEBFEDBACDCBCA*

Chapter 2

1. (a) Not possible
(b) A possible answer is:
$X_3X_2X_{11}X_{12}X_6X_{13}X_{20}X_{16}X_{17}X_{14}X_{10}X_1X_4X_9X_{15}X_{18}X_{19}X_8X_7X_5X_3$

3. Possible answers include:
(a) $X_3X_4X_2X_5X_6X_1X_3$
(b) $X_3X_2X_1X_6X_7X_8X_9X_{10}X_{11}X_{12}X_5X_4X_3$
(c) $X_3X_1X_2X_7X_6X_9X_8X_5X_4X_3$

5. (a) Yes
(b) Yes
(c) Yes

7. (a) No for (a); yes for (b); no for (c)
(b) No longer be possible to send messages between these two sites

9. Answers will vary.

11. (a) Yes for both.
(b) Answers will vary.
(c) Add edges X_2X_8, X_8X_6, X_6X_4, and X_4X_2.

13. Drawings can vary. Possible renderings include the following:

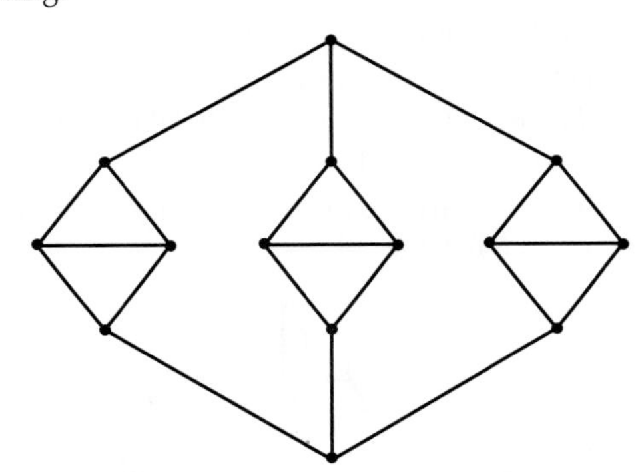

15. (a) No
(b) No

17. (a) Yes
(b) No
(c) No

19. (a) Yes
(b) No
(c) Answers will vary.

21. (a) Hamiltonian circuit: yes; Euler circuit: yes; Additional answers will vary.
(b) Hamiltonian circuit: yes; Euler circuit: yes
(c) Hamiltonian circuit: yes; Euler circuit: no
(d) Hamiltonian circuit: no; Euler circuit: yes; Additional answers will vary.

23. (a) Hamiltonian circuit: yes; Euler circuit: no
(b) Hamiltonian circuit: yes; Euler circuit: no
(c) Hamiltonian circuit: yes; Euler circuit: no
(d) Hamiltonian circuit: no; Euler circuit: no

25. (a) Drawings will vary.
(b) Drawings will vary.
(c) Answers will vary.

27. **(a)** 2520
(b) 16,807
(c) 16,800

29. **(a)** 15,120
(b) 17,576

31. Yes; 172

33. **(a)** 17,558,424
(b) Answers will vary.

35. 10,000,000; 900

37. Drawings will vary.; 6, 10, and 15 edges, respectively; $\frac{n(n-1)}{2}$ edges; 3, 12, and 60, respectively

39. **(a)** Possible drawings include the following:

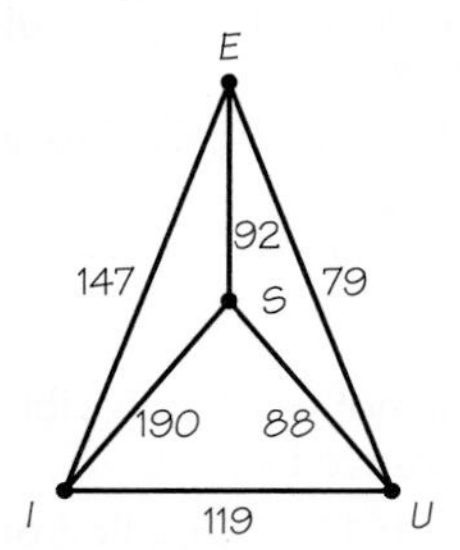

(b) Tour (1): *UISEU*: 480
Tour (2): *USIEU*: 504
Tour (3): *UIESU*: 446
(c) Tour (3)
(d) No
(e) Tour (1); yes; yes; no
(f) Tour (2); no

41. *FMCRF*

43. **(a)** *HBCAH*
(b) *HBCAH*
(c) The cheapest route costs 132 and coincides with the nearest-neighbor solution from *H* and the sorted-edges solution.

45. *MACBM*

47. A traveling salesman problem

49. Yes; Hamiltonian circuit; Chinese postman problem; Answers will vary and requires at least 9 reuses of edges.

51. Answers will vary.

53. The optimal tour is the same but its cost is now 4700.

55. Diagram (a): **(a)** There is a circuit and wiggled edges do not include all vertices. **(b)** The circuit does not include all the vertices of the graph.
Diagram (b): **(a)** The tree does not include all vertices of the graph. **(b)** Not a circuit
Diagram (c): **(a)** Not a tree **(b)** Not a circuit
Diagram (d): **(a)** Not a tree **(b)** Not a circuit

57. **(a)** 1, 2, 3, 4, 5, 8; Cost is 23.
(b) 1, 1, 1, 2, 2, 3, 3, 4, 5, 6, 6; Cost is 34.
(c) 1, 1, 1, 2, 2, 2, 2, 2, 3, 3, 3, 3, 4, 4, 4, 5, 5, 6, 7; Cost is 60.
(d) 1, 2, 2, 3, 3, 3, 4, 5, 5, 5, 6, 6; Cost is 45.

59. 27; 27; at least 26

61. Yes

63. Yes; Additional answers will vary.

65. Yes; yes

67. There are three different trees with the same cost.

69. **(a)** True
(b) False (unless all the edges of the graph have the same weight)
(c) True
(d) False
(e) False

71. **(a)** Answers will vary for each edge.
(b) 5; one less than the number of vertices in the graph
(c) No (*CD* must be included.)

73.

	A	B	C	D
A	0	16	13	5
B	16	0	19	11
C	13	19	0	8
D	5	11	8	0

75. **(a)** 22; $T_3T_2T_5$
(b) 30; $T_3T_5T_7$

77. T_1, T_5, and T_7 if shortened would reduce the earliest completion time, while shortening the other tasks will not; 29; $T_1T_4T_7$.

79. Answers will vary.

81. Drawings can vary. Possible renderings include the following:

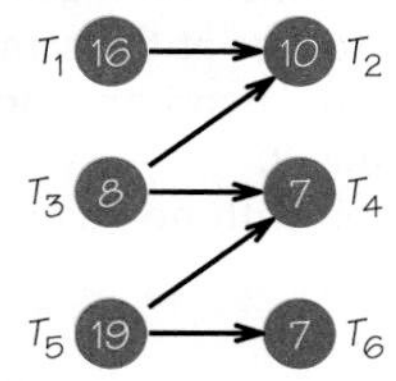

Chapter 3

1. Answers will vary.

3. Answers will vary.

5. **(a)** Processor 1: T_1, T_2, T_3, T_5, T_7; Processor 2: Idle 0 to 2, T_4, T_6, idle 4 to 5
(b) Processor 1: T_1, T_2, T_3, T_6, T_7; Processor 2: Idle 0 to 2, T_4, T_5, idle 4 to 5
(c) Yes
(d) No
(e) T_3 and T_5

7. The two vertices with no incoming edges will correspond to the only tasks that will be ready at time 0. These will be assigned to different processors at time 0, forcing the third processor to be idle for some period of time.

9. (a) (i) Processor 1: T_1 from 0 to 13, T_3 from 13 to 25, T_6 from 25 to 45; Processor 2: T_2 from 0 to 18, T_4 from 18 to 27, T_5 from 27 to 35, idle from 35 to 45
(ii) Processor 1: T_1 from 0 to 13, T_3 from 13 to 25, T_4 from 25 to 34, T_5 from 34 to 42; Processor 2: T_2 from 0 to 18, T_6 from 18 to 38, idle from 38 to 42
(b) Yes
(c) T_2, T_6, and 38; Sum of the task times divided by 2 is 40.

11. (a) Yes **(b)** No

13. Answers will vary.

15. (a) Processor 1: T_1, T_6, idle 15 to 21, T_7, idle 27 to 31; Processor 2: T_2, T_5, T_8; Processor 3: T_3, T_4, idle from 13 to 31
(b) Processor 1: T_1, T_6, idle 15 to 21, T_7, idle 27 to 31; Processor 2: T_3, T_4, idle from 13 to 21, T_8; Processor 3: T_2, T_5, idle from 21 to 31
(c) Processor 1: T_4, idle 10 to 11, T_6, idle 18 to 21, T_8; Processor 2: T_2, T_5, T_7, idle 27 to 31; Processor 3: T_1, T_3, idle 11 to 31

17. Yes

19. (a) 15
(b) 19
(c) Use the list T_1, T_4, T_3, T_2, T_5, T_6, T_7 to obtain the following schedule:

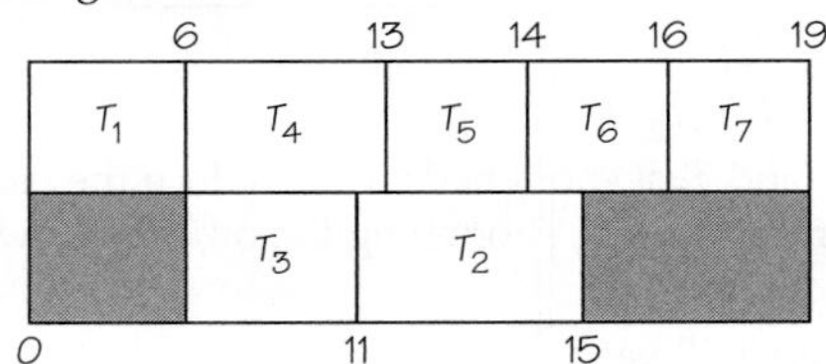

21. (a) No
(b) If T_5 could be started on machine 3 and machine 2 was also free at the same time, proper use of the list-processing algorithm should have assigned this task to machine 2.
(c) Use the digraph with no edges and the list: T_2, T_5, T_4, T_3, T_1.

23. (a) T_1, T_2, T_3, and T_6
(b) No tasks require that T_1 and T_6 be done before these other tasks can begin.
(c) T_6
(d) Processor 1: T_1, T_6; Processor 2: T_2, T_4, idle from 18 to 30; Processor 3: T_3, T_5, idle from 12 to 30
(e) No
(f) Processor 1: T_6, idle from 20 to 22; Processor 2: T_3, T_5, T_1; Processor 3: T_2, T_4, idle from 18 to 22
(g) Yes
(h) Yes

25. (a) 120
(b) No; T_1 must be assigned to the first machine at time 0.
(c) No; Because when 2 divides 31, there is a remainder of 1.
(d) No

27. Yes

29. Answers will vary.

31. No

33. (a) Task times: $T_1 = 3$, $T_2 = 3$, $T_3 = 2$, $T_4 = 3$, $T_5 = 3$, $T_6 = 4$, $T_7 = 5$, $T_8 = 3$, $T_9 = 2$, $T_{10} = 1$, $T_{11} = 1$, and $T_{12} = 3$. This schedule would be produced from the list: T_1, T_3, T_2, T_5, T_4, T_6, T_7, T_8, T_{11}, T_{12}, T_9, T_{10}.
(b) Task times: $T_1 = 3$, $T_2 = 3$, $T_3 = 3$, $T_4 = 2$, $T_5 = 2$, $T_6 = 4$, $T_7 = 3$, $T_8 = 5$, $T_9 = 8$, $T_{10} = 4$, $T_{11} = 7$, $T_{12} = 9$, and $T_{13} = 3$. This schedule would be produced from the list: T_1, T_5, T_7, T_4, T_3, T_6, T_{11}, T_8, T_{12}, T_9, T_2, T_{10}, T_{13}.

35. (a) (i) Processor 1: T_1, T_3, T_5, T_7, idle from 16 to 20; Processor 2: T_2, T_4, T_6, T_8
(ii) Processor 1: T_8, T_5, T_4, T_1; Processor 2: T_7, T_6, T_3, T_2
(b) Yes

37. Answers will vary.

39. In part (a), 33 is not exactly divisible by 4; In part (b) 56 is not exactly divisible by 5.

41. (a) (i) Machine 1: 12, 9, 15, idle from 36 to 50; Machine 2: 7, 10, 13, 20
(ii) Machine 1: 12, 13, 20; Machine 2: 7, 9, 15, 10, idle from 41 to 45
(iii) Machine 1: 20, 12, 9, idle from 41 to 45; Machine 2: 15, 13, 10, 7
(b) None of the schedules found is optimal
(c) The critical path list is T_6, T_5, T_4, T_1, T_7, T_2, T_3. This list does not lead to an optimal schedule but there is a schedule where the completion time is 43, achievable by scheduling the tasks of length 20, 13, and 10 on machine 1 and tasks of length 24 and 19 on machine 2.

43. (a) Machine 1: 129; Machine 2: 129
(b) Machine 1: 123; Machine 2: 123
(c) Yes

45. (a) Processor 1: 12, 13, 45, 34, 63, 43, 16, idle 226 to 298; Processor 2: 23, 24, 23, 53, 25, 74, 76; Processor 3: 32, 23, 14, 21, 18, 47, 23, 43, 16, idle 237 to 298
(b) Processor 1: 12, 24, 14, 34, 25, 23, 16, 16, 76; Processor 2: 23, 23, 21, 63, 43, idle 173 to 240; Processor 3: 32, 23, 53, 74, idle 182 to 240; Processor 4: 13, 45, 18, 47, 43, idle 166 to 240
(c) Three machines: Processor 1: 76, 45, 43, 24, 23, 18, 16, 13; Processor 2: 74, 47, 34, 32, 23, 21, 14, 12, idle 257 to 258; Processor 3: 63, 53, 43, 25, 23, 23, 16, idle 246 to 248
Four machines: Processor 1: 76, 43, 24, 23, 16, idle 182 to 194; Processor 2: 74, 43, 25, 23, 16, 13; Processor 3: 63, 45, 32, 23, 18, 12, idle 193 to 194; Processor 4: 53, 47, 34, 23, 21, 14, idle 192 to 194
(d) Processor 1: 84, 45, 43, 25, 23, 23, 16, 12; Processor 2: 82, 55, 34, 32, 23, 18, 14, 13; Processor 3: 71, 61, 43, 24, 23, 21, 16, idle 259 to 271

47. Answers will vary.

49. Each task heads a path of length equal to the time to do that task.

51. 9; Number of bins would not change, but the placement of the items in the bins would differ.

53. **(a)** 17
(b) 16
(c) 16
(d) 13

55. Yes, both are acceptable.

57. **(a)** Answers will vary.
(b) It is possible.

59. No; yes

61. Answers will vary.

63. Answers will vary.

65. **(a)** 152 min; 124 min
(b) 155 min; 120 min
(c) Yes; five-processor decreasing-time schedule
(d) 11
(e) NFD: 13; WFD: 11
(f) An optimal packing with 10 bins exists.

67. **(a)** Answers will vary.
(b) Answers will vary.
(c) Packing rectangles of width 1 in an $m \times 1$ rectangle is a special case of the two-dimensional problem.
(d) Answers will vary.

69. Answers will vary.

71. **(a)** Graph in: (a) No; (b) No; (c) No; (d) Yes; (e) Yes; (f) Yes
(b) Graph in: (a) Yes; (b) Yes; (c) No; (d) Yes; (e) Yes; (f) Yes
(c) Graph in: (a) 4; (b) 4; (c) 5; (d) 3; (e) 3; (f) 2

73. **(a)** Drawings can vary. Possible renderings include the following:

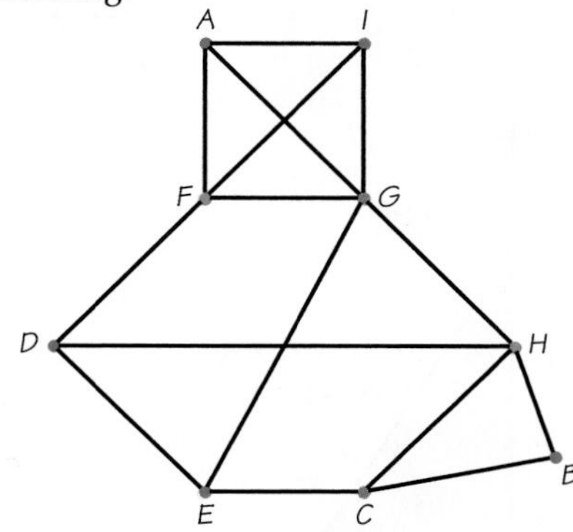

(b) 4
(c) The coloring in (a) indicates one possible arrangement.

75. **(a)** Drawings can vary. Possible renderings include the following:

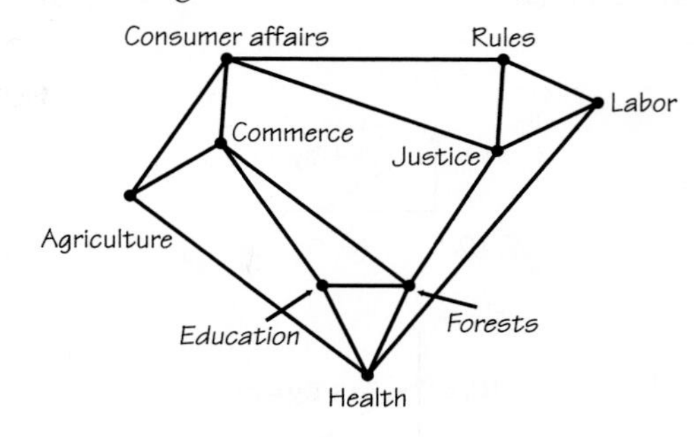

(b) 3
(c) 3; Additional answers will vary.

77. **(a)** Drawings can vary. Possible renderings include the following.

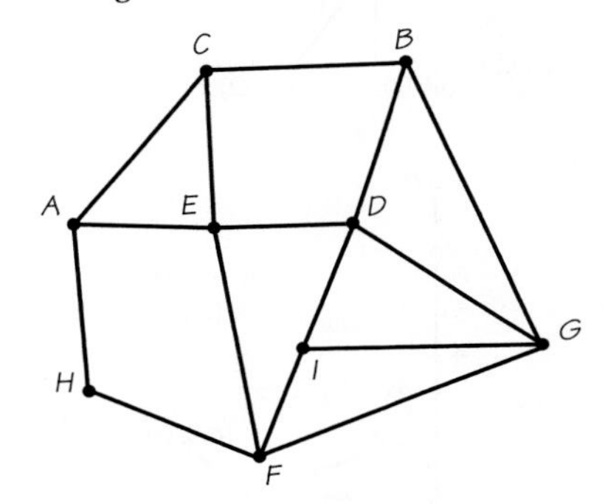

(b) 3
(c) 3

79. **(a)**

	F	E	H	P	T	A	D	C
F		×		×	×			
E	×					×	×	
H						×	×	×
P	×					×		×
T	×						×	×
A		×	×	×				
D		×	×		×			
C			×	×	×			

(b) 2
(c) 3 (Two rooms will be used for three committee meetings and one room for two committee meetings.)

81. Answers will vary.

83. Graph in: (a) 8; (b) 5; (c) 6; (d) 4; (e) 3; (f) 4; The minimum is either the largest valence of any vertex or one more than the largest valence.

85. **(a)** Graph (a) 4; (b) 2; (c) 4; (d) 4; (e) 3; (f) 3
(b) Answers will vary.

87. 3

89. 3; only if each child formed his/her own play group

Chapter 4

1. (2, 4); (2, 6); (4, 4)

3. **(a)**

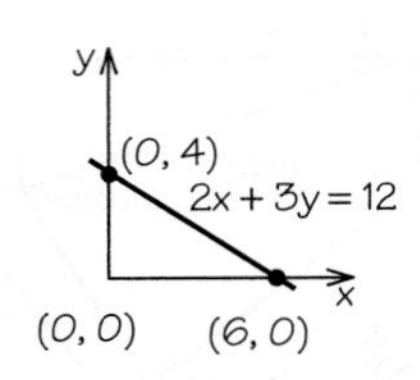

(b)

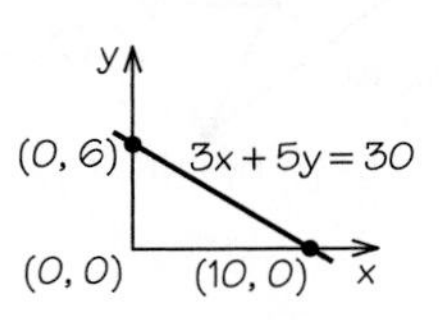

(c)

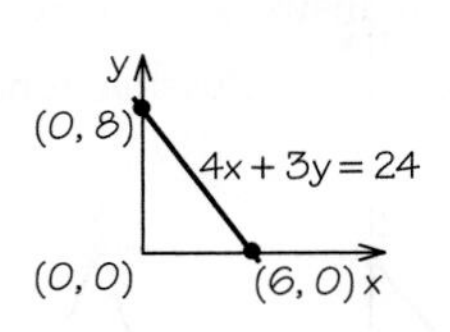

5. **(a)**

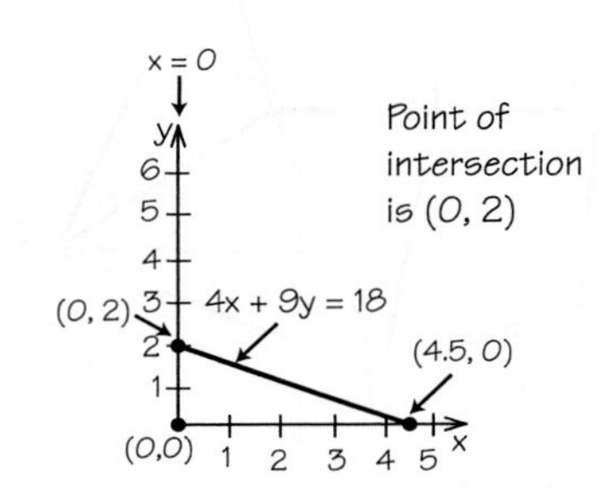

(b)

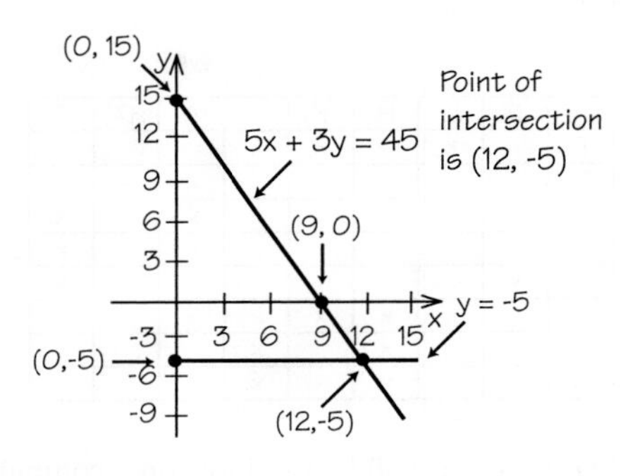

(c)

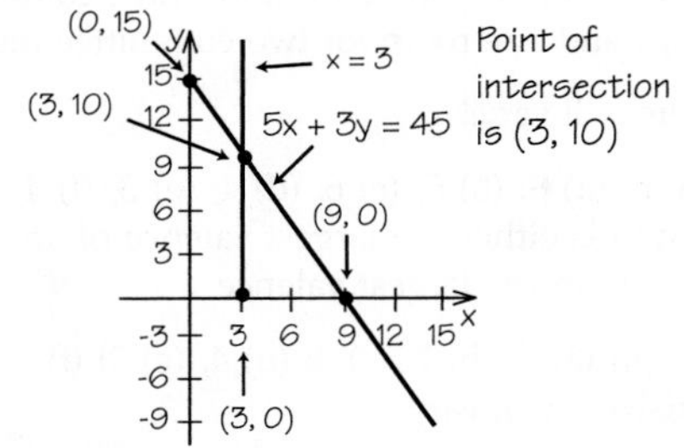

Note: For Exercises 7 and 9, first quadrant only is shown. Point of intersection is labeled.

7. **(a)**

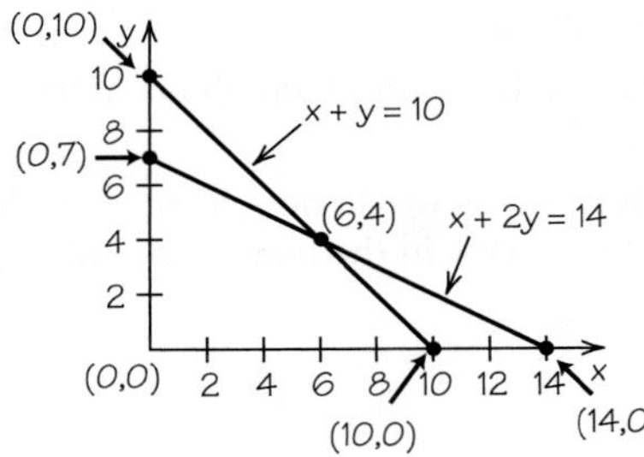

(b)

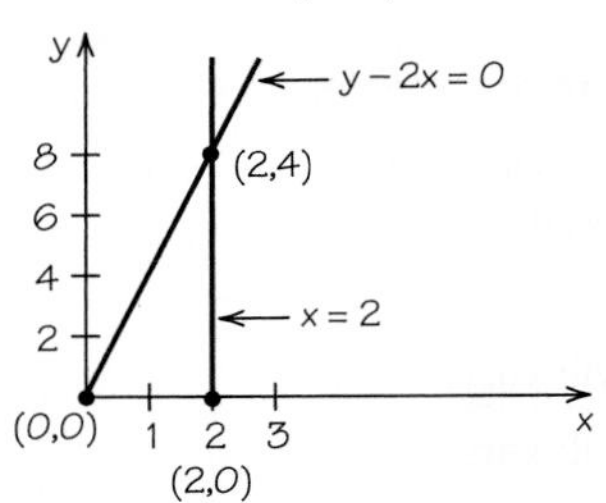

9. **(a)**

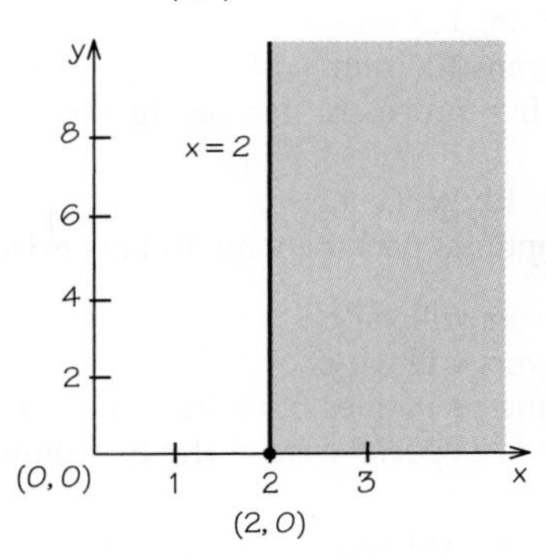

(b)

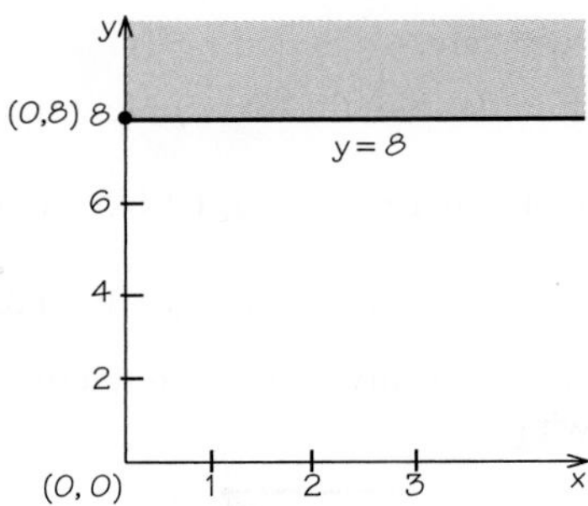

(c)

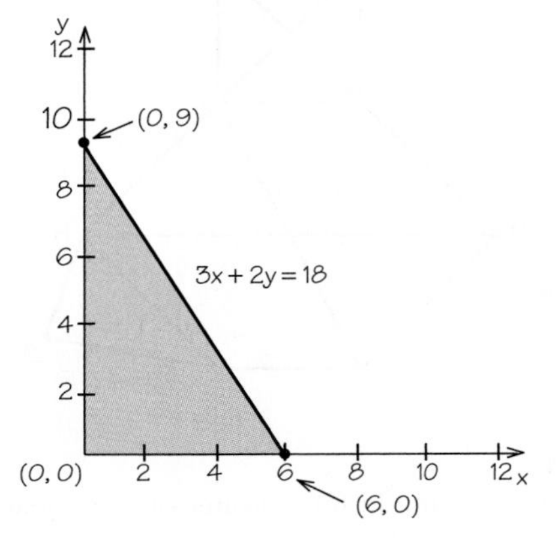

(d)

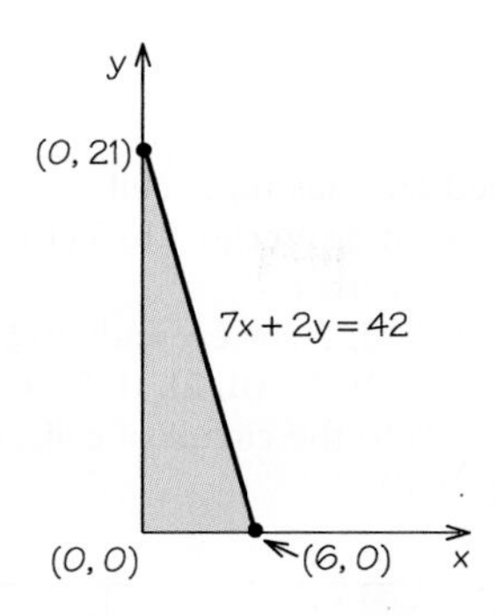

11. (a) $6x + 4y \leq 300$
(b) $30x + 72y \leq 420$

13.

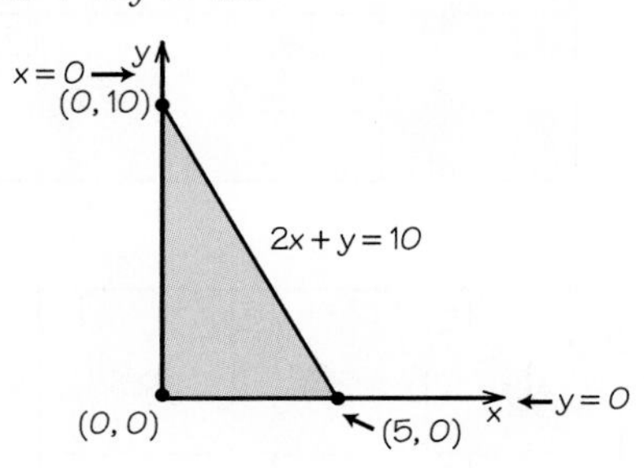

15.

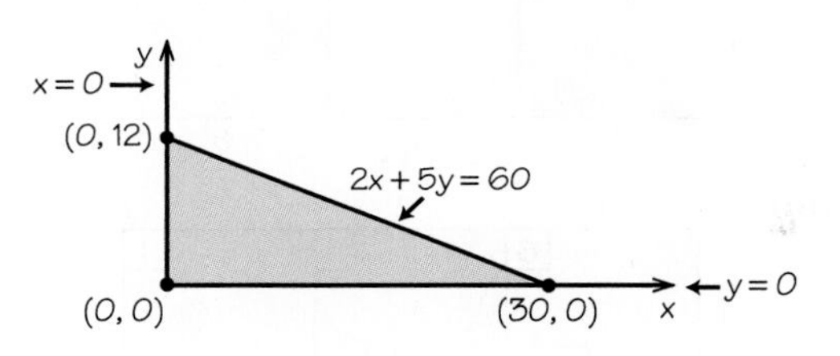

17.

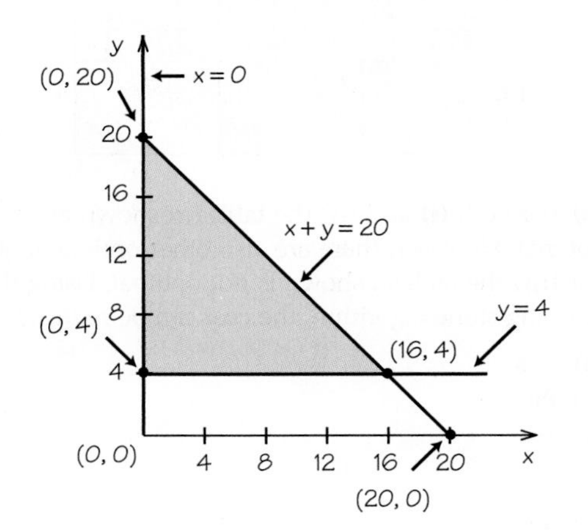

19. Exercise 13: (2, 4): yes; (10, 6): no; Exercise 15: (2, 4): yes; (10, 6): yes; Exercise 17: (2, 4): yes; (10, 6): yes

21. Make 0 skateboards and 30 dolls for a profit of $111.

23. Note: These situations are shown only for the first quadrant.

(a)

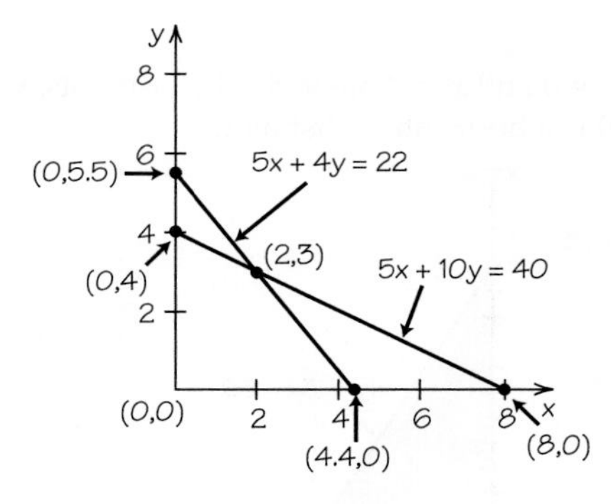

(b)

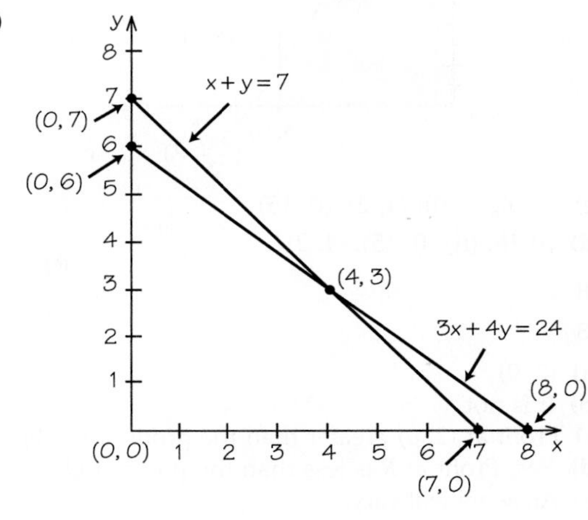

25.

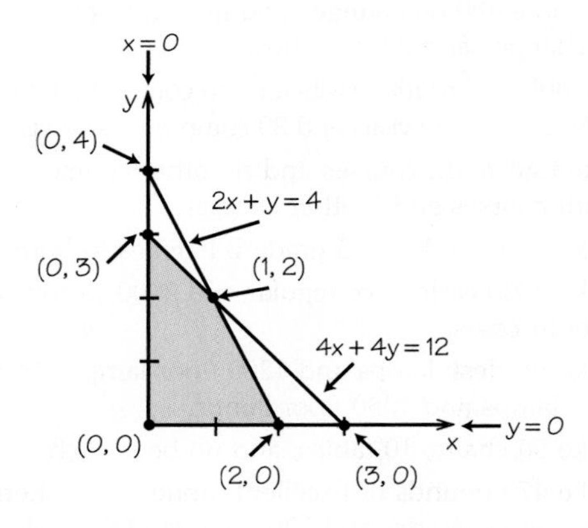

27.

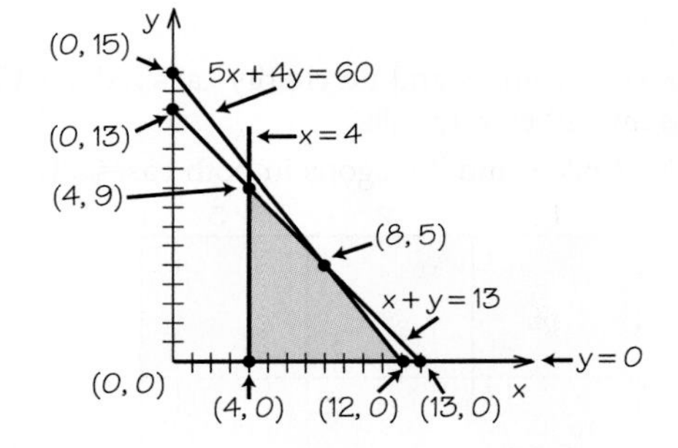

29. 29

31. **(a)** $x = 0$ and $x = 4$ are vertical constraints, while $y = 0$ is a horizontal constraint.

(b)

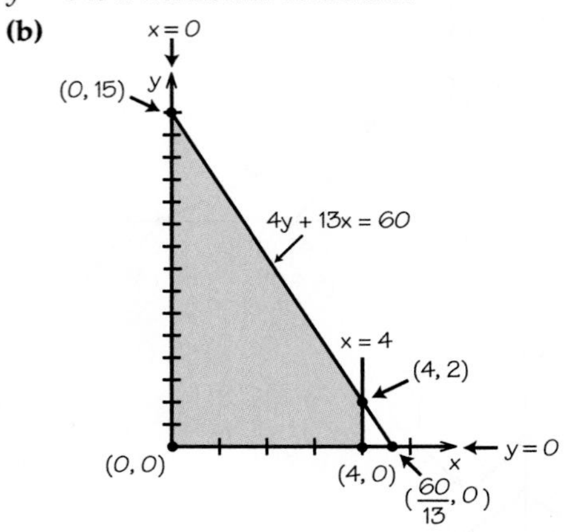

(c) (0, 0); (4, 0); (4, 2); (0, 15)

(d) (i) 16; (ii) (0, 15); (4, 2)

33. 28

35. 38

37. **(a)** (2, 0)
(b) It is not.
(c) Profit at (2, 0) greater than the profit at $(\frac{13}{7}, 0)$
(d) Yes; Profit at R is less than the profit at Q.
(e) Answers will vary.

39. Schedule 400 oil changes and no tune-ups.; Schedule 300 oil changes and 20 tune-ups.

41. Schedule 360 routine visits and no comprehensive visits.; Schedule 210 routine visits and 30 comprehensive visits.

43. Take four math courses and no other courses.; Take two math courses and 2 other courses.

45. Make 2 grade A and 5 grade B batches in both cases.

47. Make 3000 cartons of regular and 2000 cartons of diet in both cases.

49. Make no desk lamps and 1200 floor lamps.; Make 150 desk lamps and 1080 floor lamps.

51. Make 50 chairs, 10 tables, and no beds each month.

53. Make 470 pounds of Excellent, none of Southern, 2400 pounds of World, and 320 pounds of Special.

55. 43

57. Make 60 business and no charity calls.; Make 45 business and 10 charity calls.

59. Make 3 bikes and 2 wagons in both cases.

61. **(a)**

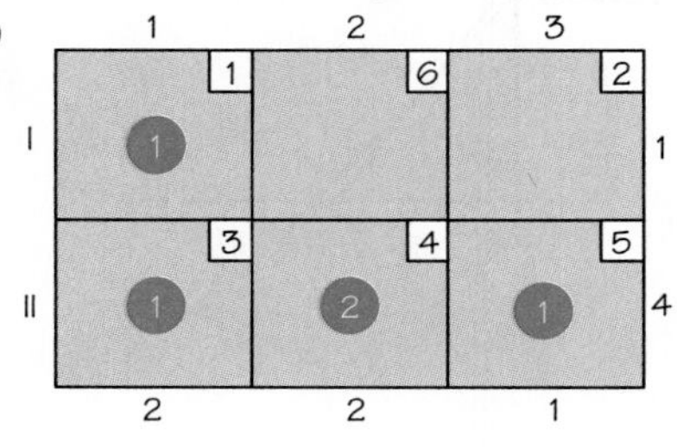

(b) 17
(c) (I, 2): 4; (I, 3): −1

63. **(a)** Connected and has no circuit
(b) Add edge joining Vertex I to Vertex 2.; Add edge from Vertex I to Vertex 3.
(c) Circuit 2, I, 1, II, 2 corresponds to the circuit of cells, (I, 2), (I, 1), (II, 1), (II, 2), (I, 2). Circuit 3, I, 1, II, 3 corresponds to the circuit of cells, (I, 3), (I, 1), (II, 1), (II, 3), (I, 3).

65. **(a)** **(i)**

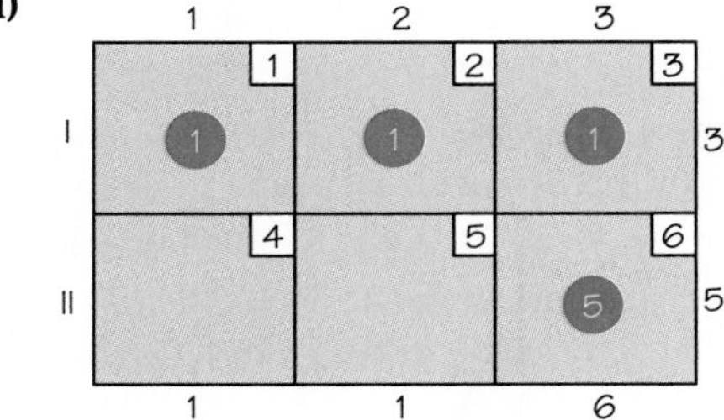

(ii)

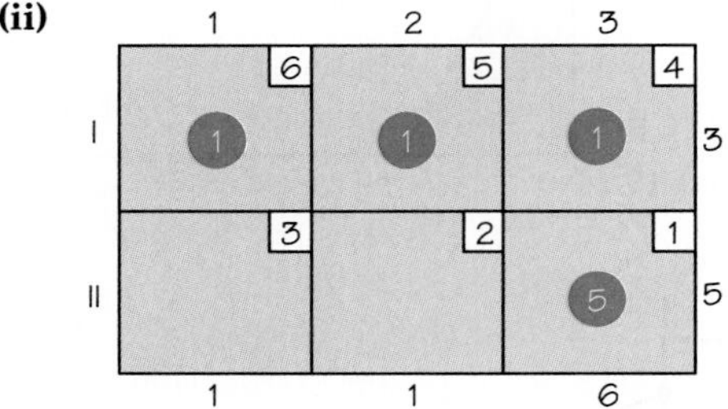

(iii)

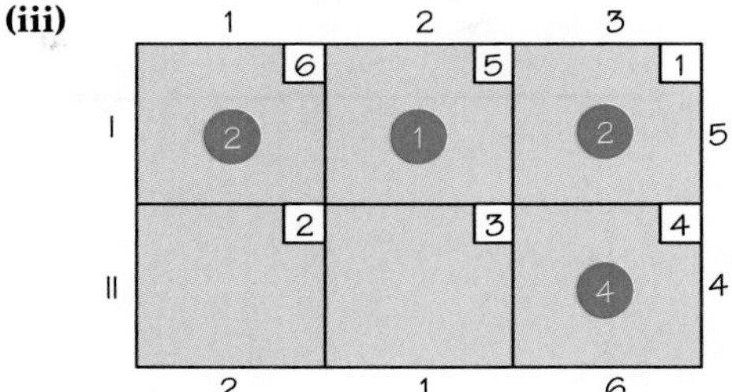

(b) For both **(i)** and **(ii)** the tableaux shown are optimal. However, there are also other optimal tableaux. For **(iii)** the tableau shown is not optimal. Using the stepping stone algorithm, the cost can be reduced to 16.

67. **(a)** Yes
(b) No
(c) Yes

Chapter 5

1. **(a)** Vehicle makes and models (i.e., the four cars).
(b) Vehicle type, transmission type, number of cylinders, city MPG, and highway MPG.
(c) Cylinders (maybe), and the two MPGs (certainly).

3. Given earlier years on the left and recent years on the right, it would be skewed to the left; draw a histogram skewed to the left because most coins in circulation were minted in recent years.

5. (a) Big countries would always top the list.
(b) Using class widths of 2 metric tons per person starting 0.0–1.9, the distribution is skewed to the right; high outliers: Canada, Australia, and the United States.

7. (a) 7.8
(b) Single-peaked and roughly symmetric, center near 13.5%, values between 7.8% and 17.4%

9. (a) There is one high outlier, 200.
(b) The center of the 17 observations other than the outlier is 137 (9th of 17). Values are between 101 and 178.

11. (a) Repeated stems break up the intervals further (e.g., 20–29 splits into 20–24 and 25–29).
(b) Reasonably symmetric.

13. (a) 141.1
(b) 137.6
(c) The high outlier pulls the mean up.

15. The distribution is strongly right-skewed, so the mean is the larger number.

17. It decreases.

19. Examples will vary. One possible answer is 1, 2, 2, 2, 3, 3, 4, 17.

21. 7.8, 12.4, 13.5, 14.3, 17.3

23. (a) 5799, 20,000, 25,942, 34,986, 36,700
(b) Two distinct clusters of values.

25. 0.0, 0.75, 3.2, 7.8, 19.9; Q_3 and the maximum are much farther from the median than are Q_1 and the minimum.

27. The median for bachelor's is greater than Q_3 for high school. The bachelor's distribution is much more spread out, especially at the high-income end, but also between the quartiles.

29. (a) A histogram or stemplot shows distribution to be unimodal and right-skewed.
(b) $\bar{x} = 48.25$, $M = 37.8$; A long right tail pulls the mean up.
(c) 2.0, 21.5, 37.8, 60.1, 204.9; Q_3 and the maximum are much farther above the M than Q_1 and the minimum are below it, showing that the right side of the distribution has more variability than the left side.

31. Arizona, California, Florida, Nevada, New Mexico, and Texas

33. (a) $\bar{x} = 5.448$, $s = 0.221$
(b) $M = 5.46$; yes

35. (a) For both datasets, $\bar{x} = 7.50$ and $s = 2.03$.
(b) The Data A set has two low outliers, and the Data B set has one high outlier.

37. (a) One possible answer is 1,1,1,1.
(b) 0, 0, 10, 10
(c) Yes. *Any* set of 4 equal numbers yields the smallest possible value for s: 0.
(d) No. Within the 0–10 constraint, numbers can't deviate any further from the mean.

39. About 27

41. Mean: A, median: B; unimodal left-skewed distribution generally has mean < median

43. 25.5%; 39.7; −11.3

45. (a) 327 to 345 days
(b) 16%

47. (a) $Q_1 = 438$, $Q_3 = 594$

49. (a) −23.94% to 45.90%
(b) a loss of at least 23.94%

51. (a) 10%; in normal distributions, median = mean
(b) 9.6% to 10.4%
(c) 9.866% to 10.134%

53. (a) 50%, 2.5%
(b) 0.37 to 0.43

55. Red: somewhat right-skewed (with no outliers); yellow: quite symmetrical (with no outliers)

57. (a) Red: $\bar{x} = 39.71$, $s = 1.799$; yellow: $\bar{x} = 36.18$, $s = 0.975$
(b) yellow distribution

59. The percent would be between 2.5% and 16%.

61. Mode, mean, and median would increase.

Chapter 6

1. (a) Latter case; study time
(b) Relationship only
(c) Latter case; rainfall
(d) Relationship only

3. (a) Life expectancy increases with GDP in a curved pattern. The increase is very rapid at first, but levels off for GDP above roughly $5000 per person.
(b) Answers will vary.

5. See answer to Exercise 31 for scatterplot; a strong, positive, straight-line relationship

7. (a) Speed
(b) The regression line for Exercise 33 is included.

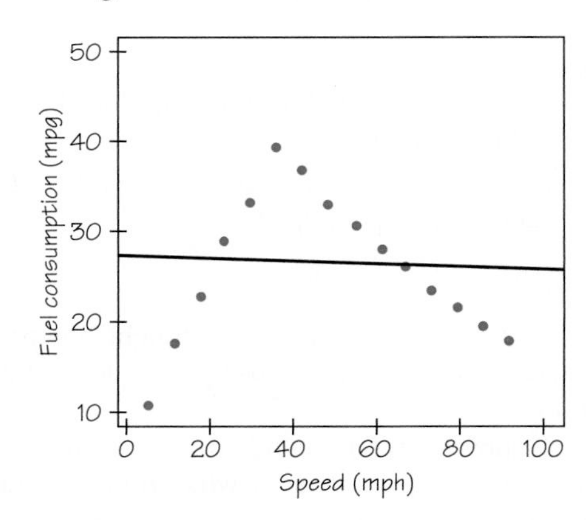

Relationship is curved; additional answers will vary.
(c) None overall
(d) Quite strong; little scatter

9. (a) Ground temperature
(b) The scatterplot is as follows; positive

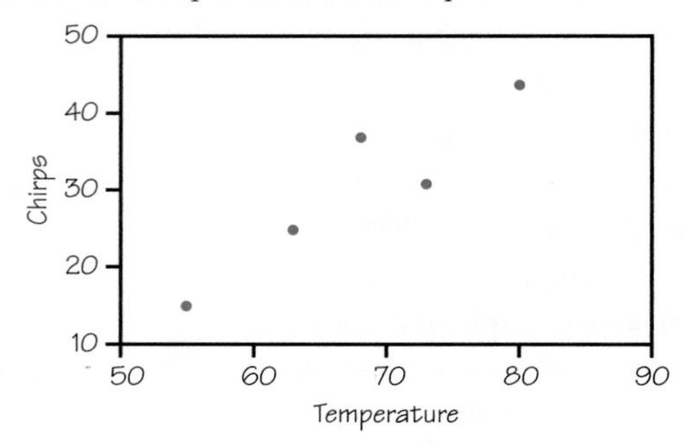

(c) Strong straight-line pattern
(d) Yes

11. 2.31

13. (a) The graph is as follows; pH decreases as the number of weeks increases.

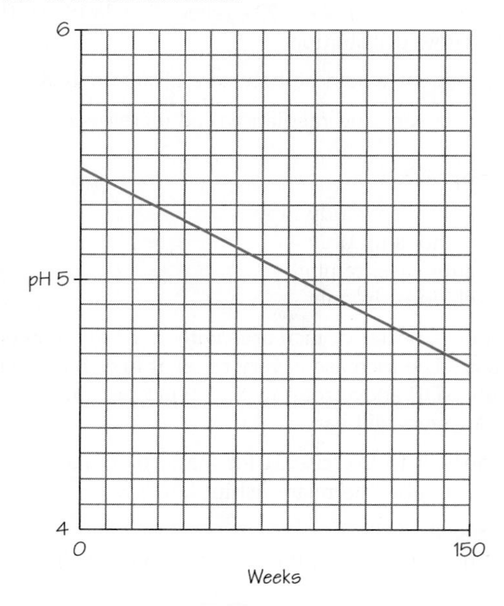

(b) 5.42 and 4.64
(c) −0.0053; on average, pH declined by 0.0053 per week during the study period.

15. (a) predicted BAC = 0.15 − 0.015 × number of hours after drinking stopped
(b) $4\frac{2}{3}$
(c) 10

17. r = 0.9507, reflecting the strong straight-line pattern in the Exercise 5 scatterplot, whose points cluster tightly about a line

19. 0.9425; correlation has slightly decreased; the point is an outlier that deviates only somewhat from the straight-line pattern

21. See the answer to Exercise 7 for the scatterplot; −0.0435; the relationship is strong, but not linear.

23. 1

25. (a) Negative
(b) Negative
(c) Positive
(d) Small

27. (a) Dividend Growth: 0.98; Small Cap Stock: 0.81; Emerging Markets: 0.35
(b) No, just that they moved in the same direction

29. (a) Predicted highway mpg = 0.21 + 1.45 × (city mpg)
(b) 24.86 mpg
(c) Yes; points follow a straight line.

31. The scatterplot with regression line and "up-and-over" method is as follows:

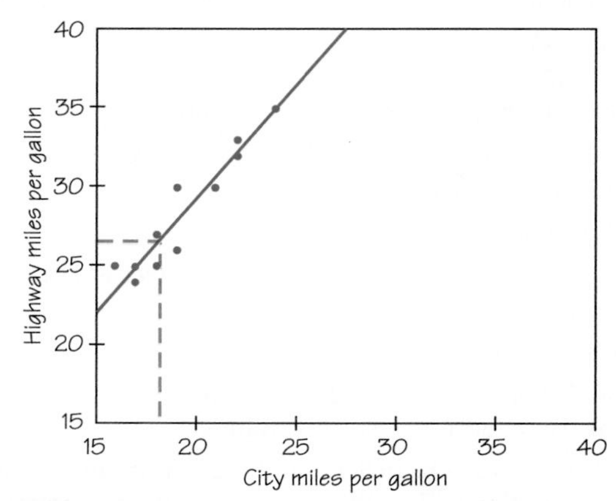

Predicted highway mileage of a car that gets 18 mpg in the city is approximately 26 or 27 mpg (26.31 mpg).

33. (a) 11.2, 37.3, and 18.3, respectively
(b) 26.96, 26.49, and 25.87, respectively
(c) See Exercise 7 answer for plot; no
(d) No; the least-squares line gives the best straight-line fit.

35. (a) Predicted height of husband = 33.67 + 0.54 × (height of woman)
(b) 69.85 in.

37. In $\hat{y} = mx + b$, substitute $b = \bar{y} - \left(r \frac{s_y}{s_x}\right)\bar{x}$, $m = r \frac{s_y}{s_x}$, and $x = \bar{x}$

39. Answers will vary.

41. (a) All four have $r \approx 0.816$ and $\hat{y} = 3.0 + 0.5x$; 8.0
(b) Here are the plots with regression lines:

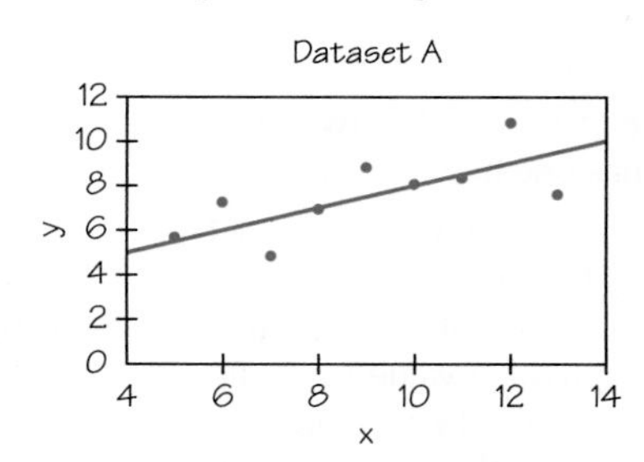

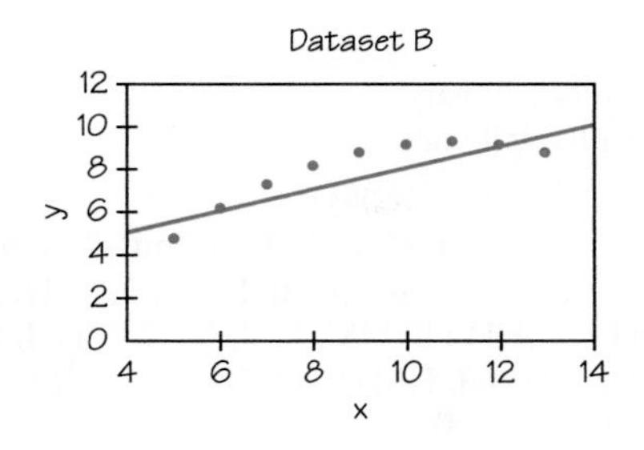

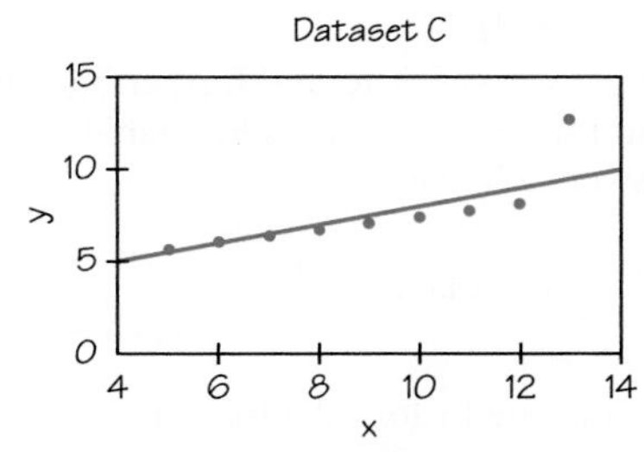

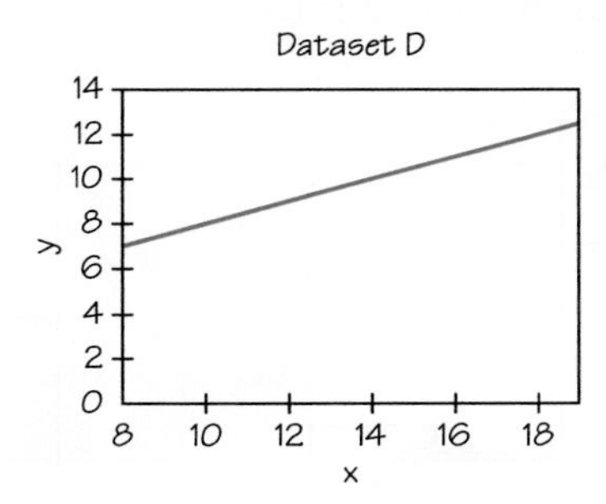

(c) Dataset A; additional answers will vary.

43. Answers will vary.

45. Answers will vary.

47. (a) Lead level and reading score, respectively
(b) Negative; answers will vary; yes

49. (a) Positive; not
(b) $r = 0.3602$

51. (a) 0.42
(b) For each additional inch of a woman's height, the height of the next person dated goes up by 0.42 inch on average.
(c) 69.22 in.

53. (a) $r = 1$; $m = 1$
(b) $r = m \approx -0.4162$

Chapter 7

1. (a) U.S. adults
(b) 1,021

3. Answers will vary.

5. (a) Answers will vary.
(b) Larger; bias due to voluntary response

7. Jessica, Lauren, Ashley, Abigail

9. (a) 001 to 371
(b) 214, 235, 119

11. Repeated samples of the same size from the same population will always be the same.

13. (a) 35, 75, 115, 155, 195
(b) Answers will vary.
(c) Answers will vary.

15. (a) All people aged 18 and over living in the United States
(b) 30%
(c) Answers will vary.

17. Answers will vary.

19. (a) Type of journal prompt
(b) Personal well-being
(c) Three groups with random assignment to groups
(d) Addresses that gratitude causes well-being

21. Classes 1, 2, 7, 8, 9, 13, 14, 15, 17, 18, 19, 22, 25, 27, and 29 receive the treatment.

23.

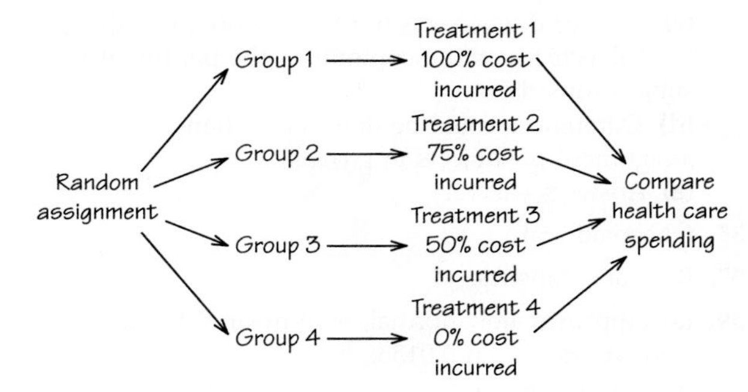

25. (a) See the treatments in the outline in part b.
(b)

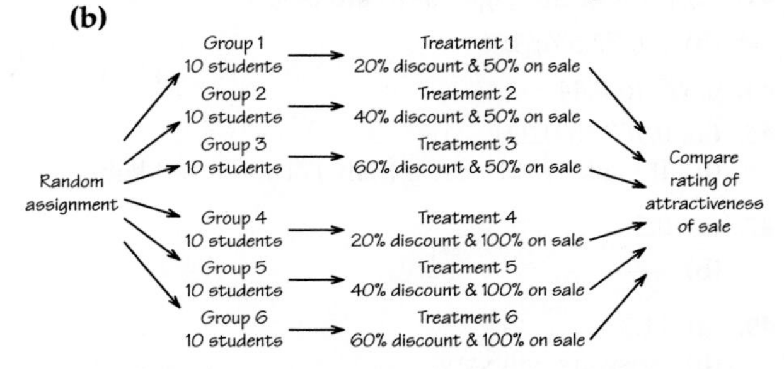

Label the subjects 01 to 60. The first group contains subjects labeled 7, 8, 10, 15, 25, 27, 54, 55, 58, and 60.

27. (a)

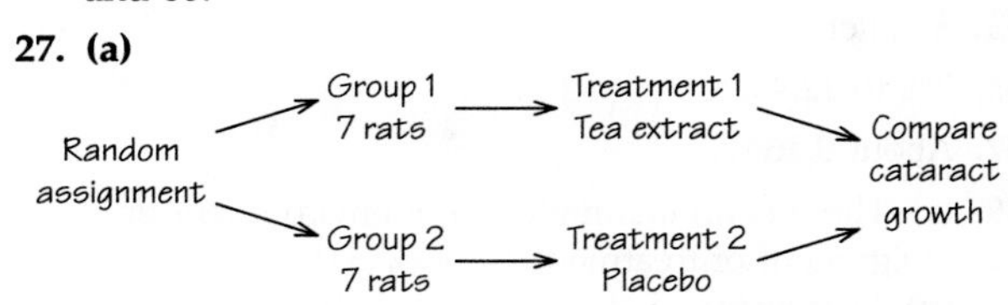

(b) Tea group contains rats 4, 6, 7, 8, 9, 11, and 12.

29. **(a)** No imposed treatment; observational

31. **(a)** Observational; additional answers will vary.
(b)–(c) Answers will vary.

33. **(a)** This is a randomized comparative experiment with four branches. Best to use groups of size 216.

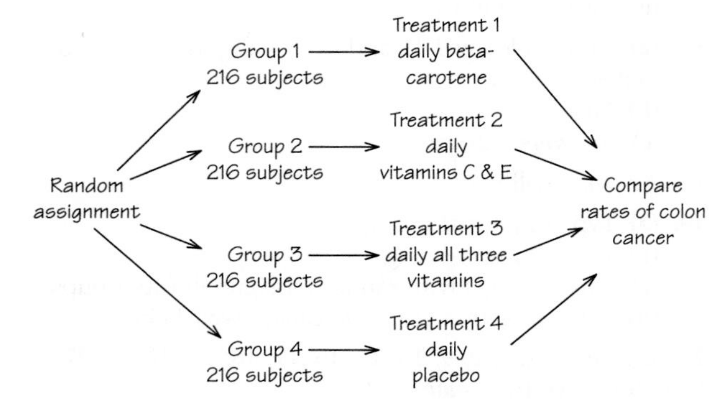

(b) 253, 296, 304, 470, 731
(c) Neither the subjects nor those working with the subjects know the contents of the pill that each subject took daily.
(d) Differences could be due to the chance assignment of subjects to groups.
(e) Answers will vary.

35. Observational

37. Both are statistics.

39. **(a)** Approximately normal, with mean 0.14 and standard deviation 0.0155.
(b) 0.109 to 0.171

41. **(a)** 60% of the digits 0–10 are 0–5.
(b) 29; 72.5%; 1; 6

43. 0.397 to 0.443

45. **(a)** 0.167 to 0.221
(b) It is likely that more than 171 ran a red light.

47. **(a)** 0.5
(b) $\frac{1}{\sqrt{n}}$

49. **(a)** 11.3%
(b) Answers will vary.

51. **(a)** No
(b) Yes

53. Smaller

55. 2% to 12.4%

57. About 0.16

59. **(a)** There is no manipulation of environment or assignment of treatment.
(b) Prospective
(c) It is not possible to manipulate or control the environment of a subject during the subject's life through age 21.

Chapter 8

1. **(a)** Results will vary.
(b) Results will vary.

3. 0.105

5. **(a)** Using H for "hit" and M for "miss", here are the 16 possible sequences: {HHHH, HHHM, HHMH, HMHH, MHHH HHMM, HMMH, MMHH, HMHM, MHHM, MHMH, HMMM, MMMH, MHMM, MMHM, MMMM}
(b) {0, 1, 2, 3, 4}

7. **(a)** It is usually easier to add further branching to a tree than further dimensions to a table.
(b) Answers will vary.
(c) $\frac{7}{8}$

9. Player A: $\frac{3}{4}$ and Player B: $\frac{1}{4}$

11. $\frac{1}{216}$

13. No; it's still true historically that under these atmospheric conditions, it rains 70 percent of the time.

15. **(a)** No.

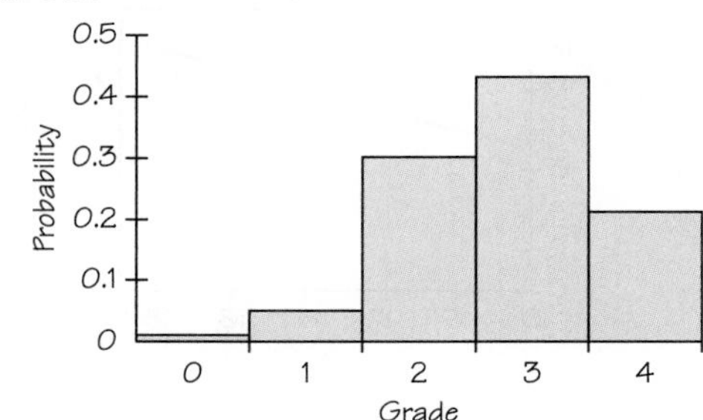

(b) 0.64

17. **(a)** No; Rule 2 was violated.
(b) Yes

19. The probability model for this pair of dice is the same as the one for regular dice.

21. $\frac{8}{15}$

23. **(a)** 1024
(b) $\frac{1}{512}$

25. **(a)** 0.919
(b) 0.377

27. **(a)** 6
(b) *asp, pas, sap, spa*
(c) 66.7%

29. **(a)** 4
(b) 2,598,960
(c) 0.00000154

31. **(a)** $\frac{1}{2} \times \text{base} \times \text{height} = \frac{1}{2}(2)(1) = 1$
(b) $\frac{1}{2}$
(c) 0.125

33. 0.25

35. 2.78; 0.8669

37. Owner-occupied units: 6.248; rented units: 4.321

39. **(a)–(b)** Both models have mean 1 because both density curves are symmetric about 1.

41. $30

43. $0.45

45. **(a)** 0; no difference
(b) $\frac{1}{16}$; helps

47. (0.11, 0.19)

49. **(a)** 5.77 mg
(b) 4; answers will vary.

51. **(a)**

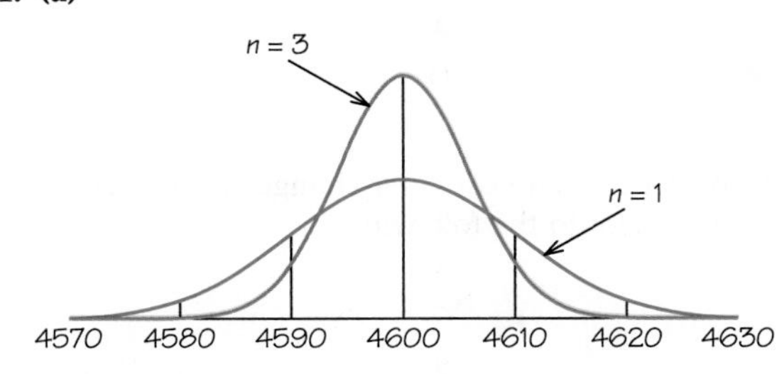

(b) (4580, 4620)
(c) (4588.46, 4611.54)

53. **(a)** About 0.16
(b) Mean: 21.0; standard deviation: 1.7
(c) About 0.0015

55. High; schools with smaller n have larger $\frac{\sigma}{\sqrt{n}}$ (i.e., more variable means)

57. **(a)** 45,697,600
(b) 2600
(c) 0.0000569

59. **(a)** 0.18
(b) 0.39

61. **(a)** 1.156 days
(b) Between 0.54 day and 1.00 day

63. **(a)** 100 percent
(b) 80 percent
(c) 12 percent

Chapter 9

1. Answers will vary.

3. Answers will vary.

5. Each one-on-one score will have a winner because there cannot be a tie.

7.

	Number of voters (4)			
Rank	1	1	1	1
First	A	B	C	D
Second	B	C	D	A
Third	C	D	A	B
Fourth	D	A	B	C

9. Adam

11. **(a)** Yes
(b) Alfonse D'Amato (*D*)

13. B wins.

15. **(a)** *C*
(b) *A*
(c) *D*
(d) *D*

17. **(a)** Five-way tie
(b) *C*
(c) Five-way tie
(d) *E*

19. **(a)** *C*
(b) *E*
(c) *E*
(d) *E*

21. **(a)** *E*
(b) Answers will vary. One possible answer is the following:

	Number of Voters (2)	
Rank	1	1
First	A	C
Second	B	B
Third	C	A

23. **(a)** If everyone prefers *B* to *D*, for example, then *D* has no first-place votes at all.
(b) Moving a winning candidate up one spot on some list neither decreases the number of first-place votes for the winning candidate nor increases the number of first-place votes for any other candidate.

25. **(a)** Condorcet winner always wins this kind of one-on-one contest.
(b) Moving a candidate up on some list only improves that candidate's chances in one-on-one contests.

27. In order to have one candidate's ranking be consistently higher than another candidate's would imply that only one candidate would be considered.

29. Answers will vary.

31. Answers will vary. One possible answer is the following:

	Number of Voters (5)	
Rank	3	2
First	A	B
Second	B	C
Third	C	A

33. **(a)** Since D has the least number of first-place votes, D is eliminated. Since B has the least number of first-place votes, B is eliminated. Thus, A is the unique winner.
(b) B

35. Answers will vary.

37. **(a)** A three-way tie with both methods
(b) That alternative is the sole winner with both methods.
(c) No; either the situation in part (a) or the situation in part (b) must occur.

39. Answers will vary.

41. **(a)** D
(b) A, B, D, and F
(c) B, D, and F
(d) A, B, D, and F

Chapter 10

1. Answers will vary. One example of two such elections is the following:

Election 1			
Rank	Number of Voters (3)		
First	A	A	B
Second	B	B	A

Election 2			
Rank	Number of Voters (3)		
First	B	A	B
Second	A	B	A

3. Answers will vary. One example of two such elections is the following:

Election 1			
Rank	Number of Voters (3)		
First	A	B	B
Second	B	A	A

Election 2			
Rank	Number of Voters (3)		
First	B	B	B
Second	A	A	A

5. **(a)** Doesn't treat all voters the same
(b) A dictatorship in which Voter #1 is the dictator
(c) A dictatorship in which Voter #2 is the dictator and a dictatorship in which Voter #3 is the dictator

7. Consider the leftmost voter changes his or her preference ballot to the following:

C
B
D
A

9. Consider the leftmost voter changes his or her preference ballot to the following:

A
C
D
B

11. Answers will vary.

13. Consider the leftmost voter changes his or her preference ballot to the following:

B
A
D
C

15. Consider the leftmost voter changes his or her preference ballot to the following:

A
C
B

17. Consider the leftmost voter changes his or her preference ballot to the following:

B
A
C

19. **(a)** Consider the agenda D, A, C, B.
(b) Consider the agenda B, D, A, C.
(c) Consider the agenda B, A, C, D.

21. A winner with the Hare system must be ranked at the top of at least one voter's ballot, or else it would be eliminated in the first round. For such a voter, there is no outcome preferred to his or her top choice being the single winner.

23. Answers will vary.

25. Consider the voters in the 7% group to change their ballots to the following:

H
J
D

27. **(a)** To go from having a unique winner to a different unique winner occurs if the winning alternative in the first election has exactly two first-place votes, and one of these two voters changed his or her ballot by moving some other alternative into first place (yielding a worse outcome for this voter).
(b) Tie in second election
(c) Answers will vary.

29. **(a)** Answers will vary.
(b) Answers will vary.

31. 1 and 4

33. Answers will vary.

Chapter 11

1. Any two of the other three voters have enough votes to carry a motion, but the weight-12 voter cannot combine his vote with any one voter to make a winning coalition. Therefore, his vote can never affect a committee decision.

3. The voters with weights 5 and 4 each have veto power, and the weight-3 voter is a dummy.

5. (a) In 1958, *B*, *G*, and *L* were dummies. In 1964, *N*, *G*, and *L* were dummies. There were no dummies in 1970 and after.
(b) 1958, 1964, and 1982

7. (a) Call the voters *A*, *B*, *C*, *D*, where *A* is the weight-3 voter. *A* is pivotal in the following permutations: *BACD, BADC, BCAD, BDAC, CABD, CADB, CBAD, CDAB, DABC, DACB, DBAC*, and *DCAB*.
(b) *ABCD, ABDC, CDBA, DCBA*
(c) $\left(\frac{1}{2}, \frac{1}{6}, \frac{1}{6}, \frac{1}{6}\right)$

9. (a) Permutations of the form 223222 and 222322
(b) 240
(c) $\left(\frac{1}{3}, \frac{2}{15}, \frac{2}{15}, \frac{2}{15}, \frac{2}{15}, \frac{2}{15}\right)$

11. No

13. (a) 3
(b) 0011, 0101, 0110, 1011, 1101, 1110

15. (12,4,4,4)

17. (a) 35
(b) 0
(c) 105
(d) 105

19. The Shapley-Shubik index of the chairperson is $\frac{2}{5}$; each of the other members has index $\frac{1}{15}$. The Banzhaf index is (840,112,112,112,112,112,112,112,112,112).

21.

	Banzhaf Index				
Year	H	N	B	G	L
1958	32	0	0	0	0
1964	32	0	0	0	0
1970	30	2	2	2	2
1976	30	2	2	2	2
1982	28	4	4	0	4

23. Nebraska's power would decrease slightly if it changed its policy.

25. No

27. (a) [4 : 2, 1, 1, 1]
(b) [9 : 3, 3, 1, 1, 1, 1, 1]

29. Faculty members each have Banzhaf index 24; administrators have index 20. Thus, faculty are slightly more powerful.

31. If a voting system has just one minimal winning coalition, then every member of that coalition has veto power. If there are just two minimal winning coalitions, then the two coalitions must overlap. Every voter who belongs to both minimal winning coalitions is essential to pass a measure—that voter has veto power.

33. (a)

Winning Coalitions	Extra Votes	Losing Coalitions	Vote Deficit
{A, B}	20	{ }	51
{A, C}	19	{A}	3
{A, D}	4	{B}	28
{B, C, D}	1	{C}	29
{A, B, C}	42	{D}	44
{A, B, D}	27	{B, C}	6
{A, C, D}	26	{B, D}	21
{A, B, C, D}	49	{C, D}	22

(b) 4 shares
(c) 19 shares
(d) 4 shares

35. (a) [8 : 6, 1, 1, 1, 1, 1, 1, 1, 1]
(b) (492, 16, 16, 16, 16, 16, 16, 16, 16)
(c) $\left(\frac{2}{3}, \frac{1}{24}, \frac{1}{24}, \frac{1}{24}, \frac{1}{24}, \frac{1}{24}, \frac{1}{24}, \frac{1}{24}, \frac{1}{24}\right)$
(d) No, there is a significant difference.

37. [9 : 4, 4, 4, 1, 1, 1, 1, 1]. (Answer is not unique).

39. Let the weight-3 voters be A_1, A_2, and A_3, and the weight-1 voters be B_1, B_2, and B_3. Then the minimal winning coalitions are $\{A_1, A_2, A_3\}$, and 9 coalitions that can be formed by including two of the *A* voters and one *B*-voter. The Banzhaf index is (30,30,30,6,6,6).

41. With three weight-1 voters, the Shapley-Shubik index is

$$\left(\frac{17}{60}, \frac{17}{60}, \frac{17}{60}, \frac{1}{20}, \frac{1}{20}, \frac{1}{20}\right)$$

and with four weight-1 voters, it is

$$\left(\frac{9}{35}, \frac{9}{35}, \frac{9}{35}, \frac{2}{35}, \frac{2}{35}, \frac{2}{35}, \frac{2}{35}\right)$$

Notice that the addition of voter B_4 causes the other weight-1 voters to gain power a bit and causes the weight-3 voters to lose power slightly in the Shapley-Shubik model. In the Banzhaf model, each of the voters loses a bit of his or her share of power when B_4 joins the system.

43. (a) Chair veto
(b) Majority rule
(c) Dictator
(d) Clique
(e) Consensus

Chapter 12

1. Assume a distribution is skewed to the left. The heavier concentration of voters on the right means that fewer voters are farther from the median. Because there are fewer voters "pulling" the mean rightward, it will be to the left of the median. Likewise, a distribution skewed to the right will have a mean to the right of the median.

3. Assume that the candidates take a common position at 0.6. Then, there are 12 voters to the left and 11 to the right of this position. If one of the candidates moves left to 0.5, then the opponent will win by 13 to 12 votes. If one of the candidates moves right to 0.7, then the opponent will win by 14 to 11 votes. Hence, neither candidate has an incentive to depart from position 0.6, making it an equilibrium position.

5. Since the districts are of equal size, the mayor's median or extended median must be between the leftmost and rightmost medians or extended medians. If, say, the left-district positions are much farther away from the mayor's median or extended median than the right-district positions, then the mayor's mean would be in the interval of the left-district positions.

7. While there is no median position such that half the voters lie to the left and half to the right, there is still a position where the middle voter (if the number of voters is odd) or the two middle voters (if the number of voters is even) are located, starting either from the left or right. In the absence of a median, less than half the voters lie to the left and less than half to the right of this middle voter's (voters') position (positions).

Hence, any departure by a candidate from a position of a middle voter to the position of a non-middle voter on the left or right will result in that candidate's getting less than half the votes—and the opponent's getting more than half. Thus, the middle position (positions) is (are) in equilibrium, making it (them) the extended median.

9. When the four voters on the left refuse to vote for a candidate at 0.6, his opponent can do better by moving to 0.7, which is worse for the dropouts.

11. Yes

13. When it is the median or the extended median; yes. Possible examples include the following:

Position i	1	2	3	4	5	6	7
Location (l_i) of position i	0.1	0.2	0.3	0.5	0.6	0.8	0.9
Number of voters (n_i) at position i	8	8	1	2	1	3	1

15. Not necessarily

17. *C* should take a position just to the left or right of *A/B*, on the side closer to the median. Thereby, *C* will receive a majority of the votes, and *A* and *B* will split the remainder.

19. If, say, *A* takes a position at *M* and *B* takes a position to the right of *M*, *C* should take a position just to the left of *M* that is closer to *M* than *B*'s position, giving *C* essentially half the votes and enabling him or her to win the election. If neither *A* nor *B* takes a position at *M*, *C* should take a position next to the player closer to *M*; the position that *C* takes to maximize his or her vote may be either closer to *M* (if the candidates are far apart) or farther from *M* (if the players are closer together), but this position may not be winning.; no

21. Following the hint, *C* will obtain $\frac{1}{3}$ of the vote by taking a position at *M*, as will *A* and *B*, so there will be a three-way tie among the candidates. Because a non-unimodal distribution can be bimodal, with the two modes close to *M*, *C* can win if he or she picks up most of the vote near the two modes, enabling *C* to win with more than $\frac{1}{3}$ of the vote.

23. *B* should enter just to the right of $\frac{3}{4}$, making it advantageous for *C* to enter just to the left of *A*, giving *C* essentially $\frac{1}{4}$ of the vote.

25. *A*: $\frac{1}{6}$; *B*: $\frac{5}{6}$; *C*: $\frac{1}{2}$; *D*: indifferent between entering just to the left of *A*, just to the right of *B*, or in between *A* and *C* at $\frac{1}{3}$, or between *C* and *B* at $\frac{2}{3}$

27. No

29. Answers will vary.

31. There is such a ban in some countries, but there is no consensus on whether this is good for boosting turnout, leads to better candidates running, induces voters to be less strategic, or leads voters to seek out more information on the quality of the candidates.

33. By definition, more voters prefer the Condorcet winner to any other candidate. Thus, if the poll identifies the Condorcet winner as one of the top two candidates, he or she will receive more votes when voters respond to the poll by voting for one or the other of these candidates. The possibility that the Condorcet winner might not be first in the poll, but win after the poll is announced, shows that the plurality winner may not be the Condorcet winner. Some argue that the Condorcet winner is always the "proper" winner, but others counter that a non-Condorcet winner who is, say, everybody's second-most-preferred candidate is a better social choice than a 51%-Condorcet winner who is ranked last by the other 49%.

35. *D*. Yes, because a poll that identifies either the top two or the top three candidates would not include *D*.

37. *A*. Answers will vary.

39. Assume a voter votes for just a second choice. It is evident that voting for a first choice, too, can never result in a worse outcome and may sometimes result in a better outcome (if the voter's vote for a first choice causes that candidate to be elected).

41. Following the hint, the voter's vote for a first and third choice would elect either *A* or *C*. If the voter also voted for *B*, then it is possible that if *A* and *B* are tied for first place, then *B* might be elected when the tie is broken, whereas voting for just *A* and *C* in this situation would elect *A*.

43. No

45. No; *D* and *C*; Class I; these voters cannot bring about a preferred outcome by voting for candidates different from *A* and *B*.

47. Without polling: (i): *A*, (ii): *D*, (iii): *B* and *D*; with polling: (i): *B*, (ii): *D*, (iii): *D*

49. On the one hand, a runoff would ensure that the winner in the runoff receives more than 50 percent of the vote. On the other hand, the need for a runoff would seem less with approval voting, because it is less likely than plurality voting to change the relative standing of the top two candidates, and it would be an added expense for a jurisdiction to incur.

51. 9.5

53. Substitute into the formula for r_i in Exercise 46 $d_i = (n_i/N)D$ and $D = R$. The proportional rule is "strategy-proof" in the sense that if one player follows it, the other player can do no better than to follow it. Hence, knowing that an opponent is following the proportional rule does not help a player optimize against it by doing anything except also following it.

55. No; yes

57. The Democrat can win the election by winning in any two states or in all three. The first three expressions in the formula for PWE_D give the probabilities of winning in the three possible pairs of states, whereas the final expression gives the probability of winning in all three states.

59. Yes

Chapter 13

1. Donald receives the Trump Tower triplex and about 87% ownership of the Palm Beach mansion. Ivana gets the rest.

3. Phil gets his way on the stereo level issue, the smoking rights issue, the phone time issue, the visitor policy issue, and about 87% of his way on the alcohol issue. Mike gets his way on the rest.

5. **(a)** Salary recommendations and external reviews
(b) Class schedules, department meetings, and calculus placement
(c) They share hiring, with Mary taking on more of the burden.

7. Answers will vary.

9. *Allocation 1:*
(a) Not proportional
(b) Not envy-free
(c) Not equitable
(d) Example: Give Bob *X*, Carol *Y*, and Ted *Z*.
Allocation 2:
(a) Not proportional
(b) Not envy-free
(c) Not equitable
(d) Example: Give Bob *Y*, Carol *X*, and Ted *Z*.
Allocation 3:
(a) Not proportional
(b) Not envy-free
(c) Not equitable
(d) It is Pareto-optimal.
Allocation 4:
(a) Not proportional
(b) Not envy-free
(c) Not equitable
Allocation 5:
(a) It is proportional.
(b) Not envy-free
(c) It is equitable.

11. Mary gets both the car and the house and pays John $43,831.25.

13. *A* receives the farm plus $7333. *B* receives $132,334. *C* receives both the house and the sculpture, while paying $139,667.

15. The allocation is as follows:

Potential Recipient	Total Points
A	16.0
B	13.5
C	17.0
D	20.5

17. Answers will vary. Possible answers include: One could use an absolute measure, awarding; for example, one point for each month a potential recipient has been waiting.

19. Carol first chooses the investments, and the final allocation has her also receiving the boat and the washer-dryer.

21. Fred first chooses the boat, and the final allocation has him also receiving the car and the motorcycle.

23. Ivana first chooses the Connecticut estate, and the final allocation has her also receiving the Trump Plaza apartment and the cash and jewelry.

25. Fred will do the calculus placement, department meetings, and class schedules. Mary will do external reviews, hiring, and salary recommendations.

27. Answers will vary. Possible ways include having Bob divide the cake into four pieces and letting Carol choose any three.

29. **(a)** Divider
(b) Bob knows the preferences of the other party. In Exercise 26, we assumed that the divider didn't know the preferences of the other party.

31. **(a)** See figures below:

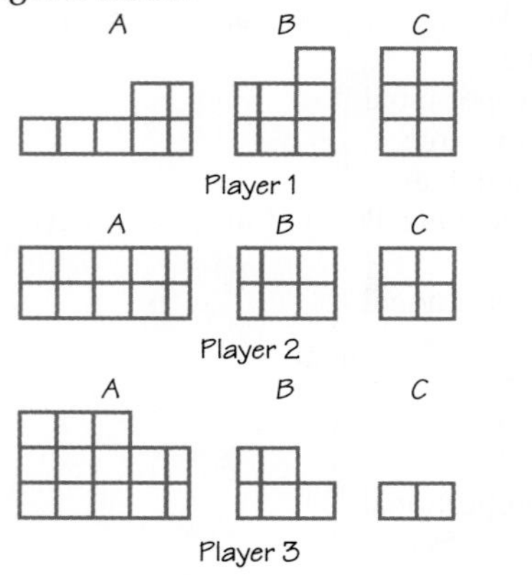

(b) Player 2 finds *A* acceptable, but not *B* or *C*. Player 3 finds *A* acceptable, but not *B* or *C*.
(c) *C*
(d) (1) Player 1: 6 square units; Player 2: 7 square units; Player 3: 10 square units; (2) Player 1: 6 square units; Player 2: $8\frac{2}{3}$ square units; Player 3: 8 square units.

33. **(a)** Ted thinks he is getting at least one-third of part of the cake (Bob's piece) plus one-third of the rest of the cake (Carol's piece).
(b) Bob gets to keep exactly two-thirds (in his own view) of the piece that he initially received and thought was at least of size one-half. Two-thirds times one-half equals one-third.
(c) Answers will vary.

35. Answers will vary.

37. **(a)** At the point where the other knife started
(b) Answers will vary.

39. If I bid, say, \$90, then Bob loses with a bid of \$80, and this is worse for him than winning with a bid of \$100 and paying \$90 for a lamp that he thinks is worth \$100.

Chapter 14

1.

	Rounded Percentage
Rent	44%
Food	23%
Transportation	10%
Gym	17%
Miscellaneous	7%
These percentages add up to 101%.	

3. Yes.

5. Straightforward rounding will not work. An apportionment method should be used, and answers may vary.

7. 0 + 2 + 1 + 2 + 3 + 2 = 10

9. The Anti-UFO and Who Cares parties are tied.

11.

Geometry	3 sections
Algebra	1 section
Calculus	1 section

13. **(a)**

Abe	36
Beth	19
Charles	23
David	20
Esther	2
Total	100

(b)

Abe	36
Beth	19
Charles	22
David	20
Esther	2
Total	100

(c) Blame the population paradox.

15.

Party	Population	Seats Apportioned		
A	5,576,330	27	28	28
B	1,387,342	7	7	7
C	3,334,241	16	17	17
D	7,512,860	37	37	38
E	310,968	2	1	1
Total	18,121,741	89	90	91

The Alabama paradox occurs when the apportionment for Party *E* decreases from 2 to 1 as the house size increases from 89 to 90.

17.

Geometry	4 sections
Algebra	canceled!
Calculus	1 section

19. Jefferson and Webster apportionments are identical:

	36 Pearls	37 Pearls
Abe	14	15
Beth	19	19
Charles	3	3

Although the Jefferson and Webster methods yield the same result, that does not mean that they are right. If there is a principle on which to choose a method, it would probably be to choose the method by which the cost per pearl is as close as possible to the same for each of the friends. The cost per pearl is the district size, so they should use the Dean method (see Exercise 48 and Writing Project 4), which minimizes absolute differences in district size. Charles might want to study up on it, because it allocates the 36th pearl to Beth, and the 37th to him!

21. (a) There will be 3 Geometry sections and 2 Pre-calculus sections, according to the following d'Hondt table:

Geometry	Pre-calculus	Calculus
~~52~~	~~33~~	15
~~26~~	~~16.5~~	7.5
~~17.3~~	11	5

(b) There will be 2 sections each of Geometry and Pre-calculus, and 1 section of Calculus.

Geometry	Pre-calculus	Calculus
~~52~~	~~33~~	~~15~~
~~17.3~~	~~11~~	5
10.4	6.6	3

23. Here is the Sainte-Laguë table. (The names of the parties are abbreviated.)

D	R	Gt	Gn	I
~~44,856~~	~~34,944~~	~~20,004~~	~~19,002~~	~~9,804~~
~~14,952~~	~~11,648~~	~~6,668~~	~~6334~~	~~3268~~
~~8,971.2~~	~~6,988.8~~	~~4000.8~~	~~3800.4~~	1960.8
~~6,408~~	~~4,992~~	2857.71	2714.57	1400.57
~~4,984~~	~~3,882.7~~	2,222.7	2,111.3	1,089.3
~~4,077.8~~	3,176.7	1,818.5	1,727.5	891.3
~~3,450.5~~	2,688	1,538.8	1,461.7	754.2
2,990.4	2,329.6	1,333.6	1,266.8	653.6

The apportionment is
Demopublicans 7
Repocrats 5
Greenocrats 3
Greenicans 3
Independents 2

25. Here is the d'Hondt table.

Whigs	Tories	Liberals	Centrists
~~5,525,381~~	~~3,470,152~~	~~3,864,226~~	201,203
~~2,762,691~~	~~1,735,076~~	~~1,932,113~~	100,602
~~1,841,794~~	~~1,156,717~~	~~1,288,075~~	67,068
~~1,381,345~~	~~867,538~~	~~966,057~~	50,301
~~1,105,076~~	~~694,030~~	~~772,845~~	40,241
~~920,897~~	~~578,359~~	~~644,038~~	33,534
~~789,340~~	~~495,736~~	~~552,032~~	28,743
~~690,673~~	~~433,769~~	~~483,028~~	25,150
~~613,931~~	385,572	~~429,358~~	22,356
~~552,538~~	347,015	386,423	20,120
~~502,307~~	315,468	351,293	18,291
~~460,448~~	289,179	322,019	16,767
~~425,029~~	266,935	297,248	15,477
394,670	247,868	276,016	14,372

The apportionment is
Whigs 13
Tories 8
Liberals 9
Centrists 0.
The Centrists will receive the 64th seat.

27. The Hill-Huntington method will apportion a seat to any party with at least 1 vote. This would obviously pose problems because splinter parties with weird agendas would be elected.

29. The Hamilton apportionment is

$$88\% + 2\% + 2\% + 8 \times 1\% = 100\%,$$

which satisfies the quota condition.

The Jefferson and Webster apportionments are

$$90\% + 10 \times 1\% = 100\%$$

The quota condition is violated.

31.

	Before Tax	After Tax
Abe	36	36
Beth	19	19
Charles	22	23
David	20	20
Esther	3	2
Total	100	100

Esther must give one of her three rare coins to her nephew.

33. **(a)** One quota will be rounded up and the other down to obtain the Webster apportionment. The quota that is rounded up will have a fractional part greater than 0.5 and will be greater than the fractional part of the quota that is rounded down. The Hamilton method will give the party whose quota has the larger fractional part an additional seat. Thus, the apportionments will be identical.
(b) These paradoxes never occur with the Webster method, which gives the same apportionment in this case.
(c) The Hamilton method, which always satisfies the quota condition, gives the same apportionment.
(d) No, each of these methods is capable of producing a different apportionment.

35. 10.77%

37. **(a)** Absolute difference, 0.413700; relative difference, 41.1%
(b) Absolute difference, 0.618695, relative difference, 44.4%
(c) 53 seats for California and 1 seat for Montana

39. 3 sections of Algebra, and 1 section each for Geometry and Calculus

41. at least 11,525

43. There are no ties with the Hamilton, Jefferson, or Webster method. The Hill-Huntington rounding of the first district's quota is 2. Hill-Huntington rounds the other quotas to 9 and 10, respectively. To adjust the quota for the first district so that it will receive 1 seat, the divisor must be greater than $\frac{100{,}000}{\sqrt{2}} = 50{,}000\sqrt{2}$. Using this divisor, the adjusted quotas would be $\sqrt{2}$ for the first district, and $6\sqrt{2} = \sqrt{8 \times 9}$ for the second district. Any divisor large enough to cause the first district's quota to be rounded down will also cause the second district's quota to be rounded down. The result would be 1 seat for the first district, 8 seats for the second district, and 10 seats for the third district, a total of 19 seats. The first and second districts are tied for the last seat.

45. **(a)** Lowndes favors small states.
(b) Yes
(c) Yes

(d)

State	Population	Quota	Lower Quota	Priority	Apportionment
KY	68,705	1.995	1	99.5%	2
RI	68,446	1.988	1	98.8%	2
DE	55,540	1.613	1	61.3%	2
VT	85,533	2.484	2	24.2%	3
SC	206,236	5.989	5	19.8%	6
CT	236,841	6.878	6	14.6%	7
NY	331,589	9.629	9	7.0%	10
MA	475,327	13.803	13	6.2%	14
PA	432,879	12.570	12	4.8%	12
NJ	179,570	5.214	5	4.3%	5
NH	141,822	4.118	4	3.0%	4
GA	70,835	2.057	2	2.9%	2
NC	353,523	10.266	10	2.7%	10
VA	630,560	18.310	18	1.7%	18
MD	278,514	8.088	8	1.1%	8
Totals	3,615,920	105	97		105

47. **(a)** No
(b) The divisor will have to be greater than the standard divisor.
(c) It favors small states.
(d) No

49. **(a)** The Jefferson method.
(b) The method of John Quincy Adams.
(c) The Hill-Huntington method, the Adams method (see Exercise 47), and the Dean method (see Exercise 48 and Writing Project 4)

Chapter 15

1. **(a), (b)** Saddlepoint at row 1 (maximin strategy), column 2 (minimax strategy), giving value 5
(c) Row 2 and column 1

3. **(a)** No saddlepoint
(b) Rows 1 and 2 are both maximin strategies; column 1 is the minimax strategy.
(c) None

5. **(a), (b)** Saddlepoint at row 3 (maximin strategy), column 3 (minimax strategy), giving value −20
(c) Column 3 dominates columns 1 and 2, so column player should avoid strategies from columns 1 and 2.

7. No. If one strategy is dominant, the strategies that dominate it may be undominated—none may dominate the others—so there may be no dominant strategy, even though one strategy is dominated.

9. Batter's optimal mixed strategy: $\left(\frac{1}{4}, \frac{3}{4}\right)$; pitcher's optimal mixed strategy: $\left(\frac{1}{4}, \frac{3}{4}\right)$; value 0.275

11. Saddlepoint is "not cheat" and "audit," giving value −\$100.

13. (a)

	Officer Does Not Patrol	Officer Patrols
You park in street	0	−\$40
You park in lot	−\$32	−\$16

(b) Your optimal mixed strategy: $\left(\frac{2}{7}, \frac{5}{7}\right)$; officer's optimal mixed strategy: $\left(\frac{3}{7}, \frac{4}{7}\right)$; value: −\$22.86
(c) It is unlikely that the officer's payoffs are the opposite of yours.
(d) Answers will vary.

15. (a) Answers will vary.
(b) Showing that your strategy is optimal involves showing that it guarantees at least a tie.

17. Always play *H*.

19. (a) "Bet, then call" should be avoided by player I.
(b) Player I: $\left(\frac{1}{3}, \frac{2}{3}, 0\right)$; player II: $\left(\frac{2}{3}, 0, \frac{1}{3}\right)$; value $-\frac{1}{12}$.
(c) Player II

21. When *A* succeeds in inducing *B* to think that the threat is real and, as a consequence, *B* defers to the threatener—without the threat being carried out.

23. (a) 50% chance of rain: leave umbrella; 75% chance of rain: carry umbrella
(b) Carry umbrella
(c) Saddlepoint at "carry umbrella" and "rain," giving value −2
(d) Leave umbrella

25. The Nash equilibrium outcomes are (4, 3) and (3, 4).

27. The Nash equilibrium outcome is (2, 4), which is the product of dominant strategies by both players.

29. Your choice will depend on whether you put more value on obtaining a payoff of 4 while avoiding a payoff of 1 by choosing your first strategy, or "playing it safe" by never doing worse than a payoff of 2, and sometimes obtaining a payoff of 3, by choosing your second strategy.

31. These choices give x as an outcome. X certainly would not want to depart from a strategy that yields a best outcome; furthermore, neither Y's departure to another outcome in the first column nor Z's departure to another outcome in the second row can improve on x for these players. It seems strange, however, that Z would choose x over z, since z is sincere and dominates x. Thus, there seem few if any circumstances in which this Nash equilibrium would be chosen.

33. Answers will vary. Here is an example: Consider the 7-person voting game in which three voters have the preference xyz (one of whom is chair), two voters have the preference zxy, and two voters have the preference zyx. For the three xyz voters, voting for both x and y dominates voting for only x; and for the two zyx voters, voting for only z dominates voting for both z and y. With the dominated strategies of x and zy eliminated, in the second-reduction matrix z dominates zy for the two zyx voters, yielding the sophisticated outcome z, which is the chair's worst outcome.

35. Answers will vary.

37. (a) If *A* hates *B*, *B* hates *C*, and *C* hates *A*, *A* does not shoot *B*, lest he be shot by *C*. *B* shoots *C*, putting her in the best position because she shoots her antagonist, though *A* also survives.
(b) Subsequent rounds would not change incentives—*C*, the player nobody hates, would be the sole survivor because nobody would shoot her on subsequent rounds.

39. *B* will shoot *C*.

41. *B* will be indifferent between shooting or not shooting *C* because whatever *B* does, he will be shot in the end by *A*.

43. The hint shows that bidding your reservation price can never be worse and, in general, will be better than bidding more or less than it. Honesty is optimal, because what you bid is not what you pay, so you can "afford" to be honest, generally paying less than your bid.

Chapter 16

1. 3

3. 3

5. 5

7. 6

9. 1

11. 3

13. 8

15. 7

17. The lead digits 97 contribute 30 to the weighted sum. Thus, the digit needed to make the sum evenly divisible by 10 is the same if you leave the 9 and 7 out of the calculation. (For example, if the sum were 162, including the 97 digits, then the sum would be 132 if it didn't include them. In either case, the check digit is 8.)

19. 6

21. Yes, it is valid.

23. No. Because 11 contributes 3 to the running total, whereas 55 contributes 16, the running total for the changed number will end with 3 instead of 0.

25. The Luhn algorithm for a valid credit card results in a running total that ends with 0, so increasing the sum by 10 would still give a total that ends with 0. Thus, the new number is valid. Increasing the sum by 12 changes the last digit of the running total to 2, which is not a valid credit number.

27. 3

29. The check digit is 7. This check digit method detects all single-digit errors.

31. In the odd-numbered positions, an error caused by replacing an odd digit by an odd digit or an even digit replaced by an even digit is not detected.

33. It is the UPC code for Kleenex. Changing exactly one digit will not result is a valid UPC number.

35. In the Codabar scheme, the digits 5, 6, 7, 8, and 9 in odd-numbered positions contribute the values 11, 13, 15, 17, and 19 to the check-digit calculation. In the method described in this exercise, these five values contribute 1, 3, 5, 7, and 9, respectively, to the calculation. Thus, both methods contribute the same amount to the last digit of the total. In the even-numbered positions, both methods use the digit itself in the weighted sum, so they agree in those positions.

37. 0-669-09325-4

39. If c_1 is the check digit for the weights 7, 3, 9, 7, 3, 9, 7, 3 and c_2 is the check digit for the weights 3, 7, 1, 3, 7, 1, 3, 7, then $c_2 = 0$ when $c_1 = 0$. Otherwise, $c_2 = 10 - c_1$.

41. The mistake of reading a 2 as a 7 is detected because the sum of the digits of the incorrect number would be odd. The mistake of reading a 2 as an 8 is not detected because the sum of the digits of the incorrect number would remain even. An error is detected when an odd digit is misread as an even one (or vice versa) because the sum of the digits changes from even to odd (or vice versa). Approximately 50% of errors are detected.

43. The computer need not know which digit is the check digit because it merely checks to see if the weighted sum is divisible by 9 for traveler's checks and divisible by 10 for the other two.

45. Because the remainder upon dividing by 9 is less than 9, 9 cannot be a remainder.

47. Because the remainder after dividing by 7 is less than 7, the digits 7, 8, and 9 cannot be a check digit.

49. Yes. The ISBN-10 scheme detects all transposition errors.

51. For the transposition to go undetected, it must be the case that the difference of the correct number and the incorrect number is evenly divisible by 11. That is, $(10a_1 + 9a_2 + 8a_3 + \cdots + a_{10}) - (10a_3 + 9a_2 + 8a_1 + \cdots + a_{10})$ is divisible by 11. This reduces to $2a_1 - 2a_3 = 2(a_1 - a_3)$ is divisible by 11. But $2(a_1 - a_3)$ is divisible by 11 only when $a_1 - a_3$ is divisible by 11 and this happens only when $a_1 - a_3 = 0$. In this case, there is no error. The same argument works for the fourth and sixth digits.

53. The combination 72 contributes $3 \cdot 7 + 2 = 23$ or $7 + 3 \cdot 2 = 13$ (depending on the location of the combination) toward the total sum, while the combination 27 contributes $3 \cdot 2 + 7 = 13$ or $2 + 7 \cdot 3 = 23$. So, the total sum resulting from the number with the transposition is still divisible by 10. Therefore, the error is not detected. Similarly, the combination 26 contributes $3 \cdot 2 + 6 = 12$ toward the total sum, whereas the combination 62 contributes $3 \cdot 6 + 2 = 20$ toward the total sum; so the new sum will not be divisible by 10. Similarly, when the combination 26 contributes $2 + 3 \cdot 6 = 20$ to the total, the combination 62 contributes $6 + 3 \cdot 2 = 12$ to the total. So, the total for the number resulting from the transposition will not be divisible by 10 and the error is detected. In general, an error that occurs by transposing *ab* to *ba* is undetected if and only if $a - b$ is 5 or -5.

55. Many single-digit errors go undetected. Substitution of b for a where $b - a$ is 5 or -5 in positions 1, 5, 7, 9, and 11 is undetected; all errors in position 3 are undetected; substitution of b for a where $b - a$ is even in position 8 is undetected.

57. Because both numbers are valid, the difference of the weighted sums is divisible by 10. That is, $(7w + 3 + 2w + 1 + 5w + 6 + 7w + 4) - (7w + 3 + 2w + 1 + 5w + 6 + 6w + 1)$ is divisible by 10. The difference simplifies to $w + 3$. So, $w = 7$.

59. **(a)** 51593-2067; 2
(b) 50347-0055; 1
(c) 44138-9901; 1

61. **(a)** 20782-9960
(b) 55435-9982
(c) 52735-2101

63. 2

65. If a double error in a block results in a new block that does not contain exactly two long bars, we know that this block has been misread. If a double error in a block of five results in a new block with exactly two long bars, the block now gives a different digit from the original one. If no other digit is in error, the check digit catches the error because the sum of the 10 digits will not end in 0. So, an error has been detected in every case. Errors of the first type can be corrected just as in the case of a single error. When a double error results in a legitimate code number, however, there is no way to determine which digit is incorrect.

67. If you replace each short bar in the bar code table by an *a* and replace each long bar in the bar code table by a *b*, the resulting strings are listed in alphabetical order.

69. Right to left

71. Wyoming, Nevada, and Alaska

73. a

75. From Spotlight 16.4, we see that the only states that have at most two three-digit codes are Montana, Delaware, South Dakota, Alaska, North Dakota, Vermont, Wyoming, Idaho, and Nevada. Of these, all but the last two have a population under 1 million.

77. $20^3 \cdot 10^3 = 8{,}000{,}000$

79. S-000, S-200, L-550, L-300, E-663, O-451

81. $26 \cdot 7^3 = 8918$

83. The digits 03 indicate that the person was born in 1903 or 2003.

85. For a woman born in November or December, the formula $40(m - 1) + b + 600$ gives a number requiring four digits.

87. August 1, 1958.

89. J-222 versus J-250, M-522 versus M-200, S versus S-000

91. Twins; sons named after their fathers (such as John L. Smith, Jr.); common names such as John Smith and Mary Johnson; states that do not include year of birth in the code

93. Because of the short names and large population, there would be a significant percentage of people whose names would be coded the same.

Chapter 17

1. No answer provided

3. **(a)** 6
(b) 3

5. 1001101

7. 000000, 100011, 010101, 001110, 110110, 101101, 011011, 111000

9. 0000000, 1000001, 0100111, 0010101, 0001110, 1100110, 1010100, 1001111, 0110010, 0101001, 0011011, 1110011, 1101000, 1011010, 0111100, 1111101. No, because 1000001 has weight 2.

11. 000000, 100101, 010110, 001011, 110011, 101110, 011101, 111000. 001001 is decoded as 001011; 011000 is decoded as 111000; 000110 is decoded as 010110; 100001 is decoded as 100101.

13. 00000000, 00010111, 00101110, 01001011, 10001101,11000110, 10100011, 10011010, 01100101, 01011100, 00111001, 11101000, 11010001, 10110100, 01110010, 11111111. The code will detect any three errors or correct any single error.

15. $2^5 = 32$; $2^8 = 256$

17. *AATAAAGCAA*

19. 111101000111001010; *AABAACAEADB*

21. T, N, and R; E

23. 13403 336 77 −49 53 −61 20 99 35 −17; we go from 59 characters to 36 characters, a reduction of almost 39 percent.

25. *A* is 0; *B* is 111; *C* is 10; *D* is 110

27. *BCEEFCDDCFF*

29. *B* is the least likely letter, *J* is the second least likely, and *G* is the third least likely.

31. UHWUHDW; ADVANCE

33. 13

35. There is no integer j such that $2j = 1$ modulo 26.

37. ROLLING STONES

39. WLZCL LZBL

41. SUN

43. 1111101, 1100011

45. 0010110, 1110110, 1000110, 1011110, 1010010, 1010100, 1010111

47. Recall that the distance between two strings is the number of positions that they differ. Observe that everywhere that u and v agree, so do $u + w$ and $v + w$. Conversely, in every position that u and v disagree, so do $u + w$ and $v + w$.

49. 23, 49, 16

51. 13

53. N converts to 14 and O converts to 15, but 14 and 77 have the greatest common divisor of 7. On the other hand, using blocks of length 4, NO converts to 1415, and the greatest common divisor of 77 and 1415 is 1.

55.

P	Q	$P \wedge Q$	$P \vee (P \wedge Q)$
T	T	T	T
T	F	F	T
F	T	F	F
F	F	F	F

Because the entries in the column for the variable P are exactly the same as the entries in the column for $P \vee (P \wedge Q)$, the two expressions are logically equivalent.

57. First, we construct the truth table for $\sim(P \wedge Q)$:

P	Q	$P \wedge Q$	$\sim(P \wedge Q)$
T	T	T	F
T	F	F	T
F	T	F	T
F	F	F	T

Next, we construct the truth table for $\sim P \vee \sim Q$:

P	Q	$\sim P$	$\sim Q$	$\sim P \vee \sim Q$
T	T	F	F	F
T	F	F	T	T
F	T	T	F	T
F	F	T	T	T

Because the last columns of the two truth tables are identical, we have shown that $\sim(P \wedge Q)$ is logically equivalent to $\sim P \vee \sim Q$.

59. First, we construct the truth table for $P \wedge (Q \vee R)$:

P	Q	R	Q ∨ R	P ∧ (Q ∨ R)
T	T	T	T	T
T	T	F	T	T
T	F	T	T	T
T	F	F	F	F
F	T	T	T	F
F	T	F	T	F
F	F	T	T	F
F	F	F	F	F

Next, we construct the truth table for $(P \wedge Q) \vee (P \wedge R)$:

P	Q	R	P ∧ Q	P ∧ R	(P ∧ Q) ∨ (P ∧ R)
T	T	T	T	T	T
T	T	F	T	F	T
T	F	T	F	T	T
T	F	F	F	F	F
F	T	T	F	F	F
F	T	F	F	F	F
F	F	T	F	F	F
F	F	F	F	F	F

Because the last columns of the two truth tables are identical, we have shown that $P \wedge (Q \vee R)$ is logically equivalent to $(P \wedge Q) \vee (P \wedge R)$.

61. The truth table for $\sim P \vee Q$ is:

P	Q	~P	~P ∨ Q
T	T	F	T
T	F	F	F
F	T	T	T
F	F	T	T

Because the last column of this truth table is identical to the one for $P \rightarrow Q$, we conclude that the two expressions are logically equivalent.

63. Let P denote "it snows" and let Q denote "there is school." Then the statement "If it snows, there will be no school" can be expressed as $P \rightarrow \sim Q$. Similarly, the statement "It is not the case that it snows and there is school" can be expressed as $\sim(P \wedge Q)$. We now construct the truth tables for each of these expressions.

A truth table for $P \rightarrow \sim Q$ is:

P	Q	~Q	P → ~Q
T	T	F	F
T	F	T	T
F	T	F	T
F	F	T	T

A truth table for $\sim(P \wedge Q)$ is:

P	Q	P ∧ Q	~(P ∧ Q)
T	T	T	F
T	F	F	T
F	T	F	T
F	F	F	T

Because the two tables have identical last columns, the two expressions are logically equivalent.

65. **(a)** 00100000
(b) 00110000

67. $s \wedge 11100000 = t \wedge 11100000$ is the same as $s \wedge 11100000 - t \wedge 11100000 = 00000000$, but in mod 2, subtraction is the same as addition because $1 + 1 = 0$.

69. s has a 0 in positions 2, 4, and 6 and a 1 in position 8.

71. 01100010, 01110010, 01101010, 01111010

73. 8.0.0.0

75. No. The network address for 172.16.17.30 with subnet mask 255.255.255.240 is 172.16.17.16; the network address for 172.16.17.15 with subnet mask 255.255.255.240 is 172.16.17.0.

Chapter 18

1. **(a)** None
(b) $4 \text{ in} \times \frac{4}{3} = 5\frac{1}{3} \text{ in}$
(c) $6 \text{ in} \times \frac{3}{4} = 4\frac{1}{2} \text{ in}$

3. **(a)** $1\frac{1}{3} \approx 1.33$
(b) 1.78
(c) 78%

5. 26% larger

7. **(a)** $\frac{1}{40} = 0.025$
(b) 64,000 times as large
(c) 400 cm = 4.00 m = 13.1 ft

9. The writer uses both multiplicative and subtractive language together. Better to say: "That would be one-fourth as many pills."

11. Nothing can decrease 150 percent without becoming negative. It's not clear what the writer meant.

13. Answers will vary.

15. 0.69

17. 32 mpg

19. **(a)** 0.00013364 ton
(b) That all parts of the scale model are made of the same materials as the real locomotive
(c) 0.267 lb
(d) 0.121 kg
(e) 0.000121 metric ton

21. \$7.70/gal

23. About 5 ft 7 in.

25. **(a)** 40 mph
(b) 28 mph
(c) 60 mph

27. **(a)** 100% (which is absurd)
(b) 50% (which makes sense)

29. **(a)** 39%
(b) 64%

31. **(a)** For Option A, yes; for Option B, no
(b) For either a loss or a gain, the absolute value of the percentage is higher for Option B.
(c) Either way, use Option B.

33. **(a)** 5.5
(b) About 33,000 lb

35. **(a)** 400,000 lb
(b) 28 lb/in^2

37. 6200 lb or 2800 kg

39. The lights are strung around the outside of the tree branches, so in effect, they cover the outside "area" of the tree (thought of as a cone). Hence, the number of strings needed grows in proportion to the square of the height: a 30-ft tree will need $5^2 = 25$ times as many strings as a 6-ft tree. However, you also could argue that a 30-ft tree is meant to be viewed from farther away, so that stringing the lights farther apart on the 30-ft tree would produce the same effect as with the shorter tree.

41. 9 ft 3 in. to 11 ft 9 in. (in modern times, there have been men over 9 ft tall); 282 cm to 358 cm

43. Less than 90.25 kg

45. 27.9

47. Answers will vary.

49. Answers will vary.

51. The square of the wingspan is proportional to the wing area, so the wing loading is proportional to weight divided by the square of the wingspan. For the 200-pounder, that ratio is $200/(50^2) = 0.080$, while for the 100-pounder, it is $100/(36^2) \approx 0.077$. Close enough.

53. **(a)** **(i)** 800
(ii) 640,000
(iii) 512,000,000
(b) One eight-hundredth.
(c) There couldn't be any such giant ants.

55. $4^4 = 256$ times as long, or $256 \times 15 = 3840$ years, so $3840 - 15 = 3825$ more

57. $A \propto d^2$ and $A \propto M^{3/4} \propto (d^2h)^{3/4} = d^{3/2}h^{3/4}$, so $d^2 \propto d^{3/2}h^{3/4}$, hence $d^{1/2} \propto h^{3/4}$ and $d \propto h^{3/2}$.

59. A small, warm-blooded animal has a large surface area–to-volume ratio. Pound for pound, it loses heat more rapidly than a larger animal, hence it must produce more heat per pound, resulting in a higher body temperature.

61. **(a)**

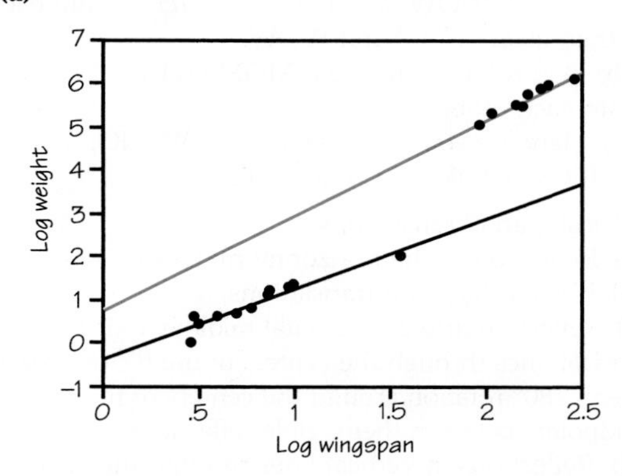

(b) Both relationships are allometric, since the results are good fits to straight lines whose slopes are not 1.
(c) The slope for birds is less steep than for planes.

Chapter 19

1. 5, 8, and 13

3. Answers will vary but will be Fibonacci numbers.

5. 1, 2, 3, 5, 8, . . . , F_{n+1}

7. The seventh number is $5m + 8n$ and the total is $55m + 88n$.

9. **(a, b)** The digits after the decimal point do not change.
(c) $\phi^2 = \phi + 1$
(d) $1/\phi = \phi - 1$

11. **(a)** 9
(b) 16

13. 6

15. **(a)** 4, 7, 11, 18, 29, 47, 76, 123
(b) 3, 1.333, 1.75, 1.571, 1.636, 1.611, 1.621, 1.617, 1.618; The ratios approach ϕ.

17. Answers will vary, but the ratio will always approach ϕ by alternating under- and over-approximations.

19. Answers will vary. Equality holds exactly when $x = y$.

21. **(a)** 1, 1, 3, 5, 11, 21, 43, 85, 171, 341, 683, 1365
(b) $B_n = B_{n-1} + 2B_{n-2}$
(c) 1, 3, 1.667, 2.2, 1.909, 2.048, 1.977, 2.012, 1.994, 2.003, 1.999
(d) $x = 2, -1$; we discard the -1 root.
$B_n = [2^n - (-1)^n]/3$.

23. Silver mean: $1 + \sqrt{2} \approx 2.414$; bronze mean: $\frac{1}{2}(3 + \sqrt{13}) \approx 3.303$; copper mean: $2 + \sqrt{5} \approx 4.236$; nickel mean: $\frac{1}{2}(5 + \sqrt{29}) \approx 5.193$. General expression: $\frac{1}{2}(m + \sqrt{m^2 + 4})$.

25. All are always true.

27. **(a)** B, C, D, E, H, I, K, O, X
(b) A, H, I, M, O, T, U, V, W, X, Y
(c) H, I, N, O, S, X, Z

29. **(a)** MOM, WOW; MUd and bUM reflect into each other, as do MOM and WOW.
(b) pod rotates into itself; MOM and WOW rotate into each other.
(c) Here are some possibilities: NOW NO; SWIMS; CHECK BOOK BOX; OX HIDE.

31. For all parts, translations
(a) Reflection in the horizontal midline
(b) None other than translations
(c) Reflection in the horizontal midline, reflections in vertical lines through the centers of the H's or between them, 180° rotation around the centers of the H's or the midpoints between them, glide reflections
(d) Reflections in vertical lines through the centers of the M's or between them

33. *d5*.

35. **(a)** *c5*
(b) *c12*
(c) *c22*

37. **(a)** *c6*
(b) (CBS) *d2*.
(c) (Dodge Ram) *d1*

39. **(a)** *c4*
(b) *d2*

41. **(a)** *p1a1*
(b) *p1m1*
(c) *p111*
(d) *p112*
(e) *pm11*
(f) *pma2*
(g) *pmm2*

43. **(a)** *pmm2*
(b) *p1a1*
(c) *pma2*
(d) *p112*
(e) *pmm2* (perhaps)
(f) *p1m1*
(g) *pma2*
(h) *p111*

45. **(a)** Patterns with vertical reflections are preferred on the Chinese pieces, while patterns with both horizontal and vertical reflections are strongly preferred on the Begho pipes.
(b) Neither culture completely excludes any strip type.
(c) **(i)** *pm11* or *pma2*: China **(ii)** *p112:* China **(iii)** *pmm2*: Begho **(iv)** *pm11*: China **(v)** *p1m1*: Begho **(vi)** *pmm2*: Begho **(vii)** *pmm2:* Begho **(viii)** *pma2*: China **(ix)** *p1a1*: China

47. (*c*) Smallest rotation is 90°, there are reflections, there are reflections in lines that intersect at 45°: *p4m*.
(*d*) Smallest rotation is 90°, there are no reflections: *p4*.

49. None of the five patterns with hexagonal symmetry can be realized, nor *p4g*, *p4*, *cm*, and *cmm*. The remaining eight can all be formed by the technique.

51. *pg*, if color is disregarded; otherwise, *p1*

53. *p6*

55. *p3*

57. Answers will vary.

59. There is no identity element.

61.

	I	F	R	T
I	I	F	R	T
F	F	I	T	R
R	R	T	I	F
T	T	R	F	I

63. Answers will vary.

65. Answers will vary.

67. Answers will vary. No, it is the cyclic group of order 4—e.g., the tire at position 1 goes to 3, then to 2, then to 4, and then back to 1.

69. There are four more derangements: CW = 2341, CCW = 4123, CWcross = 2413, and CCWcross = 3142. (CW stands for clockwise and CCW for counter clockwise.)

71. **(a)** *d2*
(b) Any two of the following: *R* (180° rotation around the center), *V* (reflection in vertical line through its center), *H* (reflection in horizontal line through its center)
(c) $\{I, R, V, H\}$

73. There are four rotational symmetries (including the identity), two reflection symmetries, and two reflections across diagonal lines: $\{I, R, R^2, R^3, H, V, RH = VR, RV = HR\}$.

75. Answers will vary.

77. As in Example 6, number the fixed positions, label with letters copies of the pattern elements in the positions, and pick a fixed position about which to make a half-turn *R*.

(a) $< T, R/R^2 = I, T \circ R = R \circ T^{-1} > = \{\ldots, T^{-1}, I, T^1, \ldots; \ldots, R \circ T^{-1}, R, R \circ T^1, \ldots\}$

(b) $< T, R, H | R^2 = H^2 = I, T \circ H = H \circ T, R \circ H = H \circ R, (R \circ T)^2 = I > = \{\ldots, T^{-1}, I, T, \ldots; \ldots, R \circ T^{-1}, R, R \circ T, \ldots; \ldots, H \circ T^{-1}, H, H \circ T, \ldots; \ldots, R \circ H \circ T^{-1}, R \circ H, R \circ H \circ T, \ldots\}$

79. $<R/R^8 = I> = \{I, R, R^2, R^3, R^4, R^5, R^6, R^7\}$, where R is a rotation by 45°

81. There are four rotational symmetries (including the identity), three reflection symmetries, and an inversion through the center.

83. The carved head is reproduced in the same shape at different scales.

85. Answers will vary.

Chapter 20

1. Exterior: 45°; interior: 135°

3. $180° - \dfrac{360°}{n}$

5. The usual notation for a vertex type is to denote a regular n-gon by n, separate the sizes of polygons by periods, and list the polygons in clockwise order starting from the smallest number of sides, so that, e.g., 3.3.3.3.3.3 denotes six equilateral triangles meeting at a vertex. The possible vertex types are 3.3.3.3.3.3, 3.3.3.3.6, 3.3.3.4.4, 3.3.4.3.4, 3.3.4.12, 3.4.3.12, 3.3.6.6, 3.6.3.6, 3.4.4.6, 3.4.6.4, 3.12.12, 4.4.4.4, 4.6.12, 4.8.8, 5.5.10, and 6.6.6.

7. 3.7.42, 3.9.18, 3.8.24, 3.10.15, and 4.5.20

9. At each of the vertices except the center one, six triangles meet, with angles (in clockwise order) of 75°, 75°, 30°, 30°, 75°, and 75°.

11. Yes

13. Three hexagonal faces meeting at a point would form a solid angle of 3 × 120° = 360°, hence they would form a flat surface. The vertex type on an ordinary soccer ball is 5.6.6. The ball has 32 faces (of which 12 are pentagons and 20 are hexagons), 60 vertices, and 92 edges.

15. (a) No
(b) No
(c) No

17. Answers will vary.

19. No; no

21. The only way to tile by translations is to fit the outer "elbow" of one tile into the inner "elbow" of another. Labeling the corners as follows works: the corners on the top A and B, those on the rightmost side C and D, the middle of the bottom E, and the middle of the leftmost side F.

23. Just label the four corners consecutively A, B, C, and D.

25. Place the skew-tetromino on a coordinate system with unit length for the side of a square and with the lower-left corner at (0, 0). Then A = (1, 2), B = (3, 2), C = (2, 0), and D = (0, 0) works.

27. Place the skew-tetromino on a coordinate system with unit length for the side of a square and with the lower-left corner at (0, 0). Then A = (0, 1), B = (2, 2), C = (3, 2), D = (3, 1), E = (1, 0), and F = (0, 0) works.

29. Answers will vary.

31. (a) Yes
(b) No
(c) No

33. See figure below.

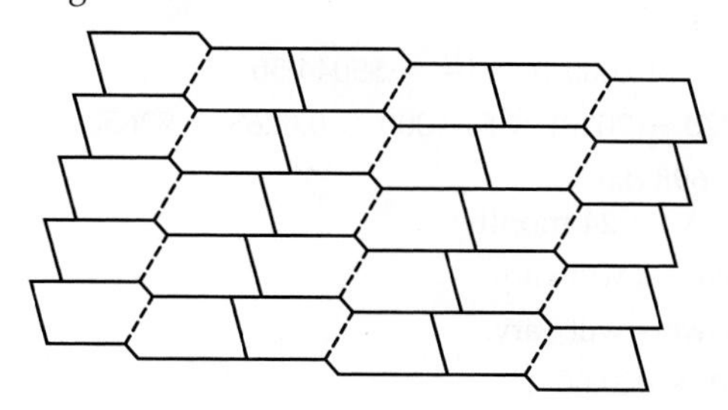

35. Answers will vary.

37. N, Z, W, P, y, I, L, V, X. See www.srcf.ucam.org/~jsm28/tiling/5–omino–trans.ps.gz.

39. Place the U on a coordinate system with unit length for the side of a square and with the lower-left corner at (0, 0). Then A = (2, 2), B = (3, 2), C = (3, 0), D = (1, 0), E = (0, 0), and F = (0, 1) works. See www.srcf.ucam.org/~jsm28/tiling/5–omino–rot.ps.gz.

41. Answers will vary.

43. (a) Consecutive segments, all of the same length; each length gives a different tiling
(b) Consecutive repetitions of any finite sequence of a's and b's

45. (a) *b; a; ab; aba; abaab; abaababa; abaababaabaab; abaababaabaababaababa*
(b) F_n segments at stage n
(c) $2m + n$

47. *ABAABABA.*

49. The two leftmost A's would have had to come from two B's in a row in the preceding month.

51. Let S_n, A_n, and B_n be the total number of symbols, the number of A's, and the number of B's at the nth stage. We note that the only B's at the nth stage must have come from A's in the previous stage, so $B_n = A_{n-1}$. Similarly, the A's at the nth stage come from both A's and B's in the previous stage, so $A_n = A_{n-1} + B_{n-1}$. Using both of these facts together, we have $A_n = A_{n-1} + A_{n-2}$. We note that $A_1 = 0, A_2 = 1, A_3 = 1, A_4 = 2, \ldots$. The A_n sequence obeys the same recurrence rule as the Fibonacci sequence and starts with the same values one step later; in fact, it is always just one step behind the Fibonacci sequence: $A_n = F_{n-1}$. Consequently, $B_n = A_{n-1} = F_{n-2}$, and $S_n = A_n + B_n = F_{n-1} + F_{n-2} = F_n$.

53. If a sequence ends in *AA*, its deflation ends in *BB*, which is impossible for a musical sequence. Similarly, if a sequence ends in *ABAB*, its deflation ends in *AA*, which we just showed to be impossible.

55. The first is part of a musical sequence; the second is not.

57. If the sequence were periodic, the limiting ratio of *A*'s to *B*'s would be the same as the ratio in the repeating part, which would be a rational number, contrary to the result of Exercise 56.

Chapter 21

1. $54 \times \$10{,}000 \times \frac{0.1121}{12} = \$5044.50$

3. $(2020 - 2010) \times \$10{,}000 \times 0.0265 = \$2650$

5. **(a)** 698 days
(b) After 24 months

7. Answers will vary.

9. Answers will vary.

11. **(a)** \$1020.00; 2%
(b) \$1020.00; 2%
(c) \$1020.15; 2.01505%
(d) \$1,020.20; 2.02008%

13. 4.8014%

15. 0.6009%; no, it will come to only \$5636.47.

17. **(a)** 480 ppm
(b) 488 ppm

19. 3.68%

21. **(a)** \$2,032.79; \$2,025.82; \$2,012.20
(b) \$1,999.00; \$1,992.56
(c) \$1,973.82; \$1,906.62; \$1,849.60
(d) For small and intermediate interest rates, the rule of 72 gives good approximations to the doubling time.

23. **(a)** 2, 2.59, 2.705, 2.7169, 2.718280469
(b) 3, 6.19, 7.245, 7.3743, 7.389041321
(c) $e = 2.718281828\ldots$; $e^2 = 7.389056098\ldots$. Your calculator may give slightly different answers because of its limited precision.

25. In both cases, \$12.07, not taking into account any rounding to the nearest cent of the daily posted interest.

27. **(a)** We seek r for which $(1 + r)^2 = (1.00)(1.10)$, so $r = (1.1)^{1/2} - 1 = 4.88\%$.
(b) We solve $e^{2r} = 1.1$, getting $r = \frac{1}{2} \ln 1.1 = 4.77\%$.

29. **(a)** $(e^{0.04} - 1) \times 100\% = 4.08108\%$
(b) The approximation for the effective rate is $r + \frac{1}{2}r^2 = 0.05 + \frac{1}{2} \times (0.05)^2 = 0.05 + 0.00125 = 0.05125$ or 5.125%, very slightly less than the true effective rate.

31. **(a)** No, it will be worth $(1 - 0.25) \times (1 + 0.25) \approx 94\%$ of the original value.
(b) Your friend's house ends up at $(1 + 0.25) \times (1 - 0.25) \approx 94\%$ of the original value, the same as your house.

33. \$81.75

35. \$67,092.02

37. \$1348.50

39. \$100,451.50

41. **(a)** \$199,149.07
(b) \$32—but you must also pay tax on that \$32 that you earn to pay that tax (so $0.32 \times \$32 = \$10.24$), and tax on that \$10.24, and so forth. All in all, you pay $\$100(0.32 + 0.32^2 + 0.32^3 + \cdots) = \$100 \frac{0.32}{1-0.32} = \$47.06$.

43. Using the Roth IRA, the entire \$199,149.07 calculated in Exercise 41a is yours tax-free at age 65. However, for the situation of Exercise 41, you will still owe tax on the interest earned: $32\% \times (\$199{,}149.07 - \$100 \times 12 \times 40) = 0.32 \times \$151{,}149.07 = \$48{,}367.70$, so your tax-free net is $\$199{,}149.07 - \$48{,}367.70 = \$150{,}781.37$. For the situation of Exercise 42, per the answer to part c, you have \$199,150.67 tax-free.

45. **(a)** $A = P(1.10)(0.75)(1.25) = 1.03125P$, so $r = (1.03125)^{1/3} - 1 \approx 1.031\%$
(b) It is the effective rate.

47. **(a)** Exponential (decay)
(b) Linear
(c) Linear
(d) Linear

49. $\$10{,}000/(1.05)^4 \approx \$8227.02$.

51. $\$150{,}000/(1.08)^{22.5} \approx \$26{,}549.53$.

53. **(a)** $\$(1.02)^3 \approx \$1.06$
(b) $\$1/1.06121 \approx \$0.94$

55. $\frac{232.0}{30.2} \times \$10.75 \approx \$82.58$; answers about price of the student's textbook will vary.

57. $\left(\frac{232.0}{179.9}\right)^{1/10} - 1 \approx 2.58\%$.

59. $(0.05 - 0.10)/(1 + 0.10) \approx -4.55\%$

61. Nowhere close; the equivalent in today's dollars is about: (a) \$65,000, (b) \$64,000.

63. \$23.3 million

65. **(a)** Nest egg: $\$100{,}000/0.072 \approx \$1{,}388{,}888.89$

Monthly deposit: $\dfrac{(\$1{,}388{,}888.89)\frac{0.072}{12}}{\left(1 + \frac{0.072}{12}\right)^{420} - 1} \approx \$735.16$

(b) $\$100{,}000/(1.03)^{35} \approx \$35{,}538.34$
(c) $\$100{,}000/(1.03)^{63} \approx \$15{,}532.98$

67. 1.60% per month, or 19.2% annual rate.

69. 4.97%. It is the effective rate. You can get the same result in a spreadsheet with =RATE(30, −10.5, 170, 0, 0.05). Since there is no exact formula for the interest rate, the spreadsheet uses a similar but more efficient method of successive approximation.

Chapter 22

1. The interest is $\$7500 \times \frac{0.068}{12} \times 15 = \$637.50$. So you owe $\$7500 + \$637.50 = \$8137.50$.

3. The first loan accumulates interest at $\$5500 \times \frac{0.068}{12} \times 51 = \$1589.50$; the second loan accumulates interest

$\$6500 \times \frac{0.068}{12} \times 39 = \$1215.50$. Your total debt is $\$5500 + \$1589.50 + \$6500 + \$1215.50 = \$14,805$, including a total of $2805 in interest.

5. $\$15,000(1.049)^5 \approx \$19,053.23$

7. $\$200,000(1.06)^{30} \approx \$1,148,698.24$

9. **(a)** $0.1724/365 \approx 0.0472329\%$
(b) $(1.0472329)^{365} - 1 \approx 18.81\%$

11. **(a)** $230.1606604, (which would be rounded up to $230.17, with the last payment less).
(b) $(120 \times \$230.1606604 - \$20,000) \approx \$7619.28$.

13. The interest for the 15 months until the start of repayment is $15 \times \$10,000 \times \frac{0.079}{12} = \$987.50$, so the starting principal when repayment begins is $10,987.50. The monthly payment will be $132.73.

15. $576.65

17. $355.96

19.

Months	0% Interest	Cash Back
36	$530.12	**$517.85**
48	$397.59	**$393.96**
60	**$318.07**	$319.67

As the term of the loan increases, the influence of the rebate declines and 0% interest becomes more desirable.

21. $536.83

23. $727.23

25. Using the amortization formula "in reverse": $8169.83. Rounding interest at each payment: $8171.48.

27. Using the amortization formula "in reverse": $27,321.51. Rounding interest at each payment: $27,322.42.

29. **(a)** $915.54

(b) $283.56

(c) 10 months

31. **(a)** $843.21.

(b) $1062.22, for an initial balance of $191,521.75

33. **(a)** Months, and billing periods, differ in their numbers of days. Also, the daily interest rate is $18\%/365 \approx 0.0493151\%$; compounded for a 30-day month, the monthly rate is then $1.000493151^{30} - 1 \approx 1.4901\%$.
(b) 179 months, or almost 15 years. The first payment is $31 with no interest due, the second payment is $77. Hint: Put the principal in column A and the interest due in column B. For the interest rounded to the nearest penny, use =ROUND(A2*.015, 2) and for the payment rounded down to the nearest dollar, use =MIN(A2+B2, MAX(25, FLOOR(0.01*A2+B2, 1))). Then adjust the last few months' interest charges by hand to be the minimum $1.50.

(c) $6873.25; $3755.42.

35. Let r be the APR, with OB the old balance (after the preceding payment) and NB the new balance (after addition of interest for this period), and let the bill be for 30 days. The daily interest rate is $i = r/365$, and we have $NB = (1 + r/365)^{30}OB$. The interest is $[(1 + r/365)^{30} - 1]$ OB.

If the interest is greater than 0.02 NB, a payment of 2% of NB will not keep up with the interest due. Solving $[(1 + r/365)^{30} - 1]OB > 0.02\ NB = 0.02(1 + r/365)^{30}OB$ gives first $[(1 + r/365)^{30} - 1] > 0.02(1 + r/365)^{30}$, then $0.98[(1 + r/365)^{30}] > 1$, and $(1 + r/365)^{30} > 1/0.98$, so that $1 + r/365 > (1/0.98)^{1/30} \approx 1.00067$, yielding finally $r > 24.59\%$. (Using a 31-day month gives $r > 23.79\%$.)

37. **(a)** 210 months, or 17.5 years. The first payment is $100 with no interest due, the second payment is $172. Hint: With no late charges, we can neglect the provision about 4%. Put the principal in column A and the interest due in column B. For the interest rounded to the nearest penny, use =ROUND(A2*.025, 2) and for the payment rounded to the nearest dollar, use =MIN(A2+B2, ROUND(MAX(35, 0.02*A2, B2 + 0.01*A2), 0)).

(b) $15,523.86

39. 3.68%

41. 391% if calculated as 14 days of 365 (390% if calculated as 2 weeks of 52)

43. 325% if calculated as 7 days of 365 (324% if calculated as 1 week of 52)

45. 4.41%

47. $319,297.70

49. **(a)** 36 months: $508,34; 60 months: $305.00
(b) Under the lease, over the lease period, you are paying for only part of the full purchase price: the full purchase price minus the value of the car at the end of the lease.
(c) Answers will vary.

Chapter 23

1. 2023

3. 5.5 billion

5. Not quite 100 percent greater than Europe's population

7. **(a)** 138 years
(b) 58 years

9. **(a)** 8.2 billion, 11.1 billion
(b) No change in growth rate, no change in death rates, no global catastrophes, etc.

11. **(a)** 17.5 years
(b) 16.2 years
(c) Answers will vary.

13. **(a)** 51 years
(b) 132 years
(c) 224 years

15. **(a)** 138 years
(b) Forever!

17. That's about 5.5 tons/plant/year ≈ 30 lb/plant/day, which is unreasonable.

19. 2.77%

21. 6.5 atoms per hour is one-fourth of 26 atoms per hour, so the sample is two half-lives (of 5730 years each) older than the sample in the text example and Table 23.1 (p. 842): 29,000 + 2 × 5730 ≈ 40,000 years. Using the formula $t = 55{,}403 - 8{,}267 \ln N$ gives 40,253 years, which again—given the imprecision of the data of 6.5 atoms per hour—should be rounded to 40,000 years.

23. Almost two half-lives have passed, so about one-fourth of strontium-90 would remain. More precisely, using $A = P\left(\frac{1}{2}\right)^{t/H}$, we get $A = P\left(\frac{1}{2}\right)^{50/28.8} \approx 0.30P$, so about 30 percent would remain.

25. 7.6 days

27. 0.2 day ≈ 5 hrs

29. 40 days

31. Answers will vary.

33. After the first year, the population stays at 15.

35. 7, 18.2, 6.6, 17.6, 8.4, 19.5, 2.0, 7.3, 18.6, 5.3

37. We must have $f(x_n) = x_n$, or $4x_n(1 - 0.05x_n) = x_n$. The only solutions are $x_n = 0$ and $4(1 - 0.05x_n) = 1$, or $x_n = 15$.

39. 10, 15.0, 11.3, 14.8, 11.6, 14.6, 11.8, 14.5, 11.9, 14.4. The population is oscillating but slowly converging to 40/3 ≈ 13.3.

41. $x_n = 0$, 40/3 ≈ 13.3

43. The red dashed line indicates the same size population next year as this year; where it intersects the blue curve is the equilibrium population size.

45. The population sizes are 11, 12.0, 12.6, 12.9, 13.1, 13.2, 13.3, 13.3, 13.3, 13.3.

47. The population sizes are 11, 15.0, 13.7, 14.9, 14.9, 15.0, 14.8, 14.3, 13.0, 9.5—and the following year the population is wiped out.

49. About 15 million pounds. The maximum sustainable yield is about 35 million pounds for an initial population of 25 million pounds.

51. **(a)** The last entry shown for the first sequence is the fourth entry of the second sequence, so the first "joins" the second, and they then both end up going through the same cycle (loop) of numbers over and over.
(b) 39, 78, 56, and we have "joined" the second sequence. However, an initial 00 stays 00 forever; and any other initial number ending in 0 "joins" the loop sequence 20, 40, 80, 60, 20, . . .
(c) Regardless of the original number, after the second push of the key, we have a number divisible by 4, and all subsequent numbers are divisible by 4. There are 25 such numbers between 00 and 99. You can verify that an initial number either joins the self-loop 00 (the only such numbers are 00, 50, and 25); joins the loop 20, 40, 80, 60, 20, . . . (the only such numbers are the multiples of 5 other than 00, 50 and 25); or joins the big loop of the other 20 multiples of 4.

53. **(a)** 133, 19, 82, 68, 100, 1,1, ... The sequence stabilizes at 1.
(b) Answers will vary.
(c) That would trivialize the exercise!
(d) For simplicity, limit consideration to 3-digit numbers. Then the largest value of f for any 3-digit number is $9^2 + 9^2 + 9^2 = 243$. For numbers between 1 and 243, the largest value of f is $1^2 + 9^2 + 9^2 = 163$. Thus, if we iterate f over and over–say 164 times—starting with any number between 1 and 163, we must eventually repeat a number because there are only 163 potentially different results. And once a number repeats, we have a cycle. Thus, applying f to any 3-digit number eventually produces a cycle. How many different cycles are there? That we leave you to work out. Hints: 1) There aren't very many cycles. 2) There is symmetry in the problem, in that some pairs of numbers give the same result; for example, $f(68) = f(86)$.

55. **(a)** 0.0397, 0.15407173, 0.545072626, 1.288978, 0.171519142, 0.59782012, 1.31911379, 0.0562715776, 0.215586839, **0.722914301**, 1.32384194, 0.0376952973, 0.146518383, 0.521670621, 1.27026177, 0.240352173, 0.78810119, 1.2890943, 0.171084847, **0.596529312**
(b) **0.723**, 1.323813, 0.0378094231, 0.146949035, 0.523014083, 1.27142514, 0.236134903, 0.777260536, 1.29664032, 0.142732915, **0.509813606**
(c) **0.722**, 1.324148, 0.0364882223, 0.141958718, 0.507378039, 1.25721473, 0.287092278, 0.901103183, 1.16845189, **0.577968093**

57. Period 2 begins at $\lambda = 3$, period 4 at $1 + \sqrt{6} \approx 3.449$, period 8 at 3.544, period 3 at $1 + 2\sqrt{2} \approx 3.828$, and chaotic behavior onsets at about 3.57. See http://www.answers.com/topic/logistic-map.

INDEX

P9-CQF-65

Effortless Elegance with Colin Cowie

EFFORTLESS ELEGANCE WITH COLIN COWIE

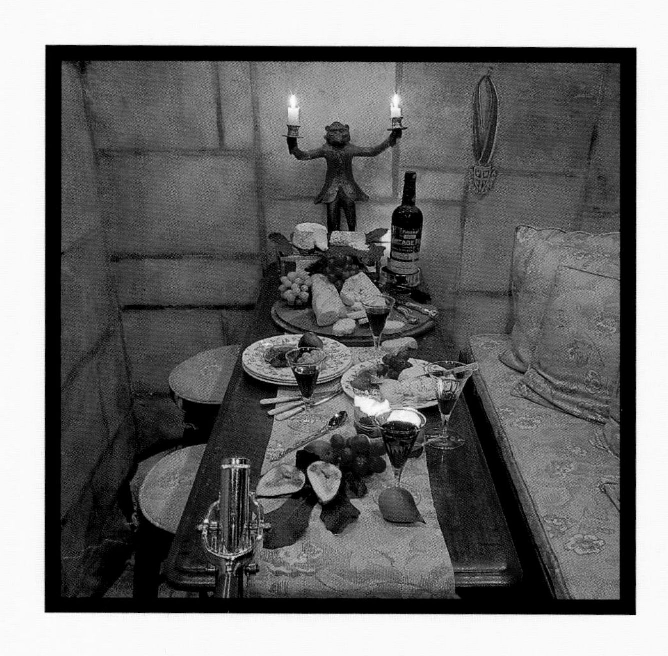

MENUS, TIPS, STRATEGIES, *and* MORE THAN 200 RECIPES
for EASY ENTERTAINING

TEXT BY MAUREEN CLANCY

HarperStyle
An Imprint of HarperCollins*Publishers*

This book is dedicated to my mother, Gloria Cowie,
who instilled in me an appreciation for beauty
and a concern for other people's well-being

Colin Cowie designs and caters parties and weddings around the world. Colin also lectures on entertaining, wedding planning, and interior design. This book is one in a series of books Colin will be authoring for HarperStyle. Colin can be contacted at:

Colin Cowie Lifestyle
P.O. Box 480228
Los Angeles, CA 90048
213-462-7183

HarperCollins books may be purchased for educational, business, or sales promotional use. For information please write: Special Markets Department, HarperCollins Publishers, Inc., 10 East 53rd Street, New York, NY 10022.

FIRST EDITION

Designed by Moritz Design

Library of Congress Cataloging-in-Publication Data

Cowie, Colin, 1962–
Effortless elegance with Colin Cowie / by Colin Cowie. — 1st ed.
p. cm.
ISBN 0-06-270152-5
1. Entertaining. 2. Dinners and dining. 3. Menus. I. Title.
TX731.C685 1996
642—dc20 96-11108

96 97 98 99 00 ❖/PIZ 10 9 8 7 6 5 4 3 2 1

CREDIT WHERE CREDIT IS DUE

This book would never have been possible without the help of my wonderful friends, associates, and colleagues.

In the kitchen: I have been inspired by Adriana Pacifici, Francis Bey, Martin Herold, Clement Bacque, Jonathon Beare, Cynthia Beare, and Sharon Richstone.

In my office: Stuart Brownstein, David Berke, Kathy Chapman, Patty Smith, and Maria Bonino.

In creating this book: Maureen Clancy and Stuart Brownstein, for writing with me; Patricia Moritz, for designing the book; Walter Hubert, for collaborating on the flowers and art direction; Alison Duke, for the food photography; Karen Gillingham, for the food styling, recipe testing, and recipe writing; Sharon Richstone, for recipe testing; Nadine Froger, for the big pictures; Jean-Jacques Pochet for the pictures of the wonderful friends who appear in this book; Leila May, for the food photographs at Don and Sharon's; and Patty Smith, for all her hard work.

For the tasteful homes and locations: Anne Archer; Jeff and Margo Barbakow; Ernie Benson; Stuart Brownstein; Maureen Clancy and Thomas Shiftan; Nick and Cathy Connor; Dusty Deyoe; Suzanne Figi; Annette and James Frehling; Kenny, Lyndie, and Max Gorelick; Lauren Gabor and Scott Goldstein; Bobbie Geller; Walter Hubert and Mark Saltzman; Donna and Mark Isham; Dr. Raj Kanodia; Cheryl Lerner; Julianne Phillips; Planet Hollywood; Don and Sharon Richstone; Andy Ross; Ted Tanaka and Diana Ho; Cheryl Tiegs; Miriam Wosk; and Caroline Yorston.

A very special thanks to: Dusty Deyoe, for the props, accessories, and friendship; Christofle, for the china and linens; Cottura, for china; Jim Block and Jerry Astorian, from Watt's Up Lighting; Cheryl Evanoff, from Regal Rents; Rick Ross, for the Moroccan lamps and rugs; Patricia Kennedy, for the Moroccan linens and cushions; Christian Navarro, for his knowledge of wines; Bruce Willis, for the black pickup truck; and Marvin Shanken from *Cigar Aficionado* and the *Wine Spectator*.

At HarperCollins: To Diane Reverand and Joseph Montebello, for bringing this book to life.

At the McBride Agency: Kim Sauer, Clare Horn, Winifred Golden, and most of all, Margret McBride, my agent who believed in my vision.

CONTENTS

INTRODUCTION

I have created a business out of parties. Besides being able to share quality time with my friends or create a moment of harmony for my clients, I enjoy orchestrating the whole affair. We are blessed with so many gifts and, when having friends over, I like to work with all the blessings. The food, the flowers, the linens, the music, the lighting, and the company.

I have written this book to inspire you to entertain. In the past decade there was the feeling we had to impress people when entertaining them. We have since come through some humbling experiences. Nowadays we entertain because it allows us to invest quality time and show one another that we care. This doesn't mean you have to spend a day or two planning a dinner. You don't even have to serve a full meal. It can be as simple as a bottle of Bordeaux, some vintage glasses, a great piece of French cheese, and a crusty baguette before dinner in a restaurant. It could be dessert, coffee, Cognac, and cigars after a show. A lazy Sunday brunch. I will show you how to prepare a dinner in less than two hours, including shopping and preparation. You will also be able to clean up before your guests arrive, spend five minutes on yourself, and be a guest at your own dinner party.

You don't have to prepare everything you serve. You get no medals. Besides, if the baker down the road bakes a better lemon tart than you, you should buy his. Your guests should always have the best. One thing that has not changed is the reason why we entertain. It's to bring people together and create an opportunity to treat every person around you like royalty, whether it be the guest, the cook, or the dishwasher.

The other reason I love to entertain is collecting. I collect for myself and find wonderful gifts for friends—china, silverware, odd and unusual serving pieces, vintage linens, first-edition cookbooks, wines, and Champagnes. I find auctions, antiques shops, swap meets, garage sales, and flea markets irresistible. It might be two vintage salad plates with a matching cloche that I will turn into a gift by placing some home-baked cookies and a favorite recipe under the cloche. Or perhaps it's a set of old bone-handled steak knives for my friend who prepares the best barbecue.

Style has changed. We all deserve to live a rich and full life, and as you will discover from this book, style is not related to money. It is about the attitude we have toward one another and ourselves. What better way to explore your own sense of style than by having friends over and entertaining them.

Enjoy!

Colin Cowie

FOREWORD

My husband and I spotted Colin at the party of a couple we knew only casually. The hostess is a wonderful artist—and dinner was to be served in her studio at her home. Jeff approached Colin, thinking he, too, was a guest, and soon learned that Colin had masterminded the entire evening. The party was delightful, with great attention paid to every detail. Although we had never worked with a party planner before, we went home that night looking forward with great anticipation to booking Colin for our next event. And the rest, as they say, is history.

We have grown to love entertaining at home. The reason is simple: Colin Cowie. He has taught us how to do it with great enthusiasm, with elegance, with taste, and with style. With budgets large and small, for business or for family and friends, for the most special of occasions, or for no reason at all. Colin has taught us how to entertain with ease, comfort, and, most important, how to enjoy our own parties.

Working with Colin has been the most fun I have ever had. I have learned so much about grace and elegance, about style and charm, about making people feel good, and about being a loving, caring person. Colin personifies all those things.

We have enjoyed our partnership and know it will endure. If this book can bring just a small flavor of what Colin has brought to our lives, I'm sure everyone will feel the same about Colin as I do.

Margo Barbakow

Effortless Elegance

with Colin Cowie

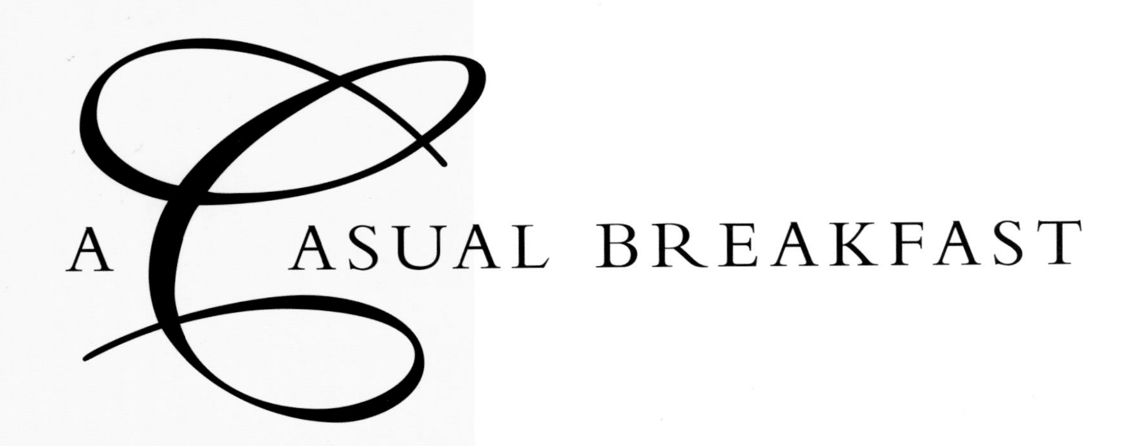

A CASUAL BREAKFAST

FRESH *grapefruit* JUICE

fried apples AND STEWED PRUNES

almond PANCAKES

scrambled eggs WITH YOGURT AND CHIVES

CHICKEN AND FENNEL *sausage*

Serves: 4

Range of Difficulty: Easy

Preparation Time: 1 hour

When I was growing up in Central Africa,

our winter breakfast was usually stewed fruits, prunes in particular, and porridge. The moment my mother's back was turned, the dog got the porridge and I ate the prunes!

How do I love breakfast? Consider this menu I did with friends who were entertaining weekend houseguests. After a wake-up call of freshly squeezed grapefruit juice, we served a compote of pan-fried apple slices and stewed prunes, which was readied the night before. Our feather-light pancakes of ground almonds were healthy and tasty.

Although the pancakes and eggs need to be made once the guests arrive, we had all the ingredients assembled ahead and invited our friends to gather in the warm kitchen while we worked.

Stewed prunes tossed in a hot pan with apple slices, brown sugar, and sweet butter. Whole wheat almond pancakes, light, tasty, and delicious under a mantle of warm maple syrup. Scrambled eggs and chives made creamy with nonfat yogurt. Chicken sausage studded with apples, fennel, and sage. Just the thought of having a delicious breakfast is reason enough to get out of bed.

On a bare wooden picnic table in the garden, we set places with assorted pieces of china from the 1950s alongside casual cloth napkins. The hostess, who is a landscape architect, had created a riot of color within the inviting space. So we picked flowers from her garden and placed them in a variety of small bottles, with one large terra-cotta potted plant to intensify the bright garden feel. It took just minutes to assemble.

To work up an appetite suggest to your friends that they join you on a morning hike in the mountains, a brisk walk along the beach, or a stroll past store windows in the neighborhood. When you return home, have a copy of the morning newspaper on hand, play some upbeat classical music, and, most important, make a freshly brewed pot of the best coffee you can find. Then sit back and enjoy a wonderfully civilized alternative to a "bagel-on-the-run."

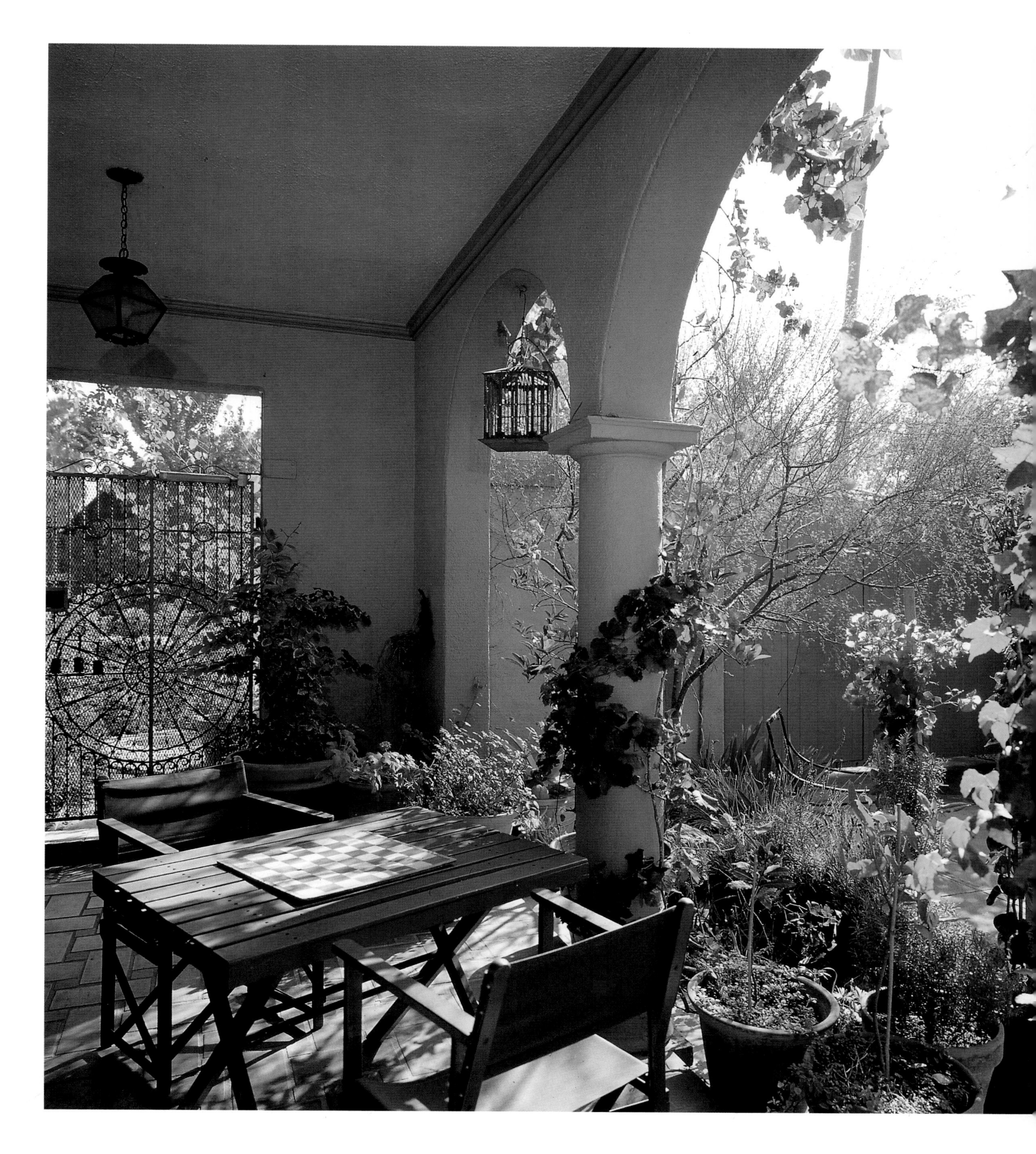

A warm, sunny garden is a lovely spot to enjoy breakfast with weekend houseguests.

Fried apples and stewed prunes make a perfect breakfast starter.

PANCAKE POSTSCRIPT

DRESS UP REGULAR PANCAKES (or, in a pinch, those made from a mix) with these additions:

- fresh berries
- chopped dates
- pecans
- orange marmalade
- sliced bananas
- maple syrup
- honey

Scrambled eggs are lightened with nonfat yogurt and teamed with savory chicken and fennel sausage.

FRIED APPLES

3 Golden Delicious apples
4 tablespoons butter
¼ cup brown sugar

1. Core and peel the apples and cut into four 1-inch slices.
2. Melt the butter in a medium skillet over medium-high heat. Add the apples and sprinkle with the sugar. Cook until caramelized, about 8 minutes per side, or until golden brown.

SERVES 4

STEWED PRUNES

10 ounces pitted prunes
2 tablespoons brown sugar
¾ cup water

In a saucepan, combine the prunes, sugar, and water. Cook over medium heat until the prunes are plump and until most of the water has drained. To serve, arrange the prunes and the fried apples in a small serving dish.

SERVES 4

Note: Stewed prunes can be prepared a day ahead, refrigerated, and served cold, at room temperature, or warmed up.

ALMOND PANCAKES

2 cups whole wheat flour
1 cup finely ground almonds
1½ teaspoons baking powder
2¼ cups milk
2 eggs, lightly beaten
4 tablespoons butter, melted

1. In a large bowl, combine the flour, almonds, and baking powder. Add the milk, eggs, and butter and stir to blend. Set the batter aside for 10 minutes.
2. Lightly grease and preheat a griddle or large skillet.
3. Using ⅓ cup of the batter for each pancake, ladle it onto the griddle and cook until holes appear in the batter. Turn the pancakes and cook the other sides until cooked through.

SERVES 3 TO 4 PANCAKES EACH

SCRAMBLED EGGS *with* YOGURT *and* CHIVES

6 eggs
¼ cup milk (nonfat optional)
1 tablespoon butter
2 tablespoons plain yogurt (nonfat optional)
2 tablespoons sliced chives
Salt and freshly ground pepper

1. Whisk the eggs and milk together in a bowl.
2. In a large skillet, melt the butter over medium heat. Pour in the eggs, reduce heat to low, and cook, stirring frequently with a wooden spoon, until soft curds form. Stir in the yogurt, chives, salt, and pepper and remove from heat. The eggs should be soft and creamy. Transfer the eggs to a serving platter and serve hot.

SERVES 4

CHICKEN *and* FENNEL SAUSAGE

1 pound ground chicken
¼ cup finely chopped onion
¾ cup peeled, cored, and chopped Granny Smith apples
¾ cup finely chopped fennel bulb
1 egg yolk, beaten
1 teaspoon crushed red pepper flakes
3 tablespoons chopped fresh parsley
Salt and freshly ground pepper
2 tablespoons vegetable oil
1 tablespoon butter

1. In a mixing bowl, combine all the ingredients except the oil and butter. Mix thoroughly with a fork. Pat the mixture into 3-inch patties.
2. Heat the oil and butter together in a pan over medium heat. Fry the patties until golden on both sides. Remove from the heat and set on paper towels to absorb the excess oil.

SERVES 4

A Lazy Sunday Brunch

FRESH *pear and ginger* JUICE

raisin MUFFINS

fruit salad WITH TARRAGON SYRUP

ASPARAGUS AND BASIL *frittata*

FISH *cakes*

fried tomatoes AND ONIONS

Serves: 8

Range of Difficulty: Easy

Preparation Time: 1 hour

THERE'S NOTHING NICER THAN A LAZY MORNING,

waking up to the smell of coffee and freshly baked muffins.

Usually, brunch guests aren't on a time schedule, so it's the perfect time to relax, to be outdoors if the weather permits, and to experiment with some imaginative menus rather than the traditional old hot cakes and sausage links.

For this Sunday brunch with eight friends, I baked a batch of my favorite raisin muffins. By timing it just right, their tantalizing aroma greeted the guests as they arrived. I don't think there's anything better than baked goods straight from the oven! Freshly squeezed juice was also waiting. A piece of fresh ginger added zip to an interesting blend of pineapple, pear, and apple juices.

Guests helped themselves to a tropical fruit salad made with oranges, papaya, strawberries, blueberries and raspberries, kiwi, grapefruit, and a subtle tarragon syrup. The herb's anise flavor brought out the sweetness of the fruits; sprigs of fresh mint added a refreshing note and vivid color.

A frittata—Italy's homey answer to the omelet—was also featured in the buffet. Because the ingredients are mixed in with the eggs instead of being folded inside, a frittata is easier than an omelet to prepare and easier to serve to large groups. If your friends love to stand around the kitchen watching you cook, you can make the frittata just before sitting down. If you'd rather unwind with your guests in front of the fireplace or out on the patio, make it in advance and serve it at room temperature the Italian way.

I made accompanying fish cakes with fish left over from the previous evening's dinner. However, you can poach a few pieces of whitefish if leftovers are not available. Fried tomatoes and onions, also prepared ahead, were savory accompaniments.

Of course, no brunch is complete without good coffee. If you have an espresso machine, offer cappuccino or whipped-cream-topped caffè mocha.

In keeping with the quiet nature of the morning, I set our table with simple white linen place mats and napkins and centered it with cacti in terra-cotta pots. For an element of color, we used mismatched Mediterranean-style china hand painted in primary colors.

Even if you live in an apartment, you can have brunch on a patio or balcony and enjoy fresh air along with good coffee, great food, and close friends.

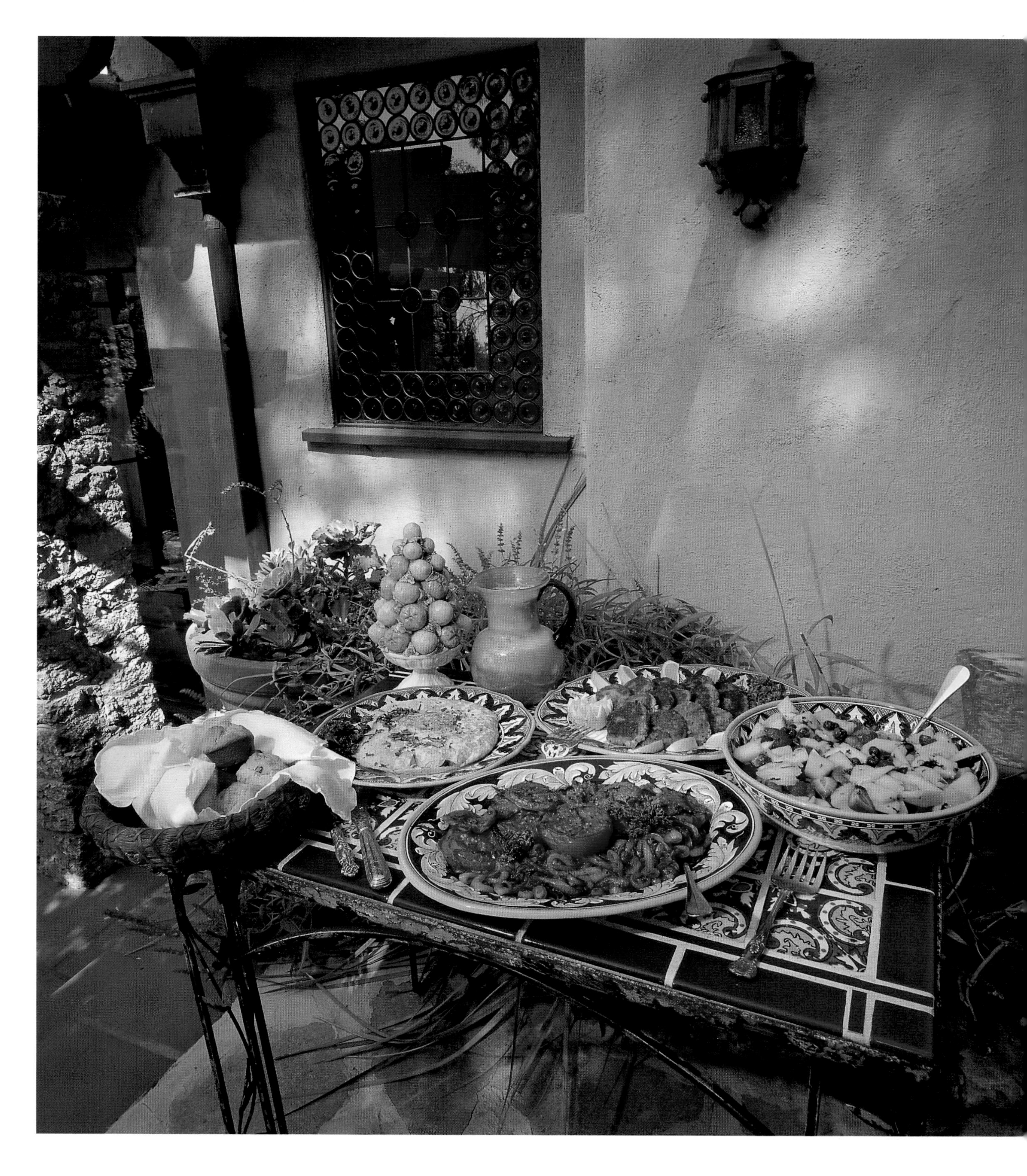

Mediterranean-style platters add to the appeal of this Sunday brunch buffet.

There's nothing more welcoming than a batch of muffins straight from the oven.

FRESH PEAR *and* GINGER JUICE

8 ripe pears, peeled, cored, and cut into chunks
8 green apples, peeled, cored, and cut into chunks
2 ripe pineapples, peeled, cored, and cut into chunks
1 ½-inch piece of fresh ginger, peeled

Pass the pear, apple, pineapple chunks, and ginger through a juicer according to manufacturer's instructions. Chill and serve.

SERVES 8

RAISIN MUFFINS

2 cups all-purpose flour
3 tablespoons sugar
1 tablespoon baking powder
½ teaspoon salt
1 cup milk
1 egg
4 tablespoons butter, melted and cooled
¾ cup raisins

1. Preheat the oven to 425 degrees.
2. Grease a muffin pan or line it with paper cups.
3. Sift the flour, sugar, baking powder, and salt into a large bowl.
4. In a small bowl, beat the milk with the egg and butter. Pour into the flour mixture and stir just until the dry ingredients are moistened.
5. Fold in the raisins, then fill the muffin cups about two thirds full with the batter. Bake for 20 minutes, or until the muffins are lightly browned on top. Remove from the pan immediately. Serve warm.

MAKES 12 MUFFINS

VARIATIONS:

Blueberry: Increase the sugar to ½ cup and substitute ¾ cup of fresh or frozen (do not thaw) blueberries for the raisins.

Cheese: Reduce the sugar to 1 tablespoon and substitute ½ to ¾ cup of shredded Cheddar cheese for the raisins.

Cornmeal: Substitute 1 cup of yellow cornmeal for 1 cup of the flour. Omit the raisins.

Whole Wheat: Substitute 1 cup of whole wheat flour for 1 cup of the all-purpose flour.

Nut: Increase the sugar to ½ cup and add ½ cup of nuts along with the raisins or in place of the raisins.

THERE'S NO NEED TO GET UP AT DAWN TO BAKE BREAKFAST MUFFINS. Just measure out all the dry ingredients the night before and place them in separate zip-topped bags. Liquids, too, can be measured, covered, and refrigerated. Pans can also be greased ahead, covered, and refrigerated, if butter is used for the greasing. In the morning, just toss the ingredients together and get ready for the fragrant pay-off!

COFFEE

Americans have grown increasingly sophisticated about coffee. Use this primer to find the right type of bean for your breakfast parties.

The two main types of coffee used today are arabica and robusta. Arabica is more complex, subtle, refined, and thus, more expensive. The cheaper robusta is hardier, with a harsher flavor. Many canned and instant coffees are made with robusta beans.

Below are the principal types of arabica beans on the market today.

Java. An aged Indonesian coffee that's full-bodied and spicy.

Kenyan. Full-bodied, acidic, and smooth; with an almost winey flavor.

Kona. From Hawaii; the real thing is hard to find but is sweet, smooth, and mellow.

Costa Rican. Strong, robust flavor with a pungent aftertaste.

Guatemalan. Mild and spicy.

Colombian. The supremo blend is smooth and light-bodied; the excelso is rich and mellow.

ASPARAGUS *and* BASIL FRITTATA

10 asparagus spears, ends trimmed and cut into 1-inch lengths
10 eggs
⅓ cup chopped fresh basil
2 tablespoons sliced chives
½ cup grated Parmesan cheese
Salt and freshly ground pepper
1 tablespoon olive oil

1. Bring a pot of salted water to a boil. Blanch the asparagus until tender and soft, about 6 minutes. Remove and set in ice water.
2. In a mixing bowl, beat together the eggs, basil, chives, Parmesan, salt, and pepper.
3. In a large ovenproof skillet heat the olive oil over medium-high heat. Pour in the egg mixture and reduce heat to medium low. Cook without stirring until the edges are set and the center is runny. Add in the asparagus. Lift the sides of the frittata to allow the uncooked eggs to run below. Continue until center of frittata is firm.
4. Place a large plate over the top of the pan and using a pot holder, invert the frittata.
5. Slide the frittata back into the pan with the uncooked side down. Allow to cook a further 2 to 3 minutes.
6. Remove and place on a serving plate. Cut into wedges and serve hot or at room temperature.

SERVES 8

FRUIT SALAD *with* TARRAGON SYRUP

½ cup water
2 tablespoons sugar
1 tablespoon chopped fresh tarragon
1 tablespoon sliced fresh mint
Pulp from 3 passion fruits
2 kiwi, peeled and cut into 1-inch cubes
1 papaya, peeled, seeded, and cut into 1-inch cubes
1 pineapple, peeled, cored, and cut into 1-inch cubes
1 mango, peeled, seeded, and cut into 1-inch cubes
1 orange, peeled and cut into 1-inch cubes
½ cup strawberry halves
½ cup raspberries
¼ cup blueberries

1. In a small saucepan, combine the water and sugar. Bring to a boil. Reduce the heat and stir in the tarragon and mint. Simmer for 15 minutes. Cool.
2. In a large bowl, combine the passion fruit pulp, kiwis, papaya, pineapple, mango, orange, strawberries, raspberries and blueberries. Pour the cooled syrup over the fruit and toss to combine thoroughly.

SERVES 8

FISH CAKES

1¼ pounds Poached Whitefish (see following recipe)
1¾ cups mashed potatoes
½ cup finely chopped onion
1 tablespoon minced parsley
Salt and freshly ground pepper
1 egg, well beaten
1 cup dry bread crumbs
4 tablespoons butter
3 tablespoons vegetable oil
2 lemons, quartered

1. In a large bowl, combine the whitefish, potatoes, onion, and parsley. Season with salt and pepper to taste. Divide the mixture into 12 equal portions and roll each into a ball. Flatten each of the balls into a round cake.
2. Dip each cake into the egg, then into the bread crumbs.
3. In a large skillet, melt the butter in the oil over medium-high heat. Add the fish cakes, four at a time, and cook until they are golden brown. Drain on paper towels. Serve the fish cakes with the lemon wedges.

MAKES 12

POACHED WHITEFISH

- 2 cups water
- ¼ white onion, sliced
- 1 celery stalk, cut into 3-inch lengths
- 3 parsley sprigs
- ½ teaspoon salt
- 1 bay leaf
- 1 cup dry white wine
- 1 pound whitefish, skin removed

1. In a large pan, combine the water, onion, celery, parsley, salt, bay leaf, and white wine. Bring to a boil, then reduce heat and simmer for 15 minutes. Carefully add the whitefish, cover the pot, and poach for about 10 minutes. (The general rule is 10 minutes for each inch of thickness.)
2. With a slotted spoon, remove whitefish and place on paper towels to drain. Let cool for 10 minutes, then transfer to a serving platter, cover with plastic wrap, and chill.

SERVES 10

FRIED TOMATOES *and* ONIONS

- 2 tablespoons butter
- 2 tablespoons olive oil
- 3 medium onions, cut crosswise into 5 slices each
- 4 large, firm tomatoes, cut crosswise into 4 thick slices each
- Salt and freshly ground pepper

Divide the butter and oil between two large skillets and place over medium-high heat to melt the butter. Add half the onions to each pan and cook the onions until browned. Add half the tomatoes to each pan and reduce the heat to medium. Cook the tomatoes about 7 minutes per side, until the juices are released, turning carefully with a spatula. Season with salt and pepper to taste, then transfer to a serving platter. Serve hot.

SERVES 8

GET CREATIVE WITH YOUR FRITTATAS. Try adding some of the following ingredients:

- tomato sauce and Parmesan cheese
- cooked fettuccine, sautéed green onions, cooked shrimp
- diced ham and sautéed mushrooms
- sautéed zucchini
- sautéed spinach
- cooked sausage and bell pepper

DRESS UP FRESH FRUIT JUICES with mint, slices of lime, or small pieces of fresh fruit frozen in ice cubes.

Or turn them into smoothies by placing the juice, a peeled banana, and lots of ice in a blender. Blend until smooth.

IF YOU DON'T HAVE A JUICER, buy several fresh squeezed juices and blend them together with fresh grated ginger.

GREAT BREAKFAST COCKTAILS

Mimosa. Top freshly squeezed orange juice with Champagne or sparkling wine. Use blood oranges when they're in season.

Bellini. Top fresh peach juice with Champagne or Asti Spumante. Use white peaches when in season.

Bloody Mary. Add a shot of vodka to good tomato juice and dress it up with Tabasco, horseradish, salt, pepper, and a celery stick.

LUNCH ON THE PACIFIC

watermelon

spicy shrimp salsa WITH TORTILLAS

poached rolled chicken WITH A PARSLEY TARRAGON SAUCE

MIXED *green salad* WITH LIGHT VINAIGRETTE

roasted potato, GREEN BEAN, AND ONION SALAD

FRANCIS BEY'S *crème brûlée*

sun tea

Serves: 4

Range of Difficulty: Easy

Preparation Time: 2 hours

1993
CACIA
ARDONNAY
CARNEROS

IT'S IMPORTANT TO BE PRACTICAL AND AWARE OF YOUR SURROUNDINGS *and other parameters when entertaining. For example, if you're working on a boat or at a picnic site, you can't take the whole kitchen with you. But you can still present your food in a beautiful, elegant, and tasteful manner.*

There's something about being out on the ocean that makes everything taste better. Maybe it's the exhilarating sense of freedom or the pure clean air and gentle breezes. Perhaps it's the quiet rhythm of the waves or the refreshing sea spray. Whatever the ingredients, these are always meals to remember.

At this informal lunch for four friends, we spent a perfect Southern California day aboard a magnificent sailboat, cruising the coast between Marina Del Rey and Point Dume, off the coast of Malibu. But our wonderful lunch, prepared the day before and packed in a picnic basket, would taste just as great under a tree, on the shore of the local neighborhood pond, or in your own backyard.

When guests arrived on board midmorning, they were served iced sun tea with sprigs of fresh mint, along with slices of juicy watermelon. Just before lunch, appetites were tempted with a spicy shrimp salsa, tortilla chips, and ice cold beers.

For lunch, there was a stuffed poached chicken breast that had been sliced to reveal a mosaic of color, served with a bright parsley sauce, and two salads: one of mixed greens; the other a medley of cold roasted potatoes, green beans, and red onions. A light, crisp Chardonnay with a tropical aroma enhanced the feast.

Dessert was a rich crème brûlée. It's easy to prepare, can be packed for a trip, and offers a sophisticated finish to a light lunch. Everyone was doubly impressed to learn everything had been entirely prepared the day before.

This informal lunch can be made ahead and served on a sailboat, in the backyard, or around the pool.

SUN TEA is made by placing tea bags in a covered jug of water and allowing the tea to steep in the heat of the sun. It takes about 1 hour to make.

SPICY SHRIMP SALSA *with* TORTILLAS

1 pound large shrimp, peeled, deveined, and poached
2 roma tomatoes, seeded and diced
¼ cup diced red onion
⅓ cup chopped fresh cilantro
½ serrano chili, seeded and minced
1 tablespoon chopped fresh parsley
Juice of 1 lime
¼ cup olive oil
Salt and freshly ground pepper

1. Cut each shrimp into 4 to 5 pieces and place them in a medium bowl. Add the tomatoes, onion, cilantro, chili, and parsley and mix to combine. Add the lime juice and oil and toss to mix thoroughly. Season with salt and pepper to taste.
2. Chill until serving time and serve with tortilla chips and beer.

SERVES 4

POACHED ROLLED CHICKEN *with a* PARSLEY TARRAGON SAUCE

7 boneless skinless chicken breast halves
8 green olives, pitted
¼ red bell pepper, seeded
¼ cup cooked spinach
1 tablespoon chopped fresh parsley
2 ounces feta cheese
Salt and freshly ground pepper
Parsley Tarragon Sauce (see following recipe)

1. Roughly chop 1 of the chicken breast halves and place in a food processor with the olives, red pepper, spinach, parsley, cheese, and salt and pepper to taste. Pulse until all the ingredients are coarsely chopped. Refrigerate.
2. Place the remaining 6 chicken breast halves between sheets of wax paper and pound until paper thin. Season all over with salt and pepper. Place the pounded breasts on a work counter.
3. Spread 1 tablespoon of the chilled chicken mixture in the center of each chicken breast and tightly roll to enclose. Wrap with plastic wrap, sealing and tying the ends to make a sausagelike casing.
4. Bring a large pot of water to a simmer. Carefully drop the wrapped chicken rolls into the water and cook about 10 minutes. Remove with a slotted spoon and set aside to cool. When cool enough to handle, unwrap the plastic and slice the rolls across the width into ¾-inch medallions. Arrange medallions on a plate and serve with the Parsley Tarragon Sauce.

SERVES 4

PARSLEY TARRAGON SAUCE

6 garlic cloves, peeled
½ cup plus 6 tablespoons water
1 bunch parsley, stems removed
3 teaspoons tarragon
¼ cup olive oil
2 tablespoons lemon juice
Salt and white pepper

1. In a small saucepan, combine the garlic and ½ cup water and poach about 15 minutes, until garlic is tender.
2. Bring another small pot of generously salted water to a boil and blanch the parsley for 2 minutes.
3. In a food processor or blender, combine the garlic, parsley, tarragon, oil, lemon juice, and the remaining water and process until puréd. Season to taste with salt and pepper and chill.

MIXED GREEN SALAD *with* LIGHT VINAIGRETTE

1 head butter lettuce, rinsed and dried, outer leaves discarded
1 head radicchio rinsed and dried
1 cup whole basil leaves
Light Vinaigrette (see following recipe)
2 tablespoons grated imported Parmesan cheese

1. Tear the butter lettuce and radicchio into bite-size pieces. Combine with the basil in a large bowl. Chill.
2. To serve, pour on the vinaigrette to coat the leaves, sprinkle with the parmesan, and toss gently.

SERVES 4

LIGHT VINAIGRETTE

1 tablespoon water
1 tablespoon red wine vinegar
1 tablespoon fresh orange juice
½ teaspoon dry mustard
3 tablespoons olive oil
Salt and freshly ground pepper

In a small bowl, whisk together the water, vinegar, orange juice, and mustard. Gradually whisk in the olive oil. Season to taste with salt and pepper.

SALSAS

Whole books are written about salsas, the rage of the health-conscious, discriminating nineties palate. These are some of my favorite combinations:

- mango, red onion, cilantro, and mild green chilies
- tomatoes, onion, herbs, and unsalted peanuts
- canned tomatillos, pineapple, onion, orange juice, and cilantro
- nectarines, fresh ginger, red onion, and lime juice
- pineapple, yellow bell pepper, red onion, and minced jalapeño
- corn kernels, red bell pepper, red onion, mango, and cilantro
- kiwi, red onion, cilantro, and jalapeño chilies
- tomatoes, garlic, red onion, and cilantro

Poached Rolled Chicken with a Parsley Tarragon Sauce.

ROASTED POTATO, GREEN BEAN, *and* ONION SALAD

¼ cup olive oil
1¼ pounds baby red potatoes, halved
½ pound green beans, trimmed
¼ red onion, very thinly sliced
1 tablespoon chopped fresh parsley
Red Wine Vinaigrette (see following recipe)
Salt and freshly ground pepper

1. In a large skillet, heat the olive oil over high heat about 2 minutes. Add the potatoes and brown all over. Reduce the heat and cook 10 minutes, until tender.
2. Bring a pot of salted water to a boil and cook the green beans until tender, 3 minutes. Drain, then immediately plunge into ice water, and drain again.
3. Combine the potatoes, beans, onion, and the parsley in a large bowl. Pour on the vinaigrette and toss lightly. Season with salt and pepper to taste.

SERVES 4

RED WINE VINAIGRETTE

1 tablespoon minced red onion
2 tablespoons red wine vinegar
1 tablespoon Dijon mustard
¼ cup olive oil
Salt and freshly ground pepper

In a small bowl, mix together the onion, red wine vinegar, and mustard. Gradually whisk in the oil. Season with salt and pepper to taste.

CRÈME BRÛLÉE VARIATIONS

Although I adore the classic mixture of heavy cream and eggs, sometimes I like to add some of the following ingredients:

- crystallized ginger
- peppermint crisp candies
- whole raspberries and blackberries
- pralines (caramelized nuts and sugar confection)
- Grand Marnier

PICNIC

roast beef SANDWICHES WITH HORSERADISH AIOLI

n SANDWICHES WITH BASIL MAYONNAISE

ahi SANDWICHES WITH SAUTÉED TOMATOES

RED CABBAGE *coleslaw*

ANET'S *potato salad*

IONS, BEETS, TOMATOES, AND *dill pickles*

chocolate chip COOKIES

ginger beer

A FAMILY PICNIC, COMPLETE WITH LAUGHING CHILDREN,

napping dogs, and favorite foods, is one of life's most memorable adventures."

I grew up in Zambia, near the Zambezi River and Lake Kafue, where I developed a great love and respect for nature and the outdoors. As a child, I loved the clean air, sudden light showers, roaming animals, and wide-open spaces that made outdoor events our favorite form of family get-together. We would sit on blankets and eat great food, sip ginger beer, and talk around a campfire.

Whether you're living in Malibu or midtown Manhattan, on a farm or high in the snow-topped mountains, a picnic is a fun departure from everyday entertaining.

Invite your guests to first come for a hike, a bike ride, or a laid-back softball game. Select the picnic site in advance and simply spread out blankets and rugs to relax on. Have all the food prepared and packed ahead so that when everyone returns with healthy appetites, you can offer them a thirst-quencher and present a hearty spread.

When I first traveled to Sun Valley, Idaho, I was taken aback by the natural beauty. For this enchanting picnic, we settled into a lovely green pasture under crystal-clear blue skies. There was a nip in the air and golden red leaves fell from the trees around us. An assortment of Pendleton blankets and old bedspreads served as our dining table. The dishes of oversize and brightly colored Fiesta ware reflected the bold, sunny colors of the great outdoors. We gathered together tight bunches of natural field flowers that had been picked while on our earlier hike and placed them in a honey pot.

While the children played and dogs napped in the warm sunshine, my friends sipped home-brewed ginger beer. Lunch included three delicious kinds of sandwiches, a colorful red cabbage coleslaw with citrusy, thyme-scented dressing, and a classic potato salad. And, of course, it was only appropriate that I had prepared my mother's favorite pickles: onions, dill cucumbers, beets, and green tomatoes.

For dessert, we had homemade chocolate chip cookies. Everyone loves them and they travel well.

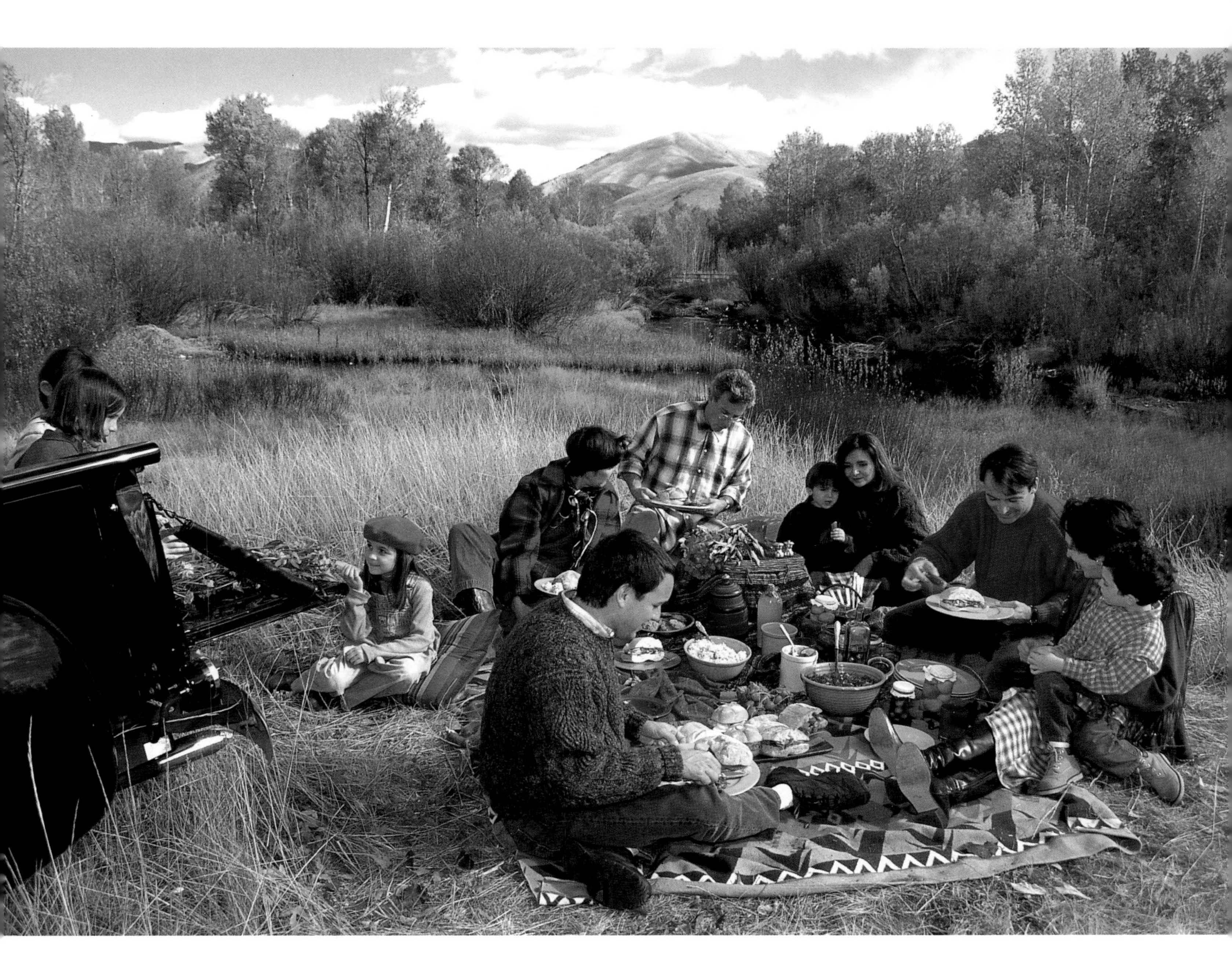

LET THE SEASONS BE YOUR GUIDE IN DECORATING YOUR TABLES. In spring, stand some slender stalks of asparagus and simple tapers in boughs of greenery or spring flowers. For summer, choose wildflowers or lush, even fully opened roses. In autumn, hollow out different-size pumpkins and fill them with fall flowers, or scatter crackly, red-orange leaves on the table. And during winter, fill copper bowls with pinecones, dried berries, and evergreen sprigs.

Sandwiches are perfect picnic fare. They can be made ahead and packed into a basket or a backpack.

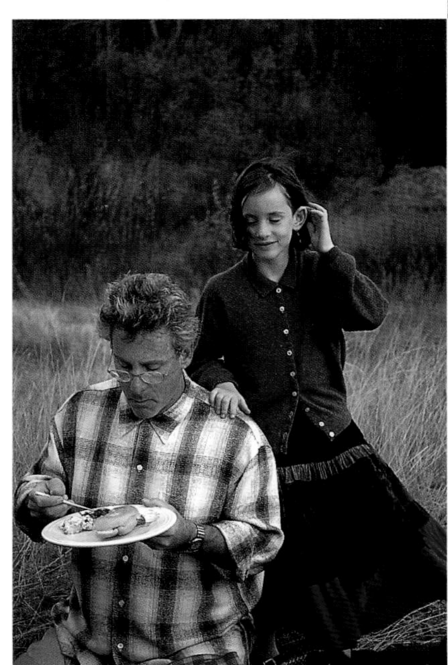

VARIATIONS ON THE CLASSIC POTATO SALAD

Dress up a traditional potato salad with some of the following:

- hard-cooked eggs and aioli (garlicky mayonnaise)
- green peas and fresh mint
- green beans, peanuts, and a spicy Thai-style dressing
- roasted potatoes with roasted garlic, red bell pepper, and fresh red onion
- crumbled blue cheese, green onions, and toasted walnuts in a walnut oil vinaigrette
- green beans, feta cheese, and a lemony vinaigrette

ROAST BEEF SANDWICHES *with* HORSERADISH AIOLI

- 1/4 cup mayonnaise
- 2 tablespoons prepared horseradish
- 1 tablespoon minced shallot
- Dash of hot pepper sauce
- 4 French rolls, split in half and toasted
- 1 pound sliced roast beef, about 1/4-inch thick
- 1 onion, sliced thinly and sautéed until golden brown
- 8 thick tomato slices
- Salt and freshly ground pepper

1. In a small bowl, combine the mayonnaise, horseradish, shallot, and hot pepper sauce. Spread the mixture on the cut surfaces of the rolls.
2. Layer some of the beef, onions, and tomato slices onto the bottom half of the rolls. Sprinkle with salt and pepper to taste. Cover the sandwiches with the roll tops.

MAKES 4 SANDWICHES

GRILLED CHICKEN SANDWICHES *with* BASIL MAYONNAISE

- 2 boneless, skinless chicken breast halves
- Salt and freshly ground pepper
- 2 tablespoons olive oil
- Basil Mayonnaise (page 120)
- 4 sandwich rolls, split in half
- 4 fresh mozzarella slices
- 8 tomato slices
- About 1 cup mixed baby greens

1. Place the chicken between pieces of wax paper and pound with a heavy object to flatten. Cut each piece in half. Season the chicken breasts with salt and pepper to taste.
2. In a large skillet, heat the oil over medium high heat. Add the chicken and cook for 1 minute on each side, or until almost cooked through. Set aside to cool.
3. Spread the Basil Mayonnaise over the cut surfaces of the rolls.
4. Place one piece of chicken on each of the roll bottoms. Add 1 mozzarella slice and 2 tomato slices with some of the greens on top. Cover the sandwiches with the roll tops.

MAKES 4 SANDWICHES

MAHIMAHI SANDWICHES *with* SAUTÉED TOMATOES

- 1 pound mahimahi fillets
- 4 to 5 tablespoons olive oil
- 1 teaspoon herbes de Provence
- Salt and freshly ground pepper
- 1/2 cup dry bread crumbs
- 2 roma tomatoes, seeded and cubed
- 2 cloves minced garlic
- 1 tablespoon chopped parsley
- 1 tablespoon balsamic vinegar
- 1 cup spinach leaves
- 4 sandwich rolls, split and toasted

1. Rub the fillets with 2 tablespoons of the olive oil, then sprinkle them with the herbs and salt and pepper to taste. Sprinkle with the bread crumbs and set aside for 1 hour.
2. In a small skillet, heat 1 tablespoon of the oil. Add the tomatoes and the garlic and sauté for 1 minute. Stir in the parsley and vinegar and season with salt and pepper to taste. Set aside.
3. In a small skillet, heat the remaining oil over medium-high heat and cook the fish for 3 minutes on each side. Set aside.
4. Arrange some of the spinach leaves over each of the roll bottoms. Spoon some of the tomato mixture onto the leaves, then top with the fillets. Cover the sandwiches with the roll tops.

MAKES 4 SANDWICHES

RED CABBAGE COLESLAW

- Juice of 1 1/2 lemons
- 1 tablespoon Dijon mustard
- 1 teaspoon honey
- 2 tablespoons olive oil
- Salt and freshly ground pepper
- 1 red cabbage, finely shredded
- 5 teaspoons chopped fresh thyme leaves

1. In a small bowl, whisk together the lemon juice, mustard, and honey. Whisk in the olive oil and season with salt and pepper to taste.
2. In a large bowl, combine the cabbage and thyme. Pour on the dressing and toss well to combine.

SERVES 8

MY SISTER JANET'S POTATO SALAD

6 cups diced cooked potatoes (about 5 large potatoes)
1 cup diced celery
½ cup diced dill pickles
2 tablespoons minced shallots
¾ cup mayonnaise
Salt and freshly ground pepper
2 hard-cooked eggs, peeled and chopped
2 tablespoons chopped chives

1. In a large bowl, combine the potatoes, celery, pickles, and shallots. Add the mayonnaise and toss to coat the ingredients evenly. Season with salt and pepper to taste.
2. Transfer the salad to a serving dish and garnish with the chopped eggs and chives.

SERVES 8

Note: Salad can be made 1 day ahead and should be garnished just before serving.

PICKLED ONIONS, BEETS, TOMATOES, *and* DILL PICKLES

Before pickling foods, sterilize the jars and their lids by boiling them in enough water to cover for 10 minutes. Leave them in the hot water until ready to use.

PICKLED ONIONS

2 pounds small white pickling onions
2 cups tarragon vinegar
1½ cups apple cider vinegar
12 whole cloves
10 peppercorns
2 bay leaves
1 cinnamon stick
3 tablespoons brown sugar, packed
1 teaspoon salt
1½ cups water
2 small dried hot red chili peppers

1. Drop the onions into a large pot of boiling water for 1 minute. Using a slotted spoon, lift the onions out of the water and, when cool enough to handle, slip off the skins. Discard the water.
2. To the pot, add the vinegars, cloves, peppercorns, bay leaves, cinnamon, brown sugar, and salt. Bring the mixture to a boil. Remove the pot from the heat and add the peeled onions. Set aside until cool.
3. Pour into a quart-size jar that has been boiled for sterilization. Add enough water to come within ¼ inch of the rim. Add 1 dried chili to the jar. Seal with the top. Store at room temperature for 2 days, then refrigerate.

MAKES 1 JAR

PICKLED BEETS

1 red onion, thinly sliced
2 pounds small beets
1½ cups cider vinegar
10 peppercorns
2 bay leaves
1 tablespoon coarse salt

1. Place the onion slices in a large bowl. Set aside.
2. Boil the beets in salted water until tender, about 15 to 20 minutes. Drain the beets over the bowl of onions, allowing the onions to wilt.
3. When cool enough to handle, slip the skins off the beets and slice into ¼-inch slices.
4. Drain the onions, reserving 1½ cups of the water. In a small saucepan, combine the reserved water, vinegar, peppercorns, bay leaves, and salt. Bring the mixture to a boil.
5. Meanwhile, pack the beets and onions into a quart-size jar. Pour in enough of the vinegar mixture to come within ¼ inch of the rims. Seal with the tops. Let the beets stand at room temperature for 24 hours, then refrigerate.

MAKES 1 JAR

PICKLED TOMATOES

2 pounds small green tomatoes, rinsed and stemmed
5 cups water
3 tablespoons coarse salt
10 garlic cloves, blanched and coarsely chopped
2 tablespoons coarsely chopped dill
2 tablespoons pickling spice
2 bay leaves

1. Pack the tomatoes into 2 quart-size sterilized jars. Set aside.
2. Boil the water and combine with the salt, garlic, dill, pickling spice, and bay leaves and pour over the tomatoes, filling the jars to within ¼ inch of the rim. Seal the jars with the tops and set in the sun for 24 hours. Store in a cool place for 3 days, then refrigerate.

MAKES 2 JARS

Note: Order green tomatoes from your market or a produce stand; even green cherry tomatoes work wonderfully for pickling.

MAKE HOMEMADE GINGER BEER and pickles to use as hostess gifts. Brew the beer and pour it into an etched glass bottle. Pack the pickles in traditional Mason jars and cover the tops with gingham fabric.

Choosing the ideal spot for a picnic is as much fun as preparing the food and enjoying the company.

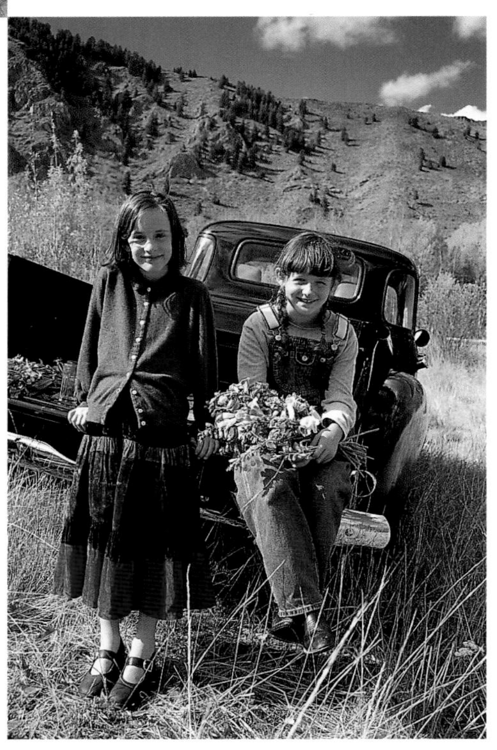

DILL PICKLES

2 pounds pickling cucumbers
5 cups water
3 tablespoons coarse salt
10 garlic cloves, blanched and coarsely chopped
2 sprigs of dill
2 tablespoons pickling spice
2 bay leaves

1. Pack the cucumbers into 2 quart-size jars. Set aside.
2. Combine the water, salt, garlic, dill, pickling spice, and bay leaves. Pour the mixture over the cucumbers, filling to within ¼ inch of the rims. Seal with the lids. Set the jars in the sun for 24 hours. Store in a cool place for 3 days before refrigerating.

MAKES 2 JARS

CHOCOLATE CHIP COOKIES

½ pound butter, softened
1 cup sugar
1 cup brown sugar
2 eggs
1 teaspoon vanilla extract
2 cups all-purpose flour
2 cups oatmeal, ground in the food processor until fine
1 teaspoon baking powder
1 teaspoon baking soda
1 12-ounce package semisweet chocolate morsels
1¾ cups chopped walnuts or pecans

1. Preheat the oven to 375 degrees.
2. Grease two cookie sheets.
3. Cream together the butter and sugars until light and fluffy. Beat in the eggs and vanilla.
4. Add the flour, oatmeal, baking powder, and baking soda. Mix until blended. Stir in the chocolate and nuts.
5. Break off tablespoons of dough and roll between the palms of your hands into small balls. Place on cookie sheets, 2 inches apart, and bake for 10 to 13 minutes, until edges are brown and bottoms set. (Do not let the tops brown or they will dry out.) Transfer to a wire rack and cool.

MAKES 60 TO 70 COOKIES

GINGER BEER

2 pounds plus ¼ cup sugar
1 pound fresh ginger, shredded
1 pack compressed yeast
1 cup black raisins
2 lemons, juiced
60–72 white raisins

1. In a large pot soak the sugar and ginger in 3 quarts water overnight. In the morning, bring the liquid to a boil and allow to simmer for 1 hour. Remove from heat and set aside to cool.
2. Pour the liquid into a clean pail and add enough water to total 2 gallons of liquid. Add the yeast and black raisins. Cover the pail and allow to stand airtight overnight.
3. In the morning, through a strainer, add the juice of 2 lemons. Cover and set aside for 6 days. Do not stir.
4. Siphon and strain all but a few inches of the ginger beer into a clean pail. Discard the remaining liquid, ginger, and raisins.
5. Bring to a boil ¼ cup sugar with 2 cups water. Add to the ginger beer and stir.
6. Pour the ginger beer into 10 to 12 750-ml. bottles, leaving 1 inch of air space at the top of each bottle. To each bottle, add 6 white raisins and cork tightly. When the raisins rise to the top, the ginger beer is ready to be served.

Note: Be careful, the corks may pop for the first few days while waiting for the raisins to rise. Place a towel on the cork when opening bottle.

A LIGHT LUNCH

chèvre mousse ON ARTICHOKE HEARTS
WITH ROASTED TOMATOES

SALAD WITH *grilled salmon,* MANGO, AND CRISPY LEEKS

lemon shortcake
WITH FRESH STRAWBERRIES

Serves: 6

Range of Difficulty: Easy

Preparation Time: 2 hours

A CARDINAL RULE WHEN CREATING FLORAL CENTERPIECES IS *to keep them low. You can have the most wonderful china and silver and food and drink in the world, but if guests can't see each other, they won't have a good time.*

A light, informal luncheon in your dining room or studio is a great way to catch up with friends or entertain business associates during the work week. Rather than running out to a restaurant, which consumes valuable time, a sophisticated yet easy-to-fix three-course lunch can be prepared. You get to set the agenda; where to sit, what to eat, when to eat, and the pace of the meal.

For this luncheon, guests were greeted with a refreshing glass of mint iced tea. During the week, I limit the alcohol served during a luncheon; otherwise, everybody will need a three o'clock nap. And the siesta is, regrettably, not an American tradition.

After a short time playing catch-up with one another's lives, we moved to the table, which I had dressed simply but elegantly with white linen place mats and a silver fruit bowl packed tightly with moonlight yellow roses.

Lunch began with a light chèvre mousse, served in an artichoke heart, with a roasted tomato and a drizzle of parsley sauce, followed by a salad of mixed greens, topped with grilled salmon and crispy leeks. The warm salmon went well with the chilled crisp vegetables and the crispy leeks. To give the salad plate a festive look, we sprinkled the rim with a "confetti" of finely cubed mango, tomato, chives, tarragon, and parsley. A smooth, assertive Champagne vinaigrette with truffle undertones brought all the components together.

An unfiltered Chardonnay is a superb accompaniment to this salad; its exceptional richness and acidity brought out the salmon's best flavor.

The entire menu is remarkably easy to prepare and can be readied in advance. Once everyone was seated, the artichoke and mousse were plated and served. It took only a minute to toss the salad with dressing and arrange the plates with the salmon and leeks.

Our delicious dessert, a lemon shortcake that can also be prepared ahead, was topped with whipped cream and fresh strawberries at the last moment. It was light, yet tasty and refreshing.

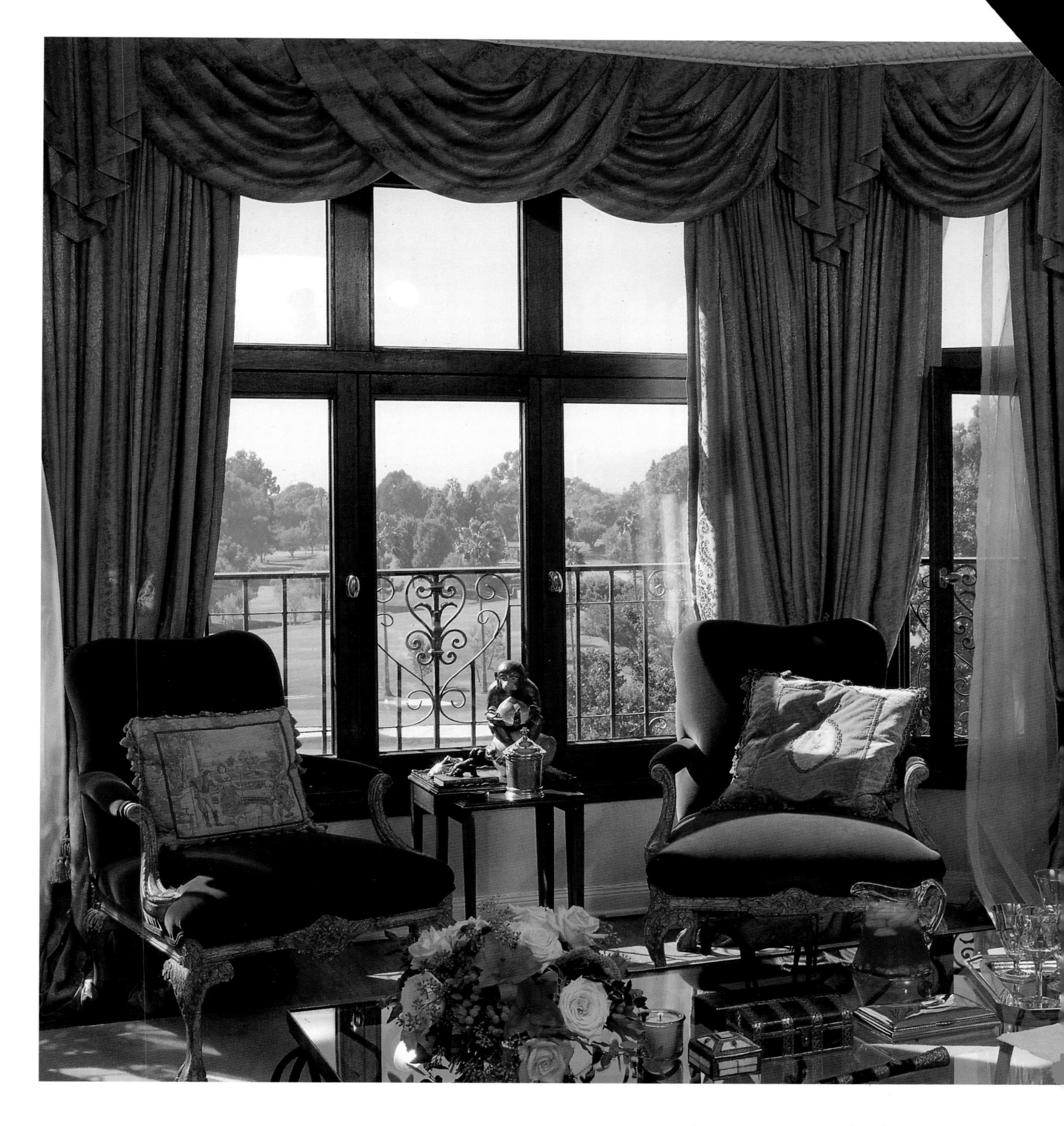

Create an inviting atmosphere for guests with a bright, airy room, fresh flowers, and a pitcher of fruity iced tea.

CHÈVRE MOUSSE *on* ARTICHOKE HEARTS *with* ROASTED TOMATOES

½ cup heavy cream, cold
3 ounces crumbled chèvre, softened
2 tablespoons sliced chives
Salt and freshly ground pepper
2 egg whites, stiffly beaten
9 roma tomatoes, peeled and halved
2 tablespoons olive oil
1 teaspoon herbes de Provence
6 cooked artichoke bottoms
Parsley Vinaigrette (see following recipe)

1. Beat the cream in a cold bowl until thick. Force the chèvre through a fine sieve and fold into the cream. Add the chives. Season with salt and pepper. Gently fold in the beaten egg whites.
2. Line a colander with muslin or several layers of cheesecloth and spoon in the mousse. Set on a tray in the refrigerator for 1 to 2 hours to set.
3. Meanwhile, preheat the oven to 375 degrees.
4. Place the tomatoes on a baking sheet. Lightly coat with olive oil and season with salt, pepper, and herbs. Bake 30 minutes and then transfer to a heated broiler. Cook until just beginning to blacken. Remove from oven and set aside.
5. To serve, place 1 artichoke bottom in the center of each of the 6 plates. Arrange 3 tomato pieces evenly around the artichoke. Place a scoop of the chèvre mousse on top of each artichoke bottom.
6. Drizzle the parsley vinaigrette all over the plate.

SERVES 6

PARSLEY VINAIGRETTE

15 parsley sprigs, stemmed
1 tablespoon sherry vinegar
4 tablespoons olive oil
2 tablespoons water
Salt and freshly ground pepper

Bring a small pot of salted water to a boil and blanch the parsley for 1 minute. Transfer the parsley to a blender, add the vinegar, oil, and water and purée. Season with salt and pepper to taste.

SALAD *with* GRILLED SALMON, MANGO, *and* CRISPY LEEKS

1 cup peanut oil
1 leek, white part only, finely julienned
½ pound mixed salad greens
Champagne Vinaigrette (see page 40)
Salt and freshly ground pepper
12 baby white potatoes, boiled, peeled, and halved
1 bunch asparagus, trimmed and blanched 5 minutes
¼ pound haricots verts, trimmed and blanched
1 ripe avocado, peeled and cubed
1 tablespoon chopped fresh parsley
1 mango, peeled and cubed
5 roma tomatoes, seeded and cubed
2 shallots, diced
18 ¼-inch strips center-cut salmon fillet

1. Preheat the broiler.
2. In a deep pot, heat the peanut oil over medium-high heat. Fry the leeks by adding to the oil a handful at a time. When golden, transfer the leeks with a slotted spoon to a paper towel to drain.
3. Place the salad greens in a large bowl and drizzle with all but 2 tablespoons of vinaigrette. Toss well and season with salt and pepper to taste.
4. Prepare the plates by dividing the salad and placing a high mound in the center of each plate. Place potatoes around the edges. Arrange the asparagus spears and haricots verts upright, leaning against the salad. Sprinkle the salad and plate with avocado, parsley, mango, tomato, shallots, and pepper to taste.
5. Season the salmon all over with salt and pepper to taste. Broil for 1 minute, being careful not to overcook. Brush the salmon lightly with the remaining vinaigrette and place 3 pieces on each salad, in an upright position. Top each with a nest of crisp leeks. Serve immediately.

SERVES 6

Confetti of finely chopped vegetables and herbs dresses up this salad of grilled salmon, mango, and crispy leeks.

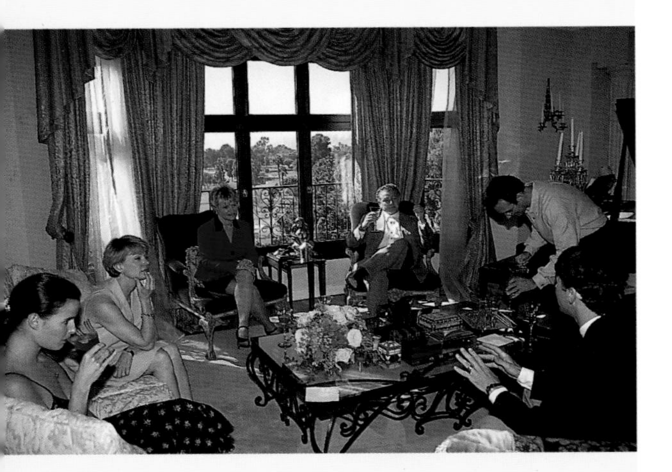

FRUITY ICED TEA

Fruit-flavored iced tea is one of my favorite lunch drinks to serve. Some are made with black teas and fruit essences; others are herbal blends of flowers, leaves, herbs, or citrus rinds. You can use one of the many tea bag products available or come up with your own blend.

I like to pour a bit of puréed mango into a glass, add a tablespoon of passion fruit, then fill the glass with iced tea or a mixture of iced tea and club soda. Garnished with a sprig of fresh mint, it's the perfect luncheon cocktail.

Another favorite cooler is made from freshly brewed iced tea served with a splash of peach nectar. And still other refreshing flavors can be made by combining bottled juices, such as raspberry or strawberry, with ordinary tea and garnishing with some mint and a few fresh berries.

CHAMPAGNE VINAIGRETTE

- 1 tablespoon Dijon mustard
- 1 teaspoon lemon juice
- ⅛ cup champagne vinegar
- ¼ cup olive oil
- ¼ cup truffle oil
- Salt and freshly ground pepper

In a metal bowl, whisk together the mustard, lemon juice, vinegar, and oils. Season with salt and pepper to taste.

Note: Salad dressings can be made up to 1 week in advance and refrigerated for later use.

LEMON SHORTCAKE *with* FRESH STRAWBERRIES

- 2 tablespoons finely grated lemon peel
- ½ cup sugar
- 4 tablespoons butter, softened
- Dash of salt
- 1⅓ cups all-purpose flour
- ½ cup heavy whipping cream (whipped)
- 48 chilled strawberries, with leaves and stems intact

1. Place the lemon peel in a small pot of water. Bring to a boil and blanch for 30 seconds. Drain and set aside.
2. Combine the sugar, butter, salt, and flour in a food processor with the metal blade in place. Run machine until a ball forms. Add the peel and run for 5 seconds. Remove the dough, pat into a disk, and cover with plastic wrap. Chill at least 1 hour.
3. To bake, preheat the oven to 400 degrees.
4. Cut the dough into 3 pieces. On a lightly floured board, roll out each to form 9 × 6-inch rectangles. Cut each rectangle in half to make six 6 × 4½-inch rectangles. With a spatula, transfer to baking sheets. Pierce all over with a fork and bake 20 to 25 minutes, until light brown. Transfer to a wire rack and allow to cool.
5. With a small spatula, spread whipped cream over each pastry shell. Arrange the berries in rows over the cream and serve immediately or chill.

MAKES 6

IF TIME IS OF THE ESSENCE, store-bought shortbread can be topped with whipped cream and strawberries.

Coffee and a second helping of dessert can be served in the comfort of a cozy den.

A Rustic Luncheon

TERRINE OF *eggplant and roasted peppers*
WITH *arugula* SAUCE

seafood paella

lemon sorbet
WITH MARINATED *grapefruit*

Serves: 6

Range of Difficulty: Easy

Preparation Time: 2 hours

Most people have great stories to tell.

You just have to create an environment for them to share them.

What could be nicer than bringing together a few old friends for lunch? Finding a couple of quiet hours stolen from the day to chat, taste some delicious food and wine, and perhaps play cards or enjoy a table game. That's a formula for a wonderful afternoon. We set our table in an old potting shed that hadn't been used in years. The patina of the walls, crumbling bricks, tangle of plants, and tarnished brass created a gentle, comfortable warmth for the occasion.

In keeping with the rustic surroundings, we dressed the table with a cotton cloth in a white-on-white print that would show off the dramatic colors of the food. The table was centered with an old metal vase filled sparsely with cascading Champagne grapes and a few blossoms from the garden. The only other elements on the table were some crusty baguettes, fragrant olive oil, bottles of Spanish wine, and a large clear pitcher of water with colorful lemon slices in it.

For the first course, we served a multilayered vegetable terrine that I have served for many years. A pungent arugula sauce gives the dish a little edge, and a small salad of baby greens shares the plate. The leftover terrine makes delicious sandwiches the following day.

The classic paella that followed, handsome and hearty in its large copper pan, left the guests speechless, but only for a moment. As they dug into the saffron rice, shrimp, chicken, clams, and mussels, and sipped a rich, dark, fruity Pesquera from Spain, there was a contented buzz of conversation. To end the meal, we served a zesty lemon sorbet with grapefruit segments that had been marinated in Grand Marnier.

The vegetable terrine, arugula sauce, sorbet, and grapefruit were prepared earlier in the morning. If time is of the essence it can be done the day before. All the components of the paella can also be readied ahead. When the guests arrive, just put the pan on the stove or in the oven to cook for 45 minutes. Paella is also a perfect choice for buffets, because it looks festive, feeds many people, and requires little "à la minute" preparation.

FLAVORED OLIVE OIL

To dress up a table and add a savory note to the meal as well, pour a good quality extra virgin olive oil in a small, attractive bowl, add a peeled garlic clove, sprigs of fresh rosemary or thyme, peppercorns, and a bay leaf. It's an authentic dip for rustic bread.

Other wonderful dipping oils can be made with either mild chili peppers (seeds removed) sliced and placed in a bottle of oil, or with fresh chopped tomatoes, ground pepper, and chopped shallots. The longer they stand, the more fragrant they become.

WINE OPTIONS

A good fruity chilled Beaujolais, served in a clear pitcher with fresh peach slices, is an ideal accompaniment to paella and other robust Spanish fare.

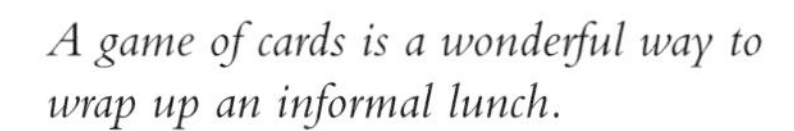

A game of cards is a wonderful way to wrap up an informal lunch.

Terrine of eggplant and roasted peppers captures the colorful spirit of the Mediterranean.

TERRINE OF EGGPLANT *and* ROASTED PEPPERS *with* ARUGULA SAUCE

1 large eggplant, peeled and cut into ½-inch slices
¼ cup olive oil
Salt and freshly ground pepper
2 tablespoons minced garlic
2 tablespoons chopped onion
2 ¼-ounce envelopes gelatin
Tomato Sauce (see following recipe)
2 each red and yellow bell peppers, roasted, peeled, seeded, and quartered
3 roma tomatoes, peeled and seeded
Arugula Sauce (see recipe)

1. Preheat the oven to 350 degrees.
2. Brush the eggplant with olive oil and season with the salt, pepper, garlic, and onion. Place in a single layer on a baking sheet and roast until golden brown, about 20 minutes. Let cool.
3. Stir the gelatin into the hot Tomato Sauce.
4. Lightly oil a 9 × 5 × 3-inch terrine and then line with plastic wrap, allowing the edges to hang over the sides of the pan. Cover the bottom and sides of the terrine with the eggplant slices, overhanging the edges. Spread a thin layer of Tomato Sauce over the eggplant. Cover with a layer of the peppers, followed by the tomatoes. Repeat the layers, ending with a top layer of eggplant. Fold over the overhanging eggplant to enclose, and then seal with the plastic liner.
5. Place a light weight or a 5-pound can on top of the terrine and refrigerate for at least 3 hours.
6. To serve, invert and remove the plastic wrap. Cut across the width with a knife and serve with Arugula Sauce. For best results, use an electric knife.

SERVES 6

TOMATO SAUCE

1 28-ounce can peeled tomatoes with basil
2 tablespoons olive oil
1 medium onion, chopped (about 1 cup)
1 garlic clove
1 bouquet garni of 10 cracked black peppercorns, 5 basil sprigs, and 1 bay leaf
Chopped fresh basil leaves or dried basil
Salt and freshly ground pepper
Sugar

1. In a blender or food processor, purée the tomatoes.
2. In a large saucepan or skillet, heat the oil over medium-high heat. Add the onion and sauté, stirring frequently, until soft and translucent. Add the garlic and sauté for about 30 seconds longer.
3. Add puréed tomatoes and bouquet garni and bring the mixture to a boil. Reduce the heat and simmer for about 5 minutes, skimming off any surface foam with a large cooking spoon. Season to taste with basil, salt, and pepper. If desired, season lightly with sugar.

ARUGULA SAUCE

3 bunches arugula, stems removed
2 tablespoons lemon juice
¼ cup olive oil
6 tablespoons water
Salt and white pepper

1. Blanch the arugula in boiling water for 3 minutes. Remove with a slotted spoon and immediately plunge into a bowl of ice water to stop the cooking process.
2. Drain the arugula and place in a blender along with the lemon juice, olive oil, and water. Blend until smooth. Season with salt and pepper to taste.

VARIATIONS ON CLASSICAL PAELLA

There are, literally, dozens of ways to prepare this beloved Spanish rice dish. Even in its homeland, there is no single official recipe. These are some of the savory interpretations I've come across:

Paella Valenciana, from the city of Valencia, is considered the original, the classic paella. It includes rabbit, chicken, assorted beans, onions, tomatoes, and, of course, short-grain rice. Snails are optional.

Shellfish paella usually features mussels, shrimp or prawns with heads and shells still on, clams, and lobster.

Vegetable paella is a kaleidoscopic feast of onions, leeks, tomatoes, cauliflower, peas, multicolored bell peppers, artichoke hearts, and tomatoes.

Other favored ingredients include pork loin, sausages such as chorizo and breakfast sausages, sea bass or other firm ocean fish, ham, and squid.

SEAFOOD PAELLA

1 teaspoon saffron threads
⅓ cup warm water
1 whole chicken breast cut up into 6 pieces
1 chicken drummettes
1½ teaspoon Cajun spice
¼ cup olive oil, or more
½ cup chopped onion
2 garlic cloves, minced
1½ cups white short grain rice
¼ teaspoon turmeric
3 cups chicken stock
1 tomato, peeled, seeded, and diced
12 jumbo shrimp, peeled and deveined
18 clams
1 pound calamari, cleaned and sliced into ½-inch slices
18 mussels
½ red bell pepper, seeded and julienned
½ yellow bell pepper, seeded and julienned
1 cup frozen peas
1 cup green beans, trimmed and sliced into 1½-inch lengths
Salt and freshly ground pepper
Juice of 1 lemon, plus 2 lemons quartered
2 tablespoons chopped parlsey

1. Dissolve the saffron in the water and set aside.
2. Season the chicken with the Cajun spice.
3. Heat the olive oil over medium heat in a very large skillet or paella pan. Cook the chicken breast pieces and drummettes for 5 minutes on each side, or until golden brown. Set aside.
4. In the same pan sauté the onion, garlic, and rice. Add the saffron mixture, turmeric, and stock. Add the chicken and cook, covered, for 15 minutes.
5. Add the tomato, shrimp, clams, calamari, mussels, bell peppers, peas, beans, salt, and pepper and cook covered for a further 10 minutes.
6. If all the stock is not absorbed, remove the lid and cook until it is absorbed.
7. Drizzle the lemon juice over the top, place the lemon wedges around the edges, and sprinkle with the chopped parsley.

SERVES 6

LEMON SORBET *with* MARINATED GRAPEFRUIT

¾ cup sugar
¾ cup water
1¼ cups freshly squeezed lemon juice, strained
½ cup dry white wine
2 egg whites
Marinated Grapefruit (see following recipe)

1. In a small saucepan, combine the sugar and water. Bring to a boil over high heat and stir just until the sugar dissolves. Cool.
2. Add the lemon juice and the wine to the cooled syrup.
3. In the bowl of an electric mixer, beat the egg whites until soft peaks form. Carefully fold them into the lemon mixture.
4. Transfer the mixture to an ice cream maker and freeze according to the manufacturer's instructions. Then transfer to a chilled bowl.
5. Serve with Marinated Grapefruit sections.

MAKES ABOUT 1 QUART

MARINATED GRAPEFRUIT

3 large grapefruits
3 tablespoons sugar
¼ cup water
⅓ cup grapefruit juice
¼ cup Grand Marnier

1. Working over a bowl to catch the juices, peel the grapefruits and, with a serrated blade, cut between the membranes to remove the sections. Set the juice and sections aside separately.
2. Combine the sugar and water in a small saucepan and cook over medium heat until the sugar just starts to caramelize. Pour in the grapefruit juice and the Grand Marnier and stir until the sugar dissolves. Pour the liquid over the grapefruit segments and set aside to cool.

DRESSING UP SORBETS

Whether you make your own sorbet or buy one of the many excellent, upscale brands, you can give this icy dessert your own unique touch with these ideas:

- grapefruit sections marinated in Grand Marnier
- fruits and berries marinated in a light syrup
- fresh strawberries puréed with a bit of sugar and Cassis
- crystallized violets, lilacs, and roses
- chopped raspberries and blackberries, tossed with sugar, fresh lime juice, and chopped mint
- mango purée thinned with a bit of orange juice
- fresh, sweet cherries, pitted and tossed with kirsch, lemon juice, and brown sugar

ENTERTAINING ADULTS AT A CHILDREN'S PARTY

POACHED *salmon* WITH CRISPY SWEET POTATOES

farfalle WITH DICED TOMATOES AND FETA CHEESE

chopped vegetable salad IN A BALSAMIC VINAIGRETTE

FRESH BAKED *tomato and oregano* BREAD

Serves: 10

Range of Difficulty: Easy

Preparation Time: 2 hours

Holden

There's nothing more inspiring than

watching the innocence on the faces of children at a birthday party. And, of course, the adults become children too.

A child's birthday celebration doesn't have to mean spending lots of money or planning an overchoreographed event. What's important is creating activities that children enjoy. That can mean something as simple as taking a group of preteens to a baseball game or a bevy of would-be prima donnas to the ballet. It can be a fanciful party in your backyard complete with pirates or ponies, magicians or kites; or a trip to a favorite restaurant or country club. Games, a birthday cake, and ice cream are the basics.

In the back garden of a Southern California home, ten adults and sixteen high-spirited four-year-olds joined their own private circus. With "clowns" popping out of colorful cardboard drums on each table; a "real" clown to paint faces, juggle, and make balloon animals; and a rousing game of Pin the Nose on the Clown, we created an exciting scene for everyone.

For the children, mini hamburgers, turkey dogs, peanut butter and jelly sandwiches, popcorn, cupcakes, and fresh fruit juices were on the menu. A towering cake, frosted with white icing and decorated with multicolored candies, met with the wide-eyed approval of all sixteen little guests.

The grown-ups sipped a refreshing Pinot Grigio, mineral water, or iced tea, and lunched on cold poached salmon, chopped vegetable salad, pasta salad, and freshly baked tomato-oregano bread.

Food was available from the moment the guests arrived—presented on a buffet table decorated with a maypole and multicolored streamers. Parents could indulge whenever their children were preoccupied with other activities.

At another birthday party, a private room at Planet Hollywood became Fire Station Number One for a very special two-year-old. It was Halloween and everyone was invited to wear costumes. The birthday boy wore firefighter garb—jacket, hat, and boots—as did his father. Other adults, too, got into the act. There were bunnies, ballerinas, French maids, bikers, lions, tigers, and cowboys in all sizes.

We worked with the restaurant to create an interesting menu for both the children and the adults, spelling out in detail what was expected, in terms of decor, food, beverages, timing of service, and costs. At this party, waiters passed trays of bite-size crab cakes, barbecued chicken sticks, and mini pizzas while guests were arriving. Later, a continuous buffet of peanut butter and jelly triangles and crunchy chicken fingers pleased the kiddies. But it was the carrot and pineapple birthday cake, shaped like a fire engine and iced in vivid primary colors, that stole the show!

A buffet of cold- and room-temperature dishes allows the adults at a child's birthday party to help themselves whenever they're not involved with the children's games and activities.

POACHED SALMON *with* CRISPY SWEET POTATOES

1 quart water
1 half white onion, sliced
2 celery stalks, cut into 3-inch lengths
6 parsley sprigs
1 teaspoon salt
2 bay leaves
2 cups dry white wine
10 6-ounce center-cut salmon fillets, skin removed
Crispy Sweet Potatoes (see following recipe)
Cucumber Yogurt Sauce (see page 55)

1. In a large pan, combine the water, onion, celery, parsley, salt, bay leaves, and white wine. Bring to a boil, then reduce heat and simmer for 15 minutes. Carefully add the salmon, cover the pot, and poach for about 10 minutes. (The general rule is 10 minutes for each inch of thickness.)
2. With a slotted spoon, remove salmon and place on paper towels to drain. Let cool for 10 minutes, then transfer to a serving platter, cover with plastic wrap, and chill. Serve with Crispy Sweet Potatoes and Cucumber Yogurt Sauce.

SERVES 10

CRISPY SWEET POTATOES

Peanut oil
2 medium sweet potatoes, peeled and julienned
Salt and freshly ground pepper

In a small saucepan heat the peanut oil over medium-high heat. Fry the potatoes, a handful at a time, until golden brown, and transfer to paper towels to drain. Season with salt and pepper to taste, and set aside.

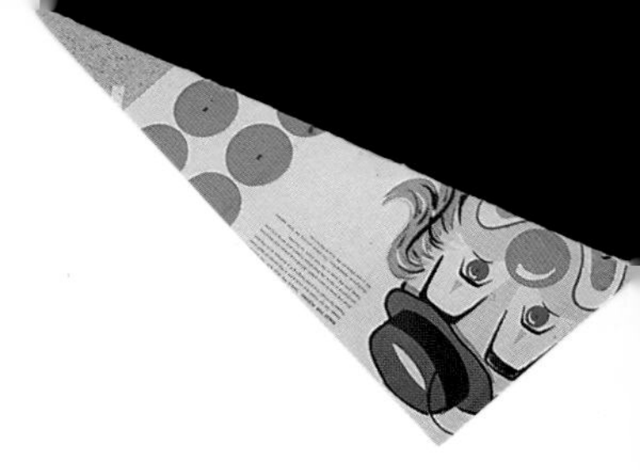

CUCUMBER YOGURT SAUCE

16 ounces nonfat yogurt
½ cucumber, peeled, seeded, and diced
2 tablespoons chopped fresh parsley
2 tablespoons chopped fresh chives
1 tablespoon fresh lemon juice
¼ teaspoon each salt and pepper

In a small bowl, combine all the ingredients.

SERVES 8

FARFALLE *with* DICED TOMATOES *and* FETA CHEESE

1½ pounds farfalle or bow tie pasta, cooked
6 roma tomatoes, seeded and diced
½ cup calamata olives, pitted
⅓ cup diced red onion
4 teaspoons drained capers
3 ounces feta cheese, crumbled
10 basil leaves, julienned
Balsamic Vinaigrette (see following recipe)

Combine the pasta, tomatoes, olives, onions, capers, cheese, and basil. Pour on the vinaigrette and toss lightly.

SERVES 8

BALSAMIC VINAIGRETTE

3 tablespoons balsamic vinegar
1 tablespoon Dijon mustard
1 garlic clove, minced
½ cup olive oil
Salt and freshly ground pepper

In a small bowl, combine the vinegar, mustard, and garlic. Add the oil in a slow steady stream, whisking constantly. Season with salt and pepper to taste.

CHOPPED VEGETABLE SALAD *in a* BALSAMIC VINAIGRETTE

1½ heads romaine lettuce
6 roma tomatoes, seeded and cubed
2 cooked artichoke bottoms, cut into cubes
1 cup canned garbanzo beans, rinsed and drained
4 ounces mozzarella cheese, cubed
2 celery stalks, cut into cubes
½ cucumber, peeled, seeded, and diced
½ white onion, diced
Balsamic Vinaigrette (preceding recipe)

Rinse and dry the lettuce, discarding outer leaves. Roughly chop the lettuce and place in a salad bowl with the tomatoes, artichoke bottoms, beans, cheese, celery, cucumber, and onion. Pour on the vinaigrette and toss lightly.

SERVES 8

FRESH BAKED TOMATO *and* OREGANO BREAD

2 packages active dry yeast
1 ½ cups warm water (105 to 115 degrees)
2 tablespoons olive oil
2 tablespoons sugar
2 teaspoons salt
2 teaspoons dried oregano
4 ½ cups all-purpose flour
⅓ cup tomato paste
Additional olive oil

1. In a large bowl, sprinkle the yeast over the warm water, stirring to dissolve. Add the olive oil, sugar, salt, and oregano. Beat in 3 cups of the flour, then beat in the tomato paste. Gradually beat in the remaining flour.
2. Rub the palms of your hands with olive oil. Gather up the dough and shape it into a ball. Place the dough ball in an oiled bowl and turn it to coat it all over. Cover the bowl with wax paper then a towel. Set the dough in a warm place to rise until doubled, about 45 minutes.
3. Oil a 9-inch pie plate.
4. Punch down the dough and knead it several times. Shape the dough into a round loaf and place it in the pie plate. Rub the dough with olive oil, then let it rise in a warm place for 30 minutes.
5. Preheat the oven to 375 degrees.
6. Bake the loaf for about 50 minutes, or until browned. Remove it from the pie plate and cool on a wire rack.

MAKES 1 LOAF

HAPPY BIRTHDAY ZACK!
HAPPY BIRTHDAY ZACK!
Raquel
Clayton
Raquel
The
LORA
Dr. Seu
Random House · New Yor
Dr. Seuss
FRESH
Popcorn

Hire a photographer to capture the fun in photos and document the occasion. Or buy disposable cameras and place them on tables for guests to snap shots of the celebration.

KID-FRIENDLY FOODS

You can't go wrong with these tried-and-true favorites:

- Hot dogs: beef, turkey, and chicken
- Mini hamburgers
- Mini turkey burgers
- Peanut butter and jelly sandwiches
- Spaghetti with meatballs
- Chicken wings
- Pizza
- Macaroni with cheese
- Cupcakes
- Popcorn
- Ice cream cones
- Cookies
- Birthday cake with ice cream
- Fresh fruit
- Soft pretzels
- Candy

WHIMSICAL TOUCHES

A child's birthday party should be one of life's sweetest moments, both for the child *and* the parents.

Let your imagination run wild with decorations, foods, activities, and gifts.

- Decorate a room around a theme that the child will enjoy.
- Paint an inexpensive tablecloth with the children's names and their hand-prints.
- Cut peanut butter and jelly sandwiches with cookie cutters shaped like favorite animals or cartoon characters.
- Be creative with place cards: Paint the children's names on shiny, yellow plastic firefighter hats, goodie bags, party hats, balloons, baseballs, or footballs.
- Be creative with "goodie bags": Use plastic beach pails, treasure chests, cowboy hats, or lunch bags.
- Use lots of balloons with streamers, letting them float to the ceiling.
- Provide plenty of games.

A NORTHERN ITALIAN BUFFET

TOMATO *lasagne*

eggplant PARMIGIANA

caprese SALAD

stuffed zucchini WITH *veal*

CHECKERBOARD OF *roasted peppers*
WITH OLIVES, CAPERS, AND ANCHOVIES

asparagus WITH CHOPPED EGG IN A WALNUT VINAIGRETTE

ADRIANA'S *chocolate torte*

raspberry WINE COOLER

Serves: 12

Range of Difficulty: Moderate

Preparation Time: 4 hours

A BUFFET IS IDEAL FOR ANY SIZE GROUP AND MOST OCCASIONS.

It contributes energy and a focus to the party. Most important, it allows the host to relax and enjoy his or her own party.

Whoever said, "The way to a man's heart is through his stomach," must have had lasagne in mind! Silky layers of homemade pasta, buffalo mozzarella, ripe tomatoes, Parmigiano-Reggiano, and a light béchamel sauce make for a heartwarming dish.

Adriana Pacifici, an Italian chef and dear friend, taught me how to make the best lasagne ever. I met Adriana in 1985 when a mutual friend invited me to her home for a cooking class. While coaxing a pot of pasta e fagioli to perfection on the stove, Adriana asked for someone to taste the hearty soup. I quickly volunteered. The soup was fabulous and a lasting friendship was born.

Adriana, a native of Rome, has been a wonderful inspiration to me, teaching me the way around the Italian kitchen and sharing recipes that have been in her family for generations. She also taught me that caring cooks can comfort, soothe, celebrate, boost morale, lift spirits, create traditions—in short, enrich the lives of everyone with whom they come in contact.

I created this menu for an informal Sunday luncheon. Though labor-intensive, everything on the menu could be made in advance.

On a sheltered terrace, surrounded by wild roses, hibiscus, and trellised greenery, we arranged a buffet table that captured the intense vitality of an Italian countryside. Topiaries were created with Styrofoam cones, fresh lemons from the garden, and Japanese moss. A collection of bottled olive oils and vinegars was both practical and decorative.

Guests sipped a refreshing Kir-like raspberry wine cooler. They settled down at two tables, which were dressed in lemon-and-lime plaid cloths with vintage terra-cotta pots packed tightly with marigolds, ivy, and ferns. Simple white plates were used, and the yellow-and-green-tinged glassware was in harmony with the lush garden setting. There was laughter and many trips to the buffet—a leisurely afternoon, designed to enjoy friends, savor good food and wine, and re-energize mind and soul for the workweek ahead.

IT'S EASY TO PREPARE A WINE COOLER.

To make enough for 3 bottles, prepare the fruit syrup by placing 2 cups of water, 6 tablespoons of sugar, and 1½ cups of fruit (strawberries, raspberries, blueberries, passion fruit, or kiwi fruit) in a saucepan. Set over medium heat for 12 to 15 minutes.

Remove the syrup from the heat and press through a fine sieve and allow to cool. From each bottle of white wine, discard ½ cup of the wine, then pour equal amounts of the mixture into the 3 bottles. Chill and serve over ice with a few pieces of the fresh fruit for garnish. For best results, use a Sauvignon Blanc and pour from a pitcher.

TOMATO LASAGNE

Tomato Sauce (see following recipe)
Lasagna Noodles (page 63)
Béchamel Sauce (page 63)
1½ cups grated imported Parmesan cheese
1½ pounds fresh mozzarella in water, drained and sliced
Additional Parmesan cheese

1. Preheat the oven to 375 degrees.
2. Spread a thin layer of Tomato Sauce over the bottom of a 15 × 10-inch baking dish. Cover the sauce with a layer of the lasagna noodles. Spread about ¾ cup of the Tomato Sauce over the noodles, then spoon about ⅔ cup of the Béchamel Sauce over the Tomato Sauce. Next, sprinkle 3 tablespoons of the Parmesan cheese over the sauces. Arrange a third of the mozzarella slices over the top. Repeat layering the noodles, the sauces, and the cheeses two more times. Add one more layer of the noodles over the top, then spread them with the remaining tomato sauce. Sprinkle with additional Parmesan cheese.
3. Bake the lasagne for 30 to 35 minutes, or until it is bubbling around the edges, the top is brown, and any exposed noodles are crisp. Remove from the oven and let the lasagne stand for at least 10 minutes before serving.

SERVES 12

Note: The lasagne can be assembled ahead, covered, and refrigerated or frozen. Before baking, let stand until at room temperature.

TOMATO SAUCE

2 tablespoons olive oil
1 medium onion, chopped (about 1 cup)
1 garlic clove
1 28-ounce can peeled tomatoes with basil
Chopped fresh basil leaves or dried basil
Salt and freshly ground pepper
Sugar

1. In a large saucepan or skillet, heat the oil over medium-high heat. Add the onion and sauté, stirring frequently, until soft and translucent. Add the garlic and sauté for about 30 seconds longer.
2. In a blender or food processor, purée the tomatoes and add to the onion in the pan. Bring the mixture to a boil. Reduce the heat and simmer for about 5 minutes, skimming off any surface foam with a large cooking spoon. Season to taste with basil, salt, and pepper. If desired, season lightly with sugar.

Note: This tomato sauce is for both the Tomato Lasagne and Eggplant Parmigiana.

BÉCHAMEL SAUCE

4 tablespoons butter
3 tablespoons all-purpose flour
2 cups hot milk
Salt and freshly ground pepper

1. In a small saucepan over medium heat, melt the butter. Add the flour and cook, stirring constantly, until the mixture begins to bubble, about 2 minutes.
2. Gradually stir in the hot milk. Continue to cook, stirring constantly, until the sauce comes to a boil. Reduce the heat and simmer, continuing to stir, 2 to 3 minutes longer, or until very thick. Season to taste with salt and pepper.

LASAGNA NOODLES

2 cups all-purpose flour
1 teaspoon salt
3 eggs
2 teaspoons vegetable oil

1. Combine the flour, salt, eggs, and oil in a food processor. Process until the dough forms a ball, adding more water, a teaspoon at a time, if dough is too dry.

2. In a large skillet, heat about 1/4 inch of the oil over medium-high heat. Add some of the eggplant slices and cook, turning once, until they are golden on both sides. Remove from the pan and place them on paper towels to drain. Repeat with the remaining eggplant slices, adding more oil as needed.
3. Preheat the oven to 375 degrees.
4. Spread a thin layer of Tomato Sauce over the bottom of a 15 × 10-inch baking dish. Arrange a third of the eggplant slices over the sauce. Scatter a third of the mozzarella cheese pieces, then a third of the Parmesan and a third of the basil over the eggplant layer. Spoon a third of the remaining Tomato Sauce over all. Repeat this process 2 more times, using half of the remaining eggplant, cheeses, basil, and Tomato Sauce for each layer.
5. Bake the casserole for about 45 minutes, or until bubbly around the edges. Remove the casserole from the oven and allow it to stand for 10 to 20 minutes before serving.

SERVES 12

Note: The casserole can be assembled, covered, and refrigerated or frozen. Before baking, let stand until at room temperature.

CAPRESE SALAD

6 large ripe roma tomatoes
6 8-ounce balls fresh mozzarella in water, drained
1/4 cup red wine vinegar
2 teaspoons Dijon mustard
1/3 cup olive oil
1 large bunch basil leaves, torn into small pieces
2 tablespoons chives, thinly cut with scissors
Salt and freshly ground pepper

1. Cut the tomatoes and the cheese into 1/4-inch-thick slices. Arrange them in alternating circles on a large flat serving dish.
2. In a small bowl, combine the vinegar and mustard. Whisk in the olive oil. Pour the dressing over the tomatoes and cheese. Sprinkle the salad with the basil and chives. Season to taste with salt and pepper.

SERVES 12

ROASTING PEPPERS

Roasting red bell peppers gives the flesh a beguilingly smoky flavor and brings out the natural sweetness of the pepper. Peppers can be prepared either over a gas burner or in a broiler. Once th charred, place the peppers in a paper bag and let them sweat f utes. The skin will peel off easi

STUFFED ZUCCHINI *with* VEAL

- 7 large zucchini (about 4 pounds)
- 3 tablespoons olive oil
- ½ medium onion, chopped (½ cup)
- 1 garlic clove, minced
- ½ teaspoon dried thyme
- 1 pound ground veal or chicken
- ¼ cup dry white wine
- 2 tablespoons whipping cream
- 3 tablespoons grated imported Parmesan cheese
- Salt and freshly ground pepper
- 2 tablespoons plain dry bread crumbs

1. Trim the ends of the zucchini, then cut 6 of them into thirds. Using a melon baller, scoop out and discard the centers of the zucchini pieces. Set aside. Chop the remaining zucchini and set aside.
2. In a large skillet heat 1 tablespoon of the oil over medium-high heat. Add the onion and cook for 2 minutes, stirring frequently. Add the chopped zucchini, garlic, and thyme and cook, stirring frequently, 5 minutes longer.
3. Transfer the mixture to a bowl and set aside.
4. In the same pan, heat the remaining oil over medium-high heat. Add the veal and cook, stirring frequently, until browned, about 5 minutes. Add the wine and bring to a boil. Return the onion mixture to the pan, cover, reduce the heat, and cook for 5 minutes.
5. Uncover the pan and continue to cook for 2 minutes to reduce the liquid. Transfer the mixture to a food processor and pulse for about 10 seconds, or until finely chopped. Add the cream and the Parmesan and process 5 seconds longer, just to blend. Season with salt and pepper to taste.
6. Preheat the oven to 375 degrees.
7. Sprinkle the inside of the zucchini pieces with salt and pepper. Fill the zucchini pieces with the veal mixture, piling it about ½ inch over the top. Arrange the stuffed zucchini pieces in a baking dish and sprinkle them with the bread crumbs. Bake for 25 minutes, or until the zucchini are tender.

SERVES 12

(OPPOSITE PAGE) *Caprese Salad, Asparagus with Chopped Egg in a Walnut Vinaigrette, Checkerboard of Roasted Peppers with Olives, Capers, and Anchovies.*

ADRIANA'S CHOCOLATE TORTE

Butter
1 cup semisweet chocolate morsels
1 cup blanched whole almonds
2½ sticks butter (½ pound plus 4 tablespoons)
1½ cups sugar
3 eggs
1 scant cup all-purpose flour
½ cup water
2 tablespoons Dutch cocoa
Dash of salt
Powdered sugar
Whipped cream

1. Preheat the oven to 400 degrees.
2. Coat a 10-inch round springform pan with butter. Line with parchment paper and then coat paper with butter.
3. Combine the chocolate morsels and the almonds in a food processor or blender and process until ground.
4. In a large mixing bowl, cream together the butter and sugar until light and fluffy. Add the eggs, one at a time, beating well after each addition. Add the chocolate mixture and beat well. Beat in the flour.
5. Bring the water almost to a simmer. Stir in the cocoa and salt until dissolved. Remove from the heat and allow to cool slightly.
6. Add the cocoa mixture to the bowl and beat until well blended.
7. Pour the batter into the prepared pan and bake for 55 to 60 minutes, until a toothpick, inserted in the center, comes out clean. Cool in the pan on a rack for 1 hour. Remove the pan sides, sprinkle the cake with powdered sugar, and serve with the whipped cream.

SERVES 12

LEMON TOPIARY

Purchase a Styrofoam cone 18 inches in height from a local arts and crafts supplies store. Using large toothpicks or bamboo skewers cut in half, pierce each lemon with a skewer and secure it to the Styrofoam. Pack lemons as closely as possible and fill any gaps with moss that has been soaked in water, then wrung out (moss is available at most florists). This centerpiece should last for more than a week if kept in a cool place. Oranges, limes, apples, and pomegranates can also be used to make attractive topiaries. Any of these fruits can be sprayed silver or gold for a festive holiday look.

MAKE AN EDIBLE TOPIARY by covering a Styrofoam cone with aluminum foil and using toothpicks to secure the food. Some of the foods from which I've made topiaries are: whole poached shrimp, strawberries, melon cubes, and pitted cherries.

Adriana's Chocolate Torte, the perfect end to a meal.

SALAD OF BUTTER LETTUCE WITH GRAPEFRUIT SEGMENTS *and grilled scallops*

pappardelle with grilled chicken AND MUSHROOMS IN A LIGHT BROTH

apricot crepes WITH CHOCOLATE SAUCE

Serves: 8

Range of Difficulty: Easy

Preparation Time: 2 hours

THERE'S NO BETTER BIRTHDAY GIFT THAN TO INVITE A FRIEND

for a special lunch at your home with six or seven of his or her best friends. It's a very sincere, generous way to say Happy Birthday.

There's something about a tray of Harvey Wallbangers that brings out the high-spirited, devil-may-care college kid in all of us!

I decided to jump start this birthday luncheon for a dear friend with this fun cocktail, served in my courtyard on a glorious autumn Sunday. For lunch, we moved inside, where I set the table with a crisply starched white cotton tablecloth, plain white china and napkins, and silver pieces from the collection I've gathered over the years. For a twist on the traditional floral centerpiece, I rested narcissus bulbs on top of six different silver pieces. The look was simple, clean, and inviting.

Midday menus are a special challenge: The food needs to be inventive and appealing yet light and refreshing. Here we started with a salad of butter lettuce, topped with sautéed sea scallops. Zesty grapefruit sections were tucked into the pale green leaves and a raspberry vinaigrette accented the fruit and seafood flavors.

For a second course, we served a warm, satisfying dish of homemade pappardelle tossed with grilled chicken, mushrooms, tomatoes, and fresh peas.

To accompany the pasta, we chose a Chardonnay on the austere side, with a brisk acidity to hold up to the sauce. And for those not drinking alcohol, serve chilled sparkling mineral water with a dash of bitters and a slice of lemon.

For dessert, homemade crepes were a welcome alternative to the traditional birthday cake.

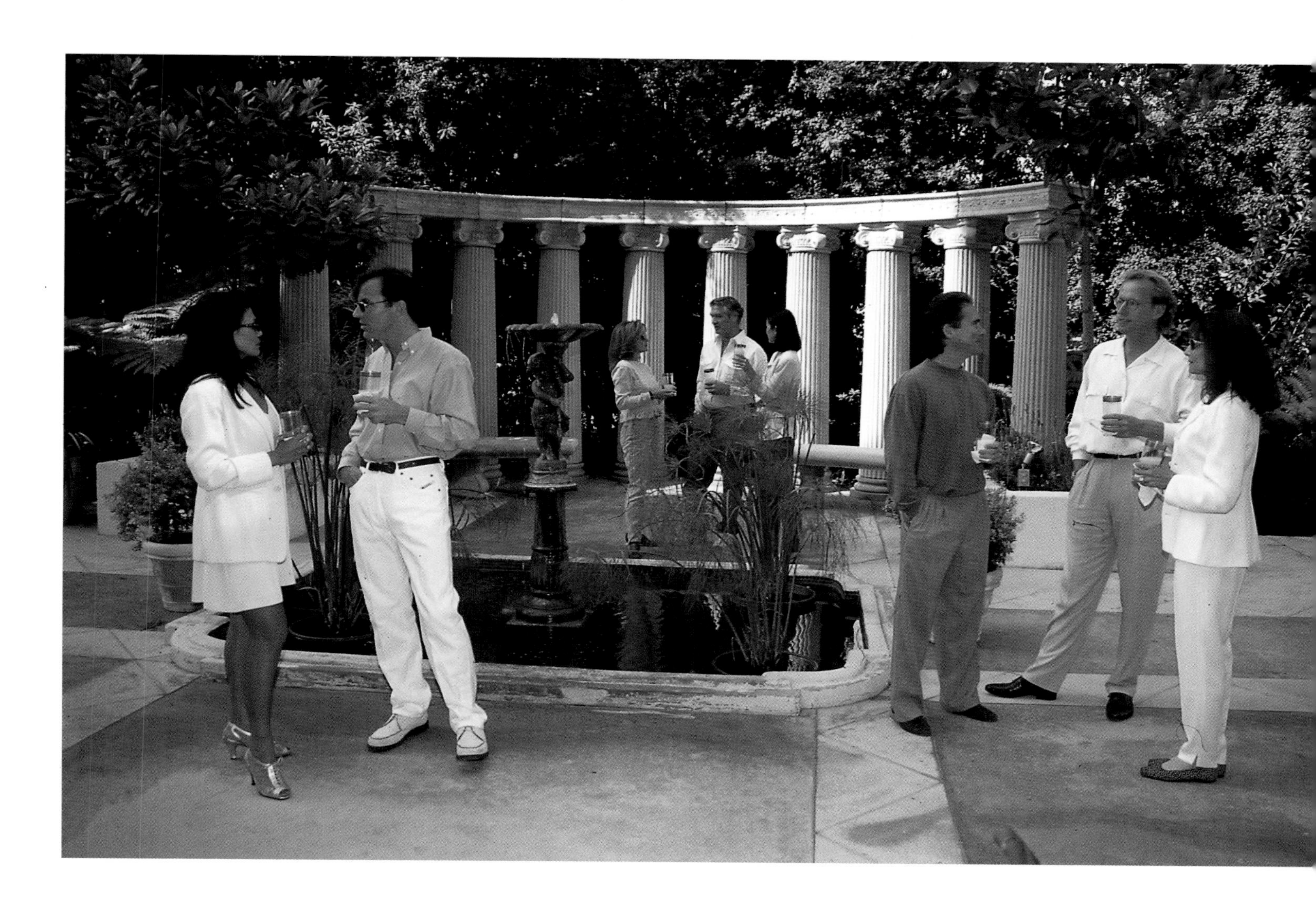

HARVEY WALLBANGER

1 ounce vodka, 4 ounces orange juice, ½ ounce Galliano. Pour the vodka and orange juice into a highball glass filled with ice cubes. Stir. Float Galliano on top.

The interplay of different textures and temperatures adds interest to this salad of butter lettuce, grapefruit, and grilled scallops. The salad is also delicious with cubed avocado and diced bell pepper.

Pasta in a light broth is ideal luncheon fare.

SALAD *of* BUTTER LETTUCE *with* GRAPEFRUIT SEGMENTS *and* GRILLED SCALLOPS

2 tablespoons flour
Salt and freshly ground pepper
24 sea scallops
2 tablespoons olive oil
3 to 4 small heads butter lettuce, rinsed and torn into pieces
Grapefruit Dressing (see following recipe)
3 grapefruits, peeled and cut into segments
2 shallots, minced

1. Season the flour with salt and pepper to taste. Sprinkle the flour over the scallops and toss to lightly coat the scallops.
2. In a large skillet, heat the olive oil over high heat. Add the scallops and cook until they are browned, about 1 minute on each side. Remove the scallops from the pan. Set aside and keep warm.
3. In a large bowl, toss the lettuce with the dressing. Pile some of the salad in the center of 8 individual salad plates. Arrange 3 scallops around the outer edge of each salad. Arrange the grapefruit segments between the scallops, then sprinkle the salad with the shallots.

SERVES 8

Note: If the grapefruit is too tart, sprinkle with sugar.

GRAPEFRUIT DRESSING

1½ tablespoons grapefruit juice
1½ tablespoons sherry vinegar
1 tablespoon Dijon mustard
1 teaspoon minced shallot
½ cup olive oil
Salt and freshly ground pepper

In a small bowl, combine the grapefruit juice, vinegar, mustard, and shallot. Gradually whisk in the oil. Season with salt and pepper to taste.

PAPPARDELLE *with* GRILLED CHICKEN *and* MUSHROOMS *in a* LIGHT BROTH

3 tablespoons olive oil
1½ pounds skinless, boneless chicken breasts
1¼ cups chopped onion
2 cups sliced mushrooms
¾ cup dry white wine
4 cups vegetable broth
1½ cups frozen peas, thawed
Salt and freshly ground pepper
1½ pounds pappardelle or fettuccine noodles, cooked and drained
2 medium tomatoes, seeded and diced
¼ cup chopped fresh tarragon leaves, lightly packed
6 tablespoons freshly grated Parmesan cheese

1. In a large skillet, heat the oil over medium-high heat. Add the chicken and cook for 2½ minutes on each side. Remove the chicken from the pan and set aside.
2. Add the onion to the pan, cover, and cook over medium heat until the onion becomes soft and has a glassy appearance. Add the mushrooms and the wine and cook for 4 minutes longer.
3. Meanwhile, slice the cooked chicken ¼ inch thick, then cut each slices in half.
4. Add the vegetable broth to the pan and bring it to a boil. Allow the broth to boil for 5 minutes to reduce slightly. Reduce the heat to medium and add the chicken pieces and the peas. Continue to cook just until chicken and peas are heated through. Season the mixture with salt and pepper to taste.
5. Divide the hot cooked pappardelle among 8 serving bowls. Top each with some of the sauce. Sprinkle each serving with some of the tomatoes, tarragon, and Parmesan cheese. Serve immediately.

SERVES 8

Homemade crepes were a welcome alternative to the traditional cake at this birthday luncheon.

APRICOT CREPES *with* CHOCOLATE SAUCE

- 1 cup all-purpose flour
- ¼ teaspoon salt
- 2 eggs
- ½ cup milk
- ½ cup water
- 4 tablespoons butter, melted
- 10 tablespoons apricot jam
- Chocolate Sauce (see following recipe)
- ½ cup heavy cream (whipped)

1. In a large bowl, combine the flour, salt, eggs, milk, and water. Whisk to make a smooth batter. Stir in 2 tablespoons of the melted butter. (Alternatively, combine the ingredients in a blender and process for 20 seconds. With a rubber spatula, scrape down the sides of the blender container and process for another 30 seconds.) Allow the batter to stand at least 30 minutes, or as long as overnight.
2. Heat a 7-inch crepe pan or skillet over medium-high heat. Brush the pan lightly with some of the remaining butter. Pour about 2 tablespoons of the batter into the pan and quickly rotate the pan so the batter coats the bottom evenly. Cook until the crepe is browned on the bottom. Turn the crepe over and brown the other side. Stack the finished crepes on a plate. Continue with the remaining batter, brushing the crepe pan with more butter as needed. Continue stacking the finished crepes on top of each other.
3. For each serving, spread 2 crepes with some of the apricot jam. Roll them jelly roll–style and place them in an oven-proof dish. Paint the crepes lightly with butter and heat in the oven for 5 minutes at 350 degrees.
4. Arrange on individual dessert plates and top with warm Chocolate Sauce and whipped cream. Serve immediately.

SERVES 8

CHOCOLATE SAUCE

- ½ cup water
- ⅓ cup sugar
- ¼ cup unsweetened cocoa powder
- 4 ounces sweetened chocolate, grated
- 2 tablespoons rum

In a small saucepan, mix the water, sugar, and cocoa powder. Set over medium heat and stir until sugar is dissolved. Reduce the heat to low and add the grated chocolate and the rum. Stir the mixture until very smooth. Keep warm.

CREPES

Use an iron frying pan, flat grill, electric crepe maker, or copper crepe pan, or simply purchase the packaged crepes sold in most supermarkets.

Here are some ideas for fillings:

- melted chocolate beaten with an egg yolk, cream, and a splash of rum
- lemon curd, either homemade or from a jar
- fresh berries marinated in Cassis or Grand Marnier
- sweetened purée of chestnuts; the French call them *marrons* and sell them canned
- crème fraîche with bananas and brown sugar
- jam or marmalade
- savory fillings might include scrambled eggs, caviar, ratatouille, goat cheese, prosciutto, or thin shavings of Gruyère cheese
- crabmeat and shrimp
- shredded chicken

An Afternoon Tea

ASSORTED *tea sandwiches*

cheese scones WITH DEVONSHIRE CREAM AND STRAWBERRY AND APRICOT JAMS

marble cake

banana nut AND DATE BREAD

madeleines

raspberry TART

Serves: 12

Range of Difficulty: Moderate

Preparation Time: 4 Hours

Entertaining is all about organization.

Whether you have help or not, you have to have a plan, a method, a way of getting to where you want to go.

I created this glorious tea party as a baby shower in a lush garden setting. To keep guests cool and comfortable on a sunny afternoon, we rented umbrellas and draped them with yards of crocheted lace to add an air of elegance. Snipping an outline of the lace pattern created a scalloped hem that hung over the edges of the umbrellas.

To dress the tables, I used a variety of vintage lace cloths that were layered over each other. On the buffet table, we swagged some of the tablecloth and secured it with a small corsage of fresh flowers. To grace the tables, roses, delphinium, and Queen Anne's lace, multihued and wildly fragrant, were gathered tightly in silver teapots, creamers, and sugar bowls.

A traditional afternoon tea might include classic sandwiches of egg salad; softened Brie with crushed walnuts and watercress; Emmenthaler cheese with Black Forest ham and very thin slices of hothouse tomatoes; and cucumber sandwiches with paper-thin slices of cucumber.

The key to the perfect tea sandwich is very thin bread. Ask your local bakery or delicatessen to slice a loaf of bread horizontally instead of vertically, or buy a loaf of unsliced bread and slice it yourself. Partially freeze the bread before attempting to cut it and use a rolling pin, pasta maker set on No. 1, or a wine bottle to further flatten the bread. Once you have assembled the sandwiches, trim the crusts off and cut into squares, rectangles, or triangles. For round sandwiches, use a cookie cutter.

A secret to tasty tea sandwiches is a spread of equal parts softened butter and cream cheese with a few drops of lemon juice, white pepper, and a dash of Tabasco. Store-bought mayonnaise can also be spiced up by adding a few drops of Tabasco.

To make elegant pinwheel sandwiches, spread a horizontal slice of bread with salmon mousse, roll it up tightly, wrap it in plastic wrap, and chill for 10 minutes before slicing. Or use slices of bread to create a savory "rainbow torte," layered with salmon mousse, egg salad, and thinly sliced cucumber. Wrap it in plastic wrap, and chill for 10 minutes before slicing and serving.

Then, of course, there's dessert. At this tea, I served a sponge cake, a colorful fruit tart, a rich, dense marble cake and my mother's Banana Nut and Date Bread. And, for me, no tea is complete without the buttery flavor of dainty madeleines.

AFTERNOON TEA IS OFTEN MISTAKENLY CALLED HIGH TEA. The ritual of afternoon, or low, tea is generally credited to the Duchess Anna of Bedford, who sipped tea and nibbled snacks in midafternoon to tide herself over to the traditional 7 o'clock dinner hour. When she started inviting her aristocratic friends to share the snack with her, word spread throughout the countryside. An elegant, more formal tea became an anticipated afternoon social engagement.

In contrast, high tea has always been the heartier early evening meal of the working class of England. In addition to the pot of tea, bread and cheese, substantial sandwiches, and perhaps a few sausages, eggs, or potatoes would be served.

A simple rental umbrella can be dressed up easily to add a whimsical note to the party scene.

CUCUMBER SANDWICHES

8 very thin slices white bread
Basic Butter Spread (page 83)
½ English cucumber, thinly sliced lengthwise

1. Flatten the bread with a rolling pin.
2. Lightly coat one side of each slice of bread with the butter spread. Place the cucumber slices in a single layer over half the bread. Cover with the remaining slices. Trim the crusts and cut into triangles.

MAKES 16 SANDWICHES

EGG SALAD SANDWICHES

5 eggs
2 tablespoons mayonnaise
1 shallot, minced
1 tablespoon lemon juice
2 dashes of Tabasco
Salt and white pepper
8 very thin slices white bread
Basic Butter Spread (page 83)

1. Cook the eggs in boiling water until hard-cooked, about 20 minutes. (A pinhole pierced in the narrow end of the egg's shell will prevent discoloring.) Rinse in cold water and peel. Roughly chop the eggs with a knife.
2. Place the eggs in a bowl. Add the mayonnaise, shallot, lemon juice, and Tabasco. Lightly mix to combine. Season with salt and pepper to taste. (Do not use the food processor.)
3. Flatten the bread with a rolling pin. Lightly coat one side of each slice of bread with the butter spread. Spread the egg mixture evenly over half the bread. Cover with the remaining slices. Trim the crusts and cut into triangles.

MAKES 16 SANDWICHES

TOMATO AND HAM SANDWICHES

8 very thin slices bread, preferably brown
Basic Butter Spread (page 83)
or Basil Mayonnaise (120)
2 hothouse tomatoes, thinly sliced
4 ounces Emmenthaler cheese, thinly sliced
4 ounces Black Forest ham, thinly sliced

Flatten the bread with a rolling pin. Lightly coat one side of the bread slices with the butter spread or Basil Mayonnaise. Top each with a single layer of tomato, cheese, and ham. Cover with the remaining slices. Trim the crusts and cut into triangles.

MAKES 16 SANDWICHES

SMOKED SALMON PINWHEELS

½ pound smoked salmon
1 tablespoon butter
1 tablespoon cream cheese
2 tablespoons lemon juice
2 dashes of Worcestershire sauce
White pepper
1 loaf white sandwich bread, thinly sliced horizontally
1 bunch chives, finely chopped

1. Combine the smoked salmon, butter, cream cheese, lemon juice, Worcestershire, and white pepper in a blender and process until puréed. Adjust the seasoning to taste with pepper.
2. Flatten the bread with a rolling pin. Coat one long slice of bread with the salmon mixture. Roll the bread lightly along the width, to create a pinwheel. Wrap well in plastic and refrigerate for at least 1 hour. Repeat with the remaining bread.
3. To serve, remove plastic and cut each log crosswise into thin pinwheels. Sprinkle with the chives and serve.

MAKES 16 SANDWICHES

BRIE, WALNUT, *and* WATERCRESS SANDWICHES

8 very thin slices bread, white or brown
4 ounces ripe Brie, softened and rind removed
¼ ounce chopped walnuts, lightly toasted
½ bunch watercress, rinsed and stems removed

1. Flatten the bread with a rolling pin.
2. Spread the cheese on one side of all the bread slices. Top half the slices with walnuts and watercress. Cover with the remaining slices. Trim the crusts and cut into triangles.

MAKES 16 SANDWICHES

ALL THESE ELEGANT SANDWICHES can be made ahead and placed in a hollowed-out *boule* or round loaf of French bread, with a layer of shredded iceberg lettuce on the bottom. Iceberg's high water content keeps the sandwiches moist throughout the party.

HOMEMADE CRÈME FRAÎCHE

If you cannot find crème fraîche in the market, you can make an acceptable substitute at home.

Combine 2 cups of heavy whipping cream with 2 tablespoons of buttermilk. Pour into a clean, warmed glass jar. Cover tightly. Keep in a warm place for 6 to 8 hours. When the mixture has thickened, refrigerate. It will thicken further in the refrigerator. Homemade crème fraîche will keep for a week in the refrigerator.

DEVONSHIRE CREAM

A product of Devonshire, England, rich clotted cream is made by heating unpasteurized milk until a dense layer of cream forms on the surface. After cooling, this dense cream layer is removed. Clotted cream or Devonshire cream can be bottled and kept refrigerated for up to 5 days.

RAINBOW SANDWICHES

1 loaf white sandwich bread, thinly sliced horizontally
1 loaf brown sandwich bread, thinly sliced horizontally
Basic Butter Spread (see following recipe)
Egg Salad (page 80)
Salmon spread (step 1 of Smoked Salmon Pinwheels, page 80)
1 bunch watercress, rinsed and stems removed

1. Flatten the bread with a rolling pin. Using 2 slices of white bread and 1 slice of brown bread, or vice versa, coat the middle slice on both sides with the butter spread. Coat the bottom and top slices, on one side only. (These are triple-decker sandwiches.)
2. Coat the buttered side of the bottom slice with egg salad and the buttered side of the top slice with salmon. Top each with watercress and insert the butter coated slice in the middle. Stack to close the sandwich, trim the crusts, and cut into small sandwiches. Refrigerate for at least 15 minutes before serving.

MAKES 16 SANDWICHES

Sandwich Notes: To keep the sandwiches fresh, cover with a damp paper towel to prevent drying. When serving, hollow out a round of bread, line it with shredded iceberg lettuce, and place the sandwiches inside. Close the top round of bread. The iceberg lettuce has a high water content and will keep the sandwiches moist and fresh.

BASIC BUTTER SPREAD

½ pound butter
8 ounces cream cheese
Juice of 1 lemon
Salt and white pepper

In a bowl, cream the butter with the cream cheese and lemon juice until completely blended. Season to taste with the salt and white pepper.

MAKES 2 CUPS

CHEESE SCONES

12 ounces self-rising flour
1¼ teaspoons baking powder
1¼ teaspoons mustard powder
Dash of salt
4 ounces mature cheddar cheese, grated
1 egg
½ pint milk

1. Preheat the oven to 425 degrees.
2. Into a medium bowl, sift the flour. Stir in the baking powder, mustard powder, and salt. Stir in the Cheddar cheese.
3. In a small bowl, lightly beat the egg and milk. Mix with the flour mixture and knead for 1 minute until soft.
4. On a floured working surface, pat out the scone mixture into a thickness of 1 inch. Stamp out 2-inch rounds or triangles and place them on a baking sheet.
5. Bake for 15 minutes or until lightly golden. Do not open the oven while baking.
6. Serve warm with butter, jam, or Devonshire cream.

MAKES 12 TO 14 SCONES

STRAWBERRY JAM

2 quarts strawberries, hulled
2½ cups sugar
1 tablespoon grated lemon peel

1. Mash 1 cup of the strawberries with a fork. Cut the remaining berries in half. Place in a large heavy pot with the sugar.
2. Bring to a boil, stirring occasionally, and then reduce the heat to medium. Simmer for 20 minutes, stirring occasionally. Stir in the peel, remove from the heat, and set aside to cool. Refrigerate.

MAKES 1½ CUPS

APRICOT JAM

2 pounds apricots (4 cups peeled and pitted)
2 cups sugar
½ fresh pineapple, peeled and cored
1 tablespoon grated orange peel

1. Bring a large saucepan of water to a boil. Blanch the apricots for 1 minute and drain. Peel, cut in half, and remove and discard the pits. Place in a heavy pot, add sugar, and let sit for 1 hour.
2. Purée the pineapple flesh in a blender or food processor to make about 2 cups of pulp.
3. Add the pineapple to the apricots in the pan and bring to a boil. Cook, stirring constantly, for 5 minutes. Reduce heat to medium and cook 15 minutes longer, stirring occasionally. Reduce heat to low and cook 30 minutes longer. Stir in the orange peel, cook 5 more minutes, and remove from heat. Set aside to cool, then refrigerate.

MAKES 2½ CUPS

Note: Since we are not using pectin, these jams will be thinner and less sweet than store-bought jams. The sugar in homemade jam should always be adjusted according to the sweetness of the fruit.

BREWING THE PERFECT POT OF TEA

Fill a kettle with fresh cold water and put it on the stove to boil. As the water heats, fill a silver, china, or porcelain teapot with hot water to warm it; discard the water. Measure tea leaves into the teapot—1 heaping teaspoon of loose tea per 6-ounce cup and for strong English tea, 1 extra for the pot. As soon as the water in the kettle comes to a rolling boil, pour into the teapot, stir quickly, cover the pot, and let it steep for 5 minutes. Pour the tea through a fine strainer into cups. Do not overboil the water or the oxygen will escape and the tea will taste flat.

A TASTE FOR TEA

Not all teas come from the tea plant. Some are infusions of flowers, herbs, or berries. Herbal teas are naturally caffeine-free.

Darjeeling. Often called the Champagne of teas, this black tea from India has a delicate and slightly winey flavor.

Earl Grey. A traditional blend of China and Darjeeling teas, with a mild flavor and smoky fragrance.

Ginger. Gingerroot gives a beguiling hint of sweetness and spice.

Formosa Oolong. Gentle brown tea with the fruity aroma of ripe peaches.

Ceylon. A mild, deep gold breakfast blend with a lovely bouquet.

Orange Pekoe. The word "orange" refers to the size of the tea leaves, not the mild flavor.

Constant Comment. A black tea scented with orange rinds.

Jasmine. Dried jasmine flowers add a lovely bouquet to green or black tea leaves.

Mint. A bright, refreshing herb tea; a national drink of Morocco.

Lemon Verbena. A soothing, slightly citrusy herbal tea.

Chamomile. Its mild flavor and calming properties make it the first choice for bedtime sipping.

MARBLE CAKE

12 tablespoons butter
¾ cup sugar
3 eggs
1 teaspoon vanilla extract
1¼ cups sifted self-rising flour
2 ounces unsweetened chocolate, melted

1. Preheat the oven to 350 degrees.
2. Grease a 9-inch springform pan and line the bottom of the pan with wax paper.
3. In a large bowl, cream the butter and sugar. Add the eggs, one at a time, beating well after each addition. Stir in the vanilla extract, then fold in the flour.
4. Transfer half of the batter to another bowl. Add the melted chocolate to one of the bowls and stir to blend.
5. Drop the batters alternately by large spoonfuls into the prepared pan, alternating between the plain and chocolate. Using the handle of a wooden spoon, make a single swirl through the batters to marbleize.
6. Bake for about 45 minutes, or until the cake tests done when a toothpick inserted near the center comes out clean. Let the cake cool in the pan for 5 minutes, then remove the cake from the pan and cool completely on a wire rack.

SERVES 8 TO 10

BANANA NUT *and* DATE BREAD

1½ cups all-purpose flour
1 teaspoon baking soda
1 teaspoon salt
1½ cups mashed bananas (about 3 medium)
1 egg
1 cup sugar
½ cup chopped dates
½ cup chopped walnuts

1. Preheat the oven to 325 degrees.
2. Butter and flour an 8 × 5-inch loaf pan.
3. Sift together the flour, soda, and salt. Set aside.
4. In a bowl, combine the bananas, egg, and sugar and blend well. Stir in the dry ingredients, then fold in the dates and nuts. Transfer the batter to the prepared pan and bake for about 60 minutes, or until a wooden toothpick inserted near the center comes out clean. Cool in the pan for 5 minutes, then turn the loaf out on a wire rack to cool completely.

SERVES 12

A POT OF EARL GREY OR PEPPERMINT TEA IS ONLY PART OF THE TEA PARTY STORY. One can also serve a fine sherry or Champagne. In warm weather, a pitcher of flavored iced tea works well; on cooler days, a few sips of port from vintage glasses is a welcome addition to the menu. All beverages, along with the appropriate glassware, should be placed on a buffet table or a side table for guests to help themselves.

MADELEINES

2 eggs, separated
½ cup sugar
¼ pound butter, melted
Grated peel and juice of ½ lemon
½ cup self-rising flour, sifted
Powdered sugar

1. Preheat the oven to 375 degrees.
2. Lightly butter 2 madeleine pans
3. In a bowl, beat the egg yolks with the sugar. Stir in the butter and the lemon peel and juice. Fold in the flour.
4. In the bowl of an electric mixer, beat the egg whites until they are stiff but not dry. Fold the whites into the batter, then spoon the mixture into the molds, filling them about two-thirds full. Bake the madeleines for 20 minutes. Cool them in the pans for 2 minutes, then remove them and cool completely, fluted side up, on a wire rack. Dust with powdered sugar.

MAKES 24

RASPBERRY TART

1 cup milk
1 vanilla bean
3 large egg yolks
4 tablespoons sugar
4 tablespoons cornstarch
Grated peel of ¼ lemon
¼ cup raspberry jam
¼ cup water
Baked Pastry Shell (see following recipe)
2 cups raspberries, picked over and washed

1. To make the filling, combine the milk and vanilla bean in a small saucepan. Cook over low heat until nearly boiling, approximately 5 minutes. Cool slightly. Remove the vanilla bean, split open lengthwise, and scrape seeds into milk, discarding bean.
2. In a bowl, whisk together the egg yolks and sugar until light and fluffy. Add cornstarch. Pour the warm milk into the egg mixture, whisking constantly.
3. Place the bowl over a pan of simmering water and cook, stirring frequently, until thick enough to coat a spoon. Remove from the heat, pour into a glass mixing bowl, and mix 1 minute longer to accelerate the cooling process. Add lemon peel and set aside to cool further.
4. Combine the jam and water in a small saucepan. Cook over low heat until melted.
5. Spoon the pastry cream into the cooked pastry shell and smooth the top. Sprinkle the raspberries over the top and brush with the raspberry jam glaze. Refrigerate until serving time.

SERVES 8 TO 10

BAKED PASTRY SHELL

2 cups all-purpose flour
Dash of salt
¼ pound butter, thinly sliced and frozen
6 tablespoons water
1 egg, beaten

1. In a food processor fitted with the metal blade, combine the flour and salt. Pulse to combine.
2. With the machine running, add the butter, 1 slice at a time. With the machine still running, add the water, 1 tablespoon at a time, until the dough forms a ball. Remove the dough and dust with a little flour. Pat into a disk, wrap well with plastic, and chill for 1 hour.
3. To bake, preheat the oven to 350 degrees.
4. Roll out the dough on a lightly floured board. Line the bottom of a 10-inch springform pan with the dough. Brush with the egg wash. Pierce the crust all over with the tines of a fork and line with foil. Fill with pie weights or rice.
5. Bake for 30 to 40 minutes, until golden brown.

AN OPEN HOUSE

SHRIMP AND SCALLOP ON *lemongrass kabobs*

bruschetta

grilled chicken WITH A TANGY TOMATILLO SALSA

grilled mahimahi WITH PESTO

CHOPPED *barbecued vegetable salad*

CLASSIC *caesar salad*

bean and red onion SALAD

blueberry cobbler WITH VANILLA ICE CREAM

Serves: 10

Range of Difficulty: Easy

Preparation Time: 3 hours

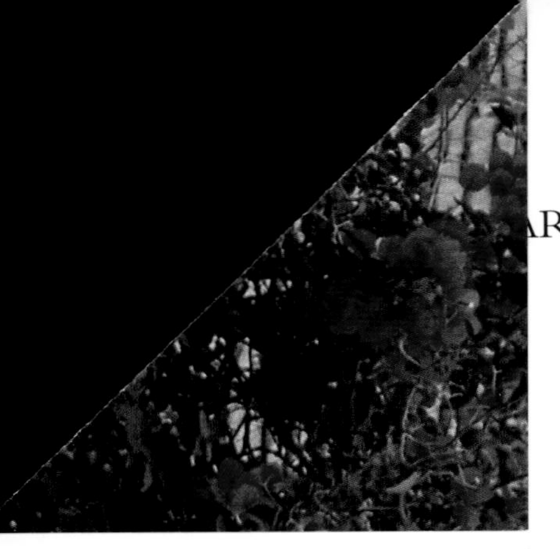

ARTY IS AN ENSEMBLE OF FOOD, MUSIC, DECOR.

No one element outshines another and guests revel in the comfort created by this balance.

An open house is a wonderful way to gather friends and neighbors, grandparents and children who love the chance to catch up on one another's lives. It's an easy way to celebrate an engagement, say bon voyage to a friend, usher in a holiday, or welcome a new baby.

For this open house, guests were invited to drop by between 5 p.m. and 9 p.m., allowing the mood of the party to flow from late afternoon, on through sunset, and into the romance of evening.

The host's stunning garden needed little dressing up for the occasion. We arranged small tables throughout the garden and pool area. We draped the tables to the ground with jute tablecloths and crisp white linen overlays. Centerpieces of whitewashed terra-cotta pots were planted with fresh rosemary. For added drama in the transition from daylight to evening, we placed groupings of votive candles on all the tables and hundreds of different size candles around the property.

The menu combined tray-passed appetizers and a hearty buffet. Appetizers from the grill were served continually until the buffet, centered around a warm, inviting barbecue, was available. Appetizers included bruschetta topped with roma tomatoes and basil, and shrimp and scallops marinated in cilantro pesto and threaded on lemongrass sticks.

Barbecued chicken with a tangy tomato salsa and barbecued mahimahi with a traditional basil pesto were served straight from the grill. A chopped salad of barbecued vegetables dressed in a citrus vinaigrette, along with a Caesar salad and a bean salad, completed the spalike menu.

For beverages, two different bars were set: one with bottles of wine, boutique beers, and a variety of bottled waters; the other with a selection of fresh fruit daiquiris made with bananas, strawberries, and mangos. Fruits piled in abundance in wicker baskets provided the bar's decor.

And for dessert, something I knew everyone would love, blueberry cobbler with homemade vanilla ice cream.

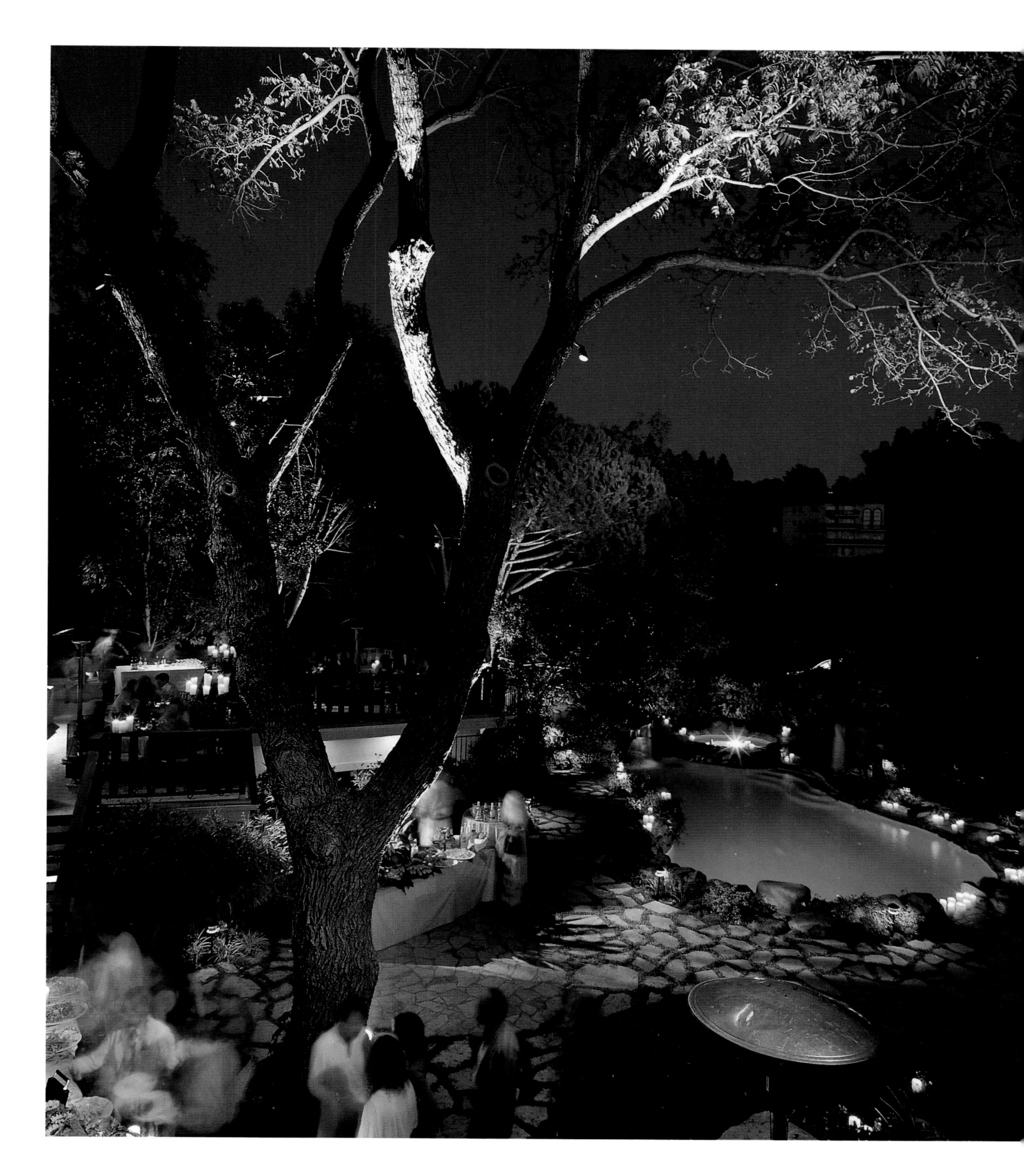

RULE OF THUMB
At an informal gathering such as an open house, a good rule of thumb is to have seating for 25 percent of the expected guests.

BRUSCHETTA

In every region where olive oil is produced, it has been traditional to celebrate the harvest and make bruschetta by grilling thick slices of good, chewy peasant bread over a wood fire, then drizzling it with pungent, fresh green olive oil.

Sometimes the slabs of bread are rubbed with a garlic clove first; sometimes the olive oil is sprinkled with a bit of coarse salt. But always, it's a morsel that satisfies the soul as well as the stomach.

Even bruschetta made in a broiler or toaster has some of the romance of the real thing. It's still a versatile, inexpensive, and savory accompaniment to cocktails.

The most popular topping for bruschetta is a simple mix of finely diced ripe tomatoes and torn basil leaves. But you can top the grilled bread with anything the season and your imagination dictates.

- Toss paper-thin slices of sweet onion with chopped fresh tomato.
- Mash avocado with a spritz of lemon juice and top with thinly sliced green onions.
- Make a purée of white cannellini beans cooked with a bit of pancetta; top the purée with arugula leaves.
- Roast red, yellow, and green bell peppers and cut into thin strips.
- Roast thin slices of eggplant.
- Warm ricotta cheese and dust with fresh cracked pepper.
- Overlap slices of chèvre, avocado, and basil.

Guests love to play the role of connoisseur. Serve bruschetta hot from the grill along with a variety of fine olive oils and invite guests to choose their favorites.

PESTO

The traditional pesto, created by the Genoese, is a blend of fresh basil leaves, garlic, pine nuts, Parmesan cheese, and olive oil. But this versatile topping for pasta, pizza, crostini, poultry, and fish lends itself to experimentation. For example:

- Use a mix of parsley, chervil, and thyme instead of basil.
- Substitute mint for basil and omit the cheese for a pesto that goes perfectly with lamb.
- Use pecans or almonds instead of the pine nuts.
- Substitute cilantro for basil and use a mixture of regular olive oil and olive oil infused with chili peppers.
- Purée oil-cured black olives and extra virgin olive oil for a quick and easy "pesto."
- Purée artichoke hearts with garlic, pine nuts, and extra virgin olive oil.

Potted plants, special bottles of wine, vintage napkins and linens, fragrant candles, and flavored oils and vinegars are much-appreciated gifts at an open house.

BEAN *and* RED ONION SALAD

- 1 cup cooked wax beans, cut into 1-inch pieces
- 1 cup cooked garbanzo beans
- 1 cup cooked Great Northern white beans
- 1 cup cooked kidney beans
- 1 cup cooked green beans, cut into 1-inch pieces
- 1 cup cooked lima beans
- Dijon Vinaigrette (see following recipe)
- 2 tablespoons thinly sliced fresh chives
- 2 tablespoons chopped red onion
- 2 tablespoons chopped fresh parsley
- Salt and freshly ground pepper

If using canned beans, rinse well. In a large bowl, combine all the beans. Add the vinaigrette, toss well, and sprinkle with the chives, onion, and parsley. Season with salt and pepper to taste.

SERVES 10

DIJON VINAIGRETTE

- ¼ cup red wine vinegar
- 1 garlic clove, minced
- 1 tablespoon Dijon mustard
- ½ cup olive oil
- Salt and freshly ground pepper

In a small bowl, whisk together the vinegar, garlic, and mustard. Gradually whisk in the oil. Season with salt and pepper to taste.

BLUEBERRY COBBLER *with* VANILLA ICE CREAM

- ¾ pound butter (1½ cups)
- 2 cups unbleached flour
- 1¼ cups whole wheat flour
- 1½ cups plus 2 tablespoons sugar
- ½ cup ice water, or more
- 12 cups blueberries
- Additional sugar
- Whipped cream or Vanilla Ice Cream

1. Cut the butter into small pieces and place in the freezer.
2. In a food processor, combine the flours and 2 tablespoons of the sugar. Scatter 1 cup of the butter pieces over the flour and process until the butter is the size of small peas. Pulsing the machine on and off, add the water, a few tablespoons at a time, just until the mixture begins to clump. Press it into a disk. Wrap it with plastic wrap and refrigerate for 30 minutes.
3. Preheat the oven to 425 degrees.
4. Roll out the dough to a rectangle roughly 20 × 14 inches. Tear off random pieces of dough and line the bottom and sides of a 15 x 10-inch baking dish, allowing some of the dough to hang over edges.
5. Fill the dough with the blueberries, then sprinkle them with the remaining sugar and butter.
6. Fold the overhanging dough back in toward the center of the dish. Tear any remaining dough into random pieces and place over the top of the blueberries. Sprinkle additional sugar over the dough. Bake for 45 to 60 minutes, or until the crust is golden brown.
7. Serve the cobbler warm with whipped cream or Vanilla Ice Cream.

SERVES 10

VANILLA ICE CREAM

- 2 cups milk
- 2 cups heavy cream
- 2 vanilla beans, cut into 3 parts
- 6 egg yolks
- ⅔ cup superfine sugar (or regular sugar placed in food processor)

1. In a heavy saucepan, bring the milk, heavy cream, and vanilla nearly to a boil. Remove from the heat. Discard the vanilla.
2. In a mixing bowl, beat the egg yolks with the sugar. Gradually add the egg mixture to the cream, whisking constantly.
3. Return the pan to the heat and cook over low heat, stirring constantly, until thick enough to coat a spoon. Transfer to a glass bowl and stir a few minutes longer to hasten the cooling. Chill until cold. Pour into an ice cream maker and freeze according to manufacturer's directions. Transfer to a chilled container, cover, and freeze.

SERVES 10

MY MOTHER'S *sausage rolls*
SERVED WITH WORCESTERSHIRE SAUCE

gravlax WITH DILL MUSTARD SAUCE

smoked salmon WITH RYE BREAD

ASSORTED *sausages* WITH DIJON AND HOT ENGLISH MUSTARD

IMPORTED *cheeses*

GREEN AND WHITE *crudité platter*

oatmeal raisin ginger COOKIES

Serves: 16

Range of Difficulty: Easy

Preparation Time: 3 hours

Create a wonderful ambience for a cocktail party

with fabulous music. My favorites are Eartha Kitt, Peggy Lee, Etta James, Oscar Peterson, and Duke Ellington. If you have time, make a tape in advance so you can control the energy of the party.

I love cocktail parties because they're a great way to entertain small or large groups of people, and a great way to introduce friends to one another. They can be relatively simple to orchestrate, thrown together at a moment's notice, or planned in advance with great attention to every detail. You can serve a variety of interesting foods in innovative or traditional ways, and at the same time enjoy the company of your guests.

The sophisticated martini—shaken, not stirred—is the perfect cocktail for any size party. Nothing looks better than a silver tray set with graceful glasses, shimmering gin or vodka, a handsome shaker, olives, and crisply starched linen cocktail napkins.

As guests arrived, hors d'oeuvres of puff pastry filled with sage-scented ground pork were passed on antique silver trays. On the cocktail table in the living room was placed a still-life of green and white crudités; snap snow peas, asparagus tips, and broccoli, with graceful curls of white daikon radish. For dipping, we offered a rich balsamic vinaigrette. Since the space was limited and the hostess had little help, we chose to create a series of self-contained vignettes in the living and dining rooms. Each vignette showcased a food as well as the accompanying beverage. The vignettes encouraged guests to move around the room and mingle.

The hostess's elegant linen-covered dining room table was set with an antique silver candelabra and a lavish arrangement of roses and hanging amaranth. It created the perfect backdrop for the home-cured gravlax and smoked salmon. Lemon halves wrapped in gauze were used as a garnish.

Cheese, too, is favorite cocktail fare. A few well-chosen cheeses make a better impression than a buffet of twenty cheeses. My simple rule of thumb: serve three ripe cheeses—a triple crème, a blue, and a goat's milk cheese. Put them on a separate table, arranged attractively on a wooden cutting board, in a flat basket, or on a bed of leaves. Provide a crusty baguette and some slices of lightly toasted walnut and raisin bread for the blue cheese. Garnish with bunches of champagne grapes or a few figs.

FLAVORED VODKAS

The easiest way to flavor this spirit is to add the desired flavoring to a bottle of vodka, cap the bottle, and allow it to marinate in the freezer for two or three days. The longer it marinates, the stronger the flavor.

Some of my favorite additions are:

- lemongrass, 3 or 4 stalks
- orange rind, ½ cup
- raspberries or blackberries, 1 cup
- tarragon, 4 or 5 stalks
- star anise, 6 pods
- grapefruit rind, ½ cup
- lemon rind, ½ cup
- chili peppers, 1 or 2 peppers (for a milder flavor, remove the seeds)

MAKING THE ICE MOLD

For a striking presentation, fill a metal champagne bucket or a large plastic bucket with water. Stand the vodka bottle in the middle of the bucket and tuck handfuls of roses, wildflowers, lemon slices, leaves, berries, or grapes around the bottle. Place in the freezer until the water is completely frozen, about 24 hours.

To remove the mold from the container, simply run the outside of the bucket under hot water until it releases the block of ice. (Metal buckets work best, since they do not expand and contract as plastic does.) Then place the iced bottle on a rimmed plate, with a towel underneath to absorb the water.

MY MOTHER'S SAUSAGE ROLLS
served with WORCESTERSHIRE SAUCE

- 12 ounces lean ground pork or veal
- ½ medium onion, chopped
- ¾ cup diced cooked potato
- 2 tablespoons chopped parsley
- 1½ tablespoons crushed dried sage leaves
- ½ teaspoon salt
- ½ teaspoon pepper
- ¼ teaspoon crushed red pepper flakes
- 1 17¼-ounce package frozen puff pastry, thawed
- 1 egg, well beaten
- ½ cup Worcestershire sauce

1. Preheat the oven to 350 degrees.
2. In a bowl, combine the pork, onion, potato, parsley, sage, salt, pepper, and pepper flakes. Mix with a fork to combine.
3. Transfer the mixture to a large pastry bag fitted with a large plain tip.
4. Unfold the puff pastry sheets and, following the fold marks, cut each sheet into three rectangles. Roll out slightly.
5. Pipe the sausage mixture along one long edge of each of the pastry rectangles. Paint the opposite edges with some of the beaten egg. Starting at the edge with the sausage mixture, roll up each rectangle, gently pressing the egg-painted edge to seal.
6. Using a sharp knife, cut the rolls into slices about 1½ inches thick. Arrange them on a broiler pan or a wire rack set on top of a baking sheet. Brush the slices with the remaining beaten egg and bake for 20 minutes, or until golden.
7. Serve the slices warm with the Worcestershire sauce on the side for dipping.

MAKES ABOUT 50 PIECES

Note: When serving the sausage rolls, make multiple batches and cook only a few at a time. They are best when served straight from the oven.

GRAVLAX *with* DILL MUSTARD SAUCE

- 1 side salmon, with skin, boned, about 4 pounds
- 2 tablespoons brown sugar, packed
- 1½ teaspoons coarse salt
- 1½ teaspoons freshly ground pepper
- ½ cup dried dill
- 1 cup chopped fresh dill
- ¼ cup vegetable oil
- Fresh dill sprigs or lemon leaves
- 1 loaf party rye bread, about 8 ounces
- Butter
- Dill Mustard Sauce (see following recipe)

1. Place the salmon skin side down on a large tray. Using tweezers, remove any small bones. Using a large cooking fork, pierce the salmon all over every few inches. Sprinkle the salmon with the sugar, salt, and pepper.
2. In a small bowl, combine the dried and fresh dill with the oil. Spread evenly over the salmon. Cover the salmon with plastic wrap and refrigerate for 3 days.
3. Remove the gravlax from the refrigerator and scrape away the dill. Slide it onto a large serving tray or board. Using a salmon knife, slice the gravlax very thin on a diagonal.
4. Serve on buttered rye bread and top generously with Dill Mustard Sauce.

SERVES 16

Note: To prevent damaging the salmon, ask the fishmonger to place it flat, without folding, on a piece of cardboard for support.

DILL MUSTARD SAUCE

- 1 8-ounce jar spicy mustard
- ¼ cup brown sugar, packed
- ¼ cup chopped fresh dill

1. In a small saucepan, combine the mustard and brown sugar. Set the pan over medium heat and stir until the sugar is dissolved. Stir in the dill, then allow the mixture to simmer for 5 minutes.
2. Remove the pan from the heat and allow the sauce to cool.

Sausage rolls are one of my favorite childhood treats.

Serve homemade gravlax with dill mustard sauce.

SAUSAGES ARE WONDERFUL AT COCKTAIL PARTIES—they're hearty, inexpensive, and filling. In this instance, I offered a selection of grilled sausages and two of my favorite mustards, hot English-style and Dijon. A good selection should include veal, turkey, chicken, beef, and pork, both spicy and mild. This vignette was set with small plates, toothpicks, and cocktail napkins, along with a bottle of Syrah from France's Rhone region. The wine's smoky style brought out the best in sausage.

SMOKED SALMON *with* RYE BREAD

1 head iceberg lettuce, leaves washed and dried
1 pound sliced smoked salmon
3 lemons, quartered
Freshly ground pepper

Arrange the lettuce leaves in a mound on a serving platter. Separate the salmon slices and drape them loosely on top of the lettuce. Garnish with the lemon wedges, season with pepper, and serve cold.

SERVES 16

ASSORTED SAUSAGES *with* DIJON *and* HOT ENGLISH MUSTARD

3 pounds assorted sausages
1 tablespoon vegetable oil
¼ cup Dijon mustard
¼ cup hot English mustard

1. Bring a large pot of water to a boil. Add the sausages and cook for 5 minutes. Remove the sausages with tongs and drain on paper towels.
2. In a large skillet, heat the oil over medium-high heat. Fry the sausages until evenly browned, using tongs to turn. (A fork would pierce the casing, allowing the juices to escape and dry the meat.) Drain on paper towels for 5 minutes.
3. Cut the sausages into 1½-inch lengths. Serve on cocktail toothpicks with the mustards for dipping.

SERVES 16

GREEN *and* WHITE CRUDITÉ PLATTER

About 20 each cauliflower florets, broccoli florets, asparagus spears, and snow peas or other fresh green and white vegetables in season
1 large daikon radish, peeled and very thinly sliced
2 tablespoons balsamic vinegar
1 tablespoon Dijon mustard
1 cup olive oil
2 teaspoons chopped parsley
2 teaspoons chopped chives
Salt and freshly ground pepper

1. Bring a large pot of water to a boil. Drop the cauliflower into the water and cook for about 2 minutes. With a slotted spoon, remove the florets and immediately plunge them into a large bowl of ice water. Drain. Repeat the process with the broccoli, asparagus, and snow peas.
2. On a serving platter or tray, arrange the blanched vegetables and the daikon slices attractively.
3. In a small bowl, whisk the balsamic vinegar and Dijon mustard. Slowly whisk in the olive oil until the dressing is thick and syrupy. Stir in the parsley and chives. Season to taste with salt and pepper.
4. Transfer the sauce to a small serving bowl and serve with the vegetables for dipping.

SERVES 16

Note: The vegetables can be varied according to taste. Slim celery, jicama, zucchini sticks, sugar snap peas, green beans, spears of Belgian endive, or fennel are just a few suggestions.

OATMEAL RAISIN GINGER COOKIES

1 cup whole wheat flour
½ cup all-purpose flour
1 teaspoon baking soda
1 teaspoon salt
1 teaspoon cinnamon
¼ pound plus 4 tablespoons unsalted butter (1½ sticks)
1 cup brown sugar, packed
½ cup granulated sugar
2 eggs
1 teaspoon vanilla extract
3 cups rolled oats
1½ cups raisins
¼ cup chopped crystallized ginger

1. Preheat the oven to 375 degrees. Grease 4 baking sheets.
2. Sift the flours with the baking soda, salt, and cinnamon and set aside.
3. In a large bowl, cream the butter and sugars until light. Add the eggs, one at a time, mixing well after each addition. Stir in the vanilla.
4. Using a wooden spoon, stir in the oats, raisins, and ginger.
5. Drop the dough by tablespoonfuls onto the prepared baking sheets. Using a moistened metal spatula, press the mounds of dough to flatten them slightly.
6. Cook for 15 to 20 minutes, until evenly brown.
7. Remove from the oven and loosen with the spatula. Allow to cool on the baking sheet for 5 minutes, then transfer to a wire cooling rack for 30 minutes.

MAKES 40 COOKIES

Add style to a crudité platter by selecting vegetables in two colors only and cutting them into unusual shapes.

chicken COOKED UNDER PRESSURE

spinach AND *risotto* CASSEROLE

tomatoes PROVENÇALE

dinner salad IN A LIGHT
CHARDONNAY VINAIGRETTE

apple tarte tatin
WITH VANILLA ICE CREAM

Serves: 8

Range of Difficulty: Easy

Preparation Time: 2 hours

I WANT TO INSTILL A PASSION FOR ENTERTAINING

and show how easy and how much fun it can be. I could give you my family's prized recipe for lemon tart, but if you don't have time to make it, I'd rather you go out and buy a really good tart and invite friends to share it.

A friend and I collaborated on this chicken cooked under pressure for an early evening family dinner. It's one of the tastiest ways I have ever discovered to prepare a chicken. The crispy golden skin, essence of rosemary and garlic, and plump, juicy meat are created by cooking the chicken in a heavy pan over high heat with very hot olive oil. The lid from a pan one size smaller is placed directly on top of the chicken and weighted down with heavy rocks, a pot of water, or even some work-out weights. The weight presses the chicken against the cooking surface of the pan and allows it to cook evenly. At the same time, the very hot oil seals in the natural juices. The chicken is partially cooked in the pan, then finished in the oven during the cocktail hour.

As an accompaniment: tomatoes Provençal style, and a delicious risotto casserole that is elegant and satisfying without chaining the hostess to the stove. All three dishes were simply popped in the oven for 45 minutes to finish cooking while guests sipped apéritifs.

On this evening, we chose to serve salad after the main course. The salad was brought to the table with the chicken, tossed with dressing, and set aside until ready to be served. As a beverage, we paired a lighter style Pinot Noir that offered good flavor and richness without overwhelming the chicken.

After, we indulged in tarte tatin, made in advance with fresh pastry crust and heated for 15 to 20 minutes before serving.

The setting, the table, the menu, the style of service—everything was the essence of simplicity. All the foods were prepared in advance and served family-style, leaving hostess and guests to dine in front of a roaring fire. On the handsome, weathered wooden country table, we placed ivory candles in silver holders and stuffed white anemones into a silver coffeepot, creamer, and sugar holder. Basic white plates with vintage napkins and heirloom silverware completed the setting.

Sampling a great wine in front of a fire is the perfect prelude to a memorable evening.

CHICKEN COOKED UNDER PRESSURE

- 4 chickens, about 3½ pounds each
- 8 garlic cloves, peeled and halved
- 2 tablespoons coarse salt
- 1 tablespoon freshly ground pepper
- 4 rosemary sprigs, leaves chopped
- ¼ cup olive oil

1. Preheat the oven to 350 degrees.
2. Cut the chickens in half along the breastbone. Trim and discard the backs and all the fat. (You can ask the butcher to prepare the chickens.) With a paring knife, make a slit in the skin near each thigh and insert a piece of garlic. Sprinkle all over with the salt, pepper, and the rosemary.
3. In a large skillet, heat the oil over high heat. (The oil should be ½ inch deep.) When the oil is very hot, add the chicken, 2 pieces at a time, skin side down. Top the chicken with a smaller pan, placed directly on the meat. Fill the pan with 30 pounds of weights, and cook until the skin is golden brown, about 7 minutes. Turn over, top with weights again, and cook 7 minutes longer. Repeat with remaining chicken pieces.
4. Transfer the chicken to a baking sheet, skin side up, and bake an additional 20 minutes. Serve hot.

SERVES 8

Notes: To catch splattering oil, cover the surrounding stove-top area with aluminum foil and line the floor near the stove with newspapers.

You will need about 20–30 pounds of weights (stones, bricks, water, or canned goods) for this dish.

SPINACH *and* RISOTTO CASSEROLE

- 2 tablespoons olive oil
- 1 medium onion, chopped
- 1½ cups arborio rice
- 1 cup dry white wine
- 3 cups vegetable broth
- 10 ounces spinach, rinsed and patted dry (2 cups, packed)
- 4 tablespoons freshly grated Parmesan cheese
- 1 teaspoon salt
- ¼ teaspoon pepper

1. Preheat the oven to 375 degrees.
2. In a large saucepan, heat the oil over medium heat. Add the onion and sauté until translucent. Add the rice and sauté for 2 minutes. Add the wine and sauté for an additional 2 minutes or until the liquid is absorbed. Stir in the broth. Bring to a boil.
3. Reduce the heat to low, stir, and allow the mixture to simmer 10 minutes, or until the liquid starts to thicken. Stir in the spinach, 2 tablespoons of the cheese, and the salt and pepper. Simmer 2 minutes longer. Transfer to a 2-quart casserole. Sprinkle the remaining cheese over the top. Bake for 35 minutes.

SERVES 8

Note: You can create your own original risotto casserole using my basic recipe and adding one or more of the following:

- mushrooms
- asparagus
- ham
- truffles
- sun-dried tomatoes
- shrimp and peas

TOMATOES PROVENÇALE

- 4 large ripe tomatoes, halved
- ⅓ teaspoon each salt and freshly ground pepper
- 1 teaspoon chopped garlic
- 2 teaspoons chopped rosemary
- 4 teaspoons coarse bread crumbs
- 4 tablespoons olive oil

1. Preheat the oven to 350 degrees.
2. In a baking dish, place tomatoes cut side up. Sprinkle with salt, pepper, garlic, and rosemary. Sprinkle bread crumbs over tomatoes, then drizzle ½ tablespoon of olive oil on each tomato. Bake for 20 to 25 minutes, until tender and browned on top.

SERVES 8

DINNER SALAD *in a* LIGHT CHARDONNAY VINAIGRETTE

2 tablespoons red wine vinegar
2 tablespoons balsamic vinegar
1 shallot, minced
½ teaspoon sugar
1 tablespoon Dijon mustard
¼ cup Chardonnay
1 cup olive oil
¼ teaspoon salt
½ teaspoon freshly ground pepper
1 pound mixed cleaned baby lettuces

1. Whisk together the vinegars, shallot, sugar, mustard, and Chardonnay in a mixing bowl. Gradually add the olive oil, whisking constantly. Stir in the salt and pepper.
2. In a bowl, toss the lettuces with enough dressing to lightly coat the leaves. The remaining dressing may be stored in the refrigerator.

SERVES 8

APPLE TARTE TATIN *with* VANILLA ICE CREAM

1 cup sugar
3 pounds small Pippin apples, halved, cored, and peeled
¼ pound butter, softened
Shortcrust Pastry (see following recipe)
Vanilla Ice Cream (see recipe) or frozen yogurt

1. Preheat the oven to 425 degrees.
2. Grease a 9-inch ovenproof skillet. Sprinkle ⅓ cup of the sugar over bottom. Pack the apples tightly in the pan in an upright fashion. Sprinkle with the remaining sugar. Cut the butter into small pieces and add to the apples.
3. Place the pan over medium heat and cook until the bottom is lightly caramelized, about 20 minutes. Transfer to the oven and bake 10 minutes longer. Remove from oven.
4. Meanwhile, on a lightly floured board, roll out the Shortcrust Pastry to form a 10-inch circle. Top the apples with the pastry, so the edges overhang and the dough settles onto the apples. Trim any excess pastry by pressing the edges with a rolling pin.
5. Return the pan to the oven and bake until the pastry is well browned, about 20 minutes.
6. To serve, invert the tart onto a serving plate and spoon on the caramel from the pan. Let cool 5 minutes and serve with Vanilla Ice Cream or yogurt.

Note: Tarte Tatin may be made entirely in advance. Refrigerate the finished tart and reheat, covered with foil, in a 350 degree oven for about 15 minutes.

SERVES 6 TO 8

SHORTCRUST PASTRY

¼ pound butter, softened
⅓ cup superfine sugar
1 egg
2 cups all-purpose flour
2 dashes of salt

1. In a medium bowl, cream the butter with the sugar. Add the egg and beat well. Add the flour and salt and beat until a smooth dough is formed.
2. Lightly knead on a floured board until the dough is elastic and smooth. Pat into a disk, wrap in plastic, and chill for at least 2 hours, or freeze as long as 2 weeks.

MAKES 1 PIE SHELL

Note: Store-bought puff pastry can be substituted for the Shortcrust Pastry.

VANILLA ICE CREAM

2 cups milk
2 cups heavy cream
2 vanilla beans, cut into 3 parts
6 egg yolks
⅔ cup superfine sugar (or regular sugar placed in a food processor)

1. In a heavy saucepan, bring the milk, heavy cream, and vanilla nearly to a boil. Remove from the heat. Discard the vanilla.
2. In a mixing bowl, beat the egg yolks with the sugar. Gradually add the egg mixture to the cream, whisking constantly.
3. Return the pan to the heat and cook over low heat, stirring constantly, until thick enough to coat a spoon. Transfer to a glass bowl and stir a few minutes longer to hasten the cooling. Chill until cold. Pour into an ice cream maker and freeze according to manufacturer's directions. Transfer to a chilled container, cover, and freeze.

SERVES 10

Apple tarte tatin can be served with vanilla ice cream, frozen yogurt, or rum-spiked crème fraîche.

As an alternative, make your tart with pears instead of apples and serve with whipped cream spiced with pear William.

Flea market finds can easily be transformed into interesting light fixtures.

A Feast from the Ocean

STEAMED *mussels* WITH VERMOUTH

STEAMED *clams*

lobster WITH RUSSIAN SALAD

jumbo shrimp WITH COCKTAIL SAUCE OR BASIL MAYONNAISE

ALASKAN *king crab* LEGS

FRESH *oysters on the half shell*
WITH A MIGNONETTE SAUCE

pistachio CHEESECAKE

Serves: 8

Range of Difficulty: Easy

Preparation Time: 1 hour

Pick your favorite food and build around it.

Then find a novel way to present it to leave your guests wide-eyed and smiling.

There are some occasions when nothing but the best will do. For this summer feast at a cozy beach house, I created an all-seafood menu. Although the ingredients are costly, the dinner makes a spectacular presentation and can be assembled in advance, leaving the host to enjoy the party.

Our table for eight was set on a spacious deck a seashell's throw from the Pacific. I covered an inexpensive white cloth with hundreds of overlapping galix leaves attached with a spray adhesive glue. In place of floral arrangements, I covered small votive candles and large hurricane lamps with seashells and scattered garden moss, succulents, and more sea shells down the middle of the table. When the candles were lighted and the stars came out, the table was in perfect harmony with its serene surroundings.

Guests were invited for seven o'clock so they could enjoy the sunset, along with steamed mussels, and crusty baguettes. To set a fun, informal tone for the evening, we served the mussels in a large bowl on the patio table. Everyone gathered around, savoring the mussels, and dipping chunks of bread into the aromatic broth.

Later, everyone sat down to the seafood feast, beginning with individual zinc buckets filled with steamed clams in a clam juice and white wine broth. Sourdough baguettes were passed around to mop up the divine broth. To complement the clams' tangy ocean flavor a steely Muscadet was served.

The main course was a spread of lobster, shrimp, Alaskan king crab legs, and oysters on the half shell. I placed three enormous platters down the center of the table so that guests could help themselves. Each platter was laden with poached lobsters, the heads filled with Russian salad and dressed with a truffle-oiled mayonnaise, poached shrimp with tails intact, steamed crab legs, and impeccably fresh oysters with dipping sauces. Guests were offered a rich Chardonnay to enhance the sweet, fragile flavor of the lobster and crab, or the Muscadet. My favorite pistachio cheesecake was a rich conclusion to the feast.

Most of this seafood feast can be poached, steamed, shucked, and assembled in advance. The clams are steamed while guests enjoy apéritifs; and the platters of seafood go on the table all at once, making for effortless service on the host's part and lots of fun for the guests.

SCENTED HAND TOWELS

After a feast of shellfish, barbecued ribs, fried chicken, or similar informal fare, it's nice to offer guests something to clean their hands. For an elegant take on the traditional finger bowl, dip small towels or washcloths into water scented with a few drops of almond essence, lemon juice, or rose water. Wring them out, roll tightly into cylinders, then place in a glass baking dish and cover with plastic wrap. Microwave the towels for 5 minutes, and serve them piping hot with tongs at the end of the meal.

HOW MUCH?

When creating a shellfish platter for 8, plan on ½ lobster per person plus 1 to spare; 12 inches of crab leg per person; 8 clams; 4 or 5 each of oysters, mussels, and shrimp per person. However, it's been my experience that no matter what you heap onto these sensational platters, they will always come back empty!

When serving clams or mussels in individual bowls as a first course or appetizer, plan on a dozen bivalves per person.

WITH THOUSANDS OF MILES OF SHORELINE, the United States and Canada offer oyster lovers a rich supply of these plump, sweet bivalves. Some of the best known varieties include:

Apalachicola. An Atlantic oyster with a teardrop-shaped shell and a sweet, coppery flavor.

Belon. Perhaps the best known, the true Belon is a European from the Belon River in Brittany in northern France. European flat oysters raised in Maine and New Hampshire are also called Belon. They have a lemony, slightly metallic taste.

Bluepoint. Although originally named for Blue Point, Long Island, where these oysters once flourished, the Bluepoint is a generic name for a small Atlantic oyster.

Bristol. A plump salty oyster from Maine.

Chincoteague. A crisp, sweet oyster from Chincoteague Bay in eastern Virginia.

Golden Mantle. A small delicate oyster with a pale gold shell from Vancouver, B.C.

Hawaiian. Raised in farms in Hawaii, these are plump, sweet, and firm with a high meat-to-shell ratio.

Kent Island. Also known as Chesapeake Bay oysters and found from Maryland to Virginia, these have a mild, clean, and sweet flavor.

Kumamoto. A tiny, deep-cupped Pacific hybrid cultivated in California and Washington; very mild with a sweet aftertaste.

Malpeque. A small oyster with a coppery taste from Prince Edward Island, Canada.

Olympia. Rarely larger than a quarter, this delicate oyster comes from Washington's Puget Sound.

Quilcene. A crisp oyster with bright, bracing flavor; from Washington.

Westcott Bay and Westcott Bay Petite. A very meaty, mild, and sweet oyster with a slightly metallic aftertaste, from San Juan Island, Washington.

WHEN STEAMING MUSSELS AND CLAMS, throw out any that don't open during the steaming. Never pry open a bivalve that doesn't open on its own.

STEAMED MUSSELS *with* VERMOUTH

4 tablespoons butter or olive oil
2 garlic cloves, minced
1 large onion, diced
1½ cups water
1½ cups white wine
¼ cup vermouth
96 small to medium mussels
2 tablespoons chopped parsley
2 loaves crusty bread, for dipping

1. In a large stockpot heat the butter over medium-high heat and add the garlic and onions. Sauté 4 to 5 minutes, until onions are glassy.
2. Pour in the water, wine, and vermouth and bring to a boil. Add the mussels and parsley, shake the pot to distribute evenly, and cover. Cook until the shells open, 4 to 5 minutes. Remove and discard any unopened shells.
3. To serve, place mussels in a serving bowl filled with the cooking broth and set on a table with crusty bread for dipping.

Note: Use the mussel shell like a tweezer to remove the mollusk.

STEAMED CLAMS

2 tablespoons olive oil
½ onion, diced (½ cup)
2 garlic cloves, minced
¼ teaspoon crushed red pepper flakes
2 cups clam juice
1 cup dry white wine
80 littleneck clams
2 tablespoons chopped fresh parsley
1 loaf crusty French bread

1. In a large stockpot, heat the olive oil over medium-high heat. Sauté the onions, garlic, and pepper flakes until soft.
2. Pour in the clam juice and the white wine and bring to a boil. Add the clams, shake the pot to distribute evenly, add the parsley, and cover. Cook until the shells open, 7 to 10 minutes. Remove and discard any unopened shells.
3. To serve, divide and place clams in 8 serving bowls. Ladle the broth over the clams and serve hot with the bread.

SERVES 8

SEAFOOD PLATTERS *for* EIGHT

5 whole lobsters
Russian Salad (see following recipe)
Juice of 2 lemons
2 teaspoons salt
2 celery stalks
32 jumbo shrimp, peeled, with tails on, and deveined
8 cooked Alaska king crab legs, about 12 inches long
24 oysters, shucked
6 bunches radishes
8 lemons, halved and wrapped in net or cheesecloth

1. Bring a very large stockpot of salted water to a rolling boil. Add the lobsters and cook for 8 minutes. Immediately transfer to ice water.
2. When cool enough to handle, pat the lobsters dry. Cut each in half lengthwise. Wipe out the head cavities and pat dry. Fill the heads with the Russian Salad. Divide and place on 3 large platters.
3. In a large pot, bring 2 quarts of water to a boil with the lemon juice, salt, and celery. Add the shrimp, reduce to a simmer, and cook for 6 minutes. Immediately transfer to ice water. Pat the shrimp dry and divide among platters.
4. Divide and arrange the crab legs and oysters on the platters. Garnish with the radishes and the lemons and serve.

Note: All of the seafood may be purchased already prepared from a fish market. Make sure you use a first-rate vendor for such high-quality products.

RUSSIAN SALAD

2 large potatoes, peeled, cubed, and cooked
2 carrots, peeled, cubed, and cooked
½ cup peas, cooked
½ cup ¼-inch green bean pieces, cooked
½ cup finely diced celery
¾ cup mayonnaise
¼ cup truffle oil or walnut oil
2 tablespoons lemon juice
½ teaspoon salt
Freshly ground pepper
2 tablespoons chopped parsley

1. In a mixing bowl, combine the potatoes, carrots, peas, beans, and celery and toss well.
2. In a small bowl, whisk together the mayonnaise, oil, lemon juice, salt, and pepper to taste. Pour the dressing on the vegetables to coat lightly. Gently toss to combine, and sprinkle with parsley.

SERVES 8

COCKTAIL SAUCE

- ¾ cup bottled catsup
- 3 tablespoons grated white horseradish
- 2 tablespoons lemon juice
- 3 dashes of hot red pepper sauce

In a bowl, combine the catsup, horseradish, lemon juice, and pepper sauce.

BASIL MAYONNAISE

- 3 cups basil leaves
- ¾ cup mayonnaise
- 3 garlic cloves, chopped
- 1½ tablespoons olive oil
- ½ teaspoon salt
- ¼ teaspoon freshly ground pepper

In a food processor, combine basil, mayonnaise, garlic, oil, salt, and pepper. Process to blend thoroughly.

PISTACHIO CHEESECAKE

- 7 tablespoons butter
- ¾ cup finely chopped pistachios
- 4 tablespoons sugar
- 2 cups graham cracker crumbs
- 1½ pounds cream cheese
- 1 cup brown sugar, packed
- 4 eggs
- 2 cups sour cream
- 1½ teaspoons vanilla extract
- Topping (see following recipe)

1. Preheat the oven to 350 degrees.
2. In a saucepan, melt 1 tablespoon of the butter over medium heat. Add the pistachios and 2 tablespoons of the sugar. Heat, stirring often, until the pistachios are coated and the sugar is caramelized. Turn out onto a foil-lined baking sheet and set aside to cool.
3. Melt the remaining butter. In a bowl, mix the cracker crumbs, the remaining 2 tablespoons of sugar, and the melted butter. Press the mixture on the bottom and up the sides of a 9-inch springform pan. Bake 10 minutes. Set aside to cool.
4. In a large bowl of an electric mixer, beat the cream cheese and the brown sugar until light. Beat in the eggs, one at a time, beating well after each addition. Add the sour cream, vanilla, and the caramelized pistachios, breaking them into small pieces, if necessary. Mix to blend thoroughly. Pour the mixture into the prepared crust and bake for 1 hour, or until the edges of the cheesecake are firm but the center is still soft and shiny. Cool completely on a wire rack.
5. Spread the Topping over the cooled cheesecake and chill for at least 4 hours.

SERVES 12

TOPPING

- 3 ounces cream cheese
- ⅓ cup brown sugar, packed
- 1 cup heavy whipping cream, whipped

In a bowl, beat the cream cheese and the brown sugar until light. Fold in the whipped cream.

MIGNONETTE SAUCE

Mignonette Sauce for oysters is easily made by combining vinegar, black pepper, and chopped shallots.

WHERE EAST TRULY MEETS WEST

gyoza

CRISPY PORK *wontons*

GRILLED *Peking duck pancakes* WITH PLUM SAUCE

MISO SOUP WITH *lobster* MEDALLIONS

STEAMED *striped bass*

JAPANESE *rice*

snow peas AND TOASTED SESAME SEEDS

kiwi MARINATED IN *passion fruit*

bananas MARINATED IN LIME JUICE

Serves: 16

Range of Difficulty: Moderate

Preparation Time: 3 hours

CHRISSI
BILL

THE IDEA OF EVERY GREAT PARTY IS TO APPEAL TO ALL THE SENSES—

with wonderful music, beautiful decor, tantalizing aromas, and, of course, delicious food.

East-meets-West, Pacific Rim cuisine, fusion food. . . . These buzz words of the decade point to our passion for the ingredients and cooking styles of the Orient. And, with a little effort and a lot of enthusiasm, anybody can savor these flavors at home.

While guests sipped chilled sake, we passed gyoza, crispy pork wontons, and bite-size duck pancakes topped with plum sauce. For dining, two mahogany tables with glass center insets were set with natural bamboo mats, earthenware plates, teacups, and small frosted glasses for sake. We added lacquered chopsticks, and marked each place setting with a place card made from hand-torn rice paper, with guests' names in a "Japanese" font.

Because arranged flowers would be an intrusion on this subtle, almost spiritual scene, we opted for whole orchid plants. They were removed from their pots and set in unstructured groupings of moss, with small black pebbles. To illuminate the table, we buried votive candles in the moss.

When guests were settled, we served a miso soup. To the delicious broth, we added oyster mushrooms, poached lobster medallions, and cubes of tofu. *Shakuhachi,* authentic Japanese flute music, drifted from the living room. Candlelight flickered and subtle aromas filled the air.

A shiny foil package of whole, steamed striped bass with mushrooms and herbs was placed between every two guests. When opened at the table, the room was enveloped in an intoxicating cloud of fresh ginger, cilantro, earthy mushrooms, and the barest whisper of pungent black bean sauce and mirin. Japanese rice and snow peas completed the dinner.

The dessert was inspired by the famous chef, Fredy Girardet. On a glazed celadon-green plate, we placed kiwi marinated in passion fruit syrup, sliced bananas marinated in lime syrup, and colorful butterflies made from strawberries, along with grapes and golden raspberries. The presentation was finished with a light raspberry coulis.

One might expect this to be a complex and time-consuming menu to prepare. However, the soup can be made early in the morning and set aside; the fish can be readied, wrapped, and refrigerated for several hours before guests arrive; and fruits can be cut and marinated several days in advance. The hostess needs only to cook the rice and place the fish package into the oven 40 minutes before serving time. The wontons and duck pancakes can also be prepared in advance and cooked once guests arrive.

Shakuhachi is the beloved flute music of Japan.

SAKE ETIQUETTE

The sake tradition demands serious drinking etiquette. In Japan, the custom of *o-shaku* dictates that you pour for your companions and keep pouring until the sake spills over the edges of the glass. One never pours for oneself; instead, allow dinner companions to do the honors. When receiving *o-shaku,* take the cup or glass with your right hand; extend your left hand, palm up, allowing fingertips to lightly support the vessel.

Sake is traditionally enjoyed in Western cultures as a hot beverage sipped from a small cup. However, there are several varieties of sake that are meant to be served cold in a small, wooden, box-shaped cup.

If orchid plants are not available, use a few stems of dendrobium orchids—as they are available year-round. Alternatively, eliminate the orchids, and double the number of candles for an equally mesmerizing look.

GYOZA

- ½ pound ground lean chicken or pork
- 1 cup finely shredded napa cabbage
- 2 green onions, chopped
- 1 garlic clove, minced
- 2 tablespoons soy sauce
- 1½ tablespoons cornstarch
- ½ teaspoon grated fresh ginger
- ½ teaspoon dark sesame oil
- 32 gyoza wrappers
- 1 tablespoon vegetable oil (optional)

1. In a large bowl, combine the ground meat, cabbage, green onions, garlic, soy sauce, cornstarch, ginger, and sesame oil. Mix thoroughly.
2. Spoon 1 teaspoon of the meat mixture onto a gyoza wrapper. With water, moisten the edges of the wrapper and fold over to seal. Repeat with the remaining meat mixture and gyoza wrappers.
3. If desired, in a large skillet, heat the oil over medium-high heat. Add the gyoza and fry until browned, about 2 minutes. Turn and brown the other sides. Alternatively, the gyoza can be dropped, a few at a time, into a large pot of boiling water and cooked for 5 minutes. Remove with a slotted spoon.
4. To serve, arrange the gyoza on a serving plate with a small bowl of soy sauce in the center for dipping.

MAKES 32

CRISPY PORK WONTONS

- Vegetable oil
- ¼ pound ground pork, beef, or chicken
- ¼ pound peeled shrimp, finely chopped
- 4 green onions, finely chopped
- 4 ounces water chestnuts, finely chopped
- 1 tablespoon cornstarch
- 1 teaspoon chili garlic sauce
- 1 teaspoon soy sauce
- 40 wonton wrappers
- 1 egg, beaten
- 2 tablespoons hot sesame oil or plum sauce for dipping

1. In a pan heat the vegetable oil over medium heat and cook the meat for 5 minutes. Allow to cool.
2. In a bowl, combine the partially cooked meat, shrimp, green onions, water chestnuts, cornstarch, chili garlic sauce, and soy sauce.
3. Place 1 teaspoon of the mixture slightly off center on each wonton wrapper. Moisten the edges of the wrapper with some of the beaten egg. Fold the wrapper over the filling to create a triangle, forcing out any air. Pull the 2 corners of the folded edge together and twist to secure.
4. Drop the wontons, a few at a time, into deep oil heated over medium heat. Fry them until golden, about 2 minutes. Drain on paper towels.
5. Arrange the wontons on a serving plate with a small bowl of hot sesame oil or plum sauce for dipping.

MAKES 40

HAVE FUN WITH TRADITION: Provide a single shiny black stone, available at a plant store, for each guest to rest his or her chopsticks on.

MOST OF THE INGREDIENTS FOR THESE MENUS are readily available in many grocery stores, gourmet markets, and ethnic markets.

GRILLED PEKING DUCK PANCAKES *with* PLUM SAUCE

10 spring roll skins
1½ cups chopped cooked Peking duck meat
¾ cup chopped green onions
1 cup plum or hoisin sauce

1. Cut each of the spring roll skins diagonally into 4 triangles.
2. Place a generous teaspoonful of duck meat in the center of each triangle, then sprinkle with ½ teaspoon of green onions and add ½ teaspoon of plum sauce.
3. With water, moisten the edges of the skins, then fold over and seal.
4. Cook the pancakes on a grill until crisp or, alternatively, steam them over simmering water until they are soft.
5. To serve, arrange the pancakes on a serving plate accompanied by a small bowl of the remaining plum sauce.

MAKES 40

Miso soup can be enhanced with lobster, shiitake mushrooms, spinach leaves, clams, or tofu cubes.

MISO SOUP *with* LOBSTER MEDALLIONS

4 10-gram sachets instant dashi powder
12 cups hot water
4 green onions, julienned into 2½-inch lengths
2 frozen lobster tails, cut into ½-inch medallions
600 grams silken tofu, cut into ½-inch cubes
12 oyster mushrooms, cleaned
½ cup red miso

1. In a saucepan, combine the dashi and hot water.
2. Fill a medium bowl with cold water and place the green onions in the water until they curl. Remove and set aside.
3. Douse the lobster medallions with cornstarch and poach in boiling water for 2 minutes. Remove from water with a slotted spoon.
4. Divide the tofu, green onions, lobster, and mushrooms between 16 bowls.
5. Heat the dashi over medium-high heat; do not boil. Place the red miso into a small sieve, lower the sieve into the dashi, and stir until the miso has dissolved.
6. Remove the sieve and discard remaining miso residue.
7. Pour the dashi into the bowls and serve immediately.

SERVES 16

STEAMED STRIPED BASS

8 whole striped bass or red snappers, about $2\frac{1}{2}$ pounds each, cleaned and scaled with heads and tails on
48 thin slices peeled, fresh ginger
1 pound shiitake mushroom caps, sliced
$\frac{1}{2}$ pound wood ear mushrooms, sliced
$\frac{1}{2}$ pound enoki mushrooms, trimmed and cleaned
16 green onions, trimmed and thinly sliced
2 bunches cilantro
4 6-ounce cans straw mushrooms, drained
1 cup mirin
1 cup prepared black bean sauce

1. With a sharp knife, make three slits, down to the bone, on one side of each fish. Stuff each slit with 2 slices ginger. Cover and refrigerate for 1 hour.
2. Preheat the oven to 350 degrees.
3. Meanwhile cut 8 pieces of heavy duty aluminum foil into 24 × 24-inch sheets. Place a fish in the center of each sheet. Divide the shiitake, wood ear, enoki, green onions, cilantro, and straw mushrooms into 8 portions. Divide each portion into thirds and place a third underneath the fish, a third inside the fish, and a third on top of the fish.
4. Divide the mirin and black bean sauce into 8 portions and drizzle over each fish. Seal the fish in their foil sheets, creating a pocket with space for air to circulate.
5. Arrange the packets on a baking sheet and bake for about 40 minutes. Transfer the hot packets to a serving tray and cut open at the table, serving the fish in its juices.

SERVES 16

Turn scallions into a floral garnish.

Red snapper also works well with this menu and is readily available year round.

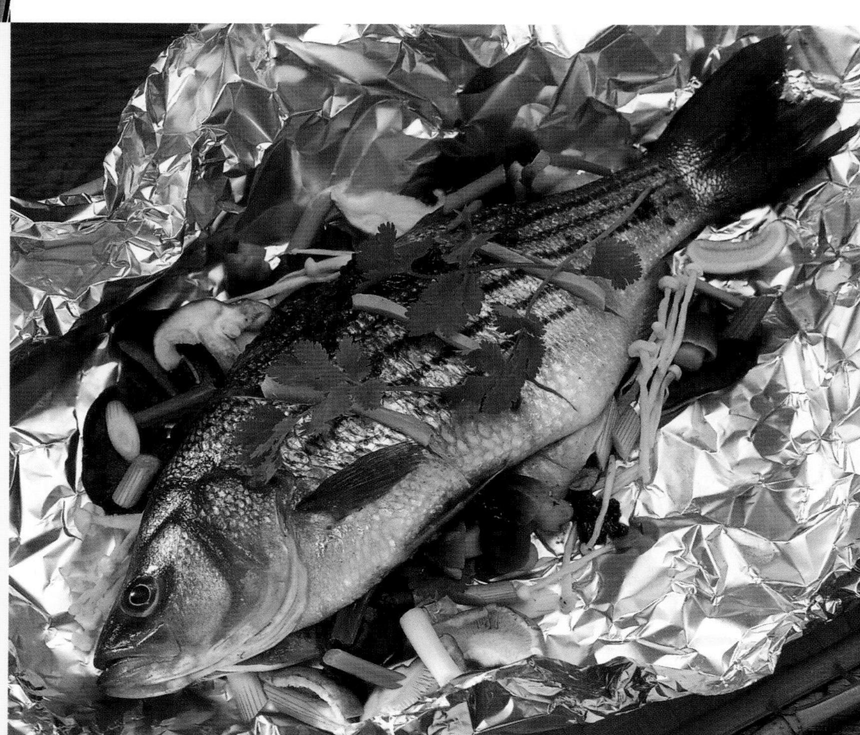

JAPANESE RICE

4 cups premium Japanese white rice
Water

Cook in a rice cooker according to the manufacturer's directions.

SNOW PEAS *and* TOASTED SESAME SEEDS

1 tablespoon sesame seeds
1½ pounds snow peas, ends trimmed
2 tablespoons sesame oil

1. Spread the sesame seeds in a dry skillet. Toast over medium heat, tossing frequently, until golden. Remove from the heat.
2. Bring a large pot of salted water to a boil. Blanch the peas for about 1 minute. Drain and immediately plunge into ice water. Drain and pat dry.
3. Just before serving, heat the oil in a large skillet over medium-high heat. Briefly sauté the snow peas, until crisp but heated through. Transfer to a platter, sprinkle with the toasted seeds, and serve immediately.

SERVES 16

KIWI MARINATED *in* PASSION FRUIT

¾ cup sugar
¾ cup fresh-squeezed orange juice
Pulp of 8 ripe passion fruits (skin should be wrinkled)
8 firm and ripe kiwi, peeled and cut into ¼-inch slices

1. In a heavy-bottom saucepan, heat sugar, orange juice, and passion fruit pulp over medium heat. When sugar has caramelized and is light golden brown in color, pour over kiwi fruit.
2. Let marinate for at least 3 hours before serving.

Note: This dish can be prepared up to 1 day in advance.

BANANAS MARINATED IN LIME JUICE

Zest of 1 orange, julienned
Zest of 1 lime, julienned
¾ cup sugar
½ cup orange juice
3 limes, pith removed and juiced
5 firm and ripe bananas, peeled and cut into ¼-inch slices

1. Blanch orange and lime zest in boiling water for 1 minute. Remove, dry with paper towels, and set aside.
2. In a heavy-bottom saucepan, heat sugar, orange juice, and lime juice over medium heat. When sugar has caramelized and is light golden brown in color, pour over the bananas.
3. Add the julienned zest and let marinate for at least 3 hours before serving.

Note: This dish can be prepared up to 1 day in advance.

For garnish you will need the following:
1½ cups raspberries
3 bunches champagne grapes
4 apples thinly sliced

Present marinated fruits in a simple Zen-like fashion.

DINNER AT THE BEACH

pissaladière

SOUPE *au pistou*

MEDITERRANEAN *sea bass*
WITH SPRING VEGETABLES

crème caramel

Serves: 8

Range of Difficulty: Easy

Preparation Time: 2 hours

CREATE A MOOD, BE IT RUSTIC, SOPHISTICATED,

opulent, whimsical, quirky, or exotic. By the time dinner is served, your guests will be transported to another land.

For this party on the Southern California coast, we created a dinner that captured the essence, the vitality, and the joie de vivre of southern France. The menu, the linens, and even the music created a relaxing mood from the moment guests arrived.

To set the stage for the light fare to come, an ordinary rectangular table was placed in the garden and covered with well-worn striped cotton cloths in bold, primary colors. Colorful hand-painted plates, green-tinged glasses, and water pitchers packed tightly with roses, hibiscus, and sunflowers added the final touch of color to the table's decor.

For guests to help themselves, wine was placed on a small table along with a pitcher filled with ice water and lemon slices and an assortment of glasses. The same striped cotton cloth was used on this table, and the decor was completed with a low basket of fruit and a tall vase of long-stemmed sunflowers. For the wine, we chose a barrel-aged Sauvignon Blanc, which provided a lightness and elegance suitable for the fare to come.

In the kitchen, everything was readied long before the first ring of the doorbell. As guests arrived, we served apéritifs and pissaladière, topped with caramelized onions and chopped marinated niçoise olives.

Once seated, guests savored soupe au pistou, a peasant-style vegetable soup made with a light broth, garden-fresh vegetables, and a dollop of fresh basil pesto. Made earlier in the day, the soup needed only to be heated and ladled into bowls.

The main course, sea bass with summer vegetables, needed minimal fussing. I sautéed the fillets long enough to seal in the flavor and give them a golden cloak, then set them aside and finished them in the oven for 6 to 8 minutes. Topped with julienned zucchini, artichoke hearts, carrots, wild mushrooms, and a final flurry of cold diced tomatoes and fresh basil, this was classic Mediterranean cuisine with an edge.

Dessert?

Crème caramel.

Classic.

What else is there to say?

It's easy. It cooks perfectly every time. And it's the quintessential comfort food with a French passport.

Soupe au pistou is a classic start to a light Mediterranean dinner.

SOUPE AU PISTOU

- ½ cup white cannellini beans
- 6 cups vegetable stock
- 1 large boiling potato, peeled and roughly diced
- 1 large carrot, peeled and roughly diced
- ¾ onion, roughly chopped
- ¾ leek, white part only, cleaned and cut into 1-inch lengths
- 2 teaspoons chopped fresh thyme
- 1 bay leaf
- 2 small tomatoes, seeded and diced
- 1 large zucchini, roughly diced
- 2 ounces green beans, trimmed and cut into 1-inch lengths
- Salt and freshly ground pepper
- Pistou (see following recipe)

1. Soak the beans in water to cover overnight.
2. Bring the stock to a boil in a large stockpot. Add the potato, carrot, onion, leek, thyme, and bay leaf. Reduce to a simmer and cook until the potato is nearly tender, about 10 minutes. Skim any scum from top of pot.
3. Drain the soaked beans and add to the stock, along with the tomatoes, zucchini, and green beans. Simmer until the green beans are tender, about 6 minutes. Again, remove any scum. Season with salt and pepper to taste. Serve in warmed bowls, with the Pistou for adding at the table.

SERVES 8

PISTOU

- 1 cup packed basil leaves
- ¾ cup olive oil
- ½ cup freshly grated imported Parmesan
- ¼ cup toasted pine nuts
- Salt and freshly ground pepper

In a blender, combine the basil, olive oil, Parmesan, and pine nuts and purée until coarsely blended. Season with salt and pepper to taste.

CARAMELIZED ONIONS

2 tablespoons butter
2 tablespoons olive oil
3 medium brown onions, thinly sliced
½ teaspoon herbes de Provence
Salt and freshly ground pepper

In a saucepan, add the butter and oil and melt the butter over medium heat. Add the onion and cook slowly for 25 to 30 minutes until caramelized and brown in color. Remove from the heat and season with herbes de Provence and salt and pepper. Spread evenly over dough.

PISSALADIÈRE

DOUGH

½ cup each warm water and milk (105 to 115 degrees)
1 envelope active dry yeast
2¼ cups all-purpose flour
1 tablespoon olive oil plus 1 teaspoon for coating
1 teaspoon salt
2 teaspoons sugar
Caramelized Onions (see preceding recipe)
½ cup niçoise olives, marinated, pitted and halved

1. Place a pizza stone on the top rack of the oven and heat for 1 hour at 500 degrees before using.
2. Place the water and milk in a small bowl and sprinkle with the yeast. Set aside until bubbly.
3. Place the flour in a food processor fitted with the plastic blade. Add the yeast mixture, 1 tablespoon oil, salt, sugar, and process until the dough forms a ball. Transfer the dough to a bowl, coat the dough, with the remaining oil, cover, and let rise in a warm place until doubled, about 45 minutes.
4. Punch down the dough then knead for 30 seconds. Allow the dough to rest for 10 minutes.
5. Divide the dough into 2 balls and roll out into 2 9-inch disks.
6. Bake the dough on the heated stone for 8 to 10 minutes, or until crisp and golden. Remove from the oven and top with Caramelized Onions and olives. Serve immediately.

SERVES 8

VARIATIONS ON PISSALADIÈRE
Thinly sliced onions, slowly cooked in olive oil with thyme, and small black niçoise olives make up the traditional pissaladière topping. Bite into this classic tart and you'll swear you're at a bougainvillea-shaded table on a sidewalk in Nice. Street vendors throughout the Riviera region bake the pizzalike pie on huge trays and sell individual portions from sidewalk stalls. You can create your own by adding one or more of these ingredients.

- multicolored bell pepper strips sautéed with garlic in olive oil
- roasted garlic purée
- sautéed wild mushrooms
- pesto
- pancetta or prosciutto
- prepared ratatouille
- sautéed, cubed new potatoes and arugula
- roughly chopped walnuts and some crumbled hard-boiled egg
- crumbled Gorgonzola cheese and fresh sage
- a mixture of chopped herbs such as parsley, oregano, basil, and rosemary

Pizza dough can be homemade or bought from your favorite local pizzeria.

Select imported European sea bass for best results.

MEDITERRANEAN SEA BASS *with* SPRING VEGETABLES

½ cup plus 2 tablespoons olive oil
1 pound assorted wild mushrooms, such as chanterelles, porcinis, or oysters, cleaned and sliced
Salt and freshly ground pepper
1 carrot, peeled and julienned
3 zucchini, julienned
5 cooked artichoke bottoms, thinly sliced
8 sea bass fillets, preferably French or Italian, about 6 ounces each
1 teaspoon herbes de Provence
2 finely chopped shallots
1 cup finely chopped fennel
3 large tomatoes, peeled, seeded, and diced
¼ cup fish stock
¼ cup dry white wine
6 fresh basil leaves, julienned
2 cloves of garlic finely chopped

1. Heat 3 tablespoons of the olive oil in a large skillet over medium-high heat. Add the mushrooms, season with salt and pepper, and sauté until golden, about 8 minutes. Transfer to a bowl and set aside.
2. Pour 4 tablespoons of the oil in the pan and return to the heat. Add the carrot and cook for 2 minutes. Add the zucchini and artichokes and cook until the vegetables are tender-crisp, about 2 minutes longer. Transfer to the bowl with the mushrooms.
3. Heat the remaining oil in another skillet over medium-high heat. Season the fillets all over with the herbes, salt, and pepper. Sauté about 3 minutes per side and set aside in a warm oven.
4. In the same pan sauté the shallots and fennel for 3 minutes. Add the tomatoes, fish stock, and white wine and reduce over medium heat for 5 minutes. Sprinkle in the basil and garlic.
5. To serve, divide the vegetable mixture and spread on 8 serving plates. Place a piece of sea bass in the center of each and spoon the sauce over the fish. Serve immediately.

SERVES 8

CRÈME CARAMEL

¼ cup water
1 cup sugar
2½ cups milk
1 vanilla bean, split
3 eggs
3 egg yolks

1. Preheat the oven to 325 degrees.
2. Combine the water and ½ cup of the sugar in a small saucepan. Cook over medium heat until the sugar has caramelized and turned a golden brown. Be careful not to stir the sugar and stay nearby as the color changes rapidly. Pour the caramelized mixture into a 10-inch ceramic or glass pie plate and swirl to coat evenly. Set aside.
3. Pour the milk into a saucepan, scrape in the vanilla seeds, and add the bean. Bring to a boil and remove from the heat.
4. In a mixing bowl, whisk together the remaining sugar, eggs, and yolks. Gradually pour the hot milk into the egg mixture, whisking constantly. Ladle off and discard the foam from top.
5. Strain into the caramelized pie plate. Place the plate in a roasting pan and fill with water halfway up the sides of the pie plate. Bake until set, about 1 hour. Let cool and then refrigerate.
6. To serve, place a plate over the pie plate and flip over. Let stand until crème releases and syrup drains from pan.

SERVES 8

Note: Custard can also be made in small ramekins for individual servings.

DINNER WHEN TIME IS SHORT

artichoke CANAPÉS

SALAD OF BABY GREENS WITH
toasted goat cheese CROUTONS

salmon with fava beans, TOMATOES, AND BASIL

flambéed cherries
WITH VANILLA ICE CREAM

Serves: 8

Range of Difficulty: Easy

Preparation Time: 1 hour

THE MOST BEAUTIFUL FLOWERS

Always do a walk-through just before guests arrive.

Check to see that music is playing, the candles are lit, the drinks are ready, the lights are dimmed, and the guest bathroom has a fragrant candle burning.

Your husband, wife, boyfriend, or girlfriend calls from the office at 5:00 p.m. and says dinner guests are arriving at 7:00 p.m. What do you do? Here's a menu and strategy that will allow you to turn out a wonderful meal *and* remain on speaking terms.

Greet guests at the door with a festive Champagne cocktail that says, "Welcome to my home!" If you don't have a great Champagne, use sparkling wine and dress it up with a splash of Cassis or Chambord to create a Kir Royale.

One of my favorite hors d'oeuvres was created for just such an impromptu event. Simply toast white bread until golden and use a cookie cutter to cut rounds the size of a half dollar. Top each with an artichoke heart and cover generously with a blend of mayonnaise, chopped shallots, and Parmesan cheese. When guests arrive, slip the canapés under the broiler and brown until golden and bubbly.

Prewashed, bagged salad greens have saved many a day. For the first course, choose a blend that includes a peppery green such as arugula or radicchio. Toss the greens with a favorite vinaigrette, then top them with what the Italians call *crostini* and the French call *croutons.* In any language, a slice of baguette, toasted, spread with goat cheese, and finished under the broiler, adds a delicious and crunchy texture to the simple salad.

Salmon is usually my answer to the question: "What do I cook when I'm in a hurry?" For this main course, I made my favorite salmon dish that presents beautifully and has an interesting combination of temperature, flavor, texture, and color. I lightly dust fillets of salmon with sea salt and pepper, then sauté them over high heat for two minutes on each side, set them aside in a glass baking dish, and finish them in the oven for 6 minutes while guests dine on salad.

When the salmon comes out of the oven, place it on a plate surrounded by fava beans and cubed tomatoes, then generously spoon a vinaigrette over everything, and sprinkle the salmon and tomatoes with fresh basil. Pour a Chardonnay with tropical fruity flavor to accompany this dish.

For dessert, we opted for cherries flambé. What can be easier than canned cherries, butter, sugar, and a shot of Armagnac? Set the contents of the pan aflame and as soon as the flames subside, spoon the cherries over ice cream.

Find a tranquil spot in your home to relax and collect your thoughts before your guests arrive.

MORE TWO-MINUTE HORS D'OEUVRES

- Roll store-bought puff pastry or phyllo dough with anchovy fillets, Stilton cheese, or sautéed mushrooms and bake.
- Hollow out cherry tomatoes and fill with mozzarella cheese, fresh basil, and balsamic vinegar.
- Spread toasted croutons with goat cheese and top with roasted peppers.
- Sauté almonds with curry powder, ground coriander, cumin, salt, and pepper.
- Marinate olives with chopped garlic, parsley, chives, and extra virgin olive oil.

MORE TWO-MINUTE DESSERTS

- Rinse two pint boxes of strawberries. Cut the berries from one box into bite-size pieces. Purée the remaining berries in a food processor with a tablespoon each of sugar and Crème de Cassis or Grand Marnier. Pour the purée over the cut berries and serve in a dessert bowl.
- Top a slice of store-bought pound cake with hot brandy syrup, fresh strawberries, and whipped cream.
- Heat a store-bought brownie, place it on a plate with raspberry coulis (frozen berries, sugar, and lemon juice puréed in a blender), and top with ice cream.

ARTICHOKE CANAPÉS

4 slices white bread, toasted, crusts removed, and quartered
1 jar marinated artichoke hearts, about 7¾ ounces
6 tablespoons mayonnaise
2 teaspoon minced shallots
3 teaspoons grated Parmesan cheese

1. Preheat the broiler.
2. With a 1½-inch round cookie cutter, cut 4 to 5 circles out of each piece of toast.
3. Cut the artichoke hearts into pieces (approximately 16 to 20 equal pieces).
4. In a bowl, combine the mayonnaise, shallots, and Parmesan cheese.
5. Place 1 artichoke piece on top of each toast round. Spoon or brush some of the mayonnaise mixture over the artichoke piece, covering it and the toast completely.
6. Arrange the canapés on a baking sheet and run under the broiler until they are golden and bubbly. Serve hot.

MAKES 16 TO 20 PIECES

SALAD *of* BABY GREENS *with* TOASTED GOAT CHEESE CROUTONS

8 ½-inch-thick slices goat cheese
¼ cup olive oil
½ teaspoon herbes de Provence
8 baguette slices, lightly toasted
10 ounces mixed baby salad greens
Tomato Juice Dressing (page 146)

1. Arrange the cheese slices in a single layer in a shallow dish. Drizzle with the olive oil, then sprinkle with the herbs. Cover and let stand for 1 to 2 hours.
2. Preheat the broiler.
3. Arrange the baguette slices on a broiler pan or heatproof tray. Top each slice with a piece of cheese. Set aside.
4. In a large bowl, toss the greens with the Tomato Juice Dressing. Pile the salad onto 8 individual salad plates.
5. Place the pan of cheese-topped baguette slices under the broiler until the cheese is bubbly and beginning to brown. Place one hot cheese toast on top of each salad and serve immediately.

SERVES 8

A salad of baby greens becomes more sophisticated with a toasted goat-cheese crouton.

VINAIGRETTE DRESSING

3 tablespoons red wine vinegar
1 teaspoon Dijon mustard
½ cup olive oil
Salt and freshly ground pepper

In a small bowl, combine the vinegar and mustard. Slowly whisk in the oil. Season to taste with salt and pepper.

SALMON *with* FAVA BEANS, TOMATOES, *and* BASIL

2 pounds fresh fava beans, shelled, or 2 10-ounce packages frozen lima beans, thawed
8 center-cut salmon fillets, about 6 ounces each, skin and bones removed
1 tablespoon coarse sea salt
1 teaspoon freshly ground pepper
2 tablespoons olive oil
8 roma tomatoes, seeded and diced (2 cups)
Vinaigrette Dressing (see following recipe)
Additional coarse sea salt
12 basil leaves, torn by hand

1. Bring a large pot of water to a boil. Drop in the shelled fava beans and cook for 2 minutes. (If using thawed lima beans, cook for 1 minute.) Immediately plunge the beans into cold water to stop the cooking. Drain the beans, then remove their outer skin. Set aside.
2. Season the salmon fillets with salt and pepper.
3. In a large skillet, heat the oil over high. Add the salmon fillets and cook for 2 minutes, skin side down. Turn and cook the other side for 2 minutes more. Remove from skillet and transfer to a shallow baking dish.
4. Five minutes prior to serving, place the salmon in a preheated oven at 375 degrees for 4 minutes.
5. Place the beans in a pot of boiling salted water for 2 minutes, or until desired tenderness. Remove beans from water and set aside.
6. Place the salmon in the center of 8 warm dinner plates. Place some of the fava beans on each side of each salmon fillet. Spoon the diced tomatoes at the top and bottom of each fillet. Drizzle vinaigrette over the salmon, beans, and tomatoes. Sprinkle lightly with additional sea salt and basil.

SERVES 8

Note: The salmon and beans can be prepared in advance, covered, and set aside for up to 2 hours. With frozen lima beans there is no shell to remove. The time can be increased in step 5 and step 1 can be removed.

TOMATO JUICE DRESSING

2 tablespoons balsamic vinegar
2 teaspoons Dijon mustard
2 tablespoons tomato juice cocktail
2 teaspoons minced shallot
⅓ cup olive oil
Salt and freshly ground pepper

In a small bowl, combine the vinegar, mustard, tomato juice cocktail, and shallot. Slowly whisk in the olive oil. Season to taste with salt and pepper.

FLAMBÉED CHERRIES *with* VANILLA ICE CREAM

¼ cup brown sugar, packed
4 tablespoons unsalted butter
2 tablespoons currant jelly
Juice of ½ lemon
2 17-ounce cans pitted sweet cherries, drained
½ cup Cognac or Armagnac
1 quart vanilla ice cream

1. In a large skillet, combine the sugar and butter. Set over medium heat and stir until the sugar is dissolved. Stir in the jelly and lemon juice. Simmer for 2 minutes to reduce slightly. Stir in the cherries and bring to a boil. Add the Cognac and heat 30 seconds longer. Standing away from the pan and using a long match, carefully ignite the cherries.
2. When flames die down, spoon the cherries and sauce over scoops of ice cream in individual dessert bowls. Serve immediately.

SERVES 8

Different colors, textures, and temperatures turn ordinary salmon into something special.

AN IMPROMPTU DINNER

FETA CHEESE AND TOMATO *canapés*

ravioli WITH RICOTTA CHEESE AND SPINACH

filet mignon ROLLED IN CRACKED PEPPER
WITH *bordelaise sauce*

roasted garlic MASHED POTATOES

spring vegetable MEDLEY

CLASSIC *pear tart*

Serves: 8

Range of Difficulty: Moderate

Preparation Time: 3 hours

Set the table using different stemware,

plates, and flatware for each person. Focus not just on the individuality of each place setting, but also on the harmony of all the elements. Not only will the table look fabulous, but every item becomes a conversation piece!

It's fun to call friends at the last minute and invite them to dinner. We dreamed up this impromptu dinner with just a day's lead time.

The evening got off to a savory start with toasted baguettes topped with sliced feta cheese, tomato, red onion, and basil, along with tall, icy glasses of Absolut Kurant vodka and cranberry juice.

In keeping with the casual nature of the event, the dinner table was set with assorted colored glasses, mismatched china collected from flea markets around the world, and jewel tone place mats found in Thailand.

For the centerpiece, rather than purchasing flowers I assembled some apples, pears, plums, and champagne grapes in a lovely old wooden fruit bowl. A couple of pineapples sprayed silver and gold were added for color, drama, and height. As a final touch, tall candle-lamps, topped with inexpensive shades glued with remnants of damask fabric, made the table decor complete.

For dinner, the robust aroma of tomato sauce filled the dining room as a bowl of ravioli was passed around the table family-style. The spinach and ricotta cheese ravioli were homemade earlier in the week and removed from the freezer just before boiling. Our sauce used fresh and canned tomatoes and took just 20 minutes to prepare.

We served the main course as a buffet in the kitchen. A gorgeous whole filet mignon was rolled in cracked pepper, then pan seared and finished in the oven for 20 minutes. A bordelaise sauce, mashed potatoes perfumed with roasted garlic, and a medley of sautéed vegetables completed the home-style dish.

With the ravioli, we poured a Byron Reserve Chardonnay known for its distinctive perfume and richness. Chateau Meyney, a full-bodied Bordeaux, handled the peppery flavors of the beef.

When serving meat as part of the main course, I often balance the meal with a light dessert. For this dessert, fresh pears in season were used to make a wonderful tart.

ORGANIZATION IS THE KEY to successful entertaining.

FETA CHEESE *and* TOMATO CANAPÉS

2 baguettes of crusty bread, cut into ¼-inch slices
16 ounces feta cheese, sliced thinly
8 roma tomatoes, sliced thinly
1 red onion, sliced thinly
2 tablespoons olive oil
24 pieces of basil

In a broiler, toast the baguette slices until golden in color. On each slice place a layer of feta, tomato, and red onion, then drizzle with olive oil. Top with a piece of basil and serve.

MAKES 24 PIECES

RAVIOLI *with* RICOTTA CHEESE *and* SPINACH

1 10-ounce package frozen spinach, thawed
1 tablespoon butter
¼ cup minced onion
1 clove minced garlic
½ cup ricotta cheese
¼ teaspoon ground nutmeg
1 egg yolk
Salt and freshly ground pepper
1¾ cups grated imported Parmesan cheese
Basic Pasta (see following recipe)
1 egg, well beaten
Cornmeal
Tomato Sauce (see recipe)
2 tablespoons butter
1 cup torn fresh basil leaves

1. Cook the spinach in boiling water for 5 minutes. Drain and squeeze out the excess water.
2. In a large sauté pan, melt the butter over medium heat. Sauté the onion and garlic until golden. Add the spinach, stir, and cook 3 minutes longer. Transfer to a mixing bowl.
3. Add the ricotta, nutmeg, and egg yolk and mix well. Season with salt and pepper to taste and stir in ¾ cup of the Parmesan. Set filling aside.
4. To form ravioli, roll out the Basic Pasta dough until very thin (No. 6 on the pasta machine). Place 1 sheet of rolled dough on a ravioli form, being careful to cover all sides evenly. Press on the top mold of the dough form to make hollows for the filling. Fill each hollow with 1 teaspoon of the filling. Brush the edges of the ravioli with the beaten egg.
5. Place another sheet of rolled pasta over the filling to cover. Run a rolling pin over the top of the rack to seal and separate the ravioli. Transfer to a baking sheet dusted with cornmeal. (If making in advance, turn the ravioli every hour to dry evenly or store in zip-top freezer bags and store in the freezer.)
6. To cook, bring a large pot of water to a boil. Add the ravioli, a handful at a time, and cook until they rise to the top.
7. Meanwhile, place 1 cup of the Tomato Sauce in the bottom of a warmed large mixing bowl. With a slotted spoon, remove the cooked ravioli and transfer to the bowl. Top the ravioli with butter, ½ cup of the basil, and the remaining Tomato Sauce. Gently toss to coat the ravioli evenly.
8. To serve, spoon the ravioli into heated pasta bowls, sprinkle with the remaining basil and 1 teaspoon Parmesan cheese. Pass the remaining Parmesan at the table.

SERVES 8

Note: Ravioli can be made in advance, wrapped in plastic after dry, and frozen for up to 1 month.

BASIC PASTA

3 cups all-purpose flour
1 tablespoon vegetable oil
4 eggs
1 teaspoon salt

1. Place the flour in the bowl of a food processor fitted with the metal blade. With the machine running, add the oil and eggs, one at a time; then add salt. When the dough forms a ball and is smooth and elastic to the touch, remove.
2. Divide the dough in half and wrap in plastic wrap.

TOMATO SAUCE

4 pounds roma tomatoes, fresh or canned
¼ cup olive oil
1 medium onion, finely chopped
2 garlic cloves, minced
½ teaspoon sugar
Salt
1 bay leaf
½ teaspoon crushed red pepper flakes
6 fresh basil leaves, torn into pieces
3 tablespoons heavy cream (optional)

1. If using fresh tomatoes, peel, seed, and purée. If using canned, seed and purée. Set aside.
2. In a large saucepan, heat the olive oil over medium heat. Sauté the onions and garlic until golden. Add the puréed tomatoes, sugar, salt to taste, bay leaf, and red pepper flakes. Simmer, uncovered, for 30 minutes. Stir in the basil and the cream, if using.

MAKES 4 CUPS

20-MINUTE TOMATO SAUCE

In a medium saucepan, sauté in 2 tablespoons of olive oil 2 chopped garlic cloves and ½ finely chopped medium onion. Add 2 cans of crushed tomatoes and season to taste with salt, pepper, fresh basil, and a bay leaf. Simmer for 20 minutes. For a sweeter sauce, add 1 tablespoon sugar. For a lighter sauce, add a 5-ounce can of tomato juice.

QUICK MUSHROOM SAUCE

In a medium saucepan, sauté 2 cups sliced white mushrooms with 1 tablespoon melted butter until tender. Add ½ cup chicken stock and ½ cup whipping cream. Season to taste with salt and pepper.

Allow to reduce over medium heat for 10 to 15 minutes. Place half the mixture in a blender and purée. Return to the saucepan and finish with 2 tablespoons of Cognac.

OTHER TOPPINGS FOR TOASTED BAGUETTES

- sliced mozzarella cheese, roma tomatoes, and fresh basil
- puréed olives and garlic
- finely chopped, marinated, roasted peppers

FOR GREAT MASHED POTATOES

- Substitute nonfat sour cream or plain yogurt for butter.
- Add sautéed lobster medallions and a bit of lobster stock, or truffle oil and sliced truffles for a very aristocratic version.
- Add a generous amount of chopped chives.

Always bring meats, fish, and chicken to room temperature before cooking.

MUSHROOM PRIMER

A variety of mushrooms can be used to enhance a Bordelaise sauce.

Porcini. Also known as cèpes, these are pale brown and have a smooth, meaty texture and a pungent, woodsy flavor. Dried porcini must be softened in warm water for 20 minutes.

Oyster. This fan-shaped mushroom, which varies in color from light gray to dark brownish-gray, has a firm texture and robust, slightly peppery flavor.

Chanterelles. A trumpet-shaped wild mushroom, the chanterelle ranges in color from bright yellow to orange-gold and has a chewy texture and slightly nutty flavor.

Wood ear. Also known as silver ear, tree ear, or cloud ear, these mushrooms are brown-black (except for the albino silver ear) and have a crunchy texture and delicate, almost bland, flavor.

Morels. This spongy, cone-shaped cap ranges in color from tan to dark brown and has an earthy, nutty, slightly smoky flavor.

FILET MIGNON ROLLED *in* CRACKED PEPPER *with* BORDELAISE SAUCE

3 whole filet mignons (1¾ to 2 pounds each)
2½ tablespoons cracked pepper
1 teaspoon coarse salt
3 teaspoons olive oil
Bordelaise Sauce (see following recipe)
¼ cup chopped parsley

1. Preheat the oven to 350 degrees.
2. Rub the meat with the pepper, pressing with your hands so it will adhere. Sprinkle the meat with the salt.
3. In a large skillet, heat the oil over high heat, add the meat, one piece at a time, and cook, turning several times, until browned all over. Transfer the filets to a roasting pan and insert a meat thermometer into the thickest part of one filet. Roast the filets in the middle of the oven until done as desired, 125 degrees for rare, 140 degrees for medium, and 160 degrees for well done.
4. Remove the filets from the oven and let them rest for 5 to 10 minutes before carving into 1-inch-thick pieces. Arrange a few slices on each dinner plate and spoon some of the Bordelaise Sauce over each, then sprinkle with the parsley.

SERVES 8

Note: The filet can be pan seared in advance and set aside for up to 2 hours before finishing in the oven.

BORDELAISE SAUCE

4 cups veal stock
½ cup chopped shallots
4 tablespoons butter
2 cups dry red wine, such as Bordeaux
2 fresh thyme sprigs
1 bay leaf
6 peppercorns
Salt and freshly ground pepper

1. In a saucepan reduce the veal stock by 50 percent over medium heat to make a demi-glace.
2. Sauté the shallots in 1 tablespoon of the butter over medium-high heat until they are soft, about 3 minutes. Add the wine, thyme, bay leaf, and peppercorns. Bring to a boil, then allow the mixture to simmer until reduced to a syrup.
3. Add the veal demi-glace and reduce for a further 10 to 15 minutes.
4. Strain the sauce and whisk in the remaining 3 tablespoons of butter, one tablespoon at a time. Season with salt and pepper.

MAKES APPROXIMATELY 1 CUP

Note: If making the bordelaise sauce in advance, reheat the strained mixture and whisk in the remaining butter just before serving.

ROASTED GARLIC MASHED POTATOES

3 pounds russet potatoes, peeled and quartered
1 tablespoon oil
5 garlic cloves, with skin on
5 tablespoons butter
2 cups hot milk
Salt and freshly ground pepper

1. Drop the potatoes into a large pot of boiling salted water. When the water returns to a boil, reduce the heat to medium and cook until the potatoes are tender when tested with a fork, about 20 minutes.
2. Meanwhile, in a small skillet, heat the oil over medium-low heat. Add the garlic and sauté, stirring frequently, until soft and browned all over. Set aside.
3. Drain the potatoes thoroughly. Add the garlic, butter, and hot milk and blend with a potato masher until almost smooth. Season with salt and pepper to taste.

SERVES 8

Note: The potatoes can be prepared 1 day ahead, covered, and refrigerated. To reheat, place the potatoes in a saucepan over medium heat. Stir in ½ cup milk and cook, stirring frequently, until heated through.

SPRING VEGETABLE MEDLEY

2½ cups asparagus tips
2½ cups fresh or frozen peas
1 cup green beans, cut in 2-inch pieces
1 cup shelled fava beans or frozen lima beans
2 tablespoons butter
1 tablespoon olive oil
Salt and freshly ground pepper

1. Drop the asparagus, peas, green beans, and fava beans into a large pot of boiling salted water and cook for about 2 minutes, or until the vegetables are tender but still crisp. Drain the vegetables and immediately plunge them into a large bowl of ice water to stop the cooking process.
2. In a large skillet, melt the butter in the oil over medium-high heat. Drain the vegetables and transfer them to the pan. Sauté the vegetables until heated through. Season with salt and pepper to taste.

SERVES 8

CLASSIC PEAR TART

TART SHELL

2 cups all-purpose flour
8 tablespoons butter
Dash of salt
6 tablespoons water

1. In a food processor mix flour, butter, and salt on pulse until pieces are the size of small peas. Add water and mix until dough turns into a ball. Remove from processor and dust with flour.
2. Wrap in wax paper and refrigerate for 30 minutes to 1 hour. If the dough is sticky, knead before refrigerating.
3. Preheat oven to 400 degrees.
4. To use, roll out on wax paper, put in tart pan, and place pie weights on top of pastry to prevent pie from rising.
5. Bake in oven for 30 to 40 minutes or until golden brown.

PASTRY CREAM

1 cup milk
1 vanilla bean
3 large egg yolks
4 tablespoons sugar
4 tablespoons cornstarch
Grated peel of ¼ lemon

1. To begin the filling, combine the milk and vanilla bean in a small saucepan. Cook over low heat until nearly boiling, approximately 5 minutes. Cool slightly. Remove the vanilla bean, split open lengthwise, and scrape seeds into milk, discarding bean.
2. In a bowl, whisk together the egg yolks and sugar until light and fluffy. Add cornstarch. Pour the warm milk into the egg mixture, whisking constantly.
3. Place the bowl over a pan of simmering water and cook, stirring frequently, until thick enough to coat a spoon. Remove from the heat, pour into a glass mixing bowl, and mix 1 minute longer to accelerate the cooling process. Add lemon peel and set aside to cool further.

FILLING

¼ cup sugar
1 tablespoon lemon juice
1 cup dry red wine
1 cup water
4 medium pears, peeled, halved, and cored
¼ cup currant jelly

1. In a large saucepan, combine the sugar, lemon juice, wine, and water and bring to a boil. Add the pears, then reduce the heat, cover and simmer until they are tender, about 10 minutes. Remove the pears from the syrup and set aside to cool. Reserve 2 tablespoons of the syrup.
2. Spread the Pastry Cream evenly over the Tart Shell.
3. With cut side down, thinly slice each pear half crosswise. With your hand or the flat side of a large knife, press each pear half to fan out the slices. Slide the knife or a thin spatula under each pear half and carefully transfer the slices to the pastry cream, retaining their pear shape and arranging them spoke fashion with the small ends toward the center.
4. In a small saucepan, combine the currant jelly and the reserved pear liquid. Stir over medium heat until the jelly is melted. Brush the mixture over the tart.

SERVES 8

An ALTERNATIVE *to the* CLASSIC PEAR TART

TART SHELL

1½ cups all-purpose flour
½ cup ground almonds
6 tablespoons sugar
8 tablespoons frozen butter, cut into small pieces
1 egg

1. Mix flour, almonds, and sugar in a food processor. Slowly add frozen butter, then add the egg.
2. When dough forms into a ball stop, the food processor. Remove the dough, wrap in wax paper, and refrigerate for 30 minutes to 1 hour. If the dough is sticky, knead once or twice on a floured board before refrigerating.
3. To use, roll out on wax paper and put in tart pan.

Note: Crust can be made ahead of time and refrigerated for up to 3 days, or wrapped tightly in wax paper, placed in a plastic bag, and frozen.

FILLING

½ cup sugar
5 small Bosc pears, peeled, cored, and halved
2 tablespoons butter, broken into small pieces
Currant Glaze

1. Preheat oven to 375 to 400 degrees.
2. Sprinkle 3 tablespoons of sugar on the bottom of the shell.
3. Slice the pears horizontally from the tip to the core across, keeping the slices in place. Arrange the pears fanlike in a circle with the smallest part of the pear pointing toward the center of the circle.
4. Sprinkle the remaining sugar over the pears.
5. Dot the butter over the pears.
6. Bake for 40 minutes or until crust turns light brown. Remove from oven, glaze, and serve.

CURRANT GLAZE

¼ cup red currant jelly
1 tablespoon sugar

In a saucepan heat jelly and sugar over high heat. Stir with a wire whip until jelly breaks down and turns into syrup (about 2 minutes). While glaze is still warm, paint pears with a very soft pastry brush.

Note: The glaze can be made while the tart is cooling.

THE POACHED PEARS, Pastry Cream, and Tart Shell can all be prepared 1 day ahead. Cover and refrigerate the pears and pastry cream.

A Wine Tasting and Dinner

wild mushroom AND SUN-DRIED TOMATO RISOTTO

insalata tricolore WITH SHAVED PARMESAN CHEESE

stuffed breast of turkey WITH WHITE WINE AND ROSEMARY

patatine

SAUTÉED *spinach and garlic*

FLOURLESS *almond torte*

WITH CHOCOLATE COFFEE SAUCE

demitasse

Serves: 8

Range of Difficulty: Easy

Preparation Time: 2 hours

THERE'S GREAT STYLE IN USING JUST *ONE* ELEMENT

and using it en masse. One type of flower or a single color. It creates a lasting impact.

I enjoy turning cocktail hour into a crash course in wine appreciation. By offering a selection of Meritage wines, Super Tuscans, or lesser-known bottles from France's Languedoc-Roussillon or California's Napa and Sonoma Valley, guests can select their favorite. The chosen wine can then accompany dinner and serve as both the beverage and conversation piece.

At this dinner party we greeted guests with three different Zinfandels of the 1993 vintage. Proclaimed the favorite, the dark ruby-purple Nalle from Sonoma, with black raspberry nose and rich, fruity flavor, was the perfect accompaniment to the robust flavors of our four-course dinner.

Our first dish, a creamy risotto with wild mushrooms and sun-dried tomatoes, instantly transported guests to the Italian countryside.

The tricolored salad of radicchio, arugula, and endive that followed was dressed with extra virgin olive oil and authentic balsamic vinegar and finished in the traditional Italian manner with thin shavings of Parmesan cheese.

Our third course was a breast of turkey, stuffed and rolled with prosciutto slices and a frittata. We prepared the plates in the kitchen, serving the turkey with fresh steamed spinach and patatine—a dish of twice-cooked potatoes with fresh rosemary.

For dessert: a feather-light flourless almond torte, drizzled with a chocolate coffee sauce and dusted with powdered sugar.

THE PURE CLASSICAL LINES of the home provided an elegant backdrop to the party. The table was set with original 1920s Arts and Crafts pottery filled with baby calla lilies, vintage pottery plates, silverware, and linens, which the hosts had collected over the years.

Enjoy a wine tasting during cocktail hour.

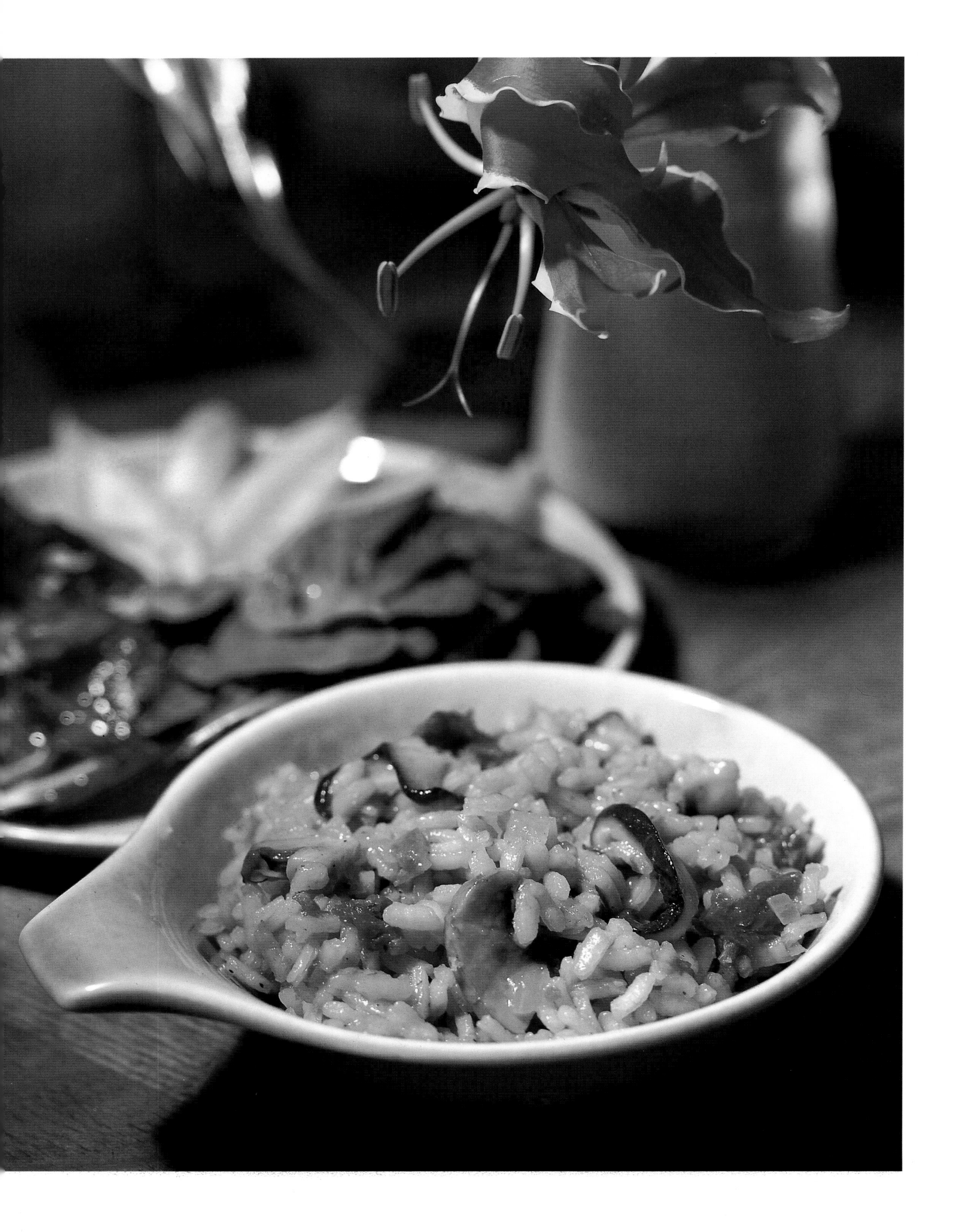

WILD MUSHROOM *and* SUN-DRIED TOMATO RISOTTO

1 ounce dried porcini mushrooms, cut into pieces
3½ cups chicken broth
2 tablespoons butter
2 tablespoons olive oil
3 tablespoons finely chopped shallots
2 cloves garlic, minced
2 cups arborio rice
1 cup dry white wine
2 tablespoons chopped sun-dried tomatoes
¼ cup grated Parmesan cheese
Salt and freshly ground pepper
Additional grated Parmesan cheese

1. Place the mushrooms in a bowl and add 2½ cups of warm water. Allow to soak for 30 minutes.
2. Drain the mushrooms in a fine sieve, reserving the soaking liquid. Set the mushrooms aside.
3. In a saucepan, bring the chicken broth and the reserved mushroom liquid to a boil. Reduce the heat to a simmer.
4. Meanwhile, in a large heavy pot, melt 2 tablespoons of the butter in the oil over medium heat. Add the shallots and garlic and sauté until translucent. Add the mushrooms and sauté for 5 minutes. Add the rice and cook, stirring, until the rice is completely coated. Stir in the white wine. Cook, stirring, for 2 minutes, or until the rice begins to absorb the liquid.
5. Begin stirring in the hot broth, about ½ cup at a time, stirring constantly after each addition until it is absorbed. When half of the broth has been added, add sun-dried tomatoes. Continue adding the remaining broth, stirring until absorbed.
6. Remove the pan from the heat and stir in the Parmesan. Season with salt and pepper to taste. Serve the risotto with additional grated Parmesan.

SERVES 8

INSALATA TRICOLORE *with* SHAVED PARMESAN CHEESE

3 tablespoons balsamic vinegar
½ teaspoon sugar
1 tablespoon Dijon mustard
1 medium shallot, minced
½ cup olive oil
Salt and freshly ground pepper
4 ounces each arugula, radicchio, and Belgian endive
6 ounces of whole Parmesan

1. In a small bowl, blend the vinegar, sugar, and mustard. Stir in the shallot. Gradually whisk in the oil. Season with salt and pepper to taste.
2. Toss each of the lettuces separately with a third of the dressing. Arrange them in sections on individual salad plates and cover with 3 shavings of Parmesan.

SERVES 8

BUY THE VERY BEST BALSAMIC VINEGAR YOU CAN AFFORD. (And look for the word *tradizionale,* which means the vinegar contains 100 percent cooked, aged, white grape must.) You'll need just a few drops of the real thing to create a taste sensation.

Use balsamic vinegar sparingly on salads or as a final touch on cooked vegetables.

FOR BEST RESULTS IN CARVING THE BREAST OF TURKEY, allow the cooked turkey to stand for 5 or 10 minutes after removing from the oven; then carve with an electric knife.

STUFFED BREAST *of* TURKEY *with* WHITE WINE *and* ROSEMARY

1 turkey breast, boned and butterflied by the butcher, about 5½ pounds
3 garlic cloves, peeled and halved
2 sprigs rosemary, or 2 teaspoons dried
Salt and freshly ground pepper
6 thin slices prosciutto
4 eggs
2 tablespoons chopped fresh parsley
½ cup olive oil
2 cups dry white wine
4 tablespoons butter
1 medium onion, chopped
1 shallot, chopped

1. With a sharp paring knife, make 6 slits in the turkey skin and insert 1 piece of garlic in each. If using fresh rosemary, insert between the skin and the meat. Sprinkle all over with salt and pepper and lay on work counter, skin side down. Place a layer of prosciutto over the turkey.
2. In a small bowl, beat eggs, parsley, and salt and pepper to taste.
3. In a large nonstick skillet, heat 1 to 2 tablespoons of the olive oil over medium heat. Add the eggs and cook until a soft omelet is formed. Do not stir or overcook the eggs. Set aside to let cool slightly and then place omelet over prosciutto.
4. Starting at one end, roll the turkey, as tightly as possible, along the length, jelly roll–style. Secure the role with string. (If using dried rosemary, rub all over the outside of the roll.) Season with salt and pepper to taste.
5. In a large covered heavy-bottomed pot, heat the remaining oil over medium-high heat. Cook the turkey roll, turning occasionally, until evenly browned. Reduce the heat to medium and sprinkle with 1 cup of the white wine. Cook until wine has evaporated. Add the butter, onion, and shallot, then reduce the heat to low. Cover and cook about 1 hour and 15 minutes, removing the cover and basting occasionally. Turn the roll once at 45 minutes. Transfer the turkey to a cutting board and cool for about 15 minutes.
6. To make the sauce, add the remaining cup of wine to the cooking pot, scrape the bottom to release any browned bits, and cook over high heat for about 10 minutes. Strain into a sauceboat.
7. To serve, cut and remove the strings. Cut the turkey roll across the width into ¼-inch slices. Arrange on a serving platter, pour some of the sauce in a thin layer, and serve with the remaining sauce.

SERVES 8

Note: The turkey can be made in advance and reheated before serving.

This menu also works well on a buffet, as it can be served hot, cold, or at room temperature.

PATATINE

4 large russet potatoes
¼ cup olive oil
2 sprigs rosemary, leaves only
Salt and freshly ground pepper

1. Peel the potatoes and cut into ½-inch cubes. (If cutting in advance, hold in a bowl of cold water to prevent discoloration.)
2. Bring a large pot of water to a boil. Add the potatoes and return to a boil. Cook for 5 minutes. Drain and pat the potatoes dry with paper towels.
3. In a large nonstick skillet, heat the oil over high heat. Add the potatoes, in a single layer, and cook until golden brown on one side, about 10 minutes. Sprinkle with the rosemary, then turn and cook about 10 minutes longer, until evenly golden and crisp. Season with the salt and pepper to taste and serve.

SERVES 8

Note: The potatoes may be prepared in advance and reheated uncovered in a hot oven on a baking sheet.

SAUTÉED SPINACH *and* GARLIC

¼ cup olive oil
6 garlic cloves, thinly sliced
4 bunches spinach, stems trimmed, rinsed and dried
Salt and freshly ground pepper

1. In a large pan, heat the olive oil over medium-high heat. Sauté the garlic until golden, being careful not to burn.
2. Add the spinach, turn the heat to high, and cook, stirring and tossing, until wilted, about 5 minutes. Season with salt and pepper to taste. Drain any excess liquid and serve.

SERVES 8

FLOURLESS ALMOND TORTE *with* CHOCOLATE COFFEE SAUCE

2½ cups blanched almonds
1½ cups sugar
7 eggs
Dash of salt
½ teaspoon vanilla extract
Grated peel of 1 lemon
Powdered sugar
Chocolate Coffee Sauce (see following recipe)

1. Preheat the oven to 350 degrees.
2. Butter a 10-inch springform pan, then line it with wax paper. Butter the paper.
3. In a food processor, combine the almonds and sugar. Process until the almonds are finely ground. Set aside.
4. Separate 4 of the eggs, placing the yolks in a large bowl and reserving the whites. Add the remaining eggs, along with the salt, vanilla, and lemon peel to the bowl with the yolks and beat to blend well. Gradually beat in the almond mixture. The mixture will become very thick.
5. In a large bowl of an electric mixer, beat the reserved egg whites until they are stiff but not dry. Gently fold them into the batter, a third at a time.
6. Transfer the batter to the prepared pan and bake for 45 minutes, or until a wooden toothpick inserted near the center of the cake comes out clean.
7. Cool the cake in the pan on a wire rack. Remove the sides of the pan and sprinkle the cake with powdered sugar. Serve with Chocolate Coffee Sauce.

SERVES 8

Note: Incidentally, this is a great cake to serve for Passover.

CHOCOLATE COFFEE SAUCE

1¼ cups milk
1 vanilla bean, split
1 tablespoon unsweetened cocoa powder
1 tablespoon finely ground coffee
2 egg yolks
2 tablespoons sugar

1. In a small saucepan, bring the milk with the vanilla bean to a simmer over medium heat. Reduce the heat.
2. Transfer 2 tablespoons of the hot milk to a bowl and stir in the cocoa powder and the coffee. Set aside.
3. In another bowl, beat the yolks and the sugar until light. Set aside.
4. Remove the vanilla bean and scrape the insides into the milk. Discard the bean.
5. Stir a small amount of the hot milk into the yolk mixture, then stir the yolk mixture along with the cocoa mixture into the pan of milk. Cook, stirring constantly, over medium heat until the sauce begins to thicken and coats the back of a metal spoon, about 10 minutes. Transfer to a serving bowl and stir for 1 minute to cool slightly.

DEMITASSE

Translated from French, demitasse means "half cup." For a rich flavor, use half espresso and half regular coffee. This works for both regular and decaffeinated.

A JAPANESE COUNTRY DINNER

SPICY *crab rolls*

YELLOWTAIL TARTAR IN *cucumber parcels*

tuna sashimi SALAD WITH SHALLOT SESAME DRESSING

EGGPLANT AND TOFU *tempura* IN MISO BROTH

litchi ICE CREAM

Serves: 8

Range of Difficulty: Easy

Preparation Time: 2 hours

I ALWAYS EQUATE FOOD TO THEATER.

When the curtain goes up, what happens in the first five minutes sets the mood of the evening. In the same way, the first course at a dinner party needs to set the mood.

The sleek black and white dining room of our host was an appropriate setting for this Japanese menu. Everything in the home is black or white; even the guests, caught up in the aura of the occasion, wore only black or white.

The only splash of color was on the table, where five colorful Siamese fighting fish swam in small acrylic boxes filled with water. In between the boxes, I nestled tea lights (votives candles with a small metal base) in a pool of coarse kosher salt. (Alternatively, this centerpiece can be created without the fighting fish and look just as mesmerizing by using twice as many candles.)

Each place was set in the minimalist manner—with a single bowl, glass, napkin, and a pair of chopsticks. The only pattern on the table was the black and white plaid of the linen napkins.

Arriving guests gathered in the kitchen to nibble spicy crab rolls and watch me prepare the first of our many courses. Rather than serving one main course, we chose to offer a variety of foods, each with a unique flavor and texture.

When seated at the table, we poured chilled sake and served a terrine of yellowtail tartar wrapped in thin cucumber slices and garnished with salmon roe and a Japanese red pepper mixture called shichimi. Accompanied by triangles of homemade melba toast, it was beautiful, tasty, and easy to make.

Our third course was inspired by chef Nobu Matsuhisa of Los Angeles and New York. Bold flavors and crisp textures unfolded in a tuna sashimi salad with baby greens, spicy daikon radishes, and a dressing of shallot, mirin, and sesame sauce.

A light, lacy eggplant and tofu tempura followed. We served it in a small bowl to savor both the broth and the tempura. Both Chardonnay and chilled sake were poured with this course.

And, to conclude: rich, homemade litchi ice cream, which can be made with fresh fruit in season or canned fruit all year round.

THE MANY SIDES OF SAKE

This pale yellow, slightly sweet wine made from fermented rice is the national alcoholic drink of Japan. Although traditionally served warm in small porcelain cups, different types of sake are now coming into their own as a chilled apéritif.

Cold sake is usually poured into a wooden box and placed on a small plate. The box is filled to the brim. Salt can be placed on the edge of the wooden box, if desired.

FOR PERFECT SUSHI RICE

Don't be intimidated by the idea of making sushi. It's a lot easier than you think! The secret is in preparing the rice with a touch of sugar and rice wine vinegar.

In some traditional Japanese cookbooks, the recipe for sushi or vinegared rice calls for a helper to stand by and fan the rice as the vinegar is added, in order to cool the rice to room temperature as soon as possible.

Follow this basic formula:

SUSHI RICE

- 1½ cups premium medium-grain Japanese rice
- 2 cups water
- 2 tablespoons rice vinegar
- 1 tablespoon sugar
- ½ teaspoon salt

1. In a medium saucepan, cook the rice with the water according to package directions. Remove from heat, cover, and set aside for 15 minutes.
2. In a small bowl, make a dressing by whisking the vinegar with the sugar and salt.
3. Transfer the rice to a wood, glass, or ceramic bowl. Gradually pour in the dressing and stir to coat evenly.

SPICY CRAB ROLLS

- ½ teaspoon shichimi togarashi (red pepper mix)
- ⅓ cup mayonnaise
- 3 tablespoons soy sauce for dipping
- 1 teaspoon wasabi (Japanese horseradish)
- 6 nori sheets
- 1½ cups cooked Sushi Rice (see sidebar)
- ½ pound lump crab meat, picked over and divided into 6 equal portions
- 1 ounce Japanese tobiko (cod roe), divided into 6 equal portions
- ½ ripe avocado, peeled and cut into strips

1. In a small bowl, combine togarashi and mayonnaise. Set aside.
2. In a small bowl, combine soy sauce and ½ teaspoon wasabi. Mix well and set aside.
3. On a flat surface, lay out the nori sheets.
4. Spread a sixth of the rice over each nori sheet, leaving a 2-inch edge along one length of each sheet.
5. Using your index fingers, make a hollow indentation in the center of the rice running the length of the nori sheet. Spread a small amount of wasabi, a portion of the crab, a generous amount of togarashi mayonnaise, and a small amount of tobiko in the indentation.
6. Place two strips of avocado across each nori sheet in a lengthwise direction.
7. Roll the nori sheets lengthwise into sausages.
8. Serve with soy and wasabi mixture.

MAKES 24 PIECES

TUNA SASHIMI SALAD *with* SHALLOT SESAME DRESSING

- ½ cup minced shallots
- 1 teaspoon freshly ground pepper
- 2 teaspoons dry mustard
- ¼ cup mirin
- 2 tablespoons soy sauce
- ½ cup sesame oil
- ½ cup vegetable oil
- 1 pound mixed salad greens
- 2 pounds tuna sashimi, cut into 32 pieces

1. In a mixing bowl, whisk the shallots with the pepper, mustard, mirin, and soy sauce. Gradually whisk in the sesame and vegetable oils.
2. In a large bowl, pour all but ¼ cup of the dressing over the salad and toss lightly. Mound the salad high in the center of 8 serving plates. Garnish each with 4 slices of sashimi and spoon the remaining dressing alongside the tuna. Serve immediately.

SERVES 8

TEMPURA

Many foods lend themselves to the tempura treatment. You can experiment with the following:

- shrimp
- potato
- sweet potato
- broccoli
- green onions
- asparagus
- carrots
- onions

Turn tuna sashimi into an elegant entrée with mixed salad greens and a Japanese-style dressing.

YELLOWTAIL TARTAR *in* CUCUMBER PARCELS

1 pound finely chopped raw yellowtail fillet
2 tablespoons diced cornichons
2 tablespoons capers, chopped
2 tablespoons Dijon mustard
2 tablespoons lemon juice
2 tablespoons chopped fresh chives
2 tablespoons chopped fresh parsley
½ cup minced red onion
2 dashes of hot pepper sauce
Salt and freshly ground pepper
1 teaspoon vegetable oil
1 English cucumber, sliced paper thin across width
1 teaspoon shichimi togarashi (red pepper mix)
1 teaspoon hot sesame oil
2 tablespoons salmon roe

1. In a bowl, combine the tuna, cornichons, capers, mustard, lemon juice, chives, 1 tablespoon parsley, red onion, and hot pepper sauce by hand. Season with salt and pepper to taste.
2. Lightly oil eight ½-cup ramekins. Line the bottoms and sides of the ramekins with the cucumber slices, allowing the side slices to overhang. Spread 2½ tablespoons of the fish mixture over the cucumbers in each ramekin and fold the slices over to enclose. Refrigerate 1 hour.
3. To serve, invert onto the center of each serving plate. Sprinkle on the shichimi togarashi, drizzle with hot sesame oil, and sprinkle with the remaining parsley and roe.

SERVES 8

EGGPLANT *and* TOFU TEMPURA *in* MISO BROTH

8 cups dashi
2½ cups soy sauce
2½ cups sweet mirin
2 cups peanut oil
2 egg yolks
1¾ to 2 cups water
2 cups all-purpose flour
8 Japanese eggplants, cut into lengthwise fans
1 pound firm tofu, cut into 2-inch cubes
2 2-inch lengths of fresh ginger, julienned
2 green onions, trimmed and thinly sliced

1. Prepare dashi with soy and mirin according to package directions. Keep warm.
2. In a deep pot or fryer, heat the peanut oil to 350 degrees.
3. Meanwhile, in a mixing bowl, blend the egg yolks and water. (For a thinner batter, which I prefer, use more water, or vice versa.) Sift the flour into the batter and stir gently to combine.
4. Dip the eggplants into the batter, one at a time, shaking off excess. Fry in the hot oil until golden brown and drain on paper towels. Repeat with the tofu pieces.
5. In each of the 8 serving bowls, place 5 small pieces of ginger. Ladle in 1 cup of the hot broth, an eggplant, and tofu. Sprinkle with the green onions and serve hot.

SERVES 8

LITCHI ICE CREAM

1 cup milk
1 cup heavy cream
1 whole vanilla bean, split
3 egg yolks
⅓ cup superfine sugar
3 11-ounce cans peeled whole litchis in heavy syrup

1. In a medium saucepan, combine the milk, cream, and vanilla bean. Place the mixture over medium heat and bring to a simmer.
2. Meanwhile, in a small bowl, whisk the egg yolks with the sugar until light. Pour a small amount of the hot milk into the yolk mixture, whisking constantly. Return the mixture to the milk in the pan and cook and stir until the mixture thickens enough to coat the back of a metal spoon. Transfer the mixture to a bowl.
3. Drain the litchis, reserving ¼ cup of the syrup. Place the litchis and the reserved syrup in a food processor or a blender. Process for 15 seconds, or until the mixture is puréed but with some pieces remaining. Stir the purée into the cream mixture until well blended. Chill.
4. Transfer the chilled mixture to an ice cream maker and freeze according to the manufacturer's directions.
5. Transfer to a chilled bowl and plate in the freezer.

SERVES 8

Note: The ice cream mixture can be prepared through step 3 up to 2 days ahead and kept refrigerated.

AN ELEGANT DINNER

caviar IN THE EGG

SALAD WITH *foie gras and asparagus tips*
WITH ARTICHOKE TRUFFLE VINAIGRETTE

breast of duck WITH GRAPE SAUCE

MEDLEY OF *rice*

CARROT *flan*

SAUTÉED *zucchini and turnips*

gratin of berries
WITH A CHAMPAGNE SABAYON SAUCE

Serves: 12

Range of Difficulty: Moderate

Preparation: 4 hours

In honor of
Robin Schwartz's Birthday
Menu
Beluga Caviar in the Egg
Frozen Ketel One Vodka
Salad of Mache
with Asparagus and Foie Gras
Chassagne Montrachet, Colin 1990
Duck Breast with Grape Sauce
Wild Rice
Carrot Flan
Baby Turnips & Zucchini
Pommard "Vignots" Domaine Leroy 1985
Gratiné of Mixed Berries
Coffee
15 September 1995
Mrs. Margo Barbakow

It's important to gear the evening to your guests.

Sometimes that means a formal dinner in the dining room, other times, it's a low-key gathering around a casual buffet in the kitchen. Most important, everyone should be made to feel special.

Inspiration can come from the most unpredictable sources. One evening, while watching a science documentary, I learned how to cut the top of the egg off with a surgeon's precision. An empty, upside-down spice bottle was placed over the small end of the raw egg, and the end of the bottle was whacked with a wooden spoon. The vibration traveled down the bottle and cut the fragile shell perfectly. A sharp knife is used to pry off the top of the egg shell. That demonstration inspired me to make French chef George Blanc's famous caviar in the egg dish as a first course for an important dinner. A simple white egg shell, perched in a collectible egg cup and filled with softly creamed eggs and glistening beluga caviar, was just the beginning of this very special evening.

Family heirlooms — vases, pitchers, and bowls — were packed tightly with flowers from the garden, and placed around the house. Two round dining tables were set with yellow damask tablecloths and pomegranate-colored overlays. Antique china fruit compotes were tightly packed with an abundance of multicolored roses to decorate the center of the tables. Dimmed lights and soft candlelight created the perfect stage for such a tasty and sophisticated menu.

When guests were seated, the caviar-crowned eggs and melba toast were waiting at the table. Vodka, from the freezer, was poured as the accompaniment.

The second course of simple mâche lettuce was topped with pâté de foie gras, plump tomatoes, and asparagus tips. The salad was dressed in light vinaigrette infused with truffle oil and a puréed artichoke heart.

Breast of duck, quickly sautéed and dressed with a refined bordelaise sauce, was served as the main course, accompanied by mixed rices, carrot flan, sautéed zucchini, and potato rounds. The 1985 Pommard wine served was full bodied and rich.

Fresh berries smothered in a rich Champagne sabayon sauce and run under the broiler until golden made a light and delicious dessert.

As extravagant as it was, dinner was prepared in four hours, marketing excluded, and served without hired help.

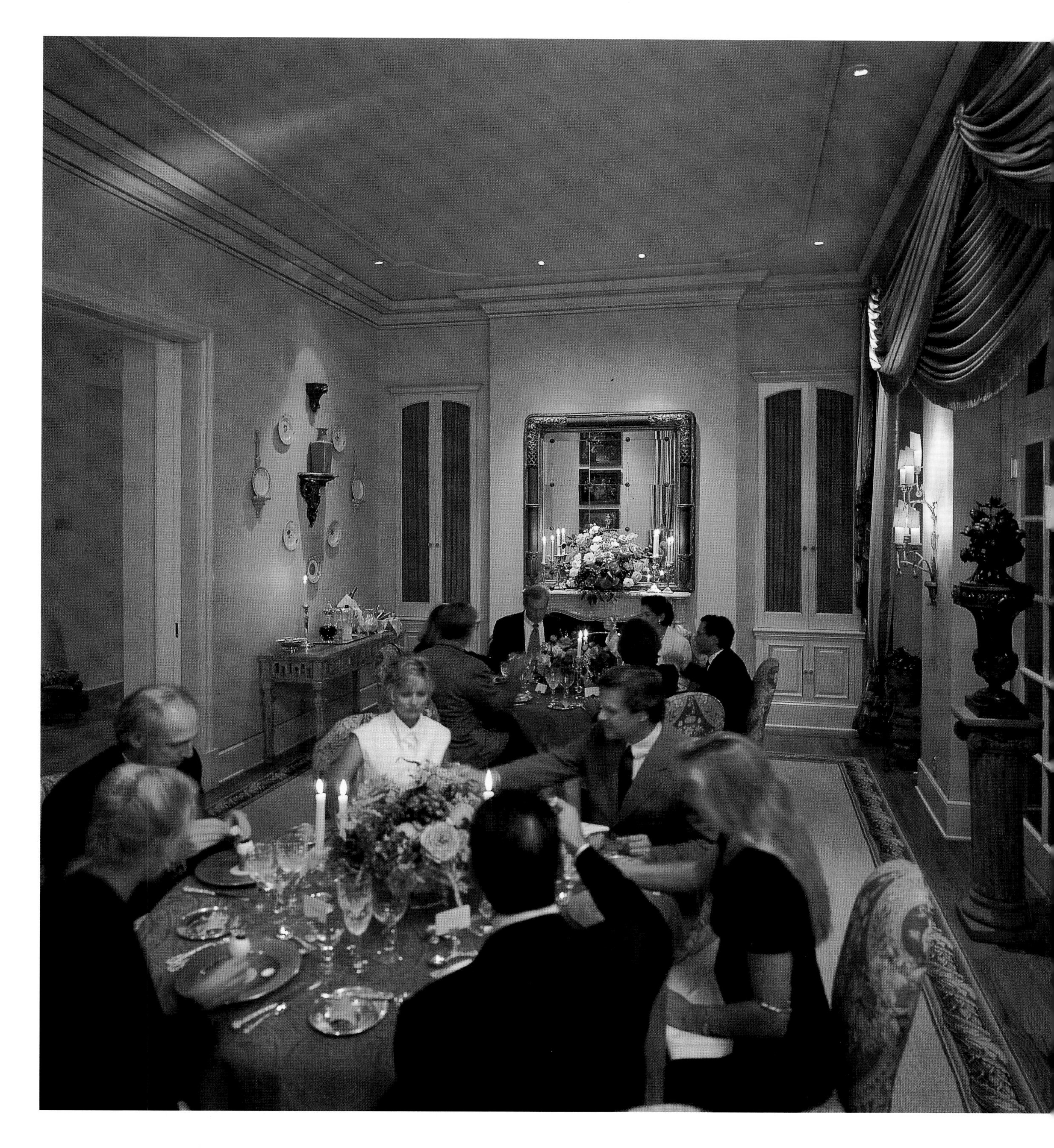

Place Cards

It is appropriate to use place cards when having more than six guests. They balance the energy of the table and alleviate anxiety from wondering "Where shall I sit?"

FOR PERFECT MELBA TOAST

1. Toast the white bread until lightly browned. Remove the crusts and with a serrated knife, cut through each slice horizontally. Then cut each slice into triangles.

2. Place the triangles on a cookie sheet, white side up, and broil until lightly browned.

3. Remove from the oven and set on a tray with the caviar and a small spoon.

Note: Watch the broiler closely, or the toast will burn within seconds. Toast can be made up to 3 days in advance and stored in an airtight container.

Fine caviar is an indulgence to treat with respect. That means a bed of crushed ice to keep it chilled, a mother-of-pearl spoon for serving, and melba toast.

CAVIAR *in the* EGG

12 jumbo eggs
36 pieces melba toast (see sidebar)
White pepper
¼ cup cream
2 tablespoons minced shallots
8 ounces beluga, osetra, or sevruga caviar

1. Remove the tops of the eggs by holding the egg, point side up, in the palm of your hand. Place an inverted empty spice bottle on top of pointed side of the egg and secure with thumb and forefinger of the same hand. With other hand, using a heavy wooden spoon, hit the bottom of the bottle. The shock will create a crack around the top of the egg. Use a sharp knife to pry open. Pour egg into a metal bowl. Set aside.
2. Rinse the egg shells and their tops under warm running water. Turn them upside down on paper towels to dry, then set the bottom shells in egg cups. Set the egg cups on small serving plates. Arrange 3 toast points around the base of each egg cup.
3. Whisk the reserved eggs in the bowl, then strain through a fine sieve. Season the eggs with white pepper to taste.
4. Just before serving, cook eggs in a metal pot over low-medium heat. Whisk continuously. When the eggs start to thicken, whisk in the cream and the shallots. Cook, whisking constantly, 30 seconds longer. Remove from heat and immediately transfer to a glass bowl.
5. Spoon some of the egg mixture in each egg shell bottom, filling about two-thirds full. Spoon the caviar on top until full, then cover each egg with the top portion of its shell.

SERVES 12

Note: This dish can be prepared 20 minutes in advance. Chopped smoked salmon can be substituted for the caviar.

JOYS OF CAVIAR

What better way to splurge than a mother-of-pearl or bone spoon heaped with glistening caviar. (Silver spoons tend to tarnish and the metal alloys affect the taste of the caviar.)

The word ***caviar*** refers only to the roe of sturgeon typically found in the Caspian Sea. Other fish roe can be called caviar legally, but only if the word is preceded by the name of the other fish, such as "salmon caviar" or "lumpfish caviar."

There are three main varieties of caviar, beluga, osetra, and sevruga. A fourth, rare caviar is sterlet.

Beluga. This is the largest species of sturgeon, yielding the largest "berries," which is a term for caviar grains or eggs. Because of its handsome appearance and very mild flavor, beluga is the most highly prized of all caviar. The color is designated by zeros with 3 zeros (000) for the lightest gray, 00 for medium gray, and 0 for the darkest black berries.

Osetra. This is the second largest species of sturgeon, producing slightly smaller caviar berries. The color of osetra ranges from golden yellow to brown; the flavor is more intense than beluga.

Sevruga. The third main species of sturgeon yields the smallest eggs with the strongest flavor. The color is usually dark gray to black. Many connoisseurs consider sevruga the most delicious of all.

Sterlet. From this type of sturgeon comes the legendary "golden caviar," which, historically, was required by law to go to the Shah of Iran if found on the Iranian side of the Caspian Sea, to the Czar if found on the Russian side. The eggs have a firm texture and slightly smoky flavor. Golden Imperial caviar is very rare.

The term ***Malossol,*** when applied to the three main types of caviar, means that the product contains less than 5 percent salt, making it of a higher quality than caviar to which more salt is added. However, the limited salt also makes it more fragile and prone to spoilage.

Accouterments. Caviar aficionados often opt to eat the elegant little eggs unadorned, straight from the tin, with a mother-of-pearl spoon.

However, for more fanfare, you can serve caviar with finely chopped hard-boiled egg white, egg yolk, finely chopped onion, lemon wedges, crème fraîche, and melba toast.

VEAL STOCK

Veal stock is a key ingredient in making delicious sauces and gravy. Using it as a base or adding it to an existing sauce will always produce a rich, intensely flavored sauce.

You may purchase one of the high-quality frozen stock products available in gourmet markets, but, if you have time to make it from scratch, make a gallon and freeze it in small containers for future dinners.

Ask your butcher for 6 pounds of veal bones. Roast the bones in a 450-degree oven until they are dark brown-black, about 15 to 20 minutes. Remove from the roasting pan and place in a large heavy pot, along with 1 pound each of coarsely chopped onion, carrot, celery, and washed leeks. Cover with water and bring to a boil. Reduce heat and simmer for 6 hours, continually skimming the grease off the stock with a large soup ladle. After 6 hours, pour the stock through a large sieve; discard the solids. Return the stock to a clean pot and allow to reduce by a further 50 percent, for a more concentrated stock.

SALAD *with* FOIE GRAS *and* ASPARAGUS TIPS *with* ARTICHOKE TRUFFLE VINAIGRETTE

- ¼ cup lemon juice
- 1 tablespoon Dijon mustard
- 1 cooked artichoke bottom, quartered
- ½ cup truffle oil
- ½ cup olive oil
- Salt and freshly ground pepper
- 1 pound mâche, frisée, or butter lettuce
- 12 slices canned foie gras (14 to 18 ounces)
- 12 cherry tomatoes, halved
- 24 asparagus tips, blanched
- 2 tablespoons chopped chives

1. In a blender, combine the lemon juice, mustard, and artichoke bottom. Process until smooth. With the blender running, gradually drizzle in the oils in a thin stream. Season to taste with salt and pepper.
2. In a large bowl, toss the mâche with the dressing. Pile the salad onto 12 individual salad plates. Top each salad with a slice of foie gras. Place the cherry tomatoes and asparagus tips decoratively on front of plate. Sprinkle with the chives.

SERVES 12

BREAST *of* DUCK *with* GRAPE SAUCE

- 6 lean boneless Mallard duck breasts
- Salt and freshly ground pepper
- 2 tablespoons olive oil
- 1 tablespoon chopped shallots
- ¾ cup red wine, such as bordeaux
- 4 cups veal stock (see sidebar)
- 2 thyme sprigs
- 1 bay leaf
- 8 peppercorns
- 1 cup green seedless grapes, peeled and halved
- 3 tablespoons of honey
- 2 tablespoons balsamic vinegar
- 1 tablespoon butter

1. Preheat the oven to 450 degrees.
2. With a sharp knife, remove excess fat from the duck and make one-inch incisions in the skin of each. Season the duck breasts with salt and pepper to taste.
3. Heat the olive oil in a large skillet over medium-high heat. Add the breasts to the pan, a few at a time, skin side down, and cook, turning once, until browned, about 5 minutes on each side. Remove from the pan and arrange on a rack in a roasting pan. Set aside.
4. In the same saucepan sauté the shallots until glassy, about 3 minutes.
5. Deglaze with ¼ cup of the wine.
6. Add the veal stock, thyme, bay leaf, and peppercorns and reduce by 50 percent to create a demi-glace. Set aside.
7. In a separate pot cook half the grapes with the honey over low heat. When the honey starts to thicken, add the balsamic vinegar and cook for an additional minute.
8. Add the remaining wine to the grape mixture and reduce for 10 to 15 minutes until syrupy.
9. Roast the breasts for 10 to 15 minutes, until medium-rare.
10. Add the demi-glace to the grape mixture and reduce for 5 to 10 minutes. Strain the sauce through a fine sieve. Add the butter and the remaining grapes. Set aside and keep warm.
11. Slice the roasted duck breasts on the diagonal, slightly less than ¼ inch thick. Arrange 6 slices of breast, fanlike, on a warm dinner plate. Spoon some of the warm sauce with the grapes over each serving.

SERVES 12

Note: Deglazing captures the natural glaze that accumulates in a cooking pan. After sautéing meats or poultry, remove the excess fat from the pan, then add a small amount of stock or wine, or in this instance bordeaux. Stir and scrape to loosen the brown bits of food at the bottom of the pan. Use as a sauce or add to prepared sauces for additional flavor.

MEDLEY *of* RICE

- 3 tablespoons butter
- 1 small onion, chopped
- 1 large carrot, peeled and finely diced
- 1 small red bell pepper, seeded and finely diced
- 2½ cups store-bought blended rice
- 2 cups chicken stock
- 2 cups vegetable stock
- Salt and freshly ground pepper

1. Heat the butter in a heavy saucepan over medium heat. Sauté the onions until soft. Add the carrot, pepper, and rice and stir well to evenly coat.
2. Pour in the chicken and vegetable stocks and season with salt and pepper. Bring to a boil, reduce to a simmer, and cook for 25 to 30 minutes. Let sit for 10 minutes, then fluff with a fork.

SERVES 12

CARROT FLAN

- 2 tablespoons butter, plus butter for greasing
- 1 onion, diced
- 3 cups chopped cleaned carrots (about 1 pound)
- 6 eggs
- 2 egg yolks
- ½ teaspoon salt
- ½ teaspoon white pepper
- ¼ teaspoon freshly grated nutmeg
- 2 tablespoons sugar
- Grated zest of ½ orange
- Juice of ½ orange
- 2 cups hot milk

1. Preheat the oven to 400 degrees.
2. Coat twelve ½-cup ovenproof ramekins with butter.
3. Melt the 2 tablespoons of butter in a small skillet over medium heat. Cook the onion until soft and set aside.
4. Bring a medium pot of water to a boil. Add the carrots and cook until tender, about 10 minutes. Drain and transfer to a food processor or blender.
5. Add the onions to the carrots and purée until smooth. Add the eggs, yolks, salt, pepper, nutmeg, sugar, zest, and juice and process to combine. Pour in the hot milk and the melted butter. Process to combine.
6. Pour into the prepared ramekins and transfer to a roasting pan. Pour hot water halfway up the sides of the ramekins. Bake about 40 minutes, until a knife inserted in the center comes out clean.
7. Cool on a rack, then invert directly onto dinner plates.

SERVES 12

SAUTÉED ZUCCHINI *and* TURNIPS

- 16 thin zucchini
- 4 turnips
- 2 tablespoons butter
- Salt and freshly ground pepper

1. With a paring knife, cut each zucchini lengthwise into thirds and each turnip into 8 pieces. Trim each zucchini and turnip into a rectangle with sharp edges and then trim the corners to form octagonal or rounded corners.
2. Melt the butter in a large skillet over medium-high heat. Sauté the zucchini and turnips until golden brown. Season with salt and pepper to taste and serve.

SERVES 12

GRATIN *of* BERRIES *with a* CHAMPAGNE SABAYON SAUCE

- 6 cups assorted berries such as strawberries, raspberries, blueberries, and blackberries
- 1 tablespoon chopped mint
- 8 egg yolks
- ½ cup sugar
- 1 cup Champagne or other sparkling wine

1. In a bowl, combine the berries and the mint. Divide the mixture evenly among twelve 4-ounce ramekins. Arrange the ramekins on a baking sheet and set aside.
2. Preheat the broiler with the rack positioned about 6 inches from the heat source.
3. In the top of a double boiler or a metal bowl, whisk the yolks and the sugar to blend well. Set the mixture over simmering water and whisk constantly until it begins to thicken. Quickly whisk in the Champagne in small amounts, then continue to whisk until the mixture is light and frothy. Spoon the mixture over the berries in the ramekins and immediately place under the broiler until they are browned and bubbly, watching carefully to prevent burning. Serve immediately.

SERVES 12

barbecued lamb kabobs WITH YOGURT AND CUMIN SAUCE

MARINATED *olives*

TOASTED MEDITERRANEAN *almonds*

baba ghanoush

hummus

CHICKEN *couscous* WITH MERGUEZ SAUSAGE AND VEGETABLES

bastilla

date nut cigar
WITH TOASTED SESAME SEEDS

Serves 10

Range of Difficulty: Moderate

Preparation Time: 3 hours

ona

The perfect party is an ensemble of

food, drink, music, decor, and energy. No one outshines the other. Guests revel in the comfort created by this balance.

Morocco travels so well! This captivating land of souks and savory foods lends itself to theme parties. Re-creating Morocco will transform an ordinary backyard into a land of vibrant colors, shimmering fabrics, and tinkling music.

It's hard to say who had more fun creating this party, the hostess or my staff and me. We started to spin the fantasy a few weeks in advance. When guests arrived, they were given gift bags and directions to a room for changing. Laughter filled the house as guests, dressed in crisp white cotton pajamas, moved to the living room, where their heads were wrapped in white turbans and their eyes decorated with black kajal pencil. With Pounana shooters of vodka, cranberry, and peach schnapps in hand, guests ventured out to the backyard and into a North African fantasyland.

The backyard was dominated by a tent, created from a canopy frame draped with yard upon yard of billowy white chiffon. The "dining table" was a simple rented banquet table, placed on milk crates, and covered with assorted printed fabrics, huge leaves, lustrous jewel tone napkins, and finally a shiny gold lamé runner. Places were set with gilded goblets and gourds painted as place cards. A spectacular arrangement of fresh fruits and flowers, chili peppers, nuts, and herbs served as the centerpiece.

The menu was a continuation of our theme. For appetizers, grilled kabobs of marinated lamb were served with a cumin-scented yogurt dipping sauce. As guests were seated, marinated olives and toasted almonds were placed on the tables. And as soon as everyone was comfortable a grand platter of couscous, chicken, vegetables, and merguez sausage was served, along with bastilla of chicken and cashews and platters of hummus and baba ghanoush.

For dessert, we offered a crispy pastry filled with dates and apricots commonly known as Cigar.

A robust, spicy red wine, such as a full-bodied Chateauneuf-du-Pape or a lusty Gigondas, marries well with this fare.

MUSIC SELECTIONS

Moroccan background music can be purchased in the International or World department of your local music store. For our evening we chose Om Kal Soum.

HOW TO CREATE A TENT

You will need to rent a tent frame from your local party rental company. Tell them how many people you'll be seating under the tent and ask them to cover the frame in white Velon. Purchase a sheer white fabric (at least 118 inches wide), to drape loosely over the tent poles. Use approximately three feet of fabric for every one foot of tent to be covered. Let the excess fabric hang down the poles, catching it with satin ribbon, strips of gold lamé, or a strong adhesive tape.

Alternatively, for a small group of people, drape some colorful ethnic fabric over a market umbrella, and use a similar piece for a tablecloth.

CREATING A
MOROCCAN MOOD

You can serve the same menu on the floor of your living room, den, or dining room. Use area rugs instead of tablecloths, centerpiece of fruits abundantly placed on the table, and ordinary pillows on the ground for sitting. Dim the lights, burn some incense, and light lots of candles.

Have your guests come dressed in white caftans or Moroccan garb, and have the fabric for making turbans waiting. It's fun and creates an energy that jump-starts the party.

FINGER BOWLS

Whenever a finger bowl is provided for guests before or after a meal, splash a few drops of rose water, lemon juice, or almond extract into the water for a fresh fragrance.

KABOBS

Just about anything can be put on a kabob for a dramatic presentation and greater ease in eating. Half-kabobs are the perfect finger food for cocktail parties.

- Consider chicken with a saté dipping sauce; shrimp with a tandoori marinade or fruit glaze; or vegetables basted with extra virgin olive oil and fresh minced herbs.
- Mix and match on each skewer: shrimp with plantains or red bell pepper, chicken with sweet onions, and lamb with cherry tomatoes and bay leaves.
- As a change to the traditional metal skewer, use bamboo skewers, rosemary branches, or lemon grass.

COUSCOUS

A thousand tiny bits of semolina, each separate, distinct, and cooked properly, having the light texture of new-fallen snow— that's couscous, the grain.

When the grain is cooked in a *cous-coussière*, above a fragrant, bubbling meat stew, then heaped on a serving platter with the stew on top, it becomes couscous, the crowning glory of Moroccan cuisine.

Typically, diners use pita bread instead of silverware to scoop the couscous from its central platter.

Bastilla is a traditional Moroccan chicken dish wrapped in phyllo pastry and dusted with powdered sugar.

CHICKEN COUSCOUS *with* MERGUEZ SAUSAGE *and* VEGETABLES

6 pounds chicken pieces
3 onions, quartered
1½ tablespoons ground cumin
1½ teaspoons ground cardamom
1½ teaspoons ground coriander
1 teaspoon ground cloves
½ teaspoon cayenne
1 teaspoon ground ginger
½ teaspoon saffron threads
10 whole black peppercorns
5 large carrots, peeled and cut into 1½-inch chunks
3 yellow squash, cut into 1½ -inch chunks
3 turnips, cut into large chunks
½ head cabbage, cut into chunks
½ bunch celery, cut into 1½-inch pieces (about 2 cups)
4 medium zucchini, cut into 1½-inch chunks (about 3 cups)
1 teaspoon harissa sauce
3 16-ounce packages couscous
20 hot grilled merguez sausages

1. Place the chicken pieces and the onion quarters in a large pot. Add the cumin, coriander, cloves, cayenne, ginger, saffron, and peppercorns. Add enough water to cover and bring to a boil. Reduce the heat and simmer for 30 minutes.
2. Skim the fat from the surface. Add the carrots, yellow squash, turnips, cabbage, and celery and continue to cook for 15 minutes. Add the zucchini and simmer for 5 minutes longer. With a slotted spoon, transfer the chicken and vegetables to a large dish and keep warm.
3. Transfer 2 cups of the broth to a small saucepan set over medium heat. (Reserve remaining broth for the couscous.) Bring to a boil. Stir in the harissa sauce, then reduce the heat and simmer, stirring occasionally, for 10 minutes, or until slightly reduced.
4. Prepare the couscous according to package directions, using the reserved cooking broth for liquid called for.
5. To serve, spoon the couscous onto a large serving platter. Arrange the chicken, vegetables, and sausage over the couscous and drizzle with some of the sauce, or pass sauce at the table.

SERVES 10

Note: Harissa sauce and merguez sausages are available at Middle Eastern markets. Chicken and vegetables, omitting the zucchini, can be made 1 day ahead. Store in the refrigerator. Before serving, return the chicken and vegetables to a boil, add the zucchini, and continue as above.

DATE NUT CIGAR *with* TOASTED SESAME SEEDS

1 pound dates, coarsely chopped
1 cup walnuts, toasted and finely chopped
½ cup apricot preserves
2 teaspoons cinnamon
1 pound fresh or frozen and thawed phyllo pastry
¼ pound butter, melted
Syrup (see following recipe)
2 tablespoons sesame seeds, toasted

1. Preheat the oven to 350 degrees.
2. In a small bowl, combine the dates, walnuts, apricot preserves, and cinnamon. Divide the mixture equally into 40 balls. With your hands, roll each ball into a log about 2 inches long. Set the logs aside.
3. Unfold the phyllo pastry on a large work surface and cover it with a barely damp kitchen towel.
4. Keeping the unused pastry covered with the towel while you work, carefully remove the top pastry sheet and brush it generously on one side with some of the butter. Set a second pastry sheet on top of the first sheet, brushing again with some of the butter. Repeat with a third sheet and more butter. Using a sharp knife, cut the stacked and buttered sheets into 8 rectangles about 7 × 4½ inches each.
5. To form "cigars," place a date log parallel to a long side of each phyllo rectangle, setting it about 1 inch from an edge and folding that edge over the log. Fold both of the short sides in over the ends of the log, then roll the log up. As you work, place the rolled pastries on a baking sheet and brush with more of the melted butter. Bake the pastries for about 30 minutes, or until golden. Remove the pastries from the oven and cool for 1 minute on baking sheets. Transfer the pastries to a wire rack and brush or drizzle them lightly with the hot syrup and sprinkle with some of the sesame seeds. Cool.
6. Repeat the process with the remaining phyllo sheets, date logs, syrup, and sesame seeds.

MAKES 40

SYRUP

⅔ cup sugar
⅔ cup water
⅔ cup honey
1 teaspoon cinnamon

1. In a heavy small saucepan, combine the sugar and water. Bring to a boil, then reduce the heat and simmer for 5 minutes.
2. Stir in the honey and the cinnamon and keep warm.

Date nut cigars can be made in advance.

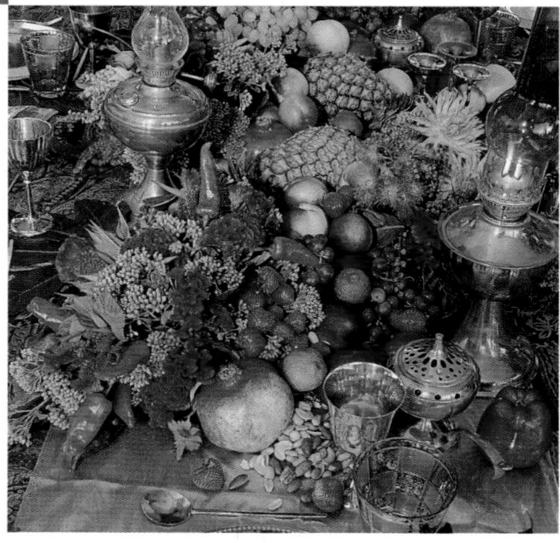

Using fruits, flowers, unscented candles, and lanterns, a stunning centerpiece can be made very easily.

A CAPE MALAY BARBECUE

CURRIED *lamb kabobs*

basmati rice WITH TOASTED CASHEWS, RAISINS, AND SULTANAS

SALAD OF *arugula,* ORANGES, AND GRILLED WHITE ONIONS

bulgur mint SALAD

mango CHUTNEY

baked bananas
WITH RUM-SPIKED CRÈME FRAÎCHE

Serves: 10

Range of Difficulty: Easy

Preparation Time: 2 hours

DRESS SHOP

Fires create a warm and cozy ambience.

One can sit in front of the fire with a glass of Armagnac and watch it forever!

You really can't go wrong when you plan a party around a barbecue. The smell of the food cooking and the informality create an easy, relaxed mood. And, of course, anything cooked over an open fire always tastes great!

For me, the pleasures of the grill are multiplied when *sosaties* are part of the menu. These exotic lamb kabobs, created by the Cape Malays on the southernmost tip of Africa, were one of my favorite treats as a child. The spicy marinade is tantalizing and food served on skewers is always fun to eat.

For this late summer barbecue in a friend's secluded garden, kabobs and fragrant basmati rice brought me back to childhood. The rice was tinged with turmeric and flavored with toasted cashews, golden raisins, and sultanas. The side dishes included crispy pappadams and mango chutney.

To balance the spicy curry flavors of the lamb, a salad of peppery arugula with grilled white onions and orange segments was served. The salad was tossed with a tart lime juice vinaigrette and finished with caraway seeds.

The dessert, cooked over cooling coals, was another childhood favorite. Whole bananas, stuffed and coated with softened butter and brown sugar, had been re-wrapped in their peels, tied with string, and placed on the barbecue to cook. When the skin was completely blackened, the bananas were ready to be served with a dollop of crème fraîche.

To make things easy, a casual bar was set in the garden, complete with a zinc tub filled with beer, root beer, sparkling mineral water, and everything needed for an old-fashioned bourbon and ginger ale.

An antique table with an old-fashioned tablecloth was used for the buffet. Napkins made from ordinary blue and white mattress ticking were stacked alongside cobalt blue plates and an assortment of pewter serving pieces from the 1950s.

Flowers were a simple complement to the look of the house. For an easy and casual look, I combined items from the garden with purchased flowers. Cockscomb, wheat berries, and Leucadia were placed in terra-cotta pots.

I like to use as many candles as possible to create an inviting atmosphere.

CURRIED LAMB KABOBS

10 6-inch wooden skewers
2 cups dried apricots
1 cup raisins
1 garlic clove, crushed
6 bay leaves
2 tablespoons brown sugar, packed
1 tablespoon ground coriander
1 tablespoon chopped chutney
2 teaspoons curry powder
2 teaspoons ground ginger
1 teaspoon salt
¾ teaspoon cayenne
2 cups white vinegar
2 large onions, quartered
5 pounds boneless lamb, cut into 1½-inch cubes

1. Prepare a charcoal fire. Soak the skewers in warm water for at least 10 minutes.
2. In a saucepan, combine 1 cup of the apricots with the raisins, garlic, bay leaves, brown sugar, coriander, chutney, curry powder, ginger, salt, cayenne, and vinegar. Bring the mixture to a boil. Reduce the heat and simmer for 5 minutes. Add the remaining apricots and the onions and remove from the heat. Allow the mixture to cool.
3. Place the lamb in a flat baking dish, just big enough to hold the meat in a single layer. Pour the cooled marinade over the lamb. Cover and refrigerate for 24 hours.
4. Remove the mixture from the refrigerator and thread 5 to 6 pieces of the lamb onto the skewers, alternating with some of the onion pieces and the apricots. Set the kabobs aside.
5. Strain marinade and transfer from the dish to a saucepan. Place the pan over high heat and bring to a boil, then reduce the heat and simmer until mixture is reduced slightly.
6. Brush the kabobs with some of the sauce and grill them over hot coals, brushing occasionally with more of the sauce, until the meat is done as desired. Serve immediately with sambals.

SERVES 10

Note: Chicken may be substituted for all or part of the lamb. The kabobs should be made the day before. The longer they marinate, the more tender and tasty they'll be.

BASMATI RICE *with* TOASTED CASHEWS, RAISINS, *and* SULTANAS

2 tablespoons butter
½ cup finely chopped onion
½ teaspoon ground turmeric
2 cups basmati rice, rinsed
2 cups chicken stock
2 cups water
⅓ cup dark raisins
⅓ cup sultanas (golden raisins)
1 cinnamon stick
½ cup toasted cashews

1. In a large saucepan, melt the butter over medium heat. Add the onion and turmeric. Cover the pan and cook until the onion is soft and has a glassy appearance.
2. Stir in the rice, chicken stock, water, raisins, and the cinnamon stick. Bring the mixture to a boil. Cover the pan, reduce the heat, and simmer for about 20 minutes, or until all the liquid has been absorbed. Remove the rice from the heat and remove the cinnamon stick. Stir in the cashews. Serve immediately.

SERVES 10

SALAD *of* ARUGULA, ORANGES, *and* GRILLED WHITE ONIONS

1 tablespoon vegetable oil
3 onions, cut into quarters
1 tablespoon water
5 ounces mixed baby lettuces
4 ounces arugula leaves
Citrus Dressing (page 202)
3 oranges, peeled and cut into segments
2 teaspoons caraway seeds

1. In a skillet, heat the oil over medium-high heat. Add the onions and toss them to coat with the oil. Carefully add the water and continue to cook, stirring frequently, until the onions begin to turn brown on the edges. Set aside to cool slightly.
2. In a large serving bowl, toss the lettuces and arugula with the dressing. Add the orange segments and the onions and toss again to combine. Sprinkle the salad with the caraway seeds.

SERVES 10

SAMBALS

Sambals are the traditional condiments served with Malaysian, Indonesian, and Indian food. They are served as a side dish or condiment to rice and curry and include chutneys (spicy fruit-based spread), achar (sweet or hot pickled relish), pappadams (crispy flatbreads flavored with fennel), and raita (yogurt salad).

BARBECUED PINEAPPLE SANDWICHES

Barbecued pineapple sandwiches are a traditional *sambal* for this dinner. For each guest, place a thin slice of cored pineapple between two pieces of buttered white bread. Cook on the barbecue until golden brown on both sides.

For an appetizer, add thinly sliced ham or cooked bacon to the sandwich, cut into triangles, and serve straight from the grill.

CITRUS DRESSING

3 tablespoons lime juice
2 tablespoons apple cider
1 tablespoon honey
1 tablespoon red wine vinegar
1 tablespoon water
1 tablespoon minced shallots
½ teaspoon dry mustard powder
½ teaspoon salt
½ teaspoon pepper
½ cup vegetable oil

In a small bowl, combine the lime juice, apple cider, honey, vinegar, water, shallots, mustard, salt, and pepper. Slowly whisk in the oil.

BULGUR MINT SALAD

2½ cups bulgur (finely cracked wheat)
3½ cups vegetable stock
2 cups chopped fresh mint
1 cup chopped red onion
½ cup chopped fresh parsley
2 tomatoes, seeded and chopped
½ cup olive oil
½ cup lemon juice
1 tablespoon ground coriander
1 tablespoon ground cumin

1. Place the bulgur in a large bowl. Bring the vegetable stock to a boil and pour over the bulgur. Let stand for 10 minutes, then fluff with a fork.
2. Add the mint, onions, parsley, tomatoes, oil, lemon juice, coriander, and cumin, mix well, and let sit for 30 minutes. Chill.

SERVES 10

MANGO CHUTNEY

4 mangos, peeled, seeded, and cut coarsely
½ cup preserved ginger, chopped
½ cup chopped candied citron
½ cup chopped candied lemon peel
1 cup golden raisins
12 whole cloves
1 cinnamon stick
1½ cups sugar
2 cups apple cider vinegar

1. In a large bowl, combine the mangos, ginger, citron, candied peel, and raisins.
2. Tie the cloves and cinnamon in cheesecloth to enclose.
3. Combine the sugar and vinegar in a medium saucepan. Bring to a boil, and add the mango mixture and cheesecloth sack. Boil for 30 minutes, stirring occasionally. Remove and discard the cheesecloth. Cool and store covered in the refrigerator. Serve with pappadams.

SERVES 10

BAKED BANANAS *with* RUM-SPIKED CRÈME FRAÎCHE

10 large firm bananas
¼ pound butter, melted
½ cup brown sugar, packed
1 cup crème fraîche
2 tablespoons dark rum

1. With a sharp knife, slit the peel open on each banana. Carefully remove peel, reserving it.
2. Roll the bananas in the melted butter and then in the brown sugar. Return each banana to its peel.
3. Using lengths of kitchen string that have been soaked in water, tie each banana. Set aside.
4. In a small bowl, blend the crème fraîche with the rum. Set aside.
5. Grill the bananas over low coals, turning frequently, for 15 minutes, or until heated through.
6. Remove the strings and peels from the bananas and serve them on individual dessert plates with some of the crème fraîche on the side.

SERVES 10

Wrapped in their skins with butter and brown sugar, bananas can be placed over dying coals to be ready for dessert when you are.

Supper After a Show

SALAD OF GREENS WITH *sautéed shrimp,* FETA CHEESE,
AND ASSORTED CRISPIES

chicken fricassee

STEAMED *rice*

chocolate soufflé
WITH FRESH WHIPPED CREAM

Serves: 14

Range of Difficulty: Easy

Preparation Time: 2 hours

I LOVE OFFERING A SPECIAL COCKTAIL AT PARTIES.

Perhaps it's a Retro Manhattan, a sophisticated martini, or an extraordinary bottle of Scotch I've discovered. I set the drinks on a silver tray with a frosty silver ice bucket, starched cloth napkins, and crystal glasses.

Whenever I have plans for the theater or a show, I always enjoy having the group to my home first, for drinks and a light appetizer. After the theater, it's fun to serve dinner and finish the evening with port, a tray of liqueurs, and a dessert or selection of champagne truffles.

For this after-theater supper, I created a menu that was easy, elegant, and light enough for late-night dining. Almost everything was prepared ahead, leaving the hostess just to shake the martinis and join in the critique of the evening's entertainment.

The glass dining table was dressed with silver and crystal candelabra with white tapers. Rather than using cut flowers, we chose an orchid plant to center the table. A black lacquered tray was used to hold the silverware, and for a slightly exotic note, napkins of an Oriental red and black fabric were tucked into a collection of mismatched sterling silver rings. Plates, wineglasses, and wines were also placed on the table.

To start, a salad of bitter greens was arranged on a large serving platter. Around the edges we tucked grilled shrimp, then sprinkled the whole platter with a "confetti" of feta cheese, red onion, and fresh mint. Just before serving, we added a nest of deep-fried julienned vegetables.

Chicken fricassee with spring vegetables served on a bed of rice was our main course. Just before serving, I like to heighten the illusion of spring by folding finely minced herbs into the fricassee.

The finale was a chocolate soufflé straight from the oven and topped with fresh whipped cream. Most of the soufflé was prepared ahead and popped into a preheated oven when the main course was served.

After dinner, the hostess offered a humidor filled with a selection of exotic cigars as we adjourned to the den for Cognac. Even the nonsmokers in the group enjoyed the novelty of the experience and puffed away contentedly, bringing an evening of good food, wine, and company to a most satisfying close.

THE MARTINI MYSTIQUE

We all know about the martini purists who insist on their favorite spirits, shaken not stirred. But the rest of us are free to experiment with the fascinating options available for making this fashionable cocktail.

At The Bar in Boston's venerable Ritz-Carlton Hotel, guests enjoy the splendor of the Boston Public Gardens while sipping from a menu of 11 martinis. Among the most popular are the Cajun Martini (3 parts Absolut Pepper vodka, 1 part Martini & Rossi dry vermouth and a green jalapeño olive); the Blue Martini (7 parts Absolut vodka, 1 part blue curaçao, and a lemon twist); the Portuguese Martini (12 parts Absolut vodka, 3 parts Dow's 30-year-old port, and a lemon twist); and the Chocolate Martini (10 parts Stolichnaya Orange vodka, 1 part crème de cacao).

Tempting martinis are also served in the Windows Lounge of the Four Seasons Hotel in Beverly Hills, including the Chevalier, made with Stolichnaya vodka and Williams pear brandy, and the Valentino, made with Skyy vodka and Midori liqueur.

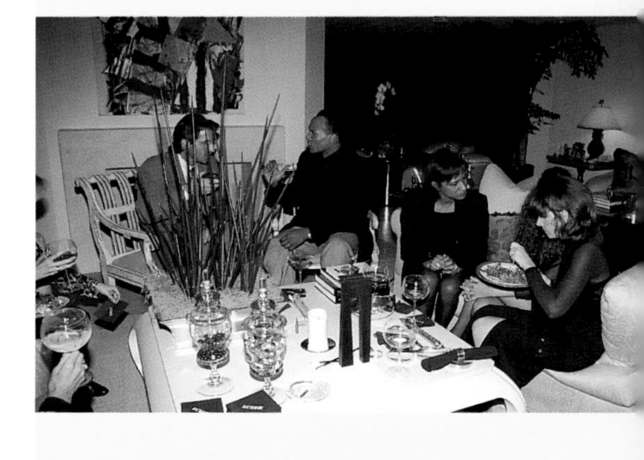

SALAD OF GREENS *with* SAUTÉED SHRIMP, FETA CHEESE, *and* ASSORTED CRISPIES

1 cup peanut oil
1 small baking potato, peeled and julienned
1 leek, trimmed and julienned
1 carrot, peeled and julienned
2 green onions, cut into 2-inch lengths and julienned
Salt and freshly ground pepper
45 large shrimp, peeled and deveined (under 15 pieces per pound)
¼ cup olive oil
2½ pounds mixed salad greens
Mint Dressing (see following recipe)
¼ cup julienned fresh mint
½ red onion, finely diced
4 ounces feta cheese, crumbled

1. In a small saucepan, heat the peanut oil over medium-high heat. Fry the potatoes, a handful at a time, until golden brown, then transfer to paper towels to drain. Repeat with the leek, carrot, and green onions, frying each vegetable separately. Toss the vegetables together in a large bowl, season with salt and pepper to taste, and set aside.
2. Season the shrimp all over with salt and pepper to taste. In a large skillet, heat the olive oil over high heat. Sauté the shrimp, being careful not to crowd the pan, until pink, about 2 minutes per side. Set aside.
3. Place the salad greens in a large bowl, add all but 2 tablespoons of the Mint Dressing, and toss well.
4. To serve, arrange the salad in the center of a serving platter. Sprinkle shrimp with the remaining dressing and arrange around the edges of platter. Top the greens with a nest of the fried vegetables, red onion and feta cheese, and serve.

SERVES 14

Note: The shrimp can be barbecued or grilled for even more flavor.

MINT DRESSING

¼ cup fresh lime juice
2 tablespoons Dijon mustard
1 teaspoon balsamic vinegar
¼ cup chopped red onion
4 sprigs fresh mint leaves
½ cup olive oil
Salt and freshly ground pepper

In a blender, combine the lime juice, mustard, vinegar, onion, and mint. With the machine running, gradually add the oil. Process until puréed. Season with salt and pepper to taste.

CHICKEN FRICASSEE

7 whole chicken breasts on the bone (14 sides)
Salt and freshly ground pepper
3 tablespoons flour
3 tablespoons olive oil
¾ pound pearl onions, peeled
4 tablespoons butter
3 tablespoons sugar
1½ pounds assorted wild mushrooms, caps and stems separated, cleaned, trimmed, and quartered
1 teaspoon minced garlic
1 teaspoon minced shallots
6 carrots, peeled
20 baby red potatoes
1 pound peas, in the shell
1 bunch asparagus, cut into 2-inch lengths
¾ pound haricots verts, trimmed and cut into 2-inch lengths
Veal Sauce (page 211)
3 tablespoons heavy cream, whipped
15 chives, sliced
3 sprigs parsley leaves, chopped
4 sprigs tarragon leaves, chopped
10 basil leaves, cut into strips with scissors
5 cups hot cooked white rice

1. Remove the skin and cut chicken meat off the bone, reserving bones for the sauce. Cut the chicken meat into 1-inch strips along the diagonal. Season all over with salt and pepper. Dust with flour.
2. In a large skillet, heat the oil over medium-high heat. Sauté the chicken until lightly browned all over, about 6 minutes.
3. Place the onions in a pot with enough water to cover. Add salt, 1 tablespoon of the butter, and the sugar. Cook over medium heat until the liquid has reduced to a syrup. Set aside.
4. In a large skillet, heat 1 tablespoon of the butter over high heat. Sauté the mushrooms for about 3 minutes. Stir in the garlic and shallots, sauté another 3 minutes, and season with salt and pepper. Remove from heat.
5. With a paring knife, cut each carrot and potato lengthwise into thirds. Trim each into a rectangle with sharp edges and then trim the corners to form octagonal or rounded corners. (Save trimmings for the Veal Sauce.) Bring a large pot of salted water to a boil and blanch the carrots, potatoes, peas, asparagus, and haricots verts separately, just until cooked. Drain and set aside.
6. In a large heavy pot, heat the remaining butter over medium heat. Add the blanched vegetables, chicken, and Veal Sauce. Cook just to heat through. Stir in the cream and add chives, parsley, tarragon, and basil and season with salt and pepper to taste.
7. Ladle into a warm soup tureen or serving bowls and garnish with the mushrooms and onions. Serve hot with steamed rice.

SERVES 14

Salads served on black plates make a stunning presentation.

ORZO AS ACCOMPANIMENT TO CHICKEN FRICASSEE

In the Italian kitchen, orzo is a member of the *pastina* family and is usually used in soups. Although often mistaken for rice, orzo is actually a tiny, rice-shaped pasta. Orzo takes just 8 or 9 minutes to cook versus 20 minutes for rice. It's tasty and just as satisfying when being served with soups, stews, and sauces.

VEAL SAUCE

3 tablespoons olive oil
Bones from the chicken
Potato and carrot trimmings
2 shallots, chopped
5 garlic cloves, chopped
½ onion, finely chopped
1 cup Veal Stock (see page 182) or chicken stock
3 tablespoons thyme vinegar
2 teaspoons molasses
2 cups white wine
1½ tablespoons tomato paste
1½ cups water
Cornstarch or arrowroot (optional)

1. In a large heavy pot, heat olive oil over high heat. Brown the bones and the potato and carrot trimmings for 6 minutes. Pour off the grease in the pan.
2. Add shallots, garlic, and onion, and cook for 4 minutes until translucent. Pour in the stock, vinegar, molasses, wine, tomato paste, and water. Bring to a boil, reduce to a simmer, and skim and discard the foam from the top. Cook until thickened as desired. (Cornstarch or arrowroot may be stirred in, if necessary, to thicken.) Pass through a strainer.

CHOCOLATE SOUFFLÉ *with* FRESH WHIPPED CREAM

14 ounces unsweetened chocolate
1 cup milk
2 tablespoons butter
1 cup plus 2 tablespoons sugar
8 egg yolks
12 egg whites
Powdered sugar
Whipped cream

1. Preheat the oven to 350 degrees.
2. Butter fourteen 4-ounce ramekins, then sprinkle them generously with sugar. Arrange them on a baking sheet.
3. In the top of a double boiler, combine the chocolate, milk, butter, and 1 cup of sugar. Set over simmering water until the chocolate is melted, stirring occasionally. Remove from the heat and let the mixture stand for 10 minutes.
4. Gradually whisk the egg yolks into the chocolate mixture.
5. In a large bowl of an electric mixer, beat the egg whites until frothy. Add the remaining 2 tablespoons of sugar and beat until the whites are stiff but not dry. Fold a third of the whites into the chocolate mixture to lighten. Carefully fold in the remaining whites. Divide the mixture equally among the prepared ramekins and bake for 20 minutes.
6. Dust the soufflés with powdered sugar and serve immediately with the whipped cream.

SERVES 14

DINNER IN AN ART STUDIO

smoked salmon canapé WITH WASABI CRÈME FRAÎCHE

caviar ON NEW POTATOES

chicken GRAND-MÈRE WITH A

shiitake mushroom SAUCE

roasted potatoes WITH GARLIC AND SHALLOTS

haricots verts SALAD WITH CHOPPED RED ONION

VANILLA *mousse meringues*

AND FRESH BERRIES

Serves: 10

Range of Difficulty: Moderate

Preparation Time: 3 hours

PUT A SUBTLE SPIN ON REALITY WHEN YOU'RE DRESSING A TABLE AND *creating an environment for a dinner party. Just look around your home; you'll be surprised by the objects you have that would enhance the evening's theme.*

"Forty people for dinner?" You gasp. "Surely that's just for professional caterers." Absolutely not. This casually elegant buffet dinner can be created by anyone with a reasonable budget and with minimal help in the kitchen. All that's needed is a fresh eye to decorating ideas, a love of good food, and a game plan that will allow the host and hostess to have as much fun as their guests.

For this dinner, we transformed an art studio into an intimate dining room. Each of the five tables was covered with an ivory floor-length cloth, then with bright vintage linen overlays. A different pattern of the hostess's china and flatware was set on each table and the centerpieces consisted of objects found around the hostess's home. Simple but unique floral accents along with elegant tapers completed the transformation.

During the short cocktail hour, guests were offered a glass of Champagne and two hors d'oeuvres to pique their appetites. Smoked salmon on puff pastry and baby potatoes partially hollowed, and refilled with crème fraîche and osetra caviar, were served on an unusual glass platter.

The simple buffet was set on the dining room table: Chicken Grand-Mère (the French term for grandmother's chicken dish), roasted potatoes tossed with garlic, shallots, and rosemary; and a vivid sunburst of pencil-thin haricots verts topped with chopped red onion and tasty dressing. Foods were served on large platters that were constantly replenished.

On a side table in the dining area, we set a bistro-style wine bar and uncorked a selection of red and white wines. This gave guests the chance to sample several wines and saved the expense of a waiter at the same time.

We offered a selection of cheeses, figs, and a decanter of port on another table, for the guests to enjoy while the hostess set out dessert in the dining room: a light vanilla yogurt mousse, tucked with fresh berries into a meringue sandwich.

COLLAGE
KURT SCHWITTERS
POPOVA

CAVIAR

Every occasion can be brightened by a bit of what I call "cosmetic caviar."

Make some wonderful little canapés and top them with a few eggs of a good quality caviar. Hollow out steamed new potatoes and refill them with a dollop of crème fraîche and a sprinkle of glistening black caviar.

The precious "berries" lend a rich flavor and an air of elegance.

SMOKED SALMON CANAPÉ *with* WASABI CRÈME FRAÎCHE

¼ cup crème fraîche or sour cream
1 teaspoon prepared white horseradish
½ teaspoon prepared wasabi (Japanese horseradish paste)
2 9-inch sheets frozen puff pastry, defrosted
3 ounces sliced smoked salmon, cut into 24 equal pieces
Fresh dill

1. Preheat the oven to 400 degrees.
2. In a small bowl, combine the crème fraîche, horseradish, and wasabi and set aside.
3. Using a 2-inch round cookie cutter or juice glass, cut out the puff pastry. Place on a baking sheet and bake for 18 minutes. Set on racks to cool.
4. Spoon a small dollop of the crème fraîche mixture on each pastry round. Top with a small mound of salmon and garnish with a tiny sprig of dill.

MAKES 24 PIECES

CAVIAR ON NEW POTATOES

12 baby red potatoes
Vegetable oil for deep frying
¼ cup crème fraîche or sour cream
1 tablespoon lemon juice
2 ounces osetra caviar

1. Bring a large pot of water to a boil. Blanch the potatoes for 2 minutes. Drain, then plunge into cold water and drain again. Pat dry with paper towels.
2. When cool enough to handle, cut the potatoes in half. Using a melon baller, scoop out and discard the centers.
3. Heat the oil in a deep pot or fryer to 350 degrees. Fry, 8 potato halves at a time, until golden brown all over. With a slotted spoon, transfer to paper towels to drain.
4. In a small bowl, mix together the crème fraîche and lemon juice. Place a dollop of the mixture into the potatoes and spoon the caviar on top.

MAKES 24 PIECES

CHICKEN GRAND-MÈRE *with* *a* SHIITAKE MUSHROOM SAUCE

2 chickens, cut into 8 pieces per chicken and trimmed of fat
Salt and freshly ground pepper
½ tablespoon herbes de Provence
2 tablespoons olive oil
8 tablespoons butter
½ cup chopped shallots
5 tablespoons flour
2 cups dry red wine
6 cups veal or chicken stock
1 bay leaf
2 sprigs fresh thyme
6 peppercorns
2 cups pearl onions
2 tablespoons sugar
4 large potatoes cut into 2-inch pieces and carved into bullets
¼ pound sliced shiitake mushrooms
2 tablespoons chopped chives

1. Season the chicken pieces with salt, pepper, and herbes de Provence.
2. In a large saucepan heat 2 tablespoons of oil and 1 tablespoon of butter until very hot. Add the chicken pieces and cook about 6 to 8 minutes on each side, or untill golden brown. Remove the chicken and set aside.
3. Remove the excess fat from the pan and add the shallots. Cook for 3 to 5 minutes, or until glassy. Add the flour and 5 tablespoons of butter and stir thoroughly.
4. Add the red wine, stock, bay leaf, thyme, and peppercorns. Bring the sauce to a boil, allow it to thicken, then reduce the heat to low and add the chicken. Cook the dark meat for 25 to 30 minutes and the white meat for 20 to 25 minutes.
5. In a small saucepan add the onions, 1 tablepoon of butter, and the sugar with enough water to half-cover the onions. Cook untill the onions are caramelized and a rich golden brown in color. Set aside.
6. In a pot with salted cold water, cook the potatoes until tender but firm. Remove from the water and sauté in a non-stick pan with 1 tablespoon of butter until the potatoes are lightly browned. Add the mushrooms and cook until tender. Add the onions and season with salt and pepper.
7. With a slotted spoon remove the chicken from the sauce and place in an ovenproof shallow roasting dish. Keep warm.
8. Bring the sauce to a boil and skim off any excess fat. Pass the sauce through a fine sieve and add the remaining tablespoon of butter with a whisk.
9. Spoon the potatoes, mushrooms, and onions over the chicken. Pour half the sauce over the chicken and serve the balance on the side. Sprinkle with the chives.

SERVES 10

Note: The sauce can be prepated 1 day ahead. Heat through before adding the chicken pieces.

MIX 'N' MATCH

There's no rule that says one must have on hand or go out and buy complete sets of china, glasses, and silver all at once, that was a trend of the past. It's more fun to buy several different sets or even one or two pieces at a time.

SERVING PLATTERS FOR A BUFFET

When preparing a dinner buffet, have *two* serving platters ready for each dish being served. While one is on the table, another can be readied in the kitchen. Don't wait until a platter is completely bare before removing it. Replacing platters before they are empty ensures that the food stays fresh and looks appetizing.

SETTING A BUFFET

The key to a successful buffet is to make the food look abundant and appetizing. Use smaller platters and containers that can be easily refilled. Never spread food out on a table that is too large. Instead, use one corner of the table and place your food, dishes, serving pieces, and buffet decor close together. When serving large groups, it's ideal to set the buffet Japanese-style, where guests can serve themselves from both sides of the table.

As a general rule, set the buffet as follows: cutlery, napkins, plates, breads, cold foods, and finally hot foods and condiments.

ROASTED POTATOES *with* GARLIC *and* SHALLOTS

½ cup olive oil
6 large russet potatoes, peeled and cut into 1½- inch pieces
12 shallots with skins
½ cup fresh rosemary leaves
Salt and freshly ground pepper
24 garlic cloves with skins

1. Preheat the oven to 425 degrees.
2. Pour the oil into a baking pan and place the pan in the oven for 15 minutes.
3. Meanwhile, in a large mixing bowl, combine the potatoes, shallots, and rosemary. Season with salt and pepper to taste. Carefully transfer the mixture to the pan with the heated oil. Bake for 1 hour, turning the potatoes occasionally. Add the garlic after 30 minutes of cooking.
4. Run the pan of potatoes under a hot broiler for 4 to 5 minutes, or until they are crisp.

SERVES 10

HARICOTS VERTS SALAD *with* CHOPPED RED ONION

1½ pounds haricots verts
3 tablespoons balsamic vinegar
2 tablespoons minced parsley
1 tablespoon minced shallot
1 tablespoon Dijon mustard
½ cup olive oil
Salt and freshly ground pepper
½ cup minced red onion

1. Drop the haricots verts into a large pot of boiling salted water. Cook just until they are tender but crisp, about 2 minutes. Immediately plunge the beans into ice water to stop the cooking process. Set aside.
2. In a small bowl, combine the vinegar with 1 tablespoon of the parsley and the shallot and mustard. Gradually whisk in the olive oil. Season with salt and pepper to taste.
3. Drain the beans and place them in a large bowl. Add about half of the dressing and toss to coat the beans thoroughly. Arrange the beans on a serving plate and sprinkle with the onion and the remaining parsley. Serve the remaining dressing on the side.

SERVES 10

CHEESE PRIMER

Strike a balance between mild and strong cheeses, old favorites, and trendy newcomers.

Roquefort. Considered the grand-père of all blue-veined cheeses, this centuries-old French cheese has a pungent, piquant flavor that tingles on the tongue in much the same way Champagne does. If this, the King of Cheeses, is a bit salty for your taste, serve it with crackers or bread and unsalted butter. Bleu de Bresse and Pipo Crem are other toothsome French blues.

Stilton. The most beloved of English cheeses, this aristocratic blue has an off-white color, pebbly consistency, and mellow flavor with subtle undertones of Cheddar. In England, it is usually enjoyed with biscuits and port, but it's also delicious with a slice of very tart apple pie.

Gorgonzola. Rich, robust, and creamy, this Italian blue cheese has a flavor all its own. Its pale green veining and very soft texture make it a popular choice for cheese trays.

Camembert. The real thing. Soft and tangy within, crusty without. It comes from the heart of Normandy in France, but several American cheese makers also offer respectable versions. Legend has it that Napoleon kissed the waitress who first served him a wedge of Camembert.

Bel Paese. This factory-made Italian cheese is semisoft and mild; it keeps well and has a reputation for exceptional consistency of quality.

St. André. Very rich, very mild, and very popular, this buttery triple crème is one of France's best-known exports.

St. Nectaire. Named for a market town in France, this popular semisoft cheese has a supple texture and mildly tangy taste that marries well with dry white wines and roses.

Montrachet. These small logs of goat cheese from the Burgundy region of France are loved for their creamy consistency and pleasantly acidic tang.

Taleggio. A soft cow's milk cheese from the Lombardy region of Italy; the flavor is rich and slightly piquant.

Explorateur. This luxurious triple crème from France has 75 percent butterfat and tastes it! (Double-crème cheeses have a minimum of 60 percent fat.)

Pont-L'Eveque. One of France's most prized cheeses, this plump, golden, square-shaped cheese has a soft texture and a deep-toned, complex flavor that goes well with full-bodied red wines, fresh cider, and the fiery apple brandy called Calvados that comes from the same region.

VANILLA MOUSSE MERINGUES *and* FRESH BERRIES

2 cups plain yogurt
1 cup sugar
2 cups heavy cream
½ teaspoon vanilla essence
3 egg whites
Rind of 1 orange, finely julienned and blanched
Meringues (see following recipe)
1½ cups mixed berries
Strawberry Coulis (see recipe)
Mint sprigs (for garnish)

1. Line a large colander with several layers of cheesecloth. Set aside.
2. In a large bowl, mix the yogurt with the sugar.
3. In a separate bowl, beat the cream and vanilla essence until stiff. Fold the cream into the yogurt.
4. In a large bowl, beat egg whites with an electric mixer until stiff but not dry. With a spatula, carefully fold in the yogurt mixture and the orange rind.
5. Transfer the mixture to the prepared colander and place over a bowl. Cover with plastic wrap and place in the refrigerator to drain for at least 4 hours.
6. When ready to serve, place one meringue on each of the 10 dessert plates. Top with some mousse and a few berries. Top each with a second meringue.
7. Drizzle some Strawberry Coulis around each meringue and garnish with a few berries and a sprig of mint.

SERVES 10

Note: Mousse can be prepared 2 days ahead.

MERINGUES

1 cup granulated sugar
1 cup superfine sugar
8 egg whites
¼ teaspoon cream of tartar
¼ teaspoon salt
1 teaspoon vanilla or almond extract

1. Preheat the oven to 250 degrees.
2. Line 2 baking sheets with parchment paper or foil. On each, draw 10 4-inch circles.
3. Mix the sugars and set aside.
4. In a bowl or electric mixer, beat the egg whites until foamy. Add the cream of tartar and salt. Continue to beat, adding 1 tablespoon of sugar at a time and beating well after each addition. This will take about 10 minutes.
5. Add the vanilla and beat at high speed until stiff and glossy.
6. Using a pastry tube fitted with a large plain tip, pipe the meringue in coils starting at the center and continuing to the outside of each circle.
7. Bake about 1 hour, until cream colored and firm. Leave the meringues in the turned-off oven for several hours or overnight without opening oven door.
8. When ready to serve, carefully peel away the paper or foil.

Note: The meringues can be made up to 2 weeks ahead. Seal them in plastic bags and store in an airtight container.

STRAWBERRY COULIS

1 cup frozen unsweetened strawberries
½ cup sugar
1 tablespoon lemon juice

1. In a medium saucepan, combine the strawberries, sugar, and lemon juice. Bring to a boil over medium-high heat. Transfer to a blender.
2. Purée until smooth, strain, and set aside. Store in a sealed container in the refrigerator for up to 1 week.

THE ULTIMATE DINNER

mille-feuille of salmon AND CAVIAR WITH CHIVE SAUCE

SALAD WITH ARTICHOKE TRUFFLE VINAIGRETTE
AND WARM *chèvre soufflé*

HERBED *rack of lamb* WITH TOMATO BORDELAISE SAUCE

spring vegetables AND BABY POTATOES

CRÈME BRÛLÉE STUFFED PEARS
WITH *raspberry coulis*
AND CRÈME ANGLAISE

Serves: 10

Range of Difficulty: Difficult

Preparation Time: 4 hours

Sometimes you want to pull out all the stops.

For me, that means caviar, foie gras, truffles, and Champagne.

Some occasions demand the very best. A no-holds-barred, no-expense-spared approach to entertaining. It will certainly leave guests wide-eyed, and be the topic of conversation for months to come.

Four round dining tables were draped with lustrous garnet and saffron French damask cloths and set with china in similar hues. Regal centerpieces of Message and velvety red roses, green hydrangeas, purple lizianthus, and burgundy and green amarynth added the finishing touches.

Our first course, a mille-feuille with the finest smoked salmon and crème fraîche was topped with 000 Imperial beluga caviar. To garnish the plates, we sprinkled the rims with minced red onion and drops of chive sauce. A rich, full-bodied Chardonnay, with fleshy tropical-fruit flavor, is a fitting partner for this grand dish.

The term "salad" took on new meaning with our next course—a very simple salad of mâche and crispy leeks. Pulling it all together was my favorite vinaigrette made with lemon juice and truffle oil and finally, a lighter-than-air cheese soufflé, straight from the oven, was spooned atop each salad. The combination of chilled lettuces, crispy leeks, and hot soufflé is wonderful.

Indeed, these first two courses were tough acts to follow. But as the main course, our rack of lamb topped with an herbed crust of bread crumbs, was up to the task. With it, we served a Canon wine from St.-Emilion, but any supple, medium-bodied Bordeaux will work very well.

For dessert, a poached pear was hollowed out and filled with a rich, crackly crème brûlée and accompanied by a glass of Demi-Sec Champagne.

Low centerpieces facilitate conversation. Use place cards to balance the energy at a dinner party.

DECORATING THE PLATE

The best and brightest chefs are turning the rims of their oversized dinner plates into their personal canvases. You can frame any appetizer, salad, or main course with a "confetti" of the following:

- minced herbs such as chervil, Italian parsley, or chives
- finely cubed peeled tomatoes and torn ribbons of basil
- chopped red onion
- cracked pepper

For dessert, plates can be decorated with the following:

- powdered sugar
- cocoa powder
- finely ground coffee
- fruit coulis
- fresh fruits
- shaved chocolate

MILLE-FEUILLE *of* SALMON *and* CAVIAR *with* CHIVE SAUCE

1 egg yolk
¼ pound butter
6 sheets phyllo pastry
Sour Cream Mixture (see following recipe)
20 ounces presliced smoked salmon, cut into twenty 1-ounce slices
10 ounces beluga, osetra, or sevruga caviar
¼ cup large diced red onion
Chive Sauce (see recipe)

1. Preheat the oven to 450 degrees.
2. In a small bowl, beat the egg with 1 tablespoon of cold water. Set aside.
3. In a small saucepan, melt the butter over medium heat until white solids separate and bubble to the top. Set aside to cool, then remove and discard the foam from the top. The butter is now clarified.
4. Place 1 sheet of phyllo on a cutting board and lightly brush with the butter. Cover with a second sheet of phyllo, coat with more butter, and repeat with the third sheet.
5. With a sharp knife, cut the phyllo into 16 equal rectangles. Transfer to a baking sheet and brush with the egg wash. Bake until golden brown, about 4 minutes. Set aside to cool.
6. Repeat the procedure with the remaining phyllo to make 32 rectangles.
7. Spread 1 heaping teaspoon of the Sour Cream Mixture over 10 phyllo pieces. Top with 1 slice of salmon and a dab of caviar. Top with another sheet of phyllo and repeat the sour cream, salmon, and caviar layers. Top with another piece of phyllo, a teaspoon of the Sour Cream Mixture and caviar.
8. To serve, place the finished pastry in the center of each of 10 serving plates. Sprinkle the plate with the onion. Drizzle with the Chive Sauce.

SERVES 10

SOUR CREAM MIXTURE

½ large red onion, diced
16 ounces sour cream
2 tablespoons thinly sliced chives
Juice of 2 lemons
½ teaspoon salt
⅛ teaspoon white pepper

In a mixing bowl, combine the red onion, sour cream, chives, lemon juice, salt, and pepper. Mix well and set aside.

MAKES 2 CUPS

CHIVE SAUCE

1 cup heavy cream
1 bunch chives
Rind of ½ lemon
Salt and freshly ground pepper

1. Place the cream in a small saucepan and bring to a boil. Cook until reduced by half.
2. Blanch the lemon in boiling water for 30 seconds and add to the cream. Season with salt and pepper. Transfer to a blender and add the chives. Blend until puréed and then strain through a fine sieve. Set aside or refrigerate.

MAKES ½ CUP

SALAD *with* ARTICHOKE TRUFFLE VINAIGRETTE *and* WARM CHÈVRE SOUFFLÉ

1 leek
1 sweet potato, peeled
Vegetable oil for frying
Salt and freshly ground pepper
1 pound baby frisée, mâche, or butter lettuce, torn into bite-size pieces
Artichoke Truffle Vinaigrette (page 110)
Chèvre Soufflé (page 229)

1. Cut the leek into 2½-inch long julienne. Using a vegetable peeler, strip off shavings of the sweet potato.
2. In a medium saucepan, heat about 2 inches of the oil over medium-high heat. Add a small handful of the leeks and fry until they are golden and crisp. Drain on paper towels. Fry the sweet potato shavings, a small handful at a time, until crisp. Drain on paper towels. Toss and season with salt and pepper.
3. In a large bowl, combine the frisée, add the Truffle Vinaigrette, and toss to combine thoroughly. Divide the salad equally among 10 individual plates, piling it high in the center. Top each salad with some of the leeks, sweet potatoes, and a scoop of the Chèvre Soufflé.

SERVES 10

Place lamb chops over a mound of vegetables for a more dramatic presentation.

CHÈVRE SOUFFLÉ

- 4 tablespoons butter
- 3 tablespoons flour
- 2 cups hot milk
- 1 cup grated imported Parmesan cheese
- ¼ teaspoon salt
- ¼ teaspoon freshly ground pepper
- 4 whole eggs, separated
- 6 ounces Chèvre or other goat cheese, at room temperature

1. In a small saucepan over medium heat, melt the butter. Add the flour and cook, stirring constantly, until the mixture begins to bubble, about 2 minutes.
2. Gradually stir in the hot milk. Continue to cook, stirring constantly, until the sauce comes to a boil. Reduce the heat and simmer, continuing to stir, 2 to 3 minutes longer, or until thick. Add the Parmesan cheese, salt, and pepper and stir until smooth. Set aside to cool until lukewarm.
3. Preheat the oven to 400 degrees.
4. Butter a 6-cup soufflé dish and then dust it with flour.
5. Force the goat cheese through a fine sieve.
6. In a large bowl, beat the 4 egg yolks with the goat cheese. Stir the mixture into the Parmesan mixture.
7. In a large mixer bowl, beat the egg whites until stiff but not dry. Fold a third of the whites into the cheese mixture to lighten it. Fold in remaining whites. Transfer mixture to the prepared soufflé dish. Bake for 35 minutes. Serve immediately.

SERVES 10

HERBED RACK *of* LAMB *with* TOMATO BORDELAISE SAUCE

- 1 bunch rosemary
- 1 bunch thyme
- 1 medium head garlic, cloves separated and chopped, plus 3 to 4 cloves, minced
- 2 tablespoons olive oil
- 4 lamb racks (1 pound each)
- 1 bunch parsley (30 sprigs)
- 2 cups dry bread crumbs
- ¼ pound butter, melted, plus 2 tablespoons
- Salt and freshly ground pepper
- 1 cup Dijon mustard
- 1 cup dry white wine
- 2 cups veal stock
- 1 cup tomato juice
- Spring Vegetables and Baby Potatoes (see following recipe)

1. Strip the leaves from the rosemary and thyme and coarsely chop them. Set aside ¼ cup of the chopped rosemary for sauce. Place the chopped herbs in a small bowl and add the chopped garlic and olive oil. Mix to blend thoroughly.
2. Rub the herb mixture all over the lamb racks. Arrange them in a shallow dish, cover, and refrigerate for 8 hours or up to 24 hours.
3. Strip the leaves from the parsley and coarsely chop them. In a medium bowl, combine the chopped parsley with the bread crumbs and melted butter. Mix to blend thoroughly and set aside.
4. Preheat the oven to 400 degrees.
5. Set a large heavy skillet over medium-high heat and add 2 tablespoons butter. When hot, add the lamb racks, two at a time, and cook for 4 to 5 minutes on each side, or until browned. Remove racks from the pan. Do not wash the pan. Sprinkle the racks with salt and pepper to taste, then brush all over with the mustard. Using your hands, press the reserved bread crumb mixture over the meat, leaving the bones exposed. Arrange the lamb racks in a large roasting pan and roast for 20 to 25 minutes or until desired doneness—125 degrees for rare, 140 degrees for medium, and 160 degrees for well done. Remove from the oven and keep warm.
6. Meanwhile, set the pan with the reserved drippings over medium-high heat. Deglaze the pan with the white wine and transfer to a pot. Add the stock, tomato juice, minced garlic, and the reserved rosemary. Bring the mixture to a boil, then lower the heat to simmer and reduce by three fourths (approximately 40 minutes). Whisk in the butter. Strain the sauce and season with salt and pepper to taste. Keep warm.
7. Carve the lamb racks into individual chops. Stand 3 chops in the center of each warm dinner plate, creating a pyramid with the chop bones at the top. Spoon some of the sauce around the chops and surround them with the vegetables.

SERVES 10

SPRING VEGETABLES *and* BABY POTATOES

- 1 pound baby red potatoes, peeled and carved into ovals
- 2 to 3 tablespoons olive oil
- Salt and freshly ground pepper
- ½ pound pearl onions, peeled
- 1½ tablespoons sugar
- ½ pound haricots verts, trimmed
- ½ pound baby carrots, peeled
- ½ pound baby turnips, peeled
- ½ pound fresh or frozen and thawed peas

1. Cook the potatoes for 5 minutes in boiling water. Remove and set aside.
2. In a large skillet, heat the oil over medium-high heat. Add the potatoes and sauté until golden brown. Season with salt and pepper to taste. Set aside and keep warm.
3. In the same skillet, add the onions, sugar, and enough water to cover the bottom of the pan by ½ inch. Sauté until glassy and caramelized. Set aside.
4. Bring a large pot of salted water to a boil. Add the haricots verts, carrots, turnips, and peas. Cook 3 minutes. Drain and season with salt and pepper to taste.
5. In the same skillet, sauté all the vegetables and season with salt and pepper to taste.

SERVES 10

1934

I HAVE AN APPRECIATION FOR GREAT CHEESES, GREAT PORT,

and wonderful cigars. While I indulge in the latter only occasionally, they provide a perfect finish for any meal.

There are many ways to warm yourself on a cold winter's night. Cashmere sweaters, shawls, and throws work well. So do a blazing fire and a couple of easy chairs. But one of my favorite strategies is to get together with friends and taste some interesting wines, preferably big, robust reds and purple-black ports with their peppery aromas and rich flavors, and fine cognacs.

For this dinner, we gathered in a friend's cozy wine cellar. Surrounded by vintage wine and port bottles, antique glasses and ice buckets, flickering candelight, and a collection of personal mementos, we savored our food and wine in a most comfortable atmosphere.

Since the cellar's table was very narrow, with little room for flowers, the food itself became the centerpiece. Fresh figs, grapes, and baby pears were placed on the table along with a few oak leaves and votives wrapped in galix leaves. The table was set with china and my best museum-quality glassware, which I had kept for special occasions up until the last big Los Angeles earthquake. The earthquake taught me that *every* day is special and the beautiful things with which we are blessed should not be left to collect dust!

Spezzatino, a wonderful, comforting Italian stew made with veal, carrots, tomatoes, and mushrooms, was served over fresh pasta. Just before bringing the stew to the table, I folded in a tablespoon of whipping cream for added richness. And for color, freshly chopped parsley was sprinkled on top. With this hearty dish we poured a robust Cabernet.

For this stew, I prepared a homemade tricolor fettuccine; spinach, tomato, and egg. However, the dish is just as good served with dried and cooked pasta, or fresh pasta from your local pasta shop.

To accompany a cheese selection of Stilton, Explorateur triple crème, and a crottin of goat cheese, we splurged on a 1934 port. The combination of creamy cheeses and rich flavor of the elegant port was all the dessert anyone could ask for. Nonetheless, we concluded our celebration with store-bought champagne truffles, good strong coffee, and cigars from the hostess's well-stocked humidor.

A wine collection can be started with just a few well-chosen bottles.

A SIMPLE RULE OF THUMB

Serve three ripe cheeses; a triple crème, a blue, and a goat cheese. My favorites are a lush Explorateur, a mellow Stilton, and a tangy crottin of ashed goat cheese. All can be served with crusty bread or crackers.

Cheese should always be served at room temperature. For an elegant look, serve a single fruit with the selection of cheese, such as figs, champagne grapes, dates, strawberries, or even kiwi.

SPEZZATINO *with* TRICOLOR FETTUCCINE

2 pounds boneless veal shoulder, cubed
Salt and freshly ground pepper
¼ cup all-purpose flour
1 tablespoon butter
2 tablespoons olive oil
½ medium onion, chopped
½ pound baby carrots
2 tablespoons tomato paste
1 bouquet garni (thyme and rosemary)
1 cup chicken broth
1 cup dry white wine
1 cup quartered mushrooms
Grated peel of ½ lemon
2 tablespoons chopped parsley
1 pound fettuccine, cooked and drained

1. Season the veal with salt and pepper to taste. Dust with the flour.
2. In a large heavy pot, melt the butter in the oil over medium-high heat. Add the veal and sauté until lightly browned. Remove the veal from the pan and keep it warm.
3. In the same pan, cook the onion over medium heat until translucent. Return the veal to the pan along with the carrots, tomato paste, and bouquet garni. Stir in the broth and wine and bring the mixture to a boil. Reduce the heat and simmer for 30 minutes, or until the veal is tender, stirring occasionally. Stir in the mushrooms and simmer for 10 minutes longer. Sprinkle with lemon peel and chopped parsley.
4. Serve over hot cooked fettuccine.

SERVES 4

Note: The stew can be made 1 day ahead. When ready to serve, bring the stew to a boil.

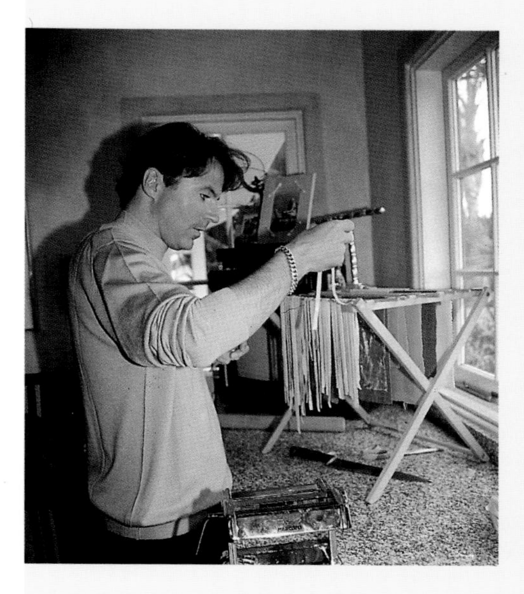

TODAY'S PASTA COMES IN A RAINBOW OF COLORS AND TASTES. You can add a purée of spinach, red bell pepper, carrots, parsley, pumpkin, squid ink, or fresh herbs to my pasta recipe:

PASTA

3 cups all-purpose flour
1 tablespoon vegetable oil
4 eggs
1 teaspoon salt

1. Place the flour in the bowl of a food processor fitted with the metal blade. With the machine running, add the oil and eggs, one at a time, then add the salt. When the dough forms a ball and is smooth and elastic to the touch, remove.

2. Divide the dough in half and wrap in plastic wrap.

3. Using a hand-cranked pasta machine, roll out the pasta from setting No. 1 through setting No. 6 for the desired thickness. Allow the pasta to dry on a pasta rack for 15 minutes before cutting into the desired type; pappardelle, fettuccine, linguine.

SPOTLIGHT ON PORT

There's no better way to conclude a wonderful meal than with a glass of mature vintage port, a sweet fortified wine that originated in Portugal's Douro Valley.

Although there are many types of port wine, there are four basic categories:

- Vintage ports are considered by most to be the best and thus, most expensive. They are made only from grapes from the best sites and a vintage is "declared" only after the best of harvests. Vintage port is bottled unfiltered, two years after the harvest. These wines of exceptional ripeness, fruit, and body can last for 50 or more years.

The most acclaimed vintages of this half-century were 1955, 1963, 1970, 1977, 1983, 1985, and 1992.

- Tawny ports are made from a blend of grapes from several different years and may be aged in wood for as long as 50 years. This is one of the least expensive ways to savor a mature port. (However, be aware, inexpensive tawny port may be merely a blend of ruby and white.)
- Ruby ports are made from lower-quality wine, aged in wood for about two years and bottled while it is still young, fruity, and bright red. The least expensive of ports, ruby ports have a straightforward sweet, grapy aroma and taste.
- White ports are produced the same way but using white grapes. Serious port aficionados don't give these pale pretenders a second look.

Some respected port "lodges" or producers include: Cockburn, Croft, Dow, Fonseca, Graham's, Quinta do Noval, Sandeman, Taylor Fladgate, and Warre.

❖

Always serve cheese at room temperature. And, if serving with fruit, serve a single fruit instead of a variety. My favorites are figs, champagne grapes, kiwi fruit, strawberries, dates, and prunes.

In France, cheese is usually served before the dessert course; in England, it is usually served after dessert.

❖

SPOTLIGHT ON COGNAC

Rather than being classified by year, cognacs are classified by name.

- VS-very special
- VSOP-very special old pale
- XO-extraordinary reserve
- Cognac is yielded from a small 20, 000-acre area between Bordeaux and La Rochelle in France.

1 2 3 4 5 6 7 8 9 10 11 12
Made by Hand in Switzerland
MARTIN

CIGARS

Wrapping up an evening in a den scented with the heady aroma of expertly aged and spiced tobacco is becoming a popular pastime.

Marvin Shanken, editor and publisher of *CIGAR Aficionado* magazine, is generally credited with fanning the fires of the premium cigar trend. The magazine offers a crash course in fine cigar appreciation. Although the subject of fine, imported, hand-rolled cigars is a complex one, there are a few basic facts to know before joining in the craze:

Brand. This is the designation given to a particular line of cigars by the manufacturer. Punch, Partagas, Macanudo, Montecristo, and Davidoff are well-known brands.

Color. This refers to the shade of the outer wrapper leaf. There are six basic shades ranging from light green and tan to brown, reddish-brown, and brown-black.

Shape. Cigars are divided into two categories: parejos, which are straight-sided, and figurados, which have irregular shapes.

Size. Length is listed in centimeters, and girth or diameter in 64ths of an inch or millimeters. A classic corona size, for example, is 6 by 42, or six inches long and 42/64ths of an inch thick.

- Coronas, Churchills, double coronas, and robustos are approximately the same size.
- Panatelas are usually longer than coronas and thinner. Lonsdales are thicker than panatelas and longer than coronas.
- Of the figurados, the Pyramid looks like a modified pyramid. A Torpedo has a pointed head, closed foot, and a bulge in the middle. A Perfecto has two closed, rounded ends and a bulge in the middle. A Diademas is a giant cigar 8 inches or longer.

Taste. Aficionados use words like acidic, salty, bitter, sweet, sour, smooth, heavy, full-bodied, rich, and balanced to describe their favorite cigars. Sight, touch, and smell also play a role in cigar smoking enjoyment.

Cutting and Lighting. A correct cut is essential to full appreciation of a cigar. Various tools are employed: single- and double-bladed guillotines, scissors, V-shaped wedges, bull's eyes, and piercers. The basic rule in lighting: Never let the cigar touch the flame. To ensure a proper light, rotate the cigar in your hand so that the foot of the cigar lights all the way around.

Humidor. A wooden box designed to store cigars in optimum conditions, by re-creating the tropical or semi-tropical environment in which most cigar tobacco is grown. A humidor must maintain a consistently tropical environment (between 68 and 70 degrees F and 70 to 72 percent humidity) over a long period of time.

Most humidors rely on some type of chemical compounds, sponges, or plain bottles to provide moisture. Cedar is the best wood for the inside because of its ability to enhance the aging process.

DINNER IN FRONT OF THE FIRE

MY SISTER ANNE'S *chicken liver pâté*

SALAD OF *mixed greens*
WITH CARAMELIZED ONIONS AND TOMATOES

pasta e fagioli

apple pie AND WHIPPED CREAM

Serves: 4

Range of Difficulty: Easy

Preparation Time: 2 hours

To create an inviting ambience, use as many candles as possible,

in groupings of different sizes, around the house. They create a warm, cozy, and romantic atmosphere.

For this autumn supper, I fixed a pot of pasta e fagioli soup to welcome my friends on a chilly evening. The tantalizing aroma of the classic Italian bean soup filled the house.

We placed the dinner table in front of a roaring fire and covered it with a kelim rug rather than a traditional tablecloth and candles in silver candelabra. Field flowers were arranged loosely in ceramic urns; their vibrant colors against the glow of the candles made the table look gorgeous and inviting.

When guests were seated, I served the soup in a rustic wooden bowl and passed a fruity green Tuscan olive oil to drizzle on top. Afterward, we had a salad of seasonal greens, dressed with a simple vinaigrette, and topped with piping hot onions and cherry tomatoes that had been caramelized in balsamic vinegar. A rich Pinot Noir with nuances of blackberry and cherry worked well with the robust nature of the menu.

A dinner scene like this one calls for a hearty dessert straight from the oven. We served my friend Sharon Richstone's foolproof apple pie with a choice of whipped cream, crème fraîche, ice cream, or frozen yogurt.

The pie, as well as the rest of the menu, was prepared in advance. When we were seated at the table, I placed the pie in the oven and it finished baking just in time.

After dinner, with feet propped up on the hearth, coffee and Cognac in hand, we once again celebrated the glorious outdoor events of the day and plotted our plans for the following day's adventures.

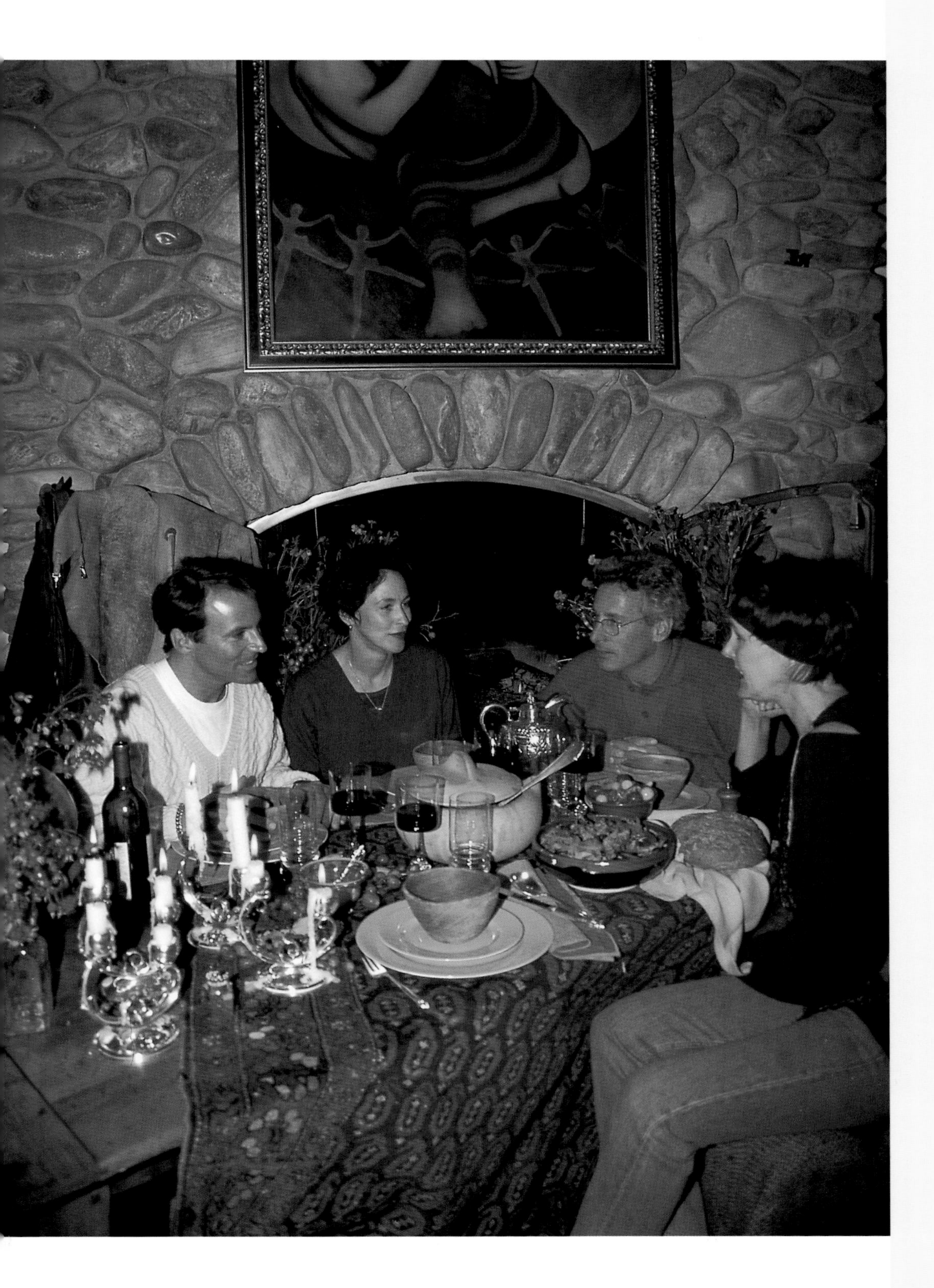

DIGESTIFS

Cozying up to a fire with an after-dinner digestif is the perfect conclusion to a perfect day. Consider the following:

Cognac. The finest of all brandies, this potent product of France has enormous finesse and a captivating aroma. It is best served in a snifter, warmed by two hands.

Armagnac. This fine French brandy has a slightly lower alcohol content than Cognac; because it undergoes a single distillation (as opposed to Cognac's double distillation), Armagnac is fuller-flavored and silky smooth.

Averna Amaro Siciliano. Created by monks some 200 years ago, this Italian digestif is made from herbs and roots and is enjoyed "neat" or with soda water.

Grappa. A colorless, high-alcohol Italian brandy; aged grappas are smooth, complex, and elegant; young ones are brash and fiery.

Fernet Branca. France's answer to the herbal digestif.

BORLOTTI (cranberry beans) are traditionally used in pasta e fagioli. However, Great Northern, cannellini, or kidney beans can also be used. Canned cannellini beans may be substituted if you don't have the time or inclination to soak dried beans.

PUTTING THE PASTA IN PASTA E FAGIOLI

You can use many different shapes of pasta in this traditional soup. These are some possibilities:

- ditalini
- orecchiette
- small shells
- elbow macaroni
- orzo

MY SISTER ANNE'S CHICKEN LIVER PÂTÉ

6 ounces streaky bacon, chopped
1 large onion, coarsely chopped (1¼ cups)
1 garlic clove, crushed
½ pound chicken livers
3 tablespoons port, sherry, or brandy
Salt and freshly ground pepper
1 parsley sprig
½ tablespoon butter
1 baguette, sliced

1. In a large skillet over medium-high heat, cook the bacon until lightly browned. Remove the bacon from the pan and set aside on paper towels.
2. Discard all but 3 tablespoons of drippings from pan. Add the onion and garlic to the reserved drippings in the pan and cook over medium-high heat until lightly browned. Return the bacon to the pan along with the chicken livers. Cook, stirring frequently, until livers are lightly browned and almost cooked through. Add the port and continue to cook for 5 minutes longer. Remove the pan from the heat and allow it to cool for 5 minutes.
3. Transfer the mixture to a food processor or blender and process until smooth. Season the mixture with salt and pepper to taste. Transfer the pâté to a small 8-ounce terrine or ramekin.
4. Garnish with a sprig of parsley. Melt the butter and pour over the top of the pâté. Refrigerate for at least 1 hour before serving.
5. Serve the pâté with baguette slices.

SERVES 4

SALAD OF MIXED GREENS *with* CARAMELIZED ONIONS *and* TOMATOES

1 tablespoon butter
1 medium onion, thinly sliced
3 to 4 cups red and/or yellow cherry tomatoes
¼ cup balsamic vinegar
Salt and freshly ground pepper
About 5 ounces of mixed greens
Mustard Dressing (see following recipe)

1. In a large skillet, heat the butter over medium heat. Add the onion and sauté until translucent. Reduce the heat to low and continue to cook, stirring occasionally, until the onion begins to caramelize, about 15 minutes.
2. Add the cherry tomatoes and, with the back of a spoon, break 4 to 5 tomatoes to release their juices. Add the vinegar and salt and pepper to taste and cook, stirring occasionally, 3 minutes longer.
3. Meanwhile, toss the greens with the Mustard Dressing. Pile the greens onto 4 individual salad plates and spoon some of the hot tomato-onion mixture alongside. Serve immediately.

SERVES 4

MUSTARD DRESSING

2 tablespoons balsamic vinegar
1½ teaspoons Dijon mustard
1 teaspoon minced shallot
⅓ cup olive oil
Salt and freshly ground pepper

In a bowl, combine the vinegar, mustard, and shallot. Slowly whisk in the oil. Season the dressing with salt and pepper to taste.

PASTA E FAGIOLI

- 8 ounces dry cannellini or Great Northern beans
- 2 medium potatoes, peeled and diced
- 3 large tomatoes, peeled and chopped
- 2 celery stalks, chopped
- 2 tablespoons minced parsley, plus 1 tablespoon, chopped, for garnish
- 6 garlic cloves, minced
- 1 teaspoon chopped thyme
- 4 fresh sage leaves
- 2 bay leaves
- 3 tablespoons olive oil
- 1 beef bouillon cube, crumbled
- 2 cups small pasta, such as pastina
- Salt and freshly ground pepper
- ½ cup thinly sliced basil
- Grated imported Parmesan cheese

1. Soak the beans in enough water to cover overnight.
2. Drain the beans, rinse with cold water, drain again. Place in a large pot with the potatoes, tomatoes, celery, parsley, garlic, thyme, sage, bay leaves, 2 tablespoons of the olive oil, and bouillon. Add enough water to cover the ingredients with 2 inches on top and bring to a boil. Reduce the heat, cover, and simmer for 2 hours, setting the pot lid ajar after 30 minutes and stirring occasionally.
3. Return the soup to a boil and add the pasta. Cook, stirring occasionally, for 8 minutes, or until the pasta is al dente. Season the soup with salt and pepper to taste. Transfer the soup to a terrine or serving bowl. Drizzle the last tablespoon of oil over soup. Sprinkle basil and parsley on soup and serve with Parmesan cheese on the side.

SERVES 4

Note: Alternatively, cook the pasta separately, then place it in the bottom of soup bowls before serving the soup.

APPLE PIE *and* WHIPPED CREAM

- 6 medium apples, cored, peeled, and sliced
- 1 tablespoon heavy cream
- ¼ cup raisins
- 1 teaspoon vanilla extract
- ½ cup brown sugar, packed
- 1½ tablespoons butter, cut into small cubes
- Pie Pastry (see following recipe)
- 1 egg, beaten with 1 tablespoon cold water
- Whipped cream or vanilla ice cream for topping

1. Preheat the oven to 350 degrees.
2. Place the apples in a large mixing bowl, add the cream, and toss to coat the apples evenly. Add the raisins, vanilla, and the brown sugar and toss well. Scatter the butter pieces over all.
3. Roll out half of the pie pastry on a lightly floured surface to ⅛-inch thickness. Line a 9-inch pie plate with the rolled pastry. Fill the pastry with the apple mixture.
4. Roll out the second half of the pastry to make a rough circle slightly larger than the pie plate and carefully place it over the apples, trimming any excess dough with scissors. Pinch the edges together to seal, then flute the edge as desired. (The pastry scraps can be rerolled and cut into shapes to use for decorating the top of the pie.) Using a sharp knife, make four 1-inch slits in the top crust for steam to escape. Brush the top with the egg mixture.
5. Bake the pie until the top is golden, 45 to 55 minutes. Cool slightly on a wire rack before serving with whipped cream or vanilla ice cream.

MAKES 1 PIE

PIE PASTRY

- 2 cups all-purpose flour
- 1 cup whole wheat flour
- 2 tablespoons sugar
- ½ pound butter, frozen, cut into slices
- ¾ cup water

1. In a food processor fitted with the metal blade, combine the flours and sugar. Pulse to combine.
2. With the machine running, add the butter slices one at a time. With the machine still running, add the water, 1 tablespoon at a time, until the dough forms a ball. Remove and divide dough into 2 equal pieces. Pat each piece of dough into a disk, wrap well with plastic, and chill for at least 30 minutes. (If pastry is very cold, let warm slightly before rolling.)

Apple pie is the quintessential American dessert.

HOUSEHOLD ESSENTIALS

A good home is a well-equipped home. In order to enjoy entertaining and be able to entertain without much time and effort, it is important to have a well-stocked, organized, and equipped kitchen. Below is a list of items to keep on hand at all times.

EQUIPMENT

COOKWARE

large saucepan
medium saucepan
small saucepan
small skillet
large skillet
non-stick skillet
stockpot
double boiler
colander
pasta pot
vegetable steamer

CUTLERY

chef's carving knife
chopping knife
boning knife
cleaver
serrated bread knife
paring knife
kitchen shears
knife block and knife sharpener

SMALL APPLIANCES

food processor
blender
coffee grinder
coffeemaker
juicer
electric beater
electric knife
electric toaster
ice cream maker
pasta machine with ravioli attachment

BAKING

glass mixing bowls
stainless steel mixing bowls
wire whisk
pastry brushes
measuring cups
kitchen scale
insulated cookie sheet
sifter
rolling pin
soufflé dishes
9-inch springform pan
muffin pans
rubber spatulas and wooden spoons
metal mixing bowls
metal whisk
ovenproof baking dishes
cutting boards
springform pan
cake pans
soufflé ramekins

MISCELLANEOUS

wine opener
garlic press
pepper grinder
pizza cutter
potato peeler
salad spinner
scissors and string
timer
candles—different shapes, sizes, and heights

SERVING ITEMS

ACCESSORIES

wood salad bowl
pasta bowl
oval serving platter
serving trays
serving utensils
pitcher
tongs
bread basket

STEMWARE AND BARWARE

12 water goblets
12 white wineglasses
12 red wineglasses
12 champagne flutes
8 highballs
8 double old-fashioneds
8 all-purpose wineglasses
6 martini glasses
1 pitcher
1 ice bucket and tongs

LINENS

dishtowels and dishcloths
apron
pot holders
napkins
placemats
tablecloths

FOOD

REFRIGERATED ITEMS

eggs
milk
butter
Parmesan cheese
fresh fruits
fresh lettuce for salad
lemons and limes
onions
tomatoes
potatoes
garlic
shallots
bottle of Champagne
two bottles of white wine
six beers
bottled water

FROZEN ITEMS

frozen yogurt
ice cream/sorbet
puff pastry
bread/rolls
regular and decaffeinated coffee
raspberries
veal stock
ice
vodka

DRY GOODS

bread
pasta
rice
tea (regular and decaffeinated)
flour
salt, pepper, sea salt
assorted herbs and spices
sugar
olive oils, peanut oil, vegetable oil
Dijon mustard

CANNED GOODS

tomato sauce, paste, and juice
chicken stock
beef stock
vegetable stock
garbanzo beans

ALCOHOLIC BEVERAGES

gin
vodka
scotch
blended whiskey
rum
brandy and cognac
Champagne
tequila
vermouth (sweet and dry)
white wine
red wine
beer
liqueurs (2–3)

COCKTAILS

Champagne Cocktail

1 sugar cube
1 splash Cognac
1 splash bitters
4½ oz Champagne or sparkling wine

Place the sugar cube in a tulip champagne glass. Add the Cognac and bitters. Top off with Champagne or sparkling wine.

Kir Royale

¼ oz crème de cassis
5 oz Champagne or sparkling wine
1 twist of lemon

Pour the liqueur into a champagne flute. Fill the glass with Champagne or sparkling wine. Garnish with a lemon twist.

Mimosa

3 oz freshly squeezed orange juice
5 oz Champagne or sparkling wine

Pour the orange juice into a champagne flute. Fill the glass with the Champagne or sparkling wine. An interesting idea is to replace the regular orange juice with blood orange juice when in season.

Bellini

3 oz Champagne or sparkling wine
1 oz fresh white peach juice

Pour the peach juice into a champagne flute. Fill the glass with the Champagne or sparkling wine.

Brandy Alexander

1½ oz brandy
1 oz crème de cacao
1 oz heavy cream
1 scoop crushed ice

Mix all the ingredients in a shaker. Strain the mixture into a chilled cocktail glass.

Vodka Gimlet

2 oz vodka
½ oz Rose's unsweetened lime juice

Combine all the ingredients in a mixing glass with several ice cubes. Stir well and strain into a chilled cocktail glass.

Sea Breeze

2 oz vodka
3 oz grapefruit juice
3 oz cranberry juice
1 scoop crushed ice

Mix all the ingredients, except the ice, in a shaker. Place the crushed ice in a highball glass and fill with the vodka mixture.

Mint Julep

6 small mint leaves
2 oz bourbon (preferably Kentucky)
1 oz lemon juice
1 oz sugar syrup
1 scoop crushed ice
1 sprig mint

Mix the mint leaves, bourbon, lemon juice, and syrup in a blender for 10 seconds. Pour over crushed ice in a double old-fashioned glass and garnish with the mint sprig.

Manhattan

1½ oz light rum
¾ oz sweet vermouth
Dash bitters
1 maraschino cherry

Mix the rum, vermouth, and bitters in a mixing glass with ice cubes. Strain the mixture into a chilled cocktail glass and garnish with a maraschino cherry.

Cosmopolitan

1¼ oz vodka
¼ oz Triple Sec
¼ oz lime juice
Splash cranberry juice
1 lime wedge

Pour all the ingredients except the lime into a mixing glass. Stir and strain into a chilled martini glass. Squeeze and garnish with the lime.

Tom Collins

3 oz gin
1½ oz lemon juice
½ oz sugar syrup
club soda
1 maraschino cherry

Mix the gin, lemon juice, and sugar syrup in a tall glass with ice. Top off with club soda and garnish with a maraschino cherry.

Sangria

20 oz fruity white wine
1½ oz gin
¾ oz sweet 'n sour mix
2 oz ginger ale
2 oz 7-Up
½ oz sliced oranges
½ oz diced apple
½ oz sliced lemon

Fill a large pitcher with ice. Add all the ingredients and stir well. Serve in a large pitcher.

Martini

2 oz vodka
¼ oz dry vermouth

Place ice in a martini shaker. Add vodka and vermouth. Shake well and strain into martini glass. Serve with an olive.

Note: Always rinse olives with water to remove any brine.

Frozen Daiquiri

1 oz light or dark rum, tequila, vodka, or gin
4 oz daiquiri mixer
¼ cup fruit (optional)

In a blender, combine all the ingredients with 1½ cups ice. Blend until smooth and serve immediately.

INDEX

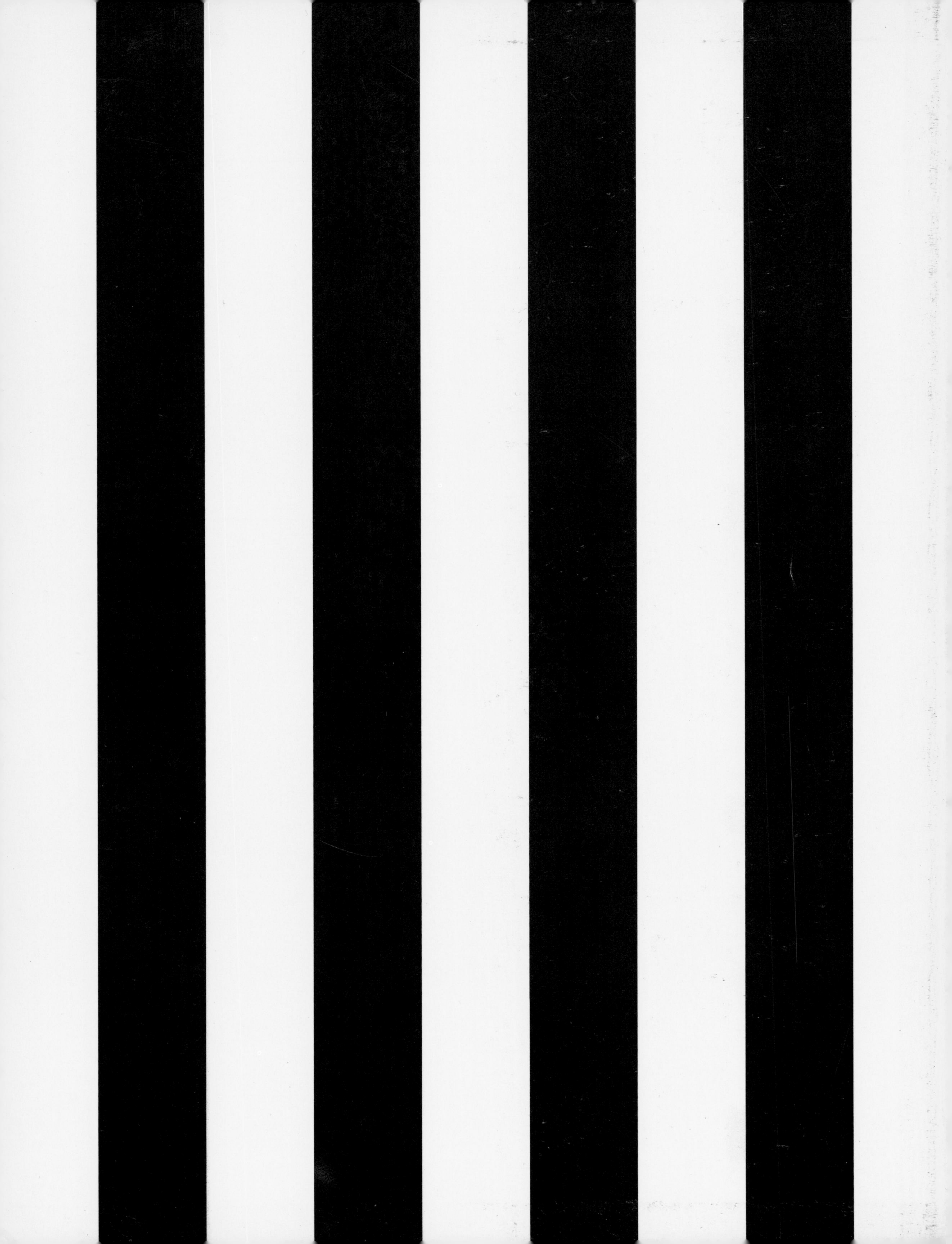